Carotenoids as Colorants and Vitamin A Precursors

FOOD SCIENCE AND TECHNOLOGY

A SERIES OF MONOGRAPHS

A complete list of titles in this series appears at the end of this volume.

Carotenoids as Colorants and Vitamin A Precursors

Technological and Nutritional Applications

Edited by

J. CHRISTOPHER BAUERNFEIND

Gainesville, Florida

1981

ACADEMIC PRESS

A Subsidiary of Harcourt Brace Jovanovich, Publishers

New York London Toronto Sydney San Francisco

ACADEMIC PRESS, INC.
111 Fifth Avenue, New York, New York 10003

United Kingdom Edition published by
ACADEMIC PRESS, INC. (LONDON) LTD.
24/28 Oval Road, London NW1 7DX

Library of Congress Cataloging in Publication Data
Main entry under title:

Carotenoids as colorants and vitamin A precursors.

(Food science and technology)
Includes bibliographical references and index.
1. Carotinoids. 2. Color of food. 3. Vitamin A in human nutrition. I. Bauernfeind, J. C.
TX553.C3C37 664'.06 79–8850
ISBN 0–12–082850–2

PRINTED IN THE UNITED STATES OF AMERICA

81 82 83 84 9 8 7 6 5 4 3 2 1

Contents

List of Contributors

Numbers in parentheses indicate the pages on which the authors' contributions begin.

C. R. Adams (563), Roche Chemical Division, Hoffmann-La Roche Inc., Nutley, New Jersey 07110

J. C. Bauernfeind* (1, 47, 319, 563), Roche Research Center, Hoffmann-La Roche Inc., Nutley, New Jersey 07110

Alan H. Brush (539), Biological Sciences Group, University of Connecticut, Storrs, Connecticut 06268

C. O. Chichester (463), Food Science and Nutrition Department, University of Rhode Island, Kingston, Rhode Island 02881

E. De Ritter (815), Food and Agricultural Product Development Department, Hoffmann-La Roche Inc., Nutley, New Jersey 07110

T. Katayama (463), Laboratory of Marine Biochemistry, Faculty of Fisheries, University of Kagoshima, Kagoshima, Japan

H. Klaüi (47), Vitamin and Nutrition Department, F. Hoffmann-La Roche and Company, Ltd., 4058 Basel, Switzerland

W. L. Marusich (319, 563), Animal Health Research Department, Hoffmann-La Roche Inc., Nutley, New Jersey 07110

Micheline M. Mathews-Roth (755), Channing Laboratory, Department of Medicine, Harvard Medical School, and Brigham and Women's Hospital, Boston, Massachusetts 02115

K. Münzel† (745), Galenical Department, F. Hoffmann-La Roche and Company, Ltd., 4058 Basel, Switzerland

A. E. Purcell‡ (815), USDA-SEA, AR, Department of Food Science, North Carolina State University, Raleigh, North Carolina 27650

*Present address: 3664 N.W. 12th Avenue, Gainesville, Florida 32605.

†Present address: Kilchgrundstrasse 8, CH-4125 Riehen, Switzerland.

‡Present address: Department of Food Science and Nutrition, Brigham Young University, Provo, Utah 84602.

K. L. Simpson (463), Food Science and Nutrition Department, University of Rhode Island, Kingston, Rhode Island 02881

J. O. Williams (787), Edward Davies Chemical Laboratories, University College of Wales, Aberystwyth, SY23 1NE, United Kingdom

Foreword

This comprehensive treatise on all aspects of carotenoid color technology will fill a gap in the carotenoid literature. J. C. Bauernfeind is to be congratulated as editor for arranging the presentation of all technological aspects of this field in ten well-documented chapters. This unique book includes about 100 figures, 250 tables, and about 4000 literature citations, and complements existing monographs such as "Carotenoids" (edited by O. Isler *et al.*, Birkhäuser, Basel, 1971); "Carotenoids" (edited by P. Karrer and E. Jucker, Elsevier, London and Amsterdam, 1950); and "The Comparative Biochemistry of the Carotenoids" (edited by T. W. Goodwin, Chapman & Hall Ltd., London, 1952).

The number of naturally occurring carotenoids with known structure increased from 80 to 300 between 1948 and 1970, and by 1976 about 400 compounds were known. About 50 of them have vitamin A activity. All vitamin A in nature originates from carotenoids. In third world countries hypovitaminosis A is a very serious deficiency disease in man. The World Health Organization (WHO) estimated in 1978 that at least 1,000,000 children principally of pre-school age were affected by xerophthalmia, a blinding disease with a death rate exceeding 25% and many permanently blind.

Carotenoids constitute a large class of well-tolerated compounds. Laboratory and biological testing under modern experimental conditions indicate that the synthetic carotenoids on the market are well utilized by man. The experts of the mixed commission of FAO and WHO classified these in the group of preferred colors with accepted toxicity tests. As members of a uniform class of compounds to which man has been constantly exposed during evolution, these synthesized carotenoids deserve special status.

O. Isler
Am Landungssteg
8592 Uttwil, Switzerland

Preface

Visual appearance, especially color, is the most important characteristic of foods in determining their selection prior to actual consumption. Very early in life the individual associates color with the flavor of food: butter, light yellow; orange, carrots; red, tomatoes; purple, boysenberry. It has been demonstrated repeatedly that a radical shift in the color of a food, even though accompanied by no change in flavor, can make it completely unacceptable. Purple butter and green sausages would be examples of foods engendering rejection because of their unfamiliar appearance. The appropriate color may enhance the appreciation of flavor.

Of the food colorants, the carotenoids are of particular importance, not only because they serve as excellent colorants but also because they represent a major dietary source of vitamin A. They are the most widespread group of naturally occurring pigments in nature, present without exception in photosynthetic tissue and occurring with no definite pattern in nonphotosynthetic tissues such as root, flower petals, seeds, and fruits. They are also found in fungi, yeast, molds, mushrooms, and bacteria, and in many cases they are the major pigment in the exoskeleton of aquatic and avian species, that is, lobster and prawn, and in the feathers of many birds. Although animals are thought to be incapable of *de novo* synthesis of carotenoid, they are able to influence their chemical structure.

The ubiquitous nature of these pigments and their conversion to vitamin A have been of great interest to scientists. The capacity to synthesize chemically both carotenoids and vitamin A and their resulting ready availability have brought about a great increase in their usage in food. The technology which governs their use has developed extremely rapidly. Major processing advances have been made in the preparation of these compounds, their stabilization in aqueous systems, and in the prevention of their oxidation during processing. The time span for these advances has been comparatively short.

Dr. Bauernfeind has been involved in scientific research that has made possible the wide application of carotenoids, and he envisioned the need for the compilation of this knowledge. How carotenoids can be utilized by man, animals, and their environment may be said to be the theme of this book now that industrial chemists have produced unlimited quantities of pure, new colorants by chemical synthesis.

Qualified authors have assembled the existing knowledge on the application of carotenoids to food and to the feed of animals, fish, poultry, and birds. The use of carotenoids in medicine, in the coloring of pharmaceutical and cosmetic products, and their unique role as photoconductors are discussed. Analytical methodology is also covered. The chapters go beyond merely supplying approaches to applications. For example, the volume illustrates the imperfect state of knowledge on the qualitative and quantitative distribution of carotenoids in natural foods, and calls for a renewed effort in applying modern analytical methodology to carotenoid determination in currently used cultivars of fruits, vegetables, and cereal grains, and in dairy, egg, fish, and poultry products so that at some future date one may have more reliable tables of individual carotenoids in the many food products around the world, one by-product of which would be an understanding of which foods contribute significant vitamin A activity and under what circumstances.

It is Dr. Bauernfeind's hope that the encyclopedic knowledge contained in this volume will serve to stimulate others to confirm, amend, and infinitely expand the existing data and thereby demonstrate, as one example, the productive partner relationship between industry and agriculture in providing a more varied food supply.

C. O. Chichester

Acknowledgment

The editor acknowledges the fine contribution of F. Hoffmann-La Roche and Company, Ltd. in printing the color inserts for the book, and the personal efforts of Dr. Franz Leuenberger and others in the guidance of this project.

J. Christopher Bauernfeind

Carotenoids as Colorants and Vitamin A Precursors

1

NATURAL FOOD COLORS

J. C. Bauernfeind

I. INTRODUCTION

Currently consumers, having become more nutrition- and health-minded in the past decade, are developing a growing interest in what is present in their food supply and particularly what is added to it in the way of food additives. Food labeling has increased this interest. Not always understanding the physiological needs of the body for certain chemicals and the body's ability to metabolize other chemical structures, consumers tend to shy away from unfamiliar compounds that comprise food additives, such as antioxidants, preservatives, and colors. In search for a solution, consumers easily adopt the concept that if the additive is in natural food it must be safe and good, since apparently it has been consumed for centuries and has withstood the test of time. On the other hand, there is also the growing awareness that nature is not infallible, as toxicants are also present in natural foods.

There has always been and always will be a desire for attractively colored foods as long as the eye signals the selection of the daily ingestion of food products for the stomach via the brain. It would seem to follow, therefore, that the food industry will

ISBN 0-12-082850-2

continue to require a wide array of acceptable, safe food colorants to satisfy consumer preferences. Worldwide, the potential market for food colors may eventually reach several hundred million dollars or more annually.

There appears to be a growing preference for natural-type colors in countries around the world. The new color list of Switzerland distinguishes between colors occurring naturally in food and colors not naturally occurring in foods, and, in Norway,* artificial colors may no longer be used. In Sweden,† the use of artificial colors has been reduced to special cases only. Iceland also has established tighter controls over color additives to foods. Botma cites the expectancy of seeing fewer permitted synthetic food colors in Europe and in turn a greater return to natural food colors.

Most food colors probably can be divided into three main categories: (a) natural organic colors of plant or animal origin, either extracted from nature or the identical compound reproduced by chemical synthesis; (b) colors of inorganic origin as taken from nature or reproduced synthetically; and (c) artificial colors, i.e., synthetic compounds, those that are unnatural or are not present in foods presently consumed, such as the azo and triphenylmethane dyes. Natural colorants can be subdivided from a chemical standpoint into other groups. There are the isoprenoid derivatives (carotenoid colors), the tetrapyrrole derivatives (chlorophyll and heme colors), the benzopyran derivatives (anthocyanin and flavonoid colors), the betalaine compounds (betanin and related colors), the flavins (such as riboflavin), and inorganic pigments. Then there is the heterogeneous group comprising products initially formed intentionally or unintentionally in food processing, including melanins, melanoidins, the Maillard reaction group such as caramel, and the carbon group such as vegetable carbon black.

Switzerland has adopted a classification‡ of colors that should be helpful as a guide to colorants around the world. It is related to a general concept that has been in the literature for several decades but has not been clearly defined or put into practice. Natural colors are colorants that naturally occur in foods and are either produced by extraction from natural sources or reproduced by chemical synthesis. This classification includes items listed in Table 1.

II. IDEAL COLOR PROPERTIES

What are the properties of an ideal food color? Some characteristics may be given as follows:

1. It should be nontoxic and noncarcinogenic at a multiple-use level; it should not contain toxic impurities.

*Godkjente Tilsetningsstoffer, (1979).

†Swedish Food Regulations, Food Additives, SLV (1975), 11: (1978), 33.

‡Lebensmittelverordnung Schweiz Änderung vom 18.10, (1978).

TABLE 1

Some Colorants Included in Natural Color Classification

Anthocyanins
Extracts of berries
Extracts of fruits such as the grape
Extracts of vegetables
Betalaines
Extracts of vegetables such as the red beet
Carotenoids
Extracts such as annatto
Apocarotenal
Canthaxanthin
β-Carotene
Xanthophylls
Chlorophyll
Coloring spices
Curcumin
Paprika
Saffron
Riboflavin
Mineral colorants
Calcium carbonate
Iron oxides
Vegetable carbon black

2. For wide usage it should have solubility and/or dispersibility characteristics for incorporation into water-base and fat-base food products.
3. It preferably should not contribute significant taste or fragrance to the food product.
4. In foods it should be stable to light in dissolved or dispersed form, over a wide pH range, preferable pH 2–8, at boiling and baking temperatures, and during shelf-life storage and preparation practices for consumption.
5. It should not react with trace elements or with oxidizing or reducing agents.
6. Its characteristics should be uniform from lot to lot and be capable of being monitored in concentrate form and in foods by suitable analytical techniques.
7. It should be widely available and relatively economical for food use.
8. It must be approved and conform to government specifications and preferably have an internationally approved status.

Weissler commented that, although the use of coloring agents to make food more attractive dates back to antiquity, the large-scale use of colorants began in the early 1800's with the development of the food processing industry. Colors that have flourished in the plant and animal kingdoms for centuries still serve in many instances as coloring substances in the food supply today. In other instances many colors in early food (taken from a nonfood source) have been deleted for safety or technological reasons. A survey of coal-tar color approval in many countries indicates that there have been many revocations of approval during the past decade, as tabulated by Botma.

III. BRIEF HISTORY OF COLORANTS

Hundreds of coal-tar dyes were synthesized by 1900, and of these seven were selected by Dr. B. C. Hesse of the United States Department of Agriculture (USDA) as physiologically harmless and suitable for food use. This list was published in 1907 under the U.S. Pure Food and Drug Act enacted the preceding year. The act also stipulated that the dye for food use would have to have a certificate filed with the Secretary of Agriculture indicating that it had been tested and found free of harmful ingredients. In later years the coal-tar dyes used in foods became known as the Food, Drug, and Cosmetic (FD&C) certified colors under the 1938 Food, Drug, and Cosmetic Act and the 1960 Color Additive Amendment. After 1907 more than the original seven FD&C certified colors were approved in the United States, as shown in Fig. 1. The volume of these approved colors through 1978 is indicated in Table 2. Of the FD&C food colors approved by the Food and Drug Administration (FDA) (Fig. 1) and in wide usage in the United States over the past 70 years, only seven remained approved in 1978. Of the original seven allowed in 1907, only two are still permitted to be in wide usage, namely, Red No. 3 and Blue No. 2. Other FD&C food colors allowed limited usage in 1978 are Citrus Red No. 2, which can be used only for coloring peels of oranges not intended or used for further processing at a level not to exceed 2 ppm by weight, and Orange B, which can be used only (tentative) for coloring casings and surfaces of frankfurters and sausages at a level not to exceed 150 ppm by weight. Currently four FD&C colors in wide usage are permanently listed, namely, Red No. 3, Red No. 40, Yellow No. 5, and Blue No. 1. The nonpermanently listed or provisional FD&C colors, Yellow No. 6, Green No. 3, and Blue No. 2, are to be reviewed in January 1981 after the completion of current chronic toxicity trials. Only one new FD&C color has been approved in the United States during the past 30 years, namely, Red No. 40 (Allura Red).

According to a U.S. tabulation done in 1967 (Table 3), many processed foods are artifically colored with the certified FD&C food colors, and a list of food groups in approximately decreasing order of volume use of color is as follows: beverages, candy and confections, dessert powders, bakery products, sausage casings, cereal products, ice cream and other dairy products, snack foods, maraschino cherries, and miscellaneous groups. The miscellaneous groups of foods using added color include dressings, gravies, jams, jellies, spices, etc. Fats and oils are not specifically mentioned. The order of importance of the food groups and of others may have changed within the intervening years. The new industry survey of food additives

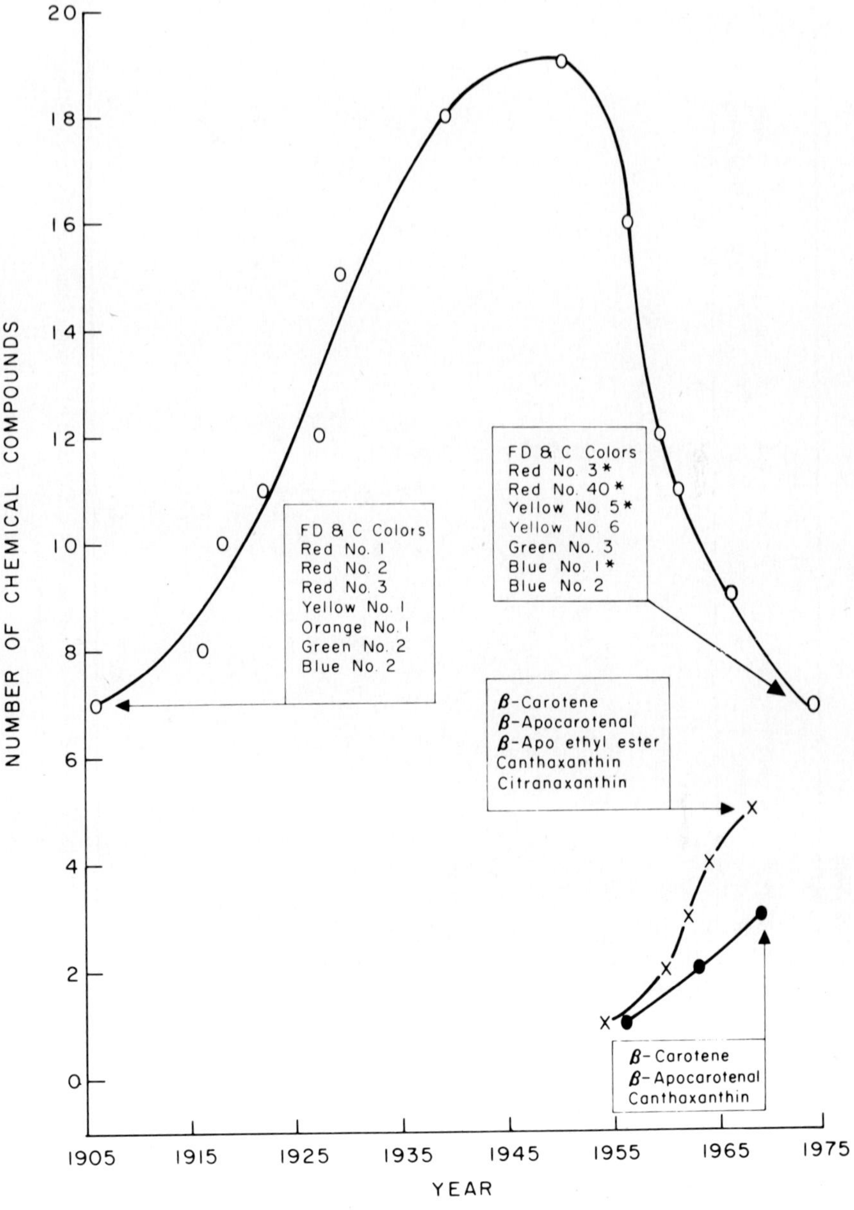

Fig. 1. Changes in the acceptance and use of FD&C colors and carotenoid food colors. The FD&C food colors currently permitted in the United States are shown. Other limited synthetic noncarotenoid permitted colors are Citrus Red No. 2, only for coloring skins of oranges not intended or used for further processing at a level not to exceed 2 ppm by weight, and Orange B (approval to be withdrawn), only for coloring casings and surfaces of frankfurters and sausages at a level not to exceed 150 ppm by weight. Legend: ○—○, U.S. permitted FD&C colors (asterisks denote permanently listed colors); ×—×, carotenoids commercially synthesized; ●—●, U.S. permitted carotenoid food colors.

TABLE 2

U.S. Certified Colors: Production Volumes by Fiscal Year[a]

Colors	FY 1974 (lb)	FY 1975 (lb)	FY 1976[b] (lb)	FY 1977 (lb)	FY 1978 (latest data) (lb)	% annual growth (FY 1974 to 1978)		1978 food usage (1b)
Primary FD&C Colors								
Blue 1	159,000	159,000	148,112	184,115	184,332			
Blue 2	89,000	85,000	79,215	98,936	85,030			
Red 2	903,000	1,378,000	239,257	—	—			
Red 3	286,000	337,000	363,875	548,558	429,552	+2.6[d]		
Red 4	28,000	35,000	4,169	7,822	6,668			
Red 40	729,000	788,000	1,500,760	1,520,648	1,799,690			
Yellow 5	1,290,000	1,391,000	1,543,764	1,165,528	1,541,179	+4.5		
Yellow 6	996,000	1,084,000	1,081,714	991,347	1,071,148	+1.8		
Citrus Red 2	496	12,000	1,752	—	—			
Green 3	5,000	9,000	6,438	4,111	5,945			
Orange B	19,060	31,000	29,045	38,909	17.788			
	4,504,000	5,310,000	4,998,101	4,559,974	5,141,302	+3.4	80	4,113,000
FD&C Lakes[c]								
Blue 1	NA	NA	49,439	96,821	74,227			
Blue 2	NA	NA	39,911	78,584	57,228			
Red 2	50,104	40,776	3,871	—	—			
Red 3	NA	NA	262,948	269,784	314,178			
Red 40	33,482	32,966	81,975	71,206	101,206			
Yellow 5	492,764	502,832	558,851	727,776	643,280			
Yellow 6	188,267	235,610	189,885	447,108	241,442			
			1,186,880	1,691,270	1,431,561		65	930,000[e]

[a] Source: U.S. Food and Drug Administration. Note: 5.04 MM lb times average price of \$6.40/lb = \$33 MM; FY denotes fiscal year; NA, not available.
[b] Five quarters; when fiscal year changed from July 1 to June 30 to October 1 to September 30.
[c] Cannot be added to primary colors since later are used to produce lakes.
[d] Actually 1.5% growth due to differences in color intensity.
[e] Only 20% average pure dye content.

TABLE 3

Food, Drug, and Cosmetic Usage of Primary Colors[a,b]

Category	FD&C Blue No. 1	FD&C Blue No. 2	FD&C Green No. 3	Orange B	FD&C Red No. 2	FD&C Red No. 3	FD&C Red No. 4	FD&C Violet No. 1	FD&C Yellow No. 5	FD&C Yellow No. 6	Total
Candy, confections	6,632	2,499	124	0	67,637	11,665	0	1,459	59,903	52,770	202,689
Beverages	15,800	2,375	301	0	282,695	1,056	0	985	78,933	181,292	563,437
Dessert powders	3,270	1,659	14	0	62,363	8,616	0	0	59,961	51,622	187,505
Cereals	843	99	0	0	15,558	1,421	0	0	52,496	35,464	105,881
Maraschino cherries	597	0	98	0	8,104	3,469	11,308	0	5,644	4,830	34,050
Pet food	1,473	6,764	0	0	67,058	1,023	0	1,278	101,743	23,226	202,565
Bakery goods	3,680	673	7	0	43,522	9,560	0	369	77,885	42,203	177,899
Ice cream, sherbet, dairy products	2,599	179	7	0	29,697	621	0	45	35,048	23,868	92,064
Sausage	647	0	0	16,890	36,084	4,970	0	0	6,502	99,605	164,698
Snack foods	305	0	0	0	3,623	766	0	2	18,456	11,409	34,561
Meat links	11	0	0	0	12	10	0	2,223	15	0	2,271
Miscellaneous	5,345	1,990	1,298	0	46,219	18,200	398	1,134	44,841	29,134	148,559
Subtotal (food use)	41,202	16,238	1,849	16,890	662,572	61,377	11,706	7,495	541,427	555,423	1,916,179
Pharmaceutical	3,250	593	220	0	21,179	12,168	1,186	347	17,275	15,938	72,156
Cosmetics	397	30	27	0	3,417	903	630	96	3,125	2,148	10,773
Total	44,849	16,861	2,096	16,890	687,168	74,448	13,522	7,938	561,827	573,509	1,999,108

[a] "Food Colors," Committee on Food Protection, Food and Nutrition Board, Division of Biology and Agriculture, National Research Council, National Academy of Sciences, Washington, D.C. (1971).

[b] In pounds. Figures represent sales for the first 9 months of 1967 and do not include exports or sales to jobbers and other manufacturers.

TABLE 4

Total Production of Processed Foods Using Certified Colors[a,b]

Category	Total color sold (lb)	Average color concentration (ppm)	Calculated total production (lb)
Candy, confections	202,689	100	2,026,890,000
Beverages	563,437	75	7,512,305,521
Dessert powders	187,505	140	1,339,348,215
Cereals	105,881	350	302,502,017
Maraschino cherries	34,050	200	170,250,000
Pet foods	202,565	200	1,012,825,000
Bakery goods	177,899	50	3,557,980,000
Ice cream, sherbet, dairy products	92,064	30	3,038,112,000
Sausage	164,698	125	1,317,584,000
Snack foods	34,561	200	172,805,000
Meat links	2,271	—	—
Miscellaneous (salad dressing, nuts, gravy, spices, jams, jellies, food packaging, etc.)	148,559	400	371,397,500
Total			20,821,999,253

[a] "Food Colors," Committee on Food Protection, Food and Nutrition Board, Division of Biology and Agriculture, National Research Council, National Academy of Sciences, Washington, D.C. (1971).

[b] During first 9 months of 1967. Calculated and based on total color sold and the reported average color concentration.

initiated in 1977 should update food color usage.* Some other statistical data related to Table 3 involving the average level of use of FD&C colors in food and the volume of food product produced are shown in Table 4.

IV. NATURAL OR NONCERTIFIED COLORS

Some naturally occurring colors approved for use in foods (Table 5) or drugs (Table 6) in the United States are exempt from FDA certification and hence are referred to as noncertified colors. Constant government surveillance is maintained for these colorants regarding specifications and use applications. A description of identity and specifications used in the United States is available for these colorants

*Food colors, 1980, Status Summary, Institute of Food Technologists, 221 N. LaSalle St., Chicago, Illinois 60601.

TABLE 5

Color Additives for Food Use That Are Exempt from Certification by the FDA[a]

Additive	Identity	Specification	Value
Algae meal, dried (permitted in chicken feed only)	Dried mixture of algae cells (genus *Spongiococcum*, separated from its culture broth), molasses, corn-steep liquor, and a maximum of 0.3% ethoxyquin; the algae cells are produced by suitable fermentation, under controlled conditions, from a pure culture of the genus *Spongiococcum*		
Annatto extract	Extract prepared from annatto seed, *Bixa orellana* L., using any one or an appropriate combination of approved food-grade extractants	Arsenic (as As)	3 ppm max
		Lead (as Pb)	10 ppm max
β-Apo-8′-carotenal		Physical state	Solid
		1% solution in chloroform	Clear
		Melting point (corrected)	136°–140°C (dec)
		Loss of weight on drying (45°C *in vacuo*)	0.2% max
		Residue on ignition	0.2% max
		Lead (as Pb)	10 ppm max
		Arsenic (as As)	1 ppm max
		Assay (spectrophotometric)	96–101%
Beets, dehydrated (beet powder)	Dark red powder prepared by dehydrating sound, mature, good-quality, edible beets	Volatile matter	4.0% max
		Acid-insoluble ash	0.5% max
		Lead (as Pb)	10 ppm max
		Arsenic (as As)	1 ppm max
		Mercury (as Hg)	1 ppm max

(Continued)

TABLE 5—*Continued*

Additive	Identity	Specification	Value
Caramel	Dark brown liquid or solid material resulting from the carefully controlled heat treatment of the following food-grade carbohydrates: dextrose, invert sugar, lactose, malt syrup, molasses, starch hydrolyzates and fractions thereof, sucrose	Lead (as Pb)	10 ppm max
		Arsenic (as As)	3 ppm max
		Mercury (as Hg)	0.1 ppm max
Carmine	Aluminum or calcium–aluminum lake on an aluminum hydroxide substrate of the coloring principles, chiefly carminic acid, obtained by an aqueous extraction of cochineal [*Dactylopius coccus costa* (*Coccus cacti* L.)]	Volatile matter (at 135°C for 3 hr)	20.0% max
		Ash	12.0% max
		Lead	10 ppm max
		Arsenic (as As)	1 ppm max
		Carminic acid	50% min
Canthaxanthin	β-Carotene-4,4′-dione	Physical state	Solid
		1% solution in chloroform	Complete and clear
		Melting point (corrected)	207–212°C
		Loss on drying	0.2% max
		Residue on ignition	0.2% max
		Total carotenoids other than transcanthaxanthin	5.0% max
		Lead	10 ppm
		Arsenic	3 ppm
		Mercury	1 ppm
		Assay	96–101%
β-Carotene	β-Carotene prepared synthetically or obtained from natural sources	Physical state	Solid
		1% solution in chloroform	Clear
		Loss of weight on drying (40–41°C, 4 hr)	0.2% max
		Residue on ignition	0.2% max
		Lead (as Pb)	10 ppm max
		Arsenic (as As)	3 ppm max
		Assay (spectrophotometric)	96–101%

Carrot oil	Liquid or solid portion of the mixture, or the mixture itself, obtained by the hexane extraction of edible carrots (*Daucus carota* L.) with subsequent removal of the hexane by vacuum distillation; the resultant mixture of solid and liquid extractives consists chiefly of oils, fats, waxes, and carotenoids naturally occurring in carrots	Hexane	25 ppm max
Corn endosperm oil (permitted in chicken feed only)	Reddish-brown liquid composed chiefly of glycerides, fatty acids, sitosterols, and carotenoid pigsments obtained by isopropyl alcohol and hexane extraction from the gluten fraction of yellow corn grain	Total fatty acids	85.0% min
		Iodine value	118–134
		Saponification value	165–185
		Unsaponifiable matter	14.0% max
		Hexane	25 ppm max
		Isopropyl alcohol	100 ppm max
Cotton seed flour, cooked, toasted, partially defatted	Product prepared as follows: food-quality cottonseed is delinted and decorticated; the meats are screened, aspirated, and rolled; moisture is adjusted, the meats heated, and the oil expressed; the cooked meats are cooled, ground and reheated to obtain a product varying in shade from light to dark brown	Arsenic (as As)	0.2 ppm max
		Lead (as Pb)	10 ppm max
		Free gossypol content	450 ppm max
Ferrous gluconate (permitted in ripe olives only)		Assay, $C_{12}H_{22}FeO_{14}$, anhydrous basis	95.0% min
		Water	6.0–9.0% max
		Arsenic (as As)	3 ppm max
		Chloride	700 ppm max
		Ferric iron	2.0% max
		Lead	10 ppm max
		Mercury	3 ppm max
		Oxalic acid	Passes test
		Reducing sugars	Passes test
		Sulfate	1000 ppm

(Continued)

TABLE 5—*Continued*

Additive	Identity	Specification	Value
Fruit juice	Concentrated or unconcentrated liquid expressed from mature varieties of fresh, edible fruits, or a water infusion of the dried fruit		
Grape skin extract (enocianina)	Purplish-red liquid prepared by the aqueous extraction (steeping) of the fresh deseeded marc remaining after grapes have been pressed to produce grape juice or wine; contains the common components of grape juice, namely, anthocyanins, tartaric acid, tanins, sugars, minerals, etc., but not in the same proportions as found in grape juice; during the steeping process, sulfur dioxide is added, and most of the extracted sugars are fermented to alcohol; the extract is concentrated by vacuum evaporation, during which practically all of the alcohol is removed; a small amount of sulfur dioxide may be present	Pesticide residues	Not more than permitted in or on grapes by regulations promulgated under section 408 of the Federal Food, Drug, and Cosmetic Act
		Lead (as Pb)	10 ppm max
		Arsenic (as As)	1 ppm max
Iron oxide, synthetic (permitted in dog and cat food only)	Any one or any combination of synthetically prepared iron oxides, including the hydrated forms; it is free from admixture with other substances	Arsenic (as As)	5 ppm max
		Lead (as Pb)	20 ppm max
		Mercury (as Hg)	3 ppm max
Paprika	Ground, dried pod of mild capsicum (*Capsicum annuum* L.)		
Paprika oleoresin	Combination of flavor and color principles obtained from paprika (*Capsicum annuum* L.) by extraction, using any one or a		

	combination of the following solvents: acetone, ethanol, ethylene dichloride, hexane, isopropyl alcohol, methanol, methylene chloride, trichloroethylene		
Riboflavin		Assay as $C_{17}H_{20}N_4O_6$, calculated on the dried basis	98% min 102% max
		Specific rotation, $[\alpha]_D^{25}$	Between $-112°$ and $-122°$, calculated on the dried basis
		Loss on drying	1.5% max
		Lumiflavin	Passes test
		Residue on ignition	0.3% max
Saffron	Dried stigma of *Crocus sativus* L.		
Tagetes (Aztec marigold) meal and extract (permitted in chicken feed only)	Tagetes (Aztec marigold) meal is the dried, ground flower petals of the Aztec marigold (*Tagetes erecta* L.) mixed with not more than 0.3% ethoxyquin	Tagetes meal shall be free from admixture with other plant material from *Tagetes erecta* L. or from plant material or flowers of any other species of plants	
	Tagetes (Aztec marigold) extract is a hexane extract of the flower petals of the Aztec marigold (*Tagetes erecta* L.); it is mixed with an edible vegetable oil, or with an edible vegetable and a hydrogenated edible vegetable oil, and not more than 0.3% ethoxyquin; it may also be mixed with soy flour or corn meal as a carrier	Tagetes extract shall be prepared from tagetes petals meeting the specifications above and shall conform to the following additional specifications:	
		Melting point	53.5°–55.0°C
		Iodine value	132–145
		Saponification value	175–200
		Acid value	0.60–1.20
		Titer	35.5–37.0°C
		Unsaponifiable matter	23.0–27.0% max
		Hexane residue	25 ppm

(Continued)

TABLE 5—*Continued*

Additive	Identity	Specification	Value
		All determinations, except the hexane residue, shall be made on the initial extract of the flower petals (after drying in a vacuum oven at 60°C for 24 hr) prior to the addition of the oils and ethoxyquin; the hexane determination shall be made on the color additive after the addition of the vegetable oils, hydrogenated vegetable oils, and ethoxyquin	
Titanium dioxide	Synthetically prepared titanium dioxide, free from admixture with other substances	Lead (as Pb)	10 ppm max
		Arsenic (as As)	1 ppm max
		Antimony (as Sb)	2 ppm max
		Mercury (as Hg)	1 ppm max
		Loss on ignition at 800°C (after drying for 3 hr at 105°c)	0.5%
		Water-soluble substances	0.3%
		Acid-soluble substances	0.5%
		Titanium dioxide after drying for 3 hr at 105°C	99.0% min
		Silicon dioxide, SiO_2, and/or aluminum oxide, Al_2O_3, as dispersing aids	2.0% total

		Lead, arsenic, and antimony shall be determined in the solution obtained by boiling 10 gm of the titanium dioxide for 15 min in 50 ml of 0.5 *N* hydrochloric acid	
Turmeric	Ground rhizome of *Curcuma longa* L.		
Turmeric oleoresin	Combination of flavor and color principles obtained from turmeric (*Curcuma longa* L.) by extraction using any one or a combination of the following solvents: acetone, ethanol, ethylene dichloride, hexane, isopropyl alcohol, methanol, methylene chloride, trichloroethylene		
Ultramarine blue (permitted only in salt intended for animal	Blue pigment obtained by calcining a mixture of kaolin, sulfur, sodium carbonate, and carbon at temperatures above 700°C; sodium sulfate and silica may also be incorporated in the mixture in order to vary the shade; the pigment is a complex sodium aluminum sulfosilicate having the approximate formula $Na_7Al_6Si_6O_{24}S_3$	Lead (as Pb) Arsenic (as As) Mercury (as Hg)	10 ppm max 1 ppm max 1 ppm max
Vegetable juice	Concentrated or unconcentrated liquid expressed from mature varieties of freshly edible vegetables		

[a] Marmion, D. M. Description and analysis of colors for foods, drugs, and cosmetics. *Encyclopedia of Industrial Chemical Analysis* **10,** 447–547 (1970).

TABLE 6

Color Additives for Drug Use That Are Exempt from Certification by the FDA[a]

Additive	Identity	Specification	Value
Alumina (dried aluminum hydroxide)	White, odorless, tasteless, amorphous powder consisting essentially of aluminum hydroxide ($Al_2O_3 \cdot xH_2O$)	Acidity or alkalinity	When 1 gm of the color additive is agitated with 25 ml water and filtered, the filtrate should be neutral to litmus paper
		Matter insoluble in dilute hydrochloric acid	0.5% max
		Lead (as Pb)	10 ppm max
		Arsenic (as As)	1 ppm max
		Mercury (as Hg)	1 ppm max
		Aluminum oxide (Al_2O_3)	
Annatto extract	Same as for food-grade additive	Same as for food-grade additive	
Calcium carbonate	Fine, white, synthetically prepared powder consisting essentially of precipitated calcium carbonate ($CaCO_3$)	Loss on drying (200°C, 4 hr)	2.0% max
		Acid insolubles	0.2% max
		Heavy metals	30 ppm
		Magnesium and alkali salts	1.0% max
		Barium	No green color when a platinum wire is dipped in 2.5% acidified sample solution and held in a non-luminous flame
Caramel	Same as for food-grade additive	Same as for food-grade additive	

Carmine	Same as for food-grade additive	Same as for food-grade additive	
β-Carotene	Same as for food-grade additive	Same as for food-grade additive	
Iron oxide, synthetic	Same as for food-grade additive	Arsenic (as As)	3 ppm max
		Lead (as Pb)	10 ppm max
		Mercury (as Hg)	3 ppm max
Pyrophyllite	Naturally occurring mineral substance consisting predominantly of a hydrous aluminum silicate, $Al_2O_3 \cdot 4SiO_2 \cdot H_2O$, intimately mixed with lesser amounts of finely divided silica, SiO_2; small amounts, usually less than 3%, of other silicates, such as potassium aluminum silicate, may be present	Lead (as Pb)	20 ppm max
		Arsenic (as As)	3 ppm max
		Lead and arsenic shall be determined in the solution obtained by boiling 10 gm of the pyrophyllite for 15 min in 50 ml of 0.5 *N* hydrochloric acid	
Talc	Finely powdered, native, hydrous magnesium silicate sometimes containing a small proportion of aluminum silicate	Lead (as Pb)	20 ppm max
		Arsenic (as As)	3 ppm max
		Lead and arsenic shall be determined in the solution obtained by boiling 10 gm of the talc for 15 min in 50 ml 0.5 *N* hydrochloric acid	
		Loss on ignition (at red heat to constant weight)	5.0% max
		Acid-soluble substances as sulfate	2.0% max
		Reaction and soluble substances	0.1% max
		Water-soluble iron	Passes test
Titanium dioxide	Same as for food-grade additive	Same as for food-grade additive	

[a] Marmion, D.M. Description and analysis of colors for foods, drugs, and cosmetics. *Encyclopedia of Industrial Chemical Analysis* **10,** 447–547 (1970).

(Tables 5 and 6). Some of these noncertified colorants have properties other than pigmenting value. For example, β-carotene and riboflavin have vitamin activity; paprika, turmeric, and saffron are spices. Others have use restrictions, such as iron oxide in pet foods, ferrous gluconate in black olives, and grape skin extract for still, carbonated, and alcoholic beverages.

The noncertified or natural-type colorants are not without problems. The disadvantages of using colorants exempt from certification cited by Soukup and Young-Maing involve (a) specific use application rather than broad application; (b) possible hygroscopicity and/or stability problems; (c) lower tinctorial intensity, water solubility, and possibly off-aroma or off-taste; (d) less reproducibility and more variability; (e) relatively greater expense; and (f) at times insufficient supply. Although the highly concentrated natural colors obtained by purification techniques or by chemical synthesis overcome many of the above objections, some remain.

Some of the problems of preparing cruder colors from natural sources arise from the difficulty of defining specifications because (a) frequently the active colorants are complex mixtures of several components, (b) the components are influenced by regional or seasonal variations, and (c) processing of the raw materials may be inherently variable. The separations in the processing of compounds from a mixture of closely related chemical components with similar physical behavior (if such purification stages are involved) usually result in lengthy and tedious processes, sometimes accompanied by significant losses or artifact development. These conditions contribute to the difficulty of arriving at guarantees of exact identity of different batches over time periods and by different manufacturers. In the meantime national and international pressures are growing for industry to overcome these problems with demands for more scientific data and more rigid specifications.

The lists of natural or noncertified food colorants do contain some pure, single compounds of synthetic origin, such as riboflavin, β-carotene, β-apo-8′-carotenal, and canthaxanthin, which are available as needed and are constant in identity and specifications. It is quite obvious that it is easier to set up more precise specifications for a pure, single compound than for a mixture. Such specifications are one of the conditions whereby government bodies or expert committees, such as the FAO/WHO, Codex Alimentarious, FDA, make judgments for colorant approval. The coal-tar dyes (azo and triphenylmethane dyes or FD&C certified colors) have experienced a decline in public confidence in their safety, as indicated by the revocation of government approval of specific colorants in the past several decades and by the appearance of isolated, adverse reports that have been widely publicized and may have influenced subsequent appraisal of data validity and significance. It is recognized today that all colorants, whether naturally or synthetically derived, must undergo elaborate safety testing and be granted approval for safety by recognized government and/or expert panel groups.

Of all the natural colorants, the synthetically produced carotenoids (identical to those in nature) have precise specifications and have been subjected to extensive toxicological testing; they have been accepted for inclusion in list A(1) by the

TABLE 7

Toxicologically Acceptable Colors with Established ADI, List A(1)[a]

EEC no.	Colorant	Established ADI[b]
E 160e	5.1 β-Apo-8′-carotenal	0–5 (as sum of total carotenoids)
E 160f	5.2 β-Apo-8′-carotenoic acid esters	See 5.1
E 160a	5.3 β-Carotene	See 5.1
E 161g	5.5 Canthaxanthin	0–25
E 150	5.6 Caramel colors (not made by the ammonia process)	Not limited
—	5.7 Chlorophyll (CI Natural Green 3)	Not limited
E 141	5.8 Chlorophyll-copper complex	0–15
—	5.10 Riboflavin	0–0.5

[a] Joint FAO/WHO Food Standards Programme. Codex Alimentarius Commission, First Series.
[b] ADI, acceptable daily intake in milligrams per kilogram body weight.

FAO/WHO on the basis of evaluations made by the Joint FAO/WHO Expert Committee on Food Additives. List A(1) (Table 7) contains those additives for which an acceptable daily intake (ADI) value has been assigned or the consumption of which has not been toxicologically limited. Colorants with temporary or provisional ADI values (Table 8) are assigned to list A(2).

Is there an ideal food color having all the attributes described earlier in this chapter? None of the compounds singularly or as a class can claim perfection; however, some colorants approach the goal.

TABLE 8

Colors in Progress with Temporary or Provisional ADI, List A(2)[a]

EEC no.	Colorant	Temporary ADI[b]
E 160b	4.1 Annatto extracts (CI Natural Orange 4)	1–1.25 (expressed as bixin)
—	4.12 Turmeric (CI Natural Yellow 3)	0–2.5
—	4.13 Curcumin	0–0.1
E 172	4.16 Iron oxides, hydrated iron oxides	Not limited
—	4.17 Beet red	Not limited
E 150	4.3 Caramel color (made by ammonia process)	0–100 (provided that level of 4-methylimidazole does not exceed 200 mg/kg)
—	4.4 Chlorophyllin-copper complex, sodium and potassium salts (CI Natural Green 3)	0–15

[a] Joint FAO/WHO Food Standards Programme. Codex Alimentarius Commission, First Series.
[b] ADI, acceptable daily intake in milligrams per kilogram body weight.

TABLE 9

Table of Permitted Natural Colors, April 1978[a,b,c,d,e]

Color	CI 1971 no.	EEC no.	Austria	Norway	Portugal	Sweden	Switzerland	Belgium	Denmark	France	Germany (West)	Ireland (Republic of)	Italy	Luxembourg	Netherlands	United Kingdom	Argentina	Australia	Brazil	Canada	Chile
General																					
Vegetable colors																		+ [13]			
Colors from edible fruits and vegetables																					
Colors from animals																		+ [10]			
Reds																					
Alkannet	75520/75530				+		+										+	+ [11]		+	
Carmine	75470	E 120	+		+		+	+		+	+	+	+	+	+	+	+	+	+	+	+
Orchil			+				+					+					+	+ [11]	+	+	
Red to yellow shades																					
Carotenoids [19]		E 160			+			+	+	+	+	+	+	+	+	+			+		+
Annatto	75120	E 160b	+	+	+	+ [2]	+	+	+	+	+	+	+	+	+	+	+	+ [12]	+	+	+
β-Carotene	75130	E 160a	+	+	+	+	+	+	+	+	+	+	+	+	+	+	+			+	+
β-Apo-8′-carotenal	40820	E 160e		+	+	+	+	+	+	+	+	+	+	+	+	+	+			+	+
Ethyl ester of β-apo8′ carotenoic acid	40825	E 160f		+	+	+	+	+	+	+	+	+	+	+	+	+	+			+	+
Canthaxanthin	40850	E 161g		+	+	+	+	+	+	+	+	+	+	+	+	+	+			+	+
Xanthophylls, other		E 161			+		+	+ [15]	+ [15]	+ [15]	+ [15]	+ [15]	+ [15]	+ [15]	+ [15]	+ [15]		+ [11]	+	+	

Red to purple shades																					
Anthocyanins		E 163	+	+	+	+	+	+	+	+	+	+	+	+	+	+		+[11]	+	+	+
Beetroot Red		E 162			+	+		+	+	+	+	+	+	+	+	+		+[11]	+	+	+
Yellows																					
Curcumin	75300	E 100	+	+	+	+	+	+	+	+	+	+	+	+	+	+	+	+[11]	+	+	+
Lactoflavin (riboflavin)		E 101	+	+[18]		+	+	+	+	+	+	+	+	+	+	+	+	+[11]	+	+	+
Greens																					
Chlorophyll	75810	E 140	+		+	+	+	+	+	+	+	+	+	+	+	+	+	+	+	+	+
Cu–chlorophyll	75810	E 141	+	+	+	+[1]		+	+	+	+	+	+	+	+	+					+
Brown																					
Caramel		E 150	+	+	+	+	+	+	+	+	+	+	+	+	+	+	+	+	+	+	+
Blacks																					
Carbon black																					
Vegetable		E 153		+	+	+		+	+	+	+	+	+	+	+	+	+[8]	+	+	+	+
Channel												+				+			+	+	
Inorganic colorings																					
Calcium carbonate					+[4]	+		+[4]	+	+[4]	+[4]	+	+[4]	+[4]	+	+	+[8]				
Iron oxides	77489																				
	77491																				
	77499	E 172		+				+[4]	+	+[4]	+[4]	+	+[4]	+[4]	+	+	+[8]	+[14]		+	
Titanium dioxide	77891	E 171		+		+		+[4]	+	+[4]	+[4]	+	+[4]	+[4]	+	+	+[8]			+	
Gold	77480	E 175			+[4]	+		+[4]	+	+[4]	+	+[9]	+[4]	+[4]	+	+[9]	+[8]				
Silver		E 174			+[4]	+		+[4]	+	+[4]	+	+[9]	+[4]	+[4]	+	+[9]	+[8]			+	
Aluminum	77000	E 173				+		+[4]	+	+[4]	+[4]	+[9]	+[4]	+[4]	+	+[9]	+[8]			+	

(Continued)

TABLE 9—*Continued*

Color	Cyprus	Egypt	Finland	Greece	Hong Kong	India	Indonesia	Iran	Israel	Japan	Kuwait	Malaya	Malta	New Zealand	Pakistan (West)	Papua New Guinea	Saudi Arabia	Singapore	South Africa	Spain	Thailand	Turkey	United States	Zambia
General																								
Vegetable colors																+								
Colors from edible fruits and vegetables	+	+			+						+	+		+			+	+	+			+		
Colors from animals																								
Reds																								
Alkannet	+	+	+	+	+		+	+	+		+	+		+			+	+				+	+ [8]	+
Carmine	+	+	+ [3]	+	+		+	+	+		+	+	+	+	+	+	+	+	+	+ [16]	+	+	+	+
Orchil	+			+	+			+		+	+	+					+	+		+ [16]		+		+
Red to yellow shades																								
Carotenoids[19]			+							+			+	+					+	+ [16]			+ [7]	
Annatto	+	+		+	+	+	+	+	+	+	+	+	+	+	+		+	+	+	+ [16]		+	+	+
β-Carotene	+	+	+	+	+	+	+	+	+	+		+	+		+		+	+	+	+ [16]	+	+	+	+
β-Apo-8′-carotenal	+	+		+		+			+			+	+						+	+ [16]	+		+	+
Ethyl ester of β-apo8′ carotenoic acid	+	+		+		+			+			+	+						+	+ [16]	+			+
Canthaxanthin	+	+		+		+			+	+		+	+						+	+ [16]	+		+	
Xanthophylls, other				+ [15]					+ [15]	+			+ [15]						+ [15]	+ [15,16]				+
Red to purple shades																								
Anthocyanins				+					+	+			+						+	+ [16]			+ [5]	
Beetroot Red				+					+	+			+						+	+ [16]			+ [6]	+
Yellows																								
Curcumin			+	+	+	+	+	+	+	+		+	+	+	+			+	+	+ [16]		+	+	+
Lactoflavin (riboflavin)				+		+			+	+			+		+				+	+ [16]	+	+	+	+
Greens																								
Chlorophyll	+	+	+	+	+	+	+	+	+	+	+	+	+	+	+	+	+	+	+	+ [16]	+	+	+ [8]	+
Cu-chlorophyll				+					+	+			+	+					+	+ [16]	+			+
Brown																								
Caramel	+	+	+	+	+	+	+	+	+	+	+	+	+	+	+	+	+	+	+	+ [16]		+	+	+

Blacks																									
Carbon black																									
Vegetable	+		+	+	+		+	+	+	+		+	+	+				+		+[16]		+			+
Channel	+				+		+	+	+			+	+	+				+							+
Inorganic colorings																									
Calcium carbonate				+[4]																					
Iron oxides																									
	+			+[4]	+		+	+	+	+[17]		+	+	+				+	+				+		+
Titanium dioxide	+	+	+	+[4]	+		+	+	+			+	+	+				+[4]	+	+[16]		+	+	+	+
Gold	+[9]											+[9]	+[9]						+[4]						
Silver	+[9]				+[9]				+			+[9]	+[9]	+				+[9]	+[4]						+
Aluminum	+[9]			+[4]	+[9]				+			+[9]	+[9]	+				+[9]	+[4]						+

[a] Courtesy of the British Industrial Biological Research Association.

[b] Not all foods may be colored even with natural colors. This table lists those colors that are generally permitted. For details of foods that may be colored and restrictions on the use of particular colors, the national legislation of the individual countries must be consulted. Plus sign indicated that a color is permitted.

[c] Only those colors that are permitted for use in more than one country are included. Paprika, turmeric, saffron, and sandalwood have not been included because they are not considered to be colors in several countries and may be used as spices in foodstuffs even though they have a coloring effect. For the purposes of this table, curcumin is taken to mean the coloring matter, not the spice turmeric. In some countries the natural colors may not be listed specifically but are permitted under the general heading of, for example, colors obtained from edible fruits and vegetables. This is indicated by the entries at the top of the table.

[d] The only "natural" colors in the Japanese permitted list are β-carotene, potassium and sodium norbixin, copper chlorophyll (chewing-gum base only), sodium copper chlorophyll (certain specified foods only), and sodium iron chlorophyll. This is because the list of permitted additives covers only those substances that are chemically synthesized. All substances obtained from natural sources by natural means are permitted, although they are not included in the additives list. The color carmine is *not* permitted in Japan because it is the aluminum lake of carminic acid and therefore not the natural form of the color. Carminic acid and cochineal may be used because they are considered to be wholly natural substances.

[e] Key to numbered superscripts: 1, max 200 mg/kg ionizable copper in copper chlorophyll; 2, color under review; 3, Ca and Al lakes of this color also permitted; 4, permitted only for surface coating of foodstuffs (West Germany and Greece: surface of sugar confectionery only); 5, only in the form of grape skin extract (enocianina); 6, as dehydrated beets only; 7, only in the form of carrot oil; 8, not generally listed but allowed in a few foods; 9, for external coloring of dragees and decoration of sugar-coated flour confectionery only (Malta: sugar-coated flour confectionery only); 10, listed specifically in Victoria only; 11, xanthophylls, but not individually specified, listed in Queensland, New South Wales only; 12, not western Australia; 13, New South Wales, named colors only; 14, in fish paste only (not western Australia); 15, specify flavoxanthin, lutein, cryptoxanthin, rubixanthin, violaxanthin, and rhodoxanthin only; 16, there is no general list of permitted natural colors, but those indicated in the table as being permitted are those listed in the regulations on specific foodstuffs; this is only a guide, since authorization is required for all additives for Spain; 17, sesquioxide only; 18, permitted only in dessert powders, max 20 ppm, of ready-to-eat product, mustard and vanilla sauce, max 20 ppm; 19, some countries specify certain carotenoids only.

V. CAROTENOIDS

There has been a very long history of practical experience with these colorants as a class, as well as more current chemical, biochemical, pharmacological, and toxicological investigations of single class members of high purity; thus, assurance of safety has been provided through a sequence of generalities and specifics. The carotenoids (see Appendix) currently marketed yield only tints of yellow to red hues. However, with the use of other colorants in the future, possibly crocin, zeaxanthan, astaxanthin, etc., and with a better understanding of carotenoid complexes, such as the carotenoproteins and other combinations, more colors of the rainbow may eventually be covered. Carotenoids possess high tinctorial properties. Although the carotenoids are oil soluble, emulsion and finely colloidal forms have been developed which disperse in water-base foods. Stability is good over the pH range of foods, and carotenoids are usually quite acceptable in food use under reducing conditions (and to light) in minimal oxygen atmospheres. Applications are best in selective product use. Oil solubility may be limiting in specific situations. Costs are sometimes comparable to those of other colorants, at times more economical, but in other cases more costly. Supplies are less limiting, and wide international acceptance has been demonstrated. β-Carotene, β-apo-8′-carotenal, and canthaxanthin are approved food colorants in the United States (Fig. 1) and other countries (Tables 9 and 10). In the form of extracts and various application forms, paprika, annatto, and saffron have had long historical use, are in current use today in the coloring of various food products, and probably will continue to meet future needs where applicable.

Another group of plant extracts are those containing chlorophylls, the pigment of green vegetation, in various concentrations. Some are stabilized by the addition of copper (replacing part of the magnesium), and others are hydrolyzed and converted to water-soluble or water-dispersible forms. The two chlorophyll structures, chlorophyll *a* and chlorophyll *b* (one oxygen atom and two hydrogen atoms less

TABLE 10

International Status of Carotenoid Food Colors, 1979[a]

Country	β-Carotene	Apocarotenal	Canthaxanthin
Argentina	+	+	+
Australia	+	+	+
Austria	+	+	+
Brazil	+	+	+
Canada	+	+	+
Chile	+	+	+
Colombia	+	+	+

(Continued)

TABLE 10—*Continued*

Country	β-Carotene	Apocarotenal	Canthaxanthin
Costa Rica	+	+	+
Cyprus	+	+	+
Czechoslovakia	+	−	−
Democratic Republic of Germany	+	−	−
EEE member countries	+	+	+
El Salvador	+	+	+
Finland	+	+	+
Greece	+	+	+
Guatemala	+	+	+
Honduras	+	+	+
Hungary	+	−	−
India	+	+	+
Indonesia	+	−	−
Iran	+	−	−
Israel	+	+	+
Japan	+	−	−
Kenya	+	+	+
Libya	+	+	−
Malaysia	+	+	+
Malta	+	+	+
Mexico	+	+	+
New Zealand	+	+	+
Nicaragua	+	+	+
Norway	+	+	+
Panama	+	+	+
Paraguay	+	+	+
Peru	+	+	+
Philippines	+	+	−
Poland	+	−	−
Portugal	+	+	+
Romania	+	−	−
South Korea	+	−	−
Spain	+	+	+
Sweden	+	+	+
Switzerland	+	+	+
Taiwan	+	+	+
Thailand	+	+	+
Turkey	+	−	−
United States	+	+	+
U.S.S.R.	+	−	−
Union of South Africa	+	+	+
Uruguay	+	+	+
Venezuela	+	+	+
Yugoslavia	+	−	−
Zambia	+	+	+

[a] +, permitted; −, not permitted.

than in chlorophyll *a*), which can be chromatographically separated, exist in plants in a fairly constant ratio. The natural chlorophylls *a* and *b* have not been used extensively in foods as yet, and further technological development is necessary to overcome stability problems and to search out favored application techniques since this class of colorants is quite reactive. Furthermore, there are legal limitations as to their use in the United States. Nevertheless, these colorants have had a history of practical experience and could probably find wider use with time.

VI. ANTHOCYANINS

Another natural class of colorants are the anthocyanins, widely spread over the plant kingdom of fruits, flowers, and vegetables, both in free form and in complexes. These water-soluble compounds, orange-red to blue, are the coloring matter in grapes, berries (cranberries, blueberries, blackberries, etc.), red cabbage, roselle flowers, and others, sources of which serve for the preparation of extracts and concentrates used industrially as food colorants. Associated compounds, the flavonoids, are pale yellow. No single pure anthocyanin or flavonoid is available today as a food colorant, either by synthesis or purified from natural sources. The extracts and concentrates contain accompanying substances, such as the tannins and polyphenols, which contribute taste or flavor. Added substances, such as sulfur dioxide or maltodextrins, may be present, depending on the product form. The anthocyanins as tinctorially strong colorants function best below pH 4. Stability may be influenced by ascorbic acid, enzymes, metal ions, oxygen, sugar, formation of complexes, degradation products, etc. Degradation by heat and light occurs. As natural colors, anthocyanins have had a very long history of practical use, particularly in the coloring of wines, and will continue to be a useful class of colorants.

VII. BETALAINES

Because of current concerns about water-soluble red colors, attention has focused more strongly on the class of betalaines. Color hues are red (betacyanines) and yellow (betaxanthines). These might be considered nitrogen-containing anthocyanin-type structures and have been known since the mid-1800's. Betanin, the main colorant of this class in the red beet and also in poke berries, yields an attractive red color. Beetroot extracts are available as liquid extracts and concentrates and can be prepared as spray-dried powders for approved uses. Although pure crystalline betanin can be produced, it has little commerical value at the current stage of development. The colorant is relatively unaffected by pH, but other variables—humidity, UV light, oxygen, and temperature—can be influencing factors. It is best used in packaged products with little exposure to light, oxygen, and humidity, in minimally processed products, and in products of short shelf life or in a dry state. Again, this is a class of product with a long history of consumption in its natural food form, and color extracts will find use in the coloring of foods that favor its best properties.

VIII. LEGAL CONSIDERATIONS

Multinational companies need colorants with a wide variety of hues and with properties that are functional in a variety of situations. More importantly, however, they require those that have a substantial safety basis and that are widely approved by government bodies in countries where they can be marketed in food products. The lists of permitted colorants vary widely from country to country. Hence, this situation causes difficulties in the shipment of colored food products among countries. Currently, attention must be given to specific allowable coloring materials, to the purity of the compounds, and to a flexible adjustment of application procedures. There is a growing number of regulations concerning the use of food colorants around the world, with some guiding international groups trying to obtain uniformity so that international trade will confront fewer barriers. Until this is achieved, food companies will have to keep track of the changing regulations in individual countries in which they conduct business. Table 9, which was prepared in England, may serve as a general guide to permitted natural colors in a given country.

Weissler states that there is no guarantee that all the present U.S. FD&C certified colors (azo and triphenylmethane dyes) will continue to be available on the same basis as they are now and suggests that the prudent technologist consider the use of the approved colorants of plant origin that are exempt from batch-to-batch certification. A more sophisticated color technology is needed, and acknowledged at times, to overcome some of the disadvantages of the natural colors.

Uneasiness among consumers and new government trends in the regulation of food colors have encouraged food manufacturers to investigate natural-type colors and to become involved in developmental studies that will result in the adaptation of these natural-type colors for specific food products. These programs include the evaluation of synthetically produced colorants that are identical with compounds naturally occurring in foods, the rationale being that colors produced in pure form should logically enjoy the same preferential status as the compounds extracted from natural origin if safety has been demonstrated by currently approved testing criteria.

The food, pharmaceutical, and cosmetic industries, ever conscious of the annual and seasonal trend of consumer preferences, will continue to develop products that have the highest appeal to the purchasers of their products and, hence, should keep alert to changing color legislation around the world. Since this volume is devoted primarily to the technology of the carotenoids in multiple-use applications, a 1979 tabulation is presented (Table 10) on the international status of carotenoid food colors. As a further guide to natural colors, a bibliography has been included at the end of this chapter. For further citations on the carotenoids, literature references in the other chapters in this volume can be examined.

IX. APPENDIX

A. β-Apo-8′-carotenal[a]

Definition	
Class	Carotenoid
Code numbers	DFG L-Orange 8, EEC no. E 160e
Chemical names	β-Apo-2′-ψ-carotenal, L-β-apo-8′-carotenal, 8′-apo-β,ψ-caroten-8′-al
Chemical formula	$C_{30}H_{40}O$
Structural formula	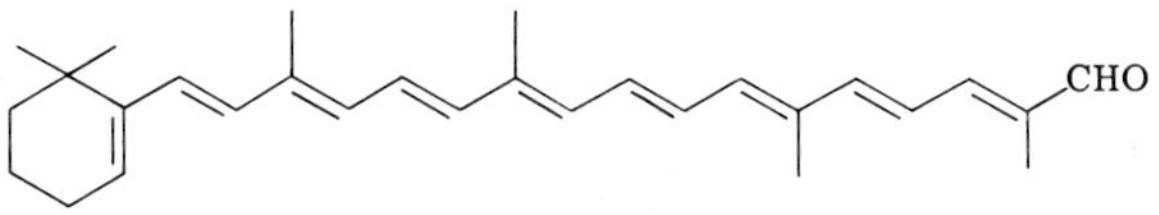
Molecular weight	416.65
Assay	β-Apo-8′-carotenal is predominantly the trans isomer and contains not less than 96% and not more than the equivalent of 101% of $C_{30}H_{40}O$ and conforms to the specifications given below
Description	Deep violet crystals with metallic luster; the articles of commerce may be solutions in oil, fat, or organic solvents, or water-dispersible forms such as powders or granules that are orange to red in color
Functional use	Food color
Characteristics	
Identifications tests	
Solubility: Insoluble in water; slightly soluble in ethanol; sparingly soluble in vegetable oils; freely soluble in chloroform	
Melting range: 136°–140°C[b]	
Determine the absorbance of sample solution B (see Method of assay) at 461 and 488 nm; the ratio A_{488}/A_{461} is between 0.80 and 0.84	
The color of a solution of apocarotenal in acetone disappears after the successive additions of a 5% solution of sodium nitrite and 1 *N* sulfuric acid	
A solution of apocarotenal in chloroform turns blue on addition of an excess of Carr–Price reagent T.S.	
Purity tests[b]	
Arsenic: Not more than 3 mg/kg	
Lead: Not more than 10 mg/kg	
Heavy metals: Not more than 20 mg/kg	
Sulfated ash: Not more than 0.1%	

(Continued)

IX. Appendix A—*Continued*

Method of assay[c,d]	Weigh accurately about 80 mg of the substance in a 100-ml volumetric flask and dissolve by shaking briefly with 20 ml pure, acid-free chloroform. Make sure that the solution is clear. Make up to volume by the addition of pure cyclohexane. Pipette 5.0 ml of the solution into a 100-ml volumetric flask and make to volume with the cyclohexane (solution A). Similarly, dilute 5.0 ml of this solution to 100 ml (solution B) and measure spectrophotometrically the extinction at the absorbance maximum (about 461 nm) against cyclohexane as a blank, using 1-cm cells. Calculate the content of apocarotenal using $E^{1\%}_{1\,cm} = 2640$. *For water-dispersible products* weigh accurately a quantity of the powder equivalent to about 10 mg free apocarotenal into a thick-walled glass tube. Add 10 ml distilled water, displace the air with nitrogen, and heat the tube on a steam bath at 60°C, stirring it from time to time with a glass stirring rod in such a way that the powder is entirely moistened and the particles do not stick to the walls of the tube. After about 10 min a homogeneous suspension of the powder should be obtained. Transfer the contents of the tube to a 250-ml volumetric flask and add 50 ml of absolute alcohol; part of the alcohol should be used to rinse the glass stirring rod and the tube. Add 100 ml of diethyl ether (free from peroxides), rinsing the stirring rod and test tube again, shake for 3 min, make up to volume with diethyl ether, and mix thoroughly. Let stand for 15 min in the absence of light so that the insoluble ingredients can precipitate. Pipette 5.0 ml of the above solution into a 100-ml round-bottom flask, replace the air with nitrogen, and evaporate to dryness at about 50°C in a rotary evaporator. After cooling, dissolve the residue in pure cyclohexane, transfer the solution quantitatively into a 100-ml volumetric flask, and adjust the volume with cyclohexane. Measure the extinction at the absorbance maximum (about 460 nm) against cyclohexane as a blank, using 1-cm

(*Continued*)

IX. Appendix A—*Continued*

	cells. Calculate the content of apocarotenal using $E^{1\%}_{1\,cm} = 2500$. This theoretical value corresponds to the proportion of isomers normally present in these water-dispersible products

[a] Supersedes the earlier specification for β-apo-8′-carotenal published in FAO Nutrition Meetings Report Series, No. 38B; WHO/Food Add./66.25 (1966).

[b] See General Methods.

[c] The assay should be carried out rapidly, avoiding exposure to air and light.

[d] In certain commercial products, such as water-dispersible powders, apocarotenal is finely dispersed in an appropriate medium, e.g., gelatin. In such preparations all the apocarotenal must be separated from the medium before the spectrophotometric assay is done, and a correction must be applied to take account of the difference in absorbance of the isomers.

B. β-Apo-8′-carotenoic Acid, Methyl or Ethyl Ester[a]

Definition	
Class	Carotenoid
Code numbers	DFG L-Orange 9, EEC No. E 160f
Chemical names	*trans*-β-Apo-8′-carotenoic acid, methyl or ethyl ester; methyl or ethyl 8′-apo-β,ψ-caroten-8′-oate
Chemical formula	$C_{31}H_{42}O_2$(methyl ester) or $C_{32}H_{44}O_2$ (ethyl ester)
Structural formula	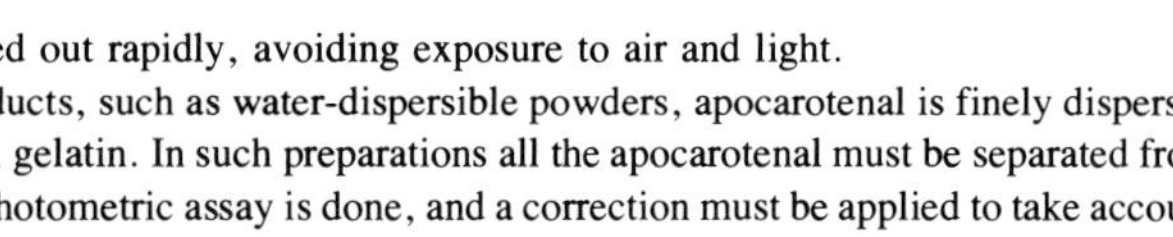 R = CH_3 or C_2H_5
Molecular weight	446.70 (methyl ester), 460.70 (ethyl ester)
Assay	β-Apo-8′-carotenoic acid, methyl or ethyl ester, is predominantly the trans isomer and contains not less than 96% and not more than the equivalent of 101% of $C_{31}H_{42}O_2$ or $C_{32}H_{44}O_2$, respectively, and both esters conform to the specifications given below; they may be further specified as to their methyl or ethyl ester content
Description	Red crystals; the articles of commerce may be solutions in oil, fat, or organic solvents, or water-dispersible forms, such as powders or granules, that are yellow to orange in color

(*Continued*)

IX. Appendix B—*Continued*

Functional use	Food color
Characteristics	
Identification tests	
Solubility: Insoluble in water; very slightly soluble in ethanol; slightly soluble in vegetable oils; freely soluble in chloroform	
Melting range: 134°–138°C[b]	
Determine the absorbance of sample solution B (see Method of assay) at 449 and 475 nm; the ratio A_{475}/A_{449} is between 0.82 and 0.86	
The color of a solution of apocarotenoic acid ester in acetone disappears after successive additions of a 5% solution of sodium nitrite and 1 *N* sulfuric acid	
A solution of apocarotenoic acid ester in chloroform turns blue on addition of an excess of Carr–Price reagent T.S.	
Purity tests[b]	
Arsenic: Not more than 3 mg/kg	
Lead: Not more than 10 mg/kg	
Heavy metals: Not more than 20 mg/kg	
Sulfated ash: Not more than 0.1%	
Method of assay[c,d]	Weigh accurately about 80 mg of the substance in a 100-ml volumetric flask and dissolve by shaking briefly with 20 ml pure, acid-free chloroform. Make sure that the solution is clear. Make up to volume by the addition of pure cyclohexane. Pipette 5.0 ml of the solution into a 100-ml volumetric flask and make to volume with the cyclohexane (solution A). Similarly, dilute 5.0 ml of this solution to 100 ml (solution B) and measure spectrophotometrically the extinction at the absorbance maximum (about 449 nm) against cyclohexane as a blank, using 1-cm cells. Calculate the content of apocarotenoic acid, methyl or ethyl ester, using $E^{1\%}_{1\,\text{cm}} = 2550$. *For water-dispersible products* weight accurately a quantity of the powder equivalent to about 10 mg free carotenoic acid ester into a thick-walled glass tube. Add 10 ml distilled water, displace the air with nitrogen and heat the tube on a steam bath at 60°C, stirring it from time to time with a

(*Continued*)

IX. Appendix B—*Continued*

	glass stirring rod in such a way that the powder is entirely moistened and the particles do not stick to the walls of the tube. After about 10 min a homogeneous suspension of the powder should be obtained. Transfer the contents of the tube to a 250-ml volumetric flask and add 50 ml of absolute alcohol; part of the alcohol should be used to rinse the glass stirring rod and the tube. Add 100 ml of diethyl ether (free of peroxides), rinsing the stirring rod and test tube again, shake for 3 min, make up to volume with diethyl ether, and mix thoroughly. Let stand for 15 min in the absence of light so that the insoluble ingredients can precipitate. Pipette 5.0 ml of the above solution into a 100-ml round-bottom flask, replace the air with nitrogen, and evaporate to dryness at about 50°C in a rotary evaporator. After cooling, dissolve the residue in pure cyclohexane, transfer the solution quantitatively into a 100-ml volumetric flask, and adjust the volume with cyclohexane. Measure the extinction at the absorbance maximum (about 449 nm) against cyclohexane as a blank, using 1-cm cells. Calculate the content of carotenoic acid ester using $E^{1\%}_{1\ \mathrm{cm}} = 2430$. This theoretical value corresponds to the proportion of isomers normally present in these water-dispersible products

[a] Supersedes the earlier specification for β-apo-8′-carotenoic acid, methyl or ethyl ester, published in FAO Nutrition Meetings Report Series, No. 38B; WHO/Food Add./66.25 (1966).

[b] See General Methods.

[c] The assay should be carried out rapidly, avoiding exposure to air and light.

[d] In certain commercial products, such as water-dispersible powders, β-apo-8′-carotenoic acid, methyl or ethyl ester, is finely dispersed in an appropriate medium, e.g. gelatin. In such preparations all the carotenoic acid ester must be separated from the medium before the spectrophotmetric assay is done, and a correction must be applied to take account of the difference in absorption of the isomers.

C. β-Carotene[a]

Definition	
Class	Carotenoid
Code numbers	DFG L-Orange 3, EEC no. E 160a

(*Continued*)

IX. Appendix C—*Continued*

Chemical name: β-Carotene (predominantly trans isomer)

Chemical formula: $C_{40}H_{56}$

Structural formula

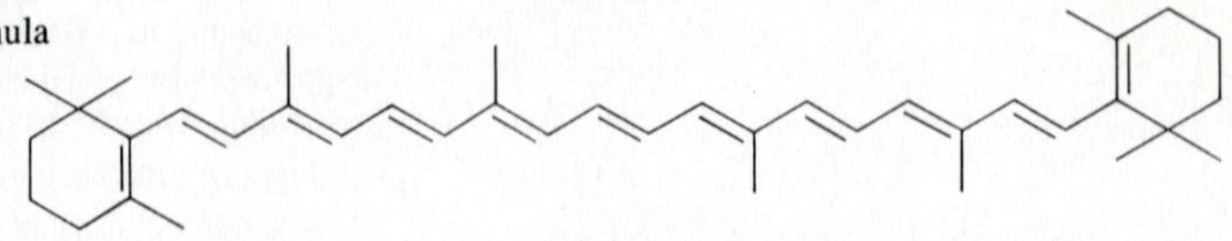

Molecular weight: 536.89

Assay: Synthetic β-carotene is predominantly the trans isomer and contains not less than 96% and not more than the equivalent of 101% of β-carotene ($C_{40}H_{56}$) and conforms to the specifications given below

Description: Red crystals or crystalline powder; the articles of commerce may be solutions in oil, fat, or organic solvents, or water dispersible forms, such as powders or granules

Functional use: Food color

Characteristics

Identification tests

Solubility: Insoluble in water; practically insoluble in ethanol; practically insoluble in methanol; slightly soluble in vegetable oil

Melting range: 178°–184°C with decomposition[b]

Determine the absorbance of sample solution B (see Method of assay) at 455 and 483 nm; the ratio A_{455}/A_{483} is between 1.14 and 1.19

Determine the absorbance of sample solution B at 455 nm and that of sample solution A at 340 nm; the ratio A_{455}/A_{340} is not lower than 1.5

Purity tests[b]

Arsenic: Not more than 3 mg/kg

Lead: Not more than 10 mg/kg

Heavy metals: Not more than 20 mg/kg

Sulfated ash: Not more than 0.2%

Method of assay[c,d]: Weigh accurately about 80 mg into a 100-ml volumetric flask and dissolve in 20 ml of pure, acid-free chloroform by swirling briefly. Check the solution for transparency against a bright surface; if necessary, complete the dissolution by allowing to stand for a short while. Dilute to volume with pure cyclohexane. Pipette 5.0 ml of this solution into a 100-ml volumetric flask and

(*Continued*)

IX. Appendix C—*Continued*

dilute to volume with cyclohexane (solution A). Dilute 5.0 ml of solution A in the same manner to 100 ml (solution B). With a spectrophotometer measure the extinction of solution B at 455–457 nm against cyclohexane as a blank, using 1-cm cells. Calculate the content of β-carotene on the basis of $E^{1\%}_{1\,\text{cm}}$ = 2500 at 455–457 nm.

For water-dispersible products weigh accurately a quantity of the powder equivalent to about 12 mg of free β-carotene into a thick-walled test tube. Add 10 ml distilled water, flush thoroughly with nitrogen, and heat the test tube in a water bath at 60°C, occasionally stirring with a glass rod. Care must be taken that the powder is thoroughly moistened and that no particles are adhering to the glass wall of the test tube. After approximately 10 min complete emulsion of the powder should be obtained. Transfer the contents of the test tube to a 250-ml volumetric flask and add 50 ml of absolute alcohol; use part of the alcohol to rinse the glass rod and the test tube. Add 100 ml of ethyl ether (peroxide-free), rinsing the glass rod and the test tube again, shake vigorously for about 3 min, dilute to volume with ethyl ether, and mix well. Allow to stand in the dark for about 15 min to allow the insoluble ingredients to precipitate and settle on the bottom. Pipette 5.0 ml of the supernatant solution into a round-bottom flask, flush with nitrogen, and evaporate in a rotary evaporator at about 50°C to dryness. After cooling, dissolve the residue in pure cyclohexane, transfer the solution quantitatively to a 100-ml volumetric flask, and dilute to volume with the same solvent. Add 4 μl of a 1% solution of iodine in absolute alcohol and allow to stand for 3 hr in the dark. With a spectrophotometer measure the extinction at the absorbance maximum (about 453 nm) against cyclohexane as a blank, using 1-cm cells. Calculate the content of β-carotene on the basis of $E^{1\%}_{1\,\text{cm}}$ = 2230

β-Carotene powders contain cis as well as trans isomers in slightly varying propor-

(*Continued*)

IX. Appendix C—*Continued*

	tions, with the trans form as the predominant component; by the addition of iodine an equilibrium is reached which has a maximum at about 453 nm with a defined $E^{1\%}_{1\,cm}$ value of 2230

[a] Supersedes the earlier specification for β-carotene published in FAO Nutrition Meetings Report Series, No. 38B; WHO/Food Add./66.25 (1966).

[b] See General Methods.

[c] Complete the determination promptly, avoiding exposure to air and light.

[d] In certain commercial products, such as water-dispersible powders, β-carotene is finely dispersed in an appropriate medium, e.g., gelatin. In such preparations all the β-carotene must be separated from the medium before the spectrophotometric assay is done, and iodine is added to establish a reproducible equilibrium between the trans and cis forms.

D. Canthaxanthin[a]

Definition	
Class	Carotenoid
Code numbers	DFG L-Orange 7, EEC no. E 161g
Chemical names	β-Carotene-4,4′-dione; *trans*-canthaxanthin; β,β-carotene-4,4′dione
Chemical formula	$C_{40}H_{52}O_2$
Structural formula	

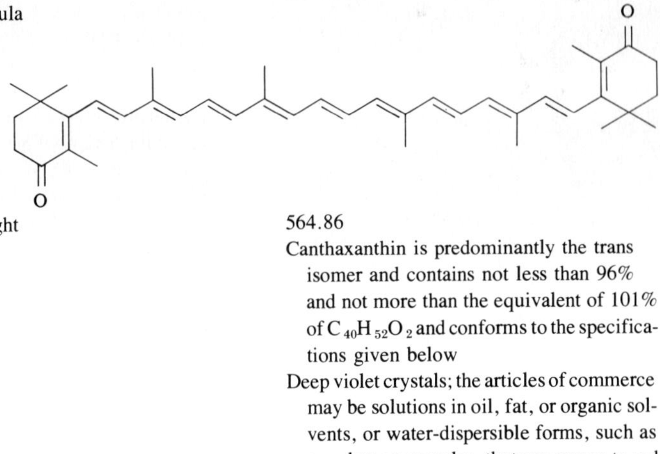

Molecular weight	564.86
Assay	Canthaxanthin is predominantly the trans isomer and contains not less than 96% and not more than the equivalent of 101% of $C_{40}H_{52}O_2$ and conforms to the specifications given below
Description	Deep violet crystals; the articles of commerce may be solutions in oil, fat, or organic solvents, or water-dispersible forms, such as powders or granules, that are orange to red in color
Functional use	Food color
Characteristics	
Identification tests	
Solubility: Insoluble in water; insoluble	

(*Continued*)

IX. Appendix D—*Continued*

in ethanol; practically insoluble in vegetable oils; soluble in chloroform	
Melting point: About 210°C with decomposition[b]	
A solution of canthaxanthin in cyclohexane has an absorbance maximum between 468 and 470 nm	
The color of a solution of canthaxanthin in acetone disappears after the successive additions of a 5% solution of sodium nitrite and 1 *N* sulfuric acid	
A solution of canthaxanthin in chloroform turns blue on addition of an excess of Carr–Price reagent T.S.	
Purity tests[b]	
Arsenic: Not more than 3 mg/kg	
Lead: Not more than 10 mg/kg	
Heavy metals: Not more than 20 mg/kg	
Sulfated ash: Not more than 0.1%	
Method of assay[c,d]	Weigh accurately approx. 10 mg of canthaxanthin in a 100-ml volumetric flask and dissolve by shaking briefly with 20 ml pure, acid-free chloroform. Make sure that the solution is clear. Make up to volume by the addition of pure cyclohexane. Pipette 5.0 ml of the solution into a 100-ml volumetric flask and make to volume with cyclohexane. Similarly, dilute 5.0 ml of this solution to 100 ml and measure spectrophotometrically the extinction at the absorbance maximum (about 470 nm) against cyclohexane as a blank, using 1-cm cells. Calculate the content of canthaxanthin using $E^{1\%}_{1\,\mathrm{cm}} = 2200$ for the pure compound. *For water dispersible products* weight accurately a quantity of the powder equivalent to about 12 mg free canthaxanthin into a thick-walled test tube. Add 10 ml distilled water (prewarmed to 60°C), flush thoroughly with nitrogen, and heat in a water bath at 60°C, stirring occasionally with a glass rod. Care must be taken that the powder is thoroughly moistened and that no particles are adhering to the wall of the test tube. After approx. 10 min, complete emulsion of the powder should be obtained in the form of a red, turbid liquid. Transfer

(*Continued*)

IX. Appendix D—*Continued*

the contents of the test tube quantitatively to a 250-ml volumetric flask with 10 ml absolute alcohol, dilute to volume with chloroform, and mix well. Filter part of the solution and immediately pipette 5.0 ml of the clear filtrate into a graduated 100-ml round-bottom flask. Flush with nitrogen and evaporate to dryness at 50°C, using a rotary vacuum evaporator. After cooling, moisten the residue with a few drops of absolute alcohol, add 50 ml pure cyclohexane, and completely dissolve by heating for a short time in a water bath at 50°C. Cool to room temperature and dilute to 100 ml with cyclohexane. With a spectrophotometer measure the extinction at the absorbance maximum (about 470 nm) against cyclohexane as a blank, using 1-cm cells. Calculate the content of canthaxanthin using $E^{1\%}_{1\,cm}$ = 1970. This theoretical value corresponds to the proportion of isomers normally present in these water-dispersible products.

[a] Supersedes the earlier specification for canthaxanthin published in FAO Nutrition Meetings Report Series, No. 38B; WHO/Food Add./66.25 (1966).

[b] See General Methods.

[c] Complete the determination promptly, avoiding exposure to air and light.

[d] In certain commercial products, such as water-dispersible powders, canthaxanthin is finely dispersed in an appropriate medium, e.g., gelatin. In such preparations all the canthaxanthin must be separated from the medium before the spectrophotometric assay is done, and a correction must be applied to take account of the difference in absorbance of the isomers.

BIBLIOGRAPHY

General

Allman, G. A. (1973). Color: The specialized approach for the food manufacturer. *Flavor Ind.* June, 256–269.

Apfelbaum, M., Custot, F., Manchon, P., Souverain, R., and Thomas, M. (1977). Les colorants alimentaires: Toujours inutiles, parfois dangereux. *Concours Med.* **99,** 236–240, 243–246.

Auslander, D. E., Goldberg, M., Hill, J. A., and Weiss, A. L. (1977). Naturally occurring colorants. *Drug Cosmet. Ind. Nov.,* 36–40, 105, 114; Dec., 55–60, 138–140.

Bayer, E., and Völter, N. (1965). Naturliche Farbstoffe. *In* "Handbuch der Lebensmittelchemie" (J. Schormüller, ed.), Vol. 1, p. 699. Springer-Verlag, Berlin and New York.

Bentley, K. W. (1960). "The Natural Pigments." Wiley (Interscience), New York.

Botma, Y. (1974). Food color control in 21 nations. *Food Eng.* **46**(5), 81–83.

Clydesdale, F. M., and Francis, F. J. (1976). Pigments. *In* "Principles of Food Sciences: Food Chemistry" (O. Fennema, ed.), pp. 385-425. Dekker, New York.
Damon, G. E., and Janssen, W. F. (1973). Additives for eye appeal. *FDA Consumer* **7**(6), 15-21.
Dinesen, N. (1974). Vegetable pigments: Toxicology. *Meet. Inst. Food Technol. 34th, New Orleans, La.*
Drake, J. J. P. (1975). Food colors: Harmless aesthetics or epicurean luxuries. *Toxicology* **5,** 3-42.
Druckrey, H. (1955). Schädliche und unschädliche farbstoffe fur lebensmittel. *Z. Krebsforsch.* **60,** 344-360.
Dufour, M., and Espriad, E. (1977). Les colorants naturels et l'industrie alimentaire. *Ind. Aliment. Agric.* **94,** 973-975.
Dupaigne, P. (1974). Les colorants rouges d'origine naturelle. *Fruits* **29,** 797-814.
FAO/WHO Expert Committee on Food Additives (1976). "Specifications for the Identity and Purity of Some Food Colors, Flavor Enhancers, Thickening Agents, and Certain Other Food Additives," WHO Food Additives Series, No. 7, pp. 4-78. World Health Organ., Geneva.
"Food Colors" (1971). Committee on Food Protection, Food and Nutrition Board, Division of Biology and Agriculture, National Research Council, National Academy of Sciences, Washington, D.C.
Francis, F. J. (1969). Color and pigment content in fruits and vegetables. *Food Technol.* **23**(1), 32-36.
Francis, F. J. (1970). Effect of color on acceptability of foods. *Proc. Int. Congr. Food Sci. Technol., 3rd* pp. 415-420.
Francis, F. J. (1974). Plant pigments. *In* "Encyclopedia of Food Technology" (M. S. Peterson and A. Johnson, eds.), pp. 694-699. Avi, Westport, Connecticut.
Goodwin, T. W. (1965). "Chemistry and Biochemistry of Plant Pigments." Academic Press, New York.
Goto, R. (1973). Natural colors for foods. *Kagaku Kogyo* **24,** 507-514.
Harborne, J. B. (1965). Chemistry of plant pigments *Nature (London)* **208,** 1051-1052.
Ikawa, F. (1970). Natural pigments and their use in food products. *Jpn. Food Sci.* **9**(2), 34-47.
Kläui, H. (1977). Färbung mit Naturfarbstoffen. "Probleme um neue Lebensmittel. Probleme der Ernährungs und Lebensmittelwissenschaft" (R. Harmer, W. Auerswald, N. Zöllner, B. Blanc, and B. Brandstetter, eds.), Vol. 5, pp. 331-351. Wilhelm Maudrich Verlag, Vienna.
König, R. (1965). Zur Färbung von Lebensmitteln verwendete Farbstoffe. *In* "Handbuch der Lebensmittelchemie" (J. Schormüller, ed.), Vol. 1. Springer-Verlag, Berlin and New York.
Kramer, A. (1978). Benefits and risks of color additives. *Food Technol.* **32**(8), 65-67.
Land, D. G. (1962). Stability of plant pigments. *In* "Recent Advances in Food Research" (J. Hawthorn and J. M. Leitch, eds.), pp. 50-66. Butterworth, London.
Little, A. C., and Mackinney, G. (1972). The color of foods. *World Rev. Nutr. Diet.* **14,** 59-84.
McClelland, C. W. (1977). Economic constraints on the use of naturally derived food colors. *In* "Current Aspects of Food Colorants" (T. E. Furia, ed.), pp. 67-76. CRC Press, Cleveland, Ohio.
Mackinney, G. (1961). The coloring matter of the orange. *In* "The Orange" (W. B. Sinclair, ed.), pp. 302-333. Univ. of California Press, Berkeley.
Mackinney, G., and Chichester, C. O. (1954). The color problem in foods. *Adv. Food Res.* **5,** 301-351.
Mackinney, G., and Little, A. C. (1972). The coloring matter of food. *World Rev. Nutr. Diet.* **14,** 85-99.
Manz, U. (1980). Farbstoffe für lebensmittel und Kosmetika. *In* "Schweizerisches Lebensmittelbuch." In preparation.
Marmion, D. M. (1970). Description and analysis of colors for foods, drugs, and cosmetics. *Encycl. Ind. Chem. Anal.* **10,** 447-547.
Mikkelsen, H., Larson, J. C., and Tarding, F. (1978). Hypersensitivity reactions to food colors with special reference to the natural color annatto extract (butter color). *Arch. Toxicol., Suppl.* No. 1, 141-143.
Noonan, J. (1968). Color additives in food. *In* "Handbook of Food Additives" (T. E. Furia, ed.), pp. 25-49. CRC Press, Cleveland, Ohio.

Ono, M. (1969). Food manufacture and natural pigments. *Kagaku To Kogyo (Osaka)* **43,** 623–629.
Packard, V. S., Jr. (1977). Food color. *In* "Processed Foods and the Consumer," pp. 79–96. Univ. of Minnesota Press, Minneapolis.
Phillip, T. (1975). Unilization of plant pigments as food colorants. *Food Prod. Dev.* **9**(3), 50–56.
Presgrave, J. E. (1971). Coloring of food and drinks. *Process Biochem.* Dec., 29–30, 40.
Riboh, M. (1977). Natural colors, what works . . . what doesn't. *Food Eng.* May, 66–68.
Richardson, G. (1973). Coloring matters: Their addition to different types of confectionery. *Manuf. Confect.* **53**(10), 34–41.
Schaffner, R. M. (1977). Regulations and future activities of food and drug administration for regulating colors. *In* "Current Aspects of Food Colorants" (J. E. Furia, ed.), pp. 85–90. CRC Press, Cleveland, Ohio.
Shimano, Y. (1972). Natural coloring agents in the future. *Shokuhin Kogyo* **15**(14), 52–59.
Somogyi, L. P. (1977). "Food Grade Colorants and Flavorings," Stanford Research Institute International Publication. Menlo Park, California.
Soukup, R. J., and Young-Maing, I. Y. (1977), The color needs of the food industry. *In* "Current Aspects of Food Colorants" (T. E. Furia, ed.), pp. 77–84. CRC Press, Cleveland, Ohio.
Specifications on Colors Resulting from Activities of the Joint FAO/WHO Expert Committee in Food Additives:
Specifications for Identity and Purity of Food Additives (Food Colors) (1963). (Fourth Report of the Expert Committee.) These specifications were subsequently revised and published as "Specifications for Identity and Purity of Food Additives, Vol. II, Food Colors." Food and Agriculture Organization of the United Nations, Rome (out of print).
Specifications for the Identity and Purity of Food Additives and Their Toxicological Evaluation (1965). "Food Colors and Some Antimicrobials and Antioxidants (Eighth Report of the Expert Committee). "FAO Nutrition Meetings Report Series, No. 38; WHO Technical Report Series, No. 309 (out of print).
Specifications for the Identity and Purity of Food Additives and Their Toxicological Evaluation (1970). "Some Food Colors, Emulsifiers, Stabilizers, Anticaking Agents, and Certain Other Substances (Thirteenth Report of the Expert Committee)." FAO Nutrition Meetings Report Series, No. 46; WHO Technical Report Series, No. 445.
Specifications for the Identity and Purity of Some Food Colors, Emulsifiers, Stabilizers, Anticaking Agents, and Certain Other Food Additives (1970). FAO Nutrition Meetings Report Series, No. 46B; WHO/Food Add/70.37.
Specifications for the Identity and Purity of Some Food Colors, Flavor Enhancers, Thickening Agents, and Certain Food Additives (1975). FAO Nutrition Meetings Report Series, No. 54B; WHO Food Additives Series, No. 7.
Specifications for Identity and Purity of Some Food Additives, Including Antioxidants, Food Colors, Thickeners, and Others (1977). FAO Nutrition Meetings Report Series, No. 57.
Specifications Resulting from the Twenty-second Report of the Joint FAO/WHO Expert Committee on Food Additives (1979). Food and Nutrition Paper, No. 7.
Thomas, B. R. J. (1959). The technology of coloring agents. *Chem. Ind. (London)* No. 15, 468–470.
Weissler, A. (1975). Regulation of food colors. *Food Technol.* **29**(5), 46.
Winston, R. P. (1965). Food colors. *In* "A Symposium on Chemical Additives in Food" (R. W. L. Goodwin, ed.), pp. 115–123. Churchill, London.
Yoshizumi, T. (1972). Natural coloring agents in the future: Properties and uses. *Shokuhin Kogyo* **15**(14), 20–27.

Carotenoids

Bauernfeind, J. C. (1972). Carotenoid vitamin A precursors and analogs in foods and feeds. *J. Agric. Food Chem.* **20,** 456–472.

Bauernfeind, J. C. (1974). Carotenoids and food technology. *In* "Encyclopedia of Food Technology" (A. Johnson and M. Peterson, eds.), pp. 163–173. Avi, Westport, Connecticut.

Bauernfeind, J. C., Rubin, S. H., Surmatis, J. D., and Ofner, A. (1970). Carotenoids and fat-soluble vitamins: Contribution to food, feed and pharmaceuticals. *Int. Z. Vitaminforsch.* **40,** 391–416.

Isler, O. (1971). "Carotenoids." Birkhaeuser, Basel.

Isler, O., Montavan, M., and Rüegg, R. (1966). Les carotenoides, colorants alimentaires naturrels. *In* "Chimie et Civilisation." Compte-Rendu du 36e. Congres International de Chimie Industrielle, Bruxelles; *Ind. Chim. Belge* **32,** 853–856 (1967).

Kläui, H. (1968). The use of synthetic carotenoids for coloring of foods. Proc. Special Color Symposium (U.K.). *Inst. Food Sci. Technol.* **1**(5), 1–8.

Kläui, H., and Münzel, K. (1965). Color coating drugs by carotenoids. *Pharm. Acta Helv.* **40,** 153–157.

Kläui, H., and Raunhardt, O. (1976), Coloring of foods with carotenoids. *Alimenta* **15,** 37–45.

Kläui, H., Capeder, A., and Münzel, K. (1963). Water-soluble carotenoids in tablet coating. *Pharm. Ind.* **25,** 173–178.

Manz, U., and Isler, O. (1976). Loebensmittel-Farbstoffe. *In* "Methodicum Chimicum" (F. Korte, H. Aebi, E. Baumgartner, H. F. Fiedler, and G. Ohloff, eds.), Vol. 10. Thieme, Stuttgart.

Anthocyanins

Adams, J. B., and Ongley, M. H. (1973). The degradation of anthocyanins in canned strawberries. *J. Food Technol.* **8,** 305–307.

Adams, J. P. (1973). Thermal degradation of anthocyanins with particular reference to the 3-glycosides of cyanindin: In acidified aqueous solution at 100°C. *J. Sci. Food Agric.* **24,** 747–762.

Akiyoshi, M., Webb, A. D., and Kepner, R. E. (1963). The major anthocyanin pigments of *Vitis vinifera* varieties. *J. Food Sci.* **28,** 177–181.

Asen, S., Stewart, R. N., and Norris, K. H. (1972). Copigmentation of anthocyanins in plant tissues and its effect on color. *Phytochemistry* **11,** 1139–1144.

Baker, C. H., Johnston, M. R., and Barber, W. D. (1974). Cherry plum pigment as a natural red colorant. *Food Prod. Dev.* **8**(3), 83.

Ballinger, W. E., Maness, E. P., and Kushman, L. J. (1970). Anthocyanins in ripe fruit of the high bush blueberry, *Vaccinium corymbosum*. *J. Am. Soc. Hortic. Sci.* **95**(3), 283–285.

Bate-Smith, E. C. (1954). Flavonoid compounds in foods. *Adv. Food Res.* **5,** 161–259.

Benk, E., Bergmann, R., and Ballendat, W. (1975). Characteristics and use of Hibiscus flowers. *Mineralbrunner* **24**(9), 300–301.

Berg, H. W., and Akiyoski, M. (1962), Color behavior during fermentation and aging of wines. *J. Am. Enol. Vitic.* **13,** 126–132.

Bourzeix, M., Dubernet, M. O., and Heredia, N. (1975). Sur l'extraction de divers constituants phenoliques des raisins et de leurs rafles. *Ind. Aliment. Agri.* **92,** 1057–1064.

Bridle, P., Stott, K. G., and Timberlake, C. F. (1973). Anthocyanins in Salix species: New anthocyanin in Salix purpurea bark. *Phytochemistry* **12,** 1103–1106.

Buckmire, R. E., 1975. Evaluation of miracle fruit (*Synsepalum dulcificum schum*) pigments as possible food colorants. *Dissert. Abstr. Int.* **35**(9) 4494. Order No. 75-5997.

Buckmire, R. E., and Francis, F. J. (1976). Anthocyanins and flavonols of miracle fruit, *synsepalum dulcificum schum*. *J. Food Sci.* **41,** 1363–1365.

Calvi, J. P., and Francis, F. J. (1978). Stability of concord grape (*V. lambrusca*) anthocyanins in model systems. *J. Food Sci.* **43,** 1448–1456.

Chandler, B. V. (1958), Anthocyanins of blood oranges. *Nature (London)* **182,** 993.

Chandler, B. V., and Clegg, K. M. (1970). Pink discoloration in canned pears. Role of tin in pigment formation. *J. Sci. Food Agric.* **21,** 315–319.

Chiriboga, C., and Francis, F. J. (1970). A recovery system for anthocyanins from cranberry pomace. *J. Am. Soc. Hortic. Sci.* **95,** 233–236.

Chiriboga, C. D., and Francis, F. J. (1973). Ion-exchange purified anthocyanin pigment as a colorant for cranberry juice cocktail. *J. Food Sci.* **38,** 464–467.

Clydesdale, F. M., Main, J. H., and Francis, F. J. (1979). Effect of anthocyanin preparation as colorants on the hygroscopicity of dry pack food products. *J. Food Prot.* **42**(3), 225–227.

Clydesdale, F. M., Main, J. H., and Francis, F. J. (1978). Roselle anthocyanins as colorants for beverages and gelatin desserts. *J. Food Prot.* **42**(3), 204–207.

Clydesdale, F. M., Main, J. H., and Francis, F. J. (1978). Cranberry pigments as colorants in selected food products. *J. Food Prot.* **42**(3), 196–201.

Clydesdale, F. M., Main, J. H., Francis, F. J., and Damon, R. H. (1978). Concord grape pigments as colorants for beverages and desserts. *J. Food Sci.* **43,** 1687–1692.

Co, H., and Markakis, P. (1968). Flavanoid compounds in the strawberry fruit. *J. Food Sci.* **33,** 281–283.

Dekazas, E. D. (1970). Anthocyanin pigments in red tart cherries. *J. Food Sci.* **35,** 237, 242.

Diaz, L. S., Gasque, F., and LaFuente, B. (1976). Stability of anthocyanin pigments in both vinifera and hybrid red grape juice. *Rev. Agroquim. Tecnol. Aliment.* **16,** 509–515.

Du, C. T., and Francis, F. J. (1973). Anthocyanins of roselle, *Hibiscus sabdariffa J. Food Sci.* **38,** 810–812.

Du, C. T., and Francis, F. J. (1975). Anthocyanins of ornamental cherries. *J. Food Sci.* **40,** 1142–1144.

Du, C. T., Wang, P. L., and Francis, F. J. (1975). Anthocyanins of pomgranate, *Punica granatum J. Food Sci.* **40,** 417–418.

Erlandson, J. A., and Wrolstad, R. E. (1972). Degradation of anthocyanins at limited water concentration. *J. Food Sci.* **37,** 592–595.

Esslen, W. B., and Sammy, G. M. (1973). Roselle: A natural red colorant for foods. *Food Prod. Dev.* **7**(1), 80–82.

Esselen, W. B., and Sammy, G. M. (1975). Applications for roselle as a red food colorant. *Food Prod. Dev.* **9**(8), 37–38, 40.

Forsyth, W. G. C., and Simmonds, N. W. (1954). A survey of anthocyanins of some tropical plants. *Proc. R. Soc. London, Ser. B* **142,** 549–564.

Francis, F. J. (1957). Color and pigment measurement in fresh cranberries. *Proc. Am. Hortic. Sci.* **69,** 296–301.

Francis, F. J. (1972). Anthocyanins of Durham and Heritage raspberries. *Hortic. Sci.* **7,** 398.

Francis, F. J. (1975). Anthocyanins as food colors. *Food Technol.* **29**(5), 52, 54.

Francis, F. J. (1977). Anthocyanins. *In* "Current Aspects of Food Colorants" (T. E. Furia, ed.), pp. 19–28. CRC Press, Cleveland, Ohio.

Francis, F. J. (1978). Natural food colors. *In* "Yearbook of Science and Technology," pp. 176–177. McGraw-Hill, New York.

Francis, F. J., and Harborne, J. B. (1966). Anthocyanins of the garden huckleberry, *Solanum guineese. J. Food Sci.* **31,** 524–528.

Francis, F. J., and Servadio, G. J. (1963). Relation between color of cranberries and color and stability of juice. *Proc. Am. Soc. Hortic. Sci.* **83,** 406–415.

Francis, F. J., Harborne, J. B., and Barker, W. G. (1966). Anthocyanins of the lowbush blueberry, *Vaccinium angustifolium. J. Food Sci.* **31,** 583–587.

Fuleki, T., and Francis, F. J. (1968). Determination of individual anthocyanins in cranberry and cranberry products. *J. Food Sci.* **33,** 471–478.

Harborne, J. B. (1965). Flavonoids: Distribution and contribution to plant color. *In* "Chemistry and Biochemistry of Plant Pigments" (T. W. Goodwin, ed.), pp. 247–278. Academic Press, New York.

Harborne, J. B. (1967). The anthocyanin pigments. *In* "Comparative Biochemistry of the Flavinoids," pp. 1–30. Academic Press, New York.

Harper, K. A. (1957). Anthocyanin pigments in foods. *CSIRO Food Preserv. Q.* **17**(4), 68–72.

Herrmann, K. (1972). The occurrence and importance of anthocyanins in food. *Z. Lebensm.-Unters. -Forsch.* **148,** 290–302.

Horowitz, R. M. (1961). The citrus flavonoids. *In* "The Orange" (W. B. Sinclair, ed.), pp. 334–372. Univ. of California Press, Berkeley.

Hrazdina, G. (1974). Reactions of anthocyanins in food products. *Lebensm.-Wiss. Technol.* **7,** 193–198.

Hrazdina, G., and Franzese, J. (1974). Structure and properties of the acylated anthocyanins from Vitis species. *Phytochemistry* **12,** 2005; **13,** 225–229.

Hrazdina, G. H., Borzell, A. J., and Robinson, W. E. (1970). Stability of the anthocyanidin 3,5-diglucosides. *Am. J. Enol. Vitic.* **21,** 201–204.

Huang, H. T. (1956). The kinetic of decolorization of anthocyanins by fungal "anthocyanase." *J. Am. Chem. Soc.* **78,** 2390–2393.

Ingalsbe, D. W., Neubert, A. M., and Carter, G. H. (1963). Colloidal blue pigments of juice. *J. Agric. Food Chem.* **11,** 263–266.

Jurd, L. (1964). Reactions involved in sulfite bleaching of anthocyanins. *J. Food Sci.* **29,** 16–19.

Jurd, L. (1964). Some benzopyrilium compounds potentially useful as color additives for fruit drinks and juices. *Food Technol.* **18**(4), 157.

Jurd, L. (1969). Polphenol condensation reactions and their possible occurrence in the aging of wines. *Am. J. Envol. Vitic.* **20,** 191–195.

Jurd, J. (1972). Some advances in the chemistry of anthocyanin-type plant pigments. *In* "Chemistry of Plant Pigments" (C. O. Chichester, ed.), pp. 123–187. Academic Press, New York.

Kulesza, J., and Wilska-Jeszka, J. (1967). Usefulness of anthocyanin dyes for coloring dessert concentrates. *Przem. Spozyw.* **21**(1), 7–8.

Li, K. C., and Wagenknecht, A. C. (1956). Anthocyanin pigments of sour cherries. *J. Am. Chem. Soc.* **78,** 979–980.

Lukton, A., Chichester, C. O., and Mackinney, G. (1956). The breakdown of strawberry anthocyanin pigment. *Food Technol.* **10,** 427–431.

McLellan, M. R., and Cash, J. N. (1978). Application of anthocyanins as colorants for Maraschino-type cherries. *J. Food Sci.* **44**(2), 483–487.

Main, J. H., Clydesdale, F. M., and Francis, F. J. (1978). Spray dried anthocyanins as food colorants. *J. Food Sci.* **43,** 169–1694, 1697.

Markakis, P. (1974). Anthocyanins and their stability in foods. *Crit. Rev. Food Technol.* **4,** 437–456.

Markakis, P., Livingston, G. E., and Fellers, C. R. (1957). Quantitative aspects of strawberry pigment degradation. *Food Res.* **22,** 117–130.

Meschter, E. E. (1953). Effects of carbohydrates and other factors on strawberry products. *J. Agric. Food Chem.* **1,** 574–579.

Morton, J. F. (1974). Renewed interest in roselle, the forgotten "Florida cranberry." *Proc. Fla. State Hortic. Soc.* **87,** 415–425.

Mourgues, J. (1977). Modification des matieres colorantes et de autres constituents des piquettes de marc, des vins rouges et des lies au cours de la distillation. *Ind. Aliment. Agric.* **94,** 273–276.

Nedelchev, N. (1975). Preparation of a colorant concentrate for food industry purposes. *Khranit. Prom.* **24**(4), 10–14.

Palamidis, N., and Markakis, P. (1975). Stability of grape anthocyanin in a carbonated beverage. *J. Food Sci.* **40,** 1047–1049.

Pankajamani, K. S., and Seshadri, T. R. (1955). Anthocyanins. *J. Sci. Ind. Res., Sect. B.* **14,** 93–98.

Patane, G. (1948). Anthocyanin extraction from oranges. *Ann. Regia Stn. Sper. Fruitticolt. Agrumicolt.* **17,** 1–12.

Peng, C. Y., and Markakis, P. (1963). Effect of phenolase on anthocyanins. *Nature (London)* **199,** 597.

Peri, G., and Bonini, V. (1976). Recovery of anthocyanins from wine distillation wastes. *J. Food Technol.* **11**(3), 283–295.

Philip, T. (1974). Anthocyanin recovery system from grape wastes. *J. Food Sci.* **39,** 859.

Pratt, D. E., Balkcom, C. M., Powers, J. J., and Mills, L. W. (1954). Fruit color stability: Interaction of ascorbic acid, riboflavin and anthocyanin pigments. *J. Agric. Food Chem.* **2,** 367–372.

Ribereau-Gayon, P. (1964). Phenolic compounds of grapes and wine. *Ann. Physiol. Veg.* **6,** 119–139.

Robinson, W. B., Weirs, L. D., Bertino, J. J., and Mattick, L. R. (1966). The relation of anthocyanin composition to color stability of New York State wines. *Am. J. Enol. Vitic.* **17,** 178–184.

Sakamura, S., and Francis, F. J. (1961). The anthocyanins of the American cranberry. *J. Food Sci.* **26,** 318–321.

Salt, F. W., and Thomas, J. G. (1957). The anaerobic corrosion of tin in anthocyanin solutions and fruit syrups. *J. Appl. Chem.* **7,** 231–238.

Sastry, L. V. L., and Tischer, R. G. (1952). Stability of the anthocyanin pigments in Concord grape juice. *Food Technol.* **6,** 264–268.

Scheffeldt, P., and Hrazdina, G. (1978). Copigmentation of anthocyanins under physiological conditions. *J. Food Sci.* **43,** 517–520.

Shewfelt, R. L., and Ahmed, E. M. (1977). Anthocyanin extracted from red cabbage shows promise as coloring for dry beverage mixtures. *Food Prod. Dev.* **11**(4), 52, 57–59, 62.

Shewfelt, R. L., and Ahmed, E. M. (1978). Enhancement of powdered soft drink mixes with anthocyanin extracts. *J. Food Sci.* **43,** 435–438.

Servadio, G. J., and Francis, F. J. (1963). Relation between color of cranberries and color and stability of sauce. *Food Technol.* **17**(5), 124–128.

Shrikhandle, A. J. (1976). Anthocyanins in foods. *Crit. Rev. Food Sci. Nutr.* **7,** 193–218.

Sistrunk, W. A., and Cash, J. N. (1970). Effect of certain chemicals on the color and polysaccharides of strawberry puree. *J. Food Technol.* **24,** 473–477.

Skalski, C., and Sistrunk, W. A. (1973). Factors influencing color degradation in Concord juice. *J. Food Sci.* **38,** 1060–1062.

Smith, B. J. (1977). Anthocyanins: The chamelons of food coloring. *Food Trade Rev.* **47**(1), 16–18.

Somaatmadja, D., and Powers, J. J. (963). Anthocyanins: Anthocyanin pigments of Cabernet Sauvignon grapes. *J. Food Sci.* **28,** 617–622.

Somers, T. C. (1971). Polymeric nature of wine pigments. *Phytochemistry* **10,** 2175–2186.

Sondheimer, E., and Kertesz, Z. I. (1953). Participation of ascorbic acid in the destruction of anthocyanin in strawberry juice and model systems. *Food Res.* **18**(5), 475–479.

Starr, M. S., and Francis, F. J. (1968). Oxygen and ascorbic acid effect on the relative stability of four anthocyanin pigments in cranberry juice. *Food Technol.* **22**(12), 91–93.

Starr, M. S., and Francis, F. J. (1973). Effect of metallic ions on color and pigment content of cranberry juice cocktail. *J. Food Sci.* **38,** 1043–1046.

Swain, T. (1965). Nature and properties of flavonoids. *In* "Chemistry and Biochemistry of Plant Pigments" (T. W. Goodwin, ed.), pp. 211–245. Academic Press, New York.

Tanchev, S. S., and Timberlake, C. F. (1969). The anthocyanins of red cabbage, *Brassica oleracea*. *Phytochemistry* **8,** 1825–1827.

Timberlake, C. F., and Bridle, P. (1968). Flavylium salts resistant to sulphur dioxide. *Chem. Ind. (London)* **43,** p. 1489.

Usova, E. M., Red'ko, E. G., and Dubkova, L. P. (1968). Use of pigments from pressed cranberry skins in the manufacture of nonalcoholic beverages. *Izv. Vyssh. Uchebn. Zaved., Pishch. Tekhnol.* **5,** 105–106.

Van Buren, P. J., Bertino, J. J., and Robinson, W. B. (1968). The stability of wine anthocyanins on exposure to heat and light. *Am. J. Enol. Vitic.* **19**(3), 147–154.

Van Buren, J. P., Hrazdina, G., and Robinson, W. B. (1974). Color of anthocyanin solutions expressed in lightness and chromaticity: Effect of pH and type of anthocyanin. *J. Food. Sci.* **39,** 325–328.

Volpe, T. (1976). Cranberry juice concentrate as a red food coloring. *Food Prod. Dev.* **10**(9), 13–14.

Wallin, B. K., and Smith, B. J. (1977). Grape anthocyanins as food colorings: Sources compared. *Int. Flavors Food Addit.* **8**(3), 102–105.

Zamorani, A., and Pifferi, P. G. (1966). Natural pigments: Determination of the composition of some anthocyanins. *Bull. Sci. Fac. Chim Ind. (Bologna)* **24**(1), 31–40.

Zapsalis, C., and Francis, F. J. (1965). Cranberry anthocyanins. *J. Food Sci.* **30,** 396–399.

Betalaines

Adams, J. P., and von Elbe, J. H. (1976). Production of a betacyanine concentrate by fermentation of red beet juice with *Candida utilis*. *J. Food Sci.* **41,** 78–81.

Aurstad, K., and Dahle, H. K. (1973). The effect of heat treatment, UV and gamma radiation on some beet root pigments. *Z. Lebensm.-Unters-Forsch.* **151,** 171–174.

Bayer, E. (1966). Komplexbildung und Blütenfarben. *Angew. Chem.* **78,** 834–841.

Bhillon, A. S., and Maurer, A. J. (1975). Evaluation of betalain pigments as colorants in turkey summer sausages. *Poult. Sci.* **54,** 1272–1277.

Driver, M. W., and Francis, F. J. (1979a). Stability of phytolaccin, betanin, and FD and C red No. 2 in dessert gels. *J. Food. Sci.* **44**(2), 518–520.

Driver, M. W., and Francis, F. J. (1970b). Purification of phytoloccanin (betanin) by removal of phytoloccatoxin from phytolacca americana. *J. Food Sci.* **44**(2), 521–523.

Georgakis, S. G., and Vareitzis, K. (1976). Verwendung von betanin an stelle von natriumnitrit für die herstellung von fleischkonserven. *Fleischwirtschaft* **56,** 1153–1154.

Kimler, L., Mears, J., Mabry, T. J., and Roesler, L. (1970). Mutual exclusiveness of betalains and anthocyanins. *Taxon* **19**, 875–878.

Köhler, K. H. (1973). Die betalaine, ein neue klasse chymochromer pigmente. *Pharmazie* **28,** 18–24.

Kopelman, I. J., and Saguy, I. (1977). Drum dried beet powder, *J. Food Technol.* **12,** 615–621.

Lempka, A. (1970). Stability of betanin in acid solutions. *Pr. Zakresu Towarozn. Chem., Wyzsza Szk. Ekon. Poznaniu, Zesz. Nauk., Ser. 1* No. 36, 121–128.

Mabry, T. J. (1966). The betacyanins and betaxanthins. *In* "Comparative Phytochemistry" (T. Swain, ed.), pp. 231–244. Academic Press, New York.

Mabry, T. J. (1970). Betalains, red-violet and yellow-alkaloids of the centrospermae. *In* "Chemistry of the Akaloids" (S. W. Pelletier, ed.), pp. 367–384. Van Nostrand-Reinhold, New York.

Mabry, T. J., and Dreiding, A. S. (1968). The betalains. *In* "Recent Advances in Phytochemistry" (T. J. Mabry, ed.), Vol. 1, pp. 145–160. North-Holland Publ., Amsterdam.

Nilsson, T. (1970). Studies into the pigments in beetroot. *Lantbrukshoegsk. Ann.* **36,** 179–219.

Nilsson, T. (1973). The pigment content of beetroot with regard to cultivar, growth, development and growing conditions. *Swed. J. Agric. Res.* **3,** 187–200.

Pasch, J. H., and von Elbe, J. H. (1975). Betanine degradation as influenced by water activity. *J. Food Sci.* **40,** 1145–1146.

Pasch, J. H., and von Elbe, J. H. (1977). Betalaines as confectionary colorants. *Candy Snack Ind.* **142**(3), 32–35.

Pasch, J. H., von Elbe, J. H., and Dinesen, N. (1975). Betalaines as natural food colorants. *Food Prod. Dev.* **9**(9), 38–40.

Pasch, J. H., von Elbe, J. H., and Sell, R. J. (1975). Betalains as colorants in dairy products. *J. Milk Food Technol.* **38**(1), 25–28.

Peterson, R. G., and Joslyn, M. A. (1958). Nature of betanin, the pigment of red beet. *Nature* (*London*) **182,** 45–46.

Piattelli, M., and Minale, L. (1964). Pigments of centrospermae: Distribution of betacyanins. *Phytochemistry,* **3,** 547–557.

Piattelli, M., Minale, L., and Prota, G. (1965). Struttura della betanina. *Ann. Chim.* (*Rome*) **54,** 955–962.

Piattelli, M., Minale, L., and Prota, G. (1965). Pigments of centrospermae: Betaxanthins from *Beta vulgaris*. *Phytochemistry* **45,** 121–125.

Powrie, W. C., and Fennema, O. (1963). Electrophoretic separation of beet pigments. *J. Food Sci.* **28,** 214–220.

Schmidt, O., Becher, T., and Hübner, M. (1960). Zur kenntnis der farbstoffe der roten rube, *Beta vulgaris*. *Chem. Ber.* **93,** 1296–1304.

von Elbe, J. H. (1975). Stability of betalaines as food colors. *Food Technol.* **42**(5), 42–44.

von Elbe, J. H. (1977). The betalaines. *In* "Current Aspects of Food Colorants" (T. E. Furia, ed.), pp. 29–41. CRC Press, Cleveland, Ohio.
von Elbe, J. H. (1977). Status of colors in the dairy industry. *Dairy Ice Cream Field* **161**(3), 60–62.
von Elbe, J. H., and Maing, I.-Y. (1973). Betalains as possible food colorants of meat substitutes. *Cereal Sci. Today* **18** 263-264, 316-317.
von Elbe, J. H., Sy, S. H., Maing, I.-Y., and Gabelman, W. H. (1972). Quantitative analysis of betacyanins in red table beets, *Beta vulgaris*. *J. Food Sci.* **37,** 932-934.
von Elbe, J. H., Klement, J. T., Amundson, C. H., (1974). Evaluation of betalain pigments as sausage colorants. *J. Food Sci.* **39,** 128–132.
von Elbe, J. H., Maing, I. Y., and Amundson, C. H. Cassens, R. G., and Lindsay, R. C. (1974). Color stability of betanine. *J. Food Sci.* **39,** 334-337.
von Elbe, J. H., Pasch, J. H., and Adams, J. P. (1974). Betalaines as food colorants. *Proc. Int. Congr. Food Sci. Technol., 4th, Madrid,* Vol. I, pp. 485–492.
Watson, W. C. (1964). The metabolism of betanin. *Biochem. J.* **90,** 3p.

Miscellaneous Colors

Aronoff, S. (1953). The chemistry of chlorophyll. *Adv. Food Res.* **4,** 134–179.
Burini, G., Damiani, P., and Avellini, P. (1975). Riboflavin content of commercial samples of alimentary pastes. *Rev. Sci. Technol. Aliment. Nutr.* **5,** 313-314.
Cardas, S. (1978), Experimental and clinical considerations with regards to tartrazine (FD&C No. 5). *J. Am. Diet. Assoc.* **77,** 696-707.
Dusinsky, G. (1970). Analytical study of natural red food dye, alkannin. *Nabrung* **4,** 356-364.
Freedman, B. J. (1977). Asthma induced by sulfur dioxide, benzoate and tartrazine contained in orange drinks. *Clin. Allergy* **7,** 407–415.
Govidarajan, V. S. (1981). Turmeric-chemistry, technology, and quality. *Crit. Rev. Food Sci. Nutr.* **12**(3), 199-301.
Humphrey, A. M. (1980). Chlorophyll. *Food Chem.* **5,** 57–67.
Ingram, J. S., and Francis, B. J. (1969). The annatto tree (*Bixa orellana L.*): A guide to its occurrence, cultivation, preparation and uses. *Trop. Sci.* **11**(2), 97–103.
Janaki, N. (1967). An improved method for the isolation of curcumin from turmeric. *J. Indian Chem. Soc.* **44,** 985-986.
Kadzhaya, M. P., Darchiashvili, T. L., and Chkheidze, N. Y. (1971). Use of riboflavin as a food dye for coloring marmalades. *Tr. Gruz. Nauchno-Issled. Inst. Pishch. Promsti.* **4,** 326–327.
Krishnamurthy, M. N., Padmabai, R., Natarajan, C. P., and Kuppuswamy, S. (1975). Color content of tumeric varities and studies on its processing. *J. Food Sci. Technol.* **12**(1), 12–14.
Lannung, A. (1962). Preparation of cochineal tincture. *Arch. Pharm. Chem.* **69,** 229-231.
Mitra, C. R. (1975). Important Indian species: *Curcuma longa* (Zingiberaceae). *Reichst. Aromeh. Korperflegm.* **25**(1), 15.
Mottier, R. (1974). Cochenille acide carminique et carmin. *Rev. Conserve Aliment. Mod.* **27,** 83, 85-86.
North, R. S. (1973). Caramel: The versatile coloring. *Flavor Ind.* **4,** 337–338.
Pelham Wright, N. (1963). A thousand years of cochineal: A lost but traditional Mexican industry on its way back. *Am. Dyest. Rep.* **52,** 635-639.
Perotti, A. G. (1975). Curcumin: A useful vegetable color not well known. *Ind. Aliment.* **14**(6), 66–68.
Preston, H. D., and Rickard, M. D. (1980). Extraction and chemistry of annatto. *Food Chem.* **5,** 47-56.
Shankaracharga, N. B., and Nataryan, C. P. (1973). Tumeric chemistry, technology and uses. *Indian Spices* **10**(4), 8-10.
Szücs, K. (1975). Cultivation and processing of paprika. Mezögazolasagi Kiado, Budapest.
Tylova, M., Dusinsky, G., and Szokalay, A. (1960). Alkanna tinctoria, the source of natural dye. *Farmacia* **29,** 274-277.
Vernon, L. P., and Seely, G. R. (1966). "The Chlorophylls." Academic Press, New York.

Fig. 1. Corn on the cob, an orange, and a tomato. Carotenoids are responsible for the colors of these foods.

Fig. 2. Carrots and green plants. Both contain the yellow carotenoid, β-carotene.

Fig. 3. Lobster, tomato, mushroom, and pepper. Red carotenoids such as lycopene are found in the tomato, canthaxanthin in a variety of mushroom, astaxanthin in the lobster, and capsanthin in the pepper.

Fig. 4. Coho salmon. Skin, flesh, and ova are pigmented when salmon consume a diet containing red carotenoids (courtesy of John Spinelli, Northwest and Alaska Fisheries Center)

Fig. 5. Cock-of-the-rock *(Rupicola rupicola).* The bird's plumage color is derived from red carotenoids.

Fig. 6. Flamingo. Red carotenoids produce the flamingo's plumage coloration.

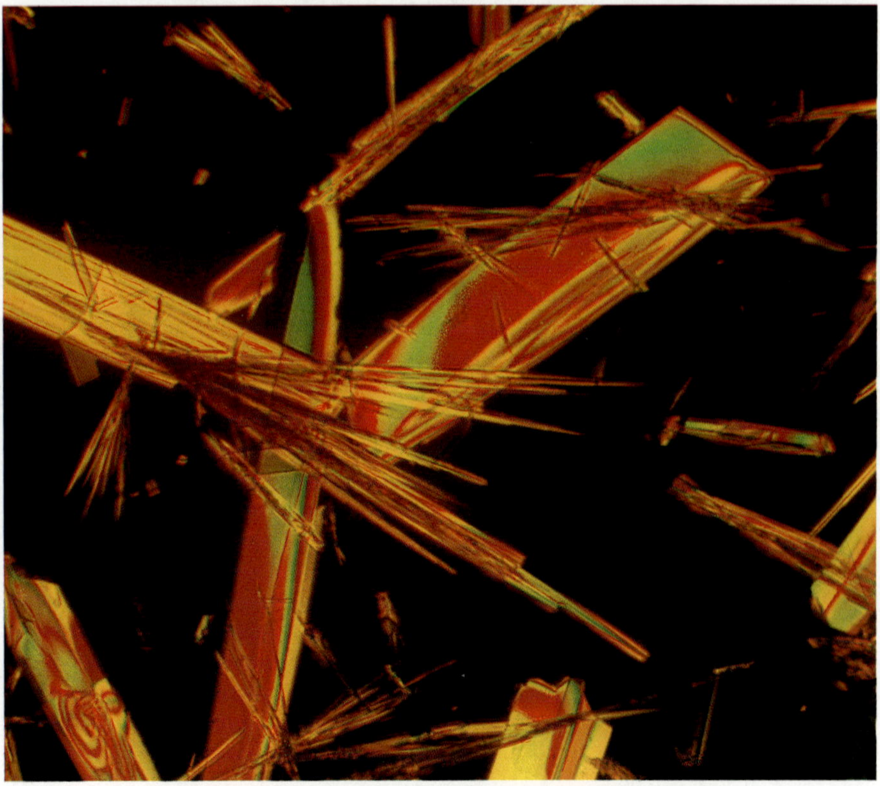

Fig. 7. Zeaxanthin. Carotenoids such as this one have characteristic crystalline structure.

Fig. 8. Drying tower and carotenoid powders. Since pure carotenoids are unstable in air, they must be converted to stable liquid and/or dry application forms such as free-flowing powders by special manufacturing processes.

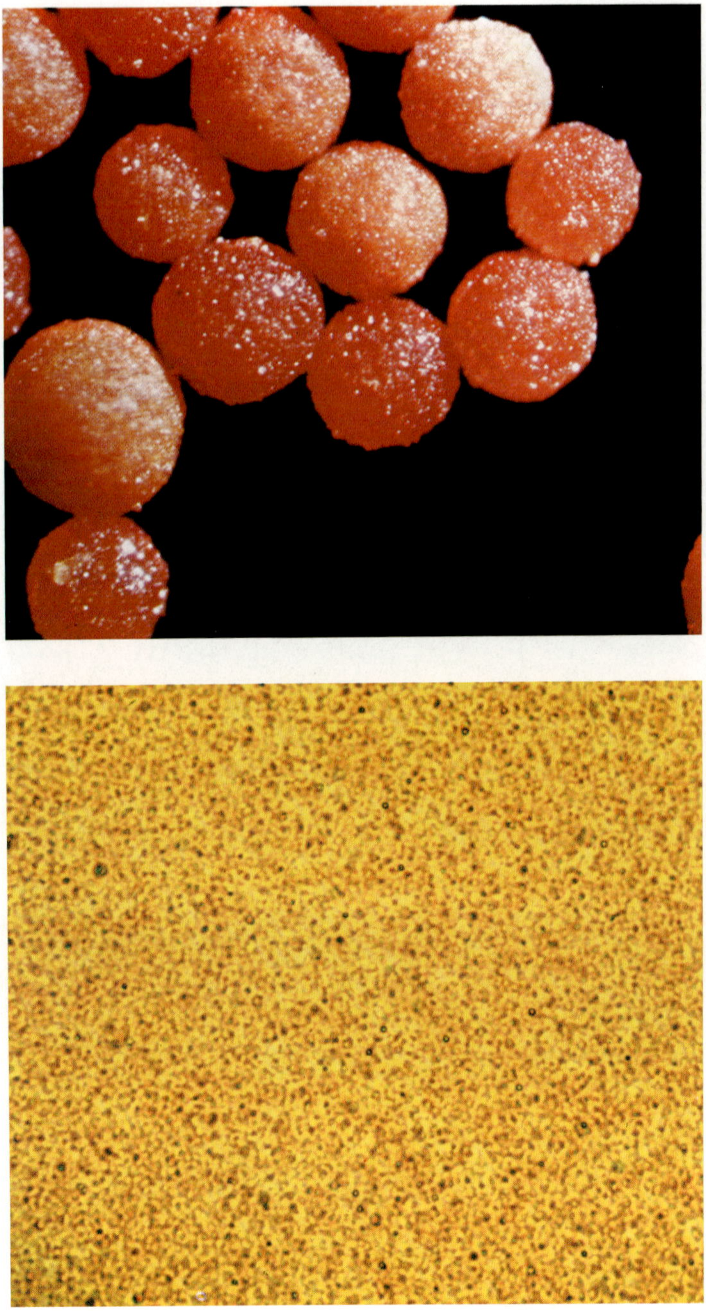

Fig. 9. Beadlets and interior view. Some dry carotenoid powders are in microbeadlet form, the carotenoid stabilized inside in either emulsion or colloidal states.

Fig. 10. Dishes of color. Carotenoids in liquid form give colors varying from yellow to red.

Fig. 11. Red birds. The plumage color of the canary can be varied by the type of carotenoid addition to the diet.

Fig. 12. Trout Fillets. Trout grown in captivity need added dietary carotenoids to duplicate natural pigmentation.

Fig. 13. Egg yolks. The golden yellow color of these yolks depends on the amount of yellow carotenoids in the diet of the chicken.

Fig. 14. Color fan. Any shade of yellow color (e.g., egg yolks) can be standardized by comparison to such a color chart.

Fig. 15. Raw chickens. The attractive yellow color results from dietary carotenoids.

Fig. 16. Broiled chickens. Again, the pleasing coloration depends on dietary carotenoids.

Fig. 17. Margarine. β-Carotene is a widely favored colorant for margarine, providing both pleasing color and vitamin A.

Fig. 18. Beverages. Juices and other drinks can be prepared with yellow or red carotenoid colorants.

Fig. 19. Snacks. The yellow color in such snacks as cakes and cookies can be provided by β-carotene.

Fig. 20. Soups. Soups, stews, sauces, etc., expected to have a yellow to red hue can be colored with added carotenoids.

Fig. 21. Cheese. Both natural and processed cheese can be colored with β-carotene or β-apo-carotenal.

Fig. 22. Candy and ice cream. Added β-carotene gives ice cream a pleasing yellow color. Some candies can be colored with yellow and red carotenoids.

Fig. 23. Tablets. Sugar-coated tablets colored with carotenoids (see Chapter 7 for more details).

Fig. 24. Suppositories. Fat-based suppositories colored with carotenoids (see Chapter 7 for more details).

2

CAROTENOIDS AS FOOD COLORS

H. Kläui and J. C. Bauernfeind

ISBN 0-12-082850-2

I. INTRODUCTION

Few properties of food are more important than color to most consumers. This interest in and overall concern for chroma, hue, shade, and tint of our visual world, and particularly our foods, follow naturally enough from the fact that sight is a very important sense. The old adage, "We eat first with our eyes," needs little amplification, as it is a part of our daily personal experience. The color of food is a significant factor in determining the acceptability of food. When we see food in its expected coloration, the response is a pleasing one. We become cautious when a food has an unexpected color, interpreting it as a possible sign of immaturity, spoilage, improper processing, or adulteration.

The expected color of food in the marketplace influences the decision of the purchaser to buy or not to buy. The proper color of a food, when served, arouses one's anticipation of food ingestion and rekindles one's recollections of past satisfaction. Color is an early signal of the inherent qualities of a food, such as freshness, wholesomeness, and readiness for consumption, thus creating an *a priori* color–taste expectancy relationship. The association of color and acceptability of foods is universal, although there may be significant differences depending on ethnic background, geographic location, religious training, social factors, and personal habits and history. What may be attractive to one individual or group may be unappetizing or even repulsive to another. When one's eyes are closed, food odor alone is a strong stimulus to one's response to food, and, when color and odor are combined, each reinforces the other favorably or unfavorably depending on the situation. The interesting relationship between colors and odors was examined by Déribéré (*249*). He generalized about certain correlations and set forth certain well-defined tendencies. He also provided spectra of color–odor relationships and associations with their combined influence on taste perception. Schultz (*847*) examined color–taste preferences and similarly illustrated the interrelationship between taste and color in food preferences.

II. COLORS OF NATURE

Nature relies on a variety of natural compounds, namely, carotenoids, chlorophylls, anthocyanins, porphyrins, quinones, etc., for the coloration of living animal and plant matter. Carotenoids and/or carotenes derive their names from the fact that they constitute the major pigment in the carrot root, *Daucus carota*, one of the first foods (*994*) observed to possess this class of pigments. Carotenoids are probably best generally described as aliphatic, aliphatic–alicyclic, or aromatic structures composed of five-carbon isoprene groups, usually eight, so linked that the two methyl groups nearest the center of the molecule are in positions 1 and 6 and all other lateral methyl groups are in positions 1 and 5; a series of conjugated C-C double bonds constitutes the chromophoric system according to the definition proposed by Karrer. The major carotenoid subgroups are carotenes and oxycarotenoids (xanthophylls). The former include the hydrocarbon carotenoids, and the latter the hydroxy, epoxy, furanoxy, and oxy derivatives of the carotenes. Oxycarotenoids are also frequently esterified as, for example, physalien (the dipalmitoyl ester of zeaxanthin). Carotenoids were named by their discoverers for some special property or for their source, e.g., carotene (from carrots), cryptoxanthin (hidden pigment), and zeaxanthin (from corn, *Zea mays*). Another classification system subdivides the carotenoids into acyclic, monocyclic, and bicyclic derivatives, and respective parent compounds of each of the above categories are lycopene, γ-carotene, and β-carotene. The prefix "neo" is is used to designate carotenoid stereoisomers containing at least one cis configuration in the double-bond chain, the prefix "pro" to designate some poly-*cis*-carotenoids, and the prefix "apo" to designate a carotenoid that has been derived from another carotenoid by loss of a structural element through degradative action. A wide range of pigments can be derived from the parent carotenoid by such changes as double-bond migration, introduction of a hydroxyl, keto, or methoxyl group, hydrogenation, cyclization, oxidative degradation, or isomerization. More than 350 carotenoids (Fig. 1) are known (*582*). A vast literature (*355,359,430–434,436,452,455,504,541,582,583,585–587,637,901,906, 928,977,1009a,1010*) on the identification, isolation, chemistry, and properties of the carotenoids has developed over the past 80 years. Well-known basic books exist on the subject (*353,360,437,489,1039*). Straub (*900*) lists some known natural carotenoids and literature references. Tables 1 and 2 present a partial history (*450*) of a few of the yellow and red carotenoids, respectively.

A wide variety of foods, for example, yellow vegetables, tomatoes, mushrooms,

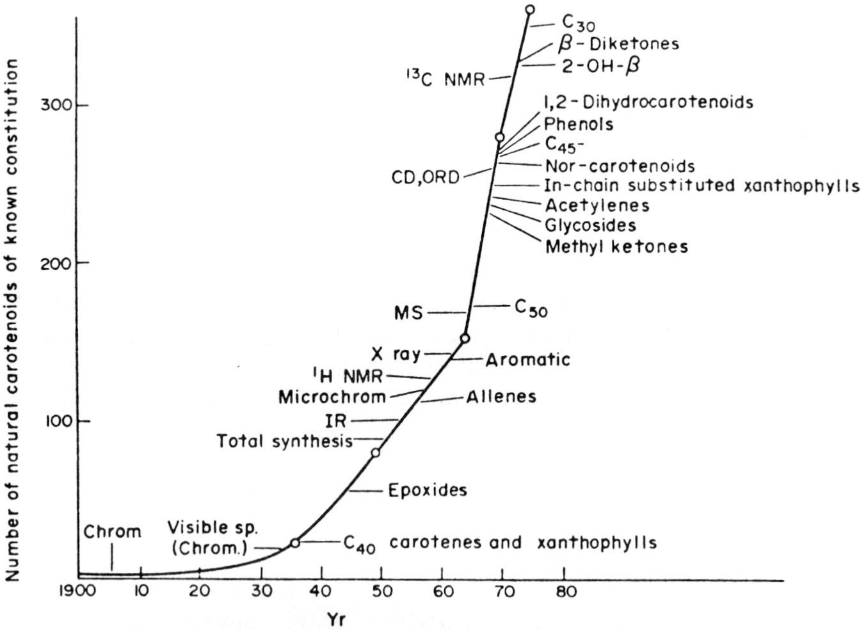

Fig. 1. Status of currently recognized carotenoids. [From Liaaen-Jensen (*582*).]

TABLE 1

Historical Development of Yellow Carotenoids[a]

Date	β-Carotene	Dihydro-β-carotene	Xanthophyll	Zeaxanthin	Cryptoxanthin	Physalien	Bixin
1830	Wachenroder (1831)[b]		Berzelius (1837)[b]				Boussingault (1825)[b]
1850				Thudichum (1869)[b]			
1870							
1890	Willstätter (1906)[c]		Willstätter (1907)[c]				
1910		Wuest (1918)[b]					Heisuschka (1917)[c]
1930	Karrer (1929–1931)[d]	Karrer (1931–1941)[d]	Karrer (1930–1933)[d]	Karrer (1929–1933)[d]	Yamamoto (1932)[b] Kuhn (1933)[d]	Kuhn (1929)[b] Kuhn (1929–1933)[d]	Kuhn (1928–1931)[d]
1950	Karrer-Inhoffen (1950)[f] Isler *et al.* (1953)[g]	Inhoffen (1950)[f]		Isler *et al.* (1955)[f]	Isler *et al.* (1956)[f]	Isler *et al.* (1955)[f]	Weedon (1951)[f]

[a] Adapted from Isler *et al.* (*450*). (These listings do not necessarily represent refs. cited in this chapter.)
[b] First mention.
[c] Molecular formula.
[d] Constitution.
[e] Isolation.
[f] Total synthesis.
[g] Commercial method.

TABLE 2

Historical Development of Red Carotenoids[a]

Date	Lycopene	Canthaxanthin	Astaxanthin	Capsorubin	Torularhodin	Capsanthin
1830						Branconnet (1817)[b]
1850						
1870	Harsten (1873)[b]		Pouchet (1876)[b]			
1890						
1910	Willstätter (1910)[c]					Zechmeister *et al.* (1927)[e]
1930	Karrer (1929–1931)[d]		Kuhn *et al.* (1933)[e]	Zechmeister *et al.* (1934)[e]	Lederer (1933)[b]	Zechmeister *et al.* (1927–1935)[d,*]
			Kuhn *et al.* (1938)[c]	Zechmeister *et al.* (1935)[c]	Karrer *et al.* (1943)[c]	Karrer *et al.* (1927–1935)[d,*]
			Kuhn *et al.* (1938)[d]	Zechmeister *et al.* (1934–1935)[d,*]	Karrer *et al.* (1946)[d,*]	
1950	Karrer (1950)[f]	Haxo (1950)[e]		Karrer *et al.* (1956)[d,*]		
	Isler *et al.* (1956)[g]	Zechmeister *et al.* (1956)[c]				
		Zechmeister *et al.* (1956)[f]				
1960				Weeden *et al.* (1960[b], 1965[i])		Weeden *et al.* (1960[b], 1965[i])

[a] Adapted from Isler *et al.* (*450*). Asterisk indicates proposed structure. (These listings do not necessarily represent refs. cited in this chapter.)
[b] First mention.
[c] Molecular formula.
[d] Constitution.
[e] Isolation.
[f] Total synthesis.
[g] Commercial method.
[h] Structure proof.
[i] Weeden synthesis.

apricots, oranges, berries, egg yolk, butter, chicken, shrimp, lobsters, salmon, and trout, owe their characteristic red, orange, or yellow color principally to carotenoids, as do certain food color extracts from natural sources, such as palm oil, paprika, annatto, and saffron. Carotenoids are widely distributed in nature, and it has been estimated that the annual natural production of these pigments amounts to about 100 million tons (*452*). Their various roles include provitamin A activity, absorption of light energy, oxygen transport, singlet oxygen quenching, light yellow to dark red food coloring, and probably other functions unknown at this time. When they are complexed with protein, green and blue colorations are achieved. The form in which carotenoids exist in all living cells and an elucidation of the various projected roles still require continuing research.

The addition of color to foods extends back in time to before the existence of written records. The addition of color by the present-day food technologist in the processing of modern refined or fabricated foods of convenience to enhance their appeal is merely a reflection or imitation of the pattern nature systematically performs in plant, animal, and mineral matter. The changing aspect in the use of food colors over the past several centuries has been the choice of colorants. β-Carotene was the first synthetic carotenoid to be marketed as a food color. Other carotenoids that have since become commercially available for food coloring are β-apo-8′-carotenal and canthaxanthin. Citranaxanthin has been developed as an indirect pigmenter for poultry products. More will probably follow in the decades ahead. Natural extracts containing carotenoids have been used for coloring foods for centuries. These include annatto (bixin, the main coloring component), saffron (crocin and other carotenoids), paprika (capsanthin and capsorubin), xanthophyll or oxycarotenoid extracts from leaves, carrot extracts, red palm oil, and others.

III. FUNCTIONS AND BIOSYNTHESIS

The functions (*357,541*) of carotenoids occurring in nature have been a subject of much speculation in the past and continue to be an area of research. It has been deduced that their functions include the following: a role as accessory pigments in energy transfer in the photosynthetic process; a role involving photoresponses such as phototropism in simple and higher plant forms; a photoprotective role, which may occur in bacteria and also in man under certain circumstances; a functional role in the oxygen cycle, such as quenching of singlet oxygen; a varied metabolic role in the physiological conversion of certain structures to vitamin A or in the conversion of other compounds to abscisic acid; a role as plant growth-regulating substances; a role involving oxidation products that function as reproduction regulators in fungi or as color attractants in flowers for insect pollination; a role in the coloration of food for mankind, from the light yellow of butter to the orange-yellow of egg yolk, the orange and red of salmon flesh, and the dark green of the live lobster, which becomes a vivid red upon heating.

It is evident that carotenoids play a number of vital roles in the physiology of the plant kingdom, and they are intimately involved with photosynthesis (*358,1016,1028a*), the fundamental life-sustaining reaction on this planet. Because of this important function, as well as the pleasing contribution they make to the human food supply, the light absorption spectra of carotenoids may be their most important characteristic. Burnett (*159*) states that carotenoids will continue for some time to astonish and frustrate investigators and observers by their variety and apparent lack of specific function. He considers it difficult to suppose that natural selection is so inefficient that this large group of compounds occurs passively and without function in so many organisms, and, as natural selection operates in diverse and unexpected ways, this appears likely to apply to carotenoid function. Krinsky (*539b,540*) considers the evolution of carotenoid function. He postulates that several billion years ago primitive anaerobic organisms needed to convert radiant energy to chemical energy to carry out a form of photosynthesis, and in the evolutionary process those organisms that gradually developed the biosynthetic mechanisms to manufacture colored carotenoid pigments enjoyed a selective environmental advantage in the competition for the utilization of visible radiation, thereby benefiting from an important function of the carotenoids. As biological evolution continued, the photosystem could use water as an electron donor, whereby oxygen was evolved. In the presence of a suitable photosensitizer and visible radiation, molecular oxygen could be converted to singlet oxygen, which carotenoids can quench and deactivate back to ground-state molecular oxygen, thus playing another major role in the oxygen evolutionary process. The metabolic role of certain carotenoids is envisioned as developing at a still later stage, such as in the requirements of visual growth and reproduction. Most visual systems require an isomer of retinol (*803*) resulting from a metabolite of certain carotenoids when dietary plant sources are consumed. Other examples could be cited. It is therefore conceivable that carotenoids have been on the planet for billions of years and even today are widely distributed in nature, serving functions other than that of food coloration.

Vallentyne (*981*) demonstrated the existence of β-carotene molecules that must be presumed to have survived for 20,000 years. This finding was made during a systematic investigation of the organic compounds in fossil sediment layers from Searles Lake, California. The sample examined was orange-green and had a sweet smell. Analysis showed about 30% dry matter, of which one-quarter was organic. Extraction with successive solvents yielded 28 mg of carotenes and 23 mg of xanthophyll (oxycarotenoids) per kilogram of the original dry matter. By the use of paper chromatography to investigate the carotene fraction, both α- and β-carotenes were identified. Another component was probably echinenone (myxoxanthin). Further extraction of the sediment showed the presence of green pigments, and the spectrum of a methanol extract suggested the presence of chlorophylls. The survival of these compounds for such a long period must be attributed to the reducing nature of their environment in the sediment.

The biosynthetic pathways of carotenoid formation have been fairly well clarified. Carotenoids are synthesized in their entirety only by plants and microorganisms. All animal carotenoids are ultimately derived, by the consumption of foods, from these two sources, although the carotenoids may be slightly altered by oxidative metabolism during digestion and absorption for accumulation in various animal tissues (*356,357*). Many papers and reviews (*97,177,179,235a,236,246, 352a,354,355,357,559,581,584,588,765–768,1023,1036*) have appeared on carotenoid biosynthesis. Acetate is considered the starting compound in the biosynthesis of the carotenoids. Two molecules of acetyl-CoA condense to form acetoacetyl-CoA, which in turn condenses with acetyl-CoA to produce β-hydroxy-β-methyl glutaryl-CoA, and through a reduction step mevalonic acid is formed. Mevalonic acid, in the presence of adenosine triphosphate (ATP), is converted to mevalonic acid phosphate and further phosphorylated to mevalonic pyrophosphate (MVAPP). In the presence of ATP and in combination decarboxylation and dehydration steps, MVAPP is converted to the important biological 5-carbon isoprene unit, isopentenyl pyrophosphate (IPP). Isopentenyl pyrophosphate is first isomerized to dimethylallyl pyrophosphate (DMP), and then IPP and DMP condense to form geranyl pyrophosphate. This 10-carbon unit, by a continued condensation with IPP, yields farnesyl pyrophosphate and by one more condensation forms the 20-carbon unit, geranylgeranyl pyrophosphate, which in turn, by dimerization, forms phytoene, the basic 40-carbon acyclic carotenoid structure. In the cyclization of the acyclic carotenoids and the introduction of the oxygen moiety, there is more uncertainty about the pathways, and hence there are different hypotheses for specific carotene or oxycarotenoid compounds.

β-Carotene production in the tomato can result from more than one pathway according to evidence presented by Raymundo *et al.* (*795*). Conversion of lycopene to β-carotene by chloroplasts was indicated by Hill and Rogers (*410*) through the use of ^{14}C-labeled lycopene. Plastids isolated from the tomato converted 3% of the radioactive carbon to β-carotene. Earlier workers also showed this and demonstrated that the reaction is reversible. Edwards and Reuter (*272*) examined the pigment contents of 11 commerical varieties of tomatoes differing appreciably in varietal and maturity factors. Pigment content was measured in the San Marzano variety during progressive maturity at six stages of ripeness from hand-picked fully green to uniformly red tomatoes. Twenty-one pigments were identified (Table 3) and evaluated quantitatively to show the dramatic pigment changes (*272*) that occur. Lycopene is the principal carotenoid, followed by β-carotene.

In most red and pink grapefruits the principal carotenoid pigments are lycopene and β-carotene, and the relative amount in the fruit changes during the maturing process. Ting and Deszyck (*940*) reported a consistent decrease in color of Ruby Red and Thompson Pink varieties on two different root stocks as measured by a decrease in lycopene content and an increase in carotene content as the season progressed. A number of chemical bioregulators of carotenoid biosynthesis have been developed which when sprayed on citrus fruit cause accumulation of carotene and lycopene and induce carotenogenesis (*192,421,976*). Because of penetration

TABLE 3

Pigments Isolated from Six Maturity Levels of the San Marzano Variety of Tomato[a,b]

Pigment	Pigment concentration in wet product (μg/gm)					
	1	2	3	4	5	6
Chlorophyll *a*	Trace	0.66	—	—	—	—
Pheophytin *a*	8.7	5.3	2.5	—	—	—
Chlorophyll *b*	5.1	0.3	—	—	—	—
Violaxanthin	—	—	—	—	—	0.05
Lycophyll	—	—	0.5	—	0.77	1.1
Lutein 5,6-epoxide	0.75	1.5	1.7	1.5	1.9	2.3
Lutein	0.27	0.15	—	0.2	—	0.5
Lycoxanthin	—	—	1.5	1.9	2.5	2.2
Mutatochrome	—	—	—	—	—	0.25
Lycopene	—	1.0	10.4	51.7	73.6	180.6
Neolycopene A	—	—	—	0.5	—	—
Tetrahydrolycopene	—	—	—	—	0.06	—
δ-Carotene	—	—	—	—	—	0.2
Poly-*cis*-lycopene II	—	—	0.1	—	0.2	0.1
Prolycopene	0.3	—	—	0.38	—	—
Poly-*cis*-lycopene V	—	—	—	—	—	0.2
Protetrahydrolycopene	—	—	—	0.2	0.36	0.38
Pro-γ-carotene	—	—	—	—	—	1.45
β-Carotene	—	2.8	4.0	3.9	4.5	5.6
Neo-β-carotene B	1.75	—	—	0.2	0.2	—
Phytofluene	4.5	3.3	2.6	3.4	3.9	7.8
Total[c]	16.8	11.7	20.7	60.3	84.0	194.9

[a] From Edwards and Reuters (*272*).
[b] Listed in approximate order of decreasing adsorptive power on magnesia/Super Cel columns.
[c] The total does not include the colorless polyenes.

difficulties, the greater influence has been on peel color. More recently the influence of synthetic bioregulators on orange endocarps has been reported (*403*).

Carotenogenesis occurs not only in the intact growing plant but in tissues separated from the plant. After the fruit or vegetable is harvested, carotenoid biogenesis does not necessarily cease. Further chromogenesis is influenced by stage of maturity, temperature, light, and environmental gases. It was observed in the yellow carrot by Booth (*128*) in 1951 that the carotenoid pigments of the freshly harvested root increased (11%) during storage, the increase being unrelated to variety, age, or initial pigmentation. The oxycarotenoid fraction increased more than the carotene fraction, both presumably from conversion of colorless polyenes. Carotenoids in the peel of harvested citrus fruits increase following treatment with ethylene, the amount being modified by temperature of storage. Improved color was observed by

Stewart and Wheaton (*894*) in oranges, resulting primarily from an increase in the orange-red or red pigment (*579a*) β-citraurin. In tangerines the same increases in pigments (Fig. 2) and to a lesser extent an increase in violaxanthin were observed after the application of ethylene (*893*). Best color is obtained with fruit held at 15°–20°C and at 1–10 ppm ethylene concentration (*893*). In a study (*1006*) of tomato varieties the average β-carotene contents of tomatoes ripened on the vine were higher than the contents of those picked green and allowed to ripen off the plant (Table 4). Exposure of green fruit to ethylene gas had only a minor effect on increased carotene content. McCollum (*641*) in 1954 showed that tomato fruit exposed to sunlight during ripening contained more β-carotene than that ripened in the shade. Watada *et al.* (*1006*) noted higher β-carotene in riper fruit that was exposed to more sunlight. Light quality and intensity are important in determining the amount of carotenoids synthesized by the tomato (*927*). Khudairi (*503*) concluded that the process of color change in tomatoes during ripening is triggered by light followed by the change of biochemical reactions. Red light absorbed by phytochrome is specific in the ripening process. The chain reactions (Fig. 3) within the color apparatus (chloroplast–chromoplast) are ethylene production, chlorophyll degradation, oxygen uptake, abscisic acid synthesis, and the enyzmatic transformation of colorless carotenes to yellow, then orange, and finally red. Light intensity as a factor has been studied (*419*).

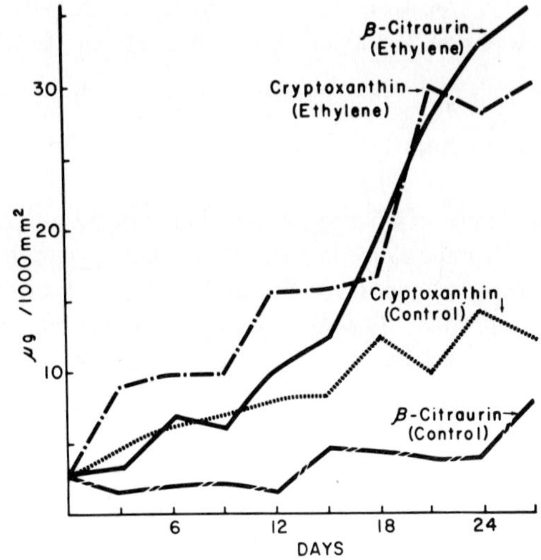

Fig. 2. Increase in cryptoxanthin and β-citraurin with time in the flavedo of ethylene-treated Robinson tangerine fruit. [From Stewart and Wheaton (*893*).] The curves for cryptoxanthin are now known to represent a mixture of cryptoxanthin and β-citraurin, with the latter predominating in the cultivar (*579*).

TABLE 4

β-Carotene Content in Ripe Tomatoes Harvested Mature Green and Ripe in 1973[a,b]

Cultivar	Stage at harvest: Mature green[c] (μg/100 gfw)	Stage at harvest: Ripe[c] (μg/100 gfw)
Fantastic	432 bcde	558 a
Double Rich	425 bcde	573 a
Rutgers	403 cde	490 abc
Homestead	396 cde	400 cde
Walter (a)[d]	394 cde	484 abc
Manapal	388 cde	408 cde
Heinz 1350	382 cde	524 ab
Cal-Ace	356 de	428 bcde
Campbell 1327	352 de	441 bcd
Walter (b)[d]	320 e	446 bcd
Average	385 m	475 n
Caro-Red	1918	4190

[a] From Watada *et al.* (*1006*).

[b] Values are average of two lots of 10 fruit each. Caro-Red was not included in statistical analysis.

[c] Values not followed by common letters are significantly different ($P = 0.05$)

[d] The two sets of Walter were harvested from different plantings.

IV. STRUCTURE IN FOODS

The question of how, and in what physical and chemical forms, carotenoids occur in nature is of particular importance in view of the challenge to the food scientist of developing and utilizing synthetic food coloring forms of carotenoids to extend nature's plan of coloring foods. In some instances carotenoids in natural foods may be present in free form dissolved in the oily or fatty phase of the food tissue. Substantial quantities of carotenoids are also present in nature in very fine physical dispersions and, in this form, are capable of coloring aqueous media. The carotenoids in oranges, tomatoes, and carrots are well-known examples. A similar situation may exist in the sweet potato (*1003*). The outstanding stability of the color of natural, water-dispersible carotenoid–protein or lipoprotein complexes is probably due to their ultrastructure. However, remarkably little is known of the nature of protein–carotenoid linkages, and this applies also to the carotenoproteins, which contain carotenoids in stoichiometric proportions and which are of extraordinary interest in view of the green and blue and other colors of these combinations.

In photosynthetic bacteria, carotenoids are concentrated together with bac-

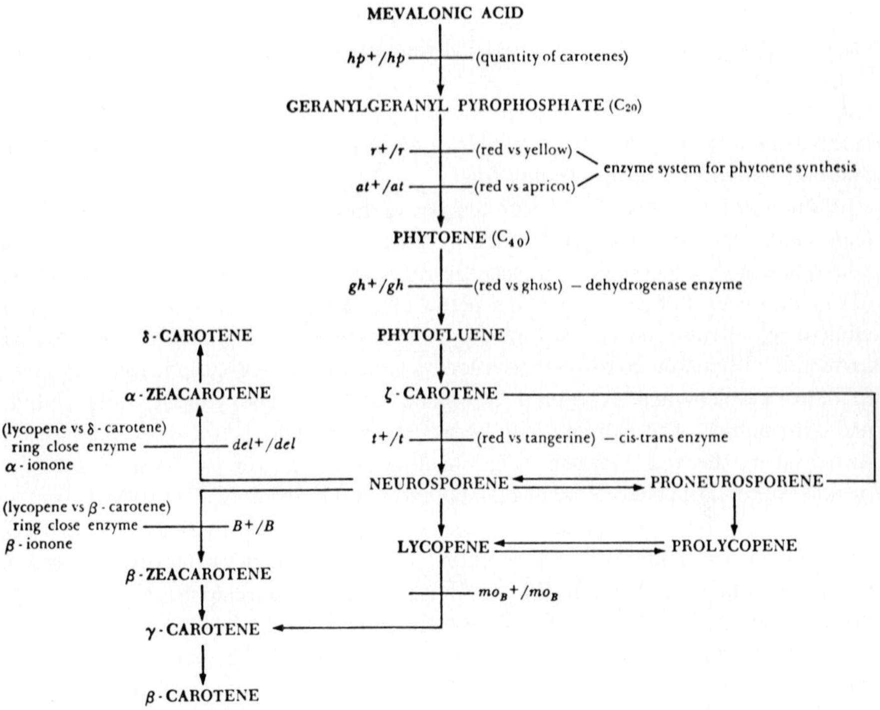

Fig. 3. Genetics of tomato carotenes. Each arrow represents a chemical reaction of a carotene with a specific enzyme system controlled by genes. [From Khudairi (*503*).]

teriochlorophyll exclusively in the chromatophores. Carotenoids in nonphotosynthetic fungi occur in oil droplets or in pericapsular fat; there has been no direct demonstration that they occur as chromoproteins. In nonphotosynthetic bacteria, however, carotenoids have never been reported in fat globules, but the pigments appear to form protein complexes or to be attached to the protoplast membrane. Carotenoids in algae occur in the chloroplasts along with the chlorophylls. The pigments are confined to the lamellae. Unlike photosynthetic bacteria, some algae can, under certain conditions, accumulate extraplastidic carotenoids (*356*), particularly in the gametes. In higher plants, carotenoids are present in the chromoplasts; as in all photosynthetic organisms, the xanthophylls are not esterified. In autumn leaves, the xanthophylls liberated from the disintegrating chloroplasts are esterified and thereby become more lipophilic (*273,274,276*).

Fruit carotenoids are attached to proteins, as is obvious from the general properties of fruit juices, but it is not known whether the carotenoids of flower petals are attached to protein. In the few roots that contain significant amounts of carotenoids, carotenes generally predominate, e.g., carrots and sweet potatoes (*361*). Carotene in carrots is located in lipidic droplets, or "globuli," in filaments and crystals. The crystalline chromatoplastids contain, in addition to protein, RNA, DNA, and phospholipids (*902*). Frey-Wyssling and Schwegler (*317*) showed by electron microscopic and polarization microscopic studies that the "crystalline" chromoplasts contain anisotropic layers of lipoprotein, in which the surplus carotene is stored. These layers have a lamellar structure and represent a particular type distinct from those that hold globuli or needle-shaped fila.

A study of the "linkages" of carotenoids to lipoproteins of chloroplasts in green leaves indicates a heterogeneity of linking strength (*203*), which is associated with a heterogeneous metabolism. The specific radioactivity of extracts of young wheat (*Triticum durum*) obtained by using petrol-ether containing increasing amounts of ethanol as solvent showed differences that obviously are related to the ease of extraction. The carotenoids were purified by column chromatography (on aluminum oxide for carotene and on calcium carbonate for lutein, epoxylutein, violaxanthin, and neoxanthin). The extractability of the carotenoids by petrol-ether was greatly increased by thermal treatment; e.g., heating of the leaves for 15 min at 100°C increases the extractable amount of total carotenoids from 12% in fresh leaves to approximately 70% in denatured leaves. Nishimura and Takamatsu (*715*) in 1957 reported on a carotene–protein complex isolated from green leaves, absorption spectra of which are shown in Fig. 4. An isolation procedure applicable to spinach, oats, wheat, and soybean plants for carotenoproteins has been described (*500*). Subbarayan and Cama (*907*) isolated and characterized a β-carotene–lipoprotein complex in the mango.

In human and animal tissues carotenoids may be present in fats in dissolved form. Goodwin (*361*) classified mammals into three groups according to their ability to metabolize carotenoids:

Group A (man): Tissues accumulate carotenoids indiscriminately

Group B (cattle and horses): Tissues accumulate mainly carotenes

Members of the A and B groups accumulate carotene in the fatty tissues of the

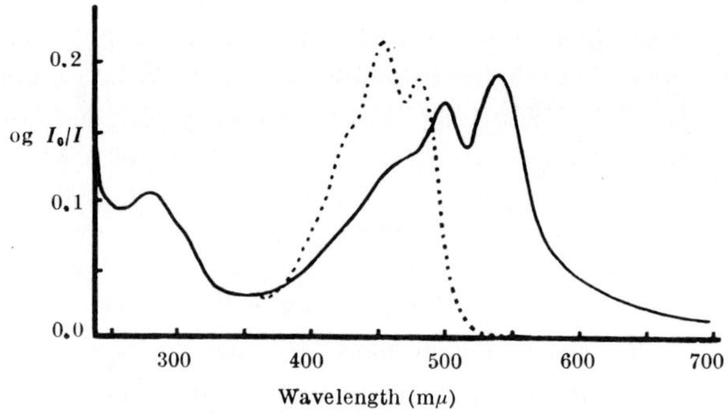

Fig. 4. Absorption spectra of a green leaf carotene–protein complex dissolved in water (solid line) and of the β-carotene separated from it and dissolved in *n*-hexane (dashed line). [From Nishimura and Takamatsu (*715*).]

body throughout life, accounting for the fact that infant fat is almost white and adult fat is yellow. This accumulation is apparently irreversible, as indicated by the fact that when body fat is mobilized, for any reason, the carotene remains in the depot. This accounts for the characteristic yellowness observed in cases of emaciation. Other members of these groups include sheep, dogs, deer, antelope, Indian buffalo, water buffalo, elephants, hedgehogs, foxes, fitchets, badgers, and roe deer.

Group C (pigs and goats): Tissues accumulate few carotenoids or none This group includes rats, rabbits, hares, giraffes, and reindeer.

Group D (birds), according to Deuel (*252*): Birds accumulate oxycarotenoids Correlations between the *β*-carotene content of blood serum (in man and various animals) and particular albumin fractions separable by electrophoresis have been found by various authors, and these have been interpreted as an indication of some linkage between carotenes and proteins (*31,326,812*). Carotenoprotein research probably was initiated by the report of Pouchet (*770*) on pigments of crustaceans. Early compounds were crustacyanin (*556*), the blue carotenoprotein of the lobster shell, ovorubin (*172*), a snail egg carotenoprotein, and ovoverdin (*1000*), the lobster egg carotenoprotein, all of which are complexes of astaxanthin. More recently Zagalsky and Herring (*1036*) studied the blue astaxanthin–proteins of the Coelenterata. Lee (*573*) described a bluish canthaxanthin–protein complex in animal tissue. Other reviews on carotenoproteins, about which very little is known, have appeared (*173,572*). Proteins in which carotenoids are present in stoichiometric proportions as prosthetic groups constitute a very interesting set of compounds, since the combination of a carotenoid with protein can extend the range of colors to green, purple, blue, and black. Despite these interesting properties, remarkably little is known of these compounds. Cheesman *et al.* (*173*) compiled a list of papers reporting the presence of carotenoproteins in invertebrates. Their table includes over 120 species of animals, and this surprisingly widespread occurrence suggests that thousands of carotenoproteins are to be found in nature.

Early observations indicated the occurrence of protein–carotenoid complexes in a wide variety of organisms. The water solubility of a large number of blue, green, and gray pigments in crustaceans and other invertebrates, coupled with the change of color to red when boiled or treated with various reagents and the increased stability of the pigments to light, was attributed to the association of the pigments with protein. Cheesman *et al.* (*173*) divided these complexes into (a) true carotenoproteins, with a strictly stoichiometric relationship between carotenoid and protein, and (b) lipoproteins, in which carotenoids are associated with the lipid component and in which a stoichiometric relationship has been proved. A characteristic property of both types of complexes is shown in an important modification of the absorption spectrum of the carotenoid. The majority of carotenoprotein complexes are blue to green, with their main absorption maxima lying between 560 and 680 nm; the others are red to purple, with their maxima between 490 and 532 nm. Rather exceptionally, their maxima are lower than those of the free carotenoid, and some green complexes appear to result from a blue stoichiometric complex in combination with

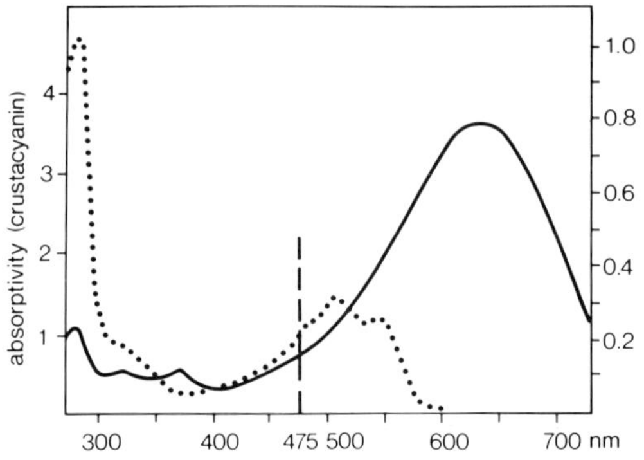

Fig. 5. Absorption spectra of crustacyanin and ovorubin. The prosthetic group of both proteins is astaxanthin, although that of ovorubin was erroneously stated [Cheesman (*172*)] to be an astaxanthin derivative. The absorption spectrum of astaxanthin in petroleum ether shows a single maximum in the visible range at 470–475 nm. Key: —, crustacyanin (1 mg/ml) in 0.2 *M* Na-K-phosphate buffer, pH 7; ••••, ovorubin (1 mg/ml) in water. [From Cheesman *et al.* (*173*).]

a lipid-dissolved free carotenoid. Typical examples (*173*) of absorption spectra are shown in Figs. 5–7.

The carotenoids found in carotenoproteins commonly have reactive keto groups in the 4 and 4′ positions of the ionone ring. Attempts at making artifical carotenoproteins with various carotenoids have succeeded only when a keto group was present in at least one of these positions. Carotenoids isolated from carotenoproteins of invertebrates include astaxanthin, canthaxanthin, astaxanthin esters, and others.

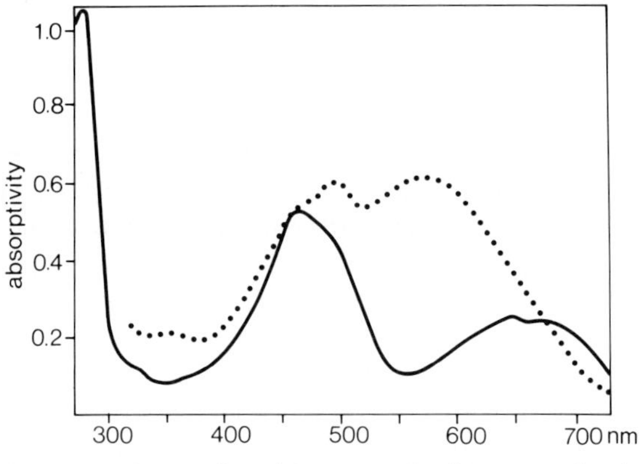

Fig. 6. Absorption spectra of ovoverdin and the egg protein of *Eupagurus bernhardus*. The prosthetic group of ovoverdin is astaxanthin; that of the *Eupagurus* protein is an ester of axtaxanthin. Both proteins are labile, and the heights and positions of the absorption maxima differ significantly in different preparations. Key: —, ovoverdin (1 mg/ml) in 0.05 *M* Na-K-phosphate buffer, pH 7; •••, *Eupagurus* protein (uncertain concentration) in 0.1 *M* Na_2HPO_4. [From Cheesman *et al.* (*173*).]

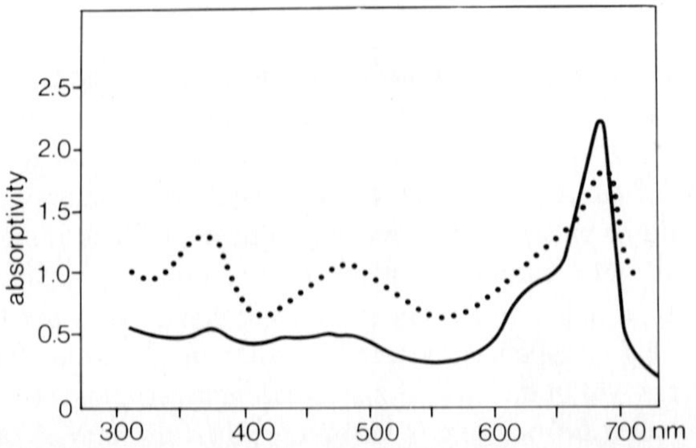

Fig. 7. Absorption spectra of two canthaxanthin–protein complexes. Both proteins are of unknown concentration in 0.05 *M* Na–K-phosphate buffer, pH 7. The prosthetic group, canthaxanthin, shows an absorption maximum in petroleum ether at 460–463 nm. The absorption maxima of the proteins in the region of 500 nm appear to be due to lipid-associated xanthophylls. Key: —, carotenoprotein from *Idothea granulosa*; •••, carotenoprotein from *Tanymastix lacunae*. [From Cheesman *et al.* (*173*).]

β-carotene, xanthophylls, anthocyanins, bile pigments, and others have been found as accompanying or associated pigments.

Since all carotenoproteins so far examined are readily split into carotenoid and apoprotein by treatment in aqueous solution with water-miscible organic solvents, it seems improbable that covalent bonds are involved in these systems. Detailed studies of the protein–carotenoid linkages, the nature of the proteins, and the conditions of interaction might open the possibility of producing complexes of any desired color shade. Stabilization of protein and carotenoid, and particularly control of protein configuration, seem to be possible functions of carotenoproteins. In a few cases, stabilization of the carotenoid against photooxidation and of the protein against denaturation by complex formation was proved. On the basis of more circumstantial evidence, Cheesman *et al.* (*173*) discussed the possible participation of carotenoproteins in a variety of functions, including protective coloration, development of photosensitivity, electron transport, and enzymatic activity. The provitamin A activity that has been clearly established for β-carotene and some other related compounds in mammals, including man, does not seem to play a part here.

V. COLORING APPROACHES

In discussing the utility of carotenoids as food colors, we must consider two possible approaches (*58*) to the end result of imparting color to a given food product. One method is the *direct addition* of the carotenoid to the food product, and the other, the *indirect route,* involves mixing the carotenoid into the daily ration of animals, poultry, or fish, which then is absorbed and deposited in the tissues. The latter is the method that nature uses in imparting color to foods.

A. Indirect Route

Egg yolks, the skin and fatty tissues of poultry, butter, and fatty tissues of cattle owe their yellow color to the absorption of natural carotenoid pigments from the ration and deposition in those tissues. Likewise, the pink and red flesh color and pigmented skin of certain trout and salmon result from natural carotenoids consumed in the wild or under modern management in artificial rearing practices.

Dairy cows on good summer pasture produce cream of a light yellow color, which results in a rich yellow butter. This color is due primarily to dietary β-carotene that has escaped conversion to vitamin A during the digestion and absorption process within the cow. Some α-carotene, cryptoxanthin, and lutein are also usually present. In the winter, however, when the cow is on grain and dry roughage and does not have access to green grass, the cream and corresponding butter are very pale; hence, winter butter must be colored during the butter-making process. Carotene can be used to color winter butter. Supplementation of the dairy cow's ration during the winter months with a high level of pure crystalline carotene in a concentrated form could be used to maintain the color of the cream and butter throughout those months.

Avian species differ from mammals such as the dairy cow since the latter convert most of their ingested carotenes to vitamin A, whereas avian species preferentially store certain ingested oxycarotenoids to provide color in eggs, body skin and fat, liver, and feathers. Yellow canaries and pink flamingos on a low-carotenoid diet lose their characteristic feather color. Normal color returns when yellow carotenoids, such as zeaxanthin, are added to the diet of the canary, and red carotenoids, such as canthaxanthin, to the diet of the flamingo. Moderately colored egg yolks from chickens or ducks are preferred for table use in many parts of the world; in certain areas consumers want highly colored yolks. Highly colored yolks are also desired for the commercial production of bakery products, noodles, mayonnaise, prepared cake mixes, and other egg products. Under natural conditions of barnyard feeding of poultry in the past, the chicken obtained the yellow oxycarotenoids from grass and maize. In modern feeding practices the dietary pigmenter must be supplied by natural, yellow, oxycarotenoid-containing ingredients and/or by the addition of oxycarotenoids of chemical synthesis or fermentation origin. The same pattern holds for skin and fatty tissues of yellow broilers and fryers for table use.

Salmonids, namely, certain trout and salmon, inherently store red oxycarotenoids acquired from their natural varied diet of larvae, insects, crustaceans, algae and other plant foods, etc., to obtain their bright skin and flesh color. Under the current system of managed rearing of large numbers of trout with specially prepared dry fish pellets, salmonids do not have the bright colors of wild trout, unless sources of these red oxycarotenoids are provided in their daily ration. It has been shown that canthaxanthin and/or astaxanthin in these diets impart a natural-appearing coloration. The choice of oxycarotenoid pigmenter or pigmenter combinations in the feed of different species is governed by the desired objective, economics, legal statutes and regulations, and current availability of the products.

B. Direct Addition

Considering the low efficiency achieved by going via the digestive system in the *indirect approach* to coloring butter, egg yolks, etc., it is obvious that the process of feeding carotenoids for pigmenting purposes is best applied to the production of fresh shell eggs for table use, yellow-fleshed broilers, fryer chickens, and fish, as nature initiated. *Direct addition* of carotenoid food colors to foods, whenever possible, is by far the preferred method from the viewpoint of flexibility, convenience, uniformity, and economics (*154,156,622*). This chapter, therefore, is concerned with the carotenoids naturally present in foods, carotenoid preparations that are commercially available for addition to foods, and the application studies that have been undertaken to extend nature's pattern by adding these carotenoids to specific foods to provide colors acceptable to consumers.

VI. NATURAL CONCENTRATES

The addition of carotenoids to food predates the commercial production of carotenoids by chemical synthesis. Annatto extracts, carrot extracts, palm oil extracts, saffron, tomato extracts, and paprika products, all containing carotenoids, have been used to color a variety of food products. One concern with some of the natural carotenoid concentrates is that flavor or odor characteristics inherent in the plant source of the concentrate are carried over into the food product to be colored and may not be compatible with it. A pretesting trial can determine whether this feature may or may not be troublesome.

Natural colors as described by the Joint FAO/WHO Expert Committee on Food Additives have been used in food for a long time and have been accepted for such use in much the same manner as vegetables and cereal products. In considering such food colors, the committee was faced with serious problems owing to the lack of published information relating to adequate identification and chemical composition. It was noted that natural colors may be available in different forms. In the case of botanical products, both the powdered plant and the extract of the powdered plant have been used as coloring agents for many years. Also, because of the differences in soil, climatic conditions, age of plant, and time of harvest, the nature and proportions of the colored and other components of the same species of the same plant may vary widely. Such products may contain a large percentage of substances that have not been defined. Morphological and histological examination of such botanicals has been used for a long time for the identification and quality evaluation of these materials, but in the opinion of the committee the information obtained in this way was inadequate. Dinesen (*262*) discussed the toxicological evaluation of plant extracts, which is complicated by the presence of many compounds besides the one under investigation. Thus, the description of the preparation of the extract becomes as important as the actual toxicological test data. Although many natural pigments have a long history of safe use for human consumption, this fact can be difficult to verify by modern toxicological testing procedures. As a result of efforts by natural plant processors, decisions taken by the Joint FAO/WHO Expert Committee include new annatto and turmeric specifications that are improved and much more realistic than those previously proposed and have become part of the official proceedings of the Eighteenth Meeting of the Joint Expert Committee on Food Additives. Some problems remain to be solved, but these now appear to be much more surmountable than before.

With certain products of other than plant origin—for example, natural and synthetic ultramarines, cochineal, and caramel—similar problems of chemical identification have been encountered. For this reason, the committee concluded that it was impossible to prepare adequate specifications for these products. The committee considered it essential that methods of analysis, including adequate identification and quality data, be provided whenever any evaluations are undertaken on coloring matter of natural origin. It is hoped that future studies will provide such information.

A. Paprika

Paprika, *Capsicum annuum,* is one of the oldest and most important natural carotenoid food colors. It is used in two forms for coloring foods: (a) paprika dry powder (micronized selected, ripe, dried pods) and (b) oleoresin of paprika (oil extract of coloring and flavoring components from the pods). Paprika powder, an insoluble red-brown powder, is used as a condiment in chili products and is added to soups, stews, and some speciality meat products such as Italian sausages. The hue of the powder is influenced by particle size.

Paprika is frequently heated to produce the desired flavor. A relationship between color and content of browning compounds has been observed in paprika powder. Ramakrishnan and Francis (*790*) reported that heating the powder at 125°C for 60 min reduced the carotenoid by 23% of the control value of 347 mg per 100gm and increased browning compounds by 306%. Conducting the heating phase under nitrogen slightly reduced the carotenoid loss. The nature and stability of paprika carotenoids have been studied and reviewed by Philip (*755*), Philip and Francis (*756*), and Philip *et al.* (*757*). The effects of sun-drying versus oven-drying of red peppers were studied by Park (*734*).

Oleoresin of paprika, unlike ground paprika, is a liquid product completely soluble in oils and therefore does not impart specks or particles of vegetable tissue to the finished food product. An advantage of oleoresin is the possibility of standardizing the color of the oil extract, since ground paprika may vary depending on such factors as geographic and climatic effects, variety of pepper, harvesting conditions, drying temperatures, oil (*261*) and moisture content (*333*), and storage conditions. A review of the red pepper pigments, their isolation and purification, color degradation, and the effects of moisture content, temperature, light, fat content, and antioxidants on color was published by Krishnamurthy and Natarajan (*545*) in 1973. The ultrastructure of the chromoplasts of different color varieties of *Capsicum* was also studied by Kirk and Juniper (*509*) and Davies *et al.* (*237*). The red fruit had the highest carotenoid content, the orange and yellow varieties possessing only 3% of the amount in the red variety. The spectra of the yellow variety were lacking in capsanthin and capsorubin. Chemical regulation of plastid development was attempted by Simpson *et al.* (*865*). Fruit dipping or on-vine treatment with CPTA [2-(4-chlorophenylthio)ethyldiethylammonium chloride] altered the ultrastructure and carotenoid composition of chromoplasts of *Capsicum annuum* cultivars. The chromoplasts lacked fibrils, synthesis of capsanthin and capsorubin was inhibited, β-carotene content was decreased, and lycopene accumulation was increased.

Various genetic studies (*663*) showed wide variation in pigment content of *Capsicum annuum*, but no interdependence was established between fruit yield and pigment content. The extractable color values of 15 selected cultivars (*700*) of red pepper and two methods of measuring color intensity were compared. Significant differences in objective and subjective color values were found among cultivars. Extractable red pigment values were much greater than yellow pigment values in all cases. As xanthophyll content increased, β-carotene content increased. All Gardner values decreased with increasing pigment content. The Gardner Color Difference Meter proved to be a better indicator of extractable pigments than the subjective evaluation. The nature and biosynthesis of the carotenoids of different color varieties (*237*) of *Capsicum annuum* have been investigated by Davies *et al.* (*237*). After harvesting of peppers, biochemical processes continue (*108*). Ripening is completed within 6–8 weeks, during which pigment (capsanthin and capsorubin) formation increases about 75–80%. Pigment increase during postripening in red and yellow carotenoids has been confirmed by others (*85,391*).

The color imparted by paprika to a food product (*13,820*) can range from a deep crimson red to pale pinkish-yellow, depending on the concentration used. The total carotenoid content of paprika extracts is approximately 4.3% (*91*). Rother (*820*) described the detection of an addition of paprika to an orange beverage based on the identification of the typical pigment, capsanthin. Oleoresin of paprika can be used in salad dressings, in sauces, and particularly in meat products, including some types of sausages (*124,836*), where it is allowed as a spice. It is forbidden as a color for meat products in most countries. For example, Klein (*525*) discussed the problem of declaring the presence of paprika in German sausage. It is often used in combination with annatto to color processed cheese. A water-dispersible form of paprika oleoresin is also manufactured. It consists of paprika oleoresin and food-grade emulsifiers and is used in baked products, beverages, cereals, and certain cheese products.

Oleoresin of paprika is available in various potencies of approximately 50,000–

120,000 so-called standard color units. Capsanthin and capsorubin (*47*) are generally considered to be the main carotenoids in paprika. However, Bunnell chromatographed a commercial sample (*155*) of oleoresin of paprika, a 50,000 color unit product, and found the following composition: capsanthin esters, 44%; zeaxanthin and cryptoxanthin esters, 23%; capsorubin esters, 19%; and carotenes, 11%. Cholnoky *et al.* (*180*) identified pigments of paprika (*Capsicum annuum* var. *lycopersiciforme rubrum*) as β-carotene, β-carotene epoxide, mutatochrome, neo-β-carotenes B and U, violaxanthin, lutein, foliaxanthin, foliachrome, antheraxanthin, aurochrome, cryptoxanthin, cryptocapsin, lutein epoxide, zeaxanthin, capsorubin, and capsanthin. Curl (*218*) reported on the carotenoids of red bell peppers, a variety of *C. annuum*; capsanthin was the principal carotenoid present (Table 5), with carotene and capsorubin as the next most important pigments.

Figure 8 gives the formulas of some carotenoid pigments in paprika, showing the complexity of the compounds present. Other workers (*594,991*) have presented analytical procedures for determining carotenoid content of paprika. A series of four papers on color measurement of *Capsicum* spices was authored by Moster and Prater (*687–690*).

A study was carried out to investigate quantitatively the changes in the carotenoid pigments of paprika caused by the bleaching process (*242*). The workers used commercial samples of bleached and normal paprika, which were extracted with acetone, separated by countercurrent distribution fractionation into monol, diol, polyol, and hydrocarbon, purified by column chromatography, and identified by spectral and chemical methods. The bleaching treatment consisted of exposure to sunlight for 3 months in aluminum-lined trays covered with Pyrex glass, and the samples were then kept in air-tight, clear-glass containers. The bleaching treatment resulted in a loss of nearly 96% of total extractable pigment expressed as β-carotene. Although most grades of paprika showed a predominance of diol and polyol carotenoids, one grade had a far greater amount of hydrocarbons. There were 54 pigments isolated from bleached and 37 from unbleached paprikas; only 33 and 31, respectively, were completely or tentatively identified. Sixteen of the known carotenoids in unbleached paprika were also observed (*242*) in the bleached sample. In different reports, Kanner and Mendel (*485*) and Kanner *et al.* (*487*) concluded that the main proteinous, carotenoid-bleaching activity of pepper extracts is due to the presence of a peroxidase-like protein; however, ascorbic acid also is involved.

It is known that the carotenoid pigments of powdered paprika possess 9–12 conjugated double bonds and hence are easily altered by the stress of heating and bleaching. Factors affecting the retention of red color in peppers have been studied by Lease *et al.* (*569*). Attempts have been made to incorporate antioxidants (*570*). In published studies the addition of ethoxyquin (an antioxidant approved for use in animals in the United States) to paprika powder (*107*) (at levels of 0.01 and 0.02%) delayed pigment loss during a storage period or delayed oxidation of extracts (*176*) of chili peppers. Extraction processes have been described by Todd (*943*), Segura Ferns (*855*), and Szabo (*917*).

TABLE 5

Carotenoids Obtained from Red Bell Peppers (Variety of *Capsicum annum*)[a,b]

Carotenoid	Spectral absorption maxima[c] (nm)	Approx. %
I-1 Phytoene	(298), 286, (276)	1.7
I-2 Phytofluene	366, 348, 332	1.1
I-3 α-Carotene	486, 456, 431, 339[d]	0.2
I-4 β-Carotene	447, 449, (424), 338	11.6
I-5 ζ-Carotene	423, 397, 377	1.5
I-6 Mutatochrome-like	451, 428, 402	0.3
IIA-1 Hydroxy-α-carotene	473, 444, 420, 332	1.0
IIA-2 Cryptoxanthin	476, 447, (422), 338	6.7
IIA-3 Hydroxy-α-carotene-like	473, 445, (423), 333	0.4
IIB-1 Cryptoflavin-like	451, 426, (402)	0.5
IIB-2 Cryptocapsin	(497), 470, (445), 353	4.3
IIIA-1 Capsolutein	486, 455, (430), 339	2.3
IIIA-2 Zeaxanthin	493, 463, (437), 346	2.3
IIIB-1 P-482, diol[e]	482, 451, 424, 336, 320	0.6
IIIB-2 Capsolutein 5,6-epoxide[f]	483, 451, 425, 336, 322	1.5
IIIB-3 Antheraxanthin	487, 456, (431), 338	1.6
IIIB-4 Capsolutein 5,8-epoxide[f]	459, 430, 406, 318, 303	0.4
IIIB-5 Mutatoxanthins	464, 437, (412), 320	1.6
IIIC-1 Violaxanthin	484, 452, 425, 336, 321	9.9
IIIC-2 Luteoxanthins	460, 431, 406, 319, 303	0.9
IIIC-3 Capsanthin	(510), 483, 363	34.7
IIID-1 P-441, tetrahydrocapsorubin[g]	441, 414, 392	0.7
IIID-2 Hydroxycapsolutein[f]	486, 456, (430), 338	1.9
IIID-3 Capsanthin 5,6-epoxide[f]	(509), 478, (455), 357, 344	0.9
IIID-4 Capsochrome[f]	(483), 456, (431), 345	0.3
IIID-5 Capsorubin	522, 487, (460)	6.4
IV-1 Hydroxycapsolutein 5,6-epoxide[d]	482, 449, 421, 336, 321	1.0
IV-2 Neoxanthin	478, 447, 421, 335, 321	0.7
IV-3 Trolliflor-like	482, 449, 423	0.1
IV-4 Carbonyl[g]	458	1.0
IV-5 Hydroxycapsanthin-like	478, 358	0.5

[a] From Curl (*218*).
[b] Values in parentheses are for shoulders or humps on spectral absorption curves.
[c] I, IIA, IIB in hexane; others in benzene.
[d] In benzene.
[e] P-482 refers to a pigment with highest wavelength spectral absorption maximum at 482 nm.
[f] Tentative identification.
[g] Spectral absorption curve with one maximum.

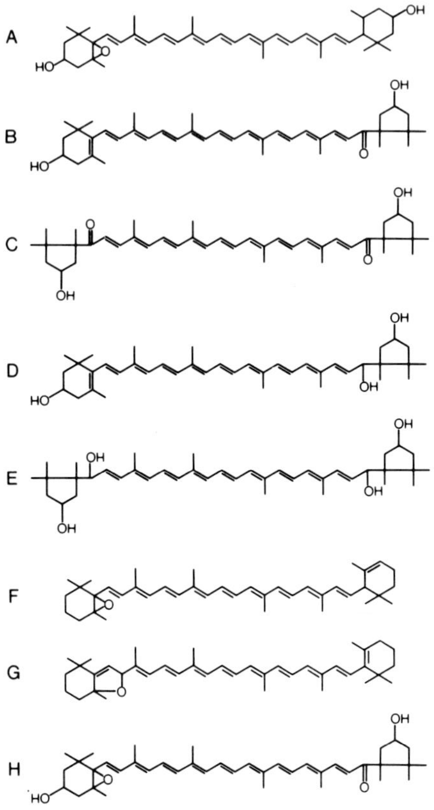

Fig. 8. Formulas of some carotenoid pigments in paprika: A, antheraxanthin; B, capsaxanthin; C, capsorubin; D, capsanthol; E, capsorubinol; F, α-carotene 5,6-epoxide; G, mutatochrome; H, capsanthin 5,6-epoxide. [From de la Mar and Frances (*242*).]

B. Annatto

Annatto is another natural carotenoid with a long history of use in coloring food products. The term "annatto" includes a whole series of coloring preparations consisting of carotenoid-type pigments, all based on extracts of the seed of the tree *Bixa orellana,* which grows abundantly in the tropics. The pigments, variable in amount (*396*), are present in the thin, resinous coating of the seeds from which the basic pigments are extracted, the crude extract being refined through saponification and steam distillation procedures to remove other unwanted fat-soluble components. The major component (*651*) consists of bixin; up to and more than 80% was found to be α-bixin, the monomethyl ester of the dicarboxylic acid α-norbixin. The stereochemistry of bixin was reported by Barber *et al.* (*46–48*). The other pigments include *cis*-norbixin, *trans*-norbixin, and an unknown pigment fraction consisting of a number of components, possibly degradation products.

Bixin is the main component of oil-soluble annatto preparations, and norbixin of the water-soluble products (*649*). Emulsions are also available. An important use of all types of annatto preparations is as a food color for the dairy industry (*42,607,808*). Annatto oil solutions of various strengths (approximately 0.2–1.5% as bixin) are available, and typical applications include coloring of butter, bakery products, and salad oil. Oil-soluble suspensions (about 2–5% bixin) have similar applications. Mixtures with paprika oleoresin yield more red colors and may be used, for example, in processed cheese; mixtures with curcumin (approximately 2–5%) are more yellow and may be used in oils and fats. It has been claimed that the color shade and stabilization of bixin can be modified by adding curcumin (*968*). Dispersions are also available as solid bars.

Emulsions of the oil-in-water type are based on the use of propylene glycol, water, and partial glycerides as emulsifiers; they contain various combinations of bixin and norbixin (approximately 0.5–3.5%) and may serve for coloring cheese and butter. Water-soluble preparations consist of solutions mainly of norbixin (about 0.5–3%) in aqueous potassium hydroxide with propylene glycol and at times polysorbate 80. Water-soluble powders or tablets contain potassium salts and have a total color value corresponding to 15% bixin. Fields of application include cheese, cereals, and ice cream. An insoluble powder is available for coloring spices and meat.

At present, many annatto preparations of widely varying quality are available on the market. Comparisons of the coloring efficacy and the price of annatto products in various countries have shown wide variations, and products can be more expensive to use in butter than is β-carotene, a carotenoid naturally present in butter (*607*).

A study of the properties of bixin and norbixin and the composition of annatto extracts was carried out by Reith and Gielen (*802*). Their work aimed at contributing to an accurate picture of the qualitative and quantitative composition of annatto butter and cheese colors; such information is necessary for drawing up specifications. The results of analytical and stability tests showed that the annatto butter color examined contained α-bixin as the principal pigment; in addition, at least eight different pigments were present. Because of complex formation of bixin and norbixin with a yellow pigment, the simple chromatographic method became rather laborious. Complex formation did not occur in annatto suspensions in oil and fat. The cheese color investigated by Reith and Geilen (*802*) was prepared by extracting the annatto seed with an aqueous alkaline solution, and the analysis showed that these extracts contained at least seven color components.

The absorption spectra (*652*) of unheated and heated bixin in corn oil are shown in Fig. 9. In several reports McKeown (*649–651*) commented on thermal degradation of bixin and listed a yellow pigment (monoethyl ester of an unsymmetric dimethyltetradecahexanedoic acid), all-*trans*-bixin, *cis*-bixin isomers, a second yellow pigment, and *m*-xylene as products. Since the main components of annatto preparations are carboxylic acids, all these products are pH sensitive. Bixin is most

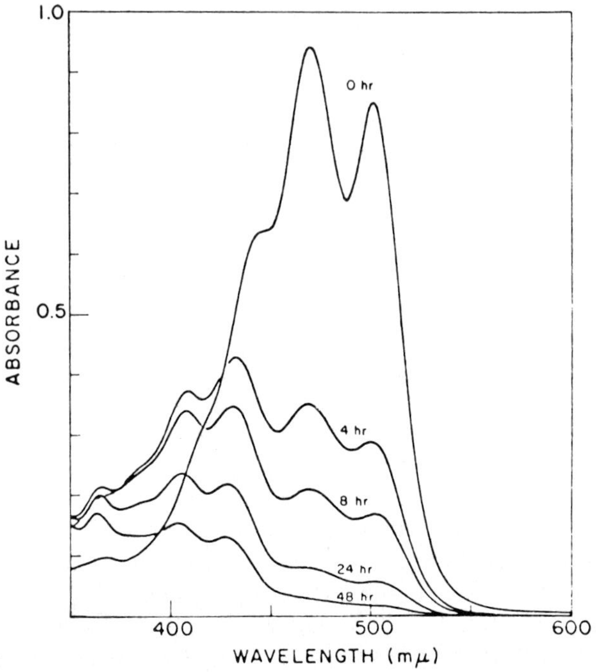

Fig. 9. Absorption spectra of a solution of bixin in corn oil, unheated and heated at 125°C for the times indicated (diluted 500 times with chloroform, 10-mm cell). [From McKeown and Mark (*652*).]

stable at pH 8, showing diminished stability in the pH 4–8 region; however, performance depends to some extent on the application forms. This sensitivity is of practical significance only in foods of low pH, in which the color may turn pink ("pinking"). Although bixin is considered to be the main component of annatto extracts, its actual relative amount may be lower than 50% of the carotenoids present. McKeown and Mark (*652*) examined the composition of commercial oil-soluble annatto food colors and obtained the results summarized in Table 6. Annatto preparations have a good shelf life, but it is generally good practice to use up the color within 3–6 months. The stability of commercial annatto butter coloring preparations was studied by Rajan *et al.* (*788*). Examination of total pigment levels during storage under prooxidant conditions showed 8.8–18.5% loss after 7 days in diffuse daylight, 15.9–29.2% loss after 30 days in diffuse daylight, and 11.5–13.3% loss after 48 hr in the presence of 4 and 5 ppm, respectively, of copper and iron at 40°C.

A method that has been developed especially for dyeing unpeeled citrus fruits consists of dissolving an annatto derivative in a hydrocarbon solvent, such as diterpene or orange oil, adding oleic acid and an amine, emulsifying the mixture with water, and applying the emulsion to the skin of citrus fruits (*963*). A thin-layer

TABLE 6

Total Pigment and Bixin Content in Commercial Annatto Preparations[a]

Color	Run	Total pigments (gm/100 gm)	Bixin (gm/100 gm)	Bixin (% of total pigments)
A	1	0.608	0.266	44.7
	2	0.608	0.277	
B	1	0.332	0.103	31.0
	2	0.332	0.103	
C	1	0.570	0.337	59.1
	2	0.575	0.340	
D	1	1.45	0.468	32.1
	2	1.45	0.462	
E	1	0.248	0.129	51.6
	2	0.252	0.128	
F	1	0.685	0.297	42.5
	2	0.689	0.288	

[a] From McKeown and Mark (*652*).

chromatographic method for identifying annatto and other food colors has been developed by Ramamurthy and Bhalerao (*791*).

C. Saffron

Saffron contains crocin, the digentiobioside of crocetin. Its pure yellow color is attractive in beverages and specialty foods (*1032*), such as risotto and curry. Montag (*678*) described the identification of saffron in bakery products and sausages. Stahl and Wagner (*885*) also published an identification method. Saffron is believed to be one of the earliest additives used to impart color to food, and until two or three generations ago it was widely used as a coloring agent in cakes and other baked goods.

Saffron is also used as a spice; it has the advantage of being widely accepted in soups, meat products, cheese, and many other foods. However, as with other natural products, there is much batch-to-batch variation in coloring as well as in flavoring strength, and, since there is no correlation between the variations of color and flavor, standardization can become a problem.

D. Tomato Extracts

Tomato extracts for coloring were found to contain 78.6–89.3% lycopene and 4.8–9.5% β-carotene (*91*). The isolation and identification of a tomato color in meat were described by Günther and Grau (*387*).

E. Carrot Extracts

Carrot extracts, carrot oil, and palm oil, and related plant food extracts originally were available on the market, and their main components are α- and β-carotenes. Processes (*304,962*) for the commercial extraction of carotenes from carrots were developed. Purified crystalline products (20% α- and 80% β-carotene), dispersions of microcrystals in oil for coloring fat-based products, and carrot oil were also available at one time. Extraction of carotene from the sweet potato was proposed in the past (*860*). However, many of these products gradually disappeared with the advent of synthetic carotenoids.

F. Xanthophyll Pastes

Xanthophyll pastes were described by Benk *et al.* (*91*). Starting materials consisted of stinging nettles, alfalfa, and broccoli, which were extracted. Two commercial pastes were examined, one containing 0.81 and the other 2.21% total carotenoids. Of the total, β-carotene represented 35.8 and 37.8%, xanthophyll esters 17.6 and 16.7%, and lutein 45.5 and 46.6%. The use of xanthophyll-containing material obtained by saponifying waxy pyrethrum residues with an alkaline agent (especially with alcoholic sodium hydroxide) has been suggested (*334*) for the pigmentation of poultry and eggs and as a coloring agent for foods and pharmaceuticals. An economic process (*970*) suitable for the commercial separation and recovery of carotenoids, phytol, and proteolytic enzymes of papaya plants also has been described.

VII. FERMENTATION PRODUCTS

The microbiological synthesis of β-carotene has received attention in recent years. *Phycomyces blakesleeanus* and *Mucor hiemalis* have been used to prepare ^{14}C-labeled β-carotene. The microorganisms *Pyronema confluens, Choanephora cucurbitarum, Rhodotorula gracilis, Sporobolomyces roseus, Candida humicola, Candida curvata, Torulopsis candida,* and *Actinomyces* no. 2426 have been examined. Various microbiological processes for producing β-carotene and other carotenoids have been studied extensively. Industrial production of β-carotene by fermentation from *Blakeslea trispora* was announced some time ago (*174*). This fungus synthesizes carotenoids when its two opposite forms are cultivated together in a special fatty medium. Ninet *et al.* (*714*) reported a two- to threefold increase in the quantity of β-carotene present in the culture media of *Blakeslea trispora* by the addition of certain amides, imides, lactams, hydrazides, or substituted pyridines. Lilly *et al.* (*592*) obtained 4 mg β-carotene per gram dry mycelium of *Phycomyces blakesleeanus.* The Northern Regional Research Laboratory has worked on developing a commercial microbiological synthesis (*185,186*). Various patents (*741*) describe methods and organisms suitable for the microbiological production of carotenoids. Some of the problems in the fermentation process are the stabilization of the crude dried carotene or carotenoid-containing mycelium (*348, 758*) and the preparation of pure crystalline carotenoids from the fermentation broths. Production of carotenoids from algae may become practical (*175,263,863*) in the future.

VIII. CHEMICAL SYNTHESIS

Carotenoids prepared by laboratory synthesis became available only relatively recently, within the last three decades. If we consider availability in terms of the volume of synthetic carotenoids produced on a commercial scale, and the period of their availability is reduced to less than the last two decades. This is a relatively short period for a new source of food colors to make inroads into extensive worldwide use, considering that other food colors have been available for many decades and some for more than a century. Yet within the past decade remarkable progress has been made in adapting pure carotenoids to the various needs of a wide range of processed, refined, and fabricated foods consumed around the world.

Some pertinent dates and events leading to commercially produced carotenoids are shown below:

Date	Event	Researcher
1831	Carotene isolated from carrots	Wackenroder
1906–1911	Column chromatography developed for carotenoids	Tswett
1907	Empirical formula of carotene established	Willstatter and Mieg
1928	Carotene gives Carr–Price reaction and cures vitamin A deficiency in rats	von Euler and co-workers
1928–1931	Structure of carotene and vitamin A established	Karrer and co-workers
1930	Carotene demonstrated to be converted to vitamin A	Moore
1950	Synthesis of β-carotene	Karrer and co-workers, Inhoffen and co-workers, and Milas and co-workers
1953–1956	Industrial synthesis of β-carotene commercial production of β-carotene	Isler and co-workers
1958–1959	Synthesis of β-apo-8′-carotenal	Isler *et al.*, Rüegg *et al.*
1959	Synthesis of β-apo-8′-carotenoic acid ethyl esters	Isler *et al.*
1960	Commercial production of β-apo-8′-carotenal	
1956–1963	Industrial synthesis of canthaxanthin	Isler *et al.*, Zeller *et al.*
1964	Commercial production of canthaxanthin	

In considering the growth of knowledge leading to carotenoid syntheses, Isler (*436*) observed that research on carotenoids can be separated into four broad periods. During the nineteenth century, the emphasis was on isolation of the pigments and their characterization by measurements of light absorption. The second period (1900–1927) centered on the determination of empirical formulas and tentative efforts to discover a role in photosynthesis. The third period (1928–1949) was dominated by the provitamin A concept, the establishment of structural formulas, and the development of synthetic methodology. The latest period (1950 to the present) is characterized by an exponential increase in the number of known carotenoids accompanied by notable advances in total synthesis and the determination of absolute configurations. The explosive growth in knowledge has been due to new methods whereby the number of individual identified carotenoids has increased greatly so that several hundred are known today.

A. Commercial Synthesis

The chemistry involved in the commercial synthesis of carotenoids is closely related to that involved in the synthesis of vitamin A now employed by several chemical manufacturers (*435a,436,438–440*). The Roche synthesis of vitamin A (*442*) and β-carotene (*447*) starts from β-ionone. The β-ionone may be prepared from citral, a constituent of lemongrass oil, or prepared synthetically from acetylene or acetone (*506,507*) as starting materials (Fig. 10). The industrial pathways (*435a*) of synthesis of vitamin A, β-carotene, and other C_{30} and C_{40} carotenoids are illustrated in Figs. 10–14. In one β-carotene procedure, β-ionone is converted to C_{14} aldehyde, which, through further chain-lengthening steps, is converted to C_{16} and then to C_{19} aldehyde. Two C_{19} aldehydes are joined through a Grignard reaction to form the C_{40} diol, which, through allylic rearrangement and dehydration, is converted to 15,15-dehydro-β-carotene. A partial hydrogenation and rearrangement

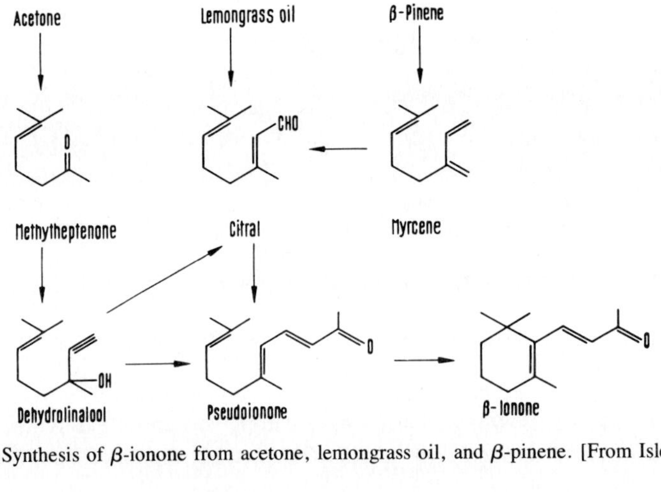

Fig. 10. Synthesis of β-ionone from acetone, lemongrass oil, and β-pinene. [From Isler (*435a*).]

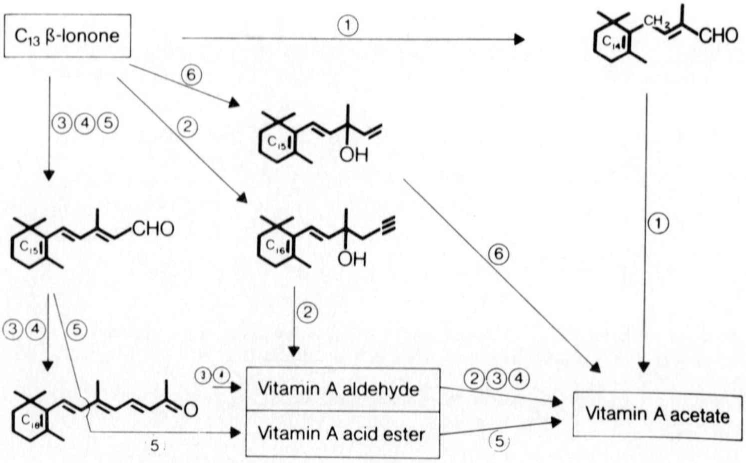

Fig. 11. Key intermediates of vitamin A manufacture. 1, Roche; 2, D.P.I./Glaxo; 3, A.E.C.; 4, Philips; 5, Sumitomo; 6, BASF. [From Isler (*435a*).]

yield *trans*-β-carotene. An alternative synthesis for β-carotene has been proposed by Surmatis and Ofner (*911*) using a C_{15} aldehyde + C_{15} aldehyde building scheme.

Synthesis (*451,826,850*) of the β-apo-8′-carotenoids, including the β-apo-8′-carotenic acid esters, are shown in Fig. 15. The synthesis of β-apo-8′-carotenal is similar to the synthesis of β-carotene through the C_{19} aldehyde; then the chain is lengthened by first condensing the aldehyde with the C_6 acetal. By a series of reactions, as used in the β-carotene synthesis for chain lengthening, the C_{30} aldehyde (β-apo-8′-carotenal) is attained. Several syntheses of canthaxanthin have been developed (*499,104,1042*). The technical synthesis of canthaxanthin may start from β-carotene. Treatment with *N*-bromosuccinimide in acetic acid and chloroform gives isozeaxanthin diacetate, which is saponified and subsequently subjected to an Oppenauer oxidation. One synthesis is shown in Fig. 13. In addition to those cited above, carotenoids that have been prepared synthetically in pure crystalline form include γ-carotene, cryptoxanthin, zeaxanthin, physalien, isozeaxanthin, lycopene, torularhodin, bixin, crocetin, astaxanthin and others, but they are not yet produced

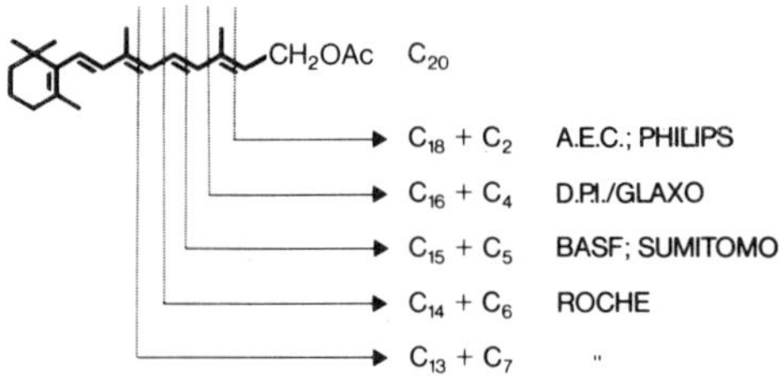

Fig. 12. Vitamin A manufacturing procedures. [From Isler (*435a*).]

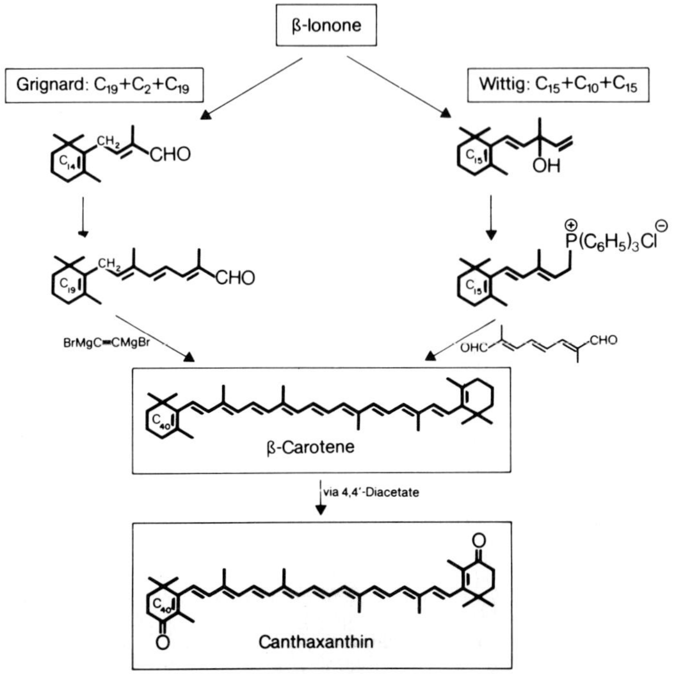

Fig. 13. Manufacture of β-carotene and canthaxanthin. [From Isler (*435a*).]

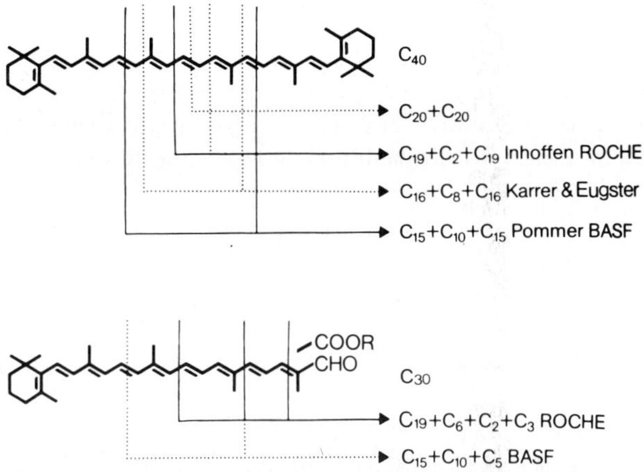

Fig. 14. Carotenoid manufacturing procedures. [From Isler (*435a*).]

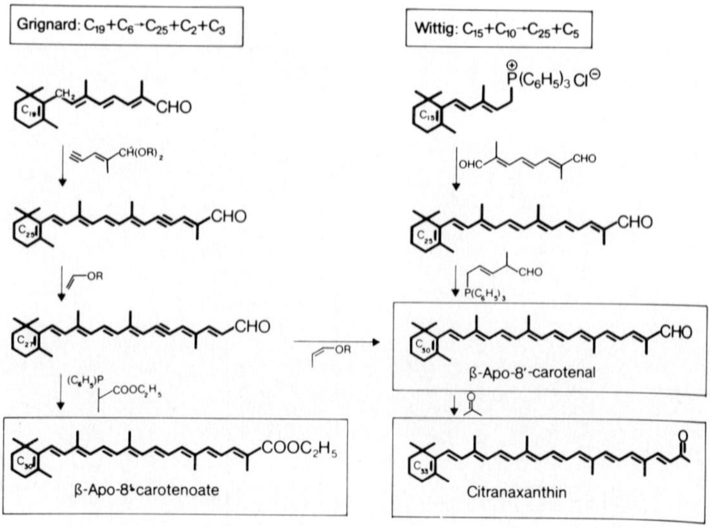

Fig. 15. Manufacture of β-apo-8′-carotenoids. [From Isler (*435a*).]

commercially. Some chemical schemes of carotenoid synthesis are indicated in Fig. 16. Surveys on the synthesis of carotenoids have been given by Isler and co-workers (*160,439,441,443–448,452–455,851*). The original carotene synthesis described by Inhoffen *et al.* (*428*), Pommer (*762*), and Sarnecki *et al.* (*336*) was adapted for commercial production by Isler and co-workers. Some new carotenoid compounds that can be used as food colorants are described in the patent literature (*973*). Reviews of current chemistry of the carotenoids have been prepared by Isler (*435a*), Mayer (*636a*), and Mayer and Isler (*637*). Other reviews (*789,1028*) have also been published. Symposia on the carotenoids have been held at intervals over the past decade (*430–434*). In Fig. 17 the structural formulas of the four carotenoids now commercially produced indicate their related structural characteristics.

Crystalline carotenoids produced by chemical synthesis are of high purity and uniform color. Technological advances have made possible the synthesis, at reasonable prices, of carotenoids with well-controlled, reproducible colors, without quality variations, and in a volume that can be scheduled to meet the needs of the food industry. The syntheses of carotenoids employ a variety of organic synthetic techniques. Readers interested in organic syntheses or details of the procedures should consult the extensive literature of primary references and reviews.

B. Physicochemical Properties

Carotenoids crystallize in a variety of forms, the crystal color varying from red through violet to almost black. The size of the crystals influences the color of the specific carotenoid. Melting points are fairly high and tend to increase with increas-

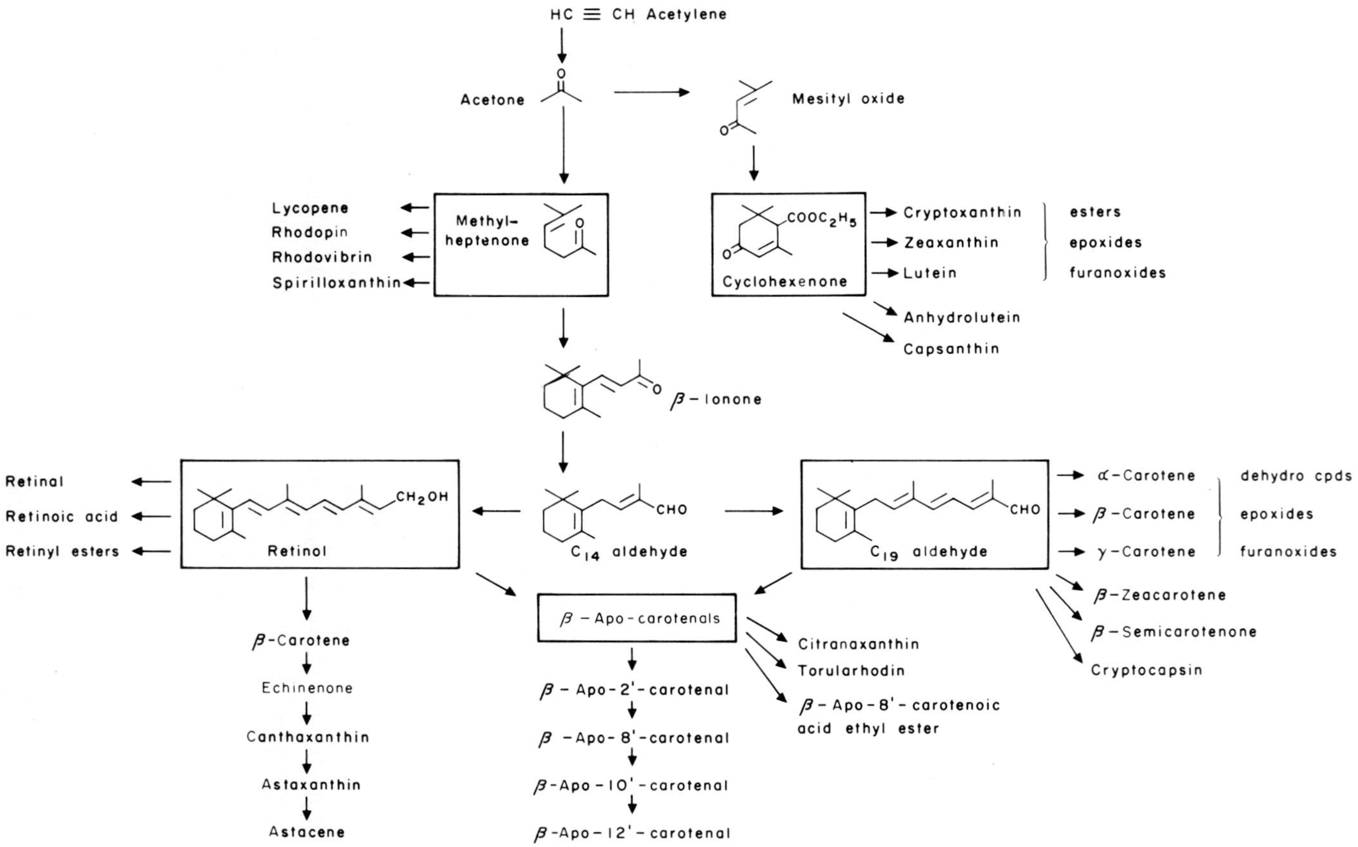

Fig. 16. Chemical schemes of carotenoid synthesis. (Courtesy of J. Surmatis.)

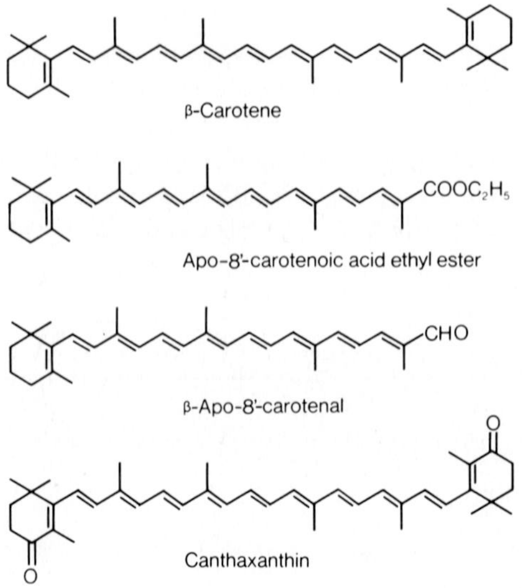

Fig. 17. Structural formulas of four commercially produced carotenoids.

ing molecular weight and functional groups. The melting points of β-apo-8′-carotenal, β-carotene, and canthaxanthin are 136°–140°C, 176°–182°C, and 208°–210°C, respectively. The physical properties of the carotenoids are discussed by Schwieter *et al.* (*849*).

Because of the conjugated double-bond structure in the molecule, the crystalline materials are very sensitive to oxidative decomposition when exposed to air. Crystals must therefore be stored in sealed containers under vacuum or inert gas at low temperatures. When micronized and suspended or dissolved in vegetable oil, however, their stability is adequate for practical use in coloring fat-base foods. Reducing the degree of unsaturation of an oil by hydrogenization increases its effectiveness in stabilizing carotene (*41*). The use of food-grade antioxidants further improves stability. Carotene stability can be enhanced as much as 20-fold by antioxidants added to the vegetable oil carrier (*1022*).

For the carotenoids used as food colors, the remarkable stabilizing effect of the tocopherols (*200,501,511,779*) and of ascorbic acid and its fatty acid esters is of particular importance. α-Tocopherol is favored as an antioxidant for carotene-in-oil solutions because of its remarkable *in vivo* properties (*71*). The oxidation of the carotenoids is accelerated by light and—in the presence of oils as a primary copper, manganese, and iron (*202,684*). Fatty acid hydroperoxides, such as linoleic hydroperoxide, initiate radical decomposition; they may also, in oil solutions, attack β-carotene directly (*634*). Gamma radiation causes the destruction of carotenoid pigments by secondary reactions. Carotenoids may also be destroyed by enzymatic oxidation, which has a definite role in the decay of vegetable matter (*377,484*), and by heat (*332,611*).

Interesting observations on the stability of carotene have been reported. Crystalline carotene in both cottonseed (Wesson) and mineral oil was examined with and without added antioxidants for stability characteristics by Morgal *et al.* (*681*). Hydroquinone (0.01; 0.1%) in cottonseed oil and phenylnaphthylamine (0.01%), α-tocopherol (0.01; 0.1%), and hydroquinone (0.01; 0.1%) in mineral oil effectively delayed carotene decomposition. Bickoff and Bickoff *et al.* found that a number of phenol (*99*), pyrogallol (*100*), bisphenol (*103*), and quinoline (*101*) anti-

TABLE 7

Effect of Various Antioxidants on the Stability of Carotene in Mineral Oil[a]

Compound tested	Time for 20% loss of carotene at 75°C (hr)
Aniline derivatives	
2-Aminobenzenethiol	32
Benzidine	6
p-Toluidine	6
o-Aminodiphenyl	5
4-Amino-1,2-dimethylbenzene	5
Aniline	3
p-Aminodiphenyl	2
Dimethylaniline	2
Dibenzylaniline	2
Benzalaniline	2
Benzylaniline	2
Naphthylamine derivatives	
1,8-Diaminoaphthalene	232
Phenyl-α-naphthylamine	193
Phenyl-β-naphthylamine	150
β-Naphthylamine	15
α-Naphthylamine	14
Diphenylamine derivatives	
4-Aminodiphenylamine	234
2,4-Diaminodiphenylamine	225
4-Isopropoxydiphenylamine	205
4-Hydroxydiphenylamine	196
4,4′-Methoxydiphenylamine	190
4-Nitrosodiphenylamine	155
4,4′-Diheptyldiphenylamine	102
4,4′-Dioctyldiphenylamine	95
4-Heptyldiphenylamine	60
Diphenylbenzidine	50
Diphenylamine	45
Benzeneazodiphenylamine	45
Carbazole	9
Acetyldiphenylamine	8
Acridone	5
Triphenylamine	4
2,4-Dinitrodiphenylamine	3

(Continued)

TABLE 7—*Continued*

Compound tested	Time for 20% loss of carotene at 75°C (hr)
Phenylenediamine derivatives	
N,N'-Diphenyl-*p*-phenylenediamine	245
N,N'-Di-*sec*-butyl-*p*-phenylenediamine	245
N,N'-Di-4-(2,6-dimethylheptyl)-*p*-phenylenediamine	226
Di-*b*-naphthyl-*p*-phenylenediamine	210
p-Phenylenediamine	106
2,4-Diaminotoluene	98
o-Phenylenediamine	78
p-Aminoacetanilide	3
Quinoline derivatives	
Polymer of 2,2,4-trimethyl-1,2-dihydroquinoline	154
6-Ethoxy-2,2,4-trimethyl-1,2-dihydroquinoline	140
Acid rearranged 2,2,4-trimethyl-1,2-dihydroquinoline	92
6-Phenyl-2,2,4-trimethyl-1,2-dihydroquinoline	51
2,2,4-Trimethyl-1,2-dihydroquinoline	44
8-Hydroxyquinoline	6
2-Hydroxylquinoline	2
Quinoline	2
Isoquinoline	2
Urea and related compounds	
sym-Diphenylcarbazide	205
1-Phenylsemicarbazide	195
1-Phenylthiosemicarbazide	190
4,4-Diphenylsemicarbazide	78
Diphenylthiocarbazide	75
Di-*o*-tolylthiourea	60
Diphenylguanidine	9
Allylthiourea	8
N,N'-Diphenylurea	6
Diphenylthiocarbazone	4
N-Phenyl-*β*-hydroxyethylthiourea	5
Urea	3
N,N'-Diphenylurea	2
Thiourea	2
Triphenylguanidine	2
Phenothiazine derivatives	
Phenothiazine	155
Phenothiazone	210
Thionol	160
Azobenzene derivatives	
p-Aminoazobenzene	8
p-Hydroxyazobenzene	5
Azobenzene	2
Diphenylmethane derivatives	
sym-Diphenylethylenediamine	14
p,p'-Diaminodiphenylmethane	6
4,4'-Tetramethyldiaminotriphenylmethane	5
Tetramethyldiaminodiphenylmethane	2

[a] From Bickoff *et al.* (*102*).

oxidants greatly reduced the destruction of carotene (Table 7) in mineral oil (*102*). Alkylation of the phenolic compounds in general enhanced this activity. In instances, an excellent stabilizing effect was obtained with 2,4-dimethyl-6-butylphenol, 3-laurylcatechol, thionol α-tocopherol, ethyl gallate, 1,5-dihydroxynaphthalene, and *N-sec*-butyl-*p*-aminophenol. Animal fat and the antioxidant 6-ethoxy-1,2-dihydro-2,2,4-trimethylquinoline were effective stabilizers in carotenoid concentrates (*597, 931*). Nordihydroguaiaretic acid (*104,105*) is an effective antioxidant for carotene, and its effect is enhanced when used with phospholipid and citric acid. The effectiveness of α-tocopherol (*104,514*) is improved by adding acid-type inhibitors. Carotenoid stability can be improved by the addition of suitable antioxidant formulations involving ascorbic acid and its fatty acid esters, tocopherol, lecithin, butylated hydroxyanisole (BHA), and butylated hydroxytoluene (BHT). Ascorbic acid has been studied in model systems (*200,486*). The tocopherols are widely distributed in food (*51*).

Another deterrent encountered in the practical application of carotenoids to foods is the relatively poor solubility of these natural pigments. Therefore, in working with carotenoids, a knowledge of their solubility characteristics is quite helpful. Carotenoids are insoluble in water, slightly soluble in vegetable oils, moderately soluble in aliphatic and aromatic hydrocarbons, and very soluble in chlorinated hydrocarbons such as chloroform. The solubilities of β-carotene, β-apo-8′-carotenal, and canthaxanthin are given in Table 8 along with other physicochemical data (*58*). The purer the compound, the more limited the solubility. Crystalline *trans*-β-carotene prepared by chemical synthesis possesses high purity since, when chromatographed on alumina or lime, one principal zone is obtained. Absorption spectra (*450,895*) of β-carotene are plotted in Figs. 18 and 19. The solubility of crystalline β-carotene in organic solvents and vegetable oil was observed by Mikhailovnina and Savinov (*665,666*). Solution velocity was reported to be very slow, and a tendency to form supersaturated solutions was noted. The variation of β-carotene solubility in edible vegetable oils is little affected by the composition of the oil (*665*). However, in spite of their relatively low solubility in oils and fat, their high tinctorial power largely overcomes this handicap. In conditions of practical use in fat-base foods, the levels of carotenoid employed as coloring are well below their solubility limits. Solubility in lipoid solvents, such as oils, can be dramatically increased by heating, and, although crystallization occurs on cooling, this property can be put to practical use both in the manufacturing of food application forms and in the coloring of certain food products.

Another characteristic of the carotenoids is the cis–trans phenomenon. The number of possible carotenoids is increased by cis–trans isomerization. Theoretically, each double bond in the β-carotene chain, including the two in the β-ionone rings, can exist in two configurations. Although 1056 *cis,trans*-β-carotene isomers are theoretically possible, in actuality methyl groups along the chain cause steric hindrance, which greatly limits the number of rearrangements (*1040*). The specific absorption curve of carotenoids is modified by isomerization from the trans to the

TABLE 8

Physicochemical Data on Carotenoid Food Colors[a]

Property	β-Carotene	Apocarotenal (C_{30})	Apo ethyl ester (C_{30})	Canthaxanthin
Color of crystals	Violet-red	Purplish-black	Orange-red	Brown-Violet
Color of oily solution	Light yellow to orange	Light orange to orange-red	Yellow to orange	Red-orange to red
Color of aqueous dispersion	Yellow-orange to orange	Orange-red to tomato red	Yellow to reddish-orange	Red shades
Melting point (°C), crystalline compounds	176–182	136–140	134–138	~210
Molecular weight	536.9	416.7	460.7	564.9
Solubility (gm 100 mg solution, 20°C)				
Fats, oils	0.05–0.08	0.17–1.5	~0.7	0.005–0.02
Orange or lemon oil	0.2–1.0	2–4	3–5	0.02–0.08
Water	Insoluble	Insoluble	Insoluble	Insoluble
Glycerol	Insoluble	Insoluble	Insoluble	Insoluble
Ethanol	Below 0.01	~0.1	Below 0.1	Below 0.01
Cyclohexane	~0.1	~0.8	~2	Below 0.01
Ether	~0.1	~1.5	~2.5	~0.03
Benzene	~2	~12	~16	~0.2
Chloroform	~3	~20	~30	~10
Acetone	~0.1	~1.8	~1.2	~0.03
Spectrophotometric data[b]				
λ_{max} (nm) (in cyclohexane)	455–456	460–462	448–450	468–472
$E^{1\%}_{1cm}$ (nm) (in cyclohexane)[c]	456 > 2400	461 > 2530	449 > 2440	470 > 2110
	485 > 2030	388 > 2030	475 > 2000	

[a] From Bauernfeind (*58*).

[b] The carotenoids isomerize in solution during storage or on heating; the predominant factor in the equilbrium mixture is the trans compound. The values given correspond to the trans isomer.

[c] Of commerical products.

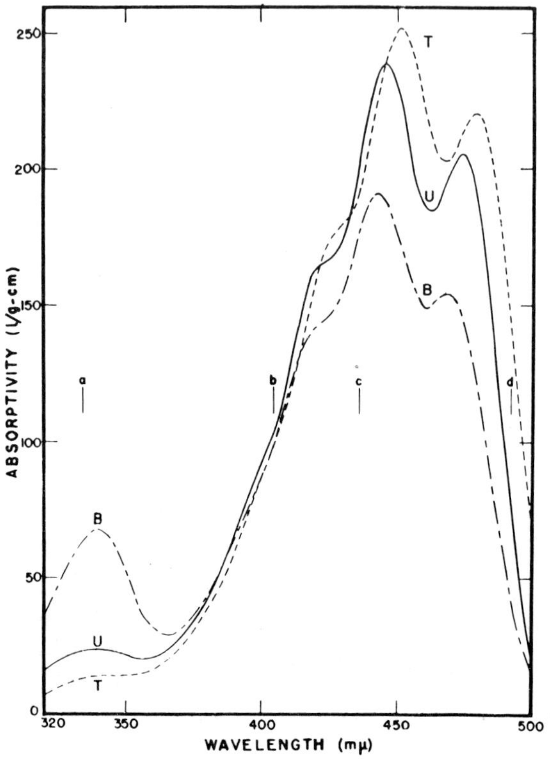

Fig. 18. Absorption spectra of three stereoisomers of β-carotene. Key: B, neo-β-carotene B; U, neo-β-carotene U; T, all-*trans*-β-carotene; a, b, c, d indicate the location of mercury arc lines 334.1, 404.7, 435.8, and 491.6 nm, respectively. [From Stitt *et al.* (*895*).]

cis form, which is characterized generally by the appearance of a cis peak in the near ultraviolet and a corresponding lowering and slight shifting toward the ultraviolet of the main absorption peaks. The majority of the carotenoids, however, occur in nature as the all-trans form, with mono-*cis* forms occurring occasionally and poly-*cis* forms rarely. The practical significance of a shift from an all-*trans* arrangement to a *cis* arrangement is a very slight loss in color strength and hue. In most situations the carotenoids in foods, whether naturally present or added, particularly if heat has been applied, are in a cis–*trans* arrangement in equilibrium and hence rarely pose problems of technical significance.

IX. APPLICATION FORMS

Market forms (*78*) of synthetic carotenoids are of several general types: oil- and water-dispersible liquid products and dry products (Fig. 20). The pure crystals are

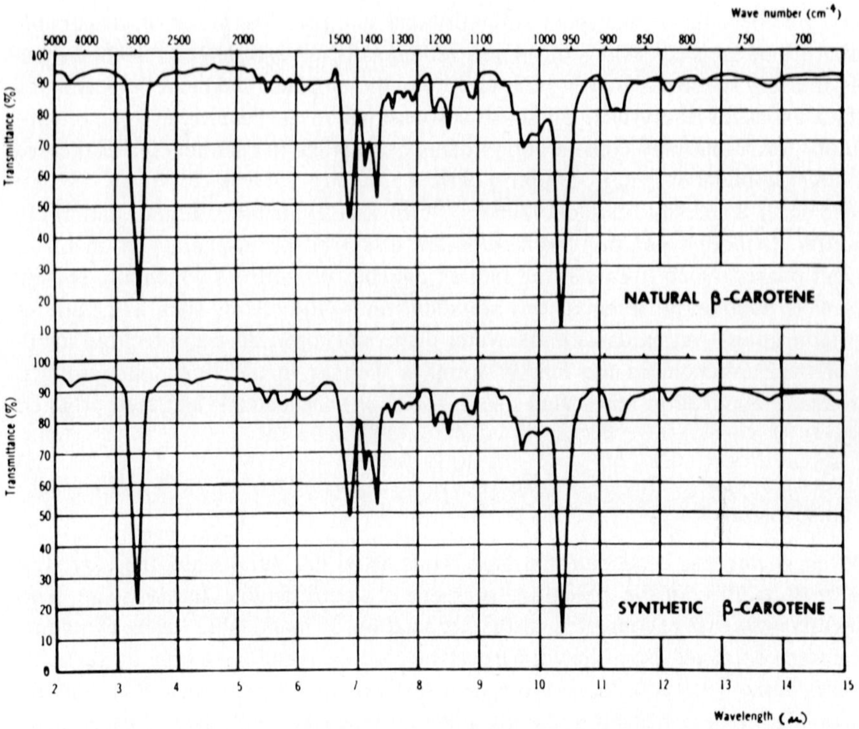

Fig. 19. Infrared absorption spectra of β-carotene. [From Isler *et al.* (*450*).]

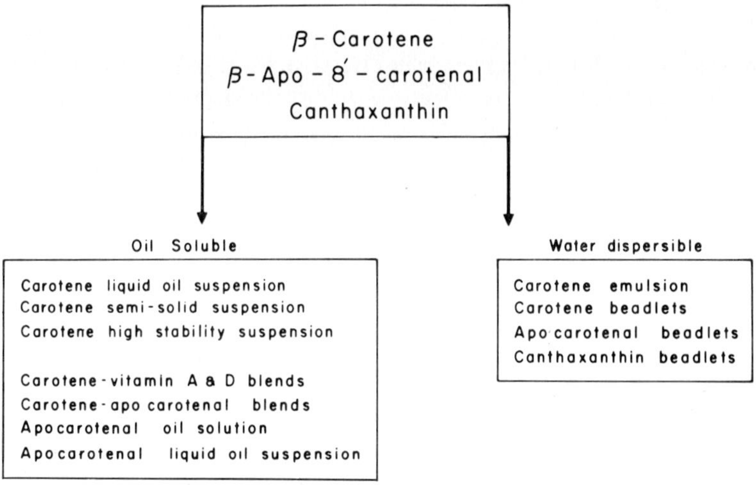

Fig. 20. Some carotenoid market application forms. [From Bauernfeind *et al.* (*78*).]

rarely used because of their poor solubility and stability. The major oil-dispersible market forms are suspensions of micronized crystals in vegetable oil. Such suspensions are very stable due to the good stability of the oil phase and the relatively small surface exposure of crystals. These oil suspensions do not require antioxidants for stability per se but may contain antioxidants to stabilize the carotenoids in the end use stress applications, e.g., popping corn.

The water-dispersible carotenoid market forms are available as liquids and as dry products. In both types the carotenoids are dissolved or suspended in an oil or solvent phase, which is emulsified into an aqueous phase from which any solvent present is removed. The emulsions are stabilized with colloids such as gelatin or vegetable gums. Antioxidants in the water-dispersible products can affect the stability of the carotenoid in the market form as well as in the final food product. Carotenoid application forms find use in food, pharmaceutical, and feed products (*78*).

A. Suspensions

When β-carotene is added to fat-base foods on an industrial scale, the very slow rate of dissolution of the crystalline pigment is a considerable disadvantage. This difficulty was first encountered in the preparation of marketable forms of natural β-carotene extracted from dehydrated carrots and has been overcome by reducing the crystal size (*960*). Reduction in crystal size hastens the rate of solution, and this principle easily adapts itself to the manufacture of marketable forms of β-carotene. Hartmann and Barnett (*960*) were the first to reduce the size of crystalline β-carotene in the preparation of a useful carotene oil paste composition for food application use. Reduction in crystal size alone is not enough, however. A dry milling operation reduces crystal size, but dry, micropulverized, crystalline *trans*-β-carotene requires an inert gas atmosphere; otherwise, it is subject to progressive oxidative destruction (*314*). Reduction in crystal size can be done within a liquid oily or aqueous environment. The vegetable oil or hydrogenated fat of suspensions of micropulverized β-carotene provides a barrier to oxygen and air and retards oxidation.

The physical form of the vegetable oil carrier is also of practical concern. Pure β-carotene suspended in an edible oil has been ground in an attritor-type mill under nitrogen to a particle size of approximately 2–5 μm. The microcrystalline β-carotene so ground and suspended in oil has a large crystal surface area. Suspensions usually contain 20–40% of the carotenoid by weight. If a liquid vegetable oil carrier is used to prepare a β-carotene suspension, the resulting product has the physical advantage of pourability, but such suspensions, upon standing for prolonged periods (weeks), may show some settling of the suspended crystals if the crystals are not very fine. Hence, bulk containers of these suspensions must be stirred or agitated to ensure homogeneity if the contents are used over widely intermittent intervals of time. By preparing this type of suspension at an optimal level of crystal content and crystal size, the settling tendency is minimized.

For example, oil suspensions (*79*) of micronized crystalline β-apo-8′-carotenal were prepared by grinding the crystals in oil suspension in a small Szegvari attritor. Micronized suspensions were made in hydrogenated coconut oil and in peanut oil at the potency range of 20–30%. The grinding of the suspension was carried out for 5–6 hr under a blanket of inert gas. In the case of the hydrogenated coconut oil, the temperature of the grind was held at 43°–49°C by circulating warm water through the jacket of the attritor in order to keep the fat in the liquid state. The suspensions were chocolate brown in color. A 30% suspension in peanut oil was just pourable at room temperature, whereas suspensions in hydrogenated coconut oil become solid at this temperature. The stability of samples of these suspensions, with and without antioxidants, was tested at different temperatures (23°, 37°, and 45°C) for various periods of time. No significant losses were observed at temperatures up to 45°C for 6 weeks, 23°C for 1 year, or 37°C for 3 months.

Dilute suspensions can be prepared by dilution with oils and the tendency of these less viscous preparations to settle out and agglomerate can be prevented by several approaches. One method, for example, involves dissolving stearic acid in the suspension and cooling, thereby forming a thixotropic system with a reversible gel structure (*62,79*). When this carotene–fatty acid–vegetable oil mixture cools, a semisolid gel forms, which is retained at room temperature until the composition is stirred before use to become a pourable liquid. Although this form of β-carotene enjoys unique characteristics for general usage, it has been particularly successful on a research basis, in small-scale trials in the coloring of butter churned by the batch process. Another method to ensure a physically stabilized product of micropulverized β-carotene is to employ a hydrogenated vegetable oil as the oil diluent since the semisolid character of the base holds back crystal settling. This type of product, at room temperature (23°C), can be readily scooped out of a bulk container and weighed or, if heated to 40°C or above, becomes pourable. Still another method is to micronize the crystals to an average size of below 1 μm. However, this approach leads to reduced pourability of the suspension.

When preparations of combined β-carotene and vitamin A (for example, for the simultaneous coloring and fortification of margarine) are used in an easily pourable composition, more dilute suspensions of micropulverized β-carotene in a solution of vitamin A in vegetable oil are employed. To ensure homogeneity, these can be packaged in custom, batch-size, sealed containers directly after manufacturing. Settling out of the β-carotene in the sealed container is of no concern since, in the use of the product, the can is flushed out with hot oil and the entire contents are used at one time.

Since the market forms for fat-base foods usually consist of micropulverized β-carotene suspended in a vegetable oil composition base and are packaged and marketed in sealed containers, stability after manufacture has not been a problem.

The vegetable oil composition acts as an oxygen barrier, and, if the headspace of the container is replaced by inert gas, CO_2, or N_2, product stability is excellent. After months of storage in closed containers, the top surface of carotene preparations may show some color fading due to slow oxidative changes caused by any residual oxygen in the sealed container. This slight surface change, held to a minimum with inert gas packing and by an aluminum foil disk on the semisolid suspensions, does not result in a significant decrease in carotenoid potency or color value of the contents. All micronized carotenoid products may be safely held at refrigerated to room-temperature (23°C) storage conditions for a 6-month storage period. However, cool storage for these carotene concentrates is advised whenever possible. Prolonged storage at temperatures of 40°C and above should be avoided. The stability performance of the three β-carotene suspension products (*79*) discussed is illustrated in Table 9.

B. Solutions

It is possible to prepare and market true solutions of certain carotenoids, such as β-apo-8′-carotenal, for food coloring applications at relatively low carotenoid content (1–4%) and corresponding high oil content. The stability characteristics (*69*) of such oil solutions have been observed (Table 10). For use in certain foods such as salad dressings these have been economically attractive to the food industry. Mi-

TABLE 9

Stability of Market Forms of β-Carotene[a]

				Retention (%)		
				6 weeks storage		6 months storage
Form	Run	Potency (% carotene)	Container	40°C	45°C	23°C
β-Carotene suspension	1	24.0	Sanitary metal can	—	97	100
	2	24.0	Sanitary metal can	—	92	100
	3	24.0	Aluminum tube	99	—	100
	4	30.0	Sanitary metal can	98	—	100
β-Carotene gel	1	1.0	Sanitary metal can	93	97	98
	2	3.5	Sanitary metal can	100	100	100
	3	3.5	Laminated plastic pouch	100	100	97
	4	7.0	Laminated plastic pouch	100	100	100
β-Carotene–vitamin A blend	1	1.4	Sanitary metal can	—	100	100
	2	1.4	Sanitary metal can	—	98	100
	3	1.6	Sanitary metal can	—	100	—
	4	3.7	Sanitary metal can	—	100	—

[a] From Bauernfeind *et al.* (*79*).

TABLE 10

Stability of β-Apo-8′-carotenal Solutions in Vegetable Oil[a]

Solution	Initial potency (%)	Retention (%) 2 months (23°F)	3 months (45°C)	3 months (7°C)	3 months (23°C)
1% apocarotenal in vegetable oil					
No antioxidant	0.97	84	76	93	93
BHT/BHA	0.97	94	76	95	92
1.5% apocarotenal in vegetable oil					
No antioxidant	1.42	91	80	97	86
BHT/BHA	1.42	94	83	95	97

[a] From Bauernfeind and Bunnell (*69*).

cronizing carotenoids increases the rate of solution, but it does not increase the total solubility of a specific carotenoid. An increase in concentration can be achieved by heating β-carotene and other carotenoids in oils or fats to above 100°C (*961*). Solutions of β-carotene in oil properly packaged, stored, and protected from light can serve, for convenience, as laboratory standards (*783*).

C. Aqueous Products

The food application forms described so far are suitable for coloring oils and fats, and in some cases other fat-base food products as well, but usually they cannot be applied to aqueous media or water-base foods. Nature, however, has not confined carotene to lipoidal systems but by means of protein complexes and colloidal structures has allowed carotene to be dispersed in primarily aqueous systems. For example, Troitskii and Tarasova (*948*) have shown that carotene forms complexes with blood proteins, which may contain as much as 21% β-carotene. The carotenoids in sweet potatoes, orange juice, tomato juice, carrot juice, etc., are most likely present in the form of complexes with proteins or lipoproteins. These are examples of water-dispersible carotenoids in nature.

At first the conversion of crystalline carotenoids with their highly unsatisfactory solubility characteristics to water-dispersible preparations was quite a problem. One early approach omits the pure crystalline compounds and uses natural carotenoid sources instead. These pigments can be obtained from carrots by extraction, the water-soluble substance being purified by the formation of a metal salt (*464*). The conversion of aqueous citrus fruit extracts to dry powders is achieved by mixing with water-insoluble swelling powders, such as cellulose, starch, or pectin (*335*). A concentrated carrot juice has been prepared by lactic acid fermentation of carrot mash (*271*). Yellow edible coloring agents are obtained by mixing plant cells of spices, such as curcumin, paprika, or saffron, with a carrier material, such as gelatin or starch, suspending the mixture in water, and drying (*337*). A process (*339*) has been described for preparing absorbates of natural pigments containing the complexes without denaturing. The process consists of centrifuging carrot juice together with starch and drying the mixture, avoiding any denaturation, precipitation, or coagulation, i.e., maintaining the product in a colloidally dispersible form. As a raw material, juices of carrots are used. Water-insoluble or swelling powders are suspended into these juices, and the resulting coloring agent is separated by filtration, sedimentation, or centrifugation. Suitable water-insoluble or swelling powders comprise starches, modified starches, and pectins. After separation, the resulting solid mass can be dried either conventionally or by adding a water-absorbing powder, such as saccharose, starch, or anhydrous dextrose.

Using pure crystalline carotenoids, it is possible to prepare water-dispersible carotenoids by (a) the formation of colloidal suspensions, (b) the emulsification of oily solutions, and (c) dispersion in suitable colloids, particularly with the addition of surface-active agents. The formation of *colloidal* suspensions is usually accomplished by dissolving carotene in a water-miscible solvent and adding this solution to water, followed by removal of the solvent. However, concentrations in these suspensions are very low, and they show, in general, an unnatural pinkish tinge. *Emulsions* of oily solutions of carotenoids in aqueous media can be prepared with the aid of emulsifying agents, but the poor solubility of carotenoids again allows, in such emulsions, only low levels of carotenoid potency (*975*).

D. Emulsion Products

Although the above-described methods can be used to produce a water-dispersible form of β-carotene, in order to be economically feasible a commercial form must have a reasonable concentration of β-carotene in a convenient form that can be used in a small amount to produce a desired color effect in a finished food product. The normal limiting factor in the preparation of useful forms of β-carotene is the relatively low order of solubility in solvents and vegetable oils (*975*). The problem of producing a dry form of β-carotene that was stable, water-dispersible, and of a practical potency was solved by the discovery that a supersaturated solution of high concentration produced by dissolving β-carotene in hot vegetable oil (above 100°C) would not crystallize from solution if it was rapidly emulsified before crystallization started. Emulsions (*74*) in liquid form are also stable in closed conditions (Table 11). The resulting emulsion was then converted to dry particle form, which retained the carotene in solution in the finely dispersed oil droplets for an indefinite period. Dispersion of the dry product in an aqueous system did not affect the stability of the dissolved β-carotene in the fine oil droplets with regard to crystallization. This discovery opened the way for the development of a low-potency, stable, dry beadlet form of carotenoids for use in coloring water-base foods. Some of the methods of preparing dry particle forms were previously described in the fat-soluble vitamin literature by Kläui *et al.* (*520*).

TABLE 11

Stability of β-Carotene Emulsions[a]

		Retention (%) with temperature and time		
Emulsion product	Initial potency (%)	45°C, 45 days	23°C, 3 months	23°C, 6 months
G-2-245	3.51	109	102	103
G-1-124	3.65	100	102	—
B-3-61	3.61	100	100	103
B-3-87	3.52	108	104	107
G-1-274	4.60	99	102	104
G-2-154	4.83	105	108	108
G-3-3	4.87	106	102	102

[a] From Bauernfeind *et al.* (*74*).

As an example, in the preparation of dry, water-dispersible β-carotene beadlets, a vegetable oil suspension of crystalline β-carotene stabilized by adding food-approved antioxidants is heated to dissolve the β-carotene crystals. As soon as the carotene is completely dissolved, the hot solution is emulsified in a gellable colloid–plasticizer composition, such as a gelatin–sugar solution. The key to the successful preparation of this product is the complete emulsification of the hot β-carotene solution before crystallization of the carotene has started. Once the emulsification is complete, no further precautions are necessary, and the emulsion is then sprayed and dried to a uniform, dry beadlet product. The carotene content of the beadlets can be varied to a certain extent, depending on variables in the formula. A potency of 2.4% was chosen as the standard β-carotene content in a product identified as dry β-carotene beadlet type 2.4 S, water-dispersible (*158*).

The dry β-carotene beadlets are spherical and have an orange-red color and a mesh size range of 30–80. The outer surface is irregular, and this property aids in keeping the beadlets uniformly distributed in a dry mix. The beadlets are very hard and capable of withstanding high pressures. The finely dispersed oil droplets containing the dissolved β-carotene can be seen throughout the interior. They disperse readily in water to give a cloudy yellow to orange solution resembling orange juice.

In the dry β-carotene beadlet, the carotene is well "sealed in" by the gelatin–sugar matrix so that the stability of the carotene in the beadlet is excellent. Stability tests (*158*) on the dry β-carotene beadlets were run. Data from these stability trials (Table 12) on several preparations of dry β-carotene beadlets demonstrate stability to be excellent. Similar data (*69*) have been collected on the β-apocarotenal beadlet (Table 12). Whereas ordinary room humidity has little effect on the beadlets, the sugar–gelatin matrix of the beadlet becomes tacky and tends to fuse at very high humidities if not protected by suitable packaging. Furthermore, under such storage conditions losses in carotenoid content result.

Some methods of preparing water-dispersible preparations will be mentioned.

TABLE 12

Stability of Beadlet Preparations

	β-Carotene[a]				
Lab no.	10 days, 45°C, 85% RH,[b] open beaker	6 weeks, 45°C, open beaker	6 months, room temp, (23°C), open beaker	12 months, room temp (23°C) Open beaker	12 months, room temp (23°C) Closed vial
75-13	88	100	98	90	96
75-14	55	100	100	79	99
75-15A	75	100	100	90	93
75-16A	46	95	92	78	92
75-24B	50	100	87	82	100

β-Apo-8′-carotenal[c]

		Retention (%)					
		23°C, 85% RH[b]		42 days, 45°C		12 months, 23°C	
Lot no.	Initial assay (%)	10 days	21 days	Open	Closed	Open	Closed
G-2-85	2.4	92	94	92	100	90	100
G-1-218	11.8	99	74	85	100	83	100

[a] Bunnell *et al.* (*158*).
[b] RH, relative humidity.
[c] Bauernfeind and Bunnell (*69*).

Ashida *et al.* (*471*) dissolved β-carotene (1.5 kg) with an antioxidant in coconut oil (3 kg) at 185°C under nitrogen gas and poured this into a hot mixture of glycerol, ethanol, and sucrose palmitate (4 kg). They subjected the mixture to homogenization, which resulted in a water-soluble food coloring. Walter and Purcell (*974*) obtained a water-dispersible product by adding an ether–ethanol solution of β-carotene to boiling solutions of soluble starch under agitation and recovering the water-dispersible precipitate. Water-dispersible carotenoid preparations were developed by Ikawa *et al.* (*473*) by homogenizing heated natural essential and plant oil solutions of the carotenoid in a gum arabic–dextrin mixture and spray-drying or by emulsifying an alcoholic solution of the carotenoid into a mixture of sucrose acetate, hexaisobutyrate, coconut oil, and gum arabic.

The solubility of *trans*-β-carotene in orange oil on heating at 65° and 100°C for 10, 30, 60, and 120 min was studied by Terasaki and Mima (*922*). Orange oil dissolved approximately 5% β-carotene, but some isomerization and decomposition occurred during heating. The optimal heating conditions were found to be 30 min at

100°C in the presence of α-tocopherol. The strong and specific odor of orange oil and most of the aromatic solvents mentioned above restricts the fields of application of such products. The abietic acid esters proved to be suitable for wider application (*923*).

Besides the application of heat, which leads to some isomerization of the dissolved carotenoid, it is also possible to increase the concentration by the use of a better solvent and thereby increase the concentration of carotenoid in the finished preparation. Solvents that dissolve relatively large amounts of carotenoids include orange oil (*463*), diterpene acid esters, e.g., abietic acid esters and its other derivatives (*472*), acetoglycerides (acetylated partial fatty acid glycerides) (*466*), aromatic aldehydes, and aromatic alcohols or their derivatives (*467*). Particularly interesting solvents are vitamin A and vitamin E (*84*). Extremely high concentrations of carotenoids, particularly apo-8′-carotenal and citranaxanthin, are soluble in vitamin A, vitamin E, and their derivatives having a melting point below 100°C. The solutions can be emulsified into an aqueous gelatin solution and converted to dry products, for example, by spray-drying or by a double dispersion.

E. Colloidal Products

Fine dispersions of carotenoids can be obtained by using solvents (*517*) in combination with surface-active agents, such as polyhydric alcohol esters of fatty acids (e.g., sucrose fatty acid esters), and emulsifying these solutions into water or aqueous dispersions of special colloids (*461*). In another process, β-carotene, polyvinylpyrrolidone, butylhydroxyanisole, and dibutylhydroxytoluene are dissolved in chloroform, and the solvent is then evaporated; the resulting solid is soluble in water (*465*). Copolymers of vinylpyrrolidone and vinyl acetate or methylcellulose and methylcellulose derivatives can also be used as colloids (*338*). Suitable solvents include chloroform, carbon tetrachloride, methylene chloride, and trichloroethylene. The carotene and antioxidant are first dissolved in the solvent, and then the macromolecular compound is added. In a subsequent process, methylcellulose and powdered water-soluble polysaccharides, sugars, or polyhydric alcohols plus antioxidants such as butylhydroxyanisole, dibutylhydroxytoluene, and others are mixed with a carotenoid dissolved in chloroform; the resulting product after evaporation of the solvent, is soluble in water (*468*). In this example, a solution of methylcellulose in chloroform containing carotene and antioxidants is added dropwise to anhydrous dextrose with stirring and then dried. The solutions obtained by means of the above-discussed procedures are quite stable, and, as in aqueous dilutions, the carotene can be further stabilized by the addition of ascorbic acid (*462,518*). However, the use of these special colloids and solvents in foods is prohibited in most countries.

A congealing process has been described (*469*). Carbowax 4000 was heated to 170°C under carbon dioxide and β-carotene, BHA and BHT were added, and finally the mixture was cooled rapidly. A flaky, water-soluble β-carotene preparation resulted which is claimed to be useful as a food dye. Similarly, poly(propylene glycol) (MW 1200) gave an oily, water-soluble product containing 30% β-carotene.

Carotenoid pigments (based on protein complexes) are obtained (*967*) by emulsifying an oily carotenoid solution in an aqueous dispersion of a protein (such as milk, soy, or peanut protein), homogenizing the resulting mixture, and maintaining it at a pH within the range of 5–9 and below the denaturing temperature until a complex of the carotenoid and the protein results. In such a carotenoid–protein complex, a substantial part of the carotenoid becomes unextractable by petroleum ether. The water-dispersible complexes have improved color stability, particularly at baking temperatures.

Stable, water-soluble, transparent compositions are obtained by heating carotenoids, such as carotene, lycopene, lutein, and cryptoxanthin, with hydrogenated castor oil–poly(oxyethylene) ether (degree of polymerization 60–100) to 100°–150°C under nitrogen (*470*). Aqueous dilutions of these preparations look quite appealing; however, nonionic emulsifiers are not permitted in foods in most countries.

Water-dispersible carotenoid preparations that disperse easily in water are obtained by emulsifying a solution of a carotenoid in a volatile solvent into an aqueous solution of a hydrophilic colloid, evaporating the solvent, and optionally converting the dispersion to a dry product (*966*). Still finer dispersions are produced by using salts of palmitoylascorbic acid as surface-active agents (*518,879,969*). The high degree of dispersibility is a characteristic feature of the coloring agents prepared by this process. Aqueous solutions possess a strong coloring power, and they show an absorption curve that is similar to that of a true solution, whereas colloidal suspensions obtained by the usual processes have a low and unspecific absorption. It is assumed that the carotenoid in these coloring agents is present as hydrophobic particles of crystalline structure and colloidal size. The particles are stabilized by a hydrophilic protective colloid. In aqueous solutions the colloidal particles protect them from flocculating. Electron microscopic examination of the residue after evaporating an aqueous solution illustrates the extreme fineness of the particles (*523*).

Water-dispersible preparations containing the carotenoid in fine dispersion in a gelatin matrix, the particles being less than 2 μm in size, may suffer from the disadvantage that, upon aging, water dispersibility decreases. This occurs particularly with carotenoids containing an aldehyde or ketone functional group such as apocarotenal, canthaxanthin, capsanthin, capsorubin, astacene, and β-citraurin. Borenstein and Bunnell (*971*) found that addition of an alkali metal bisulfite enhances the shelf life of such water-dispersible preparations considerably.

The stabilized water-soluble powders of β-carotene, β-apo-8′-carotenal, β-apo-8′-carotenoic acid ethyl ester, and canthaxanthin currently marketed contain 5–10% of the pure active substance and are brownish to violet-brown in the solid state. They dissolve in water to give slightly cloudy dispersions. β-Carotene and the ethyl ester of apocarotenoic acid give a yellow to reddish-orange solution (depending on the concentration), apocarotenal gives a yellowish-red solution, and canthaxanthin a red solution. One example can be cited. Because of the poor solubility of canthaxanthin in vegetable oil, major emphasis in market-form development was placed on water-dispersible beadlets. This form is a dry, 40- to 100-mesh beadlet containing emulsified canthaxanthin in nearly colloidal-size oil droplets in a matrix of gelatin. The process is described in the patent literature (*966*) and was adapted to the preparation of a 10% canthaxanthin beadlet. Representative batches (Table 13) were subjected to stability testing and found satisfactory at room temperature (23°C) and elevated temperatures (*157*).

The change of color shade with varying concentration is a property that water-soluble carotenoids have in common with lakes. At the same time, however, they share with soluble dyes the direct proportionality of coloring strength associated with pure dye content. The reason for this peculiar behavior is obviously connected with the fact that, although the dispersions are very fine, the carotenoids are still present in the form of insoluble particles. The characteristics of the color shade and strength, as well as the stability and chemical behavior of the color, depend not only on the chemical nature of the coloring carotenoids present in food, but to a high degree on the physicochemical distribution in the tissue as well as in the coloring preparation (*68*). The water-soluble carotenoid products may be used alone or in combination. If more intense yellows are required, free of red tints, then riboflavin, preferably in the form of the much more soluble sodium phosphate derivative, may be used in combination with lower concentrations of β-carotene. The relatively low light stability of riboflavin, however, limits the possible applications of such combinations.

TABLE 13

Stability of 10% Canthaxanthin Beadlets[a]

	Retention (%)			
Storage conditions[b]	Lot 184-61 (11.7%)[c]	Lot 184-66 (11.3%)[c]	Lot 195-21 (11.6%)[c]	Lot 195-46 (11.0%)[c]
6 weeks at 70°C	—	—	90	—
3 weeks at 55°C	96	96	98	95
6 weeks at 45°C	99	98	100	98
12 weeks at 45°C	97	94	97	99
4 months at 23°C	98	97	97	95
8 months at 23°C	94	94	94	97
12 months at 23°C	93	93	97	95

[a] From Bunnell and Borenstein (*157*).

[b] Samples stored above 23°C in open jars; samples stored at 23°C in screw-cap jars with large headspace.

[c] Initial assay.

F. Marketed Application Forms

The market forms (*58*) of synthetic carotenoids (Table 14) comprise the following groups of products:

Crystalline forms
- β-Carotene
- β-Apo-8′-carotenal
- β-Apo-8′-carotenoic acid ethyl ester
- Canthaxanthin

Oil-soluble forms
- Suspensions of microcrystals of carotenoids (20–30%) in vegetable oil
- β-Apocarotenal solution
- Carotene–vitamin A–vitamin D blends

Water-dispersible forms
- Carotenoid beadlets and powders (0.5–10%)
- Carotenoid emulsions

The oil-soluble and water-dispersible forms are used for coloring food products. The development of special products to fit specific applications continues.

G. New Product Forms

In addition to these market forms, various types of experimental products are being evaluated and may eventually lead to new market forms:

Fat-soluble beadlet forms of carotenoid crystals
Liquid products (both oil soluble and water dispersible)
Thixotropic gel (stabilized suspension of carotene microcrystals)
Cold-water dispersible products (maltodextrins, skimmed-milk powders, etc.)
Syrups (with or without protective hydrocolloid)
Water-insoluble pigment powder for use in cosmetics
Carotenoid–noncarotenoid color combinations
Water-dispersible, optically clear carotenoid products

Applications for these experimental formulations include the following:

1. Cherries. Maraschino-type cherries may be colored with canthaxanthin dispersible liquid concentrate. The cherries have an attractive red color similar to that obtained by current FD&C dyes. When placed in water, a slight bleeding of color may be observed.

2. Hard candies. Yellow colors are produced with β-carotene, and darker colors with apocarotenal and canthaxanthin. The raw materials (sugar, water, citric acid) are heated to 150°C, colors and flavors are added, and candies are formed.

3. Fruit jellies. Fruit jellies can be made similarly to candies. Raw materials are heated to 104°–106°C; flavor, sugar, and color are added before cooling. The product is molded or cut into cubes and powdered with sugar.

4. Sugar-coated tablets. The coloring of these tablets with carotenoids is achieved

TABLE 14

Possible Application Forms of Carotenoid Food Colors[a]

Form	Description	Uses	Stock solution[b]
30% β-carotene liquid suspension[c]	Micronized β-carotene suspension in food-grade vegetable oil; pourable at 75F (23°C) and above; semisolid at 40′–60°F (4°–16°C); no added antioxidants; dissolves readily in warm oil	Coloring fat and oil products, margarine, processed cheese, frozen and dried egg yolk, winter butter (continuous process), and other fat-base foods	1.5 gm suspension added to 450 ml warm vegetable oil, with stirring to dissolve; 1 ml = 1 mg β-carotene
24% β-carotene smi-solid suspension	Micronized β-carotene in hydrogenated vegetable oil; semisolid at 70°–90°F (21°–32°C); pourable at 100°F (38°C) and above; no added antioxidants; dissolves readily in warm oil	Coloring fat and oil products, margarine, and other fat-base foods	2 gm suspension added to 480 ml warm vegetable oil, with stirring to dissolve; 1 ml = 1 mg β-carotene
22% β-carotene HS liquid suspension	Micronized β-carotene suspension in food-grade vegetable oil; pourable at 23°C and above; semisolid at 4°–16°C; stabilized with added antioxidants; dissolves readily in warm oil	Coloring popping oil, heat-stressed cooking oils; fat-base foods when greater carotene stability is required	2 gm suspension added to 440 ml warm vegetable oil, with stirring to dissolve; 1 ml = 1 mg β-carotene
10% β-carotene dry beadlet[c]	Colloidal β-carotene in a gelatin–carbohydrate matrix, stabilized with added antioxidants; readily dispersible, with stirring, in warm water, 140°F (60°C)	Coloring water-base foods and beverages; dry products to be reconstituted in warm water	1 gm beadlets added to 100 ml warm water, with stirring to disperse; 1 ml = 1 mg β-carotene

(Continued)

TABLE 14—*Continued*

Form	Description	Uses	Stock solution[b]
3.6% β-carotene liquid emulsion	Soluble β-carotene in orange oil and brominated vegetable oil emulsified in a hydrolyzed protein base; pourable at 23°C; specific gravity, oil phase, 1.052 ± 0.003 (12.5–13.6°C Brix); stabilized with added antioxidants	Especially, coloring carbonated and noncarbonated orange-flavored drinks and fruit juice blends adjusted for specific gravity in clear-glass or opaque containers	3 gm emulsion added to 108 ml water, with stirring to disperse; 1 ml = 1 mg β-carotene
0.5% β-carotene CWS powder	Colloidal β-carotene in matrix consisting predominantly of malto dextrin; stabilized with added antioxidants; readily dispersible in cold water	Coloring instant beverage powders, soup powders, dessert mixes, and ice cream mix (yields an orange hue)	
1% β-carotene-type yellow powder[d]	Spray-dried emulsion of β-carotene in vegetable oil carrier maltodextrin with added antioxidants; readily dispersible in cold water	Coloring instant beverage powders, soup and pudding powders, ice cream mix (yields a yellow hue)	10 gm powder in 95 ml water, with stirring to disperse; 1 ml = 1 mg β-carotene
2.4% β-carotene dry beadlets	Soluble β-carotene in vegetable oil emulsified in a gelatin–carbohydrate matrix; stabilized with added antioxidants; readily dispersible in water (23°C)	Coloring juices, juice drinks, dry beverage bases, beverages, and other water-base liquid or dry foods to be reconstituted in water	4 gm beadlets added to 96 ml warm water, with stirring to disperse; 1 ml = 1 mg β-carotene
β-Carotene blend with vitamin A (and D)	Micronized β-carotene suspension in vegetable oil with added dissolved vitamin A (and D); packaged in custom batch-size cans; must be shaken before use	Simultaneous coloring and vitamin A fortification of margarine, mellorine, etc.	
20% apocarotenal liquid suspension[c]	Micronized apocarotenal suspension in food-grade vegetable oil; pourable at 23°C; no added antioxidants; dissolves readily in warm oil; semisolid at 4°–17°C	Coloring fat and oil products, processed cheese, and other fat-base foods	1 gm suspension added to 200 ml warm vegetable oil, with stirring to dissolve; 1 ml = 1 mg apocarotenal

Apotene liquid suspension (20% total carotenoid)	Mixture of micronized apocarotenal and β-carotene suspended in food-grade vegetable oil; pourable at 23°F and above; semisolid at 4°–16°C; stabilized with added antioxidants; dissolves readily in warm oil	Coloring popping oil, heat-stressed cooking oils, and fat-base foods when greater carotenoid stability is required	1 gm suspension added to 200 ml warm vegetable oil, with stirring to dissolve; 1 ml = 1 mg carotenoids
Carotenal solution 73	A combination of 1.4% apocarotenal and 0.6% β-carotene	Coloring processed cheese, cheese sauces; fat-base foods	
2–4% apocarotenal solution	Soluble apocarotenal in a food-grade modified vegetable oil composition; stabilized with added antioxidant	Coloring processed cheese, salad dressings, and other fat-base products; spice blends, cheese powders, salad dressings, breading products; replacement for paprika oleoresin	2.5–5 gm solution added to 100 ml vegetable oil, with stirring to dilute; 1 ml = 1 mg apocarotenal
10% apocarotenal dry beadlets[c]	Colloidal apocarotenal in a gelatin–carbohydrate matrix; stabilized with added antioxidants; readily dispersible, with stirring, in warm water (60°C)	Coloring water-base foods and beverages; dry products to be reconstituted in warm water	1 gm beadlets added to 100 ml warm water, with stirring to disperse; 1 ml = 1 mg apocarotenal
10% canthaxanthin dry beadlets[c]	Colloidal canthaxanthin in a gelatin–carbohydrate matrix; a dry beadlet; stabilized with antioxidants; readily dispersible, with stirring, in warm water (60°C)	Coloring tomato products, simulated meat products and water-base products; dry products to be reconstituted in warm water	1 gm beadlets added to 100 ml warm water, with stirring to disperse; 1 ml = 1 mg canthaxanthin
10% canthaxanthin CWS powder[d]	Spray-dried, cold-water-dispersible powder	Coloring tomato red sauces, pizza, pasta sauce; punch-type beverages; water-base foods	1 gm powder added to 100 ml water, with stirring to disperse; 1 ml = 1 mg canthaxanthin

[a] Adapted from Bauernfeind (*58*).
[b] A suggested convenience for small-scale laboratory coloring of foods for trial purposes.
[c] Usually available in other countries.
[d] Similar to beadlets, except it is readily dispersed in cold water.

by dispersing the colors in water in a ratio of 1 : 4. The color solution is added to a hot sugar syrup made with 100 parts sugar dissolved in 40 parts boiling H_2O. The sugar syrup is then used to coat the tablets.

5. Cheese wax. Yellow to red cheese wax can be obtained depending on the amount of canthaxanthin SD powder used. Two to 10 mg of the spray-dried powder is dispersed or suspended in 35 gm of hot wax. The color is very stable in the wax and does not bleed into the cheese.

Much interest has been shown in the last type of product listed above. β-Carotene, apocarotenal, and canthaxanthin, commercially synthesized carotenoids, have been used for many years to color fat- and opaque-water-base products. Recently, these carotenoids were formulated into stable, dry powders containing 2.5–10% carotenoids, which produce optically clear aqueous solutions (*282*). The powders offer a range of colors from yellow to cherry red and are useful for color preparations, such as gelatin desserts and formulated drinks, in which clarity is desired.

X. OCCURRENCE IN FOODS

How and in which chemical and physical form carotenoids occur in nature are questions of importance, particularly to food technologists involved in the processing and development of new food products. They are also important to nutritionists because fruits and vegetables have been declared the world's source of vitamins A and C (*137,550*) and also are important as sources of minerals (*550*). There exists a vast number of scientific papers, monographs, and reviews on the distribution and type of carotenoids in foods. The overall carotenoid pattern may vary from relatively simple mixtures to extremely complex ones. The simplest mixtures may be found in foods of animal origin, due to the limited ability of the animal to absorb, modify, and deposit carotenoids. The other extreme is the formidable array of carotenoids encountered in citrus products, dehydrated alfalfa meal, and paprika, for example, which are only recently succumbing to modern, sophisticated analytical chemistry. Color measurements in foods and analytical aspects have been reviewed by Francis and Clydesdale (*308,309*) and Clydesdale (*191*) and are discussed in Chapter 10 by DeRitter and Purcell.

The determination of carotenes in fruits and vegetables has been of particular interest since it is used as a measure of the provitamin A content of foods. Column chromatography has been the method used most frequently (*23,34*). The validity of the AOAC method depends on the assumption that β-carotene is the major hydrocarbon carotenoid present, since α-carotene, β-carotene, and cryptoxanthin are retained on the column and are not eluted separately. The results of this procedure are usually labeled carotene, total carotene, or β-carotene in the literature. Gebhardt *et al.* (*330*) warned that the reported calculated vitamin A values for some red and yellow fruits and vegetables may be erroneously high (Table 15A–C) due to the

TABLE 15A

Amount of Total and Effective Carotene in Raw and Canned Peaches[a]

Method	Raw (μg/100 gm)[b]	Canned (μg/100 gm)[c]	Retention (%)
Method I	1790	552	31
Method II	456	386	85
Effective carotene	321	270	84

[a] Effective carotene calculated from the results of chemical method II of Gebhardt *et al.* (*330*).
[b] Mean of three samples.
[c] Mean of 24 samples.

TABLE 15B

Amount of Carotenoid Isomers in Raw and Canned Peaches[a]

Isomer	Raw[b]		Canned[c]	
	μg/100 gm	%	μg/100 gm	%
Neo-β-carotene B	51	11	36	9
All-*trans*-β-carotene	186	41	157	41
Neo-β-carotene B	18	4	24	6
Cryptoxanthin	201	44	169	44
Total	456		386	

[a] Analyzed by chemical method II of Gebhardt *et al.* (*330*).
[b] Mean of three samples.
[c] Mean of 24 samples.

TABLE 15C

Effective Carotene in Raw Peaches[a]

Isomer	Biological activity (*4*)	Amount (μg/100 gm)	Effective carotene (μg/100 gm)
All-*trans*-β-carotene	100	186	186
Neo-β-carotene B	53	51	27
Neo-β-carotene U	38	18	7
Cryptoxanthin	50	201	101
Total			321

[a] Calculated from results of chemical method II of Gebhardt *et al.* (*330*).

inclusion of carotenoids without activity in the calculation. They suggested that a greater effort be made to determine more accurately the vitamin A activity of these products. A more complete discussion of methodology of carotenoid determination is given in Chapter 10. For current guidance, compilations (*362,638,716*) of the carotene and carotenoid content of fruits, vegetables, and other foods may be consulted (Table 16). New data with improved methodology are needed.

TABLE 16

Carotene and Retinol Content of Foods

Food	Retinol (mg/100 gm)	Carotene (mg/100 gm)[a]
Apples, Delicious, raw		0.012
Apples, Jonathan, raw		0.03
Apples, raw[b]		0.02–0.30
Apricots, raw		0.61
Apricots, raw[b]		1.5
Apricots, canned halves		0.74
Avocado, raw		0.02
Bananas, raw		0.03–0.22
Bilberries, raw[b]		0.13
Cherries, red, canned		0.03
Cherries, sour, canned[b]		0.5
Currents, black, raw[b]		0.2
Fruit salad, fresh		0.02–0.09
Gooseberries, raw[b]		0.18
Loganberries, raw[b]		0.07–0.08
Mandarins		0.13–0.28
Mangoes, raw		2.07
Mangoes, raw[b]		1.2
Melons, yellow, raw[b]		1.1–2.0
Nectarines, raw[b]		0.5
Olives, green, canned[b]		0.15
Oranges and juice, raw[b]		0.03–0.05
Orange juice, Navel		0.02
Orange and apricot juice, canned		0.33
Peaches, Alberta, raw		0.34
Peaches, raw[b]		0.5
Peaches, dried[b]		2.0
Pears, raw		0.01–0.02
Pineapples[b]		0.06
Pineapples, raw		0.03
Plums, raw[b]		0.16–0.22
Prunes, dried[b]		1.0

(*Continued*)

TABLE 16—*Continued*

Food	Retinol (mg/100 gm)	Carotene (mg/100 gm)[a]
Prunes, dried		0.48
Prunes, dried, stewed		0.42
Prunes, dried, stewed[b]		0.5
Raspberries, raw[b]		0.08
Rhubarb, raw[b]		0.05–0.06
Strawberries, raw[b]		0.03
Tangerines, raw[b]		0.07–0.1
Artichokes, boiled		0.09
Asparagus, canned		0.04
Asparagus, boiled[b]		0.5
Beans, french, boiled[b]		0.14
Beans, runner, boiled[b]		0.14
Beans, cut, boiled		0.30
Beans, frozen, boiled		0.24
Beet greens, boiled[b]		5.0
Broccoli, boiled		0.36
Broccoli, boiled[b]		2.5
Brussels sprouts, boiled[b]		0.4
Cabbage, raw or boiled[b]		0.3
Cabbage, boiled		0.02
Carrots, boiled		2.9–5.4
Carrots, boiled or canned[b]		6.0–7.0
Cauliflower, raw or boiled[b]		0.03
Celery, raw[b]		Trace
Celery, raw		0.01
Endive, raw[b]		2.0
Kale, boiled		5.0
Lettuce, raw[b]		1.0
Lettuce, raw		0.2, 0.04
Mint, raw[b]		11.0
Mustard greens, raw[b]		5.0
Parsley, raw[b]		7.0
Parsley, raw		5.0
Peas, boiled[b]		0.3
Peas, dried, boiled[b]		0.08
Peppers, raw or boiled[b]		0.2
Pumpkin, boiled, mashed		1.1–6.0
Silverbeet, boiled		1.2–3.3
Spinach, boiled[b]		6.0
Sweet corn, on cob[b]		0.24
Sweet potatoes[b]		0.01
Sweet potatoes, boiled		4.0
Tomatoes, canned, juice[b]		0.5

(*Continued*)

TABLE 16—*Continued*

Food	Retinol (mg/100 gm)	Carotene (mg/100 gm)[a]
Turnip greens, boiled[b]		6.0
Water cress, raw[b]		3.0
Yams, boiled[b]		0.01
Apple pie	0.022[c]	—
Beef sirloin, grilled	0.011	0.04
Beef vegetable soup	0–0.023	0.1–0.6
Bread, 90 gm; butter 10 gm	0.098	0.08
Butter	0.72–1.05	0.43–0.77
Butter[b]	0.75–0.97	0.35–0.65
Buttercake, chocolate, plain	0.023	—
Buttercake, plain, iced	0.027	—
Cheesecake[b]	0.28	0.11
Cheese, cheddar	0.17–0.33	0.07–0.71
Cheese, cottage[b]	0.032	0.018
Cheese, primary[b]	0.22–0.37	0.14–0.23
Cheese, processed[b]	0.24	0.120
Cheese, soufflé[b]	0.20	0.04
Cream, fresh	0.46	0.24
Cupcake, iced	0.029	0.17
Eclairs	0.20	0.07
Eggs, cooked[b]	0.13–0.19	Trace
Eggs, boiled	0.05–0.15	0.01–0.15
Egg custard, baked	0.06–0.11	0.02–0.04
Egg custard[b]	0.06	0.02
Eggs, scrambled, cheese	0.224	0.11
Fruitcake[b]	0.120	0.01
Herring, canned	0.054	0.07
Ice cream	0.04–0.11	0.05–0.07
Kidney, beef, stewed[b]	0.04–0.25	—
Kidney, pig, stewed[b]	0.07–0.14	—
Liver, calf, fried[b]	17.4	0.1
Liver, chicken, fried[b]	11.1	—
Liver, lamb, fried, bacon	0.6	0.13
Margarine[b]	0.900	0–0.66
Milk, cow, pasteurized	0.03–0.09	0.01–0.06
Milk, whole, dried[b]	0.29	0.17
Milk, skim, dried[b]	Trace	Trace
Pizza, cheese, tomato[b]	0.07	0.23
Salmon, canned[b]	0.09	Trace
Snapper filet, margarine	0.42	—
Welch rarebit	0.21	0.13
Yogurt, low fat[b]	0.008	0.05–0.280

[a] Vitamin A equivalent values can be calculated using conversion value chosen.
[b] Values from McCance and Widdowson (*638*); others from Nobile and Woodhill (*716*).
[c] Retinol values by assay (not converted from carotene column).

A. Fruits

The color of fruits and fruit juices is important for characterizing these products and is one of the criteria used for commercial standards. However, many factors affect color, including variety of fruit, maturity, place of origin, seasonal and climatic changes, and processing methods. Carotenoids in fruits, juices, and concentrates were reviewed by Bauernfeind (*63*), Goodwin and Goad (*62*), and others, dealing with the distribution of carotenoids and changes during maturity. General reviews, such as those of Goodwin (*353,360*) and Moore (*679*), may also be consulted.

Goodwin and Goad (*362*) in 1970 examined the literature relative to carotenoid distribution in fruits. The total carotenoid and β-carotene content of some fruits (Table 17) were tabulated by these researchers. They listed over 50 carotenoids occurring in fruits: antheraxanthin, β-apo-8′-carotenal, β-apo-10′-carotenal, aurochrome, auroxanthin, capsanthin, capsorubin, α-carotene, β-carotene, β-carotene 5,6-epoxide, β-carotene 5,6,5′,6′-diepoxide, γ-carotene, δ-carotene, ε-carotene, ξ-carotene, β-carotenone, celaxanthin, citranaxanthin, β-citraurin, crocetin, cryptochrome, cryptoflavin, α-cryptoxanthin, β-cryptoxanthin, chrysanthemaxanthin, eschscholtzxanthin, flavoxanthin, hydroxysintaxanthin, lutein, lutein 5,6-epoxide, luteochrome, luteoxanthin, lycopene, lycophyll, lycoxanthin, mutatochrome, mutatoxanthin, neochrome, neoxanthin, neurosporene, persicaxanthin, phytoene, phytofluene, reticulataxanthin, rhodoxanthin, rubixanthin, semi-β-carotene, sinenisiaxanthin, sintaxanthin, trollein, trollixanthin, valenciachrome, valenciaxanthin, violaxanthin, β-zeacarotene, and zeaxanthin.

The total carotene content of 54 fruits and vegetables grown in Israel has been recorded (*389*). Carotenoid analyses have been made on various European fruits, particularly wild fruits from Germany. There is a highly varying ratio between β-carotene and total carotenoids in these fruits. The carotene content of native eastern arctic plants (berries, roots, leaves) indicates that some represent a valuable reserve (*415*) in times of survival. Benk (*87*) reported on the carotenoids in wild and cultivated fruits; Lantz (*567*) also reported on wild plants. In 1969 Fonseca *et al.* (*303*) reported on the carotene content of Brazilian fruits and vegetables. The red guava, passion fruit, mango, red papaya, and peach were richest in carotene content. Kobayashi *et al.* (*529*) researched the carotenoid pigments and color of fruits in Japan. Similar studies have been made in other countries (*181,303,389,406,627, 698,716,747,793,843,859,904*). In ripening fruit the decrease in chlorophyll content is frequently accompanied by an increase in carotenoid concentration and in the ratio of carotenes to oxycarotenoids. α-Carotenes, γ-carotenes, and lycopene are the common fruit hydrocarbon carotenoids. α-Carotenes and γ-carotenes occur more frequently in fruit than in leaves. β-Carotene is quite widely distributed in some fruits, in a range of 1–60% of the total carotenoid content (*74*). Lycopene is particularly noticeable in the tomato, pink, grapefruit, pink orange, watermelon, and also some varieties of apricots. Fruit oxycarotenoids often are in an ester form and are frequently unique

TABLE 17

Total Carotenoid and β-Carotene Content of Some Fruits[a]

Fruit	Portion	Carotenoid (mg/kg fresh wt)	β-Carotene (mg/kg fresh wt)	Reference
Arum maculatum		200	22.7	Goodwin (1956)
Atropa belladonna		18	4.1	Goodwin (1956)
Avocado		5.6	0.4–0.5	Lassen *et al*. (1944)
Berberis spp.	Berries		0.17–5.9	Bubicz and Wierzchowski (1960)
Berberis barbarossa		0.25	Trace	Bubicz (1965), Goodwin (1956)
Carica papaya	Pulp	13.8	4.1	Subbrarayan and Cama (1964)
Capsicum annuum				
Pepper (green)		9.0–11.2	1.2–1.5	Curl (1964)
			0.1–377	Lanz (1943)
Pepper (red)		127–248	11.6–33	Curl (1962), Goodwin (1956)
Pepper (yellow)		234	31	Cholnoky *et al*. (1955)
		62.6	4.3	Cholnoky *et al*. (1958)
Citrus paradisi	Pulp	8.2	2.2	Curl and Bailey (1957)
			Trace to 2.07	Curl and Bailey (1953)
Ruby Red	Peel	10.4	0.75	Khan and Mackinney (1953)
Navel oranges	Pulp	23	0.115	Curl and Bailey (1961)
	Peel	67	0.11	Curl and Bailey (1961)
Nagpur oranges	Pulp	14.4	0.01	Subbarayan and Cama (1965)
	Peel	38.5	0.45	Subbarayan and Cama (1965)

Tangerines	Peel	186	0.74	Curl and Bailey (1965)
	Pulp	27	1.11	Curl (1964)
Valencia oranges	Peel	98	0.29	Curl and Bailey (1957)
	Pulp	24	0.26	Curl and Bailey (1957)
Cotoneaster spp.		1.5–16	1.2	Goodwin (1956)
Crataegus spp.		4.25	2.8–5.8	Goodwin (1956)
Musk melons (cantaloups)		20.9–61.7	0.41–5.96	Tomes *et al.* (1963)
(Cucamis melo)	Edible portion	20–2	17.1	Curl (1966)
Watermelons				
Orange		33.7	1.4	Tomes and Johnson (1965)
Red		20.9–61.7	0.41–5.96	Tomes *et al.* (1963)
Japanese persimmons		21.6–97.9		Brossard and Mackinney (1963)
(Diospyros kaki)		54	3.9	Curl (1960, 1964)
Dura nigrescens	Oil	3900	3800	Argoud (1958)
Dura virescens	Oil	930	871	Argoud (1958)
Ficus carica (figs)		8.5	0.49	Curl (1964)
Fragaria chiloensis (strawberries)			1.5	Galler and Mackinney (1965)
Hippophae rhamnoides		52	1.6	Goodwin (1956)
Lonicera spp.		18	1.3	Goodwin (1956)
Malpighia punicifolia	Pulp	57.9	26.2	Baraud (1958)
Mangifera indica		12.5	7.5	Jungalwala and Cama (1963)
Mango spp.		13–62	4.77–39.9	Sadana and Ahmad (1946)
Badami		108–154	294	Ramasarma and Banerjee (1941)
Malgoa		21.8–25.7		Ramasarma *et al.* (1946)
Physalis alkekengi	Sepals	8683	86.5	Baraud (1958)
	Berries	573	3.85	Baraud (1958)

(Continued)

TABLE 17—*Continued*

Fruit	Portion	Carotenoid (mg/kg fresh wt)	β-Carotene (mg/kg fresh wt)	Reference
Prunus spp.			0.64	Curl (1964)
(cherry)			5–11	Galler and Mackinney (1965)
Prunus (cling peaches)		27	2.7	Curl (1964)
Prunus armenica (apricots)		35	21	Curl (1964)
Prunus domestica (prunes)			3.9	Curl (1962)
Pyracantha spp.		1–15	Trace to 4.95	Goodwin (1956)
Pyrus spp. (apples)	Whole fruit	54.9–126.1	1.98–76.3	Valadon and Mummery (1967)
	Peeled mesocarp	0.9–5.4		Galler and Mackinney (1965)
	Peel (Golden Delicious)	5.6		Galler and Mackinney (1965)
Pyrus spp. (pears)	Peeled mesocarp	0.3–1.28		Galler and Mackinney (1965)
	Peel (Comice)	5.6		Galler and Mackinney (1965)
Pyrus baccata (crab apples)		19.8	4.2	Valadon and Mummery (1967)
Pyrus prunifolia (crab apples)		2.3	0.12	Valadon and Mummery (1967)
Punica granatum		0.16		Curl (1964)
Rosa spp.		25–224	4.3–24.6	Goodwin (1956)
Rubus procerus (blackberries)		5.9	0.56	Curl (1964)
Sambucus nigra		16	3.0	Goodwin (1956)
Solamun dulcamara		90	55	Goodwin (1956)
Sorbus aucuparia		120	46.8	Goodwin (1956)
Taxus baccata	Whole berry	10	0.4	Goodwin (1956)
Viburnum opulus		607	8.1	Goodwin (1956)
Vaccinium macrocarpon (cranberries)		5.8	0.3	Curl (1964)
Vaccinium spp. (blueberries)		2.7		Curl (1964)
Vitis vinifera (grapes)		1.8	0.58	Curl (1964)
Tamus communis	Berries	529	25	Baraud (1958)
		96	3.8	Goodwin (1956)

[a] From Goodwin and Goad (*362*).

structures. Intensive investigations of the type of carotenoids (*212,223,227*) of the orange have revealed that over 50% are oxycarotenoid esters, about 25% are unesterified oxycarotenoids, and about 10% are carotenes. The oxycarotenoids of fruit are often esterified. Oxygen is required for maximal carotenoid production, and the temperature range is critical. Light, however, is not required during maturation for carotenoid synthesis.

1. Apples

The peel of an apple usually contains a greater concentration of carotenoids than the flesh. In two varieties of crab apple investigated (*979*), *Pyrus baccata* and *Pyrus prunifolia,* there were more carotenes present than oxycarotenoids. The reverse was true in two varieties of eating apples, Golden Delicious and Cox's Orange Pippin. Epoxycarotenoids, such as chrysanthemaxanthin, flavoxanthin (5,8-monoepoxylutein), and auroxanthin (5,8-diepoxyzeaxanthin), predominated. Knee (*527,528*), in investigating the carotenoid content of unripe Cox's Orange Pippin apples, found the carotenoids present to be characteristic of photosynthetic tissue: β-carotene, lutein, violaxanthin, and neoxanthin. As ripening proceeded, there was a significant increase in carotenoid esters, particulary of violaxanthin. Donchev (*264*) also reported the existence of β-carotene, violaxanthin, neoxanthin, lutein, and unidentified pigments in the skin of the apple. Kvale (*560*) observed that carotenoids were synthesized during ripening, that carotenoids were positively correlated with total nitrogen in the peel tissue, and that carotenoid synthesis was highest in the fruit from high-nitrogen trees. An orange-fleshed apple, higher in carotenoid content, has been reported by Savinov and Kudritskaya (*838*).

2. Apricots and Peaches

Apricots (*220*) are excellent sources of β-carotene. Unlike peaches, they contain small amounts of oxycarotenoids. β-carotene, γ-carotene, and lycopene are the main carotenoids (*220,360*). Curl (*220*) found that the major carotenoid distribution was about 60% β-carotene, 5% γ-carotene, 4% cryptoxanthin, 5% lycopene, and 2% lutein and that the carotenes included cis and poly-cis isomers. The distribution of carotenoid in Japanese apricots as reported by Kobayashi *et al.* (*529*) was 70% β-carotene, 7% γ-carotene, 7% cryptoxanthin, 5% lycopene, 2% ϵ-carotene, 2% zeaxanthin, 1.5% lutein, and 1.5% violaxanthin. α-Carotene, mutatochrome, neurosporene, hydroxy-ϵ-carotene, rubinxanthin, persicaxanthin, luteoxanthin, flavoxanthin, and auroxanthin were present in lesser amounts. The carotenoids in canned apricots stored for 6 months were similar to those of fresh fruit except that 5,6-epoxides were converted to 5,8-epoxides in the canned preparation. Bespechal'naya and Faraiman (*96*) recorded carotenoid levels in apricots of 0.9–4.0 mg %. A total carotenoid level of 4.5–5.5 mg % (about 70% β-carotene) for apricots was reported by Aczel (*7*), the level depending on variety, storage, and other variables.

A series of papers (*4,7,9–11*) has been published by Aczel on the changes in the carotenoid content of apricots and peaches during industrial processing. Carotenoid retention is influenced by processing variables, but it is possible to attain fairly good retention performance. Mancini and Schianchi (*620*) reported 5.1–5.6 mg carotene per 100 gm in dried apricots. Foda *et al.* (*301*) found the carotenoid content to be well retained in dried apricots. Preservation of apricot juice by canning or freezing for limited periods was observed by Sarhan *et al.* (*835*). Apricot jam and juice contain about 0.5 mg carotene per 100 gm and canned apricots about 1.7 mg per 100 gm, according to Mancini and Schianchi (*620*).

Mackinney *et al.* (*613*) reported β-carotene to be 10% of the total carotenoids of peaches. Aczel (*4*) about 35 years later gave a higher figure, 20–22%, with a total content of about 2 mg %. Cling peaches were found by Curl (*222*) to contain primarily violaxanthin, cryptoxanthin, β-carotene, and persicaxanthin, but 25 other carotenoids, including neoxanthin, were identified. McCarty and Lesley (*639*) noted that peaches contain eight pigments, with five of them identified as α-carotene, β-carotene, cryptoxanthin, lutein, and zeaxanthin. Dei-Tutu *et al.* (*241*) reported a range of 1.2–5.4 mg total carotenoid pigment per 100 gm in 13 canning clingstone peach varieties grown in Australia. Cryptoxanthin values of 0.8 and 2.0 mg per 100 gm were observed in 2 different varieties. The total carotene content of 7 British Columbia tree fruits (Table 18) including 9 varieties of peaches and 10 varieties of apricots was determined byStrachan *et al.* (*898*).

Chung and Luh (*183,184*) studied ripening and peeling methods of freestone peaches. They noted increases in β-carotene during ripening and temperature influence on the color of the canned products. Patterns of the major carotenoids were followed by Katayama *et al.* (*496*) during ripening of apricots and peaches. In ripening apricots, β-carotene increased rapidly to predominate; γ-carotene, lycopene, lutein, crytoxanthin, and violaxanthin concentrations also rose. β-Carotene, lutein, and violaxanthin increased substantially during ripening of the peach, but cryptoxanthin and neoxanthin increased slightly. When ^{14}C-labeled mevalonic acid was supplied to apricots at the green-yellow stage, β-carotene, cryptoxanthin, and lutein were strongly labeled, β-carotene 10-fold more so than cryptoxanthin or lutein. Phytoene and phytofluence contained practically no radioactivity after 24 hr, suggesting that these compounds were turned over rapidly. The extent of labeling decreased as the fruit approached maturity.

3. *Avocadoes*

The avocado (*Persea americana*) is a fruit with an unusually high oil content (between 10–30%) and a relatively high concentration of chlorophyll under the peel. The fruit is consumed on a large scale in both America and Europe. In Israel many varieties are successfully cultivated both for export and for the local markets. The carotenoid composition as determined by Gross *et al.* (*381*) is shown in Table 19. Lutein, chrysanthemaxanthin, trollichrome, isolutein, cryptoxanthin, and β-carotene were the principal pigments. In a later paper Gross *et al.* (*380*) identified two pigments in the ripe pulp as α-citraurin and mimulaxanthin. Oxycarotenoids occur esterified to a major extent in the pulp but are entirely free in the peel.

TABLE 18

Carotene Content of Apricots and Peaches[a]

Variety	No. of samples	Carotene content of fresh edible portion (mg/100 gm)	
		Average	Range
Apricots			
All varieties	49	1.94	0.70–4.15
Blenheim (Royal)	10	2.23	1.55–2.80
English Moorpark	5	3.49	2.40–4.15
Kaleden	4	1.46	0.80–2.05
Leslie	1	1.85	—
Perfection	2	2.30	2.10–2.50
Reliable	1	2.45	—
Rose	2	2.20	1.75–2.65
Tilton	9	1.94	1.50–2.35
Variety no. 27-5	1	2.15	—
Wenatchee Moorpark	14	1.21	0.70–2.05
Peaches			
All varieties	23	0.79	0.31–1.46
Elberta	2	0.42	0.38–0.48
Fisher	1	0.72	—
J. H. Hale	4	0.44	0.31–0.60
Rochester	4	0.74	0.46–1.08
Spotlight	1	0.78	—
Superior	1	0.88	—
Valiant	4	1.20	1.03–1.40
Vedette	3	1.01	0.85–1.25
Veteran	3	0.78	0.34–1.46

[a] From Strachan *et al.* (*898*).

4. Bananas

Although the banana (*Musa cavendishi*) is a popular fruit, few studies have appeared on its carotenoid content. The carotenoids of banana pulp and peel were investigated by Gross *et al.* (*384*) omitting saponification so that the different xanthophyll esters could be separated. The main pattern of the pulp, which had a very low carotenoid content (0.8 μg/gm), was α-carotene (31%), β-carotene (28%), and lutein (33%), the latter appearing in equal parts of the diester, monoester, and free form. The peel, the carotenoid content of which was eightfold higher, had a pattern closer to that of chloroplasts but with all oxycarotenoids appearing fully and partially esterified or in the free form. Banana leaves were examined only for their main carotenes. The ratio of α-carotene to β-carotene (1 : 1) varied with the cultivar, being the same in chloroplasts. Maia *et al.* (*615*), in a study on storage stability and quality of freeze-dried bananas, recorded retention of the natural carotenoids present in the cold-air-freezing process.

TABLE 19

Quantitative Composition of Carotenoids from Avocado Fruit, *Persea americana*, Nabal variety[a]

Composition	Percentage of total carotenoids[b]
Carotenes	
α-Carotene	0.9
β-Carotene	4.0
ζ-Carotene	0.5
Monols	
OH-α-Carotene	1.2
Cryptoxanthin	5.2
α-Citraurin	0.7
Diols and polyols	
Lutein	25.0
Isolutein	9.0
Violaxanthin	4.0
Chrysanthemaxanthin	20.4
Luteoxanthin	2.1
Trollichrome	9.7
Neoxanthin-like	7.3
Unknown 437 (trollein?)	8.1
Neoxanthin	0.5
Unknown 437	0.5

[a] From Gross *et al.* (*381*).
[b] Expressed as β-carotene. Isomeric forms of the same pigment were determined together. Minor pigments were not recorded.

5. *Berries*

In the elderberry the principal pigments are β-carotene, cryptoxanthin, lutein, flavoxanthin, and neoxanthin (*358*). Black chokeberries produced in Latvia (*2*) showed an ascorbic acid content of 10–18 mg % and a carotene content of 0.8–3.1 mg %. Various methods of preservation were studied. Kulikova (*557*) reported cloudberries to contain 70–87 mg % of ascorbic acid and 1.8–2.0 mg % of carotenes on a fresh weight basis. The frost-resistant wild currants of the Far East contain 47–120 mg % ascorbic acid and 0.1–0.3 mg % carotenes, according to Torchinskaya (*946*). Zhamsran (*1044*) recorded that the mild, edible berries of Mongolia contain 1.9–6.7 mg % carotene. European and Siberian varieties (*372*) of black currants contain more than 0.8 mg % carotenes by dry weight. In 1975 Valadon *et al.* (*980*) investigated 11 different berries for their carotenoid component. Phytofluene and β-carotene were the only two polyenes identified in all of them; lutein was found in small quantities in only six species, and lycopene, a characteristic fruit pigment, was not always present. The distribution of carotenoids in berries appears to have no taxonomic significance, although rubixanthin was obtained only from rose hips. There was no correlation between total carotenoids and color of berries. The color depends on which major carotenoids are present and also on the presence of other noncarotenoid pigments, namely, flavonoids and chlorophyll. Earlier these workers published a paper on the carotenoids of Rowan berries (*978*). Goodwin (*358*) made a survey of the polyenes (*358*) of a number of ripe berries, some data of which are shown in Table 20.

6. *Cherries*

Galler and Mackinney (*323*) considered the carotenes, among other carotenoids, to predominate in cherries. In 1971 Schaller and von Elbe (*839*) detected the following carotenoids in Montmorency cherries: phytofluene, α-carotene, β-carotene, cryptoxanthin, cryptoflavin, lutein, zeaxanthin, mutatoxanthin, and three unknown carotenoids. The carotenoids of sweet cherries were examined by Salunkhe and Wu (*831*) during storage trials of the fruit.

TABLE 20

Total Carotenoid Content and Relative Amounts of Hydrocarbon Polyenes in Berries of Various Species[a]

Species	Carotenes (% of total carotenoids)	Amount of carotenoids (μg/gm wet wt)
Arum maculatum	92, 89	200
Atropa belladonna	23	18
Berberis barbarossa	—	0.25
Berberis sp. (Pirate King)	—	0.2
Capsicum annuum	35	127
Cotoneaster bullata	10	12
C. frigida	37	1.5
C. hebephylla	7	16
Crataegus oxyacantha	40, 41	4, 25
C. pratensis	57	5
Hippophaë rhamnoides	3	52
Lonicera periclymenum	15	18
Pyracantha flava	59, 27	9, 15
P. rogersiana	—	1
Rosa canina	26	25
R. moyesii	61	224
R. rubrifolia	49	88
Sambucus nigra	19	16
Solanum dulcamara	61	90
Sorbus aucuparia	39	120
Tamus communis	85	96
Taxus baccata	5	10[b]
Viburnum opulus	32.5	607

[a] From Goodwin (*358*).

[b] Calculated on whole berry, not just the aril.

7. *Citrus*

Many investigations have been conducted on the carotenoid content and identification of individual carotenoids in citrus products. The yellow, orange, and red colors of citrus fruit are due mainly to carotenoids. Approximately 115 different carotenoids (Table 21) have been reported (*890–892*) to occur in citrus. Because carotenoids readily form artifacts, it is suspected that many of those reported were formed during the extraction process. Few carotenoids contribute to citrus color. Some of the important ones are β-citraurin, cryptoxanthin, antheraxanthin, violaxanthin, lycopene, and β-apo-8′-carotenal. DiGiacomo (*256–257*) reviewed carotenoids in citrus fruits, their characteristics, and the influence of citrus species.

TABLE 21

Carotenoids Reported in Citrus[a]

Carotenoid	Investigator
1. Violaxanthin	Zechmeister and Tuzson (1931a, 1931b, 1936)
2. Cryptoxanthin	Zechmeister and Tuzson (1931a, 1931b, 1936)
3. Zeaxanthin	Zechmeister and Tuzson (1931a, 1931b, 1936)
4. β-Citraurin	Zechmeister and Tuzson (1931a, 1931b, 1936)
5. Phytofluene	Natarajan and MacKinney (1952)
6. α-Carotene	Natarajan and MacKinney (1952)
7. β-Carotene	Natarajan and MacKinney (1952)
8. ζ-Carotene	Natarajan and MacKinney (1952)
9. Antheraxanthin	Curl and Bailey (1954)
10. *cis*-Antheraxanthin	Curl and Bailey (1954)
11. Mutatoxanthin	Curl and Bailey (1954)
12. *cis*-Mutatoxanthin	Curl and Bailey (1954)
13. Luteoxanthin a	Curl and Bailey (1954)
14. Luteoxanthin b	Curl and Bailey (1954)
15. *cis*-Luteoxanthin	Curl and Bailey (1954)
16. Auroxanthin	Curl and Bailey (1954)
17. Valenciaxanthin	Curl and Bailey (1954)
18. Sinensiaxanthin	Curl and Bailey (1954)
19. Trollixanthin	Curl and Bailey (1954)
20. *cis*-Trollixanthin	Curl and Bailey (1954)
21. Valenciachrome	Curl and Bailey (1954)
22. Trollichrome	Curl and Bailey (1954)
23. Phytoene	Curl and Bailey (1955)
24. Lutein	Curl and Bailey (1955)
25. Trollein	Curl and Bailey (1955)
26. Hydroxy-α-carotene	Curl (1956)

(*Continued*)

TABLE 21—*Continued*

Carotenoid	Investigator
27. *cis*-Hydroxy-α-carotene	Curl and Bailey (1956)
28. Cryptoflavin-like	Curl and Bailey (1956)
29. *cis*-Luteoxanthin a	Curl and Bailey (1956)
30. *cis*-Luteoxanthin b	Curl and Bailey (1956)
31. *cis*-Sinensiaxanthin	Curl and Bailey (1956)
32. Cryptoxanthin epoxide	Curl and Bailey (1956)
33. Cryptochrome-like	Curl and Bailey (1956)
34. Capsanthin-like	Curl and Bailey (1956)
35. *cis*-ζ-Carotene	Curl and Bailey (1956)
36. Hydroxy-α-carotene furanoxide-like	Curl and Bailey (1956)
37. Sinensiachrome-like	Curl and Bailey (1956)
38. Poly-*cis*-cryptoxanthin	Curl and Bailey (1956)
39. *cis*-Cryptoflavin	Curl and Bailey (1956)
40. Cryptoflavin	Curl and Bailey (1956)
41. Flavoxanthin-like	Curl and Bailey (1956)
42. Mutatoxanthin b	Curl and Bailey (1956)
43. *cis*-Auroxanthin	Curl and Bailey (1956)
44. Trollein a	Curl and Bailey (1956)
45. Trollein b	Curl and Bailey (1956)
46. *cis*-Trollein	Curl and Bailey (1956)
47. Valenciachrome a	Curl and Bailey (1956)
48. Valenciachrome b	Curl and Bailey (1956)
49. Trollichrome-like a	Curl and Bailey (1956)
50. Trollichrome-like b	Curl and Bailey (1956)
51. Sinensiachrome	Curl and Bailey (1956)
52. γ-Carotene	Curl and Bailey (1957)
53. Lycopene	Curl and Bailey (1957)
54. Rubixanthin-like	Curl and Bailey (1957)
55. Hydroxycanthaxanthin-like	Curl and Bailey (1957)
56. Lutein 5,6-epoxide	Curl and Bailey (1959)
57. β-Apo-2′-carotenal	Winterstein *et al.* (1960)
58. β-Apo-8′-carotenal	Winterstein *et al.* (1960)
59. β-Apo-10′-carotenal	Winterstein *et al.* (1960)
60. 3,4,3′,4′-Tetradehydrolycopene	Winterstein *et al.* (1960)
61. Cryptoxanthin 5,6,′,6′-diepoxide	Curl and Bailey (1961)
62. Hydroxy-α-carotene 5,6-epoxide	Curl and Bailey (1961)
63. Cryptoxanthin 5,6-epoxide	Curl and Bailey (1961)
64. Hydroxy-α-carotene 5,8-epoxide	Curl and Bailey (1961)
65. Cryptoxanthin 5,6,5′8′-diepoxide	Curl and Bailey (1961)
66. ψ-Criptoflavin	Curl and Bailey (1961)
67. Reticulataxanthin	Curl and Bailey (1961)
68. *trans*-Neoxanthin	Curl and Bailey (1961)
69. Trolliflor-like	Curl and Bailey (1961)
70. *cis*-ψ-Cryptoflavin	Curl and Bailey (1961)
71. Tangeraxanthin	Curl (1962a)
72. Cryptoxanthin 5,8-epoxide a	(Curl (1962b)

(*Continued*)

TABLE 21—*Continued*

Carotenoid	Investigator
73. Cryptoxanthin 5,8-expoxide b	Curl (1962b)
74. Cryptoxanthin 5,6,5′,8′-diepoxide a	Curl (1962b)
75. Cryptoxanthin 5,6,5′,8′-diepoxide b	Curl (1962b)
76. Citranaxanthin	Yokoyama and White (1965a)
77. Sintaxanthin	Yokoyama and White (1965b)
78. Prolycopene	Subbarayan and Cama (1965)
79. 8′-Hydroxy-7′,8′-dihydrocitranaxanthin	Yokoyama and White (1966)
80. 3-OH-Sintaxanthin	Yokoyama and White (1966)
81. 3-OH-β-Apo-10′-carotenal	Yokoyama and White (1966)
82. β-Zeacarotene	Yokoyama and White (1966)
83. Neurosporene	Yokoyama and White (1966)
84. 8′-Hydroxy-7′,8′-dihydroreticulataxanthin	Yokoyama and White (1966)
85. η-Carotene	Yokoyama and White (1966)
86. β-Carotene 5,6-epoxide	Yokoyama and Vandercook (1967)
87. Neo-β-carotene U	Yokoyama and Vandercook (1967)
88. OH-ζ-Carotene	Yokoyama and Vandercook (1967)
89. Neoxanthin a	Yokoyama and Vandercook (1967)
90. Neoxanthin b	Yokoyama and Vandercook (1967)
91. *cis*-Zeaxanthin	Yokoyama and White (1967)
92. Apo-10′-vlolaxanthal	Curl (1967)
93. Semi-β-carotenone	Yokoyama and White (1968)
94. β-Carotenone	Yokoyama and White (1968)
95. Semi-α-carotenone	Yokoyama and White (1970)
96. Triphasiaxanthin	Yokoyama and White (1970)
97. Neo-β-carotene	Gross *et al.* (1971)
98. Poly-*cis*-cryptoxanthin a	Gross *et al.* (1971)
99. Poly-*cis*-cryptoxanthin b	Gross *et al.* (1971)
100. Rubixanthin	Gross *et al.* (1971)
101. *cis*-Lutein	Gross *et al.* (1971)
102. Isolutein	Gross *et al.* (1971)
103. *cis*-Isolutein	Gross *et al.* (1971)
104. Chrysanthemaxanthin	Gross *et al.* (1971)
105. OH-Sintaxanthin a	Gross *et al.* (1971)
106. OH-Sintaxanthin b	Gross *et al.* (1971)
107. *cis*-Violaxanthin a	Gross *et al.* (1971)
108. *cis*-Violaxanthin b	Gross *et al.* (1971)
109. *cis*-Violaxanthin c	Gross *et al.* (1971)
110. OH-Sintaxanthin 5,6-epoxide	Gross *et al.* (1971)
111. 3-Hydroxy-5,8-epoxy-5,8-dihydro-8′-apô-β-caroten-8′-a	Gross *et al.* (1975)
112. 5,6-Dihydro-β,β-carotene-3,3′,5′,6′-tetrol	Gross *et al.* (1975)
113. 5,8-Epoxy-5,8,5′6′-tetrahydro-β,β-carotene-3,3′,5′,6′-tetrol	Gross *et al.* (1975)
114. β-Citraurinene	Leuenberger and Stewart (1976)
115. β-Citraurol	Leuenberger *et al.* (1976)

[a] From Stewart (*890, 891*).

Correlations have been made between carotenoid absorbance and numerical color scores of citrus products by consumers (*750*). Umeda (*954*) applied carotenoid identification as a means of classifying citrus species and detecting added carotenoids. Yokoyama *et al.* (*1030*) reviewed the carotenoids occurring in hybrids and citrus relatives, which are highly desirable fruits for the enhancement of fruit color in mixed fruit products. Carotenoid content, like flavor, is usually considered to be improved by long-term plant breeding. Moreover, carotenoid content of citrus fruit has been found to be enhanced by the application of chemicals.

Cryptoxanthin, β-citraurin, and to a lesser extent violaxanthin accumulate in citrus cultivars following treatment (*893*) with ethylene or 2-chloroethylphosphonic acid. Good color (*1015*) in oranges and tangerines was obtained on exposure to ethylene at 15°–25°C for a number of days. Cultivars of low natural carotenoid levels (*458*), such as lemons and grapefruits, were stimulated by exposure to 2-(4-chlorophenylthio)triethylamine. Preharvest sprays of 2-chloroethylphosphonic acid were applied to tangerines to speed color development (*1033*). About 80 or more compounds have been noted to have bioregulating effects. The nature of carotenoid esterification in citrus fruit was investigated by Philip (*754*).

a. Grapefruit. Grapefruit has been examined for its carotenoid distribution. In 1953 Khan and MacKinney (*502*) studied the pigments of pink, red, and white grapefruit and reported 11 carotenoid bands by chromatographic separation. In the pink varieties β-carotene and lycopene predominated, with smaller amounts of ζ-carotene and phytofluene and traces of α- and γ-carotenes. Lycopene and β-carotene accounted for 80% or more of the pulp color of Ruby Red grapefruit. The distribution of carotenoids in Ruby Red grapefruit (Table 22) was determined by Curl and Bailey (*225*). Lime *et al.* (*593*) showed the relation of visual color to lycopene and carotene content during the maturation of Ruby Red. As the fruit reaches optimal maturity, lycopene content decreases and carotene content increases (Fig. 21). Similar phenomena were observed by Meredith and Young (*660*) in the pulp color of Redblush grapefruit, by Meredith *et al.* (*659*) in the quality of Hudson Red grapefruit, and by Ting and Deszyck (*940*) in the internal color of red and pink grapefruit in which lycopene development was influenced by temperature and storage conditions. A variety of red grapefruit, Burgundy, has been reported to possess a much deeper flesh color than Ruby Red (*721*). Philip (*754*) reported that most, if not all, of the oxycarotenoids of the Marsh Seedless grapefruit are esterified. The presence of β-citraurin as the myristate ester was observed in grapefruit by Philip.

b. Lemons. The principal carotenoid of the Meyer lemon is cryptoxanthin, according to Curl (*216*). The peel carotenoids include a number of unusual substances, including cryptoxanthin mono- and diepoxides and fractions tentatively identified as hydroxy derivatives of phytoene, phytofluene, and ζ-carotene. An unusual polyene was also found. Apparently it contains two allylic hydroxyl groups, one of which is allylic to the conjugated double-bond system. Yokoyama and

TABLE 22

Carotenoids of Citrus Pulp and Peel[a,b]

	Orange		Ruby Red grapefruit		Tangerine	
Carotenoid	Pulp (%)	Peel (%)	Pulp (%)	Peel (%)	Pulp (%)	Peel (%)
Phytoene	4.0	3.1	16.0	47.0	5.8	4.2
Phytofluene	13.0	6.1	4.4	14.0	7.2	3.5
Phytofluene-like	—	—	—	0.4	0.1	0.1
α-Carotene	0.5	0.1	—	0.1	0.3	0.2
β-Carotene	1.1	0.3	27.0	7.2	4.1	0.4
ξ-Carotene	5.4	3.5	3.5	7.2	6.9	2.0
γ-Carotene-like	—	—	0.8	0.4	0.1	—
Lycopene	—	—	40.0	11.0	0.1	0.02
Substance 377	—	—	0.2	—	—	—
Substance 335	—	—	0.6	0.5	—	—
Hydroxy-α-carotene-like	1.5	0.3	0.2	0.1	1.0	0.6
Cryptoxanthin epoxide	—	—	—	—	0.9	1.4
Cryptoxanthin epoxide-like	—	0.4	—	—	—	—
Cryptoxanthin	5.3	1.2	0.7	1.4	33.0	24.0
Cryptoxanthin-like	—	—	0.4	—	—	—
Hydroxy-α-carotene	—	—	—	—	—	0.4
furanoxide-like	—	—	—	—	—	
Cryptoflavin-like	0.5	1.2	—	1.3	0.8	3.4
Cryptochrome-like	—	0.8	—	0.2	—	0.1
Rubixanthin-like	—	—	0.5	0.3	—	0.2
Lutein	2.9	1.2	0.3	0.9	2.9	3.3
Zeaxanthin	4.5	0.8	—	—	3.3	3.5
Capsanthin-like	—	0.3	—	—	—	—
Hydroxycanthaxanthin	—	—	—	—	0.1	2.7
Antheraxanthin	5.8	6.3	0.7	—	9.7	6.2
Mutatoxanthins	6.2	1.7	0.4	0.2	2.2	2.8
Violaxanthin	7.4	44.0	0.9[c]	1.0[c]	14.0	24.0
Luteoaxanthins	17.0	16.0	0.4	1.8	3.5	9.1
Auroxanthins	12.0	2.3	0.3	1.6	0.4	1.9
Valenciaxanthin	2.8	2.2	—	—	0.2	0.4
Valenciachromes	1.0	0.7	0.2	0.3	—	—
Sinensiaxanthin	2.0	3.5	—	—	0.2	1.1
Sinensiachrome-like	—	0.2	—	—	—	—
Trollixanthin-like	2.9	0.5	—	—	1.0	2.6
Trollichrome-like	3.0	0.8	—	0.2	0.3	0.7
Trollein	—	—	—	—	0.9	—

[a] From Curl and Bailey (*225*).

[b] Assuming that all constituents have same specific extinction coefficient. Individual carotenoids expressed as percentage of the total.

[c] Plus zeaxanthin?

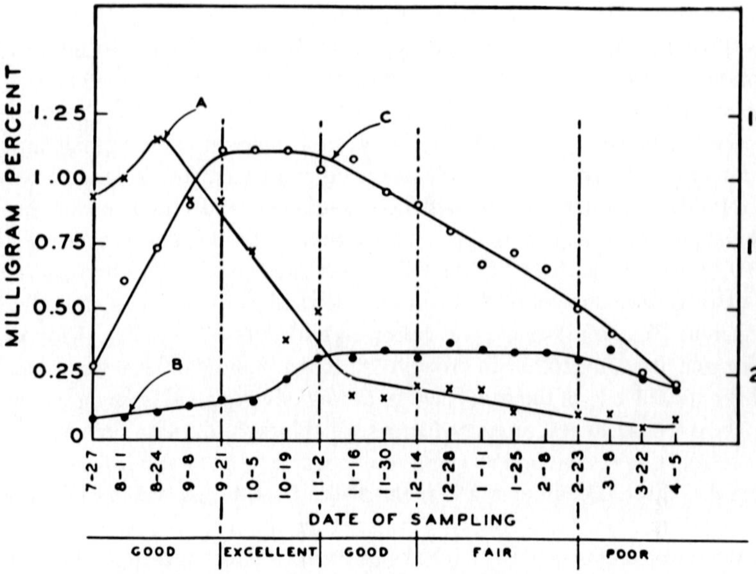

Fig. 21. Carotene and lycopene content of Ruby Red grapefruit during growth and maturity and internal coloration pattern. Key: A, lycopene; B, carotene; C, index of fading. [From Lime *et al.* (*593*).]

Vandercook (*1031*) examined both the peel and pulp of mature green and yellow lemons. As the lemons ripened and chlorophyll disappeared, α-carotene disappeared and δ-carotene and previously unreported η-carotene-like compounds appeared in pulp peel. Small but significant amounts of β-carotene 5,6-monoepoxide and its isomeric 5,8-epoxide, mutatochrome, were detected in the yellow lemon pulp. Soni (*877*) observed changes in carotenoid pigments of lemon peel in two varieties. Maximal levels of 1 mg per 100 gm of fresh peel were attained in trials on the chemical coloring of lemon fruit. El-Zeftawi (*281*) employed ethephan or cyclamic acid, which when applied in solution quickly colored the fruit and also altered the carotenoid content.

c. Oranges. The color of the orange, no doubt, has contributed to the many investigations of its carotenoid content. According to Borenstein and Bunnell (*133*), past investigators were Zechmeister and Tuzson, Karrer and Jucker, Natarajan and Mackinney, Curl and Bailey (*223*), and Curl (*212*). Investigations of carotenoids of the orange were reviewed by Mackinney (*612*) in 1961. Table 22 summarizes the carotenoids found in orange pulp and peel in 1957 by Curl and Bailey (*227*). β-Apo-8′-carotenal was isolated from orange peels by Winterstein *et al.* (*1024*) and from peel and pulp by Thommen (*930*). Thommen (*929*) reported on the detection and identification of naturally occurring ketocarotenoids and carotenals. Curl reported β-citraurin to be present also in orange peel.

As reported by Subbarayan and Cama (*908*) the pulp of the Nagpur orange contains 15 different carotenoids and the peel contains 14. Cryptoxanthin is the major carotenoid in the pulp, whereas violaxanthin forms the largest amount of total carotenoids of the peel. Significant amounts of cryptoflavin and prolycopene are also present in the pulp and comparatively large amounts of cryptoxanthin and neoxanthin are in the peel. Philip (*754*) isolated and characterized laurate esters of cryptoxanthin, violaxanthin, auroxanthan, β-citraurin, and reticulataxanthin from the peel and pulp of Valencia oranges. A 5,6-epoxide carotenoid aldehyde identified as apo-10′-violaxanthal was found in Valencia orange peel in 1967 by Curl (*211*).

The seasonal development of the external color and carotenoid content in the peel of the ripening Shamouti orange was followed by Eilati *et al.* (*277*). The external peel color and its contents of chlorophyll, total carotenoids, carotenes, and xanthophylls in the flavedo of the fruits change during development from the young and green to the overripe stages; an early decrease in chlorophyll and initial yellowing of the peel result. Later, concomitant with the appearance of the typical orange color (color break), the total carotenoid content of the flavedo, as well as its carotene/oxycarotenoid ratio, increase rapidly. Similar observations were made in the pulp of the orange by Rotstein *et al.* (*822*). The color break in the endocarp is similar to that in the peel but precedes it by more than a month. The carotene fraction of the pulp decreases with maturation from 15–20 to 3%. The monol fraction shows a decrease in the early stage and then increases with continuing maturation, reaching a final level of 16%. The diol and polyol fractions form the bulk of the total pigments (75%) in the early stage and continue to increase to 85% through maturation. The pigment mixture of the immature green fruit was analyzed and was qualitatively as complex as that of the mature fruit. A pink sport of Shamouti orange contains lycopene, according to Monselise and Halevy (*677*).

In 1974 Malachi *et al.* (*617*) observed that ripe navel oranges grown in the United States and Israel do not possess the same carotenoid distribution. The total carotenoid level in flavedo in oranges grown in Israel was 60–75 μg/gm, reflecting the rather lightly pigmented orange between grapefruit and Shamouti orange. The general pattern of the pigments and the observed ratio (1 : 11) of carotenes (8.4%) to xanthophylls (91.6%) are characteristic of citrus. The diepoxides of zeaxanthin constitute about 60% of the total carotenoids. Hydroxy-α-carotene and trollixanthin and derivatives, which are usually found in citrus flavedo grown in the United States, were absent in the Israeli fruit. Taylor and Witte (*921*) determined the total carotenoids of navel and other varieties of oranges using β-carotene as a primary standard and found a wide variation in market oranges arriving in the New York area. These differences were due not only to variety but to district of origin. Florida oranges were generally lower in carotenoids than California oranges. Usually the total carotenoid content of oranges is a satisfactory measure of total color, and variations in carotenoid types are not significant except in the case of isomerization of 5,6-epoxides to 5,8-epoxides. Seifert and Davidek (*856*) reported interactions of the carotenoids of oranges with some organophosphorus insecticides.

Orange Juice. For many years color has been an important characteristic of citrus juices and juice concentrates and is one of the criteria of present commercial standards as cited by Stewart and Leuenberger (*892*). Total color is influenced by maturity, variety, and processing methods, such as extraction and concentration. In the U.S. Department of Agriculture (USDA) and Florida Department of Citrus, standards have been set for concentrated orange juice, namely U.S. Grade A, U.S. Grade B, and substandard, based on a scoring system of 40 points for color, 20 for defects, and 40 for flavor. Orange juice having a color score of 36–40 is Grade A, 32–35 is Grade B, and 31 or less is substandard (regardless of other quality attributes). Kläui and Manz (*522*) in 1967 summarized the carotene and total carotenoid contents of some commercial citrus juices (Table 23). Carotene and total carotenoid values of California–Arizona orange juice products, obtained by Bernath and Swisher (*94*), are shown in Table 24. Quantitative values for 10 different carotenoids were determined chromatographically on seven different Floridian varieties by Stewart (*889*) in 1977 (Table 25). The carotenoid content of juice of citrus cultivars was followed by Stewart during the fruit maturation period by high-performance liquid chromatography procedures that were rapid and reproducible. Isomers of cryptoxanthin as well as the carotenes were separated. The β isomer

TABLE 23

Carotene and Total Carotenoid Contents of Some Citrus Fruit Juices[a]

Fruit juice	Total carotenoids (mg/100 ml)	β-Carotene (mg/100 ml)	β-Carotene (% of total carotenoids)
Orange juice			
Italy	1.19–1.48	0.06–0.15	5.0–10.1
Spain	0.69–1.81	0.018–0.20	2.6–11.0
South Africa	0.60–2.02	0.023–0.062	3.8–30.7
California			
Valencia oranges	0.87–1.31	0.062–0.092	7.0–7.1
Navel Oranges	0.64–0.85	0.024–0.50	5.7–5.9
Mandarin juice, Ethiopia	2.06	0.10	4.8
Clementine juice, Morocco	2.10–2.13	0.11 –0.20	5.2–9.4
Orange juice concentrates from 1961–1962 crop			
Italy	1.44–4.60	0.03–0.49	2.1–10.6
Spain	0.71–5.84	0.02–0.58	2.7–9.9
Mexico	2.44–3.41	0.04–0.49	1.6–14.4
South Africa	0.87–0.99	0.13	14.9
Greece	2.63–3.17	0.03–0.42	1.1–13.2

[a] From Kläui and Manz (*522*).

TABLE 24

Ratio of Total Carotenoids to Carotenes and β-Carotene as Percentage of Total Carotenoids in California–Arizona Sweet Orange Juice Products[a]

Variety of citrus	Total carotenoids (mg/100 gm)	Carotenes, as β-carotene (mg/100 gm)	Ratio	Carotene, (% of total) as β-carotene
Early-season Valencia conc.	7.60	0.54	14:1	7.1
Late-season Valencia conc.	9.01	0.63	14:1	6.9
Early-season navel conc.	4.53	0.22	21:1	4.8
Late-season navel conc.	5.40	0.46	12:1	8.6
Arizona sweet conc.	3.20	0.15	21:1	4.7
Arizona Valencia conc.	5.09	0.28	18:1	5.2
Yuma Valencia conc.	6.73	0.57	12:1	8.4
Bottlers Base I	6.70	0.61	11:1	9.1
Bottlers Base II	6.90	0.48	14:1	8.0
Valencia pulp	7.47	0.54	14:1	7.3
Late raw Valencia juice	2.50	0.16	16:1	6.4
Early Valencia juice	0.68	0.05	14:1	7.1
Late navel juice	0.84	0.07	12:1	8.5
Navel pulp wash	4.83	0.46	11:1	9.5
Blood orange conc.	9.03	0.36	27:1	5.2

[a] From Bernath and Swisher (*94*).

was the main source of provitamin A in citrus juice. The provitamin A content of citrus juice varied with fruit cultivar. Juice of the sweet orange cultivars had the lowest provitamin A carotenoid content, and those of Dancy and Robinson tangerines were intermediate. The provitamin A content of Murcott was the highest among all the cultivars studied. The principal provitamin A carotenoid occurring in citrus juice are α-carotene, β-carotene, and β-cryptoxanthin. Stewart (*889*) calculated the provitamin A content of seven cultivars of citrus juice (Table 26).

Natarajan and Mackinney (*706*) reported that the carotenoids of Valencia orange juice were 66% xanthophyll esters, 25% unesterified xanthophylls, and 9% epiphasic carotenoids (carotenes). The greater portion of the oxycarotenoids are epoxides (*227*). The cryptoxanthin, lutein, zeaxanthin, hydroxy-α-carotene, and xanthophyll 5,6-epoxides are mainly esterified, and the isomeric xanthophyll (5,8-epoxides) furanoxides are mainly unesterified (*229*). The carotenoid composition of fresh Valencia orange juice (*230*), investigated by countercurrent distribution, included 25 constituents (Fig. 22). The carotenoids of Italian orange juice were

TABLE 25

Carotenoid Content of Juice from Seven Citrus Cultivars[a,b]

Cultivar	Sampling dates							
	10–7	10–27	11–19	12–19	1–7	1–28	2–23	3–30
				Phytofluene				
Hamlin	0.8 ± 0.1	0.9 ± 0.0	0.7 ± 0.0	2.6 ± 0.1				
Pineapple		0.4 ± 0.0	0.8 ±0.0	2.5 ± 0.2				
Valencia				1.1 ± 0.1				
Robinson	2.6 ± 0.2	11.4 ± 0.0	16.2 ± 0.4	24 ± 0.4				
Dancy	3.8 ± 0.0	11.8 ± 0.4	11.2 ± 0.4	36 ± 0.2				
Orlando	0.6 ± 0.0	2.1 ± 0.1	2.7 ± 0.0	10.3 ± 0.2				
Murcott		4.1 ± 0.0	12.3 ± 0.4	80 ± 2.0				
				α-Carotene				
Hamlin	0.8 ± 0.0	0.4 ± 0.0	0.7 ± 0.0	0.7 ± 0.0	1.3 ± 0.1	1.7 ± 0.1	2.6 ± 0.3	2.4 ± 0.1
Pineapple	0.6 ± 0.0	0.7 ± 0.0	1.6 ± 0.1	3.0 ± 0.1	3.9 ± 0.5	4.2 ± 0.2	4.0 ± 0.3	6.5 ± 0.1
Valencia	0.8 ± 0.0	0.4 ± 0.0	0.7 ± 0.0	2.0 ± 0.1	2.5 ± 0.0	5.6 ± 0.0	5.1 ± 0.2	7.0 ± 0.0
Robinson	3.4 ± 0.0	9.0 ± 0.5	11.2 ± 0.1	9.0 ± 0.0	13.2 ± 1.2			
Dancy	4.6 ± 0.2	8.0 ± 0.0	6.2 ± 0.4	22.5 ± 1.5	13.8 ± 0.4	17.5 ± 1.5	28.2 ± 0.0	
Orlando	1.7 ± 0.1	3.4 ± 0.1	3.2 ± 0.1	9.0 ± 0.1	9.8 ± 0.7	9.9 ± 0.2	11.7 ± 0.3	
Murcott	1.4	4.0 ± 0.1	6.0 ± 0.8		11.0 ± 1.2	7.5 ± 0.0	18.0 ± 0.0	11.0 ± 0.0

(*Continued*)

TABLE 25—*Continued*

Cultivar	Sampling dates							
	10-7	10-27	11-19	12-19	1-7	1-28	2-23	3-30
				β-Carotene				
Hamlin	1.6 ± 0.1	1.1 ± 0.0	1.5 ± 0.1	1.8 ± 0.1	3.0 ± 0.1	3.9 ± 0.1	5.2 ± 0.1	5.7 ± 0.1
Pineapple	0.7 ± 0.0	1.1 ± 0	2.3 ± 0.0	6.2 ± 0.1	4.6 ± 0.3	6.1 ± 0.1	8.6 ± 0.0	11.9 ± 0.4
Valencia	0.8 ± 0.1	0.6 ± 0.1	0.6 ± 0.0	1.9 ± 0.0	3.4 ± 0.2	6.2 ± 0.2	8.1 ± 0.1	8.9 ± 0.0
Robinson	6.0 ± 0.3	24.5 ± 0.0	32.8 ± 0.0	51.0 ± 0.0	59.3 ± 0.8			
Dancy	6.6 ± 0.3	18.8 ± 0.4	18.7 ± 0.3	42.5 ± 3.5	56.7 ± 1.4	52.5 ± 2.5	66.9 ± 0.9	
Orlando	1.7 ± 0.0	3.7 ± 0.0	5.8 ± 0.2	15.9 ± 0.0	16.3 ± 1.0	24.6 ± 1.0	32.6 ± 0.1	
Murcott	2.7 ± 0.0	9.4 ± 0.6	9.4 ± 0.6	43.5 ± 0.0	64.1 ± 1.1	66.0 ± 0.8	105.8 ± 1.6	237.5 ± 7.5
				ζ-Carotene				
Hamlin	4.5 ± 0.8	2.0 ± 0.4	2.7 ± 0.2	5.7 ± 0.2	11.0 ± 0.4	9.0 ± 0.5	9.4 ± 0.8	7.7 ± 0.0
Pineapple		1.6 ± 0.1	4.0 ± 0.2	7.9 ± 0.5	8.3 ± 0.0	10.6 ± 1.4	12.9 ± 0.0	16.2 ± 0.9
Valencia				2.6 ± 0.2	6.1 ± 0.0	11.8 ± 1.1	14.9 ± 0.2	13.1 ± 0.6
Robinson	18.4 ± 0.0	55.5 ± 0.0	56.4 ± 0.0	102 ± 9.0	101.0 ± 0.0			
Dancy	24.5 ± 0.0	61.4 ± 0.0	59.0 ± 0.0	147.0 ± 0.0	180.0 ± 0.0	153.0 ± 6.0	202.0 ± 0.0	
Orlando	3.4 ± 0.1	10.8 ± 0.2	15.2 ± 0.1	53.5 ± 1.5	60.7 ± 0.0	70.5 ± 2.5	67.5 ± 1.2	
Murcott	7.6 ± 0.0	27.7 ± 0.5	61.9 ± 6.4	215.0 ± 32.0	215.8 ± 13.8	271.5 ± 0.0	285.0 ± 0.0	437.5 ± 7.5
				α-Cryptoxanthin				
Hamlin	1.2 ± 0.1	0.5 ± 0.1	1.0 ± 0.0	1.0 ± 0.1	1.8 ± 0.1	2.6 ± 0.1	4.0 ± 0.2	3.2 ± 0.0
Pineapple	0.5 ± 0.0	0.5 ± 0.0	1.9 ± 0.1	4.2 ± 0.0	4.0 ± 0.3	7.3 ± 0.3	7.5 ± 0.1	11.0 ± 0.1
Valencia		0.3 ± 0.0	0.5 ± 0.0	2.6 ± 0.0	4.0 ± 0.0	10.5 ± 0.0	10.4 ± 0.1	12.2 ± 0.2
Robinson	11.2 ± 0.5	29.3 ± 2.3	37.6 ± 0.0	29.3 ± 2.3	38.0 ± 2.0			
Dancy	8.3 ± 0.6	15.0 ± 1.2	16.9 ± 1.7	29.5 ± 0.5	32.0 ± 0.0	19.5 ± 1.5	30.0 ± 2.4	
Orlando	1.4 ± 0.1	3.6 ± 0.0	6.2 ± 0.2	12.9 ± 0.3	11.8 ± 0.6	12.0 ± 0.6	20.7 ± 0.4	
Murcott	3.2 ± 0.0	15.0 ± 1.3	10.2 ± 0.2	39.0 ± 0.0	28.2 ± 1.2	28.9 ± 1.2	54.3 ± 0.9	63.3 ± 2.0

				β-Cryptoxanthin				
Hamlin	5.4 ± 0.3	3.5 ± 0.0	3.6 ± 0.1	11.2 ± 0.3	20.6 ± 0.2	20.0 ± 0.0	20.0 ± 0.7	34.6 ± 0.6
Pineapple	1.3 ± 0.1	2.7 ± 0.1	7.7 ± 0.4	11.3 ± 0.2	8.5 ± 0.5	21.8 ± 0.8	25.7 ± 0.2	52.3 ± 1.2
Valencia	0.5 ± 0.0	0.7 ± 0.0	0.6 ± 0.0	8.6 ± 0.6	14.4 ± 0.0	24.3 ± 0.4	37.9 ± 0.0	27.3 ± 0.0
Robinson	223.8 ± 6.4	427.0 ± 3.0	511.0 ± 0.0	565.5 ± 7.5	573.0 ± 15.0			
Dancy	105.5 ± 1.5	170.5 ± 4.1	147.6 ± 6.0	189.0 ± 21.0	242.0 ± 18.0	213.5 ± 2.5	429.9 ± 0.9	
Orlando	2.9 ± 0.1	9.3 ± 0.0	10.4 ± 0.5	24.0 ± 1.1	25.0 ±.7	28.9 ± 2.0	72.2 ± 1.9	
Murcott	73.2 ± 0.0	240.5 ± 1.5	194.0 ± 16.8	426.0 ± 9.0	425.3 ± 4.8	472.5 ± 63.5	612.0 ± 9.0	1466.0 ± 40.0
				Lutein				
Hamlin	9.1 ± 0.2	6.4 ± 0.1	3.7 ± 0.2	83.0 ± 1.1	15.0 ± 0.0	20.0 ± 1.6	47.0 ± 1.4	35.8 ± 0.2
Pineapple	16.6 ± 0.2	7.3 ± 2.0	22.0 ± 0.4	32.0 ± 3.1	44.8 ± 0.0	70.8 ± 0.4	67.6 ± 0.0	89.0 ± 2.6
Valencia	9.7 ± 0.4	5.1 ± 0.1	2.7 ± 0.2	26.6 ± 1.3	34.0 ± 1.6	83.5 ± 1.5	82.0 ± 5.6	60.8 ± 0.0
Robinson	37.1 ± 0.5	124.6 ± 0.6	128.8 ± 3.6	144.9 ± 4.5	128.5 ± 1.5			
Dancy	75.5 ± 5.5	111.4 ± 5.0	117.4 ± 4.2	173.8 ± 6.2	225.5 ± 0.5	195.6 ± 6.0	266.0 ± 0.0	
Orlando	27.2 ± 0.6	51.0 ± 1.2	45.6 ± 3.4	76.2 ± 1.7	140.0 ± 0.0	186.8 ± 3.6	194.6 ± 4.0	
Murcott	19.3 ± 0.5	48.0 ± 0.8	86.8 ± 2.8	85.5 ± 4.5	131.0 ± 6.0	85.8 ± 0.6	167.4 ± 12.6	175.2 ± 4.0
				Antheraxanthin				
Hamlin	14.3 ± 0.1	8.1 ± 0.2	11.1 ± 0.0	12.0 ± 0.3	22.0 ± 0.5	33.0 ± 5.0	62.6 ± 1.0	49.8 ± 0.0
Pineapple	11.4 ± 0.2	5.6 ± 0.5	17.0 ± 0.3	20.8 ± 1.3	35.7 ± 0.0	47.4 ± 2.4	65.5 ± 3.9	87.0 ± 1.0
Valencia	5.3 ± 0.1	3.6 ± 0.6	4.5 ± 0.3	33.4 ± 2.1	24.0 ± 2.4	66.0 ± 1.5	84.0 ± 1.0	72.3 ± 3.9
Robinson	45.8 ± 1.0	114.2 ± 2.6	120.0 ± 4.0	131.1 ± 7.5	159.5 ± 10.5			
Dancy	53.0 ± 0.6	91.4 ± 2.2	60.4 ± 1.6	96.6 ± 5.0	174.0 ± 4.0	153.9 ± 4.5	212.7 ± 10.5	
Orlando	18.4 ± 0.1	60.0 ± 0.1	33.0 ± 3.3	70.2 ± 1.3	127.2 ± 0.0	173.8 ± 3.8	234.6 ± 11.8	
Murcott	14.0 ± 0.3	44.8 ± 0.0	61.0 ± 3.0	48.0 ± 3.6	139.5 ± 11.5	106.8 ± 1.2	209.6 ± 10.4	248.4 ± 10.0
				Violaxanthin				
Hamlin	14.1 ± 0.1	8.5 ± 0.4	9.5 ± 0.0	11.7 ± 1.6	18.0 ± 0.0	23.8 ± 2.4	38.7 ± 0.5	33.4 ± 1.6
Pineapple	14.0 ± 0.4	6.2 ± 0.1	10.2 ± 0.1	23.5 ± 0.1	34.9 ± 0.0	50.3 ± 0.5	49.6 ± 3.4	52.9 ± 0.5
Valencia	9.0 ± 0.4	7.6 ± 0.1	12.0 ± 0.4	59.3 ± 4.7	35.8 ± 0.9	78.6 ± 2.6	92.7 ± 3.3	74.6 ± 0.0
Robinson	63.0 ± 0.0	138.0 ± 0.8	140.0 ± 4.4	198.9 ± 9.9	183.5 ± 3.5			
Dancy	17.9 ± 0.7	29.0 ± 1.8	21.8 ± 1.8	34.2 ± 1.4	57.6 ± 0.0	60.6 ± 0.6	52.2 ± 3.0	
Orlando	12.3 ± 0.9	34.1 ± 1.1	32.0 ± 1.0	37.2 ± 1.3	67.4 ± 0.6	97.1 ± 11.9	83.4 ± 1.0	
Murcott	21.8 ± 0.4	47.4 ± 1.0	67.2 ± 3.2	44.1 ± 3.3	114.5 ± 1.5	90.0 ± 5.4	155.4 ± 0.0	165.0 ± 9.0

(Continued)

TABLE 25—*Continued*

Cultivar	Sampling dates							
	10-7	10-27	11-19	12-19	1-7	1-28	2-23	3-30
				cis-Antheraxanthin				
Hamlin	34.2 ± 1.9	27.7 ± 0.2	37.7 ± 1.7	33.7 ± 1.9	44.5 ± 1.0	53.3 ± 1.5	97.4 ± 2.2	110.8 ± 4.0
Pineapple	33.0 ± 0.5	13.8 ± 0.3	58.3 ± 1.8	80.1 ± 3.7	90.1 ± 3.0	138.3 ± 1.5	100.2 ± 3.6	240.0 ± 6.0
Valencia	10.1 ± 0.6	7.9 ± 0.4	15.4 ± 0.6	62.8 ± 1.0	72.5 ± 3.7	180.8 ± 8.0	147.8 ± 7.4	170.5 ± 7.3
Robinson	97.5 ± 4.5	199.4 ± 10.6	295.8 ± 7.0	281.1 ± 2.1	243.5 ± 6.5			
Dancy	130.5 ± 1.5	187.4 ± 3.0	154.4 ± 0.0	265.4 ± 13.4	360.0 ± 0.0	404.8 ± 5.6	378.6 ± 7.8	
Orlando	76.3 ± 1.3	143.0 ± 3.0	95.5 ± 5.5	152.0 ± 4.0	277.0 ± 0.0	354.2 ± 13.4	334.0 ± 7.2	
Murcott	57.0 ± 3.0	119.0 ± 5.0	274.4 ± 7.6	232.2 ± 6.0	250.0 ± 0.0	235.0 ± 22.0	413.5 ± 19.1	840.0 ± 32.0
				cis-Violaxanthin				
Hamlin	32.9 ± 0.2	21.4 ± 0.4	30.7 ± 1.5	27.6 ± 1.0	48.9 ± 0.2	55.8 ± 1.0	94.7 ± 4.3	110.8 ± 4.0
Pineapple	41.1 ± 0.3	16.8 ± 1.4	22.1 ± 0.1	95.1 ± 0.4	133.0 ± 3.0	177.0 ± 3.6	115.2 ± 4.0	240.0 ± 6.0
Valencia	22.4 ± 0.8	18.1 ± 0.4	36.0 ± 0.0	115.0 ± 0.0	154.0 ± 10.0	317.0 ± 1.0	216.0 ± 4.0	170.5 ± 7.3
Robinson	163.0 ± 3.0	304.0 ± 4.0	436.0 ± 12.0	452.7 ± 2.7	528.0 ± 8.0			
Dancy	41.8 ± 0.9	67.0 ± 2.2	53.2 ± 0.8	87.0 ± 5.0	206.0 ± 0.0	222.6 ± 4.8	191.4 ± 0.0	
Orlando	42.3 ± 1.8	76.3 ± 2.5	112.0 ± 7.0	67.0 ± 0.9	160.0 ± 4.0	184.8 ± 6.0	158.2 ± 2.6	
Murcott	106.0 ± 2.0	223.2 ± 0.0	414.8 ± 25.2	304.2 ± 4.8	445.0 ± 2.0	396.9 ± 17.1	620.4 ± 40.0	840.0 ± 32.0

[a] From Stewart (*889*).
[b] Expressed as micrograms per 100 ml plus or minus standard error of mean of duplicate determinations.

TABLE 26
Provitamin A Content of Citrus Juice[a,b]

Cultivar	Vitamin A activity[c] (IU)	Daily recommended percentage[c]
Hamlin	80	1.6
Pineapple	133	2.7
Valencia	83	1.7
Robinson	1142	23
Dancy	965	19
Orlando	236	4.7
Murcott	3195	64

[a] From Stewart (*889*).

[b] Values obtained from the last date sampled for each cultivar (see Table 25). Calculations based on the following: β-carotene, 1.667 IU vitamin A per microgram = 100%; α-carotene, 52.7%; β-cryptoxanthin, 57%. Conversion factor from 100 ml of juice to 6 oz, 1.77.

[c] Values based on 6 oz of juice and calculated on the dietary daily allowance of 5000 IU.

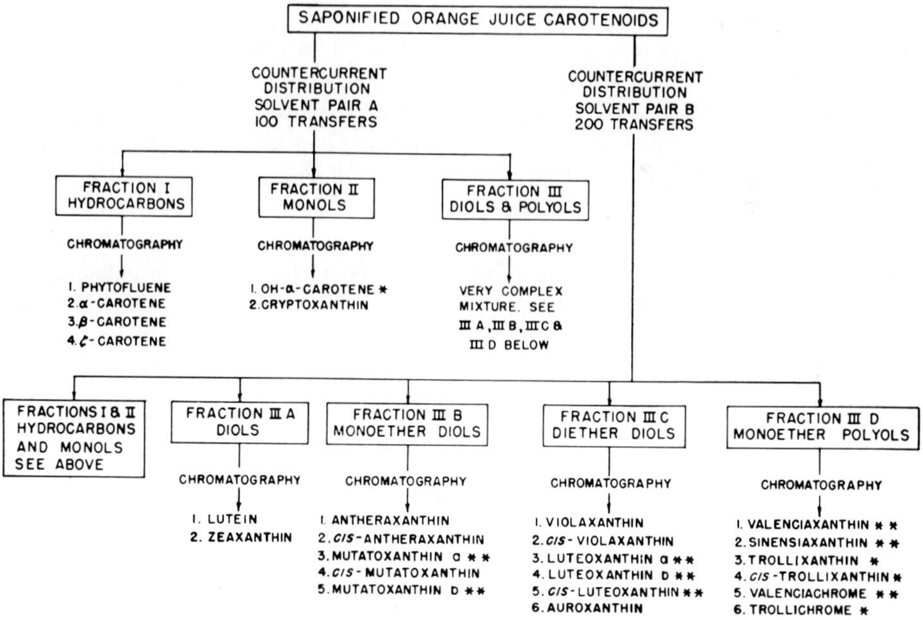

Fig. 22. Fractionation of saponified orange juice carotenoids by countercurrent distribution and chromatography. Key: *, tentative identification; **, proposed name; solvent pair A, petroleum ether–99% methanol (1.8:1); solvent pair B, benzene–petroleum ether–87% methanol (1:1:1.5). [From Curl and Bailey (*230*).]

examined by Calvarano (*161*). A study of canned orange juice by Curl and Bailey (*228*) demonstrated that 5,6-epoxides, which are prevalent in fresh juice, isomerized completely to 5,8-epoxides during storage. This loss of one double bond from the conjugated double-bond system causes both a shift in the wavelength absorption maximum and a decrease in the molar absorbance. These changes partly account for an apparent loss of 20–30% total carotenoids in canned orange juice during 1 year of storage at 70°F. Frozen sixfold-concentrated Florida Valencia orange juice contains 3.9 μg total carotenoids per 100 gm and the distribution of the carotenoids is similar to that in single-strength juice (*224*). Dehydration of this concentrate to 3% moisture caused little change in carotenoid distribution. Conditioning of the powder at 25°C for 78 days to reduce moisture further caused a large decrease in the diepoxide diols because of isomerization of 5,6-epoxides to 5,8-epoxides. On subsequent storage the remaining 5,6-epoxides isomerized. Stability of the provitamin A sources, cryptoxanthin and β- and α-carotenes, was excellent during the entire process.

Over 50 carotenoids were separated by Gross *et al.* (*383*) from the juice of Shamouti orange (*Citrus sinensis*), the principal variety cultivated in Israel. Besides those carotenoids previously reported as occurring in common orange varieties, the following were found: γ-carotene, rubixanthin, sintaxanthin and its hydroxyl derivative, and lutein 5,6-monoepoxide. The diol–polyol fraction accounted for about 70% of the total carotenoids. The light color of Shamouti juice appeared to be due to the low total concentration of carotenoids rather than to the absence of any specific colored pigments. In another study by Gross *et al.* (*382*) the carotenoid composition of juices of three varieties of oranges grown in Israel was investigated. The pigment pattern was similar in all three varieties, although some quantitative variation was observed. The differences in color between Valencia, Shamouti, and navel varieties result from these quantitative differences in carotenoid composition. The ratio of the orange pigments (λ_{max} 445 nm) to the yellow pigments (420 nm) decreased from Valencia through Shamouti to navel. Two new carbonyl pigments with unknown structure were also detected. A new apocarotenal was shown by Gross *et al.* (*379*) to be 3-hydroxy-5,8-epoxy-5,8-dihydro-8′-apo-β-caroten-8′-al. Two UV-fluorescent apocarotenols found recently in avocado were also present. For the pigments previously designated trollixanthin and trollichrome, the new structures 5,6-dihydro-β,β-carotene-3,3′,5,6-tetrol and 5,8-epoxy-5,8,5′,6′-tetrahydro-β,β-carotene-3,3′,5′,6′-tetrol are assigned, both containing a trihydroxylated ring as in heteroxanthin. Patterns of carotenoid distribution in orange juices of Japan were studied by Umeda *et al.* (*956*). Two typical patterns were found. In the Naruto orange type 50% diol diepoxides were noted, whereas in the Unshisu type 70% or more were monol carotenoids.

The color of orange juice varies from pale yellow for early season varieties to deep orange for late season types. In the past, color had been scores visually by comparing the orange juice samples with USDA plastic color standards. As a result of the work of several investigators, the Hunter Citrus Colorimeter (*422*) is used exclusively to determine the color scores of Florida processed orange juices. Petrus *et al.* in 1975 (*750*) reported on the visible and UV absorption characteristics of alcoholic solutions of orange juices, which have utility in the correlation of visible absorbance with numerical color score. Absorption curves (*749*) for each variety were similar, and visible absorption intensity increased as the Hamlin and Pineapple varieties matured. The visible absorbance of Valencia varied only slightly with maturity. Absorption intensity also increased in the order of Hamlin to Pineapple to Valencia varieties.

Researchers have been concerned with the detection of carotenoids added to orange juice and have offered analytical procedures for assay (*88,89,258,531,819*). Natural color enhancers (*941*) for juice can be prepared from orange peels, but methods of detecting adulteration of juice with peel have been published (*821*). Verma and Sastry (*989*), investigating the preparation, composition, and storage behavior of comminuted orange squashes, noted good retention of carotenoid content with storage. The carotenoid composition pattern of blended orange juices was investigated by Tada *et al.* (*918*).

d. Tangerines and Mandarins. Tangerines and mandarins are reddish-orange, loose-skinned fruits, which exhibit a complex spectrum of carotenoids and generally have a higher color intensity than the orange. Curl and Bailey (*212,226*) found that the redder color of the tangerine than of the orange was due to higher concentrations of cryptoxanthin in the pulp and peel as well as of β-carotene in the pulp and a hydroxycanthaxanthin-like substance in the peel. Minor amounts of other carotenoids not found in the orange were also reported. The peel of the tangerine has about twice the carotenoid content of the peel of a typical Valencia orange, and the pulp has about the same content as the Valencia pulp. The carotenoid constituents of the pulp and peel, as determined by Curl and Bailey, are shown in Table 22. Curl (*217*) has also characterized five carbonylcarotenoids from tangerine peel: reticulataxanthin, tangeraxanthin, β-citraurin, β-apo-8′-carotenal, and apo-12′-violaxanthal, obtained principally from the peel. The β-apo-8′-carotenal levels (*217*) in tangerine peel ranged between 0.4 and 1.9% of the total carotenoids. β-Citraurin makes an important contribution to the red color of the tangerine. DiGiacomo *et al.* (259) observed the total carotenoid, carotenoid ester, and β-carotene content of the juice of Italian tangerines and noted levels of 18.1, 11.2, and 1.4 mg/liter, respectively, at the maximal stage of ripeness.

Elahi and Shah (*279*) determined the content of total carotenoids and carotenes in peel, pulp, and juice of three varieties of mandarin (*Citrus reticulata*), three of sweet orange (*C. sinensis*), and one each of grapefruit (*C. paradisi*) and lemon (*C. limon*). Peel had the highest contents of carotenoids, with average values of 145, 132, 5.1, and 4.9 mg/kg, respectively. Carotene contents were generally higher in pulp and juice than in peel. There are a large number of carotenoids in the peel of mandarins as identified by chromatography and spectrophotometry (*554*). The main components are violaxanthin (35%), β-carotene (15%), cryptoxanthin (15%), mutatoxanthin (12%), and cynthaxanthin (9%). Prolycopene, cryptoxanthin epoxide, cryptoflavin, flavoxanthin, and apo-10′-violaxanthal are present in traces. In 1974 Gershtein and Kochurina (*340*) also investigated the Unshisu variety of mandarin and noted that the carotene and oxycarotenoid levels are influenced by time, weather conditions, and variety of fruit. Earlier in the development of the fruit the carotenoid content of the pulp is high, but as the fruit matures the peel has higher levels.

8. Mangoes

Like other fruit, mangoes vary in their ascorbic acid and carotenoid content. Iguina DeGeorge *et al.* (*425*), in investigating 30 varieties for chemical composition, reported 7 varieties (Adams, Carries, Edward, Paheri, Palmer, Sensation, and Zill) to be highest in carotenoid vitamin A precursors. Jungalwala and Cama (*479*) reported β-carotene (60%) forming the largest proportion of total carotenoids in mango fruit (Alfonso variety), with oxycarotenoids present. The Badami and Alfonso varieties were also studied by Jungalwala and Cama (*479*) and John *et al.* (*477*), and 16 different carotenoids, including β-carotene, γ-carotene, monoepoxy-β-carotene, and cryptoxanthin, were identified in the fully ripe fruit. The level of oxycarotenoids in ripe fruit was higher than in partially ripened fruit. In partially ripened fruit, the hydrocarbon carotenes (phytofluene, β-carotene, and phytoene) constitute the major part (about 85%) of the carotenoids. The isolation of carotenoid-protein complex in the mango was reported by Subbarayan and Cama (*907*). De Leon and De Lima (*243*), reporting on postharvest changes in Pico mangoes, obtained total carotenoid values of 0.05–0.9, 1.2–2.3, and 2.3–4.3 mg per 100 gm for the stages of green, yellow, and yellow-orange fruit, respectively. β-Carotene values for five varieties of mangoes ranged from 2.4 to 8.3 mg per 100 gm, as determined by Roy (*823*) in 1973. Adsule and Roy (*15*), in their study of five mango cultivars, reported the Dashehari variety to have the greatest total carotenoid content (9.1 mg per 100 gm). This variety also had the best fruit color and flavor and the best overall quality in canned or frozen form. Thomas (*926*), working with the Alfonso variety, found that storage temperature below 25°C adversely affected the development of aroma, flavor, and carotenoid formation of mangoes upon ripening. The carotenes always exceeded (60–70%) the level of oxycarotenoids of the total pulp carotenoids. Carotenogenesis in mangos was studied by Mattoo *et al.* (*635*).

9. Persimmons

The carotenoids of the Japanese persimmon have been investigated intensively (*148,221*). Forty varieties were found to contain 2–11.5 mg per 100 gm of total carotenoids. The major carotenoid was cryptoxanthin (30–35%), with an extremely variable lycopene content (0–30%). The varieties with low lycopene content showed marked increases in the mono- and diepoxydiol fraction. Curl (*221*) isolated 26 carotenoids from the Hachiya variety. The major carotenoids were cryptoxanthin

(38%) and zeaxanthin (18%). The fruit was also a good source of antheraxanthin. Minor constituents were β-carotene, lutein, ζ-carotene, violaxanthin, and neoxanthin. The distribution of carotenoids in three varieties of persimmons (Table 27) illustrates how variable the individual carotenoids are among varieties (*148*).

10. Miscellaneous Fruits

Several varieties of table grapes of the Donets Basin (*532*) were investigated for carotenoid content after harvesting and 3 months of storage. Values of 0.4–0.9 mg carotenoids per 100 gm grape skins were found. In another study of Sultana grapes grown in South Australia, Bottrill and Hawker (*135*) found carotenoid content to be influenced by drying and light conditions.

Papaya contains β-carotene (29%), cryptoxanthin (48%), and cryptoflavin (13%) in addition to 16 other carotenoids, according to Subbarayan and Cama (*909*) in a study of carotenoids in Bangalore and Carica varieties. Papaya puree prepared from varieties grown in Hawaii have a carotenoid content of 3.5–3.9 mg per 100 gm, as reported by Brekke *et al.* (*141*). Chan *et al.* (*168*) studied carotenoid patterns in papaya puree during processing.

Major components of passion fruit are β-carotene and unesterified and esterified xanthophylls, according to Pruthi and Lal (*774*). Leuenberger and Thommen (*578*) investigated the pasteurized juice of passion fruit (*G. edulis*) for its carotenoid

TABLE 27

Percent Distribution of Carotenoids in Three Varieties of Persimmons[a,b]

Fraction	Component	Fuyu	Honan Red	Tamopan
Hydrocarbon	Phytofluene	4.2	2.6	0.5
	α-Carotene	0.6	1.5	0.8
	β-Carotene	4.6	6.4	7.5
	ζ-Carotene	7.6	3.8	Trace
	γ-Carotene	0.5	1.2	Trace
	Lycopene	29.0	19.0	0.2
Monols[c]	Hydroxy-α-carotene	2.3	0.1	2.0
	Cryptoxanthin	31.0	38.0	28.0
Diols	Lutein	2.2	2.0	4.5
	Zeaxanthin	6.8	11.0	12.0
Monoepoxydiols[c]	Antheraxanthin	5.0	9.2	21.0
	Mutatoxanthin	2.0	0.5	1.2
Diepoxydiols[c]	Violaxanthin	1.6	1.8	14.3
	Luteoxanthin	—	—	1.7
Polyols[c]	Neoxanthin	1.0	1.0	2.0

[a] From Brossard and Mackinney (*148*).

[b] Fuyu, Honan Red, and Tamopan contained, respectively, 65, 80, and 61 mg of carotenoid per kilogram.

[c] Miscellaneous unidentified components are not included.

content. Besides the already known carotenoids, β-carotene, ζ-carotene, and phytofluene, β-apo-12′-carotenal, β-apo-8-carotenal, cryptoxanthin, auroxanthin, and mutatoxanthin could be identified. The three main carotenoids, β-carotene, ζ-carotene, and phytofluene, were traced quantitatively in the various parts of the fruit, as well as in pasteurized and deep-frozen specimens of the fruit juice.

The carotenoid composition (Table 28) of fresh pineapples was examined by

TABLE 28

Carotenoids Obtained from Pineapples[a]

Fraction	Probable identity[b]	Solvent[c]	Absorption maxima (nm)[d]	Approx. %[e]	
I-1	Phytofluene	H	367, 347	0.5	
I-2	α-Carotene	H	474, 445, 421	0.5	
I-3	β-Carotene	H	476, 449, (427)	6.7	
I-4	ζ-Carotene	H	426, 400, 377	0.9	
I-5	Neurosporene	H	468, 439, 416	0.7	
II-1	Hydroxy-α-carotene	H	472, 445, 420	0.9	
II-2	Cryptoxanthin	H	475, 449, (426)	0.9	
II-3	Cryptoxanthin-like	H	470, 445, (427)	0.4	} 0.7
II-4	Cryptoxanthin-like	H	470, 445, (424)	0.3	
IIIA-1	Lutein	B	483, 453, 426	2.5	
IIIA-2	*cis*-Lutein	B	480, 445, (426)	1.0	
IIIB-1	Lutein 5,6-epoxide	B	481, 450, 426	0.3	
IIIB-2	Antheraxanthin	B	482, 454, (432)	1.0	} 1.7
IIIB-3	*cis*-Antheraxanthin	B	480, 452, (430)	0.7	
IIIB-4	Flavoxanthin	B	456, 432, 408	0.3	
IIIB-5	Mutatoxanthin	B	462, 434, 410	0.6	
IIIC-1	*cis*-Violaxanthin	B	477, 447, 422	25.0	} 52.0
IIIC-2	Di-*cis*-violaxanthin	B	471, 441, 416	27.0	
IIIC-3	Luteoxanthins	B	460, 432, 407	6.2	} 13.4
IIIC-4	*cis*-Luteoxanthins	B	458, 430, 406	7.2	
IIIC-5	Auroxanthins	B	436, 410, 386	3.4	
IV-1	Neoxanthin-like	B	480, 449, 426	0.7	} 1.5
IV-2	Neoxanthin-like	B	480, 450, 426	0.8	
IV-3	Neoxanthin	B	480, 450, 425	8.1	
IV-4	Trollixanthin-like	B	480, 451, 427	1.1	} 2.1
IV-5	Trollixanthin-like	B	480, 450, 428	1.0	
IV-6	Neochromes	B	457, 431, 408	0.7	} 1.5
IV-7	Neochromes	B	458, 430, 406	0.8	
IV-8	Substance 407[f]	B	432, 407, 385	0.8	

[a] From Morgan (*683*).
[b] Tentative identification indicated by absorption spectra etc.
[c] H, hexane; B, benzene.
[d] Values in parentheses represent shoulders on spectral absorption curves.
[e] Based on absorbance at principal wavelengths.
[f] Substance 407 refers to that pigment with absorption maximum at 407 nm.

TABLE 29

Carotenoid Constituents of Italian Prunes[a]

Fraction	Constituent	Approx. % of carotenoid mixture[b]
I	Phytoene	1.3
	Phytofluence	1.1
	α-Carotene	0.5
	β-Carotene	18.7
	ζ-Carotene	0.2
	Mutatochrome[c]	0.4
II	Hydroxy-α-carotene	0.2
	Cryptoxanthin 5,6,5′,6′-diepoxide	1.2
	Cryptoxanthin 5,6-epoxide[c]	3.5
	Cryptoxanthin	7.3
	Cryptoxanthin 5,6,5′,8′-diepoxide[c]	0.4
	Cryptoxanthin 5,8-epoxide	0.6
	P 374[d]	1.2
IIIA	Lutein	15.5
	Zeaxanthin	0.4
IIIB	Lutein 5,6-epoxide	0.4
	Antheraxanthin	2.1
	Flavoxanthin	0.04
	Mutatoxanthins	0.2
IIIC	Carbonyl[a]	0.2
	Carbonyl[b]	0.2
	Violaxanthin	35.0
	Luteoxanthins	2.3
IV	Persicaxanthin	2.9
	Neoxanthin[a]	0.4
	Neoxanthin[b]	0.4
	Trolliflor[c]	0.6
	Taraxanthin[c]	0.07
	P 399[d]	0.3
	Persicachromes	0.1

[a] From Curl (*215*).

[b] Percentages based on total absorbance of each constituent at its principal spectral absorption maximum.

[c] Tentative identification.

[d] Pigments absorbing at 374 and 399 nm unidentified as also for pigments a and b.

Morgan (*683*). The major carotenes were violaxanthin (50%), luteoxanthin (13%), β-carotene (9%), and neoxanthin (8%) plus smaller amounts of ζ-carotene, hydroxy-α-carotene, cryptoxanthin, lutein, auroxanthin, and neochrome.

Italian prunes (Table 29) contain violaxanthin, β-carotene, and lutein as the primary carotenoids (*215*) and, less abundantly, cryptoxanthin, mono- and diepoxides, persicaxanthin, luteoxanthin, antheraxanthin, phytoene, and phytofluene. More recently, Moutounet (*692*) published a paper on the carotenoids of the plum and prune and changes during processing. Troyan and Gol'yan (*950*) also published on pigment changes during growth and ripening of plums.

Peterson *et al.* (*748*) reported 0.16 mg % carotene content in cranberries (*Oxycoccus microcarpus* and *O. quadripetalus*). Galler and Mackinney (*323*) discussed the carotenoid distribution of pears, strawberries, and cherries. Strawberry pigments were also investigated by Woodward (*1026*) during the ripening process. Oxycarotenoids, mainly esterified diepoxydiols and diepoxypolyols, predominate in pears and strawberries. The carotenoids of several fruits of low carotenoid content have been determined by Curl (*213*) (Table 30).

The total carotenoid content of watermelon (*Citrullus ovulgaria*) was examined by Morgan (*682*). The total pigment of pulp samples expressed as lycopene was 24.6 μg/gm. This consisted of 90% hydrocarbons, 6.7% monols, and 3.3% diols plus polyols. The contents of various carotenes separated by chromatography were as follows: phytoene, 2.1%; phytofluene, 1.4%; α-carotene, 0.06%; β-carotenes, 4.1%; δ-carotenes, 1.6%; pigments with maxima at 429, nm 432, and 434 nm, 0.01% of each; proneurosporene, 0.01%; mutatochrome-like pigment, 0.03%; pro-γ-carotene, 0.03%; poly-*cis*-lycopenes, 0.01%; ζ-carotenes, 0.36%; neurosporene, 0.14%; neolycopene, 7.6%; and lycopene, 73.7%. Carotenoid changes in developing muskmelons were studied by Curl and Reid *et al.* (*799*).

B. Vegetables

Green and yellow vegetables are on the usual recommended lists of healthful foods to be consumed by adults and children for their carotenoid (provitamin A), ascorbic acid, fiber, and mineral contents. Consequently, there have been quite a few studies on the carotenes of vegetables produced in various countries (*181,300, 303,389,406,600,627,698,716,747,793,837,843,859,904*).

1. Carrots

A process for the commercial extraction of carotenes from carrots has been described (*962*), as cull carrots once served as a source of a natural carotenoid concentrate. According to early work, garden varieties of carrots averaged 5.4 mg of total carotenes per 100 gm, mainly α- and β-carotenes, with ζ-carotene prevalent and lycopene present in some orange varieties (*392*). Malewski and Michalik (*618*) reported that carrot varieties adaptable to freezing and canning contained 6.9–12.2 mg of total carotenes per 100 gm; Macedonian carrots were reported to contain

TABLE 30

Carotenoid Constituents Obtained from Low-Carotenoid Fruits by Chromatography[a]

Constituent	Approx. % carotenoid mixture[b]				
	Cran-berry	Black fig	Seedless grape	Black-berry	Blue-berry
Hydrocarbons (I)					(11.0)[c]
Phytoene	4.7	2.4	+	1.5	
Phytofluene	0.9	0.7	0.2	0.2	
α-Carotene	0.14	0.24	0.2	1.1	
Unidentified	—	—	0.4	—	
β-Carotene	5.3	9.9	32.3	9.5	
ζ-Carotene	0.6	—	—	—	
Mutatochrome[d]	0.6	—	—	—	
Monols (II)		(2.1)[c]	(2.1)[c]	(3.2)[c]	(2.2)[c]
Hydroxy-α-carotene-like	0.4				
Hydroxy-α-carotene	1.1				
Cryptoxanthin	1.8				
Diols (IIIA)					
Lutein	31.3	28.8	22.0	44.3	39.2
Zeaxanthin	2.8	1.1	1.1	2.3	2.1
Monoepoxide diols (IIIB)		(2.0)[c]	(6.0)[c]	(1.9)[c]	(13.7)[c]
Carbonyl[d]	0.18				
Lutein 5,6-epoxide	10.5				
Antheraxanthin (cis)	4.6				
Lutein 5,8-epoxide	0.7				
Mutatoxanthins	0.33				
Diepoxide diols (IIIC)				(22.0)[c]	(19.4)[c]
Carbonyl[d]	0.37	—	—		
Violaxanthin	20.7	40.0	15.7		
Luteoxanthins	1.8	2.5	1.6		
Polyols (IV)				(14.1)[c]	(11.4)[c]
Valenciaxanthin[d]	1.1	0.4			
Sinensiaxanthin[d]	0.25	1.0			
Neoxanthin	6.9	7.1	(12.2)[e]		
Sinensiaxanthin-like	1.8	1.0			

[a] From Curl (*213*).

[b] Percentages based on total absorbance of each constituent at its principal spectral absorption maximum.

[c] Percentage of traction from countercurrent distribution; fractions not chromatographed.

[d] Tentative identification.

[e] Spectral absorption curve of entire fraction very similar to that of neoxanthin.

1.4–8.2 mg per 100 gm (*188*). The effect of variety and planting date on carotenoid content (*138*) is illustrated in Table 31. Investigators have reported 5–46% α-carotene by weight of total carotenes in carrots (*360*). Small amounts of γ-, δ-, and ζ-carotenes and lycopene are also present. The oxycarotenoid fraction is usually only 5–10% of the total carotenoid content (*129*), but yellow varieties contain 75–93% oxycarotenoids. The distribution of carotenoids in different parts of the carrot has been examined by Booth (*127*). The phloem has a higher concentration of carotenoids than the xylem, as is obvious from the relative color of the two parts of the root (*127*).

Comparisons of several varieties of European and Asiatic carrots showed that the European carrot contained more carotene than did the Asiatic varieties and also contained greater amounts of total soluble solids and total solids. The larger core size and higher percentage of moisture was responsible for the lower nutritive value of the Asiatic carrots (*342*). Genetic studies on the carrot have resulted in an improved carotene content (*577*). Consistency of seasonal performance is not always obtained, as indicated by the investigation of Bradley and Rhodes (*139*). Eleven of 29 varieties and lines gave good canned slice color in both unusually warm and unusually cool growing seasons for carrots. A number of lines showed

TABLE 31

Effect of Variety and Planting Date on Color and Carotenoid Content of Carrots[a]

	Hunter *a/b* ratio	β-Carotene	α-Carotene	β- plus α-carotene	β-/α-carotene ratio
Planting date[b]					
March 11	0.60	5.0	4.6	9.6	1.1
March 25	0.57	4.2	4.4	8.6	1.0
July 22	0.68	5.7	2.2	7.9	2.6
August 3	0.63	5.2	2.1	7.3	2.5
L.S.D. at 0.05	0.05	0.5	0.4	0.5	0.3
Variety[c]					
Danvers 126	0.70	6.0	3.3	9.3	1.8
Ch. Red Core	0.56	4.8	2.7	7.5	1.8
Royal Chantenay	0.54	4.5	2.5	7.0	1.8
Scarlet Nantes	0.58	5.0	2.6	7.6	1.9
Pioneer	0.60	5.2	3.5	8.7	1.5
Waltham Hi-Color	0.83	4.7	3.6	8.3	1.3
L.S.D. at 0.05	0.08	0.9	0.9	0.9	0.5

[a] From Bradley and Dyck (*138*). Values are given as mg/100 gm fresh wt.
[b] Average of 10 varieties or strains.
[c] Average of four planting dates. Ten varieties or strains were used, but the strains did not differ significantly, so only one is reported. There was no variety × planting date interaction in any of the measured components.

seasonal variation in color, with better color in the cool season, whereas others gave consistently fair or poor color in both growing seasons. No single pigment was highly correlated with color across the range of environmental and genetic diversity encountered. β-Carotene was the only single component that had a significant correlation with color. The carotene content of recent carrot cultivars has been reported by Blattna *et al.* (*114*). Inheritance of color and associated carotenoid composition has been a subject of interest to several groups of workers (*426,561,957*). Fertilization and/or herbicide treatment may influence the carotene content of the carrot root. Salminen *et al.* (*829*) reported that herbicide treatment retarded significantly and systematically carotene formation in the carrot, but Sweeney and Marsh (*914*) found higher carotenoid values. A third report by Zechalko *et al.* (*1038*) showed inconsistent findings.

Pigment concentration changes during growth and storage in the carrot. Changes in the isomeric composition of the carotene fraction were studied during growth and storage. An increase in carotene concentration was found during growth. Different curves were obtained from leaves and roots. During storage, the concentration of carotenes increases initially, reaches a maximum, and decreases again, and the α-/β-isomeric component changes (*539a*).

The carotene content of carrots is also influenced by conditions of storage. Among various tested methods (*792*) for storage of Nante variety carrots at 10° to −2°C and 72–98% relative humidity for 7 months, the best appeared to be storing in polyethylene bags. When carrots are stored alone in a cellar and in sand, losses of weight, dry substance, sugar, and carotene become greater as the duration of storage increases.

2. *Corn*

Corn (maize) consumed on the cob or cut from the cob after cooking is usually considered a vegetable, but when dried, ground, whole, or refined, in cooked or baked foods, it is usually categorized as a cereal grain. There are red, yellow, white, and mixed varieties. Yellow corn is important for its carotenoid vitamin A precursor value. Benk (*86*) studied the carotene levels of corn. According to Zimmerman *et al.* (*1046*) Golden Cross Bantam sweet corn contained about 1 μg/gm total carotene by chemical assay and bioassay. Sweet golden corn varieties contained 1.2–5.3 μg/gm total carotene (*852*). Evertender C and Royal Gold sweet corn contained 8–9 μg/gm total carotenoids, including 0.9–1.3 μg/gm β-carotene and 2.3–3.1 μg/gm cryptoxanthin (*938*), as influenced by year of planting and growing cycle (Fig. 23). β-Carotene in these varieties was stable in refrigerated and frozen storage. Diol and polyol carotenoids comprised about 50% of the total carotenoids of both varieties. A relatively recently discovered carotene in corn is zeacarotene (*752*). It has provitamin A activity and is present at lower levels than cryptoxanthin. It appears to be an all-*trans*-monohydroxy-α-carotene. Typical carotenoid analyses of yellow corn have been reported by Quackenbush *et al.* (*781*). Considerable work has been done to breed yellow corn with high carotenoid content (*118*).

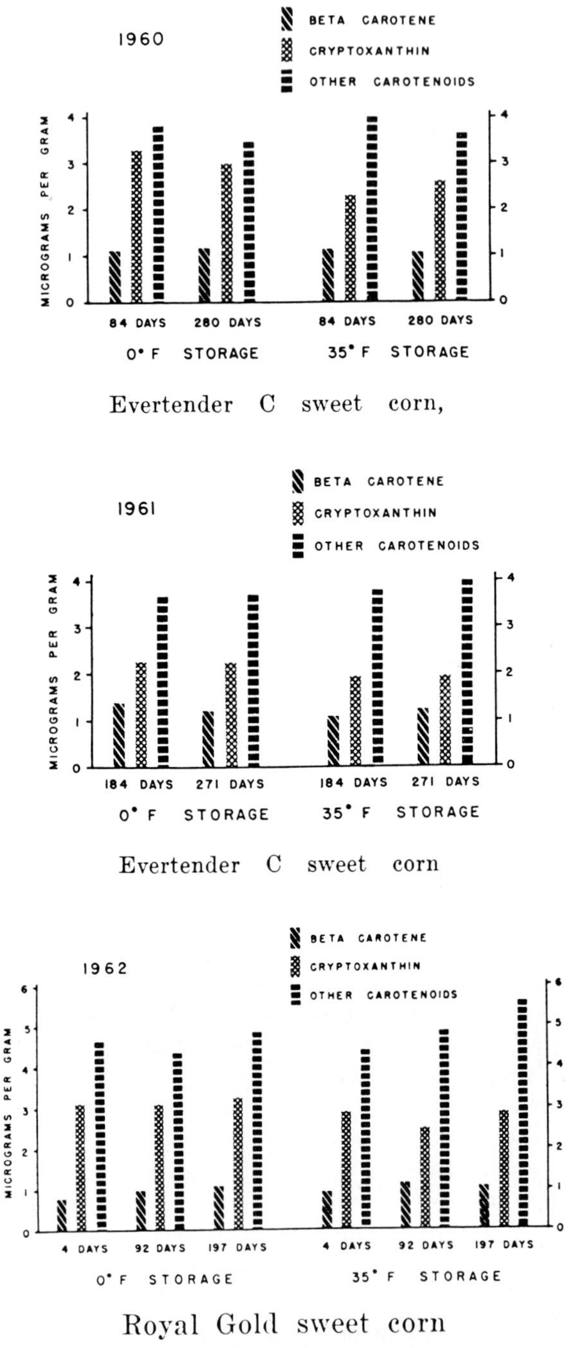

Fig. 23. Carotenoids of frozen and refrigerated sweet corn. [From Tichenor *et al.* (*938*).]

3. *Peppers*

As mentioned earlier in this chapter, red peppers are a source of natural carotenoid concentrates. Garden-grown green peppers are considered more as a source of ascorbic acid. In a 1976 study, peppers (*Capsicum annuum*) were harvested at the mature green and red ripe stages for ascorbic acid (*632*) and carotene determinations. Green fruit of Yolo Wonder L Early Calwonder, Resistant Florida Giant, and Florida 136 varieties contained at harvest 100–117 mg ascorbic acid per 100 gm. Yolo Wonder L fruit harvested at the red stage had significantly higher ascorbic acid levels (153 mg per 100 gm) than green fruit. The carotene content of green bell peppers was reduced after 14 days of storage at 45°F and after 7 days at 70°F. Green bell peppers contain 9–11 μg/gm total carotenoids. α-Carotene, β-carotene, ζ-carotene, hydroxycarotene, cryptoxanthin, lutein, zeaxanthin, violaxanthin, luteoxanthin, auroxanthin, neoxanthin, and neochrome were reported in green peppers by Curl (*214*). Red bell peppers have a complex carotenoid mixture, including capsanthin, capsorubin, and at least six other pigments containing cyclopentane rings (*218*). Total carotenoid concentration is 127–248 μg/gm.

4. *Potatoes*

White potatoes, *Solanum tuberosum,* as expected, are very low in total carotenoid content. A mixture of carotenoids is found in the Katahdin Irish potato, the total carotenoid content being about 3 μg/gm, dry weight (*151*). Lutein (0.5–0.8 μg/gm), β-carotene (0.3 μg/gm, and ξ-carotene (aurochrome) (0.1 μg/g) were isolated among several unidentified carotenoids. Lepage (*575*) reported that lutein constitutes 50% of the carotenoid content of white potato tubers, the remainder being α-carotene, β-carotene, an unidentified pigment, and lutein epoxide. β-Carotene, β-carotene 5,6-monoepoxide, cryptoxanthin 5,6-diepoxide, lutein, *cis*-violaxanthin, *cis*-antheraxanthin 5,6-monoepoxide, *cis*-neoxanthin, and an unknown carotenoid were noted in British varieties (*744*). During growth, the epoxides are abundant; at maturity, the nonepoxides are also present in equal ratio; and during storage, nonepoxides increase, and epoxides decrease. In a German study (*491*), the following carotenoids were isolated from 50 kg of raw, unpeeled tubers: α-carotene, β-carotene 5,6:5′,6′-diepoxide, lutein, lutein 5,6-epoxide, violaxanthin, and, presumably, neoxanthin. The pigments were identified by light absorption curves, paper chromatography, and rearrangement of 5,6-epoxides to 5,8-epoxides.

Sweet potatoes or yellow potatoes, extensively grown in the southern United States, are a significant source of carotenoid provitamin A. A good yellow flesh color is a reliable indicator of the provitamin A value of sweet potatoes. Sweet potatoes as well as carrots maintain their carotene content during cool storage in root form (*287*). Sweet potatoes, in addition to having significant quantities of β-carotene, also contain ζ-carotene, hydroxy-ζ-carotene, β-carotene furanoxide, γ-carotene, and *cis*-cryptoxanthin epoxide. The total carotene content, 0–151 μg/gm, depends greatly on variety (*672*) and varies more than 400% within a variety (*288*). The effects of variety, season, planting and harvest time, curing, and storage of sweet potatoes have been intensively studied at the Georgia Experimental Station. The effect of variety (*881*) is shown in Table 32. Season and field location as well as storage conditions are other highly influencing variables (*631*). Colorless polyenes, namely, phytofluene and phytoene, have been detected in the sweet potato, and their significance in carotenoid biosynthesis has been considered (*1034*). Purcell and Walter (*776*) in 1968 studied the carotenoids of the centennial variety of sweet potato, *Ipomoea batatas,* and reported the total carotene content of sweet potato to be 0.13 mg/gm for fresh matter and 0.45 mg/gm for dry matter. The composition of carotenes and oxy derivatives was determined; some values were α-carotene, 0.9%; β-carotene, 86.4%; γ-carotene, 0.8%; ζ-carotene, 1.8%; phytoene, 2.6%; and phytofluene, 2.0% of total carotenoids. Philippine sweet potatoes are regarded (*329*) as lower in starch than others. As the depth of the color of the sweet potato flesh increases, the concentration of the carotenoid provitamin A increases and the starch content decreases.

TABLE 32

Mean Carotene Content at Time of Harvest of 15 Varieties or Lines of Sweet Potatoes Grown in Two or More Plantings[a]

Variety or line	Number of plantings	Carotene content (mg/100 gm dry basis) Unadjusted	Adjusted[b]
Allgold	19	44.4	44.6
Goldrush	9	42.2	40.0
Heartogold	7	25.6	26.1
L-37	12	25.2	25.3
Maryland Golden	6	19.6	21.9
Nancy Gold	15	20.2	20.5
Australian Canner	14	20.4	20.2
Ranger	12	19.2	19.9
Queen Mary	11	19.7	19.3
Unit I Porto Rico	46	12.0	12.0
Bunch Porto Rico	14	11.3	10.7
L-132	4	8.2	9.6
B-703	2	8.2	7.3
B-2934	7	5.4	5.2
Little Stem Jersey	3	1.0	0.5
Overall mean	181		18.9

[a] From *Southern Cooperative Series Bulletin (881).*

[b] Adjusted for variation due to plantings at different locations and in different years on the basis of the performance of Unit I Porto Rico variety. Least difference (approx.) required for significance in comparisons involving less than six plantings, 8 mg; for 6 or more plantings, 6 mg.

5. *Spinach and Other Green Leaves*

Spinach contains 53–58 μg/gm total carotene, 57–74 μg/gm lutein, and some other carotenoids (*920*). Another report (*412*) indicates that fresh spinach leaves contain appreciable amounts of β-carotene. Other oxycarotenoids, lutein, zeaxanthin, violaxanthin, and neoxanthin are also present, which is the pattern in most higher plant tissue. Hirayama and Oida (*412*) found 100 gm of fresh spinach leaves to contain 4 mg β-carotene, 7 mg lutein, 1 mg zeaxanthin, 1 mg violaxanthin, and 2 mg neoxanthin. Storage at 45°C for 25 hr resulted in 43–75% decomposition of the total carotenoids.

Leafy vegetables are an important source of vitamin A activity in the diet of the peoples of tropical and warm-temperate countries. Begum and Pereira (*82*) investigated 32 varieties of edible green leaves for β-carotene content during different seasons of the year and divided them into foods having a high, moderate, and low pigment content (Table 33) and showed some plant leaves, such as the amaranths, to

TABLE 33

β-Carotene Content of Edible Green Leaves[a]

Edible green leaves[b]	β-Carotene content[c] (μg/100 gm)
High content of β-carotene	
Tribulus terrestris (5)	7365 ± 856
Neeringi keerai	
Amaranthus viridis (7)	7160 ± 2760
Kuppai keerai	
Solanum nigrum (5)	6990 ± 3730
Manathakkali keerai	
Sesbania grandiflora (5)	6610 ± 2160
Agathi keerai	
Euphorbia hirta (5)	6215 ± 640
Pacharisi keerai	
Oxalis corniculata (7)	6050 ± 2250
Puliara keerai	
Amaranthus spinosus (6)	5735 ± 2815
Mullu keerai	
Raphanus sp. (6)	5570 ± 1245
Kasini keerai	
Amaranthus polygonoides (4)	5260 ± 1410
Sirukeerai	
Alternanthera sessilis (5)	5245 ± 490
Poonanganni keerai	
Moringa oleifera (6)	5220 ± 2180
Muringa keerai	
Alternanthera sp. (4)	4630 ± 810
Seemai poonanganni	

(*Continued*)

TABLE 33—*Continued*

Edible green leaves[b]	β-Carotene content[c] (μg/100 gm)
Moderate content of β-carotene	
Amaranthus sp. (5)	3880 ± 320
Chakravarthy keerai	
Amaranthus tristis (6)	3445 ± 1010
Arai keerai	
Portulaca wightiana (7)	3570 ± 1050
Nadirsan keerai	
Cardiospermum helicacabum (4)	3535 ± 680
Modakathan keerai	
Brassica oleracea var. *caulorapa* (5)	3470 ± 685
knol khol keerai	
Celosia sp. (5)	3405 ± 980
Panna keerai	
Hibiscus cannabinus (5)	3395 ± 2051
Pulichai keerai	
Amaranthus gangeticus (4)	3250 ± 1100
Thandu keerai	
Amaranthus sp. (4)	3065 ± 880
Molai keerai	
Digera arvensis (4)	2730 ± 980
Toyya keerai	
Boerhaavia repens (4)	2630 ± 1080
Mukarrate keerai	
Balanites roxburghi (4)	2530 ± 1350
Nachu kottai keerai	
Low content of β-carotene	
Raphanus sativus (5)	2280 ± 615
Mulangi keerai	
Petroselinum sp. (4)	2180 ± 490
Pain pasalai keerai	
Colocasia antiquorum (5)	2020 ± 725
Seppam keerai	
Trigonella foerman graecum (5)	1970 ± 442
Vendia keerai	
Portulaca oleracea sp. (4)	1850 ± 885
Paruppu keerai	
Spinacia oleracea (5)	1840 ± 440
Pasalai keerai	
Rumex vesicarius (4)	1810 ± 240
Ambut chukka keerai	
Tamarindus indicus (4)	1180 ± 125
Puli elai	

[a] From Begum and Pereira (*82*).

[b] The names in Tamil are given below the botanical names. The number of analyses is given in parentheses.

[c] Mean values ± 1 SD.

be significant sources of β-carotene. There are changes in the carotene content of green leafy vegetables during cooking, which should be recognized (*878*).

Amaranths belong to a small group of plants in which photosynthesis is exceptionally efficient. The sunlight they capture is utilized more effectively than in most plants. Vigorous and tough, amaranths have been termed self-reliant plants, germinating and growing well under adverse conditions. They are easily cultivated, adapt well to rural small plots, can be harvested by hand, and are easy to cook. Amaranths are annuals with large leaves, and they are cereal-like plants producing full, fat seed heads. In addition to their seeds, some species have edible and nutritious leaves that are widely eaten as a boiled vegetable. The Department of Food Science and Technology of the University of Nairobi has collaborated with UNICEF in a program of research and extension aimed primarily at increasing the utilization and availability of local green leafy vegetables as sources of vitamin A. It has been observed that many indigenous wild plants are excellent sources of carotene (Table 34) as well as other micronutrients.

6. *Squash*

The term "squash" is applied to a variety of yellow and green curbits, some with no keeping or storage capacity and others that can be stored for months at low temperatures in fresh form. Many of these are good sources of carotenoid provitamin A. Flesh color serves as a crude guide to carotenoid content. Seven varieties (*416*) of the nonstorage or summer squashes (*Cucurbita pepo* L. and *Cucurbita*

TABLE 34

β-Carotene Content of Wild and Cultivated Green Leafy Vegetables from the Central Province of Kenya

Vegetable	Local name	β-Carotene content (μg/100 gm wet wt)	
		Range	Mean
Amaranthus sp.	Terere	3,750–4,750	4,416
Crotalaria brevidens	Mito	6,250–7,750	7,000
Gynandropsis gynandra	Sageti	15,500–16,750	15,916
Vigna unguiculata	Kunde	6,250–8,250	7,416
Cucurbita pepo		7,750–9,000	8,375
Solanum nigrum	Managa	7,500–7,750	7,625
Curcurbita sp.		7,875–9,125	8,291
Beta vulgaris		4,875–7,375	6,125
Manihot utilissima	Muhage	12,125–16,750	14,437
Phaseolus vulgaris		9,500–11,225	10,665
Brassica oleracea		7,625–9,000	8,312
Chorchoris olitorius	Murenda	8,250–9,000	8,750
Solanum tuberosum		10,825–11,520	11,260

TABLE 35

Carotenoids in Butternut Squash during Storage[a,b]

Weeks of storage	AOAC[c] carotene	Partition chromatography		
		α-Carotene	β-Carotene	Xanthophylls
0	2.09	0.37	0.44	0.88
1	2.71	0.45	0.64	0.98
2	3.28	0.37	0.86	1.29
3	3.90	0.68	0.82	1.50
4	4.24	0.86	0.79	1.44
7	4.59	0.51	0.80	2.14

[a] From Lewis and Merrow (*580*).
[b] Fresh basis, mean of eight replications expressed as milligrams β-Carotene in 100 gm.
[c] Association of Official Analytical Chemists.

moschata L.) contained from 0.6 μg/gm total carotene for Early White Bush Scallop to 6 μg/gm for Golden Cusham. Lewis and Merrow (*580*) demonstrated that Butternut squash (storage type) has approximately equal quantities of α- and β-carotenes and considerably higher quantities of mono- and dihydroxycarotenoids. The influence of storage after harvesting on carotenoid content (*580*) is illustrated in Table 35. The total carotene content of six varieties of winter squash (storage type) determined at harvest and at 5-week intervals ranged from 4 to 29 μg/gm at harvest. After 10 weeks of storage, total carotene content ranged from 6 to 36 μg/gm on an adjusted fresh weight basis.

7. *Tomatoes*

Lycopene is by far the predominant carotenoid of tomatoes; β-carotene, γ-carotene, and various oxycarotenoids (a minor fraction) make up the remainder (*219,769*). Total carotenoid content is 20–60 mg/kg (dry basis). The outer pericarp of tomatoes (Table 36) has the highest total carotenoid concentration, and the locular contents have the highest carotene content (*640*). Regarding carotene content, Cirkova-Gorgievska *et al.* (*187*) reported values of 0.33 and 0.29% for two varieties; 0.64, 0.74, and 0.85 mg % were cited for three varieties of ripe tomatoes by Bagdasaryan *et al.* (*38*). Caro-Red, a variety of tomato high in provitamin A carotenoids, was reported by Tomes and Quackenbush (*944*). Curl (*219*) determined the xanthophyll distribution (Table 37) of tomatoes: 15% were monols, 49% diols, 4% monoepoxide diols, 22% diepoxide diols, and 11% polyols. Phytoene 1,2-oxide and related epoxy derivatives have more recently been recognized in the tomato (*144*). Krishna Mallia *et al.* (*543*) studied the carotenoid distribution in tomatoes grown in India.

The carotenoid composition of tomato fruit and the inheritance of fruit color of many species and cultivars (*488,949*) have been studied extensively since the early

TABLE 36

Carotenoids of the Regions of the Stokesdale Tomato at Different Stages of Ripeness[a]

	Number of days ripened				
Region of fruit	4	8	12	16	20
			Total carotenoids (mg/100 gm)		
Outer pericarp	4.40	7.84	8.64	8.64	8.40
Inner pericarp	2.96	4.40	4.40	4.00	4.40
Locular contents	4.80	4.16	4.00	6.56	6.32
			Carotene (mg/100 gm)		
Outer pericarp	0.23	0.23	0.18	0.18	0.26
Inner pericarp	0.22	0.12	0.15	0.18	0.39
Locular contents	0.69	0.71	0.60	0.57	0.55
			Total carotenoids/carotene		
Outer pericarp	15	34	47	47	23
Inner pericarp	14	37	30	22	11
Locular contents	7	6	7	12	12

[a] From McCollum (*640*).

1930's. Ulrich *et al.* (*952*) recognized all-*trans*-lycopene to be the major carotenoid of red-fruited tomatoes but in the genotype tangerine found poly-*cis*-lycopene, all-*trans*-ζ-carotene, and poly-*cis*-neurosporene. Yellow tomatoes accumulate mostly neurosporene, β-carotene, and lutein and during ripening reach a final carotenoid level of approximately 40 mg/kg dry weight of fruit, according to Jen (*474*). Ripe tangerine tomatoes (variety of *Lycopersicon esculentum*) have orange-colored pulp, which contains prolycopene as the main carotenoid (*1041*). In hybrids studied by Grigorovskaya *et al.* (*371*) large qualitative and quantitative differences in carotenoid pigments were found. Another factor influencing carotenoid development is light intensity. Khudairi (*503*) concluded that the process of color change in tomatoes is triggered principally by red light.

Many investigations have involved color and carotenoid development during the ripening process. Carotenoid content influenced by stage of ripeness (*1014*) and by temperature and light is shown in Table 38. During the growth and maturation process, carotenoid metabolism is localized at the center of the fruit and later in the external portions, according to Laval-Martin *et al.* (*568*). Other publications (*535,595,673*) deal with the influence of temperature and diurnal change, stage of ripeness, etc., on the pigment composition of the tomato. Von Euler *et al.* (*993*) found that green tomatoes ripened normally at 20°C but became yellow rather than red at 30°C. Verma *et al.* (*990*) showed that vine-ripened tomatoes contained less acid but more reducing sugars, ascorbic acid, and carotene than tomatoes ripened

TABLE 37

Xanthophylls Obtained from Tomatoes[a]

Fraction	Identity	Spectral absorption maxima[b] (nm)	Percentage of total xanthophylls[c]
II-1	Monol 487	487, 455, 428	6.8 } 9.8
II-2	Poly-*cis*-monol 487[d]	453, 431	3.0
II-3	Poly-*cis*-rubixanthin[d]	(490),458, 437	4.4
II-4	Poly-*cis*-lycoxanthin a[d]	(497), 468, (447)	2.1
II-5	Poly-*cis*-lycoxanthin b[d]	(496), 469, (448)	1.6 } 4.7
II-6	Lycoxanthin	504, 473, 445	1.0
IIa-1	Substance 398	398, 376, 357	0.8
IIa-2	Substance 378	378, 357, 344	2.0
IIIA-1	Lutein	487, 455, 431	33.5
IIIA-2	Zeaxanthin	490, 461, (435)	1.3
IIIB-1	Lutein 5,6-epoxide	483, 452, 425	2.0
IIIB-2	*cis*-Antheraxanthin[d]	481, 453, 429	1.7
IIIB-3	Flavoxanthin	459, 430, 406	0.8
IIIB-4	*cis*-Mutatoxanthin[d]	461, 434, 411	0.9
IIIC-1	Violaxanthin	484, 453, 426	12.5
IIIC-2	Luteoxanthin a	461, 432, 408	0.6
IIIC-3	Luteoxanthin b	460, 431, 407	1.0 } 2.1
IIIC-4	*cis*-Luteoxanthin a	456, 427, 405	0.2
IIIC-5	*cis*-Luteoxanthin b	456, 427, 404	0.3
IV-1	Neoxanthin	479, 447, 423	11.7
IV-2	Trolliflor[d]	482, 449, 423	0.4
IV-3	Neochromes[d]	457, 431, 408	0.9

[a] From Curl (*219*).
[b] In petroleum ether for II and IIa; benzene for remainder.
[c] Based on total absorbance of each constituent at the principal spectral absorption maximum.
[d] Tentative identification.

artificially after harvesting at the green mature stage, a finding confirmed by Al-Vagab *et al.* (*20*) and Watada *et al.* (*1005*) and generally believed by tomato consumers. In the investigations of Meredith and Purcell (*658*) on the Homestead tomato, the concentration of α- and β-carotenes did not change significantly during the ripening process. Proper potassium nutrition is needed for carotenoid development in the tomato, as several studies have revealed carotenoid synthesis to be reduced in potassium deficiency (*951*).

Tomato Juice. A good percentage of the tomato crop is converted to processed tomato products of a high degree of color (*33*), such as juice and paste. Zubeckis (*1050*) examined the tomato juice produced in Canada. In 13 samples of raw tomatoes the carotene content varied from 0.23 to 0.74 mg per 100 gm. Commercial

TABLE 38

Carotenoid Content[a] and Some Growth Rates of Tomatoes Grown Under Various Environmental Conditions[b]

	Environmental conditions[c]								
	A	A	A	B	B	C	D	E	F
Temp (°C), day	26.5	26.5	26.5	20.0	20.0	20.0	26.5	26.5	20.0
Temp (°C), night	20.0	20.0	20.0	20.0	20.0	20.0	20.0	26.5	26.5
Total pigment (as lycopene)	386	286	302	219	365	485	615	512	708
Xanthophyll	4.6	3.8	3.8	3.4	5.8	7.1	8.7	5.2	7.8
Lycoxanthin	21.2	12.1	11.4	9.8	16.5	20.3	26.6	26.8	30.7
Lycopene	270	180	188	138	220	308	375	351	442
Neolycopene A	17.7	16.5	19.5	12.1	21.7	14.8	20.6	16.3	25.5
Neolycopene B	1.5	—	—	1.8	0.4	2.7	3.2	6.7	1.5
Unidentified carotene	1.7	—	—	1.0	—	1.5	2.5	—	1.2
γ-Carotene + neo-γ-carotene	1.2	—	—	1.1	0.4	2.0	1.5	3.6	1.4
Unidentified	0.5	0.1	0.2	0.5	0.2	0.5	0.3	2.2	0.3
Isomer of β-carotene	2.1	0.7	0.6	1.7	0.5	2.8	1.6	1.4	1.9
β-carotene	6.2	3.1	2.5	4.7	1.3	8.9	4.6	7.6	7.8
Neo-β-carotene	1.4	0.7	—	1.7	0.6	3.5	0.5	1.4	2.3

[a] Expressed as milligrams per 100 gm dry fruit.

[b] From Went *et al.* (*1014*).

[c] Conditions: A and B, low night temperature without additional light; C and D, illuminated at night, E and F, high night temperature.

brands of tomato juice varied from 0.38 to 0.55 mg %, with a mean value of 0.44 mg %. Large variations were observed both among brands and among samples from the same manufacturer. In a study (*596*) on the effect of harvest maturity on carotenoids in tomato paste, phytoene, phytofluene, β-carotene, and lycopene increased in content as the maturity of the tomato fruit advanced. The effects of variety (*609*), heat in processing (*238,669,670*), and storage (*8*) on the color and carotenoid content of tomato juice and tomato puree have been investigated.

8. Miscellaneous Vegetables

Cantharellus cinnebarinus, an edible pink mushroom most commonly recognized in Europe, contains canthaxanthin as its principal pigment (*401*). β-Carotene is present to a lesser degree. *Phyllotopsis nidulans* is composed of 58% β-carotene, 29% α-carotene, 8% echinenone, and 4% astaxanthin (*296*). The commonly used U.S. mushroom, *Agaricus campestris* L., has few if any carotenoids. The carotenoids of the rutabaga root or swede (*478*) include a number of carotenes, neurosporene, and lycopene and are unusual in the proportion of poly-*cis*-

carotenoid isomers. Fresh broccoli contains 6–46 μg/gm total carotene. The total carotene content of asparagus is 5.8–7.9 μg/gm (*1045*). Lima beans, fresh or frozen, contain about 2.3 μg/gm total carotene, two-thirds of which is β-carotene (*1045*); frozen baby lima beans contain 1.2 μg/gm total carotene (*196*). Snap beans contain 5.5 μg/gm (*643*), of which two-thirds is β-carotene (*1045*). Soybeans, cow peas, lima beans, Thomas Laxton peas, and French (snap) beans contain 2–7 μg/gm total carotene. The content of soybeans is influenced by maturity, sprouting, processing, and storage (*49*). The total carotene content of 34 early and late varieties or strains of peas was found to be 3.4–7.1 μg/gm (*405*). The carotenes of the pumpkin have been studied by Berger *et al.* (*93*) and Walczak and Gronowska-Senger (*999*).

C. Grains

Because of the volume of cereal grains consumed in the diet of most of the populations of the world, some examination of grain products is justified.

1. Maize (Corn)

As early as 1933–1934 zeaxanthin, cryptoxanthin, and β-carotene were isolated from dried corn (*555*). Petzold and Quackenbush (*751*) also reported zeinoxanthin in corn. A chromatographic separation of carotenoids of corn seedlings by Moster *et al.* (*691*) is shown in Fig. 24. The carotene and oxycarotenoid contents of yellow

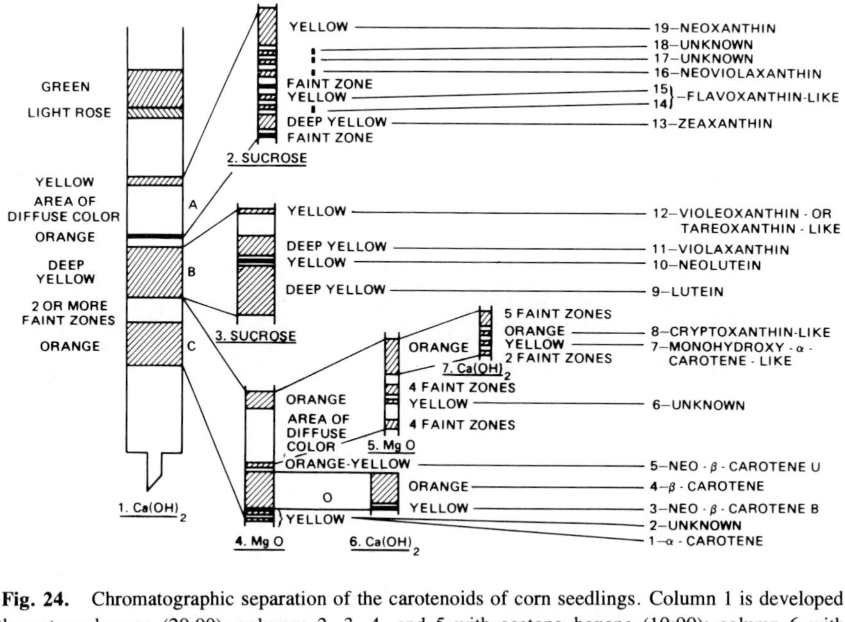

Fig. 24. Chromatographic separation of the carotenoids of corn seedlings. Column 1 is developed with acetone–hexane (20:90); columns 2, 3, 4, and 5 with acetone–hexane (10:90); column 6 with acetone–hexane (0.7:99.3); and column 7 with acetone–hexane (3.97). [From Moster *et al.* (*691*).]

TABLE 39

Carotenoids of Corn and Sorghum[a]

Sample	Nonsaponified		Saponified		
	Carotenes (ppm)	Xanthophylls (ppm)	Carotenes (ppm)	Xanthophylls (ppm)	Xanthophyll esters (ppm)
High-xanthophyll corn	3.0	35.9	2.6	36.3	0.4
Hybrid yellow corn	2.0	20.1	1.6	20.5	0.4
Corn gluten (60% protein)	16.9	201.6	13.2	205.3	3.7
Red sorghum	0.3	1.8	0.2	1.9	0.1
Sorghum gluten (60% protein)	0.8	6.5	0.6	6.7	0.2

[a] From Blessin (*115*).

corn and sorghum, as determined by Blessin (*115*), are shown in Table 39. Corn should be analyzed for β-carotene, cryptoxanthin, and other provitamin forms (*141a*). Quackenbush *et al.* (*781*) in 1961 separated 11 carotenoid fractions in five corn strains and compared theoretical provitamin A activity with animal bioassay results. Lutein and zeaxanthin predominated; the content of total colored carotenoids ranged from 18 to 55 μg/gm. Similar studies were made by Grogan and Blessin (*373*) in 1968, Mihajlovic *et al.* (*667*) in 1969, and Zetea and Bodea (*1043*) in 1971. Hungarian-grown *Zea maize* L. was studied at two stages of maturity by Zsolt *et al.* (*1049*). Carotenes, with β-carotene predominating, were approximately one-third of the total carotenoids. The total carotenoid content ranged from 10 (white corn) to 24 μg/gm (red corn). In examining 20 samples of hybrid Spanish maize or maize flour from different regions, Carballido *et al.* (*163*) found a range of carotenoid content of 8.2–28.0, with an average value of 16.1 μg/gm. Quackenbush *et al.* (*780*) completed a study in 1963 dealing with the carotenoid content of some 125 corn inbreds. The carotenoid provitamin A content of yellow inbreds varied from a trace to 7.3 μg/gm, including one-third of the β-zeacarotene, one-half of the cryptoxanthin, and all of the β-carotene as the vitamin A precursor forms. The content of oxycarotenoid pigments also varied widely (2–33 μg/gm of corn) and bore no close relationship to the provitamin A pigments. In a third paper, Quackenbush (*778*) reported that 11%-moisture corn lost 75% of initial carotenes when stored for 3 years at 25°C and 50% at 7°C. Losses of total carotenoids were most rapid during the early part of the storage period. The oxycarotenoids were more stable than the carotenes. Climatic conditions influence carotene and oxycarotenoid content in corn strains. The distribution of carotenes and oxycarotenoids in hand-dissected and dry-milled fractions of yellow dent corn was determined by Blessin *et al.* (*117*). Benk (*86*) cited the carotenoid value of yellow maize starch for mayonnaise production.

2. *Wheat*

Since wheat products, such as flour, farina, and semolina, are widely used in cooked and baked food products, two aspects of carotenoid content are important: either its intensification, as is desired in pasta, or its elimination, as is desired in refined bread or white cake flour. The carotenoid contents of four types of wheat were measured (Table 40) by Parker and Harris (*735*). The total carotenoid content of whole wheat was 1.8–2.3 μg/gm, and the embryo contained the highest concentration (4–11 μg/gm). Regarding carotene content, whole wheat contained 0.15–0.25 μg/gm, and the embryo 0.5–1.13 μg/gm. The major oxycarotenoids, as found by Irvine and Anderson (*435*), were lutein and taraxanthin. Lepage and Sims (*576*) studied the carotenoids of flour from two varieties of wheat. The total carotenoid contents were 3.7 and 2.8 μg/gm, and in both varieties the major components were lutein and its esters. Likewise, Lier (*590*) and Lier and Lacroix (*591*) examined the carotenoids of durum wheat. The carotene and total carotenoid contents in 39 samples of semolina and flour were 0.05 and 5.3 mg/kg, respectively (*1018*). Shkvarkina and Soshina (*861*) cited the carotenoid content of flour and bread as

TABLE 40

Carotenoid Content of Wheat and Its Fractions[a]

	Endosperm		Embryo		Bran		Whole wheat	
Class of wheat	μg/gm	%	μg/gm	%	μg/gm	%	μg/gm	%
Hard red spring								
Carotene	0.09	5.7	0.72	10.0	0.04	4.3	0.18	10.0
Xanthophyll	0.84	53.5	5.78	80.3	0.42	45.1	0.99	55.0
Xanthophyll ester	0.64	40.8	0.69	9.6	0.47	50.5	0.63	35.0
Total	1.57		7.19		0.93		1.80	
Hard red winter								
Carotene	0.11	5.5	0.80	10.2	0.02	2.1	0.21	10.3
Xanthopyll	0.77	38.5	5.98	76.2	0.33	34.8	0.79	38.7
Xanthophyll ester	1.12	56.0	1.06	13.5	0.60	63.2	1.04	51.0
Total	2.00		7.84		0.95		2.04	
Soft white								
Carotene	0.21	9.6	1.13	10.2	0.03	3.4	0.25	10.8
Xanthophyll	1.18	54.2	9.70	87.8	0.32	36.4	1.33	57.8
Xanthophyll ester	0.79	36.3	0.21	1.9	0.53	60.2	0.72	31.4
Total	2.18		11.04		0.88		2.30	
Durum								
Carotene	0.08	4.1	0.50	12.2	0.10	4.5	0.15	7.6
Xanthophyll	1.78	90.8	2.93	70.8	1.31	59.0	1.67	84.8
Xanthophyll ester	0.10	5.1	0.70	17.0	0.81	36.5	0.15	7.6
Total	1.96		4.13		2.22		1.97	

[a] From Parker and Harris (*735*).

0.05–0.08 mg % and 0.03–0.07 mg %, respectively, and reported significant loss upon transportation and storage. The loss of carotenoids during intensive kneading in the baking process have been attributed to lipoxidase activity stimulated by aeration of the dough (*265*). Factors affecting the stability of carotene were discussed by Dahle (*234*). Carotenoid oxidation is inhibited during dough mixing carried out in a vacuum (*708*). Since the preferred color of pasta products is yellow, many papers have appeared on the carotenoids of pasta products and factors influencing color (*17,623*). It is well recognized that there are losses (30–60%) during the processing of pasta (*260,480*) and that carotenoids are added when egg or spinach is included in the formulation (*709*).

3. *Miscellaneous Grains*

The grain of common varieties of sorghum contains 1.5–2.0 μg/gm total carotenoids; some yellow strains contain 8–9 μg/gm (*120*). Major carotenoids were identified as lutein, zeaxanthin, and β-carotene. Yellow sorghum carotenoid losses (35–53%) during outdoor weathering periods were reported (*119*). Blessin (*115*) compared the total carotenes, xanthophylls, and xanthophyll esters of corn and sorghum (Table 39). Additional data on the carotenoid content of sorghum grains were developed by Wall and Blessin (*1001*) in 1969. The carotenoids of sorghum have also been studied by Deostali *et al.* (*248*). Flour preparations of six Russian rye varieties were demonstrated to contain β-carotene, poly-*cis*-lycopene, lutein epoxide, and taraxanthin (*526*). The carotenoid compositions of barley seed oil (*245*), rye seed oil, and rice seed oil (*834*) were also investigated.

D. Vegetable Oils

Of the vegetable oils that are widely consumed, palm oil (*90*) has by far the highest concentration of carotenoids, usually 500 mg/kg or more occurring in the crude, unbleached oil. Red palm oil is obtained from the fruit of the palm tree *Elaeis guineensis,* of which a large number of subspecies are known (*106*). Palm forests grow in West Africa, but the tree is also cultivated in East Africa, Java, Malay, and South America. Palm oils in the Far East and from the Belgian Congo contain 0.05–0.08% total carotenes and those from the Ivory Coast (especially Dahomey) contain 0.1–0.16%, but oil yield is less (*112*). The carotene composition of palm oil was published by Pinto (*759*). α-Carotene and β-carotene are the major carotenoids present (*423,695*), usually in a 3 : 2 ratio, but small amounts of γ-carotene, lycopene, neolycopene, neolutein, and probably neo-γ-carotene are detected in some specimens. The oil content of the mesocarp and also the carotene content of the oil appear to increase with the maturity of the fruit. The oil (*424*) is remarkable for the high proportion of α-carotene (Table 41). In Africa and elsewhere, palm oil can be a significant food source of provitamin A (*589,933*). If the oil is decolorized (*598*) and transesterified, as is usually done if the oil is to be used as a commercial food oil, its vitamin A value is substantially reduced. Carotenoid

TABLE 41

Carotene in West African Red Palm Oil[a]

Variety	Maturity	Percentage of oil in mesocarp	α- + β-carotene (%)	α-carotene (% of total)	Provitamin activity of oil (IU/100 gm)
Afia oku	Very immature	15.8	0.0473	43	62,000
	Unripe	37.6	0.0597	30	85,000
	Ripe	37.5	0.0634	41	84,000
	Overripe	55.8	0.0785	27	113,000
Lisombe	Very immature	22.6	0.0165	42	21,000
	Unripe	27.5	0.0496	43	65,000
	Ripe	56.2	0.0726	40	98,000

[a] From Hunter *et al.* (*424*).

separation studies have been carried out (*598*), and the total carotene (α and β) contents of palm oil differ slightly as judged by samples from seven countries (*694*). Crude oil of yellow maize (corn) is a fairly good source (*668*) of the provitamin A carotenoids. The carotenoid pigments in peanut oil have been studied during maturation (*742,743*). Carotenoid esters in soybean and rapeseed oils have been researched by Froehling *et al.* (*318*).

E. Dairy Products

In milk and colostrum, almost all the vitamin A present occurs in the ester form (*170,645,883*) and only a small percentage as the alcohol (*169,325,740,936*). The breed of cattle and stage of lactation have little influence on the ester–alcohol ratio, although dietary changes such as carotenoid intake may be a factor (*170,646*). Vitamin A activity of milk is due not only to true vitamin A but to the presence of carotenoid vitamin A precursors. The carotenoid content of milk is dependent largely on the daily intake (*126,393,475,548,644*) of carotenoids by the cow and to a minor extent on breed. β-Carotene makes up the greatest percentage of the carotenoids in milk (*343,899*). Other carotenoids observed in milk are α-carotene (*400*), neo-β-carotenes (*1048*), γ-carotene (*1048*), cryptoxanthin (*344*), all of which are provitamin A structures, and nonprovitamin A carotenoids such as lycopene (*344*), lutein (*345,782*), and zeaxanthin (*310*). Since milking cows usually are on fresh green pasture in late spring, summer, and early fall, the vitamin A value is much higher (*408,800,853,1004,1019*), in milk produced at these times than in that produced during winter (Fig. 25). The carotenoid content of cow milk shows little change between the second and fortieth weeks of lactation. Total carotenes are 4.3–6.0 μg/gm, 65–85% of the total carotenoids, according to

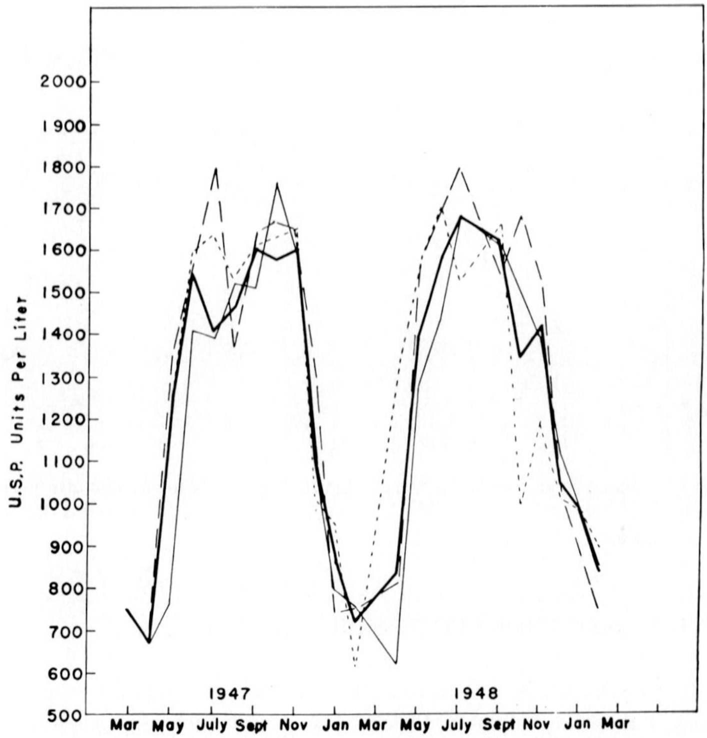

Fig. 25. Seasonal and regional variation of the vitamin A potency of Ohio Market milk. Key: —, average; ---, central; ——, north; ----, south. [From Washburn *et al.* (*1004*).]

Chandra (*169*). New Zealand butterfat contains 9–14 μg/gm total carotenoids, depending on the season of the year (*647*). The variation in the carotenoid pigments in the butterfat of milk according to different investigations has been charted (Fig. 26) by Reinart and Brown (*800*). In an extensive U.S. survey, butter (80% fat) was found to contain 2.7 μg carotene and 5.9 μg vitamin A per gram during the winter and 6.1 μg carotene and 7.5 μg vitamin A during the summer (*166*). In addition to butter, cheese made from whole milk also can be a significant source of carotenes and vitamin A since 80–100% of the milk content is incorporated into the cheese (*810*). It is important to recognize that, in the removal of the fat from milk such as occurs in the production of liquid or dry nonbutterfat dairy products, the carotenes and fat-soluble vitamins are removed from these products. Reviews (*270,370,397,398,508,533,648,784,932,935*) on the carotenoid and vitamin A content of dairy products are available.

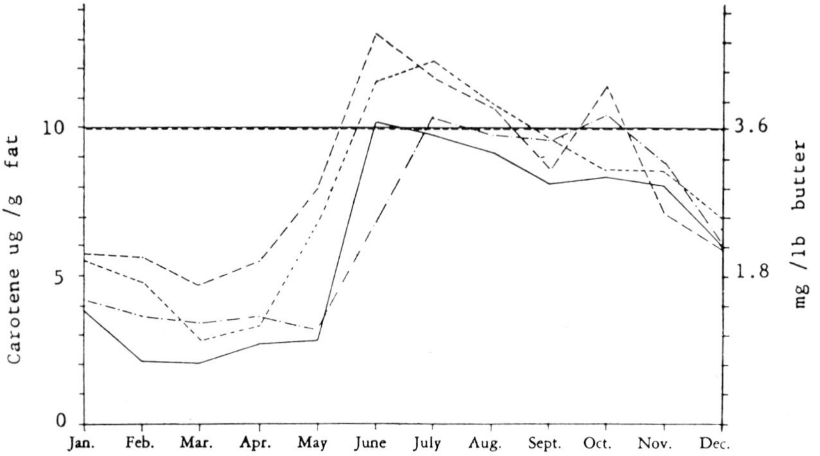

Fig. 26. Variations in the natural pigments in butterfat according to different investigations. Key: --, Manitoba; ---, Idaho; ——, Nebraska; -·-·, Sweden; horizontal line, suggested standard color of butter. [From Reinart and Brown (*800*).]

XI. INFLUENCE OF FOOD PROCESSING

As indicated earlier, factors influencing carotenoid production in ripening fruit are oxygen, light, and temperature (*360*). Carotenoids possess carbon–carbon double bonds, a system subject to oxidation, which is accelerated by light, metals, and peroxides. The main factors affecting carotenoids during food processing and storage are oxidation by oxygen of the air and structural changes by heat. Heat also isomerizes the all-*trans*-carotenoids to cis forms. Although not a usual type of degradation process, exposure to oxidases under conditions favorable to their activity results in the destruction of carotenoids, as can occur in unblanched vegetables (*871*) or bread making (*266*). The naturally occurring carotenoprotein complexes and the protein-absorbed carotenoids generally are believed to be more stable than the free carotenoids. Destroying the natural complexed structure by heating or treating with water-soluble solvents, such as alcohol and acetone, usually leads to a decrease in carotenoid stability. At the same time the carotenoids become more easily extractable by nonpolar solvents, and this behavior is also an indication that the embedding protein matrix is changed in such a manner that its protective effect on the carotenoids is decreased.

The stability of food carotenoids during and after food processing has received considerable study, such as the report of Strimska and Holukova (*903*). Carotenoids are stable in systems with minimal oxygen content. During processing it is advisable to minimize the oxygen content, particularly of clear-glass items, which is normally done by hot packing, vacuum packing, and oxygen scavenging with ascorbic acid. The common unit operations of food processing, in general, have relatively minor effects on carotenoid content. Blanching and freezing generally cause little or no degradation. Packaged frozen foods and heat-sterilized foods sealed from air in containers with few exceptions exhibit good to excellent carotenoid stability throughout their normal shelf life. In general, dehydrated and powdered fruits and vegetables have poor carotenoid stability unless carefully processed and promptly placed in a hermetically sealed, inert atmosphere for storage (*888,938*).

A. Freezing

Blanching had little effect on the total carotene content of asparagus, snap beans, peas, and corn (*643*). Blanching and freezing also had little effect on asparagus and lima bean carotenoids (*1045*). In frozen corn there was little change in β-carotene, cryptoxanthin, or total carotenoids during 9 months of storage at 0°F (*938*). Frozen broccoli had no loss in total carotene during storage at 0°F for 61 weeks (*625*). The biochemistry and physiology, including color stability, involved in the freezing preservation of commercially important fruits was reviewed by Finkle (*298*) in 1971. Carotenoids properly added to food products and stored in frozen condition are quite stable physically and chemically (*57*).

B. Heating, Cooking

Carotenoids are stable to heat in systems with a minimal oxygen content. Oil solutions can be heated under nitrogen to 150°C with only small losses. Day and Erdman (*239*), Mader (*611*), Nakagawa *et al.* (*701*), and Lotspeich *et al.* (*601*) observed the degradation of β-carotene and identified some of the breakdown products. Before 1970, cooking was believed to have little or no effect on carotenoids, even though isomerization as a result of heating was recognized. For carotenoids without provitamin A value, this general statement still holds for short cooking periods. Cooking had little effect on carotenoids in lima beans, as concluded by Cook *et al.* (*196*). Cole and Kapur (*194*) observed a 2% lycopene loss in heating tomato pulp after 1 hr at 65°C and a 15% loss after 3 hr. Martin *et al.* (*625*) reported that retention of total carotene was almost 100% when fresh or frozen broccoli was cooked. Losses were small or nil in carrots (*925*) and broccoli (*171*) that were microwave-cooked. Microwave cooking had no effect on total carotene content in peas (*275*). Cooking carrots in distilled, tap, or salt water at atmospheric pressure or in a pressure cooker had no effect on total carotene content, according to Lantz (*565*). Purcell *et al.* (*777*) in 1970 observed that some high-carotenoid vegetables showed a distinct color shift when heated. The absorption spectra of fresh and heated carrots, sweet potatoes, squash, and tomatoes were determined. The spectra of fresh vegetables having well-defined chromoplasts were nearly duplicated by aqueous suspensions of the major carotene. The color shift was attributed to the degradation of chromoplasts and solutions of carotenes in other cellular lipids. Spencer (*882*) briefly discussed carotenoid changes during the cooking of foods.

When one considers carotenoid provitamin A structures in food processing, stereoisomer form becomes important as it affects vitamin A value. Carotene solutions heated above 60°C undergo cis–trans isomerization, resulting in a mixture of stereoisomers consisting principally of all-*trans*-β-carotene, neo-β-carotene B, and neo-β-carotene U, with minor amounts of other isomers (*253*). Cis–trans isomerization also occurs slowly at room temperature but has relatively little effect on tinctorial potency or hue. cis-Carotene isomers are usually lower in vitamin A activity than the corresponding trans form. In nature, the carotenoids exist primarily in their all-trans form; however, cis forms do exist and may even be abundant, as in the case of pro-γ-carotene (*360*). As shown in the Sweeney and Marsh studies (*915,916*), the principal cis-β isomer in cooked broccoli is neo-β-carotene U, a mono-cis isomer. Cooking of broccoli (25 min) caused increases in neo-β-carotene U (24%) but not neo-β-carotene B (5%), a di-cis isomer. Similar results were obtained with all green vegetables studied (*915*), including brussels sprouts, spinach, collards, kale, beet greens, and endive (Table 42). Carotene in sweet potatoes is chiefly all-*trans*-β-carotene. Cooking (40 min) produced large amounts of neo-β-carotene B (24%) and small amounts of neo-β-carotene U (5%). Similar results (*915*) were obtained with carrots, red peppers, pumpkins, and squash (Table 43). It seems obvious that the principal β-carotene isomer formed during the cooking of green vegetables is neo-β-carotene U, whereas that formed during the cooking of red or yellow vegetables is neo-β-carotene B. α-Carotene, when present, performed in the same fashion as β-carotene. The reason for the difference between green and nongreen vegetables in carotene isomer formation is not known. Although the chemical determination of "crude carotenes" in fresh and cooked vegetables may not demonstrate a very noticeable crude value change, analyses of individual carotenoids show that significant changes do occur. Panalaks *et al.* (*732*) discussed the effect of processing on carotene isomers. Determinations of crude carotene in fresh canned, dehydrated, and cooked sweet potatoes, carrots, peas, and green beans and in fresh and canned tomatoes and peaches were compared with those of individual carotenoids that constituted the crude pigments. The pigments α-carotene, neo-β-carotene B, neo-β-carotene, and all-*trans*-β-carotene were separated chromatographically. The carotenoids possess vitamin A activities calculated as 63–95% of the corresponding crude carotene values.

When cooked vegetables were kept hot (approximately 90°C) for 30–45 min, carrots showed losses of β-carotene of 21–38%, and Savoy cabbage lost approximately 50% of its β-carotene content after having been kept hot for 3 hr (*842*). The detrimental effects are increased when larger quantities are being cooked, since the periods of time involved are correspondingly longer. El-Wakeil and Morse (*280*) noted greater carotenoid losses during baking of sweet potatoes than encountered during the blanching process. Observable differences in the color of carrots from the same lot cooked in different ways prompted Borchgrevink and Charley (*130*) to study the effects of cooking on both the color and the carotenoid content of cooked carrots. Carrots cooked in a saucepan had the highest concentrations of crude carotene and of all-*trans*-β-carotene and were more typically red-orange and bright, whereas those cooked for 2 min in a pressure saucepan were more yellow and dull and were lowest in β-carotene content. Neo-β-carotene B was highest in the pressure-cooked carrots, but the amount did not entirely account for the lower concentration of the all-trans isomer. Baloch *et al.* (*43*) found β-carotene more susceptible to heat damage than α-carotene in carrot processing.

Mudambi and Rajagopal (*693*) studied the effect of heat on palm oil. Fresh palm oil samples were obtained from Nigerian local markets and analyzed for their β-carotene content. Portions of the palm oil samples were also heated to 11 temperature levels between 138° and 258°C with an interval of 12°C, and the β-carotene content was determined. The amount of β-carotene decreased with an increase in temperature. The destruction of β-carotene was greater when the oil was heated continuously for 30 min at any given temperature. When palm oil is used as a frying medium, only a very small fraction of the β-carotene is retained.

C. Canning, Retorting

Fellers (*293*) in 1940 indicated that canning and subsequent storage at ordinary temperatures were not believed to affect the total carotenoid content of processed foods. Booher and March (*125*), by means of a rat bioassay, even reported an increase in the vitamin A activity of green beans, collard, endive, kale, iceberg lettuce, peppers, frozen spinach, and turnip greens on canning. Glass packing of peas, spinach, carrots, and tomato juice, with subsequent exposure to diffused light, was reported to have little effect on total carotene content (*294*), and asparagus, snap beans, peas, and corn had excellent carotene stability according to McConnell *et al.* (*643*) in 1945. Mechanical hulling of lima beans has been reported to cause buising and significant losses of carotene and ascorbic acid (*404*).

Whereas total carotene content in spinach is stable on canning, lutein undergoes substantial losses when spinach is processed at 240°F and minimal losses at 270°–280°F (*920*). Gstirner and Saad (*385*) reported losses of 33% carotene in 44 days in light-exposed, glass-packed spinach puree, but in those experiments sealed test tubes were only one-third filled with product, resulting in unrealistically large headspaces compared with commercial glass packing. Tan and Francis (*920*) also reported on the effect of processing on the carotenoids and color of spinach, as did Kaur and Manjrekar (*497*). A comparison of carotenoids in frozen and canned carrots in a 4-year study by Weckel *et al.* (*1009*) indicates a higher carotenoid concentration in the canned products: 852–1352 versus 534–831 μg/gm dry weight, apparently due to a leaching of the carrot soluble solids into the canning liquid. The canning process caused losses in provitamin A activity of 7–12% because of cis–trans isomerization of both α- and β-carotenes (Table 44). The influences of canning and drying of carrots on carotenoid content were also studied by Lantz (*566*). Observations of color and carotenoid content (Table 45) were included in a study of aseptic canning of strained sweet corn puree sterilized at 285°F for 24 sec versus

TABLE 42

Stereoisomers of β-Carotene in Fresh and Frozen Green Vegetables as Affected by Cooking Time[a,b]

Vegetable sample	Cooking time (min)	Neo-β-carotene E (%)	Neo-β-carotene B (%)	All-*trans*-β-carotene (%)	Neo-β-carotene U (%)	Biological value[c] (%)	Xanthophylls (% of total carotenoids[d])	Total carotene as β-carotene[e] μg/100 gm)	Effective carotene[f] (μg/100 gm)
Broccoli									
Fresh	0	—	5.7	81.0	13.3	89.1	54	1080	960
	15	—	4.7	74.5	20.8	84.9			920
	25	—	5.4	70.4	24.2	82.5			890
Frozen									
Heads	0	—	6.6	83.8	9.6	90.9	68	950	860
	8	—	8.6	78.2	13.2	87.8			830
	15	1.0	7.0	76.4	15.6	86.5			820
Stems	0	—	5.4	81.1	13.5	89.1	70	270	240
	8	1.2	11.2	73.0	14.6	85.1			230
	15	0.7	9.0	73.5	16.8	85.0			230
Brussels sprouts									
Fresh	0	—	6.8	82.5	10.7	90.2	74	480	320
	10	—	6.1	75.3	18.6	85.6			310
	20	—	6.2	70.6	23.2	82.7			300
Frozen	0	—	5.7	82.0	12.3	89.7	71	370	330
	10	—	5.7	74.6	19.7	85.1			310
	20	—	4.4	70.4	25.2	82.3			300
Collards									
Fresh	0	—	6.5	80.9	12.6	89.1	70	5400	4810
	7	—	7.3	77.5	15.2	87.2			4710
	15	—	6.7	77.5	15.8	87.1			4700
Frozen	0	—	7.8	81.0	11.2	89.4	77	5510	4930
	15	—	8.3	78.3	13.4	87.8			4840
	25	—	9.0	76.4	14.6	86.7			4780

Kale									
Fresh	0	—	3.8	85.7	10.5	91.7	65	4720	4330
	20	—	5.4	77.0	17.6	86.6			4090
	30	—	6.6	70.6	22.8	82.8			3910
Frozen	0	—	9.6	80.6	9.8	89.4	73	2840	2540
	20	—	8.7	75.5	15.8	86.1			2450
	30	—	12.7	70.3	17.0	83.5			2380
Spinach									
Fresh	0	—	4.9	82.6	12.5	90.0	59	3300	2970
	5	—	7.8	76.4	15.8	86.5			2860
	10	—	7.7	76.9	15.4	86.9			2860
Fresh[g]	0	—	9.4	80.2	10.4	89.2	69	4840	4320
Fresh[h]	0	—	9.5	79.9	10.6	88.9	65	3770	3350
Frozen	0	—	5.6	84.2	10.2	91.1	48	4940	4510
	5	0.3	8.0	80.5	11.2	89.2			4410
	10	0.4	8.2	78.6	12.8	88.1			4360
Beet greens, fresh	0	—	7.0	80.7	12.3	89.1	67	2560	2240
	10	—	5.8	77.4	16.8	86.9			2190
	20	—	6.8	74.8	18.4	85.4			2150
Endive, fresh	0	—	5.2	80.8	14.0	88.9	65	960	850
	7	—	6.5	76.5	17.0	86.4			830
	15	—	4.8	76.5	18.7	86.1			830

[a] From Sweeney and Marsh (*915*).
[b] Vegetables obtained at retail markets, except as noted. All values are mean of four replications on at least two sample lots.
[c] Percentage of value if all carotene were assumed to be all-*trans*-β-carotene.
[d] Xanthophyll values determined only where indicated.
[e] Calculation based on assumption that carotene is entirely all-*trans*-β-carotene.
[f] Corrected for lower vitamin A values of carotenes other than all-*trans*-β-carotene.
[g] Bounty variety, grown at Agricultural Research Center, Beltsville, Maryland.
[h] 668.9 variety, grown at Agricultural Research Center, Beltsville, Maryland.

TABLE 43

Stereoisomers of α- and β-Carotenes in Fresh and Frozen Red and Yellow Vegetables as Affected by Cooking Time[a,b]

Vegetable	Variety	Cooking time (*min.*)	Neo-α-carotene B (%)	All-trans-α-carotene (%)	Neo-α-carotene U (%)	Neo-β-carotene E (%)	Neo-β-carotene B (%)	All-trans-β-carotene (%)	Neo-β-carotene U (%)	Biological value[c] (%)	Xanthophylls (% of total carotenoids[d]	Total carotene/as β-carotene[e] (μg/100 gm)	Effective carotene[f] (μg/100 gm)
Carrots													
Fresh	—	0	1.0	21.9	—	—	2.9	72.6	1.6	86.5	18	15,180	13,130
		15	1.1	22.4	0.4	—	9.0	65.9	1.2	83.3			12,640
		25	3.9	19.0	0.5	—	10.2	65.4	1.0	82.0			12,450
Fresh	Danvers[g]	0	1.2	38.6	0.7	—	5.3	52.4	1.8	76.7		14,160	10,860
	Chantenay[g]	0	1.8	37.4	0.9	—	3.2	54.6	1.1	76.9		6,270	4,820
Frozen	—	0	0.4	29.0	0.8	—	5.0	63.0	1.8	82.1	7	11,090	9,070
		10	1.8	27.5	0.5	0.9	12.9	54.2	2.2	77.3			8,570
		15	2.2	25.2	0.8	0.8	15.9	51.5	3.6	75.7			8,390
Squash													
Fresh		0	0.6	19.0	0.6	—	4.2	71.6	4.0	85.6	61	5,700	4,880
	Butternut	20	2.9	16.1	1.0	—	14.9	60.5	4.6	79.2			4,510
		40	4.8	14.0	0.9	—	21.2	53.5	5.6	75.1			4,280
Fresh	Butternut[g]	0	1.2	32.5	1.2	—	6.6	53.1	5.4	76.3	66	2,060	1,570
	Hubbard[g]	0	—	—	—	—	8.6	84.2	7.2	91.5	83	820	740

Frozen	—	0	—	4.0	—	—	20.8	71.0	4.2	85.7	85	2,780	2,380
		10	—	3.7	—	—	23.2	68.1	5.0	84.3	85		2,340
Sweet potatoes													
Fresh		0	—	—	—	—	0.1	96.7	3.2	97.9	12	7,620	7,470
		20	—	—	—	—	24.9	69.7	5.4	85.0			6,480
		40	—	—	—	5.2	23.6	66.0	5.2	83.3			6,350
Frozen	—	0	—	—	—	—	13.3	83.1	3.6	91.5	12	6,220	5,690
		20	—	—	—	3.6	16.0	76.2	4.2	88.2			5,460
		40	—	—	—	3.7	18.9	74.2	3.2	87.4			5,430
Pumpkin, fresh		0	—	3.1	—	—	5.6	86.0	5.3	92.6	71	1,100	1,020
		10	—	3.0	—	—	16.2	75.3	5.5	87.6			970
		20	—	3.0	—	—	14.3	75.5	7.2	87.4			970
Red peppers, fresh	—	0	—	2.6	—	—	6.3	83.5	7.6	91.1	85	2,280	2,080
		5	—	2.3	—	—	24.8	64.6	8.3	82.1			1,870
		10	—	3.1	—	—	31.6	58.3	7.0	79.4			1,810

[a] From Sweeney and Marsh (*915*).
[b] All values are mean of four replications on at least two purchase lots.
[c] Percentage of value if all carotenes were assumed to be all-*trans*-β-carotene.
[d] Xanthophyll values determined only where indicated.
[e] Calculation based on assumption that carotene is entirely all-*trans*-β-carotene.
[f] Corrected for lower vitamin A values of carotenes other than all-*trans*-β-carotene.
[g] Vegetable grown at Agricultural Research Center, Beltsville, Maryland. All others obtained at retail markets.

TABLE 44

Average Content of Total Carotenoids and Carotene Components of Canned Carrots[a]

Year	Total carotenoids (μg/gm dry weight)	Percentage of total						
		Neo-α-B	All-*trans*-α	Neo-α-U	Neo-β-B	All-*trans*-β	Neo-β-U	Loss in pro-vitamin A activity due to isomerization
1957	879 ± 243	2.1	30.9	2.7	6.0	42.0	3.8	7.3
1958	927 ± 179	6.4	24.8	1.6	13.1	41.5	4.5	11.7
1959	852 ± 149	5.4	24.2	2.4	8.5	44.9	4.7	9.7
1960	1352 ± 335	2.3	26.6	2.5	10.7	46.4	4.2	8.3

[a] From Weckel *et al.* *(1009)*.

TABLE 45

Effect of Processing on Carotene Retention of Strained Corn[a,b]

Process	Total solids (%)	Absorbance at 438 nm	Carotene retention (%)[c]
Frozen corn	24.35	0.425	100
Strained, before processing	22.10	0.370	95.8
Aseptic	21.90	0.227	59.4
Retort	21.93	0.282	73.6

[a] From Daoud *et al.* (*235*).
[b] Moisture content was considered in calculations.
[c] Percent retention was based on raw material (frozen corn = 100% retention).

240°F for 42 min (*235*). The product heated for a shorter time was judged to be superior.

Carotene has been declared to be relatively stable during the canning of sweet potatoes (*32*). Total carotene content appears to be relatively stable but suffers from isomerization (*571*). Sweet potatoes showed maximal retention of β-carotene during canning when subjected to the still retort process at 116°C or to the agitating heat process at 127° and 132°C. Poorest retention occurred at 121°C in either the still or agitating process. However, neo-β-carotene was found in the highest amounts in sweet potatoes processed at 121°C, suggesting that β-carotene was converted to neo-β-carotene during heat processing. Neo-β-carotene U was not affected by the heat processing. The shelf life of canned, precooked, dehydrated sweet potato flakes packaged in air was extended by the use of BHA and BHT and mixtures with citric acid or sodium acid pyrophosphate (*247*).

Noble (*718*) in 1975 investigated color and carotenoid pigment changes as a result of heat concentration of tomato pulp using juices reconstituted from pastes of 10, 15, and 20% soluble solids. With increasing heat treatment, the color measured by reflectance, using a Hunter Color Difference Meter, decreased in redness or hue, in saturation, and in lightness. Total carotenoid content and isolated lycopene decreased with the heat concentration, resulting in losses of 11, 22, and 57% lycopene relative to the unconcentrated control tomato pulp. In a study by Aczel (*8*) β-carotene and lycopene decreased during the storage of tomato puree for 1 year at normal temperature to about half the initial values (β-carotene, 1522 mg per 100 gm; lycopene, 31–44 mg per 100 gm). Smaller losses were achieved under cooler storage conditions. Vladimirov (*992*) reported the commercial processing of some tomato varieties on which carotenoid observations were made.

Sweeney and Marsh (*915*) observed that canning or pressure cooking tended to increase the number of stereoisomers formed (Table 46) because of the high temperatures involved. Neo-β-carotene B was detected in only a few of the vegetables

TABLE 46

Carotene Stereoisomers in Canned or Pressure-Cooked Foods[a,b]

Vegetable	Treatment	Neo-α-carotene B (%)	All-*trans*-α-carotene (%)	Neo-α-carotene U (%)	Neo-β-carotene (%)	Neo-β-carotene B (%)	All-*trans*-β-carotene (%)	Neo-β-carotene U (%)	Neo-β-carotene V (%)	Biological value[c] (%)	Xanthophylls (% of total carotenoids[d])	Total carotene as β-carotene[e] (μg/100 gm)	Effective carotene[f] (μg/100 gm)
Carrots													
Brand 1	Canned	4.7	29.5	0.3	2.8	4.5	55.9	2.3	—	76.4		8,100	6,180
Brand 2	Canned	1.8	22.8	0.9	1.7	3.6	65.2	4.0	—	82.0	5	7,800	6,400
Brand 3	Canned	8.2	24.2	0.7	4.8	13.9	43.8	4.4	—	69.6		8,370	5,830
Carrots	15 lb, 5 min	0.8	30.0	0.3	2.3	12.4	50.4	3.8	—	75.6		14,460	10,920
Squash	Canned	—	1.8	—	—	15.4	73.2	9.6	—	85.9	75	940	810
Squash	15 lb, 5 min	—	4.0	—	—	23.8	67.0	5.2	—	83.7		2,920	2,450
Spinach													
Brand 1	Canned	—	—	—	—	5.4	72.5	21.6	6.5	86.1	53	5,200	4,360
Brand 2	Canned	—	—	—	1.6	6.9	62.5	29.0	—	78.0	57	4,430	3,460
Spinach	15 lb, 5 min	—	—	—	2.1	9.6	72.0	16.3	—	84.4		4,960	4,180
Pumpkin													
Brand 1	Canned	5.3	35.1	3.3	2.0	6.2	43.2	4.2	0.7	68.9	20	9,880	6,800
Brand 2	Canned	—	3.3	—	—	20.5	62.4	13.8	—	80.2	62	580	460
Sweet potatoes													
Brand 1	Canned	—	—	—	3.3	10.2	79.5	7.0	—	89.4	17	4,640	4,150
Brand 2	Canned	—	—	—	5.7	6.2	82.9	5.2	—	91.2	17	6,600	6,020
Kale													
Brand 1	Canned	—	—	—	—	6.4	69.1	24.5	—	81.8	48	3,810	3,110
Brand 2	Canned	—	—	—	—	5.2	63.9	30.9	—	78.4	54	3,930	3,080
Apricots													
Brand 1	Canned	—	—	—	1.1	8.0	83.0	7.4	0.5	90.8	23	1,320	1,200
Brand 2	Canned	—	—	—	1.8	10.0	74.7	13.5	—	86.1	16	1,710	1,470

[a] From Sweeney and Marsh (*915*).
[b] All values are mean of four replications on at least two purchase lots.
[c] Percentage of values if all carotene were assumed to be all-*trans*-β-carotene.
[d] Xanthophyll values determined only where indicated.
[e] Calculation based on assumption that carotene is entirely all-*trans*-β-carotene.
[f] Corrected for lower vitamin A values of carotenes other than all-*trans*-β-carotene.

cooked without pressure. For the canned or pressure-cooked vegetables, it was detected in all except kale and squash. Also neo-β-carotene V, which was not detected in foods cooked without pressure, was found in canned spinach, some samples of canned pumpkin, and canned apricots. These isomers were not present in amounts large enough to significantly affect biological value percentages. In general, these values were similar to those obtained for vegetables cooked without pressure. Processing of vegetables, by cooking or canning, converts all-*trans*-carotenes to isomers with lower vitamin A activity. This conversion lowers the vitamin A values of green vegetables by 15–20% and those of yellow vegetables by 30–35%.

Aczel (*10*) observed decreases in the carotenoid content of apricots during the canning process and storage. In the course of storage, the total carotenoid content decreased to 85% of its original value, and the β-carotene content was 74% of its original value. Destruction was directly related to pasteurizing temperatures and storage in the processing of apricot compote. Peaches lost 50% of their initial total carotenoids on canning in one study by Mitchell *et al.* (*674*). Since the carotenoid distribution of peaches is similar to that of orange juice and the peach pH is low, one may conjecture that isomerization of 5,6-epoxides is a factor in the large losses of carotenoids in the canning of peaches. The decomposition of carotenoids and β-carotene during the processing and storage of peaches (*25*) as well as apricots was studied by Aczel (*4,7,11*). The rate of decomposition of total carotenoid in both fruits was the same, namely, to about 90% of the fresh fruit level. Losses of β-carotene were 37% for peaches and about 20% for apricots. In an examination of papaya puree and concentrates during processing, Chan *et al.* (*168*) noted that changes in carotenoids were characterized by a hypsochromic shift which intensified during the processing sequence. The shift was attributed to the acid-catalyzed isomerization of 5,6-monoepoxycryptoxanthin to 5,8-monoepoxycryptoxanthin. Concentration did not affect flavor quality and had little or no effect on aroma quality. Biochemical and physiological changes in the heat processing of canned citrus products, which include changes occurring in the carotenoid pigments, were studied by Blundstone *et al.* (*122*).

Three classes of pigments of commercially important fruits, namely, carotenoids, flavonoids, and chlorophylls, were reviewed by Chichester *et al.* (*178*). Individual pigments influenced by processing techniques were considered rather than the various classes of pigments. The content of carotenoids in citrus fruit as studied by DiGiacomo (*257*) was influenced by methods of concentration, pasteurization, and storage in juice production. Fang-Yung and Aliev (*290*) reported some losses of carotene in pulpy fruit juices during the finishing stage in the processing cycle. Pulp-containing juice was prepared by Kozenko and Kleshchunova (*536*) from quinces by crushing the fruit, passing through a sieve, heating to 85°C for 5 min, packing, and pasteurizing. The juice was of satisfactory organoleptic quality and showed good retention of carotene, ascorbic acid, pectin, flavonols, and other substances. DiGiacomo reported on carotenoid variation of citrus fruits as influenced by technological processing (*257*).

D. Beverage Manufacture

Fruit juices and beverages containing carotenoids should be stored so that they are protected from air; gassing, preferably with carbon dioxide, and reduction of headspace volume to a minimum should be observed. Flavonoids, such as hesperidin, and ascorbic acid function as natural protective agents and retard the destruction of carotenoids (*811*). Rother (*817*) studied the influence of light on the stability of soft drinks containing orange juice. Damage is caused by oxidation of coloring and aromatic substances, leading to loss of color and flavor. The degree depends on the amount of oxygen present and the duration of exposure. The addition of liposoluble antioxidants provides no protection, but by adding ascorbic acid the stability of beverages of this type can largely by ensured, and this form of stabilization has been shown to depend on the concentration of ascorbic acid. Schara and Tsoumanis (*840*), in a similar study, came to the same conclusions. Liposoluble antioxidants in general have a favorable effect on carotenoids (perhaps via stabilization of the essential oil). Oxygen is the cause of bleaching in orange soft drinks, and bleaching is associated with a loss of ascorbic acid and the development of a soapy flavor. The size of the headspace and the partial pressure of oxygen are a direct indication of the tendency to bleach. The authors pointed out that these conditions are not restricted to aqueous systems, but apply also to orange oils and dried products.

The use of ascorbic acid in fruit juices and beverages represents one of its main fields of application in the food industry. Ascorbic acid is of particular interest as a stabilizer in beverages containing extremely delicate flavors such as grapefruit and tangerine, but it is also used in large quantities for citrus-type beverages. Ascorbic acid decomposes if oxygen remains in the product and in the headspace, and under these conditions, when the ascorbic acid is exhausted, oxidation is responsible for the usual color and flavor changes. However, oxygen removal and exclusion by the use of sealed containers and a sufficient amount of ascorbic acid enable added ascorbic acid to act as an oxygen acceptor, thus creating an anaerobic environment. It has been reported that under such conditions it is possible to achieve an inhibition of the growth of thermoresistant molds in apple juices by the addition of approximately 500 mg of ascorbic acid per liter (*746*), although ascorbic acid is not a general antimicrobial agent (*268*).

The mechanism of coupled oxidation leads to oxidative breakdown of components that, in the absence of easily oxidizable products, would not react directly with oxygen (*599*). Therefore, an excess of easily oxidizable substances cannot guarantee complete protection of the product. The detrimental effect of traces of heavy metals, particularly copper and iron, and their catalytic effect on the nonenzymatic oxidation in carbonated beverages were studied by Urion *et al.* (*959*). They pointed out the importance of avoiding contamination with traces of these metals. Iron, brass, bronze, and copper equipment should be avoided and replaced by stainless steel, aluminum, or plastic equipment. Furthermore, the processing time necessary for comminuting the fruits should be as short as possible, and the oxygen should be removed and any further exposure to air be avoided as completely as possible. Only when these processing precautions are followed strictly can maximal stability and efficacy of ascorbic acid be achieved (*514*).

E. Dehydration

Dehydration of food products, particularly fruits and vegetables, is widely used for increasing shelf life. Dehydration and increase of surface area (in powdered or lyophilized fruit and vegetable products) generally lead, however, to very poor stability unless the products are protected from air by storing in an inert atmosphere or a vacuum (*574*). The influence of heat treatment on some plant carotenoids, including those of the swede, apricot, peach, and tomato, were reviewed by Hanson (*390*). Concurrent with carotenoid destruction, there is a deterioration of culinary quality.

Popovskii *et al.* (*763*) reported that lyophilization preserved the quality of strawberries, apricots, and peaches, although some vitamin C was lost from strawberries and a small amount of carotenes from apricots. The quality of lyophilized fruitberry puree was excellent, and there were minimal losses of vitamin C and carotenes. The losses of ascorbic acid and β-carotene during freeze-drying and further storage up to 7 months were studied by Fonseca *et al.* (*302*) in two papaya varieties, passion fruit, and strawberries. The losses of ascorbic acid during dehydration varied between 0 and 7.0%, except for strawberries, where they reached values of 41.6 and 36.7%. No losses were observed for carotene, except in yellow pulped papaya, which lost 4.8%. During storage, the mean monthly losses of carotene varied between 1.6 and 3.0%, except in passion fruit, where 75% was lost in the first month. Changes in the vitamin content of freeze-dried fruit and vegetables were also observed by Lempka and Prominski (*574*). Some of the products were heated or blanched before freeze-drying. When necessary they were diced, chopped, or pulped. Heat treatment caused considerable loss of ascorbic acid and riboflavin, ascribed mainly to the duration of the treatment rather than to the blanching solution or temperature. The carotenoid content was not affected. In freeze-dried products stored for different lengths of time without a vacuum, losses of carotenoids were high. The advantages of freeze-drying, compared with other drying processes at normal pressure or under vacuum, were that most products did not change chemically, retained their normal appearance, flavor and aroma, and could be reconstituted easily. During the production and storage of sun-dried apricot juice, a brown color developed, as noted by Salem and Hegazi (*828*). Sodium metabisulfite in different concentrations was used to inhibit the Maillard reaction, which leads to this off-color. The results showed that the ascorbic acid, total carotenoid, and total protein contents of the apricot fruit were greatly decreased by processing. Bolin and Stafford (*123*) also observed that drying methods influenced carotene retention in apricot halves and sheets, with little or no losses occurring during dehydration but up to 30% during sun-drying. Sulfured dried apricot sheets retained 13% more carotene and 300% more vitamin C after 13 weeks of storage at 32°C compared to the unsulfured samples.

Changes in carotenoid pigments were noted by Curl and Bailey (*224*) in the preparation of Florida Valencia orange juice powder (Table 47). A minor part of the carotenoid 5,6-epoxides was isomerized to 5,8-epoxides, a conversion that in-

TABLE 47

Carotenoid Composition of Florida Valencia Orange Juices[a]

Constituent	Approx. % of carotenoid mixture[b]					Ratio to amount found in freshly prepared powder	
	Single strength	Concentrate	Fresh powder	Conditioned powder	Aged powder	Conditioned	Aged
Phytoene	1.4	—[c]	2.1	2.5	3.2	1.02	1.00
Phytofluene	0.6	0.6	0.6	0.8	0.9	1.08	0.98
α-Carotene	1.0	1.4	1.1	1.5	1.7	1.09	1.01
β-Carotene	1.2	1.8	1.6	2.1	2.4	1.11	0.98
ζ-Carotene	1.1	1.5	1.7	2.2	2.4	10.7	0.91
γ-Carotene-like	0	0.2	0.2	0.2	0.5[d]	0.88	—
Hydroxy-α-carotene	2.9	2.8	2.5	2.5	2.8	0.85	0.73
Cryptoxanthin	5.4	5.6	4.4	4.9	6.6	0.93	0.98
Lutein	7.1	7.1	7.2	8.7	9.2	1.03	0.85
Zeaxanthin	9.2	9.0	8.3	9.9	11.3	1.01	0.90
Lutein 5,6-epoxide	0.3	0.3	0.3	0	0.04	0	0.10
Antheraxanthin	11.6	13.4	12.2	5.4	0.7	0.38	0.007
Flavoxanthin	0.14	0.3	0	0.3	0.4	—	—
Mutatoxanthins	7.3	8.0	9.4	16.7	22.9	1.50	1.63
Violaxanthin	14.7	18.8	13.6	1.5	0	0.09	0
Luteoxanthins	12.7	11.5	15.0	10.6	0.7	0.60	0.03
Auroxanthins	3.9	2.8	5.0	14.8	19.7	2.50	2.63
Valenciaxanthin	1.8	1.5	1.5	1.3	0.14	0.73	0.06
Sinensiaxanthin	1.6	1.6	1.5	0.4	0	0.24	0
Trollixanthin-like	10.4	7.6	7.1	6.1	0	0.73	0
Trollein	—[e]	—[e]	—[e]	—[e]	4.4	—	—
Valenciachromes	0.5	0.2	0.9	1.6	3.1	1.46	2.20
Trollichrome-like	3.4	1.7	2.1	4.5	6.4	1.87	1.99
Sinensiachrome	0.8	0.5	0.4	0.8	0	1.72	0

[a] From Curl and Bailey (*224*).

[b] Calculated from sum of absorbances at principal maximum on Cary spectrophotometer curve for each constituent.

[c] Probably present but obscured by impurity.

[d] Spectrophotometer curve indicated that band was quite impure.

[e] Probably present but obscured by trollixanthin-like pigment.

creases during storage. Nonepoxide carotenoids may also undergo trans–cis isomerization. No significant losses in cryptoxanthin or α- and β-carotenes were observed during conditioning or storage. In the dehydration of orange and tangerine flavedo (*136*) during the preparation of citrus pulp feed for cattle, carotenoid pigment losses occur, which are lowered if an antioxidant is added before the drying application.

Schillinger and Zimmermann (*842*) studied the stability of vitamins and β-carotene during storage in dried snap beans, carrots, red cabbage, white cabbage, savoy cabbage, sauerkraut, potatoes, parsley, leeks, and spinach. Parsley, leeks, and spinach were freeze-dried and stored *in vacuo*, and the other vegetables were heat-dried and stored under nitrogen. All were stored for up to 3 years at ambient temperatures of 5°–28°C, with relative humidities of 35–63%. β-Carotene showed large losses of 38–55% in all samples except carrots and parsley. Ascorbic acid showed a similar loss, but in general it was significantly less stable than β-carotene, particularly in the presence of air and higher levels of humidity. The effects of dehydration and refrigeration on the carotene of squash were investigated by Garcia and Geuco (*328*). Nutting *et al.* (*720*) showed that dried samples of curled or plain leaf parsley contained a higher percentage of cis isomers and neo-β-carotene but a lower percentage of all-*trans*-β-carotene. Pordab *et al.* (*764*) observed carotene losses in prepared powdered purees made from cereals mixed with either vegetables or fruit. For maximal carotene content, it was concluded that carrots should be packed only in nitrogen in cans or in laminated bags with low gas and steam permeability because the presence of oxygen results in rapid deterioration of the product during storage. The influence of oxygen content is illustrated by the graphic data (Fig. 27) of Spiess *et al.* (*884*). Packing in cans in air was thought to be suitable for spinach, green pea, and apple powdered puree when a shelf life of 12–18 months was sufficient.

Fresh carrots have between 75 and 114 mg α-carotene per 100 gm dry matter. After blanching they had somewhat more. From 40 to 50% was lost in air-drying

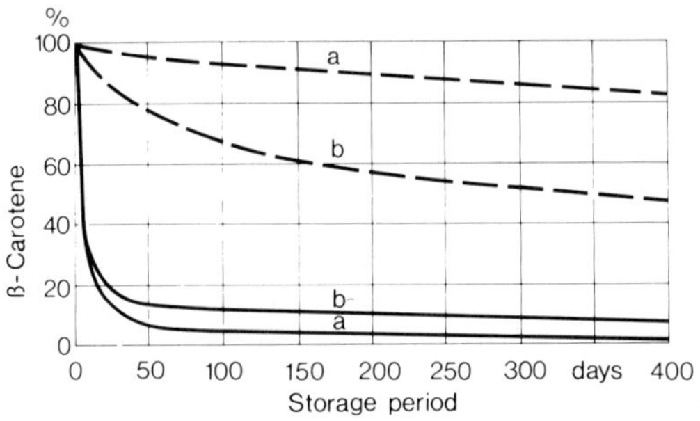

Fig. 27. Freeze-drying of fruits and vegetables. Retention of the β-carotene content during storage. Oxygen concentration in packaging atmosphere—*P*, 1 mg/gm in mg O_2/gm dry substance. Key: —, 0.1 mg/gm; --, 5 mg/gm; a, paprika; b, carrots. [From Spiess *et al.* (*884*).]

and 20% in vacuum-drying, but if the vacuum-drying was done under nitrogen the loss was only about 7%. The loss of carotene depended more on the amount of oxygen present than on the drying temperature in the studies of Pazarincevic-Trajkovic and Baras (*733*). Tomioka *et al.* (*945*) found 10 carotenoids in fresh carrots, and 93% of the total carotenoids consisted of carotenes. The carotenoids were stable when the carrots were freeze-dried to 10% moisture but were unstable at 0% moisture. The carotenoids of microwave-treated, freeze-dried carrots were more stable than those treated with hot water. The stability of α-carotene was similar to that of β-carotene during freeze-drying and storage. Carrots were prepared by Sweeney and Marsh (*915*) for freeze-drying by grating, slicing, or formation of a slurry. Mean values (Table 48) indicated that stereoisomers of carotene are not formed during freeze-drying. However, total carotene values for freeze-dried carrots were somewhat lower than those for fresh carrots.

Carotene oxidation and off-flavor development in dehydrated carrots have been correlated by Falconer *et al.* (*289*). Off-flavor can be avoided by packaging in an inert atmosphere. Dehydrated carrots stored under high vacuum have good carotenoid stability according to trials of Mackinney *et al.* (*614*). If the carrots are blanched or sulfitized before drying, according to Androsova and Skrobanskii (*22*), the loss of carotene is smaller owing to the inactivation of enzymes. With storage of dry carrots in the air, carotene is quickly decomposed. Tajiri *et al.* (*919*) found that blanching eliminated 6% of the lycopene of carrots during 10 days of storage. Added L-ascorbic acid was a good antioxidant. L-Ascorbic acid addition after blanching was very effective; lycopene content decreased only 15–16% during 300 days of storage, and no discoloration appeared. If lycopene decreased to 45–50% of its original level, some discoloration and disagreeable odor were detected. Lantz (*565*) reported excellent stability of total carotene content in carrots stored 17 weeks at 42°F. Cooking under different conditions had little effect on total carotene content. The total β-carotene content of an explosive puffed carrot product compared

TABLE 48

Carotene Stereoisomers in Carrots as Affected by Freeze-Drying[a]

Carotene	Fresh	Freeze-Dried
Neo-α-carotene B (%)	1.8	1.9
All-*trans*-α-carotene (%)	42.9	42.1
Neo-α-carotene U (%)	0.8	0.7
Neo-β-carotene B (%)	5.1	5.0
All-*trans*-β-carotene (%)	48.2	49.2
Neo-β-carotene U (%)	1.2	1.0
Total carotene (μg/100 gm)[b]	9140	8020

[a] From Sweeney and Marsh (*915*).

[b] Calculated on fresh weight basis.

quite favorably with that of the vacuum freeze-dried product, both before and after cooking (*244*). The conventional air-dried product had a lower total β-carotene content. Of the trans isomer available in the blanched fresh carrot, approximately 60% was retained by both the explosive puffed and conventional dried products, whereas about 80% was retained by the freeze-dried products. After proper reconstitution, however, the retention of *trans*-β-carotene was 76% for freeze-dried carrots, 64% for explosive puffed, and 52% for conventional dried carrots.

Blanching has a destructive influence on carotene content in the dehydration (Table 49) of sweet potatoes (*860*). Commercially dehydrated sweet potatoes lost 10% of their total carotene in 18 weeks at 40°–50°F, as noted by Mallette *et al.* (*619*). Hoover (*417,418*) reported that sodium acid pyrophosphate improved natural color retention in precooked frozen sweet potato products and in sweet potato flakes. The destruction of carotene and fatty acids during autoxidation of precooked, dehydrated sweet potato flakes was studied in relation to oxygen uptake by Walter and Purcell (*1002*). Major fatty acids were identified and determined by gas chromatography. Carotene was determined photometrically. Surface carotene was lost about 100-fold more rapidly than bound carotene. In addition, surface fatty acids were not destroyed nearly as rapidly as carotene but much faster than bound fatty acids. An antioxidant added to the flakes (*775*) and storage under nitrogen and at low temperature decreased losses of carotenoids in storage.

Concentration of tomato solids in the manufacture of tomato puree causes lycopene degradation, but this is minimized by deaeration and high-temperature, short-time heat treatment, according to Monselise and Berk (*676*). The degradation of lycopene in centrifuged, washed tomato pulp under different heating conditions was studied by Cole and Kapur (*194*). Heating for 3 hr at 100°C in daylight under a current of oxygen caused the greatest degradation. Lovric *et al.* (*602*), in studying cis–trans isomerization of lycopene and color stability of foam-mat-dried tomato powder during storage up to 1 year, noted that color changes depend on the interplay

TABLE 49

Destructive Effect of Blanching on Carotene Content during Dehydration of Sweet Potatoes[a]

	Moisture (%)		Carotene content (mg/100 gm dry basis)		Destruction (%)	
Sweet potato sample	Unblanched	Blanched	Unblanched	Blanched	Unblanched	Blanched
Fresh	74.20	74.20	14.02	14.02	—	—
Sun-dried	6.75	8.30	10.69	7.68	23.8	45.2
Dry heat (electric oven)	6.05	6.35	10.73	7.15	23.4	48.9
Controlled humidity (fuel oil)	24.90	—	14.02	—	—	—
	4.06	6.47	9.11	5.80	35.0	58.6

[a] From Sherman and Koehn (*860*).

of isomerization and oxidation. Excessive desiccation strongly promoted oxidation, whereas storage in an inert atmosphere improved color retention.

Fresh tea leaves contain carotenoids. Total carotenoids and individual compounds were investigated by Blanc (*113*) by thin-layer and column chromatography, and the main constituents of this fraction were identified in green, withered, fermented, and black tea. Values for β-carotene, lutein, violaxanthin, a fraction resembling auroxanthin, neoxanthin, α-cryptoxanthin, and epoxy-5,6-α-carotene were compiled for the four types of tea. Tirimanna and Wickremasinghe (*942*) were able to identify 9 of the 14 carotenoids found as β-carotene, phytoene, phytofluene, lycopene, γ-carotene, cryptoxanthin, violaxanthin, lutein, and zeaxanthin. The remainder included two red carotenoids, which may have been keto compounds, and two 5,6-epoxides. During wilting, the amounts of carotene and neoxanthin decrease, whereas the amounts of lutein and violaxanthin increase, as cited by Nikolaishvili and Adeishvili (*713*). During fermentation, the carotene content increases slightly, whereas the content of the other pigments decreases. There is thermal degradation of β-carotene in green tea (*499*). Drying is accompanied by a significant decrease in the content of yellow pigments. Yellow pigments are believed to participate actively in the biochemical process occurring in tea leaves during processing. Yamamoto and Muraoka (*1029*) reported on the carotenoids in fresh and fermented tea leaves; likewise, Venkatakrishna *et al.* (*986,987*) observed the effect of processing on carotenoid composition. Black tea aroma and the role of carotenes in black tea formation were studied by Sanderson and Gonzalez (*832*) and Sanderson and Graham (*833*).

F. Milling and Macaroni Processing

Milling and processing of seeds or grains may destroy or remove a considerable portion of the carotenoid content (*116*). Flours, such as wheat flour, are frequently low in carotenoids (*360*). Furthermore, the bleaching stage in the processing of flour for white bread destroys the residual carotenes. Chemical oxidizing and reducing agents in milling and baking and their influence on carotenoid pigments have been reviewed by Alexander (*18*).

When a special yellow color is desired, such as in pasta products like macaroni, special wheats (durum) are grown and carefully processed to produce semolina and farina. The factors affecting oxidative stability of carotenoid pigments of milled durum products (semolina and flours) were investigated by Dahle (*234*). Carotenoid pigments are mainly responsible for the yellow color of pasta products, macaroni, and spaghetti, and since a bright color is a mark of quality for the consumer, the stability of the carotenoids is a matter of concern to the macaroni processor. Processing losses in pasta of 30–60% of the original color have been reported (*841*). The application of a vacuum during dough mixing as avocated by Nazarov and Egorova (*708*) resulted in a reduction in the oxidation of carotenoids and a considerable improvement in the color of the product. The carotenoids in macaroni products undergo oxidative changes during manufacture, and Menger (*657a*) studied the influence of various factors on the destruction of pigments. Monoglycerides increased the losses, whereas sodium chloride, lecithin, and ascorbic acid decreased them. The study showed that free polyunsaturated fatty acids, which increase with the aging of wheat and flour, play a significant role in stability. Tocopherols, if present in sufficient quantity, can improve oxidative pigment stability.

G. Irradiation

The effects of γ radiation on carotenoids (*610*) were studied in a variety of systems, including β-carotene and lycopene in the solid and dissolved states, and carotenoids in tomato purees, prawns, green tomatoes, canned apricot nectar, peach nectar, peach halves, carrots, sweet potatoes, green beans, broccoli, sweet corn, and hard red spring and winter wheat. The intensity of radiation, processing, and storage conditions had a significant influence. By irradiation and storage in an inert atmosphere, retention of carotenoids in the food product is improved. Furthermore, radiation is less likely to cause damage to β-carotene protected in a food environment than to β-carotene in a solvent (*558,827*).

Irradiation of green tomatoes retarded the synthesis of lycopene, and at high levels prevented it. The red color of tomatoes faded at doses of from 5×10^5 to 1×10^6 rad. Irradiation of fully ripened fruit with γ rays and storage at $20° \pm 2°C$ were effective in preserving the fruit for up to 1 week, according to Kim *et al.* (*505*). Doses of up to 3.72×10^6 rad had no apparent effect on the carotenoids of canned apricot nectar, peach nectar, or peach halves, according to Salunkhe *et al.* (*830*). Pablo *et al.* (*728*) irradiated mangoes, which delayed their decay without causing carotenoid destruction. Similar findings were noted by Jiravatana *et al.* (*476*) with papayas and by Umeda (*955*) with Japanese persimmons.

Franceschini *et al.* (*307*) studied the effects of γ irradiation (Table 50) on the carotenoids of carrots, sweet potatoes, green beans, and broccoli. Green bean carotenoids were unstable when irradiated at 1.86 Mrad after freezing but reasonably stable when irradiated at room temperature. The carotenoids of sweet potatoes showed relatively little destruction on irradiation at 1.86 Mrad; however, visual color changes were greater than pigment changes in storage. Carotenoid destruction of broccoli was 25–50% at 1.86 Mrad. Carotenoid destruction of carrots was moderate at 1.86 Mrad except when the carrots were irradiated in an air atmosphere. In trials of Tichenor *et al.* (*938*), packing in nitrogen improved retention of carotenoid pigments of irradiated sweet corn. Retention of β-carotene, cryptoxanthin, and other carotenoids is good at 1.0 Mrad but falls off at 3 and 5 Mrad. Irradiation of deep-frozen spinach with γ rays (50–500 krad) and storage at $-18°C$ for 1 month had little effect on carotenoid content (*45*).

Lai *et al.* (*562*) γ-irradiated both hard red spring wheat and hard red winter wheat. Total carotenoids per 100 gm of flour decreased from 13.7 to 10.5 mg at 1.0×10^6 rad. These values for carotenoids in flour are higher than those reported by

TABLE 50

Carotenoid Content of Four Vegetables Packed Dry in Enameled Cans and Stored at Room Temperature for 11 Months[a,b]

	Irradiated at room temperature			Irradiated frozen		
Vegetable[c]	Vacuum	Nitrogen	Air	Vacuum	Nitrogen	Air
Green beans						
Carotene	52	30	7	39	11	3
Xanthophyll	84	44	9	60	13	7
Broccoli						
Carotene	95	81	78	103	111	92
Xanthophyll	155	157	129	174	213	179
Sweet potatoes						
Total carotenoids	359	393	374	396	417	371
β-Carotene	325	354	339	363	383	338
Xanthophyll	1.72	1.75	1.65	2.04	2.03	1.67
Reflectance	19.7	18.9	18.2	20.4	18.1	17.0
Hue	34.8	34.3	29.9	34.0	33.0	29.6
Chroma	27.2	26.8	24.8	27.0	25.2	23.0
Carrots						
Total carotenoids	574	496	344	585	415	318
β-Carotene	352	326	222	377	259	208
α-Carotene	109	84	58	115	75	53
Xanthophyll	6.3	5.35	6.43	5.38	4.61	3.93
Reflectance	12.0	12.5	12.9	12.9	14.5	15.3
Hue	41.3	40.8	37.6	40.9	42.2	39.5
Chroma	30.4	30.2	28.7	31.4	32.5	30.4

[a] From Franceschini *et al.* (*307*).

[b] Pigment contents in micrograms per gram dry weight. Each datum was obtained from an average of 10 cans, i.e., two at each of five storage periods.

[c] The reflectance, hue, and chroma were represented by *Rd*, $\tan^{-1} a/b$, and $(a^2 + b^2)^{1/2}$, respectively.

other workers. Bancher *et al.* (*44*), observing carotene destruction in γ-irradiated palm oil, proposed that this preservation technique not be applied to palm oil. Irradiation products of β-carotene have been examined (*44*). Fluorescent light irradiation of fluid milk in clear bottles decomposed β-carotene in the studies of Sinha (*870*).

H. Oil Refining

Crude plant or vegetable oil extracts, particularly palm oil, obtain their color from dissolved carotenoid and other plant colors. The significance of these coloring agents in oils and the influence of acids, leaches, enzymes, heat, light, oxygen, hydrogen, and adsorbents, during the refining of the oils have been frequently reviewed. A few references are cited (*349,547,626,953*). For example, Franzke *et al.* (*311,312*) reported that raw rapeseed oils contained 40–70 mg/gm of carotenoids and 3–5 mg/gm of carotenes, whereas corresponding values for refined oils were 0.3–0.7 and 0.4 mg/gm, respectively. Needless to say, crude oils frequently contain significant levels of carotenes as in the case of palm oil, which in areas where it is produced and available, can contribute markedly to the vitamin A status of the consuming public, if used as the principal edible and cooking fat. Highly refined oils contain little or no carotenoids unless they are added back after the refining process.

I. Storage

Although storage of some plant and tree foods after harvesting, a period sometimes referred to as maturation, causes increases in provitamin A activity, such as in most varieties of sweet potatoes, the carotenoids in fresh foods in general tend to deteriorate upon storage. The type of storage conditions is significant in controlling the stability of carotenoids in fresh food finished products.

Carrots can be successfully frozen and canned or dried as a food source of carotene (*244,888,1009*). Carotenogenesis in stored carrots was investigated by Booth (*128*). The studies by Ezell and Wilcox (*287*) included the influence of storage temperature on sweet potatoes. Carotenoid changes were very sensitive to differences in storage temperature after harvesting. At 60°F there was an increase in carotene content of distinct nutritional importance in most varieties. The variety is very important in determining carotene content during storage. A suitable genetic background is necessary for the ability to synthesize carotene when environmental conditions are favorable. Quite late in the storage period, after 5 months, great losses of carotene occur. The postharvest increase in carotene in sweet potatoes appears to be analogous to the postharvest increase in lycopene that occurs in tomato fruits harvested before fully ripe. Veloso de Almeida *et al.* (*985*), in studying the β-carotene and lycopene content and fluctuations in stored tomatoes, reported pronounced heterogeneity among cultivars in content and storage performance. Genetic lines for good storage and processing performance were also reported by Mohr (*675*). Segal (*854*) reviewed the effects of temperature and humidity on fruit and vegetables during storage and changes in carotenoids, carbohydrates, pectins, organic acids, tannins, ascorbic acid, and volatile compounds in stored products.

The storage of leafy green vegetables at 0°–20°C causes carotenoid losses (*287*). Wilting cannot be used as a measure of the rate of carotenoid destruction; 4 days at 0°C caused 0–13% losses in carotene in turnip, kale, collards, and rape; at 10°C, losses were 17–41%.

With processed foods, pretreatment methods before packing as well as storage conditions influence carotenoid content. Korobkina (*534*) discussed the ascorbic acid and carotene content of fresh and processed fruits during storage. Aczel recorded changes in carotenoid composition of apricots and peaches (5–7,9,11), tomato juice, and puree (*8,12*) during storage. Fang-Yung and Aliev (*290*) observed carotene losses in heat-processed juices of 17% when stored at 20°C and 22% at 36°C toward the end of a year. Canned tomato juice retained over 95% of the original total carotene during 2 years at 10°–27°C in studies by Sheft *et al.* (*859a*). Miki and Akatsu (*669*) reported no difference in lycopene content in tomato juice under the following conditions of storage: at room temperature (20°–37°C) for 6 months; heating at 95°, 110°, 120°, or 130°C for 10 min; exposure to sunlight for 5 months inside of a window. The lycopene content decreased by 2% in sunlight and by 5–15% after heating. Lycopene was slightly more stable on heating in the presence of Fe^{3+}, Cu^{2+}, or Sn^{2+} than in the absence of these elements.

XII. NUTRITIONAL CONTRIBUTION

Because of the interrelationship of the carotenes and vitamin A as a nutrient source for man, considerable research has been conducted on the metabolism and nutrition of the carotenes and other carotenoid vitamin A precursors. For general vitamin A activity in mammals, the structure of a carotenoid provitamin A compound must include at least one unsubstituted β-ionone ring and a polyene side chain. The other end of the molecule may have a cyclic or an acyclic structure and be lengthened but not shortened to less than an 11-carbon polyene chain. Chain lengthening decreases activity. β-Carotene possesses two β-ionone rings, one at either end of a long polyene chain, and is a provitamin A with high activity. α-Carotene and γ-carotene are also biologically active at approximately half the β-carotene value.

Oxygen may be introduced into the molecule in certain locations and at certain oxidation states. The β-apocarotenals were postulated by Glover (*346*) and Redfearn (*347*) to be degradation products if terminal oxidation of the β-carotene molecule is one pathway for the conversion to vitamin A; they are biologically active when given orally or by parenteral injection. More currently, Singh and Cama (*868*) suggested that they are directly attacked enzymatically at the 15,15′ bond. The associated carotenals, carotenoic acids, and some acid esters are also active (*869*). The 5,6-monoepoxide and the 5,8-monofuranoxide are vitamin A precursors, but less so the diepoxides (*490,868,910*). It is assumed that *in vivo* these compounds are converted back to β-carotene. Karrer and Jucker (*490*) were early investigators of the distribution of the carotenoid epoxides. Other oxygen-containing carotenoids are biologically active, such as cryptoxanthin, isocryptoxanthin, 5,6-dihydroxy-β-carotene, citranaxanthin and torularhodin. Oxidation that opens both rings of β-carotene destroys activity, but when only one ring is attacked, as in the case of β-semicarotenone, for example, activity remains. Dehydrogenation of the β-ionone ring, as in the case of 3,4-dehydro-β-carotene, still allows the compounds to be active. Anhydrolutein (*152,153*) has shown activity for the chick, mouse, rat, and fish, and astaxanthin is active for fish (*365,378*) but only to a small extent in the rat (*164*). Activity is lost when hydroxyls and ketones are attached to both rings (positions 3 and 3′, 4 and 4′) or when both β-ionone rings are removed and the molecule become acyclic. Hydrogenation of both rings of β-carotene completely destroys its activity.

A. Vitamin A Precursors

A list of carotenoids (*60,867*) that are currently regarded as having vitamin A activity is as follows: β-carotene, α-carotene, 3,4-dehydro-β-carotene, 3,4,3′,4′-bisdehydro-β-carotene, 2,2′-dimethyl-β-carotene, 1,1′-bisdemethyl-1,1′-bisethyl-β-carotene, γ-carotene, 7′,8′-dihydro-γ-carotene (β-zeacarotene), β-carotene 5′,6′-monoepoxide, α-carotene 5,6-monoepoxide, β-carotene 5′,8′-monofuranoxide (mutatochrome, citroxanthin, flavacin), 4-keto-β-carotene (4-oxo-β-carotene, echinenone, aphanin, myxoxanthin), 3-keto-β-carotene (3-oxo-β-carotene), 3,4-dehydro-4′-keto-β-carotene, β-semicarotenone, 3,4-diketo-β-carotene (euglenanone), 3-hydroxy-β-carotene (cryptoxanthin), 3-hydroxy-β-carotene 5′,6′-monoepoxide (cryptoxanthin monoepoxide), 4-hydroxy-β-carotene (isocryptoxanthin), isocryptoxanthin methyl ether, 3,4-dehydro-3′-hydroxy-β-carotene (anhydrolutein, deoxylutein), hydroxydihydro-γ-carotene, 5,6-dihydroxy-β-carotene, 3-hydroxy-4-keto-β-carotene (hydroxyechinenone), β-apo-2′-carotenal, β-apo-4′-carotenal, β-apo-8′-carotenal, β-apo-10′-carotenal, β-apo-12′-carotenal, β-apo-14′-carotenal, β-apo-8-carotenol, β-apo-4′-carotenoic acid (neurosporaxanthin), β-apo-8′-carotenoic acid, β-apo-12′-carotenoic acid, 3′,4′-dehydro-17′-oxo-γ-carotene (oxotorulene), β-apo-8′-carotenoic acid ethyl ester, citranaxanthin, sintaxanthin, torularhodin, and cryptocapsin.

B. Metabolism

The conversion of carotenoid precursor to vitamin A can be accomplished either by (a) symmetric or asymmetric fission or (b) terminal oxidation (Fig. 28). The biosynthesis of vitamin A from β-carotene and probably other carotenoids takes place mainly during absorption in the intestinal mucosa. An *in vivo* reaction sequence occurs: (a) the cleavage of β-carotene into two molecules of retinal and (b) the reduction of retinal to retinol. A carotenoid-cleavage enzyme (15,15′-dioxygenase) has been purified (*563*) and demonstrated by Olson and co-workers in a number of species of animals and is believed to exist in man (*724,725*). Studies by Goodman *et al.* (*351*) with doubly labeled β-carotene on two rats with mucosa *in vitro* showed that β-carotene was converted to vitamin A without removal of the hydrogen atoms attached to the central carbon atoms.

Goodman *et al.* (*351*) gave β-[15,15′$-^3$H]carotene by mouth in oil emulsified into skim milk to two human subjects, and lymph was collected and analyzed.

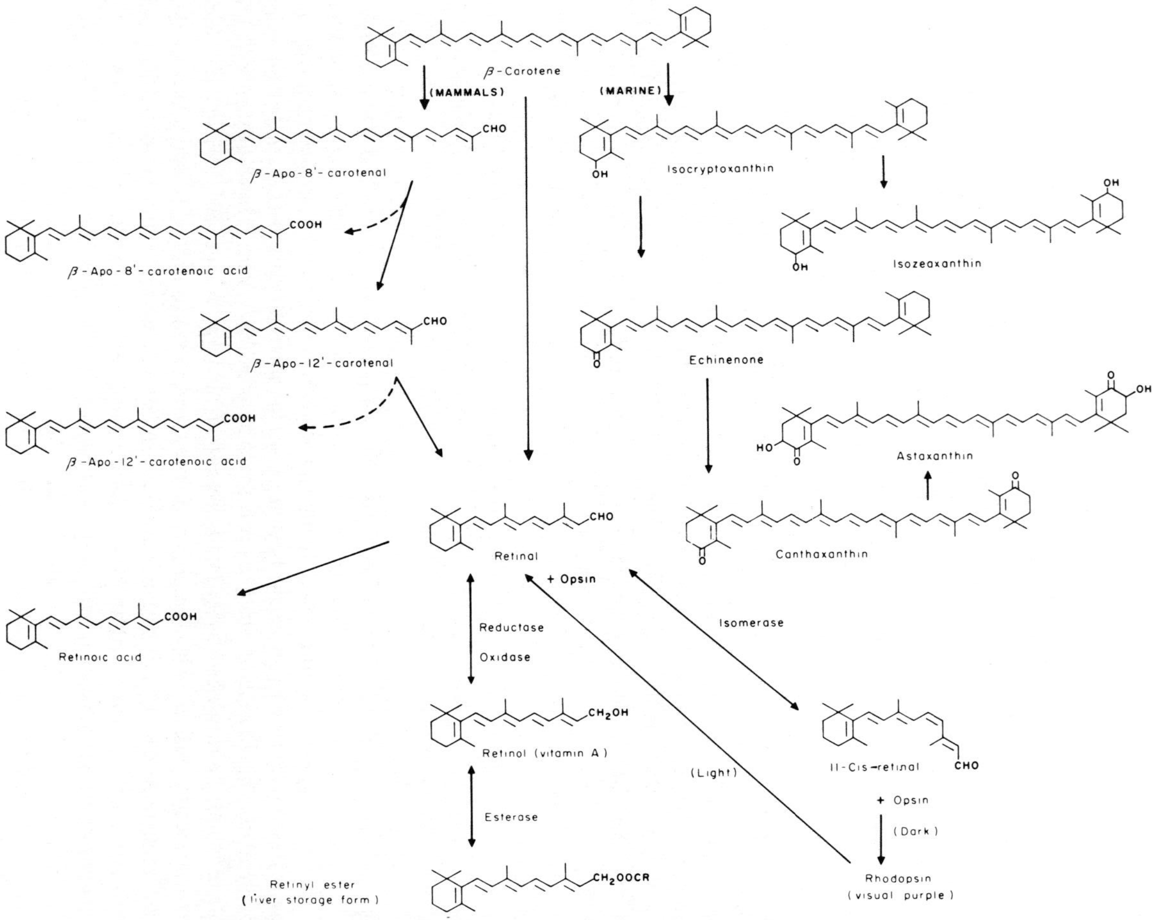

Fig. 28. Metabolism schemes for β-carotene. [From Bauernfeind (*60*).]

β-[15,15'-^{3}H]carotene was absorbed at the same rate as the other lipids in the diet. About 60–70% of the radioactivity was absorbed after between 3 and 10 hr, and absorption was complete by 12 hr but only 9–17% of the radioactivity was recovered in the lymph. β-Carotene (Table 51) was present in lymph mainly as retinyl ester (*351*), but some 20–30% was unchanged. The retinyl esters were palmitic, stearic, oleic, and linoleic. It was concluded that the human intestine possesses only a limited ability to absorb unchanged dietary β-carotene into the lymph. Similar studies were run by Blomstrand and Werner (*121*). β-[15,15'-^{14}C]Carotene and β-[15,15'–^{3}H]carotene were given to four human patients in whom polyethylene cannulas had been inserted in the thoracic duct in the neck. The lymph lipids were extracted and chromatographed on columns and on thin-layer plates of alumina. Only 9–17% of the labeled β-carotene was normally absorbed by way of the lymphatic pathway and recovered mainly as retinyl esters. Unchanged, labeled β-carotene comprised 2–28% of the radioactivity after ingestion of the labeled β-carotene. Absorption of radioactivity into the lymph followed the appearance of chylomicron triglycerides. Labeled retinyl esters, mainly retinyl palmitate, represented 60–75% of the absorbed radioactivity after the administration of labeled β-carotene. In one patient given β-[15,15'-^{14}C]carotene, 47% of the radioactivity was recovered as β-carotene, suggesting the presence of an enzymatic block in the conversion mechanism. Cornwell *et al.* (*198*) suggested portal absorption as part of the transport pattern. McLaren (*654*) reported in 1971 a young Lebanese-Arab girl possessing vitamin A deficiency symptoms but high concentrations of total carotenoids and β-carotene in the plasma. Repeated blood sampling after large oral dose of β-carotene failed to produce any retinyl ester in the plasma, but a positive response resulted from preformed vitamin A administration, thus suggesting failure in the β-carotene enzymatic conversion mechanism.

TABLE 51

Distribution of Radioactivity in Human Lymph after β-[15,15'-^{3}H] Carotene Oral Administration[a]

	Percent distribution of absorbed ^{3}H			
	Study I		Study II	
Alumina column fraction	Chylomicrons	Bottom fraction[b]	Chylomicrons	Bottom fraction[b]
1, β-carotene	23	22	30	27
2, retinyl esters	67	71	61	65
3, retinal	3	2	3	2
4, retinol	3	3	3	3
5 + 6, polar; acids	4	2	3	3

[a] From Goodman *et al.* (*351*).

[b] Lymph from which chylomicrons had been removed by centrifugation.

Goodman (*350*) concluded that biosynthesis of vitamin A from β-carotene takes place mainly in the intestinal mucosa during the absorption of dietary β-carotene. The normal sequence involves (a) the cleavage of β-carotene and (b) the reduction of retinal to retinol. The newly formed retinol is then mainly esterified with long-chain fatty acids, and the retinyl esters are incorporated into lymph chylomicrons and transported from the intestines via the intestinal lymphatics. The distribution of vitamin A structures (Table 52) in human lymph after ingestion of β-carotene was demonstrated by Goodman *et al.* (*351*).

Man differs from the rat in that the human intestine is able to absorb a small amount of unchanged dietary β-carotene into the lymph. In man the composition of the lymph retinyl esters is remarkably constant regardless of fatty acid composition of the diet. Retinyl palmitate predominates. After entering the vascular compartment, the newly formed retinyl esters are mainly taken up by the liver, where they undergo turnover and where vitamin A is stored mainly in ester form. The transport of β-carotene in human blood was examined by Bjornson *et al.* (*111*), and their results confirmed earlier studies of Krinsky *et al.* (*542*). The major portion of β-carotene in most subjects was present in the low-density lipoproteins, irrespective of the lipid distribution in the lipoproteins. Plasma and red blood cell (RBC) carotene concentrations showed a much slower and nonparallel increase in response to carotene administration, and, when administration was stopped, elevated plasma levels persisted longer than elevated RBC levels. Conversion of β-carotene to vitamin A in man is variable (*189,864*) and uncertain because of various factors, particularly in acute and chronic disease (*552*). The enzymatic breakdown of a number of carotenoids that have been examined by *in vitro* studies would lead one to believe that they might perform similarly in man.

TABLE 52

Labeled Retinyl Ester Compositions in Human Lymph[a]

Study	Labeled compound fed	Oil in test meal	Sample analyzed	Percent distribution of labeled retinyl esters			
				Palmitate	Stearate	Oleate	Linoleate
I	β-[^{3}H] Carotene	Olive oil	Chylomicrons	57	25	12	6
			Bottom	54	25	14	7
II	β-[^{3}H] Carotene	Kronolja	Chylomicrons	53	27	13	7
			Bottom	57	26	12	5
III	[^{14}C] Retinol	Kronolja	Chylomicrons	60	22	10	8
			Bottom	60	24	10	6
IV	[^{3}H] Retinyl acetate	Olive oil	Whole lymph	56	19	16	9

[a] From Goodman *et al.* (*351*).

C. Factors Influencing Metabolism

Some factors that affect the biological activity of the carotenoid vitamin A precursors, as determined by animal studies, are as follows: the physical form of the carotenoid; its oil-phase solubility; the state of isomerization; the dietary level of the carotenoid administered; the dietary fat level, its degree of unsaturation, and its state of oxidation; the presence of dietary antioxidant (tocopherol, ascorbic acid, and others) or prooxidants (nitrites); the presence of dietary absorption inhibitors (for example, mineral oil); the adequacy of bile production in the tract; added dietary emulsifiers; the type and adequacy of dietary protein; the role of thyroid-active compounds; the surface area of the digestive tract; the enzymatic digestion efficiency; and the presence of disease and/or internal parasites. Environmental temperature may be a factor. Complete insolubility in aqueous media, low solubility in oils, and a very slow rate of dissolution generally are characteristic properties of all the carotenoids, and the purer the compounds, the more limited is their solubility. Solubility and other physicochemical data for some carotenoids, namely, β-carotene, β-apo-8′-carotenal, β-apo-8′-carotenoic acid ethyl ester, and canthaxanthin are presented in Table 8.

β-Carotene, β-apo-8′-carotenal, and β-apo-8′-carotenoic acid ester are poorly absorbed from the gastrointestinal tract (*150,552*). A factor overlooked by most nutritionists estimating the vitamin A activity of carotenoid vitamin A precursors is the fact that vitamin A value is highly influenced by the amount consumed. The efficacy of conversion to vitamin A decreases as the intake increases. This is easily illustrated by β-carotene feeding trials (Table 53) with animals such as the rat (*1011*).

TABLE 53

Vitamin A Activity of β-Carotene as Influenced by Level of Feeding[a]

Dosage of β-carotene	Relative vitamin A activity (%)
Curative growth assay	
1–3 IU; 0.6–1.8 μg	100
Vaginal smear assay	
125 IU; 75 μg	>95
250 IU; 150 μg	95
500 IU; 300 μg	68
1000 IU; 600 μg	50
2000 IU; 1200 μg	30
Liver storage assay	
2000 IU; 1200 μg	27
4000 IU; 2400 μg	15
8000 IU; 4800 μg	10
16,000 IU; 9600 μg	5
32,000 IU; 19,200 μg	4

[a] From Weiser and Studer (*1011*).

A little more than a decade ago, Canadian workers researched the literature and issued the document entitled "Biological Values of Provitamins A for Man, Domestic Mammals and Birds," a report of the National Committee on Animal Nutrition of the National Coordinating Committee on Agricultural Services. The authors (*140*) of the general conclusions and recommendations, Branion and Eslie, declared that carotene from a variety of sources in typical animal rations should have a biological effect similar to that of the following fractions of its mass as preformed vitamin A: for cattle, sheep, horses, and pigs one-seventh, and for chickens and turkeys one-third. Furthermore, they declared that the appropriate relationship in human dietetics appears to be similar to that for domestic animals, i.e., one-seventh and not the one-half observed for the rat under very restrictive conditions. In an FAO/WHO report (*292*), the absorption of carotene was discussed. The FAO/WHO group recommended that, in the absence of more specific data for foods, the availability of β-carotene be taken as one-third and that the efficiency of conversion in the body be accepted as one-half of the available β-carotene; hence, the utilization efficiency in human beings is taken as one-sixth. Thus, in human beings, 1 mg β-carotene in the diet is taken to have the same biological activity as 0.167 mg retinol. Other reviews show a similar conclusion. Furthermore, there is no constancy in the conversion ratio of β-carotene to vitamin A. The vitamin A activity of β-carotene decreases with increasing daily dose and increasing excess above the average requirement. Conversion ratios vary with (a) the criteria used for evaluation (plasma level, liver storage, growth, etc.) and (b) the level of carotene consumed. In man, the absorption of carotene varies widely, depending on the source. Pure crystalline β-carotene administered orally at low levels in an oil solution is almost completely absorbed, whereas carotene from natural food sources is usually absorbed less readily. With higher intakes of carotenoids, a higher percentage of that ingested is excreted in the feces.

Again by studying the vitamin A content of tissues in animals fed simultaneously β-carotene and various levels of dietary protein, a number of workers (*92,456, 459,630,896*) have concluded that either the absorption of β-carotene or its conversion to vitamin A is directly dependent on a normal protein intake. About a decade ago, Deshmukh and Ganguly (*251*) indicated that in rats fed β-carotene the conversion process is affected by dietary protein level. The 1970 data of Gronowska-Senger and Wolf (*376*) on the rat support this concept and demonstrate that the activity of the enzyme carotene dioxygenase, which converts β-carotene to retinal, is dependent on dietary protein intake. It is well appreciated that children consuming a poor diet or suffering from protein–calorie deficiency do not adequately convert carotenoid vitamin A precursors to preformed vitamin A. Other studies support the role of protein in carotene absorption (*411,482,712*).

The level of dietary fat in some situations seems to be important in the absorption of carotenoids (*679*). Roels *et al.* (*814*) traced the fairly widespread vitamin A deficiency in Ruanda Urundi in central Africa to the very low fat content of the diet of the population. There fat represents less than 7% of the total caloric intake. When supplements of fat were given to vitamin A-deficient boys in this region, the absorption of dietary carotenoids increased quite remarkably from less than 5% to about 50%, with resulting increases in serum levels of vitamin A. Similarly, in vitamin A-deficient Indonesian boys, Roels *et al.* (*813*) demonstrated significantly increased absorption of dietary carotenoids when oil supplement was given. Some authors have examined the influence of the quality of fat on the absorption of dietary provitamin A from the intestinal tract. Dietary antioxidants, in particular vitamin E (α-tocopherol), apparently contribute in some manner either to the efficient absorption of carotene or to the protection of carotene or vitamin A from oxidative breakdown during storage in the body (*375*). Differences in absorption may be the consequence of different tocopherol contents of oils, and some of the contradictory results published in the literature may be due to the fact that the effect of the tocopherol content of oils on the absorption rate was neglected. The influence of triglycerides containing various amounts of linoleic acid on the absorption of β-carotene was examined by Kasper and Ernst (*493*) in rats and man. Corn (maize) oil was more effective in promoting liver storage of vitamin A in rats than was olive oil (lower in tocopherol), but was improved on tocopherol addition. Corn oil was superior to olive oil in raising serum carotene levels in man and was less influenced by tocopherol addition than in the rat trials. Corn oil, in addition to having a higher linoleic acid content, also has a lower viscosity than olive oil, which may have contributed to its better solvent characteristics, and this may have favored its dispersibility in aqueous media. The improvement of the absorption of β-carotene by using emulsifiers such as phosphatides was reported by Kasper and Krennrich (*495*) and Wagner (*995*). The latter author found that fat alone did not improve the absorption but that the addition of deoxycholic acid significantly increased the β-carotene blood levels. The highest values were obtained after administration of an aqueous dispersion of carotene. The utilization of β-carotene by man has been a topic of various discussions (*208,285,537,726,983*). The efficiency of carotene absorption has been used as a screening test for the detection of steatorrhea in man (*182*).

D. Tissue Levels

Carotenes and other carotenoids are usually present in human blood and body fat. In man, excessive consumption of carotene causes the deposition of carotene in the skin, creating a yellow coloration known as hypercarotenemia. Carotenoids have been reported to appear in human blood and organs in approximately the same ratio in which they occur in the diet. The level of carotenoids in blood plasma of man and the type and quantity stored in depot fat reflect primarily human eating habits. The carotene and carotenoid blood levels of healthy subjects have been examined by Kübler (*551*). The carotenoid content of human plasma is higher in summer, coinciding with a greater abundance of fresh vegetables and fruits and thus serving as an indicator of their dietary intake. In an individual on a diet free of carotenoids, the blood levels become low in or free of carotenoids in weeks or months. The limited capacity of the mucosa of the small intestine to absorb carotene and carotenoids has long been known, since the work on animals by Tiews (*939*) and on human beings by Fujita and Morimoto (*320*) and Kübler (*552,553*). Ross and Parker (*816*) reported that the normal range for serum carotene in 227 apparently healthy individuals on a normal mixed diet was 70–300 μg per 100 ml, with a mean of 153 and a standard deviation of ±57. No statistical variation was found according to sex, age, or season. Serum carotene values reported by a number of investigators are shown in Table 54, as compiled by Ross and Parker (*816*). Serum values are regularly reported in the literature (*1,14,50,142,255,295,322,314,492,494,629,642,730,745*).

The only mammal that absorbs carotenoids in a relatively unselective fashion is man, according to Goodwin (*360*). The presence of both carotenes and many oxycarotenoids in human blood plasma and milk was first demonstrated by Palmer and Eckles (*731*) in 1914, although the presence of lipochromes was first noted in 1869 by Thudichum (*937*). As reviewed by Goodwin the carotenoid concentration

TABLE 54

Serum Carotene Values in Humans[a,b]

Range	Mean	Standard deviation	Comment	Investigator
100–348	199	—	25 males	Murill *et al.*
136–420	227	—	20 females	
80–370	210	± 31	70 subjects	Harris *et al.*
49–202	120	± 34	40 females 15–22 years of age	Haworth, Moschette, and Tucker
50–300	166	± 59	30 males	Kimble
90–340	187	± 77	34 females	
117–241	158	—	11 subjects, summer	Yudkin
50–118	85	—	12 subjects, winter	
70–282	123	± 47	110 subjects	Wenger, Kirsner, and Palmer
18–278	138	± 62	71 healthy persons, chiefly college students on ordinary diet	Caveness, Satterfield, and Dann
34–422	131	—	197 subjects 5 months to 68 years	Yarbrough and Dann
200–320	—	—	155 middle-aged and old individuals; 47 younger subjects	Kirk and Chieffi
110–370	207	—	20 European males	Highman
113–340	244	—	20 European males	
100–310	162	—	11 Indian males	
30–260	131	—	8 Bantu males	

[a] From Ross and Parker (*816*).
[b] Expresses as micrograms per 100 ml of serum.

decreases rapidly as the colostrum (*483*) changes into mature milk. Milks higher in fat have a higher carotenoid content, and the concentration is subject to seasonal variations according to the ingestion of green and yellow plant foods. Based on literature reports, human tissue other than fat, blood, and milk which have carotenoids are the adrenals, seminal vesicles, placenta, heart, liver, pancreas, nerves, bone marrow, corpora lutea, and retina. Carotenoids accumulate in the fatty depots throughout life, and this accounts for the fact that adult fat is yellow and infant fat is almost white. The corpora lutea and human milk (*331,483*) contain carotenes in higher proportion. Life cycle studies have been made of the plasma carotene and/or carotenoid content of tissues in health and in a variety of infectious diseases and altered metabolic situations in man. As reviewed by Goodwin, lowered carotenoid plasma values may occur in cases of scarlet fever, celiac disease, acute stage of jaundice, glomerular nephritis, hyperthermia, pneumonia, etc. Increased levels (or carotenemia) may be associated with hyperlipemia, myxoedema, etc., and in instances of very large daily consumption of foods containing high levels of carotenoids. The carotenoid content of serum of healthy Danish (*366*) subjects has been reported, with men averaging 60 μg per 100 ml (range 21–1231) and women 72 μg per 100 ml (range 30–143). The carotenoid content (*231*) of 17 Ghanains (5 months to 75 years of age) was as follows: body fat, mean 2.8 mg per 100 gm (range 0–9.8); liver, mean 5.9 mg per 100 gm (range 0.4–15). Wright and Milner (*1027*) observed a lowering of serum levels after the age of 40.

Carotenemia due to excessive consumption of high-carotene foods is a yellowish coloration of skin caused by the gradual deposition of excess pigment (*3,19, 35,193,232,324,369,399,407,413,680*) in the epithelial tissue. It disappears with time when the ingestion of excess carotene is discontinued. Lycopenemia is a similar situation caused by excessive consumption of lycopene-rich foods (*798*). Carotenemia is not considered harmful and does not produce hypervitaminosis A, as shown by the high consumption of volunteers ingesting carotene for prolonged periods (*40,232,413*). Serum vitamin A values do not increase greatly. In human beings (*719*) subject to photosensitization in erythropoietic photoporphyria, oral β-carotene given daily (60–180 mg/day) appears to prevent photosensitivity. Mathews-Roth *et al.* (*633*) reported the liver carotene values of two patients consuming large amounts of carotene (A, a 4-year-old consuming 30 mg/day β-carotene, and B, a 22-year-old consuming 180 mg/day for about 1 year), finding that abnormal amounts of β-carotene and vitamin A do not accumulate in the liver as a result of the ingestion of pure β-carotene in substantial amounts (Table 55).

Administration of 0.1% β-carotene by weight of diet to four generations of rats for periods including the entire life span of some groups produced no adverse manifestations (*410*), supporting earlier data (*1037*). Safe oral levels in dogs exceed 8000 mg/kg body weight. Doses of 100 mg β-carotene per kilogram body weight 5 days per week for 13 weeks evidenced no influence on the growth of dogs or any adverse reactions (*410*). Greenberg *et al.* (*368*) administered 60 mg β-carotene per day to 15 human subjects for 3 months. Serum β-carotene levels were increased

TABLE 55

Carotene and Vitamin A in the Liver[a]

Investigator/source	Carotene (mg/kg wet wt) Mean	Carotene (mg/kg wet wt) Range	Vitamin A (kIU/kg wet wt) Mean	Vitamin A (kIU/kg wet wt) Range
Ralli *et al.*				
Diabetic patients	26.30	1.20–96.80	364.56	19.05–613.80
Nondiabetic patients	10.08	1.50–37.50	305.90	88.00–1126.40
Dzialoszynski and Tomaszewski	5.40	0–14.00	504.00	0–2500.00
Jensen and With	20.42	1.12–42.59	201.42	20.16–483.20
Ralli *et al.*	14.40	3.30–39.00	1210.00	34.00–6390.00
Blankenhorn	7.64	2.31–27.30		
Hoppner *et al.*				
Males	8.40	0–35.50	341.28	0–2222.35
Females	11.20	0–43.40	321.22	0–3852.32
Matthews-Roth *et al.*				
Patient A	21.88		57.57	
Patient B	11.92		96.29	
	3.06		22.10	

[a] From Matthews-Roth *et al.* *(633)*.

from 128 to 308 μg per 100 ml, but there were no clinical signs of vitamin A toxicity. A 2-year feeding study with rats, plus a second 2-year study with the F_1 generation, showed no adverse effect on feeding β-apo-8′-carotenal at 0.1% except that changes in the epithelium of the seminiferous tubules were slightly more frequent ($P = 0.17$) in the treated animals. This was apparently due to the fact that high-level feeding of β-apo-8′-carotenal depresses vitamin E tissue stores (*149*), which was confirmed in subsequent carotenoid feeding trials (*133*). Bagdon *et al.* (*39*) fed dogs a daily dose of 1000 mg β-apo-8′-carotenal for 14 weeks with no adverse manifestations. Canthaxanthin in rats was studied in a three-generation test similar to that of β-apo-8′-carotenal described above. No adverse indications were noted. Four grams per day for 15 weeks had no adverse effects on dogs.

Since the rate of absorption of carotene (percent absorbed) by mammals decreases as ingested intakes increase, the fecal excretion of carotene by man increases accordingly. Depending on the source and amount of carotene consumed, fecal excretion of intact carotene by man has been reported to range from 2 to 99% of the ingested quantity (*292,794*). Carotene is not usually found in urine, and no carotenoids have been detected in human sweat (*361*). Seifter and Rettura (*857*) reported that supplemental β-carotene in the diet (90 mg/kg) had a theraupeutic action on two tumor systems in mice and suggested its further study in tumor inhibition.

XIII. HUMAN CONSUMPTION PRACTICES

In societies that consume both plant and animal foods, the vitamin A needs of the population are met by the combined consumption of carotenoid vitamin A precursors and true vitamin A, assuming that a selection of plant and animal foods has been ingested. Murray and Campbell (*699*) declared that provitamin A is a major source of vitamin A activity, but the value of dietary intake of such carotenoid vitamin A precursors is governed by (a) the nature and amount of biologically active carotenoids, (b) the trans–cis isomer content of the carotenoids, (c) their stability in the gastrointestinal tract, and (d) their digestibility. They noted that, in practice, biologically active carotenoids can be separated and determined chemically, and correction is usually made for incomplete absorption, but the trans–cis isomer content and instability in the gastrointestinal tract are not usually taken into account. Different food sources of carotene vary so greatly in digestibility that no single factor can be applied to individual foods. When evaluating mixed diets, it is desirable to have a digestibility factor, and so selected factors have been used by different investigators or organizations. The FAO/WHO (*292*) recommendation of 1967 is that one-sixth of dietary carotenes is metabolically available as vitamin A. Hence, if one were to arrive at total international units of vitamin A activity in a diet containing both retinol and retinyl esters given in international units and total carotenes given in micrograms, one would use the following formula:

$$\text{IU of preformed vitamin A} + \left(\frac{\mu\text{g carotene}}{6} \times 1.67 \right) = \text{total IU}$$

In the 1974 NAS–NRC Recommended Dietary Allowances (*796*) the proposal is included that in addition to the international unit activity expression, vitamin A also be given in terms of retinol equivalents defined as follows:

$$\begin{aligned} 1 \text{ retinol equivalent} &= 1\ \mu\text{g retinol} \\ &= 6\ \mu\text{g } \beta\text{-carotene} \\ &= 12\ \mu\text{g other carotenoid vitamin A precursors} \\ &= 3.33 \text{ IU vitamin A activity from retinol} \\ &= 10 \text{ IU vitamin A activity from } \beta\text{-carotene} \\ &= 20 \text{ IU vitamin A activity from other carotenoid vitamin A precursors} \end{aligned}$$

Hence, to convert international units of vitamin A activity from retinol plus micrograms of β-carotene plus micrograms of other carotenoid vitamin A precursors to retinol equivalents, the following equation would be used:

$$\frac{\text{IU preformed vitamin A}}{3.33} + \frac{\mu\text{g } \beta\text{-carotene}}{6} + \frac{\mu\text{g other carotenoid vitamin A precursors}}{12} = \text{total retinol equivalents}$$

In the 1974 NAS–NRC Recommended Dietary Allowance (*796*) it was stated that the usual foods available to the consumer are estimated to provide about half of the total vitamin A activity as preformed vitamin A and half as carotenoid vitamin A precursors. Witsch *et al.* (*1024a*) in a dietary study of 156 American adults of both sexes and various ages during different stages of activity and covering a 6- to 12-month period, found that approximately equal amounts of carotenes and preformed vitamin A made up their total vitamin A activity intake. Fruits and vegetables accounted for two-fifths of the total vitamin A activity. About 60% of the subjects studied met or exceeded the total vitamin A activity daily allowance of 5000 IU or 1000 retinol equivalents. Greaves and Tan (*367*) and Thompson (*933*) reviewed the relative sources of vitamin A activity for American and British diets and concluded that the British receive about one-third of their vitamin A requirements in the form of carotenoid vitamin A precursors (Fig. 29). Kübler (*552,553*) reported on the carotenes in the nutrition of infants.

Thompson (*934*) pointed out the outstanding source, when consumed, in Africa of provitamin A (Table 41), namely, red palm oil. In western Nigeria the daily intake of 100–200 gm of this oil supplies 60,000–120,000 IU, causing the blood

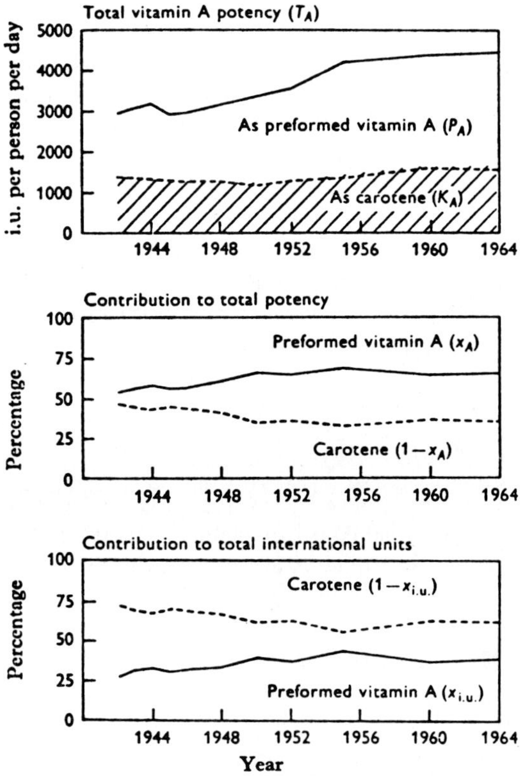

Fig. 29. Vitamin A and carotene in the British diet. Results of the National Food Survey: 1942–1949, urban working-class households; 1950–1964, all households. [From Greaves and Tan (*367*).]

serum to be unusually rich in carotene (*165*). Other important sources are edible green leaves, red paprika, cape gooseberries, paw-paws, and mangoes plus foods also eaten in the Western Hemisphere, such as carrots, yellow corn, and sweet potatoes. For various reasons these rich sources of provitamins are not always consumed in adequate amounts, and vitamin A deficiency, including blindness, is prevalent in many parts of Africa and other parts of the Far and Middle East and Central and South America (*1017*). In Africa and the Far East, little fish and milk and very few vegetables and fruits are consumed. The diets consist mainly of cereals, such as millet, rice, tapioca, and wheat, and supply less vitamin A activity than the recommended allowance for protection against xerophythalmia and other symptoms of vitamin A deficiency. A similar situation can be found in many parts of Central and South America, as described by Flores *et al.* (*299*), where not only does the diet often supply far less than the recommended allowance of vitamin A for adults, but the majority of it comes in the form of carotenoid vitamin A precursors. Steps are being taken in many of these countries to augment diets with carotenoid vitamin A precursors in the form of naturally grown foods and to nutrify certain basic food ingredients with added formed vitamin A esters. In sharp contrast to inhabitants of these warmer climate countries, Eskimos in 1855, according to Sinclair (*866*), were mainly carnivorous, ingesting about 10,000 IU of vitamin A daily, mainly from seal flesh. Berries supplied about 15 IU of vitamin A in carotenoid vitamin A precursor form.

Most of the older chemical methods for determining β-carotene content measured only total carotene in foods, no allowance being made for the difference in activity of the various provitamin A carotenoids, α-, β-, and γ-carotenes, cryptoxanthin, etc., that might be present. The problem is further complicated by current knowledge that each of the carotenoids in processed foods can exist in stereoisomeric forms, which may have widely different biological properties. Since most of the "carotene" values of foods were established some time ago, they do not necessarily represent the true vitamin A value of present-day foods. Gebhardt *et al.* (*330*) have shown concern that reported values of vitamin A activity for some red and yellow fruits (such as peaches) and vegetables may be erroneously high due to the inclusion of carotenoids without activity in the vitamin A calculation of chemically derived figures. The newer analytic methodology needs further development, testing, and extending to current varieties of vegetables, fruits, processed foods, and fabricated foods, so that eventually reliable food composition tables of vitamin A activity values can be established for currently produced foods.

A better understanding of the loss of biological activity of carotenoids during the processing of vegetables came about during some studies in the early 1970's. Sweeney and Marsh (*915*) studied the effects of cooking and various processing procedures on the carotenoids in a number of different vegetables. High-pressure liquid chromatography methods (*916*) were used, which permitted the separation and determination of the various stereoisomers. Samples of vegetables with high concentrations of provitamin A were purchased fresh in the retail market, and two brands of frozen and canned samples of the vegetables were also taken for study. The stereoisomers were measured in the canned vegetables, in the fresh vegetables as purchased and after two periods of cooking, and in the frozen vegetables after thawing and cooking. Samples of the spinach, carrots, and squash were also analyzed after pressure cooking, and carrots after freeze-drying.

In the case of the fresh and uncooked frozen green vegetables the isomeric patterns were very similar, showing that neither blanching nor frozen storage had any appreciable effects. Cooking resulted in a change in the distribution of isomers, the most significant being conversion of the all-*trans*-β-carotene form to neo-β-carotene U, which has a lower biological activity (about 38% compared with the all-trans form). The loss was greater with the longer cooking time, and in most of the green vegetables the calculated biological activity was reduced by between 15 and 20% on cooking.

The fresh and uncooked frozen yellow and red vegetables contained both α- and β-carotenes, with the exception of sweet potato and Hubbard squash. Considerable variation was observed in the proportion of α-carotene in the carrots studied, which ranged from 20 to 40% of the total. Cooking these vegetables resulted in a greater reduction of calculated biological activity because of the large amount of α-carotene present and because of a similar isomerization of the all-trans β-isomer to neo-β-carotene B. The decrease due to cooking was of the order of between 30 and 35%.

Canning and pressure cooking tended to increase the number of isomers formed, probably because of the higher temperatures involved. In general, however, the losses due to the isomerization were comparable to those after conventional cooking procedures, as the amounts of the less active isomers formed were small. Freeze-drying of carrots had no effect on the isomeric patterns observed, although the total carotene values found were a little lower than those found in fresh carrots.

The refined separation procedures used by these authors show that it is no longer possible to hold the view that the vitamin A activity of the carotenes in vegetables is unaffected by cooking or processing, and, although this adds one more difficulty to the prediction of the vitamin A activity of a diet containing cooked vegetables, at the same time its wider application should permit better estimates of the provitamin activity in foods.

The most important assignment of biological activity to the various carotenoids is unsatisfactory because it rests on relatively few studies, particularly with man. For the many countries where carotenoid provitamins are the major source of vitamin A activity in the diet, the application of these refined analytical procedures in studies of the availability and vitamin A activity of the isomers has nutritional significance.

A. Availability to Man

Ciusa and Nebbia (*189*) stated that carotene is a poor source of vitamin A for man. As reviewed by Goodwin, the absorption of carotenoids from plant food sources by man is generally fair to poor and poorer if the foods are uncooked or not finely divided. Absorption is facilitated somewhat by the presence of fat or dispersing agents such as lecithin. Fraps and Meinke (*315*) found that carotene fed in the form of vegetable was less effective than when given as a component of butterfat. Between 25 and 45% of carotene from raw spinach or grated carrots can be absorbed by man (*983*), whereas the absorption from the cooked material is considerably less, i.e., 2–12%. Van Zeben and Hendriks (*984*) suggested that the poor absorption of carotene from cooked carrots is attributable to locked-in carotene within undigested plant cells, as consumption of homogenized cooked carrots results in higher serum carotene values. It was also reported by Kreula and Virtanen (*539*) that, in the case of man, carotene is biologically more effective when given in butter than when administered as a component on carrots. Kreula (*538*) found that carotene administered in olive oil was absorbed to a greater extent than that from raw, finely grated carrots in three human subjects on a fat-free diet. The utilization of carotene from sweet potatoes by young human adults was reported to be 54%, determined by adequate balance studies by James and Hollinger (*460*), with three women and five men as subjects. The absorption and nutritional value of carotene were examined by Bernhard (*95*) and Simonnet (*864*) in 1963 Thompson (*934*) in 1964, Krondl (*546*) in 1966, and Kübler (*551*) in 1970.

The Japanese have been interested in the absorption of carotene in man since their diets have a high content of plant foods. Kawaguchi and Fujita (*498*), using one female human subject, observed the following percentages of carotene absorption: cooked carrots, 21; cooked pumpkin, 35–53; cooked spinach, 33–35; dried seaweed soaked in hot water, 8. The absorption of carotene from food by a healthy male subject was studied by Fujita and Morimoto (*320*) and Morimoto (*685*). The absorption of carotene from food based on observations with cooked squash, carrots, and spinach was 31–55%. In Indonesia, Oey *et al.* (*722*) reported a study in which young children were given palm oil (4 ml/day) as a provitamin A source, which was judged effective in decreasing xerophthalmia and increasing serum vitamin A values. Roels *et al.* (*814*) found that African boys receiving 200 gm carrots in addition to their normal diet absorbed only 5% of the carotene, but a supplement of 18 gm olive oil increased the absorption to 25% and produced increases in serum content of carotene and vitamin A.

Pirie (*760*) reported on the value of carotene in the prevention of xerophthalmia in man. The absorption of carotene from green leafy vegetables (Table 56) was observed by Lala and Reddy (*564*). Indian children were given cooked amaranth (40 gm/day), which was judged to yield a carotene absorption between 57 and 93%. The daily consumption of this gave an increase in serum vitamin A values and if followed would, in practice, prevent vitamin A deficiency blindness and xerophthalmia. Muhilal and Karyadi (*697*) fed a mixture of vegetables (amaranth, swamp cabbage, cassava leaves, green yardlong bean, *Sauropus androginus* leaves, and yellow sweet potato) to 12 children (3–5 years old), analyzing carotene intake and excretion. They recorded absorption percentages of from 11.9 to 65.3%, with an average value of 34.3%. They questioned whether vegetable carotene could be

TABLE 56

Effect of Feeding Green Leafy Vegetables on Serum Vitamin A Level[a,b]

Initial level of serum vitamin A (μg/100 ml)	Number of children		Initial values of serum vitamin A (μg/100 ml)		Final values of serum vitamin A (μg/100 ml)		Differences between initial and final values of serum vitamin A	
	Experimental	Controls	Experimental	Controls	Experimental	Controls	Experimental	Controls
<25	17	3	15.4 ± 1.41	17.5 ± 3.04	28.0 ± 2.26	17.9 ± 1.33	12.6 ± 1.83 ($P < 0.001$)	0.3 ± 2.63 (NS)
>25	12	3	31.2 ± 1.98	31.8 ± 1.28	37.5 ± 4.47	27.0 ± 1.92	6.2 ± 4.27 (NS)	−−.8 ± 1.72 (NS)
All children	29	6	21.9 ± 1.83	24.7 ± 3.51	31.9 ± 2.39	22.5 ± 2.31	10.0 ± 2.11 ($P < 0.001$)	−2.2 ± 1.82 (NS)

[a] From Lala and Reddy (*564*).

[b] Amaranth, 40 gm, providing 1200 μg β-carotene, fed daily for 15 days.

TABLE 57

Absorption of Carotene from Different Sources[a,b]

Source	Ingested (μg) Total carotene	Ingested (μg) β-Carotene	Absorption (%) Total carotene	Absorption (%) β-Carotene
Amaranth	5151	2926	58.1 (44.1–79.6)	75.7 (52.4–99.8)
Papaya	9840	3319	45.8 (19.3–63.5)	90.3 (73.3–99.4)
Carrot	7288	1381	35.5 (15.9–67.3)	81.1 (73.2–96.8)
Crystalline β-carotene	4500	3679[c]	97.7 (93.2–100.0)	98.8 (95.4–100.0)
Diet with mixed sources[d]	4253		33.4 (15.7–59.2)	

[a] From Rao and Rao (*794*).
[b] Values are means of four subjects with range of values in parentheses.
[c] Determined by column chromatography on calcium hydroxide.
[d] Contained yellow maize, spinach, orange, and peas as sources of carotene.

used to achieve optimal vitamin A status in children. Rao and Rao (*794*), studying four healthy men, reported carotene absorption (Table 57) from amaranth of 58–76%; from papaya, 46–90%; from carrots, 36–81%, and from mixed sources, 33%. Rajalakshmi *et al.* (*787*) studied the availability of carotene in leafy vegetables in Indian adult men, and spinach and amaranth were shown to be available sources of carotenes for man. Krishnamurthy *et al.* (*544*) also reported on the biological utilization of β-carotene from amaranth. Employing two healthy men, Van Eekelen and Pannevis (*982*) found carotene from cooked spinach and carrots to be poorly absorbed compared to carotene dissolved in oil. A 1967 FAO/WHO report (*292*) summarized the results of 19 reports (Table 58) on the availability of carotene from various sources.

B. Availability Judged by Animal Studies

The availability of vitamin A from carotenoids of local fruits and vegetables was studied by Giasuddin *et al.* (*341*) in 1973 by the bioassay method. Three local fruits and vegetables that showed relatively high carotenoid content were selected, and their ability to produce vitamin A in albino rats was measured. With pure β-carotene as reference standard, it was found that the carotenoids of palang sak (*Spinacia oleracea* Linn.), ripe papaya (*Carica papaya* Linn.), and mistikumra (*Cucurbita maxima* Duch) were only 42.7, 31.3, and 24.3% (active, respectively, to produce) vitamin A. The availability and biological utilization of carotene from four commonly eaten Philippine leafy vegetables were investigated by Ortaliza *et al.* (*727*);

TABLE 58

Availability of Carotene from Various Sources[a]

Number of subjects	Source of carotene	Intake (mg/day)[b]	"Absorbed" percentage of intake[c]
33 infants	Carotene in oil	36–54	72 (31–95)
3 infants	Carotene in oil	0.4–2.6	55 (30–70)
1 infant	Carrot meal	3.8	50
7 infants	Carotene in oil	1.1–1.5	36 (25–40)
2 infants	Carrot, cooked, mashed	0.9–1.9	2, 3
2 infants	Carrot, raw, mashed	0.4, 0.8	.26, 33
2 infants	Spinach, cooked, mashed	0.3, 0.7	12, 2
2 infants	Spinach, raw, mashed	0.3, 0.4	25, 45
4 infants	Tomato juice	0.26	8 (7–11)
5 infants	Carotene in oil	6–14	68 (59–81)
10 infants	Carrot, cooked, pureed	26.7[d]	60 (46–75)
4 children	Carotene in oil	8–15	52 (30–75)
1 child	Carrot puree	8	30
1 child	Carrot meal	38	45
1 child	Spinach, cooked, minced	5	30
1 child	Carotene in oil	0.9	52
6 children	Sweet potato, cooked, mashed	3.6	28 (5–58)
5 children	Carrot, raw, grated	19	Less than 5
5 children	Above + oil	19	About 25
4 children	Carotene in oil	28	About 45
1 adult	Carrot, raw	5.5, 15.5	78, 87
1 adult	Spinach, cooked	2.6	88
1 adult	Above, fat-free diet	2.6	52
2 adults	Carotene in oil	1.1	59
2 adults	Carrots, cooked	5.7	1
2 adults	Spinach, cooked	1.1	6
2 adults	Above + fat	1.7	5
2 adults	Carrot, raw, finely grated	7.9	21 (20, 22)
2 adults	Carrot, raw, coarse	10	2 (2.3, 2.4)
2 adults	Carrot, cooked, cubed	26	4 (1.6, 5.4)
5 adults	Carotene in oil	3	52 (46–56)
2 adults	Carotene in oil	2	55 (40, 70)
2 adults	Carrot, puree	11, 14	55 (50, 60)
1 adult	Spinach, minced	2	35
1 adult	Carrots, raw	20	1
1 adult	Carrots, cooked	13	19

(Continued)

TABLE 58—*Continued*

Number of subjects	Source of carotene	Intake (mg/day)[b]	"Absorbed" percentage of intake[c]
1 adult	Spinach, raw	32	45
1 adult	Spinach, cooked	33	58
2 adults	Carotene in oil	1.9–19.6	40 (30–51)
3 adults	Carotene in fat	1.6–2.7	38 (35–41)
3 adults	Carrot, raw, finely grated	6.7	10 (8–13)
3 adults	Carrot, raw, blended in oil	7.2	47 (32–50)
3 adults	Carrots, fresh, raw	8.3	13 (11–15)
2 adults	Carrots, fresh, cooked	8.6	9 (8, 9)
1 adult	Above + oil	8.6	13
3 adults	Carrots, stored, raw	7.6–30.3	4 (2–7)
1 adult	Carrots, stored, cooked	32	1.4
1 adult	Above + oil	8	4
2 adults	Spinach, fresh	9.6	9 (9, 10)
2 adults	Spinach, dried	6.3	7 (7, 8)
1 adult	Above + oil	27	7
1 adult	Above, extracted, in oil	6	45
2 adults	Beet greens	4.5	8 (5, 11)
1 adult	Above, extracted, in oil	22	46
5 adults	Carotene in oil or fat	3.1–3.6	73 (69–82)
2 adults	Carrots, canned, sliced	2.2	24 (16, 31)
4 adults	Carrots, cooked, puree	3.0	25 (6–36)
4 adults	Above + fat	3.0	33 (0–60)
2 adults	Carrots, cooked, homogenized	3.0	56 (46, 63)
3 adults	Spinach, canned, puree	2.5	41 (28–62)
2 adults	Spinach, canned, homogenized	4.0	43 (35–47)
3 adults	Cabbage, dried, outer leaves	1.9	41 (13–60)
2 adults	As above	3.8	28 (28, 27)
3 adults	Carrots, fresh, raw	2.9–6.9	42 (37–46)
2 adults	Carrots, stored, raw	9.2	43 (36, 49)
3 adults	Carrot meal, coarse	6.0–7.5	37 (34–42)
2 adults	Carrot meal, fine	1.2	55 (54, 56)
8 adults	Sweet potato, cooked, mashed	3.5	54 (41–50)
1 adult	Carotene in oil	1.7	98
1 adult	Carrot, cooked	2.6	29
1 adult	Carrot, grated, raw	2.1	21
1 adult	Carrot, grated, cooked	2.1	48
1 adult	Above + fat	2.1	46
1 adult	Pumpkin, strained	0.9	53
1 adult	Above + fat	0.9	66
1 adult	Squash, strained	3.2	35
1 adult	Above + fat	3.2	38

(*Continued*)

TABLE 58—*Continued*

Number of subjects	Source of carotene	Intake (mg/day)[b]	"Absorbed" percentage of intake[c]
1 adult	Spinach, strained	2.4	33
1 adult	Above + fat	2.5	42
1 adult	Spinach, strained	3.4	37
1 adult	Above + fat	3.2	37
1 adult	Seaweed, dried	2.1	8

[a] From FAO/WHO Joint Expert Committee on Food Additives (*292*).

[b] When intakes were expressed in international units, they were recalculated on the basis of 1 IU = 0.6 μg β-carotene. In some instances, the methods employed did not differentiate between the various carotenes.

[c] For present purposes the "absorption" is taken as the difference between intake and fecal content (often corrected for carotene excretion on carotene-free diets). Values in parentheses indicate the reported range.

[d] Value represents total dose administered; number of days not known.

mustasa (mustard), gabi leaves (taro, *Calocasiai esculentum*), saluyot (jute, *Corchorus olitorius*), and kalabasa tops (squash, *Cucurbita maxima*) were studied by rat liver vitamin A storage tests. Relative to the control, β-carotene in oil, carotene availability of the four vegetables was 53, 38.7, 21.4, and 18.8%, with significant differences in carotene utilization among the vegetables. When percent availability and total β-carotene content were considered, amounts of available carotene from the four vegetables were 12.7, 10.5, 4.25, and 2.5 mg per 100 gm. The data for all the vegetables showed statistically significant differences in carotene utilization. The availability of carotene from the tomato as judged by rat trials was reported by Premachandra *et al.* (*771,772*). Sweeney and Marsh (*913*) in 1974 observed availability of carotene to rats. Comparisons were made with all-*trans*-β-carotene dissolved in cottonseed oil. The results indicated that carotene in carrots, as measured by storage of vitamin A in liver and kidney, was equal in availability to all-*trans*-β-carotene dissolved in cottonseed oil. Despite similar liver and kidney storage values, fecal excretions of carotene for rats receiving carrots were much higher than for those receiving carotene in cottonseed oil. The disruption of carrot cells by freezing, blending, freeze-drying, or ultrasonic treatment did not increase carotene availability.

Rajalakshmi *et al.* (*786*) reported on experiments conducted between 1967 and 1973 in growing rats. The effects of naturally occurring carotene were compared with those of synthetic retinyl acetate with regard to the total quantity of vitamin A in the liver, expressed as a percentage of the amount ingested. This proportion ranged from 9% for mint leaves to 50% for spinach, with intermediate values for fenugreek, radish tops, yellow pumpkin, coriander leaves, and mustard greens.

These animal studies indicated about the same range of carotene utilization as reported in section XIII,A.

XIV. APPLICATIONS IN COLORING FOODS

Hues obtainable with carotenoids added to foods are dependent on both the market form used and the type of food involved. Carotene produces yellow to orange colors in oil solution; 1–10 mg/liter produces a yellow hue, and 30–50 mg/liter approaches the orange range. β-Apo-8′-carotenal is more orange than β-carotene in oil solution and becomes red as concentration increases. The solubility of carotenoids in fats and oils (triglycerides) is low, and it is frequently necessary to heat oils to achieve the desired concentration. The solubility of canthaxanthin in triglycerides is so low that practical applications in fat-based foods have not yet been achieved, despite its very high tinctorial potency. In certain food uses the low solubility of canthaxanthin may be advantageous, e.g., in coloring formulated tomato-based foods—spaghetti sauce, tomato soup, pizza sauce. A water-dispersible form of canthaxanthin has been developed which simulates the natural color of lycopene-rich tomatoes very well. When added to tomato products, the canthaxanthin remains in the aqueous phase, producing an attractive red hue, whereas the oil phase has a slight yellow hue, which is not readily observed. Natural carotenoid concentrates, as previously mentioned, exist in a variety of oil-soluble, liquid, and dry forms adapted to specific coloring needs.

In aqueous foods it is necessary to use water-dispersible forms of carotenoids. The only commercially available water-soluble carotenoid is norbixin, which is a dicarboxylic acid. Norbixin is soluble in alkaline aqueous solutions and is available in this form. Other carotenoids can be solubilized by preparation of oil-in-water emulsions in which the carotenoid is in the oil phase. Such products are commercially available in liquid and dry forms. Water dispersion of carotenoids frequently produces colors that differ from the colors obtained from an equivalent concentration of carotenoids in vegetable oil. β-Carotene dispersions are yellow-orange to orange rather than yellow. β-Apo-8′-carotenal produces peach-orange hues at low concentration and more red shades at higher concentrations. Canthaxanthin is reddish at all concentrations in aqueous dispersion and resembles tomato juice or whole blood. Most water-dispersible carotenoid preparations do not produce the sparkling clarity of truly water-soluble compounds and hence are limited in applications where clarity is necessary, unless especially prepared water-clear dispersible products (*282*) are used. The opacity resulting from the use of higher levels of water-dispersible carotenoids in aqueous foods simulates the insoluble form of the natural pigments in fruit tissue and therefore produces a more natural appearance.

A guide (*58*) to the potential level of use of these carotenoids in a variety of foods is shown in Fig. 30. Since the pure carotenoids have been available for only a

Fig. 30. Natural occurrence, structural formulas, and coloring guide for carotenoid food colors. This guide is intended as a starting point for those who wish to color goods with carotenoid food colors. Trial studies with these carotenoids (singularly or in combination) in formulated food products with tests under intended holding, storage, and use conditions are recommended as a prerequisite to final color and level selection. [From Bauernfeind (*58*).]

Formula	Natural occurrence	Guide to potential use level[a]
β-Carotene 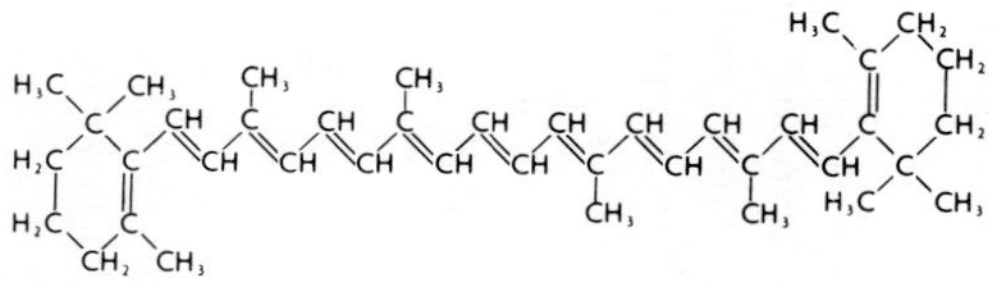	Butter, cheese, milk, ice cream, egg yolk, fish, crustaceans, bivalves, cauliflower, asparagus, broccoli, lettuce, fenugreek, spinach, tomatoes, watermelons, carrots, sweet potatoes, squash, peppers, pumpkins, beans, peas, wheat, corn, sorghum, pasta products, palm oil, grass, alfalfa, hay, silage, pineapples, oranges, cranberries, figs, grapes, apricots, peaches, prunes, apples, pears, strawberries, nectarines, mango fruit	(mg/lb or 0.5 kg) Butter and butter sauce, 1–4; cheese, 2–8; mellorine or ice cream, 2–5; sherbet, 3–8; regular margarine, 3–4; whipped margarine, 5–7; salad oil, shortening or lard, 3–6; frying oil[b] (for french fries, 6–15; popping oil[b] (for popcorn), 30–60; oil spray[c] (for popcorn), 50–200; toppings, 1–10; egg yolk, bulk,[c] 20–50; baked goods, 2–10; frostings, fillings, sauces, or chicken dishes, 1–6; gelatin mix[d] (dessert or salad use), 5–20; confections or candy, 3–15 (mg/qt or liter) Imitation fluid milk, 0.5–2; coffee whitener or cream, 1–5; eggnog, 3–4; juices, drinks, carbonated beverages, 1–10; orange concentrate[c], 10–40; soups, 2–8; honey, 2–5

β-Apo-8′-carotenal (apocarotenal) 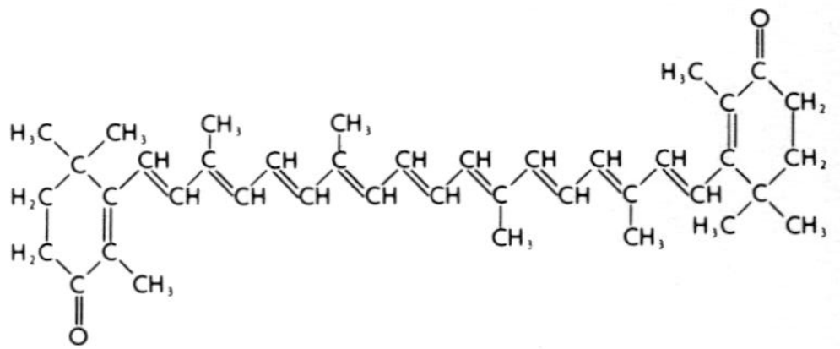	Oranges, tangerines, other citrus fruits, grass, spinach, alfalfa, marigolds, other green plants, animal tissue	(mg/lb or 0.5 kg) Salad dressing, 4–15; cheese, 3–6; mellorine, ice cream, or sherbet, 2–8; baked goods, 4–10; gelatin mix[d] (dessert or salad), 5–15; confections or candy, 3–15 (mg/qt or liter) Drinks or beverages, 3–15; chowders or stews, 4–10
4,4′-Diketo-β-carotene (canthaxanthin) 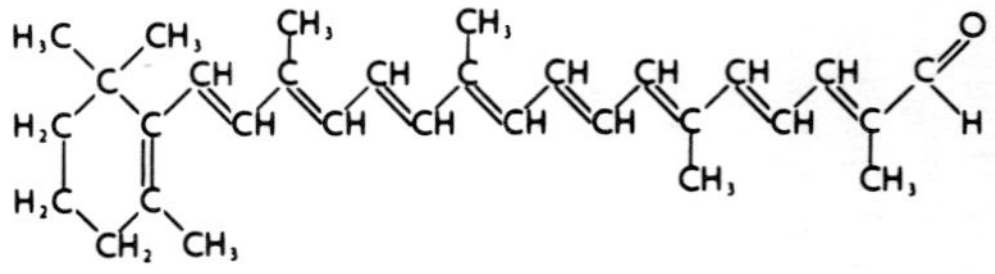	Edible mushrooms, sea trout, brine shrimp, crustaceans, algae, microorganisms, avian tissues of the flamingo and other fed-feathered birds	(mg/lb or 0.5 kg) Salad dressings, 10–30; cheese, 2–6; simulated meat products, 4–10; sausage or frankfurters, 3–6; spaghetti, pizza, or barbecue sauce, 3–15; baked goods, 2–6; gelatin mix[d] (dessert or salads), 10–20; confections or candy, 3–15 (mg/qt or liter) Drinks or beverage, 3–15; tomato soup or cocktails, 2–10

[a] Based on pure compound.
[b] A cooking fat, not consumed directly.
[c] Food ingredients, not consumed directly.
[d] A dry mix, to be reconstituted with water.

relatively short time compared to some other classes of coloring compounds, the science and technologies involved in the synthesis, preparation, and incorporation in food products are not complete and in certain instances warrant further developments. Reviews of various aspects are available (*53,57,58,60,78,81,133,134,156, 199,206,283,319,363,364,388,450,452,453,513,515,518,521,524,622*).

A. Shortening, Lard

The desire for yellow color in processed food products prepared with hydrogenated vegetable oil, lard, and liquid vegetable oils leads to another application whereby the homemaker may improve the appearance of baked goods. Shortening in particular seems to be an excellent vehicle (*79,707*) for the addition of carotene to food products, from the standpoint of both color and color retention during baking. Proponents of carotene addition point out that the attraction of yellow fat, such as lard, is not simply a matter of passing novelty; the consumer finds that yellow lard enhances and glamorizes many foods, since it gives them an especially attractive hue. Pie crust made with yellow lard takes on a golden brown color. Fried chicken is also an attractive dish when prepared with a yellow fat. Consumers are pleased to find that many other foods are similarly glamorized.

In the process of coloring fat or oil, a β-carotene suspension (20–30% β-carotene) can be used. Usually about 3.5–5.0 mg of β-carotene colors 1 lb or 500 gm of product. This level of color may be obtained by using between 24 and 32 gm (about $\frac{7}{8}$–$1\frac{1}{8}$ oz) of 30% β-carotene or between 30 and 40 gm (about 1–$1\frac{1}{2}$ oz) of 24% β-carotene or between 36 and 48 gm (about $1\frac{1}{4}$–$1\frac{3}{4}$ oz) of 20% β-carotene per ton or 1000 kg of product. Higher levels may be used depending on the use of the colors of product and consumer preference. Large manufacturers may prefer to use a batch-size container holding the exact amount of β-carotene to color a given quantity of product. Before color addition the fat product is first clarified and deodorized. If the fat or oil is not deodorized, the coloring should be added after it has settled and been filtered. The usual point of addition is in the makeup tank before chilling and packaging. If the entire product is not heated at any phase in the usual operation, an aliquot or portion of it should be removed and warmed and the carotenoid added with stirring until dissolved, cooled, and returned to the production batch (*79*). If food-grade antioxidants are added for protection against oxidative rancidity of the fat, these likewise will provide an added stabilizing influence on the carotenoids. β-Carotene suspension in the correct amount is added tc the warm fat product (100°–130°F; 38–56°C) with agitation, which is continued until complete solution is achieved. Subsequent treatment of fat, such as chilling and packaging, is accomplished in the usual manner. It should be remembered that the coloring will not improve the original quality of the processed fat. All steps in the manufacturing should follow the most highly recommended procedures scrupulously, so that the product has a high smoke point and a bland odor. Data from some plant trial runs (Table 59) illustrate the stability (*79*) of added β-carotene under commercial pack-

TABLE 59

Color Observations on β-Carotene-Colored Shortening and Vegetable Oil[a]

Product	Form of carotene	Type of packaging	Appearance	Color as carotene (mg carotene/kg) Initial	6 months, 5°C	6 months, 23°C
Hydrogenated vegetable oil	β-Carotene suspension	1-lb metal can	Evenly colored	8.4	8.2	8.2
Pure lard	β-Carotene suspension	2-lb carton	Evenly colored	7.7	7.4	7.7
Hydrogenated vegetable oil	β-Carotene suspension	2-lb metal can	Evenly colored	11.7	—	—
Hydrogenated vegetable oil	β-Carotene suspension	1-lb metal can	Evenly colored	11.1	—	—
Liquid vegetable oil	β-Carotene suspension	1-qt amber bottles, air headspace	Evenly colored	8.0	7.5[b]	6.9
Liquid vegetable oil	β-Carotene suspension	1-qt amber bottles, nitrogen head-space	Evenly colored	7.9	7.8[b]	—
Liquid vegetable oil plus antioxidants	β-Carotene suspension	1-qt amber bottles, air headspace	Evenly colored	7.8	7.8[b]	7.8

[a] From Bauernfeind *et al.* (79).
[b] Three months, 23°C.

aging conditions. The variation in the color of the shortening that may be obtained is shown below.

Added β-carotene (mg/kg)	Added β-apo-8′-carotenal (mg/kg)	Shortening color
10	0	Yellow
15	0	Bright yellow
20	0	Deep yellow
5	2.5	Light orange
7.5	2.5	Orange-yellow
10	2.5	Yellow-orange

Tests were run to obtain data on the behavior of carotene-fortified shortening in cooking trials. For these tests, shortenings were fortified at three arbitrary levels: shortening A, 9.2 mg of β-carotene per kilogram; shortening B, 19.8 mg of β-carotene per kilogram; shortening C, 26.4 mg of β-carotene per kilogram; the control, uncolored shortening. Four types of food products were prepared in the tests: cookies, cake, pie crust, and uncooked frosting. In planning these baking trials, attention was given to the ingredients containing natural carotene in an attempt to eliminate or standardize the natural carotene sources wherever possible. A control sample was prepared to determine the amounts present in the initial unbaked product. All baking trials were carried out in a home-style kitchen and baked in the oven of an electric home range. The three β-carotene additions resulted in evenly colored yellow shortenings of three definite shades. Baked food products containing these shortenings were evenly colored. Retention of carotene in the baked foods, varying from 75 to 95% of the added β-carotene, was found by analysis. The favorable carotene retention data reported here are comparable to vitamin A retention values found in baking studies with vitamin A-fortified margarine. As would be expected, excellent retention was found in the uncooked frosting.

If specialty oil products, such as popping and deep-frying oils, are to be colored, the conditions of high heat put added stress on the carotenoids (*332*). Stabilized carotenoid suspension application forms must be used for these oils. Popcorn and french-fried potatoes (*549*) to be frozen for home and institutional use are two food products colored in this fashion.

B. Margarine

Synthetic β-carotene was first used to color margarine and butter, and successful industrial trials were carried out by Schuchardt (*844,845*) in the year of its commercial introduction. Earlier Matzko (*636*) proposed improving the vitamin A content of margarine by the addition of a carotene preparation. In 1938 Stoltz and Sutton (*897*) published a paper on carotene in the coloring of butter. Wagner (*998*) also reported on color and vitamin studies in margarine. As early as 1913–1915 the yellow color of summer butter was associated with vitamin A activity.

The technology of coloring fat-based foods has been described in detail by Bauernfeind and co-workers (*79*). The stability of the suspensions as well as of the added β-carotene in margarine is very good, and margarine is still considered an important commercial food product for the application of β-carotene. Suspensions in vegetable oil containing 20–30% micronized β-carotene are the most suitable formulations for coloring margarine. Approximately 3–9 gm pure β-carotene per 1000 kg may be added depending on the color shade desired and the type of margarine to be produced. The addition of β-carotene to margarine is widely practiced around the world (*395,520*). Margarine is a water-in-oil emulsion of partially hydrogenated vegetable oils and nonfat fluid milk with flavors and with or without salt. Like butter as a table spread, margarine requires a rich yellow color for ready consumer acceptability. Coloring materials for margarine as used in the United States as well as in other parts of the world are described in the literature. Espoy and Barnett (*286*) and Luckmann *et al.* (*608*) presented methods for the detection and determination of the coloring materials, carotene, annatto extract, etc., in margarine. The latter authors (*608*), in their analytical paper on the determination of vitamin A in margarine, pointed out that carotene, in addition to imparting a yellow color, acts as provitamin A. For this reason, this vitamin A contribution must be taken into account in margarine analyses, and steps must be taken to prevent erroneously high vitamin A estimates from being made in the assay of margarine colored with other materials. A treatise on the technology of margarine manufacture by Crump (*210*) claims that carotene is the ideal color for margarine as well as butter and foretells the possibility that carotene will eventually replace all other coloring materials in margarine.

According to Harris (*395*), the nutrification of margarine with vitamins A and D is mandatory in Denmark, the United Kingdom, Italy, Norway, Sweden, and India (Vanaspati). It is nutrified on a voluntary basis in Australia, Austria, Belgium, Brazil, Canada, Germany, Finland, Greece, Holland, Newfoundland, Portugal, Russia, South Africa, Switzerland, Turkey, and the United States. In 1957, in the United States, about 50% of the margarine produced was colored with carotene, contributing one-third of the vitamin A value of the product. In 1970, Morton (*686*) reviewed the vitamin A nutrification of margarine (Table 60). In the United States margarine must contain at least 15,000 IU of vitamin A activity per pound, with vitamin D as an optional additive. The Federal Definitions and Standards of Identity for margarine specifically permit the vitamin A activity of β-carotene used for color to be included as part of the required vitamin A level. The majority of margarines are colored with 3.0–3.3 mg of crystalline β-carotene per pound (6–7 mg/kg). A few brands have a slightly higher concentration. β-Carotene supplies about one-third of the total vitamin A requirement of margarine besides contributing color. The air or gas in the whipped form of margarine tends to dilute the color. Thus,

TABLE 60

Amount of Vitamins A and D Commonly Added to Margarine[a,b]

Country	Vitamin A (IU/kg)	Vitamin D (IU/kg)
Australia	30,000	4000
Austria	20,000	1000
Belgium	20,000	1000
Brazil	15,000–50,000	500–2000
Canada	33,000	
Chile	14,000–18,000	1000
Colombia	20,000–30,000	2000–4000
Denmark	20,000	625
Finland	20,000	2500–3500
Germany	20,000–30,000	1000
Greece	25,000	1500
Israel	30,000	3000
Japan	30,000–40,000	
Mexico	20,000	2000
Netherlands	20,000	2000
Norway	20,000	2500
Philippines[c]	22,000	1100
Portugal	20,000–35,000	875–1000
South Africa	20,000	1000
Sweden	30,000	1500
Switzerland	30,000	3000
Turkey	20,000	1000
United Kingdom	30,000–33,000	2900–3500
United States	33,000	4400

[a] From Morton (*686*).

[b] When color is added in the form of β-carotene, it provides 10,000–12,000 IU of vitamin A activity per kilogram.

[c] One hundred milligrams vitamin B_1 also added.

higher levels of β-carotene (10–12 mg/kg) are required to provide a color equal to that of regular margarine. In both regular and whipped margarine, the balance of the vitamin A requirement is provided by added vitamin A, which has no effect on the margarine color.

Margarine with added β-carotene is prepared in commercial plants in batch-size churning units of 500 kg or more. To the weighed amounts of warm margarine oil (100°–130°F), micropulverized β-carotene in the form of β-carotene–vitamin A blend is added and thoroughly mixed until a clear, uniform solution is obtained before the emulsification step, after which the usual remaining production stages are followed. In some cases it is advantageous to add the vitamin A and β-carotene to stainless steel bulk tanks of margarine oil having a capacity in the range of 10,000 kg or more. Tanks must have adequate mixing facilities. Either built-in mixers or a circulation system to pass the tank contents through a pump of sufficient size to provide rapid mixing may be used. The advantage is that, when the bulk oil is colored and nutrified, each churn of margarine does not have to be treated separately. Maximal holding time of the bulk oil is 4–5 days, during which the temperature of the oil should be controlled.

Samples of 18 lots of margarine were subjected to vitamin A and carotene analyses initially and after storage for 2 and 6 months at 7° and 23°C. The stability results (Fig. 31) on β-carotene in margarine are in agreement with the excellent stability findings of Melnick *et al.* (*656*) from studies in which carotene of natural origin was added to margarine manufactured on a commercial scale and stored at 7°C. Stability performance was also confirmed by the study of Geminder and MacDonough (*332*), wherein vegetable oil suspensions of β-carotene were declared to have excellent stability and to be readily adaptable for coloring margarine. In all the plant trial runs, no flavor defects were observed in the β-carotene-colored margarine product. Coloring was uniform without mottling or speckling. Although the data (Fig. 31) are based on chemically determined values (*79*), an earlier report (*628*) showed excellent correlation of chemical values with biological values as determined by the rat curative growth method applied to β-carotene-colored margarine (Table 61). Hence, by biological assay, the added β-carotene in margarine has been observed to be fully biologically available (*628*), and chemical assays of nutrified margarines made by different manufactures (Table 62) (Fig. 31) show good stability under simulated use conditions (*628*).

Schuchardt (*844*) described successful margarine coloring trials with β-carotene in which the result was a pure yellow-orange color, without a greenish shade, which did not change during a 6-month storage period. Stability, taste, and odor were not

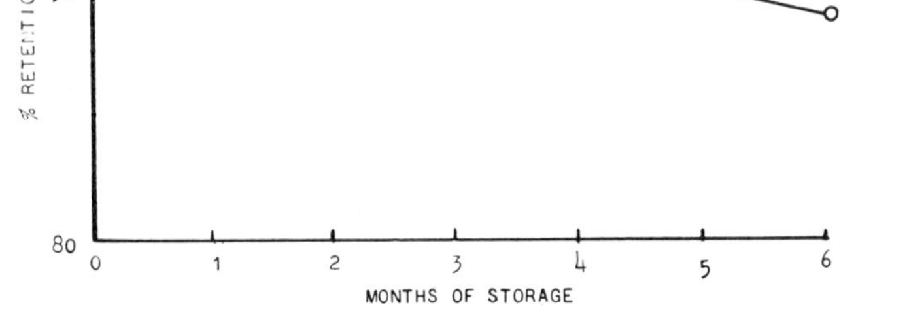

Fig. 31. Stability of β-carotene and vitamin A in 18 commercial lots of margarine. Key: •, vitamin A values; ○, carotene values. [From Bauernfeind *et al.* (*79*).]

TABLE 61

U.S.P. Rat Curative Growth Assays of Commercial Margarines with Added Vitamin A and β-Carotene[a]

Material fed[b]					Vitamin A activity per lb margarine					
					Biological total		Colorimetric			
Brand of margarine	Type of carotene	No. of rats	Dosage per rat/4 days IU	Gain in weight after 28 days (gm)	IU	± SE (%)	Vitamin A (IU)	Carotene[c] (IU)	Total (IU)	Biological colorimetric × 100 (%)
Assay 1										
U.S.P. reference		12	3.19	37.6						
standard vitamin A		11	4.69	59.7						
		10	4.69	57.0						
		13	6.91	83.2						
			Margarine lb + 10^{-4}							
A	Natural	12	2.76	61.5	18,400	5.8	11,000	5700	16,700	110
		12	4.06	91.3						
A	Synthetic	12	2.76	55.2	16,600	5.1	10,300	5600	15,900	104
		11	4.06	84.7						
B	Sunthetic	12	2.11	61.2	26,000	5.8	18,600	7800	26,400	99
		12	3.12	96.0						
Assay 2			IU							

U.S.P. reference standard	8	3.00	33.9						
	9	3.00	39.8						
Vitamin A	8	4.41	61.6						
All contain synthetic β-carotene	10	4.41	59.4						
		Margarine lb × 10^{-4}							
C	8	1.79	36.1	17,100	4.9	11,900	5800	17,700	97
	10	2.64	64.0						
D	8	1.79	32.4	15,000	5.7	10,700	5800	16,500	91
	9	2.64	51.7						
E	10	1.79	45.7	18,500	6.0	12,900	6200	19,100	97
	9	2.64	63.4						
F	10	1.79	39.7	18,200	5.2	12,900	6500	19,400	94
	10	2.64	68.7						
G	9	1.79	36.7	17,100	5.5	12,300	5500	17,800	96
	9	2.64	63.1						
H	8	1.79	37.6	17,800	5.8	11,200	5000	16,200	110
	10	2.64	64.4						
I	9	1.79	39.1	18,100	5.0	12,400	5400	17,800	102
	10	2.64	67.3						Average 100

[a] From Marusich *et al.* *(628)*.

[b] Each letter refers to a different manufacturer.

[c] 0.6 μg β-carotene = 1 IU, or 1 gm = 1,667,000 IU.

TABLE 62

Percent Retention of β-Carotene in Commercial Margarine[a]

Manufacturer	Initial assay provitamin A (IU/kg)	After 2 months storage 5°C	After 2 months storage 5° and 23°C[b]	After 2 months storage 23°C	After 6 months storage 5°C	After 6 months storage 5° and 23°C[b]	After 6 months storage 23°C
J	14,080	105	98	94	95	92	86
	13,420	97	97	97	—	—	—
K	11,660	98	96	93	—	—	—
	11,660	96	96	93	—	—	—
L	13,420	—	—	—	100	93	85
M	11,440	—	—	—	92	—	92
	11,440	—	—	—	96	—	102
	11,440	96	100	92	102	91	—
N	13,090	—	—	—	91	91	89
	14,080	98	91	91	105	98	83
O	12,870	101	104	99	101	99	94
	13,310	103	103	98	94	84	86
P	12,650	99	97	94	97	96	85
	13,200	95	95	90	102	98	—
	12,210	96	94	92	103	99	94
	11,110	97	97	95	107	100	89
Q	12,100	102	102	98	96	95	86
	14,300	99	97	91	97	95	83
Average		99	98	94	99	95	89

[a] From Marusich *et al.* *(628)*.
[b] Alternating temperature exposures; more typically in use conditions.

unfavorably influenced. Some literature reports indicate that some discoloration can occur in colored fats under conditions conducive to such developments. Hayes and Steele (*402*) reported a hitherto unknown type of deterioration of hydrogenated arachis oil containing vitamin A and β-carotene at the World Fat Congress in Hamburg in 1964; after several months of storage at −9°C the sample had turned green. The authors found that the characteristic deterioration occurred very rapidly in a hydrogenated palm-kernel oil, which was used for further investigations. The discoloration occurred most rapidly at temperatures between −3° and −6°C. Washing the fat with aqueous sodium hydroxide solution increased the stability of β-carotene. The formation of anhydrovitamin A during this deterioration process suggests that the fat contained an acid catalyst. The more rapid rate of reaction of free retinol compared with the palmitate is further evidence in favor of this hypothesis. McWeeny (*655*), continuing the work on this phenomenon, was able to isolate a number of β-carotene oxidation products. He suggested that the green discoloration was due to oxidation of β-carotene by traces of peroxy acids to epoxide derivatives, which were converted to green-blue ionized forms by the action of an acidic material present in the fat. A study of the factors affecting the rate at which the green discoloration occurred indicated possible methods of avoiding the problem. These methods include a suitable choice of oil, and the use of certain additives, adequate "tempering" of the finished fat, and care in the use of cool storage. One reason that the green discoloration has not been more widely encountered, according to McWeeny, is that commercial products are rarely based on one type of fat. The susceptibility of certain types of fat to green discoloration may be greatly reduced when they are blended with other fats, particularly if a liquid oil is included in the blend.

In an attempt to find stabilizing additives, a whole series of products was tested (*621*). Citric acid and ascorbyl palmitate, at 20 and 50 ppm, respectively, significantly retarded the green discoloration. The influence of various antioxidants on the color retention values of peanut oil containing 10 mg/kg β-carotene and stored at 65°C was studied. Rapid oxidation occurred when the samples were exposed to air, but a slight stabilizing effect was obtained by using an ascorbyl palmitate–α-tocopherol mixture. Samples protected from air under nitrogen were stabilized effectively with a series of antioxidants (*510*), and mixtures with ascorbyl palmitate gave a significant synergistic improvement (Fig. 32), particularly ascorbyl palmitate plus gallate.

Heating peanut oil containing 300 ppm β-carotene at 100° and 135° for 10, 30 and 60 min leads to increasing degradation (*621*) of the β-carotene, as shown by the absorption curves (Figs. 33 and 34). The losses occurring during the first 10 min of heating at 100°C are caused only to a minor degree by oxidative degradation; they represent mainly the heat-induced isomerization to *cis*-β-carotenes, which have a lower extinction. However, at 135°C rapid degradation occurs within a relatively

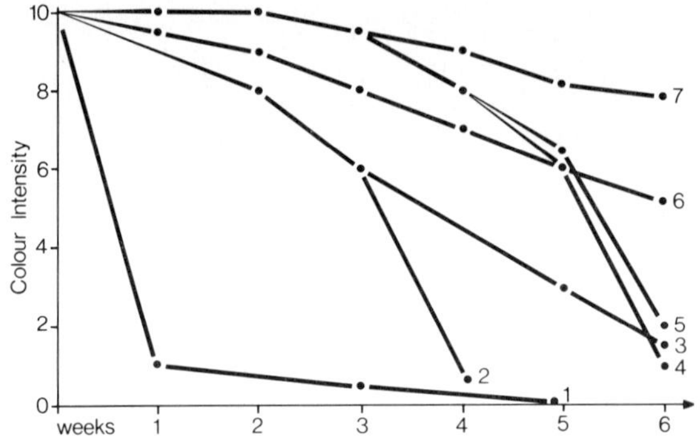

Fig. 32. Color stability of β-carotene in peanut oil at 65°C in the absence of air. Key: 1, control without antioxidant, NDGA, or α-tocopherol; 2, ascorbyl palmitate; 3, BHA; 4, α-tocopherol + ascorbyl palmitate, NDGA + ascorbyl palmitate; 5, BHA + ascorbyl palmitate; 6, propyl gallate; 7, propyl gallate + ascorbyl palmitate. [From Kläui (*510*).]

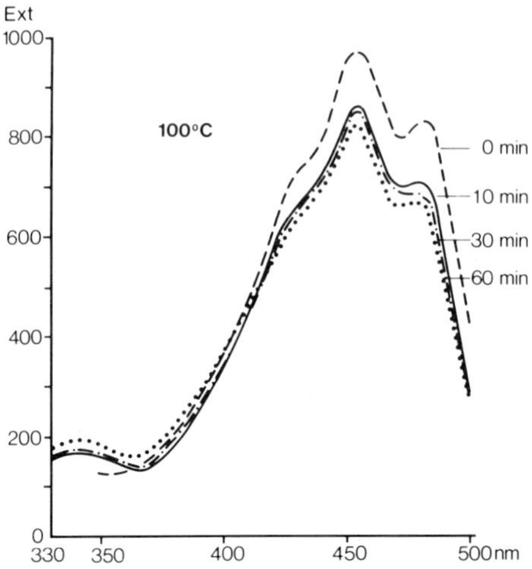

Fig. 33. Absorption curves of β-carotene heated in peanut oil at 100°C. [From Manz (*621*).]

short time. At 135°C β-carotene in coconut fat is much more stable than in peanut oil; after 60 min practically no decomposition, apart from isomerization, is visible, whereas practically complete destruction occurs in peanut oil (Fig. 35).

The decolorization of carotene occurring in fats in solution under oxidizing conditions, which is strongly accelerated by light and particularly heat, was studied by Venkataro and Achaya (*988*). The authors offered an explanation of the observed results based on the following hypothesis. (a) Carotene is decolorized by hydrogen

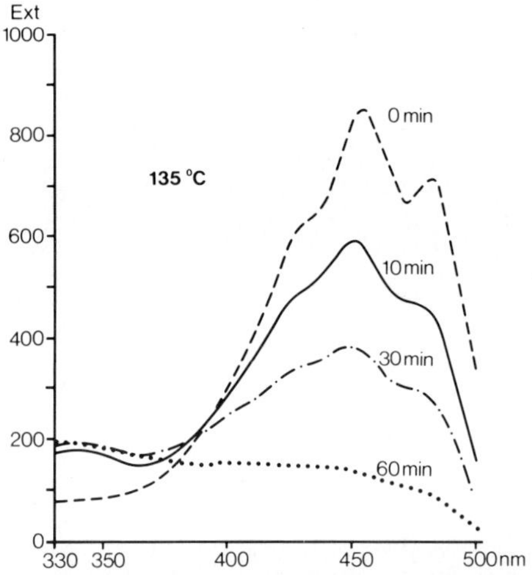

Fig. 34. Absorption curves of β-carotene heated in peanut oil at 135°C. [From Manz (*621*).]

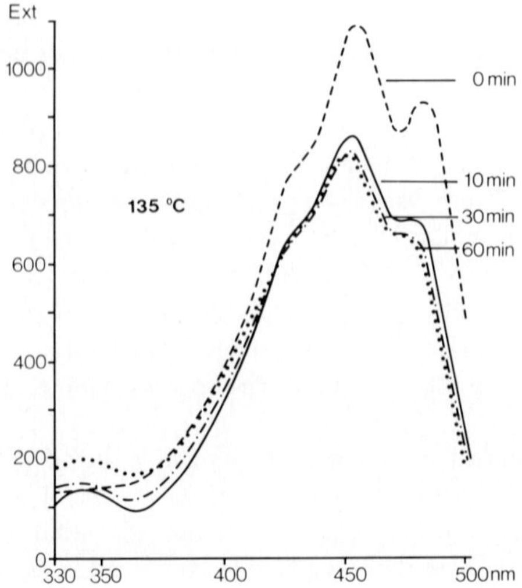

Fig. 35. Absorption curves of β-carotene heated in coconut fat at 135°C. [From Manz (*621*).]

abstraction, which upsets the entire conjugated chromophoric system, and (b) the degree of internal stability of the substrate-derived peroxyfree radicals, which are the hydrogen-abstracting agents, is inversely related to the life of carotene in that substrate.

C. Butter

The natural color of butter varies with the season of the year and dairy cattle feeding practices. In midwinter under typical winter conditions of dairy cattle feeding, butter is almost white (if no artifical butter color is added), very yellow during the weeks of fresh, luxuriant pastures, and, between these extremes, an intermediate color. Since this extreme change in color is decidedly objectionable to the consumer, the practice was started years ago, about 1886, in the United States, of using added artificial color to produce a uniform shade of color throughout the year. Unlike American, Canadian, and European butter, which is low in color in the winter, the butter of New Zealand and Australia is the lowest in carotene content during the summer season (February), even though the cattle are grazing on good pastures; hence, a seasonal effect causing some decreased availability of carotene from the pasture is believed to be operative (*647*) in addition to grazing practices (*313*). Not only does the natural color of butter vary over the year, but these changes are correlated with variable vitamin A concentration. Part of this variable concentration is due to variation in the concentration of true vitamin A in the product, but a very significant part is due to variation in natural carotene content (*313,647,801,815,1004,1021*).

Whereas the margarine industry quickly introduced the color that also nourishes, namely, β-carotene, after it became available, this has not been true in the butter industry. This older industry had a history of coloring winter butter with coal-tar dyes or annatto extracts, (*286,606*) and so today some of the latter coloring substances are still in wide use. Butter-coloring problems have been discussed by Yver and Luguet (*1035*). There is, however, a growing appreciation for the use of β-carotene as a butter-coloring agent, older studies having been done first with carotene extracts (*606,607,804,805,825,897,912*) and later with synthesized β-carotene (*36,110,606,607,737,785,807,809,846,905,1012,1013*). Winter butter can be successfully colored with β-carotene both by the continuous and batch processes. In the United States, color added to winter butter is not declared on the label.

In the continuous butter process, after the cream is heated (50°C) and separated as butterfat, β-carotene can be easily added in the form of a 20–30% β-carotene suspension (Table 63), at which stage the micronized crystals of β-carotene quickly dissolve in the warm butterfat, and processing continues as usual (*62,1013*). Most of the butter in the world is still made by the batch process of placing cold cream in wooden or metal churns of various capacities and churning it into butter. In the batch process method, β-carotene dissolved in oil (or as a β-carotene gel) can be added (Table 63) to the churn (*62,79*). However, a preferred method is the addition of β-carotene beadlets (*606,607,1013*) previously dissolved in a small amount of warm water (Table 10). Butter, because of its fluctuating natural β-carotene content (*62,647,800,801,805,1021*) over the year, as influenced by the diet of the cow, may range in vitamin A activity from as low as 12,000 to over 40,000 IU/kg. Adding β-carotene (4–6 mg/kg) as a coloring agent to winter butter could provide approximately 6000–10,000 IU of vitamin A activity per kilogram, which would be a significant improvement in the nutritive value of butter produced during the winter or dry feeding period of the dairy cow (*62,807*). During those months in which the carotene added for color uniformity did not supply enough vitamin A to attain the goal of 30,000 IU per kilogram (15,000 IU/lb) of butter, use could be made of a β-carotene–vitamin A blend to attain the objective. Uniformity of color and vitamin A activity in butter is desirable throughout the year. Butter colored with chemically produced β-carotene is highly acceptable in hue, color stability, flavor, and keeping qualities and also with regard to chemical evaluation, acid degree, and peroxide and aldehyde values (*606,607,807,809*).

Richardson and Konar (*804*) summarized the earlier history of carotene coloring trials with butter and observed that, when high-quality carotene concentrates were used, no deleterious effects were observed, nutritive value increased, and coloring costs were not prohibitive. Early reports include the studies of Richter (*806*). Riel and Johns (*809*) reported successful batch-type butter coloring trials with crystalline β-carotene manufactured by the Roche process. Twelve churnings of butter were made using both first-grade and second-grade cream, wherein β-carotene was added to cold, 32% fat cream (5°C) in the churn previously pasteurized (77°C, 10 min).

TABLE 63

β-Carotene Stability in Colored Butter[a]

Type of churn	Form of carotene	Color appearance	Carotene added	Assay values after storage (mg carotene/kg butter)				
				Initial	2 months, 5°C	6 months, 5°C	2 months, 5° and 23°C[b]	2 months, 23°C
Batch type								
New Jersey, January	β-Carotene gel added to the cream (5°C) before churning	Uniform	0	7.7	8.0	—	7.8	1.9
			6.6	13.5	14.0	—	14.0	14.0
Iowa, February	β-Carotene gel added to the cream before churning	Uniform	2.2	7.7	7.3	—	7.7	6.8
Pennsylvania, January	β-carotene gel added to the cream before churning	Uniform	0	6.9	7.4	—	7.4	7.3
			6.6	13.2	14.0	—	13.8	14.0
Continuous type Canada, November	β-Carotene suspension added to the warm butter fat	Uniform	2.2	7.9	—	7.9	—	—

[a] From Bauernfeind *et al.*(79).
[b] Alternately subjected to 5° and 23°C for 24- to 60-hr periods.

Prints of the salted (2%) butter, wrapped in parchment paper and boxed, were stored and observed (*809*) for color, flavor, and Vitamin A potency (Table 64). The authors concluded that butter colored with chemically produced β-carotene was acceptable in color and flavor as well as by chemical evaluation, acid degree, and peroxide and aldehyde values. Later trials of coloring butter with β-carotene were described by Bélanger and Giroux (*83*). Aust (*36*) also reported favorable color and flavor findings for β-carotene-colored butter in a study in which the colored butter was examined by the Austrian Dairy Commission. Furthermore, reddish and greenish tints usually associated with other types of butter colors were not observed in the β-carotene-colored butter. Favorable German and Dutch butter trials with β-carotene have also been described (*785,845*).

In cooperative studies between the Roche Laboratories and other institutions, β-carotene-colored butter samples were assayed to determine the uniformity of color and stability of the added β-carotene (Table 63). Yver and Luguet (*1035*) showed that β-carotene, mixed with the cream at the beginning of fermentation, yielded a homogeneous color in the final butter made by the modern continuous process. Thus, the butter color in winter can be increased to equal that normally produced in the spring and early summer. Parman (*737*) cautioned that all present forms of β-carotene may not be suitable for coloring butter. For example, in a process in which fast dissolution of β-carotene in the cream is critical, a diluted oily solution would seem to be the answer. However, a universal method of application has not yet been chosen by the industry, since commercial butter production is in a changing period of production methodology.

Butter makers have always been interested in a quick, simple method of estimating the natural color of the fat in cream for calculating the amount of color to add to each batch. Richardson and Konar (*804*) suggested a final butter color equivalent to 500 mg of carotene per 100 lb of fat (10–12 μg per gram of fat). Reinart and Brown

TABLE 64

Flavor and Color Observations on β-Carotene-Colored Butter[a]

	Color as β-carotene			
Storage	(μg/gm fat)	mg/kg butter	Flavor[b] (official grade)	Vitamin A[c] (IU/kg)
Initial	10.6	8.47	36.5 (37.3)	27,940
2 weeks, 10°C	10.7	8.58	36.7 (30.7)	28,820
6 months, −20°C	11.2	8.95	32.5 (31.2)	29,700
6 months, −20°C, plus 2 weeks, 10°C	10.7	8.58	22.8 (22.7)	28,160

[a] From Riel and Johns (*809*).
[b] Figures in parentheses are for a noncarotene-colored butter as a control.
[c] Includes vitamin A from carotene and true vitamin A.

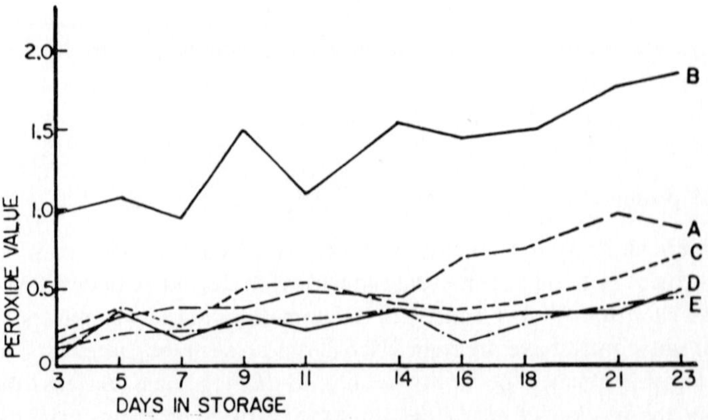

Fig. 36. Peroxide values determined for butter colored with different annatto and β-carotene solutions. Key: A, annatto A; B, annatto B; C, β-carotene standard solution; D, β-carotene water solution; E, blank. [From Lück *et al.* (*607*).]

(*800*) prefer that butter be colored to a total carotene concentration of 10 μg of carotene per gram of fat. Since currently there is a lack of uniform coloring methods, the use of β-carotene as the butter color and the adoption of solutions of β-carotene as butter standards would bring this standardization to the butter industry. Because the color that β-carotene imparts to butter is identical to the natural color of butter and added β-carotene restores the vitamin A activity lost to winter butter due to seasonal effects or feeding practices, β-carotene is a preferred butter color. Cowgill (*207*) believes that a more uniform vitamin A butter is a worthy objective of the butter industry.

In an African study, Lück *et al.* (*607*) compared the use of annatto and β-carotene for coloring butter. They found that the different annatto extracts sold to the South African butter industry varied considerably in quality and price per color unit (price per color unit, based on $E^{1\%}_{1cm}$ at 468 nm, was lowest for β-carotene, 10% water dispersible). Butter colored with annatto showed a higher degree of oxidative deterioration (*607*) during storage at 20°C than butter colored with oil suspensions of β-carotene (Fig. 36). Butter colored with 10% water-dispersible β-carotene did not differ in oxidation development from uncolored butter manufactured from the same cream. The authors concluded that, on the basis of price and practical advantages, β-carotene is an ideal coloring matter for dairy products. Ten percent water-dispersible β-carotene was preferred.

The stability of added β-carotene in butter has been shown to be very good (*79*). The method of measuring butter color by means of a reflectance spectrophotometer was described in detail by Lück and Novello (*606*). In France, color is an important factor of quality, according to Casalis *et al.* (*167*), and the same seasonal variation has been found there as in other countries. Studies were undertaken to add β-carotene to cream to be churned into butter to achieve a standardized color.

Consumer preference was shown for β-carotene-colored butters. Stavrova *et al.* (*886*) also successfully investigated β-carotene as a coloring agent for winter butter. β-Carotene of sufficient purity from a fermentation source was reported to be suitable for coloring butter by Protsenko *et al.* (*773*).

D. Butter Sauce

The American Dairy Association sponsored research at the University of Wisconsin, Department of Food Science and Industries, to develop a better flavored line of vegetables canned in butter sauce. In vacuum-can-packed vegetables with butter sauce, the sauce must have the right viscosity to prevent the butter fat from separating and to give the sauce the ability to cling to the vegetable. Successful work was carried out by Weckel *et al.* (*1008*) to solve various problems, among them the addition of fat-soluble yellow colors that would be completely dispersible, and would provide a sauce of uniform color. The authors found that β-carotene water-dispersible beadlets satisfied these requirements. The color must be adjusted to the type of product, and specific formulas (Table 65) of suitable amounts of β-carotene and methods have been developed (*1008*) for processing. The amount added depends on the vegetable and the butter sauce formula.

β-Carotene-colored butter sauce also finds application in frozen specialty vegetable packs. Butter as normally prepared does not contain sufficient depth of color to capture consumer appeal. As a result it is customary to add color to the butter sauce before freezing. A level of 2–4 mg of β-carotene per pound of butter is recommended. Some frozen food processors use water-dispersible β-carotene beadlets for this purpose. In these instances there is a preference to add the color after the butter that they purchase has passed internal bacterial control. This enables them to reject any butter failing to meet their tests, something they could not do if the butter were deeply colored to specifications at the creamery. The beadlets are more convenient because of their lower concentration and ease of handling. The beadlets can be dispersed in a small quantity of warm water used in preparing the sauces. Other processors have been equally successful in coloring sauces with fat-soluble β-carotene suspension added to the heated butter. β-Carotene can be mixed into the warm, molten butter during the preparation of the sauce.

The reasons for using β-carotene in food sauces are (a) bright, attractive, and natural-appearing color, (b) uniformity and purity of color and no variation from batch to batch, (c) convenience, and (d) safe and nutritious color that can be used in butter sauce products made with the coloring nature provided in butter.

E. Eggs

Coloring of egg yolks is normally achieved by the indirect process of feeding chickens rations rich in pigmenting carotenoids. However, in instances in which insufficient coloring is obtained by the indirect process, particularly with industri-

TABLE 65

Suggested Formulas and Process Conditions for Full-Pack Canning of Vegetables in Butter Sauce[a]

	Formulas for butter sauce[b]							Process conditions[c]		
Vegetable variety	Water (%)	Salt (%)	Sugar (%)	Starch (%)	Vis-carin (%)	Butter (%)	β-Carotene (gm/cwt)	Fill-in vegetable to sauce (gm)	Calculated minimum process (min at 255°F)	Recommended process (min at 255°F)
For 307 × 306 (12-oz) size cans										
Peas	76.8	3.0	6.0	1.25	0.4	12.5	3.5	280/120	13.5	16.9
Green snap beans	80.0	2.6	5.2	0.80	0.6	10.8	3.5	240/140	9.5	11.9
Whole kernel corn	76.8	3.0	6.0	1.25	0.4	12.5	2.6	280/120	19.0	23.8
Sliced carrots	75.4	3.0	6.0	1.37	0.4	13.8	3.5	280/110	8.5	10.6
Lima beans	81.5	2.4	4.8	0.93	0.4	10.0	1.8	250/150	11.5	14.4
For 303 × 406 (16-oz) size cans										
Peas	76.8	3.0	6.0	1.25	0.4	12.5	3.5	325/140	14.5	18.1
Green snap beans	79.6	2.6	5.2	0.80	0.6	11.2	3.5	279/163	10.2	12.8
Whole kernel corn	76.8	3.0	6.0	1.25	0.4	12.5	2.6	325/140	19.0	23.8
Sliced carrots	76.1	3.0	6.0	1.37	0.4	13.1	3.5	325/130	10.2	12.8
Lima beans	81.5	2.4	4.8	0.93	0.4	10.0	1.8	297/178	16.8	21.0

[a] From Weckel *et al.* (1008).
[b] Conform to amended standards of identity for canned vegetables in butter sauce.
[c] These values not confirmed by inoculated pack studies.

ally used yolk, the desired color may be achieved by the direct addition of β-carotene or other natural-type carotenoids in the batch or continuous mixing or churning processing of egg yolk before freezing, drying, or further treatment. Direct addition of carotenoid to eggs has the advantage of economy and flexibility. Several market forms of β-carotene can be used.

As a result of the confirmatory research by Ganguly *et al.* (*327*), β-carotene is now recognized as a coloring component of egg yolk, the major components being the hydroxycarotenoids typified by lutein and zeaxanthin. Every country dweller obtaining fresh eggs from a small poultry flock knows that the yolk color varies, the exact shade depending on the feeds consumed and other variables. Modern indoor management of large flocks and the extensive use of uniform commercial feeds have brought greater constancy of yolk color. The trend has been toward the lighter or yellow yolk desired by the fresh egg market and a gradual decrease in the volume of dark or orange yolks required by food processors for the manufacture of properly colored noodles, mayonnaise, sponge cake, sweet goods, speciality breads, cookies, pie crusts, and other products (*306*). Dark egg yolk and yolk products used by the food industry are usually in the form of frozen egg yolk (plain, salted, or sugared), frozen whole egg, frozen whole egg and yolk blends, dried egg yolk solids, dried whole egg solids, and dried blends of yolk and whole egg. They are bought and sold at a premium price according to basis U.S. color specification (Table 66). The color (*481*) of these yolk products has been expressed in terms of carotene concentration.

In the United States some of the shell eggs (about 10%) produced from laying hens are processed into frozen yolk, frozen whole egg, or dried yolk, bulk packed, for use in mayonnaise and noodle manufacture and for bakery use. These yolks are intended to be dark orange-yellow (*761*) (NEPA scale no. 5 or 120 ppm β-carotene standard), in contrast to yolks of shell eggs for table use (NEPA scale no. 2 or 40

TABLE 66

Natural Yolk Color Expressed as Carotene Units[a]

NEPA yolk color	Color equivalents in terms of carotene (ppm)
1 (very light yellow)	15
2	40
3	70
4	90
5	120
6 (orange)	150

[a] From Kahlenberg (*481*).

[b] Technical Committee of the National Egg Products Association.

ppm β-carotene standard). In most instances the hens producing these highly colored eggs are fed a ration of high carotenoid content (*306*), as previously arranged by contract with the poultry feeder to meet the specification of the eggs produced. In terms of present-day technology, this is an inefficient and relatively costly method of adding carotenoids to bulk frozen and dried yolks.

A technology, assay techniques, and carotenoid application forms exist today for the direct addition of β-carotene to bulk yolk products, which would produce the dark yolk color more economically and in quantities as needed so there would be no oversupply or deficiency of the product (*862*). It would be analogous to the permitted coloring of butter and cheese in the United States or the addition of vitamin D to fluid milk to standardize the vitamin potency in the product. Since β-carotene is already naturally present in egg yolk and since solutions of it are regarded as possessing color comparable to extracts of natural egg yolk, the approach of commercially adding β-carotene to weighed batches of yolk in the churning or mixing process before freezing or drying to produce the desired natural color grade has the advantages of flexibility, economy, and absence of foreign coloring.

Various market forms of β-carotene have been added to frozen egg yolk products to provide additional color. The necessity of maintaining low temperatures, 5–13°C (40°55°F), in the yolk products during mixing and before freezing makes complete solution of the β-carotene suspensions difficult. However, if the β-carotene suspension is first uniformly dispersed in a small portion of yolk warmed quickly to room temperature, this warm premix of dissolved carotenoids can be incorporated into the cold batch of yolk product immediately. It is recommended that this final blended product be pasteurized by high-capacity, multiple-plate heat exchangers at approximately 60°C (140°F), followed by cooling, packaging, and freezing. The pasteurization step ensures complete solution of the added β-carotene suspension. Frozen egg yolks have been prepared on a pilot-plant scale (Table 67) with a water-dispersible form of β-carotene, such as β-carotene beadlets, which is easier to apply (*158*). The beadlets are added to a small volume of lukewarm water and stirred or shaken. The resulting dispersion is added to the cold egg yolk in the mixing churn. A chart can be developed, once the natural color level is established, to determine

TABLE 67

Stability of β-Carotene in Frozen Egg Yolks[a]

Storage period at −20°C	Total carotenoid content (mg/100 gm yolk)
Initial	16.9
3 months	16.5
12 months	15.0

[a] From Bunnell *et al.* (*158*).

TABLE 68

Stability of β-Carotene in Dried Egg Yolks[a,b]

	Storage temperature			
Storage period	7°C	23°C	30°C	37°C
Initial	35.2	—	—	—
3 months	35.0	33.0	34.0	33.0
4 months	—	—	—	33.0
12 months	29.2	28.2	31.0	—

[a] From Bunnell *et al.* (*158*).
[b] Data expressed as milligrams carotenoids per 100 gm yolk.

how much β-carotene is to be added to each batch to bring it to a desired color (NEPA no. 5) level. The carotene-fortified egg yolks are then frozen.

β-Carotene suspension can also be uniformly mixed into liquid egg yolk to be dried; the heat before and during the spray-drying solubilizes the carotene, and a uniformly colored dried product is obtained. Pilot-plant trials (*79*) with spray-dried yolks containing added β-carotene have been described. Various trials indicated that β-carotene losses were not great in either frozen egg yolk or dried egg yolk when β-carotene was added during processing using β-carotene suspensions. Excellent performance with a water-dispersible form of β-carotene for coloring dried egg yolk products (*158*) has been reported (Table 68). With the water-dispersible beadlet, warming of the egg yolk before mixing is not needed to dissolve the added β-carotene.

Dried yolks fortified with β-carotene are relatively stable at various storage temperatures for up to 3 months and show less than 20% loss at various storage temperatures for up to 1 year. Frozen yolks fortified with β-carotene are stable when stored for up to 3 months and show approximately a 12% loss for over a year's storage. Frozen egg yolks were used to prepare mayonnaise and found to give a uniform, rich, natural-appearing color. Both the frozen and dried yolks were also satisfactory in the preparation of noodles. Most market forms of β-carotene added to yolk products performed equally well, and the added β-carotene, with storage time, showed a higher retention (lower losses) than the natural carotenoids in the nonfortified product.

F. Baked Goods

Yellow or golden color can be introduced into baked goods, such as cakes, sweet dough products, pastry, special breads, doughnuts, cookies, crackers, fillings, and frostings, through the use of colored shortenings, margarine, butter, or eggs. It can also be provided by the direct addition of yellow carotenoids in a proper application form (*661*) during the dough-making procedure, assuming that there are no legal restrictions on direct color addition.

For direct addition, two commercial β-carotene products, one oil soluble, the other water dispersible, are especially suitable for use in bakery operations. The levels of β-carotene required to color bakery products vary somewhat depending on the specific product being colored and the color intensity desired by the manufacturer. Some small losses may occur during the mixing and/or baking operations. Prolonged mechanical aeration of nonyeast dough product batters can cause a decrease in color. Color loss depends on the severity of aeration. In yeast-raised doughs, there is little destruction of β-carotene during mixing and proofing. The typical loss of β-carotene in the baking of yeast-raised products is approximately 10%. Yeast dough was retained in a frozen state for 3 months or more before thawing, and baking revealed no loss of yellow color during this storage period. Dry mix blends (doughnut mix, cake mixes, etc.) have excellent color stability. In cookie-type products, color stability is good to excellent. Replacement of the presently used yellow colors with carotene will not alter the flavor, texture, volume, or other organoleptic characteristics of baked goods. Some baking trials with indirect and direct addition of β-carotene are discussed in subsequent paragraphs.

Since β-carotene-colored fats behave similarly during the dough-making and baking processes, one may take, for example, trials with β-carotene-colored shortening as indicative of the behavior of carotene-colored fats. In cooking studies conducted in the Roche Product Development test kitchen, shortening was colored at three levels with micronized β-carotene as an oily suspension. Baked food products (*79*) containing these colored shortenings were evenly colored, and the deepened coloration was striking in comparison with control products containing uncolored or white shortening (Table 69). The products tested included cookies, cakes, and pie crust.

Cooking trials were also run on egg yolk with added β-carotene suspension. Frozen yolk was thawed and incorporated in a baking trial of cookies (vanilla drop). The frozen β-carotene-colored egg yolk was used in the dough formula in place of regular egg yolk. Data were obtained showing excellent retention of carotene during baking; 94% of the carotene in the unbaked dough (by assay) was recovered in the baked cookies after correcting for moisture losses. Frozen as well as dried β-carotene-colored egg yolk was also used to prepare home-style egg noodles from wheat flour, water, salt, and salad oil. When the frozen yolk product was used, a carotene recovery of 94% was obtained in the finished noodle product; when the dried yolk product was used, a recovery of 96% was obtained. A third food product, homemade mayonnaise, was prepared from β-carotene-colored frozen egg yolk, salad oil, white vinegar, and salt. Although the final product was not assayed, a uniformly colored, creamy-yellow mayonnaise resulted.

Direct addition of yellow color during baking is not a novel practice. Historically, many bakery products in the United States have been colored yellow with FD&C Egg Shade, a combination of dyes; FD&C Yellow No. 5; and/or FD&C Yellow No.

TABLE 69

Stability of β-Carotene in Colored Shortening in Test Kitchen Trials[a]

Food product	Color	Assay values (mg carotene/kg) Natural carotene	Added β-carotene	Unbaked	Baked	Carotene retained
Vanilla drop cookies						
Control	Creamy white	0.55	0	0.64	0.73	100+
Colored A	Pale yellow-orange	0.55	2.07	2.84	2.71	95
Colored B	Pale yellow-orange	0.55	2.11	5.48	5.19	95
Colored C	Medium yellow-orange	0.44	6.38	—	6.05	—
Two-egg cake[b]						
Control	Pale yellow	1.21	0	1.12	0.90	80
Colored A	Pale yellow-orange	1.21	1.12	2.13	1.76	82
Colored B	Pale yellow-orange	1.21	1.96	3.32	2.68	81
Colored C	Bright yellow-orange	0.77	2.84	3.45	3.12	90
Pie crust						
Control	—	0	0	0	—	—
Colored A	—	0	2.90	3.01	2.66	88
Colored B	—	0	6.29	6.62	5.63	85
Colored C	—	0	8.56	7.85	6.80	87

[a] Adapted from Bauernfeind *et al.* (*79*).

[b] A pound-cake series was also run with longer baking time, which resulted in a 62–76% carotene retention.

6 (*305*). Water-dispersible forms of the carotenoids, particularly β-carotene, have been used both experimentally and in commercial production to impart yellow color to bakery products. When β-carotene beadlets are used, the color of the baked product is maintained during market shelf life without fading (*305*). The use of β-carotene eliminates concern (*197*), real or imagined, for allergic reactions (*98, 1047*) or behavioral problems that have been alleged to occur with FD&C Yellow No. 5.

The dry, free-flowing beadlets are easy to handle in the bakery. Usually a stock color solution of beadlets in warm water is prepared for use over a period of a few hours for batch-to-batch bakery operation. A measured quantity of stock color solution as part of the water is added to the formula, or it may be incorporated into doughs at the slurry mixer containing the dry ingredients. In sweet dough production, the stock solution is added to the proofed sponge in the second mix. Formula water levels should be adjusted to compensate for the water in the coloring solution. For example, where light creamy to yellow bread can be legally produced by direct color addition, one can disperse 35–165 gm ($1\frac{1}{4}$–6 oz) of 10% β-carotene beadlets in 2 qt of warm water, 55°–66°C (130°–150°F), to make a stock solution. In dough preparation, 120 ml (4 fl oz) of the stock solution per 45 kg (100 lb) of flour are used as part of the water. Stock solution not used can be refrigerated for overnight use.

Bakers making only sweet goods may prefer to use an oil suspension of β-carotene that is dissolved directly in the formula shortening. Because of the small amounts of β-carotene suspension needed, it is usually convenient to prepare a "premix" or stock solution of the β-carotene in fat or oil, which can be added in more easily measured quantities. For example, dissolve 30–140 gm (1–5 oz) of 30% β-carotene suspension in 10 lb of shortening or vegetable oil by heating shortening to 53°–66°C (130°–150°F). Stir without beating in air. This stock solution should be stored in a wide-mouth container for easy removal. Use 120 ml (4 fl oz) of stock solution per 45 kg (100 lb) flour. Stock solution must be added to the dough as part of the shortening in dough.

Although the carotenoids are limited to the yellow–orange–red end of the color spectrum, they may be combined with other approved colorants to obtain desired colors. To establish formulas that include these carotenoids, suggested starting levels (based on pure compound) are the following: β-carotene in baked goods, 4.4–22 mg/kg; β-carotene in fillings and frostings, 2.2–13.2 mg/kg; apocarotenal in baked goods, 8.8–22 mg/kg; canthaxanthin in baked goods, 4.4–13.2 mg/kg. Whereas β-carotene provides a yellow color in baked foods, β-apocarotenal simulates a carrot color or orange color, and canthaxanthin gives a peach color to the crumb of the cake. These carotenoid coloring agents are not affected by reducing substances such as ascorbic acid and are relatively stable in the pH ranges normally found in foods. Freezing of products colored with carotenoids results in little or no color damage.

A few trial runs can be cited involving test kitchen studies with a dry white cake mix. β-Carotene beadlets were added and blended into commercial white cake mix of the self-rising type, and, although the dry mix itself was not colored, the batter and finished cake were colored to an even golden yellow. As the first portion of warm water was added and blended with the dry cake mix, the color appeared quickly and evenly throughout. One set of cakes was baked at the normally recommended oven temperature of 177°C (350°F) and another set was baked at 205°C (400°F) to see whether the higher temperature would be more destructive to β-carotene, but little or no difference was noted (Table 70).

TABLE 70

Stability of 10% β-Carotene Preparation in Cake[a]

Baking temp	Addition rate (mg/kg cake)	Retention (%)
350°F (177°C)	4.4	94
	6.6	92
	22	94
400°F (205°C)	4.4	95
	6.6	90
	22	93

[a] From B. Borenstein and F. Jahns (unpublished).

TABLE 71

Stability of 10% Canthaxanthin Preparation in Cake[a]

	(Canthaxanthin) (mg/kg)	Retention (%)
Amount added	38.9	—
Baked	37.2	96
Baked, 4 days room temp	30.6	82

[a] From Bunnell and Borenstein (*157*).

The stability of the dry β-carotene beadlet in cake mix was investigated by storing samples of mix under various conditions. There was no loss of carotene after exposure in open containers for 42 days at 113°F or 3 months at room temperature (23°C). The stability of the dry β-carotene beadlet in the dry cake mix is excellent. Similarly successful trials have been made with water-dispersible forms of β-apocarotenal and canthaxanthin in dry cake mixes (*157*) and during the baking process (Table 71). Carotene-colored breads have a particularly pleasing color when they are sliced or toasted. Because of lipoxygenase action during bread making, some destruction of β-carotene occurs (*266*). This type of destruction of β-carotene can be demonstrated using soybean lipoxygenase (*1007*). Drobot *et al.* (*267*) described the use of a carotene concentrate of fermentation origin as an additive to bread.

Flour, sugar, or starch colored with oily forms of β-carotene or other carotenoids can also be used as a premix that blends easily with bakery products. However, such absorbates may have only a limited color stability and sometimes show a tendency to decrease the dispersibility of the carotenoid. Thus, the even and stable coloring of dry mixes must still be considered an unsolved problem. A suitable product would have to be highly diluted in order to color a dry mix evenly; it would also have to be stable and quickly soluble in cold water.

Bread crumbs used, for example, in coating food products such as fish fingers can be colored with carotenoids. Canthaxanthin in its water-soluble form can be used alone or in combination with other coloring material. Bauernfeind *et al.* (*79*) also mentioned the possibility of blending an oil solution of β-carotene with the dry bread crumbs or adding a water-dispersible form to the batter before use. The color persists during precooking, freezing, storage, and subsequent reheating. Snack foods of the baked or heat-treated type can be colored with carotenoids (*872*).

G. Macaroni, Pasta Products

Macaroni products prepared from durum wheat have a light yellow color, whereas the corresponding products from soft wheat are practically colorless. Because the available quantity of durum wheat is not sufficient to cover all of the annual requirements, it is often necessary to use soft wheat for the production of macaroni products. In Italy, the flour of soft wheat is usually colored with β-carotene to get the same color as that obtained with durum wheat. This procedure is tolerated because of current needs. Some countries, recognizing the naturalness of β-carotene, permit carotene coloring of semolina when durum wheat is in short supply. Several β-carotene market forms can be used. An oily solution of carotene containing about 0.4–0.5% β-carotene is thoroughly mixed with a certain amount of flour. The resulting highly colored premix is afterward added to the flour used for the preparation of macaroni products. A flow sheet of macaroni coloring with an oil solution of β-carotene is given in Fig. 37. Also, it is technically feasible to add β-carotene in the form of water-dispersible emulsions and colloidal beadlet products in warm water to the dough during the mixing and blending operations (79).

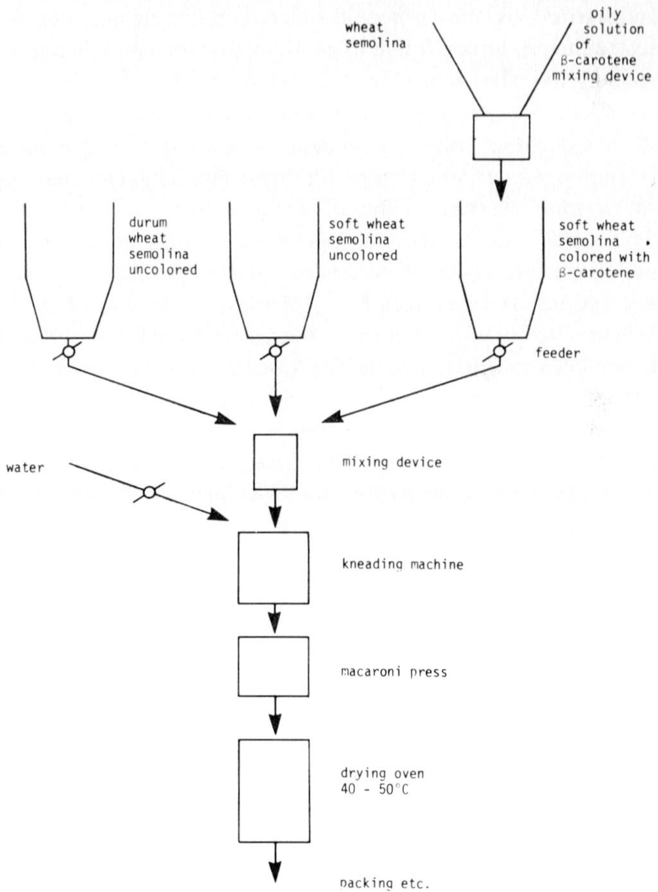

Fig. 37. Flow sheet of macaroni coloring with β-carotene. The coloring of soft wheat semolina with β-carotene is carried out batchwise; other steps of the production are a continuous process.

β-Carotene has been used successfully in producing well-colored macaroni products from pale-colored semolina. Added β-carotene shows good stability in these products. About 4 gm of β-carotene are used per ton (1000 kg) of final pasta product. Several commercial forms of β-carotene can be used.

H. Salad Oils, Dressings, Spreads

Vegetable oils (corn, cottonseed, olive, peanut, soybean, etc.) used in salads are easily colored, if desired, by the use of β-carotene oily suspension application forms. After the vegetable oil has been clarified and deodorized, the carotenoid suspension is added to the warm oil product, 44°–50°C (110°–120°F), with agitation, which is continued until complete solution is obtained. If the entire product is not heated at any phase in the usual operation, an aliquot or portion of it should be removed and warmed and the carotenoid added with stirring until dissolved, cooled, and returned to the production batch (*79*). If food-grade antioxidants are added for protection against oxidative rancidity of the fat, these likewise will provide an added stabilizing influence on the added carotenoid. Good storage stability of color has been noted in carotenoid-colored vegetable oil (Table 72). If exposed to direct sunlight for prolonged periods, a more light resistant container may be required.

The color of salad dressing of the oil–water mixture (unstabilized) type or the emulsified type (stabilized) containing vegetable, eggs or emulsifying agent, spices, flavors, condiments, etc., is usually a concern of the consumer. Mayonnaise, one of the earliest dressings, is recognized by its pleasing color, consistency, and flavor. β-Carotene added to the oil of an oil–water mixture makes an attractive product (10–25 mg per kilogram of product). The so-called French and Russian dressings have an orange or red hue, thus requiring addition of carotenoid other than β-carotene. The color of "French" dressing is attributable in part to the condiments, paprika, vinegar, etc. Both β-apocarotenal (*69*) and canthaxanthin (*157*) qualify for this usage. Apocarotenal seems to be more light stable (Table 73) and

TABLE 72

Laboratory Storage Data for Salad Dressing with Added β-Carotene in Closed Clear-Glass Bottles[a]

	Calculated carotene		3 weeks, 5°C		3 weeks, 23°C		3 weeks, 28°C, sunlight		3 weeks, 45°C	
Sample no.	In oil	In dressing	In oil	In dressing	In oil	In dressing	In oil	In dressing	In oil	In dressing
1	15.0	5.0	16.4	6.5	15.1	6.0	13.2	5.2	13.2	5.2
2	7.3	3.0	8.4	3.3	7.4	3.0	6.9	2.8	6.9	2.8
3	9.0	5.6	9.8	6.1	9.0	5.7	9.0	5.7	8.0	5.0
4	11.2	5.6	12.8	6.4	11.2	5.6	12.3	6.1	10.9	5.5

[a] Data in milligrams of β-carotene per 100 gm.

TABLE 73

Bleaching of French Dressing in Direct Sunlight[a]

Bleaching time/(days)	Added β-Apo-8′-carotenal	Added oleoresin paprika
1	None	Moderate
2	Slight	Severe
3	Moderate	Complete

[a] From F. Jahns (unpublished).

can be used to replace paprika and/or oleoresin of paprika in a variety of dressing products. Apocarotenal has been used commercially in a variety of nonstandardized salad dressings, including low-calorie dressings, creamy-type (egg-base) dressings, sandwich spreads, tartar sauces, Thousand Island dressings, and "Russian" dressings. A wide range of color in the "French" dressing-"Russian" dressing collection can be produced in products by using 30 mg of apocarotenal (pure basis) per kilogram dressing.

Nonseparating dressings are readily colored with apocarotenal by using apocarotenal oil suspension. The required level of this product is dissolved in the oil phase of the dressing before the addition of the other ingredients. After the apocarotenal is dissolved in the oil, the product is processed exactly as usual. However, in the nonseparating dressings, there is an alternative; namely, the carotenoid water-dispersible beadlets, singly as in the case of apocarotenal or in combination, can be added to the aqueous or water phase. The beadlet products should be dispersed in warm water before addition to the main water phase of the formula. Homogenization of the product follows.

Similarly, apocarotenal oily suspension can be used to color the oil phase of separating-type dressings. An apocarotenal product more convenient to use in coloring both types of dressing is a carotenal solution in oil currently manufactured commercially. The advantages of using apocarotenal in coloring nonstandardized dressings and spreads include economy and uniformity of color distribution in the product.

Canthaxanthin at 15–30 mg/kg has a desirable color range for dressings in the "Russian"-type group. The water, vinegar, and alginate are mixed, and the corn oil is added gradually, with constant stirring. The calculated amount of canthaxanthin beadlet suspension is added and mixed thoroughly. Canthaxanthin has excellent light stability and good stability at room temperature (Table 74). Bunnell and Borenstein (*157*) found that canthaxanthin imparts a good color range to dressings and has better light stability than paprika.

Finberg (*297*) reported the results of a study carried out to determine the comparative color stability of ground paprika (standard quality), oleoresin of paprika (40,000 CV), and apocarotenal (20% suspension in oil). The three colorants were

TABLE 74

Stability of Canthaxanthin in Russian Dressing[a]

Storage conditions	Canthaxanthin (mg/kg)	Retention (%)
Initial	32.1	—
1 month at 23°C	32.8	100
3 months at 23°C	31.7	99
6 months at 23°C	30.0	93

[a] From Bunnell and Borenstein (*157*).

added to dressings under various commerical plant production conditions using "French"- and "Russian"-type dressing formulations in which no paprika, tomato products, or other color-bearing ingredients were used. The "French" dressing formulation contained 35% oil, and the "Russian" dressing formulation contained 63% oil. The initial color substitution equivalents were for ground paprika, 100 (parts by weight); oleoresin of paprika (40,000 CV), 6.25; and β-apo-8′-carotenal (20% suspension), 0.78. The conclusion drawn from the color stability comparisons made at seven participating companies was that apocarotenal provides an effective, economical, and uniform color for use in dressings. The color of the oleoresin of paprika and ground paprika was observed to change in two ways: (a) surface fading and/or (b) a deepening of the brownish tones. With the increasing use of plastic containers, the improved color uniformity and stability of apocarotenal was judged to be of major value for dressings packed in plastic.

I. Cheese

The performance of added β-carotene, β-apo-8′-carotenal, and canthaxanthin has been studied in both natural and processed cheese (*69,79,157*). Usually the carotenoids were added to primary or natural cheese in the form of an aqueous solution prepared from water-dispersible beadlets or water-dispersible emulsions in amounts of pure carotenoid equivalent to 3–12 mg/lb. For processed cheese, carotenoid suspensions in oil were used. Amounts of pure carotenoid ranging from 3 to 6 mg per pound or 0.5 kg of cheese were of interest.

In the manufacture of natural cheese, the major portion of the fat of the whole fluid milk from which it is prepared is entrapped in the coagulated protein; hence, the seasonal variation (*409*) in the fat-soluble color component of primary cheese, such as cheddar, is a direct reflection of the variation in the color of the fat of the original milk (*801,1004*) produced over the seasons. The coloring of cheese with water-dispersible carotenoid preparation has been discussed by Lück (*603,605*). Added carotenoids can be incorporated by dispersing carotenoid beadlet products in warm water and then mixing evenly throughout the milk before the addition of the starter and acidity adjustment. Rennet is added, the milk allowed to set, the curd cut and hardened, the whey separated, and the processing continued as usual. The precipitated curd absorbs the color if given adequate time, leaving none in the whey. In some cases, and depending on the process and the conditions used, absorption may be incomplete. Means of overcoming this difficulty must be adapted to the particular circumstances involved. Trials conducted with natural cheese (*69*) were run (Table 75) to indicate the excellent stability of color provided by the added carotenoids. Organoleptic evaluations were favorable in carotenoid-colored cheeses. In a trial in which β-carotene (10 mg/kg) as a water-dispersible form was added to cheddar cheese before curd formation, color stability (98% retention) was excellent during a 14-month storage period at 5°C (40°F). Added β-carotene gives cheese a rich yellow color between that of butter and egg yolk. A range of interesting colors can be created by using the other carotenoids singly and in combination for specialty cheeses. A blend of one or two parts of β-carotene to one part of apocarotenal yields a most acceptable color (*69*).

β-Carotene in water-dispersible form and an annatto cheese color were compared by Schwarz *et al.* (*848*). Three types of cheese were produced: Tilsit, Edam, and Romadur. After a storage time of 10 weeks, annatto was no longer detectable, whereas β-carotene, particularly when added in the form of water-dispersible beadlets, showed good stability.

Process or processed cheese prepared by comminuting primary or natural cheeses and heating them in a jacketed kettle with added citrate and salt can be colored with apocarotenal suspension in oil or apocarotenal oil solution to yield the orange-tan color most frequently associated with this product (*69,513,738*). The coloring preparation is added directly to the melted premix cheese batch or, if desired, can first be mixed with warm melted cheese or butter oil and the premix added to the kettle. The amount of the carotenoid depends slightly on the natural color already present. Variation in color can be obtained, if desired, by using, for example, β-carotene (*79*), an apocarotenal-β-carotene blend, or canthaxanthin (*157*). After the carotenoid is uniformly blended into the cheese mass, further processing continues as usual. The most satisfactory color in the processed cheese samples was achieved with about 8 mg of β-apo-8′-carotenal per kilogram. Quite good color stability (*69,79*) is obtained with added β-carotene (Table 75) or β-apocarotenal (Table 75) under usual packaging and storage conditions and even when exposed to incandescent or fluorescent light (*69*). Cheese spreads are also ideal foods for coloring with added apocarotenal. Canthaxanthin produces an attractive cheese color (*157*) that is different in hue from that of β-apo-8′-carotenal. It resembles the hue of cheese colored with a large amount of annatto or annatto–paprika mixtures. The preliminary data shown in Table 76 indicate fair stability.

Problems in coloring cheese wax coatings without color migration into the cheese proper have plagued cheese processors from time to time. A finely ground canthaxanthin water-dispersible beadlet evenly suspended in the melted wax appears to be one solution for a red color cheese wax coating.

TABLE 75

Stability of Carotenoids in Cheese[a]

Sample	Initial assay (mg/kg)	Retention (%) Sunlight at 23°C		Darkness at 23°C		Light at 5°C	Darkness at 5°C	
		48 hr	144 hr	48 hr	144 hr	1 month	1 month	1 month
Natural cheese								
β-Carotene beadlets, type 2.4-CWS	11.9	95	90	96	91	90	100	100
Apocarotenal beadlets, type 4.8-CWS	14.3	85	99	99	100	88	92	100
β-Carotene beadlets, type 2.4-CWS, and apocarotenal beadlets, type 4.8-CWS	22.4	77	87	92	92	96	94	100
Control[b]	0.88	50	100	75	100	100	100	100
Process cheese								
30% β-carotene suspension	11.9	92	93	93	100	93	100	100
30% apocarotenal suspension	13.4	92	94	90	92	92	100	94
30% apocarotenal suspension	6.4	97	97	100	100	100	100	100
Control[b]	6.4	100	97	100	100	93	100	100
30% apocarotenal suspension	8.6	69	95	100	100	100	100	95

[a] From Bauernfeind and Bunnell (*69*).
[b] Total color measured against a *trans*-β-carotene standard.

TABLE 76

Retention of Canthaxanthin in Process Cheese[a]

	Canthaxanthin (mg/kg)	Retention (%)	Canthaxanthin (mg/kg)	Retention (%)
Amount added	7.3	—	10.9	—
Initial	6.8	—	10.7	—
3 months refrigerator	5.4	79	7.4	69
9 months refrigerator	5.1	75	7.0	65

[a] From Bunnell and Borenstein (*157*).

J. Ice Cream

Ice cream is a popular and universally pleasing food. As a dairy product, ice cream contains appreciable butterfat, and it is the carotene and vitamin A contents that would be expected to vary in ice cream over the seasons, although no detailed study has been published on this point. Dahlberg and Loosli (*233*) acknowledged vitamin A variation, and on the basis of data collected from June–July cream they gave an average value of 548 IU vitamin A per 100 gm for ice cream. Other frozen products are mellorine, parvine, frozen custard, ice milk, fruit sherbet, and water ice. All can be easily colored with carotenoids by the use of carotenoid water-dispersible forms.

A given trial will be described. Vanilla ice cream was colored with β-carotene to produce a "French" vanilla type or golden vanilla ice cream. Dry, water-dispersible β-carotene (type yellow, 1 or, 2.4% preparations, equivalent to 0.15 gm pure carotene) was mixed with 120 ml (4 oz) of warm tap water and the beadlets dispersed by stirring. The amount of β-carotene used may be larger or smaller depending on the degree of color desired in the final product. The mixture was shaken until a uniform dispersion was obtained. The carotenoid water-dispersible preparation was then added to about 5 gal (20 liters) of the ice cream mix at a temperature of 5°C. Vanilla extract was added, and the mix placed in a 10-gal capacity Emory Thompson freezer operated for 7 min at a 90% overrun. The ice cream was filled into 66-pt containers and frozen at −20°C (−5°F). The carotene-enriched ice cream and the control without carotene were tasted initially and after both a 3- and a 6-week storage period. No difference in flavor could be detected after manufacture and after storage. The pleasing creamy color of the carotene-colored lot of ice cream was noted by members of the taste panel.

The stability (*158*) of β-carotene in ice cream is good (Table 77). β-Carotene retention during normal storage has been found to be about 97%.

For powdered ice cream mixes, water-dispersible beadlets or powder can be added directly to the dry mix, provided that the composition and its consistency ensure stable distribution and that the quantity of water or milk and the length of

TABLE 77

Stability of β-Carotene in Ice Cream/Ice Cream Mix[a]

	Carotene content (mg/kg)
Control mix	0.97
Carotene-fortified mix	7.7
Churned and frozen ice cream, initial	7.5
3 weeks storage at −20°C	7.4
6 weeks storage at −20°C	6.6

[a] From Bunnell *et al.* *(158)*.

time and temperature used in the preparation of the ice cream mix are appropriate for the carotene product to disperse. Carotenoids in a cold-water-soluble application form are particularly well suited for coloring ice cream mixes. Surfactants, hydrocolloids, and other stabilizers may influence the dispersibility of the carotenoid preparation.

While β-carotene yields a golden vanilla product, apocarotenal gives an appetizing peach color (*69*); single and combinations with fruit-containing products can be tried for other effects. Apocarotenal addition to a sherbet formulation contributes a desirable strong orange hue; canthaxanthin gives a tomato red hue. The carotenoid colors do not cover the whole range of hues commonly used in frozen desserts. Mellorine, a frozen vegetable fat dessert, like ice cream, can be colored with carotene to provide both color and vitamin A activity. Mellorine producers should bear in mind that both β-carotene and apocarotenal, when used as colors in this product, also contribute substantially to the vitamin A activity required in the product, a condition of sale in a number of states requiring vitamin A nutrification.

K. Yogurt

Quite low concentrations of carotenoids, 2–8 mg/kg, in water-dispersible forms, are adequate to produce an attractive range of colors in yogurt, from a lemon yellow through apricot and peach to redder tints. For trials on yogurt, which can be carried out quite easily, 1–2% of skim milk powder is added to whole milk. The end product contains 3.5% fat. The milk is incubated in large tanks at 42–45°C, and, after the fermentation is complete, sugar, flavor, and the colorant are added with stirring (*204a*).

L. Dairy Product Substitutes

Although dairy products are frequently thought of as being pure white, the natural β-carotene content of butterfat provides a creamy, off-white appearance, which is obviously absent when vegetable fats are used in conjunction with nonfat dry milk solids, sodium caseinate, or other proteins to produce imitation dairy products. Nature does not provide milk-based products with a consistent color. Cows ingest more of nature's colorants from summer pasture than from their dry winter feeds. The products made from milk throughout the seasons reflect the variations. Added β-carotene has been used almost from the inception of filled-milk products to restore the natural creamy appearance. Because of the extremely low levels required (0.1–10 μg/gm), the attractive appearance, and the low cost of using β-carotene in these products, it has been used almost exclusively as the source of color in imitation milk, whipped toppings, both fluid and dry coffee whiteners, and sour cream. Whipped cream and toppings, in which a creamy color is characteristic, are stable vehicles for β-carotene. For a natural color, low β-carotene levels are recommended in these products.

Filled milk, a product made from a milk-base protein ingredient (usually skim milk or nonfat dry milk) with a nonanimal fat, may easily be colored with β-carotene. There are two methods of adding β-carotene to imitation fluid milk: (a) oil-phase addition and (b) aqueous-phase addition.

Oil-phase addition is recommended for processors using a liquid oil phase. One dissolves the required quantity (usually 1.0–1.6 gm per 100 gal or 400 liters of milk) of 30% β-carotene suspension in oil (equivalent to 0.35–0.55 gm pure) in 500 ml or 1 pt or pound of warm vegetable oil, 50°–60°C (120°–150°F). Then the β-carotene preparation is added to the remainder of the vegetable oil required per 400-liter batch of fluid milk. The colored vegetable oil is homogenized into the aqueous milk phase as usual. If colored imitation fluid milk is regularly prepared, as an added convenience a quantity of β-carotene stock solution can be prepared that is 10–20 times the 500-ml quantity mentioned. Oil-phase addition is not recommended for processors who add coconut oil flakes directly to the aqueous phase, since it is essential that the oil-soluble β-carotene be uniformly dissolved in the vegetable oil before homogenization.

Aqueous-phase addition in filled-milk production uses β-carotene water-dispersible forms. A stock solution of the required quantity of β-carotene beadlets (equivalent to 1.5 gm pure) is first prepared with 1 liter or quart of warm water. Fresh stock solution should be prepared daily. The required volume (about 4 fl oz or 120 ml) of this stock solution is then added to the warm fluid milk (400 liters). If a deeper color is desired, additional stock solution is added at a rate of 1 fl oz (30 ml) at a time. The stock solution should be added as early as possible during the preparation of the milk so that the β-carotene can be dissolved in the vegetable oil phase during the mixing and homogenizing stages. Consequently, it is important to standardize mixing time, homogenization, etc., when evaluating the level of β-carotene required. Color should be evaluated immediately, after processing, and after overnight storage during the initial trials.

Synthetic or imitation milk is made from nondairy protein and fat ingredients. At present, such products are being marketed only in insignificant quantities. As in the case of the other emulsion-type products mentioned, coloration would pose no technical problems. A product based on the use of whey and containing orange juice has been offered as a breakfast beverage (*671*).

Coffee whiteners are available in various countries, mostly in the form of a spray-dried powder. They are also marketed as liquid and frozen products. These products consist of an emulsion of vegetable (or animal) fat in an aqueous phase containing sodium caseinate, carbohydrates, emulsifier, thickener, stabilizers, flavor, and color. Compared with conventional dairy products, the new products are cheaper and have a longer shelf life but may or may not have a nutrient distribution pattern similar to that of dairy products. β-Carotene (about 0.05–01 gm pure per 400 liters of whitener) dissolved in the oil phase is a suitable coloring agent and also imparts nutritional value as provitamin A. Some processors prefer the combined use of β-carotene dissolved in the oil phase and riboflavin in the aqueous phase (1 μg/gm pure of each) to provide color.

M. Sauces, Gravies, Soups

Tomato products prepared from field tomatoes picked fully ripened under ideal sunlight, exposure, etc., possess full color and need no further color supplementation. The trend today, however, is toward a one-time mechanical picking of the crop because of labor, personnel limitations, economics, and a desire to reduce microbial content often present in overripe fruit; hence, at times full field color may not be present in all the fruit. Tomato products for soup, vegetable, cocktail sauce, pasta sauce, barbecue sauce, pizza sauce, etc., may be benefited by the addition of canthaxanthin. Canthaxanthin beadlets are dispersed in warm water and added in the required amount (*157*). The hue produced by canthaxanthin is partly dependent on the opacity, color, and fat content of the food; in most applications it produces a tomato red hue. Furthermore, canthaxanthin beadlets alone or blended with apocarotenal beadlets can be added directly to dry mixes, such as dehydrated spaghetti sauce, dry soup mixes, and spice mixes, provided that the dry mixes are reconstituted in warm water or used in final food products that have a heat preparatory step before consumption. Carotene beadlets are useful in chicken soup products, chicken gravy, chicken–vegetable soup, and dry soup mixes. Some dry soup mixes have a fat component. In this case, instead of the dry beadlet being added to the dry mix soup proper, the fat may be colored with fat-soluble forms of the carotenoid before packaging of the fat component. When carotenoid beadlets are added to a dry soup mix, they can be seen as specks of color resembling spice particles.

Chowders, soups, and stews are candidates for water-dispersible forms of the carotenoids, β-carotene being more suitable for chicken and egg soups and dishes, and canthaxanthin for tomato or beef–vegetable preparations. Apocarotenal has been of interest as an additive to "Manhattan" clam chowder. Carotenoids can also be predispersed into one of the ingredients of a liquid soup and then converted to a dry product by the usual dehydration process. This procedure may help to homogenize the color both into the dry soup mix and upon reconstitution or dispersion in water. Quantitively, about 2–10 mg of β-carotene (pure basis) are required per 500 gm or 1 pound of dry soup mix or 1–3 mg per pound of chicken gravy sauce. For pizza or barbecue sauce or tomato soup, 2–10 mg of canthaxanthin (pure basis) are used per pound; for tomato ketchup or tomato pastes, 5–20 mg/lb may be needed.

Bunnell and Borenstein (*157*) added canthaxanthin beadlets dispersed in water to canned tomato soup. The heat-processed cans were then stored and assayed for canthaxanthin; approximately 90% retention (Table 78) was found after 6 weeks at 45°C or after 6 months at room temperature (23°C). The minimal concentration required for visual detection was 6 ppm canthaxanthin. The same authors studied the performance of canthaxanthin in spaghetti sauce. Three batches of sauce were prepared: one control and two levels of canthaxanthin. The tomato solid contents of the three samples were varied. The procedure used was to add water slowly to the tomato paste with constant stirring. All dry ingredients were mixed thoroughly. The dry ingredient mix was then added to the diluted paste. The final product was heated to 82°C (180°F) and filled into 211 × 400 cans. The cans were processed in boiling water for 1 hr and cooled in running water to 37°C. After heat processing, the canned product was stored for 6 weeks at 45°C and up to 12 months at room temperature; in both instances the canthaxanthin retention (Table 79) was well above 90% of the initial assay. The tomato solid concentrations were varied in this study, and the results showed that 20 mg/kg canthaxanthin (by weight of spaghetti sauce) made up for a reduction of ripe tomato solids (from 7.7 to 4.7%), and the product containing canthaxanthin was markedly superior in color to the control or standard formula after 14 months of storage at room temperature.

Similar results were obtained with canthaxanthin beadlets in ravioli sauce and barbecue sauce.

TABLE 78

Retention of Canthaxanthin in Tomato Soup[a]

Storage conditions	(Canthaxanthin) (mg/kg)	Retention (%)
Claim	6.0	—
Initial	6.1	—
3 weeks at 45°C	5.9	97
6 weeks at 45°C	5.4	89
2 months at 23°C	5.5	90
5 months at 23°C	5.5	90
6 months at 23°C	5.7	93

[a] From Bunnell and Borenstein (*157*).

TABLE 79

Retention of Canthaxanthin in Spaghetti Sauce[a]

	Canthaxanthin concentration			
	10 μg/gm		20 μg/gm	
Storage	mg/kg	Retention (%)	mg/kg	Retention (%
Claim	10	—	20	—
Initial	9.2	—	18.1	—
6 weeks at 45°C	9.4	100	17.6	97
2 months at 23°C	9.1	99	17.7	98
6 months at 23°C	9.3	100	18.1	100
14 months at 23°C	—	—	18.1	100

[a] From Bunnell and Borenstein (*157*).

N. Meat Products, Fish Products

Meat products in which there is an insufficient red color from nitrosomyoglobin level (or in which one wishes a substitute for nitrosomyoglobin) may be benefited (and if legally permitted) by the incorporation of canthaxanthin beadlets previously dispersed in warm water to enhance the level of red color. Another application is the fabrication of simulated meat products from plant or fermentation protein sources to be used as meat extenders. Pet and other animal foods are other types of products to be considered. Trials run on products so far indicate an encouraging technical potential for canthaxanthin in these applications (*157*). The hue obtained is dependent on the fat content and physical structure of the specific product. Canthaxanthin is the carotenoid with the relatively deepest color and is best suited for coloring meat products. Bunnell and Borenstein (*157*) reported that 11 μg/gm canthaxanthin by weight of meat produced an attractive color in sausages. A typical frankfurter formulation was used, except that the meat component was 100% pork to obtain a low heme pigment content. The formulation was chopped in a laboratory-model silent cutter until a good meat emulsion was obtained (8 min). Sodium ascorbate was added (10% solution), and chopping continued. A sample was taken as a control. Canthaxanthin aqueous dispersion was added (equivalent to 10 mg pure per kilogram meat), and the emulsion mixed for 2 min. Samples were stuffed in cellophane casings and cured in an electric oven at 80°–90°C. The cellophane casings were removed, and the frankfurters stored at 5°C for analytical studies and color comparisons. The control was almost colorless; the colored sample had a pink color similar to cured beef, indicating that canthaxanthin has a promising potential for coloring meat substitutes. Concentrations as low as 7 μg/gm in sausage products give colors similar to that of cured beef (*157*).

According to Furia (*321*), a situation in which the majority of artificial food dyes (azo food dyes) show instability occurs when they are combined with protein and heated, such as in the retorting sterilization process. In canned meat pet foods, for example, only erythrosine was mentioned as being a stable color. Carotenoids, such as canthaxanthin, are suitable as a red color for meat dog foods, the hue of which is influenced by the fat content of the product.

The pigment most commonly found in ocean fish and crustaceans is astaxanthin (3,3′-dihydroxy-4,4′-diketo-β-carotene), which with its esters is the principal pigment in the flesh of the salmon, giving it a pink to orange-pink color. Canthaxanthin is closely related structurally to these compounds and may be considered a biogenetic intermediate. The color hue, its intensity, and uniformity are among the most important quality factors in salmon, including the canned product. There is a need and a desire to increase and standardize color in the processing operation. Since most salmon are caught in streams, the indirect pigmentation approach is not feasible.

Direct pigmentation studies (*109*) were undertaken at the Technological Station of the Fisheries Research Board of Canada on the University of British Columbia campus at Vancouver and, more recently, at new facilities in West Vancouver. The method initially employed was to dip cut salmon for 5–10 min in a solution containing canthaxanthin, β-carotene, L-ascorbic acid, and sodium chloride. Aerosol sprays were also tried. Pigment concentrations varied as follows: canthaxanthin, 0.05–0.25%; β-carotene, 0.05–0.10%; ascorbic acid, 0.05–0.20%; and sodium chloride, 0–20%. Although details have not been fully worked out, a multineedle injection approach with multihole needles combined with a spray of fillets or steaks may be an interesting approach. Improved color can be obtained, but adaptation to commercial processing equipment has not been fully worked out.

Shrimp and lobster products, particularly peeled, canned, or deep-frozen shrimp, may be colored with water-soluble canthaxanthin. Dip treatments for coloring fresh peeled shrimp were studied using water and 3% brine solutions containing 0, 20, or 40 μg/gm canthaxanthin (in aqueous dispersion made from beadlets). Dip times were 0, 5, and 15 min. The shrimp were then cooked for 4 min in boiling water. Upon cooking the relatively small areas of the shrimp surface containing protein-bound carotenoids became red-orange. Dipping for 15 min in brine containing 40 μg/gm canthaxanthin changed these ring areas to a distinctly different, darker red color. Canthaxanthin is deposited or concentrated along the rings because they are richer in fat. Other types of fish products that might benefit from canthaxathin addition are lobster, shrimp, and crawfish dishes, but application studies are needed (*157*).

O. Toppings, Icings

Toppings, icings, fillings, and frostings are easily colored with carotenoids. The color can vary from the faintest touch of yellow to deep yellow-orange, to orange, and to tomato red. Although there are many formulations for these products, the usual method of adding β-carotene depends on whether fat or oil is used in the formula. When fat or oil is present, a 20–30% carotenoid oil suspension may require dilution with shortening or oil for certain operations. The required amount is added to all or a portion of the fat or oil and stirred until the β-carotene is completely dispersed. During this step, the fat should preferably be at least at room temperature. Heating the fat phase to 37°–50°C speeds the rate of solution. The colored fat phase is mixed into the formulation in the usual manner. When the formula contains no fat or oil or only a trace, carotenoid water-dispersible products previously dispersed in a small amount of lukewarm water are used. In making the dispersion, about four times the volume of water for each volume of dry carotenoid is used. The water dispersion is mixed into the formula, and processing is completed in the usual manner.

The amount of carotenoid needed to produce the desired color is influenced to some extent by the other ingredients in the formula and the physical characteristics of the product. For example, air beaten into a topping tends to lighten color, and more color will be needed. A few trials usually establish the required level of β-carotene for any particular product. As a guide, a range of from 0.5 to 2 mg of carotenoid (pure basis) per pound or 500 gm of product will provide a trace of color; 6–10 mg/lb will provide a pronounced color. The color stability of carotenoids in most toppings, icings, frostings, and fillings is good. In some products, unusual conditions may cause some fading. These can usually be controlled by the judicious use of appropriate food antioxidants.

P. Gelatin Desserts, Puddings

Gelatin desserts consist mostly of gelatin, sugar, fruit acids, flavor, color, and water and are usually colored with water-soluble colors, giving complete clarity in the prepared gelatin dessert. The amount of carotenoid that can be incorporated into gelatin desserts, however, is limited, since at high levels these carotenoid preparations produce an undesirable clouding (*157*).

Only recently developed research-scale carotenoid powders employing improved formulations and highly sophisticated technology are compatible with products that are optically clear. These new "microcolloidal dispersions" of carotenoids in spray-dried form meet the following requirements: (a) yield optically clear aqueous preparations without introducing a haze; (b) disperse easily in hot or cold water; and (c) produce colors in the yellow to cherry red range (*282*). One must await the commercialization of this development to see whether a gelatin dessert mix containing optically clear carotenoid preparations can be marketed.

In other types of desserts, such as puddings, that contain skimmed-milk powder, starch, gums, gelatin, etc., or that are prepared with fresh milk, yielding an opaque mixture with spoonable final consistency, the rate of dissolution of the water-dispersible carotenoid beadlets may be a limiting factor during the preparation of the product (heat is necessary). In some of the newer cold-soluble pudding formulations (made with cold water), the regular carotenoid beadlets are not always dispersed completely; therefore, new and more diluted cold-water-soluble carotenoid products are being developed. Such a highly diluted product would also have the advantage of imparting an even, smooth color to the dry mix, whereas the regular beadlets are too coarse and concentrated. Fruit preserves and syrups can be colored with carotenoids. However, a cloud is produced with present forms of water-dispersible carotenoid products, which in some cases may limit the amount of carotenoid that can be added. The newly developed, research-scale, optically clear products may also have application here.

Q. Juices, Beverages

Fruit juices are particularly rich sources of vitamin C, and many of them also contain appreciable amounts of carotenoids, including β-carotene; these and various other ingredients contribute to the nutritional value of these products (*154,530*). The natural β-carotene of fruit can vary from 1 to 60% of the total carotenoid content (*63*). The color of citrus fruit is due to carotenoids, and a wide variety of these pigments have been found in the peel and juice. The seasonal trend in carotenoid content for California–Washington navel oranges is plotted in Fig. 38 by averaging the data of Taylor and Witte (*921*). Shown on the same graph are data (*158*) obtained by the Roche laboratories on samples of Florida frozen orange

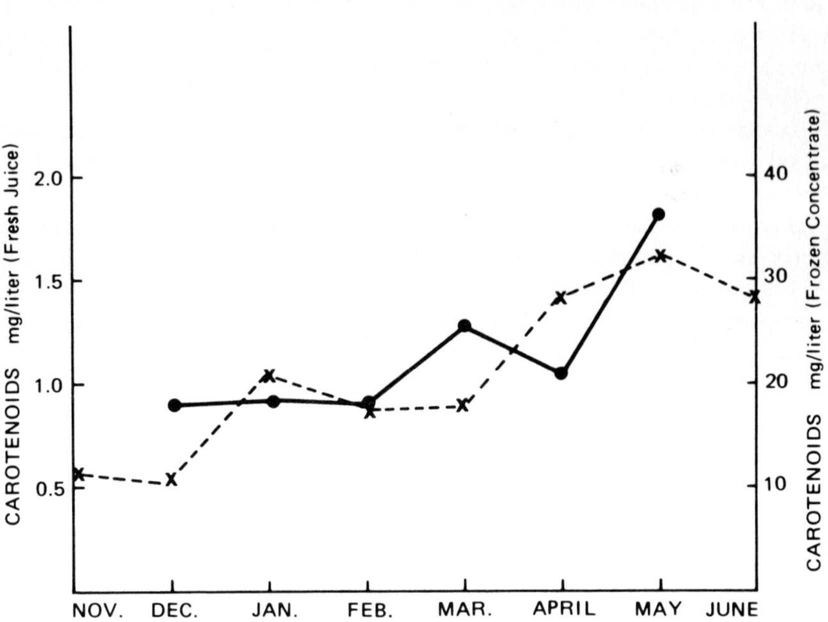

Fig. 38. Seasonal variations in carotenoid content of oranges. Key: ●—●, California-Washington navel oranges (data from Taylor and Witte); x---x, Florida frozen orange juice concentrate (data from Roche). [From Bunnell *et al.* (*158*).]

concentrate from different seasons of the year. These data illustrate the difficulties encountered by packers of citrus juice when attempting to produce a product of reasonably uniform color throughout the year.

Analyses of fruit consumption figures per capita for the past decade reveal an increasing trend for a greater percentage of the total fruit crop to be processed into juices and juice drinks because of the current demand for convenient foods and greater variety. This trend also offers the producer a market for fruit of irregular size, fruit with slight defects, overripe fruit, and small-lot varieties. The 1959 Yearbook of Agriculture (*887*), entitled "Food," recommends "A Daily Food Guide" as an aid in selecting foods for adequate nutrition and food variety. One of the four major food classifications, vegetables and fruits, is declared by the authors to furnish a substantial share of the vitamin A (carotene) and nearly all of the vitamin C (ascorbic acid) available to us from food. Therefore, the combination of added β-carotene and vitamin C in fruit juice products would seem to be physiologically, dietetically, and technologically sound. The addition of carotenoids to fruit juices appeals to consumers (*429*).

The possibility was studied of blending apple, tomato, and apricot juices with a natural source of ascorbic acid and carotenoids, sweet red pepper juice, in order to enrich the ascorbic acid and carotene content of the juices. The addition of 15% of pepper juice to fruit juices and 30% to tomato juice contributed to the nutritional value, vitamin content, color, and taste of the juices (*291*). Added β-carotene in the form of β-carotene beadlets or emulsion imparts a natural orange appearance, whereas apocarotenal beadlets alone or in combination with β-carotene, dispersed in warm water, give a deeper reddish-orange color. Apocarotenal matches very well the orange color often associated with and desired in carbonated orange drinks. Added ascorbic acid or naturally present ascorbic acid has a stabilizing effect on β-carotene and apocarotenal in storage tests in beverages (*74,158,522*). This is in contrast to the instability of FD&C Yellows No. 5 and No. 6, which show fading in the presence of ascorbic acid in light. Procedures for adding nutrients to juice have been discussed by Parman (*736*) and Kläui (*522*). Apocarotenal, however, is not quite as stable as β-carotene on prolonged exposure to sunlight. Canthaxathin added in beadlet form, alone or in combination with β-carotene, is very stable in beverages. Under normal processing and storage conditions, the stability of the carotenoids appears to be quite adequate. When possible, if the fruit beverage contains natural juice or fruit components, a brief homogenizing step often brings out the brilliance of the color and ensures physical homogeneity of the product. Carotenoid-colored beverages can be processed in plain tin containers without loss of color on storage (*158*). The addition of β-carotene and/or apocarotenal and ascorbic acid to juices and fruit beverages is in keeping with the nutritive heritage of these products.

A small laboratory trial involved adding β-carotene to a selection of juices and fruit drinks. The fruit juices and drinks were purchased as heat-processed products in 46-oz cans. The products were opened, blended, colored with β-carotene emulsion, then filled into sterilized, clear, flint glass bottles, sealed, and heat-processed for 20 min at 180°F. β-Carotene analyses were made initially and following storage. In most instances β-carotene retention values (*74*) exceeded 90% of the initial value (Table 80) following short-time accelerated aging trials. Data were also collected on commercial runs of juices, juice drinks, and concentrates colored with either dry β-carotene beadlets or β-carotene emulsion. β-Carotene was added to the warm, liquid fruit products in the tanks in the quantity required to give the desired color effect (10–25 mg β-carotene per gallon). After blending, the β-carotene-colored juice was filled into enamel-lined or occasionally plain tin containers, sealed, and heat-processed. β-Carotene analyses were run initially and after storage. Commercial trial runs of juices, juice drinks, and juice concentrates colored with β-carotene in emulsion or in dry beadlet form revealed high carotene retention values (*74*) after 6 and 12 months at 23°C (Table 81).

TABLE 80

Accelerated Aging of β-Carotene Heat-Processed Fruit Juices Packed in Glass[a]

Product	Amount added[b] (mg/8 fl oz)	β-Carotene (mg/8 fl oz or 237 ml)[c]		
		Initial assay	3 weeks, 40°C	3 months, 30°C
Pineapple–grapefruit drink				
Control A	—	0.06	0.06	0.06
Added β-carotene	1.38	1.38	1.35 (98)	1.38 (100)
Control B	—	0.05	0.06	0.04
Added β-carotene	1.38	1.44	1.24 (86)	1.44 (100)
Grapefruit drink				
Control	—	0.02	0.03	0.02
Added β-carotene	1.38	1.35	1.33 (99)	1.35 (100)
Orange juice				
Control A	—	1.21	1.21	1.18
Added β-carotene	1.38	2.60	2.65 (102)	2.04 (79)
Control B	—	0.80	0.80	0.63
Added β-carotene	1.38	2.34	2.15 (92)	1.99 (85)
Control C	—	0.02	0.02	0.02
Added β-carotene	1.38	1.60	1.52 (95)	1.60 (100)
Orange–pineapple drink				
Control	—	0.44	0.44	0.42
Added β-carotene	1.38	1.88	1.84 (98)	1.80 (96)
Orange–lemon drink				
Control	—	0.09	0.133	0.08
Added β-carotene	1.38	1.44	1.38 (96)	1.44 (100)
		Average	(96)	(95)

[a] From Bauernfeind *et al.* (*74*).
[b] β-Carotene added as an emulsion as milligrams per 8 fl oz, or 237 ml.
[c] Figures in parentheses indicate percent retention from initial assay.

TABLE 81

Stability of Added β-Carotene in Juices, Juice Drinks, and Juice Concentrates[a]

Product	Size of can (oz)	Type of process-ing	β-Carotene product	β-Carotene (mg/8 fl oz or 473 ml)[b] Initial assay	6 months, 23°C	12 months, 23°C
Juice						
Orange juice	46	Heat	Beadlets	1.04	0.87(83)	—
Juice drinks						
Juice blend	29	Heat	Beadlets	1.28	1.16(90)	1.29(100)
Orange drink						
I. Enamel-lined can	10	Heat	Beadlets	1.12	1.01(90)	—
Plain tin can				1.10	1.04(94)	—
II. Enamel-lined can[c]	10	Heat	Solution	0.19	0.20(103)	0.23(121)
Enamel-lined can[c]			Solution	0.25	0.28(112)	0.33(132)
Enamel-lined can			Beadlets	1.12	1.08(96)	1.07(95)
Plain tin can			Beadlets	0.96	104(108)	1.13(117)
III. Enamel-lined can	12	Heat	Beadlets	1.26	1.12(89)	1.07(85)
Plain tin can			Beadlets	1.01	1.05(104)	1.07(106)
Enamel-lined can			Beadlets	0.89	0.77(86)	0.76(85)
Enamel-lined can			Emulsion	0.80	0.80(100)	0.73(91)
IV. Enamel-lined can	46	Heat	Emulsion	1.23	1.23(100)	1.24(100)
Orangeade	6	Heat	Beadlets	2.20	—	2.12(96)
Pineapple drink	46	Heat	Beadlets	0.31	0.39(125)	0.40(129)
Pineapple–orange drink	46	Heat	Beadlets	0.63	0.53(84)	—
Juice blend	46	Heat	Emulsion	1.32	1.32(100)	1.20(90)
Concentrates						
Orangeade base	6	Heat	Beadlets	7.87	7.30(92)	—
Orangeade base	6	Heat	Beadlets	7.73	7.20(93)	—
Orange juice	32	Heat	Beadlets	9.99	10.40(104)	—
Orange drink[d]	6.4	Frozen	Beadlets	0.99	0.85(85)	0.83(83)
				Average	(97)	(102)

[a] From Bauernfeind *et al.* (*74*).
[b] Figures in parentheses indicate percent retention from initial value.
[c] Assays indicate total β-carotene and lycopene content.
[d] Assays for this sample were conducted on the reconstituted product.

As described in Section XIV,B (Table 61), β-carotene prepared by chemical synthesis, put in a fat-base application form, and added to commercially processed margarine was demonstrated to possess vitamin A activity (*628*) equivalent to that of β-carotene from natural sources. Biological trials (*74*) with animals have also been conducted with synthetic β-carotene in water-base application forms; emulsions and dry beadlets, employed in coloring juices and beverages, have also been shown to be fully biologically active as vitamin A sources (Table 82).

The color and concentration of carotenoids required for coloring various types of drinks or beverage products have been studied. Since blue and crimson are missing from the carotenoid palette, it is not possible to match all the direct shades. However, acceptable yellow, orange, and orange-red colors can be achieved, for example, with β-carotene and canthaxanthin, and the amounts needed range from 1 to approximately 10 mg/liter (1–10 ppm). In general, quantities below 5 ppm are used, and these yield appealing colors at reasonable costs. Experience has shown that neither apo-8′-carotenal nor canthaxanthin should be used alone in simulated orange juice-type products, that is, in low proportions, since (a) they would give excessively red shades and impart an unnatural appearance, and (b) combinations with β-carotene give more fullness of color, or "body," to the beverage and are more economical to use. For a lemon-colored drink one method involves a combination of dilute chlorophyll extracts (5 μg/gm as pure chlorophyll) with β-carotene suspension (0.2 ppm of pure β-carotene). In a trial, physical and chemical stability appeared to be satisfactory. In instances in which it is desirable for the reddish hue of carotene in aqueous dispersion to be suppressed, riboflavin may be added. Another suggestion for color variation includes the combined use of carotenoids and anthocyanins. The possibility of imparting color to beverages by coloring the flavoring oil with carotenoids before emulsification was studied from the onset.

The very low solubility of β-carotene in essential oils and the heat sensitivity of these oils, precluding higher dissolution temperatures, present a difficult problem. The much higher solubility of the apocarotenoids allows their inclusion in essential oils in tinctorially significant amounts. A 20% suspension of the micronized carotenoids in vegetable oil may be used for this purpose. The solubilities of the four carotenoids (without heating) in an orange oil and a lemon oil (Table 83) were determined by Manz (*621*). β-Carotene and canthaxanthin are obviously unsuitable for processing via the essential oil, since the amounts that can be incorporated are far too low. However, in conjunction with the two apocarotenoids, suitable coloring effects can be obtained. The use of a combination is also preferable, since the stability of β-carotene and canthaxanthin is better than that of the apo compounds.

β-Carotene and β-apo-8′-carotenal have been used to color orange drinks since the commercial availability of water-dispersible forms of these food-approved colors renders this feasible. The major technological problem is in the creaming of the flavor oil emulsion and the carryover of this problem to added fat-soluble colors such as the carotenoids. Two types of emulsions are used in the production of carbonated and noncarbonated fruit juice beverages or imitation fruit juice drinks:

TABLE 82
U.S.P. Rat Curative Growth Assays of β-Carotene Beadlets and Emulsions[a]

Assay	Percent total carotene by assay	No. of rats	Total μg carotene fed per rat	Gain in weight after 28 days	Vitamin A potency: Biological (IU/gm)	Vitamin A potency: Chemical[b] (IU/gm)	$\frac{\text{Bio.}}{\text{chem.}} \times 100$	Range at $P = 0.05$
Assay no. 1[c]								
Beadlet, 75-1B	1.31	8	33.3	55	22,000	21,800	101	86–116
		11	51.0	79				
Beadlet, 75-24B	2.47	8	31.1	48	38,000	41,200	92	76–108
		8	48.4	73				
Assay No. 2[d]								
Beadlet, P-004	2.59	14	24.7	62	43,000	43,200	100	83–117
		24	40.4	80				
Beadlet, P-005	2.51	23	22.2	51	41,000	41,800	98	83–113
		23	37.9	85				
Beadlet, P-006	2.65	18	25.8	68	51,000	44,200	115	98–132
		23	43.0	93				
Assay No. 3[d]								
Emulsion, P-031	3.72	14	13.1	22	59,000	62,000	95	79–111
		14	19.2	42				
		16	28.3	71				
Emulsion, G-2-64	5.12	17	14.8	29	84,000	85,000	99	86–112
		16	21.7	50				
		18	31.9	83				
Emulsion, A-2-119 adsorbed on starch	0.035	15	17.1	33	530	584	91	70–112
		18	27.1	62				
		20	45.4	82				
							Av. 99	
Orange drink[c] Lot no. 209-4	0.62 mg/100 ml	15	13.6	16	920	1,034	89	75–105
		20	27.2	45				
		21	40.8	56				
							Av. 98	

[a] From Bauernfeind *et al.* (74).
[b] Chemical assay multiplied by biological value factor; 1 gm β-carotene = 1,666,667 IU.
[c] All supplements fed in ration; calculated from feed intake.
[d] Oral dosing by syringe.

TABLE 83

Solubility of Pure Crystalline Carotenoids[a]

Carotenoid	Lemon oil ("lemon concentrate," Givaudan-Esrolko) (%)	Orange oil ("orange concentrate," Givaudan-Esrolko) (%)
Crystalline compounds		
β-Carotene	~0.3	~0.3
β-Apo-8′-carotenal	~4.5	~3.2
β-Apo-8′-carotenoic acid ethyl ester	~5.0	~4.4
Canthaxanthin	~0.1	~0.04
20% suspension of micronized crystals in oil		
β-Carotene	~0.2	~0.2
β-Apo-8′-carotenal	~3.9	~3.1
β-Apo-8′-carotenoic acid ethyl ester	~4.0	~3.4
Canthaxanthin	~0.08	~0.05

[a] From Manz (*621*).

flavoring emulsions and cloudifying agents are usually used in fruit pulp or juice-free beverages. The production of these emulsions is done by standard methods including the emulsification of the oil in an aqueous phase and homogenization, which in general includes a two-step high-pressure homogenization. The aqueous phase usually contains a hydrocolloid such as gum acacia, tragacanth, or one of the recently developed starch derivatives used to replace gum acacia, such as Steri Mul and Capsul, and sugars, sugar alcohols, preserving agents, etc., may be added if desired or necessary. The oily phase may contain, in addition to the flavoring oil, a weighted oil or another substance that increases the specific weight of the oil phase and thus stabilizes the system by retarding or preventing creaming and ring formation in the finished beverage. Until recently the classic weighted oils were brominated vegetable oils (BVO); however, BVO are no longer tolerated in most countries. The FDA restrictions on brominated BVO as a flavor oil weighting agent and the industry switch from high-quality, high-cost flavor emulsifiers (gum acacia) to modified starch products have made it more difficult to make physically stable emulsions. This is evident in the commonly found white ring of flavor ingredients in glass-packed drinks. The ring may not be objectionable to the consumer, but it does demonstrate instability of the flavor emulsion. Flavor oil droplets should be 2 μm in diameter, or smaller, to retard creaming or ringing and should be weighted as closely as possible to the specific gravity of the aqueous phase of the drink. Substances used for adjusting the specific weight of the oil phase to that of the aqueous phase, i.e., to approximately 1.02–1.03, are insoluble in water and therefore also act as clouding agents. Efficient emulsification requires homogenizer pressures in the 4000 psi range in many emulsion systems. Products used for control are abietic and dihydroabietic acid derivatives, such as the glycerol esters, Ester Gum, natural resins, such as dammar resin, and synthetic compounds, such as sucrose acetate-isobutyrate. For the legal acceptance of these products, food additive lists of specific countries have to be consulted. The use of carotenoids to color yellow to orange-red drinks necessitates attention to the flavor oil emulsion.

Water-dispersible carotenoid preparations yield a slightly cloudy solution, imparting a pulplike appearance to some products. Homogenizing the color solution with the flavoring and fruit concentrate is of great importance in obtaining a physically stable product with a brilliant color. Some water-dispersible carotenoid preparations contain gelatin as a carrier, and this could react with some components of the fruit juice, leading to precipitation or flocculation. Homogenization delays or prevents this by finely dispersing the reaction products and imparting the same charge and surface characteristics to them. A similar reaction may occur with flavoring oil emulsions and cloud emulsions, depending on the type of hydrocolloid used as an emulsifier. The pH conditions prevailing in beverages favor precipitation or coacervation. Homogenizing the whole mixture with a sufficient amount of fruit juice or concentrate or with an excess of anionic hydrocolloid emulsifier may overcome this difficulty.

Carotenoids as coloring agents in beverage systems can be applied by dissolving the carotenoid in the oily phase of the emulsion before emulsification. Although some clouding agents, such as BVO and abietates, are relatively good solvents for carotenoids, the total amount of carotenoid that can be added in this way is limited due to the poor solubility of the carotenoids, on the one hand, and the limited amount of oil phase, on the other. Heating increases the amount dissolved, but the sensitivity of the flavoring oils limits the temperature and length of heating. Furthermore, heating for too long at high temperatures may lead to excessive isomerization of the carotenoid, which is undesirable.

It has been found that many difficulties can be overcome by predissolving the carotenoid in a vegetable oil. Since vegetable oils are quite stable, heating can be done at relatively high temperatures, e.g., at 120°–160°C, and for correspondingly short times, e.g., 20 min down to less than 1 min. The sensitive flavoring ingredients may then be added at lower temperatures, followed by immediate emulsification. Thus, products of improved quality and containing flavorings, cloud agents, and color at the same time can be prepared. The relatively small amount of additional vegetable oil introduced by this process usually necessitates a corresponding adjustment of the specific weight by increasing the proportion of the heavier component or components. Food technologists preparing juice beverages of the above type must have an appreciation of the phenomena mentioned in the above paragraphs.

The natural vitamin C content of fruits and vegetables and the standardization of fruit juices with added ascorbic acid to delay oxidative or enzyme-catalyzed oxidation (*818*) and to serve as a uniform source of vitamin C have been described by Bauernfeind (*64*) and Bauernfeind and Pinkert (*76*). Tests have been run on beverages containing both added carotenoid (β-carotene) and L-ascorbic acid. A 1-year storage stability and flavor trial was completed in a canned orange drink beverage. β-Carotene emulsion had been added on an experimental basis at the level of 1.2 mg of carotene (pure) per 9 fl oz (250 ml) of drink. Vitamin C was added at a level to permit a claim of some 30 mg per 9-oz serving. Samples were set up under normal (23°C) and accelerated storage conditions (30°, 37°, and 45°C) to determine the stability of flavor and color and the carotene and vitamin C content. Retention of added ingredients (80–100%) was quite acceptable; color and flavor evaluations were favorable. Water-dispersible carotenoid preparations can also be used to color frozen orange drink products.

In canned beverages colored with azo dyes and containing ascorbic acid, the color is protected from light and the products have a reasonable shelf life. Another problem may arise here, since azo dyes appear to have a causative effect on can corrosion. The rate of corrosion is directly proportional to the azo dye concentration (*240,624*). The type of acid used may also influence the results; phosphoric acid significantly increases the expected percentage of perforations, according to Hotchner and Poole (*420*). These authors also studied the rate of corrosion by various dyes. The nonazo dye Brilliant Blue FCF had practically no effect on perforations, whereas Amaranth was corrosive and Sunset Yellow had a very high corrosivity. β-Carotene in water-dispersible form was also examined by Hotchner and Poole. Their results show clearly that β-carotene is much less corrosive than azo-type dyes.

The injurious effects of light on soft drinks containing orange juice can be summarized as a light-catalyzed oxidation of coloring and flavoring substances, and Rother (*817*) has shown that, by adding ascorbic acid, stability can largely be ensured. The prooxidative effect of residual oxygen in the headspace of bottles, which leads to the deterioration of orange-based beverages, including bleaching, has been studied by Schara and Tsoumanis (*840*). Ascorbic acid is often used in similar conditions as an oxygen acceptor. However, a major source of color fading is contact with reducing agents such as ascorbic acid (*321,604*). This is of particular practical importance when artificial food colors, including azo and triphenylmethane dyes, are added to supply color, since they are easily reduced to colorless compounds. Color fading in carbonated and still beverages is caused by ascorbic acid, which may be naturally present in the fruit drink or beverage or may have been added either as a flavor antioxidant or as a vitamin source. The reductive action of ascorbic acid is strongly catalyzed by light as has been demonstrated by Kläui and Manz (*522*). The excellent stability of added carotenoids in juices and drinks following processing and storage has been reported by Bauernfeind (*63*), Bunnell *et al.* (*158*), Bauernfeind *et al.* (*74*) Kläui (*518*), Kläui and Manz (*522*), and Manz and Kläui (*622*). These colorants can easily be added in carbonated beverage production.

Carbonated beverages colored with carotenoids retain their color for several months even when exposed to light. The hue of canthaxanthin, apocarotenal, and β-carotene is independent of pH. No change of hue is observed in the pH range 2–7.

The carotenoid colors are not easily reduced in the presence of metals such as zinc, copper, tin, and aluminum. Therefore, no problem of loss of color may occur if products containing these colors are packaged in cans. Currently, marketed water-dispersible carotenoids, however, do not yield complete clarity in clear or transparent aqueous drinks.

The production of a 1000-liter batch of a simple orange beverage colored with 1 μg/gm each of β-carotene and canthaxanthin was described in detail (*522*). Twelve kilograms of orange concentrate was thoroughly mixed with 100 liters of sugar solution (65%), 4 liters of citric acid solution (50%), and 35 gm of added ascorbic acid (vitamin C). This relatively small quantity of ascorbic acid, together with the natural ascorbic acid already present in the concentrate, had the purpose of stabilizing the beverage and improving the life of the flavoring substances. Then 1.0 gm of canthaxanthin and 1.0 gm of β-carotene, each in the form of 10 gm of a 10% water-dispersible preparation, were dispersed in 200 ml of warm water and stirred into the orange concentrate–sugar solution. The mixture was intimately homogenized with the aid of a pressure homogenizer, pasteurized at 60°C for 30 min, and deaerated. All the apparatus and pipework were of stainless steel. One-liter bottles with lever stoppers were each filled with 114 ml of the colored orange syrup and topped up with cold, carbonated mineral water. The bottles were stoppered and inverted several times to mix the orange syrup evenly with the water.

Bottles containing the carbonated orange beverage colored with 1 mg each of β-carotene and canthaxanthin per liter were kept (a) in the dark, (b) in diffuse daylight, and (c) in direct sunlight. For comparison purposes, sample bottles containing a beverage made with the same raw materials but colored with a mixture of tartrazine and Sunset Yellow were similarly exposed. The beverage colored with azo dyes and exposed to direct sunlight faded completely within less than 2 weeks, and in diffuse daylight the color retention after 6 months was only 40% (Fig. 39).

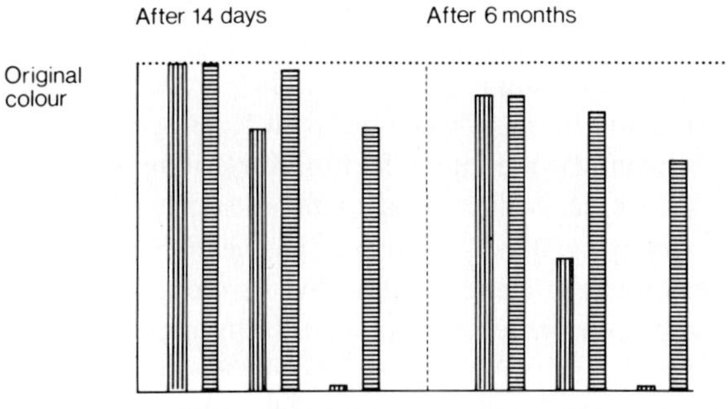

Fig. 39. Stability of azo dyes and carotenoids in carbonated orange beverages. Key: vertical stripes, azo dyes; horizontal stripes, carotenoids; 1, darkroom; 2, diffuse daylight; 3, direct sunlight. [From Kläui and Manz (*522*).]

After 6 months of direct sunlight, the beverage colored with carteonoids still showed a higher color retention.

A commercial trial was set up to study the performance of β-carotene and β-apo-8′-carotenal in carbonated beverages with and without added ascorbic acid. The carotenoids were added in emulsion (water-dispersible) form. The trial was set up with the following variables (based on the beverage): (a) 3.6 mg β-apo-8′-carotenal without ascorbic acid, (b) 3.6 mg β-apo-8′-carotenal with 100 mg ascorbic acid, (c) 6 mg β-carotene without ascorbic acid, and (d) 6 mg β-carotene with 100 mg ascorbic acid. Flavor, sweetness, and acidity were controlled by added orange-flavored beverage base and 32°C Baume sugar syrup. All ingredients were dispersed in 96 fluid oz of syrup and beverage base. Four and one-half ounces of final concentrate blend was added to each 29-oz bottle filled with carbonated water, and the bottles were capped on the regular production line. β-Carotene imparted a color to beverages simulating natural orange juice, whereas β-apo-8′-carotenal, and especially combinations of the latter with β-carotene, gave a deeper reddish-orange. β-Apo-8′-carotenal gave the deep orange color often desired by carbonated and orange drink manufacturers. Performance data (Table 84) show that ascorbic acid had a stabilizing effect on β-carotene and β-apo-8′-carotenal in these aging studies (*69*). β-Apo-8′-carotenal, however, is not as stable as β-carotene when exposed to direct sunlight. Canthaxanthin in carbonated beverages is fairly stable (*621*) if not exposed to sunlight for lengthy periods (Table 85). It therefore appears advisable that exposure to continuous direct sunlight for prolonged periods be controlled, even when carotenoids are used as the coloring agent. Under normal manufacturing and shelf-life storage conditions, the stability of carotenoids in beverages appears to be adequate.

In carbonated beverages or still beverages in which clarity is required, azo dyes or anthocyanins have been the main colorants. However, investigations of the pharmacological effects of these coal-tar type of dyes have led to the delisting of FD&C Violet No. 1 and Red No. 2 from the certified list of colors permitted in the United States for use in foods. United States beverage manufacturers are concerned that other FD&C colors (Yellow No. 5, Red No. 40) may eventually follow. These factors would present a real problem for the industry as the number of potential color substitutes becomes narrowed to a very few. It is in this area that new experimental application forms of the carotenoids may offer a potential solution. Most recently, chemically stable 2.5–10% spray-dried carotenoid powders were developed using a gelatin–carbohydrate matrix. The powders yield optically clear aqueous solutions in the color range of yellow to bright red, according to Emodi *et al.* (*282*). These preparations are currently classified as research-scale products. These carotenoid preparations may not offer an exact match of colors obtainable with FD&C colors, but they do produce a wide array of colors ranging from yellow to red. Philip (*753*) discussed the application of new colors, including the carotenoids, in the coloring of soft drinks of the future.

TABLE 84

Performance Data for Carotenoids in Carbonated Beverages[a]

Sample	Amount added[b]	Initial assay[b]	Retention (%)		
				Indirect light	
			1 month, south window	2 months, 30°C	3 months, 23°C
Apocarotenal, no ascorbic acid	3.6	3.51	0	83	78
Apocarotenal + 100 mg ascorbic acid	3.6	3.69	16	94	90
β-Carotene, no ascorbic acid	6.0	5.87	68	97	85
β-Carotene + 100 mg ascorbic acid	6.0	5.62	78	100	95

[a] From Bauernfeind and Bunnell (*69*).
[b] Milligrams per 29 fl oz (860 ml).

TABLE 85

Stability of Canthaxanthin in Carbonated Soda[a,b]

Storage conditions	Canthaxanthin (mg/bottle)	Retention (%)
Initial determination	12.0	—
Direct sunlight, room temp.		
1 month	11.2	93
2 months	10.0	83
6 months	9.0	75
12 months	8.4	70
Dark room, room temp.		
6 months	11.4	95
12 months	9.5	79

[a] From Manz (*621*).
[b] Packed in flint bottles, no ascorbic acid added.

R. Confectionary Products, Candy

Carotenoids in their water-dispersible form are well suited for coloring candies, frostings, decorative coatings, fruit gums, fruit drops, and pastry fillings. Confections, candy, fillings, and sugar frostings usually have little or no added fat, and hence the carotenoid beadlet products dispersed in warm water are the preferred vehicles for providing color in these products (*157,457,515,516,662*). β-Carotene, β-apocarotenal, and canthaxanthin can impart, respectively, an intense yellow, yellowish-orange, orange, and redish-orange hue to the above products. Blends of these compounds are also possible. Riboflavin, or vitamin B_2, is not a carotenoid. Riboflavin has been used in amounts of about 2 gm per 100 kg of candy without effect on flavor. With the carotenoids, 0.3–1.5 gm (pure compound) are used per 100 kg of candy. To minimize degradation in making cooked hard candy, the carotenoids are added as an aqueous dispersion after cooking of the sugar syrup (*457*). In using the water-dispersible forms of carotenoids such as beadlets, a stock solution can be prepared and added to the base formula late in the process. To color high-boiled goods, hard carmels, etc., premix of color, sugar, or citric acid may be stirred into the liquid melt on the cooling slab or folded into the batch before pulling. The rate of use of these pigments in individual products would have to be determined by the manufacturer. Some adaptation trials may be required to determine the best procedure for adding these colors to individual confectionary products as emphasized by Metzner (*662*).

Fruit jellies can be attractively colored with carotenoids, using β-carotene for yellows and apocarotenal and canthaxanthin for the darker colors. Colored syrups of the beadlets can be prepared by dissolving the beadlets in water, adding the appropriate amount of sugar, and stirring until solution is complete. For use in coloring sugar-coated candies, the colored syrups should have a high sugar concentration to facilitate application of a uniform, color-stable coating. Suggested amounts (milligrams per candy) of pure carotenoids added to yield certain colors are shown in the tabulation below. The candies colored with β-carotene showed essentially no color change when stored for several months at elevated temperatures or when exposed to direct light for 3 months.

Amount added (mg)	β-Carotene	Apocarotenal	Canthaxanthin
0.15	Deep orange	Red-brown	Red
0.045	Orange	Orange-red	Orange-red
0.015	Yellow-orange	Orange-yellow	Orange-yellow
0.0045	Yellow	Peach	Red-peach

Manz (*622*) colored soft sugar candies under factory conditions with aqueous dispersions of four carotenoids. The paper-wrapped candies were kept at room temperature in diffuse daylight and assayed periodically (Table 86) for their carotenoid content (*622*). The stability of carotenoids dispersed in sucrose syrup (sirupus simplex) is good (*519*) if not heated at too high a temperature or for too long a period.

S. Popcorn, Fish Sticks, Potato Sticks or Chips

When high heat is applied to food cooked in oil, stress is put on carotenoid stability. Stabilized carotenoid products containing a relatively high amount of antioxidants have been developed especially for preparing yellow popcorn and similar products (*964*) undergoing high heat treatment. Color-stable preparations suitable for making yellow popcorn on the basis of natural or synthetic carotenes have also been patented (*162*). Other processes reported for stabilizing carotenoids in oils, e.g., xanthophyll oils, involve mixing the oils with soybean lecithin so that the mixture contains 25–75% of lecithin (*965*). Particularly effective synergistic antioxidant mixtures for carotenoids comprise an antioxidant and a primary lower alkanolamine (*143*) or an antioxidant with a secondary amine as a synergist (*880*). Tocopheramines have also been found experimentally to have a strong stabilizing effect in foodstuffs containing carotenoids, and, moreover, they have significant vitamin E activity (*711*) but may not be on accepted food additive lists.

β-Carotene is recommended for its ability to provide a pleasant and nutritive yellow color to popcorn. Application can be made by popping the corn in β-carotene-colored oil and/or shortening. A special stabilized β-carotene suspension in vegetable oil has been developed to meet the needs of popcorn-seasoning manufacturers. This product is characterized by outstanding stability under the severe heat conditions of popcorn processing. About 5–15 gm of β-carotene (pure) are used per 100 kg of popping oil. The β-carotene suspension is dissolved into the popping oil and, with continued heating, stirred. Since the usual levels are well

TABLE 86

Stability of Carotenoids in Soft Candies[a,b]

	β-Carotene		Apocarotenal		Apocarotenoic ester		Canthaxanthin	
	mg/kg	Retention (%)	mg/kg	Retention (%)	mg/kg	Retention (%)	mg/kg	Retention (%)
Initial content								
Quantity added	60.8		60.8		60.8		60.8	
Initial assay	54.3	100	70.5	100	60.6	100	60.0	100
After storage								
5 months	56.0	100	72.1	100	58.8	97	63.9	100
12 months	47.2	87	69.0	98	52.8	87	54.5	91
Initial content								
Quantity added	15.2		15.2		15.2		15.2	
Initial assay	15.1	100	16.3	100	15.2	100	17.7	100
After storage								
5 months	13.4	89	15.7	96	16.8	100	16.3	92
12 months	11.5	76	15.0	92	14.4	95	13.6	77

[a] From Manz (*621*).

[b] The quantities of carotenoids refer to the pure compound. Storage conditions: room temperature and diffuse daylight.

below the maximal solubility of β-carotene in vegetable oil, there is no difficulty in dissolving the carotene by manual or electrical stirring. After oil or shortening is evenly colored, it is used to pop corn in the usual manner. Darker colors can be obtained with proportionately more β-carotene.

Another approach to coloring popcorn is to use noncolored popping oil and to spray the popcorn with a solution of β-carotene-colored and butter-flavored oil after popping. About 10–40 gm of β-carotene (pure) are used for 100 kg of warm oil sprayed onto the popcorn after the popping operation. The amount of spray oil used depends on the minimum needed to get the desired color effect.

Stability data on film-packed carotene-colored popcorn have been good. After 2 months of storage under various conditions, retention values were high; nearly 100% of the carotene color found on initial assay was retained when the samples were stored in the dark at 23° and 30°C. Normal light exposure at room temperature storage showed 90% color retention during the same length of time.

The golden brown color desired in the breading of food products, such as fish sticks, is produced by blending stabilized oil suspensions of β-carotene with dry breading or by adding a water-dispersible form (*62*) to the batter before use. The color persists during the precooking, freezing, storage, and subsequent reheating of the breaded product. β-Carotene has been observed (*549*) to be effective in imparting color to frozen french-fried potatoes and gives a golden brown color to the product. Frying oils colored with β-carotene offer convenience in processing, handling, and reducing frying time.

Stabilization of β-carotene on the surface of frozen fried potato products is achieved by adding β-carotene to the frying oil containing antioxidants (*972*). Following this method, sticks of potatoes suitable for the manufacture of french-fried potatoes were submerged in a vat containing hot hydrogenated cottonseed oil, to which had been added a suspension of β-carotene containing antioxidants (BHA and propyl gallate). About 1.5–3.0 gm (pure) of β-carotene were added per 100 kg of oil. The potato sticks were fried for 3 min and then frozen at −20°F. After storage for 6 months at 0°F, the original β-carotene content of the sticks remained substantially the same. Color loss of the oil solution is a function of time and temperature; consequently, the amount of β-carotene to be added is dependent on individual processing conditions.

Another process (*621*) yielded evenly, lightly colored chips even when different types of potatoes were used. This procedure involved a blanching step followed by rinsing combined with coloring of the chips in an aqueous 0.0025–0.05% β-carotene solution. After frying, the potato chips had an attractive, homogeneous, light color. A critical factor was the thickness of the chips: in order to diminish or avoid excessive browning reactions (Maillard), the thickness of the chips was reduced from the usual 1.10–1.25 mm to 0.85–1.00 mm.

Processes also exist for coloring foods by applying, for example, a spray. Whenever this approach can replace the application of carotene in a hot oil cook, less destruction of the carotenoid can be expected and better control can be exercised over the coloring formulation. An aqueous dispersion of carotenoids in a large amount of dextrins or sugar syrups has been applied to food (*278*). This process can be applied to color, e.g., breakfast cereals or other dried protein–cereal products, infant foods, and snack preparations. A spray operation allows a carotene application either before a heat processing step or immediately thereafter.

T. Miscellaneous Products

Eggnog is a seasonal specialty item. The color of eggnog is first dependent on the color of the egg yolks used. If there is insufficient color from the yolks, β-carotene beadlet products, first dissolved in warm water, can be added to obtain the desired creamy or yellow appearance (*158*). The use of β-carotene makes it possible to standardize the creamy appearance of eggnog regardless of the color of the egg yolks employed in its preparation.

The addition of carotenoid pigments to soy milk (*190*) before coagulation produces soybean curd varying in color from yellow to orange, to reddish-orange, to tan, and to red, depending on the carotenoid or blend of carotenoids used and the method of addition. In the Orient, soybean curd (tofu) is a large-volume, well-known item of diet. Tofu has the advantages of low cost, high protein and calcium content, and desirable texture. However, it is devoid of vitamin A activity, which is a major nutritional deficiency among the world's undernourished populations. Therefore, Clausen and Nakayama studied the nutrification of soybean curd with carotenoids which combine provitamin A activity with color (*190*). β-Carotene β-apo-8′-carotenal, and, for coloring purposes only, canthaxanthin were used in the form of water-dispersible beadlets, solutions in vegetable oil, and suspensions of

TABLE 87

Color of Curds as Affected by Type and Form of Carotenoid Addition[a]

Carotenoid	(%)	Color
β-Carotene		
Beadlet	0.01	Orange
Beadlet	0.003	Light yellow
Oil solution	0.01	Yellow
Microcrystals	0.01	Pink
Microcrystals	0.003	Colorless
β-Apo-8′-carotenal		
Beadlet	0.01	Orange
Beadlet	0.004	Orange
Oil solution	0.01	Reddish-orange
Microcrystals	0.01	Tan
Microcrystals	0.004	Light tan
Canthaxanthin, beadlet	0.01	Reddish

[a] From Claussen and Nakayama (*190*).

microcrystals. The addition of the carotenoid pigments in these forms was made at the milk stage before coagulation (by salts). Recovery of the added carotenoids in tofu was essentially quantitative in the case of the beadlets. Addition of oil solutions resulted in leaching and lack of color uniformity. The stability (*190*) of the colors (Table 87) under the usual conditions of frying and boiling was reported to be excellent.

β-Carotene in the water-dispersible form has a use in the coloring of mustard, either by itself or in combination with chlorophyll or riboflavin. Since the choice of final color is subjective, the only approach is trial and error until a pleasing formula is found.

U. Pilot-Type Production Trials

A tabulation of results of small commercial or pilot-type commercial trials made with prevailing equipment on a number of food products containing added carotenoids as food colors is given in Table 88 (*57,58*). In general, added carotenoids in the direct coloring of food products are quite stable when prepared by good technological practice. If the added carotenoid is in an oil phase, the addition of food-grade, oil-soluble antioxidants, such as the tocopherols, is helpful. Heating carotenoids above 60°C (104°F) induces isomerization but has little effect on the tinctorial potency. However, around 200°C (400°F) or deep-frying temperatures, carotenoid stability is poor, although it is improved somewhat by antioxidant addition.

Considering that the first synthetic carotenoid was permitted for food use in 1956, the commercial use of carotenoids is still relatively new. Although there has been great progress in the development of stable, convenient application forms, there remains more to be accomplished with market forms and their application to food products. The rapid growth of industry technology, new product development, and changing needs preclude any abatement of the developmental work that must be pursued by interested food processors. As with any ingredient, the carotenoids as food colors may have some shortcomings, but their strengths (*55*) as listed below encourage their wider use in the food industry.

1. They possess high tinctorial potency as colorants. On a weight basis, the carotenoids impart a more intense color to food than many of the azo and triphenylmethane dyes.
2. They are fat-soluble colorants, but emulsion and colloidal forms exist for water applications.
3. They are relatively stable under reducing conditions.
4. They are relatively stable to light, particularly in a minimal-oxygen atmosphere.
5. They are relatively stable over the acid and neutral pH range of food products.

6. They are available in a constant supply of a predetermined quality and potency.
7. They are not foreign substances but exist widely in nature.
8. Some possess vitamin A activity and hence have a dual function.

XV. VITAMIN A

A. Technology

When vitamin A activity is desired, it may be more practical and economical to use vitamin A rather than a carotenoid vitamin A precursor, particularly if added color is of low priority. Vitamin A, retinol ($C_{20}H_{29}OH$), is an isoprenoid polyene alcohol, 3,7-dimethyl-9-(2′,6′,6′-trimethyl-1′-cyclohexen-1′-yl)-2,4,6,8,-nonatetraen-1-ol; it is also known as axerophthol, the antixerophthalmic vitamin, and the antiinfective vitamin. With a trimethylcyclohexenyl ring and a methylated side chain containing four conjugated double bonds (Fig. 40), it can exist in different isomeric forms, each with different biological activities. The all-trans isomer, possessing the highest biological activity, is primarily produced commercially by chemical synthesis. Vitamin A complexed with protein (retinol-binding protein) is the principal form of transport of the vitamin in blood from the liver. Vitamin A esters are the liver storage form, and vitamin A aldehyde (retinal) in the retina combines with protein to form the essential photosensitive pigment, rhodopsin, in the visual process of man and animals. Vitamin A palmitate, propionate, or acetate, also of the all-trans type manufactured by chemical synthesis, are employed most frequently in food and animal applications. The chemical properties of vitamin A compounds (*70*) are fairly well known (Table 89), and a technology of application has been well developed (*59*) as has occurred with other vitamins (*54,205*). All industrial syntheses (*435a,438*) of vitamin A are based on β-ionone.

B. Lability and Stability

The stability characteristics of vitamin A (a labile structure) and its derivatives are influenced by a number of factors, such as oxygen or ozone, other oxidizing agents, acids, and heat. Vitamin A is fairly stable when heated to moderate temperatures in an inert atmosphere in the absence of light, but it is unstable in the presence of oxygen or air or when exposed to ultraviolet light. In the presence of alkali, vitamin A is quite stable, so that alkaline saponifications of vitamin A-containing materials may be carried out without serious loss of vitamin A. The presence of trace metals may also accelerate the oxidation of vitamin A. Therefore, in the handling of vitamin A, some precautions are the following: (a) an inert atmosphere, (b) subdued light, and (c) avoidance of trace metals and a strongly acidic environment. Stability performance in food (*250*) and other products is difficult to predict because the

TABLE 88

Stability of Carotenoids in Food Products[a]

Product	Packaging	Added carotenoid	Method of addition	Carotenoid level[b]					
					After storage 70°–75°F (23° C)				
				After processing	2 months	3 months	4 months	6 months	12 months
Apricot drink	Metal can	β-Carotene	Before canning	1.21 mg/6 fl oz		1.21		1.21(100)	1.16(96)
Apricot–orange drink	Metal can	β-Apo-8′-carotenal (beadlets)	Before canning	0.44 mg/6 fl oz		0.42		0.43(98)	
Butter	Paper, waxed carton	β-Carotene (emulsion)	At salting stage	4.7 mg/lb		4.2[a]		4.2[c] (89)	
	Paper, waxed carton	(beadlets)	At salting stage	5.2 mg/lb		4.4[c]		4.6[c] (89)	
	Paper, waxed carton	(gel)	At salting stage	7.9 mg/lb		7.2[c]		7.4[c] (94)	
	Paper, waxed carton	(gel)	In cream	7.8 mg/lb		7.2[c]		7.3[c] (94)	
				3.23 mg/lb	3.29[c]			3.41[c] (100)	
	Paper, waxed carton	(gel)	In continuous processing	3.60 mg/lb				3.60[c] (100)	
Cheese spread, canned	Sealed can	β-Carotene (beadlets)	To the milk	1.7 mg/2 oz		1.8		1.7(100)	1.8(100)
									1.7[d] (100)
Cheddar	Waxed wedge	β-Carotene (beadlets)	To the milk	5.0 mg/lb		5.1[c]		5.4[c] (100)	4.9[c] (98)
Primary	Waxed wedge	β-Carotene (beadlets)	To the milk	5.4 mg/lb		5.6[c]			5.7[c,e] (100)
		β-Apo-8′-carotenal	To the milk	6.5 mg/lb		6.7[c]			7.8[c,e] (100)

Primary	Waxed wedge	β-Carotene (A) and β-apo-8′-carotenal (B)	To the milk	A, 2.4 mg/lb B, 1.7 mg/lb A, 4.3 mg/lb B, 1.7 mg/lb				2.2[c] (92) 1.6[c] (99) 4.2[c] (98) 1.6[c] (94)
Processed	Film	β-Apo-8′-carotenal	To warm cheese	1.09 mg/lb	1.03[c]		1.06[c] (97)	1.03[c] (94)
Processed	Film	β-Carotene (A) and β-apo-8′-carotenal (B)	To warm cheese	A, 2.1 mg/lb B, 1.4 mg/lb A, 3.9 mg/lb B, 1.6 mg/lb				2.0[c] (95) 1.3[c] (93) 3.8[c] (97) 1.4[c] (88)
Processed	Paper and box	β-Apo-8′-carotenal (suspension)	To warm cheese	4.9 mg/lb		5.0[e]	5.1[c] (100)	
Fruit juice drink	Metal can	β-Carotene (emulsion)	Before canning	1.32 mg/fl oz		1.38	1.32(100)	1.20(91)
	Metal can	β-Carotene	Before canning	0.96 mg/6 fl oz		0.92	0.87(91)	0.97(100)
Fruit juice base, chilled	Metal can	β-Carotene (emulsion)	Before canning	2.9 mg/8 fl oz		2.98[e]	2.86[c] (96)	2.90[c] (100)
Lard	Waxed paper carton	β-Carotene (suspension)	To warm fat	3.24 mg/lb			3.25(100)	
Margarine	Waxed paper carton	β-Carotene	To warm oil before churning	3.84 mg/lb	3.60		3.30(86) 3.66[c] (95)	
	Waxed paper carton	β-Carotene	To warm oil before churning	3.51 mg/lb	3.48		3.30(94) 3.54[c] (100)	
	Waxed paper carton	β-Carotene	To warm oil before churning	3.63 mg/lb	3.54		3.12(86) 3.42[c] (94)	
	Waxed paper carton	β-Carotene	To warm oil before churning	3.33 mg/lb	3.06		3.12(94) 3.14[c] (100)	
	Waxed paper carton	β-Carotene	To warm oil before churning	3.90 mg/lb	3.54		3.24(83) 3.78[c] (97)	

(*Continued*)

TABLE 88—*Continued*

Product	Packaging	Added carotenoid	Method of addition	Carotenoid level[b]					
					After storage 70°–75°F (23° C)				
				After processing	2 months	3 months	4 months	6 months	12 months
Orange beverage, carbonated	Glass bottle	β-Carotene (emulsion)	Before bottling	0.78 mg/6 fl oz				0.81(100)	0.66(85)
	Glass bottle	β-Apo-8′-carotenal alone	Before bottling	3.5 mg/29 fl oz		2.73			
		With added ascorbic acid	Before bottling	3.70 mg/29 fl oz		3.31			
	Glass bottle	β-Carotene alone	Before bottling	5.87 mg/29 fl oz		5.00			
		With added ascorbic acid		5.62 mg/29 fl oz		5.32			
Orange drink, low calorie	Glass bottle	β-Carotene	Before canning	0.96 mg/10 fl oz	0.93	0.96		0.94(98)	0.92(96)
Orange drink	Metal can	β-Carotene and β-apo-8′-carotenal (3:1)	Before canning	2.75 mg/12 fl oz		2.71		2.64(97)	
				2.71 mg/12 fl oz		2.68		2.64(97)	
	Enamel-lined can	β-Carotene (beadlets)	Before canning	1.73 mg/12 fl oz				1.67(97)	1.60[f](92)
	Plain tin can	β-Carotene (beadlets)	Before canning	1.52 mg/12 fl oz				1.58(100)	1.60[d](92)
	Enamel-lined can	β-Carotene (beadlets)	Before canning	0.56 mg/100 ml				0.51(91)	
	Plain tin can	β-Carotene (beadlets)	Before canning	0.55 mg/100 ml				0.52(94)	

Orange juice, concentrate	Metal can	β-Carotene (beadlets)	Before canning	2.51 mg/16 fl oz		2.64		2.60(100)	2.88(100)
Orangeade base	Metal can	β-Carotene (beadlets)	Before canning	5.90 mg/6 fl oz		5.30		5.46(93)	
Pie filling	Metal can	β-Apo-8′-carotenal (beadlets)	Before canning	2.24 mg/kg		2.07		2.10(94)	2.00(89)
Pineapple–orange drink	Metal can	β-Carotene	Before canning	0.92 mg/6 fl oz		0.85		0.85(92)	
Popcorn oil	Pint bottles	β-Carotene (suspension)	To warm oil	13.6 mg/lb		12.7		13.1(96)	12.9(95)
Shortening	Sealed can	β-Carotene	To warm fat	3.80 mg/lb				3.71[c] (98)	
Whole yolk									
Dry	Film	β-Carotene (beadlets)	In liquid yolk	5.5 mg/100 gm	5.0		4.7	4.7(85)	
	Film	β-Carotene (suspension)	In liquid yolk	7.7 mg/100 gm	7.5		7.5	6.8(88)	
	Film	β-Carotene (beadlets)	In liquid yolk	11.7 mg/100 gm	10.4		9.4	9.5(81)	
	Film	β-Carotene (emulsion)	In liquid yolk	14.1 mg/100 gm	12.8		12.1	11.3(80)	
Frozen	Sealed containers	β-Carotene (beadlets)	In liquid yolk before freezing	16.9 mg/100 mg		16.5[g]			15.0[g](89)
Yolk product, dry	Film	β-Carotene	In liquid yolk	5.8 mg/100 gm	5.8		5.4[h]		5.1(88)

[a] From Bauernfeind (*57,58*).
[b] Values in parentheses are percent retention during storage.
[c] Storage temperature 5°C.
[d] Storage period of 21 months.
[e] Storage period of 24 months.
[f] Storage period of 18 months.
[g] Storage temperature −20°C.
[h] Storage temperature 30°C.

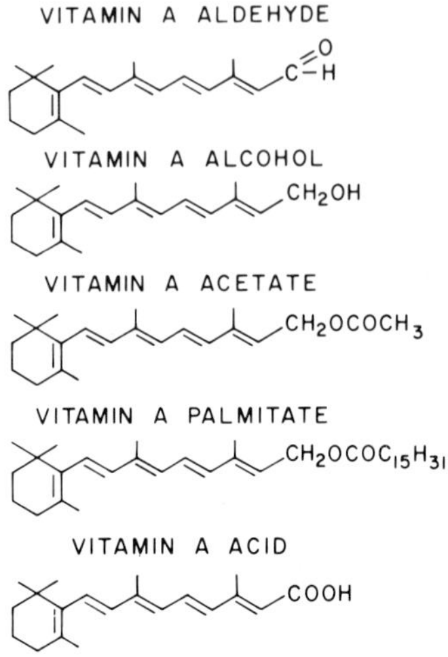

Fig. 40. Structural formulas of vitamin A compounds.

details of pH, moisture content, packaging, storage conditions, and other significant variables are usually insufficient.

C. Application Forms

Vitamin A is offered in a number of application forms (*71,520*). The principal methods of stabilizing vitamin A involve (a) sealing under vacuum or inert gas, (b)

TABLE 89

Physicochemical Data on Vitamin A[a]

Property	Vitamin A palmitate	Vitamin A acetate	Vitamin A (alcohol form)
Color of crystals (pure)	Semisolid, yellow	Light yellow	Light yellow
Melting point (°C)	28–29	57–58	62–64
Molecular weight	524.9	328.5	286.4
Solubility (%), 20°C			
Fats, oils, ether, chloroform	>50	>50	>50
95% ethanol, isopropanol	<10	<10	<10
Spectrophotometric data			
Max at 325 nm			
$E^{1\%}_{1\,\mathrm{cm}}$ (ethanol)	975	1550	1835
Biological activity, vitamin A value (IU/gm)	1.82×10^6	2.9×10^6	3.33×10^6

[a] From Bauernfeind and Cort (*70*).

storage at low temperatures, (c) the addition of antioxidants, (d) absorbing and/or sealing with a protective matrix, such as a gelatin, vegetable gums, or wax composition, and (e) chemically complexing with other compounds. The first two are common methods of storing any unstable compound, but they are not always convenient and practical. The use of antioxidants to conserve vitamin A is quite common. One is confined in this respect, however, to those antioxidants that are completely acceptable in a given country for food or pharmaceutical use. This list may include tocopherols, ascorbic acid and derivatives, butylated hydroxyanisole, butylated hydroxytoluene, and propyl gallate. In the preparation of vitamin A in an acceptable, dry, particle form, the selected matrix or absorbent must be nontoxic, edible, digestible, impervious to oxygen, relatively resistant to moisture, easy to handle, relatively economical, and available. Much effort (*520*) has gone into the development of dry, stable, fat-soluble vitamin compositions, as detailed by Kläui *et al.* (Table 90). In the preparation of dry forms of vitamin A, powders, granulates, microspheres, beadlets, agglomerates, etc., have been prepared by a variety of processing methods involving absorption, granulation, spray-congealing, encapsulating, and chemical complexing procedures.

Basically, the application forms of vitamin A fall into two categories: *liquid* and *dry* products (Fig. 41). Specialty liquid vitamin A formulated products in emulsion form are produced when water dispersibility or spray-on characteristics are desired in food applications. Dry vitamin A products prepared from the esters have been developed in various particle sizes and shapes with digestible coatings. In some instances the product is consumed in dry form; in others, the dry form becomes water dispersible at a given temperature before it is consumed, and thus special characteristics are required for the intended end use. In addition to physical characteristics, the application forms of micronutrients have certain chemical and physiological properties that contribute to, for example, adequate performance (*131,520*). When liquid or dry vitamin forms are used in the nutrification (*704*) of foods for human use, they should not influence significantly the color, appearance, and flavor of the nutrified food, should be fully available biologically, should be reasonably stable, and should be relatively economical to use (*28,30,146,209,875*). To monitor the vitamin A content in foods (*739*) and tissues (*958*), qualitative and quantitative methods of analysis have been published. To be successfully nutrified most foods require centralized processing locations, followed by food distribution programs that will ensure regular access to and consumption of the nutrified foods.

1. Use of Application Forms

For the past 30 years a continuous effort (Fig. 41) has been made to meet the demands (*78*) of the food and pharmaceutical industries for suitable vitamin A formulations. In practice, vitamin A can be given as an oral supplement, or it can be administered parenterally. It can also be added to a variety of foods by nutrification. Nutrification is the incorporation of micronutrients into foods for nutritional im-

TABLE 90

Stabilized Dry Forms of Fat-Soluble Vitamins and Processes[a]

Products	Form or type	Processes
Adsorbates and mixtures	Powders	Mixing, air-mixing, spray-mixing
	Granulates	Granulating
		Crushing
Fat-based powders	Microspheres (“fat beadlets”)	Spray-congealing
	Flakes	Drum-cooling
	Powders	Spraying or dropping into liquids
		Comminuting (crushing of chilled mass), deep freezing, grinding
		Dispersion in aqueous liquids
		Mixing, air-mixing, spray-mixing
Dried emulsions and suspensions	Spherical or spheroidal beadlets	Spray-drying (spraying into “hot gas”)
	Flakes	Spraying into liquids (“liquid catch”)
	Powders	Spraying into powders (“powder catch”)
		Spraying onto moving or stationary surface
		Double dispersion process (dual emulsion)
		Drum-drying
		Air suspension coating
		Comminuting sheets of dried emulsion
Microcapsules	Spherical or spheroidal beadlets	Coacervation
	Agglomerates	Air suspension coating (Wurster/WARF, United States)
		Multiorifice encapsulation (Southwest Research Institute, United States)
		Vacuum encapsulation (National Research Corporation, United States)
Inclusion compounds and various products (including chemical derivatives, precipitates, and coagulates)	Crystals	Crystallization
	Powders	Precipitation and coagulation
	Granules	Chemical reaction

[a] From Kläui *et al.* (*520*).

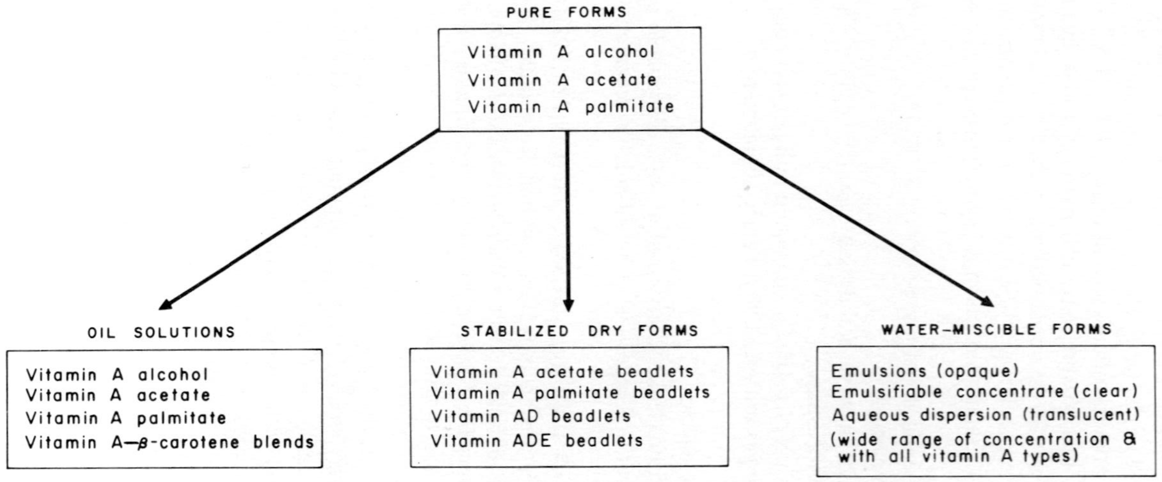

Fig. 41. Some chemical and physical forms of vitamin A in commercial production. [From Bauernfeind (*52*).]

provement. In addition to vitamin A products, variations include combination with vitamin D and/or vitamin E in stabilized form. Premixes of vitamin A may include, if desired, the water-soluble vitamins (e.g., folic acid and ascorbic acid) and minerals (e.g., iron and iodine) as well. Thus, many application market forms are possible with their respective specifications and uses (*59,71,73,512*).

Vitamin A can either (a) be consumed daily in foods in amounts that meet physiological needs, or (b) be consumed intermittently in amounts greatly beyond daily needs and the excess vitamin stored in the liver for future needs. The delivery systems dealing with vitamin A of chemical synthetic origin can be divided into two principal categories:

I. Prevention of vitamin A deficiency
   A. Short-term measures
      1. Intermittent massive oral dosing
   B. Longer-term measures
      1. Nutrification of regularly consumed indigenous foods
      2. Regular ingestion of a daily dietary supplement

II. Treatment of vitamin A deficiency
   A. Emergency measure
      1. Initial intramuscular injection of a water-dispersible preparation
      2. Subsequent oral ingestion of an oil-soluble preparation

a. Intermittent Oral Dosing (*28,56,73,386,653,723,797,1017*). For intermittent massive oral dosing of young children (or adults), a technology exists for the manufacture of sealed, soft gelatin capsules of vitamin A ester, usually palmitate (with vitamin E). These can be swallowed directly, or the contents can be squeezed out onto the tongue of the child not yet old enough to swallow the capsules. Soft gelatin capsules are made and filled continuously on rotary die process machines by skilled operators from ribbons of gelatin. Content variation among capsules is very small. Vitamin A palmitate (200,000 IU) plus α-tocopherol (40 IU) in oil formulation is filled into a 6-minim, one-piece, clear, soft gelatin capsule (15 mm long) with a nipple or nozzle tip (*52*). The tip can be cut off if the contents are to be squeezed out. After the capsule is swallowed, dissolution occurs within 10 min. Capsules containing less or more vitamin A can be prepared; however, due to the lengthy preparatory steps in machine adjustments and operations, commercial runs of less than 0.5 million capsules are difficult to arrange. The stability (*52*) of vitamin A in the capsules is excellent, as determined by assay data (Table 91). Field studies (*56,73,386,723,1017*) indicate that one capsule should be given every 6 months, as a minimal program, to young children as an xerophthalmia-preventing measure. If cost and distribution systems permit, one capsule given every 3–4 months would give better assurance, particularly to older children, whose vitamin A requirements are higher; this would also ensure that young children missed in the previous dosing would receive a capsule. Objections to

TABLE 91

Retention of Vitamin A in Stored Soft Gelatin Capsules[a]

	Vitamin A value (IU/capsule)		
Capsulation run	Initial[b]	After storage[c]	
A	214,000[a]	—	211,000 (20 mo)
B	205,000	—	219,000 (25 mo)
C	209,000	—	206,000 (29 mo)
D	214,000		207,000 (31 mo)
E	303,000	294,000 (17 mo)	296,000 (42 mo)
F	295,000	313,000 (17 mo)	300,000 (42 mo)
G	300,000	293,000 (17 mo)	292,000 (42 mo)
H	290,000	280,000 (17 mo)	297,000 (42 mo)

[a] From Bauernfeind (*52*).
[b] All values by duplicate chemical assay.
[c] Numbers in parentheses indicate months of storage at 23°C in closed containers.

the capsule approach are (a) the skipped child, (b) the possibility of regurgitation of contents after swallowing or not swallowing the capsule, and (c) the misconception that one has to take ''medicine'' to prevent nutritional blindness (faulty diet is one cause of blindness. Intermittent massive oral dosing has been used intensively in India and Bangladesh and to some extent in other countries, such as Haiti, Indonesia, Philippines, and El Salvador. In certain instances vitamin A oil solutions in containers, dispensed to children by spoon or dropped dispenser, have been used in place of capsules.

b. Parenteral Dosing (*56,72,73,386,723,1017*). When a patient (child or adult) is recognized as having severe signs of corneal ulceration with xerosis and there is danger of keratomalacia and irreversible loss of sight, one should apply vitamin A as effectively as possible. Intramuscularly (IM) injected oil solutions of vitamins have been found to be relatively ineffective compared to water-dispersible vitamin A palmitate parenterals, which are quick acting. A water-base formulation of vitamin A palmitate (100,000 IU) containing appropriate emulsifiers, antioxidants, and preservatives, heat-sterilized in 1- or 2-ml glass ampules for hospital use, may be administered IM in a sterile syringe. A vitamin A formulation of this type administered IM moves quickly into the bloodstream, and the excess is stored in the liver (*56,72,73,723*), as shown by animal experiments. Liver stores maintain blood levels as long as such stores remain and protein foods are consumed. Properly formulated vitamin A parenterals are quite stable (*52*) with respect to vitamin A potency, as monitored by storage assay data (Table 92).

TABLE 92

Retention of Vitamin A in Stored Water-Dispersible Vitamin A Parenterals[a,b]

Storage conditions	Retention (%)
6 weeks, 45°C	91
3 months, 37°C	94
6 months, 23°C	98
12 months, 23°C	94
18 months, 23°C	90
12 months, 5°C	100
24 months, 5°C	97

[a] From Bauernfeind (*52*).
[b] Initial assay 120,000 IU/ml.

The schedule of treatment currently recommended by WHO (*1017*) and IVACG (*386*) for severe xerophthalmia is tabulated below.

Treatment schedule[a]	Preparation
Immediately upon diagnosis	Inject IM 50,000–100,000 IU water-dispersible vitamin A
Second day	Give orally 100,000–200,000 IU oil solution vitamin A
Before discharge	Give orally 100,000–200,000 IU oil solution vitamin A

[a] Lower levels to very small children and those less than 1 yr.

All hospitals in countries with recognized cases of vitamin A deficiency should keep in stock at all times ampules of water-dispersible vitamin A and oral capsules of oil solutions of vitamin A.

D. Nutrified Foods (*59,61,67,71,77,131,209,395,512,657,703*)

Vitamin A has been successfully added (*71*) to many foods, such as salad oils, margarine, peanut butter, liquid skim milk, canned evaporated or condensed milk, nonfat dry milk, ice cream, mellorine, parevine, butter, cheese, white flour, bread, cakes, cookies, corn grits, corn meal, white rice, breakfast cereals, snacks, potato chips, potato flakes, potato granules, fruit juice beverages and powders, infant and geriatric liquid and dry formulas, confectionaries, tea dust, tea leaves, refined crystalline sugar, and such seasonings as refined white fine crystalline salt and monosodium glutamate.

1. Oils and Fats (254,628,656,686,996,1025)

Oils and fats (shortening, butter, margarine, ghee, vanaspati, etc.) are dietary ingredients of most populations. Liquid vegetable oils (cottonseed, maize, olive, peanut, soybean, etc.) used in light cooking or consumed with leafy vegetables or other foods can be nutrified with added vitamin A. After the liquid vegetable oil in a bulk container during processing has been clarified, a liquid vitamin A ester concentrate in a proper amount is added and agitated for uniformity before the nutrified oil product is placed in sealed containers. Although nutrification of vegetable oils is not practiced to a large extent in commerce, it is feasible if the oils are packaged in opaque or darkly colored glass containers with closures to eliminate the effect of light, a stimulant to oxidative processes. Edible antioxidants may also be added to protect both the vegetable oil and the added vitamin A ester from oxidation. An inert gas replacement for the air in the headspace of the container is advisable at the time of initial packaging, if practical.

Margarine, commonly used as a bread spread interchangeable with butter, is generally nutrified with vitamin A ester (20,000–50,000 IU/kg) to equal or exceed the average levels of this nutrient in butter. Nutrification of margarine with vitamin A is now carried out in many countries (Table 60). Nutrification of margarine with vitamin A or vitamins A, D, and E is relatively simple. Vitamin A, premeasured in amounts required for the batch size of the margarine oil tanks used in a particular margarine factory, is added to one tank of oil, a simple act and an easy factory control operation. If β-carotene is also to be used, both color and vitamin A can be added simultaneously. A safe color, β-carotene provides color and much of the required vitamin A activity. Available data indicate that added vitamin A is quite stable (*254,628,656*) during the margarine manufacturing operations and during the use of the nutrified product in the home. Furthermore, in biscuits, cakes, and breads prepared with vitamin A-nutrified fats, 80–100% of the added vitamin survives the baking process. Lard, shortening, ghee, vanaspati, and other fats likewise can be easily nutrified with added vitamin A.

2. Milk (65,75,195,204,997,1020)

When fat is removed from whole fluid milk, all fat-soluble vitamins including vitamin A are likewise removed. That the resulting *skim milk* can serve as a vehicle for added fat-soluble vitamins has been well documented. All skim milk or low-fat liquid or dry milk products should have vitamin A added and, if given to infants, should probably have vitamins D and E and possibly C added as well. Two techniques have been developed for the nutrification of nonfat dry milk (skim-milk powder).

A liquid form of pure vitamin A palmitate (preferably with added antioxidants) with or without vitamin D is diluted in hydrogenated coconut oil. This warm oil dilution is homogenized into liquid condensed skim milk to make a high-fat premix. The required amount of the premix is blended into a known quantity of condensed skim milk and mixed thoroughly. The mixture is spray-dried and then, if desired, instantized in the usual manner. The use of coconut oil has been found desirable for good stability of the added vitamins. The amount used increases the total fat content of nonfat dry milk (about 0.1–0.2%). Blending of the premix into the condensed skim milk may be done on a batch basis or on a continuous basis, the latter requiring controlled metering of the introduced vitamin A premix with the flow rate of the condensed milk.

A dry, stable, water-dispersible form of vitamin A ester (type 250-S, 250-CWS, or 325-L) is blended with a weighed quantity of nonfat dry milk to form a premix. An aliquot or proper portion of the premix is added to a given quantity of nonfat dry milk and mixed in, to produce the final nutrified product. Blending may be done batch style in a suitable dry blender, such as a ribbon mixer, or continuous blending may be employed, metering in the dry vitamin A or the prepared premix in proportion to the flow of the dry milk product.

Both the wet and dry methods are practical and proven. The dry technique provides slightly better stability but is slightly more costly than the wet technique because the dry forms of vitamin A are somewhat more costly than the liquid vitamin concentrates and also because of the extra labor involved in the dry mixing operation. Problems to be aware of in the nutrification of dry skim milk are (a) uniformity of introduction of vitamin A in the continuous process, (b) good blending in both the continuous and batch systems, and (c) avoidance of overprocessing, which leads to off-flavors. Moisture content of the dry milk product should not exceed 4%. The stability (*75*) of synthetic vitamin A added to nonfat dry milk is good to excellent (Table 93). Data also exist to show that synthetic vitamin A added to dry milk is fully available biologically (*65*). Proper packaging influences the vitamin A stability of the dry product; packages with high moisture resistance are preferred. When reconstituted (*75,1020*), nutrified nonfat dry milk retains its added vitamin A content during a time interval satisfactory for ready consumption.

Nonfat dry milk exported from the United States under the Food for Peace program has been nutrified with vitamins A and D for the past decade. After adequate experiments had demonstrated vitamin A to be stable in nonfat dry milk, the product was nutrified with 5000 IU of vitamin A and 500 IU of vitamin D per 100 gm for the P.L. 480 overseas distribution in May 1965. Some years thereafter, these levels were reduced to 2200 IU of vitamin A and 440 IU of vitamin D per 100 gm to coincide with the same levels in nutrified nonfat dry milk initiated in the United States in 1968. The World Health Organization recommends nutrification of milk for developing countries at 5000 IU of vitamin A and 500 IU vitamin D per 100 gm, which has been adopted by the EC in Brussels. The vitamin A nutrification of nonfat dry milk has been long favored by WHO. Over a decade ago UNICEF (*65*) encouraged studies that resulted in the development of technology for incorporating vitamin A in the product. Once again, in 1977 WHO (*284,729*) stressed the desirability of nutrifying nonfat dry milk with vitamin A as one means of improving vitamin A intake and hence preventing xerophthalmia blindness in children. To

TABLE 93

Retention of Added Vitamin A in Stored Nonfat Dry Milk[a]

			Retention (%)																	
			Polyethylene bags						Kraft foil bags						Tin cans					
			Weeks at 37.5°C			Weeks at 45°C			Weeks at 37.5°C			Weeks at 45°C			Weeks at 37.5°C			Weeks at 45°C		
Sample no.	Vitamin A added (IU/kg)	Initial assay (IU/kg)	3	6	12	3	6	12	3	6	12	3	6	12	3	6	12	3	6	12
1	22,000	22,200	93	97	100	101	88	89	104	99	100	111	100	86	99	100	95	92	88	85
2	22,000	22,800	100	91	95	97	88	86	99	100	98	109	100	87	90	88	94	82	91	82
			Months at 23°C			Months at 5°C			Months at 23°C			Months at 5°C			Months at 23°C			Months at 5°C		
1	22,000	22,200	108	87	69	100	76	85	104	79	78	109	74	84	111	82	85	115	85	96
2	22,000	22,800	93	82	85	106	80	84	103	78	77	121	79	107	108	85	89	115	78	94

[a] From Bauernfeind and Parman (*75*).

assure consumers of vitamin A-nutrified nonfat dry milk that vitamin A is actually present, a simple qualitative test has been devised which can be used in the field (*269*).

3. Cereal-Grain Products (21,37,132,145,146,201,394,395,702,705,824)

Cereal-grain products usually constitute an important calorie source of many populations. In Mexico, India, and Africa, they make up 55–65% of the daily caloric intake (*37*). In the United States it is estimated that about 26% of the daily caloric intake (*705*) comes from products based on cereal grains (wheat, corn, rice, etc.) and their by-products. In 1974 a proposed nutrification policy (*705*) for U.S. cereal-grain products was set forth by the Panel of the Committee on Food Standards and Fortification Policy of the Food and Nutrition Board, National Research Council, National Academy of Science, for 10 nutrients, namely, vitamin A, thiamin, riboflavin, niacin, pyridoxine, folacin, iron, calcium, magnesium, and zinc. The studies of Anderson *et al.* (*21*), Cort *et al.* (*201*), Rubin *et al.* (*824*), and others attest to the technical feasibility of cereal-grain product nutrification. More recent commercial-scale (*702*) testing of nutrification of cereal-grain products has demonstrated the practicality of adding vitamin A, other vitamins, and minerals to these products. Nutrification of cereal-grain products with added thiamin, riboflavin, niacin, and iron has been fairly widely adopted in the United States and has spread to other countries (*66*).

Weaning food, principally with a cereal-grain base and usually mixed with a vegetable protein or milk and nutrified with micronutrients, has been tried or considered in Egypt, Algeria, India, Iran, Morocco, Tunisia, Turkey, Guyana, Jamaica, and other countries as a nutrient delivery system for infants. Various degrees of success have been observed. Extruded cereal grain–vegetable protein product mixtures are of current interest in Sri Lanka, Bolivia, Costa Rica, Honduras and other countries.

a. Flour, Maize Meal. The development in the last decade of stable dry forms of vitamin A ester of smaller particle size (*132*) has greatly broadened the type of food products amenable to vitamin A nutrification. Foremost among these food products are cereal-grain products, such as white flour, corn or maize meal, and white rice. The need for the nutrification of cereal-grain products with vitamin A for developing countries was first suggested by UNICEF. The small-particle, dry vitamin A ester product (type 250–SD), introduced into cereal-grain products in premix form in laboratory trial runs, has shown good stability in corn (maize) meal (15% moisture), namely, 86% retention of initial values after 6 months at room temperature (23°C), and in wheat flour (11% moisture), 90% after the same time interval. Furthermore, when the dry vitamin A ester form was incorporated into nutrified cereal-grain products and baked into chappaties, breads, and tortillas, vitamin A retention values between 87 and 95% were observed. Vitamin A premixes for white wheat flour are designed to be metered into the continuous flow of

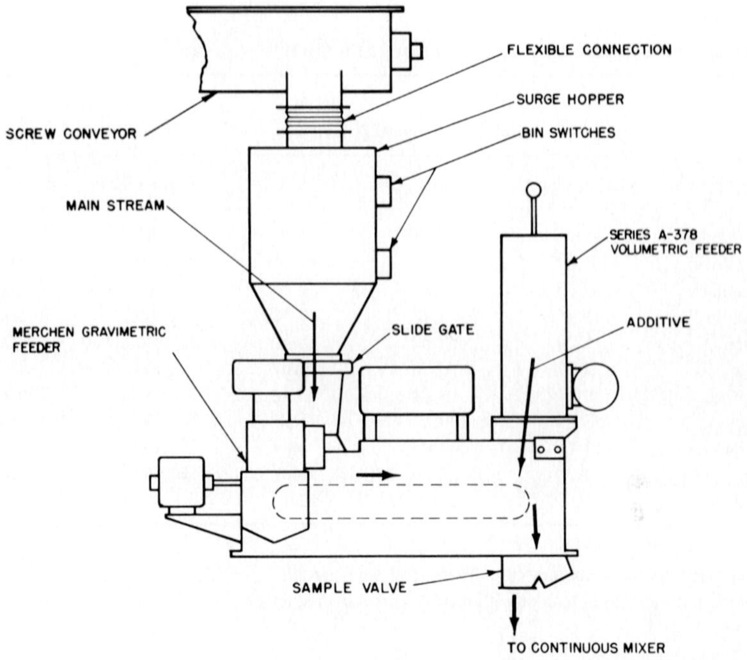

Fig. 42. Combination of gravimetric feeder for food product and volumetric feeder for miconutrient premix before the mixing unit in continuous food processing line. (Courtesy Wallace & Tiernan Division, Pennwalt Corp., Bellville, N.J.).

flour with the aid of a mechanical feeder (Fig. 42) at the rate of 0.25 oz (7.09 gm) per 100 lb (45.5 kg) of flour without problems of bridging, compression, lumping, or segregation. The fine particle size of the dry vitamin A form developed for flour permits screening with a coarse screen without excessive loss or physical separation of the vitamin A particles from the flour.

Recently, the stability of vitamin A when incorporated into white flour, corn meal, and white rice as a multimicronutrient premix was retested as part of the feasibility studies of the 1974 proposed NAS–NRC–FNB nutrification program for cereal-grain products. Two premixes were prepared. Premix X contained all 10 micronutrients, including vitamin A, and premix Y was like premix A but without calcium and magnesium. Nutrified flour was prepared in commercial bakery facilities with premixes X and Y, and breads were baked from each nutrified flour. Nutrified flour samples were stored and assayed for nutrient content, as were the baked breads. Tabular data (Tables 94 and 95) show excellent retention (*824*) of the added vitamin A in the multinutrient nutrified flour and good to excellent retention in bread baked with such flours. Bread characteristics, such as color, grain score, and flavor, although slightly modified when closely compared with those of unnutrified controls, were suitable, and, with slight modification of the magnesium

TABLE 94

Stability of Vitamin–Mineral Premix in Flour (12% H_2O)[a]

	Initial assay (mg/100 gm)			Retention, 6 months at 23°C(%)	
Supplement	Label claim	Premix X[b]	Premix Y[c]	Premix X[b]	Premix Y[c]
Vitamin A[d]	1600[e]	1730[e]	1800[e]	101	100
Vitamin B_1	0.64	0.76	0.71	99	101
Vitamin B_2	0.40	0.39	0.40	100	100
Vitamin B_6	0.44	0.48	0.56	100	100
Niacin	5.29	5.60	5.60	100	100
Folic acid	0.07	0.063	0.07	100	96
Iron	8.81	9.34	9.46	97	100
Calcium	198.2	209.0	NR[f]	NR[f]	—
Magnesium	44.1	47.0	NR[f]	NR[f]	—
Zinc	2.2	2.3	NR[f]	NR[f]	NR[f]
Moisture		11.6	12.1	11.8	12.5

[a] From Rubin *et al.* (824).
[b] Fortified with the complete vitamin–mineral supplement.
[c] Vitamin–mineral supplement without calcium and magnesium.
[d] Type 250-SD.
[e] International units per 100 gm.
[f] NR, not run.

TABLE 95

Retention of Added Vitamins and Minerals in Baked Bread (38% H_2O)[a]

		Retention (%)			
		After baking		5 days at 23°C	
Nutrient	Level added (mg/lb, or 454 gm)	Premix X[b]	Premix Y[c]	Premix X[b]	Premix Y[c]
Vitamin A	992[d]	83	95	83	95
Vitamin B_1	1.8	101	101	101	100
Vitamin B_2	1.1	105	101	108	101
Vitamin B_6	1.2	100	105	100	105
Niacin	15.0	100	102	100	106
Folic acid	0.19	94	105	80	102
Iron	24.8	104	105	NR[e]	NR[e]
Calcium	558.0	NR[e]	—	NR[e]	—
Magnesium	124.0	107	—	NR[e]	—
Zinc	6.2	NR[e]	NR[e]	NR[e]	NR[e]

[a] From Rubin *et al.* (824).
[b] Fortified with the complete vitamin–mineral supplement.
[c] Vitamin–mineral supplement without calcium and magnesium.
[d] International units per pound.
[e] NR, not run.

and/or calcium levels, the products were judged to be very acceptable with all 10 added nutrients present. In some of these studies, vitamin E (as α-tocopheryl acetate) was added in the micronutrient premix, and subsequent flour and bread studies demonstrated excellent stability of added vitamin E (*201*). Yellow corn (maize) meal and grits were likewise nutrified with a multinutrient premix, and the nutrified products were observed (*824*) for nutrient retention after storage and when cooked for consumption (Tables 96 and 97). Although the added vitamin A showed some losses, 80% or more was retained in the storage tests and 70% or more in the cooking tests of the nutrified corn products (*201,824*).

Vitamin A nutrification of corn meal and flour became a part of P.L. 480 purchase specifications for food exported from the United States on December 1, 1968, and May 1, 1969, respectively. The range specified was 4000–6000 IU/lb. Currently, additional nutrified foods sent abroad from the United States under the U.S. Agency for International Development (AID) or Food for Peace program have the following specifications: soy-fortified corn meal, 2–3 mg thiamin, 1.2–1.8 mg riboflavin, 16–24 mg niacin, 4000–6000 IU vitamin A, 13–26 mg iron, and 500–750 mg calcium per pound; soy–wheat flour blend, 2–2.5 mg thiamin, 1.2–1.5 mg riboflavin, 16–20 mg niacin, 4000–6000 IU vitamin A, and 500–1107 mg (for 6% soy) or 750–1364 mg (for 12% soy) calcium per pound; other foods have other specifications. Programs have been considered for nutrified flour in such countries as India, Brazil, Chile, Indonesia, Tunisia, Pakistan, Jordan, and Iran. In the past Guatemala and South Africa have shown an interest in nutrified maize meal. The technology exists for the incorporation of vitamin A into wheat and maize products in amounts capable of making substantial improvement in the diets of consumers

TABLE 96

Stability of Vitamin–Mineral Premix in Yellow Corn Meal (6.5% H_2O) and Corn Grits (11.4% H_2O) at Room Temperature (23°C)[a,b]

	Corn meal			Corn grits		
Nutrient	Initial	3 months	6 months	Initial	3 months	6 months
Vitamin A[c]	6000[d]	5820[d]	5880[d]	6000[d]	5700[d]	4850[d]
Thiamin	3.17	3.25	3.07	3.83	3.81	3.86
Niacin	26.0	25.7	NR[e]	36.0	36.0	NR[e]
Pyridoxine-HCl	4.5	4.0	4.5	4.4	4.3	4.5
Folic acid	0.6	0.5	0.5	0.49	0.45	0.5
Riboflavin	2.02	1.81	1.96	2.82	2.64	2.58
Iron	41.0	39.0	40.0	55.0	52.1	56.0

[a] From Rubin *et al.* (*824*).
[b] Expressed as millgrams per pound (454 gm) unless otherwise indicated.
[c] Type 250-SD.
[d] International units per pound.
[e] NR, not run.

TABLE 97

Retention of Vitamin A in Cooked Grits and Cooked Yellow Corn Meal[a]

	Retention (%) after cooking				
Product	4 min	5 min	6 min	10 min	30 min
Quick grits	80	—	75	70	—
Regular grits	—	—	—	—	66–75
Yellow corn meal	—	87	—	—	—

[a] From Rubin *et al.* (*824*).

without impairing significantly the acceptance of the nutrified products. Other cereal-grain products made from barley, sorghum, rye, etc., can be nutrified.

b. Rice (*201,664,824*). Rice is consumed as a whole kernel product, which makes nutrification more difficult than it is for a meal or flour product. The problem is compounded by the custom of washing rice before cooking. The added nutrients must be placed on rice in a rinse-resistant coating, yet in one that dissolves during the cooking process, releasing the nutrients. White rice nutrification requires, therefore, an entirely different approach from that of flour or meal nutrification. Rice nutrification is accomplished by mixing 1 nutrified grain or kernel with 199 kernels of regular white rice. The nutrified kernel can be made either by coating white rice kernels with the nutrients or by producing simulated kernels made from a cereal dough composition containing the nutrients. Commercially, only the former is being carried out at present. There are two procedures for manufacturing the nutrified rice grains. One of these, referred to as the HLR–Mickus procedure, was developed at Roche many years ago and described by Mickus (*664*). This procedure essentially entails soaking the rice in an acidified mixture of water-soluble vitamins, then adding various coatings, followed by a powdered iron source, vitamin A, and calcium, in a Trumbol mixer, and finally whitening and polishing the nutrified grains. A second variation is the Wright process, in which all the vitamins and minerals are introduced in powder form with suitable coating agents. Both procedures yield a stable, coated product with good vitamin A retention (*201*) (Table 98) and rinse resistance, and the nutrients are biologically available in the cooked rice. With any nutrified rice product, excess water should not be used in cooking, since water poured off after cooking will mean some loss of nutrients.

4. Sugar (16,24,26–30,427,947)

A food carrier that is to be nutrified must be one or several foods that are consumed on a regular basis throughout the year by a given population. In Central America, crystalline white sucrose is such a food carrier for micronutrients. The

TABLE 98

Stability of HLR-Mickus Nutrified Rice Premix and Retention after Cooking[a]

	Per lb (454 gm) of premix			Per lb of rice	
Nutrient	Initial	6 months, room temp (23°C)	4 weeks (45°C)	Uncooked rice with premix diluted 1:200	Rice cooked nutrified (1380 gm)[b]
Vitamin A[c]	1,200,000	1,050,000	1,050,000	6,000	5,800
Pyridoxine (mg)	400	400	390	2.1	2.5
Folic acid (mg)	46	44	41	0.23	0.27
Vitamin E (IU)	3,240	3,210	3,200	16	15.9
Thiamin (mg)	550	530	NR[d]	2.6	NR[d]

[a] From Cort *et al.* (*201*).
[b] One pound of uncooked rice = 1380 gm of cooked rice.
[c] Type 250-SD.
[d] NR, not run.

amount of vitamin A added to sugar provides the FAO–WHO recommended daily allowance for an adult man, namely, 750 μg or 2500 IU.

A project was initiated at INCAP with the cooperation of Roche in 1969 to test and implement the nutrification of sugar with a suitable form of vitamin A. Of the vitamin A types on the market, vitamin A palmitate, types 250-SD and 250-CWS, were found to be adaptable to the project. Product type 250-CWS was finally chosen. It is a pale yellow beadlet product with particles approximately 100–450 μm in size. Separation of the vitamin A particles from the sugar crystals is prevented by providing an edible bonding agent when the dry vitamin A is blended with the crystalline sugar. In the nutrification of sugar, a vitamin A–sugar premix is first prepared. This premix is added in a set ratio, by means of a mechanical feeder, to the flow of nutrified sugar, followed by adequate mixing to ensure uniformity of the final vitamin A-nutrified product. At the concentration of vitamin A in nutrified sugar (50 IU/gm), no significant color or flavor problems were encountered when the sugar was consumed in hot coffee, tea, orange, pineapple, and lemon beverages, and other food products. Biological effectiveness of the added vitamin A in sugar was demonstrated by animal assay and by increased vitamin A levels in blood serum of mothers, babies, and children and in the milk of nursing mothers after the consumption of vitamin A-nutrified sugar. Stability data (Table 99) demonstrate adequate retention of added vitamin A (*52*) in the product for time periods involved in manufacturing, marketing, and consumer use. Lengthy storage of vitamin A-nutrified sugar is not necessary since sugar and vitamin A are regularly available for the nutrifying operations. Nutrification of sugar is commercially practiced in Costa Rica, Guatemala, Panama, and Honduras and is expected to occur in El

TABLE 99

Stability of Vitamin A in Nutrified Sugar[a]

			Retention of vitamin A (%)						
			45°C			23°–25°C			
Product	Trial	Type of vitamin A	1 month	2 months	3 months	1 month	2 months	3 months	6 months
Premix[b]									
A	1	250-CWS	92	83	—	99	98	—	99
B	1	250-SD	87	79	—	97	95	—	96
C	2	250-CWS	—	—	86	—	—	92	94
D	2	250-SD	—	—	77	—	—	93	89
E	3	250-CWS	93	—	81	97	—	90	90
F	3	250-SD	92	—	74	100	—	95	93
Nutrified sugar[c]									
G	4	250-CWS	91	—	76	96	—	96	92
H	4	250-SD	90	—	73	100	—	88	85

[a] From Bauerenfeind (*52*).
[b] Vitamin A–sugar premix (50,000 IU/gm).
[c] Vitamin-nutrified sugar (50–70 IU/gm).

Salvador and Nicaragua, Interest has been shown in Brazil, Haiti, Indonesia, and Chile.

5. *Salt (NaCl)*

Technical problems exist in the nutrification of salt. If the salt is of pure grade, low moisture content, and uniform small particle size and is packaged in moisture-resistant containers, it can be successfully nutrified with a dry form of vitamin A ester, as is evidenced by data obtained with white fine crystalline salt. This salt was nutrified with a premix containing vitamin A palmitate (type 250-SD), thiamin, riboflavin, pyridoxine, niacin, and ascorbic acid. Over 90% of the vitamin A was retained after 1 year of storage. In the United States vitamin-nutrified uniform crystal salt is employed by some bakers for the nutrification of bread, buns, and rolls with thiamin, riboflavin, niacin, and iron. Another nutrified salt has been developed for toppings on snacks such as potato chips to provide supplementary vitamin A, thiamin, riboflavin, niacin, and ascorbic acid. The general experience with nutrified, high-grade, uniform crystal salt, therefore, is similar to that with nutrified sucrose. Salt must be centrally produced, have some degree of standardized quality, and be subjected to proper controls in the nutrification process and in subsequent packaging and distribution programs.

Salt from several countries, among them India and Java, received for the purpose of nutrification trials has been damp, of high moisture content (as high as 4%), of highly variable crystal size (up to 12 mm or more), and containing dissolved trace mineral impurities. As mentioned earlier, vitamin A is a labile compound subject to these stress factors found in impure salt. These stress conditions overcome the stabilizing characteristics built into the dry vitamin A products. There is no known technology that can bring about the successful uniform nutrification of these cruder salt products.

6. *Monosodium Glutamate (873–876)*

There are situations with some seasoning items wherein small-volume packaging is practiced and daily or uniform use of such packages is employed. Such is the case in the Philippines,where surveys sponsored by the Cebu Institute of Medicine indicated that a high percentage of the children consumed monosodium glutamate (MSG) in seasoned food prepared in the home, that little variation existed in per capita consumption, and that MSG was centrally produced and marketed by two local firms. Monosodium glutamate is produced as a pure crystalline substance of uniform physical characteristics and of low moisture content. With the cooperation of Philippine personnel and Cornell consultants, a study was initiated at Roche to nutrify MSG since all indications showed MSG to be a promising vehicle for conveying vitamin A to the Philippine population. A technology was developed to bring about uniformity of particle distribution of a vitamin A-nutrified MSG. Packets (15,000 IU) were prepared and stored to determine vitamin A stability, which was found to be good to excellent over a 6-month period. Since MSG is locally available in the Philippines, there is no need for lengthy storage. Biochemical and clinical studies showed that vitamin A-nutrified MSG is an effective vitamin A delivery system. Other countries consuming MSG on a regular basis are considering it as a potential carrier of micronutrients.

More recently there has been interest in incorporating both added vitamin A and iron into MSG. A relationship between serum vitamin A and hemoglobin iron or anemia in human beings has been observed (*414,616,696*). On the basis of laboratory trials in progress, conducted by Roche, it appears feasible technically to add both dry vitamin A palmitate and either micronized ferric orthophosphate or coated ferrous sulfate to MSG (*80*).

7. Tea (147,924)

Tea is widely consumed in some parts of the world. The idea of using tea for conveying vitamin A to an entire population was conceived at the New Delhi Mission of AID as early as 1967.

It was determined at Roche (*147*) that tea dust could be nutrified by dry-mixing the dust with a fine powdered vitamin A ester product (type 250-SD). Tea dust so nutrified retained 85% of the vitamin A activity after storage for 1 year at room temperature (23°C). Tea leaves were sprayed with an emulsion of vitamin A palmitate. Satisfactory stability resulted for as long as 1 year. Vitamin A palmitate applied both in the powdered form to the dust and as an emulsion form sprayed onto leaves showed 100% retention even after 1 hr of boiling in the tea-brewing process.

Trials on a commercial scale in a tea factory in south India have demonstrated the feasibility of nutrifying large quantities of tea during its processing, either in individual factories or in tea-blending plants. Presumably tea leaves would be sprayed after the fermentation stage, near the final drying stage. Tea has been nutrified with both vitamin A and iodine (*924*). Some interest in nutrifying tea has been shown in the past in India and Pakistan.

E. Pilot-Type Production Trials

In order to provide guidelines relative to the stability of vitamin added to a variety of processed foods, a tabulation of results of commercial or pilot-type trials (*70*) made with prevailing equipment has been prepared (Table 100). In most food development projects, the appropriate market form of the vitamin is chosen after the stage and mode of addition are decided. The product is prepared and packaged simulating the anticipated commercial process as closely as possible. Samples freshly prepared and others previously stored are used to determine organoleptic effects on the food and the stability of the added vitamin A. In almost all nutrification practices, an input or overage above the label claim is required because of (a) possible imperfect distribution of the added nutrient, (b) analytical assay variation, and (c) potential degradation of the nutrient during processing and storage. The overages required vary with (a) the stability characteristics of vitamin A form, (b)

TABLE 100

Stability of Added Vitamin A in Food Products[a]

Product	Packaging	Type of vitamin A added	Method of addition	Vitamin A assay after processing	Vitamin A level (%) after storage, 70°–75°F (23°C)				
					2 months	3 months	4 months	6 months	12 months
Breakfast cereal	Liner box	Palmitate PIMO	Before extrusion	4600 IU/oz	I	89	—	83	—
	Liner box	Palmitate emulsion[b]	Spray	2000 IU/oz	100[c]	100[c]	—	97	95
Canned bread	Can	Palmitate 250-S	In water–flour mix	2245 IU/1¾ oz	—	—	—	99	100
	Can	Palmitate 250-S[b]	In water–flour mix	2260 IU/1¾ oz	—	—	—	100	100
Chocolate bars	Film, paper	Palmitate PIMO[b]	In chocolate	3600 IU/1½ oz	—	—	—	100	100
Chocolate drink, dry									
Milk additive	Can	Palmitate PIMO[b]	In cocoa	2600 IU/¾ oz	—	98	—	96	75
	Can	Palmitate 250-S[b]	Dry mix	2725 IU/¾ oz	—	97	—	94	76
	Can	Palmitate 250-S[b]	Dry mix	2150 IU/¾oz	—	—	—	93	92[d]
	Can	Palmitate 250-S[b]	Dry mix	2200 IU/¾ oz	—	—	—	89	90[d]
Instant type	Film, pouch	Palmitate 250-S	Dry mix	2200 IU/¾ oz (reconstituted)	—	96	—	87	87
	Liner box	Palmitate 250-S	Dry mix	2670 IU/qt (reconstituted)	—	100	—	90	90
	Liner box	Palmitate 250-CWS	Dry mix	2600 IU/qt (reconstituted)	—	100	—	96	100
Chocolate drink, liquid	Bottle	Palmitate 250-S[b]	Disperse in liquid	392 IU/100 gm	—	93	—	93	92
Chocolate milk shake, dry (milk additive)	Liner box	Palmitate 250-CWS[b]	Dry mix	5280 IU/30 gm	—	100	—	100	—
Chocolate milk powder	Foil pouch	Palmitate 250-S[b]	Dry mix	23800 IU/lb	—	84[c]	—	92	96
Chocolate powder (milk additive)	Liner box	Palmitate 250-CWS[b]	Dry mix	6350 IU/20 gm	—	—	—	98	88
	Liner box	Palmitate 250-CWS[b]	Dry mix	6200 IU/20 gm	—	—	—	90	82
	Liner box	Palmitate 250-CWS[b]	Dry mix	6400 IU/20 gm	—	—	—	94	87

(*Continued*)

TABLE 100—*Continued*

Product	Packaging	Type of vitamin A added	Method of addition	Vitamin A assay after processing	Vitamin A level (%) after storage, 70°–75°F (23°C)				
					2 months	3months	4 months	6 months	10 months
Chocolate syrup	Aerosol can	Palmitate Pl.8[b]	Wet blend	4200 IU/oz	—	100	—	100	83
	Aerosol can	Palmitate Pl.8[b]	Wet blend	7200 IU/oz	—	88	—	78	76
	Can	Palmitate PIMO[b]	Wet blend	2800 IU/56 gm	—	—	—	95	88
	Can	Palmitate PIMO	Wet blend	2940/24 gm	—	100	—	100	—
	Can	Palmitate PIMO	Wet blend	3150/24 gm	—	96	—	96	100
Cocoa powder	Can	Palmitate 250-S	Dry mix	45,000 IU/100 gm	—	90	—	89	—
Cookies	Liner box	Palmitate 250-S[b]	Dough stage	5050 IU/lb	82	—	—	—	—
Cookie filling									
Butterscotch	Glass jar	Palmitate 250-CWS[c]	In sugar–water mix	1730 IU/gm	100	—	—	100	—
Fudge	Glass jar	Palmitate 250-CWS[c]	In sugar	1890 IU/gm	100	—	—	97	—
Corn meal									
15% moisture	Liner box	Palmitate 250-S	Dry mix	7700 IU/lb	—	94	—	86	—
12% moisture	Liner box	Palmitate 250-S	Dry mix	7200 IU/lb	—	94	—	92	—
Dietary candy	Liner box	Palmitate PIMO[b]	Blend in end of heat stage	35 IU/gm	—	100	—	89	100
Evaporated milk	Can	Palmitate Pl.8[b]	Homogenized in fat phase	6400 IU/qt (reconstituted)	—	—	—	100	—
Flour									
White	Liner bag	Palmitate 250-SD	Dry mix	5000 IU/lb	94	—	93	90	—
Premix	Liner bag	Palmitate 250-SD	Dry mix	73,000 IU/gm	100	—	100	90	87
Fruit punch	Can	Palmitate 250-CWS[c]	Disperse in water phase	5370 IU/9 fl oz	—	100	—	98	80
Graham crackers	Liner bag	Palmitate 250-S[b]	Dough stage	4400 IU/lb	—	85[f]	—	73	85[c]
Grapefruit drink, conc. (frozen)	Can	Palmitate 250-S[c]	Disperse in warm concentrate	6250 IU/6 fl oz	—	100[h]	—	89	—
Infant formula, liquid	Can	Palmitate Pl.8	Homogenize in fat phase	2860 IU/13 fl oz	—	—	100	87	90
	Can	Palmitate PIMO[c]	Homogenize	2570 IU/13 fl oz	—	99	—	93	86
	Can	Palmitate PIMO	Homogenize	2850 IU/qt (reconstituted)	—	—	—	98	98 89[g]
Infant formula, liquid (with added iron)	Can	Palmitate PIMO	Homogenize	2900 IU/qt (reconstituted)	—	—	—	98	96 95[g]

Infant formula, liquid (soy base)	Can	Palmitate 250-S[b]	Disperse in warm base	8300 IU/liter	—	98	—	94	90
Infant formula, dry	Can	Palmitate PIMO	Blend in condensed stage	2800 IU/160 gm	—	—	—	100	100 100[g]
Infant formula, dry (with added iron)	Can	Palmitate PIMO	Blend in condensed stage	2850 IU/160 gm	—	—	—	98	98 96[g]
Infant formula, dry	Can	Palmitate Pl.8	Blend in condensed stage	19,400 IU/lb	—	—	—	—	85
Infant formula, dry (soy base)	Can	Palmitate PIMO	Blend in condensed stage	34,000 IU/lb	—	100	—	91	93
Instant milk mixer (dry product)	Film pouch	Palmitate 250-CWS	Dry mix	1400 IU/25 gm	—	—	—	100	100
Instant snack (dry milk additive)	Film envelope	Palmitate 250-CWS[e]	Dry mix	1600 IU/30 gm	90	91	—	93	—
	Film envelope	Palmitate 250-CWS[e]	Dry mix	4800 IU/30 gm	94	85	—	85	—
Jam	Glass jar	Palmitate 250-S[c]	In water, end of heat stage	1220 IU/oz	100	—	—	—	—
Juice concentrate (frozen)	Can	Palmitate 250-CWS[c]	End of heat cycle	1090 IU/4 fl oz (reconstituted)	—	—	—	—	95[h]
Liquid diet food	Can	Palmitate 250-S	Wet-blending	7100 IU/qt	—	98	—	97	95
Lollipops	Film	Palmitate PIMO[b]	End of heat cycle	3500 IU/pop	—	—	100	96	—
Lollipops	Film	Palmitate PIMO[b]	End of heat cycle						
Pineapple				4100 IU/pop	—	—	—	—	100
Lemon				4500 IU/pop	—	—	—	—	97
Lime				2800 IU/pop	—	—	—	—	100
Low-calorie milk drink	Can	Palmitate 250-S[c]	Wet-blending	7000 IU/qt (reconstituted)	—	100	—	100	93
Low-calorie dietary product	Film package	Palmitate 250-CWS[c]	Dry mix						
Coffee				5200 IU/8 oz	—	88	—	88	77
Vanilla				5500 IU/8 oz	—	90	—	84	77
Chocolate				5450 IU/8 oz	—	91	—	81	72
Low-calorie diet mix, dry	Film packet	Palmitate 250-CWS	Dry mix	2065 IU/2 oz	—	88	—	86	84
Margarine	Film box	Palmitate PIMO	In oil blend	17,900 IU/gm	—	100[i]	—	97	—
	Film box	Acetate AIMO	In oil blend	18,000 IU/lb	—	100[i]	—	97	—
	Film box	Fish liver oil	In oil blend	18,800 IU/lb	—	100[i]	—	89	—
	Film box	Acetate AIMO	In oil blend	17,200 IU/lb	—	100[i]	—	98	—
	Film box	Acetate AIMO	In oil blend	17,700 IU/lb	—	99[i]	—	98	—
	Film box	Palmitate PIMO	In oil blend	19,400 IU/lb	—	100[i]	—	98	—

(*Continued*)

TABLE 100—*Continued*

Product	Packaging	Type of vitamin A added	Method of addition	Vitamin A assay after processing	Vitamin A level (%) after storage, 70°–75°F (23°C) 2 months	3months	4 months	6 months	10 months
Milk amplifier, dry mix	Film pouch	Palmitate 250-S	Dry mix	1120/14 gm	—	85	—	76	73
	Glass jar	Palmitate 250-CWS	Dry mix	1680 IU/37 gm	—	85	—	90	87
	Film pouch	Palmitate 250-CWS	Dry mix	1690 IU/37 gm	—	100	—	100	97
Milk amplifier, dry mix (with minerals)	Glass jar	Palmitate 250-CWS	Dry mix	1720 IU/37 gm	—	96	—	82	79
	Film pouch	Palmitate 250-CWS	Dry mix	1620 IU/37 gm	—	96	—	96	84
Nonfat dry milk	Liner box	Palmitate 250-S	Dry mix	6200 IU/100 gm	94	—	94	92	89
	Liner box	Palmitate PIMO in coconut oil	Homogenize in condensed stage	94 IU/gm	—	94[c]	—	88[c]	84[f]
	Can	Palmitate PIMO in coconut oil	Homogenize in condensed stage	97 IU/gm	—	92[c]	—	86[c]	—
	Film box	Palmitate coconut oil emulsion	Wet-blending	7400 IU/100 gm	—	100	—	98	95
	Film box	Palmitate coconut oil emulsion	Wet-blending	8200 IU/100 gm	—	100	—	98	96
	Liner box	Palmitate 250-CWS	Dry mix	12,600 IU/qt (reconstituted)	—	99	—	100	84
Orange beverage									
Dry mix	Glass jar	Palmitate 250-CWS[b]	Dry mix	1970 IU/16.25 gm	—	100	—	—	87
Powder	Can	Palmitate 250-CWS[e]	Dry mix	1500/4 fl oz (reconstituted)	—	100	—	100	100
Powder	Glass jar	Palmitate 250-CWS	Dry mix	3365 IU/oz	—	94	—	96	92
						89[c]	—	88[c]	83[c]
Orange drink									
Granulate	Glass jar	Palmitate 250-CWS[e]	Dry mix	3450 IU/oz	—	97	—	99	97[f]
								90[c]	
Dry mix	Glass jar	Palmitate 250-CWS[e]	Dry mix	3230 IU/oz	—	96	—	85	80

Frozen concentrate	Can	Palmitate 250-S[b]	Disperse in warm concentrate	6500 IU/6 fl oz (reconstituted)	—	96[h]	—	88[h]	—
				6900 IU/6 fl oz	—	96[h]	—	85[h]	—
Frozen concentrate	Can	Palmitate 250-CWS[e]	Disperse in warm concentrate	1900 IU/4 fl oz	—	100[h]	—	100[h]	87[h]
Frozen concentrate	Can	Palmitate Pl.8[e]	Homogenize in warm concentrate	2110 IU/4 fl oz	—	100[h]	—	100[h]	84[h]
Orange concentrate, frozen	Can	Palmitate 250-S[e]	Disperse in warm concentrate	41,000 IU/kg	—	100[h]	—	100[h]	100[h]
									100[h,j]
		Palmitate 250-S[e]	Disperse in warm concentrate	39,900 IU/kg	—	100[h]	—	—	100[h]
Orange oil, emulsion	Glass jar	Palmitate Pl.8	Homogenize in oil phase	870 IU/gm	—	—	99	—	—
Peanut butter	Glass jar	Palmitate PIMO	Blend after grinding	16,000 IU/gm	—	—	—	—	91
	Glass jar	Palmitate 250-S[b]	Salt premix	16,000 IU/lb	—	100	—	100	—
Pineapple juice	Can	Palmitate	Disperse in warm juice	525 IU/4 fl oz				90[f]	
		250-CWS[e]			—	97	—	75	—
Pineapple–grape fruit drink	Bottle	Palmitate emulsion	Disperse in drink	5200 IU/8 fl oz					
	Bottle	Palmitate	Disperse in drink	4400 IU/8 fl oz	—	78	—	67	—
		250-CWS			—	80	—	68	—
Potato chips	Film bag	Palmitate 250-SD	Salt premix	5000 IU/lb	100	98	—	—	—
Potato flakes	Can	Palmitate	Dry mix						
Air packed		250-CWS[b]		9900 IU/3 oz	84[c]	81	—	75	—
Nitrogen packed				10,900 IU/3 oz	85[c]	97	—	97	—
Potato flakes	Can	Palmitate 250-SD	Dry mix	16,700 IU/lb	100	—	100	90	—
				1,000 IU/lb	100	—	100	92	—
Potato granules	Liner box	Palmitate 250-S	Dry mix	1350 IU/25 gm	—	97	—	98	91
Rice premix	Liner box	Palmitate 250-SD	Added in coating process	3450 IU/gm	100	—	—	92	92
	Film pouch	Palmitate 250-S	Simulated kernel	2500 IU/gm	98	—	97	96	92
Salt, India	Glass jar	Palmitate 250-SD	Dry mix coating						
0.1- to 1-cm crystals				4440 IU/15 gm	9—	93	—	82	85
Crushed				4550 IU/15 gm	—	75	—	—	70
Salt, U.S. (fine crystal)	Glass jar	Palmitate 250-SD	Dry mix coating	4300 IU/15 gm	—	98	—	96	94
Seasoning mix	Film pouch	Palmitate 250-S[b]	Dry mix	8500 IU/27.2 gm	96	96	95	97	—
Soda drink (grape)	Bottle	Palmitate emulsion	Disperse in drink	4500 IU/8 fl oz	—	78	—	57	—
	Bottle	Palmitate 250-CWS	Disperse in drink	4100 IU/8 fl oz	—	74	—	60	—
Sugar premix (El Salto)	Paper box	Palmitate 250-SD	Dry mix	26,000 IU/gm	97	—	93	85	—
Sugar premix	Paper box	Palmitate 250-SD	Dry mix	7000 IU/gm	100	—	—	92	—

(*Continued*)

TABLE 100—*Continued*

Product	Packaging	Type of vitamin A added	Method of addition	Vitamin A assay processing	Vitamin A level (%) after storage, 70°–75°F (23°C)				
					2 months	3months	4 months	6 months	10 months
Sugar	Glass jar	Palmitate 250-SD	Dry mix	85 IU/gm	97	—	93	85	80
	Paper box	Palmitate 250-SD	Dry mix	70 IU/gm	100	—	—	95	—
	Paper box	Palmitate 250-SD	Dry mix	70 IU/gm	100	—	—	100	—
Syrup milk amplifier	Glass jar	Palmitate 250-CWS[b]	Disperse in warm syrup						
Root beer flavor				2580/46 gm	—	97	—	86	79
Berry flavor				2580 IU/46 gm	—	100	—	88	81
Tea dust	Glass jar	Palmitate 250-S	Dry mix	200 IU/gm	100	—	—	100	98
	Glass jar	Palmitate 250-SD	Dry mix	200 IU/gm	96	—	—	94	90
Tea leaves	Glass jar	Palmitate or acetate emulsion in 50% sucrose	Spray	1000 IU/5 gm	—	—	—	90	—
Tomato soup (condensed)	Can	Palmitate 250-S	Disperse in warm soup	1535 IU/8 fl oz (reconstituted)	—	100	—	83	—
Vegetable oil	Amber bottle	Palmitate Pl.8[k]	Blend in oil						
Air pack				11,400 IU/lb	—	—	—	89[l]	—
Air pack and anti-oxidants				11,000 IU/lb	—	—	—	95[l]	—
Vitamin mix packet	Foil packet	Palmitate 250-CWS[b]	Dry mix	2100 IU/½ oz	96	—	92	90	—

[a] From Bauernfeind and Cort (*70*).
[b] Other vitamins or vitamins and minerals were also added.
[c] 98°F storage.
[d] 9-month storage.
[e] Ascorbic acid also added.
[f] 86°F storage.
[g] 24-month storage.
[h] −10°F storage.
[i] Alternate storage at 75°F and 45°F for periods of 24 hours.
[j] 18-month storage.
[k] Beta-carotene also added for color and vitamin A value.
[l] 5 months indirect light exposure, 75°F.

TABLE 101A
Recommended Daily Dietary Allowances[a,b]

	Age (years)	Weight (kg)	Weight (lbs)	Height (cm)	Height (in)	Protein (gm)	Fat-soluble vitamins: Vitamin A (μg R.E.)[c]	Vitamin D (μg)[d]	Vitamin E (mg α T.E.)[e]	Water-soluble vitamins: Vitamin C (mg)	Thiamin (mg)	Riboflavin (mg)	Niacin (mg N.E.)[f]	Vitamin B_6 (mg)	Folacin[g] (μg)	Vitamin B_{12} (μg)	Minerals: Calcium (mg)	Phosphorus (mg)	Magnesium (mg)	Iron (mg)	Zinc (mg)	Iodine (μg)
Infants	0.0–0.5	6	13	60	24	kg × 2.2	420	10	3	35	0.3	0.4	6	0.3	30	0.5[h]	360	240	50	10	3	40
	0.5–1.0	9	20	71	28	kg × 2.0	400	10	4	35	0.5	0.6	8	0.6	45	1.5	540	360	70	15	5	50
Children	1–3	13	29	90	35	23	400	10	5	45	0.7	0.8	9	0.9	100	2.0	800	800	150	15	10	70
	4–6	20	44	112	44	30	500	10	6	45	0.9	1.0	11	1.3	200	2.5	800	800	200	10	10	90
	7–10	28	62	132	52	34	700	10	7	45	1.2	1.4	16	1.6	300	3.0	800	800	250	10	10	120
Males	11–14	45	99	157	62	45	1000	10	8	50	1.4	1.6	18	1.8	400	3.0	1200	1200	350	18	15	150
	15–18	66	145	176	69	56	1000	10	10	60	1.4	1.7	18	2.0	400	3.0	1200	1200	400	18	15	150
	19–22	70	154	177	70	56	1000	7.5	10	60	1.5	1.7	19	2.2	400	3.0	800	800	350	10	15	150
	23–50	70	154	178	70	56	1000	5	10	60	1.4	1.6	18	2.2	400	3.0	800	800	350	10	15	150
	51+	70	154	178	70	56	1000	5	10	60	1.2	1.4	16	2.2	400	3.0	800	800	350	10	15	150
Females	11–14	46	101	157	62	46	800	10	8	50	1.1	1.3	15	1.8	400	3.0	1200	1200	300	18	15	150
	15–18	55	120	163	64	46	800	10	8	60	1.1	1.3	14	2.0	400	3.0	1200	1200	300	18	15	150
	19–22	55	120	163	64	44	800	7.5	8	60	1.1	1.3	14	2.0	400	3.0	800	800	300	18	15	150
	23–50	55	120	163	64	44	800	5	8	60	1.0	1.2	13	2.0	400	3.0	800	800	300	18	15	150
	51+	55	120	163	64	44	800	5	8	60	1.0	1.2	13	2.0	400	3.0	800	800	300	10	15	150
Pregnant						+30	+200	+5	+2	+20	+0.4	+0.3	+2	+0.6	+400	+1.0	+400	+400	+150	[i]	+5	+25
Lactating						+20	+400	+5	+3	+40	+0.5	+0.5	+5	+0.5	+100	+1.0	+400	+400	+150	[i]	+10	+50

[a] See Food and Nutrition Board, National Academy of Sciences—National Research Council (1979).

[b] The allowances are intended to provide for individual variations among most normal persons as they live in the United States under usual environmental stresses. Diets should be based on a variety of common foods in order to provide other nutrients for which human requirements have been less well defined.

[c] Retinol equivalents. 1 retinol equivalent = 1 μg retinol or 6 μg β-carotene.

[d] As cholecalciferol. 10 μg cholecalciferol = 400 IU vitamin D.

[e] α tocopherol equivalents. 1 mg d-α-tocopherol = 1 α T.E.

[f] 1 N.E. (niacin equivalent) is equal to 1 mg of niacin or 60 mg of dietary tryptophan.

[g] The folacin allowances refer to dietary sources as determined by *Lactobacillus casei* assay after treatment with enzymes (''conjugases'') to make polyglutamyl forms of the vitamin available to the test organism.

[h] The RDA for vitamin B_{12} in infants is based on average concentration of the vitamin in human milk. The allowances after weaning are based on energy intake (as recommended by the American Academy of Pediatrics) and consideration of other factors such as intestinal absorption.

[i] The increased requirement during pregnancy cannot be met by the iron content of habitual American diets nor by the existing iron stores of many women; therefore the use of 30–60 mg of supplemental iron is recommended. Iron needs during lactation are not substantially different from those of nonpregnant women, but continued supplementation of the mother for 2–3 months after parturition is advisable in order to replenish stores depleted by pregnancy.

TABLE 101B

Estimated Safe and Adequate Daily Dietary Intakes of Additional Selected Vitamins and Minerals[a,b]

		Vitamins			Trace elements[c]						Electrolytes		
	Age (years)	Vitamin K (μg)	Biotin (μg)	Pantothenic Acid (mg)	Copper (mg)	Manganese (mg)	Fluoride (mg)	Chromium (mg)	Selenium (mg)	Molybdenum (mg)	Sodium (mg)	Potassium (mg)	Chloride (mg)
Infants	0–0.5	12	35	2	0.5–0.7	0.5–0.7	0.1–0.5	0.01–0.04	0.01–0.04	0.03–0.06	115–350	350–925	275–700
	0.5–1	10–20	50	3	0.7–1.0	0.7–1.0	0.2–1.0	0.02–0.06	0.02–0.06	0.04–0.08	250–750	425–1275	400–1200
Children and adolescents	1–3	15–30	65	3	1.0–1.5	1.0–1.5	0.5–1.5	0.02–0.08	0.02–0.08	0.05–0.1	325–975	550–1650	500–1500
	4–6	20–40	85	3–4	1.5–2.0	1.5–2.0	1.0–2.5	0.03–0.12	0.03–0.12	0.06–0.15	450–1350	775–2325	700–2100
	7–10	30–60	120	4–5	2.0–2.5	2.0–3.0	1.5–2.5	0.05–0.2	0.05–0.2	0.1 –0.3	600–1800	1000–3000	925–2775
	11+	50–100	100–200	4–7	2.0–3.0	2.5–5.0	1.5–2.5	0.05–0.2	0.05–0.2	0.15–0.5	900–2700	1525–4575	1400–4200
Adults		70–140	100–200	4–7	2.0–3.0	2.5–5.0	1.5–4.0	0.05–0.2	0.05–0.2	0.15–0.5	1100–3300	1875–5625	1700–5100

[a] Recommended dietary allowances; revised 1979. Food and Nutrition Board, National Academy of Sciences–National Research Council.
[b] Because there is less information on which to base allowances, these figures are not given in Table 101A and are provided here in the form of ranges of recommended intakes.
[c] Since the toxic levels for many trace elements may be only several times usual intakes, the upper levels for the trace elements given in this table should not be habitually exceeded.

TABLE 102

Tentative List of Countries and Territories in Which Vitamin A Deficiency Is a Public Health Problem[a]

Afghanistan	Iran	Sao Tome and Principe
Antiqua	Iraq	Senegal
Benin	Jordan	Socialist Republic of Viet Nam
Bangladesh	Kenya	Somalia
Bolivia	Lao People's Democratic Republic	South Africa
Brazil	Lebanon	Southern Rhodesia
Burma	Libyan Arab Republic	Sri Lanka
Cape Verde	Malawi	St. Kitts—Nevis and Anguilla
Chile	Malaysia	St. Lucia
Colombia	Maldives	St. Vincent
Comores	Mali	Sudan
Democratic Kampuchea	Mauritania	Syrian Arab Republic
Democratic People's Republic of Korea	Mozambique	Thailand
Dominica	Nepal	Togo
Ecuador	Nicaragua	Tonga
Egypt	Niger	Tunisia
El Salvador	Nigeria	Turkey
Ethiopia	Pakistan	Tuvalu
Fiji	Panama	United Republic of Tanzania
Ghana	Papua New Guinea	Upper Volta
Grenada	Peru	Uganda
Haiti	Philippines	Uruguay
Honduras	Republic of Korea	Zaire
India	Rwanda	Zambia
Indonesia		

[a] Adopted from PAG Bulletin (*729*). Prepared by WHO for guidance in world food programs. The list includes countries in which the existence of xerophthalmia has been reported, even if only a small area of the country.

the specific food application, and (c) the food plant process and type of packaging. Recommended daily allowances are indicated in Table 101.

XVI. VITAMIN A DEFICIENCY AROUND THE WORLD

Xerophthalmia occurs in infants and young children in certain geographic areas of the world, such as Asia, the Middle East, Africa, and Central and South America, where intakes of vitamin A food sources are inadequate. It is estimated that more than 250,000 children around the world may go blind annually as a result of this nutritional deficiency. In addition to ocular symptoms, other outward symptoms of vitamin A deficiency are growth depression, suspected greater susceptibility to infectious and parasitic diseases, and, in many instances, death (see Table 102).

Through the cooperative efforts of AID of the U.S. State Department, WHO, UNICEF, international blindness societies, and other international and national groups, a concerted effort is being made to wipe out xerophthalmia throughout the world. An International Vitamin A Consultative Group (IVACG) was formed in 1975 by representatives of the above groups for this purpose. Programs at various stages and of various types have been carried out, are underway, or are planned in India, Indonesia, Bangladesh, Pakistan, Sri Lanka, Haiti, Guatemala, Costa Rica, El Salvador, Panama, Brazil, the Philippines, and other countries.

A number of intervention programs have been developed to overcome the vitamin A deficiency problem, one approach of which is the vitamin A nutrification of food. A food to be nutrified should be one that is regularly consumed by the segment of the population to be protected against the deficiency disease. The food also must be one that can be centrally processed by an existing technology or one that can be developed. Obviously, the nutrified product should not have a changed appearance or taste, and the added vitamin should be physiologically available to the body. Last, there should be some monitoring system to gauge the intake of the nutrified product in the population.

Nutritional interventions exist to minimize xerophthalmia or eradicate it entirely, but a strong continuous effort is required with full government commitment to the program.

REFERENCES

1. Abboud, I. A., Osman, H. G., and Massoud, W. H. (1968). *Med. J. Cairo Univ.* **36**(1), 1-13.
2. Abolina, I. (1970). *Latv. Lauksaimn. Akad. Raksti* **25,** 26-28.
3. Abrahamson, I. A., and Abrahamson, I. A., Jr. (1962). *arch. Ophthalmol.* **68,** 4-7.
4. Aczel, A. (1977). *Lebensm.-Wiss. + Technol.* **10**(2), 120-
5. Aczel, A. (1976). *Ind. Obst.- Gemeuseverwert.* **61**(8), 203-205.
6. Aczel, A. (1974). *Ind. Obst.- Gemueserverwert.* **59**(18), 491-496.
7. Aczel, A. (1973). *Lebensm.-Wiss. + Technol.* **6**(1), 36-37.
8. Aczel, A. (1972). *Ind. Obst.- Gemueseverwert.* **57**(16), 436-440.
9. Aczel, A. (1971). *Ind. Obst.- Gemueseverwert.* **56**(16), 465-467.
10. Aczel, A. (1970). *Elelmiszervizsgalati Kozl.* **16**(4/5), 217-223.
11. Aczel, A. (1969). *Konzerv-Paprikaip.* **3,** 60-62.
12. Aczel, A. (1969). *Elelmiszervizsgalati Kozl.* **14,** 246-251; **15**(1), 35-39.
13. Adachi, K. (1970). *Jpn. Food Sci.* **9**(2), 48-52.
14. Adam, W., and Körner, W. F. (1968). *Int. Z. Vitaminforsch.* **38,** 160-165.
15. Adsule, P. G., and Susante, K. (1974). *J. Food Sci. Technol.* **11**(6), 269-273.
16. Aguilar, J. R., Arroyave, G., and Gallardo, C. (1977). *Publ. INCAP* **E-913.**
17. Alause, J., and Feillet, P. (1970). *Tec. Molitoria* **21**(18), 511-517.
18. Alexander, J. (1971). *Lik. Milling* **153**(5), 48-50, 52, 56; (6), 32-33, 36-37; (7), 33-34, 36-38; (8), 30, 32-34, 36.
19. Almond, S., and Logan, R. F. L. (1942). *Br. Med. J.* **2,** 239-241.
20. Al-Vagab, K. M. (1971). *Visn. Sil's'kogospod. Nauki* **9,** 56-58.
21. Anderson, R. H., Maxwell, D. L., Mulley, A. E., and Fritsch, C. H. (1976). *Food Technol.* **30**(5), 110-114.
22. Androsova, O. G., and Skrobanskii, G. G. (1967). *Tr. Khar'k. S-kh. Inst.* **61,** 129-135.
23. A.O.A.C. (1975). "Methods of Analysis." Assoc. Off. Anal. Chem., Washington, D.C.
24. Araujo, R. L., Souza, M. S. L., Mata-Machado, A. J., Mata-Machado, L. T., Lourdes Mello, M., Costa Cruz, T. A., Vieira, E. C., Souza, D. W. C., Palhares, R. D., and Borges, E. L. (1978). *Nutr. Rep. Int.* **17**(3), 307-314.

25. Arnrich, L., and Morgan, A. F. (1954). *J. Nutr.* **54**(1), 107–119.
26. Arroyave, G. (1977). *Int. Vitam. A. Consult. Group Meet.*, Geneva.
27. Arroyave, G. (1972). *West. Hemisphere Nutr. Congr., Proc., 3rd, 1971* pp. 68–79.
28. Arroyave, G., Aguiler, J. R., and Flores, M. (1978). *Nutr. Transition, Proc., West. Hemisphere Nutr. Congr., 5th, 1977* p. 46–56.
29. Arroyave, G., Aguilar, J. R., and Portela, I. E. (1975). *Pub. INCAP* **E-853.**
30. Arroyave, G., Beghin, I., Flores, M., Stoto de Guido, C., and Ticas, J. M. (1974). *Arch. Latino-am. Nutr.* **24,** 485–512.
31. Arroyave, G., Wilson, D., Contreras, G., and Behar, M. (1963). *J. Pediat.* **62,** 920–925.
32. Arthur, J. C., Jr., and McLemore, T. A. (1957). *J. Agric. Food Chem.* **5,** 863–867.
33. Asselbergs, E. A., Wyszecki, G. W., and Mohr, W. P. (1961). *Food Technol.* **15,** 156–159.
34. Association of Vitamin Chemists (1966). "Methods of Vitamin Assay," pp. 97–122. Wiley (Interscience), New York.
35. Auckland, G. (1952). *Br. Med. J.* **2,** 267–268.
36. Aust, H. (1955). *Milchwiss. Ber.* **5,** 85–92.
37. Austin, J. E. (1978). *Cereal Foods World* **23**(5), 229–233, 265.
38. Bagdasaryan, E. G., Tarosova, E. O., and Ter-Karapetyan, M. A. (1971). *Biol. Zh. Arm.* **24**(10), 38–43.
39. Bagdon, R. E., Impellizzeri, C., and Osadca, M. (1962). *Toxicol. Appl. Pharmacol.* **4**(4), 444–456.
40. Bagdon, R. E., Zbinden, G., and Studer, A. (1960). *Toxicol. Appl. Pharmacol.* **2**(2), 225–236.
41. Bailey, A. C., Oliver, G. D., Singleton, W. S., and Fisher, G. S. (1943). *Oil Soap (Chicago)* **20,** 251–255.
42. Baldwin, T. T., and Armstrong, J. G. (1969). *Can. Inst. Food Technol. J.* **1**(4), 154–159.
43. Baloch, A. K., Buckle, K. A., and Edwards, R. A. (1977). *J. Food Technol.* **12**(3), 285–293.
44. Bancher, E., Washuettl, J., and Riederer, P. (1971). *Seifen, Oele, Fette, Wachse,* **97**(5), 121–126; *Monatsh. Chem.* **103**(21), 464–472 (1972).
45. Bancher, E., Washuettl, J., and Riederer, P. (1970). *Lebensm. Ernaehr.* **23**(1/2), 4–7.
46. Barber, M. S., Hardisson, A., Jackman, L. M., and Weedon, B. C. L. (1961). *J. Chem. Soc.* pp. 1625–1630.
47. Barber, M. S., Jackman, L. M., Warren, C. K., and Weedon, B. C. L. (1960). *Proc. Chem. Soc. London,* pp. 19–20.
48. Barber, M. S., Jackson, L. M., and Weedon, B. C. L. (1960). *Proc. Chem. Soc., London* pp. 23–24.
49. Bates, R. P., and Mathews, R. F. (1976). *Proc. Fla. State Hortic. Soc.* **88,** 266–271.
50. Batista, F. M., and Gomes, S. M. (1969). *Hospital (Rio de Janeiro)* **76,** 489–494.
51. Bauernfeind, J. C. (1980). *In* "Vitamin E" (L. Machlin, ed.) pp. 99–167.
52. Bauernfeind, J. C. (1978). *Proc. Int. Cong. Nutr. 11th,* Rio de Janeiro.
53. Bauernfeind, J. C. (1978). *In* "The Encyclopedia of Food Science" (M. S. Peterson and A. H. Johnson, eds.), pp. 113–126. Avi Publ. Co., Westport, Connecticut.
54. Bauernfeind, J. C. (1978). *In* "The Encyclopedia of Food Science" (M. S. Peterson and A. H. Johnson, eds.), pp. 788–843. Avi Publ. Co., Westport, Connecticut.
55. Bauernfeind, J. C. (1975). *Food Technol.* **29**(5), 48–49.
56. Bauernfeind, J. C. (1974). *Jt. WHO/US-AID Meet. Control Vitam. A Defic.,* Jakarta.
57. Bauernfeind, J. C. (1974). *In* "The Encyclopedia of Food Technology" (A. H. Johnson and M. S. Peterson, eds.), pp. 163–173. Avi Publ. Co., Westport, Connecticut.
58. Bauernfeind, J. C. (1973). "The IFT World Directory and Guide," pp. 96–109. Inst. Food Technol., Chicago, Illinois.
59. Bauernfeind, J. C. (1973). "Vitamin A, Xerophthalmia and Blindness," Vol. 3. Office of Nutrition, Technical Assistance Bureau Agency for International Development, U.S. Dept. of State, Washington, D.C.
60. Bauernfeind, J. C. (1972). *J. Agric. Food Chem.* **20**(3), 456–473.

61. Bauernfeind, J. C. (1971). *Proc. SOS/70, Int. Congr. Food Sci. Technol., 3rd, 1970* pp. 217-232.
62. Bauernfeind, J. C. (1959). *Food Can.* **19,** 35-40.
63. Bauernfeind, J. C. (1958). *In* "Symposium on Fruit Juice Concentrates" (Scientific and Technical Commission, International Federal Fruit Juice Producers), pp. 265-290. University of Bristol, Bristol, England.
64. Bauernfeind, J. C. (1953). *Adv. Food Res.* **4,** 359-431.
65. Bauernfeind, J. C., and Allen, L. E. (1963). *J. Dairy Sci.* **46**(3), 245-254.
66. Bauernfeind, J. C., Benevenga, N. J.,and Mertz, W. (1980) *In* "Chemistry and the Food Systems" (ACS), pp. 13 & 88-94.
67. Bauernfeind, J. C., and Brooke, C. L. (1973). *Food Eng.* **45**(6), 91-97, 100.
68. Bauernfeind, J. C., Brubacher, G. H. Kläui, H. M., and Marusich, W. L. (1971). *In* "Carotenoids" (O. Isler, ed.), pp. 743-770. Birkhaeuser, Basel.
69. Bauernfeind, J. C., and Bunnell, R. H. (1962). *Food Technol.* **16**(12), 76-81.
70. Bauernfeind, J. C., and Cort, W. M. (1974). *Crit. Rev. Food Technol.* **4,** 337-375.
71. Bauernfeind, J. C., and Cort, W. M. (1974). *In* "Encyclopedia of Food Technology" (A. Johnson and M. Peterson, eds.), pp. 891-899. Avi Publ. Co., Westport, Connecticut.
72. Bauernfeind, J. C., Marusich, W. L., and Roncalli, R. A. (1972). *Can. Vet. Med. Assoc. Meet., 1972* Quebec.
73. Bauernfeind, J. C., Newmark, H., and Brin, M. (1974). *Am. J. Clin. Nutr.* **27,** 234-253.
74. Bauernfeind, J. C., Osadca, M., and Bunnell, R. H. (1962). *Food Technol.* **16**(8), 101-108.
75. Bauernfeind, J. C., and Parman, G. K. (1964). *Food Technol.* **18**(2), 52-57.
76. Bauernfeind, J. C., and Pinkert, D. M. (1970). *Adv. Food Res.* **18,** 219-315; *in* "The Encyclopedia of Food Technology" (A. Johnson and M. Peterson, eds.), pp. 67-75. Avi Publ. Co., Westport, Connecticut, 1974.
77. Bauernfeind, J. C., Rokosny, A. D., and Siemers, G. F. (1953). *Food Eng.* **25**(6), 81, 82, 85, 87.
78. Bauernfeind, J. C., Rubin, S. H., Surmatis, J. D., and Ofner, A. (1970). *Int. Z. Vitaminforsch.* **40,** 391-416.
79. Bauernfeind, J. C., Smith, E. G., and Bunnell, R. H. (1958). *Food Technol.* **12**(10), 527-535.
80. Bauernfeind, J. C., and Timreck, A. (1978). *Int. Vitam. A Consult. Group Meet.,* Rio de Janeiro, *Int. Anemia Consult. Group Meet., 1978* Campinas, Brazil.
81. Baumgartner, E. (1976). *Alimenta* **15**(2), 31-32.
82. Begum, A., and Pereira, S. H. (1977). *Trop. Geogr. Med.* **29,** 47-50.
83. Bélanger, G., and Giroux, R. N. (1972). *Que. Lait Aliment.* **31**(9), 4-7.
84. Belgian Patent 721, 167/67; 1967.
85. Benedek, L. (1972). *Acta Aliment. Acad. Sci. Hung.* **1**(2), 187-202.
86. Benk, E. (1973). *Gordian* **73**(4), 140.
87. Benk, E. (1967). *Riechst., Aromen, Koerperflegem.* **17**(3), 96, 98.
88. Benk, E. (1964). *Riechst., Aromen, Koerperflegem.* **14,** 143-147; *Ind. Obst.- Gemueseverwert.* **56**(23), 692-694 (1971).
89. Benk, E. (1961). *Dtsch. Lebensm.-Rundsch.* **57,** 324-329.
90. Benk, E., and Brixius, L. (1965). *Fette, Seifen, Anstrichm.* **67,** 65-69.
91. Benk, E., Walff, I., and Bergmann, R. (1963). *Gordian* **63,** 466-468.
92. Berger, S., Rechcigl, M., Jr., Loosli, J. K., and Williams, H. H. (1962). *J. Nutr.* **77,** 174-178.
93. Berger, S., Wertheim, A., and Krzyzanowska, M. (1952). *Przem. Spozyw.* **6,** 459-463.
94. Bernath, P., and Swisher, H. E. (1969). *Food Technol.* **23,** 815-818.
95. Bernhard, R. (1963). *Wiss. Veroeffentl. Dtsch. Ges. Ernaehr.* **9,** 169-178.
96. Bespechalnaya, V. V., and Fraiman, I. A. (1969). *Sadovod. Vinograd. Vinodel, Mold.* **24**(8), 16-18.
97. Beytia, E. D., and Porter, J. W. (1976). *Annu. Rev. Biochem.* **45,** 911-942.
98. *BIBRA Information Bulletin* (1971). **10**(4), 174-176.

99. Bickoff, E. M. (1951). *J. Am. Oil Chem. Soc.* **28**(2), 65–68.
100. Bickoff, E. M., Coppinger, G. M., Livingston, A. L., and Campbell, T. W. (1952). *J. Am. Oil Chem. Soc.* **29,** 51–53.
101. Bickoff, E. M., Livingston, A. L., Guggolz, J., and Thompson, C. R. (1954). *J. Agric. Food Chem.* **2**(24), 1229–1231.
102. Bickoff, E. M., Livingston, A. L., Guggolz, J., and Thompson, C. R. (1952). *J. Am. Oil Chem. Soc.* **29**(11), 445–446.
103. Bickoff, E. M., Livingston, A. L., and Thompson, C. R. (1954). *J. Am. Oil Chem. Soc.* **32,** 64–68.
104. Bickoff, E. M., and Williams, K. T. (1946). *Oil Soap (Chicago)* **23,** 65–68.
105. Bickoff, E. M., Williams, K. T., and Sparks, M. (1945). *Oil Soap (Chicago)* **22,** 128–131.
106. Bienaymé, A. (1955). "Etudes sur de caroténe de l'huile de palme," Ser Sci. No. 8 Inst. Rech. Huiles et Oléagineux, Paris.
107. Biliczky, L., and Csiszar, F. (1975). *Elelmiszervizsgalati Kozl.* **21**(1–2), 47–52.
108. Biliczky, L., and Csiszar, F. (1973). *Hrana Ishrana* **14**(7–8), 304–312.
109. Billings, F. L. (1971). *Pigm. Canned Salmon Canthaxanthin, 2nd Roche Meet.*, London.
110. Birglof, A., and Kjell, S. (1963). *Sven. Mejeritidn.* **55,** 531.
111. Bjornson, L. K., Kayden, H. J., Miller, E., and Moshell, A. N. (1976), *J. Lipid Res.* **17,** 343–352.
112. Blaizot, P., and Cuvier, P. (1953). *J. Am. Oil Chem. Soc.* **30,** 586–587.
113. Blanc, M. (1972). *Lebensm.-Wiss. + Technol.* **5**(3), 95–97.
114. Blattna, J., Krcova, H., and Blattny, C. (1976). *Sb. Vys. Sk. Chem.-Technol. Praze, Potraviny* **E45,** 87–96.
115. Blessin, C. W. (1962). *Cereal Chem.* **39,** 236–242.
116. Blessin, C. W., Brecher, J. D., and Dimler, R. J. (1964). *Cereal Chem.* **41,** 543–548.
117. Blessin, C. W., Brecher, J. D., and Dimler, R. J. (1963). *Cereal Chem.* **40,** 582–586.
118. Blessin, C. W., Brecher, J. D., Dimler, R. J., Grogan, C. O., and Campbell, C. M. (1963). *Cereal Chem.* **40,** 436–442.
119. Blessin, C. W., Dimler, R. J., and Webster, O. J. (1962). *Cereal Chem.* **39,** 389–392.
120. Blessin, C. W., Van Etten, C. H., and Wiebe, R. (1958), *Cereal Chem.* **35,** 359–365.
121. Blomstrand, R., and Werner, B. (1967). *Scand. J. Clin. Lab. Invest.* **19,** 339–345.
122. Blundstone, H. A. W., Woodman, J. S., and Adams, J. B. (1971). *In* "Biochemistry of Fruits and Products" (A. C. Hulme, ed.), pp. 543–572. Academic Press, New York.
123. Bolin, H. R., and Stafford, A. E. (1974). *J. Food Sci.* **39**(5), 1034–1036.
124. Bono, G. D., and Sivieri-Buggiani, S. (1971). *Atti Soc. Ital. Sci. Vet.* **25,** 398–401.
125. Booher, L. E., and Marsh, R. L. (1941). *U.S. Dep. Agric., Tech. Bull.* **802.**
126. Booth, R. G., Kon, S. K., Dann, W. J., and Moore, T. (1933). *Biochem. J.* **27,** 1189–1196.
127. Booth, V. H. (1951). *J. Sci. Food Agric.* **2,** 350–352.
128. Booth, V. H. (1951). *J. Sci. Food Agric.* **2,** 353–358.
129. Booth, V. H. (1945). *J. Soc. Chem. Ind.* **64,** 194–199.
130. Borchgrevink, N. C., and Charley, H. (1966). *J. Am. Diet. Assoc.* **49,** 116–121.
131. Borenstein, B. (1971). *Crit. Rev. Food Technol.* **2**(2), 171–186.
132. Borenstein, B. (1969). *Northwest. Miller* **276**(2), 18–19.
133. Borenstein, B., and Bunnell, R. H. (1967). *Adv. Food Res.* **15,** 195–176.
134. Borenstein, B., and Gordon, H. T. (1972). *Abstr. Pap., 164th Meet. Am. Chem. Soc.* INDE25.
135. Bottrill, D. E., and Hawker, J. S. (1970). *J. Sci. Food Agric.* **21**(4), 193–196.
136. Braddock, R. J., and Kesterson, J. W. (1973). *Proc. Fla. State Hortic. Soc.* **85,** 214–216.
137. Bradley, G. A. (1972). *Hort Science* **7**(2), 141–143.
138. Bradley, G. A., and Dyck, R. L. (1968). *Proc. Am. Soc. Hortic. Sci.* **93,** 402–407.
139. Bradley, G. A., and Rhodes, B. B. (1969). *J. Am. Soc. Hortic. Sci.* **94**(1), 63–65.
140. Branion, N. D., and Emslie, A. R. G. (1966). *Can. Dep. Agric., Publ.* **1238.**
141. Brekke, J. E., Chan, H. T., Jr., and Cavaletto, C. G. (1972). *Food Prod. Dev.* **6**(6) 36–37.

141a. Bressani, R., Duhtimdhse, N. S., and Squibb, R. L. (1957). *Food Res.* **22**(3), 303–307.
142. Briggs, M., and Bennum, M. (1972). *Contraception* **6,** 275–280.
143. British Patent 990, 902 (1965).
144. Britton, G., and Goodwin, T. W. (1969). *Phytochemistry* **8**(11), 2257–2258.
145. Brooke, C. L. (1969). *Proc. West. Hemisphere Nutr. Congr., 2nd, 1968* pp. 137–139.
146. Brooke, C. L. (1968). *J. Agric. Food Chem.* **16,** 163–167.
147. Brooke, C. L., and Cort, W. M. (1972). *Food Technol.* **26**(6), 50–52, 58.
148. Brossard, J., and Mackinney, G. (1963). *J. Agric. Food Chem.* **11**(6), 501–503.
149. Brubacher, G., Schärer, K., Studer, A., and Wiss, O. (1965). *Z. Ernaehrungswiss.* **5,** 190–202.
150. Brubacher, O., Gloor, U., and Wiss, O. (1960). *Chimia* **14,** 19–20.
151. Brunstetter, B. C., and Wiseman, H. G. (1947). *Plant Physiol.* **22,** 421–436.
152. Budowski, P., Ascarelli, I., Gross, J., and Nir, I. (1963). *Science* **142,** 969–971.
153. Budowski, P., and Gross, J. (1965). *Nature (London)* **206,** 1254–1255.
154. Bunnell, R. H. (1968). *J. Agric. Food Chem.* **16**(2), 177–183.
155. Bunnell, R. H., Hoffmann-La Roche Inc., Nutley, New Jersey, (unpublished work).
156. Bunnell, R. H., and Bauernfeind, J. C. (1962). *Food Technol.* **16**(7) 36–43.
157. Bunnell, R. H., and Borenstein, B. (1967). *Food Technol.* **21**(3), 13–16A (331–334).
158. Bunnell, R. H., Driscoll, W., and Bauernfeind, J. C. (1958). *Food Technol.* **12,** 536–544.
159. Burnett, J. H. (1965). *In* "Chemistry and Biochemistry of Plant Pigments" (T. W. Goodwin, ed.), pp. 381–400. Academic Press, New York.
160. Businger, A., Isler, O., Rüegg, R., and Zeller, P. (1958). *J. Sci & Ind. Res. (India)* **17A,** 502–510.
161. Calvarano, M. (1962). *Essenze Deriv. Agrum.* **32**(2), 92–100.
162. Canadian Patent 737,963; 1966.
163. Carballido, A., Lorente, V., and Centeno, M. C. (1972). *An. Bromatol.* **23**(3), 291–309.
164. Carricaburu, P., Tayebi, and Chardenot, P. (1970). *C.R. Hebd. Seances Acad. Sci., Ser. D* **270,** 1282–1283.
165. Carter, R. A., and Cook, G. C. (1963). *Br. J. Nutr.* **17,** 515–522.
166. Cary, C. A., Geddes, W. F., Guilbert, H. R., Hathaway, I. L., Peterson, W. H., Salmon, W. D., Snedecor, G. W., and Zscheile, F. P., Jr. (1947). *U.S., Dep. Agric., Misc. Publ.* **636.**
167. Casalis, J., Chardon, Y., Luquet, F. M., Mainguy, P., and Yver, M. (1972). *Lait* **52,** 28–42.
168. Chan, H. T., Jr., Kuno, M. T. H., Cavaletto, C. S., and Nakayama, T. O. M. (1975). *J. Food Sci.* **40**(4), 701–703.
169. Chanda, R. (1952). *Biochem. J.* **52,** 2P.
170. Chanda, R., Clapham, H. M., and Owen, E. C. (1954). *Biochem. J.* **56,** 453–458.
171. Chapman, V. J., Putz, J. O., Gilpin, G. L., Sweeney, J. P., and Eisen, J. N. (1960). *J. Home Econ.* **52,** 161.
172. Cheesman, D. F. (1954). *Biochem. J.* **58,** 38P.
173. Cheesman, D. F., Lee, W. L., and Zagalsky, P. F. (1967). *Biol. Rev. Cambridge Philos. Soc.* **42,**131–160.
174. Chemical Engineering News (1966). **44**(8) 44–45.
175. Chemical Week (1978). May 3, p. 42.
176. Chen, S. L., and Gutmanis, F. (1968). *J. Food Sci.* **33**(3), 274–280.
177. Chichester, C. O. (1967). *Pure Appl. Chem.* **14,** 215–226.
178. Chichester, C. O., and McFeeters, R. (1971). *In* "Biochemistry of Fruits and Their Products" (A. C. Hulme, ed.), Vol. 2, pp. 707–719. Academic Press, New York.
179. Chichester, C. O., and Nakayama, T. O. M. (1967). *In* "Biogenesis of Natural Compounds," (P. Bernfeld, ed.) 2nd ed., pp. 641–678. Pergamon, Oxford.
180. Cholnoky, L., Györgyfy, K., Nagy, E., and Panczel, M. (1955). *Acta Chim. Acad. Sci. Hung.* **6,** 143–171.
181. Chong, Y. H., and Soh, C. C. (1969). *Med. J. Malays.* **3,** 282–287.
182. Chow, T. C., and Marlatt, A. L. (1953). *J. Nutr.* **51,** 305–315.
183. Chung, J. I., and Luh, B. S. (1972). *Korean J. Food Sci. Technol.* **4**(1), 6–12.

184. Chung, J. I., and Luh, B. S. (1971). *Confructa* **16,** 275–280.
185. Ciegler, A. (1965). *Adv. Appl. Microbiol.* **7,** 1–34.
186. Ciegler, A., Lagoda, A. A., Sohns, V. E., Hall, N. H., and Jackson, R. W. (1963). *Biotechnol. Bioeng.* **5,** 109–121.
187. Cirkova-Georgievska, M., and Pesevska-Sotirova, V. (1977). *Hrana Ishrana* **18**(3/4), 172–176.
188. Cirkova-Georgievska, M., and Pesevska-Sotirova, V. (1967). *Skopje, Zemjod.* **21,** 67–75.
189. Ciusa, W., and Nebbia, G. (1954). *Acta Vitaminol.* **8,** 257–262.
190. Claussen, E. E., and Nakayama, T. O. M. (1971). *J. Food Sci* **36,** 632–634.
191. Clydesdale, F. M. (1977). Color measurement. *In* "Current Aspects of Food Colorants" (T. E. Furia, ed.), pp. 1–18. CRC Press Inc., West Palm Beach, Florida.
192. Coggins, C. W., Jr., Henning, G. L., and Yokoyama, H. (1970). *Science* **168,** 1589–1590.
193. Cohen, L. O. B. (1958). *Ann. Intern. Med.* **48**(2), 219–227.
194. Cole, E. R., and Kapur, N. S. (1957). *J. Sci. Food Agric.* **8,** 366–368.
195. Conochie, J., and Wilkinson, R. A. (1956). *Proc. Int. Dairy Cong,* [*Proc.*], *14th, 1956* Vol. I (2), 357–364 (1956).
196. Cook, B. B., Gunning, B., and Uchimoto, D. (1961). *J. Agric. Food Chem.* **9,** 316–321.
197. Cordas, S. (1978). *J. Assoc. Off. Anal. Chem.* **77,** 696–707.
198. Cornwell, D. G., Kruger, F. A., and Robinson, H. B. (1962). *J. Lipid Res.* **3**(1), 65–70.
199. Corral, A. (1971). *Afinidad* **28**(288), 747–749.
200. Cort, W. M. (1974). *J. Am. Oil Chem. Soc.* **51**(7), 321–325.
201. Cort, W. M., Borenstein, B., Harley, J. N., Osadca, M., and Scheiner, J. (1976). *Food Technol.* **30**(4), 52–62.
202. Cort, W. M., Mergens, W., and Greene, A. (1978). *J. Food Sci.* **43,** 797–798.
203. Costs, C., and Deroche, M. E. (1965). *C.R. Hebd. Seances Acad. Sci., Ser. C* **261,** 4893–4896.
204. Coulter, S. T., and Thomas, E. L. (1968). *J. Agric. Food Chem.* **16,** 158–162.
204a. Counsell, J. N. (1979). Private communication.
205. Counsell, J. N. (1971). *Food Process. Ind.* **40**(1180), 39–41.
206. Counsell, J. N., and Webb, G. K. (1971). *Flavour Ind.* **2**(9), 519–524.
207. Cowgill, G. R. (1951). *In* "Handbook of Nutrition," 2nd ed. Council of Food & Nutr., Am. Med. Assoc., Blakiston Press, Philadelphia, Pennsylvania.
208. Crain, F. D., Lotspeich, F. J., and Krause, R. F. (1965). *Proc. Soc. Exp. Biol. Med.* **119**(3), 606–608.
209. Crowley, P. R. (1974). *WHO/US-AID Meet. Control Vitam. A Defic.* Jakarta.
210. Crump, G. B. (1958). *Prog. Chem. Fats Other Lipids* **5,** 151.
211. Curl, A. L. (1967). *J. Food Sci.* **32**(2), 141–143.
212. Curl, A. L. (1965). *J. Food Sci.* **30**(1), 13–18.
213. Curl, A. L. (1964). *J. Food Sci.* **29** 241–245.
214. Curl, A. L. (1964). *J. Agric. Food Chem.* **12,** 552–524.
215. Curl, A. L. (1963). *J. Food Sci.* **28,** 623–626.
216. Curl, A. L. (1962). *J. Food Sci.* **27**(2), 171–176.
217. Curl, A. L. (1962). *J. Food Sci.* **27,** 537–543.
218. Curl, A. L. (1962). *J. Agric. Food Chem.* **10**(6), 504–509.
219. Curl, A. L. (1961). *J. Food Sci.* **26**(2), 106–111.
220. Curl, A. L. (1960). *Food Res.* **25,** 190–196.
221. Curl, A. L. (1960). *Food Res.* **25**(5), 670–674.
222. Curl, A. L. (1959). *Food Res.* **24,** 413–422.
223. Curl, A. L., and Baily, G. F. (1961). *J. Food Sci.* **26,** 442–447.
224. Curl, A. L., and Bailey, G. F. (1959). *J. Food Technol.* **13**(7), 394–398.
225. Curl, A. L., and Bailey, G. F. (1957). *Food Res.* **22,** 63–68.
226. Curl, A. L., and Bailey, G. F. (1957). *J. Agric. Food Chem.* **5,** 605–608.
227. Curl, A. L., and Bailey, G. F. (1956). *J. Agric. Food Chem.* **4,** 156–159.
228. Curl, A. L., and Bailey, G. F. (1956). *J. Agric. Food Chem.* **4,** 159–162.

229. Curl, A. L., and Bailey, G. F. (1955). *Food Res.* **20,** 371-376.
230. Curl, A. L., and Bailey, G. F. (1954). *J. Agric. Food Chem.* **2,** 685-690.
231. Dagadu, J. M. (1967). *Br. J. Nutr.* **21,** 453-456.
232. Dagadu, M., and Gillman, J. (1963). *Lancet* **1,** 531-532.
233. Dahlberg, A. C., and Loosli, J. K. (1948). *J. Am. Diet. Assoc.* **24,** 20-22.
234. Dahle, L. (1965). *J. Agric Food Chem.* **13**(1), 12-15.
235. Daoud, H. N., Luh, B. S., and Seehafer, M. E. (1970). *Confructa* **15**(1), 4-20.
235a. Davies, B. H. (1979). *Pure Appl. Chem.* **51**(3), 623-630.
236. Davies, B. H. (1973). *Pure Appl. Chem.* **35,** 1-28.
237. Davies, B. H., Mathews, S., and Kirk, J. T. (1970). *Phytochemistry* **9,** 797-805.
238. Davis, R. B., and Gould, W. A. (1955). *Food Technol.* **9,** 540-547.
239. Day, W. C., and Erdman, J. G. (1963). *Science* **141,** 808-810.
240. Dean, G. E. (1966). *Am. Soft Drink J.* **36,** 63.
241. Dei-Tutu, J., Edwards, R. A., and Buckle, K. A. (1970). *J. Food Technol.* **22**(6), 272-277.
242. de La Mar, R. R., and Francis, F. J. (1969). *J. Food Sci.* **34,** 287-290.
243. De Leon, S. Y., and De Lima, L. S. (1970). *Philipp. J. Sci.* **97**(4), 337-347.
244. Della Monica, E. S., and McDowell, P. E. (1965). *Food Technol.* **19,** 1597-1599.
245. Demchenko, A.I. (1969). *Izv. Vyssh. Uchebn. Zaved., Pishch. Tekhnol.* **5,** 18-20.
246. Dempsey, M. E. (1974). *Annu. Rev. Biochem.* **43,** 967-996.
247. Deobald, H. J. (1964). *Food Technol.* **18,** 739-742, 1970-1975.
248. Deostale, Y. G., Mohan, V. S., and Radhakrishnan, M. R. (1969). *J. Nutr. Diet.* **6**(3), 224-228.
249. Déribéré, M. (1971). *Parfum. + Kosmet.* **52,** 369-377.
250. DeRitter, E. (1976). *Food Technol.* **30**(1), 48-54.
251. Deshmukh, D. S., and Ganguly, J. (1964). *Indian J. Biochem.* **1,** 204-207.
252. Deuel, H. J., Jr. (1957). "The Lipids, Their Chemistry and Biochemistry," Vol. III. Wiley (Interscience), New York.
253. Deuel, H. J., Jr., (1951). "The Lipids," Vol. I, Wiley (Interscience), New York.
254. Deuel, H. J., and Greenberg, S. M. (1953). *Food Res.* **18,** 497-503.
255. Dietl, B., Warantowicz, M., Carzenska, J., and Ziemlanski, S. (1973). *Rocz. Panstw. Zakl. Hig.* **24,** 605-615.
256. Di Giacomo, A. (1970). *Essenze Deriv. Agrum.* **40**(1), 5-32.
256a. Di Giacomo, A. (1970). *Essenze Deriv. Agrum.* **40**(2), 105-128.
257. Di Giacomo, A. (1970). *Essenze Deriv. Agrum.* **40**(3), 255-274.
258. Di Giacomo, A. (1966). *Essenze Deriv. Agrum.* **36**(3), 167-176.
259. Di Giacomo, A., Rispoli, G., and Tita, S. (1968). *Aromi-Saponi, Cosmet.-Aerosol.* **50**(2), 64-67, 80.
260. Di Lieto, A. (1968). *Tec. Molitoria* **19**(20), 557-565.
261. Dimitrov, D., and Tenov, R. (1969). *Nauchni Tr., Nauchnoizsled. Inst. Konservna Prom-st. Plovdiv* **6,** 79-100.
262. Dinesen, N. (1975). *Food Technol.* **29**(5) 40-46.
263. Dolev, A., Nativ, A., Evron, M., and Ben-Amotz, A. (1978). *Proc. Int. Congr. Food Sci. Technol., 5th,* Kyoto, *1978* Abstract, 1a-26.
264. Donchev, K. (1977). *Gradinar. Lozar. Nauka* **14**(4) 25-32.
265. Drapron, R., Beaux, Y., Cormier, M., and Geffroy, J. (1971). *C.R. Seances Acad. Agric. Fr.* **57**(3), 245-251.
266. Drapron, R., Beaux, Y., Cormier, M., Geffroy, J., and Adrian, J. (1974). *Ann. Technol. Agric.* **23**(3), 353-365.
267. Drobot, V. I., and Stabnikova, E. V. (1976). *Izv. Pishch. Tekhnol.* **3,** 57-59.
268. Duden, R., and Siddiqui, J. R. (1966). *Z. Lebensm.-Unters. -Forsch.* **132,** 1-4.
269. Dustin, J. P. (1977). *Food Nutr.* **3**(1), 7.
270. Dzhurov, D. G. (1962). *Mezhdunar. Konf. Vitam. Dokl. Suobsht., 1960* p. 423.
271. East German Patent 43,064; 1966.

272. Edwards, R. A., and Reuters, F. N. (1961). *J. Food Technol.* **19**(8), 352–357.
273. Egger, K. (1968). *Ber. Dtsch. Bot. Ges.* **81**(3–4), 153–156.
274. Egger, K., and Schwenker, U. (1966). *Z. Pflanzenphysiol.* **54,** 407–416.
275. Eheart, M. S., and Gott, C. (1964). *J. Am. Diet. Assoc.* **44**(2), 116–119.
276. Eichenberger, W., and Grob, E. C. (1962). *Helv. Chim. Acta* **45,** 974–981.
277. Eilati, S. K., Monselise, S. P., and Budowski, P. (1969). *J. Am. Soc. Hortic. Sci.* **94**(4), 346–348.
278. Eire Patent 554/63; 1963.
279. Elahi, M., and Shah, M. A. (1971). *Pak. J. Sci. Ind. Res.* **14**(4/5), 353–355.
280. El-Wakeil, F. A., and Morsi, M. K. (1976). *Agric. Res. Rev.* **54**(3), 99–109.
281. El-Zeftawi, B. M. (1973). *J. Food Technol.* **25**(3), 151–153.
282. Emodi, A. (1978). *Food Technol.* **32**(5), 38–42, 78.
283. Emodi, A., Scialpi, L., and Antoshkiw, T. (1976). *Food Technol.* **30**(7), 58–60.
284. Enrichment of Dried Skim Milk; With Special Reference to Vitamin A (1977). *Food Nutr.* **3**(1), 1–7.
285. Eriksen, B., and Hoygaard, A. (1941). *Klin. Wochenschr.* **20,** 200–201.
286. Espoy, H. M., and Barnett, H. M. (1955). *Food Technol.* **9,** 367–372.
287. Ezell, B. D., and Wilcox, M. S. (1952). *Plant Physiol.* **27,** 81–94.
288. Ezell, B. D., and Wilcox, M. S. (1946). *Science* **103,** 193–194.
289. Falconer, M. E., Fishwick, M. J., Land, D. G., and Sayer, E. R. (1964). *J. Sci. Food Agric.* **15,** 897–901.
290. Fang-Yung, A. F., and Aliev, I. A. (1971). *Lh. Klim.* Abstract 21R170.
291. Fang-Yung, A. F., Solvid, M. E., and Khotivari, A. V. (1973). *Izv. Vyssh. Uchebn. Zaved., Pishch. Tekhnol.* **1,** 157–158.
292. FAO/WHO Joint Expert Committee on Food Additives (1967). *W. H. O. Tech. Rep. Ser.* **362.**
293. Fellers, C. R. (1940). *Proc. Inst. Food Technol.* **97,** 107.
294. Fellers, C. R., and Buck, R. E. (1941). *Food Res.* **6,** 135.
295. Ferrari, G., and Depetris, M. P. (1969). *Minerva Pediatr.* **21,** 1421–1425.
296. Fiasson, J. L. (1969). *C.R. Hebd. Seances Acad. Sci., Ser. D* **266**(5), 786–788.
297. Finberg, A. J. (1971). *Food Prod. Dev.* **5**(2) 46–47.
298. Finkle, B. J. (1971). *In* "Biochemistry of Fruit and Their Products" (A. C. Hulme, ed.), Vol. 2, pp. 653–686. Academic Press, New York.
299. Flores, M., Garcia, B., Flores, Z., and Lara, M. Y. (1964). *Br. J. Nutr.* **18,** 281–293.
300. Flores, M., Menchu, M., Marta, Y. L., and Arroyave, G. (1969). *Arch. Latinoam. Nutr.* **19**(3), 311–324.
301. Foda, Y. H., Hegazi, S. M., and Salem, S. A. (1972). *Sudan J. Food Sci. Technol.* **4,** 57–63.
302. Fonseca, H., Nogueira, J. N., and Leme, J., Jr. (1972). *Solo* **64**(2), 53–59.
303. Fonseca, H., Nogueira, J. N., and Marcondes, A. M. S. (1967). *Arch. Latinoam. Nutr.* **19**(1), 9–16.
304. *Food Engineering* (1961). May, pp. 89–91.
305. *Food Processing* (1976). October, p. 54.
306. Forsythe, R. (1954). *Macaroni J.* **46,** July issue.
307. Franceschini, R., Francis, F. J., Livingston, G. E., and Fagerson, I. S. (1959). *Food Technol.* **13,** 358–365.
308. Francis, F. J., and Clydesdale, F. M. (1975). "Food Colorimetry: Theory and Practice." Avi Publ. Co., Westport, Connecticut.
309. Francis, F. J., and Clydesdale, F. M. (1971). *Food Prod. Dev.* **4**(4), 91–92.
310. Frankel, E. N., Smith, L. M., and Jack, E. L. (1958). *J. Dairy Sci.* **41,** 483–491.
311. Franzke, C., Grunert, K. S., and Kroschel, H. (1972). *Nahrung* **16**(8), 873–890.
312. Franzke, C., Grunert, K. S., and Petzold, H. (1971). *Nahrung* **15**(1), 1–9.
313. Fraps, G. S. (1947). *Tex., Agric. Exp. Stn., Bull.* **690.**

314. Fraps, G. S., and Kemmerer, A. R. (1941). *Am. Chem. Soc., News Ed.* August 10.
315. Fraps, G. S., and Meinke, W. W. (1945). *Food Res.* **10,** 187–196.
316. French Patent 1,456,569; 1966.
317. Frey-Wyssling, A., and Schwegler, F. (1965). *J. Ultrastruct. Res.* **13,** 543–559.
318. Froehling, P. E., Van Den Bosch, G., and Boekenoogen, H. A. (1972). *Lipids* **7**(7), 447–449.
319. Fujii, T. (1972). *Shokuhin Kogyo* **15**(14), 60–66.
320. Fujita, A., and Morimoto, H. (1960). *J. Vitaminol.* **6,** 278–290.
321. Furia, T. E. (1968). "Handbook of Food Additives." CRC Press, West Palm Beach, Florida.
322. Gal, I., and Parkinson, C. E. (1972). *Int. J. Vitam. Nutr. Res.* **42,** 565–575.
323. Galler, M., and Mackinney, G. (1965). *J. Food Sci.* **30,** 393–395.
324. Gandy, D. T. (1964). *Med. Rec. Ann.* **57,** 274–275.
325. Ganguly, J., Kon, S. K., and Thompson, S. Y. (1947). *Br. J. Nutr.* **1,** iii.
326. Ganguly, J., Krinsky, N. I., Mehl, J. W., and Deuel, H. J., Jr. (1952). *Arch. Biochem. Biophys.* **38,** 275–282.
327. Ganguly, J., Mehl, J. W., and Deuel, H. J., Jr. (1953). *J. Nutr.* **50,** 59–72.
328. Garcia, E. H., and Geuco, C. (1968). *Philipp. J. Plant Ind.* **33**(1/2), 85–93.
329. Garcia, E., Querido, J., and Cahanap, A. (1970). *Philipp. J. Plant Ind.* **35**(3/4), 203–213.
330. Gebbardt, S. E., Elkins, E. R., and Humphrey, J. (1977). *J. Agric. Food Chem.* **25**(3), 629–632.
331. Gebre-Medhin, M., Vahlquist, A., Hofvander, Y., Uppsäll, L., and Vahlquist, B. (1976). *Am. J. Clin. Nutr.* **29**(4), 441–451.
332. Geminder, J. J., and MacDonough, E. E. (1957). *J. Am. Oil Chem. Soc.* **34,** 314–318.
333. Gerhardt, V. (1974). *Fleischwirtschaft* **54**(3), 469–470, 1132–1134.
334. German Patent 1,077,448; 1967.
335. German Patent 1,096,177; 1960.
336. German Patent 1,158,505; 1963.
337. German Patent, 1,173,780; 1964.
338. German Patent 1,258,527; 1968.
339. German Patent 1,290,794; 1969.
340. Gershtein, L. A., and Kochurina, A. P. (1974). *Fiziol. Biokhim. Kul't. Rast.* **6**(4), 423–429.
341. Giassuddin, A. S. M., Ul-Alam, A., and Ahmad, K. (1973). *Dacca Univ. Stud.* **21**(2), Pt. B, 65–69.
342. Gill, H. S., and Kataria, A. (1974). *Curr. Sci.* **43**(6), 184–185.
343. Gillam, A. E., and El-Ridi, M. S. (1937). *Biochem. J.* **31,** 251–253.
344. Gillam, A. E., and Heilbron, I. M. (1935). *Biochem. J.* **29,** 834–836.
345. Gillam, A. E., Heilbron, I. M., Morton, R. A., Bishop, G., and Drummond, J. C. (1933). *Biochem. J.* **27,** 878–888.
346. Glover, J. (1960). *Vitam. Horm. (N.Y.)* **18,** 371–386.
347. Glover, J., and Redfearn, E. R. (1954). *Biochem. J.* **58,** 15P–16P.
348. Gogek, C. J. (1968). *J. Agric. Food Chem.* **16,** 730–734.
349. Goldovskii, A. M., and Nikitinskaya, V. N. (1973). *Maslob.-Zhir. Prom-st.* **6,** 9–14.
350. Goodman, D. S. (1969). *Am. J. Clin. Nutr.* **22,** 963–965.
351. Goodman, D. S., Bloomstrand, R., Werner, B., Huang, H. S., and Shiratori, T. (1966). *J. Clin. Invest.* **45**(10), 1615–1623.
352. Goodman, D. S., Huang, H. S., and Shiratori, T. (1966). *J. Biol. Chem.* **241,** 1929–1932.
352a. Goodwin, T. W. (1979). *Pure Appl. Chem.* **51**(3), 593–569.
353. Goodwin, T. W., ed. (1976). "Chemistry and Biochemistry of Plant Pigments," 2nd ed. Academic Press, New York.
354. Goodwin, T. W. (1971). *Biochem. J.* **123**(3), 293–329.
355. Goodwin, T. W. (1971). *In* "Carotenoids" (O. Isler, ed.), pp. 577–636. Birkhaeuser, Basel.
356. Goodwin, T. W. (1963). *In* "Carotine und Carotinoide" (K. Lang, ed.), pp. 1–25. Steinkopff, Darmstadt.

357. Goodwin, T. W. (1963). *In* "Biosynthesis of Vitamins and Related Compounds," pp. 270–319. Academic Press, New York.
358. Goodwin, T. W. (1961). *Annu. Rev. Plant Physiol.* **12,** 219–243.
359. Goodwin, T. W. (1955). *Annu. Rev. Biochem.* **24,** 497–522.
360. Goodwin, T. W. (1954). "Carotenoids, Their Comparative Biochemistry." Chem. Publ. Co., New York.
361. Goodwin, T. W. (1952). "The Comparative Biochemistry of the Carotenoids." Chapman & Hall, London.
362. Goodwin, T. W., and Goad, L. J. (1970). *In* "The Biochemistry of Fruits and Their Products" (A. C. Hulme, ed.), pp. 305–327. Academic Press, New York.
363. Gordon, H. T. (1977). *In* "Current Aspects of Food Colorants" (T. E. Furia, ed.), pp. 41–48. CRC Press Inc., West Palm Beach, Florida.
364. Gordon, H. T. (1972). *J. Food Technol.* **26**(5), 64, 66.
365. Grangaud, R., Nicol, M., LeGall, J., and Soussy, A. (1964). *Arch. Sci. Physiol.* **18,** 235–251.
366. Gravesen, K. J. (1967). *Scand. J. Clin. Lab. Invest.* **20**(1), 57–62.
367. Greaves, J. P., and Tan, J. (1966). *Br. J. Nutr.* **20,** 819–825.
368. Greenberg, R., Cornbleet, T., and Jeffay, A. I. (1959). *J. Invest. Dermatol.* **32,** 599–604.
369. Greene, C. H., and Blackford, L. M. (1926). *Med. Clin. North Am.* **10,** 733–744.
370. Gregory, M. E. (1975). *J. Dairy Res.* **42**(1), 197–216.
371. Grigorovskaya, M. V., and Litvinova, V. S. (1973). *Izv. Akad. Nauk. Mold. SSR, Ser. Biol. Khim. Nauk* **5,** 1824.
372. Gritsishin, I. S., and Snezhko, V. L. (1973). *Fiziol. Biokhim. Kul't. Rast.* **5**(1), 87–89.
373. Grogan, C. D., and Blessin, C. W. (1968). *Crop Sci.* **8,** 730–732.
374. Groh, J., Krch, V., Bartos, V., Bastercky, J., Jr., Dite, P., and Kozak, J. (1966). *Int. Congr. Gerontol., Proc., 7th, 1966* Vol. 2, pp. 231–236.
375. Gronowska-Senger, A. (1973). *Postepy. Biochem.* **19,** 325–330.
376. Gronowska-Senger, A., and Wolf, G. (1970). *J. Nutr.* **100,** 300–308.
377. Grosch, W., Laskawy, G., and Kaiser, K. P. (1977). *Z. Lebensm.-Unters.-Forsch.* **165,** 77–81.
378. Gross, J., and Budowski, P. (1966). *Biochem. J.* **101,** 747–754.
379. Gross, J., Carmon, M., Lifschitz, A., and Sklarz, B. (1975). *Phytochemistry* **14**(1), 249–252.
380. Gross, J., Gabai, M., and Lifshitz, A. (1973). *Phytochemistry* **12**(9), 2259–2263.
381. Gross, J., Gabai, M., and Lifshitz, A. (1972). *J. Food Sci.* **37**(4), 589–591.
382. Gross, J., Gabai, M., and Lifshitz, A. (1972). *Phytochemistry* **11**(1), 303–308.
383. Gross, J., Gabai, M., and Lifshitz, A. (1971). *J. Food Sci.* **36,** 466–473.
384. Gross, J., Michaels, C., and Lifshitz, A. (1976). *Lebensm.-Wiss. + Technol.* **9**(4), 211–214.
385. Gstirner, F., and Saad, S. N. I. (1959). *Z. Lebensm.-Unters.-Forsch.* **109,** 483–487.
386. Guidelines for the Eradication of Vitamin A Deficiency and Xerophthalmia (1976). A report of the International Vitamin A Consultative Group (VACG). Nutrition Foundation, Inc., New York.
387. Günther, H., and Grau, R. (1962). *Fleischwirtschaft* **14**(1), 34–35.
388. Hadert, H. (1972). *Alimenta* **11**(6), 224.
389. Halevy, S., Koth, H., and Guggenheim, K. (1957). *Br. J. Nutr.* **11,** 409–413.
390. Hanson, S. W. (1954). *In* "Symposium: Color in Foods" (K. T. Farrell, J. R., Wagner, M. S. Peterson, and G. Mackinney, eds.), p. 136–158. Adv. Board Quartermaster Res. Dev. Comm. Foods, Natl. Acad. Sci., Natl. Res. Counc., Washington, D.C.
391. Harkay-Vinkler, M. (1974). *Acta Aliment. Acad. Sci. Hung.* **3**(3), 239–249.
392. Harper, R. H., and Zscheile, F. P. (1945). *Food Res.* **10,** 84–97.
393. Harris, P. L., Swanson, W. J., and Hickman, K. C. D. (1947). *J. Nutr.* **33,** 411–427.
394. Harris, R. S. (1963). *J. Agric. Food Chem.* **16**(2), 149–152.
395. Harris, R. S. (1959). *J. Agric. Food Chem.* **7**(2), 88–102.
396. Hart, G. (1964). *Papua New Guinea Agric. J.* **11**(1), 8–11.

397. Hartman, A. M., and Dryden, L. P. (1974). *In* "Fundamentals of Dairy Chemistry" (B. H. Webb and A. H. Johnson, eds.), 2nd ed., pp. 325-401. Avi Publ. Co., Westport, Connecticut.
398. Hartman, A. M., and Dryden, L. P. (1965). "Vitamins in Milk and Milk Products," pp. 1-116. Am. Dairy Sci. Assoc., Chicago, Illinois.
399. Hashimoto, H. (1922). *J. Am. Med. Assoc.* **78,** 1111-1112.
400. Hauge, S. M. (1942). *Food Ind.* **14**(2), 48-49, 98-99.
401. Haxo, F. (1950). *Bot. Gaz. (Chicago)* **112,** 228-232.
402. Hayes, E., and Steele, K. A. (1964). *Abstr., World Fat Congr., 1st, 1964,* p. 161.
403. Hayman, E. P., Yokoyama, H., and Poling, S. M. (1977). *J. Agric. Food Chem.* **25,** 1251-1253.
404. Head, M. K., and Giesbrecht, F. G. (1976). *J. Am. Diet. Assoc.* **69**(6), 640-644.
405. Heinze, P. H., Hayden, F. R., and Wade, B. L. (1974). *Plant Physiol.* **22,** 548.
406. Herrmann, K. (1975). *Ernaehr.-Umsch.* **22**(2), 45-49.
407. Hess, A. F., and Myers, V. C. (1919). *J. Am. Med. Assoc.* **73,** 1743-1745.
408. Hibbs, J. W., and Krauss, W. E. (1946). *J. Anim. Sci.* **5,** 401.
409. Higuchi, K., and Peterson, W. H. (1946). *J. Dairy Sci.* **29,** 789-793.
410. Hill, H. M., and Rogers, L. J. (1969). *Biochem. J.* **113**(3), 31-32.
411. Hirayama, M. (1962). *Annu. Rep. Natl. Inst. Nutr., Tokyo* pp. 36-37.
412. Hirayama, O., and Oida, H. (1969). *J. Agric. Chem. Soc.* **43**(7), 423-428.
413. Hoch, H. (1943). *Biochem. J.* **37,** 430-433.
414. Hodges, R. E., Sauberlich, H. E., Canham, J. E., Wallace, D. L., Rucker, R. B., Majia, L. A., and Mohanram, M. (1978). *Am. J. Clin. Nutr.* **31,** 876-885.
415. Hoffmann, I., Nowosad, F. S., and Cody, W. J. (1967). *Can. J. Bot.* **45,** 1859-1862.
416. Holmes, A. D., Spelman, A. F., and Jones, C. P. (1945). *Food Res.* **10,** 489-496.
417. Hoover, M. W. (1964). *Food Technol.* **18,** 1793-1796.
418. Hoover, M. W. (1963). *Food Technol.* **17,** 636-638.
419. Horvath, G. (1972). *Phytochemistry* **11**(1), 183-187.
420. Hotchner, S. J., and Poole, C. J. (1967). *Proc. Int. Congr. Canned Foods, 5th, 1967.*
421. Hsu, W. J., Poling, S. M., DeBenedict, E., Rudash, C., and Yokoyama, H. (1975). *J. Agric. Food Chem.* **23,** 831-834.
422. Huggart, R. L., Barron, R. W., and Wenzel, F. W. (1966). *Food Technol.* **20**(5), 109-111 (667-679).
423. Hunter, R. F., and Krakenberger, R. M. (1946). *Biochem. J.* **40,** 492-493.
424. Hunter, R. F., Scott, A. D., and Williams, N. E. (1944). *Biochem. J.* **38,** 211-213.
425. Iguina DeGeorge, L. M., Collazo DeRivera, A. L., Benero, J. R., and Pennock, W. (1969). *J. Agric. Univ. P. R.* **53,** 100-105.
426. Imam, M. K., and Gabelman, W. H. (1968). *Proc. Am. Soc. Hortic. Sci.* **93,** 419-428.
427. INCAP (1974). *Publ. INCAP* **V-36** (Engl. transl., Jan. 1975).
428. Inhoffen, H. H., Bohlman, F., Bartram, K., Rummert, G., and Pommer, H. (1950). *Justus Liebigs Ann. Chem.* **750,** 54-69.
429. *Int*ernational Fruit World (1975). **33**(3), 81.
430. International Symposium on Carotenoids (1978). *Pure & Applied Chem.* **51**(3, 434-675 (1979).
431. International Symposium on Carotenoids (1975). *Pure & Applied Chem.* **47**(2/3), 96-243 (1976).
432. International Symposium on Carotenoids (1972). *Proc. Carotenoids Other Than Vitam. A, Pure & Applied Chem.* **35**(1), 1-130 (1973).
433. International Symposium on Carotenoids (1969). *Proc. Carotenoids Other Than Vitam. A, Pure & Applied Chem.* **20**(4), 365-553 (1970).
434. International Symposium on Carotenoids (1967). *Proc. Carotenoids Other Than Vitam. A, Pure & Applied Chem.* **14**(2), 215-278 (1967).
435. Irvine, G. N., and Anderson, J. A. (1949). *Cereal Chem.* **26,** 507-512.
435a. Isler, O. (1979). *Pure Appl. Chem.* **51**(3), 447-462.
436. Isler, O. (1971). *In* "Carotenoids" (O. Isler, ed.), pp. 11-27. Birkhaeuser, Basel.
437. Isler, O., ed. (1971). "Carotenoids." Birkhaeuser, Basel.

438. Isler, O. (1970). *Experientia* **26,** 225–240.
439. Isler, O. (1967). *Chim. Ind. (Milan)* **49,** 1317–1333.
440. Isler, O. (1950). *Chimia* **4,** 103–118.
441. Isler, O., Guex, W., Lindlar, H., Montavon, M., Rüegg, R., Ryser, G., Saucy, G., and Schwieter, U. (1958). *Chimia* **12,** 89–90.
442. Isler, O., Huber, W., Ronco, A., and Kofler, M. (1947). *Helv. Chim. Acta* **30,** 1911–1927.
443. Isler, O., Klaui, H., and Solms, U. (1957). *In* "The Vitamins" (W. H. Sebrell, Jr. and R. S. Harris, eds.), 2nd ed., Vol. 1, pp. 101–113. Academic Press, New York.
444. Isler, O., Lindlar, H., Montavon, M., Rüegg, R., Saucy, G., and Zeller, P. (1957). *Helv. Chim. Acta* **40,** 456–467.
445. Isler, O., Lindlar, H., Montavon, M., Rüegg, R., Saucy, G., and Zeller, P. (1956). *Mitt. Geb. Lebensmittelunters. Hyg.* **47,** 183–190; *Chem. Soc., Spec. Publ.* **4,** 47–63.
446. Isler, O., Lindlar, H., Montavon, M., Rüegg, R., Saucy, G., and Zeller, P. (1956). *Helv. Chim. Acta* **39,** 2041–2053.
447. Isler, O., Lindlar, H., Montavon, M., Rüegg, R., and Zeller, P. (1956). *Helv. Chim. Acta* **39,** 249–259.
448. Isler, O., Montavon, M., and Rüegg, R. (1967). *Ind. Chim. Belge* **32,** Spec. No. Pt. 3, 853–856.
449. Isler, O., Montavon, M., Rüegg, R., Saucy, G., and Zeller, P. (1956). *Vehr. Naturforsch. Ges. Basel* **67,** 379–391.
450. Isler, O., Ofner, A., and Siemers, G. F. (1958). *Food Technol.* **12,** 520–526.
451. Isler, O., Guex, W., Rüegg, R., Ryser, G., Saucy, G., Schwieter, U., Walter, M., and Winterstein, A. (1959). *Helv. Chim. Acta* **42,** 864–871.
452. Isler, O., Rüegg, R., and Schwieter, U. (1967). *Pure Appl. Chem.* **14,** 245–263.
453. Isler, O., Rüegg, R., and Schudel, P. (1961). *Chimia* **15,** 208–226.
454. Isler, O., and Schudel, P. (1963). *In* "Carotine and Carotenoide" (K. Lang, ed.), pp. 54–104. Steinkopff, Darmstadt; *Wiss. Veroeff. Dtsch. Ges. Ernaehr.* **9,** 54–104.
455. Isler, O., and Zeller, P. (1958). *Vitam. Horm. (N.Y.)* **15,** 31–71.
456. Jagannathan, S. N., and Patwardhan, V. N. (1960). *Indian J. Med. Res.* **48,** 775–784.
457. Jäger, A. (1963). *Zucker-Suesswaren-Wirtsch.* **16**(1), 10–14; (2), 63–68.
458. Jahn, O. L., and Young, R. (1975). *J. Am. Soc. Hortic. Sci.* **100**(3), 244–246.
459. James, W. H., and El-Gindi, I. M. (1963). *J. Nutr.* **51,** 97–108.
460. James, W. H., and Hollinger, M. E. (1954). *J. Nutr.* **54,** 65–74.
461. Japanese Patent Appl. 23,165/57; 1957.
462. Japanese Patent 9,535/60; 1960.
463. Japanese Patent 21,476/61; 1961, and 8,532/62; 1962.
464. Japanese Patent 1,646/63; 1963.
465. Japanese Patent 8,476/63; 1963.
466. Japanese Patent 21,760/64; 1964.
467. Japanese Patent 18,018/65; 1965.
468. Japanese Patent 10,978/66; 1966.
469. Japanese Patent 14,428/66; 1966.
470. Japanese Patent 6,986/69; 1969.
471. Japanese Patent 7,225,220; 1972.
472. Japanese Patent 7,284/63; 1963, U.S. Patent 3,227,561; 1966.
473. Japanese Patent 75,104,271.
474. Jen, J. J. (1974). *J. Agric. Food Chem.* **22**(5), 908–910.
475. Jenness, R., and Palmer, L. S. (1945). *J. Dairy Sci.* **28,** 473–494.
476. Jiravatana, V., Cuevas-Ruiz, J., and Graham, H. D. (1970). *J. Agric. Univ. P.R.* **54**(2), 314–319.
477. John, J., Subbarayan, C., and Cama, H. R. (1970). *J. Food Sci.* **35**(3), 262–265.
478. Joyce, A. E. (1954). *Nature (London)* **173,** 311–312.
479. Jungalwala, F. B., and Cama, H. R. (1963). *Indian J. Chem.* **1**(1), 36–40.

480. Jurkovic, N., and Lackovic, M. (1971). *Kem. Ind.* **20**(8), 379-381.
481. Kahlenberg, O. J. (1949). *Food Ind.* **21**(4), 101-104, 467-470.
482. Kamath, S. K., MacMillan, J. B., and Arnrich, L. (1972). *J. Nutr.* **102**(2), 1579-1584.
483. Kaneda, K. (1956). *J. Vitaminol.* **2**(3), 177-180.
484. Kanner, J., and Budowski, P. (1978). *J. Food Sci.* **43**(2), 524-526.
485. Kanner, J., and Mendel, H. (1977). *J. Food Sci.* **42**(6), 1549-1551.
486. Kanner, J., Mendel, H., and Budowski, P. (1977). *J. Food Sci.* **42**(1), 60-64.
487. Kanner, J., Mendel, H., and Budowski, P. (1976). *J. Food Sci.* **41,** 183-185.
488. Kargl, T. E., Quackenbush, F. W., and Tomes, M. L. (1960). *Proc. Am. Soc. Hortic. Sci.* **75,** 574-578.
489. Karrer, P., and Jucker E. (1948). *In* "Carotenoide." Birkhaeuser, Basel (Engl. transl., Elsevier, Amsterdam, 1950).
490. Karrer, P., and Jucker, E. (1944). *Helv. Chim. Acta* **31,** 802-803.
491. Kasim, M. (1967). *Nahrung* **11**(5), 405-409.
492. Kasper, H. (1968). *Int. Z. Vitaminforsch.* **38,** 142-148.
493. Kasper, H., and Ernst, B. (1969). *Int. Z. Vitaminforsch.* **39**(1), 23-29.
494. Kasper, H., and Hospach, R. (1974). *Dtsch. Med. Wochenschr.* **99,** 198-200.
495. Kasper, H., and Krennrich, O. (1965). *Z. Gesamte Exp. Med.* **139,** 644-650.
496. Katayama, T., Nakayama, T. O. M., Lee, T. H., and Chichester, C. O. (1971). *J. Food Sci.* **36**(5), 804-806.
497. Kaur, B., and Manjrekar, S. P. (1975). *Indian Food Packer* **29**(4), 9-14.
498. Kawaguchi, T., and Fujita, A. (1956). *J. Vitaminol.* **2,** 115-127.
499. Kawashima, K., and Yamanishi, T. (1973). *J. Agric. Chem. Soc. Jpn.* **47**(1), 79-81.
500. Ke, B. (1971) *In* "Methods in Enzymology" (A. San Pietro, ed.), Vol. 23, Part A, pp. 624-636. Academic Press, New York.
501. Kesterson, J. W., and McDuff, O. R. (1949). *Am. Perfum. Essent. Oil Rev.* Oct., 285-287.
502. Khan, M., and MacKinney, G. (1953). *Plant Physiol.* **28**(3), 550-552.
503. Khudairi, A. K. (1972). *Am. Sci.* **60**(16), 696-707.
504. Kienzle, F., and Isler, O. (1978). *In* "The Chemistry of Synthetic Dyes" (K. Venkataraman, ed.), Vol. 8, pp. 389-414. Academic Press, New York.
505. Kim, H. S., Kim, Y. S., and Park, K. T. (1969). *J. Nucl. Sci.* **9**(1/2), 125-132.
506. Kimel, W., Sax, N. W., Kaiser, S., Eichman, G. G., Chase, G. O., and Ofner, A. (1958). *J. Org. Chem.* **23,** 153-157.
507. Kimel, W., Surmatis, J. D., Weber, J., Chase, G. O., Sax, N. W., and Ofner, A. (1957). *J. Org. Chem.* **22,** 1611-1618.
508. Kirchgessner, M., Friesecke, H., and Koch, G. (1967). "Nutrition and the Composition of Milk." Crosby Lockwood, London.
509. Kirk, J. T. O., and Juniper, B. E. (1967). *In* "Biochemistry of Chloroplasts" (T. W. Goodwin, ed.), Vol. 2, pp. 691-701. Academic Press, New York.
510. Kläui, H., F. Hoffmann-La Roche & Co., Ltd., Basel (unpublished research).
511. Kläui, H. (1976). "Symposium on Oil Soluble Food Additives." Institute of Food Science & Technology, Norwich, U.K., 1976; *Flavours,* July/August, 1976.
512. Kläui, H. (1975). *Proc. Int. Congr. Food Sci. Technol., 4th, 1974* Vol. 1, pp. 740-762.
513. Kläui, H. (1973). *In* "Functional Properties of Fats in Foods" (J. Solms, ed.), pp. 133-169. Forster Verlag, Zurich.
514. Kläui, H. (1971). *In* "University of Nottingham Residential Seminar on Vitamins" (M. Stein, ed), pp. 110-143. Churchill-Livingston, Edinburgh and London.
515. Kläui, H. (1969). *J. Food Technol.* **21**(12), 620.
516. Kläui, H. (1968). *Inst. Food Sci. Technol.* **1**(5), 1-8.
517. Kläui, H. (1967). "Karotinoider," SIK Rep. No. 224. Svenska Institutet för Konserveringsforskning.

518. Kläui, H. (1963). *In* "Carotine und Carotinoide" (K. Lang, ed.), pp. 390-400. Steinkopff, Darmstatt; *Wiss. Veroeff. Dtsch. Ges. Ernaehr.* **9,** 390-400 (1963).
519. Kläui, H., Capeder, A., and Münzel, K. (1963). *Pharm. Ind.* **25,** 173-178.
520. Kläui, H., Hausheer, W., and Huschke, G. (1970). *In* "International Encyclopedia of Food and Nutrition" (R. A. Morton, ed.), pp. 113-159. Pergamon, Oxford.
521. Kläui, H., (1969). *J. Food Technol.* **21**(12), 620-627.
522. Kläui, H., and Manz, U. (1967). *Beverages* **8**(1), 16-20.
523. Kläui, H., and Münzel, K. (1965). *Pharm. Acta Helv.* **40,** 153-167.
524. Kläui, H., and Raunhardt, O. (1976). *Alimenta* **15**(2), 37-45.
525. Klein, D. (1978), *Fleischwirtschaft* **58**(2), 227-228.
526. Klyushkina, Y. F., Denisenko, Y. I., Nechaev, A. P., and Yantovskii, M. T. (1970). *Prikl. Biokhim. Mikrobiol.* **6**(1), 95-98.
527. Knee, M. (1972). *J. Exp. Bot.* **23**(74), 184-196.
528. Knee, M. (1971). *J. Sci. Food Agric.* **22**(7), 371-377.
529. Kobayashi, K., Miyamoto, K., and Akuta, S. (1970). *Yamaguchi Daigaku Nogakubu Gakujutsu Hokoku* **21,** 139-146.
530. Koch, J. (1972). "Fruchtsaefte." Werner Verlag, Hofheim.
531. Koch, J., and Hess, D. (1971). *Dtsch. Lebensm.-Rundsch.* **67**(6), 185-195.
532. Kolensnik, A. A., and Dmitrieva, A. B. (1968). *Tovarovedenie* **3,** 10-13.
533. Kon, S. K. (1973). "Milk Products in Human Nutrition," 2nd ed. Unipub., New York.
534. Korobkina, Z. V. (1968). *Tr. Vses. Semin. Biol. Akiv. Veshehest vam. Plodov Yagod 3rd 1968,* 384-388. *Zh. Khim.* Abstr. No. 14R81 (1969).
535. Koskitalo, L. H., and Ormrod, D. P. (1972). *J. Food Sci.* **37**(1), 56-59.
536. Kozenko, S. I., Kleschunova, G. A., and Korastileva, N. N. (1973). *Konservn. Ovoshchesush. Prom-st.* **11,** 34-35.
537. Kreula, M. S. (1950). *Ann. Acad. Sci. Fenn., Ser. A.* **38,** 7-23.
538. Kreula, M. S. (1947). *Biochem. J.* **41**(2), 269-273.
539. Kreula, M. S., and Virtanen, A. I. (1939). *Laekarefoeren. Foerh.* **45,** 355-362.
539a. Kriggre, G. (1971). *Latv. Lauksaimn. Akad. Raksti* **29,** 118-125.
539b. Krinsky, N. I. (1979). *Pure Appl. Chem.* **51**(3), 649-660.
540. Krinsky, N. I. (1972). *Abstr. Commun. Int. Symp. Carotenoids Other Than Vitam. A, 1972* pp. 70-72.
541. Krinsky, N. I. (1971). *In* "Carotenoids" (O. Isler, ed.), pp. 669-716. Birkhaeuser, Basel.
542. Krinsky, N. I., Cornell, D. G., and Oncley, J. L. (1958). *Arch. Biochem. Biophys.* **73,** 233-246.
543. Krishna Mallia, A., Subbarayan, C., and Cama, H. R. (1967). *Indian J. Nutr. Diet.* **4,** 277-280.
544. Krishnamurthy, M. N., Geetha, S., and Devadas, R. P. (1976). *Indian J. Nutr. Diet.* **13**(9), 293-295.
545. Krishnamurthy, M. N., and Natarajan, C. P. (1973). *Indian Food Packer* **21**(1), 39-44.
546. Krondl, A. (1966). *Vnitr. Lek.* **12,** 1092-1097.
547. Kroschel, H., Grunert, S., and Franzke, C. (1973). *Lebensm-Ind.* **20**(7), 317-321.
548. Kruisheer, C. I., and den-Herder, P. C. (1952). *Neth. Milk Dairy J.* **6,** 109-126.
549. Krukar, R. J., and Schnyder, W. A. (1966). *Quick Frozen Foods,* June.
550. Kübler, W. (1975). *Landwirtsch. Forsch.* **32**(2), 211-221.
551. Kübler, W. (1970). *In* "Vitamins," pp. 21-30. F. Hoffman-La Roche & Co., Ltd., Basel.
552. Kübler, W. (1963). *Wiss. Veroeff. Dtsch. Ges. Ernaehr.* **9,** 222-234.
553. Kübler, W. (1963). *Dtsch. Med. Wochenschr.* **88**(6), 1319-1325.
554. Kudritskaya, S. E., Savinov, R. G., and Fishman, G. M. (1973). *Subtrop. Kul't.* **4,** 30-34.
555. Kuhn, R., and Grundmann, C. (1934). *Ber. Dtsch. Chem. Ges.* **67,** 593-595.
556. Kuhn, R., and Sörensen, N. A. (1938). *Angew. Chem.* **51**(27), 465-466.
557. Kulikova, N. T. (1972). *Tr. Vses. Semin. Biol. Akt. (Lech.) Veshchestvam Plodov Yagod, 4th, 1970* pp. 91-96.

558. Kumta, V. S., Saivant, P. L., and Ramakrishnass, T. V. (1970). *Radiat. Bot.* **10,** 161-174, 395-399.
559. Kushwaka, S. C., Susue, G., Subbarayan, C., and Porter, J. W. (1970). *J. Biol. Chem.* **245**(18), 4708-4717.
560. Kvale, A. (1971). *Acta Agric. Scand.* **21**(3), 207-213.
561. Laferriese, L., and Gabelman, W. H. (1968). *Proc. Am. Soc. Hortic. Sci.* **93,** 408-418.
562. Lai, S., Finney, K. F., and Milner, M. (1959). *Cereal Chem.* **36,** 401-408.
563. Lakshmanan, M. R., Chansang, H., and Olson, J. A. (1972). *J. Lipid Res.* **13**(4), 477-482.
564. Lala, V. R., and Reddy, V. (1970). *Am. J. Clin. Nutr.* **23,** 110-113.
565. Lantz, E. M. (1949). *N.M., Agric. Exp. Stn., Bull.* **350.**
566. Lantz, E. M. (1946). *N.M., Agric. Exp. Stn., Bull.* **327.**
567. Lantz, E. M., and Smith, M. (1944). *N.M., Agric. Exp. Stn., Bull* **989.**
568. Laval-Martin, D., Quennemet, J., and Moneger, R. (1972). *C.R. Hebd. Seances Acad. Sci., Ser. D* **274**(21), 2879-2882.
569. Lease, J. G., and Lease, E. J. (1956). *Food Technol.* **10,** 368-373.
570. Lease, J. G., and Lease, E. J. (1956). *Food Technol.* **10,** 403-405.
571. Lee, W. G., and Ammerman, G. R. (1974). *J. Food Sci.* **39**(6), 1188-1190.
572. Lee, W. L. (1977). "Carotenoproteins in Animal Coloration." Academic Press, New York.
573. Lee, W. L. (1966). *Comp. Biochem. Physiol.* **18,** 17-36.
574. Lempka, A., and Prominski, W. (1967). *Nahrung* **11,** 267-276.
575. Lepage, M. (1968). *Lipids* **3**(6), 477-481.
576. Lepage, M., and Sims, R. P. A. (1968). *Cereal Chem.* **45,** 600-604.
577. Lesunova, N. G. (1972). *Sel'sk. Khoz. Biol.* **7**(5), 673-677.
578. Leuenberger, F. J., and Thommen, H. (1972). *Z. Lebensm.-Unters. -Forsch.* **149**(5), 279-282.
579. Leuenberger, U., and Stewart, I. (1976). *Phytochemistry* **15,** 227-229.
580. Lewis, E. P., and Merrow, S. B. (1962). *J. Agric. Food Chem.* **10,** 53-56.
581. Liaaen-Jensen, S. (1977). *In* "Marine Natural Products Chemistry" (D. J. Fauerner and W. H. Fenical, eds.), pp. 239-259. Plenum, New York.
582. Liaaen-Jensen, S. (1976). *Pure Appl. Chem.* **47,** 129-149.
583. Liaaen-Jensen, S. (1973). *Pure Appl. Chem.* **35,** 81-112.
584. Liaaen-Jensen, S. (1972). *Annu. Rev. Microbiol.* **26,** 225-248.
585. Liaaen-Jensen, S. (1971). *In* "Carotenoids" (O. Isler, ed.), pp. 61-199. Birkhaeuser, Basel.
586. Liaaen-Jensen, S. (1970). *In* "Biochemistry of Terpenoids" (T. W. Goodwin, Ed.), pp. 223-254. Academic Press, New York.
587. Liaaen-Jensen, S. (1967). *Pure Appl. Chem.* **14,** 227-244.
588. Liaaen-Jensen, L. (1965). *Annu. Rev. Microbiol.* **19,** 163-182.
589. Lian, O. K., Tie, L. T., Rose, C. S., Prawiranegara, D. D., and Gygörgy, P. (1967). *Am. J. Clin. Nutr.* **20**(12), 1267-1274.
590. Lier, J. B. (1975). *Diss. Abstr. Int. B* **35**(9), 4369.
591. Lier, J. B., and Lacroix, L. J. (1974). *Cereal Chem.* **51**(2), 188-194.
592. Lilly, V. G., Barnett, H. L., and Krause, R. F. (1960). *W.V., Agric. Expt. Stn., Bull.* **441T.**
593. Lime, B. J., Stephens, T. S., and Griffiths, F. P. (1954). *Food Technol.* **8,** 566-569.
594. Lippert, L. F., and Hall, M. O. (1973). *HortScience* **8**(1), 38-40.
595. Lipton, W. J. (1970). *J. Am. Soc. Hortic. Sci.* **95**(6), 680-684.
596. Liu, Y. K., and Luh, B. S. (1977). *J. Food Sci.* **42,** 216-220.
597. Livingston, A. L., Bickoff, E. M., and Thompson, C. R. (1955). *J. Agric. Food Chem.,* **3**(5), 439-441.
598. Loncin, M. (1976). *Oleagineux* **30**(2), 77-80.
599. Loncin, M., and Jacobsberg, B. (1964). *Fette, Seifen, Anstrichm.* **66,** 910-911.
600. Lopez, H., Navia, J. M., Clement, D., and Harris, R. S. (1963). *J. Food Sci.* **28**(5), 600-610.
601. Lotspeich, F. J., Krause, R. F., Lilly, V. C., and Barnett, H. L. (1961). *Fed. Proc. Fed. Am. Soc. Exp. Biol.* **20,** 269.

602. Lovrič, T., Sablek, Z., and Boskovič, M. (1970). *J. Sci. Food Agric.* **21**(12), 641-647.
603. Lück, H. (1973). *S. Afr. J. Dairy Technol.* **5**(4), 181-185.
604. Lück, H. (1965). *Z. Lebensm.-Unters.-Forsch.* **126,** 193-201.
605. Lück, H., and Downes, T. E. H. (1972). *S. Afr. J. Dairy Technol.* **4**(4), 207-213.
606. Lück, H., and Novello, J. C. (1970). *S. Afr. J. Dairy Technol.* **2**(2), 91-96.
607. Lück, H., Novello, J. C., and Van Zyl, D. J. (1969). *S. Afr. J. Dairy Technol.* **1**(1), 15-23.
608. Luckmann, F. H., Melnick, D., and Vahlteich, H. W. (1952). *J. Am. Oil Chem. Soc.* **29,** 321-330.
609. Luh, B. S., and Daoud, H. N. (1968). *Fruchtsaft-Ind.* **13**(5), 204-211.
610. Lukton, A., and Mackinney, G. (1956). *J. Food Technol.* **10,** 630-632.
611. Mader, I. (1964). *Science* **144,** 533-534.
612. Mackinney, G. (1961). *In* "The Orange: Its Biochemistry and Physiology" (W. B. Sinclair, ed.), p. 302. University of California, Davis.
613. Mackinney, G., Aronoff, S., and Bornstein, B. T. (1942). *Ind. Eng. Chem., Anal. Ed.* **14,** 391-395.
614. Mackinney, G., Lukton, A., and Greenbaum, L. (1958). *Food Technol.* **12,** 164-166.
615. Maia, G. A., and Luh, B. S. (1970). *Confructa* **5**(4), 216-230.
616. Majia, L. A., Hodges, R. E., Arroyave, G., Viteria, F., and Torun, B. (1977). *Am. J. Clin. Nutr.* **30,** 1175-1184.
617. Malachi, T., Gross, J., Lifshitz, A., and Sklarz, B. (1974). *Lebensm.-Wiss. + Technol.* **7**(6), 330-334.
618. Malewski, W., and Michalik, H. (1974). *Biul. Warzywniczy* **16,** 389-404.
619. Mallette, M. F., Dawson, C. R., Nelson, W. L., and Gortner, W. A. (1946). *Ind. Eng. Chem.* **38,** 437-441.
620. Mancini, F., and Schianchi, G. F. (1963). *In* "Carotine and Carotinoide" (K. Lang, ed.), pp. 401-406. Steinkopff, Basel.
621. Manz, U., F. Hoffmann-La Roche & Co., Ltd., Basel, (unpublished work).
622. Manz, U. (1967). *Chimia* **21**(7), 329-335.
623. Marine Font, A., and Alvarez Reverter, M. (1973). *Circ. Farm.* **31** 383-402.
624. Martin, L. E. (1965). *Am. Soft Drink J.* March issue.
625. Martin, M. E., Sweeney, J. P., Gilpin, G. L., and Chapman, V. J. (1960). *J. Agric. Food Chem.* **8,** 387-390.
626. Martinenghi, G. B. (1972). *Oleagineux* **27**(11), 553-555.
627. Maruna, R. F. L., (1966). *Z. Ernaehrungswiss.* **7**(24), 127-131.
628. Marusich, W., DeRitter, E., and Bauernfeind, J. C. (1957). *J. Am. Oil Chem. Soc.* **34,** 217-211.
629. Massoud, W. H., Bird, A. C., and Perkins, E. S. (1975). *Br. J. Ophthalmol.* **59,** 200-204.
630. Mathews, J., and Beaton, G. H. (1963). *Can. J. Biochem. Physiol.* **41,** 543-549.
631. Matsuzaua, K., Innami, S., and Kuga, T. (1958). *Eiyogaku Zasshi* **16,** 145-148.
632. Matthews, R. F., Locascio, S. J., and Ozaki, H. Y. (1976). *Proc. Fla. State Hortic. Sci.* **88,** 263-265.
633. Matthews-Roth, M. M., Abraham, A. A., and Gabuzda, T. G. (1976). *Clin. Chem.* **22**(6), 922-924.
634. Mattill, H. A. (1927). *J. Am. Med. Assoc.* **89,** 1505-1508.
635. Matto, A. K., Modi, V. V., and Reddy, V. V. R. (1968). *Indian J. Biochem.* **5**(3), 111-114.
636. Matzko, S. N. (1936). *Z. Unters. Lebensm.* **72,** 143-148.
636a. Mayer, H. (1979). *Pure Appl. Chem.* **51**(3), 535-564.
637. Mayer, H. J., and Isler, O. (1971). *In* "Carotenoids" (O. Isler, ed.), pp. 325-575. Birkhaeuser, Basel.
638. McCance, R. A., and Widdowson, E. M. (1978). "The Composition of Foods" (A. A. Paul and D. A. T. Southgate, eds.), Elsevier, Amsterdam.
639. McCarty, C. D., and Lesley, J. W. (1954). *Proc. Am. Soc. Hortic. Sci.* **64,** 289-292.
640. McCollum, J. P. (1955). *Food Res.* **20,** 55-59.

641. McCollum, J. P. (1954). *Food Res.* **19,** 182–189.
642. McConaghey, R. M. S. (1952). *Lancet* **2,** 714–715.
643. McConnell, J. E. W., Esselen, W. B., Jr., and Guggenberg, N. (1945). *Fruit Prods. J.* **24**(5), 133–135.
644. McDowell, A. K. R., and McDowall, F. H. (1953). *J. Dairy Res.* **20,** 76–100.
645. McGillivray, W. A. (1957). *J. Dairy Res.* **24,** 95–101.
646. McGillivray, W. A. (1957). *J. Dairy Res.* **24,** 102–108.
647. McGillivray, W. A. (1956). *N. Z. J. Sci. Technol.* **38,** 466–472.
648. McGillivray, W. A., and Gregory, M. E. (1962). *J. Dairy Res.* **29,** 211–232.
649. McKeown, G. G. (1965). *J. Assoc. Off. Agric. Chem.* **48,** 835–837.
650. McKeown, G. G. (1963). *J. Assoc. Off. Agric. Chem.* **46,** 790–796.
651. McKeown, G. G. (1961). *J. Assoc. Off. Agric. Chem.* **44,** 347–351.
652. McKeown, G. G., and Mark, E. (1962). *J. Assoc. Off. Agric. Chem.* **45,** 761–766.
653. McLaren, D. S. (1964). *Nutr. Rev.* **22,** 289–291.
654. McLaren, D. S., and Zekian, B. (1971). *Am. J. Dis. Child.* **121,** 278–280.
655. McWeeny, D. J. (1968). *J. Sci. Food Agric.* **19,** 250–265.
656. Melnick, D., Luckmann, F. H., and Vahlteich, H. W. (1953). *Food Res.* **18,** 504–510.
657. Menden, E., and Cremer, H. G. (1958–59). *Food Manuf.* July, August, September, October, November issues (1958); February, March issues (1959).
657a. Menger, A. (1964). *Dtsch. Lebensm.-Rundsch.* **60,** 387.
658. Meredith, F. I., and Purcell, A. E. (1966). *Proc. Am. Soc. Hortic. Sci.* **89,** 544–548.
659. Meredith, F. I., Redman, G. H., and Griffith, F. R. (1974). *J. Rio Grande Val. Hortic. Soc.* **28,** 149–153.
660. Meredith, F. I., and Young, R. H. (1971). *HortScience* **6**(3), 233–234.
661. Metzner, A. A. (1978). *Cereal Foods World* **23**(3), 120–123.
662. Metzner, A. A. (1976). *Manuf. Confect.* July.
663. Michna, M. (1968). *Hodowa Rosl., Aklim. Nasienn.* **12**(3), 313–328.
664. Mickus, P. R. (1955). *Food Eng.* **27,** 91–94.
665. Mikhailovnina, A. A., and Savinov, B. G. (1950). *Ukr. Khim. Zh.* **16,** 183–187.
666. Mikhailovnina, A. A., and Savinov, B. G. (1949). *Ukr. Khim. Zh.* **15,** 285–291.
667. Mihajlovic, M., Hadzi-Taskovic, V., Sukalovic, V., and Markovic, G. (1971). *Eucarpia, Proc. Maize Sorghum Sect., 5th, 1971,* pp. 82–84.
668. Mikhitaryan, S. S., Nechaev, A. P., Denisenko, Y. I., and Lyubushkin, V. T. (1966). *Izv. Vyssh. Uchebn. Zaved., Pischch. Tekhnol.* **6,** 29–30.
669. Miki, N., and Akatsu, K. (1971). *Nippon Shokuhin Kogyo Gakkai-Shi* **18**(7), 309–312.
670. Miki, N., and Akatsu, K. (1970). *J. Food Sci. Technol.* **17**(5), 175–181.
671. Milchwissenschaft (1970). **25**(1), 55–56.
672. Miller, J. C., and Covington, H. M. (1942). *Proc. Am. Soc. Hortic. Sci.* **40,** 519–522.
673. Minamide, T., and Ogata, K. (1972). *J. Jpn. Soc. Food Nutr.* **25**(4), 333–338.
674. Mitchell, J. H., Van Blaircom, L. O., and Roderick, D. B. (1948). *S.C., Agric. Exp. Stn., Bull.* **372.**
675. Mohr, W. P. (1976). *J. Can. Inst. Food Sci. Technol.* **9**(1), 20–24.
676. Monselise, J. J., and Berk, Z. (1954). *Isr. Res. Counc., Bulletin* **4,** 188–191.
677. Monselise, S. P., and Halevy, A. H. (1961). *Science* **133,** 178.
678. Montag, A. (1962). *Z. Lebensm.-Unters. -Forsch.* **116,** 413–420.
679. Moore, T. (1957). "Vitamin A," pp. 168–170. Elsevier, Amsterdam.
680. Moran, A. M. (1961). *Rev. Sanid. Milit. (Peru)* **34,** 32–42.
681. Morgal, P. W., Byers, L. W., and Miller, E. J. (1943). *Ind. Eng. Chem.* **35,** 794–797.
682. Morgan, R. C. (1967). *J. Food Sci.* **32**(3), 275–278.
683. Morgan, R. C. (1966). *J. Food Sci.* **31**(2), 213–217.
684. Mori, Y. (1973). *Kaseigaku Zasshi* **24,**(2), 99–103.
685. Morimoto, H. (1960). *Vitamins* **20**(4), 524–531.

686. Morton, R. A. (1970). *J., R. Soc. Health* **90**(11), 21–28.
687. Moster, J. B., and Prater, A. N. (1957). *Food Technol.* **11**(3), 146–148.
688. Moster, J. B., and Prater, A. N. (1957). *Food Technol.* **11**(4), 222–225.
689. Moster, J. B., and Prater, A. N. (1957). *Food Technol.* **11**(4), 226–229.
690. Moster, J. B., and Prater, A. N. (1952). *Food Technol.* **6**(12), 459–463.
691. Moster, J. B., Quackenbush, F. W., and Porter, J. W. (1942). *Arch. Biochem. Biophys.* **38,** 287–296.
692. Moutounet, M. (1976). *Ann. Technol. Agric.* **25**(1), 73–84.
693. Mudambi, S. R., and Rajagopal, M. V. (1977). *J. Food Sci.* **42**(5), 1414–1415.
694. Mueller-Mulot, W. (1976). *Fette, Seifen, Anstrichm.* **78**(1), 18–22.
695. Mueller-Mulot, W. (1975). *Rev. Fr. Corps Gras* **22**(3), 131–136.
696. Muhilal and Karyadi, D. (1978). *Report of Nutrition Research and Development Center,* Bogor, Indonesia.
697. Muhilal and Karyadi, D. (1977). *Int. Vitam. A Consult. Group Meet., 1977,* Geneva.
698. Munsell, H. E., Williams, L. O., Guild, L. P., Kelley, L. T., McNally, A. M., and Harris, R. S. (1950). *Food Res.* **15,** 439–453.
699. Murray, T. K., and Campbell, J. A. (1966). *Can., Dep. Agric., Publ.* **1238.**
700. Nagle, B. J., Villalon, B., and Burns, E. E. (1978). *38th Meet. Inst. Food Technol.* Abstract No. 335, p. 151.
701. Nakagawa, H., Kushida, T., Ogura, N., and Takehana, H. (1971). *J. Food Sci. Technol.* **18**(6), 259–263.
702. NAS-NRC-FNB Workshop on Cereal Grain Enrichment (1977). Washington, D.C.
703. NAS-NRS (1975). "Technology of Fortification of Foods." Food and Nutrition Board, National Research Council, National Academy of Science, Washington, D.C.
704. NAS-NRS (1974). "General Policies in Regard to the Improvement of Nutritive Quality of Foods." Food and Nutrition Board, National Research Council, National Academy of Science, Washington, D.C.
705. NAS-NRC (1974). "Proposed Fortification Policy for Cereal-Grain Products." Food and Nutrition Board, National Research Council, National Academy of Sciences, Washington, D.C.
706. Natarajan, C. P., and Mackinney, G. (1952). *J. Sci. Ind. Res., Sect. B* **11,** 416–418.
707. National Provisioner (1957). Jan. 26.
708. Nazarov. N. I., and Egorova, N. I. (1972). *Khlebopek. Konditer. Prom-st.* **5,** 24–25.
709. Nazarov, N. I., Kondratenko, S. S., and Shebershneva, N. N. (1968). *Khlebopek. Konditer. Prom-st.* **12**(7), 23–24.
710. Neamtu, G. (1970). *Natura (Bucharest)* **22**(1), 62–67.
711. Netherlands (NL) Patent 6,414,112; 1965.
712. Newsome, R. S. (1961). *Diss. Abstr.* **22,** 997–998.
713. Nikolaishvili, D. K., and Adeishvili, N. I. (1966). *Byull. Vses. Nauchno-Issled. Inst. Chain. Prom-sti.* **2,** 57–60.
714. Ninet, L., Renault, J., and Tissier, R. (1969). *Biotechnol. Bioeng.* **11,** 1195–1210.
715. Nishimura, M., and Takamatsu, K. (1957). *Nature (London)* **180,** 699–700.
716. Nobile, S., and Woodhill, J. M. (1973). *J. Food Technol.* **25**(2), 1–22.
717. Nobile, S., Woodhill, J. M., and Rosevear, P. E. J. (1975). *J. Food Technol.* **27,** 342–347.
718. Noble, A. C. (1975). *J. Agric. Food Chem.* **23**(1), 48–49.
719. *Nutrition Reviews* (1943). *1,* 242; **6,** 112 (1948); **7,** 288 (1949); **9,** 217 (1951); **10,** 265 and 329 (1952); **11,** 19 (1953); **12,** 19, 23, 59, 179, 191 (1954); **13,** 123 (1955); **14,** 157 (1956); **19,** 334 (1961); **20,** 63, and 282 (1962); **24,** 16 (1966).
720. Nutting, M. D., Neumann, H. J., and Wagner, J. R. (1970). *J. Sci. Food Agric.* **21**(4), 197–202.
721. Oberbacher, M. F., Ting, S. V., and Deszyck, E. J. (1960). *Proc. Am. Soc. Hortic.* **75,** 262–265.
722. Oey, K. L., Liem, T. T., Rose, C. S., Prawiranegara, D. D., and György, P. (1967). *Am. J. Clin. Nutr.* **20,** 1267–1274.

723. Olson, J. A. (1972). *Isr. J. Med. Sci.* **8**(8–9), 1199–1206.
724. Olson, J. A., and Hayaishi, O. (1965). *Proc. Natl. Acad. Sci. U.S.A.* **54**(5), 1364–1370.
725. Olson, J. A., and Lakshmanan, M. R. (1970). *In* "The Fat-Soluble Vitamins" (H. F. DeLuca and J. W. Suttie, eds.), pp. 213–226. Univ. of Wisconsin Press, Madison.
726. Onsteed, G. R., and Zieve, L. (1972). *J. Am. Med. Assoc.* **22**(7), 677–679.
727. Ortaliza, I., Del Rosario, I. F., Santos, M. H., Aguilar, C. G., and Dumada-Ug, L. M. (1971). *Philipp. J. Sci.* **100**(2), 95–102.
728. Pablo, I. S., Manalo, J. A., and Cardeno, V. A. (1971). *I.A.E.A., Bull.* pp. 101–111. Disinfectation Fruit Irradiat. Proc. Panel, 1970, Vienna.
729. *PAG Bulletin* (1976). **6**(4), 1–7.
730. Palamarchuk, K. I. (1977). *Vestn. Dermatol. Venerol.* **3,** 23–27.
731. Palmer, L. S., and Eckles, C. H. (1914). *J. Biol. Chem.* **17,** 223–236.
732. Panalaks, T., and Murray, T. K. (1970). *Can. Inst. Food Technol. J.* **3**(4), 145–151.
733. Pararincevic-Trajkovic, J., and Bara, J. (1970). *Hrana Ishrana* **11,** 325–330.
734. Park, C. R. (1975). *Henguk Yongyang Hakhoe Chi* **8**(4), 167–172.
735. Parker, H. K., and Harris, M. C. (1964). *In* "Wheat; Chemistry and Technology" (I. Hlynka, ed.), pp. 435–462. Am. Assoc. Cereal Chem., St. Paul, Minnesota.
736. Parman, G. K. (1964). *Milk Dealer,* Apr. 32, 90.
737. Parman, G. K. (1964). *Manuf. Milk Prod. J.* **55**(5), 5.
738. Parman, G. K., and Borenstein, B. (1964). *Food Eng.* **36**(5), 77–78.
739. Parrish, D. B. (1977). *Crit. Rev. Food Sci. Nutr.* **9**(4), 375–394.
740. Parrish, D. B., Wise, G. H., and Hughes, J. S. (1947). *J. Biol. Chem.* **167,** 673–678.
741. Patent References: U.S. Patents 2,974,044; 1961. 3,079,380; 1963. 3,095,357; 1963. 3,128,236; 1964. 3,191,701; 3,226,302; 1965. 3,280,502; 1966. British Patent 1,003,546; Canadian Patent 696,273; Belgian Patents 633,200; 644,957; 647,485; 651,382; 1964. 702,235; French Patents 1,325,656, 1963; 1,344,264, 1963; 1,371,272; 1964. 1,403,839, 1965. 1,412,506, 1965. 1,499,879; 1966; 1,449,880; 1966. NL Patent 6,411,184; 1965. Russian Patents 169,749; 1965. 179,427; 1966.
742. Pattee, H. E., and Purcell, A. E. (1969). *J. Am. Oil Chem. Soc.* **46,** 629–631.
743. Pattee, H. E., and Purcell, A. E. (1967). *J. Am. Oil Chem. Soc.* **44,** 328–330.
744. Pendlington, S., Dupont, M. S., and Trussell, F. J. (1965). *Biochem. J.* **94,** 25p.
745. Perlin, C. (1970). *Cesk. Gastroenterol. Vyz.* **24**(8), 364–367.
746. Peter, A. (1966). *Fruchtsaft-ind.* **11,** 53–59.
747. Peters, F. E., Tomono, M., and Wills, P. A. (1960). *Food Res.* **25**(2), 211–228.
748. Petersone, M., Melgalve, L., Kozule, V., Vimba, B., Lapsa, E., and Zukauska, I. (1976). *Latv. Lauksaimn. Akad. Raksti* **88,** 74–77.
749. Petrus, D. R., and Daugherty, M. H. (1973). *J. Food Sci.* **38,** 659–664.
750. Petrus, D. R., Huggart, R. L., and Daugherty, M. H. (1975). *J. Food Sci.* **40**(5), 922–924.
751. Petzold, E. N., and Quackenbush, F. W. (1960). *Arch. Biochem. Biophys.* **86,** 163–165.
752. Petzold, E. N., Quackenbush, F. W., and McQuistan, M. (1959). *Arch. Biochem. Biophys.* **82,** 117–124.
753. Philip, M. V. (1978). *Food Prod. Dev.* Sept., 24–25.
754. Philip, T. (1973). *J. Agric. Food Chem.* **21**(6), 963–966.
755. Philip, T. (1970). *Diss. Abstr. Int. B* **31**(5), 2752 (Order No. 70-22789).
756. Philip, T., and Francis, F. J. (1971). *J. Food Sci.* **36**(5), 823–827.
757. Philip, T., Nawar, W. W., and Francis, F. J. (1971). *J. Food Sci.* **36**(1), 98–100.
758. Piasecki, M., and Szebiotko, K. (1975). *Przem. Ferment. Rolny* **19**(3), 16–21.
759. Pinto, G. L. (1960). *Arq. Port. Bioquim.* **4,** 270–275.
760. Pirie, A. (1975). *Baroda J. Nutr.* **2**(2), 79–84.
761. Pohle, W. D., and Mehlenbacher, V. C. (1955). *Food Technol.* **9,** 565–568.
762. Pommer, H. (1960). *Angew. Chem.* **72,** 911–915.

763. Popovskii, V. G., Ivasyuk, N. T., and Kreponosova, A. N. (1968). *Tr. Mold. Nauchno-Issled. Inst. Pishch. Prom-sti.* **8,** 20–35.

764. Pordab, Z., Piechanowski, J., and Maik, L. (1971). *Pr. Inst. Lab. Badaw. Przem. Spozyw.* **21**(1), 109–118.

765. Porter, J. W. (1969). *Pure Appl. Chem.* **20,** 449–496.

766. Porter, J. W., and Anderson, D. G. (1967). *Annu. Rev. Plant Physiol.* **18,** 197–228.

767. Porter, J. W., and Anderson, D. G. (1962). *Arch. Biochem. Biophys.* **97,** 520–528.

768. Porter, J. W., and Spurgeon, S. L. (1979). *Pure Appl. Chem.* **51**(3), 609–622.

769. Porter, J. W., and Zscheile, F. P. (1946). *Arch. Biochem. Biophys.* **10,** 537–546.

770. Pouchet, G. (1872). *J. Anat. Physiol., Norm. Pathol. Homme Anim.* **8,** 401–407.

771. Premachandra, B. R., Vasantharajan, V. N., and Cama, H. R. (1978). *Indian J. Exp. Biol.* **16**(4), 468–472.

772. Premachandra, B. R., Vasantharajan, V. N., and Cama, H. R. (1978). *Indian J. Exp. Biol.* **16,** 473–490.

773. Protsenko, A. L., Frolova, L. S., and Leonenko, S. G. (1978). *Molochn. Prom-st.* **1,** 31.

774. Pruthi, J. S., and Lal, G. (1958). *Food Res.* **23,** 505–510.

775. Purcell, A. E. (1962). *J. Food Technol.* **16,** 99–102.

776. Purcell, A. E., and Walter, W. M., Jr. (1968). *J. Agric. Food Chem.* **16,** 769–770.

777. Purcell, A. E., Walter, W. M., Jr., and Tompkins, W. T. (1969). *J. Agric. Food Chem.* **17**(1), 41–42.

778. Quackenbush, F. W. (1963). *Cereal Chem.* **40,** 266–274.

779. Quackenbush, F. W., Cox, R. P., and Steenbock, H. (1942). *J. Biol. Chem.* **145,** 169–177.

780. Quackenbush, F. W., Firch, J. G., Brunson, A. M., and House, L. R. (1963). *Cereal Chem.* **40,** 250–259.

781. Quackenbush, F. W., Firch, J. G., Raburn, W. J., McQuistan, M., Petzold, E. N., and Kargl, T. E. (1961). *J. Agric. Food Chem.* **9,** 132–135.

782. Quackenbush, F. W., Steenbock, H., and Peterson, W. H. (1938). *J. Am. Chem. Soc.* **60,** 2937–2941.

783. Rabourn, W. J., and Schall, E. D. (1956). *Anal. Chem.* **28,** 1197–1198.

784. Radema, L. (1958). *Off. Org. K. Ned. Zuivelband.* **50,** 216–217.

785. Radema, L. (1956). *Ned., Zuivelband. Off. Org.* **48,** 185–190; *Misset's Zuivel* **62,** 191 (1956).

786. Rajalakshmi, R., Chari, K. V., Advani, M., Chary, T. M., Patel, M. A., Mutdlik, A. M., Bhat, T. K., and Vyas, A. D. (1975). *Baroda J. Nutr.* **2**(2), 103–119.

787. Rajalakshmi, R., Sail, S. S., and Ramachandran, K. (1974). *Baroda J. Nutr.* **1**(1), 29–35.

788. Rajan, T. S., Wood, F. W., and DeMan, J. M. (1965). *J. Dairy Sci.* **48,** 1134.

789. Ramage, R. (1972). *In* "Chemistry of Terpenes and Terpenoids" (A. Newman, ed.), pp. 288–336. Academic Press, New York.

790. Ramakrishnan, T. V., and Francis, F. J. (1973). *J. Food Sci.* **38,** 25–28.

791. Ramamurthy, K. M., and Bhalerao, V. R. (1964). *Analyst* **89**(1064), 740–746.

792. Ramane, I. (1972). *Nauka—Zhivotnovod.* **11,** 113–119.

793. Rao, C. N. (1967). *J. Nutr. Diet.* **4**(1), 10–16.

794. Rao, C. N., and Rao, B. S. N. (1970). *Am. J. Clin. Nutr.* **23**(1), 105–109.

795. Raymundo, L. C., Griffiths, A. E., and Simpson, K. L. (1970). *Phytochemistry* **9,** 1239–1245.

796. Recommended Dietary Allowances (1974). 8th ed., pp. 20–54. Food and Nutrition Board, National Research Council, National Academy of Sciences, Washington, D.C.

797. Reddy, V. (1977). *Int. Vitam. A Consult. Group Meet., 1977* Geneva.

798. Reich, P., Shwachman, H., and Craig, J. M. (1960). *N. Engl. J. Med.* **262**(6), 263–269.

799. Reid, M. S., Lee, T. H., Pratt, H. K., and Chichester, C. O. (1970). *J. Am. Soc. Hortic. Sci.* **95**(6), 814–815.

800. Reinart, A., and Brown, R. W. (1953). *Proc. Int. Dairy Congr., 13th, 1953* Vol. 3, pp. 1238–1240.

801. Reinart, A., and Nesbitt, J. M. (1956). *Int. Dairy Congr.* [*Proc.*], *14th, 1956* Vol. 1, pp. 934–945.
802. Reith, J. F., and Gielen, J. W. (1971). *J. Food Sci.* **36**(6), 861–864.
803. Report of a Joint WHO/AID Meeting (1976). *W.H.O., Tech. Rep.* **590.**
804. Richardson, G. A., and Konar, M. R. (1949). *Proc. West. Div., Am. Dairy Sci. Assoc., 30th Meet.* pp. 105–119.
805. Richardson, G. A., and Long, M. L. (1947). *J. Dairy Sci.* **30,** 533–534.
806. Richter, F. (1935). *Milchwirtsch. Forsch.* **17,** 72–86.
807. Riel, R. R. (1957). *Can. Dairy Ice Cream J.* **36**(4), 61, 66.
808. Riel, R. R., and Gibson, C. A. (1958). *J. Dairy Sci.* **41**(5), 624–629.
809. Riel, R. R., and Johns, C. K. (1957). *J. Dairy Sci.* **40,** 192–199.
810. Rifaat, I. D., El-Koussy, L. A., and Sokkary, A. M. (1969). *Agric. Res. Rev.* **47**(6), 245–250.
811. Ripoli, G., and DiGiacomo, A. (1962). *Essenze Deriv. Agrum.* **32,** 187–192.
812. Rödel, W. (1966). *Nahrung* **10,** 213–223.
813. Roels, O. A., Djaeni, S., Trout, M., Lauw, T. G., Heath, A., Poey, S. H., Tarwotjo, M. S., and Suhadi, B. (1963). *Am. J. Clin. Nutr.* **12,** 380–387.
814. Roels, O. A., Trout, M., and Dujacquier, R. (1958). *J. Nutr.* **65,** 115–128.
815. Ronnenberg, P. (1945). *Yearb. R. Vet. Agric. Coll.* pp. 47–48.
816. Ross, G., and Parker, J. G. (1962). *N. Y. State J. Med.* **62,** 3584–3589.
817. Rother, H. (1963). *Naturbrunner* **13,** 4.
818. Rother, H. (1963). *Naturbrunner* **13,** 94.
819. Rother, H. (1962). *Riechst. Aromen* **12**(2), 33–41.
820. Rother, H. (1962). *Naturbrunnen* **12,** 298–299.
821. Rother, H. (1961). *Naturbrunnen* **11,** 204.
822. Rotstein, A., Gross, J., and Lifshitz, A. (1972). *Lebensm.-Wiss. + Technol.* **5**(4), 140–143.
823. Roy, S. K. (1973). *J. Food Sci. Technol.* **10,** 45.
824. Rubin, S. H., Emodi, A., and Scialpi, L. (1977). *Cereal Chem.* **54**(4), 895–904.
825. Rubowicz, K., and Jarmolowicz, H. (1955). *Prace Glownego Inst. Przemyslu Rolnego i Spozyvczego* **3**(3), 1–12.
826. Rüegg, R., Montavon, M., Ryser, G., Saucy, G., Schwieter, U., and Isler, O. (1959). *Helv. Chim. Acta* **42,** 854–864.
827. Saivant, P. L., Ramakrishnan, T. V., and Kumta, U. S. (1970). *Radiat. Bot.* **10,** 169–174.
828. Salem, S. A., and Hegazi, S. M. (1973). *J. Sci. Food Agric.* **24**(2), 123–126.
829. Salminen, K., Karinpaeae, A., Koivistoinen, P., and Mukula, J. (1970). *Acta Agric. Scand.* **20**(1), 49–57.
830. Salunkhe, D. K., Gerber, R. K., and Pollard, L. H. (1959). *Proc. Am. Soc. Hortic. Sci.* **74,** 423–429.
831. Salunkhe, D. K., and Wu, M. T. (1973). *J. Am. Soc. Hortic. Sci.* **98**(1), 113–116.
832. Sanderson, G. W., Co, H., and Gonzalez, J. G. (1971). *J. Food Sci.* **36**(2), 231–236.
833. Sanderson, G. W., and Graham, H. N. (1972). *Abstr. Pap., 164th Meet. Am. Chem. Soc.* AGFD69.
834. Sandler, Z. Y., Denisenko, Y. I., Nechaev, A. P., and Yanotovskii, M. T. (1969). *Izv. Vyssh. Uchebn. Zaved., Pishch. Tekhnol.* **3,** 23–26.
835. Sarhan, M. A. I., El-Wakeil, F. A., and Morsi, M. K. S. (1971). *Sudan J. Food Sci. Technol.* **3,** 41–47.
836. Sato, T. (1966). *Nippon Shokuhin Kogyo Gakkai-shi* **13**(11), 488–491.
837. Satyanarayana, M. N., and Kadhol, S. B. (1963). *Indian J. Med. Res.* **51,** 764–767.
838. Savinov, B. G., and Kudritskaya, S. E. (1977). *Vitam. Rastitel'n. Resursy Ikh Ispolz, 1977* pp. 178–181.
839. Schaller, D. R., and von Elbe, J. H. (1971). *J. Food Sci.* **36**(4), 712–713.
840. Schara, A., and Tsoumanis, A. (1963). *Mineralwasserztg.* **16,** 610–614, 817–820.
841. Schettino, O., and Di Lieto, A. (1968). *Tec. Molitoria* **19,** 557–565.

842. Schillinger, A., and Zimmermann, G. (1965). *Dtsch. Lebensm.-Rundsch.* **61**(45), 144–147.
843. Schmandke, H., and Guerra, O. O. (1969). *Nahrung* **13**(6), 523–530.
844. Schuchardt, W. (1955). *Oleagineux* **10,** 259–264.
845. Schuchardt, W. (1954). *Fette, Seifen, Anstrichm.* **56,** 986–989.
846. Schuller, W. (1957). *Milchwiss. Ber.* **7**(1), 1–26.
847. Schultz, H. G. (1954). *In* "Symposium on Color in Foods" (K. T. Farrell, J. R. Wagner, and M. S. Peterson, eds., pp. 16–23. Quartermaster Food and Container Institute, Chicago, Illinois.
848. Schwarz, G., Mumm, H., and Woerner, F. (1958). *Molk. Kaeserei-Ztg.* **9**(43), 2–12.
849. Schwieter, U., Bolliger, H. R., Chopard-Dit-Jean, L. H., Englert, G., Kofler, M., Koenig, A., Planta, C. V., Rüegg, R., Vetter, W., and Isler, O. (1965). *Chimia* **19,** 294–302.
850. Schwieter, U., Gutmann, H., Lindlar, H., Marbet, R., Rigassi, N., Rüegg, R., Schaeren, S. F., and Isler, O. (1966). *Helv. Chim. Acta* **49,** 369–390.
851. Schwieter, U., and Isler, O. (1967). *In* "The Vitamins" (W. H. Sebrell, Jr. and R. S. Harris, eds.), 2nd ed., Vol. 1, pp. 5–100. Academic Press, New York.
852. Scott, G. C., and Belkengren, R. O. (1944). *Food Res.* **9,** 371–376.
853. Searles, S. K., and Armstrong, J. G. (1970). *J. Dairy Sci.* **53**(2), 150–154.
854. Segal, B. (1971). *Biochim. Prod. Aliment.* pp. 361–380.
855. Segura Ferns, R. (1961). *Galenica Acta B* **13,** 381–389.
856. Seifert, J., and Davidek, J. (1973). *Pestic. Sci.* **4**(3), 261–263.
857. Seifter, E., and Rettura, G. (1978). *Abstr. Pap., 175th Meet. Am. Chem. Soc.* AGFD 45.
858. Sexton, E. L., Mehl, J. W., and Deuel, H. J. (1946). *J. Nutr.* **31,** 299–319.
859. Shah, M. A., Qureshi, A. K., and Elahi, M. (1969). *Pak. J. Ind. Res.* **12,** 43–46.
859a. Sheft, B. B., Griswald, R. M., Tarlowsky, E., and Halliday, E. G. (1949). *Ind. Eng. Chem.* **41,** 144.
860. Sherman, W. C., and Koehn, C. J. (1948). *Ind. Eng. Chem.* **40**(8), 1445–1448.
861. Shkvarkina, T. I., and Soshina, V. K. (1973). *Khlebopek. Konditer. Prom-st.* **3**(3), 7–8.
862. Siemers, G. F., and Parman, G. K. (1958). *Macaroni J.* **39**(4), 27.
863. Silerio, V. F., Mendoza de Flores, C., and Avila, G. E. (1976). *Tec. Pecu. Mex.* **31,** 47–54.
864. Simonnet, H. (1963). *Dtsch. Ges. Ernaehr.* **9,** 205–221.
865. Simpson, D. J., Rahmen, F. M. M., Buckle, K. A., and Lee, T. H. (1974). *J. Plant Physiol.* **1**(1), 135–147.
866. Sinclair, H. M. (1953). *Proc. Nutr. Soc.* **12,** 69–82.
867. Singh, H., and Cama, H. R. (1975). *J. Sci. Ind. Res.* **34,** 219–230.
868. Singh, H., and Cama, H. R. (1974). *Biochem. Rev.* **45,** 77–88.
869. Singh, H., Krishna Mallia, A., and Cama, H. R. (1972). *Biochem. J.* **128**(1), 11P.
870. Sinha, S. P. (1963). *Int. Z. Vitaminforsch.* **33,** 262–268.
871. Sistrunk, W. A., and Cash, J. N. (1970). *J. Food Sci.* **35**(5), 645–648.
872. *Snack Food* (1976). **65**(10), 24–25.
873. Solon, F., Fernandez, T. L., Latham, M. C., and Popkin, B. M. (1979). *Am. J. Clin. Nutr.* **32,** 1445–1453.
874. Solon, F. S., Fernandez, T. L., Latham, M. C., and Popkin, B. M. (1978). *Proc. Int. Cong. Nutr., 11th, 1978,* Rio de Janeiro.
875. Solon, F. S., Fernandez, T. L., Latham, M. C., and Popkin, B. M. (1974). Research to Determine the Cost & Effectiveness of Alternate Means of Controlling Vitamin A Deficiency. Cebu Institute Medicine, Cebu, Philippines.
876. Solon, F. S., Popkin, B. M., Fernandez, T. L., and Latham, M. C. (1978). *Am. J. Clin. Nutr.* **31,** 360–368.
877. Soni, S. L. (1969). *Indian J. Hortic.* **26**(1/2), 21–26.
878. Sood, R., and Bhat, C. M. (1974). *J. Food Sci. Technol.* **11**(3), 131–133.
879. South Africa Patent, R61/1660; 1960.
880. South Africa Patent, 63/5232; 1964.
881. *Southern Cooperative Series* (1953). Bulletin 30. Georgia Exp. Stn, Athens.

882. Spencer, M. (1973). *Nutr. Food Sci.* **32,** 11–14.
883. Spielman, A. A., Thomas, J. W., Loosli, J. K., Whiting, F., Norton, C. L., and Turk, K. L. (1947). *J. Dairy Sci.* **30,** 343–350.
884. Spiess, W. E. L., Sole, P., Askar, C., and Askar, A. (1968). *Therapiewoche* **29,** 1165–1168.
885. Stahl, E., and Wagner, C. (1969). *J. Chromatogr.* **40**(2), 308.
886. Stavrova, E. R., Kremneva, L. S., and Veiner, A. G. (1969). *Izv. Vyssh. Uchebn. Zaved., Pishch. Tekhnol.* **5,** 68–70.
887. Stefferud, A. (1959). *In* "The Yearbook of Agriculture" (USDA Staff, ed.), pp. 132, 159, 226–268. US Govt. Printing Office, Washington, D.C.
888. Stephen, T. S., and McLemore, T. A. (1969). *Food Technol.* **23,** 1600–1604.
889. Stewart, I. (1977). *J. Agric. Food Chem.* **25**(5), 1132–1137.
890. Stewart, I. (1977). *Proc. Int. Soc. Citricult.* **1,** 308–311.
891. Stewart, I. (1973). *In* "Congreso Mundial de Citricultura" (O. Carpena, ed.), pp. 325–330. Int. Soc. Citricult.
892. Stewart, I., and Leuenberger, U. B. (1976). *Alimenta* **15**(2), 33–36.
893. Stewart, I., and Wheaton, T. A. (1972). *J. Agric. Food Chem.* **20**(2), 448–449.
894. Stewart, I., and Wheaton, T. A. (1971). *Proc. Fla. State Hortic. Soc.* **84,** 264–266.
895. Stitt, F., Bickoff, E. M., Bailey, B. F., Thompson, C. R., and Friedlander, S. (1951). *J. Assoc. Off. Agric. Chem.* **34,** 460–471.
896. Stoecker, B., and Arnich, L. (1973). *J. Nutr.* **103**(8), 1112–1118.
897. Stoltz, R. B., and Sutton, T. S. (1938). *J. Dairy Sci.* **21,** A49.
898. Strachan, C. C., Moyls, A. W., Atkinson, F. E., and Britton, J. E. (1951). *Can., Dep. Agric., Publ.* **862.**
899. Strain, H. H. (1939). *J. Biol. Chem.* **127**(1), 191–201.
900. Straub, O. (1976), "Key to Carotenoids, List of Natural Carotenoids." Birkhaeuser, Basel.
901. Straub, O. (1971). *In* "Carotenoids" (O. Isler, ed.), pp. 771–850. Birkhaeuser, Basel.
902. Straus, W. (1956). *Exp. Cell Res.* **11,** 289–296.
903. Strmiska, F., and Holcikova, K. (1970). *Fluess. Obst* **37**(2), 46–53.
904. Strmiska, F., and Holcikova, K. (1969). *Fluess. Obst* **36**(11), 140–142, 144–146.
905. Stussi, D. (1957). *Schweiz. Milchztg.* **83**(54), 430; (55), 433.
906. Subbarayan, C., and Cama, H. R. (1968). *J. Sci. Ind. Res.* **27**(8), 306–315.
907. Subbarayan, C., and Cama, H. R. (1966). *Indian J. Biochem.* **3**(4), 225–227.
908. Subbarayan, C., and Cama, H. R. (1965). *Indian J. Chem.* **3,** 463–465.
909. Subbarayan, C., and Cama, H. R. (1964). *Indian J. Chem.* **2**(1), 451–454.
910. Subbarayan, C., Lakshmanan, M. R., and Cama, H. R. (1966). *Biochem. J.* **99,** 308–311.
911. Surmatis, J. D., and Ofner, A. (1961). *J. Org. Chem.* **26,** 1171.
912. Swartling, P. (1945). *Sven. Mejeritidn.* **37,** 103–105, 115–118.
913. Sweeney, J. P., and Marsh, A. C. (1974). *J. Nutr.* **104**(9), 1115–1123.
914. Sweeney, J. P., and Marsh, A. C. (1971). *J. Agric. Food Chem.* **19**(5), 854–856.
915. Sweeney, J. P., and Marsh, A. C. (1971). *J. Am. Diet. Assoc.* **59,** 238–243.
916. Sweeney, J. P., and Marsh, A. C. (1970). *J. Assoc. Off. Anal. Chem.* **53,** 937–940.
917. Szabo, P. (1969). *Konzerv- Paprikaip.* **1,** 7–11.
918. Tada, M., Umeda, K., Ifuku, Y., and Shiroishi, M. (1976). *J. Food Sci. Technol. (Tokyo)* **23**(3), 113–118.
919. Tajiri, T., Matsumoto, K., and Hara, K. (1973). *Ergei Gakkai Zasshi* **42**(1), 81–88.
920. Tan, C. T., and Francis, F. J. (1962). *J. Food Sci.* **27,** 232–241.
921. Taylor, A. L., and Witte, P. J. (1938). *Ind. Eng. Chem.* **30,** 110–111.
922. Terasaki, M., and Mima, H. (1964). *J. Jpn. Soc. Food Nutr.* **17,** 111–114.
923. Terasaki, M., and Mima, H. (1964). *J. Jpn. Soc. Food Nutr.* **17,** 195–198.
924. "The Technical Feasibility of Fortifying Tea." Reports of International Venture Research, Chaska, Minnesota.
925. Thomas, M. H., Brenner, S., Eaton, A., and Craig, V. (1949). *J. Am. Diet. Assoc.* **25,** 39–45.

926. Thomas, P. (1975). *J. Food Sci.* **40**(4), 704–706.
927. Thomas, R. L., and Jen, J. J. (1975). *J. Food Sci.* **40,** 566–568.
928. Thommen, H. (1971). *In* "Carotenoids" (O. Isler, ed.), pp. 637–668. Birkhaeuser, Basel.
929. Thommen, H. (1967). *Int. Z. Vitaminforsch.* **37**(2), 175–183.
930. Thommen, H. (1962). *Naturwissenschaften* **49,** 517–518.
931. Thompson, C. R., Bickoff, E. M., Van Atta, A. R., Kohler, G. O., Guggolz, J., and Livingston, A. L. (1960). *U.S. Dep. Agric., Tech. Bull* **1232.**
932. Thompson, S. Y. (1968). *J. Dairy Res.* **35,** 149–169.
933. Thompson, S. Y. (1965). *Proc. Nutr. Soc.* **24,** 136–146.
934. Thompson, S. Y. (1964). *Exp. Eye Res.* **3,** 392–404.
935. Thompson, S. Y., Henry, K. M., and Kon, S. K. (1964). *J. Dairy Res.* **31,** 1–25.
936. Thompson, S. Y., and McGillivray, W. A. (1957). *J. Dairy Res.* **24,** 108–114.
937. Thudichum, J. L. W. (1869). *Proc. R. Soc. London* **11,** 253–256.
938. Tichenor, D. A., Martin, D. C., and Wells, C. E. (1965). *Food Technol.* **19**(3), 106–109.
939. Tiews, J. (1963). *Wiss. Veroeff. Dtsch. Ges. Ernaehr.* **9,** 235–261.
940. Ting, S. V., and Deszyck, E. J. (1958). *Proc. Am. Soc. Hortic. Sci.* **71,** 271–277.
941. Ting, S. V., and Hendrickson, R. (1969). *Food Technol.* **23**(7), 947–950.
942. Tirimanna, A. S. L., and Wickremasinghe, R. L. (1971). *Qual. Plant. Mater. Veg.* **20**(4), 341–346.
943. Todd, E. (1957). *Food Eng.* **29**(8), 82–83.
944. Tomes, M. L., and Quackenbush, F. W. (1958). *Econ. Bot.* **12,** 256–260.
945. Tomioka, K., Imaya, R., Endo, K., and Yamamoto, Y. (1973). *Kaseigaku Zasshi* **24**(8), 688–693.
946. Torchinskaya, M. M. (1972). *Tr. Vses. Semin. Biol. Akt. (Lech.) Veshchestvam Plodov Yagod., 4th, 1970* pp. 97–99.
947. Toro, O., de Pablo, S., Aguayo, M., Gattas, V., Contreras, I., and Monckeberg, F. (1975). Prevention of Vitamin A Deficiency by Enrichment of Sugar. Dept. of Nutrition and Food Technology, University of Chile, Santiago.
948. Troitskii, G. V., and Tarasova, L. S. (1955). *Biokhimiya* **20,** 19–30.
949. Trombly, H. H., and Porter, J. W. (1953). *Arch. Biochem. Biophys.* **43,** 443–457.
950. Troyan, A. V., and Gol'yan, D. S. (1973). *Izv. Vyssh. Uchebn. Zaved., Pishch. Tekhnol.* **1,** 22–24.
951. Trudel, M. J., and Ozbun, J. L. (1971). *J. Am. Soc. Hortic. Sci.* **96**(6), 763–765.
952. Ulrich, J. M., and Mackinney, G. (1968). *Photochem. Photobiol.* **7**(3), 315–318.
953. Umanskaya, A. N., and Klyuchkin. V. V. (1972). *Tr., Vses. Nauchno-Issled. Inst. Zhirov* **29,** 101–103.
954. Umeda, K. (1973). *Eiyo To Shokuryo* **26**(2), 81–90.
955. Umeda, K. (1971). *Nippon Shokuhin Kogyo Gakkai-Shi* **18**(4), 161–166.
956. Umeda, K., Tanaka, Y., and Ohira, K. (1971). *Nippon Shokuhin Kogyo Gakkai-Shi* **18**(11), 519–523.
957. Umill, N., and Gabelman, W. H. (1972). *J. Am. Soc. Hortic. Sci.* **97**(4), 453–460.
958. Underwood, B. A. (1974). *World Rev. Nutr. Dis.* **19,** 123–172.
959. Urion, E., Chapon, L., and Creff, R. (1960). *Boissons Fr.* February/March issue.
960. U.S. Patent 2,477,928; 1948.
961. U.S. Patents 2,830,908; 1958. 2,861,891; 1958; 2,874,058; 1959. 3,227,561; 1966. Japanese Patent 21,476/61; 1961.
962. U.S. Patent 2,848,508; 1958.
963. U.S. Patent 2,943,943; 1960.
964. U.S. Patent 3,039,877; 1962.
965. U.S. Patent 3,081,171; 1963.
966. U.S. Patent 3,110,598; 1963.
967. U.S. Patent 3,125,451; 1964.

968. U.S. Patent 3,162,538; 1964.
969. U.S. Patent 3,206,316; 1965.
970. U.S. Patent 3,248,301; 1964.
971. U.S. Patent 3,316,101; 1965.
972. U.S. Patent 3,423,213; 1969.
973. U.S. Patents 3,403,184; 1968. 3,408,406; 1968. 3,432,566; 1968. 3,446,801; 1969.
974. U.S. Patent 3,790,688; 1974.
975. U.S. Patent 2,876,160; 1959.
976. Valadon, L. R. G., and Mummery, R. S. (1978). *Phytochemistry* **17,** 818-819.
977. Valadon, L. R. G., and Mummery, R. S. (1977). *Phytochemistry* **16,** 613-614.
978. Valadon, L. R. G., and Mummery, R. S. (1972). *Ann. Bot. (London)* [N.S.] **36,** 471-474.
979. Valadon, L. R. G., and Mummery, R. S. (1967). *Ann. Bot. (London)* [N.S.] **31,** 495-503.
980. Valadon, L. R. G., Sellens, A. S., and Mummery, R. S. (1975). *Ann. Bot (London)* [N.S.] **39,** 785-790.
981. Vallentyne, J. R. (1957). *Arch. Biochem. Biophys.* **70,** 29-34.
982. Van Eekelen, M., and Pannevis, W. (1938). *Nature (London)* **141,** 203-204.
983. Van Zeben, W. (1946). *Maandschrift. Kindergeneeskd.* **14,** 45.
984. Van Zeben, W., and Hendriks, T. F. (1948). *Int. Z. Vitaminforsch.* **19,** 265-266.
985. Veloso de Almeida, P. G., Brune, W., and Kling de Moraes, G. H. (1977). *Rev. Ceres* **24**(132), 187-199.
986. Venkatakrishna, S., Premachandra, B. R., and Cama, H. R. (1977). *Tea Q.* **47,** 28-31.
987. Venkatakrishna, S., Premachandra, B. R., and Cama, H. R. (1976). *Agric. Biol. Chem.* **40**(12), 2367-2371.
988. Venkataro, P., and Achaya, K. T. (1969). *Fette, Seifen, Anstrichm.* **71,** 270-272.
989. Verma, A. K., and Sastry, L. V. L. (1969). *Indian Food Packer* **23**(6), 5-11.
990. Verma, A. N., Srivastava, D. C., Misha, H. R., and Sharma, R. K. (1970). *Poona Agric. Coll Mag.* **60**(1/2), 72-74.
991. Vinkler, M., and Kiszel-Richter, M. (1972). *Acta Aliment. Acad. Sci. Hung.* **1**(1), 41-58.
992. Vladimirov, B. M. (1967). *Izv. Vyssh. Uchebn. Zaved., Pishch. Tekhnol.* **3,** 89-91.
993. von Euler, H., Karrer, P., Krauss, E. V., and Walker, O. (1931). *Helv. Chim. Acta* **14,** 154-162.
994. Wackenroder, H. (1831). *Mag. Pharm.* **33,** 144.
995. Wagner, K. H. (1954). *Klin. Wochenschr.* **32,** 87-89.
996. Wagner, K. H. (1954). *Fette, Seifen, Anstrichm.* **56,** 681-687.
997. Wagner, K. H. (1952). *Milchwissenschaft* **7,** 250-253.
998. Wagner, K. H. (1950). *Int. Z. Vitaminforsch.* **22,** 289-308.
999. Walczak, B., and Gronowska-Senger, A. (1975). *Technol. Rolno-Spozyw. Warshaw Szk. Gospod. Wiejsk.* **10,** 129-135.
1000. Wald, G., Nathanson, N., Jencks, W. P., and Tarr, E. (1948). *Biol. Bull. (Woods Hole, Mass.)* **95**(2), 249-250.
1001. Wall, J. S., and Blessin, C. S. (1969). *Cereal Sci. Today* **14**(8), 264-270.
1002. Walter, W. M., Jr., and Purcell, A. E. (1974). *J. Agric. Food Chem.* **22**(2), 298-302.
1003. Walter, W. M., Jr., and Purcell, A. E. (1971). *J. Agric. Food Chem.* **19**(1), 175-178.
1004. Washburn, R. G., Hibbs, J. W., and Krauss, W. E. (1950). *Ohio, Agric. Exp. Stn., Res. Circ.* **8.**
1005. Watada, A. E., Norris, K. H., Worthington, J. T., and Massie, D. R. (1976). *J. Food Sci.* **41,** 329-332.
1006. Watada, A. E., Aulenbach, B. B., and Worthington, J. T. (1976). *J. Food Sci.* **41,** 856-858.
1007. Weber, F., Laskawy, G., and Grosch, W. (1974). *Z. Lebensm.-Unters. -Forsch.* **115**(1), 142-150.
1008. Weckel, K. G., Huang, T. T., and Ziemba, J. V. (1967). *Food Eng.* **39**(5), 68-77.
1009. Weckel, K. G., Santos, B., Herman, E., LaFerriere, L., and Gabelman, W. H. (1962). *Food Technol.* **16**(8), 91-94.

1009a. Weedon, B. C. L. (1979). *Pure Appl. Chem.* **51**(3), 435–446.
1010. Weedon, B. C. L. (1971). *In* "Carotenoids" (O. Isler, ed.), pp. 267–324. Birkhaeuser, Basel.
1011. Weiser, H., and Studer, W., unpublished data.
1012. Wells, J. L. (1959). *Can. Dairy Ice Cream J.* **38**(8), 56, 58.
1013. Wells, J. O. (1962). *Food Can.* **22**(4), 36–38.
1014. Went, F. W., LeRosen, A. L., and Zechmeister, L. (1942). *Plant Physiol.* **17,** 91–100.
1015. Wheaton, T. A., and Stewart, I. (1973). *J. Am. Soc. Hortic. Sci.* **98**(4), 337–340.
1016. Whittingham, C. P. (1965). *In* "Chemistry and Biochemistry of Plant Pigments" (T. W. Goodwin, ed), pp. 357–380. Academic Press, New York.
1017. WHO/US-AID Report (1976). Vitamin A deficiency and xerophthalmia. *W.H.O., Tech. Rep. Ser.* **590.**
1018. Wildfeuer, I., and Acker, L. (1968). *Mitt. Geb. Lebensmittelunters. Hyg.* **59,** 392–400.
1019. Wilkinson, H. (1939). *Analyst* **64,** 17–23.
1020. Wilkinson, R. A., and Conochie, J. (1958). *Aust. J. Dairy Technol.* **13,** 29–31.
1021. Willard, H. S. (1947). *Wyo., Agric. Exp. Stn., Bull.* **279.**
1022. Williams, K. T., Bickoff, E., and Lowrimore, B. (1944). *Oil Soap (Chicago)* **21,** 161–164.
1023. Williams, R. J. H., Britton, G., and Goodwin, T. W. (1967). *Biochem. J.* **105,** 90–105.
1024. Winterstein, A., Studer, A., and Ruegg, R. (1960). *Chem. Ber.* **93,** 2951–2965.
1024a. Witschi, J. C., Houser, H. B., and Littell, A. S. (1970). *J. Am. Diet. Assoc.* **57,** 13–16.
1025. Wodsak, W. (1952). *Dtsch. Lebensm.-Rundsch.* **48,** 208.
1026. Woodward, J. R. (1972). *J. Sci. Food Agric.* **23**(4), 465–473.
1027. Wright, W. G., and Milner, A. R. (1965). *Gerontol. Clin.* **7,** 120–127.
1028. Yamaguchi, M. (1974). *Yukagaku* **23**(10), 624–632.
1028a. Yamamoto, H. Y. (1979). *Pure Appl. Chem.* **51**(3), 639–648.
1029. Yamamoto, R., and Muraoka, T. (1973). *Sci. Pap., Inst. Phys. Chem. Res. (Jpn.)* **19,** 127–131.
1030. Yokoyama, H., Guerrero, H. C., and Boetiger, H. (1972). *Adv. Food Res. Suppl.* **3,** 1–7.
1031. Yokoyama, H., and Vandercook, C. E. (1967). *J. Food Sci.* **32**(1), 42–48.
1032. Yoshimuzi, T. (1974). *Shokuhin Kogyo* **17**(20), 52–62.
1033. Young, R., and Jahn, O. (1972). *Proc. Fla. State Hortic. Soc.* **85,** 33–37.
1034. Yuichiro, T. (1967). *Mem. Fac. Agric., Kagoshima Univ.* **6**(1), 31–38.
1035. Yver, M., and Luguet, F. M. (1973). *Ind. Aliment. Agric.* **90**(1), 23–25.
1036. Zagalsky, P., and Herring, P. J. (1977). *Philos. Trans. R. Soc. London, Ser. B* **279,** 289–326.
1037. Zbinden, G., and Studer, A. (1958). *Z. Lebensm.-Unters. -Forsch.* **108,** 113–134.
1038. Zechalko, A., Rola, J., Jasinska, M., and Biernat, J. (1971). *Pamiet. Pulawski* **46,** 227–242.
1039. Zechmeister, L. (1962). "Cis-trans Isomeric Carotenoids, Vitamin A and Arylpolyenes," Springer-Verlag, Berlin and New York.
1040. Zechmeister, L. (1944). *Chem. Rev.* **34,** 267–344.
1041. Zechmeister, L., Le Rosen, A. L., Went, F. W., and Pauling, L. (1942). *Proc. Natl. Acad. Sci. U.S. A* **27,** 468–474.
1042. Zeller, P., Bader, F., Lindlar, H., Montavon, M., Müller, P., Rüegg, R., Ryser, G., Saucy, G., Schären, S. F., Schwieter, U., Stricker, K., Tamm, R., Zuercher, P., and Isler, O. (1959). *Helv. Chim. Acta* **42,** 841–847.
1043. Zetea, F., and Bodea, C. (1971). *Stud. Cercet. Biochim.* **14**(2), 211–214.
1044. Zhamsran, Z. (1971). *BNMAU Shinzhlekh Ukhaany Akad. Medee* **1,** 53–59.
1045. Zimmerman, W. I., Tressler, D. K., and Maynard, L. A. (1941). *Food Res.* **6,** 57–68.
1046. Zimmerman, W. I., Tressler, D. K., and Maynard, L. A. (1940). *Food Res.* **5,** 93–101.
1047. Zlotlow, M. J., and Settipane, G. A. (1977). *Am. J. Clin. Nutr.* **30,** 1023–1025.
1048. Zscheile, F. P., Henry, R. L., White, J. W., Jr., Nash, H. A., Schrewsbury, C. L., and Hauge, S. M. (1944). *Chem. Anal. Ed.* **16,** 190–193.
1049. Zsolt, J., Schneider, G., and Matkovics, B. (1963). *Can. J. Biochem. Physiol.* **41,** 481–486.
1050. Zubeckis, E. (1968). *Can. J. Public Health* **59,** 193–196.

3

OXYCAROTENOIDS IN POULTRY FEEDS

W. L. Marusich and J. C. Bauernfeind

CAROTENOIDS AS COLORANTS
AND VITAMIN A PRECURSORS

ISBN 0-12-082850-2

I. INTRODUCTION

Nature relies on a variety of compounds for the pigmentation of living matter, with carotenoids probably representing the most widespread group of naturally occurring pigments. Reviews by Bauernfeind (*62,63*), Bauernfeind *et al.* (*69*), Borenstein and Bunnell (*106*), Nadai and Brubacher (*599*), and Weedon (*847,848*) summarize their occurrence and properties in nature. Carotenoids are universally present in plant photosynthetic tissues, and, although their role in photosynthesis is not yet completely defined, they may transfer absorbed energy to chlorophyll through electron transport, as proposed by Bodea (*99*). They can also be found in nonphotosynthetic tissues of higher plants (fruit, flower petals, pollen, roots), in fungi and nonphotosynthetic bacteria, and in some species of all animal phyla. The chemistry and biosynthesis of carotenoids have been reviewed by Mackinney *et al.* (*526*), Britton (*119a*), Moss and Weedon (*594a*), Porter and Anderson (*644*), Davies (*202,203a*), Grob (*334*), and Goodwin (*326a–328*), their metabolism by Thommen (*796*), and their functions by Krinsky (*474*). Carotenoids are synthesized *de novo* only by higher plants and protista; all animal carotenoids must come from a dietary source, although they may be slightly altered by oxidative metabolism before being accumulated in animal tissues, as shown by Goodwin (*326*) and Olson (*615*). Today, large numbers of carotenoid structures have become available through total synthesis, as described in detail by Mayer and Isler (*563*) and tabulated by Straub (*762*).

Carotenoids are aliphatic or aliphatic alicyclic structures composed of five-carbon isoprene groups, usually eight, so linked that the two methyl groups nearest the center of the molecule are in positions 1 and 6 and all other lateral methyl groups are in positions 1 and 5; a series of conjugated C-C double bonds constitutes the chromophoric system, as depicted by Karrer and Jucker (*443*). Carotenoids can be divided into two main groups: (a) hydrocarbons, containing only hydrogen and carbon, which are called carotenes, and (b) oxygen-containing derivatives, which are called xanthophylls by Davies (*203*) and oxycarotenoids by Deuel (*229*). In this chapter the term "oxycarotenoids" is synonymous with "xanthophylls," and both terms are used interchangeably.

From the functional viewpoint as avian tissue pigmenters, Marusich (*542*) has categorized the carotenoids as follows: (a) vitamin A precursors that do not pigment, such as α- and β-carotenes; (b) vitamin A precursors that pigment, such as cryptoxanthin, β-apo-8′-carotenal, and β-apo-8′-carotenoic acid ethyl ester; (c) nonvitamin A precursors that do not pigment or pigment poorly, such as violaxanthin and neoxanthin; and (d) nonvitamin A precursors that pigment, such as lutein, zeaxanthin, and canthaxanthin. Carotenoids of type (b) and (d) are of significant economic interest in the pigmentation of egg yolks and/or the skin and fatty tissues of growing chickens for meat production (broilers) or heavier-meat birds (roasters and capons). Avian species preferentially store these orally ingested oxycarotenoids in the liver, eggs, body fat, skin, feathers, beak, and shanks.

β-Carotene is not an effective *in vivo* pigmenter for avian species since they convert β-carotene efficiently to vitamin A, as noted by Goodwin (*325*). With the dairy cow, however, ingested β-carotene is both a pigmenter and a potent provitamin A source. Ingested β-carotene that is not converted to vitamin A is stored in both the fatty tissues of the body and the butterfat of milk and milk by-products, as detailed by Goodwin (*324*). Morgan and Everitt (*591*) also found that cattle deposit β-carotene in their fat. Soliman *et al*. (*732*) reported that cattle deposit from 10 to 20 times the amount of β-carotene in body fat as the buffalo, camel, and sheep.

II. SIGNIFICANCE OF YELLOW COLOR IN POULTRY PRODUCTS

The term "pigmenter" refers to carotenoids that when present in, or added to, an animal feed (ration) will color body tissues, such as skin and fat, or animal products, such as eggs, butter, and cheese. This indirect method of coloring food is as old as the animal kingdom. Pigmentation of poultry involves primarily the skin coloration of meat chickens (broilers and roasters) and the degree or shade of egg yolk color. A secondary aspect is the pigmentation of newly hatched chicks, as cited by Bauernfeind *et al*. (*65*) and Bauernfeind and Marusich (*67*). Douglas *et al*. (*239*) showed that the intensity of visual pigmentation in 1-day-old hatched chicks is related to the concentration of oxycarotenoids in the feed fed to their dams. As noted by Bauernfeind *et al*. (*65*), the rations of the breeder hen or dam should be supplemented with adequate amounts of xanthophylls so that the newly hatched chicks will have sufficient yellow color, a visual quality indicator, acceptable to purchasers and producers. Chicks with pale shanks and down are less desirable to poultry growers in the business of producing market-weight broilers and roasters.

Market-weight broilers with uniformly well pigmented skin and shanks are generally preferred by consumers as indicated by numerous studies (*7,32,36,51,67,112, 142,181,182,265,290,352,375,381,386,581,582,596,661,875*). Lambert and Knox (*488*) reported a close relationship between the intensity of yellow skin and shank coloration. A bright yellow or yellow-orange color is often associated with quality and health. Wagner and Mitchell (*841*) suggested that the capability to deposit a single oxycarotenoid or group of oxycarotenoids by chickens is quantitatively inherited. A premium may be paid to the producer of well-pigmented broilers (*76,258,264,265,384,875*). In specific geographic areas of the United States (i.e., the Northeast) a premium of up to 10 cents/lb is routinely obtained in the retail market for branded, heavily pigmented whole roasters, broilers, and cut-up parts as compared to less pigmented, branded, and nonbranded sources. On the other hand, in some geographic areas lightly pigmented chickens may be preferred (*72,258,289, 359,779*) depending on the breeds of chickens involved, poultry management practices, or consumer preference.

With table eggs (whole shelled eggs) a uniformly moderate light to deep yellow yolk color is generally preferred by most consumers (*7,23,27,28,85,107,124,219, 225,247,490,698,714,775,783,809,811,836,875,887*). In addition, highly colored egg yolks (fresh, frozen, or dried) are required for the commercial production of bakery products, noodles, mayonnaise, prepared cake mixes, and other egg-containing products (*25,67,140,145,258,271,287,440,524,605,698,777,869*). A premium price may be paid for dark-colored yolks that are intended for the food processing industry (*25, 53, 140, 219, 287, 490, 698, 745, 777, 816*). This outlet represents a significant percentage of all the eggs produced in the United States. These eggs are referred to as "breaker eggs." Carlson (*141*) in 1961 stated that about 10% of the eggs produced in the United States are broken for processing purposes. Today, this percentage of the total is probably substantially higher.

Just as the majority of consumers prefer red tomatoes over yellow ones, red salmon over pink salmon, red radishes over white ones, white cabbage over red cabbage, and green beans over yellow beens, there is a preference for yellow or well-pigmented poultry products over white or less well pigmented broilers and eggs.

III. HISTORICAL ASPECTS OF PIGMENTED POULTRY PRODUCTS

Skin and yolk pigmentation presented no problems when the principal source of poultry and poultry products was the barnyard flock, as was the case in the United States in the early part of this century. Under these conditions the liberal feeding of whole or cracked yellow corn, free access of chickens to green grass and the opportunity to forage by themselves, coupled with a markedly slower growth rate for broilers and a low rate of egg production in hens, allowed for adequate pigmentation in poultry and poultry by-products. Snyder (*730*) in 1945 showed graphically the increase in yolk color when confined hens were allowed free access to green pasture. Current methods, namely, intensive confined management practices, selecting genetically fast-growing broiler strains, marketing at an earlier age, and utilizing high-energy, low-fiber feeds, have created a distinct problem in the production of adequately pigmented broilers, as pointed out by Williams *et al.* (*875*) and Ratcliff *et al.* (*665*). High-energy feeds are formulated with little or no dehydrated alfalfa meal, a good source of xanthophylls but high in fiber. The feed conversions (grams of feed consumed per gram of weight gain) have been constantly reduced so that broilers actually take in less feed and thus fewer xanthophylls per unit of body weight. Wildfeuer and Acker (*864*) noted that, with the laying hen, genetic improvements have allowed for markedly increased egg production with less feed consumed per dozen eggs laid. Thus, the total xanthophylls ingested per day, to be transferred and deposited in the yolk, are also reduced, resulting in yolks of lighter color. Ratcliff *et al.* (*664*) stated that the problem is especially acute for the "breaker egg" producer. Levels of dehydrated alfalfa meal ranging from 15 to

25% had to be included in hens' rations to attain the intense yolk pigmentation required for "breaker eggs" (*144,177,664,777*). Scott *et al.* (*713*) noted that, when environmental temperatures are high and feed consumption low, it is sometimes necessary to feed as much as 30–40% dehydrated alfalfa meal to produce deeply colored (NEPA no. 5) yolks. Couch (*170*), Couch *et al.* (*179*), and Farr *et al.* (*261*) also reported a decrease in egg yolk color amounting to about 33% during hot summer weather in eggs produced by hens on the same dietary program as that during the cooler seasons of the year due to a marked reduction in feed intake.

The pigmenting burden in modern commercial feeds falls on yellow corn, which alone cannot produce adequate results without the addition of corn gluten meal and/or dehydrated alfalfa meal (*172,382,449,485,575,666*). The increased usage of milo in the United States, as noted by Couch and Crawford (*174*), Harms *et al.* (*359*) and Elrod *et al.* (*252*), and of wheat, barley, and rice in other parts of the world, as cited by Bornstein and Bartov (*108,109*) and Bauernfeind *et al.* (*65*), as yellow corn substitutes, further accentuates the difficulty of producing adequately pigmented eggs and/or broilers. In a series of papers, Terpstra (*788–790*) treated the problem of yolk pigmentation and some possible approaches to its solution. A number of additional oxycarotenoid sources have been developed over the years in an attempt to augment the xanthophylls in yellow corn, corn gluten meal, and dehydrated alfalfa meal. These include alga meals (*588,630*), marigold meal (*115,529*) and concentrates made from marigold meal, higher protein–lower fiber dehydrated alfalfa meals, a 45% protein alfalfa concentrate (*347,735,890*), and pure oxycarotenoids in dry stabilized form including β-apo-8′-carotenal, β-apo-8′-carotenoic acid ethyl ester, canthaxanthin (*546,547*), and citranaxanthin.

IV. EGG AND POULTRY TISSUE CAROTENOIDS

A. Egg Yolks

The first attempts to identify the yellow pigment in egg yolk, by Stadeler (*743*) in 1867, were unsuccessful. In 1869, Thudichum (*808*) reported the pigment in egg yolk to be nonsaponifiable and exclusively soluble in fat solvents and classified it as a "lutein" or "lipochrome." In 1903, Schunck (*709*) isolated the pigment from egg yolk and considered it identical with one of the xanthophylls found in flowers. Willstätter and Escher (*878*) first isolated what they considered to be a singular pure pigment and named it lutein. Karrer and Helfenstein (*441*) demonstrated that the yolk pigment, although possessing the same melting point as leaf xanthophyll (lutein), differed in its optical rotation. Kuhn *et al.* (*482*) were able to separate the pigments in the egg yolk into two components identified as lutein and zeaxanthin. Their results agreed with those of Kuhn and Smakula (*481*), who found the yolk pigments to be composed of 70% lutein and 30% zeaxanthin. Subsequently, Gillam and Heilbron (*316*) identified cryptoxanthin and carotene as being present in small quantities in the yolk besides zeaxanthin, which they reported to be the major component. The presence of traces of cryptoxanthin and carotene in egg yolk was also reported by Kennedy (*452*), Ganguly *et al.* (*302*), Kuhn and Brockmann (*478*), and Kuhn and Grundmann (*479*). Peterson *et al.* (*632*) found the presence of esterified xanthophylls and the occasional occurrence of isolutein, neoxanthin, and very small quantities of compounds of the flavoxanthin type in yolks, besides lutein, cryptoxanthin, and carotene (mainly β-carotene with only traces of α-carotene). Van Koetsveld (*828*) was able to confirm the presence of α-carotene only when it was specifically included in hens' feed; however, he also identified lycopene in the yolk. Nash and Zscheile (*602*) reported the presence of trace amounts of ζ-carotene in yolks.

As the carotenoids comprising the pigments in the yolk were being identified, the question arose as to their origin in the yolk. Numerous trials were run to ascertain whether they were synthesized by the hen or absorbed from feed ingredients. Stewart and Atwood (*753*) and Dryden (*240*) appear to have been the first American workers to study the effect of feed on yolk color. Palmer (*620*) and Palmer and Kempster (*623*) conducted more extensive studies demonstrating that the entire carotenoid content of the egg yolk was derived from ingested feed. A flock of White Leghorn pullets and cockerels was raised from hatching to maturity on rations so devoid of carotenoids that the mature birds showed only the slightest traces of pigment in adipose tissue, no demonstrable amounts in the blood serum or skin, and none in the yolks of the eggs laid by mature hens. Feeds rich in xanthophylls (yellow corn) brought about a rapid coloration in all parts of the body of roosters and nonlaying hens and effectively colored the yolks of eggs from laying hens. In turn, feeding carotene in the form of highly colored (colostrum) butterfat had little effect on the amount of pigment carried by the blood serum and deposited in the egg yolk. Others have shown that carotenes are inefficiently deposited in the yolk (*369,725,870*). Thus, the hen selectively deposits only xanthophylls in both the egg yolk and the body fat, as shown by Palmer and Eckles (*622*), in contrast to the cow, which stores β-carotene in various fat depots.

Palmer's pioneering work (*621*) proved that this class of carotenoids (xanthophylls) also occurs in the fat, skin, shanks, and liver of poultry. Hughes and Payne (*415*) studied the effect of feeding graded levels of yellow corn to Leghorn hens. They separated and measured by solvent extraction the xanthophylls (zeaxanthin and lutein) in the methanol phase and the carotenes plus cryptoxanthin in the petroleum ether phase. They determined the rates of deposition for the two groups of carotenoids by relating the quantities fed daily to the hen and the quantities recovered in the yolks. The lutein plus zeaxanthin fraction accounted for the majority of pigment, with the carotene plus cryptoxanthin fraction being a minor component.

Kuhn and Brockmann (*478*) studied the effect of administering pure physalien (zeaxanthin dipalmitate) to hens depleted of oxycarotenoids. Measurement of the yolk pigments showed a 5:1 ratio of free zeaxanthin to zeaxanthin ester, even though only the ester was ingested orally. These data indicated that the hen was able to hydrolyze the ester. Brockmann and Völker (*120*) reported on the results of administering purified lutein, zeaxanthin, carotene, lycopene, and violaxanthin to hens. Lutein and zeaxanthin gave the most intense color to the yolks, compared to about 0.5% deposition for carotene and only trace amounts of lycopene and no violaxanthin. Wenland (*857*) also could not find violaxanthin in the yolk.

Peterson *et al.* (*632*) orally administered samples of purified lutein, zeaxanthin, cryptoxanthin, and carotene to laying hens. Neither carotene nor cryptoxanthin was found in the yolks; however, substantial quantities of lutein and zeaxanthin were measured. Strain (*759*) observed that eggs from White Leghorns fed 20% fresh alfalfa or an equivalent quantity of freshly dried alfalfa contained minute traces of neoxanthin, traces of flavoxanthin-like compounds, and considerable quantities of zeaxanthin, isolutein, and cryptoxanthin. Although the hens had consumed large quantities of carotene from the alfalfa samples, only traces were found in the eggs. The transfer of dietary cryptoxanthin to the yolk along with β-carotene was later reported by Ganguly *et al.* (*302*) and Grimbley and Black (*332,333*), but both were transferred in relatively small amounts. Van Koetsveld and Grimbergen (*829*) found from 2 to 10 μg of β-carotene per yolk. Rauch (*669*) reported pure β-carotene to have no effect on yolk pigmentation, and Wildfeuer *et al.* (*865*) showed it to be very ineffectively deposited in yolk (0.23%). Thumim (*809*) and Tiews and Zucker (*810*) found lutein and zeaxanthin to be the major oxycarotenoids associated with pigmentation resulting from feeding corn, corn gluten meal, and lucerne (alfalfa) meal. Williams *et al.* (876) fed purified β-carotene, zeaxanthin, and lutein to laying hens and measured the relative concentration of these carotenoids in serum, adipose tissue, liver, and egg yolk. The results (Table 1) show that β-carotene was poorly deposited whereas zeaxanthin and lutein were well deposited and in about equal amounts.

Benk *et al.* (*73*) determined the total oxycarotenoid content and β-carotene

TABLE 1

Deposition of Carotenoids in Serum, Abdominal Adipose Tissue, Liver, and Egg Yolk of Laying Hens Fed Purified Carotenoids[a]

Supplement to basal ration	Level fed (mg/lb)	Carotenoid concentration: Serum (μg/ml)	Adipose tissue (μg/gm)	Liver (μg/gm)	Egg yolk (μg/gm)
None	—	Trace	Trace	Trace	0.41
β-Carotene	8.6	0.08	0.10	Trace	Trace
Zeaxanthin	8.6	1.28	1.62	1.98	8.10
Lutein	8.6	1.30	1.75	2.01	8.21

[a] From Williams *et al.* (*876*).

portion in 55 samples of egg yolks and found that the level of β-carotene, as a percentage of total oxycarotenoids, fluctuated between 0.8 and 4.6%. Thus, the oxycarotenoid portion represented from 95 to 99% of the total. Brockmann and Völker (*120*) considered that only xanthophylls containing two hydroxyl groups were utilized. However, Titus *et al*. (*812*) concluded that xanthophylls containing one, two, three, or possibly more hydroxyl groups were readily transferred to the yolk. Karrer and Krause-Voith (*444*) claimed that epoxides were not transferred to the yolk, whereas Strain (*758*) found that a monoepoxide (flavoxanthin) was transferred, but, at the same time, the diepoxide (violaxanthin) was not.

As a result of numerous experiments with a multitude of carotenoid sources, it can be stated that the color of yolk results from a relatively nonspecific deposition of oxycarotenoids. One requirement for deposition is that the oxycarotenoid possess functional groups containing oxygen such as hydroxyl, keto, or ester groups, which impart moderately polar characteristics. The hydrocarbon carotenoids, such as carotene, do not possess functional groups and are not deposited in the yolk in any significant quantities. The oxycarotenoids are deposited in egg yolks in direct proportion to their distribution in the ration, based on the findings of Schrenk *et al.*

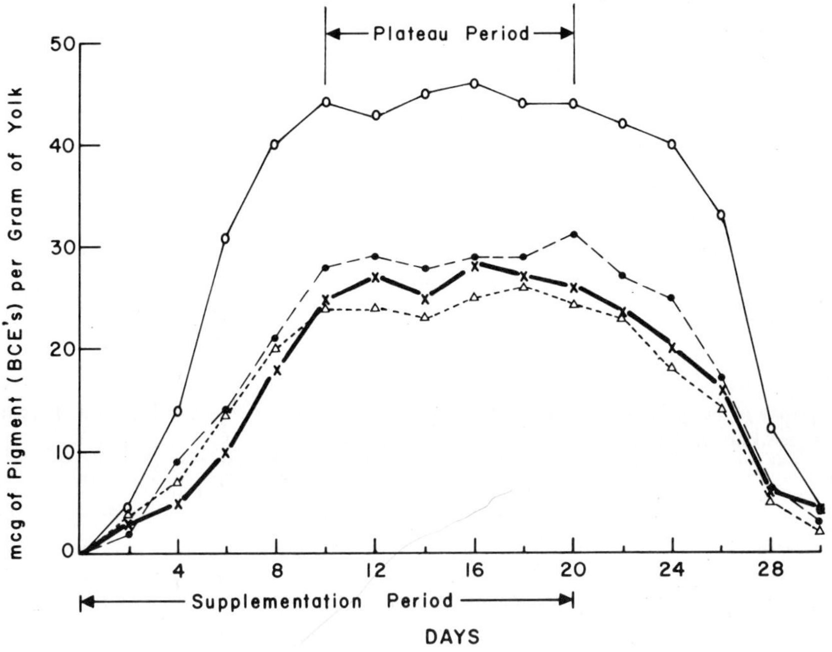

Fig. 1. Deposition of pigment in yolks of eggs from hens fed a low-pigment ration supplemented with 10 mg oxycarotenoids per kilogram of feed using several pigmentation sources. Key: ●, canthaxanthin; △, apocarotenal (APO); ○, apocarotenoic acid ethylester (APO-EE); x, corn + alfalfa. All are in stabilized beadlet form.

(*708*), Thompson *et al.* (*806*), Smith and Perdue (*729*), and Osadca and DeRitter (*619*). Thus, hens transfer to the yolk at least part of the many oxycarotenoids fed to them. Some are efficiently absorbed and transferred, others are altered and deposited, and still others are inefficiently absorbed or not absorbed at all.

The deposition of xanthophylls in eggs is a rapid process. Hens previously maintained on an oxycarotenoid-free feed will transfer dietary xanthophylls to the egg yolk within 48 hr after being fed an oxycarotenoid-rich feed, as shown by Albright and Thompson (*8*), Heiman and Wilhelm (*385*), and Titus *et al.* (812). A period of 9–11 days or more must elapse before uniform maximal color deposition (a plateau level) is obtained throughout the yolk, as demonstrated by Elrod *et al.* (*252*), Marusich *et al.* (*548*) and Williams *et al.* (*876*). Typical yolk pigment deposition data have been obtained (Fig. 1) for corn and alfalfa oxycarotenoids as compared to three pure oxycarotenoids in dry stabilized beadlet form fed to laying hens. After withdrawal of the oxycarotenoid source from the ration, a rapid decrease occurs in yolk oxycarotenoid content.

B. Poultry Tissue

Palmer and Kempster (*623*) demonstrated that cockerels fed an oxycarotenoid-free ration showed only trace amounts of oxycarotenoids in body tissues, but, when a ration high in xanthophylls was fed, the birds exhibited xanthophylls in the body fat, blood plasma, and skin (especially the shank and toe-web skin areas). Lutein, the main pigment found in these tissues, was in an esterified form. The presence of lutein esters in these tissues has been confirmed (*120,230,561*). This is in contrast to the findings of Peterson *et al.* (*632*), which showed that the xanthophylls stored in egg yolk were mainly in the "free form," with only about 8% of the pigments being esterified. Guilbert and Hinshaw (*344*) found lutein to be the main pigment of body fat and liver. Goodwin (*325*) concluded that the oxycarotenoids in the skin are in the form of esters. He further stated that these esters are mobilized in the free state into the eggs during the laying season. Bohren *et al.* (*102*) reported that this transfer of ester forms into the yolk in the free form occurs even if the hens are maintained on an oxycarotenoid-poor ration during the laying season, provided that they had access to oxycarotenoid-rich feed before the start of laying. Quackenbush *et al.* (*656*) also reported that the oxycarotenoids occurred principally as esters in the skin, with lutein, zeaxanthin, and cryptoxanthin being the effective and predominant pigments. They also showed that the skin oxycarotenoid profile paralleled the oxycarotenoid composition of the ration fed. The results of Smith and Perdue (*729*) were not in agreement; these researchers fed corn and alfalfa meal or corn and alga meal and tentatively identified four of the oxycarotenoids as α-carotene-like, cryptoxanthin-like, lutein, and zeaxanthin, with α-carotene-like constituting 35% of the total in shank skin. Later work by Livingston *et al.* (*518*) confirmed the earlier findings of Quackenbush *et al.* (*656*). Livingston *et al.* (*518*) observed that, during saponification of broiler shank extracts, a large amount of waxy material was present which took up considerable amounts of alkali, and, unless excess alkali was present, incomplete saponification occurred and esters remained, suggesting the presence of carotene-like and monohydroxy-like oxycarotenoids. They suggested that this might account for the differences in results reported by Smith and Perdue (*729*), on the one hand, and their results and those of Quackenbush *et al*. (*656*), on the other.

Livingston *et al*. (*518*) fed pure all-*trans*-lutein in the ration and identified lutein and its mono- and diesters in broiler shank skin. They further reported that lutein comprised from 62 to 93% of the total depending on whether corn gluten meal or alfalfa meal was the major oxycarotenoid source. When alfalfa was fed, cryptoxanthin comprised less than 1% of the total oxycarotenoids and zeaxanthin comprised about 5%. With corn gluten meal, lutein, zeaxanthin, and cryptoxanthin comprised 62, 29, and 7%, respectively.

In a collaborative broiler pigmentation trial reported by Marusich (*541*), it was shown that the ratio of lutein to zeaxanthin, as they occur in dehydrated alfalfa meal or corn gluten meal, is approximately the same after these pigments are deposited in the shank skin. Similarly, when the ratio of dihydroxycarotenoids to monohydroxycarotenoids is determined in these feed ingredients, essentially similar ratios are seen in the shank skin (Table 2).

Some oxycarotenoids may be deposited in broiler tissues in a form that differs from that administered. Kuzmicky *et al*. (*486*) administered neoxanthin or violaxanthin to chickens. Following neoxanthin administration, skin analysis showed the presence of both neoxanthin and neochrome, a 5,8-epoxide of neoxanthin; after violaxanthin administration, violaxanthin along with luteoxanthin and auroxanthin (5,8-epoxide isomers of violaxanthin) were found. The amounts of all

TABLE 2

Ratio of Lutein to Zeaxanthin in Diet versus Deposition in Shank Skin and Ratio of Dihydroxy to Monohydroxy Pigments in Diet versus Deposition in Shank Skin[a]

	Dehydrated alfalfa meal[b]				Corn gluten meal[c]			
	Lutein/ zeaxanthin		DHP/MHP[d]		Lutein/ zeaxanthin		DHP/MHP[d]	
Collaborator	Diet	Shank	Diet	Shank	Diet	Shank	Diet	Shank
1	6.8	6.3	9.1	9.9	2.7	1.8	4.3	10.3
2	7.5	5.1	10.7	7.2	1.8	2.1	5.6	6.6
3	7.0	6.1	12.4	9.1	2.1	1.4	6.8	7.0

[a] From Marusich (*541*).
[b] 3.55% dehydrated alfalfa meal fed in ration.
[c] 3.60% corn gluten meal fed in ration.
[d] DHP, dihydroxy pigments; MHP, monohydroxy pigments.

oxycarotenoids found were extremely small since neither neoxanthin nor violaxanthin is an effective pigmenting agent. Kuzmicky *et al.* (*486*) observed that neoxanthin and violaxanthin were ineffective for broiler pigmentation, whereas lutein, zeaxanthin, and cryptoxanthin were found Livingston *et al.* (*518*) and Kuzmicky *et al.* (*485*) to be the predominant and effective pigmenters in skin. Seaweed meals, which are added to poultry layer rations in Scandinavia, contain fucoxanthin as the principal oxycarotenoid. Layers have the ability to convert fucoxanthin to isofucoxanthin, which is deposited to a considerable extent in the yolk, according to Jensen (*436,437*).

As with egg yolk pigmentation, the intensity of broiler skin and shank pigmentation is determined primarily by the level of oxycarotenoids present in the ration, as cited by Palmer (*620*), Heiman and Tighe (*384*), Fritz *et al.* (*290*), Goodwin (*325*), and Fritz and Wharton (*289*). The oxycarotenoids are also rapidly deposited in the breast skin and shanks of birds; increases in color intensity can be seen after 5 days of feeding, as noted by Heiman and Tighe (*384*). These authors also showed that the colorimetrically determined pigment content continues to increase as the birds are maintained on the oxycarotenoid-rich ration, rather than plateauing as is seen in egg yolk (Fig. 1). Similar findings were reported by Day and Williams (*211*), Bartov and Bornstein (*58*), Marusich and Bauernfeind (*547*) and Harms *et al.* (*356*). Both Harms *et al.* (*356*) and Herrick *et al.* (*391*) found skin pigment concentrations to be repleted and depleted more rapidly than shank pigment concentrations.

V. MEASUREMENT OF PIGMENTATION AND OXYCAROTENOID CONTENT

Pigmentation reponses to various supplements differ, and therefore some mode of measuring differences among treatments is necessary. Generally, these fall into four categories: (a) visual observation by means of a color-graduated visual aid, (b) visual observation and ranking from deepest to lightest pigmentation, (c) extraction of the oxycarotenoids from specific tissues, such as serum (plasma), yolk, or skin (toe web, shank, breast feather tract), followed by quantitative colorimetry, and (d) use of special photoelectric instruments that measure reflectance.

It must be recognized that depth of color as perceived by the eye in both egg yolk and broiler shanks or skin tissues is not directly proportional to the pigment deposited as determined by colorimetry at all concentration ranges. The deposition of pigment in broiler tissues from the feeding of one or more oxycarotenoids increases linearly as the level of oxycarotenoid intake increases over a wide range of supplementation, whereas the visual observation of color increase is logarithmically related to the colorimetric extract reading, based on the findings of Heiman and Tighe (*384*), Heiman (*382*), Fritz *et al.* (*290*), and Kuzmicky *et al.* (*485*). Thus, at low levels of pigmentation, the eye readily detects color differences with small changes in concentration of oxycarotenoids, but, as the intensity of color increases,

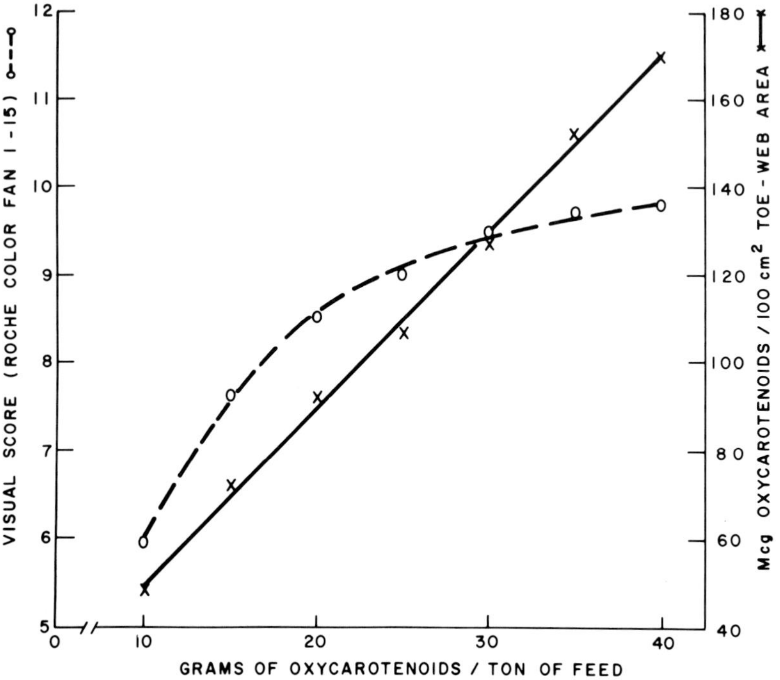

Fig. 2. Relation of visual score to pigment content in toe-web skin of broilers fed graded levels of dietary oxycarotenoids from feed ingredients (○, visual scores; x, colorimetry). [From Marusich and Wilgus (*558*).]

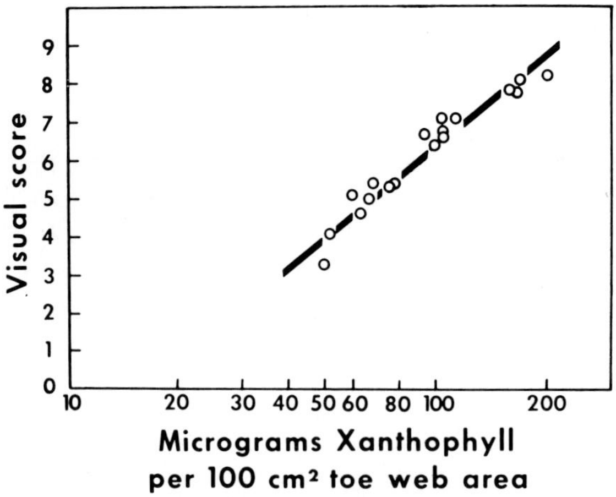

Fig. 3. Relation of visual score (Roche Yolk Color Fan, 1–15) to pigment content in toe-web skin ($y=8.76 + 7.57X$; $r=0.954$). [From Kuzmicky *et al.* (*485*).]

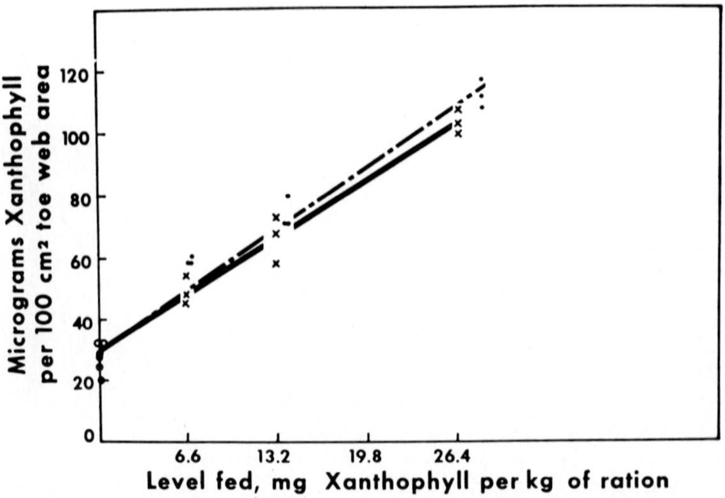

Fig. 4. Relation of dietary oxycarotenoid level to pigment content in toe-web skin. Key: ●, alfalfa meal, $y=30.15 + 3.00X$; x, corn gluten meal, $y=30.15 + 2.77X$. [From Kuzmicky *et al.* (*485*).]

progressively much larger amounts of pigment are required to produce a detectable difference to the eye (Figs. 2, 3, and 4).

The same phenomenon occurs with yolk pigmentation, as shown by numerous studies. Yolk pigment concentration (in terms of colorimetric measurements) is linearly dependent on the dietary levels of xanthophylls (*12,56,107,132,287,527, 529,548,783*). The color increase seen by the eye is also logarithmically related to the colorimetric reading. Thus, for both broiler shank and skin and egg yolk, each successive increase in pigment content results in a smaller increase in coloring effect visible to the eye. In practical terms, this means that, when there is a considerable degree of pigmentation already present, it will take a larger amount of added dietary oxycarotenoids to produce a small-increment color increase discernible to the eye.

The colorimeter does not interpret what the eye sees, but records only the concentration of pigment in the solution extracted from the particular tissue. Figure 5 illustrates this relationship. Based on colorimetric values, dehydrated alfalfa and β-apo-8′-carotenal feeding result in comparable pigment content in yolks. However, by visual scoring of yolks, the β-apo-8′-carotenal-pigmented yolks average about two scores higher at all levels of supplementation.

A. Visual Observation and Visual Aids

Ridgeway (*678*) in 1912 devised a scale for yolk pigmentation ranging from 27 for light cream to 13 for dark orange-red yolks. Parker *et al.* (*624*) in 1926 set forth other color standards intended specifically for yolk color measurement, but these

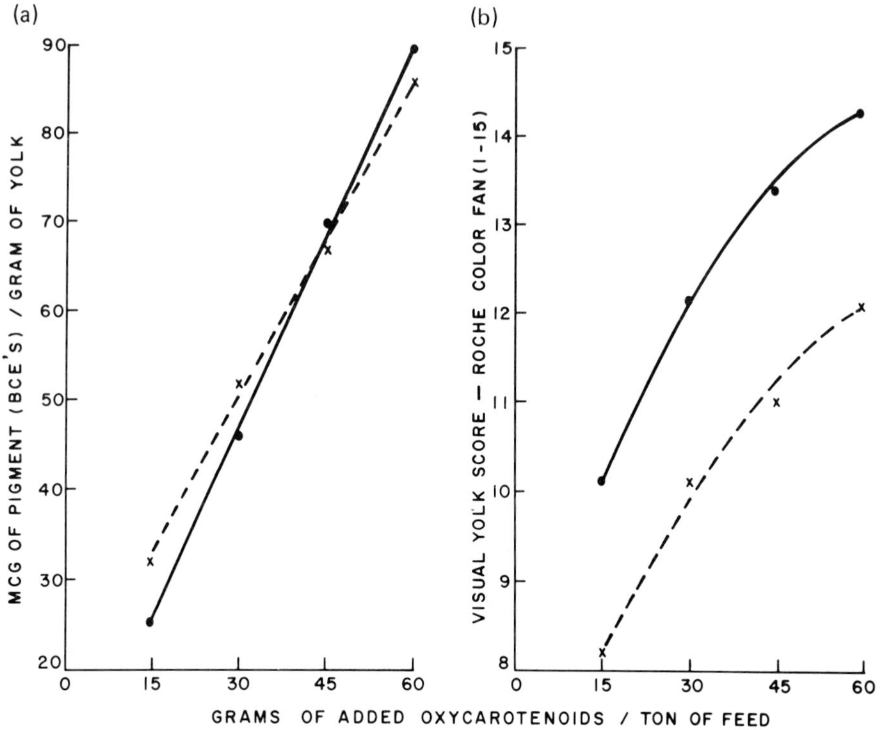

Fig. 5. Comparative yolk pigmentation [colorimetric assay (a) versus visual scores (b)] resulting from feeding β-apo-8′-carotenal (●—●) or dehydrated alfalfa meal (x---x). [From Marusich (*538*).]

never came into general use. Sharp and Powell (*715*) also prepared a set of yolk standards, and these were duplicated and used by several laboratories. Brown (*121*) used the Bradley Color Top to measure shank color.

Some years later (1935) Heiman and Carver (*383*) published a description of a yolk color rotor, which was manufactured in limited quantity and enjoyed moderate usage and acceptance among egg quality researchers. It consisted of 24 concave watch glasses painted on the undersurface and ranging from off-white to deep orange-red in color; these were mounted on a circular board, which could be rotated. Although it was designed for yolks, the rotor served for both egg yolks and broiler skin measurements. It represented a major step forward in the development of visual scoring aids. Today, it is difficult to locate rotors, and those still available have suffered color degradation. Titus *et al.* (*812*) published a chart for converting readings on the Sharp and Powell (*715*) color scale to readings on the Heiman–Carver Yolk Color Rotor. Other workers, such as Maw (*562*), determined the color of body fats by comparing them with the color of carotene solutions. Ringrose *et al.* (*679*) compared shank color with glass tubes colored with various shades of yellow.

Poultry investigators, namely, Culton and Bird (*185*) and Bird (*88*), used the Heiman–Carver Yolk Color Rotor to score shank pigmentation. Others, such as Hammond and Harshaw (*352*), Collins *et al*. (*154*), Elrod *et al*. (*252*), Fritz *et al*. (*290*), Waldroup *et al*. (*844*), Marusich *et al*. (*548*), and Bletner *et al*. (*96*), devised their own visual aids for broiler or egg yolk color evaluation. Elrod *et al*. (*252*) devised a series of 11 increasing colors ranging from creamy white to deep orange-yellow for broilers. Marusich *et al*. (*548*) filled 15 test tubes with increasing concentrations of sodium dichromate and activated alumina to simulate increasing yolk color. Bornstein and Bartov (*107*) prepared similar tubes with increasing concentrations of potassium dichromate to score yolk color; however, they omitted the alumina because it settles out. Waldroup *et al*. (*844*) arbitrarily scored shanks from 1 to 5 and yolks from 1 to 4 after selecting typical examples of each and scoring all others against these. Bletner *et al*. (*96*) made a color chart, ranging from almost white to deep orange with 13 gradations, to score shank color. In 1962, Ashton and Fletcher (*47*) published a yolk color standard consisting of 15 concave colored aluminum disks, each with a center hole for viewing the egg yolk under test.

In 1957, Steinegger and Zanetti (*748*) described the use of the first Roche Yolk Color Fan. It consisted of 12 shades of color ranging from light cream-yellow to deep orange in the form of a folding fan. The use of a color fan to measure yolk pigmentation was actually pioneered by Kupsch (*483*). In 1963, a revised and improved Roche Yolk Color Fan with 15 color gradations became available. It is currently in use and has been extensively employed by researchers throughout the world for yolk color scoring. It has also been used successfully for shank skin scoring by numerous investigators, including Lipstein *et al*. (*508*), Kuzmicky *et al*. (*485*), Bartov and Bornstein (*58*), Harms *et al*. (*356*), and Twining *et al*. (*825*). It has been described in detail by Vuilleumier (*836*) and Streiff (*769*). The latter workers also related the color readings of the newer fan (with color grades 1–15) directly to the older (1957) fan with color grades 1–12, to the Heiman–Carver Yolk Color Rotor, and to the Ashton–Fletcher Color Ring numerical values. Marusich and Bauernfeind (*546*) also presented the relationship between the old (1–12 grades) and new (1–15) Roche Yolk Color Fans in a conversion chart and related both to the Heiman–Carver Yolk Color Rotor and the Ashton–Fletcher Color Rings (Table 3) on the basis of scores assigned to egg yolks with the four visual aids. Variable results were obtained, depending on whether the scales were compared directly with each other or whether egg yolk colors were assessed independently against the four scales.

The Roche Yolk Color Fan evolved from a study of the basic properties of color. Judged subjectively, every color is distinguished by three characteristics: (a) hue (e.g., yellow or red); (b) saturation (the proportion of spectral colors in a mixture of spectral light and white, such as pale yellow or saturated yellow); and (c) brightness (the intensity of light reflection—hence, dull brownish-yellow or luminescent yellow). In order to define a color objectively, reflected light is measured at several wavelengths; the corresponding tristimulus values (X, Y, Z) of the CIE (Commis-

TABLE 3

Relationship between Several Visual Aids for Scoring Egg Yolk Color[a]

Roche Yolk Color Fan (1-15)	Roche Yolk Color Fan (1-12)		Heiman-Carver Color Rotor (1-24)		Ashton-Fletcher Color Rings (1-15)	
	a	b	a	b	a	b
1	1-2	2	4	6	1	1
2	2-3	2	6	7	1-2	2
3	3	3	7	8	2-3	3
4	4	3	8	10-12	4	4-5
5	4-5	4	8-9	12	4-5	6-7
6	5	5	9	13	5	8-9
7	5-6	6	9-10	14	5-6	9
8	6		10-11	15	6-7	10
9	6-7	7	10-12	16	7-8	11-12
10	7	7	12-14	16	8-9	13
11	7.5	7.5	15-16	18	10-11	
12	8	8	17	19-20	12	14
13	9	9	18-19	21-22	13	14
14	9-10	9	19-20	22	13-14	15
15	10	10	20-21	23	14	
	11		21		15	
	12		22			
			23			
			24			

[a] Data in columns a from Marusich and Bauernfeind (*546*); data in columns b from Vuilleumier (*836*).

sion Internationale de 'Eclairage) colorimetric system are then calculated. These values completely characterize a color. When the tristimulus values are converted to a two-dimensional system of chromatic coordinates (x, y), a color triangle results (Fig. 6), which contains all colors of the same brightness; every color is represented by a point inside this triangle. To evaluate the range of color in yolks, representative eggs were taken at random from a number of locations. The main sources of pigments in the feed, such as maize, dried lucerne, and grass meal, were present in amounts and grades that varied widely, so that the color of the eggs produced ranged from light green-yellow through golden yellow to reddish-orange. The objective measurement of the colors of these yolks and the determination of their position in the CIE color triangle produced relationships that were used as the basis for developing the color fan. The points of color of the yolks under test all fell within a sharply defined area of the color triangle termed the "yellow yolk strip" (Fig. 7). This area begins at pale yellow and passes into the deeper yellow region with an increasing degree of saturation before bending into the orange zone. For further discussion and

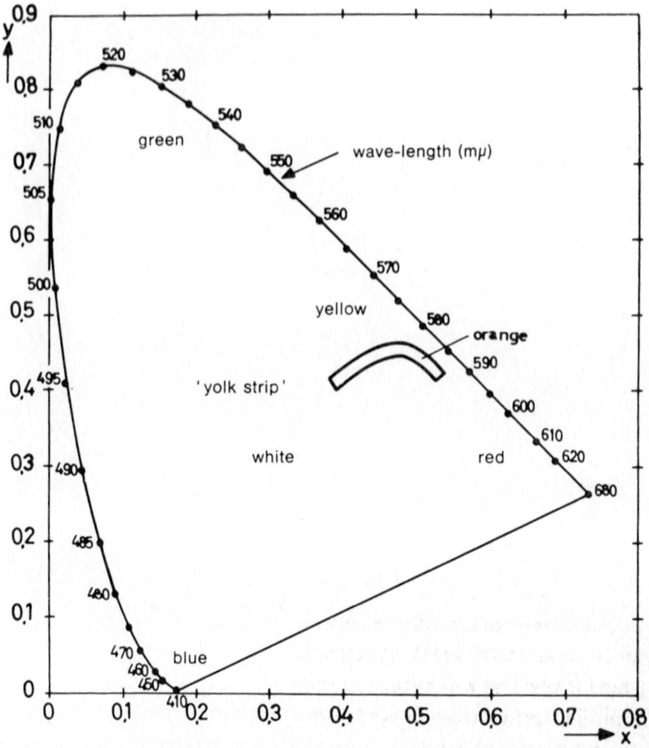

Fig. 6. Location of the "yolk strip" in the CIE chromaticity diagram. [From Vuilleumier (*836*).]

explanation of the CIE colorimetric system, the reader is referred to other papers (*163,248,283,439,520–522,892,893*).

Harms *et al.* (*356*) and Twining *et al.* (*825*) reported on the use of the Purina Skin Pigmentation Guide to measure breast feather tract color. This visual aid consists of a series of blades or paddles ranging from eggshell white to deep orange (0–13) in a fan design similar to that of the Roche Yolk Color Fan. However, it is designed specifically for breast feather tract scoring and also attempts to simulate the texture of the feather tract skin surface. Mainguy and Rouques (*530,531*) and Vuilleumier (*836*) have reviewed many of the devices that have evolved over the years for visual scoring of color.

B. Visual Observation and Ranking

Another approach to visual evaluation of broiler pigmentation is the use of the so-called rank test (*119,827*). With this technique, all the broilers in one experiment, regardless of the group to which they belong, are visually ranked according to

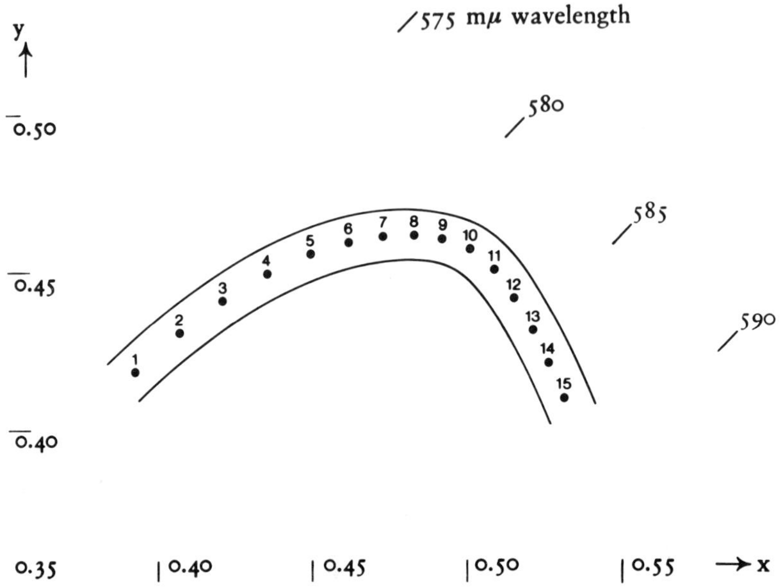

Fig. 7. Location of the colors of the Roche Yolk Color Fan (1–15) in the "yolk strip." [From Vuilleumier (*836*).]

the intensity of their pigmentation. Each bird is given a rank number according to its position in the sequence of increasing or decreasing intensity of color. Differences between average group numbers for particular treatments indicate a disparity of pigmentation. The significance of the differences can be evaluated by statistical methods. There is one disadvantage: The method does not permit the results of two individual experiments to be compared when run at different times.

C. Tissue Extraction of Oxycarotenoids and Colorimetry

1. Poultry Tissues

Heiman and Tighe (*384*) in 1943 were the first to report on a chemical extraction method for shank skin and subsequent colorimetric measurement of the amount of pigment in the shank relative to the visual evaluation of shank color. They peeled off the shank skin, removed all adhering tissue, spread the shank skin over a wooden block, and cut out circles of skin with a ⅝-in. punch. The disks were placed in acetone and allowed to stand for 48 hr in the dark with frequent shaking. The acetone solution was then read in a Klett–Summerson photoelectric colorimeter with a no. 42 filter. The scale readings on the colorimeter were directly proportional to the concentration of the pigment extracted from the skin. The relation of the concentration of pigment in the shank skin to the visual gradation of color seemed to be logarithmic. Thus, each successive quantity of pigment in the shank skin exerted a smaller coloring effect visible to the eye (Fig. 2).

This basic solvent extraction procedure for shank skin as originally devised by Heiman and Tighe (*384*) has been modified by selecting toe-web skin or breast feather tract skin to obtain disk samples, the tedious operation of obtaining shank skin thus being eliminated. The use of cork borers or paper-punch-like tools has been introduced, and the area of the disk sample obtained is calculated, care being taken that each disk is uniform in area. The concentration of pigment is then expressed in terms of the area of the disk surface (micrograms pigment per 100 square centimeters), the fact that each toe-web disk has two surfaces being taken into account. Techniques for handling samples along with extraction procedure modifications have been described by Wilgus (*867*), Potter *et al*. (*646*), Day and Williams (*211*), Mitchell *et al*. (*584*), Quackenbush *et al*. (*656*), and Kuzmicky *et al*. (*485*). Dua *et al*. (*242*) showed that the concentration of xanthophylls in the skin increased as the dietary level of oxycarotenoids increased (Fig. 8).

Serum and/or plasma levels of carotenoids have also been measured to evaluate pigmentation status. Wilson (*879*) in 1955 observed in nonlaying hens that the level of serum carotenoids increased rapidly, to a maximum in about 6 days. A high level of serum carotenoids in hens could be used to identify nonlaying hens. It seemed desirable to Davis and Kratzer (*208*) to investigate the changes in serum carotenoids resulting from changes in dietary oxycarotenoid intake to determine whether this measure could be used to predict the pigmentation changes that would occur later. They found that the level of serum carotenoids of chicks fed a low-pigment ration for 1 week and then a test ration with adequate oxycarotenoids for 1 week was highly correlated with the degree of pigmentation of the shank. Serum and plasma

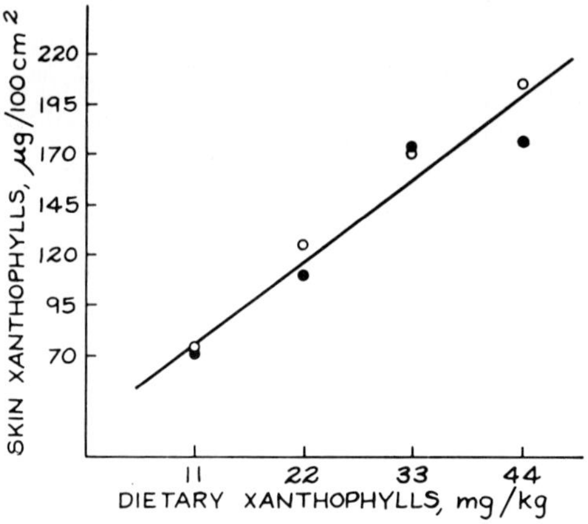

Fig. 8. Relationship between dietary level of xanthophylls and shank skin xanthophylls. Key: ●, alfalfa meal; ○, corn gluten meal; $\hat{Y} = 35.0 + 3.68X$; $r = 0.925$ ($p < 0.01$). [From Dua *et al*. (*242*).]

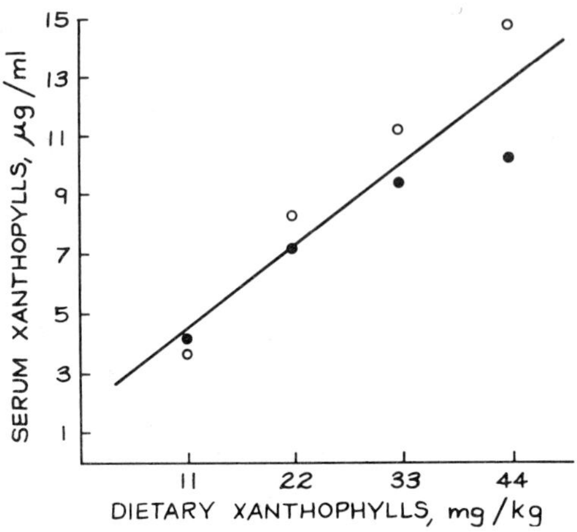

Fig. 9. Relationship between dietary level of xanthophylls and serum xanthophylls. Key: ●, alfalfa meal; ○, corn gluten meal; $\hat{Y}, = 1.39 + 0.259X$; $r = 0.841$ ($p < 0.01$). [From Dua *et al.* (*242*).]

carotenoid concentrations have been shown to be correlated with the level of oxycarotenoids in the ration and the carotenoid content of skin when extracted and measured chemically on the basis of reports by Combs and Nicholson (*160*) and Dua *et al.* (*242*). Figure 9 depicts the serum response to the same dietary levels of oxycarotenoids as those shown in Fig. 8 for skin. Similar data are shown in Fig. 10 from Twining *et al.* (*825*); good correlation of plasma carotenoid levels was shown with shank visual scores using the Roche Yolk Color Fan and with breast skin using the Purina Skin Pigmentation Guide (Table 4). Monitoring of plasma or serum carotenoid concentrations in broilers before marketing enables producers to increase

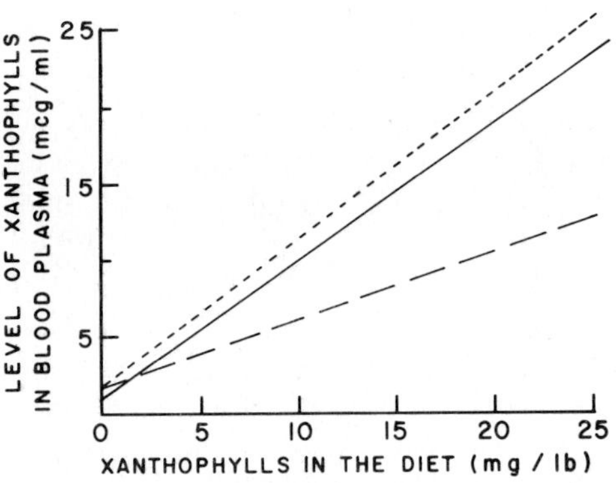

Fig. 10. Effect of dietary xanthophylls from corn gluten or marigold meal on plasma levels of xanthophylls. Curves based on 80% availability assumed for xanthophylls in corn gluten meal (---); 100% availability assumed for xanthophylls in corn gluten meal (———); marigold meal (----). [From Twining *et al.* (*825*).]

TABLE 4

Relative Visual and Colorimetric Values Obtained by Various Methods of Rating Broiler Pigmentation[a]

Treatment	Shank score[b]	Skin score[c]	Breast color[d]	Visual rating[e]	Plasma (μg/ml)
ANRC reference standard (10 mg/lb)[f]	5.14	4.86	16.32	3.17	11.55
Basal ration	<1.00	0.14	5.86	5.00	0.57
Corn meal (5 mg/lb)	2.00	3.88	13.39	4.00	5.17
Corn and corn gluten meal standard curve					
5 mg/lb	2.38	3.75	13.69	3.63	5.10
10 mg/lb	5.43	6.00	15.60	2.86	10.86
15 mg/lb	7.00	8.67	20.43	1.83	16.26
20 mg/lb	6.43	8.43	22.11	1.86	17.19
25 mg/lb	8.25	9.88	24.48	1.13	23.46
Marigold meal curve (sample F)					
5 mg/lb	1.50	2.50	10.83	4.00	3.96
10 mg/lb	3.57	3.71	15.21	3.57	7.88
15 mg/lb	3.50	3.88	15.66	3.13	8.35
20 mg/lb	4.71	5.86	17.39	2.57	10.80
25 mg/lb	4.38	5.38	16.40	2.88	11.57
Marigold meal samples at 15 mg/lb					
Sample A	3.88	4.75	16.91	2.88	8.89
Sample B	4.40	5.80	16.77	3.17	7.42
Sample C	4.86	5.00	18.24	2.57	10.81
Sample D	2.75	3.75	13.70	3.63	7.12
Sample E	4.43	5.86	18.71	2.57	10.96

[a] From Twining *et al.* (*825*).
[b] Roche Yolk Color Fan.
[c] Purina Skin Pigmentation Guide.
[d] Hunterlab Color Difference Meter.
[e] With this scale, 1 is highest and 5 is lowest degree of pigmentation.
[f] β-APO-8′-carotenoic acid ethyl ester in stabilized gelatin beadlet form.

or decrease total dietary oxycarotenoids to achieve a constant, uniform pigmentation level in their marketed birds. Plasma or serum carotenoid concentration is also an excellent indicator of relative flock health. This parameter is more sensitive to changes occurring than is visual observation of shank color in housed growing birds.

Thomas *et al.* (*791*) reported that chickens held on a pigment-free diet and then fed graded levels of xanthophylls for only 2 days gave the same relative differences in serum levels as those attained when birds were held on the respective two rations for a 4-week period. Middendorf *et al.* (*575b*) presented results obtained using an even more rapid bioassay procedure to compare oxycarotenoid availability from various pigmentation sources. Oxycarotenoid-depleted, fasted birds (24 hr) are intubated with equal amounts of oxycarotenoids from different sources. The level of serum xanthophylls rises rapidly between 6 and 14 hr following intubation. The increases in serum levels that occur between 14 and 24 hr following intubation are not statistically different. Thus, blood samples for serum analysis may be obtained at any time during the latter time span, but the time must be the same for all birds.

The serum oxycarotenoids are read at 474 nm and expressed as β-carotene equivalents (BCE's). Results are calculated as the increase in serum oxycarotenoids (micrograms per milliliter) over the intubated controls per milligram oxycarotenoid intubated per kilogram of body weight. This method offers a dual advantage of rapidity and requiring extremely small quantities of test material. In the case of pure, crystalline oxycarotenoids, which may be available from chemical synthesis in only milligram quantities, this procedure would be ideal.

2. *Egg Yolks*

a. NEPA Method. Methods for the extraction of yolk pigments, similar to the procedures for broiler tissues, have been developed and used to classify yolks into different categories of color intensity. A procedure widely used in the United States is the National Egg Products Association (NEPA) method, as described by Kahlenberg (*440*). An acetone extract of yolk pigments is compared with a series of standard aqueous potassium dichromate solutions of increasing concentrations, which are assigned arbitrary numbers of 1 throuth 10, indicating increasing color from light yellow to deep orange. Another method, which is used by the American Association of Cereal Chemists, is that of Dalby (*196*). Both methods use acetone for extraction and potassium dichromate standards, but the calculations are somewhat different. Both are adaptations of the methods originally developed by Turner and Conquest (*824*).

b. AOAC Method. Pohle and Mehlenbacker (*640*) revised the NEPA method by replacing the potassium dichromate standard with a β-carotene standard. This revised method is referred to as the Association of Official Analytical Chemists (AOAC) method (*296*). It was evaluated by Forsythe (*272–274*) and Campbell *et al.* (*139*) and found to be preferable since β-carotene has a light absorption curve similar to that of the yolk carotenoids, whereas those of the potassium dichromate solutions are different from the yolk carotenoid curves. The AOAC (*48*) method expresses the concentration of pigment in egg products in terms of BCE's since increasing concentrations of β-carotene provides the standard curve against which the egg product is compared. The method is further described by Vuillemier *et al.* (*837*) and Wildfeuer *et al.* (*865*). Table 5 relates the NEPA and AOAC values to several visual aids previously discussed (Table 3).

A modification of the AOAC (*48*) method was proposed by Marusich (*538*). It involved the use of a 1:1 mixture of acetone and chloroform for more efficient extraction of the pigments from the yolk than could be achieved by the use of acetone alone as in the AOAC procedure. The modified procedure consistently

TABLE 5

Relation of Several Visual Aids to NEPA Number and AOAC β-Carotene Equivalents for Scoring Egg Yolk Color[a]

	Colorimetric values		Visual scoring aids			
	Potassium dichromate solution	AOAC β-carotene equivalents	Roche Yolk Color Fans		Heiman–Carver Color Rotor	Ashton–Fletcher Color Rings
NEPA	(%)	(μg/gm yolk)	1–15	1–12	(1–24)	(1–15)
1	0.005	15–25	10	7	12	8–9
2	0.010	30–45	11	7–7.5	15–16	10–11
3	0.015	50–60	11–12	7.5–8	16–17	11–12
4	0.020	65–75	12	8	17	12
5	0.025	80–100	13	9	18	13
6	0.030	105–125	14	9–10	19–20	13–14

[a] From Marusich and Bauernfeind (*546*).

extracted more pigment from the yolk, as shown by the results of a collaborative trial with eight participants. Scott *et al.* (*713*) obtained improved results with this procedure compared to those obtained by the AOAC method (*48*) and a method previously used in their laboratory. Eggs for fresh market use (table eggs) approximate a NEPA number of 1 to 2, (BCE's of 20–40 μg per gram of yolk), whereas those for use in the processed food industry (breaker eggs) are desired at an NEPA number of 4–6 (BCE's of 80–120 μg per gram of yolk) or higher. The approximate relation of visual scores to colorimetric values are shown in Table 5.

D. Reflectance Instruments

Specialized reflectance instruments have been utilized in an attempt to quantitatively measure skin and yolk pigmentation. The instrumental approach has the advantage of objectivity, compared to the more subjective visual scoring, and of rapidity, compared to the colorimetric extraction procedures. Davies *et al.* (*205*) used a Hunterlab D-20 Color Difference Meter to study skin color in processed broilers. Foot pad and shank color were not correlated with breast or thigh skin color. The correlation was good between the instrumental readings and the visual measurements on breast skin, breast feather tract, and thigh skin color. The authors concluded that the instrumental technique was useful for the rapid, objective measurement of skin color. Hunter (*418*) described the principle and properties of the Color Difference Meter. Twining *et al.* (*825*) also reported on the use of the Hunterlab Color Difference Meter to measure the skin pigmentation of dressed broilers in comparison with visual scores of shanks and breast skin and plasma carotenoid concentrations. All parameters correlated well (Table 4 and Fig. 11).

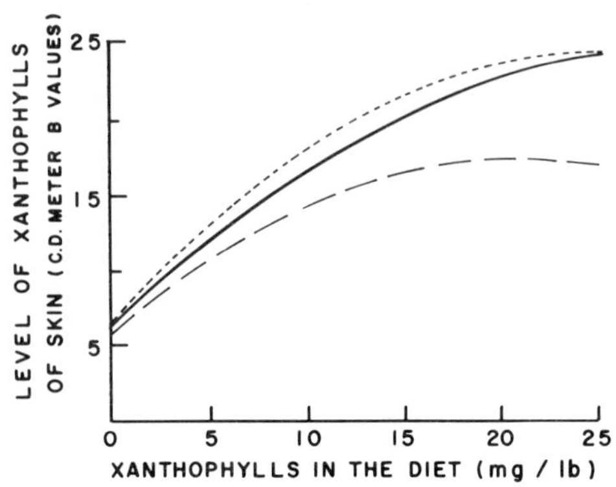

Fig. 11. Effect of dietary xanthophylls from corn gluten or marigold meal on deposition of pigment in the breast skin. Curves based on 80% availability assumed for xanthophylls in corn gluten meal (---); 100% availability assumed for xanthophylls in corn gluten meal (———); marigold meal (— —). [From Twining *et al.* (*825*).]

Dansky (*198,199*) compared visual dressing plant observations on broilers with representative samples evaluated by the Hunterlab D-20 and obtained good agreement by the two methods. Arnold (*46*), Livingston *et al.* (*519*), and Yacowitz *et al.* (*894a*) also reported favorably on its use.

Philips *et al.* (*636a*) found the Hunter D25 Color and Color Difference Meter to be well suited for the measurement of yolk color. Excellent correlation was obtained between the instrumental approach and visual scores based on the Roche Yolk Color Fan.

Fry *et al.* (*294,297–299*), Herrick *et al.* (*390*), Fry and Harms (*295*), Fry (*293*), and Harms *et al.* (*356a,b*) reported on the use of another instrument, the IDL (Instrument Development Laboratories) Color Eye to measure skin pigmentation. Harms *et al.* (*356*) described the properties of the IDL Color Eye and reported significant correlation coefficients for the pigmentation measurements when breast skin values were compared to shank skin values, and both were related to the chemically extracted pigment content in toe-web samples. They concluded that this instrument was satisfactory for the objective evaluation of pigmentation. It has also been used for the measurement of yolk pigmentation intensity by Hinton *et al.* (*400,402,403*), Fry *et al.* (*300*), Fry and Harms (*296*), and Fletcher *et al.* (*270a,b,c*). Rauch (*667,668,671*), Scholtyssek *et al.* (*707*), Livingston *et al.* (*519*), Boguth and Czernicki (*100*), and Hartfiel and Smitten (*362*) have also published papers on a photoelectric shank color evaluation method employing color reflectance.

The instrumental approach offers a solution to some of the problems encountered in pigmentation evaluation. Those who have employed these instruments concede that further work on either instrument improvement, sample preparation, or conditions for reading samples or all three is necessary before the instruments can be considered totally effective tools for the quantitative comparative evaluation of pigmentation. With either the instrumental approach or the solvent extraction of tissues followed by colorimetry, visual observations are still necessary to ensure that the color being produced is an acceptable color.

E. Oxycarotenoid Assay of Feed Ingredients

The separation and quantitative determination of the xanthophylls and carotenes evolved over the years after the initial observation by Borodin (*111*) in 1883 that the carotenoids could be separated into alcohol-soluble and petroleum ether-soluble fractions. The progression of improved analytical techniques involving extraction, characterization, and measurement of the carotenoids encompassed the efforts of numerous investigators, including Arnaud (*45*), Euler *et al.* (*255*), Coward (*183*), Schertz (*701*), Kuhn and Brockmann (*478*), Russell *et al.* (*694*), Guilbert (*342*), Peterson *et al.* (*631*), and Fraps (*284,285*). The development of and improvements in column chromatography for the separation of pigments as originally reported by Tswett (*817*) involved the contributions of Palmer and Eckles (*622*), Kuhn and Lederer (*480*), Karrer and Walker (*446*), Kuhn *et al.* (*482*), Kuhn and Brockmann (*478*), Strain (*757*), Mackinney (*525*), Winterstein and Stein (*884*), Winterstein (*880*), and Peterson *et al.* (*632*).

Methodology for the separation, identification, and quantitative measurement of oxycarotenoids and carotenes gradually improved, as highlighted by the procedures of Strain (*761*), Karrer *et al.* (*445*), Thompson and Bickoff (*803*), Quackenbush (*650*), Moster *et al.* (*595*), Gugliemelli (*341*), and Bickoff *et al.* (*80*). The Bickoff *et al.* (*80*) procedure was developed in 1954, in which two alternate assay methods for alfalfa meal were applied. These investigators noted changes in the oxycarotenoid profile of fresh meal compared to that of the same meal after dehydration. Fresh alfalfa meal contained five xanthophylls (lutein, violaxanthin, cryptoxanthin, zeaxanthin, and neoxanthin), comprising 99% of the total, plus seven minor bands. The same five oxycarotenoids still comprised 87% of the total in dehydrated alfalfa meal; however, an additional 35 minor bands were present, comprising 13% of the total. The Bickoff *et al.* (*80*) method involved basically extraction, chromatographic separation, and spectrophotometric reading of the components. All subsequent improvements have involved alterations in these basic steps.

1. CIRF Method

In 1961, Tennyson *et al.*, as detailed in the review by Rathmann and Daggy (*666*), developed a method for determining the oxycarotenoid and carotene content of corn products, including corn, corn gluten, corn gluten meal, and corn gluten oil. This procedure was accepted by the member companies of the Corn Industries Research Foundation (CIRF) and is generally identified as the CIRF method. It is still being used in many laboratories to determine the oxycarotenoids and carotenes in specific corn products used in prepared poultry feeds. Data from a limited collaborative trial comparing the CIRF procedure for corn gluten meal with biological responses as reported by Marusich (*541*) indicated that the CIRF method may overestimate the effective xanthophylls. Livingston *et al.* (*519*) have treated this aspect in detail since their laboratory was one of the collaborators. They and Knowles *et al.* (*457*) have also stressed that further research in this area should be conducted to assess the validity of these chemical procedures in relation to biological effectiveness.

2. *WRRL Method*

In 1967, Kohler *et al.* (*466*) published a procedure for the determination of both carotenes and oxycarotenoids in fresh and dehydrated alfalfa meal and Bermuda grass. The procedure, termed the WRRL (Western Regional Research Laboratory of the U.S. Department of Agriculture) method, was a relatively simple chromatographic method that could be used in control laboratories. Quantitative yields were possible, the chlorophyll error was minimized, and the solvent system eluted almost all the oxycarotenoids including lutein, zeaxanthin, and violaxanthin and their cis isomers after the carotenes had been removed. The method underestimated neoxanthin and carotenoid oxidation products; however, these have little or no biological activity. Kohler *et al.* (*466*) stated that the xanthophylls in alfalfa meal constitute a very complex mixture of chemical entities, and most occur as the all-trans isomers in fresh meal but are partially converted to cis isomers during dehydration. Because of the multiplicity of components, any procedure for analysis of the total xanthophylls represents a compromise and must be followed exactly in order to yield reproducible results. This procedure is an improvement over a previous method, which had been accepted as the official method of analysis of the AOAC (*612*).

3. *AOAC Method*

Most recently, Quackenbush (*652*) and Quackenbush *et al.* (*653*) have developed a chemical assay method for the quantitative measurement of the biologically effective oxycarotenoids in dehydrated alfalfa meal, corn gluten meal, and finished feeds. The method provides for the separation and determination of the monohydroxy pigments (zeinoxanthin and cryptoxanthin), the dihydroxy pigments (lutein, zeaxanthin, and their isomers), and the polyoxy pigments (violaxanthin and neoxanthin). The dihydroxy pigments predominate in dehydrated alfalfa meal and corn gluten meal and are the active components responsible for pigmentation. By taking the total amount of dihydroxy pigments plus half the amount of monohydroxy pigments, a good chemical estimate of the biologically effective oxycarotenoids is obtained; this value has been termed the dihydroxy pigment equivalent by Quackenbush *et al.* (*653*). The method was adopted by the AOAC as "official-first action" in 1970, as reported by Deutsch (*233*). Table 6 summarizes the results obtained for alfalfa and/or corn gluten meal by the CIRF, WRRL, and AOAC methods.

TABLE 6

Relative Oxycarotenoid Potency of Dehydrated Alfalfa and Corn Gluten Meal Determined by Several Colorimetric Assays[a]

		Chemical analysis of dehydrated alfalfa meal		Chemical analysis of corn gluten meal	
Collaborator	Time in refrigerated storage	AOAC[b] (DHPE)[c]	WRRL[d] (total xanthophylls)[e]	AOAC[b] (DHPE)[c]	CIRF[f] (total xanthophylls)
1	Initial	138.4	148.9	146.9	183.3
	2 Weeks	133.1	—	136.4	—
	4 Weeks	141.9	—	137.1	—
2	Initial	136.5	150.1	134.4	199.0
	4 Weeks	138.2	142.8	134.9	196.0
3	Initial	144.3	142.5	142.0	213.0
	2 Weeks	147.0	140.0	152.0	209.4
	4 Weeks	150.0	144.5	153.0	200.5
Average		141.2	144.8	142.1	200.2

[a] From Marusich (*541*).

[b] Quackenbush *et al.* (*653*).

[c] Dihydroxy pigment equivalents (DHPE) include all dihydroxy pigments (lutein, zeaxanthin, and their isomers) plus one-half the monohydroxy pigments (zeinoxanthin and cryptoxanthin).

[d] Western Regional Research Lab (*466*).

[e] Includes the monohydroxy and dihydroxy pigments plus violaxanthin and part of neoxanthin content.

[f] Corn Industries Research Foundation (Rathman and Daggy, Corn Products Co. Tech. Brochure, 1961).

4. Chemical Method versus Biological Effectiveness

As cited previously, Marusich (*541*) reported on the results of a limited collaborative study with six participating laboratories to compare the AOAC chemical method with biological response. It was concluded that the procedure measures the biologically effective xanthophylls in dehydrated alfalfa meal and corn gluten meal. This method and the WRRL procedure yielded essentially equivalent values for the oxycarotenoids in dehydrated alfalfa meal. However, when the values obtained for corn gluten meal were compared with the values obtained by the CIRF procedure, the CIRF method gave higher values. These higher values were not substantiated by the biological responses. Knowles *et al.* (*457*) have reviewed these chemical procedures in great detail and also have presented a new method for measuring oxycarotenoids in alfalfa products (WRRL - 1971). Basically it encompasses the WRRL method (*466*) modified by the addition of 0.5 ml of water to the WRRL extracting solvent before overnight soaking (*515*). To reiterate, further concerted research efforts should be made in this area to perfect a better chemical method and to correlate chemical methodology with biological response for the numerous feed ingredient oxycarotenoid sources.

Ideally, the chemical method should be relatively simple. The equipment should be readily available, the time commitment necessary to perform the analysis reasonable, and the techniques involved performable by technicians with relatively brief training. It should be accurate. The procedure must measure the true content of the oxycarotenoids present. Chlorophyll contamination of eluates, improper analytical wavelength or absorptivity, and incomplete band separation on the absorption column must be avoided. Test results must be reproducible with adequate precision and be obtainable routinely so that a true variation in the samples tested can be noted with confidence. For effective pigmenting oxycarotenoids, the chemical analyses must give a measure of the relative pigmenting value of the test material. This chemical value may be an empirical value that is consistently related to the biological pigmenting potency of the sample, or it may be an absolute value for the content of effective pigmenting oxycarotenoids as determined by extensive collaborative testing.

VI. FACTORS INFLUENCING PIGMENTATION

Pigmentation involves more than having the right oxycarotenoids in the feed or ration, since numerous other factors are involved. Thus, any level of oxycarotenoids supplied by the feed ingredients in the ration does not always produce the same pigmentation results.

A. Breed, Strain, and Sex

Breed, strain, and sex differences influence pigmentation. Bateson (*59*), Hurst (*419*), and Dunn (*245*) investigated the genetics of skin color and established that white skin (*W*) is autosomal and dominant to yellow (*w*) skin. Hutt (*420*) concluded that in chickens the dominant gene for white skin restricted entry of oxycarotenoid pigments into the skin but not into other tissues. Bohren *et al*. (*102*) and Kaufman and Baczkowska (*451*) also showed that different breeds vary in their ability to deposit pigment in the yolk. White Rocks were reported by Bunnell and Bauernfeind (*130*) to be less efficient than sex-linked crosses in the utilization of dietary oxycarotenoids on the basis of relative pigment deposition in the yolk measured quantitatively. Using reflectance colorimetry (IDL Color Eye), Fletcher *et al*. (*270b*) compared the color of yolks from 12 layer strains. They found that yolk pigmentation varies among strains and housing conditions. Several strains produced more highly pigmented yolks regardless of housing conditions; however, the pigmentation of some strains appeared to vary with housing conditions (caged versus floor rearing).

Collins *et al*. (*154*) reported on breed differences in shank pigmentation. New Hampshire broilers were significantly darker than White Plymouth Rocks, with Barred Plymouth Rocks intermediate. Hammond and Harshaw (*352*) also studied the effect of breed and diet on skin and shank color. Light Sussex chicks were found to have almost no yellow pigment in either skin or shanks regardless of the diet fed. White Wyandottes, White Leghorns, and Rhode Island Reds were reported to have approximately the same shank and skin color on any given dietary level of oxycarotenoids. Moyer and Collins (*596*), Jaap (*433*), Zervas *et al.* (*900*), Henderson (*387*), and Stone *et al.* (*755*) all showed that shank pigmentation intensity was an inherited characteristic. Two experiments were reported by Harms *et al.* (*356b*) evaluating shank pigmentation from different strains and crosses of broilers. In the first study, shanks from 11 strains and crosses from a commercial breeder and in the second study shanks from 8 different crosses resulting from two male lines and four females lines were compared. Pigmentation was measured by the IDL Color Eye (reflectance colorimetry). Significant differences were seen among different strains and crosses. The authors estimated that the strains with the least amount of shank color would require about 4 gm more xanthophylls per ton to have pigmentation equal to that of the strain with the highest color intensity. They further observed that broilers from one male line had significantly better pigmentation than broilers from a second male line, indicating that degree of pigmentation is a heritable characteristic.

Genetic factors also influence the ability of hens to absorb and deposit xanthophylls in the egg yolk, and there are differences even among individual hens within a single strain, according to Scott *et al.* (*713*). Hens vary in their ability to assimilate oxycarotenoids from the ration even when they are all of the same breed and strain since they differ in their physiological capacity to transfer the oxycarotenoids from the feed to the egg yolk (*416,632,680,681*). Genetic studies by Bird (*86*) involving only yellow-skinned (*ww*) chickens indicated that genotype and sex influence the subjective score of pigmentation intensity of the skin.

Stone *et al.* (*756*) determined the heritability of blood carotenoid concentrations for each sex and concluded that sex-linked genes appeared to influence the variation in blood carotenoid concentration. Bird (*86*) also noted that pigmentation intensity varied from one broiler strain to another, with males being more intensely pigmented than females. Our own observations show that males are more strongly pigmented than females when both are maintained on the same ration for the same period of time.

Torges (*814*) selected laying hens over five generations for several egg quality characteristics. Positive results were obtained in selection for albumen height and shell breaking strength; however, no success could be demonstrated in selection for egg yolk color.

B. Diseases

Poultry diseases, such as coccidiosis, chronic respiratory disease (CRD), and Newcastle disease, depress shank and skin pigmentation. Mitchell *et al.* (*584*) and Bletner *et al.* (*96*) found that broilers inoculated with sporulated *Eimeria maxima* oocysts exhibited reduced shank pigmentation and lower concentration of carotenoids in toe-web tissue and serum when either natural ingredients or an oxycarotenoid concentrate served as the dietary pigmentation source. Douglas (*238*) observed that infection in broilers with *E. maxima* and *E. necatrix* reduced pigmentation, based on serum carotenoid levels and visual shank color. The inferior pigmentation was due not to a reduction in feed intake and thus a lowered oxycarotenoid intake, but to impaired intestinal absorption since the adverse effect was not counteracted by providing extra oxycarotenoids in the feed, but was negated when xanthophylls were injected into the breast muscle. Kowalski and Reid (*471*) also reported lower plasma oxycarotenoid levels after artificial infection with *E. brunetti*.

Marusich *et al.* (*556,557*) established subclinical *E. maxima* and *E. acervulina* infections in broilers and noted slightly reduced growth and impaired feed conversions. However, feed intake and thus oxycarotenoid intake were not affected by either infection. Shank visual scores, plasma and toe-web tissue carotenoid concentrations, and liver vitamin A levels were depressed by both infections, again suggesting an impairment in the efficiency of absorption as a result of the two mild coccidial infections. These results (Table 7) agree with the earlier findings of Douglas (*238*). Marusich *et al.* (*556*) extended these studies to include mild infections with a nonpathogenic species (*E. praecox*) and a pathogenic species that inhabits the cecum (*E. tenella*), from which site nutrient absorption is nil. It was found that neither of these mild infections affected oxycarotenoid absorption on the basis of visual scoring of the shanks or breast feather tract and colorimetric determination of plasma and toe-web skin concentrations of pigment. Nor was vitamin A absorption, as measured by liver storage of the vitamin, affected by either infection (Table 8). Thus, neither a nonpathogenic coccial species located in the intestinal lumen, where active absorption takes place, nor a mild (subclinical) infection with a pathogenic species located in a nonabsorptive section of the gut adversely influenced the absorption of oxycarotenoids and vitamin A.

Yvore and Mainguy (*898*) reported a marked reduction in plasma carotenoids when chickens were exposed to a very mild *E. acervulina* infection. Yvore *et al.* (*897*) also noted a depigmentation effect on the broiler carcass with a mixed low-grade infection of *E. tenella, E. maxima,* and *E. acervulina*. In a follow-up study, Yvore (*896*) found only a transient and minor effect on pigmentation with a mild *E. tenella* infection, as contrasted to a severe and persistent effect on the absorption capabilities of the chicken in the presence of a mild *E. acervulina* infection. *Eimeria acervulina, E. brunetti,* and *E. mivati* all caused a decrease in plasma carotenoids (with *E. acervulina* causing the most severe drop), according to the studies of Stephen *et al.* (*750*). Ruff *et al.* (*689*) reported that any one of the six major species of chicken coccidia induced lower blood carotenoid levels. The maximal decrease in plasma levels occurred 7 or 8 days postinfection in all species; it was least with *E. tenella* (49.8%), and 62.4–74.0% with the other five intestinal species (*E. acervulina, E. mivati, E. maxima, E. necatrix,* and *E. brunetti*).

TABLE 7

Effect of *Eimeria maxima* and *Eimeria acervulina* Infections on Broiler Pigmentation Parameters and Liver Vitamin A Deposition[a]

Ration	Infection[b]	Oxycarotenoid intake (mg/bird)[c]			Pigmentation parameters			Total Vitamin A intake (IU/bird)			Liver Vitamin A[e]	
		Feed ingredient	Canthaxanthin	Total	Shank visual score[d]	Plasma ($\mu g/ml$)	Toe-web ($\mu g/100\ cm^2$)	As vitamin A	From β-carotene	Total	IU	Deposition (%)
Finisher 1												
(20 gm oxycarotenoids from feed ingredients + 3 gm canthaxanthin per ton)	None	45.8	6.9	52.7	8.6	12.6	160.0	20,160	18,730	38,890	13,580	34.9
	E. maxima	45.1	6.8	51.9	7.9[g]	11.3	136.9[g]	19,870	18,450	38,320	12,330	32.2[f]
	E. acervulina	45.4	6.8	52.2	8.0[g]	11.9	122.8[g]	19,990	18,570	38,560	11,400	29.6[g]
Finisher 2												
(26 gm oxycarotenoids from feed ingredients per ton)	None	58.3	—	58.3	7.8	12.8	163.8	19,760	26,980	46,740	15,330	32.7
	E. maxima	60.7	—	60.7	6.7[g]	12.7	153.5	20,560	27,420	47,980	12,250	25.5[g]
	E. acervulina	58.6	—	58.6	6.9[g]	12.1	134.0[g]	19,840	27,090	46,930	12,290	26.2[g]

[a] From Marusich *et al.* (*557*).

[b] Birds infected at 5 weeks by daily oral dosing for 5 days with 6000 sporulated oocysts of *E. maxima* or 30,000 sporulated oocysts of *E. acervulina*; birds were observed 16 days after the last dose (total of 21 days from initiation of dosage).

[c] Average data for 30 birds (15 females and 15 males) per treatment group.

[d] Roche Yolk Color Fan (1–15).

[e] Significance of difference from respective control (one-sided test).

[f] $p < 0.05$.

[g] $p < 0.005$.

TABLE 8

Lack of Effect of *Eimeria praecox* and *Eimeria tenella* Infections on Broiler Pigmentation Parameters and Liver Vitamin A Deposition[a]

		Oxycarotenoid intake (mg/bird)[c]			Pigmentation parameters				Total vitamin A intake (IU/bird)			Liver vitamin A		
Ration	Infection	Feed ingredients	Cantha-xanthin	Total	Shank visual score[d]	Breast visual score[e]	Plasma μg/ml[f]	Toe web μg/100 cm²	From vit. A	From β-carotene	Total	IU per liver	IU per liver (corrected)[g]	Deposition (%)
Finisher 1	None	42.8	5.4	48.2	10.4	10.9	10.6*	200.7	25,750	17,160	42,910	15,200	9,400	21.9
(20 gm oxycarotenoids	*E. praecox*	43.1	5.4	48.5	9.7	10.4	11.2*	188.8	25,890	17,260	43,150	15,500	9,700	22.5
from feed ingredients + 2.5 gm canthaxanthin per ton)	*E. tenella*	40.9	5.1	46.0	10.1	10.6	11.3*	214.6	24,600	16,400	41,000	16,400	10,000	25.8
Finisher 2	None	62.9	—	62.9	9.6	10.6	14.2†	207.0	25,190	16,790	41,890	14,200	8,400	20.0
(30 gm oxycarotenoids	*E. praecox*	63.5	—	63.5	9.6	10.6	14.4†	194.2	25,440	16,960	42,400	13,500	7,700	18.2
from feed ingredients per ton)	*E. tenella*	63.8	—	63.8	9.6	10.6	13.7†	229.4	25,530	17,020	42,550	15,600	9,800	23.0

[a] From Marusich *et al.* (556).
[b] Birds infected at 5 weeks by daily oral dosing for 5 days with 5000 sporulated oocysts of *E. praecox* or 10,000 sporulated oocysts of *E. tenella*, birds were observed 16 days after the last dose (total of 21 days from initiation of dosage).
[c] Average data for 30 birds (15 females and 15 males) per treatment group.
[d] Roche Yolk Color Fan (1–15).
[e] Purina Skin Pigmentation Guide (fan).
[f] Values with different superscripts are statistically different ($p < 0.05$).
[g] Corrected for liver vitamin A content of birds when 3-week study was initiated.

Ruff and Fuller (*688*) observed that plasma carotenoids were markedly reduced in broilers infected with *E. acervulina* or *E. tenella*. They found that the mechanism for this depigmentation differs between the two species. With *E. acervulina* it is an interference with absorption of xanthophylls from the intestinal lumen. With *E. tenella* it is leakage of the xanthophylls through the damaged wall of the cecum. It was also observed that determining plasma carotenoid concentrations is a more rapid and sensitive means of measuring changes in pigmentation levels than is determining visual skin scores or skin carotenoid concentrations.

Ruff *et al*. (*690*) showed good correlation between plasma carotenoid concentrations and visual observations of breast skin color after both severe and mild exposuure to mixed *Eimeria* species. As the degree of protection improved by increasing the anticoccidial level in the ration, the magnitude of pigmentation improved similarly.

Impairment of intestinal absorption of vitamin D due to coccidiosis was reported by Stafseth (*744*), who noted that rickets appeared earlier or was more severe in chickens suffering from coccidiosis than in uninfected chickens. Davies (*201*), Erasmus *et al*. (*254*), and Singh and Donovan (*723*) had also shown reduced absorption of vitamin A on the basis of lowered liver vitamin A content in the presence of coccidiosis. Turk and Stephens (*819–823*) concluded from a series of experimental infections that the absorption of zinc and oleic acid could be impaired in broilers, depending on the species of *Eimeria* used. Chickens infected with *E. acervulina* exhibit reduced plasma levels of protein, carotene, and vitamin A, as shown by Kouwenhoven and van der Horst (*470*). They observed a strongly acidic pH in the intestinal tract in the presence of infection, which apparently influences protein uptake in the intestinal villi, thereby bringing about less efficient absorption of xanthophylls. Littlefield *et al*. (*510*) surgically removed the portion of the jejunum–ileum normally inhabited by *Eimeria maxima* and, on the basis of blood carotenoid concentration and visual observation of pigmentation after a 2-week feeding period of a ration high in oxycarotenoids, found no significant absorption of xanthophylls. This finding, coupled with the impairment of pigmentation by coccidiosis, strongly suggests that the capabilities of the active absorption sites in the intestinal tract are adversely affected by the presence of pathogenic coccidial oocysts.

Bird (*87*) stated that the earliest sign of CRD or ''air sac'' is loss of pigmentation. This loss of color occurs before the producer is aware that his flock is sick. Couch *et al*. (*179*) found that pigmentation of broilers is directly related to the health of the birds and noted that this is quite apparent from the depigmentation of the skin and shanks seen in flocks experiencing outbreaks of coccidiosis and CRD. Bletner (*95*) reported that chickens infected with either fowl cholera, respiratory diseases, large roundworms, or capillaria worms will depigment rather rapidly. Squibb *et al*. (*738*) also noted that blood serum vitamin A and carotenoid levels were depressed in New Hampshire and male White Leghorn (*737*) chickens infected with Newcastle disease. They (*738*) also observed a similar depression with coryza and fowl cholera infections (Table 9). Treat (*816*) concluded that any disease condition, such as coccidiosis, enteritis, or worm infestation, would have a detrimental effect on yolk color. Berg (*74*) and Berg *et al.* (*75*) postulated that diseased birds in general may not absorb the pigmenting components from the intestine or deposit those absorbed as efficiently into the yolk as noninfected hens.

There is a paucity of published data on the effect of mycotic infections on pigmentation; nevertheless, a few papers have appeared recently. Tung and Hamilton (*818*) showed that a dietary level of 1.25 μg of aflatoxin per gram (1.25 ppm) or above lowers significantly ($p < 0.05$) the concentration of plasma carotenoids in broiler chickens. Similarly, Huff and Hamilton (*413*) reported that dietary ochrato-

TABLE 9

Effect of Disease on Vitamin A and Carotenoid Blood Levels in Chickens[a]

Disease	Flock no.	No. of birds	Sex[b]	Age (months)	Stage of disease	Vitamin A (μg %)	Total carotenoids (μg %)
Coryza	1	15	F	3	Control	42	265
		15	F	3	Moderate	24[c]	158[d]
	2	15	M	3	Control	37	275
		15	M	3	Moderate	21[c]	209
	3	12	F	5	Control	50	425
		12	F	5	Moderate	31[c]	364
	4	10	M	10	Control	25	277
		10	M	10	Moderate	17	240
	5	10	F	10	Control	35	279
		10	F	10	Moderate	8[c]	241
Average					Control	38	303
					Infected	21[c]	237[d]
Fowl cholera	6	9	ND	0.75[e]	Moderate	42	366
		9	ND	0.75	Advanced	26[c]	157[d]
	7	8	M	6	Moderate	35	171
		7	M	6	Advanced	26	169
	8	8	F	6	Moderate	26	—
		11	F	6	Advanced	24	251
Average					Controls[f]	38	270
					Infected	28[c]	177[d]
Newcastle disease	11	15	ND	6[g]	Controls	28.8	361
	12	15	ND	6[g]	Highly infected	18.1[d]	148[c]

[a] From Squibb *et al.* (*738*).
[b] ND, not determined.
[c] Statistically significant from respective controls at $p < 0.01$.
[d] Statistically significant from respective controls at $P < 0.05$.
[e] 3 weeks.
[f] Average data from controls in flock 5.
[g] Weeks.

xin A at levels of 4.0 and 8.0 μg/gm also lowered significantly ($p < 0.05$) the plasma carotenoid concentrations in broiler chicks. Huff *et al.* (*414*) induced aflatoxicosis in laying hens by adding from 0 to 10.0 μg aflatoxin per gram of layer ration. Egg production and egg size were decreased; yolk and plasma carotenoid concentrations were elevated. The authors concluded that plasma and yolk lipids reflect the inhibition of lipid synthesis and transport from the liver by aflatoxin, but plasma and yolk carotenoids, which are dietary in origin, necessarily increase when egg production decreases during aflatoxicosis. It has been noted by other workers (*26*) that copper sulfate in the ration occasionally improves broiler pigmentation, possibly because of its ability to partially control mycotoxins.

C. Market Preparation Practices

Regardless of the level of pigment deposition desired, everyone likes a uniformly pigmented chicken. Each chicken should have a uniform color over its skin area, and there should be uniformity among chickens. Difficulties in feather removal have caused many processors to increase scald temperature and/or time to remove all feathers. The pick time has also been altered in some cases. This has resulted in "barked" chickens and cases in which skin color uniformity is affected. This lack of uniformity is caused by the fact that some skin areas, such as feather tracts, differ in color from other areas. Also, some chickens have pigment loss whereas others do not. To gain more information about the effect of scald conditions and pick time on the color of broiler skin, a series of experiments was conducted by Heath (*375*), Heath and Thomas (*377–379*), and Heath and Wabeck (*380*). Their studies showed that the pigmentation of broilers can be drastically affected during processing, the main factors being (a) the temperature of the scald water, (b) the scald time, (c) the kind of plucker, and (d) the plucking time. A reduction in the loss of oxycarotenoids was achieved by reducing the scald temperature to below 128°F (53°C) and by reducing the scalding time interval to 60 sec or less. Scald-water temperature is probably most important since these workers reported no effect of exposure time when chickens were scalded for 60, 120, or 240 sec at 52°C (126°F). A decrease in xanthophylls of breast skin and breast feather tract area and yellow color of the skin was attributed to an increase in temperature from 52°C (126°F) to 55°C (131°F). Another factor in processing involves the feather pickers. If these are set too close, the outer tender skin (cuticle) may be abraded, resulting in poorly pigmented chickens.

D. Ration Ingredients

Components or ingredients of the feed may have an effect on pigmentation intensity. Culton and Bird (*185*) reported that soybean meal, meat scraps, and fish meal in the ration had a suppressing effect on broiler pigmentation. Bird (*88*) found one sample of meat and bone meal to depress pigmentation and another sample to have no adverse effect. Mann (*532*) noted the same effect with fish meal. Hammond and Harshaw (*352*) reported fortified cod liver oil to be a potent source of a pigment-suppressing factor. Ordinary cod liver oil was about one-half as potent and ordinary sardine oil about one-third as potent as the fortified cod liver oil in depressing shank and skin pigmentation. Mattson and Deuel (*561*) and Goldhaber *et al.* (*323*) obtained similar effects with cod liver oil on shank pigmentation. Hammond *et al.* (*353*) observed that large quantities of cod liver oil or small quantities of sulfur interfered with the transfer of pigment from the ration through the egg to shank of the 1-day-old chick. The vitamin A content of fortified cod liver oils probably is primarily responsible for the depigmentation effect. Whereas cod liver oil suppresses pigmentation in the chick, vegetable oils do not, according to Carver (*147*). Heath and Shaffner (*376*) reported improved breast skin pigmentation in broilers as the soybean oil level was increased from 4 to 7–10%. Tissue lipids also rose. Thus, more oxycarotenoids were deposited as the tissue lipid content increased as a result of higher dietary soybean oil levels.

1. Vitamin A

When fed at high levels in the feed (generally in excess of 10,000 IU/lb) vitamin A tends to depress yolk and skin pigmentation, according to the studies of Deuel *et al.* (*230*), Mattson and Deuel (*561*), Rubin and Bird (*685*), Vavich and Kemmerer (*830*), Sunde (*778*), Berg (*74*), Donovan *et al.* (*236*), Donovan and Luneau (*237*), Farr *et al.* (*261*), Madiedo and Sunde (*529*), and Dua *et al.* (*243*). The normal levels of vitamin A (2000–6000 IU/lb) routinely added to layer feeds do not markedly affect yolk pigmentation.

Bauernfeind (*61*) reported that the addition of 10,000 IU vitamin A per pound of feed had no adverse effect on yolk pigmentation when two pure oxycarotenoids (β-apo-8′-carotenal and canthaxanthin) in dry stabilized form were fed, as determined by quantitative measurement of yolk pigment content. The addition of 100,000 IU/lb reduced relative yolk pigment content by 5 and 10% for β-apo-8′-carotenal and canthaxanthin, respectively. Studies were also run with other dietary stresses and feed additives (Table 10). Hayes (*372*) studied the effect of feeding layer rations with 12,000, 24,000 and 36,000 IU vitamin A per pound of feed and noted depressed egg yolk pigmentation at the 24,000 IU/lb level and above. When 3% beef tallow was also added to the ration, the pigment-depressing effect of the vitamin A was eliminated even at the highest level (36,000 IU/lb). Mehner (*570*) fed three groups of laying hens with rations containing 8000, 32,000 and 64,000 IU vitamin A per kilogram of feed for 10 weeks and noted slightly reduced pigmentation, especially at the two higher levels. Sunde (*778*) showed that increasing levels of supplemental vitamin A (25,000, 50,000, 75,000, 100,000 and 125,000 IU/kg) progressively decreased yolk pigmentation, with the highest (125,000 IU/kg) group containing only half the yolk pigment content seen in those eggs from hens receiving the lowest level (25,000 IU/kg). In turn, the group receiving the lowest vitamin A level (25,000 IU/kg) showed an approximately 25%

TABLE 10

Relative Yolk Pigmenting Efficacy of Stabilized Oxycarotenoids in Contact with Feed Additives and Ingredients

Ration additive or ingredient and concentration	Stabilized oxycarotenoid beadlets: Apocarotenal[a]	Stabilized oxycarotenoid beadlets: Canthaxanthin[b]
Basal ration alone (control)	100[c]	100[c]
Trace minerals,[d] 0.015%	95	100
Fish oil, 0.44%	88	109
Fish meal, 5.0%	98	99
Meat scraps, 4.0%	92	98
Trithiadol, 0.1%	118	90
Unistat, 0.1%	117	103
Sulfaquinoxaline, 0.0125%	108	105
Nicarbazin, 0.0125%	103	98
Zoalene, 0.0125%	86	98
Amprolium, 0.0125%	111	102
Chlortetracycline, 50 gm/ton	118	102
Arsanilic acid, 0.01%	110	—
Roxarsone,[e] 0.005%	104	106
Roxarsone,[e] 0.005% + Zn bacitracin, 10 gm/ton + pro. penicillin, 8 gm/ton	100	97
Zn bacitracin, 10 gm/ton + pro. penicillin, 8 gm/ton	99	—
NFZ,[f] 0.0055% + NF−180,[g] 0.008%	—	96
Butylated hydroxytoluene, 0.0125%	—	107
Vitamin A, 10,000 IU/lb[h]	98	98
Vitamin A, 100,000 IU/lb[h]	95	91

[a] Two milligrams fed per hen daily in beadlet form in the ration. [From Bauernfeind (*61*).]

[b] One milligram fed per hen daily in beadlet form in the ration. [From Bauernfeind (*61*).]

[c] Arbitrarily set at 100 as a comparison standard. Criterion was pigment deposited in yolk as determined chemically. Basal ration was a low-pigment ration containing 3500 IU vitamin A (as beadlets) per pound. Additives and ingredients were mixed into the basal ration and fed as indicated. No additive or ingredient was unduly suppressing.

[d] Manganese, zinc, iron, copper, iodine, and cobalt.

[e] 3-Nitro-4-hydroxyphenylarsonic acid.

[f] Nitrofurazone.

[g] Furazolidone.

[h] As stabilized vitamin A palmitate beadlets.

depression in yolk oxycarotenoid content compared to that receiving the control ration without supplemental vitamin A. The rations all contained 12 gm oxycarotenoids per ton supplied by 55% yellow corn and 5% dehydrated alfalfa meal. Dua *et al*. (*243*) found that feeding of 220,000 IU vitamin A per kilogram for 14 days significantly depressed plasma and egg yolk carotenoids, whereas 22,000 IU/kg did not when added to a basal ration containing 4500 IU/kg and fed to White Leghorn hens in their eighth month of production.

The data available on the effect of high-level feeding of vitamin A on broiler pigmentation are more limited than those for egg yolk pigmentation. However, it again appears that the levels required to depress broiler pigmentation are substantially above those levels routinely used for optimal growth, feed conversion, and accumulation of vitamin A in the liver. Tagwerker *et al.* (*783*) found that about 80,000 IU vitamin A per kilogram of feed depressed skin pigmentation. Dua (*241*) and Dua *et al.* (*244*) noted that the carotenoid content of serum, liver, and skin of chickens was decreased when 100,000 IU vitamin A per kilogram were fed compared to when 5000 IU/kg were fed. The depression was seen with vitamin A palmitate, aldehyde, alcohol, and acid, with retinoic acid manifesting the greatest depressing effect when all were fed at 100,000 IU/kg. Deuel *et al.* (*230*) found the carotenoid content in the body fat of laying hens to be suppressed to 25% of the normal levels when fed 100,000 UI vitamin A per pound of ration for 6½ months. However, when 200,000 IU/lb were fed for 3 months, no suppression was noted. The authors considered that this discrepancy was the result of the sluggish turnover of the carotenoids in the body fat and that 3 months on the higher level of vitamin A was not sufficient time to complete this turnover. According to Rubin and Bird (*685*), 6-week-old chickens showed considerably less pigmentation when 40,000 IU vitamin A per pound were fed in a basal ration containing 40% yellow corn, 10% corn gluten meal, and 3% alfalfa. In another trial two groups of baby chicks were maintained on an oxycarotenoid-deficient diet for 6 weeks, one group receiving a normal vitamin A level, the other group receiving 454,000 IU/lb. At 6 weeks of age, and for 36 days thereafter, both groups were given a diet containing 40% yellow corn, 10% corn gluten meal, and 3% alfalfa meal. The shanks were scored with the Heiman–Carver Color Rotor three times each week, and it was observed that the chickens receiving the high vitamin A level accumulated pigment in the shanks at a considerably lower rate.

2. *Vitamin E*

Vitamin E (α-tocopherol) enhances pigmentation. Since one of its biological functions is that of an *in vivo* antioxidant, its positive influences on pigmentation are understandable. Pitt and Morton (*639*), in a review of fat-soluble vitamins, suggested that the main function of vitamin E is to act as a physiological antioxidant. Hislop and Whittle (*406*) noted that α-tocopherol is an effective antioxidant protecting substances sensitive to oxidation during storage, both in the feed (*in vitro*) and in the digestive tract (*in vivo*). According to Mecchi *et al.* (*569*) and Phillips and Williams (*637,638*), it is the only antioxidant absorbed in quantity through the intestinal walls and stored in organs of the body and in the body fat, where it plays an important role in maintaining the stability of the fat of poultry. Its role as a tissue antioxidant was reviewed by Marusich (*543*). Wiss *et al.* (*888*) presented comparative data on the uptake of vitamin E and a synthetic antioxidant (ethoxyquin) in rats. Two milligrams of C^{14}-labeled *dl*-α-tocopheryl acetate or 2 mg of ^{14}C-labeled ethoxyquin were administered orally per rat. Both substances were absorbed and deposited in the vital organs; however, ethoxyquin was rapidly eliminated. After 96 hr, the concentrations of vitamin E were, on the average, about 10-fold higher than those of ethoxyquin. The authors concluded that the absorption, distribution, and elimination of ethoxyquin follow the pattern of nonphysiological substances that the body tries to excrete rapidly. On the other hand, the increase of vitamin E in specific tissues and the maintenance of a certain level over a period of several days reflects its biological function in these tissues.

Adamstone (*4*) stated that the extreme paleness of shanks, indicating low concentrations of xanthophylls in the body, is a characteristic of vitamin E deficiency. Pope *et al*. (*641*) found absorption of xanthophylls to be increased with concurrent administration of α-tocopherol. The addition of vitamin E to a natural ingredient ration was reported by Goldhaber *et al*. (*323*) to result in a 42% increase in plasma levels of xanthophylls. Day and Williams (*211*) noted a reduction in broiler pigmentation when the supplementary vitamin E was removed from the ration. Russo *et al*. (*695*) found that vitamin E enhanced yolk pigmentation when added each day at a level of 3.5 mg per bird to layer rations containing three levels of xanthophylls from yellow corn and/or corn gluten mean. The pigment content of the yolk was directly related to the three levels of oxycarotenoids fed and was higher for each group supplemented with vitamin E. Vitamin E was found by Saunders *et al*. (*699*) to exert a positive effect on yolk pigmentation when the ration contained 3% sunflower oil. Without vitamin E in the ration, the sunflower oil exerted a negative effect on pigmentation. Some negative reports have also appeared, including those of Runnels *et al*. (*692*), who noted no enhancement of shank pigmentation with supplemental vitamin E or chemical antioxidants. They noted, however, that the depth of shank coloration was excellent in all cases, leaving little room for improvement. Williams *et al*. (*875*) did not obtain a positive pigmenting effect with added vitamin E in broiler rations. Couch (*168*) observed that the addition of 20,000 IU vitamin E per ton of feed increased the liver storage of vitamin A in broilers but had no effect on pigmentation.

3. *Antioxidants*

Antioxidants are routinely added to poultry feeds or to a specific ingredient, such as dehydrated alfalfa meal, during its processing. In the United States ethoxyquin or 1,2-dihydro-6-ethoxy-2,2,4-trimethlquinoline (EMQ), butylated hydroxyanisole (BHA), or butylated hydroxytoluene (BHT) is used, with primary emphasis on EMQ. Another very effective antioxidant, diphenyl-*p*-phenylenediamine (DPPD), is not currently permitted in commercial feeds but was extensively tested in the past in research trials. The oxycarotenoids effective as pigmenters are structurally similar to the carotenes and vitamin A, and all are very susceptible to oxidative destruction in the ration. Thus, the reason for including an antioxidant in the ration of broilers and layers is that it will reduce the rate of oxidative destruction of these unsaturated molecules including unsaturated fats during feed or ingredient storage and possibly in the gastrointestinal tract following ingestion. The *in vivo* effect has not been clearly demonstrated; attempts to measure the effect of antioxidants by enhanced pigmentation of poultry when the *in vitro* stability aspect has been eliminated have produced conflicting results. Wilgus (*867*), Potter *et al.* (*646*) and Fritz *et al.* (*290*) obtained improvement in broiler pigmentation with DPPD included in the ration. Potter *et al.* (*646*) found both DPPD and BHT to be effective, although BHT was less effective than DPPD. The influence of DPPD and BHT was compared by Matterson *et al.* (*559*) in the utilization of carotenoid pigments from dehydrated alfalfa meal, and both antioxidants were about equivalent in increasing plasma and skin carotenoid levels. Fritz *et al.* (*290*) reported equivalence for BHT and DPPD. Elrod *et al.* (*252*) compared the effect of BHT and EMQ alone and combined. Both improved shank pigmentation slightly, but the two combined were no better than either fed alone. Romoser *et al.* (*682*) observed that the feeding of EMQ in the ration increased skin pigmentation in broilers. A positive effect was also noted by Waldroup *et al.* (*844*) with EMQ on the skin and shanks of broilers. Ratcliff *et al.* (*663*) found EMQ to enhance broiler pigmentation significantly in three of four groups. They also reported that BHT significantly improved pigmentation in one instance, but it significantly depressed pigmentation in another group. Runnels *et al.* (*692*) reported carcass color to be visibly improved by the addition of either BHT or EMQ, with both being about equivalent. However, no enhancement of shank pigmentation was noted with the use of either BHT or EMQ, as noted previously in the section on vitamin E.

Improvement in yolk pigmentation with both DPPD and BHT was reported by Shellenberger *et al.* (*716*), with EMQ by Anjaneyalu *et al.* (*13,14*), Madiedo and Sunde (*528,529*), Nakaue *et al.* (*600*), Lautner (*495*), Bartov and Bornstein (*56*), Bushong *et al.* (*135*), and with BHT by Saunders *et al.* (*699*).

The antioxidants have also been reported to have no effect or actually have a depressing effect. Harms *et al.* (*358*), in an effort to determine the effect of feeding DPPD, found that it significantly depressed skin pigmentation in broilers. It was noted by Day and Williams (*211*) that BHT reduced broiler pigmentation somewhat in most instances, but the differences were not generally significant. Pope *el al.* (*641*) reported that BHT was partially effective. Williams *et al.* (*875,877*) and Couch (*168*) failed to show any beneficial effect of DPPD, BHT, or EMQ on broiler pigmentation as measured by serum and skin xanthophylls. Bletner *et al.* (*96*) reported that BHT had no ameliorating effect on the depression of pigmentation due to coccidiosis. Ratcliff *et al.* (*663*), as noted previously, obtained both positive and negative results with BHT. Carlson *et al.* (*145*) reported that EMQ did not improve the utilization of xanthophylls by laying hens as measured by both visual observation and chemical determination of yolk pigment content. Bartov and Bornstein (*56*) observed no effect of EMQ on yolk pigment concentration *in vivo* when both egg and feed samples were collected and assayed for their respective pigment content. They suggested that the action of antioxidants is restricted to the stabilization of dietary xanthophylls. On the other hand, Anjaneyalu *et al.* (*14*) considered the positive effect of EMQ to be on yolk pigmentation *in vivo*. Streiff (*768*) failed to obtain any effect with EMQ when added directly to beadlets containing β-apo-8′-carotenoic acid ethyl ester or directly to the feed, as determined by yolk pigment deposition measured visually and colorimetrically.

The evidence is greatly in favor of antioxidants functioning as *in vitro* protectors of xanthophylls and carotenes, rather than actually having a protective effect *in vivo*. Numerous reports attest to the *in vitro* antioxidant effects of DPPD, BHT, and EMQ. Thompson (*802*), Bickoff *et al*. (*80*), Tiews and Zucker (*811*), Livingston *et al*. (*511*), Knowles *et al*. (*458*), and Witt *et al*. (*890,891*) all showed the superior stability of carotenes and xanthophylls in dehydrated alfalfa meals in the presence of EMQ. A similar effect was observed by Livingston *et al*. (*517*) with dehydrated turf grass and EMQ. Siedler *et al*. (*719*) reported EMQ to be slightly more effective than DPPD and BHT in preventing vitamin A, carotene, and fat oxidation in commerical poultry rations; however, all were effective. Ratcliff and Day (*662*) noted BHT to be effective in improving retention of the oxycarotenoids in dehydrated alfalfa meal and two samples of corn gluten meal held at 100°F for 8 weeks.

Potter *et al*. (*646*) found a relationship between the vitamin B_{12} content of the ration and the effect of DPPD on pigmentation. They reported that vitamin B_{12} alone did not promote consistent increases in vitamin A or carotenoids in the liver and blood plasma but did consistently increase the amount of carotenoids in the skin. The addition of DPPD to this ration produced small increases in vitamin A and carotenoid storage, whereas the combination of DPPD and vitamin B_{12} produced substantial increases.

4. *Fats and Oils*

Fat content and type of fat ingested have been reported to have both a positive and negative influence as well as no influence on the pigmentation of broilers and egg yolks. The fat content of feeds and resulting changing ratio of calories to protein may be inseparably tied together in some of the effects observed. Maw (*562*) and Hill and Dansky (*399*) reported a positive correlation between carcass fat and pigment deposition. In turn, Donaldson *et al*. (*235*) showed that the amount of carcass fat is increased by increasing the calorie–protein ratio of broiler rations. Day and Williams (*211*) noted that increasing the calorie–protein ratio in feed during the last 2- or 3-week feeding period before marketing the broilers significantly improved pigmentation. They also showed that the addition of 5% fat (stabilized beef tallow) significantly improved pigmentation. According to Slinger *et al*. (*726*) better finished birds can be obtained by increasing the energy content of the ration and simultaneously reducing the protein content for the last 4-week feeding period before marketing. Quackenbush *et al*. (*656*) found that the addition of 2 or 10% lard improved pigmentation efficiency on the basis of extracted pigment concentration in shank skin. As cited previously, Heath and Shaffner (*376*) reported improved breast skin pigmentation in broilers as the level of soybean oil was increased from 4 to 7–10%. Biedermann and Prabucki (*84*) and Wiemann and Marquering (*861*) also obtained positive effects with fat.

Carver (*147*) reported that certain samples of yellow grease, no. 1 tallow, hydrolyzed animal and vegetable fat, and methyl esters of vegetable fat reduce pigmentation in broilers, as determined by both shank visual color and colorimetric measurement of extracted toe-web pigment. It was further demonstrated that the unknown factors that exist in certain fat supplements to depress pigmentation vary in concentration since some samples of yellow grease and hydrolyzed animal and vegetable fat did not depress pigmentation.

Ratcliff *et al.* (*665*) determined that the addition of 4% fat or 2% lecithin to broiler rations did not significantly influence pigmentation. Similarly, varying the calorie–protein ratio of the rations (35:1 through 60:1) had no significant influence on broiler pigmentation. The efficiency of utilization of the xanthophylls was lower when the calorie–protein ratio was below 40:1 in all instances. Williams (*874*) found that fish oil, soybean lecithin, and various grades of oils and fats added to broiler diets depressed the deposition of xanthophylls in all cases.

Williams *et al.* (*875*) concluded that the effect of fat on broiler pigmentation depends on the type of fat. Animal fat at a level of 20% significantly depressed serum carotenoid levels, pigmentation, and liver vitamin A storage. The same results were obtained with 20% corn oil, soybean oil, or cottonseed oil, but the depression of serum carotenoid levels and skin pigmentation was greater than with the 20% animal fat. On the other hand, poultry oil at 20% did not significantly reduce serum carotenoid levels or pigmentation but did elicit a significant increase in liver vitamin A. Couch (*168*) showed that the use of 2.5, 5, and 10% added fat had no effect on broiler pigmentation or vitamin A storage in the liver. The addition of 20% fat significantly depressed both parameters. Couch *et al.* (*179*) also reported that the type and level of fat may definitely affect the deposition of skin carotenoids in growing broilers based on colorimetric determination of toe-web skin extracted pigment content and serum carotenoids. The latter workers found that a fat high in free fatty acids (90%) was more deleterious in its effect on the deposition of skin carotenoids than a fat lower in free fatty acids (53%). Hammond (*351*) found that mineral oil caused chicks to have pale shanks and severe rickets. The feces of these chickens contained more yellow pigments than those of chicks fed the same ration without mineral oil, strongly suggesting an adverse effect on intestinal absorption. Herrick *et al.* (*389*) reported 5% added fat to have no effect on broiler pigmentation, as measured by shank color or skin color changes. Heath (*374*) presented data to suggest that added dietary fat enhanced skin deposition of xanthophylls.

Fat has also been reported as having either a positive, negative, or no effect on egg yolk pigmentation. Fritz *et al.* (*291*) reported that a readily digestible fat may improve the utilization of xanthophylls by the hen. On the other hand, Fritz (*287*) stated that if the energy content of the ration is increased substantially by the addition of fat, the overall effect is to reduce the quantity of feed consumed and thereby reduce oxycarotenoid intake and thus yolk color. Mackay *et al.* (*523*) reported that the addition of 4% animal fat to the diet increased the utilization of dietary xanthophylls and egg yolk color when paprika extract was used as the oxycarotenoid pigmentation source. In studies by Sullivan and Holleman (*777*) the addition of 4% animal fat to feed did not influence egg yolk color. Madiedo and Sunde (*529*) reported that 5% safflower oil, which is rich in linoleic acid, did not affect yolk pigmentation when yellow corn provided the oxycarotenoids. Saunders *et al.* (*699*) found that 3% sunflower oil had either no effect or a depressing effect when an antioxidant such as BHT or vitamin E was omitted from the ration. If the ration contained BHT or vitamin E and 3% sunflower oil, yolk pigmentation was significantly enhanced. Biedermann and Prabucki (*84*) observed improved yolk color with increased fat levels ranging from 5 to 15%. In a second trail, different fats were fed, all at a 10% level. Those fats with lower melting points gave higher carotenoid levels in yolks. Blum and Leclercq (*98*) added tallow, olive oil, or sunflower oil at 4% to layer rations containing 20 gm xanthophylls per ton extracted from lucerne meal. Only tallow gave a positive effect on yolk carotenoid content, as determined by colorimetric assays.

A study was run in our laboratory (*677*) to determine the effect of feeding polyunsaturated fatty acids (PUFA) on yolk pigmentation. A low-pigment ration devoid of all supplemental vitamin E and antioxidants was used. A special gelatin beadlet containing 1% β-apo-8′-carotenal, also free of antioxidants, was made. This beadlet was fed in the following rations for 21 days: (a) the low-pigment ration, free of antioxidants; (b) ration plus 10% stripped safflower oil (approximately 70% linoleic acid), providing the high level of PUFA; (c) ration plus 10% hydrogenated

TABLE 11

Yolk Pigmentation Data from Hens Fed β-Apo-8′-carotenal in the Presence of Saturated and Polyunsaturated Fatty Acids[a,b]

Supplement to basal ration	Visual score[c] (Roche Yolk Color Fan)	AOAC BCE[d] content		Percentage deposited
		μg/gm	μg/yolk	
None	13.0	49.5	941	31.4
10% saturated fat	13.1	44.9	853	28.4
10% stripped safflower oil	13.1	45.6	866	28.9
5% saturated fat and 5% stripped safflower oil	13.3	48.6	923	30.8

[a] From Reese (*677*).

[b] β-Apo-8′-carotenal (3.0 mg) fed daily per hen in a stabilized gelatin beadlet form for 20 days with six hens per treatment.

[c] Data based on all eggs laid from ninth through twentieth day.

[d] BCE, β-carotene equivalent.

coconut oil (saturated fat); and (d) ration plus 5% stripped safflower oil plus 5% hydrogenated coconut oil. On the basis of yolk visual scores and colorimetric determination of pigment content it was concluded that none of the treatments affected pigment deposition when β-apo-8′-carotenal was the pigment source (Table 11).

5. *Arsenicals*

Arsenicals, including 3-nitro-4-hydroxyphenylarsonic acid (roxarsone) and *p*-aminophenylarsonic acid (arsanilic acid) and its sodium salt (sodium arsanilate), enjoy widespread usage in the United States as additives in broiler rations to increase growth, improve feed conversion (*468*), and enhance pigmentation. Roxarsone is the most commonly used. It is permitted at levels of 0.0025–0.005%, whereas sodium arsanilate and arsanilic acid are added at 0.005–0.01%. Although general field observations support the contention that arsenicals enhance pigmentation in broilers, laboratory trials run in batteries and/or in floor pens are not conclusive. Patrias (*627*) stated that both roxarsone and arsanilic acid were effective but did not present experimental data. Couch (*166,167*) also reported positive effects with roxarsone and arsanilic acid, respectively. Couch (*167*) orally administered corn oil containing xanthophylls to broilers with and without an arsenical in the ration. Shanks were visually scored and showed superior pigmentation in the group receiving the arsenical. Similar improvements due to an arsenical were obtained when yellow corn and dehydrated alfalfa meal served a the oxycarotenoid sources. Smidt *et al*. (*728*) observed that organic arsenicals had no effect alone but enhanced pigmentation when fed with xanthophylls; roxarsone had a greater effect than arsanilic acid. Kowalski and Reid (*471*) also obtained positive pigmentation in broilers with roxarsone in both the presence and absence of coccidiosis.

Negative responses (no effect due to feeding of arsenicals) were reported by Tarver *et al*. (*785*), Elrod *et al*. (*252*), West (*858*), and Marusich *et al*. (*552*) with roxarsone and by Frost *et al*. (*292*) with arsanilic acid. Johnson (*438*) and Marusich *et al*. (*551*) failed to obtain any effect on broiler pigmentation with either roxarsone or arsanilic acid. Others, such as Milligan *et al*. (*578*) and Ratcliff *et al*. (*665*), reported both improvement and no effects with roxarsone in the same studies as measured by several broiler pigmentation parameters. Marusich *et al*. (*552*) studied the effect of feeding continuously for 8 weeks 0.005% roxarsone added to rations containing from 20 to 40 gm oxycarotenoids per ton on the performance of broilers reared in cages. The arsenical had no effect on rate of growth, feed conversion, or pigmentation. Pigmentation parameters included visual shank scores determined using the Roche Yolk Color Fan and colorimetric assay of total carotenoid concentrations in plasma, toe-web skin, and shank skin (Table 12).

The bulk of the literature (Table 13) evaluating the effect of arsenicals on growth, feed conversion, and pigmentation in broilers has been summarized. The data show the responses to both roxarsone and arsanilic acid to be quite variable. Morehouse (*587*), in a review of the effects of roxarsone on pigmentation, noted that the

TABLE 12

Effect of Roxarsone and Oxycarotenoids on Broiler Pigmentation Parameters[a]

Oxycarotenoid level (gm/ton)[b]			Total oxycarotenoid intake[c] (mg/bird)		Visual score (Roche Yolk Color Fan, 1–15)		Tissue carotenoid concentration: Plasma (μg/ml)		Toe web (μg/100 cm²)[c]		Shank skin (μg/gm)	
Feed ingredients	Canthaxanthin	Total	A[d]	B[e]	A[d,f]	B[e,f]	A[d,g]	B[e,g]	A[d,g]	B[e,g]	A[d,g]	B[e,g]
20	0	20	75.7	73.3	8.3	8.8	11.6	12.0	208	195	43.0	41.5
20	2	22	80.0	77.9	10.0	9.6	12.7	11.1	208	213	45.0	43.0
20	4	24	84.8	83.7	10.7	10.5	13.0	13.8	257	253	52.0	59.5
40	0	40	120.0	116.7	9.6	9.6	16.6	16.0	279	301	73.0	65.0
40	4	44	126.6	125.3	10.8	10.9	15.8	17.0	309	310	74.5	69.5
Average for all levels of oxycarotenoid supplementation			97.4	95.4	9.9	9.9	13.9	14.0	252	254	57.5	55.7

[a] From Marusich *et al.* (*553*).
[b] Fed from 5 weeks; first 5 weeks all birds on same starter ration with 20 gm oxycarotenoids from feed ingredients per ton.
[c] From 0 to 8 weeks + 5 days.
[d] Group A, no roxarsone.
[e] Group B, fed 0.005% roxarsone for 8 weeks.
[f] Average for 20 birds.
[g] Average for 10 birds.

TABLE 13

Summary of Published Responses for Roxarsone or Arsanilic Acid Fed to Growing Chickens[a,b]

Roxarsone			Arsanilic acid			
Growth	Feed conversion	Pigmentation	Growth	Feed conversion	Pigmentation	References
+	*	*	*	*	*	Morehouse and Mayfield *(590)*
+	*	*				Bird *et al.* *(89)*
+	*	*				Morehouse *(586)*
−	−	*				Scott and Glista *(711)*
+	*	*				Stokstad and Jukes *(754)*
+	±	*	+	±	*	Couch *(166)*
±	*	+	±	*	+	Patrias *(627)*
+	*	*				Combs and Laurent *(158)*
			+	*	*	Elam *et al.* *(250)*
+	−	*				Milligan *et al.* *(577)*
			+	*	*	Abbott *et al.* *(3)*
+	*	*	+	*	*	Carlson *et al.* *(143)*
			±	±	*	Combs *et al.* *(162)*
*	*	+	*	*	+	Couch *(167)*
+	±	*	−	−	*	Morrison *et al.* *(594)*
+	+	*				Pepper *et al.* *(629)*

–	–	–				Tarver *et al.* *(785)*
			+	–	–	Frost *et al.* *(292)*
+	*	*	+	*	*	Libby and Schaible *(506)*
			+	–	*	Libby *et al.* *(505)*
			–	–	*	McDonald *(565)*
+	±	±				Milligan *et al.* *(578)*
+	+	–				West *(858)*
+	+	*				Day *et al.* *(210)*
–	–	–				Elrod *et al.* *(252)*
			–	–	*	Heywang *(397)*
+	*	*				Wharton *et al.* *(859)*
			±	*	*	Wisman *(887)*
–	–	*				Sibbald and Slinger *(718)*
*	*	+	*	*	+	Smidt *et al.* *(728)*
+	+	*	+	+	*	Chang and Butters *(148)*
–	–	–	–	–	–	Marusich *et al.* *(551)*
*	*	+				Kowalski and Reid *(471)*
+	+	+				Olson *et al.* *(614)*
			–	–	*	Kolar and Seymour *(468)*
*	*	+				Fry and Harms *(295)*
*	*	+				Fry *et al.* *(297)*
*	*	+				Fry *et al.* *(298)*

[a] From Marusich *et al.* *(552)*.
[b] Key: +, positive improvement; –, no improvement; ±, both positive and no improvements reported; *, data not reported; no notation, not tested.

literature showed variable results, since many of the studies were based on quantitative determination of the xanthophylls in the skin. In his work with roxarsone, he was concerned with qualitative differences in pigmentation between roxarsone-fed birds and those receiving no roxarsone. He stated: "It is possible that this arsonic acid does not produce quantitative differences in xanthophyll content of the skin of birds. However, it appears that its use in some way enhances the optical appearance of the yellow pigments present." Patrick (*628*) also noted that the arsonic acids probably make broilers appear to have greater pigmentation by enhancing visual appearance but the true effect may be increased peripheral circulation, which makes the skin have a "lively" appearance.

Another possible explanation relates to the weak anticoccidial and weak antibacterial properties of roxarsone and arsanilic acid against both pathogens and nonpathogens, as noted by Frost *et al.* (*292*) A mild or subclinical coccidiosis may adversely affect broiler pigmentation; thus, in the presence of a subclinical or mild disease state the arsenical may exert a beneficial pigmenting effect. Similarly, in an environment conducive to the establishment of an intestinal microflora that exhibits no overt signs of disease but prevents the bird from manifesting its full potential for growth and efficiency of feed conversion, the arsenical may exert a positive effect similar to that attained by feeding low levels of antibiotics. Under these conditions, oxycarotenoid absorption may also be enhanced, resulting in better pigmentation. Marrett *et al.* (*533*) showed that the addition of erythromycin thiocyanate at 21.6 gm/ton but not at 4 gm/ton significantly increased yolk pigmentation, as determined by colorimetric measurement of extracted yolk pigments. Perdue and Smith (*630*) had previously reported positive effects on yolk pigmentation with either 5 or 10 gm of erythromycin thiocyanate added per ton of layer feed. Similarly, coccidiosis, CRD, and other infections interfere with pigmentation responses, as discussed in Section VI, B. Arsenicals have slight coccidiostatic properties. For example, sodium arsanilate has an approved claim "for prevention of coccidiosis" in chickens when fed at the 0.04% level for 8 days. Roxarsone at 0.005% significantly reduces fecal oocyst counts when broiler chicks are infected with *Eimeria tenella* sporulated oocysts. This activity is also measurable when roxarsone is fed in combination with an approved prophylactic anticoccidial.

The anticoccidial properties of arsenicals probably explains the improvement generally seen in the field when birds are reared on the floor and constantly exposed to *Eimeria* oocysts, in contrast to the lack of effects seen in laboratory studies in which birds are raised in cages on raised wire floors in the absence of oocysts.

Bunnell *et al.* (*132*) did not obtain any effect with 0.01% arsanilic acid or 0.005% roxarsone added to a hen layer ration when β-apo-8′-carotenal served as the source of oxycarotenoids, as determined by colorimetric measurement of pigment content in the yolk. Bauernfeind (*61*) also failed to obtain a positive response with roxarsone when canthaxanthin served as the single oxycarotenoid in a hen layer ration, again using colorimetric measurement of yolk pigments.

E. Oxycarotenoid Instability

The instability of the oxycarotenoids in feed ingredients is relative and depends on the source, storage conditions, length of storage, addition of fat and/or an antioxidant, and probably other less well defined factors. During the years, as stability data have accumulated and improvements have been made in handling and storing the products, better retention of either carotenes or xanthophylls or both has been achieved.

1. Corn Products

Both corn and corn gluten meal exhibit problems of xanthophyll and carotene instability. The carotenes represent only about 10% of the total carotenoids, and the xanthophylls comprise 90%. Watson (*846*) reported that corn lost 50% of its oxy carotenoids when stored for 1 year at 25°C (77°F). Quackenbush (*651*) found that corn lost 25% of its pigment content after being subjected to drying to reduce its mosture content from 11 to 3% (Table 14). He also noted that corn stored for 1 year at 25°C lost 50% of its pigment content. The oxycarotenoid content of corn varies with the time of year it is received at a wet-milling plant to be processed to obtain corn gluten meal. Lower oxycarotenoid potencies are seen during the summer and early fall. These data have been presented in detail by Watson (*846*) and Heiman (*382*). Marusich and Wilgus (*558*) showed the average monthly potencies for corn gluten meal over a 3-year period (Table 15), which correlates well with the data for corn presented by Watson (*846*). Quackenbush (*651*) showed the losses of carotenes to be proportionally greater than the losses of xanthophylls in corn (Table 14) and similar to the losses of xanthophylls in dehydrated alfalfa (Tables 16–19). Thompson and Maclay (*807*) stated that corn loses its xanthophylls under ordinary storage conditions in approximately the same way as dehydrated alfalfa meal.

Marusich (*542*) pointed out that the corn gluten meals available from the fall of 1970 through the fall of 1971 were substantially lower in oxycarotenoids than those available in previous years. This was related to the fact that large quantities of stored corn moved into corn gluten meal production due to the severity of the southern corn leaf blight that reduced the corn crop for 1970. Harms and Goff (*357*) had also shown that mold-damaged corn did not pigment broilers as well as no. 1 or no. 2 yellow corn free of mold. This could be related to its aflatoxin content.

Although ordinary sorghum (milo) varieties are highly deficient in carotenes and oxycarotenoids, plant geneticists and breeders have recently developed true yellow-endosperm varieties containing one-fifth to one-half the normal carotenoid content of yellow corn (*94*). A major problem is the rapid loss of carotenoids (both carotenes and oxycarotenoids) from the sorghum grain in the field (*683*). Typical stability data are shown in Fig. 12 for carotenes and xanthophylls in two typical strains of yellow-endosperm sorghum.

TABLE 14

Changes in the Carotenoid Fractions of Corn with Different Moisture Contents under Different Storage Conditions[a,b]

Moisture (%)	Storage temp (°C)	Storage period (months)	Carotenes	Zeinoxanthin	Cryptoxanthin and esters	Lutein	Zeaxanthin	Polyoxypigments	Sum of pigments
11	25	0	4.8	3.4	5.5	19.8	16.9	3.4	53.8
		4	3.6	2.1	5.1	17.5	16.5	2.0	46.8
		8	2.5	2.2	3.3	16.0	10.4	2.8	37.2
		12	1.8	1.9	2.4	10.8	6.7	1.6	25.2
		24	1.7	1.8	3.2	11.3	7.7	1.9	27.6
		36	1.0	1.3	2.8	7.6	5.2	1.8	19.5
11	7	0	4.8	3.4	5.5	19.8	16.9	3.4	53.8
		4	3.1	2.2	5.0	20.5	12.5	2.6	45.9
		8	3.5	2.4	4.5	23.2	15.1	2.7	51.4
		12	3.4	2.9	4.0	14.6	14.2	2.3	41.4
		24	3.4	2.4	6.1	19.3	12.4	2.5	46.1
		36	2.6	2.5	5.6	15.8	12.6	2.3	41.1
3	25	0	3.9	2.3	5.0	15.7	9.9	2.8	39.6
		4	3.3	1.8	5.0	14.4	10.1	2.3	36.9
		8	2.6	1.7	3.4	12.0	7.8	2.0	29.5
		12	2.7	2.1	3.1	8.4	6.3	1.7	24.2
		24	1.4	1.4	3.1	8.6	5.8	2.0	22.3
		36	1.1	1.3	2.5	6.5	4.4	1.6	17.2
3	7	0	3.9	2.3	5.0	15.7	9.9	2.8	39.6
		4	3.1	1.4	5.0	13.2	10.7	2.5	35.9
		8	3.5	2.1	4.0	17.9	12.4	2.2	42.1
		12	3.5	2.5	4.0	11.5	8.8	2.5	32.8
		24	3.0	2.3	5.5	15.0	10.2	2.0	38.0
		36	2.2	2.1	4.5	10.3	8.6	2.0	29.5

[a] From Quackenbush (*651*).
[b] All values are averages of two analyses, expressed as γ per gram of corn.

TABLE 15

Stability of the Oxycarotenoids in Corn and Corn Gluten Meal

A. Stability of lutein and zeaxanthin in corn[a,b]			B. Losses of oxycarotenoids from corn gluten meal[c,d]								
	Retention (%) stored at		Oxycarotenoids			Oxycarotenoids			Oxycarotenoids		
Storage period (months)	7°C	25°C	Month	mg/lb	%[e]	Month	mg/lb	%[e]	Month	mg/lb	%[e]
0	100	100	November	178	90	May	169	85	July	147	75
4	89	93	December	191	95	June	160	80	August	146	75
8	103	72	January	189	95				September	131	65
12	79	48	February	196	100				October	137	65
24	85	52	March	179	90						
36	78	35	April	178	90						
			Average	185			165			140	

[a] From Quackenbush (*651*).
[b] Values combined because they are essentially alike.
[c] From Marusich and Wilgus (*558*).
[d] Average of monthly values for three crop years: 1963–1964, 1964–1965, 1965–1966.
[e] Oxycarotenoid content for Feburary taken as 100%.

TABLE 16

Loss of Carotenes and Xanthophylls during Storage of Alfalfa Meal with Different Moisture Contents and Two Types of Dehydrators[a,b]

	Content before storage (mg/kg)					Loss at 6 weeks (%)					Loss at 12 weeks (%)				
Alfalfa preparation and % moisture	C	N	V	L	Total xanthophylls	C	N	V	L	Total xanthophylls	C	N	V	L	Total xanthophylls
Arnold dehydrator															
Freeze-dried control, 3 trials	362	91	130	494	788	33	14	66	26	31	47	22	84	28	38
Dehydrated meal, 9.2%	342	55	15	346	508	64	56	100	54	53	78	56	100	65	64
Stearns–Roger dehydrator															
Freeze-dried control, 3 trials	364	97	174	493	825	25		77		28	40	21	82	25	37
Ethoxyquin-treated	282	66	135	353	598	39	44	77	14	32	43	57	82	22	39
Dehydrated meal															
12.2%	353	82	46	350	574	72	65	100	53	60	84	70	100	69	71
7.1%	342	64	33	306	484	71	62	100	53	57	84	65	100	67	68
2.5%	293	35	20	207	346	59	44	100	29	38	76	44	100	48	52
2.5%	309	40	22	232	388	25	39	100	7	19	37	61	100	17	31

[a] From Livingston *et al.* (*515*).

[b] Key: C, total carotenes; N, neoxanthin; V, violaxanthin; L, lutein.

TABLE 17

Loss of Carotenes and Xanthophylls during Dehydration plus Storage of Alfalfa Meals with Different Moisture Contents and Outlet Temperatures[a]

		Carotenoids in fresh alfalfa (mg/kg)			Carotenoids after dehydration						Carotenoids after dehydration plus 12 weeks storage					
					Remaining (mg/kg)			Loss (%) in fresh			Remaining (mg/kg)			Loss (%) in fresh		
Moisture in meal (%)	Outlet temp (°F)	Carotenes	Lutein	Total xanthophylls	Carotenes	Lutein	Total xanthophylls	Carotenes	Lutein	Total xanthophylls	Carotenes	Lutein	Total xanthophylls	Carotenes	Lutein	Total xanthophylls
Arnold dehydrator																
9.2	300	377	550	834	342	346	508	9	37	39	75	121	185	80	78	78
7.8	310	355	425	735	269	222	368	24	48	50	86	119	176	76	72	76
2.3	330	353	508	796	291	159	286	18	69	64	97	75	136	73	85	83
Stearns–Roger dehydrator																
12.2	250	342	480	800	353	350	574	+3	27	28	53	108	167	85	78	79
7.1	250	412	517	862	342	306	484	17	41	44	57	101	156	86	80	82
2.5	250	337	482	814	293	207	346	13	57	58	68	108	165	80	78	80

[a] From Knowles *et al.* (458).

TABLE 18

Loss of Xanthophylls and Carotenes during Dehydration of Alfalfa Meal[a]

Dryer	Outlet temp of dryer (°F)	Moisture of meal (%)	Carotenes[b]			Xanthophylls[b]		
			Fresh	Meal	Loss (%)	Fresh	Meal	Loss (%)
Industrial Arnold	300	9.2	378	344	9	831	508	39
Trial 1	310	7.8	353	267	24	736	368	50
	330	2.3	353	290	18	798	286	64
Pilot Arnold	270	2.8	329	290	12	784	363	53
Trial 1	300	1.6	334	290	13	750	295	61
Stearns–Roger	330	1.5	348	281	19	821	220	73
Trial 1	240	8.3	411	387	6	876	552	37
	270	9.5	407	353	13	905	542	40
	300	9.9	411	363	12	861	552	36
	330	5.9	363	353	3	822	450	45
Trial 2	250	12.2	349	353	0	784	568	28
	250	7.1	411	339	17	861	484	44
	250	2.5	339	290	14	808	329	59
	275	7.1	445	377	15	972	556	43
	275	3.1	445	300	33	1005	397	60
	275	1.5	363	286	21	847	310	63

[a] From Livingston *et al.* *(516)*.
[b] Dry basis, average of duplicate analyses (mg/kg).

TABLE 19

Stability of Individual Xanthophylls during Storage[a] of Dehydrated Alfalfa–Effect of Dehydration Conditions[b]

Dryer temp			Neoxanthin			All-*trans*-lutein			Isomer I (lutein)			Isomer II (lutein)			Deoxylutein			Violaxanthin		
				Retained in meal			Retained in meal			Retained in meal			Retained in meal			Retained in meal			Retained in meal	
At inlet (°F)	At outlet (°F)	Moisture of Meal (%)	Content in fresh alfalfa (mg/kg)	Stored 4 weeks (%)	Stored 10 weeks (%)	Content in fresh alfalfa (mg/kg)	Stored 4 weeks (%)	Stored 10 weeks (%)	Content in fresh alfalfa (mg/kg)	Stored 4 weeks (%)	Stored 10 weeks (%)	Content in fresh alfalfa (mg/kg)	Stored 4 weeks (%)	Stored 10 weeks (%)	Content in fresh alfalfa (mg/kg)	Stored 4 weeks (%)	Stored 10 weeks (%)	Content in fresh alfalfa (mg/kg)	Stored 4 weeks (%)	Stored 10 weeks (%)
1600	220	11.0	70	50	19	297	50	35	37	50	30	18	50	38	29	47	40	29	47	40
1400	220	9.0	68	53	19	352	53	40	35	53	34	15	55	52	31	58	32	40	30	25
1200	220	8.6	31	50	50	275	67	41	24	67	48	13	34	49	18	64	62	57	25	13
1600	270	4.0	44	70	40	242	80	65	22	71	54	13	67	45	22	70	54	59	25	21
1400	270	3.2	48	75	55	245	82	68	31	72	36	7	80	100+	22	80	60	33	54	40
1200	270	1.6	22	94	80	216	99	69	18	100	100	4	100+	100+	15	100+	70	20	80	67
1600	320	1.2	22	90	35	117	74	62	18	75	55	11	72	60	59	63	48	31	72	55
1400	320	0.7	18	65	60	148	66	62	18	54	70	9	52	100+	66	51	52	37	55	52
1200	320	0.3	22	70	45	145	67	63	15	60	60	4	52	100+	42	53	55	22	70	70

[a] Stored at 90°F.

[b] From Livingston *et al.* *(514)*.

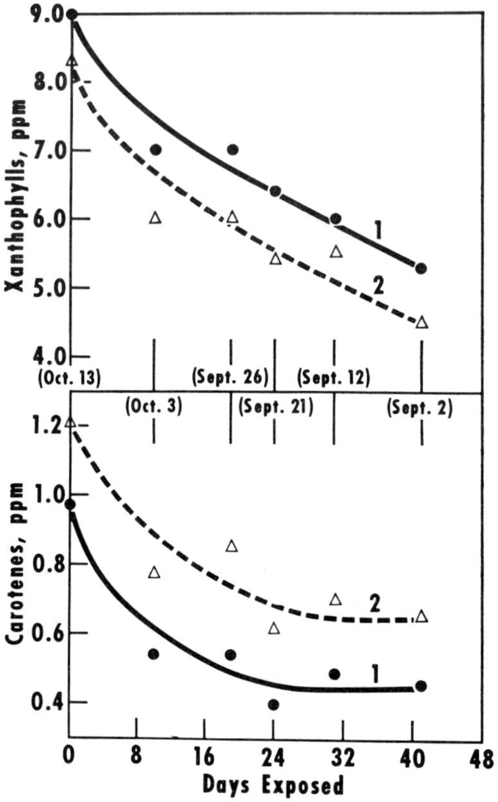

Fig. 12. Loss of carotenes and xanthophylls in representative strains of yellow-endosperm sorghum grain during weathering. [From Blessin *et al.* (*93*).]

2. *Alfalfa Meal*

The use of dehydrated alfalfa meal as a major contributor of xanthophylls in broiler rations has steadily declined in the United States. However, its worldwide usage (lucerne meal and related forms) is still significant. Thus, its stability is treated in great detail in this section. The rate of loss of xanthophylls and carotenes in dehydrated alfalfa meal is accelerated by heat, as determined by Wiseman *et al.* (*886*), and by both heat and light, as reported by Hauge and Aitkenhead (*368*), Kon and Thompson (*469*), Guilbert (*343*), and Dolge *et al.* (*234*). Xanthophylls, carotenes, and fat-soluble vitamins are all subject to oxidation losses during storage, as shown by Wilder and Bethke (*862*), Silker *et al.* (*720*), Griffith and Thompson (*331*), Thompson and Maclay (*807*), Kohler *et al.* (*463*), Bartov and Bornstein (*57*), and Thompson *et al.* (*805*). Both xanthophylls and carotenes form stereoisomers during the dehydration process, with a resulting loss of the provitamin A activity of the carotenes as well as a possible loss of the pigmenting potency of the xanthophylls, according to the findings of Bickoff *et al.* (*80*), Bickoff and Thompson (*81*), Thompson *et al.* (*804*), and Livingston *et al.* (*513*). The complexity of this isomerization and the effect of storage on the oxycarotenoid content are emphasized in Table 19 for all-*trans*-lutein, lutein isomer I, lutein isomer II, deoxylutein, violaxanthin, and neoxanthin. Livingston *et al.* (*514,516*) reported losses of xanthophylls ranging from 28 to 73% during pilot and industrial-scale alfalfa dehydration. These losses were correlated with the moisture content of the meal (see Tables 16–19 and Fig. 12 and 13).

Although the instability characteristics of the xanthophylls of alfalfa meal have not been examined as thoroughly as those of the carotenes, Thompson and Maclay (*807*), Livingston *et al.* (*511*), Elrod *et al.* (*252*), Knowles *et al.* (*458*), and Witt *et al.* (*890*) have all reported that the xanthophylls are destroyed to a lesser extent than the carotenes during storage. Nakaue *et al.* (*600*) found greater losses of xanthophylls than carotenes in feeds stored at room temperature for 4 weeks that contained ethoxyquin-protected alfalfa meal. Without EMQ, the losses for both were about equivalent (24 and 26%). Thompson and Maclay (*807*) reported on the results from nine samples of dehydrated alfalfa obtained from several different dehydrators. The carotenes and xanthophylls were determined initially and after storage at 77°F for 16 weeks. Losses of 42 and 76% were observed for the carotenes and losses of 25–54% for the xanthophylls. Elrod *et al.* (*252*) noted that under

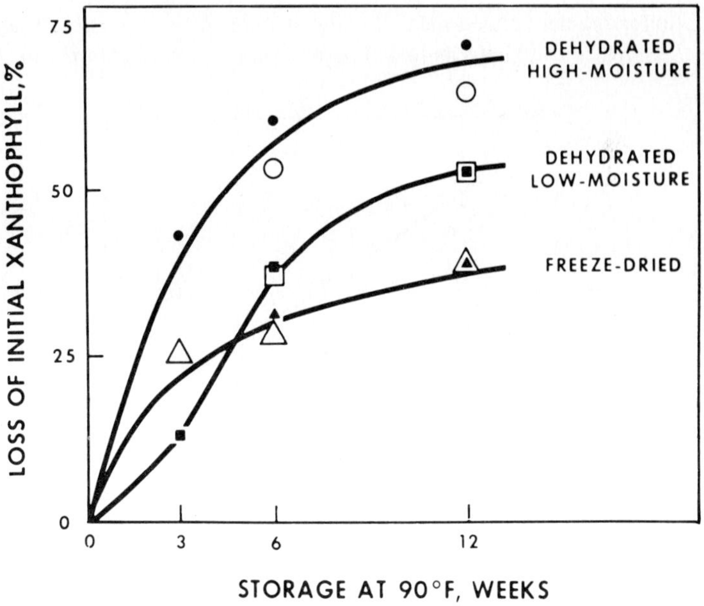

Fig. 13. Loss of xanthophylls during storage of dehydrated alfalfa meal. Key: Stearns–Roger dehydrator trials: ●, ▲, ■; Arnold dehydrator trials: ○, △, □. [From Knowles *et al.* (*458*).]

certain storage conditions about half the xanthophylls of ingredients in mixed feeds are lost after 6 months at room temperature. Fassler *et al.* (*263*) investigated the stability of the carotenes and oxycarotenoids in four different samples of formulated feeds. After 3 months storage at normal temperatures, losses of total carotenes plus oxycarotenoids ranged from 40 to 75%; losses of the pigmenting oxycarotenoids ranged from 17.5 to 56%. Feed ingredient losses are quite serious; however, since formulated feeds are consumed quickly after manufacture, long-term stability data are of less practical significance here.

Due to the excessive loss of carotenes coupled with isomerization losses and loss of xanthophylls, numerous attempts have been made to solve these problems. Inert gas storage has aided the retention of carotenes (*407*), as have treatment with an antioxidant, (*83,270,340,458,511,579,600,802,890*), and addition of animal fats (*82,580*). Typical stability data for dehydrated alfalfa meal obtained utilizing an antioxidant under controlled dehydration conditions including mosture content of the meal and temperature were presented by Knowles *et al.* (*458*) (Tables 16 and 17) and Witt *et al.* (*890*) (Fig. 12). Additional data are shown in Tables 18 and 19 and Fig. 14. Even when stabilized with ethoxyquin, losses for xanthophylls of 20 and 40% are obatined after 4 and 12 weeks of storage at 100°F, respectively. Problems still exist in this area. Knowles *et al.* (*458*) found that alfalfa meals that showed the lowest losses during dehydration suffered the highest losses during storage. Livingston *et al.* (*513*) called attention to the almost constant xanthophyll–carotene ratio of 2.1:1 in fresh alfalfa meal compared to widely varying ratios in the dehydrated meal. In nine samples of fresh meal, all rations fell within 2.1–2.3 parts xanthophylls to 1.0 part carotene. After dehydration, the ratio

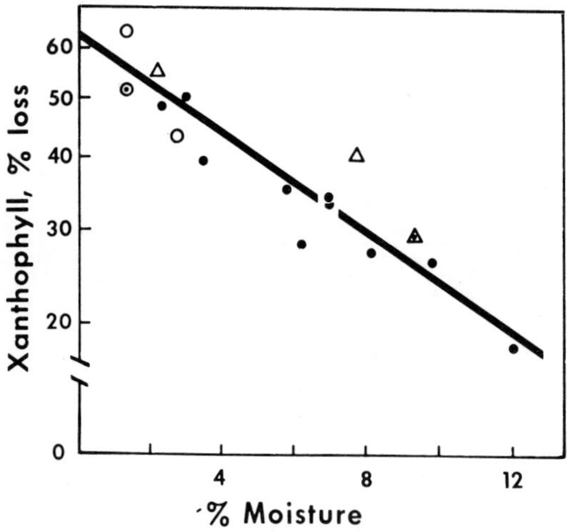

Fig. 14. Correlation of loss of xanthophylls with three dehydrators and moisture content of dehydrated alfalfa meal. Key: ●, Stearns-Roger; △, Arnold; ○, pilot Arnold dehydrator. [From Livingston *et al.* (*516*).]

was 0.9 to a high of 2.0 for xanthophylls to 1.0 for carotenes depending on the dehydration conditions. The authors stated that one must analyze each sample of dehydrated meal for its true oxycarotenoid content rather than assume a potency on the basis of its carotene content.

VII. PIGMENTATION SOURCES

A. Natural Feed Ingredients

The principal natural feed ingredient sources of oxycarotenoids utilized during the past several decades include yellow corn, corn gluten meal, and dehydrated alfalfa meal (*46,242,382,485,654,656,664,666,846,872*). Reliance on these natural ingredients to provide a sufficient amount of oxycarotenoids for adequate pigmentation of broilers and egg yolks is not always possible. Poultry rations are in a continuing state of change, as are other phases of agriculture. Many new ingredients and special oxycarotenoid concentrates have been conceived, developed, and evaluated. Of all the ingredients tested, a few have found a place in poultry rations in the United States to augment the oxycarotenoids of yellow corn, corn gluten meal, and dehydrated alfalfa meal. These include alga meals, marigold meals and concentrated extracts from marigold meals, and concentrated meals made from specialized alfalfa processing (Pro-xan process). These special concentrates are discussed below. Major problems in the use of natural ingredients for pigmentation include (a) variation in oxycarotenoid content, (b) bioavailability of the oxycarotenoids, (c) instability of the oxycarotenoids in the feed ingredient and finished feed, and (d) poor correlation between chemical assay of oxycarotenoid content and active biological performance.

1. Yellow Corn

Wide variations in oxycarotenoid content among strains, hybrids, and varieties of corn have been reported. Quackenbush *et al.* (*654*) compiled results on 125 corn inbreds and showed a range for oxycarotenoid content of from 1 to 15 mg/lb (2.2–33.0 μg/gm). They found that the provitamin A content of yellow inbreds also varied from a trace to 7.3 μg/gm. These workers noted that each inbred had its own characteristic carotenoid content and distribution. Rubin and Shillinger (*686*) analyzed 28 corn varieties and found the oxycarotenoid content to range from 20.0 to 48.6 μg/gm. The oxycarotenoid content of corn is controlled by genetic factors, and thus its total content or ratio of oxycarotenoid components can be altered, as reported by Brunson and Quackenbush (*127*), Ratcliff *et al.* (*665*), Blessin *et al.* (*92,93*), Quackenbush *et al.* (*654,656*), Grogan and Blessin (*335*), Grogan *et al.* (*336*), Watson (*846*), and Dua *et al.* (*242*). Ratcliff *et al.* (*664*) and Rubin and Shillinger (*686*) in studies on broilers, Day *et al.* (*209*) in studies on both broilers and yolk, and Hartung *et al.* (*364*) in studies on yolk showed that specially developed strains of corn high in oxycarotenoid content were equally as effective as regular corn in improving pigmentation; however, less of the high-oxycarotenoid strains was required to produce the same pigmentation effects. The corn oxycarotenoid profile as a typical composition has evolved over the years. Blessin *et al.* (*94*) in 1958 listed in order of percent occurrence lutein, zeaxanthin, and cryptoxanthin. In 1961, Quackenbush *et al.* (*655*) cited the principal oxycarotenoids of corn to be lutein (54%), zeaxanthin (23%), zeinoxanthin, and cryptoxanthin (8%). they also showed that different corn samples varied greatly in both the absolute and relative amounts of lutein and zeaxanthin (*655,656*). Ratios of 1:1 to 6:1 were found for lutein–zeaxanthin contents.

Grogan and Blessin (*335*) characterized the major carotenoids in 25 lines and crosses of corn, including polyoxycarotenoids, zeaxanthin, lutein, cryptoxanthin,

TABLE 20

Major Carotenoid Fractions of 25 Maize Samples Differing in Total Carotenoid Content[a]

Sample no.	Polyoxy		Zeaxanthin		Lutein		Cryptoxanthin		Zeinoxanthin		Carotenes		Total carotenoids (ppm)
	ppm	%	ppm	%	ppm	%	ppm	%	ppm	%	ppm	%	
50	0.4	7	0.8	13	1.9	31	0.8	13	0.7	11	1.5	25	6.1
49	1.9	12	3.3	21	4.3	28	2.4	16	0.7	5	2.8	18	15.4
41	1.4	7	7.3	35	7.9	38	1.6	7	0.7	3	2.1	10	21.0
11	1.8	7	10.6	43	8.1	33	2.1	9	0.4	2	1.5	6	24.5
35	1.8	6	12.9	47	6.7	24	2.2	8	0.5	2	3.6	13	27.7
34	3.2	11	18.5	63	2.2	8	2.5	8	0.5	2	2.4	8	29.3
42	2.2	7	13.3	44	6.2	20	4.1	14	0.7	2	3.9	13	30.4
16	2.0	7	15.3	50	4.1	13	4.5	15	1.0	3	3.8	12	30.7
37	1.9	6	16.1	50	4.4	14	5.6	17	0.8	2	3.5	11	32.3
44	3.6	11	12.5	39	7.7	24	4.0	12	1.1	3	3.5	11	32.4
8	1.8	6	9.5	29	12.6	38	2.3	7	2.7	8	3.9	12	32.8
38	2.0	6	17.7	54	4.1	12	4.8	15	0.7	2	3.7	11	33.0
9	1.9	6	18.1	55	4.3	13	4.1	12	0.5	1	4.3	13	33.2
32	2.4	7	16.9	51	6.2	18	3.6	11	0.7	2	3.6	11	33.4
14	2.9	9	17.7	52	2.7	8	6.9	20	0.6	2	3.0	9	33.8
31	1.5	4	13.7	40	9.8	29	4.2	12	2.0	6	3.2	9	34.4
17	2.5	7	20.6	56	5.7	15	5.0	14	0.8	2	2.2	6	36.8
36	1.9	5	23.8	64	3.0	8	4.5	12	0.4	1	3.5	10	37.1
48	1.7	4	24.6	65	3.9	10	3.8	10	0.7	2	3.3	9	38.0
7	3.9	10	16.9	44	8.2	21	5.3	14	0.7	2	3.7	9	38.7
24	2.8	7	15.3	38	12.4	31	4.3	11	1.6	4	3.7	9	40.1
23	3.6	8	21.8	48	8.6	19	4.4	10	2.8	6	4.3	9	45.5
22	3.5	8	21.4	45	10.4	22	4.9	10	2.5	5	4.8	10	47.5
13	2.3	5	33.1	67	2.9	6	6.2	13	0.6	1	4.0	8	49.1
12	2.8	6	33.1	67	3.7	8	5.2	10	0.7	1	4.0	8	49.5

[a] From Grogan and Blessin (*335*).

zeinoxanthin, and the carotenes. These are shown in Table 20. Thus, the oxycarotenoid profile is not a rigidly fixed entity but a highly variable property of corn strain and line genetics.

2. *Corn Gluten Meal*

The oxycarotenoid content of corn gluten meal is related to the protein content, with 41% protein meal averaging 60 mg/lb and 60% protein meal averaging about 120 mg/lb. It is also directly related to the oxycarotenoid content of the corn used in its manufacture, as shown by Watson (*846*) and Heiman (*382*). The average oxycarotenoid content and range of values found in the literature are summarized in Table 21 for corn, corn gluten meals, dehydrated alfalfa meals, as well as other sources. Livingston *et al.* (*518*) showed the oxycarotenoid profile for corn gluten meal to be comprised of lutein (53%), zeaxanthin (29%), cryptoxanthin (10%), and zeinoxanthin (5%). The oxycarotenoid profile of corn gluten meal, as determined by Livingston *et al.* (*518*), is in good agreement with that determined for whole ground corn by Quackenbush *et al.* (*655*). Lutein and zeaxanthin were shown to be good skin pigmenters by Quackenbush *et al.* (*656*), with cryptoxanthin less effective and zeinoxanthin least effective.

The oxycarotenoids of yellow corn have generally been reported to be more efficiently utilized than those contained in corn gluten meal and dehydrated alfalfa meal for both broiler skin and egg yolk pigmentation. Some workers have expressed this relative effectiveness qualitatively; other have attempted to quantitate the relationships. Day and Williams (*211*), Blessin *et al.* (*94*), Ratcliff *et al.* (*665*), Waldroup *et al.* (*844*), Rathmann and Daggy (*666*), Ratcliff *et al.* (*664*), Combs *et al.* (*159*), Combs and Nott (*161*) and Dua *et al.* (*242*) all reported that the corn xanthophylls were more effective than those in corn gluten meal and that, in turn,

TABLE 21

Oxycarotenoid Content of Feed Ingredient Pigmentation Sources

Ingredient	Average (mg/lb)	Range (mg/lb)
Yellow corn	8	4–14
Corn gluten meal (41% protein)	60	40–80
Corn gluten meal (60%)	120	80–180
Dehydrated alfalfa meal (17% protein)	110	70–130
Dehydrated alfalfa meal (20% protein)	140	90–170
Dehydrated alfalfa meal (22% protein)	170	120–220
Dehydrated alfalfa meal (25% protein)	220	160–240
Alfalfa concentrate (pro-xan) (45% protein)	350	200–450
Alga meal	1,000	600–2,000
Marigold meal	4,000	2,000–6,000
Marigold meal concentrate	23,000	20,000–25,000

the corn gluten meal xanthophylls were more effective than those in dehydrated alfalfa meal for broiler pigmentation where the three-way comparisons were made. Kuzmicky *et al.* (*485*) and Arnold (*46*) reported equivalence for the three feed ingredients, and Wilkinson and Barbee (*872*) found that corn gluten meal was slightly less effective (92%) than dehydrated alfalfa meal (arbitrarily assigned 100%). They did not compare these two sources with yellow corn. A wide range of relative availabilities has been reported over the years (Table 22). The earlier the work, generally, the lower the relative responses compared to corn. Many factors probably have influenced these values, especially as they pertain to dehydrated alfalfa meal, including specific oxycarotenoid profile, isomerization, instability, overestimation of oxycarotenoid content due to chemical assay methodology employed, and biological variation associated with the response parameters. Kohler *et al.* (*467*)

TABLE 22

Effectiveness of Oxycarotenoids in Corn Gluten Meal and Alfalfa Meal Relative to Yellow Corn for Pigmentation[a]

Corn gluten meal	Dehydrated alfalfa meal	Reference
Skin and shank		
36	36	Day and Williams (*211*)
75	75	Day and Williams (*211*)
50	36	Blessin *et al.* (*94*)
47	31	Ratcliff *et al.* (*665*)
—	50	Waldroup *et al.* (*844*)
74	53	Rathmann and Daggy (*666*)
—	74	Ratcliff *et al.* (*665*)
80	65	Combs *et al.* (*159*)
80	80	Day *et al.* (*209*)
80	65	Combs and Nott (*161*)
65[b]	60[b]	Dua *et al.* (*242*)
100	100	Kuzmicky *et al.* (*485*)
92	100[c]	Wilkinson and Barbee (*872*)
100	100[c]	Marusich (*541*)
Egg yolk		
—	61	Hughes and Payne (*415*)
86	77	Morehouse (*588*)
96	80	Morehouse and Hansen (*589*)
80	80	Day *et al.* (*209*)
—	58–69	Bartov and Bornstein (*56*)
92	88	De Groote (*221*)
95	80	Marusich and Wilgus (*558*)
80	60	De Groote (*225*)

[a] Yellow corn = 100%.
[b] Calculated from authors' data.
[c] Authors assumed 100%, not compared to yellow corn.

emphasized that these factors contribute to the wide range of availabilities reported for the oxycarotenoids in alfalfa meals. Marusich (*541*) reported essential equivalence, biologically, for the oxycarotenoids in corn gluten meal and dehydrated alfalfa meal in a limited collaborative trial, based on shank color and toe-web pigment content when specific chemical assay procedures were utilized and the rations were mixed frequently and refrigerated to ensure full oxycarotenoid potency. As Heiman (*382*) stated, it is difficult to explain why the biological availability of corn gluten meal is lower than that of the corn from which it is made. He specula ted that changes occur in the oxycarotenoids during processing such that the analytical method continues to show the expected values, but the chicken finds the pigments not fully available.

Lower relative values have also been obtained for dehydrated alfalfa meal based on the efficiency of oxycarotenoid deposition in egg yolk. Bartov and Bornstein (*56*) reported dehydrated alfalfa meal to be about 65% as effective as corn; Morehouse and Hanson (*589*) and Marusich and Wilgus (*558*) found alfalfa about 80% as effective as corn. In turn, Morehouse and Hanson (*589*) reported corn gluten meal to be more effective than corn (111%), and Marusich and Wilgus (*558*) gave a value of 96% for corn gluten meal relative to corn. Earlier, more qualitative comparisons by Gillam and Heilbron (*316*) and Peterson *et al.* (*632*) showed that the xanthophylls from yellow corn were better deposited in the egg yolks than were those from green plants when fed on an equivalent basis. The relative performance of corn, corn gluten meal, and dehydrated alfalfa meal oxycarotenoids for yolk pigmentation as reporded in the literature, has been summarized (Table 22).

3. *Alfalfa Meal*

The oxycarotenoid content of dehydrated alfalfa meals can also vary widely due to such factors as stage of maturity, dehydration conditions, storage conditions, and protein content of the meal. The oxycarotenoid content generally increases as the protein content increases. Meals with 17% protein average 100 mg/lb compared to 25% protein meals, which contain about 180 mg oxycarotenoids per pound. These properties of dehydrated alfalfa have been reviewed by Wilgus (*868*), Middendorf (*575*), Marusich and Wilgus (*558*), and Halloran (*347*). Zscheile *et al.* (*901*) obtained lutein by chromatography from alfalfa leaves in 1942. In 1954, Gugliemelli *et al.* (*341*) reported lutein to be the predominant oxycarotenoid in fresh alfalfa, with violaxanthin and zeaxanthin less abundant and cryptoxanthin, flavoxanthin, and neoxanthin constituting a minor portion of the total oxycarotenoids. In the same year, Bickoff *et al.* (*80*) observed that fresh alfalfa meal had five major bands (lutein, zeaxanthin, cryptoxanthin, violaxanthin, and neoxanthin), which comprised 99% of the total oxycarotenoids. In addition, seven minor bands were present in the remaining 1%. In dehydrated alfalfa meal, the same five pigments still comprised 87% of the total oxycarotenoids; however, more than 35 additional bands were now shown to be present due to the dehydration process. Bickoff *et al.* (*80*) listed 46% lutein, 4% zeaxanthin, 16% violaxanthin, 7% cryptoxanthin, and 14% neoxanthin in dehydrated alfalfa meal. Livingston *et al*. (*518*) noted that lutein comprised 75% of the total oxycarotenoids, with 4% zeaxanthin, 7% violaxanthin, 1% cryptoxanthin, and 3% neoxanthin, in dehydrated alfalfa meal. Livingston *et al*. (*516*), utilizing 20 separate samples, showed marked differences in the relative proportions of the principal xanthophylls of fresh and dehydrated samples. They also demonstrated that the dehydrated meal contained approximately 10% neoxanthin and 7–10% violaxanthin, with the remainder being all-*trans*-lutein along with lutein isomers I, II, and III (*514,516*). Kuzmicky *et al*. (*486*) found neoxanthin to be about 8% as effective as lutein and violaxanthin essentially ineffective as skin pigmenters. Marusich *et al*. (*548*) had previously reported violaxanthin to be ineffective as a yolk pigmenter. Lucerne (alfalfa) meal was fed at 6% for 10 weeks, and the various tissues were analyzed for total and specific oxycarotenoids by Grabowski (*329*); these values were compared to those found after feeding 7% grass meal. Shank skin, skin, subcutaneous fat, and visceral fat were assayed. Lutein was the predominant oxycarotenoid, followed by zeaxanthin. Violaxanthin was present in far smaller concentrations, and β-carotene had the lowest concentration.

Typical oxycarotenoid proviles for corn, corn gluten meal, and dehydrated alfalfa meal, along with other pigmentation sources, are shown in Table 23.

TABLE 23

Oxycarotenoid Profile of Feed Ingredients Used for Pigmentation

	Percent composition of oxycarotenoids[a]								
	Yellow corn	Corn gluten meal	Dehydrated alfalfa meal		Marigold meal		Alga meal		
Oxycarotenoid	*729*	*518*	*729*	*518*	*7*	*657*	*729*	*630*	*630*
Lutein	54	53.4	46	75.6	64	88.0	86	78.4	89.6
Zeaxanthin	23	29.3	4	4.4		3.6	2	5.1	
Violaxanthin[b]			16	7.0			4	8.1	1.4
Neoxanthin[c]			14	3.0		0.1			
Cryptoxanthin	8	10.0	7	1.0	4	0.6			
Antheraxanthin[d]					31	0.1			
Zeinoxanthin		5.1		1.6		0.7			
Polyoxyxanthophylls[e]		2.2		7.4					
Others	15		13		<1	6.9	8	8.4	9.0

[a] Italic numbers are reference numbers.
[b] Ineffective: Kuzmicky *et al. (486)*.
[c] 8% as effective as lutein: Kuzmicky *et al. (486)*.
[d] Efficacy unknown.
[e] Ineffective: Quackenbush *et al. (656)*.

4. Coastal Bermuda grass

Coastal Bermuda grass (*Cynodon dactylon*) is a hybrid characterized by a high leaf-to-stem ratio, rapid growth, fast regrowth after cutting, good drought resistance, and good response to fertilization. It is grown primarily in the southern and southeastern part of the United States, with about 6 million acres under cultivation (*872*). The product of commerce is generally a 17% protein dehydrated meal, which may be pelletized and then crumblized before addition to feeds. It may contain up to 200 mg total xanthophylls per pound of feed. It suffers from the same drawbacks that plague dehydrated alfalfa meals. The samples vary widely in oxycarotenoid content depending on time of cutting during the growing season, dehydration conditions, storage conditions and length of storage (*733*). It has local geographic utility for the poultry egg and broiler industry, as indicated by a number of trials (*206*). Studies by Davies *et al*. (*207*) indicate that dehydrated coastal Bermuda grass could be economically utilized in southern poultry rations with a ptoential market of 250,000 tons by 1985. The greater use of Bermuda grass in layer rations than in broiler rations is predicated on the layer's lower requirement for energy and amino acids and greater tolerance of fiber. Its major competitor is dehydrated alfalfa meal, and its levels of use are in line with those of 17% protein alfalfa meal. Wheller and Turk (*860*) and Wilkinson and Barbee (*872*) obtained essentially equivalent shank pigmentation in broilers with coastal Bermuda grass compared to dehydrated alfalfa. Barnett and Morgan (*54*) and Marusich and Bauernfeind (*546*) reported approximate equivalence for Bermuda grass and alfalfa meal in yolk pigmentation.

B. Special Concentrates

1. Oil Concentrates

Thompson and Maclay (*807*) prepared limited amounts of xanthophyll-in-oil concentrates from by-products in the processing of corn for zein and of dehydrated alfalfa meal for chlorophyll. Corn endosperm oil is prepared by continuous solvent extraction of corn gluten, which is a necessary step in the manufacture of zein. Corn endosperm oil (corn oil concentrate) is essentially the fat of corn gluten together with the xanthophylls and other carotenoids contained therein, which constitute the natural yellow coloring matter in corn. These commercial concentrates of xanthophylls have had limited usage in poultry feeds to augment the usual feed ingredients. Results obtained with these products for both broiler and yolk pigmentation have not been particularly satisfactory (*87, 146, 172, 211, 251, 262, 289, 290, 388, 583, 606, 665, 777, 877*). It is not used to any great extent in current feed formulations due to its poor stability and availability of better products.

2. Alga Meals

Alga meals have been extensively researched for both broiler and yolk pigmentation in the United States, primarily with two products obtained from two sources.

Perdue and Smith (*630*) and Smith and Perdue (*729*) studied the properties of alga meal from *Chlorella pyrenoidosa*, and Morehouse (*588*) and Morehouse and Hanson (*589*) studied those of alga meal from *Spongiococcum excentricum*. Fritz (*286,287*), Couch *et al*. (*172,179*) Madiedo and Sunde (*529*), Combs and Nicholson (*160*), Carlson *et al*. (*145*). Madiedo and Sunde (*529*), Madiedo *et al*. (*527*), and Marusich and Bauernfeind (*546,547*) all reported that alga meals are effective and about equivalent biologically to the oxycarotenoids in dehydrated alfalfa meals. One of the alga meals (*Spongiococcum excentricum*) was produced in commercial quantities and marketed in the United States for several years (20); however, it is no longer available. Alga meals are products of higher potency, containing from 1 to 2 gm oxycarotenoids per pound of meal, according to the data of Perdue and Smith (*630*). Smith and Perdue (*729*) showed that lutein is the major oxycarotenoid (86%) with about 2% zeaxanthin, 4% violaxanthin, and 8% classified as "others." The stability pattern of alga meal "as is" and in mixed feeds and its biological effectiveness essentially follow the properties of dehydrated alfalfa meals. Couch *et al*. (*172*) reported that alga meal produced a desirable egg yolk color indistinguishable from that obtained by feeding dehydrated alfalfa meal. It was also determined that the pigmenting compounds in alga meal carry over well into cakes, salad dressings, and other processed food products via the egg yolks from hens fed alga meal supplement. These relatively high potency alga meals could supply the necessary levels of oxycarotenoids when added at modest levels (0.25–2%).

In Europe, Lautner (*492*) reported on an alga source (*Scenedesmus quadricauda*) and found it to be highly effective for yolk pigmentation. Another alga source (*Streptomyces mediolani*) was also extensively evaluated by Olivetti and Meregalli (*613*), Auxilla and Vassura (*49*), and Quilici (*660*) for yolk pigmenting properties. By means of both the Roche Yolk Color Fan and colorimetric measurement of yolk carotenoid content, it was shown that alga meals are effective pigmenters when compared to several other feed ingredient oxycarotenoid sources, including Plata maize, yellow maize, Brazilian semiflinty maize, hybrid floury maize, or lucerne meal. Høie and Sannan (*408*) noted significant pigmenting effects on egg yolks with the use of other alga meals (seaweed meals) consisting of the species *Fucus serratus, Fucus vesiculosus, Ascorphyllum nodosum, Laminaria hyperborea,* and *Laminaria digitata*. When higher levels (10–15%) of these alga meals were used, the egg yolk colors became reddish, and the laying birds developed diarrhea, especially with *Laminaria digilata*. Olsson (*617*), Black (*90*), and Nelson and Baptist (*605*) observed little or no effect with their samples of seaweed meal.

Jensen (*436*) found that the addition of 10–15% alga meals made from the species *Ascorphyllum nodosum, Fucus vesiculosus,* and *Fucus serratus* to a pigment-deficient basic feed produced effective egg yolk pigmentation. Assays showed that the predominant carotenoids in these meals were violaxanthin and fucoxanthin. However, they were not deposited in the yolks as such but were partly converted to furanoid and other epoxidic derivatives of the original pigments. Another experiment reported from Australia showed that, of three different alga species, the species *Ulva lactuca* produced the most marked pigmentation. A level of 12% by weight in the feed was required, however, to produce the yolk color preferred by local housewives (*30*). The complexity of the oxycarotenoid profiles in some alga meals was shown by Jeffrey (*435*) with paper chromatography and Bunt (*133*) with thin-layer chromatographic techniques.

3. Marigold Meal and Concentrates

Marigold petals (Aztec marigold meal, *Tagetes erecta*) and whole marigold plants are dehydrated, ground, and marketed for use as both broiler and yolk pigmentation sources to augment yellow corn, corn gluten meal, and alfalfa meal. Many researchers have investigated marigold meals (*6, 7, 113–115, 164, 172, 179, 198, 199, 282a, 286, 350, 527, 540, 546, 547, 575a, 713, 776, 825, 843*). Generally, the performance of marigold meals is inferior to that of dehydrated alfalfa meals for both broiler and yolk pigmentation.

Marusich (*540*) reported on the results of a collaborative trial in which six out of seven participants found marigold meal to be less effective than dehydrated alfalfa meal (average of 50% availability), based on shank visual scores. Twining *et al.* (*825*) also found marigold meal oxycarotenoids to be less available. Six samples were assayed using skin pigmentation and plasma carotenoid concentration as the response parameters. The authors found the average availability for the six samples to be 50%, with a range of 43–65%, when compared to corn gluten meal and using 80% availability for the oxycarotenoids in the corn gluten meal. If corn gluten meal were assigned as arbitrary value of 100% availability, the six marigold samples averaged 58% availability, with a range of 39–76% (see Table 4, Figs. 10 and 11). Dansky (*199*) found alfalfa meals better than Ecuadorian marigold meal for broiler pigmentation. He also reported corn gluten meal to be equivalent to dehydrated alfalfa meal products containing 25 or 45% protein. Witt *et al.* (*889*) noted that the oxycarotenoids in marigold meal were only 50% as effective as those in an alfalfa concentrate (pro-xan). Streiff (*771*) reported that the combination of 80 mg oxycarotenoids from marigold meal plus 4 mg canthaxanthin per kilogram of broiler feed produced a deeper orange color than feeding 160 mg oxycarotenoids per kilogram from marigold meal alone. He also showed that 30 mg β-apo-8′-carotenoic acid ethyl ester (APO-EE) per kilogram is equivalent to 160 mg from marigold meal for comparable broiler pigmentation. In broilers, marigold meal concentrate derived by extraction from marigold petal meal was reported to be more effective than the equivalent oxycarotenoid levels supplied by marigold meal (*40*). Three commercial sources of esterified xanthophylls from *Tagetes* petals were compared by Middendorf *et al.* (*575a*) using the serum oxycarotenoid levels 18–24 hr after a single intubation (rapid bioassay method): ground petals, a concentrated extract, and an extract that was partially saponified. All three had the same biological value, which was about 70% that of alfalfa and 40% that of corn gluten meal. In another study with broilers, good correlations were obtained with shank color, as scored with the Roche Yolk Color Fan, breast skin color, as scored with the Purina

Skin Pigmentation Guide, and serum oxycarotenoid content when a commercial concentrate derived by extraction from the marigold petal meal (*38*) was used.

Marigold petal meal concentrates were fed to layers (*39*) in two trials. In one trial 45 gm of xanthophylls per ton of feed were added to a ration already containing 10 gm/ton from natural feed ingredients. This oxycarotenoid concentration produced eggs with yolks that had an NEPA number of 4–5 and a Roche Yolk Color Fan value of 13. In a second trial 45 gm of xanthophylls from the marigold concentrate were added to a ration containing 26 gm of mixed oxycarotenoids from natural feed ingredients. This level (71 gm total) produced yolks with an average NEPA number of 5 and contained 108 μg of oxycarotenoids per gram of yolk. Quantitative comparisons were made by Sullivan (*776*), Marusich and Bauernfeind (*546*), and Madiedo *et al.* (*527*) for marigold meal as a yolk pigmenter, with results paralleling those for dehydrated alfalfa meal. Scott *et al.* (*713*) compared β-apo-8′-carotenal (APO) with marigold meal for yolk pigmentation. When both were fed to provide 50 mg oxycarotenoids per kilogram of layer mash, APO-fed hens deposited 85 μg pigment per gram of yolk in contrast to 60 μg per gram of yolk for hens fed the marigold meal. In a series of egg yolk studies reported by Guenthner *et al.* (*339*) in which corn gluten meal, alfalfa meal, marigold meal, APO, and APO-EE were compared, the marigold product oxycarotenoids were least utilized and APO-EE was most efficiently utilized. McNaughton *et al.* (*568*) studied the performance of a concentrated extract derived from marigold meal in yolk pigmentation. They found the oxycarotenoids from the marigold meal concentrate to be least efficiently utilized when compared to the yolk pigmentation resulting from feeding corn gluten meal, APO-EE, and combinations of APO-EE and canthaxanthin.

Halloran (*348*) and Halloran *et al.* (*350*) evaluated marigold meal against both a 45% protein alfalfa concentrate and a 20% protein dehydrated alfalfa meal for yolk pigmentation efficacy. The marigold xanthophylls showed only 54% availability compared to those from the alfalfa products based on visual score of the yolks using the Roche Yolk Color Fan and 58% availability based on colorimetric determination of the pigment in the yolks (β-carotene equivalents per gram of yolk). Other trials by Halloran (*349*) with broilers demonstrated that the xanthophylls of marigold meal were only 52% biologically available compared to those in corn gluten meal based on relative visual scores at 8 weeks for breast feather tracts. Kuzmicky *et al.* (*487*) found the xanthophylls of marigold meal to be less effective than those of dehydrated alfalfa meal, alfalfa leaf protein concentrate (pro-xan), and freeze-dried pro-xan based on both broiler and egg yolk studies. They reported both pro-xan products to be 1.7 times more available then the xanthophylls in dehydrated alfalfa meal and 3 times those in marigold meal. The availability of the xanthophylls in dehydrated alfalfa meal was about half that of pure lutein and twice that of xanthophylls in marigold meal for the broiler. Similar availabilities were obtained with laying hens.

Another problem with marigold meal is the resolution of its oxycarotenoid profile. Kuhn *et al.* (*482*) indicated the presence of lutein in *Tagetes erecta*, but a complete analysis of the flower petal pigments was not conducted. Alam *et al.* (*6*) reported that the ground petals of *Tagetes erecta* contained 64% lutein, 31% antheraxanthin, and 3.5% α-cryptoxanthin. Antheraxanthin has not been reported to be present in any other major poultry feed ingredient, and no data could be found in the literature assessing its pigmenting efficacy for poultry. Antheraxanthin, an epoxide carotenoid, was reported to be present in certain strains of an alga (*Euglena gracilis*) by Krinsky (*472*), Krinsky and Goldsmith (*475*), and Krinsky *et al.* (*476*) and in fresh orange peel and tangerine pulp and peel by Curl and Bailey (*190,192*). Couch (*173*) reported the xanthophylls to be present in esterified form. Saponification of the marigold meal oxycarotenoids increased the pigmentation of egg yolks by approximately 30%. Thus, the meal as it would normally be added to the ration containing esterified xanthophylls would not be completely available. More recently, Quackenbush and Miller (*657*) reported on the oxycarotenoid profile of six different marigold flower petals including *Tagetes erecta*. They found that lutein comprised from 80 to over 90% of the total oxycarotenoids in all six samples. The remaining 10–20% of the oxycarotenoids present were represented by up to 17 identifiable oxycarotenoids, with no one in particular predominating. Antheraxanthin was one of the 17 but was present in a concentration of about 1% of the total. The lutein in marigold meal is present in ester form and may not be completely hydrolyzed by the chicken. Thus, this could explain the lesser availability of marigold meal oxycarotenoids when the chemically determined content is compared to biological performance (see Tables 4 and 24 and Figs. 10, 11, and 15). In contrast, the rapid bioassay procedure described by Middendorf *et al.* (*575a*) showed no significant differences in serum carotenoid levels and relative activity () when ground petals (41%), a concentrated extract (43%), and an extract that was partially saponified (43%) were compared. All were inferior to alfalfa (58%) and corn gluten meal (assigned 100%).

Stability data, presented by the manufacture of marigold concentrate (*41*), show an approximately 10% loss of potency in 24 days in mash when stored at 38°–40°C. Little has been published on the stability of marigold meal per se. Coon and Couch (*164*) reported a 28% loss of potency when the meal was stored for 10 months at 55°F. In general, it appears to have stability characteristics similar to those of stabilized dehydrated alfalfa meal.

4. *Alfalfa Concentrate*

A recent development has been the production of a high-protein, low-fiber, high-carotene, and high-oxycarotenoid product derived from alfalfa meal juice (pro-xan). Its method of manufacture from fresh alfalfa meal has been described by Bickoff (*78*), Bickoff *et al.* (*79*), Kohler (*462*), Kohler *et al.* (*465*), Lazar (*496*), Miller *et al.* (*576*), Knuckles *et al.* (*459,460*), and Spencer *et al.* (*734,735*). Its stability characteristics have been detailed by Witt *et al.* (*889–891*). Its nutritional properties have been described by de Fremery (*213*), Booth *et al.* (*104*), Kuzmicky *et al.* (*484*), and Kohler and Bickoff (*464*). Its pigmentation properties have been reviewed by Halloran (*347–349*), Knuckles *et al.* (*459*), Booth *et al.* (*104*), Arnold

TABLE 24

Summary of Visual Color Scores Obtained by Eight Collaborators in ANRC[a] Broiler Pigmentation Study[b]

Oxycarotenoid source	Level (gm/ton)	Visual color scores (Roche Yolk Color Fan, 1–15)								
		No. 1[c]	No. 2	No. 3	No. 4	No. 5	No. 6	No. 7[c,d]	No. 8[c,d]	Average
APO-EE	10	2.7	4.6	4.2	1.3	1.5	1.3	4.6	4.3	3.1
APO-EE	20	4.5	7.6	7.1	4.7	3.6	2.7	7.2	6.9	5.5
APO-EE	30	5.8	8.0	9.6	7.0	4.8	3.7	8.6	—	6.8
APO-EE	40	7.9	9.7	10.5	8.8	5.5	4.8	—	—	7.9
Yellow corn	10	2.8	4.2	4.3	1.7	1.6	1.5	3.5	3.7	2.9
Yellow corn + alfalfa (10 + 10)	20	4.5	5.9	6.6	3.7	3.4	3.3	6.0	5.2	4.8
Yellow corn + alfalfa + corn gluten meal (10 + 10 + 10)	30	5.0	6.3	7.3	4.5	3.7	3.8	7.0	—	5.3
Yellow corn + marigold meal (10 + 10)	20	3.6	—	5.4	4.0	1.6	2.4	—	—	3.5
Yellow corn + canthaxanthin (10 + 2.5)	12.5	4.2	6.2	7.0	5.8	2.5	2.9	—	—	4.8
Yellow corn + alfalfa + canthaxanthin (10 + 10 + 2.5)	22.5	6.0	8.1	8.0	8.3	3.3	4.0	—	—	6.3

[a] Animal Nutrition Research Council.
[b] From Marusich (*540*).
[c] Results of 0 to 4-week study; all others fed supplements for last 3 weeks (35–56 days).
[d] Results for alfalfa and corn gluten meal combined.

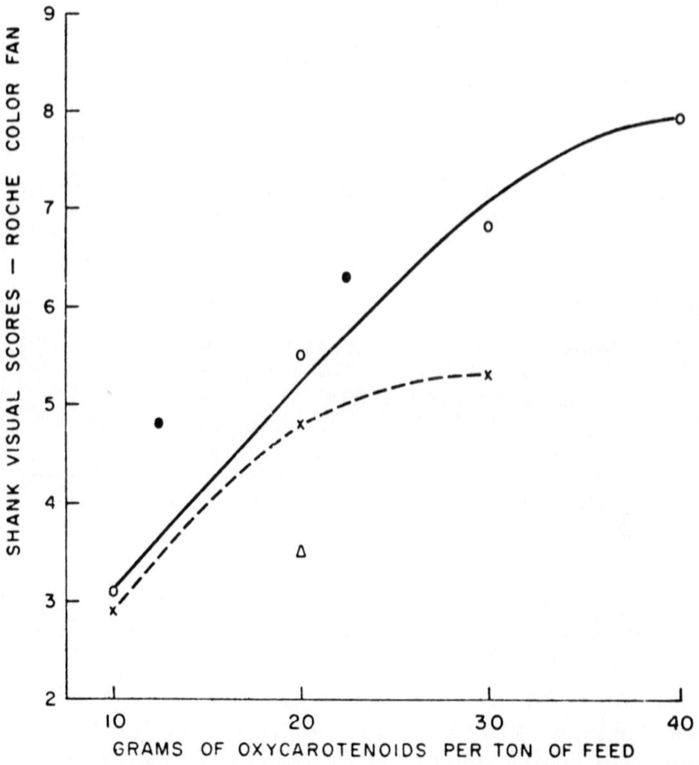

Fig. 15. Relative effectiveness of several oxycarotenoid sources based on shank pigmentation of broilers; composite response for eight collaborators. Key: x, natural ingredients (corn, alfalfa, corn gluten meal); ○, APO-EE; △, marigold meal; ●, natural ingredients plus canthaxanthin. [From Marusich (*540*).]

(*46*), and Witt *et al.* (*891*) and in a technical brochure (*894*) distributed by the manufacturer of the commercial product (X-Pro). The product contains an average of 45% protein, 400 mg oxycarotenoids per pound and 260 mg carotene per pound. Its stability characteristics (Table 25A,B) and biological efficacy are similar to those of representative samples of high-quality 20 or 22% protein dehydrated alfalfa meal. Dansky (*198,199*) found it comparable to 25% protein dehydrated alfalfa meal and corn gluten meal for broiler pigmentation.

Arnold (*46*) compared pro-xan with 25% protein dehydrated alfalfa meal, corn gluten meal, and Ecuadorian marigold petals in broilers. On the basis of pigmentation of the breast feather tract determined with the Hunterlab D-20 Color Difference Meter, corn gluten meal was more effective than the marigold petals, but the difference among chickens treated with corn gluten meal, 25% protein dehydrated alfalfa meal, and pro-xan was not statistically significant. Halloran (*348*) and Halloran *et al.* (*350*) found the xanthophylls in pro-xan to be comparable to those in 20%

TABLE 25A

Stability of Xanthophylls in Air-Dried pro-xan, Drum-Dried pro-xan, and Dehydrated Alfalfa Meal[a] Stored at 100°F for 4 Weeks[b]

Sample	Initial xanthophylls (mg/lb)	Xanthophylls retained (%)	
		Untreated	Treated[c]
Air-dried pro-xan	440	78	—
	317	84	90
	396	86	96
Drum-dried pro-xan	473	70	79
	294	56	79
Dehydrated alfalfa meal	178	56	79
	171	61	85
	138	60	86

[a] Stored in open containers.
[b] From Witt *et al.* (*891*).
[c] Samples treated with 0.125% ethoxyquin.

protein dehydrated alfalfa meal for yolk pigmentation and equal to those in 19% protein meal for broiler pigmentation. The oxycarotenoids in pro-xan were shown by Halloran (*349*) to be 91% as available as alfalfa meal xanthophylls on the basis of visual examination of foot pads and 104% on the basis of toe-web content of BCE's in broilers. These values should be considered equal to 100% availability when related to alfalfa meal.

TABLE 25B

Stability of Xanthophylls in pro-xan Drum-Dried to Different Moisture Contents and Stored at 100°F[a,b]

Sample[c] moisture (%)	Xanthophylls retained (%)			
	4 weeks		12 weeks	
	Untreated	Treated[d]	Untreated	Treated[d]
7	69	79	42	68
9	70	79	46	69
11	69	78	47	68
13	70	77	48	68

[a] From Witt *et al.* (*891*).
[b] Stored in open containers.
[c] Samples prepared from unwashed concentrate.
[d] Samples treated with 0.125% ethoxyquin.

Guenthner *et al.* (*339*) found no consistent differences in utilization efficiencies among corn gluten meal, 20% protein dehydrated alfalfa meal, and 40% protein alfalfa concentrate (pro-xan) when evaluated for egg yolk pigmentation in both caged and floor-pen-reared layers.

The availability of the oxycarotenoids in pro-xan was evaluated in two trials with broilers and one trial with layers by Kuzmicky *et al.* (*487*). The pro-xan products were shown to have oxycarotenoid availabilities about 1.7 times that of dehydrated alfalfa meal and 3 times that of marigold meal. Comparable results were obtained in the layer trial.

5. *Paprika Products*

The fruits of plants of the genus *Capsicum* (pimiento, chili, paprika) have been researched for their yolk and broiler skin pigmenting capabilities. The active components in these sources are capsanthin ($C_{40}H_{56}O_3$) and capsorubin ($C_{40}H_{56}O_4$). Morgan and Woodroof (*593*) fed ground, dried pimiento peppers, Barry (*77*) fed chili powder, and Benedek (*71*) fed paprika to laying hens. All reported yolks to have a reddish tinge. Brown (*122*) showed that capsanthin was the carotenoid in pimiento peppers, and he (*123*) found capsanthin in the yolks after feeding red peppers. However, it was absorbed by the hen and deposited in the yolk with a transfer efficiency of only about 1%. Titus *et al.* (*812*) and Fritz and Wharton (*289*) also noted that paprika extracts were inefficiently deposited in the yolk and imparted a red color. Mackay *et al.* (*524*) reported that paprika extract produced both acceptable and reddish-tinge yolks in relation to the levels fed. The latter workers reported that sponge cake made with the reddish yolks had acceptable yellow color with red overtones. In contrast, Carlson *et al.* (*145*) reported acceptable yolk color with the feeding of paprika; however, mayonnaise produced with these yolks did not have the degree of yellow color desired but exhibited a pink tinge. Although the paprika extract imparts an intense color to yolks and would appear to produce yolks acceptable for the processed food industry, the pigment is masked in mayonnaise and sponge cakes made from these yolks and actually appears as a pinkish hue, which is undesirable according to Carlson *et al.* (*145*) and Deethardt *et al.* (*212*).

Generally, paprika has been found to produce a reddish yolk color, as reported by Rauch (*671*), Czernicki and Weiser (*194*), Scott *et al.* (*713*), Tiews and Zucker (*811*), Wildfeuer *et al.* (*866*), and Streiff (*768*). Tortuero (*815*) concluded that the red xanthophylls, capsanthin and capsorubin, obtained by extraction of paprika could be used for yolk pigmentation in combination with the oxycarotenoids in dehydrated alfalfa meal. Several other workers have reported similar effects of blending the red carotenoids with the yellow carotenoids, including Scott *et al.* (*713*), De Groote and Reyntens (*226*), De Groote (*215,222*), Haertel (*361*), Scholtyssek *et al.* (*706*), and Poppe (*642*).

Streiff (*765*) reported that a commercially available paprika concentrate (oxycarotenoid concentration 10-fold higher than in paprika per se) had no value in broiler pigmentation. A similar sample was evaluated for yolk pigmentation by

Marusich (*537*) and found to be far inferior to β-apo-8′-carotenal. Splittgerber *et al.* (*736*) replaced 0.5% alfalfa meal with 0.5% Hungarian feed paprika, which assayed 2.29 gm capsanthin per kilogram, and observed no improvement in broiler pigmentation. Using the same source and level of paprika, the authors observed a positive increase in yolk pigmentation. Streiff (*766*) also reported on work comparing paprika to canthaxanthin for yolk pigmentation. The addition of 300 gm of paprika per ton of layer feed had no discernible effect on yolk color. Supplementation iwth 5.4 kg of paprika per ton of feed produced an increase in color comparable to the addition of 1 gm of canthaxanthin per ton of feed.

6. *Annatto Products*

Bixin ($C_{25}H_{30}O_4$), the principal oxycarotenoid in the seed obtained from the annatto tree (*Bixa orellana*), has also been investigated as a pigmentation source. Palmer and Kempster (*623*) noted that annatto did not influence the color of the adipose tissue of fowl. Landagora (*490*) stated that in Puerto Rico deeply pigmented yolks are preferred; thus, it was of interest to determine the effect of adding achiote seeds (annatto seeds) to layer rations. When 1% ground achiote seed meal was added to the feed or drinking water of layers, no change in yolk color was observed. However, when 2% whole seeds were sprinkled on the feed, increased yolk color was observed by the third day of feeding, which increased with time until a plateau level of deeper color was sustained as long as the bixin source was fed.

Landagora and Montalvo (*491*) reported that the addition of 0.5% whole achiote seeds to broiler rations resulted in brighter yellow to orange color in beaks, shanks, body fat, skin, and base of the feathers compared to nonsupplemented controls. They concluded that the seeds could serve as a source of pigmentation where the seeds are grown locally. Prohaszka (*648*) fed 10 or 100 mg carotenoids from annato seeds (ground) to White Leghorn hens and noted that the 10 mg/kg level had no effect on yolk color and the 100 mg/kg was slightly effective. The intensity of color of the yolks was less than that obtained with 5 mg of an effective oxycarotenoid (β-apo-8′-carotenal). The latter pigment source is discussed in Section VII, D, 1.

A newer bixin-type product has been developed in Brazil as a by-product of the annato extraction process. It contains bixin either alone or in combination with other oxycarotenoids, such as capsanthin, lutein, and/or zeaxanthin. It is called cortegg. It was reported to be effective in improving egg yolk pigmentation by Silveira and Kronka (*721*). In broiler trials, Silveira *et al* (*722*) compared cortegg to another native product (protenose) for improved skin pigmentation. Protenose was better than cortegg, and the responses to cortegg were quite variable.

7. *Tomato Products*

Lycopene ($C_{40}H_{56}$) is not an oxycarotenoid but a hydrocarbon carotene (acyclic carotene). It is a common component of vegetables (tomatoes, carrots, green peppers) and fruit (pink citrus, apricots, watermelons) (*186,187,191,249,592,702*). *Suarez* (*774*) reported that lycopene fed in the form of dried crushed tomatoes had variable effects on yolk color and concluded that lycopene is metabolized to some other carotenoid and then deposited in the yolk. Agrawal and Talapatra (*5*) observed that 20% tomato pomace (another lycopene source) imparted a brilliant reddish-yellow color to egg yolks comparable to that obtained with corn.

8. Crustacean Products

Astaxanthin ($C_{40}H_{52}O_4$) is 3,3′-dihydroxy-4,4′-diketo-β-carotene, or 3,3′-dihydroxycanthaxanthin. It is a common pigment of crustaceans (shrimp, crabs, crayfish, and lobsters), algae, fish, and oranges (*64a,94,150,190,231,443,447, 448,607,608*). The results of Wald and Zussmann (*842*) support the findings that the hen can accumulate astaxanthin in the yolk as a result of feeding on crab and lobster shells. The authors noted that these dietary additives caused yolks to be deeply pigmented, which was not acceptable in the marketplace. Shrimp meal oxycarotenoids were reported by Chawan and Gerry (*149*) to be valuable as a complementary source of pigment in broiler feeds when ground wheat replaced part of the yellow corn. The shrimp meal must be fed with a source of yellow xanthophylls. Crustacean meals derived from processing plant wastes (shells, head, viscera, etc.) may become a limited leed ingredient for poultry in areas of close proximity to the plants, especially with the development of "aquacultures" or marine life farm ponds (*573–574*).

Meyers (*573a*) reviewed crustacean meals and their use in feed formulations. He stated that shrimp meal has long been used by fish nutritionists in diets to lend desired flesh coloration to trout and salmon. It is not unusual for as much as 15% shimp meal to be used in pond trout formulations. The red or yellow color of the fins, skin, and flesh of wild rainbow trout as well as other kinds of salmonid fish is due to the carotenoids. Those that occur most commonly are astaxanthin, canthaxanthin, lutein, and, to a lesser extent, β-carotene.

Meyers (*573a*) listed pigment concentration (in micrograms per gram) for shrimp meal from brown shrimp, *Penaeus aztecus*, as 10; for shrimp meal from white shrimp, *Penaeus setiferus*, as 6; for sun-dried shrimp meal as 2–3; and for vacuum-dried shrimp meal, *Pandalus borealis*, as 104.

Leuenberger (*502*) developed a procedure for isolating astaxanthin from crabs, crab eggs, salmon eggs, and lobster eggs. About 400 mg of crystalline astaxanthin per kilogram (400 ppm) could be isolated from lobster eggs. Lambertsen and Braekkan (*489*) developed a chromotographic procedure sensitive to 0.1 μg/gm for astaxanthin that is applicable to crustacean samples, fish oils, and organs of rainbow trout.

C. Single Oxycarotenoids

The majority of research on avian carotenoids has been with "total oxycarotenoids" or "total xanthophylls" as assayed or estimated in feed ingredients or with very crude extracts and concentrates. Some research has been carried out with pure oxycarotenoids, obtained mainly by extraction, concentration, and purification from feed ingredient sources.

Lutein and zeaxanthin are two isomeric dihydroxycarotenoids having the empirical formula $C_{40}H_{51}(OH)_2$. Lutein is 3,3′-dihydroxy-α-carotene, and zeaxanthin is 3,3′-dihydroxy-β-carotene. Lutein and zeaxanthin have been isolated from similar sources since they quite frequently occur together in nature, with lutein generally the predominant oxycarotenoid and zeaxanthin occurring to a lesser extent [Quackenbush *et al.* (*655*) for yellow corn and Livingston *et al.* (*518*) for corn gluten meal]. Table 23 shows the relative proportions of each in the common feed ingredients used for pigmentation.

1. Lutein

Lutein is a constant component of milk and butter (*317*), where it varies in concentration depending on the amount of green feeds fed. It has also been found in the serum of cattle (*314*) but not in the body fat of cattle. It has been identified in chicken fat by Zechmeister and Tuzson (*899*) and in the feathers of the wild Madiera canary (*Serinus canarius*) and of the yellowhammer (*Emberizia citrinella*) by Brockmann and Völker (*120*). Titus *et al.* (*812*) worked with carefully purified leaf xanthophyll (lutein) and obtained yolk color comparable to that obtained by feeding graded levels of yellow corn.

Nelson (*604*) administered graded levels of lutein obtained by acetone extraction from a crude concentrate to Japanese quail. The pigment content of the yolks was determined as β-carotene equivalents per gram of yolk and showed a straight-line function when plotted against the dose. In turn, the yolks, when visually scored using the Roche Yolk Color Fan, showed a straight-line function when plotted against the logarithm of the dose fed.

Lutein–fatty acid esters were reported by Philip *et al.* (*636*) to be better utilized by laying hens than crystalline lutein. The higher solubility of the fatty acid esters in lipids compared to lutein may be responsible for their increased utilization. The egg yolk oxycarotenoid from different diets containing lutein–fatty acid esters was free lutein, indicating that the laying hen deacylated the lutein–fatty acid esters before deposition into the yolk.

Williams *et al.* (*876*) presented both yolk deposition and skin deposition data after the administration of pure lutein, zeaxanthin, and β-carotene. Lutein and zeaxanthin were equally well utilized by laying hens and chicks when fed at comparable levels, with both being equally deposited in the serum, egg yolk, liver, and adipose tissue of layers (Table 1) and in the serum, toe-web skin, liver, and adipose tissue of chickens (Table 26). β-Carotene was found in all tissues examined but in only trace amounts. In a comparison of the skin pigmenting efficiency of isolated oxycarotenoids during a 5-week feeding period with male chickens, Quackenbush *et al.* (*656*) found lutein and zeaxanthin to be superior to all other pigments tested. Although lutein was deposited to the same extent as zeaxanthin, the depth of color produced was slightly less for lutein than for zeaxanthin. Cryptoxanthin was less

TABLE 26

Deposition of Carotenoids in Blood Serum, Abdominal Adipose Tissue, Toe-Web Skin, and Liver of Chicks Fed Purified Carotenoids[a]

Supplement to basal ration	Carotenoid concentration			
	Serum (μg/ml)	Adipose tissue (μg/gm)	Liver (μg/gm)	Toe-web (μg/100 cm^2)
None	Trace	Trace	Trace	Trace
β-Carotene, 10 mg/lb	0.01	0.07	Trace	0.45
Zeaxanthin, 10 mg/lb	4.51	1.14	1.64	10.77
Lutein, 10 mg/lb	5.10	1.20	1.75	12.12

[a] From Williams *et al.* (*876*).

effective than either lutein or zeaxanthin but about twice as effective as zeinoxanthin. The hydrocarbon, β-carotene, was ineffective.

Fritz *et al.* (*290*) reported on feeding purified lutein to broilers over a range of 0–100 mg/lb and found 12.4 mg/lb to impart satisfactory color. Lutein levels of 25–40 mg/lb gave maximal color to the shanks. Sullivan and Holleman (*777*) worked with purified xanthophylls in concentrate form but did not obtain satisfactory yolk pigmentation compared to graded levels of alfalfa meal. Instability of the concentrate was probably a factor. Kuzmicky *et al.* (*485*) studied the effect of pure lutein (obtained from alfalfa meal) on broiler skin pigmentation compared to solvent extracts of mixed xanthophylls from alfalfa meal as compared to yellow corn, corn gluten meal, and dehydrated alfalfa meal per se. They reported that the three feed ingredient oxycarotenoid sources were equal, the solvent-extracted mixed oxycarotenoids were better utilized, and the pure lutein was most efficiently utilized. Some of these data are shown in Table 27. They further emphasized that the lesser absorption of the oxycarotenoids from the dehydrated alfalfa meal compared to the results obtained with the solvent extracts or pure lutein was due to the lesser availability of the xanthophylls contained in the intact meal. Kohler *et al.* (*467*) had shown that pelleting of alfalfa followed by regrinding produced an increase in broiler pigmentation. The pelletizing process and regrinding affect the integrity of the cell walls containing the oxycarotenoids, facilitating their absorption through the intestinal tract.

2. *Zeaxanthin*

Zeaxanthin has also been isolated and identified from the feathers of the wild canary *Serinus canarius* (*120*) and from the yolk of hens' eggs (*442,482,760*). It is widespread in higher plants (*128,354,355,603*), in blue-green algae (*1,396,684*),

TABLE 27

Broiler Pigmentation (Colorimetric Assay of Toe-Web Skin) from Several Oxycarotenoid Sources Fed to Chicks from 2 to 4 Weeks of Age[a,b]

Supplement	Oxycarotenoid level (mg/kg ration)			Relative skin pigmentation potency[c]	Range of potency (p=0.05)
	6.6	13.2	26.4		
Alfalfa meal	51.8	75.8	104.8	1.00	
Corn gluten meal	50.2	68.6	104.3	0.96	0.83–1.09
Pure crystalline lutein extracted from alfalfa meal	78.0	113.7	200.7	2.17	1.95–2.43
Mixed oxycarotenoids extracted from alfalfa meal[d]					
88:10:2	66.3	104.8	168.4	1.79	1.61–2.01
84:10:5	63.0	93.7	163.3	1.67	1.50–1.88
80:10:10	59.7	99.8	158.3	1.65	1.47–1.85

[a] From Kuzmicky *et al.* (*485*).
[b] Each value (micrograms carotenoids per 100 cm^2 toe-web area) is an average of four replicates.
[c] Calculated by slope-ratio assay.
[d] Numbers represent ratio of hexane–acetone–methanol used to elute from column.

and in certain bacterial species (*2,180,647,848a*). Zeaxanthin was obtained by Strain (*760*) and Euler and Gard (*256*) in pure form from egg yolk by chromatographic methods. Brockman and Völker (*120*) showed that purified lutein and zeaxanthin were deposited in egg yolks to a greater extent than cryptoxanthin and carotene. Peterson *et al.* (*632*) reported that purified samples of zeaxanthin were deposited in egg yolk approximately twice as efficiently as purified lutein.

Crystalline zeaxanthin was dissolved in acetone, cottonseed oil with antioxidants (BHA, BHT, EMQ) was added, the acetone was evaporated under CO_2 at 40°C, and the resulting oil solution of zeaxanthin was evaluated as an egg yolk pigmenter by Marusich *et al.* (*548*). This procedure prevented destruction and stereoisomerization. The oil solutions of zeaxanthin were added daily to the feed and fed daily. Relative efficiency was measured by visual scoring of yolk color and by colorimetric assay of the total pigment in the yolk. Zeaxanthin was found to be a highly effective oxycarotenoid providing good yolk color. Weiser and Manz (*851*) reported zeaxanthin to be highly effective for egg yolk pigmentation when fed with lutein alone or in combination with canthaxanthin.

Norman *et al.* (*610,611*) found crystalline zeaxanthin to be rapidly absorbed and transferred to the plasma of laying hens (*611*) and very efficiently transferred to the yolk (*610*) after daily oral administration.

Zeaxanthin in stabilized gelatin beadlet form was evaluated by Marusich *et al.* for both egg yolk (*554*) and broiler (*555*) pigmentation. In the egg yolk trial, a 1%

zeaxanthin stabilized beadlet was compared to APO-EE, marigold meal, and a concentrated extract of marigold meal. All were fed in a low-pigment ration at graded concentrations (0.5, 1.0, and 2.0 mg oxycarotenoids per hen daily) for 20 days. Eggs were collected twice daily for 40 days (5 days before and for 15 days after the 20-day supplementation period). All yolks were visually scored with the Roche Yolk Color Fan and an aliquot taken for colorimetry. The results are summarized in Table 28 and show zeaxanthin to be an excellent yolk pigmenter on the basis of both the colorimetric assay data and the visual score. The efficiency of oxycarotenoid transfer from feed to yolk was somewhat lower for zeaxanthin than for APO-EE but was markedly higher than for the marigold meal and the concentrated extract from the marigold meal.

Zeaxanthin in a stabilized gelatin beadlet formulation was evaluated as a broiler pigmenter in two trials (*555*). In the first experiment, low-oxycarotenoid basal broiler starter and grower rations were fed continuously for 7 weeks with zeaxanthin added to both of the rations at concentrations of 11, 22, and 33 mg/kg. Zeaxanthin was compared to three other pigmentation sources: APO-EE, dehydrated marigold petal meal, and a concentrated extract from marigold petal meal. The birds from both experiments were processed under simulated commercial conditions. Shank color was visually scored by three or four persons using the Roche Yolk Color Fan, and breast feather tract pigmentation was measured using the Purina Skin Pigmentation Guide. The shank and breast skin visual scores of the broilers fed zeaxanthin were markedly higher than those of birds fed APO-EE and the two marigold petal meal products over the entire 11–33 mg/kg feeding range. These data are shown in Table 29.

In the second experiment, zeaxanthin was added at a concentration of 2.75–11 mg/kg to broiler grower rations containing <1.0, 5.5, 11, 16.5, or 22 mg oxycarotenoids per kilogram from feed ingredients and fed from the twenty-ninth through the forty-ninth day. Zeaxanthin proved to be highly effective, as indicated by the visual scores from the shank, foot pad, and breast feather tract skin. The data, summarized in Table 30 and plotted in Fig. 16, suggest that 1 gm of zeaxanthin can replace an average of up to 2 gm of mixed oxycarotenoids from feed ingredients. The deep yellow to yellow-orange color associated with zeaxanthin as contrasted with the light yellow of lutein, which is the predominant oxycarotenoid of feed ingredients and marigold meal products, accounts for its greater potency (*349,519,713*). Thus, these trials (*554,555*) show that zeaxanthin in stabilized form imparts a very desirable color over a wide range of feeding concentrations when added to rations (layer, broiler) also varying widely in oxycarotenoid content. For broilers, it is efficacious when fed continuously for only the last several weeks before marketing.

Weiser and Manz (*854,855*) reported zeaxanthin to be highly effective as a broiler skin pigmenter, with activity slightly better than that of APO-EE. Like vitamin A and β-carotene, the oxycarotenoids are all prone to oxidative deterioration, as discussed earlier in relation to the oxycarotenoids from natural ingredient

TABLE 28

Relative Yolk Pigmentation Properties (Colorimetric and Visual) of Four Oxycarotenoid Sources Fed for 20 Days to Laying Hens[a]

	Oxycarotenoid content			Colorimetric assay data[b]				
		Level fed		Plateau area (avg. 11–20 days)		Total for 1–35 days		
Oxycarotenoid source added to low-pigment basal ration	Potency (mg/gm)	(mg/hen per day)	(mg/kg of feed)	μg BCE per whole yolk	(%)[c] Deposition	μg BCE for all yolks	(%)[d] Deposition	Visual scores, plateau area (Roche Yolk Color Fan)
None				50		600		<1.0
Zeaxanthin stabilized beadlets	10.5	0.5	5	230	46.0	4,802	48.0	9.0
		1.0	10	669	66.9	9,852	49.3	11.6
		2.0	20	1,108	55.4	16,722	41.3	12.3
β-Apo-8′-carotenoic acid ethyl	84.0	0.5	5	329	65.8	5,805	58.0	11.2
ester stabilized beadlets		1.0	10	675	67.5	11,054	56.3	12.7
(Carophyll-Yellow)		2.0	20	1,430	71.5	22,631	55.6	14.4
Mexican marigold meal—	5.4	0.5	5	169	33.8	2,870	28.7	6.9
ground *Tagetes* petals		1.0	10	397	39.7	6,302	31.5	8.9
(Florafil)		2.0	20	552	27.6	9,321	23.3	10.2
Concentrated extract of dried	47.4	0.5	5	265	53.0	3,579	35.8	7.7
ground *Tagetes* petals		1.0	10	377	37.7	6,549	32.7	8.9
(Gold'n Bloom)		2.0	20	610	30.5	11,320	28.3	10.7

[a] From Marusich *et al.* (*554*).

[b] Corrected for micrograms of pigment found in yolks from hens maintained continuously on the low-pigment basal ration alone.

[c] Average micrograms of pigment per yolk in plateau area divided by daily hen oxycarotenoid intake and multiplied by 100. This value would be representative of yolks produced when supplements were fed continuously.

[d] Total micrograms of pigment in all yolks laid over entire 35-day period divided by total milligrams of oxycarotenoids fed per hen for 20 days multiplied by 100 is the percentage of total dose accounted for in all yolks.

TABLE 29

Shank and Breast Feather Tract Visual Scores and Relative Potencies of Four Oxycarotenoid Sources Fed to Broiler Chickens for 7 Weeks[a]

				Potency and range based on			
		Visual scores[b]		Shank		Breast	
Oxycarotenoid source	Level (mg/kg of feed)	Shank[c]	Breast feather tract[d]	Relative potency	95% confidence interval	Relative potency	95% confidence interval
None (low-pigment)	<1.0	<1	<1				
β-Apo-8′-carotenoic acid ethyl	11	2.4	3.7				
ester stabilized beadlets[e]	22	6.3	7.4	1.00		1.00	
	33	8.6	8.8				
Zeaxanthin stabilized beadlets[f]	11	8.4	10.0				
	22	11.4	12.1	1.856	1.380–2.892	2.019	1.408–3.781
	33	11.8	12.2				
Dehydrated marigold petal	11	2.2	3.8				
meal[g]	22	4.6	6.8	0.625	0.253–1.008	0.963	0.532–1.689
	33	5.9	8.8				
Concentrated extract of mari-	11	1.9	3.8				
gold meal[h]	22	4.0	6.3	0.641	0.272–1.028	1.025	0.595–1.801
	33	6.6	9.5				

[a] From Marusich *et al.* (555).
[b] Average values for 20 broilers scored by three people.
[c] Roche Yolk Color Fan, 1965 edition (1–15); the higher the number, the deeper the color.
[d] Purina Skin Pigmentation Guide (0–13); the higher the number, the deeper the color.
[e] Contained 84.0 mg APO-EE per gram beadlets.
[f] Contained 10.5 mg zeaxanthin per gram beadlets.
[g] Contained 5.4 mg oxycarotenoids per gram.
[h] Contained 47.4 mg oxycarotenoids per gram.

TABLE 30

Growth, Feed Conversion, Oxycarotenoid Intake, Shank, Foot Pad, and Breast Feather Tract Visual Scores of Broiler Chickens Fed Five Basal Rations for 7 Weeks and Supplemented with Zeaxanthin for Last 3 Weeks[a]

Modification of basal ration feed ingredients	Oxycarotenoid content (mg/kg feed): Feed ingredients	Oxycarotenoid content (mg/kg feed): Zeaxanthin	Weight gain (gm)	Feed intake (gm)	Feed conversion	Oxycarotenoid intake (mg): 1–28 days, feed ingredients	Oxycarotenoid intake (mg): 29–49 days, Feed ingredients	Oxycarotenoid intake (mg): 29–49 days, Zeaxanthin	Visual scores[b]: Shank[c]	Visual scores[b]: Foot pad[c]	Visual scores[b]: Breast feather tract[d]	Relative potency of zeaxanthin to feed ingredient oxycarotenoids: Shank	Relative potency: Foot pad	Relative potency: Breast feather tract
Low-pigment milo	<1	0	1490	2965	1.99	<1.2	<1.8	0	<1	<1	<1	1.343	1.187	1.329
basal ration	<1	2.75	1501	3002	2.00	<1.2	<1.8	5.0	1.0	1.0	1.9			
	<1	5.5	1456	2999	2.06	<1.1	<1.9	10.2	1.7	1.9	4.3			
	<1	11.0	1506	3102	2.06	<1.2	<1.9	20.7	5.2	4.4	7.1			
Yellow corn[e]	5.5	0	1563	3110	1.99	6.5	10.6	0	2.0	1.8	3.1			
	5.5	2.75	1436	3073	2.14	6.4	10.5	5.2	3.0	3.0	4.8	1.601	1.532	1.273
	5.5	5.5	1531	3001	1.96	6.3	10.2	10.2	4.2	4.0	6.6			
	5.5	11.0	1549	3052	1.97	6.5	10.3	20.6	6.0	5.8	9.5			
Yellowcorn + corn	11.0	0	1524	3185	2.09	12.9	22.1	0	3.2	3.2	4.8			
gluten meal	11.0	2.75	1482	3008	2.03	12.4	20.7	5.2	5.0	4.9	7.3	3.000	1.923	1.561
(CGM)[f]	11.0	5.5	1496	3082	2.06	12.8	21.1	10.5	5.7	5.2	9.5			
	11.0	11.0	1478	3030	2.05	12.2	21.1	21.1	7.3	6.5	10.2			
Yellowcorn + CGM	16.5	0	1438	2991	2.08	19.0	30.4	0	4.8	4.6	7.9			
+ alfalfa meal	16.5	5.5	1524	3018	1.98	18.8	31.0	10.3	5.8	5.6	9.8	2.305	2.140	2.202
(AM)[g]	16.5	11.0	1498	2951	1.97	19.5	29.2	21.4	7.8	6.9	10.8			
Yellowcorn + CGM	22.0	0	1444	2960	2.05	24.5	40.6	0	5.2	5.4	8.8			
+ AM +	22.0	5.5	1532	3064	2.00	25.8	41.6	10.4	6.4	6.0	10.6			
concentrated	22.0	11.0	1519	2992	1.97	24.8	41.0	20.5	8.6	7.6	11.3	2.201	2.377	2.323
extract from	33.0[i]	0	1521	2996	1.97	25.1	61.2	0	6.7	6.2	10.0			
marigold meal[h]														
Average												2.090	1.832	1.738

[a] From Marusich *et al.* (*555*).
[b] Average values for 20 broilers scored by four people.
[c] Roche Yolk Color Fan, 1965 edition (1–15); the higher the number, the deeper the color.
[d] Purina Skin Pigmentation Guide (0–13); the higher the number, the deeper the color.
[e] Yellow corn contained 13.8 mg oxycarotenoids per kilogram and provided 5.5 mg/kg finished feed.
[f] Corn gluten meal contained 275 mg oxycarotenoids per kilogram and provided 5.5 mg/kg finished feed.
[g] Dehydrated alfalfa meal contained 290 mg oxycarotenoids per kilogram and provided 5.5 mg/kg finished feed.
[h] Concentrated extract from marigold meal contained 51.63 gm oxycarotenoids per kilogram and provided 5.5 mg/kg finished feed.
[i] 22.0 mg/kg fed for first 4 weeks, then 11.0 mg/kg from the concentrated extract of marigold meal fed for last 3 weeks.

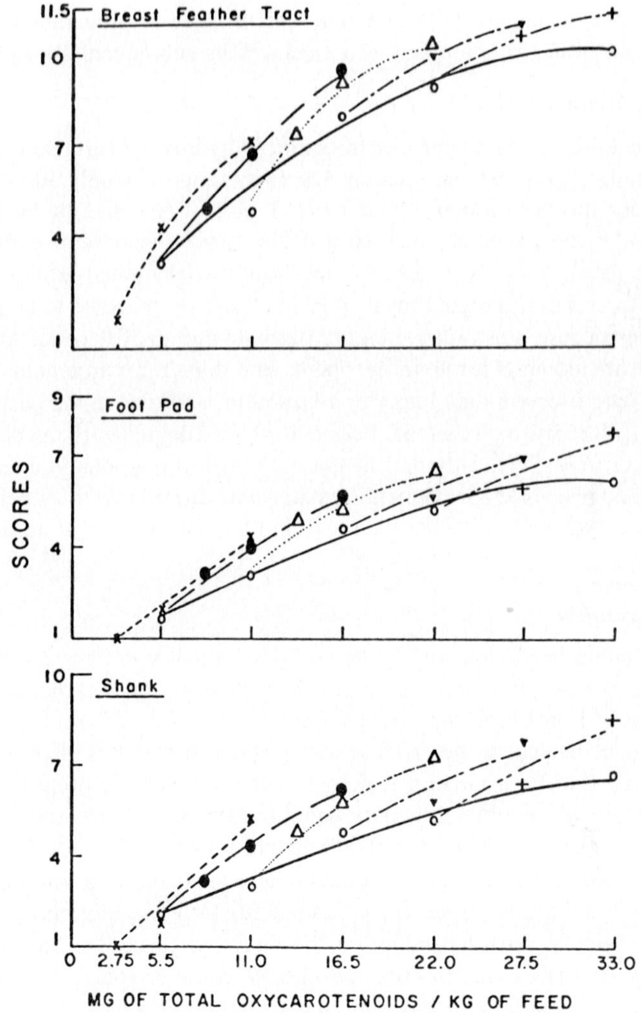

Fig. 16. Visual pigmentation scores of broilers fed graded concentrations of feed ingredient oxycarotenoids for 7 weeks and zeaxanthin supplemented for only the last 3 weeks. Key: ○, feed ingredient oxycarotenoids: 0, 5.5, 11.0, 16.5, 22.0, 33.0 mg/kg. Zeaxanthin added (mg/kg): x, 0; 5.5; △, 11.0; ▼, 16.5; +, 22.0. [From Marusich *et al.* (*555*).]

sources. Thus, to ensure that the oxycarotenoid supplement is present to the fullest extent, the pure oxycarotenoids are added in a stable form to the feed, as described by Kläui (*455*). Briefly, the oxycarotenoid is added to a gelatin–sucrose emulsion, and a dry beadlet product results which consists of minute spherical particles containing the oxycarotenoid pigment very finely distributed in a matrix of gelatin. One gram of product contains approximately 100,000 beadlets, most of which have a diameter within a range of 0.15–0.4 nm. This method of preparation protects the oxycarotenoid from oxidation and also enhances its subsequent absorption.

3. *Cryptoxanthin*

Cryptoxanthin is a monohydroxycarotenoid (3-hydroxy-β-carotene) with the empirical formula $C_{40}H_{55}OH$ and vitamin A activity (approximately 900,000 IU/gm). Cryptoxanthin has been found in butter (*315*), eggs (*316*), and the blood serum of cattle (*314*). It represents about 8–10% of the oxycarotenoids of corn (*106*) and corn gluten meal (*518*). It is also a component of dehydrated alfalfa meal (*518, 729*) and is present in marigold meal (*6,657*). It has been shown to be present in 8 of 11 species of rose hips (*320*), ranging from about 5 to 30% of the total carotenoid content (including β-carotene, lycopene, and dihydroxycarotenoids). Gram for gram, it is less effective than lutein or zeaxanthin, and thus in the chemical determination of effective oxycarotenoid content of feed ingredients, as developed by Quackenbush (*652,657*), only half of the monohydroxycarotenoid content is used in the calculations. Its lesser pigmenting potency is due mainly to its being partially metabolized to vitamin A, rather than to being poorly absorbed from the intestinal lumen (*325*).

4. *Citranaxanthin*

Citranaxanthin is a carotenoid ketone ($C_{33}H_{44}O$) that was first isolated from the peel of a trigeneric citrus hybrid (*Sinton citrangequat*) by Yokoyama and White (*895*) in 1965. It has been reported by Bauernfeind *et al.* (*65*) to be an effective yolk pigmenter but not to be efficiently deposited in the skin of broilers. Hoppe (*409*) showed it to have vitamin A activity equal to about 700,000 IU/gm. In the curative rat growth test it had activity equal to 900,000 IU/gm, and in the rat vaginal smear assay a potency of 430,000 IU/gm was found (*849*). Citranaxanthin has been synthesized (*895*) and is commercially available in a stabilized form. It cannot be fed as the sole source of oxycarotenoids for yolk color because it induces off-color (orange to red) depending on the level fed. It can be blended with the yellow oxycarotenoids and produces acceptable color in yolks. Thus, it has pigmenting properties similar to those of canthaxanthin; however, it is less effective than canthaxanthin when fed under the same conditions (*125,319,772,773*). It has also been shown to be a poor broiler skin and shank pigmenter (*772*) compared to APO-EE. These observations were confirmed by the findings of Hinton *et al.* (*404*) using a commercial source of citranaxanthin (Avizant). They reported that canthaxanthin had a relative utilization of 81.8 and 54.5% for shanks and skin, respectively, whereas the relative utilization of Avizant was only 2.8 and 3.4% for shanks and skin, respectively. These data indicate very poor broiler utilization of citranaxanthin.

In other trials with laying hens, Weiser and Manz (*850*) found canthaxanthin to be three-fold more active than citranaxanthin on the basis of visual scoring of egg yolks and five-fold more active on the basis of colorimetric assay of the yolks for oxycarotenoid content. In other yolk studies, canthaxanthin produced twice the coloring effect that citranaxanthin produced, as determined by visual observations. When colorimetrically measured for oxycarotenoid yolk deposition, canthaxanthin was from 2.7 to 3.1 times more effective (*852,853*). In other trials, Streiff (*771*) reported canthaxanthin to be deposited in yolks three-fold more efficiently than citranaxanthin; for broilers he found that 4 mg of canthaxanthin per kilogram were more effective than 16 mg citranaxanthin per kilogram. Weiser and Manz (*854*) found that the combination of 20 mg/kg of APO-EE plus 2.5 mg/kg of canthaxanthin was significantly more effective than 20 mg/kg of APO-EE plus either 5 or 10 mg/kg of citranaxanthin for broilers.

5. *Bixin*

Bixin ($C_{25}H_{30}O_4$) derived in pure form from the annatto seed has been evaluated as both a yolk and skin pigmenter. Williams *et al.* (*876*) reported that crystalline bixin was deposited in the egg yolk to a greater extent than was lutein on zeaxanthin. However, bixin was not deposited to any appreciable extent in the serum, adipose tissue, toe-web skin, or livers of chicks and laying hens. They could not explain this apparent discrepancy in bixin performance. Lutein and zeaxanthin, on the other hand, were deposited in the serum, adipose tissue, and livers of chicks and laying hens and also in the egg yolks and toe-web skin of chicks. Wildfeuer *et al.* (*865*) fed 5, 15, and 45 mg bixin per kilogram of feed for 22 days and observed no effect on color or carotenoid content of the yolk. In another trial, Wildfeuer *et al.* (*866*) also failed to show activity for bixin in egg yolk coloration. Berning (*76*) evaluated bixin for both yolk and skin pigmentation and concluded that annatto had no significance as a pigment source for poultry.

Rauch (*674*) found bixin to be ineffective for yolk pigmentation; its performance was similar to that of β-carotene. Previously, Rauch (*672*) studied the influence of different bixin preparations on yolk color using a reflectance instrument and plotting reflectance curves for different color tones in the yolks. Bixin, as artificial esters that do not occur in nature, was shown to be an efficient yolk pigmenter, with the bixin monoester being much more effective than the bixin diester. The work of McKeown (*566*) with oil-soluble annatto food colors may provide a partial explanation for the conflicting results cited in the literature on bixin efficacy as a yolk pigmenter. Oil-soluble annatto food colors are usually prepared by a high temperature extraction (about 125°C) of the seeds of the annatto tree (*Bixa orellana*) with a vegetable oil, propylene glycol, or some other suitable solvent. McKeown and Mark (*567*) had shown that bixin was unstable under the conditions of the extraction and underwent various transformations to give a number of pigments not originally found in the annatto seeds. The main product of the thermal degradation of bixin at 125°C is a monomethyl ester of an unsymmetric dimethyltetradecahexaenedioic acid. This degradation pigment comprises up to 52% of the total pigment content on a molar basis. Other pigments formed during the degradation include all-*trans*-bixin, cis isomers of bixin, *m*-xylene, and another unidentified yellow pigment.

Thus, the pigmentation properties of these altered forms of bixin may be a partial explanation for the contradictory results cited. Depending on the source and method of purification of bixin, positive or negative results could be obtained.

6. *Neoxanthin*

Neoxanthin was discovered by Strain (*759*), who assigned to it the formula $C_{40}H_{56}O_4$ in 1938. Its structure is tentatively identified as that shown by Weedon (*848*) based on the work of several investigators (*55,151,189*). Kuzmicky *et al.* (*486*) compared pure neoxanthin and violaxanthin with pure lutein as skin pigmenters; all were derived from extraction of fresh alfalfa. On the basis of toe-web analysis and visual scoring, neoxanthin had about 8% of the pigmenting potency of lutein, and violaxanthin was essentially ineffective.

7. *Violaxanthin*

Violaxanthin, or zeaxanthin diepoxide, has the same empirical formula ($C_{40}H_{56}O_4$) as neoxanthin. Violaxanthin was reported to be essentially ineffective as a yolk pigmenter by Marusich *et al.* (*548*) and ineffective as a broiler pigmenter by Kuzmicky *et al.* (*486*).

8. *Lycopene*

Lycopene ($C_{40}H_{56}$) is a hydrocarbon carotene that was extracted from tomato paste by Nelson and Baptist (*605*). They found it to produce pink yolks; however, when small quantities are combined with lutein, enhanced yolk color was obtained.

9. *Astaxanthin*

Astaxanthin ($C_{40}H_{52}O_4$), or 3,3′-dihydroxycanthaxanthin, was extracted from lobster shells by Nelson and Baptist (*605*). They reported that the combination of astaxanthin in small amounts with lutein produced deep yellow yolks similar to those obtained by feeding high levels of lutein alone. When fed alone, astaxanthin produced pink-colored yolks. A sample of mayonnaise made with egg yolks containing the mixed red (astaxanthin) and yellow pigments (lutein) had the same color as a sample made with yolks containing a higher level of only the yellow pigment. No pink color was evident in the sample prepared from the mixed pigments. The authors further stated that astaxanthin produced pink egg yolks when fed alone. Enough of the yellow pigment must be present to impart an initial yellow color to the egg yolks. Too much astaxanthin causes pinkish yolks and too little results in no intensifying effect on lutein. The optimal proportions of the red and yellow pigments were not determined. Marusich and Bauernfeind (*546*) previously had added crystalline astaxanthin dissolved in oil to a layer feed devoid of pigments and found deposition in the yolk to be relatively poor, with an off-color produced.

10. *Zeinoxanthin*

Zeinoxanthin, a monohydroxycarotenoid with the empirical formula $C_{40}H_{55}OH$ (3-hydroxy-α-carotene), was isolated from ground, freeze-dried alfalfa by

Livingston and Knowles (*512*), from corn by Petzold and Quackenbush (*635*), and from red peppers (*152,782*). It is not an efficient pigmentation source, as shown by Quackenbush *et al*. (*656*).

11. Myxoxanthin (myxoxanthophyll)

Myxoxanthophyll, a $C_{46}H_{66}O_7$ rhamnoside, is a common pigment in blue-green algae (*395,504,601,605*). It was fed to Leghorn hens by Rauch (*675*) over a range of 2–60 mg per kilogram of diet, using rations with and without 45% maize. Up to 3 mg myxoxanthophyll per kilogram of diet was suitable for yolk pigmentation if fed with the 45% corn ration; higher levels produced red yolks. Nelson and Baptist (*605*) reported that a qualitative assay of alga meal containing myxoxanthophyll resulted in light brown egg yolks. Because of the off-color, no quantitative studies were conducted.

12. Echinenone

Echinenone ($C_{40}H_{56}O$) is 4-keto-β-carotene. It has been isolated from algae, sea urchins, Daphniae, Hydra, brine shrimp, and crustaceans (*204,473,475,477,795*). It has vitamin A activity of about 900,000 IU/gm (*497*). It is an effective broiler pigmenter when fed with feed ingredient sources of oxycarotenoids (*536*). Alone, echinenone induces a pink tone at low levels of addition and an orange-red tone at high levels of feeding. When fed in conjunction with 60% yellow corn at levels of 1.5, 3, and 12 gm per ton of feed for the last 4 weeks, positive increases in pigmentation were seen in the shanks, skin, and viseral fat of 8-week-old broilers (*536*).

13. Fucoxanthin

Leuenberger (*501*) isolated fucoxanthin from the brown seaweed *Fucus serratus* in crystalline form. Feeding of the fucoxanthin to laying hens had no effect on yolk color. Fucoxanthin could not be found in the yolk, liver, or feces. In contrast, a dispersion in gelatin of fucoxanthin extracted from the same seaweed *Fucus serratus* enhanced the color of egg yolks when given to hens depleted of body pigments, according to Bolton *et al*. (*103*).

14. Capsanthin

Capsanthin ($C_{40}H_{56}O_3$) in crystalline form was dissolved in oil and added to hens' ration by Marusich *et al*. (*548*). They found that it was poorly deposited in the yolk and induced a weak off-orange color.

D. Oxycarotenoids from Industrial Chemical Synthesis

Pure carotenoids obtained by chemical synthesis have been extensively investigated as poultry pigmenters, as described by Isler and Zeller (*432*) and Isler *et al*. (*423–426,428–430*). Steinegger *et al*. (*746,747*) and Steinegger and Zanetti (*748,749*) determined the effect of feeding pure zeaxanthin, canthaxanthin, isozeaxanthin, physalien, β-apo-4′,6′,8′-,* and 10′-carotenals, β-apo-8′-carotenyl acetate, β-apo-8′-carotenoic acid, β-apo-8′-carotenoic acid ester, isozeaxanthin dimethyl ether, and isozeaxanthin diacetate on the pigmentation of egg yolk. All influenced the color of egg yolk but with different efficiencies and color characteristics. Steinegger *et al.* (*747*) also reported that canthaxanthin was effective for broiler pigmentation. Early studies with some of the same oxycarotenoids and a few additional ones, such as capsanthin, violaxanthin, and methylbixin, were reported by Bunnell and Bauernfeind (*130*) and Marusich *et al.* (*550*). The yolk color and efficiency of deposition again varied widely depending on the oxycarotenoid. Marusich *et al.* (*550*) also reported canthaxanthin and isozeaxanthin dimethyl ether to be effective for broiler pigmentation. The multipurpose of the pure oxycarotenoid studies was (a) to test individual carotenoids as ingredients in typical ration formulations, (b) to measure quantitatively the relative efficiencies of transfer of the dietary pigment, (c) to note the color characteristics of the transferred pigment, and (d) to determine which would have practical applications as poultry pigmenters. Due to numerous factors, including off-color characteristics in yolks (*isozeaxanthin, isozeaxanthin diacetate, methylbixin*), poor transfer efficiency into the yolk (violaxanthin, capsanthin), and extreme difficulty in chemical synthesis (zeaxanthin, lutein), three oxycarotenoids were selected and extensively tested for pigmentation properties. These included APO for yolks (it is not an effective broiler pigmenter) and APO-EE and canthaxanthin for both yolk and broiler pigmentation.

1. β-Apo-8′-carotenal for Yolk Pigmentation

β-Apo-8′-carotenal, a derivative of β-carotene ($C_{30}H_{40}O$) containing an aldehyde group, was first prepared by Karrer and Jucker (*443*). Its total synthesis was accomplished by Isler *et al.* (*421*), Rüegg *et al.* (*687*), and Isler and Zeller (*432*).The synthesis of APO through the C_{19} aldehyde is the same as that for β-carotene; then the chain is lengthened by first condensing the aldehyde with the C_6 acetal. By a series of reactions, as used in the β-carotene synthesis for chain lengthening, the C_{30} aldehyde (APO) is attained. This compound had been known previously only as one of the degradation products of β-carotene, but with its synthesis and a better knowledge of its physiochemical and biological properties, it has been found in a variety of natural sources. β-Apo-8′-carotenal has been isolated from orange and tangarine peel by Curl (*188*), from orange peel by Winterstein *et al.* (*885*), and from orange peel and pulp (together with β-apo-2′- and 10′-carotenals) by Thommen (*793,795*). Thommen (*793*) found APO in the peel of six varieties and in the juice of four varieties of oranges. It has also been reported in bovine intestines and mucous membrane of horses (*885*) and in green vegetables and alfalfa meal by Winterstein and Hegedus (*882*), Thommen (*792*), Winterstein (*881*), Winterstein *et al.* (*883*), and Thommen and Wiss (*801*). Thommen and Leuenberger (*798*) found APO in the

*In the early literature this was termed β-apo-2′-carotenal.

pasteurized fruit juice of the passion fruit (*Passiflora edulis*) along with β-carotene, cryptoxanthin, and β-apo-12′-carotenal.

Euler *et al*. (*257*) reported APO to be biologically active as a vitamin A source in the rat. Daily supplements of 5 μg fed to vitamin A-deficient rats stimulated growth. Glover and Redfearn (*322*) prepared β-apo-8′-, β-apo-10′- and β-apo-12′-carotenals and fed them to vitamin A-deficient rats. All were transformed into vitamin A. The vitamin A activity of pure APO was determined quantitatively by Marusich *et al*. (*549*) using the classic U.S.P XIV rat curative growth assay. It was found to be 72% ± 8% as active as all-*trans*-β-carotene, or the equivalent of about 1,100,000 IU vitamin A activity per gram. It was shown to have activity as a vitamin A source in chickens by Ferrando (*267*) and in rats (*266*) using the liver storage assay for vitamin A. Haslach (*365*) calculated a potency of 658,000 IU/gm using the chick growth assay. It was also shown to have vitamin A activity in the Japanese quail (*Coturnix coturnix japonica*) by Al-Hasani and Parrish (*9,10*). They noted that β-apo-8′-, β-apo-10′- and β-apo-12′-carotenals were all metabolized to vitamin A by the Japanese quail. Al-Hasani and Parrish (*9*) reported that significant quantities of vitamin A were formed from the three β-apocarotenals and carried in the serum, stored in the liver, and transferred to the egg. They concluded that the storage of vitamin A by quail receiving the β-apocarotenals as the only vitamin A source demonstrated a substantial vitamin A activity for the β-apocarotenals. The metabolism of APO, and the apocarotenals in general, to vitamin A was reviewed by Glover and Redfearn (*322*), Glover (*321*), Moore (*585*), Brubacker *et al*. (*126*), Olson (*615*), and Thommen (*796*).

The extreme safety of APO was demonstrated by Bagdon *et al*. (*50*) in dogs, by Ferrando (*266*) and Schärer and Studer (*700a*) in rats, and by Marusich *et al*. (*534*) in poultry. It is approved for use as a food color in the United States and most other countries (*62,66,69,106,626*). It is not, however, an approved *feed* color additive in the United States. Most other countries do not distinguish between food and feed color additives. Steinegger *et al*. (*747*) were the first to report on APO as an effective yolk pigmenter. Steinegger and Zanetti (*550*) reported that 0.6 mg APO fed per hen daily in a layer feed poor in feed ingredient oxycarotenoids produced yolks of good yellow color comparable to that obtained when the feed contained grass meal. They also observed no influence of APO on the color of the shank, skin, or beak of the hen. Bunnell and Bauernfeind (*130*) reviewed this work and that of Marusich *et al*. (*550*) and concluded that APO could serve as an effective yolk pigmenter. Rauch (*667*) compared the effect of lutein, zeaxanthin, APO, β-carotene, and alfalfa meal on yolk color using reflectance curves obtained by means of a spectrophotometer. He observed that the yolks of eggs laid by hens fed the same ration varied greatly, causing considerable scatter of the reflectance values. Zeaxanthin, APO, and alfalfa meal effectively pigmented the yolks, whereas carotene was ineffective.

A screening program was initiated by Marusich *et al*. (*548*) to measure quantitatively the relative efficiency of transfer of the various oxycarotenoids into the yolk;

the results were obtained in conjunction with a qualitative visual scoring technique. It was determined that, with a 16-day feeding period, peak pigment content was obtained in eggs laid from the tenth to eighteenth day after supplementation began. This was termed the plateau period (see Fig. 1). By acetone extraction and reading of the color in an Evelyn Colorimeter with a 440-nm filter and using a β-carotene standard curve, the micrograms of pigment (expressed as β-carotene equivalents) were calculated. This yolk content in micrograms divided by the daily dose yielded the average percent deposition. This parameter plus the actual visual score led to the selection of APO as a promising yolk pigmenter (see Fig. 5), confirming previously reported observations (*550,747*). Sunde and Madiedo (*780*) and Sunde (*778*) reported that 3 mg of APO per kilogram of a white corn ration did not yield enough yolk color, whereas 9 mg/kg gave results comparable to those obtained with a 78% yellow corn ration (Heiman–Carver Yolk Color Rotor value of 15). The addition of 6 mg/kg was about equal to 5% alfalfa meal (Heiman–Carver value of 13). Sunde (*778*) also observed that APO yielded a yolk color stable for a 6-week storage period. This confirmed the earlier observations of Steinegger and Zanetti (*550*) covering a 3-month storage period. Other U.S. workers (*296,346,403*) and investigators from other countries have reported on APO as an effective yolk pigmenter [Ward (*845*) in England, Ferrando (*266*) in France, Thumin (*809*) in Israel, Edmunson (*246*) in New Zealand, Buvanendran (*236*) in Ceylon, Tello and Urbina (*787*) in Peru, Coles (*153*) in New Zealand, Burton (*134*) in Australia, Mazziotti Di Celso (*564*) and Galvano and Lanza (*301*) in Italy, Lautner (*493*) in Czechoslovakia, De Groote (*224*) in Belgium, and DaSilva *et al.* (*200*) in Portugal.]

As stated in Section VII, C, 2, the pure oxycarotenoids have all been stabilized by incorporation into a gelatin–sugar beadlet form. Table 31 depicts the stability characteristics of APO in beadlet form "as is" and when added to poultry mashes. Bunnell *et al.* (*132*) presented data on yolk pigmentation with APO in stabilized gelatin beadlet form when fed in conjunction with numerous feed additives, including possible stress agents, such as trace mineral mix, meat scraps, and fish meal, as well as nicarbazin, zoalene, roxarsone, arsanilic acid, bacitracin–penicillin combination, and the later combination plus roxarsone. None of the feed additives adversely influenced the effectiveness of APO as a yolk pigmenter as shown by the efficiency of pigment transfer from diet to yolk (Table 10). Couch (*179*) and Couch and Farr (*178*) noted that APO could be used as the sole yolk pigmenter in the absence of all feed ingredients oxycarotenoids since the color produced is indistinguishable from that resulting when hens are fed feed ingredient sources of xanthophylls. It could also be effectively combined with yellow corn and dehydrated alfalfa meal to obtain staisfactory yolk color.

Boguth and Czernicki (*101*) reported that APO yielded a deeper color by visual inspection than the equivalent level of mixed xanthophylls from natural ingredients. Czernicki and Weiser (*194*) recommended APO because of its favorable coloring properties and vitamin A activity. Braünlich (*116*), Bauernfeind (*60*), Carlson (*140*), Carlson *et al.* (*144,146*), Tagwerker *et al.* (*347*), Farr *et al.* (*259,261*),

TABLE 31

Stability of β-Apo-8′-carotenal 10% Beadlets Used in ANRC Collaborative Study[a]

Storage conditions[b]	β-Apo-8′-carotenal (%)	Retention (%)
As is		
Initial	10.1	100
3 weeks, 45°C	9.7	96
6 weeks, 45°C	10.2	101
3 months, RT	10.2	101
6 months, RT	10.1	100
12 months, RT	10.0	99
19 months, RT	10.1	100
In chick mash		
3 weeks, 45°C	10.0	99
6 weeks, 45°C	10.2	101
3 weeks, RT	10.0	99
6 weeks, RT	10.0	99
12 weeks, RT	10.2	101

[a] From Marusich (*538*).
[b] RT, room temperature (23°C).

Poppe *et al.* (*643*), Sullivan (*776*), Marusich *et al.* (*544*), Bauernfeind and Marusich (*67*), Rauch (*667,668*), Hall *et al.* (*346*), Couch and Farr (*175,176*), Guenther *et al.* (*338*), Giavarini (*310*), Scott (*712*), Scott *et al.* (*713*), Vollrath (*834*), Livingston *et al.* (*519*), and Bauernfiend (*61*) reported favorable results with APO. Carlson *et al.* (*145*) reported the combination feeding of 11% of a high-quality 20% protein dehydrated alfalfa meal and APO at 6 mg/lb or 18–20% of the alfalfa meal alone to yield a satisfactory deeply pigmented egg yolk that scored 19–20 on the Heiman–Carver Yolk Color Rotor, had a β-carotene equivalent pigment content of 100–110 μg/gm, and had satisfactory carryover in mayonnaise.

Marusich (*538*) summerized the results of a collaborative trial with eight participating laboratories whereby APO in the form of a 10% stabilized beadlet was evaluated as the potential Animal Nutrition Research Council (ANRC) Yolk Pigmentation Reference Standard. Comparisons were made between APO and the oxycarotenoids of dehydrated alfalfa meal fed at 15, 30, 45, and 60 gm/ton. Both sources were equally well deposited in the yolk over the entire feeding range. Yolk visual scores averaged about two fan values higher with APO using the Roche Yolk Color Fan at all levels of feeding. These comparisons are shown graphically in Fig. 5. On the basis of the uniformity of response, satisfactory yolk color characteristics, and stability properties, the 10% preparation of APO was accepted as the ANRC Yolk Pigmentation Standard (*288,538*) in 1966.

TABLE 32

Hunter *L*, *a*, and *b* Values and Roche Yolk Color Fan Visual Scores of Egg Yolks[a]

	Instrumental values						
	Raw yolk			Hard boiled yolk			Visual scores
Treatment	*L*	*a*	*b*	*L*	*a*	*b*	(Roche fan)
Control	52.5	−4.5	19.9	88.5	−5.1	26.7	2.9
Lutein, 2.5 mg/lb	52.4	−3.3	26.3	87.0	−4.1	35.7	6.2
Lutein, 5.0 mg/lb	52.1	−2.6	28.2	85.7	−3.3	40.2	7.3
Lutein, 10.0 mg/lb	49.4	0.0	29.2	84.1	−1.2	45.4	8.8
Lutein, 20.0 mg/lb	47.9	1.3	29.1	82.3	−0.5	45.7	9.9
Apocarotenal, 2.5 mg/lb	50.8	−0.9	26.4	85.3	−1.7	37.9	8.4
Apocarotenal, 5.0 mg/lb	47.6	2.9	27.3	82.4	1.9	42.0	10.9
Apocarotenal, 10.0 mg/lb	46.3	6.4	27.8	79.4	7.3	44.9	12.5
Apocarotenal, 20.0 mg/lb	43.6	11.7	26.3	74.9	14.3	46.2	14.7
Canthaxanthin, 2.5 mg/lb	44.8	11.1	21.1	78.8	15.3	31.8	14.4
Canthaxanthin, 5.0 mg/lb	43.2	16.6	21.8	76.3	19.7	32.0	14.8
Canthaxanthin, 10.0 mg/lb	40.6	20.0	21.5	72.3	29.6	35.5	14.9
Canthaxanthin, 20.0 mg/lb	36.5	23.0	20.0	67.2	33.3	37.0	15.0

[a] From Philip *et al.* (*636a*).

Philip *et al* (*636a*) compared the pigmenting value of crystalline lutein, APO, and canthaxanthin for egg yolks by means of both reflectance colorimetry (Hunter D25 Color and Color Difference Meter) and visual scoring using the Roche Yolk Color Fan. They fed a low-pigment layer ration (69% milo, 14.25% soybean meal) supplemented with 2.5, 5.0, 10.0, and 20.0 mg/lb of each of the oxycarotenoid sources. Good agreement was obtained between the instrumental values and the visual scores. The authors found that 20 mg lutein per pound of feed produced light orange yolks similar in color to yolks produced by APO at 2.5 mg/lb. In consumer preference study, 95% of the judges preferred table egg yolk colors having a Roche fan score of 7, 8, or 9. These colors corresponded to feeding from 10 to 20 mg/lb of lutein and 2.5 mg/lb of APO. Canthaxanthin gave the highest readings at all four levels; however, the color produced was unacceptable (orange to red yolks). These data are shown in Table 32.

2. *β-Apo-8′-carotenoic Acid Ethyl Ester for Yolk Pigmentation*

β-Apo-8′-carotenoic acid ethyl ester ($C_{32}H_{44}O_2$), an analogue of β-carotenal. It has been reported by Baraud *et al.* (*52*) to have been isolated from yellow maize. It was synthesized by Isler *et al.* (*422*) and, like APO, has vitamin A activity. Ferrando (*266*) reported it to have less activity than APO in the rat liver storage assay. In the rat curative growth assay, Bauernfeind *et al* (*65*) and Kläui (*456*) reported APO-EE to have about one-third the activity of APO or about 420,000 IU of vitamin activity per gram. Hoppe (*409*) reported a value of 1,300,000 IU/gm using male broiler-type chicks in a curative growth assay. It is not an approved food or feed color additive in the United States, but is widely used throughout the rest of the world (*62*). Like APO, it is available in stabilized gelatin beadlet form. It has stability characteristics comparable to those of APO (Table 31). It has been extensively tested as a yolk pigmenter by Bayley and Newlands (*70*), Stevens and Lewis (*751*), Stevens *et al.* (*752*) and Portsmouth (*645*) in England, Ferrando (*266*) in France, Streiff (*764*) in Switzerland, Scholtyssek and Bohm (*705*) and Vollrath (*835*) in Germany, Blommaert and Steenis (*97*), De Groote (*221*), and De Groote and Reyntens (*226*) in Belgium, Kivimae *et al.* (*454*) in Sweuen, Soliman (*731*) in Egypt, Quilici (*659*) in Italy, Saunders *et al* (*699*) in South Africa, Prohaszka and Toth (*649*) in Hungary, and DeJongh (*227*) in Mexico. All reported APO-EE to be a highly effective yolk pigmenter producing color comparable to that produced by the oxycarotenoids of feed ingredients, but with less APO-EE required than mixed oxycarotenoid sources. Portsmouth (*645*) stated that APO-EE is 1.5 times as effective as feed ingredient xanthophylls; thus, 3 mg of the latter can be replaced in layer rations with 2 mg APO-EE. Marusich and Bauernfeind (*546*) measured pigment content in yolks after feeding a diverse array of pigmenting sources and found that hens fed APO-EE produced eggs with the highest pigment content (two- to threefold higher then other sources). Increased visual sources were also related to the increased pigment measured colorimetrically. In a symposium on yolk color held in London, Comben (*156*), Streiff (*770*), and Braünlich (*117,118*) presented data substantiating the utility of APO-EE for egg yolk pigmentation compared to a number of other oxycarotenoid sources.

Numerous evaluations of APO-EE and APO have also been made to determine the relative pigmenting efficacy of each. β-Apo-8′-carotenal was originally developed as a commercial product but was rapidly replaced by APO-EE due to a generally greater potency relationship reported by numerous investigators (*61,100, 194,214,216,218–220,223,247,266,310,362,541,609,668,670,673,674,676,707, 763,783,809,863,864,866*). Figure 17 compares the yolk responses obtained with APO and APO-EE when added to several different layer rations (white corn, milo, or yellow corn). In Fig. 18, APO-EE is compared to a mixture of feed ingredient sources and APO and a combination of APO and canthaxanthin.

β-Apo-8′-carotenoic acid ethyl ester continues to be reported as the most active single oxycarotenoid for yolk pigmentation (*31,119,339,554,568,832,873*). Guenthner *et al.* (*339*) compared the pigmenting potential of the xanthophylls in corn gluten meal (60% protein), alfalfa meal (20% protein), pure lutein isolated from alfalfa meal, lutein extracts from corn gluten meal and alfalfa meal, an alfalfa concentrate (pro-xan), marigold petal meal, APO, and APO-EE using both caged and floor-pen-reared laying hens and producing eggs for both the "table" and "breaker" egg market. Of all the materials tested APO-EE was the most potent. The authors

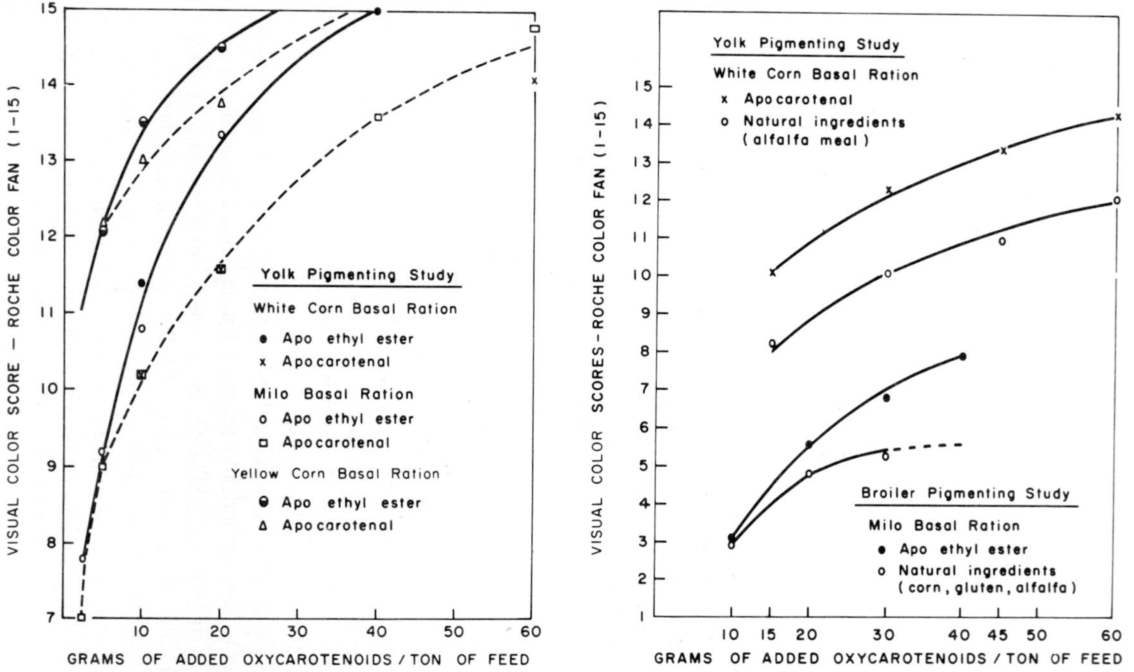

Fig. 17. Efficacy of APO and APO-EE in pigmenting egg yolk and APO-EE in pigmenting broilers as supplements to different types of rations compared to mixed oxycarotenoids from feed ingredients. [From Bauernfeind (*63*).]

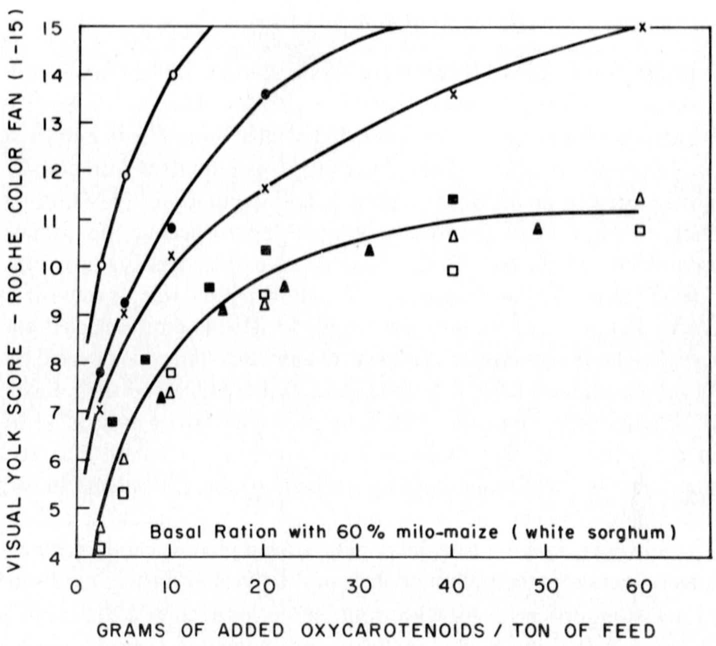

Fig. 18. Visual yolk response after oxycarotenoids from different sources were added to a milo-maize ration determined using the Roche Yolk Color Fan. Key: ●, APO-EE; x, APO; ○, APO-EE + Canthaxanthin (4:1), Natural ingredients: △, marigold meal; □, alga meal; ▲, alfalfa meal; ■, corn gluten meal. [From Marusich and Bauernfeind (*546*).]

concluded that to obtain an intensely pigmented yolk, i.e., NEPA No. 5, containing approximately 100–135 μg of BCE per gram of yolk, the level of xanthophylls required from these products would be 80–100 ppm. About one-half this amount (40–50 ppm) of APO-EE would be required, whereas if the source were marigold meal about one-third more xanthophylls (105–135 ppm) would be needed.

Marusich *et al* (*554*) compared APO-EE (Carophyll-Yellow) to Mexican marigold meal (Florafil), a concentrated extract of marigold meal (Gold'n Bloom), and a 1% stabilized zeaxanthin gelatin beadlet experimental formulation. This study is described in Section VII, D, 2, and the detailed data are shown in Table 28. The APO-EE was markedly superior to the other three products, and zeaxanthin was decidedly better than the two marigold products, as indicated by both BCE's in the yolk and visual score.

In studies designed to measure the partial or total replacement of corn oxycarotenoids, it was shown that 2.4 mg APO-EE per kilogram equaled 3.6 mg of oxycarotenoids from corn for yolk pigmentation, or that there was an effective ratio of 1.5:1 for APO-EE to corn (*42*).

3. Canthaxanthin for Yolk Pigmentation

Canthaxanthin (4,4′-diketo-β-carotene) was unknown until isolated from an edible mushroom, *Cantharellus cinnabarinus,* in 1950 by Haxo (*371*). Saperstein and Star (*696*) and Saperstein *et al.* (*697*) isolated canthaxanthin from mutant strains of *Corynebacterium michiganese*. They determined its empirical formula, $C_{40}H_{52}O_2$, and recognized its ketone character. Petracek and Zechmeister (*634*), in a study of the products of reaction of β-carotene with *N*-bromosuccinimide, isolated 4,4′-diketo-β-carotene, which they (*633*) established to be identical to canthaxanthin from *Corynebacterium michiganese*. Canthaxanthin was synthesized from β-carotene by Petracek and Zechmeister (*634*) in 1956 and by Entschel and Karrer (*253*) in 1958. The first complete synthesis of canthaxanthin was devised by Isler *et al.* (*427*) and described in detail by Isler and Zeller (*432*) and then developed by Isler and Schüdel (*431*) into the first industrial synthesis. Surmatis *et al.* (*781*) presented a synthesis of canthaxanthin from vitamin A alcohol. The absence of provitamin A activity was noted both in synthetic canthaxanthin and in canthaxanthin isolated from *Corynebacterium michiganese* by Petracek and Zechmeister (*633,634*). However, in retinol-depleted guppies and platies, canthaxanthin, as well as β-carotene, increased the retinol content of the liver, according to Olson (*616*).

Some interesting observations with trout and shrimp suggest that canthaxanthin may play a role in fertility: Sexual maturity and spawning occurred earlier in trout that received a diet supplemented with canthaxanthin than in controls without this carotenoid (*232*). Analyses have shown that shrimp, which form part of the natural diet of trout, also contain oxycarotenoids. These can be recovered from the roe and the spawn and exert a chemotactic attraction on the spermatozoa (*363*). Increasing the oxycarotenoid content of a feed by the addition of canthaxanthin results in distinct improvements in the fertilization rate. The proportion of unfertilized eggs was reduced from 4.1% in controls to only 0.1% by the addition of supplementary canthaxanthin to the diet.

In flamingoes, as in trout, oxycarotenoid intake appears to influence fertility and reproduction (*330,838*). In the species *Phoenicopterus ruber*, kept in captivity, appropriate supplementation of the diet resulted in previously pale feathers being replaced after the next molt by brilliantly colored feathers and the nest-building drive being manifested for the first time. Eggs laid thereafter hatched regularly, and this was also considered to be the result of the improvement in diet (*839*).

It is interesting that, during the first few weeks after hatching, flamingoes feed their young a red juice, which flows from the beak of the older birds into that of the young. Samples of this juice have been found to contain significant quantities of canthaxanthin, with values of 5.4, 4.9, and 1.6 μg/ml (*840*) being recorded. Gilchrist and Lee (*313*) noted the possible role of the carotenoid pigments in reproduction in the sand crab *Emerita analoga*.

The biosynthesis of 4-keto- and 4,4′-diketocarotenoids in *Brevibacterium* sp. KY-4313 was studied by Hsieh *et al.* (*411*). Echinenone and canthaxanthin were isolated from the cultures grown on a medium containing several *N*-alkanes. When glutathione was added to the bacterial cultures, the formation of canthaxanthin was inhibited, whereas β-carotene and its hydroxy derivatives accumulated. It was suggested that these 4-hydroxy compounds, isocryptoxanthin, isozeaxanthin, and 4-hydroxy-4′-keto-β-carotene are intermediates in the biosynthesis of canthaxanthin.

Cooney *et al.* (*165*), Ungers and Cooney (*826*), and Schwartzel and Cooney (*710*) reported canthaxanthin to be the major pigment in *Micrococcus roseus*. Canthaxanthin has its widest distribution in marine organisms. It is the major carotenoid of *Daphnia*, according to Thommen and Wackernagel (*800*). It has been isolated from algae (*Chlorella*) by Czygan (*195*) and Hertzberg and Liaaen-Jensen (*393,394*), hydra by Krinsky and Lenhoff (*477*), the marine isopods *Idothea granulosa, Idothea resecata,* and *Idothea montereyensis* by Lee (*498–500*), freshwater anostracens by Gilchrist (*311*), *Carcinus meanas* (shore crab) by Gilchrist and Lee (*312*), brine shrimp, *Artemia salina*, by Krinsky (*473*), California *Artemia* by Hsu *et al.* (*412*), crayfish by Czeczuga and Czerpak (*193*), and sea trout, *Salmo trutta,* by Thommen (*794*) and Thommen and Gloor (*797*). Lee (*500*) stated that there are strong indications that the canthaxanthin utilized for overt color by the idoteid isopods is derived from dietary β-carotene through a stepwise oxidation, probably involving monohydroxy-β-carotene, echinenone, and 4-hydroxy-4′-keto-β-carotene.

In an excellent review of crustacean meals and oxycarotenoid-fortified diets, Meyers (*573a*) focused on the increasing importance of crustacean mea.. The rapidly expanding crawfish industry in Louisiana produces over 3000 tons of dry crawfish waste yearly. Preliminary data on pigment extractions from the large claw of the crawfish *Procambarus clarkii* have shown concentrations of total carotenoids as high as 196 μg per gram dry tissue.

In Japan, consumer acceptance of the prawn *Penaeus japonicus* is based largely on color, supplied in the diet either as astaxanthin or as an oxycarotenoid that can be converted to astaxanthin by the animal. Biosynthesis of astaxanthin by *P. japonicus* has been extensively examined in Japan by Tanaka (*783a*). It was demonstrated that pure pigments, such as β-carotene, zeaxanthin (from the Chinese lantern), and canthaxanthin, can be converted to astaxanthin. Astaxanthin was biosynthesized from zeaxanthin, thus indicating the existence of a second pathway to astaxanthin in the prawn. Canthaxanthin and astaxanthin from crab waste were metabolized to and absorbed as body astaxanthin. In addition, pigmented preparations from corn gluten, alfalfa, and *Spirulina* increased the level of body astaxanthin to various degrees. Results of feeding prawn on an artificial diet supplemented with 10% corn gluten, 10% alfalfa, and 10% *Spirulina*, and 3 mg % canthaxanthin are shown in Table 33.

Spirulina, a blue-green alga rich in β-carotene, and synthetic canthaxanthin yielded a greater concentration of astaxanthin than either corn gluten or alfalfa. Prawn fed *Spirulina* and canthaxanthin were also visibly redder than other animals.

TABLE 33

Contents of Astacin[a] in Prawn (*Penaeus japonicus*) Fed Corn Gluten, Alfalfa, Canthaxanthin, and *Spirulina*[b,c]

	Artificial diet (control)	Artificial diet containing 10% corn gluten	10% alfalfa	3 mg % canthaxanthin	10% *Spirulina*
Concentration of Astacin (μg/gm body wt)	3.4	5.1	5.5	10.4	18.4

[a] Obtained by the oxidation of astaxanthin.
[b] From Tanaka *et al.* (*783a*).
[c] *Spirulina* is a blue-green alga.

In tests with ^{14}C-labeled β-carotene, it was shown that β-carotene can be converted to astaxanthin by the prawn. Zeaxanthin was more effective as a pigment source than β-carotene in being converted to astaxanthin (Table 34).

It is interesting that Fletcher *et al.* (*270c*) considered the possibility of oxy carotenoid changes being stimulated by natural light. They observed that broilers raised with constant low-intensity lighting in windowless houses exhibited a significantly lower dominant wavelength (less orange) than birds reared in open-type houses. With a natural diurnal light cycle present, broilers reared in pens along open walks had a higher dominant wavelength (more orange) than broilers reared in interior pens where light intensity was lower. The authors concluded that light affects broiler pigmentation on the basis of instrumental measurement of skin color using the IDL Color Eye. They stated that one or both of the xanthophyll pigments (lutein and zeaxanthin) in the ration fed to the broilers may have been changed to redder pigments such as astaxanthin, thus accounting for the more orange present in the broilers exposed to natural light.

The high concentration of canthaxanthin in various crustaceans is partly responsi-

TABLE 34

Incorporation of Carotene and Zeaxanthin into Astaxanthin in the Prawn *Penaeus japonicus*[a]

Pigment	Astaxanthin isolated (μg/gm body wt)
None	3.3
β-Carotene, 20 mg/gm	21.0
Zeaxanthin, 20 mg/gm	55.0

[a] From Tanaka *et al.* (*783a*).

ble for the widespread occurrence of canthaxanthin in water birds that feed on crustaceans. Canthaxanthin is the major pigment of several flamingo species. It predominates in the feathers, leg skin, egg yolk, blood plasma, and liver of the flamingoes *Phoenicopterus ruber* and *P. chilensia,* according to Fox (*276*). Canthaxanthin and astaxanthin appear to be elaborated by the bird, either in the intestinal flora or at cellular sites from carotenoid precursors in the diet (*276*). Thommen and Wackernagel (*799*) reported canthaxanthin to be the main pigment of the feathers, skin, heart, kidney, and liver of the lesser flamingo *Phoeniconaias minor*. Wackernagel (*839*) isolated 4.34 mg of canthaxanthin from 34 gm of liver from *phoeniconaias minor*. Canthaxanthin is also present in significant amounts in the Andean flamingo *Phoenicoparrus andinus* and two other species, *Phoeniconaias jamesi* and *P. ruber*, (*278,279,282*), in the roseate spoonbill *Ajaia ajaja* (*277,281*), in the scarlet ibis *Guara rubra* (*275,280*), and in the feathers of the cardinal *Cardinalis cardinalis,* the trogon, *Pharomachrus mocino, Calochaetes coccineus,* and *Pyrocephalus rubineus* (*833*).

Hsu *et al*. (*412*) studied the metabolism of β-carotene in the brine shrimp *Artemia salina* L. and found pure β-carotene to be converted to echinenone (4-keto-β-carotene) and canthaxanthin (4,4′-diketo-β-carotene). After the feeding of β-carotene, the kinetics of changes in pigment composition infer that echinenone is an intermediate in the conversion of β-carotene to canthaxanthin. Canthaxanthin is the final product, with no intermediates found, other than echinenone. In turn, the eggs and freshly hatched nauplii of *Artemia salina* have only echinenone and canthaxanthin. Krinsky (*473*) studied the oxycarotenoid pigments of the brine shrimp *Artemia salina* and found only canthaxanthin and echinenone in a ratio of 19:1. Davies *et al*. (*204*) proposed a mechanism for the conversion of β-carotene to canthaxanthin by the brine shrimp with echinenone as an intermediate. Hata and Hata (*367*) also demonstrated that β-carotene is converted mainly to echinenone and canthaxanthin by *Artemia salina*.

Fox *et al*. (*282*) first depleted the American flamingo (*Phoenicopterus ruber*) of its carotenoids and then fed a surplus of β-carotene daily in the diet. They noted that the birds voided some excess β-carotene and considerable zeaxanthin in their feces, while accumulating some unchanged β-carotene and some echinenone. However, canthaxanthin was the predominant pigment in the liver. When the blood carotenoid levels were restored with β-carotene in the diet, the only oxycarotenoids detected were echinenone (3–9%) and canthaxanthin (90%+).

Gloor (*318*) found canthaxanthin in yam tubers (*Dioscorea*). He noted that this was the first time canthaxanthin had been discovered in a higher plant other than the edible mushroom *Cantharellus cinnabarinus*. More recently, canthaxanthin was also found in the Colorado potato bettle *Leptinotarsa decemlineata* by Leuenberger and Thommen (*503*). These workers showed that the Colorado potato bettle is able to change carotenes and xanthophylls taken up from the leaves of the potato plant into echinenone and canthaxanthin, respectively. Arroya (*46a*) found β-carotene and canthaxanthin to be the two predominant pigments in Philippine duck egg yolks. Lower levels of echinenone, cryptoxanthin, and zeaxanthin were also identified.

Canthaxanthin is approved for use as a food colorant in the United States and most other countries (*62,69,106,131*). A recent tabulation shows it to be approved in 36 countries (*64*). Like APO, it is not permitted to be added to poultry feeds in the United States; however, approval as a broiler pigmenter is expected early in 1981. It is also dispensed in a stabilized beadlet form. Table 35 shows typical stability data. Osadca *et al.* (*618*) provided a method for its determination in concentrates and finished feeds. Canthaxanthin is efficiently deposited in yolks and is an effective yolk pigmenter (see Fig. 1) but its use may result in pink to deep orange or orange-red yolks when used a the entire or major source of oxycarotenoid, as reported by Marusich *et al.* (*548*), Philip *et al.* (*636a*), and Sullivan (*776*). Canthaxanthin is entirely satisfactory for the pigmentation of egg yolks for table use when fed in combination with yellow xanthophylls from feed ingredient sources or APO or APO-EE, according to Czernicki and Weiser (*194*), DeGroote (*214,217–220,222,223,225*), Nobel (*609*), Carlson *et al.* (*145*), Giavarini (*309,310*), Tasaki and Myoga (*786*), Rauch (*671,673,674*), Scott *et al.* (*713*), Wildfeuer *et al.* (*863,866*), Gawecki and Potkanski (*306*), Vollrath (*834*), Marusich and Bauernfeind (*546*), Couch and Farr (*176,177*), Biedermann and Prabucki (*84*), and Vollrath (*835*).

TABLE 35

Stability of 10% Canthaxanthin Beadlets "as Is," in Mash, and in Pelleted Feed

Storage conditions[a]	Retention (%)
Stabilized gelatin beadlets	
2 months, RT (75°F)	99
4 months, RT	98
6 months, RT	96
12 months, RT	96
3 weeks, 45°C (113°F)	99
6 weeks, 45°C	99
9 weeks, 45°C	99
12 weeks, 45°C	96
In chick mash	
3 months, RT (75°F)	97–100
3 months, 37°C (98°F)	98–100
In pelleted feed	
3 months, RT (75°F)	92–93
3 months, 37°C (98°F)	81–88

[a] RT. room temperature. [From Bauernfeind (*68*).]

The results of Scott *et al.* (*713*) are typical of the type of response obtained when canthaxanthin is added to a ration containing moderate levels of oxycarotenoids from feed ingredient sources. These investigators compared the effect of adding 2, 4, 5, or 10 gm canthaxanthin per ton of feed to that obtained with 10, 20 or 40 gm APO per ton when added to a basal ration containing 65.75% yellow corn and 2.5% dehydrated alfalfa meal (17% protein). A marked increase in yolk pigmentation was seen by visual inspection with increasing levels of APO. The highest level (40 gm/ton) produced a deep yellow-orange yolk with an average BCE content of 81 μg per gram of yolk. In turn, supplementation with only 2 gm canthaxanthin per ton also produced an orange-colored yolk, which appeared by visual inspection to be almost as deeply pigmented as those of hens receiving 40 gm APO per ton. However, colorimetric determination of pigment content of the yolks gave BCE values of 49–57 μg per gram of yolk, compared to 45 μg/gm for the yolks of hens on the basal ration alone. Thus, by visual inspection, these yolks were deeply colored but, in reality, contained far less additional pigment. These workers also noted the effect of these yolks on the color of cakes. Cakes made with eggs from hens on the basal ration showed only the slighest tinge of yellow color. Ten grams of APO per ton produced a definite yellow color and 40 gm/ton produced eggs so deeply pigmented that the cake was a golden yellow. Cakes made with eggs from hens receiving 10 gm canthaxanthin per ton were not as well pigmented as those made with eggs from hens receiving 10 gm APO per ton, although by visual inspection the egg yolks from canthaxanthin-fed hens had a much deeper orange color. Similar observations were reported by Carlson *et al.* (*145*) and Deethardt *et al.* (*212*).

Couch and Farr (*177*) reported the results of a comprehensive study whereby three layer rations with 5, 10, or 20% dehydrated alfalfa and 31.4, 28.5, or 22.8% yellow corn, respectively, were each supplemented with canthaxanthin at levels of 4–8 gm/ton. The most dramatic improvement in yolk color was seen when canthaxanthin was added to the ration with 20% alfalfa. The 20% alfalfa meal ration alone gave yolks with NEPA values of 4.5, which were increased to an average of 5.5 with the canthaxanthin additions. Carlson *et al.* (*145*) showed that canthaxanthin when added at 8 gm/ton to a ration already containing 16 gm oxycarotenoids per ton from 70% yellow corn and 2% dehydrated alfalfa meal (17% protein) produced yolks with adequate Heiman–Carver Yolk Color Rotor numbers for breaker egg color. Although canthaxanthin addition adequately enhanced visual yolk color, the color of mayonnaise produced with these yolks was comparable to the light color of mayonnaise made with the control yolks except for a pink tinge. Paprika extract (oleoresin paprika) added at 1 and 2 lb/ton produced similar effects when yolks containing it were used for mayonnaise. The addition of canthaxanthin at 4 gm/ton to a ration containing 11% dehydrated alfalfa (20% protein) and 60% yellow corn providing 30 gm oxycarotenoids per ton produced dark yolks that were satisfactory for breaker egg use. They also scored 19–20 on the Heiman–Carver Yolk Color Rotor and contained from 100 to 110 μg of BCE's per gram of yolk. Deethardt *et al.* (*212*) reported that sponge cakes made with yolks colored with canthaxanthin when added at both 4 and 8 gm/ton, as described by Carlson *et al.* (*145*), induced a pink color in the sponge cake.

Vollrath (*835*) showed that the addition of both canthaxanthin and APO-EE to a 30% yellow corn ration represented a good combination for achieving desirable yolk color in table eggs. Similar results were reported by McNaughton *et al.* (*568*) for combinations of canthaxanthin with APO-EE. Thus, canthaxanthin and other red oxycarotenoids can be suitably blended with yellow oxycarotenoids from feed ingredient sources or combined with APO or APO-EE to produce highly acceptable, deeply colored orange table-use yolks, which are preferred in many parts of the world, with much lower total oxycarotenoid levels being fed. However, these yolks may not be acceptable as breaker eggs for use in specialized processed foods, acceptance depending on the relative contribution of the yellow and red oxycarotenoids. Braünlich (*119*) treated this subject thoroughly.

4. *β-Apo-8′-carotenoic Acid Ethyl Ester for Skin Pigmentation*

β-Apo-8′-carotenoic acid ethyl ester, in addition to having yolk pigmenting properties, is a highly effective pigmenter for both shanks and skin of broilers, in contrast to APO, which is a very ineffective shank and skin pigmenter, as reported by Steinegger and Zanetti (*550*), Ewing (*258*), Rauch (*671*), Kohler (*461*), and Quackenbush *et al.* (*656*). Its conversion to vitamin A probably accounts for this lesser pigmenting effect in the broiler, as with cryptoxanthin (see Section VII, D,2).

Rauch (*671*) studied the effect of adding APO-EE at 12 and 24 gm/ton to an almost pigment-free feed utilizing an optical technique for the quantitative determination of shank color in pullets. Samples of skin were secured and the reflectance measured in a Beckmann DU spectrophotometer with a reflectance attachment. The shanks were also visually scored by taking colored photographs at intervals and comparing these colors to the color tables developed by Hickethier (*398*). Rauch (*671*) found APO-EE to be highly effective at both levels of supplementation on the basis of both visual observation and reflectance readings. Streiff (*767*), in a review of broiler pigmentation problems and solutions, stated that APO-EE is the "oxycarotenoid of choice" for broiler pigmentation due to its properties of excellent absorption through the intestinal tract and deposition of pigment in the tissues coupled with its golden yellow color and attributes of stability in a gelatin–sugar matrix. The feed formulator can predict the magnitude of pigmentation to be produced each time it is added with greater certainty than when conventional feed ingredients are depended on solely. Gazo *et al.* (*308*) showed increased pigmentation in shanks and breast skin which resulted from adding 12 and 17 mg APO-EE per kilogram to a ration already containing 8 mg oxycarotenoids per kilogram. The responses to APO-EE were quite dramatic.

Marusich (*539*) reported a study comparing the effect of graded levels of oxycarotenoids from yellow corn, corn gluten meal, and dehydrated alfalfa meal ranging from 3 to 29 gm/ton to the pigmenting effect of APO-EE added at 5–25 gm/ton. Although the shank visual scores, determined using the Roche Yolk Color Fan, increased with increasing levels of both oxycarotenoid sources (feed ingredients or APO-EE), the trend suggested superiority in the respons to APO-EE at equivalent levels (see Table 36 and Fig. 19). Resulting from this trial was a follow-up collaborative trial authorized by the ANRC to evaluate the potential of APO-Ee as the ANRC Reference Broiler Pigmentation Standard, similar to the use of APO as the ANRC Reference Yolk Pigmentation Standard. Marusich (*540*) reported on the results of this trial with eight collaborators including both university and industrial laboratories. Oxycarotenoids from feed ingredients were fed at 10 gm/ton (yellow corn), 20 gm/ton (yellow corn plus dehydrated alfalfa), and 30 gm/ton (yellow corn plus dehydrated alfalfa plus corn gluten meal) and compared to APO-EE fed at 10, 20, 30, and 40 gm/ton. The same general dose-response curves were obtained by all eight collaborators. The composite response of all investigators showed the response of APO-EE at 10 and 20 gm/ton to essentially equal the feed ingredient oxycarotenoid response. However, above these levels APO-EE increased shank pigmentation, as shown by visual scores determined using the Roche Yolk Color Fan, to a greater degree than the equivalent supplement derived from feed ingredients. As a result of the uniformity of response to APO-EE up to 20 gm/ton and its excellent stability characteristics both as a 10% concentration in a gelatin–sucrose beadlet alone and in finished feeds, it was accepted as the ANRC Reference Broiler Standard in 1969 (*540,717*). These data are detailed in Table 24 and shown graphically in Fig. 15.

Extensive comparative data on APO-EE, marigold meal (*Tagetes erecta*), alga meals (*Chlorella pyrenoidosa, Spongiococcum excentricum*), dehydrated alfalfa meals (*20, 22,* 25% protein), corn gluten meals, and combinations of APO-EE with canthaxanthin and canthaxanthin with feed ingredients were presented by Marusich and Bauernfeind (*547*). These trials included adding the above supplements to both low-pigment (milo) rations and rations containing low to moderate levels of oxycarotenoids (25–60% yellow corn base). All pigmentation sources were effective in increasing the color of shanks, as measured by the Roche Yolk Color Fan and the Heiman–Carver Yolk Color Rotor, as dietary levels of supplementation were increased. At the lower levels of addition APO-EE equalled the pigmentation obtainable with the other sources. However, as the levels increased, especially above 15 gm/ton, APO-EE was decidely more effective, providing shank visual scores of 9–10 (Roche) or 12–14 (Heiman–Carver). Selected data are shown in Tables 37 and 38 and Fig. 20.

In another trial, Marusich *et al.* (*555*) tested the relative potency of APO-EE, zeaxanthin, and two marigold products when fed at levels of 11, 22, and 33 mg/kg in a low-oxycarotenoid basal ration for 7 weeks. Shanks were visually scored with the Roche Yolk Color Fan, and breast feather tracts with the Purina Skin Pigmentation Guide. Zeaxanthin in stabilized beadlet form was the most effective, followed by APO-EE. The marigold meal and concentrated extract of marigold meal were least effective and about equal in effect (Table 29).

Herrick *et al.* (*390,392*), Hinton *et al.* (*404*), Halloran (*349*), Garlick *et al.*

TABLE 36

Shank Pigmentation of Broilers Fed Several Oxycarotenoid Sources for Last 4 Weeks before Marketing[a]

Basal ration	Oxycarotenoid level (gm/ton feed) Feed ingredients	Pure	Total	Shank visual score (Roche Fan, 1–15)
25% yellow corn[b]	3.0	—	3.0	0.9
50% yellow corn	6.0	—	6.0	2.2
50% yellow corn + 3% corn gluten[c]	11.0	—	11.0	3.4
50% yellow corn + 6% corn gluten	16.0	—	16.0	4.4
50% yellow corn + 6% corn gluten + 1.25% alfalfa[d]	23.0	—	23.0	5.3
50% yellow corn + 6% corn gluten + 2.5% alfalfa	29.0	—	29.0	5.8
60% milo	1.0	—	1.0	0.5
60% milo + apo ethyl ester[e]	1.0	5.0	6.0	2.0
	1.0	10.0	11.0	4.0
	1.0	15.0	16.0	6.3
	1.0	20.0	21.0	7.0
	1.0	25.0	26.0	7.4
60% milo + apo ethyl ester + canthaxanthin[f] (6:1)	1.0	4.3 + 0.7	6.0	2.1
	1.0	8.6 + 1.4	11.0	4.5
	1.0	12.2 + 2.8	16.0	6.5
	1.0	17.1 + 2.9	21.0	8.5
	1.0	21.5 + 3.5	26.0	10.1
60% milo + apo ethyl ester + canthaxanthin (4:1)	1.0	4.0 + 1.0	6.0	2.4
	1.0	8.0 + 2.0	11.0	4.7
	1.0	12.0 + 3.0	16.0	8.3
	1.0	16.0 + 4.0	21.0	9.8
	1.0	20.0 + 5.0	26.0	11.2
60% milo + apo ethyl ester + canthaxanthin (2:1)	1.0	3.3 + 1.7	6.0	2.8
	1.0	6.7 + 3.3	11.0	5.4
	1.0	10.0 + 5.0	16.0	9.5
	1.0	13.3 + 6.7	21.0	10.7
	1.0	16.7 + 8.3	26.0	11.7
50% yellow corn + 6% corn gluten + canthaxanthin	16.0	2.0	18.0	7.8
	16.0	4.0	20.0	8.7

[a] From Marusich and Bauernfeind (*547*).
[b] 25 and 50% yellow corn rations provided 3.0 and 6.0 gm oxycarotenoids per ton, respectively.
[c] Corn gluten meal (60% protein) contained 165 gm oxycarotenoids per ton (0.18 gm/kg).
[d] Alfalfa meal (25% protein) contained 408 gm oxycarotenoids per ton (0.45 gm/kg).
[e] Stabilized beadlet of β-apo-8′-carotenoic acid ethyl ester (115 gm/kg).
[f] Stabilized beadlet of canthaxanthin (100 gm/kg).

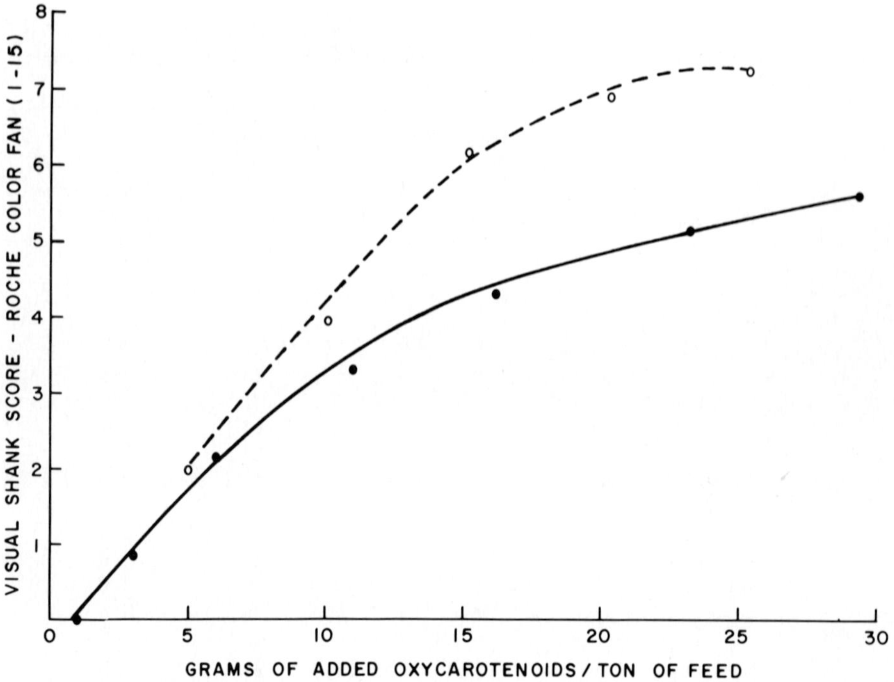

Fig. 19. Broiler shank pigmentation resulting from feeding graded levels of oxycarotenoids from feed ingredients (●—●) or APO-EE (○-○). [From Marusich (*539*).]

(*303,304*), and Harms *et al.* (*356*) concluded that the use of APO-EE offers excellent promise in broiler pigmentation studies. In a study of pigment repletion and depletion rates in broiler shanks and skins, Harms *et al.* (*356*) and Herrick *et al.* (*391*) utilized APO-EE as one of the pigmentation sources. They observed that, when birds were switched from high- to low-oxycarotenoid diets and from low- to high-oxycarotenoid diets at 0, 5, 10, 15 and 20 days before slaughter at 8 weeks of age, pigmentation values for the depleting and repleting birds equalized between the sixth and seventh day for the skin and at the tenth day for the shanks (Fig. 21). Heiman and Tighe (*384*) showed that with feed ingredient sources of oxycarotenoids, the repletion and depletion time for shanks also equalized at about the tenth day. Thomas *et al.* (*791*) observed that, when diets containing from 5 to 25 mg APO-EE per kilogram were fed for only 2 days to broilers previously held on a low-oxycarotenoid ration, serum carotenoid levels ranged from 4.7 to 19.8 μg/ml. Ferrando and Mainguy (*269*) reported that APO-EE has the highest pigmenting activity of the pigmenters on the market that can be used as the sole pigmentation source. Livingston *et al.* (*519*) also reported APO-EE to be an effective broiler pigmenter.

TABLE 37

Shank Pigmentation of Broilers Fed Several Oxycarotenoid Sources for the Last 3 Weeks before Marketing[a]

	Oxycarotenoid level (gm/ton feed)			
Basal ration	Feed ingredients	Pure	Total	Shank visual score (Roche fan, 1–15)
60% yellow corn[b]	11.0	—	11.0	2.4
60% yellow corn + 1.4% alfalfa[c]	17.5	—	17.5	3.9
60% yellow corn + 2.4% alfalfa	22.0	—	22.0	3.9
60% yellow corn + 3.5% alfalfa	26.0	—	26.0	4.3
60% milo[d] + apo ethyl ester[e] + canthaxanthin[f] (4:1)	1.5	8.0 + 2.0	11.5	4.6
	1.5	12.8 + 3.2	17.5	7.5
	1.5	20.0 + 5.0	26.5	9.6
60% milo[d] + apo ethyl ester[e] + canthaxanthin (3:1)	1.5	7.5 + 2.5	11.5	5.9
	1.5	12.0 + 4.0	17.5	7.9
	1.5	18.75 + 6.25	26.5	10.6
60% milo[d] + apo ethyl ester + canthaxanthin (2:1)	1.5	6.7 + 3.3	11.5	5.9
	1.5	10.7 + 5.3	17.5	8.6
	1.5	16.7 + 8.3	26.5	10.9
60% milo[d] + apo ethyl ester[e] + canthaxanthin (1:1)	1.5	5.0 + 5.0	11.5	9.3
	1.5	8.0 + 8.0	17.5	10.8
	1.5	12.5 + 12.5	26.5	12.6

[a] From Marusich and Bauernfeind (*547*).
[b] 60% yellow corn ration provided 11.0 gm oxycarotenoids per ton.
[c] Alfalfa meal (20% protein) contained 456 gm oxycarotenoids per ton (0.50 gm/kg).
[d] 60% milo ration provided 1.5 gm oxycarotenoids per ton.
[e] Stabilized beadlet (112 gm/kg).
[f] Stabilized beadlet (110 gm/kg).

Middendorf *et al.* (*575b*) determined the relative bioavailability of several feed ingredients oxycarotenoid sources relative to APO-EE (100%) using the rapid serum bioassay procedure described in SectionV, B, 1. Samples of corn gluten meal ranged from 64 to 89%, one marigold meal sample assayed 29%, one marigold meal extract assayed 63%, and dehydrated alfalfa meal showed 47% availability. There was also great variation among different lots of corn gluten meal, dehydrated alfalfa meal, alfalfa meal concentrates (pro-xan), marigold meal, and marigold meal extracts. Thus, the overall availability of the oxycarotenoids from these sources is relative and variable.

In trials designed to measure the partial or total replacement of corn oxycarotenoids by APO-Ee for broiler pigmentation, it was observed that 4.3 mg of APO-EE could replace 6.5 mg of oxycarotenoids from corn per kilogram. At higher levels, 7.3 mg APO-EE replaced 11.0 mg oxycarotenoids from corn, for an effective re-

TABLE 38

Shank Pigmentation of Broilers Fed Several Oxycarotenoid Sources for the Last 3 Weeks before Marketing[a]

Basal ration	Oxycarotenoid level (gm/ton feed) Feed ingredients	Pure	Total	Shank visual score (Roche fan, 1–15)
60% yellow corn[b]	10	—	10	3.9
60% yellow corn + 2.3% alfalfa[c]	20	—	20	6.3
60% yellow corn + 4.6% alfalfa	30	—	30	7.5
60% yellow corn + 6.9% alfalfa	40	—	40	8.5
60% yellow corn + apo ethyl ester[d]	10	10	20	7.3
60% yellow corn + apo ethyl ester[d]	10	20	30	9.2
60% yellow corn + apo ethyl ester[d]	10	30	40	10.2
60% yellow corn + 2.3 alfalfa +	20	2.5	22.5	8.2
canthaxanthin[e]	20	4.0	24.0	9.9
60% yellow corn + 4.6% alfalfa +	30	2.5	32.5	8.9
canthaxanthin[e]	30	4.0	34.0	10.1

[a] From Marusich and Bauernfeind (*547*).
[b] 60% yellow corn ration provided 10.0 gm oxycarotenoids per ton.
[c] Alfalfa meal (22% protein) contained 440 gm oxycarotenoids per ton (0.48 gm/kg).
[d] Stabilized beadlets of β-apo-8′-carotenoic acid ethyl ester (115 gm/kg).
[e] Stabilized beadlets of canthaxanthin (104 gm/kg).

placement ratio of 1: 1.5 for APO-EE to the oxycarotenoids in yellow corn (*42*).

Hinton *et al.* (*404*) fed 5, 10, 20, 30, and 40 mg APO-EE per kilogram of feed. All birds were evaluated visually by the Roche Yolk Color Fan and The Purina Skin Pigmentation Guide and objectively by the IDL Color Eye (a reflectance colorimeter). The authors found good correlation with the three scoring tools; however, the colorimeter was superior to the visual scoring aids. They found that APO-EE served as an excellent broiler skin pigmentation standard. They also constructed a relationship between the Roche Yolk Color Fan values for shanks and the excitation purity (EP) reading of the IDL Color Eye and between the Purina Skin Pigmentation Guide values and the EP readings for the same skin as recorded by the IDL Color Eye using the birds fed APO-EE. The authors hope that a guide such as this will enable researchers from different experiment stations to have a common point of reference for data analysis.

Since the stabilized gelatin–sugar beadlet formulations used for APO-EE and canthaxanthin are water soluble, the pigmenters can also be added to drinking water with excellent results. A daily water intake of 10–15 mg APO-EE for 8 days or a total of 80–120 mg per bird has been shown to be quite effective (*865*).

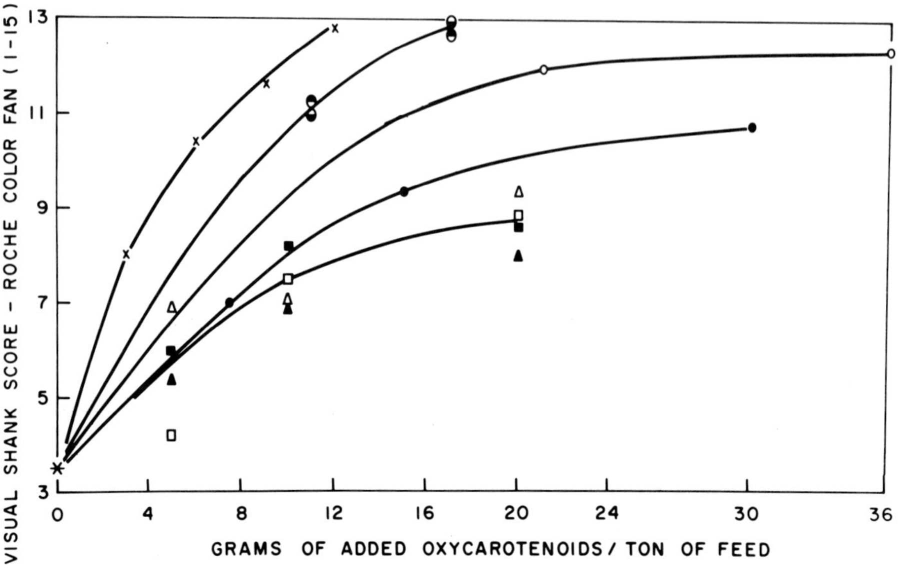

Fig. 20. Shank color response to feed ingredient oxycarotenoid sources and pure oxycarotenoids added to a yellow corn ration. Key: *, basal ration with 60% yellow corn (maize), 9.5 gm oxycarotenoids per ton; x, conthaxanthin; ●, APO-EE; ○, canthoxanthin + APO-EE; ◓, Canthaxanthin + alfalfa meal; ◒, canthaxanthin + gluten meal. Natural ingredients: △, marigold meal; □, corn gluten meal; ▲, alga meal; ■, alfalfa meal. [From Marusich and Bauernfeind (*547*).]

5. *Canthaxanthin for Skin Pigmentation*

Steinegger *et al.* (*747*) were the first to note the broiler pigmentation properties of canthaxanthin. It was also observed early that canthaxanthin cannot serve as the sole source of oxycarotenoids for broiler pigmentation. If added at low levels to a pigment-free ration, pink overtones are evident; if added at high levels to a ration containing moderate levels of feed ingredient oxycarotenoids, deep orange overtones are produced. A number of studies have shown that canthaxanthin can serve well as a pigmenter adjuvant when added at low levels to a ration reasonably adequate in other sources of pigmentation. The addition of a few grams induces increased color in shanks and skin that cannot be obtained with the equivalent amount of any other oxycarotenoid source.

Marusich *et al.* (*550,552,556*), Couch *et al.* (*171,172,179*), Camp *et al* (*173,178*), Bauernfeind (*60*), Marusich and Bauernfeind (*545,547*), Combs (*157*), Combs and Nicholson (*160*), Matterson *et al.* (*560*), Landagora and Montalvo (*491*), Marusich (*540*), Gawecki *et al.* (*305*), Scott (*712*), Bauernfeind (*61*), Ferrando and Henry (*268*), Farr and Couch (*260*), Halloran (*348*), Livingston *et al.* (*519*), and Hinton *et al.* (*404*) all have studied and reported on the pigmenting attributes of canthaxanthin.

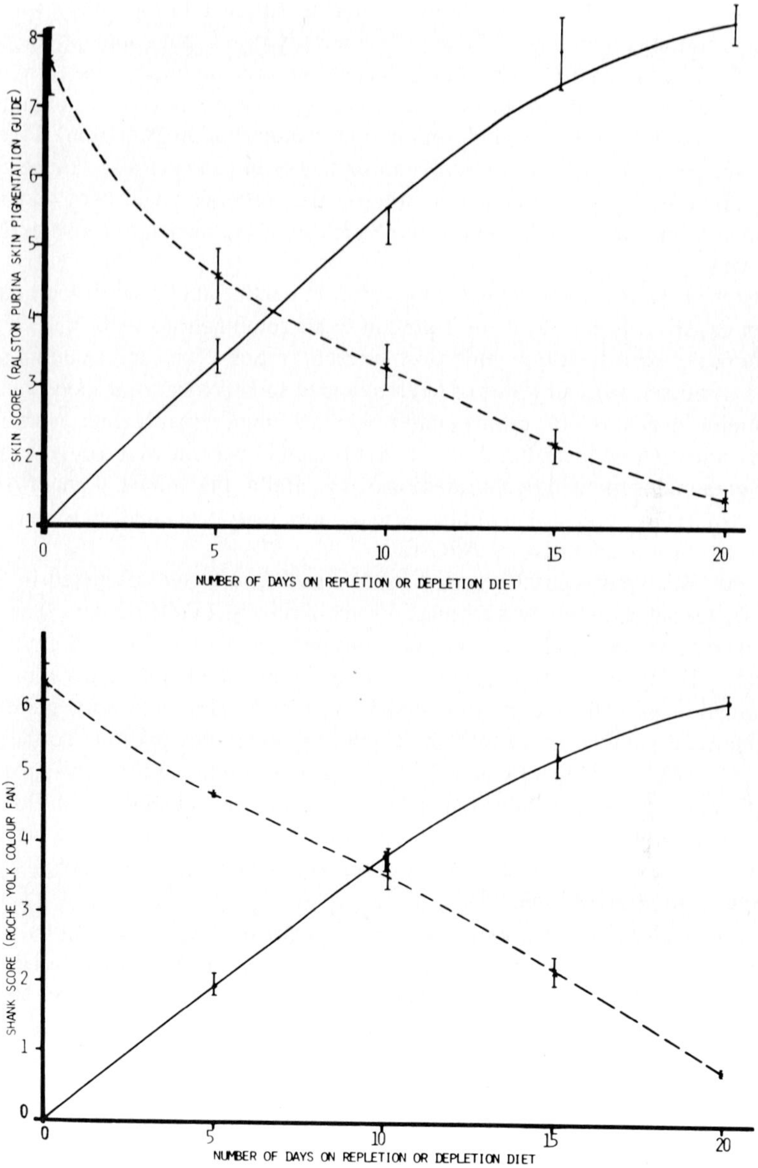

Fig. 21. Broiler pigmentation repletion (———) and depletion (---) in skin and shanks after feeding feed ingredient oxycarotenoids. [From Herrick *et al.* (*391*).]

Combs and Nicholson (*160*) presented the results of a broiler study whereby a practical ration containing 60–70% yellow corn with 8.7 gm xanthophylls per ton was fed; 12 gm of additional xanthophylls per ton were supplied in the form of corn gluten meal only, alfalfa meal only, corn gluten meal (4 gm/ton) plus alga meal (8 gm/ton), and corn gluten meal (4 gm/ton) plus canthaxanthin (8 gm/ton). Pigmentation responses were judged by visual score and by serum and toe-web skin carotenoid concentrations. By all pigmentation criteria, the combination of corn gluten meal providing 4 gm xanthophylls per ton plus 8 gm canthaxanthin per ton gave the highest responses.

Matterson *et al.* (*560*) evaluated the effect of canthaxanthin when fed in a block design experiment at 0, 2, 4, or 8 gm/ton in all combinations with 0, 1, 2, or 4% alfalfa using a practical-type milo (low-pigment) ration. The canthaxanthin supplements combined with all levels of alfalfa yielded satisfactory color; however, when the alfalfa level was 0%, canthaxanthin produced unacceptable pink overtones. In the combination with alfalfa, 2 gm of canthaxanthin per ton were equivalent to the oxycarotenoids provided by the addition of 1% alfalfa. In contrast, 4 gm of APO-EE were required to replace 1% alfalfa. Thus, under these test conditions, canthaxanthin was twice as effective as APO-EE.

In the collaborative trial reported by Marusich (*540*) and discussed in Section VII, D, 4, canthaxanthin was included by six of the eight collaborators. The yellow corn alone contributed 10 gm oxycarotenoids per ton. An additional 10 gm/ton were provided by alfalfa meal to provide a total of 20 gm xanthophylls per ton. A third combination of yellow corn and alfalfa meal with additional corn gluten meal providing 10 gm/ton gave a total of 30 gm oxycarotenoids per ton. To the yellow corn ration (10 gm/ton) were added 2.5 gm canthaxanthin per ton. This group gave visual scores for shank color equal to the visual score obtained with 20 gm/ton utilizing yellow corn plus alfalfa based on the average values for the six participants. Thus, 2.5 gm canthaxanthin per ton equalled 10 gm xanthophylls per ton derived from alfalfa (Table 37).

Halloran (*348*) and Halloran *et al.* (*350*) supplemented broiler rations, containing either a 45% protein alfalfa concentrate or 19% protein dehydrated alfalfa meal, with canthaxanthin and observed improved visual pigmentation scores using the Roche Yolk Color Fan. The gizzard fat was evaluated at the same time as the carcass color and yielded equal color fan values. Halloran (*348*) stated that this technique is especially useful when broilers are processed at ''scalding'' temperatures (temperatures far above 128°F), which may drastically affect shank and skin color by actual removal of the outer skin layer.

Marusich *et al.* (*556,557*) presented data indicating that 1 gm of canthaxanthin replaced more than 2 gm of oxycarotenoids from feed ingredients (see Tables 7 and 8). corn alone, from a combination of corn and corn gluten meal, or from a combination of corn and dehydrated alfalfa meal. Similarly, combinations of canthaxanthin and APO-EE were reported as highly effective. Typical responses are shown in Table 36 and Fig. 22. For maximal efficiency, the yellow carotenoids should be present in

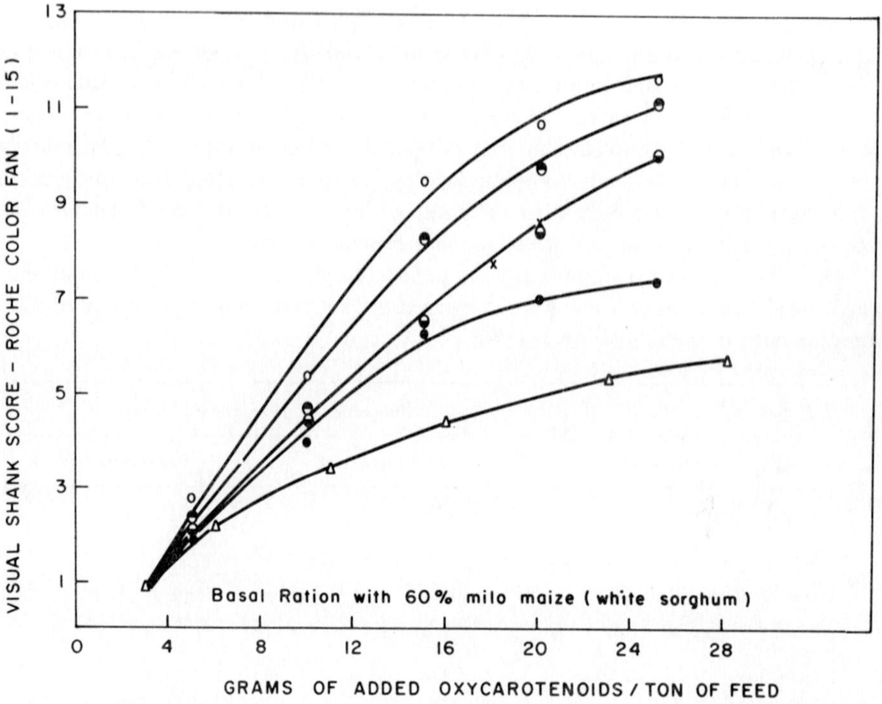

Fig. 22. Shank color response to single and blended pure oxycarotenoids added to a milo ration (basal ration with 60% milo–maize (white sorghum). Key: ●, APO-EE; ○, APO-EE + canthaxanthin (2:1); ◓, APO-EE + canthaxanthin (4:1); ◒, APO-EE + canthaxanthin (6:1); △, natural ingredients; x, natural ingredients + canthaxanthin. [From Marusich and Bauernfeind (*547*).].

ratios of 4 to 5 parts or higher to 1 part of canthaxanthin. This will ensure improved color without eliciting orange overtones, which are acceptable or even desirable in some geographic areas but unacceptable in others.

Marusich *et al.* (*556,557*) presented data indicating that 1 gm of canthaxanthin replaced more than 2 gm of oxycarotenoids from feed ingredients (see Tables 7, 8). Later, the same workers (*553*) showed that equivalent breast skin and shank skin visual scores were obtained with 20 gm of feed ingredient oxycarotenoids plus 2.5 gm canthaxanthin added per ton of feed as were obtained with 30 gm of feed ingredient oxycarotenoids added per ton when fed for the last 3 weeks before marketing. This is an effective replacement ratio of 1 gm canthaxanthin for 4 gm of yellow oxycarotenoids (see Table 12).

In another study, Halloran (*349*) reported that 1 gm of canthaxanthin could replace 3 gm of xanthophylls provided by pro-xan in a high-energy broiler ration, as shown by visual scores for breast, thigh, and foot pad skin using the Roche Yolk Color Fan. Bauernfeind and Marusich (*68*) found that 1 gm of canthaxanthin could replace up to 4 or more grams of the yellow oxycarotenoids to provide the expected yellow-orange range of pigmentation (Roche Yolk Color Fan values of 5–10). In geographic areas where highly pigmented birds with orange overtones are desired or even preferred, canthaxanthin may comprise 15–20% of the total oxycarotenoid content of the broiler ration. In areas where lighter shades of yellow to yellow-orange are desired, canthaxanthin may provide 5–10% of the total oxycarotenoids in the ration. The wide range of pigmentation potential resulting from the use of canthaxanthin as an adjuvant or enhancer of yellow oxycarotenoids enables the broiler producer to meet any color preference of consumers.

Feed ingredient oxycarotenoids from corn, corn gluten meal, dehydrated alfalfa, alga meal, and marigold meal are compared to canthaxanthin as broiler pigmentation sources in the tabulation below.

Factor or characteristic influencing pigmentation	Properties of oxycarotenoid source	
	Feed ingredient	Canthaxanthin
Bread and strain	Same	Same
Sex	Same	Same
Coccidiosis	Same	Same
Length of feeding	Same	Same
Dose–response	Same	Same
Stability	Variable	Stable
Availability	Variable	Uniform
Oxycarotenoid profile	Complex	Single, pure
Deposited as fed	Variable	Yes
Concentration	Wide range (0.001–1.0%)	10% stabilized product
Absorption/deposition efficiency	Variable; dependent on specific oxy-carotenoid profile	Uniform Uniform

E. Variety of Oxycarotenoid Sources

Although the principal sources of oxycarotenoids for both egg yolk and broiler pigmentation today are still yellow corn, corn gluten meal, and dehydrated alfalfa meal, as noted by Scott (*712*), Lautner (*494*), Hartel (*360*), Groza (*337*), Bauernfeind (*61*), Comben (*155*), Kingan and Sullivan (*453*), Hunter (*417*), and Kathe (*450*), many other sources of oxycarotenoids have been considered. Numerous feed ingredients, forages, grasses, and even exotic sources of oxycarotenoids have been evaluated for their pigmentation potential. The following are those that have been reported to result in some favorable pigmentation response or at least to have had a positive effect: Hughes and Payne (*415*) with green barley grass; Titus *et al.* (*812*) with chili pepper meal and pimiento pepper meal; Sipe and Polk (*724*) with green mustard, rye, oat, and kikuyu grass; Tomhave and Hoffmann (*813*) with broccoli leaf meal, carrot top meal, pea vine meal, lima bean vine meal, and turnip top meal; Damron and Janky (*197*) with peanut leaf meal; Squibb et al. (*739–742*) with ramie meal, kikuyu grass, desmodium meal, ixbut meal, teosinte meal, bamboo leaf meal, banana leaf meal, sweet potato meal, achiote meal (annatto seed meal—source of bixin), and zempa meal; Garlich *et al.* (*303*) with sweet potato vine meal; Fritz *et al.* (*290*) with carrot meal, red pepper powder, chili pepper meal, paprika, paprika oleoresin, sweet potato meal, sunflower blossom meal, annatto seed and hull meal, and stinging nettle meal; Runnels *et al.* (*693*) with dehydrated brocolli leaf meal; Schaible *et al.* (*700*) with dehydrated lettuce meal; Madiedo and Sunde (*528*) with alga meal and dried lake weed meal; Creger *et al.* (*184*) with aquatic flowering plants; Madiedo *et al.* (*527*) with marigold petals (*Tagetes erecta*); Carlson *et al.* (*146*) with carrot meal and paprika; Landagora (*490*) and Landagora and Montalvo (*491*) with achiote seeds; Ratcliff *et al.* (*663,664*) with clover meal; Head (*373*) with pyrethrum extract from chrysanthemums; Scholfield and Dutton (*704*) with soybean lecithin; Bornstein and Lipstein (*110*), Budowski *et al.* (*129*), Lipstein *et al.* (*508,509*), Menge (*571*), and Menge and Beal (*572*) with by-products from the refining of soybean oil (acidulated soybean soap stocks); Scott *et al.* (*713*) with pollen from beehives; Wilkinson and Barbee (*872*) with pearl millet and coastal Bermuda grass; Høie and Sannon (*408*), Jensen (*436*), and Booth (*105*) with seaweed meals; Meyers and Rutledge (*574*) with shrimp meal wastes; Blattna *et al.* (*91*) with celandine (*Chelidonium majus*); and others with dehydrated turf grass clippings (*33–35,37,43,571*). More recently (*44*) a leafy green vegetable from the highlands of Ethiopia, *Brassica carinata,* a relative of collard and mustard greens, has been shown to be effective. Fritz *et al.* (*290*) presented an extensive list of items tested for broiler pigmentation, and those that were somewhat effective have been cited above. They also listed a number of additives that had very little or no effect on broiler pigmentation, including corn distillers' dried grains, dried tomato pomace, fresh tomatoes, palm oil or palm oil foots, cottonseed meal, and turmeric. Runnels (*691*) studied the effect of 2.5, 5.0, and 7.5% corn distillers' dried solubles (CDDS) on broiler performance, including pigmentation. He reported that the addition of 2.5% CDDS resulted in increased shank color. However, the further addition of 5.0 or 7.5% did not increase the color intensity of the shanks over that attributable to the 2.5% level of CDDS. To our knowledge, there are no other reports of CDDS contributing to pigmentation. De Groote (*219*) and Scott *et al.* (*713*) have also compiled lists of sources of oxycarotenoids for yolk pigmentation.

F. Current Worldwide Oxycarotenoid Sources

In the United States, the poultry industry depends on yellow corn, corn gluten meal, dehydrated alfalfa meal, marigold meal or extraction of marigold petals, providing a more concentrated product, and special high-protein alfalfa-derived products, such as pro-xan (X-Pro). Table 21 lists these major pigmentation sources as well as average content and range of oxycarotenoid content found by chemical assay.

Worldwide, numerous items are utilized, some quite effectively and others not so effectively. These include grass, lucerne, nettles, and similar green crops, which may be dried, ground, and used without further processing in the form of a meal or extracted, with the pigment extract being absorbed on a cereal carrier. Xantalfa 10 Br is one of the most important in this category; it is manufactured in France and used primarily for broilers. Another stabilized lucerne extract, Probione, has been shown to be very effective for yolks in Belgium (*97*).

Commerically available raw paprika varies widely in quality depending on its origin, growing conditions, and period of storage. Paprika and its derivatives are used mainly for egg yolks and to a lesser extent for broiler pigmentation, since the deposition rates of capsanthin and capsorubin in the skin are poor. An extract of paprika is sold under the trade name Duphacolor. The oxycarotenoid content of this product is almost 10 times that of raw paprika, and it is stabilized by gelatin coating. Its pigmenting properties per unit of oxycarotenoids are comparable to those of raw paprika.

Combination products are also available. In principle, these products are combinations of yellow and red oxycarotenoids. Lutenal in Germany is a combination of lucerne extracts plus paprika or paprika extracts. Pigmentene is also thought to be a combination of a paprika-like product and an extract from marigold petals.

Alga meals and fermentation products have been used in the United States (A-Zanth) and in Europe, Australia, Norway, and South Africa. A-Zanth had been withdrawn from the U.S. market, apparently for economic reasons. It had properties similar to those of dehydrated alfalfa meals.

Annatto products, which contain bixin and other oxycarotenoids from the seeds of the annatto tree in the form of ground seeds as well as extracts, have been used but with basically poor results. Products include Extracolor in France, Plantengeel in Belgium, and Bixinol in Germany.

Acidulated soybean soapstock may contain from 160 to 250 mg oxycarotenoids per kilogram; however, the pigmenting efficiency is less than that of corn xanthophylls. The product has been used in Israel for both broiler and yolk pigmentation.

An extract of lucerne meal and orange peel and possibly other materials is sold in Israel and Portugal as Plantexine. Another is the residue from the refining of pyrethrum extract. The waxy material obtained is absorbed onto a carrier of plant origin and is available in Italy as H. T. Pigmenter. Citranaxanthin in a stabilized beadlet form is available (Avizant) in Europe primarily for yolk pigmentation. It is also sold in combination with lucerne meal.

Canthaxanthin and APO-EE are sold worldwide as 10% concentrations in a gelatin stabilized form as Carophyll-Yellow and Carophyll-Red, respectively. A combination product (Carophyll-Orange) contains 5% APO-EE plus 5% canthaxanthin in the gelatin matrix. All have been well received for both yolk and broiler pigmentation, as documented in Section VII,D,1,3–5.

We have thoroughly treated poultry pigmentation in the preceding pages. The last area to be covered is achieving maximal effect at the least possible cost. In geographic areas and markets where broiler skin color is important to the producer because of consumer perference, millions of dollars are spent by the industry on pigment sources to ensure the depth and uniformity of color desired.

VIII. MODE OF FEEDING OXYCAROTENOIDS

A. Broiler Production

In order to achieve proper pigmentation of shanks and skin of broilers one need not feed oxycarotenoid sources throughout the life of the broiler (essentially a 7- to 8-week period in the United States). The broiler starter ration, which is routinely fed for the first 3–4 weeks of life, may contain substantially lower levels of oxycarotenoids than the broiler grower and finisher rations, which are fed for the last 3–4 weeks and contain higher levels of oxycarotenoids to provide adequate pigmentation for the marketed chickens. Bird (*88*) stated that excellent shank pigmentation was induced in a period of only 10 days by proper feeding. Day and Williams (*211*) reported that the utilization of xanthophylls was greater during the latter part of the gorwing period. Fritz *et al.* (*290*) and Mitchell *et al.* (*581,584*) found that desirable pigmentation resulted from feeding oxycarotenoid-containing rations for the last 4 weeks of the growing period. Combs and Nicholson (*160*) concluded that supplementation of the ration during the last 4 weeks was as effective as feeding for the entire period. It was reported by Couch *et al.* (*179*) that skin and shank pigmentation could be accomplished within 2.5–3 weeks and that it was not necessary to use high-oxycarotenoid rations except during the finishing period. Conversely, Heiman (*382*) stated that any given level of xanthophylls produced the highest degree of pigmentation when fed for the longest time. Bartov and Bornstein (*58*) also reported that the best results were obtained when oxycarotenoids were fed throughout the entire growing period. Higher oxycarotenoid levels (approximately two-fold higher) are required (Table 39) when supplements are fed for a shorter time but lesser amounts of total oxycarotenoids are required (Fig. 23), as shown by Marusich *et al.* (*547*). The pigment content of broiler tissues increased or decreased at a similar rate when the oxycarotenoid level was either increased or decreased, as reported by Heiman and Tighe (*384*), Heiman (*382*), Harms *et al.* (*356*), and Herrick *et al.* (*391*) (Fig. 24). Gazo *et al.* (*307*) reported on the improving of pigmentation in ducks during the growing period with results comparable to those of broilers.

B. Egg Production

The transfer of oxycarotenoids from the feed to the egg yolk is also a very rapid process. Heiman and Wilhelm (*385*) reported that dietary xanthophylls were transferred within 48 hr after feeding began. Rates of repletion and depletion in yolks

TABLE 39

Effect of Graded Levels of Canthaxanthin Fed for Variable Periods on Shank Color[a]

Canthaxanthin supplement to basal ration[b] (gm/ton feed)	Duration of feeding (weeks)	Oxycarotenoid intake (mg) Feed ingredients	Canthaxanthin	Ratio	Total	Visual score (Roche Yolk Color Fan)
0	8	61	0	—	61	5.4[c]
3	8	62	14	4.4:1	76	8.7
6	8	63	30	2.1:1	93	11.6
3	6	62	11	5.6:1	73	8.5
6	6	59	23	2.6:1	82	10.8
3	4	62	8	7.8:1	70	8.2
4.5	4	59	12	4.9:1	71	10.0
6	4	61	17	3.6:1	78	10.4
9	4	59	22	2.7:1	81	11.6
6	3	65	14	4.6:1	79	10.0
9	3	64	20	3.2:1	84	10.4
12	3	64	28	2.3:1	92	11.2
Basal + 2.5% alfalfa[d] + 2.5% corn gluten meal[e,f]	8	126	—	—	126	10.2

[a] From Marusich and Bauernfeind (*547*).
[b] 60% yellow corn ration provided 13 gm oxycarotenoids per ton.
[c] Average for 50 birds; all others, average for 10 birds.
[d] Alfalfa meal (20% protein) contained 360 gm oxycarotenoids per ton (0.40 gm/kg).
[e] Corn gluten meal (60% protein) contained 180 gm oxycarotenoids per ton (0.20 gm/kg).
[f] Ration provided 26.4 gm oxycarotenoids per ton.

were shown by Marusich *et al.* (*548*). Yolk color intensity reaches a plateau (constant concentration) in from 8 to 14 days depending on the sources of oxycarotenoids fed. The plateau begins between the ninth and eleventh day (Fig. 1). Couch *et al.* (*179*) stated that 9–10 days are required for egg yolks to reach the maximal level of pigmentation when alfalfa meal or yellow corn is used in the ration as the oxycarotenoid source. Williams *et al*. (*876*) also found 8–10 days as the time for the egg to reach a constant value. Hinton *et al.* (*401,405*) demonstrated in pullets maintained on an oxycarotenoid-free ration until the first egg was laid that comparably colored yolks could be produced within 2 weeks after hens were placed on a ration rich in oxycarotenoids when scored against yolks of eggs from pullets continuously fed the oxycarotenoid-rich ration.

Figure 24 shows the relative efficiency of pigment deposition in the egg yolk for various oxycarotenoid sources. The values are based on average data obtained in numerous trials in which the oxycarotenoid sources were fed for 20 days to hens previously maintained on a low-oxycarotenoid layer ration. Eggs were collected

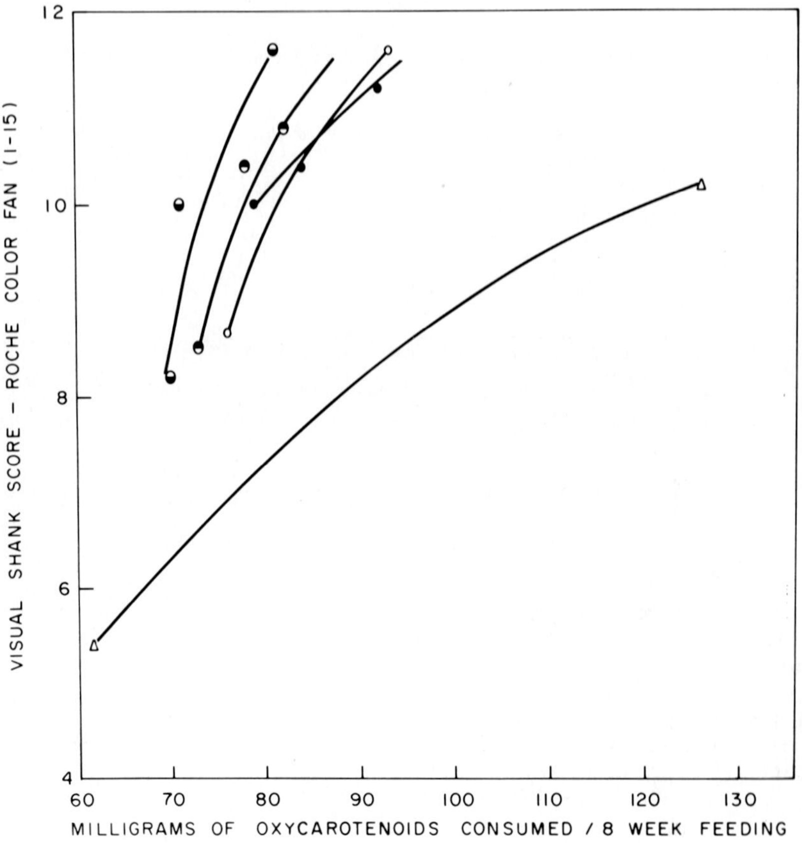

Fig. 23. Shank color response to feed ingredient oxycarotenoids and canthaxanthin consumed for different time periods. Key: ○, canthaxanthin + natural ingredients (8 weeks); △, natural ingredients (8 weeks); ◓, natural ingredients (2 weeks) + canthaxanthin (4 weeks); ●, natural ingredients (4 weeks) + cathaxanthin (4 weeks); natural ingredients (5 weeks) + canthaxanthin (3 weeks). [From Marusich and Bauernfeind (*547*).]

from the tenth through the twentieth day. The yolk was separated from the albumen, and the pigment content was extracted using a 1:1 mixture of acetone and chloroform and read against a β-carotene standard, as described by Marusich (*538*). The percent pigment deposition was calculated by dividing the total pigment content in the yolk by the average daily intake of oxycarotenoid and multiplying by 100.

Due to the rapidity of oxycarotenoid depletion and repletion, Lifschitz *et al.* (*507*) used this property of yolks to study the ovule maturation process and the possible causes of interruptions in the oviposition cycle of hens. They concluded

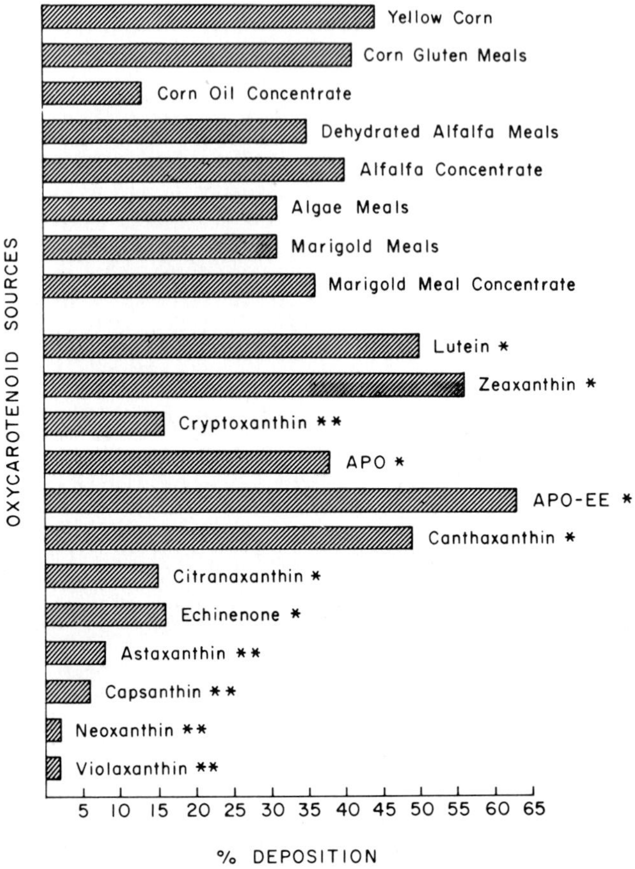

Fig. 24. Relative deposition of pigment in egg yolks after feeding oxycarotenoids from feed ingredients and single sources (*, in stabilized beadlets; **, in oil solution).

that a change in feeding conditions every 7 days (alternating low pigment to high pigment content) produced effects in the yolk pigment content that coincided with the period of fast maturation. The results also suggested that the yolk pigment content could allow for a study of the way in which a pause in laying is manifested.

Nelson (*604*) studied the deposition of oxycarotenoids in the yolk of Japanese quail (*Coturnix coturnix japonica*). Increased yolk color was observed within 2 days after supplementation began and reached a maximum after 7 days. When the quail were placed on a low-pigment ration, 8 days were required to deplete the yolk pigments. The Japanese quail could serve as a substitute for the laying hen in the evaluation of pigment sources in the laboratory, especially when small samples of oxycarotenoid are to be tested.

IX. QUANTITATIVE REQUIREMENTS FOR OXYCAROTENOIDS

A. Broilers

The level of oxycarotenoids required in a ration to provide adequate pigmentation can vary widely depending on the intensity of pigmentation desired by a particular market. Various levels, therefore, have been cited by numerous investigators as necessary for "adequate" pigmentation. House (*410*) reported that there is a gradual increase in the degree of shank color as one progresses from 0.8 to 12.0 mg of oxycarotenoids per pound of feed. He concluded that 9–10 mg/lb were required to obtain acceptable shank color using yellow corn, corn gluten meal, and alfalfa meal. Fritz *et al.* (*290*) fed graded levels of xanthophyll (lutein) from 0 to 100 mg/lb and determined that the addition of 12.5 mg lutein per pound produced good average pigmentation. Around 25–40 mg/lb gave the maximal intensity of color that could be obtained with this pigment based on visual shank scores. Day and Williams (*211*) found that 6.25 mg oxycarotenoids from yellow corn and other natural ingredient sources per pound provided adequate color; however, a level of 5.0 mg/lb was deemed inadequate. Mitchell *et al.* (*583*) cited 6.5 mg oxycarotenoids from yellow corn and alfalfa per pound as producing desirable color. Hastings (*366*) suggested levels between 6.25 and 12.5 mg/lb depending on feeding practices. If fed continuously for 8 weeks, 6.25 mg/lb; if fed in broiler grower feed for only 2–3 weeks, 12.5 mg/lb. Couch *et al.* (*179*) considered levels of 8 to 12 mg/lb to be adequate. However, they qualified these by pointing out that many factors, such as stability, source of xanthophylls, and state of health of the flock, influence these values.

Current levels may range from a low of 1–5 mg/lb to a high to 20–25 mg/lb depending on the intended use of the broiler and the consumer market preference. In areas in which very lighly pigmented broilers are preferred (so-called milk-fed broilers) or when broilers are sold as prepared foods, such as breaded, frozen, or breaded take-home items, the rations may be formulated without any corn gluten meal oı alfalfa meal, and even milo (low in xanthophylls) may be substituted for yellow corn for economy of feed formulation. On the other hand, certain areas demand, and producers insist on, a highly pigmented bird as a sign of premium quality. These producers formulate grower and finisher rations with the highest levels of oxycarotenoids that are economically feasible. The feed ingredients and concentrates supplying oxycarotenoids contribute substantially to the cost of a ration. Worldwide, consumer preferences as to the degree of pigmentation of broilers vary widely. For example, in Mexico, feeds are formulated to contain as much as 50 mg oxycarotenoids per pound of feed, or even higher, in order to meet the demands for highly pigmented birds. Consumers in Germany and Italy desire dressed birds with deep yellow skin, whereas in England and France the preference is for white skin (*169*).

B. Egg Yolks

The quantity of oxycarotenoids required for egg yolk pigmentation falls into two groups depending on the intended use of the egg. For table eggs, low levels are required, whereas high levels are required for eggs utilized in noodles, bakery products, mayonnaise, and other prepared foods in which a yellow color is desired and must be provided by egg yolk rather than by the direct addition of colorants. The latter eggs are "breaker eggs" and represent a sizable percentage of the total egg production, which was estimated to be 10% in 1961 by Carlson (*141*) for the United States. Hastings (*366*) found that 10 mg xanthophylls per pound was required to produce eggs with a medium-colored yolk that were acceptable as fresh table eggs. The degree of color in table eggs is also subject to consumer preferences, as shown by studies conducted by Jasper and Cray (*434*), Slocum and Swanson (*727*), Sunde and Madiedo (*780*), and Parkhurst (*625*). Regional and nationality preferences have been reported, for example, by Blommaert and Steenis (*97*) for Belgium, in which deeply colored yolks with a definite orange tinge are preferred. Quilici (*658*) stated that in Italy consumers also prefer deeply colored yolks. Similar findings were reported by Villavicencio (*831*) in Peru, Thumin (*809*) in Israel, Hagemann (*345*) in Switzerland, Coles (*153*) in Great Britain, Buvanendran (*136*) in Ceylon, Saunders (*698*) in South Africa, Schmidt (*703*) in Germany, and by others in Australia (*21,22,24,27*), New Zealand (*18*), Northern Ireland (*15*), Colombia (*29*), South Africa (*19*), and the United States (*16,17*).

In some of these countries, such as Australia (*21,22,27*), Belgium (*97*), Colombia (*29*), and New Zealand (*18*), the desired color for table eggs is related to a visual color guide (Roche Yolk Color Fan) with the actual intensity equal to or greater than a reading of 8 on the fan in order to obtain a premium price. This color fan is discussed in detail in Section V,A.

Uniformity of yolk color in the United States is probably as important as, if not more so than, the intensity of yellow color in table eggs, according to Fritz (*287*). Furthermore, producers, probably more than consumers, actually prefer light yolks to more deeply pigmented ones. Lightly colored yolks help to improve candling grade because yolk shadow scores are improved, as documented by Mueller (*598*) and Tarver (*784*). Egg yolk color and visibility of the yolk shadow during candling are closely correlated, as reported by Parker *et al.* (*624*), Almquist (*11*), Wilhelm and Payne (*871*), and Mueller (*597*). Albumen quality, determined by candling, is reduced when the yolk is highly pigmented (*598,784*). Albumen quality, as measured by Haugh units (*370*) on broken-out eggs, however, was equivalent regardless of the degree of yolk pigmentation (*598*). Tarver (*784*) obtained lower Haugh units as the level of oil supplying xanthophylls was increased. Hinton *et al.* (*400,403*) reported that candled grade of eggs is inversely related to oxycarotenoid content and that yolk color per se can affect candling grade. In these trials, Haugh units were not adversely affected. Fry and Harms (*296*) found candled quality to be negatively correlated with yolk hue (dominant wavelength) and intensity (excitation purity).

All eggs were graded on the basis of yolk shadow only by a federal state grader, and the yolk color was evaluated with a reflectance colorimeter (IDL Color Eye). Others recognized that the depth of yolk color is a modifying influence on candle grade (*11,624,778,784*). To obtain the darkly colored egg yolks desired by the food processing industry, 10–30% alfalfa meal is required in the ration, according to Carlson (*140*) and Carlson *et al.* (*146*). The levels of xanthophylls required are 25 mg/lb or higher. Farr *et al.* (*261*) and Couch *et al.* (*179*) also reported that 25 mg of xanthophylls from alfalfa meal, yellow corn, and other feed ingredients were required per pound to produce darkly colored yolks. Similar levels of addition were also reported by Carlson *et al.* (*144*) and Sullivan and Holleman (*777*) to produce "breaker eggs" of a satisfactory depth of color. As noted earlier, Scott *et al.* (*713*) pointed out that 30–40% dehydrated alfalfa meal may be required when temperatures are high and feed intake reduced. Similarly, Couch (*170*), Couch *et al.* (*179*), and Farr *et al.* (*261*) showed that hot summer weather may decrease yolk pigmentation by 33%.

X. LINEAR PROGRAMMING (COMPUTER LEAST-COST FORMULATION)

The use of computers in animal agriculture has provided a rapid means of formulating nutritionally balanced rations at least cost. The savings per ton of feed may range from minimal to substantial.

In the broiler industry, computers are used extensively to calculate least-cost rations at frequent intervals when significant changes in ingredient costs have occurred since the previous calculation. In some regions, to obtain the desired pigmented product, not only are the minimal energy, protein, amino acids, calcium, and phosphorus levels required for broiler rations calculated; the xanthophyll or oxycarotenoid levels and sources are also cost-calculated by nutritionists to select the most economical source.

A market research study carried out in 1974 indicated different minimal levels of xanthophylls in broiler grower and finisher rations throughout the broiler-producing areas of the United States (Table 40). Four geographic regions—California, parts of North Carolina, Delmarva Peninsula (Delaware–Maryland–Virginia) and Maine—require high levels of xanthophylls in their broiler grower and finisher feeds. The levels of oxycarotenoids in regional broiler grower rations are dictated by consumer preference for light or well-pigmented broilers.

In the United States, starter rations are generally fed for the first 3 or 4 weeks, a grower ration is fed from 3 to 6 weeks or 4 to 7 weeks, and a finisher ration is fed for the last week before marketing. The oxycarotenoids are substantially increased in the grower compared to the starter ration and may be kept the same or increased in the finisher ration compared to the grower when pigmentation is of prime concern.

The commonly used pigmentation sources, besides yellow corn, are corn gluten

TABLE 40

Requirements of Oxycarotenoids in Broiler Rations in Various Areas of the United States

Location	Growers sampled	Oxycarotenoids (mg/lb feed)		
		Starter ration	Grower or finisher ration	Withdrawal ration
California	2	6–8	17–18	17–18
Texas	2	4	8.5–10	9.5–10
Georgia	4	8	7–9	7–10
Arkansas	6	6	7–10	8–12
North Carolina	1	6–8	10–20	10–20
South Carolina	2	—	8–14	8–14
Alabama	2	5	6–10	7–12
Mississippi	4	8	6–10	7–11
Delaware–Maryland–Virginia (eastern shore)[a]	5	8–12	14–22	15–24
Maine	2	19	16–20	16–20

[a] Delmarva Peninsula.

meal (60% protein) dehydrated alfalfa meal (17% protein), concentrates from marigold meal, and marigold meal per se. The xanthophylls in corn gluten meal are assumed to be utilized by the broiler at the rate of 100%, whereas the xanthophylls in 17% dehydrated alfalfa meal, the concentrate from marigold meal, and marigold meal are assigned availabilities of 80, 80, and 60%, respectively.

The biological values of these oxycarotenoid levels for each ingredient are important because the computer incorporates the pigmenting source or sources to meet the required oxycarotenoid level for a broiler grower and finisher ration on the basis of *per unit cost*. The computer brings in a certain amount of 60% protein corn gluten meal for its nutrient contribution to the overall ration; however, after the amino acid requirements have been met, its further use depends on its unit cost for xanthophylls compared to the unit cost ($/gm) of other oxycarotenoid sources.

Broiler grower or finisher rations requiring from 4 to 12 mg xanthophylls per pound generally do not need supplemental oxycarotenoid sources. Over the 4–6 mg/lb range, the requirement is met with yellow corn alone and up to the 12-mg level, yellow corn and corn gluten meal generally are sufficient to meet the oxycarotenoid requirement. At levels higher than 12 mg/lb of feed, additional oxycarotenoid sources will be required. Here least-cost calculations will decide the choice of pigment supplement.

XI. FD&C FOOD COLORS

Numerous FD&C colors have been evaluated for both yolk and broiler pigmentation. None that theoretically could induce the correct color have been effective.

Denton (*228*) administered from 10 to 50 mg of FD&C Red No. 1 (Ponceau 3R), FD&C Red No. 2 (Amaranth), FD&C Orange No. 1 (Orange 1), and FD&C Yellow No. 3 (Yellow AB) in capsules per hen daily. These daily doses would be equivalent to adding FD&C colors to the feed at from 90 to 450 gm/ton. No change in color of the egg yolk or albumen was observed with any FD&C color.

Fritz *et al*. (*290*) evaluated FD&C Yellow No. 3 and FD&C Yellow No. 4 at 0.05% in the ration (454 gm/ton) for broilers and found neither to have any effects on pigmentation. Marusich (*535*) tested four more FD&C colors—Yellow No. 5 (Tartrazine), Yellow No. 6 (Sunset Yellow FCF), Red No. 3 (Erythrosine), and Red No. 4 (Ponceau SX), for their ability to pigment egg yolks and/or albumen. The FD&C colors were added to a low-pigment hen layer ration at 0.10% (908 gm/ton), equivalent to feeding about 100 mg FD&C color per hen daily, and fed for 10 days. No color change was observed in either the albumen or the egg yolk.

From these three studies, it can be concluded that these FD&C artificial coloring agents are not absorbed and deposited in poultry tissues as are the distinctly different oxycarotenoid structures. Thus, the FD&C artificial colors are not contenders as poultry pigmenters for inclusion in poultry feeds.

REFERENCES

1. Aasen, A.J., and Jensen, S.L. (1967). *Acta Chem. Scand.* **20,** 2322–2342.
2. Aasen, A.J., Jensen, S.L., and Borch, G. (1972). *Acta Chem. Scand.* **26**, 404–405.
3. Abbott, O.J., Bird, H.R., and Cravens, W.W. (1954). *Poult. Sci.* **33**, 1245–1253.
4. Adamstone, F.B. (1942). *Am. J. Med. Technol.* **8**, 76–83.
5. Agrawal, D.K., and Talapatra, S.K. (1969). *Indian Vet. J.* **46,** 692–696.
6. Alam, A.U., Couch, J.R. and Creger, C.R. (1968). *Can. J. Bot.* **46**, 1539–1541.
7. Alam, A.U., Creger, C.R., and Couch, J.R. (1968). *J. Food Sci.* **33**, 635–636.
8. Albright, W.P., and Thompson, R.B. (1935). *Poult. Sci.* **14**, 373–375.
9. Al-Hasani, S.M., and Parrish, D.B. (1968). *J. Nutr.* **94**, 402–406.
10. Al-Hasani, S.M., and Parrish, D.B. (1972). *J. Nutr.* **102**, 1437–1440.
11. Almquist, H.J. (1933). *Calif., Agric. Exp. Stn., Bull.* **561.**
12. Anjaneyalu, Y.V. (1962). PhD. Thesis, University of Arizona, Tucson.
13. Anjaneyalu, Y.V., Kurnick, A.A., and Reid, B.L. (1961). *Poult. Sci.* **40**, 1372.
14. Anjaneyalu, Y.V., Kurnick, A.A., and Reid, B.L. (1963). *Proc. Soc. Exp. Biol. Med.* **113**, 275–278.
15. Anonymous, (1954). Leaflet No. 57. Ministry of Agriculture, Government of Northern Ireland.
16. Anonymous (1953). *Ohio, Agric. Exp. Stn., Res. Bull.* **736.**
17. Anonymous (1954). *Wash., Agric. Exp. Stn., Bull.* **556**.
18. Anonymous (1961). *N. Z. Egg Mark. Auth., Circ.* **57**.
19. Anonymous (1962). Circular Letter No. 43. Egg Control Board, Ireland.
20. Anonymous (1962). *Feeds Illustrated* **13**(11), 24–25.
21. Anonymous (1963). *Australas. Poult. World* Sept. 10. 1–2.
22. Anonymous (1963). *Poultry, Sidney Aust.* Aug. 29. 1–4.
23. Anonymous (1963). *Poultryman* **33**(9), 23.
24. Anonymous (1963). Newsletter No. 16. Western Australian Egg Marketing Board.
25. Anonymous (1964). *Feedstuffs* **36**(48), 55.
26. Anonymous (1964). *Poult. Dig.* **23**(266), 158–160.
27. Anonymous (1965). Information Bulletin 1-3, Feb. 10. Western Australian Egg Marketing Board.

28. Anonymous (1965). *Poultryman* **35**(2), 7.
29. Anonymous (1966). "Marketing Eggs in Bogata," Part 5. Instituto Latinomericano de Mercadeo Agricola.
30. Anonymous (1966). *Poult. Int.* **5**(5), 62.
31. Anonymous (1967). *Gefluegelhof* **30**(25), 516–518.
32. Anonymous (1969). "Noraid Papers," Vol. 2(1). Norwich Pharmacal Co., Norwich, New York.
33. Anonymous (1972). *Agri-News* **3**(1), 4. Ciba-Geigy Corp. Ardsley, New York.
34. Anonymous (1972). *Anim. Nutr. Health* **27**(7), 14.
35. Anonymous (1972). *Broiler Ind.* **35**(9), 50.
36. Anonymous (1972). *Food Technol. (Chicago)* **26**(8), 106.
37. Anonymous (1972). *Poult. Dig.* **31**(362), 194.
38. Anonymous (1973). Technical Bulletin No. 73, Feb. 1. S. B. Penick & Co.
39. Anonymous (1973). Technical Bulletin No. 74, Feb. 1. S. B. Penick & Co.
40. Anonymous (1973). Technical Bulletin No. 75A, July 30. S. B. Penick & Co.
41. Anonymous (1973). Technical Bulletin No. 76, July 1. S. B. Penick & Co.
42. Anonymous (1973). Technical Information, pp. 1–5. F. Hoffmann-La Roche & Co., Basel.
43. Anonymous (1975). *Poult. Dig.* **34**(403), 370.
44. Anonymous (1976). *Agric. Res.* **24**(10), 7.
45. Arnaud, A. (1887) *C. R. Habd. Seances Acad. Sci.* **104**, 1293–1296.
46. Arnold, W. L. (1972). *Proc. 11th Tech. Alfalfa Conf.*, pp. 66–70.
46a. Arroya, P.T. (1975). Personal communication.
47. Ashton, E.H., and Fletcher, D.A. (1962). *Poult. Sci.* **41**, 1903–1909.
48. Association of Official Agricultural Chemists (1960). "Official Methods of Analysis," 9th ed., pp. 654–655. A.O.A.C., Washington, D.C.
49. Auxilla, M.T., and Vassura, G. (1968). *Riv. Zootec.* **41**, 522–530.
50. Bagdon, R.E., Impellizzeri, C., and Osadca, M. (1962). *Toxicol. Appl. Pharmacol.* **4**, 444–456.
51. Banks, Q.D., and Voss, L.A. (1962). *Mo. Agric. Exp. Stn., Bull.* **813**.
52. Baraud, J., Benitez, F., Genevois, L., and Maurice, A. (1965). *C. R. Habd. Seances Acad. Sci.* **260**(26), 7045–7046.
53. Barnett, B.D. (1968). Research Report No. 83. Hoffmann-Taff, Inc., Springfield, Missouri.
54. Barnett, B.D., and Morgan, C.L. (1963). *Poult. Sci.* **38**, 1188.
55. Bartlett, L., Klyne, W., Mose, W.P., Scopes, P.M., Galasko, G., Mallams, A.K., Weedon, B.C.L., Szabolcs, J., and Toth, Gy. (1969). *J. Chem. Soc. Cp.* 2527.
56. Bartov, I., and Bornstein, S. (1966). *Poult. Sci.* **45**, 297–305.
57. Bartov, I., and Bornstein, S. (1967). *Poult. Sci.* **46**, 796–805.
58. Bartov, I., and Bornstein, S. (1969). *Poult. Sci.* **48**, 495–504.
59. Bateson, W. (1902). *Rep. Evol. Comm. R. Soc.* **1**, 87–124.
60. Bauernfeind, J.C. (1962). *Proc. Tex. A&M Nutr. Conf.* pp. 110–123.
61. Bauernfeind, J.C. (1970). *World Rev. Anim. Prod.* **5**(21), 22–50.
62. Bauernfeind, J.C. (1971). *Abstr. Pap., 161st Meet. Am. Chem. Soc.* Abstr. no. 21.
63. Bauernfeind, J.C. (1972). *Agric. Food Chem.* **20,** 456–473. Institute of Food Technologists, Chicago.
64. Bauernfeind, J.C. (1973). "The IFT World 1973 Directory and Guide," pp. 96–109.
64a. Bauernfeind, J.C. (1976). *Prog. Fish Cult.* **38,** 180–183.
65. Bauernfeind, J.C., Brubacher, G.B., Kläui, H.M., and Marusich, W.L. (1971). *In* "Carotenoids" (O. Isler, ed.), pp. 743–770. Birkhaeuser, Basel.
66. Bauernfeind, J.C., and Bunnell, R.H. (1962). *Food Technol. (Chicago)* **16**(12), 76–82.
67. Bauernfeind, J.C., and Marusich, W.L. (1959). *Proc. Poult. Sci. Conv. (Aust.)* pp. 55–57.
68. Bauernfeind, J.C., and Marusich, W.L. (1974). *Proc. World's Poult. Cong. 15th, 1974* pp. 309–310.
69. Bauernfeind, J.C., Rubin, S.H., Surmatis, J.D., and Ofner, A. (1970). *Int. Z. Vitaminforsch.* **40**(3), 391–416.

70. Bayley, H.S., and Newlands, G. (1966). *Proc. World's Poult. Cong., 13th, 1966* p. 212.
71. Benedek, L. (1937). *Z. Lebensm.-Unters.-Forsch.* **74**, 297–302.
72. Benjamin, E.W., and Pierce, H.C. (1937). *In* "Marketing Poultry Products," p. 80. Wiley, New York.
73. Benk, E., Brixuis, L., and Dietl, R. (1967). *Arch. Geflüegelkd.* **31**(4), 263–268.
74. Berg, L.R. (1963). *Feedstuffs* **35**(2), 26.
75. Berg, L.R., Merrill, L.H., and Bearse, G.E. (1963). *Poult. Sci.* **42**, 492–498.
76. Berning, H.J. (1965). Thesis, University of Munich.
77. Berry, L.N. (1936). N. M., Agric. Exp. Stn., *Bull.* 785.
78. Bickoff, E.M. (1972). *Proc. Tech. Alfalfa Conf. 11th,* p. 9.
79. Bickoff, E.M., Knuckles, B.E., Spencer, R.R., Lazar, M.E., and Kohler, G.O. (1970). *J. Agric. Food Chem.* **18**, 1086–1089.
80. Bickoff, E.M., Livingston, A.L., Bailey, G.F., and Thompson, C.R. (1954). *J. Agric. Food Chem.* **2**, 563–567.
81. Bickoff, E.M.,and Thompson, C.R. (1949). *J. Assoc. Off. Agric. Chem.* **32**, 775–780.
82. Bickoff, E.M., Thompson, C.R., Livingston, A.L., Van Alta, G.R., and Guggolz, J. (1955). *J. Agric. Food Chem.* **3**, 67–69.
83. Bickoff, E.M., and Williams, K.T. (1944). *Ind. Eng. Chem.* **36**, 320–323.
84. Biedermann, R., and Prabucki, A.L. (1969). *Arch. Gefluegelkd.* **33**, 125–132.
85. Bilz, R. (1963). *In* "Carotene and Carotenoids" (K. Lang-Mainz, ed.), pp. 150–168. Steinkopff, Darmstadt.
86. Bird, F.H. (1952). *Pac. Poultryman* **27**, 8–9, 34–35.
87. Bird, F.H. (1952). *Proc. Tex. A&M Nutr. Conf.* pp. 16–22.
88. Bird, H.R. (1943). *Poult. Sci.* **22**, 205–208.
89. Bird, H.R., Groschke, A.C., and Rubin, M. (1949). *J. Nutr.* **37**, 215–226.
90. Black, W.A.P. (1953). *Agriculture (London)* **60**, 126–130.
91. Blattna, J., Blattny, C., and Pozdena, J. (1960). *Vorlaeufige Mitt. Nahr.* **4,** 816–824.
92. Blessin, C.W., Brecher, J.D., Dimler, R.J., Grogan, C.O., and Campbell, C.M. (1963). *Cereal Chem* **40**, 436–442.
93. Blessin, C.W., Dimler, R.J., and Webster, O.J. (1962). *Cereal Chem.* **39**, 389–392.
94. Blessin, C.W., Van Elten, C.H., and Wiebe, R. (1958). *Cereal Chem.* **35**, 359–365.
95. Bletner, J.K. (1972). *Poult. Dig.* **31**(364), 290.
96. Bletner, J.K., Mitchell, R.P., Jr., and Tugwell, R.L. (1966). *Poult. Sci.* **45**, 689–894.
97. Blommaert, M., and Steenis, G. (1963). *Vlaams Diergeneesdi. Tiudschr.* **32**(11), 306–317.
98. Blum, J.C., and Leclercq, B. (with Guillaumin, S.) (1970). *Ann. Biol. Anim., Biochim., Biophys.* **10**, 73–79.
99. Bodea, C. (1969). *Proc. Int. Symp. Carotenoids Other Than Vitam. A, 1969* pp. 523–529; *Pure Appl. Chem* **20**(4), 523–529 (1969).
100. Boguth, W., and Czernicki, B. (1962). *Zentralbl. Veterinaermed.* **9**(8), 779–786.
101. Boguth, W., and Czernicki, B. (1962). *Zentralbl. Veterinaermed.* **9**(9), 899.
102. Bohren, B.B., Thompson, C.R., and Carrick, C.W. (1945). *Poult. Sci.* **24,** 356–362.
103. Bolton, W., Booth, E., Dewar, W.A., and Wells, J.W. (1971). *J. Sci. Food Agric.* **22,** 349–353.
104. Booth, A.N., Saunders, R.M., Connor, M.A., and Kohler, G. O. (1972). *Proc. Tech. Alfalfa Conf., 11th, 1972,* pp. 51–57.
105. Booth, E. (1964). *Vet Rec.* **76**(48), 1410.
106. Borenstein, B., and Bunnell, R.H. (1967). *Adv. Food Res.* **15**, 195–276.
107. Bornstein, S., and Bartov, I. (1966). *Poult. Sci.* **45**, 287–296.
108. Bornstein, S., and Bartov, I. (1967). *Br. Poult. Sci.* **8**, 213–221.
109. Bornstein, S., and Bartov, I. (1967). *Br. Poult. Sci.* **8**, 223–230.
110. Bornstein, S., and Lipstein, B. (1963). *World's Poult. Sci. J.* **19**, 172–184.
111. Borodin, L. (1883). *Bull. Acad. Imp. Sci. St.-Petersbourg* **9**, 512; Abstr. in *Bot. Ztg.* **41**, 577–579. (1884).

112. Bougon, M. (1969). *Bull. Agric. Exp. Stn. Ploufragan* **9**, 14.
113. Brambila, S., Pino, J.A., and Mendoza, C. (1962). *Poult. Sci.* **41**, 1629.
114. Brambila, S., Pino, J.A., and Mendoza, C. (1963). *Poult. Sci.* **42,** 294–300.
115. Brambila, S., Pino, J.A., and Mendoza, C. (1963). *Poult. Sci.* **42,** 689–694.
116. Braünlich, K.(1962). *Gefluegelhof* **25**(26), 448.
117. Braünlich, K.(1970). *Proc. Symp. Yolk Color Egg Qual. Factor, 1970,* pp. 39–43.
118. Braünlich, K.(1970). *Proc. Symp. Yolk Color Egg Qual. Factor, 1970,* pp. 53–56.
119. Braünlich, K.(1974). *Proc. World's Poult. Cong., 15th, 1974* pp. 236–241.
119a. Britton, G. (1976). *In* "Chemistry and Biochemistry of Plant Pigments" (T.W. Goodwin, ed.), 2nd ed., Vol. 1, pp. 262–327. Academic Press, New York.
120. Brockmann, H., and Völker, O. (1934). *Hoppe-Seyler's Z. Physiol. Chem.* **224,** 193–243.
121. Broun, W.L. (1930). *Ga., Agric. Exp. Stn., Bull.* **160**.
122. Broun, W.L. (1934). *Ga., Agric. Exp. Stn., Bull.* **183**.
123. Broun, W.L. (1938). *J. Biol. Chem.* **122**, 655–659.
124. Brown, C.C. (1970). *Proc. Symp. Yolk Color Egg Qual. Factor, 1970,* pp. 15–22.
125. Brubacher, G. (1969). F. Hoffmann-La Roche & Co., Ltd., Basel (unpublished data).
126. Brubacher, G., Gloor, U., and Wiss, O. (1960). *Chimia* **14**, 19–20.
127. Brunson, A.M., and Quackenbush, F.W. (1962). *Crop Sci.* **2**, 344–347.
128. Bubicz, M. (1965). *Bull. Acad. Pol. Sci., Ser. Sci. Biol.* **13,** 251–255.
129. Budowski, P., Ascarelli, I., Gross, J., and Nir, I. (1963). *Science* **142,** 969–971.
130. Bunnell, R.H., and Bauernfeind, J.C. (1958). *Proc. World's Poult. Cong., 11th, 1958* p. 183.
131. Bunnell, R.H., and Borenstein, B. (1967). *Food Technol.* **21**, 331–334.
132. Bunnell, R.H., Marusich, W.L., and Bauernfeind, J.C. (1962). *Poult. Sci.* **41**, 1109–1115.
133. Bunt, J.S. (1964). *Nature (London)* **203**, 1262–1263.
134. Burton, H.W. (1963). *Queensl. Agric. J.* **89**, 585–587.
135. Bushong, R.D., Jr., Dilworth, B.C., and Day, E.J. (1972). *Poult. Sci.* **51**, 1768–1770.
136. Buvanendran, V. (1961). *Ceylon Vet. J.* **9**(4), 141–143.
137. Camp, A.A., Creger, C.R., and Couch, J.R. (1963). *Poult. Sci.* **42**, 1258–1259.
138. Camp, A.A., Creger, C.R., Farr, F.M., and Couch, J.R. (1962). *Inform. Poult. Nutr. Conf., 1962.*
139. Campbell, L., Lally, M., and Forsythe, R.H. (1960). *Cereal Sci. Today* **5**(6), 180, 189.
140. Carlson, C.W. (1961). *Feedstuffs* **33**(5), 52.
141. Carlson, C.W. (1961). *Poult. Process Mark.* **5**, 14, 33–34.
142. Carlson, C.W. (1967). Research Report No. 77. Hoffmann Taff, Inc., Springfield, Missouri.
143. Carlson, C.W., Guenther, E., Kohlmeyer, W., and Olson, O.E. (1954). *Poult. Sci.* **33**, 768–774.
144. Carlson, C.W., Halverson, A.W., Deethardt, D., and Aspelin, J. (1962). *Poult. Sci.* **41**, 1634.
145. Carlson, C.W., Halverson, A.W., and Kohler, G.O. (1964). *Poult. Sci.* **43**, 654–662.
146. Carlson, C.W., Shinnick, F.L., Sneller, M.A., and Halverson, A.W. (1961). *S. Dako. Farm Home Res.* **12**(2), 20–23.
147. Carver, D.S. (1959). *Poult. Sci.* **38**, 71–76.
148. Chang, T.S., and Butters, H.E. (1967). *Poult. Sci.* **46**, 1243.
149. Chawan, C.B., and Gerry, R.W. (1974). *Poult. Sci.* **53**, 671–676.
150. Cheesman, D.F., and Prebble, J. (1966). *Comp. Biochem. Physiol.* **17**, 929–935.
151. Cholnoky, L. Györgyfy, K., Ronai, A., Szabolcs, J., Toth, Gy., Galasko, G., Mallams, A.K., Waight, E.S., and Weedon, B.C.L. (1969). *J. Chem. Soc. C* pp. 1256–1263.
152. Cholnoky, L., Szabolcs, J., and Nagy, E. (1958). *Justus Liebigs Ann. Chem.* **616**, 207–218.
153. Coles, R. (1962). *Agric. Merchant (Engl.)* **42**(3), 103–106.
154. Collins, W.M., Thayer, S.C., and Skoglund, W.C. (1955). *Poult. Sci.* **34**, 223–228.
155. Comben, N. (1970). *Proc. Symp. Yolk Color Egg Qual. Factor, 1970,* pp. 9–10.
156. Comben, N. (1970). *Proc. Symp. Yolk Color Egg Qual. Factor, 1970,* pp. 45–51.
157. Combs, G.F. (1962). "Poultry Nutrition Short Course," pp. 22–25. University of Delaware, Newark.

158. Combs, G.F., and Laurent, C.K. (1953). *Md., Agric. Exp. Stn., Misc. Publ.* **151**.
159. Combs, G.F., Milligan, J.L., and Martin, J.L. (1963). *Feedstuffs* **35**(14), 44–48.
160. Combs, G.F., and Nicholson, J.L. (1963). *Feedstuffs* **35**(1), 36–38.
161. Combs, G.F., and Nott, H. (1967). *Feedstuffs* **39**(42), 36–38.
162. Combs, G.F., Romoser, G.L., and Bishop, R.W. (1954). *J. Nutr.* **53**, 511–522.
163. Commission Internationale de l'Eclairage (1931). *Proc. 8th Sess. 1931*. pp. 3–12.
164. Coon, C.N., and Couch, J.R. (1976). *Poult. Sci.* **55**, 841–847.
165. Cooney, J.J., Marks, H.W., Jr., and Smith, A.M. (1966). *J. Bacteriol.* **92**, 342–345.
166. Couch, J.R. (1953). *Broiler Growing* **4**(12), 60–62.
167. Couch, J.R. (1954). *Feed Age* **4**(1), 33, 50.
168. Couch, J.R. (1959). *Feedstuffs* **31**(7), 42.
169. Couch, J.R. (1961). Research Report No. 5. Hoffman-Taff, Springfield, Missouri.
170. Couch, J.R. (1962). *Feedstuffs* **34**(5), 38.
171. Couch, J.R., Camp, A.A., and Farr, F.M. (1971). *Br. Poult. Sci.* **12**, 205–211.
172. Couch, J.R., Camp, A.A., Farr, F.M., and Creger, C.R. (1962). *Proc. Texas A&M Nutr. Conf.* pp. 124–128.
173. Couch, J.R., and Coon, C.N. (1972). *Feedstuffs* **44**(14), 18–21, 31.
174. Couch, J.R., and Crawford, W.P. (1957). *Feeds Illustrated* **8**(12), 40.
175. Couch, J.R., and Farr, F.M. (1970). *Poult. Sci.* **49**, 1377.
176. Couch, J.R., and Farr, F.M. (1970). *Poult. Sci.* **49**, 1377–1378.
177. Couch, J.R., and Farr, F.M. (1971). *Br. Poult. Sci.* **12**, 49–55.
178. Couch, J.R., and Farr, F.M. (1971). *Br. Poult. Sci.* **12**, 87–93.
179. Couch, J.R., Farr, F.M., and Camp, A.A. (1963). *Distill. Feed Res. Counc., Conf., Proc.* **18**, 31–39.
180. Courington, D.P., and Goodwin, T.W. (1955). *J. Bacteriol.* **70**, 568–571.
181. Courtenay, H. (1962). *Poult. Dig.* **21**(250), 711–714.
182. Courtenay, H.V., and Branson, R.E. (1962). *Tex., Agric. Exp. Stn., Bull.* **989**.
183. Coward, K.H. (1924). *Biochem. J.* **18**, 1114–1122.
184. Creger, C.R., Farr, F.M., Castro, E., and Couch, J.R. (1963). *Poult. Sci.* **42**, 1262.
185. Culton, T.G., and Bird, H.R. (1941). *Poult. Sci.* **20**, 432–436.
186. Curl, A.L. (1960). *J Food Res.* **25**, 190–196.
187. Curl, A.L. (1964). *J. Agric. Food Chem.* **12**, 522–524.
188. Curl, A.L. (1965). *J. Food Sci.* **30,** 13–18.
189. Curl, A.L. (1965). *J. Food Sci.* **30**, 426–432.
190. Curl, A.L., and Bailey, G.F. (1956). *J. Agric. Food Chem.* **4,** 156–162.
191. Curl, A. L., and Bailey, G. F. (1957). *Food Res.* **22,** 63–68.
192. Curl, A.L., and Bailey, G.F. (1957). *J. Agric. Food Chem.* **5,** 605–608.
193. Czeczuga, B., and Czerpak, R. (1969). *Hydrobiologia* **33,** 379–384.
194. Czernicki, B., and Weiser, H. (1962). *Zentralbl. Veterinaermed.* **9**, 899–904.
195. Czygan, F.C. (1964). *Experientia* **20**, 573–574.
196. Dalby, G. (1948). *Cereal Chem.* **25**, 413–417.
197. Damron, B.L., and Janky, D.M. (1976). *Poult. Times* Mar. 29, p. 9.
198. Dansky, L.M. (1971). *Poult. Sci* **50**, 1569.
199. Dansky, L.M. (1972). *Feedstuffs* **44**(5), A3, A14.
200. DaSilva, J., Salgueiro, V., Vieira, L., and Carvalko, M.C. (1966). *Bol. Pecu.* **34**(3), 189–203.
201. Davies, A.W. (1952). *Nature (London)* **170**, 849.
202. Davies, B.H. (1962). *Biochem. J.* **81.**
203. Davies, B.H. (1965). *In* "Chemistry and Biochemistry of Plant Pigments" (T.W. Goodwin, ed.), 1st ed., Part IV, pp. 489–532. Academic Press, New York.
203a. Davies, B.H. (1976). *In* "Chemistry and Biochemistry of Plant Pigments" (T.W. Goodwin, ed.), 2nd ed., Vol. 2, pp. 38–165. Academic Press, New York.
204. Davies, B.H., Hsu, W.J., and Chichester, C.O. (1970). *Comp. Biochem. Physiol.* **33**, 601–615.

205. Davies, R.E., Jones, M.L., and Yacowitz, H. (1969). *Poult. Sci.* **48**, 1800.
206. Davies, U. (1972). *Feedstuffs* **44**(19), 20.
207. Davies, Y., Trotter, W.K., and Burdick, D. (1976). *Feedstuffs* **48**(22), 17–19, 38.
208. Davis, P.N., and Kratzer, F.H. (1958). *Poult. Sci.* **37**, 851–854.
209. Day, E.J., Ratcliff, R.G., and Dua, P. (1963). *Miss., Agric. Exp. Stn., Inf. Sheet* No. 810.
210. Day, E.J., West, J.W., and Hill, J.E. (1958). *Poult. Sci.* **37**, 275–281.
211. Day, E.J., and Williams, W.P., Jr. (1958). *Poult. Sci.* **37**, 1373–1381.
212. Deethardt, D.E., Burrill, L.M., and Carlson, C.W. (1965). *Food Technol.* **19,** 75–77.
213. de Fremery, D. (1972). *Proc. Tech. Alfalfa Conf., 11th, 1972,* pp. 41–50.
214. De Groote, G. (1963). *Bondsblad* **10**(5), 8–9, 12–13.
215. De Groote, G. (1963). *Landbouwkd. Tijdschr.* **16**, 1487–1505.
216. De Groote, G. (1963). *Landbouwkd. Tijdschr.* **16**, 1613–1623.
217. De Groote, G. (1963). *Belg. Rev. Agric.* **16**(10), 1491–1512.
218. De Groote, G. (1963). *Belg. Rev. Agric.* **16**(11), 1617–1627.
219. De Groote, G. (1964). *J. Agric. Vet. Chem* **5**, 9–12.
220. De Groote, G. (1964-1965). *Land Tuinbouw. Jaarb.* **19**, 149–154.
221. De Groote, G. (1966). *Belg. Rev. Agric.* **19**(2), 73–89.
222. De Groote, G. (1967). *Belg. Rev. Agric.* **20**(3), 435–453.
223. De Groote, G. (1968). *Proc. Meet. Eur. Fed. World's Poult. Sci. Assoc., 5th, 1968,* pp. 16–23.
224. De Groote, G. (1970). *G. Allevatori* **20**(2), 62–66.
225. De Groote, G. (1970). *Worlds Poult. Sci. J.* **26**, 435–441.
226. De Groote, G., and Reyntens, N. (1963). *Belg. Rev. Agric.* **16**, 249–284.
227. De Jongh, R. (1966). Ph.D. Thesis, University of Monterrey, Mexico.
228. Denton, C.A. (1940). *Poult. Sci.* **19**, 281–285.
229. Deuel, H.J., Jr. (1955). *In* "The Lipids," Vol. I, pp. 4–46. Wiley (Interscience), New York.
230. Deuel, H.J., Jr., Hrubetz, M.C., Mattson, F.H., Morehouse, M.G., and Richardson, A. (1943). *J. Nutr.* **26**, 673–685.
231. Deuel, H.J., Jr., Meserve, E.R., Johnston, C.H., Polgar, A., and Zechmeister, L. (1949). *Arch. Biochem.* **23**, 242–245.
232. Deufel, J. (1965). Staatl. Inst. Seenforsch Seenbewirtsch. Langenargen/Bodensee. Publ. no. 6, pp. 22–33.
233. Deutsch, M.J. (1970). *J. Assoc. Off. Anal. Chem.* **53,** 349.
234. Dolge, K.L., Dembiczak, C.M., Rousseau, J.E., Jr., Eaton, H.D., Beall, G., and Moore, L.A. (1955). *Conn., Agric. Exp. Stn., New Haven, Bull.* **314**.
235. Donaldson, W.E., Combs, G.F., and Romoser, G.L. (1956). *Poult. Sci.* **35**, 1100–1105.
236. Donovan, G.A., Henderson, D.C., and Luneau, R.N. (1961). *Poult. Sci.* **40,** 1395–1396.
237. Donovan, G.A., and Lueau, R.N. (1962). *Feedstuffs* **34**(3), 38, 54.
238. Douglas, C.R. (1966). *Diss. Abstr. B* **27**, 656B.
239. Douglas, C.R., Davis, E.A., Goff, O.E., and Bletner, J.K. (1964). *Poult. Sci.* **43**, 1314.
240. Dryden, J. (1905). *Utah, Agric. Exp. Stn., Bull.* **92**.
241. Dua, P.N. (1968). *Diss. Abstr. B* **28**, 3117B.
242. Dua, P.N., Day, E.J., Hill, J.E., and Grogan, C.O. (1967). *J. Agric. Food Chem.* **15**, 324–328.
243. Dua, P.N., Dilworth, B.C., Day, E.J., and Hill, J.E. (1967). *Poult. Sci.* **46**, 530–531.
244. Dua, P.N., Dubal, V.L., Day, E.J., and Hill, J.E. (1968). *Proc. Soc. Exp. Biol. Med.* **128**, 262–265.
245. Dunn, L.C. (1925). *Anat. Rec.* **31**, 343–344.
246. Edmondson, J. (1961). *N. Z. J. Agric.* **102**, 526,527,529,531.
247. Edmondson, J. (1962). *N. Z. J. Agric.* **104**, 299–300.
248. Edwards, R.A., and Reuter, F.H. (1964). *Food Technol. Aust.* **16**(9), 534–544.
249. Edwards, R.A., and Reuter, F.H. (1967). *Food Technol Aust.* **8**, 352–357.
250. Elam, J.F. Jacobs, R.L., Tidwell, W.L., Gee, L.L., Couch, J.R. (1953). *J. Nutr.* **49**, 307–317.

251. Elrod, R. (1956). *Feedstuffs* **28**(3), 20a.
252. Elrod, R.C., Robajdek, E.S., Gledhill, R.H., Witz, W.M., Diser, G.M., and Hayward, J.W. (1958). *Feedstuffs* **30**(33), 26-27, 30, 32.
253. Entschel, R., and Karrer, P. (1958). *Helv. Chim. Acta* **41**, 402-413.
254. Erasmus, J., Scott, M.L., and Levine, P.P. (1960). *Poult. Sci.* **39**, 565-572.
255. Euler, H., Demole, V., Karrer, P., and Walker, O. (1930). *Helv. Chim. Acta* **13**, 1078-1083.
256. Euler, H., and Gard, U. (1931). *Ark. Kemi, Mineral. Geol.* **B1**°(19), 1-6.
257. Euler, H., Karrer, P., and Solmssen, U. (1938). *Helv. Chim. Acta* **21**, 211-222.
258. Ewing, W.R. (1963). "Poultry Nutrition." Ray Ewing Co., Pasadena, California.
259. Farr, F.M., Castro, E., and Couch, J.R. (1963). *Poult. Sci.* **42**, 1268.
260. Farr, F.M., and Couch, J.R. (1970). *Poult. Sci.* **49**, 1384.
261. Farr, F.M., Creger, C.R., and Couch, J.R. (1962). *Poult. Sci.* **41**, 1643.
262. Farr, F.M., Hulett, B.J., Davies, R.E., and Couch, J.R. (1961). *Poul. Sci.* **40**, 1401.
263. Fässler, C., Vuilleumier, J.P., and Brubacher, G.B. (1962). *Int. Z. Vitaminforch.* **32**(4), 454-460.
264. Ferguson, T.M. (1958). *Feeds Illustrated* **9**(9), 26-28.
265. Ferguson, T.M., and Couch, J.R. (1958). *Feedstuffs* **30**(11), 82-85, 87.
266. Ferrando, R. (1962). *Rec. Med. Vet.* **138**, 547-562.
267. Ferrando, R. (1963). *Rec. Med. Vet.* **139**, 547-555.
268. Ferrando, R., and Henry, N. (1970). *Ind. Aliment. Anim.* **219**, 33-43.
269. Ferrando, R., and Mainguy, P. (1970). Roche Documentation Paper available on request from F. Hoffmann-La Roche & Co., Paris.
270. Fischer, D.R. (1962). *Feedstuffs* **34**(5), 35.
270a. Fletcher, D.L., Harms, R.H., and Janky, D.M. (1978). *Poult. Sci.* **57,** 624-629.
270b. Fletcher, D.L., Janky, D.M., Christmas, R.B., Arafa, A.S., and Harms, R.H. (1977). *Poult. Sci.* **56,** 2061-2063.
270c. Fletcher, D.L., Janky, D.M., Voitle, R.A., and Harms, R.H. (1977). *Poult. Sci.* **56,** 953-956.
271. Forsythe, R.H. (1957). *Cereal Sci. Today* **2**(8), 211-216.
272. Forsythe, R.H. (1957). *J. Assoc. Off. Agric. Chem.* **40**, 535-542.
273. Forsythe, R.H. (1958). *J. Assoc. Off. Agric. Chem.* **41**, 274-276.
274. Forsythe, R.H. (1960). *J. Assoc. Off. Agric. Chem.* **43**, 540-542.
275. Fox, D.L. (1962). *Comp. Biochem. Physiol.* **5**, 31-43.
276. Fox, D.L. (1962). *Comp. Biochem. Physiol.* **6**, 1-40.
277. Fox, D.L. (1962). *Comp. Biochem. Physiol.* **6**, 305-310.
278. Fox, D.L., and Hopkins, T.S. (1965). *Nature (London)* **206**, 301-302.
279. Fox, D.L., and Hopkins, T.S. (1966). *Comp. Biochem. Physiol.* **17**, 841-856.
280. Fox, D.L., and Hopkins, T.S. (1966). *Comp. Biochem. Physiol.* **19**, 267-278.
281. Fox, D.L., Hopkins, T.S., and Zilversmit, D.B. (1965). *Comp. Biochem. Physiol.* **14**, 641-649.
282. Fox, D.L., Wolfson, A.A., and McBeth, J.W. (1969). *Comp. Biochem. Physiol.* **29**, 1223-1229.
282a. Francis, C., Janky, D.M., and Harms, R.H. (1978). *Poult. Sci.* **57,** 1136.
283. Francis, F.J., and Clydesdale, F.M. (1972). *Food Prod. Dev.,* **6**, 47-48, 55.
284. Fraps, G.S., and Kemmerer, A.R. (1937). *Tex. Agric. Exp. Stn., Bull.* **557**.
285. Fraps, G.S., and Treichler, R. (1933). *Tex. Agric. Exp. Stn., Bull.* **477**.
286. Fritz, J.C. (1962). *Proc. Tex. A&M Nutr. Conf.* pp. 102-109.
287. Fritz, J.C. (1962). *Feedstuffs* **34**(19), 44-48.
288. Fritz, J.C. (1966). *Feedstuffs* **38**(43), 7, 66.
289. Fritz, J.C., and Wharton, F.D., Jr. (1957). *Poult. Sci.* **36**, 1118.
290. Fritz, J.C.Wharton, F.D., Jr., and Classen, L.J. (1957). *Feedstuffs* **29**(42), 18,20,22,24.
291. Fritz, J.C., Wharton, F.D., Jr., and Classen, L.J. (1958). *Proc. World's Poultry Congr., 11th, 1958,* pp. 77-85.
292. Frost, D.V., Overby, L.R., and Spruth, H.C. (1955). *J. Agric. Food Chem.* **3**, 235-243.
293. Fry, J.L. (1970). *Broiler Ind.* **33**(12), 48.

294. Fry, J.L., Ahmed, E.M., Herrick, G.M., and Harms, R.H. (1969). *Poult. Sci.* **48**, 1127–1129.
295. Fry, J.L., and Harms, R.H. (1974). *Poult. Sci.* **53**, 1404–1407.
296. Fry, J.L., and Harms, R.H. (1975). *Poult. Sci.* **54**, 1094–1101.
297. Fry, J.L., Harms, R.H., and Moeller, M.W. (1976). *Poult. Sci.* **55**, 744–747.
298. Fry, J.L., Harms, R.H., Moeller, M.W., and Kling, H.F. (1976). *Poult. Sci.* **55**, 1112–1116.
299. Fry, J.L., Herrick, G.M., Harms, R.H., and Ahmed, E.M. (1970). *Proc. World's Poult. Congr., 14th 1970* pp. 1168–1169.
300. Fry, J.L., Hinton, C.F., and Harms, R.H. (1973). *Sunshine State Agric. Res. Rep.* **18**(1-2), 12–15.
301. Galvano, G., and Lanza, A. (1972). *Avicoltura* **41**, 67–77.
302. Ganguly, J., Mehl, J.W., and Deuel, H.J. (1953). *J. Nutr.* **50**, 59–72.
303. Garlich, J.D., Bryant, D.M., Covington, H.M., Chamblee, D.S., and Purcell, A.E. (1974). *Poult. Sci.* **53**, 692–699.
304. Garlich, J.D., Covington, H.M., Chamblee, D.S., and Purcell, A.E. (1972). *Proc. 26th Annu. N. C. Nutr. Conf.* pp. 1–6.
305. Gawecki, K., Lipinska, H., and Bonczol, B. (1968). *Postepy Drobiarstwa Poznan* **10**(1), 13–20.
306. Gawecki, K., and Potkanski, A. (1968). *Postepy Drobiarstwa Poznan* **10**(1), 21–28.
307. Gazo, M., Sladka, O., and Stasko, J. (1968). *Ved. Prace Vysk. Ustavu Chov Hydiny, Ivanka Dunaji* pp. 141–147.
308. Gazo, M., Sladka, O., Stasko, J., and Koci, S. (1968). *Wiss. Arb. Inst. Gefluegelforsch,* CSR 6.
309. Giavarini, I. (1965). *Proc. Eur. Poult. Conf. 2nd 1964* pp. 281–284.
310. Giavarini, I. (1966). *Proc. World's Poult. Congr., 13th, 1966* 564. USSR.
311. Gilchrist, B.M. (1968). *Comp. Biochem. Physiol.* **24**, 123–147.
312. Gilchrist, B.M., and Lee, W.L. (1967). *J. Zool.* **151**, 171–180.
313. Gilchrist, B.M., and Lee, W.L. (1972). *Comp. Biochem. Physiol.* **42,** 263–294.
314. Gillam, A.E., and El Ridi, M.S. (1935). *Biochem. J.* **29**, 2465–2468.
315. Gillam, A.E., and Heilbron, I.M. (1935). *Biochem J.* **29**, 834–836.
316. Gillam, A.E., and Heilbron, I.M. (1935). *Biochem. J.* **29**, 1064–1067.
317. Gillam, A.E., Heilbron, I.M., Morton, R.A., Bishop, G., and Drummond, J.C. (1933). *Biochem. J.* **27**, 878–888.
318. Gloor, U. (1967). F. Hoffmann-La Roche, & Co., Ltd., Basel (unpublished data).
319. Gloor, U. (1969). F. Hoffmann-La Roche, & Co., Ltd., Basel (unpublished data).
320. Gloor, U., and Leuenberger, F. (1969). F. Hoffmann-La Roche, & Co., Ltd., Basel (unpublished data).
321. Glover, J. (1960). *Vitam. Horm. (N.Y.)* **18**, 371–386.
322. Glover, J., and Redfearn, E.R. (1954). *Biochem. J.* **58**, XV.
323. Goldhaber, P., Zacharias, L., and Kinsey, V.E. (1950). *J. Nutr.* **42**, 453–462.
324. Goodwin, T.W. (1952). "The Comparative Biochemistry of the Carotenoids." Chapman & Hall, London.
325. Goodwin, T.W. (1954). "Carotenoids, Their Comparative Biochemistry." Chem. Publ. Co., New York.
326. Goodwin, T.W. (1955). *Annu. Rev. Biochem.* **24,** 497–522.
326a. Goodwin, T.W. (1965). *In* "Chemistry and Biochemistry of Plant Pigments" (T.W. Goodwin, ed.), 1st ed., Part I, pp. 127–173. Academic Press, New York.
327. Goodwin, T.W. (1971). *In* "Carotenoids" (O. Isler, ed.). pp. 577–636. Birkhaueser, Basel.
328. Goodwin, T.W. (1976). *In* "Chemistry and Biochemistry of Plant Pigments" (T.W. Goodwin, ed.), 2nd ed., Vol. 1, pp. 225–261. Academic Press, New York.
329. Grabowski, T. (1969). *Postepy Drobiarstwa* **11**, 147–164; *Food Sci. Tech. Abstr.* **4**, 45449 (1972).
330. Green, J. (1957). *Proc. R. Soc. London, Ser. B.* **147,** 392.
331. Griffith, R.B., and Thompson, C.R. (1949). *Bot. Gaz. (Chicago)* **111**, 165.

332. Grimblely, F.H., and Black, D.J.G. (1950). *Br. J. Nutr.* **4**, 323–331.
333. Grimblely, F.H., and Black, D.J.G. (1952). *Br. J. Nutr.* **6**, 393–397.
334. Grob, E.C. (1963). *In* "Carotine and Carotinoide" (K. Lang, ed.), pp. 26–53. Steinkopff, Darmstadt.
335. Grogan, C.O., and Blessin, C.W. (1968). *Crop. Sci.* **8**, 730–732.
336. Grogan, C.O., Blessin, C.W., Dimler, R.J., and Campbell, C.M. (1963). *Crop Sci.* **3**, 213–214.
337. Groza, I. (1968). *Rev. Zooteh. Med. Vet.* **18**(7), 44–46.
338. Guenthner, E., Carlson, C.W., and Olson, O.E. (1968). *Poult. Sci.* **47**, 1677.
339. Guenthner, E., Carlson, C.W., Olson, O.E., Kohler, G.O., and Livingston, A.L. (1973). *Poult. Sci.* **52**, 1787–1798.
340. Gugliemelli, L.A., and Mitchell, H.L. (1958). *Agric. Food Chem.* **6**, 126–128.
341. Gugliemelli, L.A., Mitchell, H.L., and Silker, R.E. (1954). *Trans. Kans. Acad. Sci.* **57**(3), 368–373.
342. Guilbert, H.R. (1934). *Ind. Eng. Chem., Anal. Ed.* **6**, 452–453.
343. Guilbert, H.R. (1935). *J. Nutr.* **10**, 45–62.
344. Guilbert, H.R., and Hinshaw, V.R. (1934). *J. Nutr.* **8**, 45–56.
345. Hagemann, P. (1955). *Gefluegelhof* **18**(51), 859.
346. Hall, G.M., Waldroup, P.W., Fry, J.L., Ammerman, C.B., and Harms, R.H. (1966). *Poult. Sci.* **45**, 639–641.
347. Halloran, H.R. (1970). *Feedstuffs* **42**(41), 31–33.
348. Halloran, H.R. (1972). *Proc. Tech. Alfalfa Conf., 11th, 1972,* pp. 93–94.
349. Halloran, H.R. (1974). *Proc. World's Poult. Congr., 15th, 1974* pp. 241–244.
350. Halloran, H.R., Lyle, J.B., and Trana, W.T. (1971). *Poult. Sci.* **50**, 1583.
351. Hammond, J.C. (1943). *Off. Publ., Assoc. Am. Fd Control Off.* pp. 55–61.
352. Hammond, J.C., and Harshaw, H.M. (1941). *Poult. Sci.* **20**, 437–444.
353. Hammond, J.C., Miller, D., and Whitson, D. (1942). *Poult. Sci.* **21**, 525–527.
354. Harashima, K., and Yajima, Y. (1969). *Agric. Biol. Chem.* **33**, 1092–1094.
355. Hargreaves, K.R. (1971). *Phytochemistry* **10,** 898–899.
356. Harms, R.H., Ahmed, E.H., and Fry, J.L. (1971). *Proc. Md. Nutr. Conf.* pp. 81–86.
356a. Harms, R.H., Fletcher, D.L., and Janky, D.M. (1977). *Poult. Sci.* **56,** 1720.
356b. Harms, R.H., Fry, J.L., and McPherson, W.N. (1977). *Poult. Sci.* **56,** 86–90.
357. Harms, R.H., and Goff, O.E. (1957). *Tenn., Agric. Exp. Stn., Bull.* **269**.
358. Harms, R.H., Quisenberry, J.H., and Couch, J.R. (1958). *Poult. Sci.* **37**, 143–146.
359. Harms, R.H., Raskopf, B.D., Hochreich, H.J., Kidd, I.H., and Douglas, C.R. (1958). *Poult. Sci.* **37**, 359–363.
360. Härtel, H. (1970). *Arch. Geflüegelkd.* **34,** 109–115.
361. Härtel, H. (1970). *Proc. World's Poult. Congr., 14th, 1970* pp. 825–834.
362. Hartfiel, W., and Schmitten, F. (1965). *Arch. Geflüegelkd.* **29**(4), 367–386.
363. Hartmann, M., and Medem, F. (1947). *Naturwissenschaften* **34**, 25–26.
364. Hartung, T.E., Sullivan, T.W., and Froning, G.W. (1968). *Poult. Sci.* **47**, 1679.
365. Haslach, H. (1964). Ph.D. Thesis, University of Munich.
366. Hastings, W.H. (1961). *Feedstuffs* **33**(46), 30–31, 34–35.
367. Hata, M., and Hata, M. (1969). *Comp. Biochem. Physiol.* **29**(3), 985–994.
368. Hauge, S.M., and Aitkenhead, W. (1931). *J. Biol. Chem.* **93**, 657–665.
369. Hauge, S.M., Zscheile, F.P., Carrick, C.W., and Bohren, B.B. (1949). *Ind. Eng. Chem.* **36**, 1065–1068.
370. Haugh, R.R. (1937). *U.S. Egg Poult. Mag.* **43**, 552–555, 572–573.
371. Haxo, F. (1950). *Bot. Gaz. (Chicago)* **112,** 228–232.
372. Hayes, J.P. (1966). *S. Afr. J. Agric. Sci.* **9**, 461–467.
373. Head, S.W. (1966). *Poult. Sci.* **45**, 1269–1275.
374. Heath, J.L. (1971). *Proc. Md. Nutr. Conf.* pp. 74–77.

375. Heath, J.L. (1972). *Proc. Md. Nutr. Conf.* pp. 41–47.
376. Heath, J.L., and Shaffner, C.S. (1972). *Poult. Sci.* **51,** 502–506.
377. Heath, J.L., and Thomas, O.P. (1973). *Poult. Sci.* **52**, 967–971.
378. Heath, J.L., and Thomas, O.P. (1974). *Poult. Sci.* **53**, 291–295.
379. Heath, J.L., and Thomas, O.P. (1974). *Poult. Sci.* **53**, 1880–1885.
380. Heath, J.L., and Wabeck, C.J. (1975). *Poult. Sci.* **54**, 1288–1292.
381. Heffner, J., Roy, E.P., Davis, B.H., and Hilton, W.B. (1964). *La., Agric. Exp. Stn., Bul.* **586.**
382. Heiman, V. (1966). *Feedstuffs* **38**(16), 34–35, 38–40.
383. Heiman, V., and Carver, J.S. (1935). *U.S. Egg Poult. Mag.* **41**, 40–41.
384. Heiman, V., and Tighe, L.W. (1943). *Poult. Sci.* **22**, 102–107.
385. Heiman, V., and Wilhelm, L.A. (1937). *Poult. Sci.* **16**, 400–403.
386. Helbacka, N.V. (1958). *Md. Poultryman* Jan. 14, pp. 2–5; *Feedstuffs* **30**(7), 4 (1958).
387. Henderson, E.W. (1960). *Poult. Sci.* **39**, 53–56.
388. Henderson, P.D. (1953). Technical Report X-37. Bowman Feed Products, Inc. Holland, Michigan.
389. Herrick, G.M., Fry, J.L., Damron, B.L., and Harms, R.H. (1970). *Poult. Sci.* **49**, 222–225.
390. Herrick, G.M., Fry, J.L., and Harms, R.H. (1970). *Poult. Sci.* **49**, 1396.
391. Herrick, G.M., Fry, J.L., and Harms, R.H. (1971). *Poult. Sci.* **50**, 1467–1475.
392. Herrick, G.M., Fry, J.L., and Harms, R.H. (1972). *Poult. Sci.* **51**, 428–435.
393. Hertzberg, S., and Liaaen-Jensen, S. (1966). *Phytochemistry* **5**, 557–563.
394. Hertzberg, S., and Liaaen-Jensen, S. (1966). *Phytochemistry* **5**, 565–570.
395. Hertzberg, S., and Liaaen-Jensen, S. (1969). *Phytochemistry* **8**, 1259–1280.
396. Hertzberg, S., Liaaen-Jensen, S., and Siegelman, H.W. (1971). *Phytochemistry* **10**, 3121–3127.
397. Heywang, B.W. (1959). *Poult. Sci.* **38**, 1468–1469.
398. Hickethier, A. (1952). "Farbenordnung" (Color Classification). Verlag H. Osterwald, Hanover, Germany.
399. Hill, F.W., and Dansky, L.M. (1951). *Proc. Cornell Nutr. Conf.* pp. 27–32.
400. Hinton, C.F., Fry, J.L., and Harms, R.H. (1972). *Pap., Assoc. South. Agric. Workers, 1972; Feedstuffs* **44**(14), 18–21, 31 (1972).
401. Hinton, C.F., Fry, J.L., and Harms, R.H. (1972). *Poult. Sci.* **51,** 1818.
402. Hinton, C.F., Fry, J.L., and Harms, R.H. (1972). *Poult. Sci.* **51,** 1819.
403. Hinton, C.F., Fry, J.L., and Harms, R.H. (1973). *Poult. Sci.* **52,** 360–364.
404. Hinton, C.F., Fry, J.L., and Harms, R.H. (1973). *Poult. Sci.* **52**, 2169–2180.
405. Hinton, C.F., Fry, J.L., and Harms, R.H. (1974). *Poult. Sci.* **53**, 223–226.
406. Hislop, R.I., and Whittle, T.E. (1967). *World's Poult. Sci. J.* **23**(2), 133–150.
407. Hoffman, E.J., Lum, F.G., and Pitman, A.L. (1945). *J. Agric. Res.* **71**, 361–366.
408. Høie, J., and Sannan, F. (1960). *Sci. Rep. Agric. Coll. Norway* **39**(19), 1–76.
409. Hoppe, P. (1970). *Proc. World's Poultry Congr., 14th, 1970* pp. 845–850.
410. House, W.B. (1057). *Feedstuffs* **29**(4), 29a–31a.
411. Hsieh, L.K., Lae, T.C., Chichester, C.O., and Simpson, K.L. (1974). *J. Bacteriol.* **118,** 385–393.
412. Hsu, W.J., Chichester, C.O., and Davies, B.H. (1970). *Comp. Biochem. Physiol.* **32**, 69–79.
413. Huff, W.E., and Hamilton, P.B. (1975). *Poult. Sci.* **54**, 1308–1310.
414. Huff, W.E., Wyatt, R.D., and Hamilton, P.B. (1975). *Poult. Sci.* **54**, 2014–2018.
415. Hughes, J.S., and Payne, L.S. (1937). *Poult. Sci.* **16**, 135–138.
416. Hunter, J.A., Van Wagenen, A., and Hall, G.O. (1936). *Poult. Sci.* **15**, 115–118.
417. Hunter, J.E. (1969). *Feedstuffs* **41**(4), 32.
418. Hunter, R.S. (1958). *J. Opt. Soc. Am.* **48**, 985–995.
419. Hurst, C.C. (1905). *Rep. Eval. Comm. R. Soc.* **11**, 131–154.
420. Hutt, F.B. (1949). "Genetics of the Fowl." McGraw-Hill, New York.
421. Isler, O., Guex, W., Lindlar, H., Montavon, M., Rüegg, R., Ryser, G., Saucy, G., and Schwieter, U. (1958). *Chimia* **12**, 89–90.

422. Isler, O., Guex, W., Rüegg, R., Ryser, G., Saucy, G., Schwieter, U., Walter, M., and Winterstein, A. (1959). *Helv. Chim. Acta* **42,** 864–871.
423. Isler, O., Gutmann, H., Montavon, M., Rüegg, R., Ryser, G., and Zeller, P. (1957). *Helv. Chim. Acta* **40,** 1242–1249.
424. Isler, O., Lindlar, H., Montavon, M., Rüegg, R., Saucy, G., and Zeller, P. (1956). *Helv. Chim. Acta* **36**, 2041–2053.
425. Isler, O., Lindlar, H., Montavon, M., Rüegg, R., Saucy, G., and Zeller, P. (1957). *Helv. Chim. Acta* **40**, 456–457.
426. Isler, O., Lindlar, H., Montavon, M., Rüegg, R., and Zeller, P. (1956). *Helv. Chim. Acta* **39**, 449–454.
427. Isler, O., Montavon, M., Rüegg, R., Saucy, G., and Zeller, P. (1956). *Verh. Naturforsch. Ges. Basel* **67**, 379–388.
428. Isler, O., Montavon, M., Rüegg, R., and Zeller, P. (1957). *Justus Liebigs Ann. Chem.* **603**, 129–144.
429. Isler, O., Ofner, A., and Siemers, G.F. (1958). *Food Technol. (Chicago)* **12**, 520–526.
430. Isler, O., Rüegg, R., and Schudel, P. (1961). *Chimia* **15**, 208–226.
431. Isler, O., and Schudel, P. (1963). *Wiss. Veroff. Dtsch. Ges. Ernaehr.* **9**, 54–104.
432. Isler, O., and Zeller, P. (1957). *Vitam. Horm. (N.Y.)* **15**, 31–71.
433. Jaap, R.G. (1955). *Poult. Sci.* **34**, 389–395.
434. Jasper, A.W., and Cray, R.E. (1953). *Ohio Agric. Exp. Stn., Bull.* **736.**
435. Jeffrey, S.W. (1961). *Biochem. J.* **80**, 336–342.
436. Jensen, A. (1963). *Poult. Sci.* **42**, 912–915.
437. Jensen, A. (1966). *Acta Chem. Scand.* **20**, 1728–1730.
438. Johnson, J.R. (1962). Thesis, West Virginia University, Morgantown.
439. Judd, D.B. (1933). *J. Opt. Soc. Am.* **23**, 359–374.
440. Kahlenberg, O.J. (1949). *Food Ind.* **21**, 467–470.
441. Karrer, P., and Helfenstein, A. (1930). *Helv. Chim. Acta* **13**, 86–88.
442. Karrer, P., and Jucker, E. (1948). "Carotenoide." Birkhaeuser, Basel.
443. Karrer, P., and Jucker, E. (1950). "Carotenoids." Am. Elsevier, New York.
444. Karrer, P., and Krause-Voith, E. (1948). *Helv. Chim. Acta* **31,** 802–803.
445. Karrer, P., Krause-Voith, E., and Steinlin, K. (1948). *Helv. Chim. Acta* **31,** 113–120.
446. Karrer, P., and Walker, O. (1933). *Helv. Chim. Acta* **16**, 641–643.
447. Katayama, T., Yokoyama, H., and Chichester, C.O. (1970). *Int. J. Biochem.* **1**(4), 438–444.
448. Katayama, T., Yokoyama, H., and Chichester, C.O. (1970). *Bull. Jpn. Soc. Sci. Fish.* **36**(7), 702–708.
449. Kathe, R.L. (1962). *Proc. Tex. A&M Nutr. Conf.* pp. 157–172.
450. Kathe, R.L. (1967). *Feeds Illustrated* **18**(5), 18–19.
451. Kaufman, L., and Baczkowska, H. (1938). *Mem. Inst. Natl. Pol. Econ. Rurale* **17**, 176–187.
452. Kennedy, W.C. (1935). *Poult. Sci.* **14**, 301–306.
453. Kingan, J.R., and Sullivan, T.W. (1964) *Poult. Sci.* **43**, 1205–1209.
454. Kivimae, A., Hellström, V., and Adalberth, B. (1964). *Arch. Geflüegelkd.* **29** (1), 26–34.
455. Kläui, H. (1963). *Wiss. Veroetf. Dtsch. Ges. Ernaehr.* **9**, 390–400.
456. Kläui, H. (1968). *Proc. Spec. Colour Symp. Inst. Food Sci. Technol.* **1** (5).
457. Knowles, R.E., Livingston, A.L., and Kohler, G.O. (1972). *Proc. Tech. Alfalfa Conf., 11th, 1972,* pp. 71–83.
458. Knowles, R.E., Livingston, A.L., Nelson, J.W., and Kohler, G.O. (1968). *J. Agric. Food Chem.* **16**, 654–658.
459. Knuckles, B.E., Lazar, M.E., Bickoff, E.M., and Kohler, G.O. (1972). *Proc. Tech. Alfalfa Conf., 11th, 1972,* pp. 18–25.
460. Knuckles, B.E., Spencer, R.R., Lazar, M.E., Bickoff, E.M., and Kohler, G.O. (1970). *J. Agric. Food Chem.* **18**, 1086–1089.
461. Kohler, G.O. (1965). *Proc. Tech. Alfalfa Conf., 9th, 1965,* pp. 50–59.

462. Kohler, G.O. (1972). *Proc. Tech. Alfalfa Conf., 11th, 1972,* pp. 10–17.
463. Kohler, G.O. Beier, E., and Bolze, C.C. (1955). *Poult. Sci.* **34**, 468–471.
464. Kohler, G.O., and Bickoff, E.M. (1974). *Proc. Md. Nutr. Conf.* pp. 61–64.
465. Kohler, G.O., Bickoff, E.M., Spencer, R.R., Witt, S.C., and Knuckles, B.E. (1968). *Proc. Tech. Alfalfa Conf., 10th, 1968,* pp. 71–79.
466. Kohler, G.O., Knowles, R.E., and Livingston, A.L. (1967). *J. Assoc. Off. Anal. Chem.* **50**, 707–711.
467. Kohler, G.O., Livingston, A.L., Kuzmicky, D.D., Knowles, R.E., and Nelson, J.W. (1966). *Feedstuffs* **38** (7), 16,67.
468. Kolar, J.A., and Seymour, E.W. (1972). *Poult. Sci.* **51**, 1825.
469. Kon, S.K., and Thompson, S.Y. (1940). *J. Agric. Sci.* **30**, 622–638.
470. Kouwenhoven, B., and van der Horst, C.J.G. (1969). *Z. Parasitenkd.* **32**, 347–353.
471. Kowalski, L.M., and Reid, W.M. (1972). *Poult. Sci.* **51**, 1586–1589.
472. Krinsky, N.I. (1964). *Biochim, Biophys. Acta* **88**, 487–491.
473. Krinsky, N.I. (1965). *Comp. Biochem. Physiol.* **16**, 181–187.
474. Kinsky, N.I. (1971). *In* "Carotenoids" (O. Isler, ed.), pp. 669–716. Birkhaeuser, Basel.
475. Krinski, N.I., and Goldsmith, T.H. (1960). *Arch. Biolchem. Biophys.* **91**, 271–279.
476. Krinsky, N.I., Gordon, A., and Stern, A.I. (1964). *Plant Physiol.* **39**, 441–445.
477. Krinsky, N.I., and Lenhoff, H.M. (1965). *Comp. Biochem. Physiol.* **16**, 189–198.
478. Kuhn, R., and Brockmann, H. (1932). *Hoppe-Seyler's Z. Physiol. Chem.* **206,** 41–64.
479. Kuhn, R., and Grundmann, C. (1933). *Ber. Dtsch. Chem. Ges.* **66**, 1746–1750.
480. Kuhn, R., and Lederer, E. (1931). *Naturwissenschaften* **19** 306.
481. Kuhn, R., and Smakula, A. (1931). *Hoppe-Seyler's Z. Physiol. Chem.* **197,** 161–166.
482. Kuhn, R., Winterstein, A., and Lederer, A. (1931). *Hoppe-Seyler's Z. Physiol. Chem.* **197,** 141–160.
483. Kupsch, N. (1934). *Arch. Gelfluegelkd.* **8**, 97–115.
484. Kuzmicky, D.D., Kohler, G.O., and Bickoff, E.M. (1972). *Proc. Tech. Alfalfa Conf., 11th, 1972,* pp. 58–65.
485. Kuzmicky, D.D., Kohler, G.O., Livingston, A.L., Knowles, R.E., and Nelson, J.W. (1968). *Poult. Sci.* **47**, 389–397.
486. Kuzmicky, D.D., Kohler, G.O., Livingston, A.L., Knowles, R.E., and Nelson, J.W. (1969). *Poult. Sci.* **48**, 326–330.
487. Kuzmicky, D.D., Livingston, A.L., Knowles, R.E., and Kohler, G.O. (1977). *Poult. Sci.* **56**, 1504–1509.
488. Lambert, W.V., and Knox, C.W. (1927). *Poult. Sci.* **7**, 24–30.
489 Lambertsen, G., and Braekkan, O.R. (1971). *J. Sci. Food Agric.* **22**, 99–101.
490. Landagora, F.T. (1962). *J. Agric. Univ. P. R.* **46**, 91–96.
491. Landadora, F.T., and Montalvo, P.A. (1963). *J. Agric. Univ. P. R.* **47**, 91–97.
492. Lautner, V. (1965). *Biol. Chem. Vyz. Zvirat* **1**, 265–274.
493. Lautner, V. (1967). *Biol. Chem. Vyz. Zvirat* **3** (6), 573–575.
494. Lautner, V. (1967). *Zivocisna Vyroba* **12**, 919–926.
495. Lautner, V. (1969). *Zivocisna Vyroba* **14**, 615–622.
496. Lazar, M.E. (1972). *Proc. Tech. Alfalfa Conf., 11th, 1972,* pp. 26–34.
497. Lederer E., and Moore, T. (1936). *Nature (London)* **137**, 996.
498. Lee, W.L. (1966). *Comp. Biochem. Physiol.* **18**, 17–36.
499. Lee, W.L. (1966). *Comp. Biochem. Physiol.* **19**, 13–27.
500. Lee, W.L. (1972). *Abstr. Commun., 3rd, Int. Symp. Carotenoids Other Than Vitam. A, 1972* pp. 74–75.
501. Leuenberger, F.J. (1970). F. Hoffmann-La Roche & Co., Ltd., Basel (unpublished data).
502. Leuenberger, F.J. (1974). F. Hoffmann-La Roche & Co., Ltd., Basel (unpublished data).
503. Leuenberger, F.J., and Thommen, H. (1970). *J. Insect. Physiol.* **16** 1855–1858.

504. Liaaen-Jensen, S. (1969). *Pure Appl. Chem.* **20**, 421–448.
505. Libby, D.A., Evans, R.J., Bandemer, S.L., and Schaible, P.J. (1955). *Poult. Sci.* **34**, 972–980.
506. Libby, D.A., and Schaible, P.J. (1955). *Science* **121**, 733–734.
507. Lifschitz, E., Manso, F., Swarez, D., and Favret, E.A. (1971). *Poult. Sci.* **50**, 1017–1021.
508. Lipstein, B., Bornstein, S., and Budowski, P. (1967). *Poult. Sci.* **46**, 626–636.
509. Lipstein, B., Budowski, P., and Bornstein, S. (1970). *Poult. Sci.* **49**, 449–458.
510. Littlefield, L.H., Bletner, J.K., Shirley, H.V., and Goff, O.E. (1972). *Poult. Sci.* **51**, 1721–1725.
511. Livingston, A.L., Bickoff, E.M., and Thompson, C.R. (1955). *J. Agric. Food Chem.* **3**, 439–441.
512. Livingston, A.L., and Knowles, R.E. (1969). *Phytochemistry* **8**, 1311–1312.
513. Livingston, A.L., Knowles, R.E., Israelsen, M., Nelson, J.W., Mottola, A.C., and Kohler, G.O. (1966). *J. Agric. Food Chem.* **14**, 643–644.
514. Livingston, A.L. Knowles, R.E., and Kohler, G.O. (1970). *U.S., Dep. Agric., Tech. Bull.* **1414.**
515. Livingston, A.L., Knowles, R.E., and Kohler, G.O. (1971). *J. Assoc. Off. Anal. Chem.* **54**, 981–982.
516. Livingston, A.L., Knowles, R.E., Nelson, J.W., and Kohler, G.O. (1968). *J. Agric. Food Chem.* **16**, 84–87.
517. Livingston, A.L., Knowles, R.E., Page, J., and Kohler, G.O. (1971). *J. Agric. Food Chem.* **19**, 951–953.
518. Livingston, A.L., Kuzmicky, D.D., Knowles, R.E., and Kohler, G.O. (1969). *Poult. Sci.* **48**, 1678–1683.
519. Livingston, A.L., Kuzmicky, D.D., Kohler, G.O., and Knowles, R.E. (1972). *Proc. Tech. Alfalfa Conf., 11th, 1972,* pp. 84–92.
520. MacAdams, D.L. (1942). *J. Opt. Soc. Am.* **32**, 247–274.
521. MacAdams, D.L. (1943). *J. Opt. Soc. Am.* **33**, 18–29.
522. MacAdams, D.L., and Brown, W.R.J. (1949). *J. Opt. Soc. Am.* **39,** 454–462.
523. Mackay, E., Moutney, G.R., and Naber, E.C. (1962). *Poult. Sci.* **41**, 1660.
524. Mackay, E., Mountney, G.J., and Naber, E.C. (1963). *Poult. Sci.* **42**, 32–37.
525. Mackinney, G. (1935). *J. Biol. Chem.* **111**, 75.-84.
526. Mackinney, G., Chandler, B.V., and Lukton, A. (1960). *Proc. Int. Congr. Biochem., 4th, 1960* Vol 15.
527. Madiedo, G., Richter, E.F., and Sunde, M.L. (1964). *Poul. Sci.* **43**, 990–994.
528. Madiedo, G., and Sunde, M.L. (1962). *Poult. Sci.* **41**, 1660.
529. Madiedo, G., and Sunde, M.L. (1964). *Poult. Sci.* **43**, 1056–1061.
530. Mainguy, P., and Rouques, A. (1965). *Aliment. Vie* **53**, (4–6), 83–116.
531. Mainguy, P., and Rouques, A. (1965). *Aliment. Vie* **53**, 194–212.
532. Mann, T.B. (1945). *J. Agric. Sci.* **35**, 101–107.
533. Marrett, L.E., Nivas, S.C., and Sunde, M.L. (1968). *Poult. Sci.* **47**, 1692.
534. Marusich, W.L. (1962). Hoffmann-La Roche Inc., Nutley, New Jersey (unpublished data).
535. Marusich. W.L. (1963). Hoffmann-La Roche Inc., Nutley, New Jersey (unpublished data).
536. Marusich, W.L. (1964). Hoffmann-La Roche Inc., Nutley, New Jersey (unpublished data).
537. Marusich, W.L. (1964). Hoffmann-La Roche Inc., Nutley, New Jersey (unpublished data).
538. Marusich, W.L. (1967). *Feedstuffs* **39** (4), 48–49.
539. Marusich, W.L. (1969). *Feedstuffs* **41** (3), 26.
540. Marusich, W.L. (1970). *Feedstuffs* **42** (10), 30–31.
541. Marusich, W.L. (1971). *Feedstuffs* **43** (8), 32–34.
542. Marusich, W.L. (1971). *Proc. N. C. State Univ. Anim. Nutr. Conf.* **25**, 50–68.
543. Marusich, W.L. (1980). *In* "Vitamin E" (L. Machlin, ed). Dekker, New York, pp. 445–472.
544. Marusich, W.L., Ackerman, G.L., and Bauernfeind, J.C. (1962). *Poult. Sci.* **41**, 1664.
545. Marusich, W.L., and Bauernfeind, J.C. (1962). *Poult. Sci.* **41**, 1664.
546. Marusich, W.L., and Bauernfeind, J.C. (1970). *Poult. Sci.* **49**, 1555–1566.
547. Marusich, W.L., and Bauernfeind, J.C. (1970). *Poult. Sci.* **49**, 1566–1579.

548. Marusich, W.L., DeRitter, E., and Bauernfeind, J.C. (1960). *Poult. Sci.* **39**, 1338–1345.
549. Marusich, W., DeRitter, E., Vreeland, J., and Krukar, R. (1960). *J. Agric. Food Chem.* **8**, 390–393.
550. Marusich, W., Kadin, H., and Bauernfeind, J.C. (1958). *Poult. Sci.* **37**, 1224.
551. Marusich, W.L., Ogrinz, E., Brand, M., and Mitrovic, M. (1969). *Poult. Sci.* **48**, 217–222.
552. Marusich, W.L., Ogrinz, E.F., Brown, P.R., and Mitrovic, M. (1971). *Poult. Sci.* **50**, 1602.
553. Marusich, W.L., Ogrinz, E.F., Brown, P.R., and Mitrovic, M. (1972). *Br. Poult. Sci.* **14**, 23–30.
554. Marusich, W.L., Ogrinz, E.F., Camerlengo, N., McCambley, J., and Mitrovic, M. (1974). *Proc. World's Poult. Congr., 15th, 1974* pp. 310–312.
555. Marusich, W.L., Ogrinz, E.F., Camerlengo, N., McCambley, J., and Mitrovic, M. (1976). *Poult. Sci.* **55**, 1486–1494.
556. Marusich, W.L., Ogrinz, E.F., Schildknecht, E., Brown, P.R., and Mitrovic, M. (1973). *Br. Poult. Sci.* **14**, 541–546.
557. Marusich, W.L., Schildknecht, E., Ogrinz, E.F., Brown, P.R., and Mitrovic, M. (1972). Br. Poult. Sci. **13**, 577–585.
558. Marusich, W.L., and Wilgus, H.S. (1968). *Proc. Arkansas Nutr. Conf.* pp. 9–22.
559. Matterson, L.D., Potter, L.M., Pudelkiewicz, W.J., Carlson, D., and Singsen, E.P. (1956). *Poult. Sci.* **35**, 1156.
560. Matterson, L.D., Pudelkiewicz, W.P., and Potter, L.M. (1962). *Poult. Sci.* **41**, 1664.
561. Mattson, F.H., and Deuel, H.J., Jr. (1943). *J. Nutr.* **25**, 103–112.
562. Maw, W.A. (1939). *Proc. World's Poult. Congr. Expo. 7th 1939* pp. 177–178.
563. Mayer, H., and Isler, O. (1971). *In* "Carotenoids" (O. Isler, ed.), pp. 325–575. Birkhaeuser, Basel.
564. Mazziotti Di Celso, P. (1964). *Aliment, Anim.* **8** (7/8), 401–407.
565. McDonald, M.W. (1955). *Poult. Sci* **34**, 55–56.
566. McKeown, G.G. (1963). *J. Assoc. Off. Agric. Chem.* **46**, 790–796.
567. McKeown, G.G., and Mark, E. (1962). *J. Assoc. Off. Agric. Chem.* **45**, 761–766.
568. McNaughton, J.L., Day, E.J., and Dilworth, B.C. (1976). *Poult. Sci.* **55,** 2065.
569. Mecchi, E.P., Pool, M.F., and Klose, A.A. (1953). *Poult. Sci.* **32**, 915–916.
570. Mehner, A., (1965). *Arch. Gefluegelkd.* **29**, 356–363.
571. Menge, H. (1972). *Poult. Dig.* **31** (367), 431.
572. Menge, H., and Beal, R.E. (1973). *Poult. Sci.* **52**, 219–222.
573. Meyers, S.P. (1971). *Feedstuffs* **43** (8), 27, 30.
573a. Meyers, S.P. (1977). *Feedstuffs* **49**(19), 26–28.
574. Meyers, S.P., and Rutledge, J.E. (1971). *Feedstuffs* **43**(49), 31–32.
575. Middendorf, D. (1967). *Feedstuffs* **39**(5), 28–30, 32.
575a. Middendorf, D.F., Childs, G.R., and Cravens, W.W. (1978). *Poult. Sci.* **57,** 1157–1158.
575b. Middendorf, D.F., Childs, G.R., and Cravens, W.W. (1978). *Poult. Sci.* **57,** 1158.
576. Miller, R. E., Bickoff, E.M., and Kohler, G.O. (1972). *Proc. Tech. Alfalfa Conf., 11th, 1972* pp. 35–40.
577. Milligan, J.L., Marr, J.E., Wilcke, H.L., Bethke, R.M., and McKinney, A.W. (1953). *Poult. Sci.* **32** 916.
578. Milligan, J.L., Wilcke, H.L., Marr, J.E., and Bethke, R.M. (1955). *Poult. Sci.* **34**, 794–799.
579. Mitchell, H.L., Beauchene, R.E., and Silker, R.E. (1954). *J. Agric. Food Chem.* **2**, 939–941.
580. Mitchell, H.L., and Silker, R.E. (1955). *J. Agric. Food Chem.* **3**, 69–71.
581. Mitchell, R.P., Jr., Bletner, J.K., and Goff, O.E. (1961). *Tenn., Agric.Exp. Stn., Prog. Rep.* **40**, 6–7.
582. Mitchell, R.P., Jr., Bletner, J.K., and Goff, O.E. (1962). *Feed Bag* **38** (2), 7–9.
583. Mitchell, R.P., Jr., Bletner, J.K., and Goff, O.E. (1962). *Feedstuffs* **34**(14), 28.
584. Mitchell, R.P., Jr., Bletner, J.K., and Tugwell, R.L. (1961). *Poult. Sci.* **40,** 1432.
585. Moore, T. (1957). *In* "Vitamin A," pp. 180–191. Elsevier, Amsterdam.

586. Morehouse, N.F. (1949). *Poult. Sci.* **28**, 375–384.
587. Morehouse, N.F. (1960). *Feed Age* **10** (5), 43–44.
588. Morehouse, N.F. (1961). *Poult. Sci.* **40**, 1432.
589. Morehouse, A.L., and Hanson, A.W. (1961). *Inform. Poult. Nutr. Conf., 1961.*
590. Morehouse, N.F., and Mayfield, O.J. (1946). *J. Parasitol.* **32** 20–24.
591. Morgan, J.H.L., and Everitt, G.C. (1969). *N. Z. Agric. Sci.* **4,** 10–18.
592. Morgan, R.C. (1967). *J. Food Sci.* **32,** 275–278.
593. Morgan, W.A., and Woodroff, J.G. (1927). *Ga., Agric. Exp. Stn., Bull.* **147.**
594. Morrison, A.B., Hunsacker, W.G., and Aitken, J.R. (1954). *Poult. Sci.* **33**, 491–494.
594a. Moss, G.P., and Weedon, B.C.L. (1976). *In* "Chemistry and Biochemistry of Plant Pigments" (T.W. Goodwin, ed.), 2nd ed., Vol. 1, pp. 149–224. Academic Press, New York.
595. Moster, J.B., Quackenbush, F.W., and Porter, J.W. (1952). *Arch. Biochem. Biophys.* **37**, 287–296.
596. Moyer, S.E., and Collins, W.M. (1960). *Poult. Sci.* **39**, 662–663.
597. Mueller, W.J. (1956). *Poult. Sci.* **35**, 1074–1078.
598. Mueller, W.J. (1957). Pa., Agric. Exp. Stn., Prog. Rep. **174.**
599. Nadai, J., and Brubacher, G. (1969). *In* "International Encyclopedia of Food and Nutrition" (R.A. Morton, ed.), Vol. 9, pp. 449–487. Pergamon, Oxford.
600. Nakaue, H.E., Kurnick, A.A., Hulett, B.J., and Reid, B.L. (1966). *Poult. Sci.* **45**, 478–483.
601. Nakayama, T.O.M. (1962). *In* "Physiology and Biochemistry of Algae" (R.A. Lewis, ed.), pp. 409–420. Academic Press, New York.
602. Nash, H.A., and Zscheile, F.P. (1945). *Arch. Biochem.* **7**, 305–311.
603. Neamtu, G., Tamos, V., and Bodea, C. (1965). *Stud. Cerect. Biochim.* **8**, 411–444.
604. Nelson, T.S. (1966). *Poult. Sci.* **45**, 747–753.
605. Nelson, T.S., and Baptist, J.N. (1968). *Poult. Sci.* **47**, 924–931.
606. Newlands, G.M. (1966). *J. Inst. Corn Agric. Merchants* **14** (3), 100–102.
607. Nicola, M.D. (1956). *Exp. Cell Res.* **10**, 441–446.
608. Nicola, M.D., and Goodwin, T.W. (1954). *Exp. Cell Res.* **7**, 23–31.
609. Nobile, S. (1964). *Proc. Australas. Poult. Sci. Conv.* pp. 62–65.
610. Norman, G.M., Sykes, A.H., and Bayley, H.S. (1973). *Br. Poult. Sci.* **14**, 507–519.
611. Norman, G.M., Sykes, A.H., and Bayley, H.S. (1973). *Br. Poult. Sci.* **14**, 521–530.
612. "Official Methods of Analysis" (1965). 10th ed., Sects. 39.014–39.017. Assoc. Off. Anal. Chem., Washington, D.C.
613. Olivetti, A., and Meregalli, A. (1968). *Aliment. Anim.* **12**, 433–443.
614. Olson, G., Cobb, W.R., McManus, A.D., Van Iderstine, A., and Ott, W.H. (1972). *Poult. Sci.* **51**, 1846.
615. Olson, J.A. (1964) *J. Lipids Res.* **5**, 281–299.
616. Olson, J.A. (1967). *Pharmacol. Rev.* **19** (4), 559–596.
617. Olsson, N. (1941). *Lantbrukshoegsk., Husdjursfoersoeksanst., Sartryck Foerhandsmeed.* No. 10, pp. 1–9.
618. Osadca, M., Araujo, M., and DeRitter, E. (1972). *J. Assoc. Off. Anal. Chem.* **55**, 110–113.
619. Osadca, M., and DeRitter, E. (1965). Hoffmann-La Roche Inc., Nutley, New Jersey (unpublished data).
620. Palmer, L.S. (1915). *J. Biol. Chem.* **23**, 261–279.
621. Palmer, L.S. (1922). "Carotenoids and Related Pigments," Am. Chem. Soc. Monogr. Ser. Wiley (Interscience), New York.
622. Palmer, L.S., and Eckles, C.H. (1914). *J. Biol. Chem.* **17**, 191, 211, 223, 237, 245.
623. Palmer, L.S., and Kempster, H.L. (1919). *J. Biol. Chem.* **39**, 331–337.
624 Parker, S.L. Gossman, S.S., and Lippincott, W.A. (1926). *Poult. Sci.* **5**, 131–145.
625. Parkhurst, R.T. (1937). *Flour Feed* **12** (5), 10.
626. Parman, G.K., and Borenstein, B. (1964). *Food Eng.* **36** (5), 77–80.

627. Patrias, G. (1952). *Feed Age* **2** (12), 32–33.
628. Patrick, H. (1964). *Feed Age* **14** (3), 46–47.
629. Pepper, W.F., Slinger, S.J., and Bergey, J.E. (1954). *Poult. Sci.* **33**, 422–423.
630. Perdue, H.S., and Smith, I.D. (1962). *Proc. Tex. A&M Nutr. Conf.* pp. 77–84.
631. Peterson, W.J., Hughes, J.S., and Freeman, H.F. (1937). *Ind. Eng. Chem., Anal. Ed.* **9**, 71.
632. Peterson, W.J., Hughes, J.S., and Payne, L.F. (1939). *Kan., Agric. Exp. Stn., Bull.* **46**.
633. Petracek, F.J., and Zechmeister, L. (1956). *Arch. Biochem. Biophys.* **61**, 137–139.
634. Petracek, F.J., and Zechmeister, L. (1956). *J. Am. Chem. Soc. 78*, 1427–1434.
635. Petzold, E.N., and Quackenbush, F.W. (1960). *Arch. Biochem. Biophys.* **86**, 163–165.
636. Philip, T., Weber, C.W., and Berry, J.W. (1976). *J. Food Sci.* **41**, 23–25.
636a. Philip, T., Weber, C.W., and Berry, J.W. (1977). *Poult. Sci.* **56**, 1305–1309.
637. Phillips, H.J., and Williams, I.L. (1950). *Poult. Sci.* **29**, 776.
638. Phillips, H.J., and Williams, I.L. (1952). *Food Technol.* **6**, 74–76.
639. Pitt, G.A.J., and Morton, R.A. (1962). *Annu. Rev. Biochem.* **31**, 500–506.
640. Pohle, W.D., and Mehlenbacker, V.C. (1955). *Food Technol.* **9**, 565–568.
641. Pope, C.W., Schaible, P.J., and Evans, R.J. (1957). *Poult. Sci.* **36**, 1149.
642. Poppe, S. (1965). *Arch. Gefluegelzuecht Kleintierkd.* **13**, 137–141.
643. Poppe, S., Moldenhauer, H., and Meier, H. (1964). *Arch. Gefluegelzuecht Kleintierkd.* **13** (4/5), 251–257.
644. Porter, J.W., and Anderson, D.G. (1962). *Arch. Biochem. Biophys.* **97**, 520–528.
645. Portsmouth, J. (1970). *Fert., Feed & Pestic. J.* **66** (11), 4–5.
646. Potter, L.M., Bunnell, R.H., Matterson, L.D., and Singsen, E.P. (1956). *Poult. Sci.* **35**, 452–456.
647. Prebble, J.N. (1962). *Biochem. J.* **85**, 13P–14P.
648. Prohaszka, L. (1968). *Rev. Cubana Cienc. Agric.* **2**, 113–118.
649. Prohaszka, L., and Toth, M. (1967). *Zentralbl. Veterinaermed.* **14**, 119–128.
650. Quackenbush, F.W. (1950). *J. Assoc. Off. Agric. Chem.* **33**, 647–651.
651. Quackenbush, F.W. (1963). *Cereal Chem.* **40**, 266–269.
652. Quackenbush, F.W. (1970). *J. Assoc. Off. Anal. Chem.* **53**, 186–189.
653. Quackenbush, F.W., Dyer, M.A., and Smallidge, R.L. (1970). *J. Assoc. Off. Anal. Chem.* **53**, 181–185.
654. Quackenbush, F.W., Firch, J.G., Brunson, A.M., and House, L.R. (1963). *Cereal Chem.* **40**, 250–259.
655. Quackenbush, F.W., Firch, J.G., Rabourn, W.J., McQuistan, M., Petzold, E.W., and Kargl, T.E. (1961). *J. Agric. Food Chem.* **9**, 132–135.
656. Quackenbush, F.W., Kvakovszky, S., Hoover, T., and Rogler, J.C. (1965). *J. Assoc. Off. Agric. Chem.* **48**, 1241–1244.
657. Quackenbush, F.W., and Miller, S.L. (1972). *J. Assoc. Off. Agric. Chem.* **55**, 617–621.
658. Quilici, R. (1963). *Avicoltura* No. 6, pp. 135–140.
659. Quilici, R. (1966). *G. Pollicolt.* **16** (9), 37–39.
660. Quilici, R. (1970). *Riv. Zootec.* **43**, 113–118.
661. Raskopf, B.D., Kidd, I.H., and Goff, O.E. (1961). *Tenn., Agric. Exp. Stn., Bull.* **324**.
662. Ratcliff, R.G., and Day, E.J. (1958). *Miss., Agric. Exp. Stn., Inf. Sheet* **605**.
663. Ratcliff, R.G., Day, E.J., and Hill, J.E. (1961). *Poult. Sci.* **40**, 716–720.
664. Ratcliff, R.G., Day, E.J., Grogan, C.O., and Hill, J.E. (1962). *Poult. Sci.* **41**, 1529–1532.
665. Ratcliff, R.G., Day, E.J., and Hill, J.E. (1959). *Poult. Sci.* **38**, 1039–1048.
666. Rathmann, D.M., and Daggy, E.E. (1961). "Corn Zanthophylls in Poultry Nutrition," Tech. Brochure. Corn Products Co.
667. Rauch, W. (1959). *Arch. Gefluegelkd.* **23**, 319–331.
668. Rauch, W. (1960). *Arch. Gefluegelkd.* **24**, 417–431.
669. Rauch, W. (1961). *Arch. Gefluegelkd.* **25**, 494–502.

670. Rauch, W. (1963). *In* "Carotine and Carotinoide" (K. Lang, ed.), pp. 300–326. Steinkopff, Darmstadt.
671. Rauch, W. (1965). *Arch. Gefluegelkd.* **29**, 1–26.
672. Rauch, W. (1965). *Arch. Gefluegelkd.* **29**, 340–346.
673. Rauch, W. (1967). *Arch. Gefluegelkd.* **31**, 444–448.
674. Rauch, W. (1968). *Arch. Gefluegelkd.* **32**, 53–74.
675. Rauch, W. (1968). *Arch. Gefluegelkd.* **32**, 261–268.
676. Rauch, W. (1968). *Z. Tierphysiol., Tierernaehr. Futtermittelkd.* **23** (5), 295–296.
677. Reese, W.C. (1965). Hoffmann-La Roche Inc., Nutley, New Jersey (unpublished data).
678. Ridgeway, R. (1912). "Color Standards and Color Nomenclature." Published by the Author, Washington, D.C.
679. Ringrose, R.C., Norris, L.C., and Heuser, G.F. (1939). *Cornell Univ., Agric. Exp. Stn., Bull.* **725**.
680. Romanoff, A.L. (1960). "The Avian Embryo." Macmillan, New York.
681. Romanoff, A.L., and Romanoff, A.J. (1949). "The Avian Egg." Wiley, New York.
682. Romoser, G.L., Machlin, L.J., Maddy, K.H., and Gordon, R.S. (1959). *Poult. Sci.* **38**, 1241–1242.
683. Ross, W.M., and Webster, O.J. (1960). *U.S., Dep. Agric., Agric. Inf. Bull.* **218**.
684. Rotfarb, R.M. (1970). *Fiziol. Biokhim. Issled. Rast.* pp. 3–5.
685. Rubin, M., and Bird, H.R. (1946). *Science* **103**, 584–586.
686. Rubin, M., and Shillinger, J.A. (1973). *Proc. Md. Nutr. Conf.* pp. 110–116.
687. Rüegg, R., Montavan, M., Ryser, G., Saucy, G., Schwieter, U., and Isler, O. (1959). *Helv. Chim. Acta* **42**, 854–864.
688. Ruff, M.D., and Fuller, H.L. (1975). *J. Nutr.* **105**, 1447–1456.
689. Ruff, M.D., Reid, W.M., and Johnson, J.K. (1974). *Poult. Sci.* **53**, 1801–1809.
690. Ruff, M.D., Reid, W.M., and Rahn, A.P. (1976). *Am J. Vet. Res.* **37**, 963–967.
691. Runnels, T.D. (1957). *Distill. Feed Res. Counc., Conf., Proc.* **12** 54–60.
692. Runnels, T.D., Kobrin, B.R., and Ely, C.M. (1966). *Poult. Sci.* **45**, 1283–1290.
693. Runnels, T.D., Waller, E.F., Kelley, E.G., and Krulick, S. (1951). *Poult. Sci.* **30**, 930.
694. Russell, W.C., Taylor, M.W., and Chichester, D.F. (1935). *Plant Physiol.* **10**, 325–340.
695. Russo, V. (1966). *Prod. Anim.* **5**, 65–71.
696. Saperstein, S., and Starr, M.P. (1954). *Biochemistry* **57**, 273–275.
697. Saperstein, S., Starr, M.P., and Filfus, J.A. (1954). *J. Gen. Microbiol.* **10**, 85–92.
698. Saunders, C. (1964). *Proc. Soc. Adv. Anim. Nutr. South. Afr.* **1**, 81–98.
699. Saunders, C.F., Hayes, J.P., Erasmus, J., Jooste, J.P., DuPlessis, P.H.C., and Venter, E.L. (1967). *S. Afr. J. Agric. Sci.* **10**, 373–385.
700. Schaible, P.J., Libby, D.A., Evans, R.J., and Bandemer, S.L. (1954). *Mich., Agric. Exp. Stn., Bull.* **37** (2), 273–277.
700a. Schärer, K., and Studer, F. (1960–1962). F. Hoffmann-La Roche & Co., Ltd., Basel (unpublished data).
701. Schertz, F.M. (1923). *J. Agric. Res.* **26**, 383–400.
702. Schettino, O., and Di Lieto, A. (1968). *Tec. Molitoria* **19**, 557.
703. Schmidt, L. (1963). *Dtsch. Gefluegelvirlsch.* **15**, 1962–1965.
704. Scholfield, C.R., and Dutton, H.J. (1954). *J. Am. Oil Chem. Soc.* **31**, 258–260.
705. Scholtyssek, S., and Böhm, U. (1962). *Kraftfutter* **45** (2), 69–71.
706. Scholtyssek, S., Kirsammer, R., Hauser, E., and Kurmann, B. (1965). *Arch. Gefluegelkd.* **29**, 249–257.
707. Scholtyssek, S., Lorenz, G., and Erfani, H. (1961). *Arb. Landw. Hochsch. Hohenbeim* **1**, 77–91.
708. Schrenk, W.G., Chapin, D.S., and Conrad, R.M. (1964). *Ind. Eng. Chem., Anal. Ed.* **16**, 632–634.
709. Schunk, C.A. (1903). *Proc. R. Soc. London* **72**, 165–176.

710. Schwartzel, E.M., and Cooney, J.J. (1972). *J. Bacteriol.* **112**, 1422–1424.
711. Scott, H.M., and Glista, W.A. (1950). *Poult. Sci.* **29**, 921–923.
712. Scott, M.L. (1967). *Cornell Poult. Pointers* **19** (2), 11–14.
713. Scott, M.L., Ascarelli, I., and Olson, G. (1968). *Poult. Sci.* **47**, 863–872.
714. Seim, R., and Braun, R. (1963). *Farm J.* **87** (6), 26, 78.
715. Sharp, P.F., and Powell, C.K. (1930). *Ind. Eng. Chem.* **22**, 908–910.
716. Shellenberger, T.E., Parrish, D.B., and Sanford, P.E. (1957). *Poult. Sci.* **36**, 1313–1316.
717. Sheppard, A.J. (1969). *Feedstuffs* **41** (45), 9, 58, 59.
718. Sibbald, I.R., and Slinger, S.J. (1963). *Poult. Sci.* **42,** 1325–1332.
719. Siedler, A.J., Enzer, E., Schweigert, B.S., and Riemenschneider, R.W. (1956). *J. Agric. Food Chem.* **4,** 1023–1029.
720. Silker, R.E., Schrenk, W.G., and King, H.H. (1944). *Ind. Eng. Chemm.* **36,** 831–835.
721. Silveira, J.J.N., and Kronka, R.N. (1972). *Bol. Ind. Anim.* **29** 393–400.
722. Silveira, J.J.N., Kronka, R.N., Becker, M., and Kalil, E.B. (1971). *Bol. Ind. Anim.* **28**, 449–462.
723. Singh, S.P., and Donovan, G.H. (1966). *Poult. Sci.* **45**, 1124.
724. Sipe, G.R., and Polk, H.D. (1941). *Poult. Sci.* **20**, 406–412.
725. Sjollema, B., and Donath, W.F. (1940). *Biochem. J.* **34**, 736–748.
726. Slinger, S.J., Pepper, W.F., Hill, D.C., and Synder, E.S. (1952). *Poult. Sci.* **31**, 193–201.
727. Slocum, W.L., and Swanson, H.S. (1954). *Wash., Agric. Exp. Stn., Bull.* **556.**
728. Smidt, M.J., Classen, L.J., Pagoria, F., and Fritz, J.C. (1965). *Fed. Proc., Fed. Am. Soc. Exp. Biol.* **24** 687.
729. Smith, I.D., and Perdue, H.S. (1966). *Poult. Sci.* **45**, 577–581.
730. Snyder, E.S. (1945). *Bull. Ont. Dep. Agric.* No. 446.
731. Soliman, M.K. (1964). *Zentralbl. Veterinaermed.* **11**(6), 517–522.
732. Soliman, M.K., Elmossalami, E., and Elnawawy, F.A. (1967). *Zentralbl. Veterinaer med., Reihe B* **14**, 288–291.
733. Spencer, R.R., Austin, W.K., and Burdick, D. (1972). *Feedstuffs* **44** (14), 18–21, 31.
734. Spencer, R.R., Bickoff, E.M., Kohler, G.O., Witt, S.C., Knuckles, B.E., and Mottola, A. (1970). *Trans ASAE* **13**, 198–200.
735. Spencer, R.R., Mottola, A.C., Bickoff, E.M., Clark, J.P., and Kohler, G.O. (1971). *J. Agric. Food Chem.* **19**, 504–507.
736. Splittgerber, H., Wein, F.K., and Arhelger, R. (1971). *Dtsch. Gefluegelwirtsch.* **23**, 790.
737. Squibb, R.L., Beisel, W.R., and Bostian, K.A. (1971). *Appl. Microbiol.* **22** (6), 1096–1099.
738. Squibb, R.L., Braham, J.E., Guzman, M., and Scrimshaw, N.S. (1955). *Poult. Sci.* **34**, 1054–1058.
739. Squibb, R.L., Braham, J.E., and Scrimshaw, N.S. (1957). *Poult. Sci.* **36**, 1241–1244.
740. Squibb, R.L., Guzman, M., and Scrimshaw, N.S. (1953). *Poult. Sci.* **32**, 953–957.
741. Squibb, R.L., Guzman, M., and Scrimshaw, N.S. (1953). *Poult. Sci.* **32**, 1078–1083.
742. Squibb, R.L., Mendez, J., and Scrimshaw, N.S. (1953). *Turrialba* **3,** 163–168.
743. Stadeler, G. (1867). *J. Pract. Chem.* **100**, 148.
744. Stafseth, H.J. (1931). *J. Am. Vet. Med. Assoc.* **78**, 793–816.
745. Steele, H.L., Barnett, B.D., von Tungeln, G.R., and Boles, P.P. (1965). S. C., Agric. Exp. Stn., Bull. **518**.
746. Steinegger, P., Menzi, M., and Berger, R. (1956). *Gefluegelhof* **19** (13) 215–220.
747. Steinegger, P., Streiff, K., and Zeller, P. (1957). *Mitt. Geb. Lebensmittelunters. Hyg.* **48**, 445–449.
748. Steinegger, P., and Zanetti, G. (1957). *Arch. Gefluegelkd.* **21**, 236–245.
749. Steinegger, P., and Zanetti, G. (1959). *Arch. Gefluegelkd.* **23**, 166–173.
750. Stephens, J.F., Borst, W.J., and Barnett, B.D. (1974). *Poult. Sci.* **53**, 1735–1742.
751. Stevens, J., and Lewis, J. (1961). Report of the School of Agriculture, pp. 75–80. University of Nottingham, England.

752. Stevens, J., Payne, C.G., and Lewis, D. (1964). *Br. Soc. Anim. Prod.* **6** (2), 257.
753. Stewart, J.S., and Atwood, H. (1903). *W. Va., Agric. Exp. Stn., Bull.* **88**.
754. Stokstad, E.L.R., and Jukes, T.H. (1950). *Proc. Soc. Exp. Biol. Med.* **73**, 523–528.
755. Stone, H.A., Collins, W.M., and Urban, W.E., Jr. (1971). *Poult. Sci.* **50**, 675–681.
756. Stone, H.A., Collins, W.M., and Urban, W.E., Jr. (1971). *Poult. Sci.* **50**, 1859–1862.
757. Strain, H.H. (1934). *J. Biol. Chem.* **105**, 523–535.
758. Strain, H.H. (1936). *Science* **83**, 241–242.
759. Strain, H.H. (1938). *Carnegie Inst. Washington, Publ.* **490**.
760. Strain, H.H. (1945). "Chromatographic Adsorption Analysis." Wiley (Interscience), New York.
761. Strain, H.H. (1948). *J. Am. Chem. Soc.* **70**, 1672.
762. Straub, O. (1971). *In* "Carotenoids" (O. Isler, ed.), pp. 771–850. Birkhaeuser, Basel.
763. Streiff, K. (1959). F. Hoffmann-La Roche & Co., Ltd., Basel (unpublished data).
764. Streiff, K. (1961). *Gefluegelhof* **24** (38), 645–648.
765. Streiff, K. (1963). F. Hoffmann-La Roche & Co., Ltd., Basel (unpublished data).
766. Streiff, K. (1964). F. Hoffmann-La Roche & Co., Ltd., Basel (unpublished data).
767. Streiff, K. (1967). "Carotenoids in Broiler Pigmentation." Lecture presented at the working meeting on "Intensive Breeding and Nutritional Problems" in Roche Hause, Vienna, April 6. F. Hoffmann-La Roche & Co., Ltd., Basel.
768. Streiff, K. (1969). F. Hoffmann-La Roche & Co., Ltd., Basel (unpublished data).
769. Streiff, K. (1970). *Proc. Symp. Yolk Color Egg Qual. Factor, 1970* pp. 29–31.
770. Streiff, K. (1970). *Proc. Symp. Yolk Color Egg Qual. Factor, 1970* pp. 33–38.
771. Streiff, K. (1976). F. Hoffmann-La Roche & Co., Ltd., Basel (unpublished data).
772. Streiff, K., Thommen, H., and Weiser, H. (1966). F. Hoffmann-La Roche & Co., Ltd., Basel (unpublished data).
773. Streiff, K., and Weiser, H. (1970). F. Hoffmann-La Roche & Co., Ltd., Basel (unpublished data).
774. Suarez, D. (1969). *Poult. Sci.* **48**, 733–735.
775. Sullivan, T.W. (1964). *Feedstuffs* **36** (4), 42, 44.
776. Sullivan, T.W. (1965). *Poult. Sci.* **44**, 1418.
777. Sullivan, T.W., and Holleman, K.A. (1962). *Poult. Sci.* **41**, 1474–1478.
778. Sunde, M.L. (1962). *Poult. Sci.* **41**, 532–541.
779. Sunde, M.L. (1962). *Feed Age* **12** (12), 71–72.
780. Sunde, M.L., and Madiedo, G. (1960). *Agric. Trop.* **16**, 725–730.
781. Surmatis, J.D., Walser, A., Gibas, J., Schwieter, U., and Thommen, R. (1970). *Helv. Chim. Acta* **53**, 974–990.
782. Szabolcs, J., and Ronai, A. (1969). *Acta Chim. Acad, Sci. Hung.* **61**, 309–315.
783. Tagwerker, F.J., Streiff, K., and Brubacher, G. (1962). *Proc. World's Poult. Congr., 12th, 1962* pp. 177–181.
783a. Tanaka, Y., Matsuguchi, H., Katayama, T., Simpson, K.L., and Chichester, C.O. (1976). *Bull. Jpn. Soc. Fish.* **42,** 197–202.
784. Tarver, F.R., Jr. (1961). *Poult. Sci.* **40,** 987–991.
785. Tarver, F.R., Jr., Patrick, H., and Goff, O.E. (1954). *Poult. Sci.* **33**, 1085.
786. Tasaki, I., and Moyoga, K. (1966). *Jpn. J. Zootech. Sci.* **37**, 236–245.
787. Tello, A.G., and Urbina, R.V. (1962). *Rev. Fac. Med. Vet. Univ. Nac. Mayor San Marcos* **16/17**, 79–87.
788. Terpstra, K. (1962). *Veeteelt- Zuivelber* **5**(11). 416–422.
789. Terpstra, K. (1962). *Veeteelt- Zuivelber.* **5**(12), 465–471.
790. Terpstra, K. (1963). *Veeteelt- Zuivelber.* **6**(1), 18–26.
791. Thomas, O.P., Twining, P.V., Jr., Bossard, E.H., and Lund, P.G. (1971). *Poult. Sci.* **50**, 1636.
792. Thommen, H. (1961). *Chimina* **15**, 433–434.
793. Thommen, H. (1962). *Naturwissenschaften* **22**, 517–518.
794. Thomen, H. (1962). F. Hoffmann-La Roche & Co., Ltd., Basel (unpublished data).

795. Thommen, H. (1967). *Int. Z. Vitaminforsch.* **37**, 175–183.
796. Thommen, H. (1971). *In* "Carotenoids" (O. Isler, ed.), pp. 637–668. Birkhaeuser, Basel.
797. Thommen, H., and Gloor, U. (1965). *Naturwissenschaften* **52**, 161–162.
798. Thommen, H., and Leuenberger, F. (1971). F. Hoffmann-La Roche & Co., Ltd., Basel (unpublished data).
799. Thommen, H., and Wackernagel, H. (1963). *Biochim. Biophys. Acta* **69**, 387–396.
800. Thommen, H., and Wackernagel, H. (1964). *Naturwissenschaften* **51**, 87–88.
801. Thommen, H., and Wiss, O. (1963). *Z. Ernaehrwiss., Suppl.* **3**, 17–23.
802. Thompson, C.R. (1950). *Ind. Eng. Chem.* **42**, 922–925.
803. Thompson, C.R., and Bickoff, E.M. (1951). *J. Assoc. Off. Agric. Chem.* **34**, 219–224.
804. Thompson, C.R., Bickoff, E.M., and Maclay, W.D. (1951). *Ind. Eng. Chem.* **43**, 126–129.
805. Thompson, C.R., Bickoff, E.M., Van Alta, G.R., Kohler, G.O., Guggolz, J., and Livingston, A.L. (1960). *U.S., Dep. Agric., Tech. Bull.* **1232**.
806. Thompson, C.R., Ewan, M.A., Hauge, S.M., Bohren, B.B., and Quackenbush, F.W. (1946). *Ind. Eng. Chem., Anal. Ed.* **18**, 113–115.
807. Thompson, C.R., and Maclay, W.D. (1952). *Feed Age* **2** (10), 22–25.
808. Thudichum, J.L.W. (1869). *Proc. R. Soc. London* **17**, 253–256.
809. Thumim, A. (1962). *Proc. World's Poult. Congr., 12th, 1962* pp. 552–528.
810. Tiews, J., and Zucker, E. (1962). *Z. Tierphysiol., Tierernaehr. Futtermittelkd.* **17** (5), 302–308.
811. Tiews, J., and Zucker, H. (1963). *Tieraerztl. Umsch.* **18**, 590–597.
812. Titus, H.W., Fritz, J.C., and Kauffman, W.R. (1938). *Poult. Sci.* **17**, 38–45.
813. Tomhave, A.E., and Hoffman, E. (1944). *Del., Agric. Exp. Stn., Bull. 247.*
814. Torges, H.G. (1964). *Arch. Gefluegelkd.* **28**, 75–82.
815. Tortuero, F. (1968). *Poult. Sci.* **47**, 376–383.
816. Treat, C. (1963). *Feedstuffs* **35**, (52), 18.
817. Tswett, M. (1906). *Berl. Bot. Ges.* **24**, 384.
818. Tung, H.T., and Hamilton, P.B. (1973). *Poult. Sci.* **52**, 80–83.
819. Turk, D.E., and Stephens, J.F. (1966). *J. Nutr.* **88**, 261–266.
820. Turk, D.E., and Stephens, J.F. (1967). *Poult. Sci.* **46**, 775–777.
821. Turk, D.E., and Stephens, J.F. (1967). *Poult. Sci.* **46**, 939–943.
822. Turk, D.E., and Stephens, J.F. (1967). *J. Nutr.* **93**, 161–165.
823. Turk, D.E., and Stephens, J.F. (1969). *Poult. Sci.* **48**, 586–589.
824. Turner, A.W., and Conquest, V. (1939). *U.S. Egg Poult. Mag.* **45,** 668–670.
825. Twining, P.V., Jr., Bossard, E.H., Lund, P.G., and Thomas, O.P. (1971). *Proc. Md. Nutr. Conf.* pp. 90–95.
826. Ungers, G.E., and Cooney, J.J. (1967). *J. Bacteriol.* **96**, 234–241.
827. Van der Waerden, B.L., and Nievergelt, E. (1956). "Tables for Comparing Two Samples by X-Test and Sign Test," pp. 1–34. Springer-Verlag, Berlin and New York.
828. Van Koetsveld, E.E. (1963). *In* "Carotine and Carotinoids" (K. Lang-Mainz, ed.), pp. 194–204. Steinkopff, Darmstadt.
829. Van Koetsveld, E.E., and Grimbergen, A.H.M. (1958). *Voeding* **19** (10), 615–624.
830. Vavich, M.G., and Kemmerer, A.R. (1950). *Arch. Biochem.* **28**, 295–298.
831. Villavicencio, R.U. (1961). Ph.D. Thesis, University of San Marco, Lima, Peru.
832. Vogt, H., Krieg, R., and Rauch, W. (1972). *Ber. Landwirtschaftswiss.* **50**, 734–746.
833. Völker, O. (1961). *Naturwissenschaften* **48**, 581.
834. Vollrath, W. (1969). *Kraftfutter* **52** (1), 602–604.
835. Vollrath, W. (1969). *Dtsch. Gefluegelwirtsch.* **21** 2234–2241.
836. Vuilleumier, J.P. (1969). *Poult. Sci.* **48**, 767–779.
837. Vuilleumier, J.P., Probst, H.P., and Brubacher, G. (1967). *In* "Handbuch der Lebensmittelchemie" (L. Acker, K.G. Bergner, W. Diemair, W. Heimann, F. Kiermeier, J. Schormüller, and S.W. Souci, eds.), Vol. 2, pp. 669–876. Springer-Verlag, Berlin and New York.

838. Wackernagel, H. (1959). *Agric. Mag.* **65**, 20–21.
839. Wackernagel, H. (1963). *Wiss. Veroeffentl, Dsch. Ges. Ernaehr.* **9**, 293–299.
840. Wackernagel, H. (1964). *Int. Z. Vitaminforsch.* **34**, 141–143.
841. Wagner, R.P., and Mitchell, H.K. (1964). "Genetics and Metabolism." Wiley, New York.
842. Wald, G., and Zussmann, H. (1938). *J. Biol. Chem.* **122,** 449–460.
843. Waldroup, P. (1972). *Feedstuffs* **44** (39), 38.
844. Waldroup, P.W., Douglas, C.R., McCall, J.T., and Harms, R.H. (1960). *Poult. Sci.* **39**, 1313–1317.
845. Ward, A.H. (1964). *Milling* **132** (12), 297–300.
846. Watson, S.A. (1962). *Proc. 17th Annu. Hybrid Corn. Ind. Res. Conf.* pp. 92–100.
847. Weedon, B.C.L. (1965). *In* "Chemistry and Biochemistry of Plant Pigments" (T.W. Goodwin, ed.), 1st ed., Part I, pp. 75–125. Academic Press, New York.
848. Weedon, B.C.L. (1971). *In* "Carotenoids" (O. Isler, ed.), pp. 29–59. Birkhaeuser, Basel.
848a. Weeks, O.B., and Garner, R.J. (1967). *Arch. Biochem. Biophys.* **121,** 35–49.
849. Weiser, H., and Manz, U. (1970). F. Hoffmann-La Roche & Co., Ltd., Basel (unpublished data).
850. Weiser, H., and Manz, U. (1972). F. Hoffmann-La Roche & Co., Ltd., Basel (unpublished data).
851. Weiser, H., and Manz, U. (1972). F. Hoffmann-La Roche & Co., Ltd., Basel (unpublished data).
852. Weiser, H., and Manz, U. (1974). F. Hoffmann-La Roche & Co., Ltd. Basel (unpublished data).
853. Weiser, H., and Manz, U. (1974). F. Hoffmann-La Roche & Co., Ltd., Basel (unpublished data).
854. Weiser, H., and Manz, U. (1974). F. Hoffmann-La Roche & Co., Ltd., Basel (unpublished data).
855. Weiser, H., and Manz, U. (1975). F. Hoffmann-La Roche & Co., Ltd., Basel (unpublished data).
856. Weiser, H., and Manz, U. (1975). F. Hoffmann-La Roche & Co., Ltd., Basel (unpublished data).
857. Wenland, G. (1944). *Vitam Horm.* **5**, 46–53.
858. West, J.W. (1956). *Poult. Sci.* **35**, 835–842.
859. Wharton, F.D, Fritz, J.C., Freed, M., and Classen, L.J. (1959). *Poult. Sci.* **38**, 1258.
860. Wheeler, H.O., and Turk, D.E. (1963). *S. C., Agric. Exp. Stn., Tech. Bull.* **1007**.
861. Wiemann, H., and Marquering, B. (1969). Arch. Gefluegelkd. **33**, 344–350.
862. Wilder, O.H.M., and Bethke, R.M. (1941). *Poult. Sci.* **20**, 304–312.
863. Wildfeuer, I. (1969). *Gordian* **69,** 469–499.
864. Wildfeuer, I., and Acker, L. (1967). Z. Lebensm-Unters.-Forsch. **133**, 341–352.
865. Wildfeuer, I., Acker, L., Mehner, A., and Rauch, W. (1968). *Z. Lebensm.-Unters.-Forsch.* **136.**, 129–136.
866. Wildfeuer, I., Acker, L., Mehner, A., and Rauch, W. (1968). *Z. Lebensm.-Unters.-Forsch.* **136,** 203–214.
867. Wilgus, H.S. (1954). "Exp. B-854 Xanthophylls." Peter Hand Foundation, Chicago, Illinois.
868. Wilgus, H.S. (1955). *Feedstuffs* **27** (5), 46.
869. Wilhelm, L.A. (1963). *Proc. Tech. Alfalfa Conf., 8th, 1963,* pp. 28–31.
870. Wilhelm, L.A., and Heiman, V. (1937). *Poult. Sci.* **16**, 416–418.
871. Wilhelm, L.A., and Payne, L.F. (1934). *U.S. Egg Poult. Mag.* **40** (3), 28–29, 42–43.
872. Wilkinson, W.S., and Barbee, C. (1968). *Poult. Sci.* **47**, 1579–1587.
873. Williams, N.S., and Overfield, N.D. (1970). *Proc. Symp. Yolk Color Egg Qual. Factor, 1970,* pp. 11–14, 23–27.
874. Williams, W.P. (1962). *Feedstuffs* **34** (4), 34–35, 52.-54.
875. Williams, W.P., Jr., Davies, R.E., and Couch, J.R. (1959). *Proc. Tex. A&M Nutr. Conf.* pp. 85–95.
876. Williams, W.P., Jr., Davies, R.E., and Couch, J.R. (1963). *Poult. Sci.* **42**, 691–699.
877. Williams, W.P., Jr., Davies, R.E., Ferguson, T.M., and Couch, J.R. (1960). *Poult. Sci.* **39**, 1307.
878. Willstätter, R., and Escher, H.H. (1911–1912). *Hoppe-Seyler's Z. Physiol. Chem.* **76,** 214–225.
879. Wilson, W.O. (1956). *Poult. Sci.* **35**, 226–227.
880. Winterstein, A. (1933). *In* "Handbuch des Pflanzenanalyse" (G. Klein, ed.), Vol. 4, pp. 1403–1437. Springer-Verlag, Berlin and New York.

881. Winterstein, A. (1960). *Angew. Chem.* **72**, 902–910.
882. Winterstein, A., and Hegedus, B. (1960). *Chimia* **14**, 18–19.
883. Winterstein, A., Rüegg, R., and Studer, A. (1960). *Angew. Chem.* **72**, 46.
884. Winterstein, A., and Stein, G. (1933). *Hoppe-Seyler's Z. Physiol. Chem.* **220**, 247–263.
885. Winterstein, A., Studer, A., and Ruegg, A. (1960). *Chem. Ber.* **93**, 2951–2965.
886. Wiseman, H.G., Kane, E.A., and Cary, C.A. (1936). *J. Dairy Sci.* **19**, 466.
887. Wisman, E.L. (1960). *Poult. Sci.* **39**, 1140–1148.
888. Wiss, O., Bunnell, R.H., and Gloor, U. (1962). *Vitam. Horm. (N.Y.)* **22**, 441–455.
889. Witt, S.C., Bickoff, E.M., and Kohler, G.O. (1972). *Feedstuffs* **44** (41), 26–27.
890. Witt, S.C., Spencer, R.R., Bickoff, E.M., and Kohler, G.O. (1971). *J. Agric. Food Chem.* **19**, 162–165.
891. Witt, S.C., Spencer, R.R., Bickoff, E.M., and Kohler, G.O. (1972). *Proc. Tech. Alfalfa Conf., 11th, 1972,* pp. 95–102.
892. Wright, W.D. (1940). *Nature (London)* **146**, 155–158.
893. Wright, W.D. (1941). *Proc. Phys. Soc., London* **53**, 93–99.
894. X-Pro (1970). Tech. Bull. No. 3. Batley-Janss Enterprises, Brawley, California.
894a. Yacowitz, H., Davies, R.E., and Jones, M.L. (1978). *Poult. Sci.* **57**, 443–448.
895. Yokoyama, H., and White, M.J. (1965). *J. Org. Chem.* **30**, 2481–2482.
896. Yvore, P. (1974). *Folia Vet. Lat.* **4**, 408–425.
897. Yvore, P., Lesur, J., Mainguy, P., Tan Hung, N., and Paquin, J. (1972). *Ann. Rech. Vet.* **3**, 389–398.
898. Yvore, P., and Mainquy, P. (1972). *Ann Rech. Vet.* **3**, 381–387.
899. Zechmeister, L., and Tuzson, P. (1934). *Hoppe-Seyler's Z. Physiol. Chem.* **225**, 189–195.
900. Zervas, N.P., Collins, W.M., and Skoglund, W.C. (1962). *Poult. Sci.* **41**, 1247–1254.
901. Zscheile, F.P., White, J.W., and Roach, J.R. (1942). *Plant Physiol.* **17**, 331–337.

4

CAROTENOIDS IN FISH FEEDS

K. L. Simpson, T. Katayama, and C. O. Chichester

I. INTRODUCTION

The beautiful yellow, orange, and red colors in the skin, shell, or exoskeleton of aquatic animals are due to a group of pigments called carotenoids. In common with other animals, these organisms must be provided with carotenoids in their diet. Although unable, apparently, to form the initial pigments, these aquatic animals nevertheless rival bacteria in the originality of the final products. Astaxanthin and a

CAROTENOIDS AS COLORANTS
AND VITAMIN A PRECURSORS

ISBN 0-12-082850-2

few other pigments were once thought to be the main pigments formed. However, with the use of more sophisticated purification and analytical techniques, new carotenoids have been found that are peculiar to aquatic animals.

Bicyclic carotenes, such as β-carotene (β,β-carotene) (**I**), or xanthophylls, such as lutein (β,ϵ-carotene-3,3′-diol) (**II**), are ingested from plants and converted to such compounds as astaxanthin (3,3′-dihydroxy-β,β-carotene-4,4′-dione) (**III**). Or-

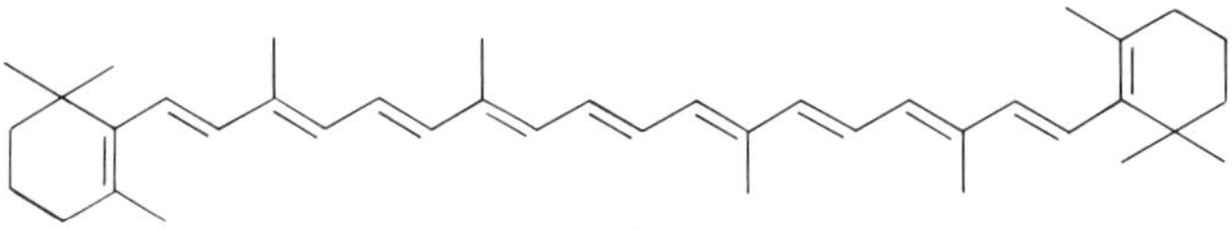

(I)

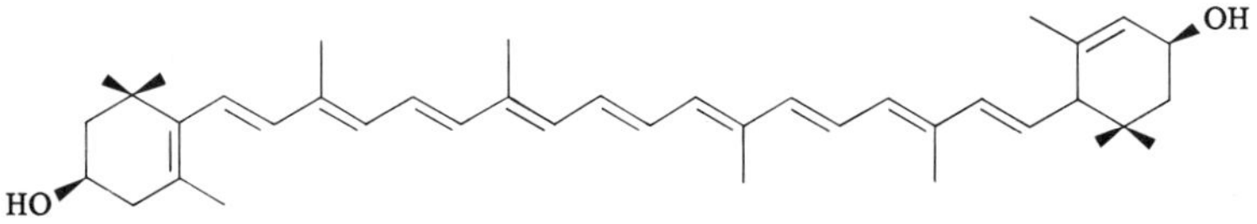

(II)

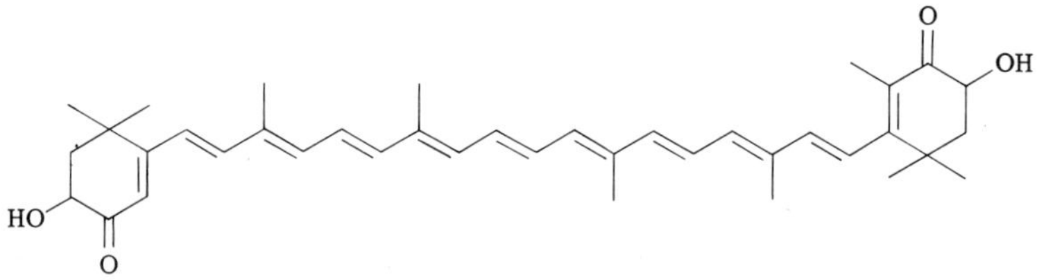

(III)

ganisms that feed on lower forms are able to absorb possible intermediate pigments, such as canthaxanthin (β,β-carotene-4,4′-dione) (**IV**) or astaxanthin, and make

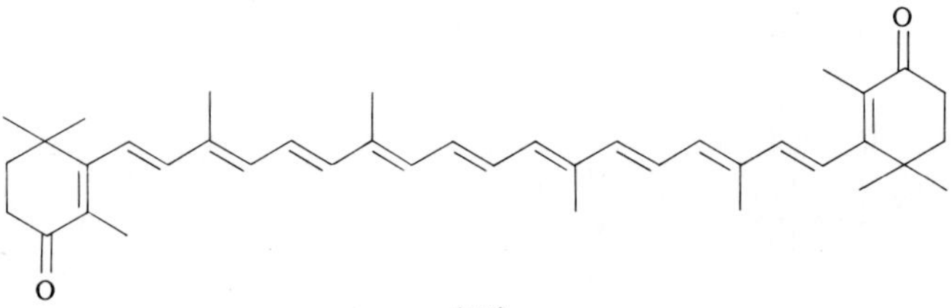

(IV)

further changes or deposit them unaltered. Carotenoids may be esterified with short- or long-chain fatty acids. Carotenoproteins occur widely among invertebrates, especially in Crustacea, and are responsible for the blue or purple color of some crustaceans. This complex (blue) can be broken on heating, liberating the red pigments involved (astaxanthin or canthaxanthin) and the protein.

Although the carotenoids are perhaps the most spectacular, they are certainly not the only pigments found in aquatic animals (*102,103,123,127*). The melanins are a diverse group of pigments which are the end products of phenolic oxidation and polymerization. These pigments are responsible for the brown to black pigmentation found in the eyes, peritoneal lining, and exposed skin (*103,123,311*). Other pigments, although less conspicuous, include the porphyrins (*103,311*), flavines (*103,123,311*), pterins (*103,123,215*), quinones (*30,311*), and ommochromes (*123,240*).

Review articles on the carotenoids of aquatic animals are generally dated prior to 1960 (*102,103,121–123,168*). More recent work (*124,161,228*) covers carotenoids in general.

The rapid increase in the world's population and the limitation of the world's land resources have caused people to look to the sea for increased harvest. Greater emphasis is being placed on culturing shrimp, lobster, trout, carp, catfish, oysters, etc., in impoundments rather than tracking and capturing them with boats. This aquaculture or mariculture has been generally limited to "luxury" products in the developed countries. The worldwide shortage of animal protein will undoubtedly justify aquaculture on the basis of the high quality of protein that can be produced. If the operation is to be efficient in terms of land use, feed must be supplied to the fish. The feed must provide all of the essential elements of the natural diet if the product is to compete with the wild harvest. Carotenoids must therefore be supplied to aquatic animals that deposit them. They must be supplied in a form that can be used by the particular organism, i.e., the right kind of carotenoid (such as astaxanthin) in a nondegraded form.

It is evident that commercial suppliers have not, in many cases, produced feeds adequate in carotenoid content. It is well known, for instance, that fishermen can readily tell the difference between hatchery trout and wild trout on the basis of color and possibly flavor.

II. GENERAL METHODS IN CAROTENOID FISH FOOD CHEMISTRY

General methods in carotenoid chemistry have recently been published by Liaaen-Jensen (*208*) (isolation and reactions), Vetter *et al.* (*310*) (spectroscopic methods), Mayer and Isler (*228*) (total syntheses), Britton and Goodwin (*18*), and Davies (*81*). This section covers generally some of the methods that are specific to the subject area of this chapter. A more rigorous treatment of the subject is to be found in Chapter 10 by DeRitter and Purcell.

A. General Precautions

Care must be exercised in the extraction, purification, and storage of carotenoids in any product. Chemical alteration may lower their value as vitamin A precursors or as absorbed pigmenters [see Chichester and Nakayama (*33*) and Simpson *et al.* (*277*)].

1. Protection from Light, Heat, and Oxygen

Light has two effects on carotenoids. The first is the formation of cis–trans double bonds, which cause a shift to the blue end of the electromagnetic spectrum of 2–10 nm depending on the number and position of the cis double bonds. The second is oxidation of the carotenoid chain. Interruption of the chromophore causes a large shift into the ultraviolet portion of the spectrum and generally a loss of color. These changes are much more pronounced in pure than in crude preparations of carotenoids.

The destructive effect of light on a solution of pure β-carotene can be seen in Fig. 1. There is a loss of color as well as qualitative changes in the spectrum. The addition of butylated hydroxyanisole (BHA) and butylated hydroxytoluene (BHT) can be seen to retard these changes.

Any pigment subjected to elevated temperatures during solvent evaporation, drying, or storage is subject to losses. Likewise, storage in an oxygen atmosphere in acetone or peroxide-containing diethyl ether causes destruction of pigments.

Knowles *et al.* (*189*) studied the stability of carotenoids and xanthophylls subjected to accelerated storage conditions. It was found that commercially dehydrated meals stored for 12 weeks had a total xanthophyll loss of 52–71%. Even freeze-

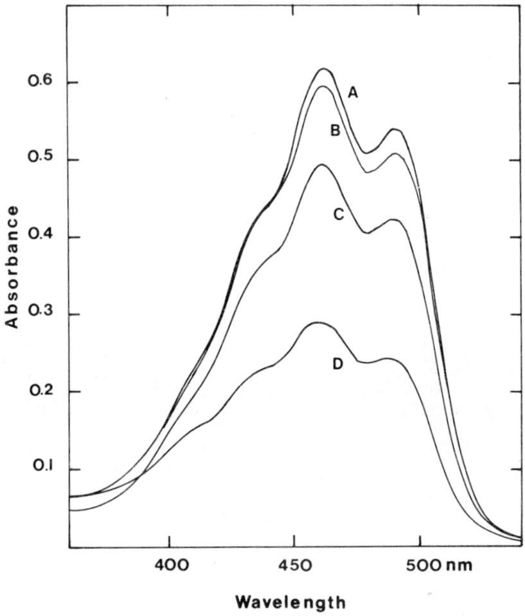

Fig. 1. Irradiation of β-carotene with and without antioxidants. (A) β-Carotene in toluene; (B) β-carotene with 0.1% BHA irradiated for 30 min with a 275-W sunlamp 8–12 inches away; (C) same with 0.1% BHT; (D) same without antioxidant.

TABLE 1

Effect of Storage on Stability of Various Xanthophyll Preparations[a]

Xanthophyll preparation	Original assay (mg)	Assay (mg)	Weeks of aging	Conditions
Crayfish extract	1186	960	4	Freezer, in CH_2Cl_2, in dark
Crayfish feed	22/lb	8/lb	5	Room temp, in dark
	22/lb	15/lb	5	Room temp, in dark, 30 mg Sustane per lb of feed
	22/lb	23.6/lb	11	Freezer, 30 mg Sustane per lb of feed
Paprika	107/lb	92/lb	5	Room temp, in dark
pepper feed	107/lb	66/lb	18	Room temp

[a] From Peterson *et al.* (*249*).

dried samples showed a 38% loss. This study showed that losses could be minimized by the addition of an antioxidant (ethoxyquin) and storage under an inert atmosphere.

Peterson *et al.* (*249*), in the course of their experiments on the use of various xanthophyll preparations for feeding trout, measured the loss of the pigments from their stored feed. Table 1 shows that a combination of darkness, lowered temperature, and an antioxidant (Sustane) provided the best storage conditions.

Pure canthaxanthin as the dry, stabilized beadlets showed a 96% recovery after 12 weeks at 45°C. When incorporated into a pelleted feed and stored at 37°C for 12 weeks, 81–88% of the canthaxanthin was retained (*12*).

2. *Protection from Acids and Bases*

As a general rule, acid reagents or acidic chromatographic adsorbents should be avoided if possible. Acidic conditions can cause isomerization of epoxycarotenoids, cis–trans isomerization, and oxidative or hydrolytic decomposition. These reactions usually lead to a loss of pigment quality and quantity. Hydrochloric acid in chloroform can cause dehydration of xanthophylls containing allylic alcohols:

$$-CH_2CHOHCH{=}CH- \rightarrow -CH{=}CHCH{=}CH-$$

Saponification treatment is a normal step in most carotenoid purifications. Alkali treatment of astaxanthin (**III**) causes the formation of astacene (3,3′-dihydroxy-2,3,2′,3′-tetradehydro-β,β-carotene-4,4′-dione) (**V**), as shown in Scheme 1. Actinioerythrin (3,3′-dihydroxy-2,2′-dinor-β,β-carotene-4,4′-dione 3,3′-diacylate) (**VI**), the main carotenoid of the red variant of the sea anemone *Actinia equina* (*150*), is converted to the blue pigment violerythrin (**VII**) on treatment with alkali (*150*) (Scheme 2). Fucoxanthin (5,6-epoxy-3,3′,5′-trihydroxy-6′,7′-didehydro-

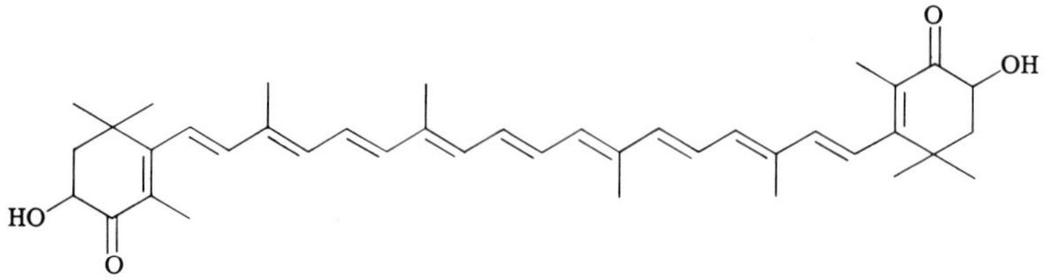

(III)

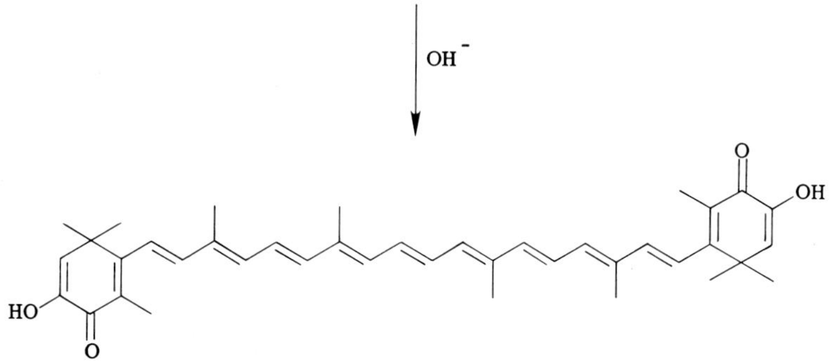

(V)

Scheme 1

5,6,7,8,5′,6,′-hexahydro-β,β-caroten-8-one 3′-acetate) **(VIII)**, a common pigment

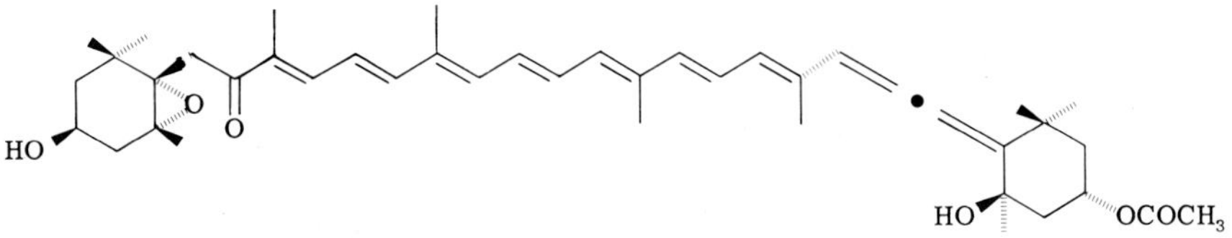

(VIII)

of algae and recently isolated from sea urchins (*115*) and Crustacea (*148*), has also been shown to be unstable to alkali. Dilute acid is used after saponification to extract some pigments (e.g., astacene) from the aqueous into the organic layer. Cold acetic acid has also been used as an eluting solvent for astaxanthin from chromatographic columns (*65,147*).

3. *Enzymatic Discoloration*

The carotenoids have long been known to be substrates for lipoxidase-type enzymes (*33*). It has also been suggested that the oxidation of fats in animal tissues is due to hemin (*297*).

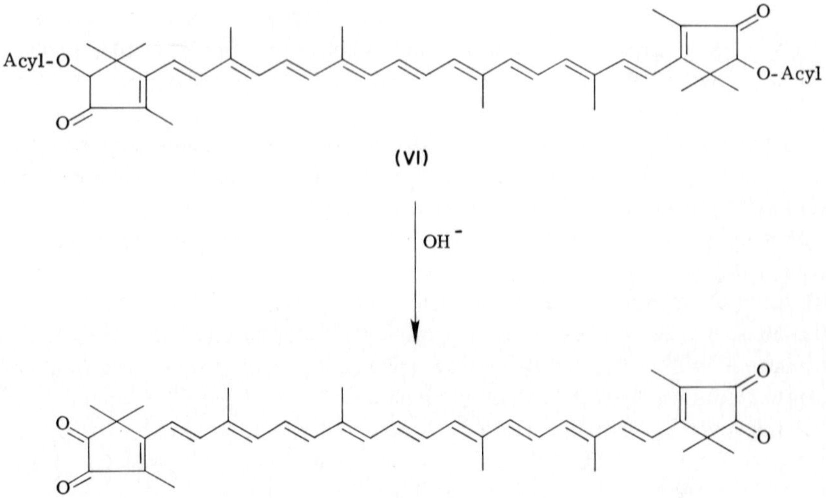

(VI)

(VII)

Scheme 2

It was observed (*304*) that discoloration of red fish occurred at refrigeration temperatures in the dark and that homogenates from the skin, muscle, and liver were able to degrade astaxanthin, tunaxanthin, and β-carotene to colorless compounds. Tsukuda (*302,305*) isolated and partially purified a heat-labile, lipoxidase-like enzyme from the skin of red fish (*Sebastes thompsoni* and *Chelidonichthys kumu*). The partially purified enzyme decolorized tunaxanthin (ϵ,ϵ-carotene-3,3′-diol) (**IX**) in the presence of linoleic and linolenic acids. As with other preparation and storage procedures, care should be exercised in protecting carotenoid-containing feed from oxidative enzymes.

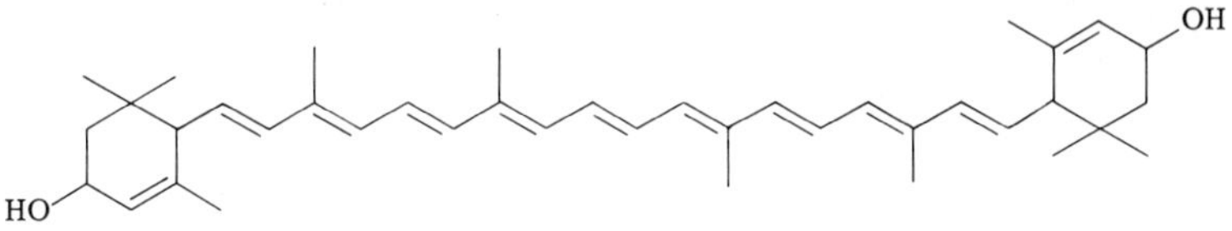

(IX)

B. Extraction

Acetone is the most commonly used solvent for the extraction of carotenoid-containing tissues. Saito and Regier (*261*) have given a detailed procedure for the extraction with acetone of a tissue homogenate from trout and salmon.

The pigments are washed free of acetone into petroleum or diethyl ether, and the latter solution is dried over sodium sulfate or *in vacuo*. When whole animals are used, it may be necessary to rid the animal of residual ingested pigments by feeding a carotenoid-free diet (*147*).

The pigments from goldfish can be extracted (*138*) by steeping the whole fish in acetone until the skin and fins are decolorized. This method was reported not to affect the pigments of other tissues, muscle, or viscera.

Carotenoids not extracted with ethyl alcohol (e.g., from cuticle) might be extracted with pyridine.

In order to liberate astaxanthin from hydrocoral (*112*), it was necessary to first either dissolve the complex in acid, treat with disodium ethylenedinitrilotetraacetate, or treat with nearly neutral oxalateoxalic acid. The pigments were then readily extracted from the residue with ethyl alcohol.

Dichloromethane can also be used for the extraction of dried tissue or pulverized crustacean shells. Lewis (*206*) obtained the mucus lipids of 10 species of marine fish by wiping the fish with paper tissue. The tissues were extracted with chloroform–methanol (2:1 v/v), and β-carotene was identified along with other lipids.

The carotenoproteins (*29,31,324*) from soft tissues can be extracted with water or dilute salt solutions. Solvents such as acetone usually break up the carotenoid–protein complex. The extraction of carapace carotenoproteins requires that the material be finely ground and decalcified by the use of pH 5.0 citrate buffer (*312*), pH 7.5 sodium ethylenediamine tetraacetate (EDTA) (*21,22,322*), ammonium sulfate (*28,322*), or hexametaphosphate at pH 5 (*192*). Traditional techniques have been utilized in the further purification of carotenoproteins (*22,322,325*). The carotenoids from fish feed should be extracted easily with solvents such as acetone, when added from such sources as soft tissue or purified preparations. The extraction of carotenoids from feeds containing ground crustacean parts may require more rigorous conditions (*195*).

C. Purification and Pigment Identification

Purification procedures may involve no more than washing an extract when an estimate of the amount of pigment present is required. For pigment identification some form of chromatographic separation is required. The purification techniques must, of necessity, be rigorous when they are followed by physical or radiochemical analysis.

It is possible to estimate the amount of pigment present by determining the light absorption of a known volume. If canthaxanthin were the major pigment present and the value of $E^{1\%}_{1\text{cm}} = 2200$ at 466 nm in hexane were used, the calculation could be made as follows:

$$\text{mg} = \frac{A \times \text{ml} \times \text{dilution}}{220}$$

Thus, if a 50-ml hexane extract had to be diluted 1 : 1 to give an absorbance (*A*) value of 0.66, the total number of milligrams would be as follows:

$$\frac{0.66 \times 50 \times 2}{220} = 0.3 \text{ mg}$$

When a spectrophotometer is not available, a visual estimation can be made between standard solutions of 1, 2, 3, 4, and 5 μg canthaxanthin per milliliter and the extract in a similar tube (*261*).

There is no universal column or thin-layer procedure for the separation of carotenoid pigments. A procedure for the pigments from one organism may fail to separate the pigments from another. Reference is made to original papers, and these should be consulted. Figures 2 and 3 show representative column separations to give some indication of the type of adsorbent and elution solvents used.

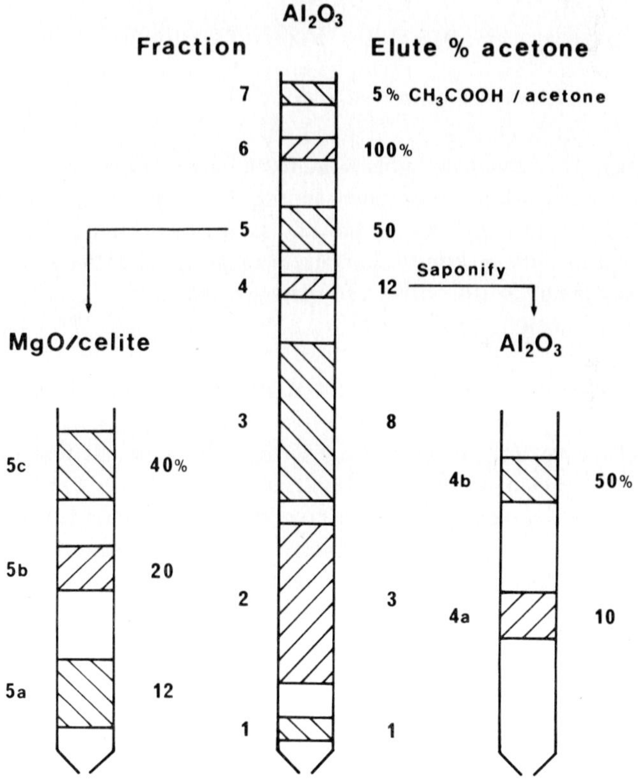

Fig. 2. Chromatographic separation of carotenoids from *Branchipus stagnalis* on columns of alumina and magnesium oxide/celite. Fractions eluted with acetone in light petroleum. Fractions were identified as follows: 1, β-carotene; 2, echinenone; 3, canthaxanthin; 4a, isocryptoxanthin; 4b, lutein; 5a, "monohydroxymonoketo-β-carotene"; 5b, lutein; 5c, "monohydroxydiketocarotenoid"; 6, xanthophyll epoxide; 7, astacene. From Gilchrist (*118*).

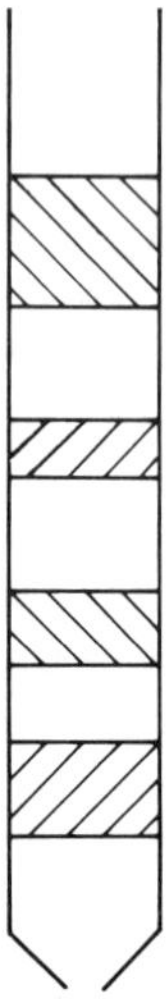

Fig. 3. Carotenoids from goldfish as they appear on MicroCel-C when developed with 2.2% acetone in petroleum ether. From Katayama *et al.* *(185)*.

Identifications of individual pigments are sometimes reported only on the basis of wavelength maxima of their absorption spectra. To properly identify a pigment, more rigorous methods must be employed, such as cochromatography with an authentic sample and the making of derivatives and physical analysis of their spectra with such instruments as IR, NMR, and mass spectrometers.

III. CAROTENOID DISTRIBUTION IN AQUATIC ANIMALS

Karrer and Jucker (*168*) in 1948 (English translation and additional material by Braude in 1950), Goodwin (*122*) in 1952, and Fox (*102*) in 1953 reviewed the literature on animal carotenoids up to the respective dates. Much recent information was given by Weedon in 1971 (*314*). This listing of the carotenoids is according to the chemical type of pigment and gives the principal organisms in which particularly the more recently identified carotenoids may be found.

More specialized reviews have also been published: Goodwin, carotenoids in fish, 1951 (*121*); Fox (*103*) pigments of fishes, 1957; Vevers (*311*) and Fox and Hopkins (*108*) pigments of echinoderms, 1966; Goodwin (*123,127*), Pigments of Crustacea, 1960, 1971; Goodwin (*128*), algal carotenoids, 1971; Goodwin (*125*), pigments of Porifera, 1968; Goodwin (*126*), pigments of Coelenterata, 1968; Fox (*104*), pigmentation of molluscs, 1966; Goodwin (*129*), pigments of Mollusca.

No attempt is made here to present a comprehensive review of material that has been covered elsewhere, nor is an attempt be made to cover any but the more recent papers. We have included animals that are now of little economic importance since they may be involved in a food chain or could be used as a source of carotenoids.

A. Protozoa

The protozoans, or "first animals," may be divided into two groups, namely, those that, like plants, make their food through the chlorophyll they possess and those that, like animals, capture and ingest other organisms. The algae may represent the bottom link in food chains. These organisms, which often appear as a bloom or colored tide, may be involved in a commensal relationship with higher animals, such as the Porifera, Coelenterata, Echinodermata, and Mollusca. Pigments isolated from higher forms may have been initially synthesized by protozoans.

The carotenoids have been shown to protect plants from photooxidation and thus are always associated with chlorophyll. When the protozoans contain chlorophyll, similar plant-type carotenoids are also found, with some exceptions.

The usual carotenoids of algae include β-carotene (**I**), lutein (**II**), neoxanthin (5′,6′-epoxy-6,7-didehydro-5,6,5′,6′-tetrahydro-β,β-carotene-3,5,3′-triol) (**X**), zeaxanthin (β,β-carotene-3,3′-diol) (**XI**), and violaxanthin (5,6,5′,6′-diepoxy-5,6,5′,6′-tetrahydro-β,β-carotene-3,3′-diol) (**XII**).

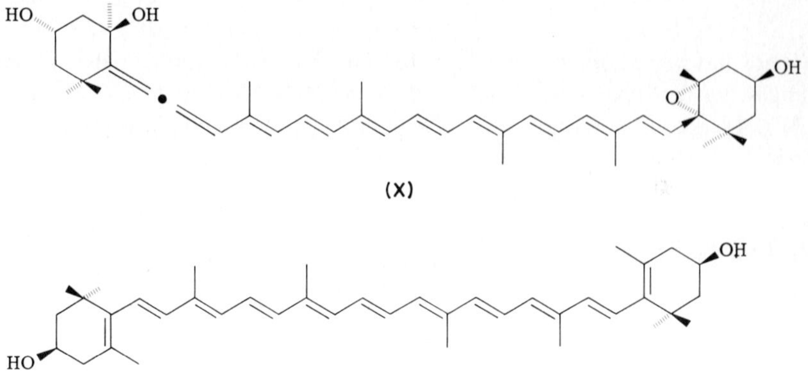

(X)

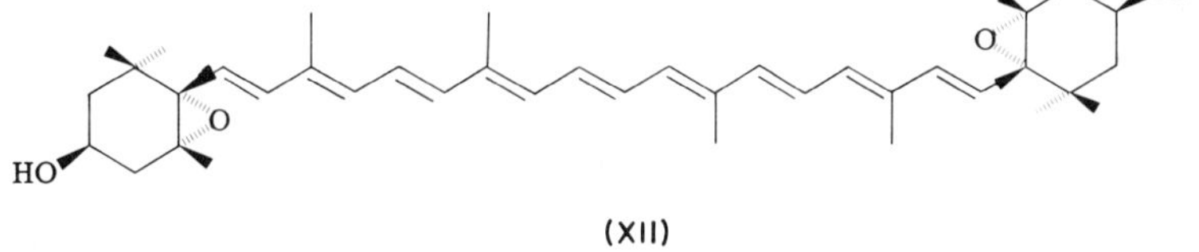

(XI)

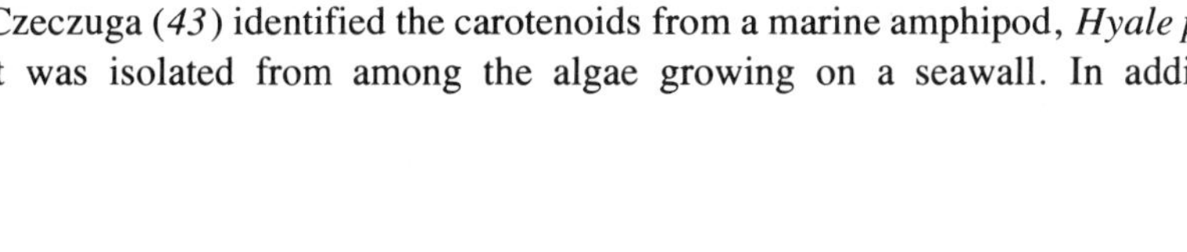

(XII)

Czeczuga (*43*) identified the carotenoids from a marine amphipod, *Hyale perieri,* that was isolated from among the algae growing on a seawall. In addition to astaxanthin, canthaxanthin, and β-carotene, the algal pigments lutein, probably violaxanthin, and the rather rare animal pigment, neoxanthin, were isolated. The freshwater filter feeding Diaptomidae, *Hemidiaptomus* and *Diaptomus*, were similarly found to contain violaxanthin in addition to typical crustacean ketocarotenoids (*63*). Ketocarotenoids are also found in bacteria (*263*), algae (*128*), and algae under nitrogen starvation (*76*).

Peridinin (5,6′-epoxy-3,5,3′-trihydroxy-6,7-didehydro-5,6,5′,6′-tetrahydro-10,11,20-trinor-β,β-caroten-19′,11′-olide) (**XIII**), a characteristic dinoflagellate carotenoid (*288*), was isolated from a natural bloom or red tide of *Conyaulax polyedra* and also from the zooxanthellae of the Pacific coast sea anemone *Anthopleura xanthogrammica* (*64*).

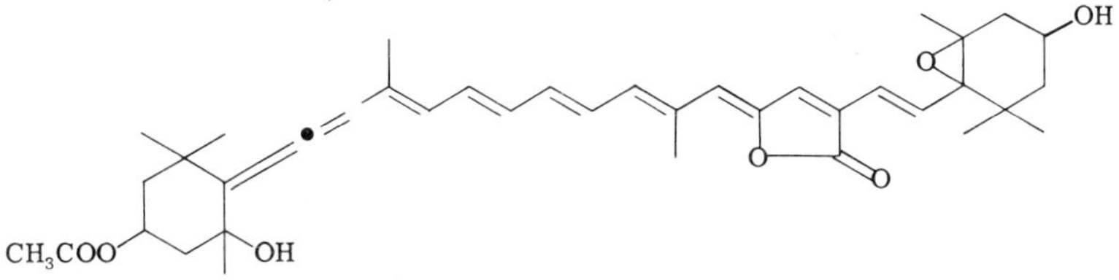

(XIII)

B. Porifera

There have been relatively few investigations of the carotenoids of the sea sponges. Earlier reports (*122*) indicated the presence of γ-carotene (β,ψ-carotene) (**XIV**), lycopene (ψ,ψ-carotene) (**XV**), torulene (3′,4′-didehydro-β,ψ-carotene) (**XVI**), and lutein.

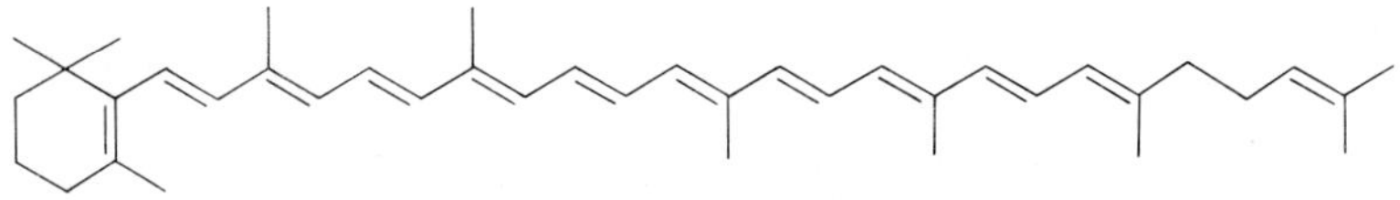

(XIV)

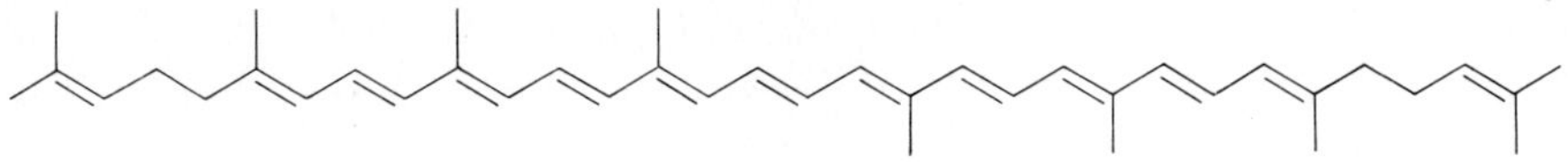

(XV)

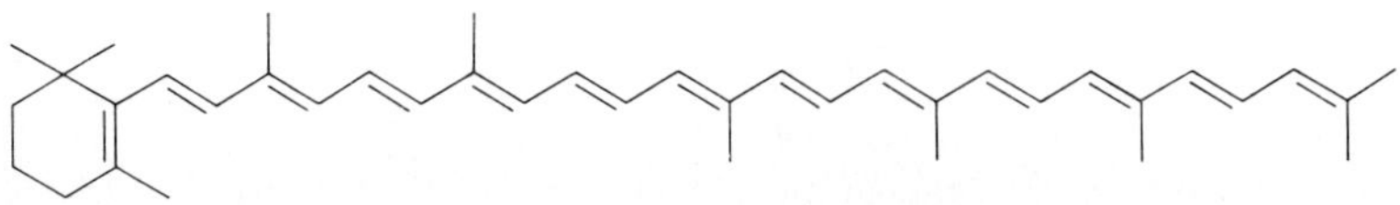

(XVI)

The finding of torulene (*197*) is of interest since this pigment is a characteristic pigment of yeasts of the genus *Rhodotorula*. It is also of interest that monocyclic and acyclic carotenoids were reported, whereas the usual animal pigments are bicyclic.

The arylcarotenoids renieratene (ϕ,χ-carotene) (**XVII**), isorenieratene (leprotene) (ϕ,ϕ-carotene) (**XVIII**), and renierapurpurin (χ,χ-carotene) (**XIX**) were

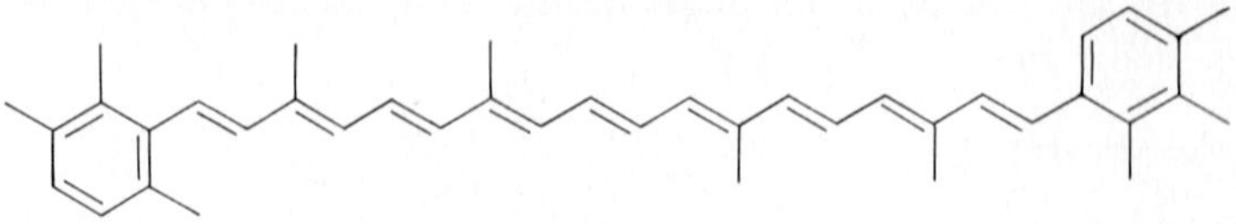

(XVII)

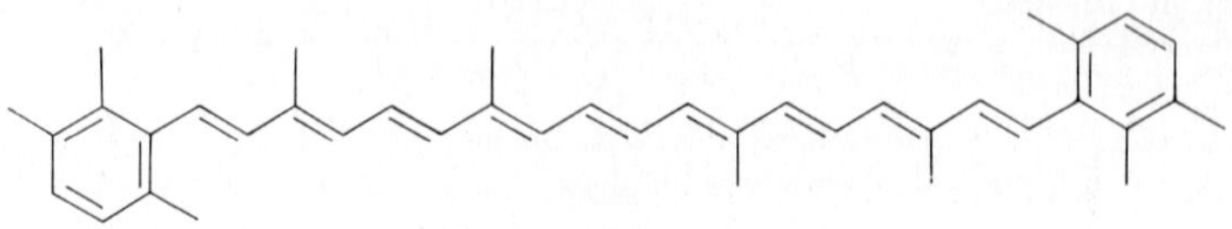

(XVIII)

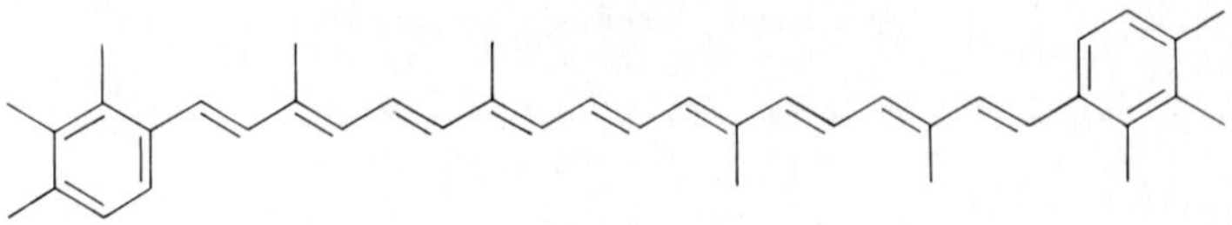

(XIX)

first isolated from the sea sponge *Reniera japonica* (*318–322*) and have now been isolated from various bacteria (*95*). Czeczuga (*45*) reported the isolation of β-carotene, γ-carotene, isozeaxanthin, lutein, zeaxanthin, neoxanthin, and astaxanthin from *Verongia aerophoba*.

Czeczuga and Ktyszejko (*72*) later reported that isocryptoxanthin was the major pigment in *V. aerophoba* caught off the west coast of Africa. This pigment was also the major carotenoid isolated from the freshwater sponge *Lubomirska baicalensis*. Aguilar-Martinez and Liaaen-Jensen (*1*) and Tanaka *et al.* (*293*) reported the isolation of the aromatic ketocarotenoid trikentriorhodin (3-hydroxy-7,8-dihyro-κ,χ-carotene-6,8-dione) from *Trikentrion helium*. The compound was also isolated from *Clathria frondifera* and *Tedania digitata* (*293*).

A new carotenoid, tedaniaxanthin (3-hydroxy-2,3-didehydro-β′,χ-carotene), was isolated from *T. digitata* by Tanaka *et al.* (*294*). A second new carotenoid, clathriacine (3-hydroxy-2,3-didehydro-β,χ-caroten-4-one), was isolated from *C. frondifera* (*289*).

It remains to be seen whether the finding of torulene and lycopene in the sponge will be verified since these hydrocarbons are rarely, if ever, encountered in marine animals (*125*).

C. Coelenterata

The name of the phylum Coelenterata is derived from *coel* (= hollow) and *entron* (= gut), which refers to the fact that the main cavity is the digestive cavity. "In this phylum we meet with the greatest variety and brightness of animal pigmentation" (*102*). The colors are often intense and may vary greatly within the same species living side by side. The phylum includes freshwater hydra and marine jellyfish, sea anemones, and corals.

TABLE 2

Distribution of Carotenoids in Coelenterates

Type, genus, and species	Carotenoid	Reference
Hydra		
Hydra littoralis	4-Hydroxy-4′-keto-β-carotene, echinenone, canthaxanthin	*(191)*
H. pirardi	Canthaxanthin, diketopirardixanthin, ketohydroxypirardixanthin, dihydroxypirardixanthin	*(191)*
H. vulgaris	Astaxanthin, "pirardixanthin", canthaxanthin	*(57)*
Jellyfish		
Aurelia aurita	β-Carotene, echinenone, canthaxanthin, astaxanthin, isocryptoxanthin, zeaxanthin	*(42)*
Corneous coral		
Paragorgia aborea	Astaxanthin	*(307)*
Primnoa resedae formis	Astaxanthin	*(307)*
Stony coral		
Lophelia petusa	Astaxanthin	*(307)*
Paramuricea sp.	Canthaxanthin, astaxanthin (?), zeaxanthin	*(307)*
Soft coral		
Alcyonium digitatum	Astaxanthin, 7,8-didehydroastaxanthin, 7,8,7′,8′-tetradehydroastaxanthin	*(307)*
	Astaxanthin, isozeaxanthin phoeniconone, zeaxanthin, canthaxanthin, phoenicoxanthin	*(50)*
Pennatula phosphorea	Canthaxanthin, astaxanthin, isozeaxanthin, zeaxanthin	*(50)*
Hydrocoral		
Allopora californica	Astaxanthin, phoenicoxanthin	*(112)*
Distichopora coccinea	Astaxanthin	*(105)*
Stylaster elegans	Astaxanthin, zeaxanthin	*(105)*
S. roseus	Astaxanthin	*(105)*
S. sanguineus	Astaxanthin	*(105)*
Sea anemone		
Actina equina	Astaxanthin, actinioerythrin	*(79, 90, 150)*
Gyrostoma sp.	Actinioerythrin	*(253)*
Metridium senile	Astaxanthin, diketopirardixanthin	*(107, 110, 307)*

Sea anemones, in particular, feed on and are eaten by various crustaceans. Table 2 shows the pigmentation pattern of some representative genera.

In studies concerned with the metabolism of dietary pigments of coelenterates, both *Hydra littoralis* and *H. pirardi* (*191*) were grown on brine shrimp, *Artemia salina*. *Hydra littoralis* apparently laid down its dietary pigments without change since echinenone (β,β-caroten-4-one) (**XX**) and canthaxanthin are also found in brine shrimp. *Hydra pirardi* contained canthaxanthin and carotenoids of unproved structure, dihydroxypirardixanthin, ketohydroxypirardixanthin, and hydroxypirardixanthin.

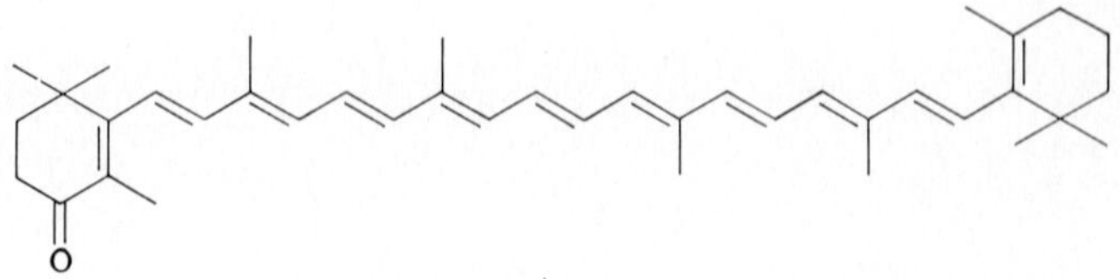

(XX)

Hydra vulgaris specimens were cultured on *Ceriodaphnia reitculata* for a 45-day period. Not all of the carotenoids found in the live food were found in the hydra, and "pirardixanthin," not found in *C. reticulata,* was found in the hydra (*56*).

The jellyfish *Aurelia aurita* contains, in addition to the usual pigments (β-carotene, echinenone, canthaxanthin, and astaxanthin), isocryptoxanthin (β,β-caroten-4-ol) (**XXI**) and zeaxanthin, which are possible intermediates from β-carotene.

Only a few representative species of the sea anemones and corals have been studied with respect to their carotenoids. Those that have been studied have yielded a series of unusual carotenoid pigments. The acetylenic pigments 7,8-didehydroastaxanthin (3,3′- dihydroxy- 7,8- didehydro- β,β- carotene- 4,4′- dione) (**XXII**) and 7,8,7′8′- tetradehydroastaxanthin (3,3′- dihydroxy- 7,8,7′,8′- tetradehydro- β,β- carotene- 4,4′- dione) (**XXIII**) (*150,307*) and the noncarotenoid actinioerythrin (**VI**) were isolated from soft coral. The same soft coral (*Alcyonium digitatum*) isolated from the Gullmar fjord, Sweden, contained "common" carotenoids (*50*). In addition, the brown variety yielded phoenicoxanthin (3-hydroxy- β,β- carotene- 4,4′- dione) (**XXIV**), and the yellow variety yielded phoeniconone (3- hydroxy- 2,3- didehydroβ,β- carotene- 4,4′- dione) as a possible artifact. Phoenicoxanthin has been reported in the hydrocoral *Allopora californica* (*112*).

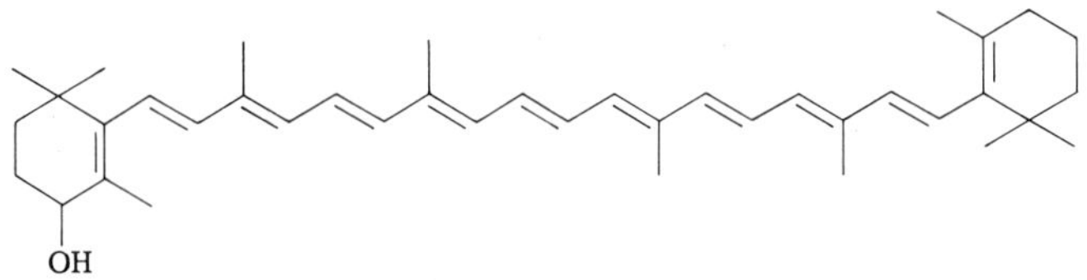

(XXI)

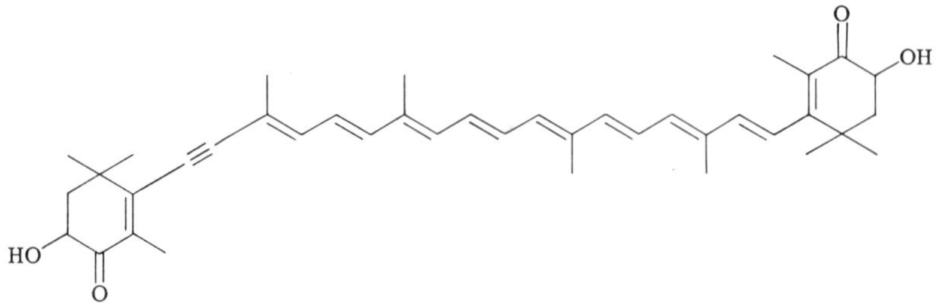

(XXII)

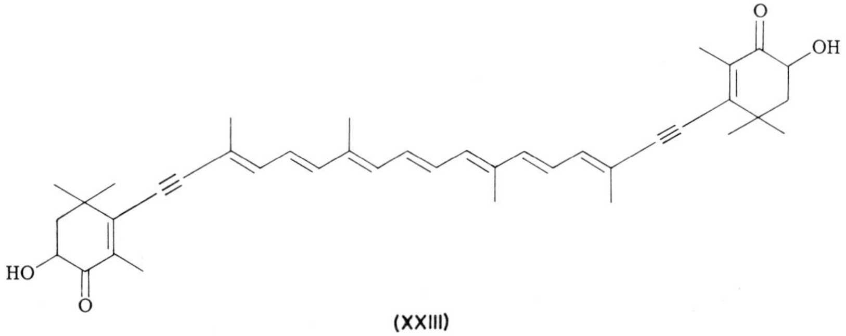

(XXIII)

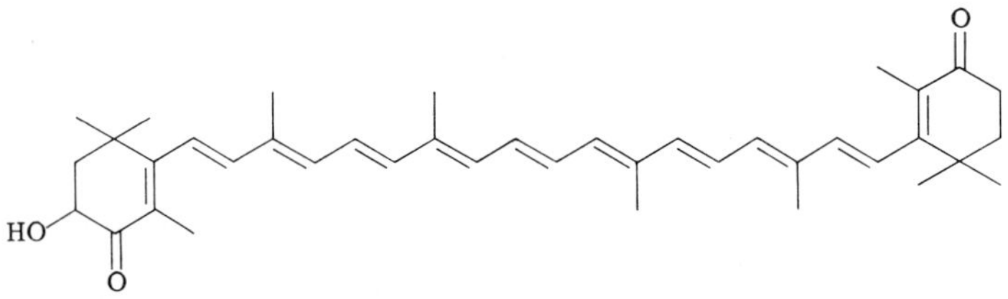

(XXIV)

Fox (*105*) studied the calcareous skeletons of several hydrocorals and found, in addition to astaxanthin and zeaxanthin, several fatty acids and fluorescent compounds. He discussed possible methods of carotenoid ester binding in the hydrocorals.

D. Annelida

The free-swimming marine worms, the sabellied and serpulid polychaetes, have crowns pigmented with carotenoids. The colors can include various shades of purple, red, orange, yellow, blue, brown, and green. The actual pigments are restricted to a few types, although only a few reports on these worms have appeared. The pigments reported to be present in the polychaetes are similar to those isolated from other invertebrates. The findings of three reports are presented in Table 3.

TABLE 3

Carotenoid Distribution in Polychaetes

Genus and species	Carotenoid	Reference
Branchiomna vesiculosum	β-Carotene, astaxanthin	*(80)*
Chone infundibuliformis	β-Carotene, astaxanthin	*(80)*
Eudistylia polymorpha	β-Carotene, astaxanthin, zeaxanthin (?)	*(80)*
E. vancoureri	β-Carotene, astaxanthin, zeaxanthin (?), lutein (?)	*(80)*
Megalomma vesiculosum	β-Carotene, echinenone, astaxanthin, isocryptoxanthin, 4-hydroxy-4′-keto-β-carotene	*(204)*
Myxicola infundibulum	β-Carotene, echinenone, canthaxanthin, astaxanthin, isocryptoxanthin, 4-hydroxy-4′-keto-β-carotene	*(80, 204)*
Nereis zonata	α-Carotene, β-carotene, cryptoxanthin, lutein, zeaxanthin, neoxanthin, astacene	*(48)*
Pomatoceros triqueter	β-Carotene, astaxanthin	*(80)*
Protula intestinum	β-Carotene, astaxanthin	*(80)*
Pseudopotamilla occelata	Astaxanthin	*(80)*
Sabella pavonina	β-Carotene, astaxanthin, lutein	*(80)*
S. penicillus	β-Carotene, isocryptoxanthin, echinenone, 4-hydroxy-4′-keto-β- carotene, canthaxanthin	*(204)*
S. spallanzanii	β-Carotene	*(80)*
Schizobranchia insignis	β-Carotene, astaxanthin, lutein	*(80)*

E. Echinodermata

The echinoderm (''spiny-skinned'') phylum is exclusively marine and is divided into five classes: starfish (Asteroidea), serpent stars (Ophiuroidea), sea urchins (Echinoidea), sea cucumbers (Holothuroidea), and sea lilies (Crinoidea). The carotenoids occur mainly in the integument, ovaries, eggs, or liver of starfish; in the sex glands of sea cucumbers and sea urchins; and in the skin and gonads of serpent stars and sea lilies. The recently reported distribution of carotenoids in echinoderms is given in Table 4. The mono- and diacetylenic derivatives of astaxanthin (*70,302*) isolated in the soft coral *Alcyonium digitatom* (*307*) have also been reported to occur in the starfish *Asterias rubens* (*113*).

Tanaka and Katayama (*290*) investigated four starfish genera (Table 4) and found 7,8-didehydroastaxanthin (**XXII**) to be present in concentrations of greater than 50% in each. The diacetylenic pigment 7,8,7′,8′-tetradehydroastaxanthin (**XXIII**) was isolated from *Acanthaster planci*. Both of these pigments had previously been found in soft coral (*150,307*). The authors proposed possible metabolic pathways from these starfish.

Three possible intermediates in the synthesis of astaxanthin have been isolated from sea urchins: β-cryptoxanthin (β,β-caroten-3-ol) (**XXV**), 3′-hydroxyechienone

TABLE 4

Distribution of Carotenoids in Echinoderms

Type, genus, and species	Carotenoid	Reference
Starfish		
Acanthaster planci	7,8-Didehydroastaxanthin, β-carotene, echinenone, α-cryptoxanthin, canthaxanthin, lutein, zeaxanthin, diatoxanthin, astaxanthin, 7,8,7′,8′-tetradehydroastaxanthin	*(290)*
Asterias amurensis	7,8-Didehydroastaxanthin, zeaxanthin, lutein, β-carotene, echinenone	*(290)*
A. rubens	β-Carotene, 7,8-didehydroastaxanthin, 7,8,7′8′-tetradehydroastaxanthin	*(85, 113)*
Asterina panceri	β-Carotene, echinenone, astaxanthin, 3′-hydroxyechinenone, β-doradexanthin	*(86, 89)*
Asteropectin auranticus	β-Carotene, echinenone, astaxanthin, β-cryptoxanthin, 3′-hydroxyechinenone, β-doradexanthin	*(87, 89)*
Coscinastrias acutispina	7,8-Didehydroastaxanthin, 7,8,7′,8′-tetradehydroastaxanthin, zeaxanthin, lutein, β-carotene	*(290)*
Echinaster sepositus	β-Carotene, astaxanthin, echinenone (?)	*(290)*
Leiaster leachi	7,8-Didehydroastaxanthin, zeaxanthin, lutein, echinenone, β-carotene	*(290)*

(*Continued*)

(3′-hydroxy-ββ-caroten-4-one) (**XXVI**), and β-doradexanthin, or adonixanthin (3,3′-dihydroxy-β,β-caroten-4-one) (**XXVII**).

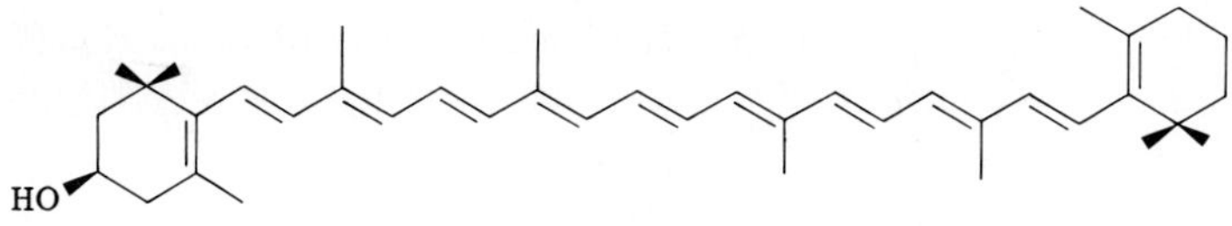

(XXV)

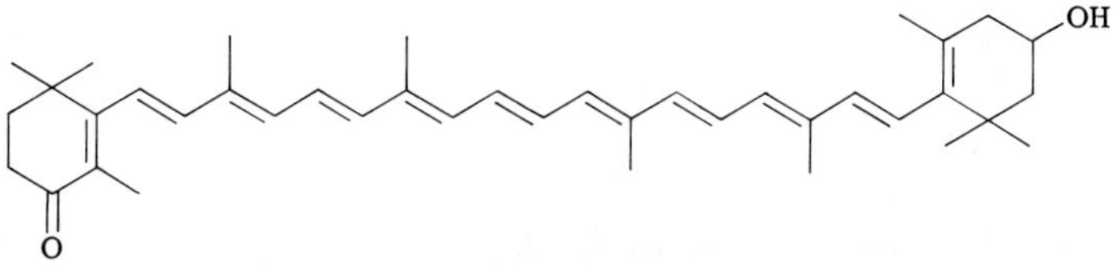

(XXVI)

TABLE 4—*Continued*

Type, genus, and species	Carotenoid	Reference
Marthasterias glacialis	β-Carotene, echinenone, astaxanthin, β-cryptoxanthin, 3′-hydroxyechinenone, β-doradexanthin	*(88, 89)*
Ophidiaster ophidianus	β-Carotene, astaxanthin, β-cryptoxanthin, 3′-hydroxyechinenone, β-doradexanthin	*(88, 89)*
Sea urchins		
Paracentrotus lividus	β-Carotene, echinenone, fucoxanthin, fucoxanthinol, paracentrone	*(91, 115, 157)*
Sphaerechinus granularis	β-Carotene, echinenone	*(90)*
Strongylocentrotus purpuratus	β-Carotene, echinenone, zeaxanthin	*(131)*
S. droebachiensis	β-Carotene, echinenone, zeaxanthin, isocryptoxanthin	*(132)*
Sea cucumber		
Psolus fabrichii	β-Carotene, echinenone, canthaxanthin, astaxanthin, phoenicoxanthin	*(23)*
Sea lily		
Florometra serretissima	Echinenone, canthaxanthin astaxanthin, β-carotene, zeaxanthin, lutein	*(243)*
Lamprometra klunzingeri	β-Carotene, ketocarotenoids, acetylenic-like pigments	*(133)*

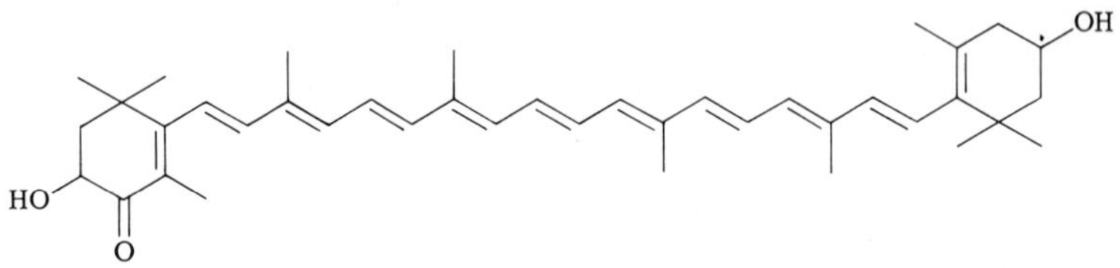

(XXVII)

The main carotenoids in the ovary and eggs of the sea urchin *Strongylocentrotus droebachiensis* were found by Griffiths and Perrott (*132*) to be echinenone, β-carotene, and a significant amount of isocryptoxanthin. The gut wall and gut contents contained about 80% fucoxanthin and isofucoxanthin. These authors suggested that the source of the echinenone in this animal is ingested β-carotene through isocryptoxanthin, with the site of conversion being the ovary. The content of echinenone in the ovary showed seasonal variations since this pigment was transported to the eggs.

Fucoxanthin (**VIII**), a common pigment of brown algae, has been isolated from the sea urchin *Paracentrotus lividus* (*115*) *Fucoxanthinol* (5,6-epoxy-3,3′,5′-trihydroxy-6′,7′-didehydro-5,6,7,8,5′,6′-hexahydro-β,β-caroten-8-one) (**XXVIII**), previously prepared from fucoxanthin, has now been shown to occur in *P. lividus* (*115*). Paracentrone (3,5-dihydroxy-6,7-didehydro-5,6,7′,8′-tetrahydro-7′-apo-β-caroten-8′-one) (**XXIX**) a C_{31} carotenoid was also isolated from *P. lividus* (*115*). Both paracentrone and fucoxanthinol are probably products of ingested fucoxanthin (*115*).

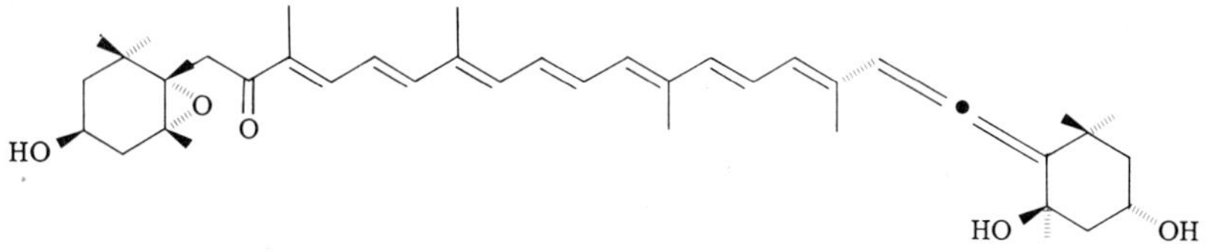

(XXVIII)

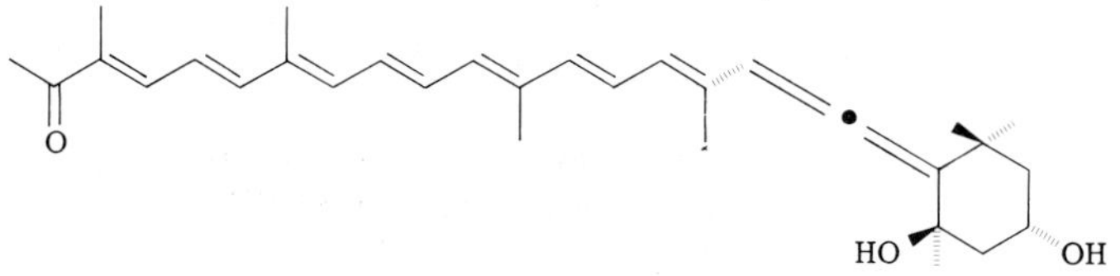

(XXIX)

Capsaxanthin (capsanthin?) was reported (*72*) to have been isolated from *Stichopus regalis* (44.8%).

There are few reports on the sea lilies. One report (*133*) states that the pigmentation in *Lamprometra klunzingeri* is due mainly to anthraquinones.

F. Mollusca

The phylum Mollusca includes such important classes as the gastropods [snails, slugs, limpets and whelks, pelecypods or bivalves (clams, oysters, scallops), etc.] and cephalopods (nautiluses, squid, octopuses, etc.). The carotenoids are usually encountered in the reproductive parts of these animals and less frequently in the body and other organs. The ink from the octopus contains xanthophylls.

Shimizu and Uchida (*273*) investigated the muscle and viscera of the short-necked clam *Veneus japonica* and identified β-carotene from the 12 isolated fractions. Shimizu and Monma (*270*) identified β-carotene, lutein, and zeaxanthin-like pigment from the hard clam *Meretrix lusoria*. Shimizu and Ohta (*272*) studied the pigments of the corb shell *Corbicula japonica* and found, among their six fractions, lutein, β-carotene, and a taraxanthin-like pigment. Shimizu and Oda (*271*) identified lutein and β-carotene from *Laqueatus sowerby*.

Alloxanthin (7,8,7′,8′-tetradehydro-β,β-carotene-3,3′-diol) (**XXX**) (*25*) isolated from various *Cryptomonas* algae has been shown to be identical to pectenoxanthin (from the giant scallop *Pecten maximus*) and cynthiaxanthin (from the tunicate *Halocynthia papillosa*). Alloxanthin was also isolated from the mussel *Mytilus edulis* (*25*), from the red-crange foot of the shellfish *Anadara broughtoni* and *A. suberenata*, and from the red-orange ovaries of *Pecten yessoensis* (*246*). The acetylenic carotenoid pectenolone (3,3′-dihydroxy-7′,8′-didehydro-β,β-caroten-4-one) (**XXXI**) was also isolated from *Pecten maximus*.

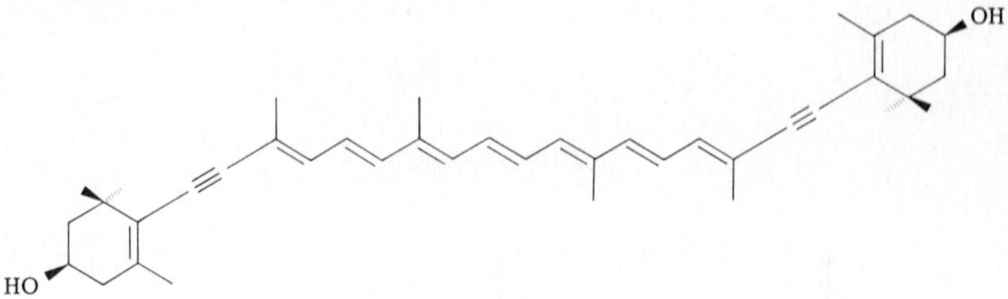

(XXX)

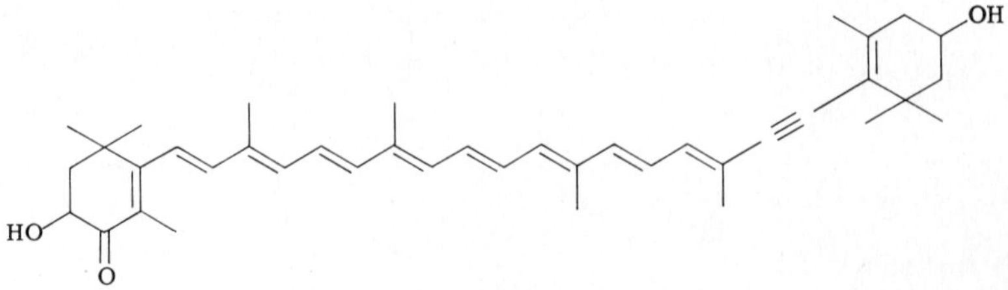

(XXXI)

McBeth (*212*) isolated α-carotene, β-carotene, and isorenieratene from the nudibranchs *Anisodaris nobilis, Dendrodoris fulva,* and *Doriopsilla albopunctata*. The arylcarotenoid isorenieratene (**XVIII**) was also isolated from sea sponges (*320*), which are the probable source of arylcarotenoids in nudibranchs. Only astaxanthin was isolated from *Flabellinopsis iodinea* (*212*). McBeth (*211*) isolated and partially characterized an acetylenic apocarotenoid ($C_{31}H_{38}O_3$), ''Hopkinsiaxanthin,'' from *Hopkinsia rosacea*. Goodwin and Fox (*130*) isolated, but did not identify, the carotenoids from *Hopkinsia rosacea, Archidoris monterevensis,* and *Flabellina iodinea*.

Czeczuga and Ktyszejko (*72*) investigated the carotenoid content of several invertebrates from the west coast of Africa. *Archidoris tuberculata* was reported to contain 20% α-carotene and *Octopus vulgaris,* 58% tunaxanthin (**IX**).

Lee and Kim (*199*) isolated β-carotene, lutein, zeaxanthin, and astaxanthin (astacene) from the sea mussel *Mytilus edulis*. The presence of mytiloxanthin, an unidentified carotenoid of possible acetylenic structure, was suggested. These authors concluded that xanthophylls and their esters, rather than carotenes, are the major pigments in the sea mussel.

More recently, mytiloxanthin (**XXXII**) was shown to be a rather novel enolized β-diketone with a cyclopentane end group (*187*). A second compound, isomytiloxanthin (**XXXIII**), was also isolated from this sea mussel (*187*).

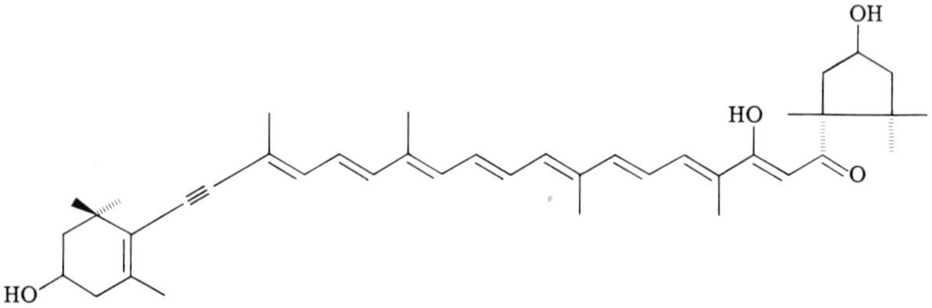

(XXXII) Mytiloxanthin

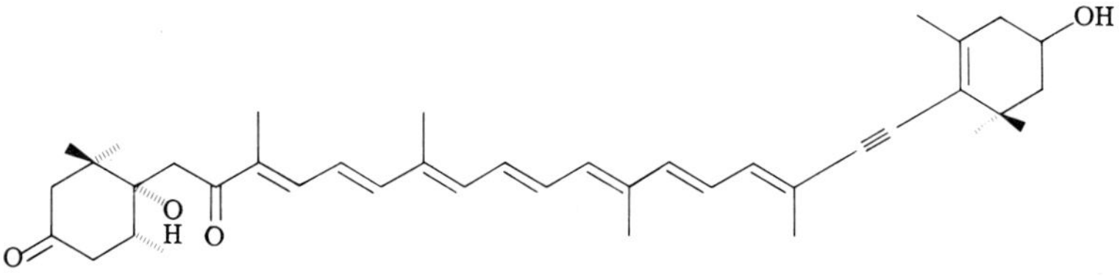

(XXXIII) Isomytiloxanthin

G. Crustacea

Three-quarters of the million or so known species of animals belong to the phylum Arthropoda, which includes the insects and crustaceans. The color of the crustaceans (crayfish, lobsters, shrimps, crabs, etc.) and the fin fish discussed in the next section is of economic importance. This fact is borne out by the relatively greater number of papers on the occurence of pigments in various species and the metabolism of these pigments (*123,127*).

Most crustaceans contain carotenoids in some combination of the eyes, blood, eggs, carapace, and various organs such as the hepatopancreas and ovary. An example of the quantitative distribution of carotenoids in the various tissues of the crayfish is given in Table 5 (*58*). Table 6 gives a listing of the distribution of carotenoids in Crustacea [see Goodwin (*127*)].

Astaxanthin (**III**) is found in most crustaceans either free, esterified, or as carotenoproteins and usually represents the major carotenoid from these sources. β-Carotene, lutein, echinenone, and canthaxanthin are the other pigments usually encountered.

Pigments, other than possible intermediates between β-carotene and astaxanthin, rarely occur in crustaceans, and when they do their origin is probably diet-related. Neoxanthin (**X**) was isolated from the marine amphipod *Hyale perieri* (*43*). This

TABLE 5

Tissue Distribution of Carotenoids in *Oreonectes rusticus*[a]

Carotenoid	Exoskeleton	Carcass	Eggs	Eye stalks
β-Carotene	1.0	25.2	32.5	2.7
Cryptoxanthin ester	5.0[b]	0.4	3.1	—
Lutein ester	5.0[b]	3.0	—	—
Astaxanthin ester	47.7	14.8	—	43.6
Ketocarotenoid	1.6	—	—	—
Lutein	6.0	45.7	27.9	0.4
Astaxanthin			24.4	
	38.7[c]	11.0[c]		53.2[c]
Astacene			12.0	

[a] From Wolfe and Cornwell (*317*). Data given as percentage of total carotenoids.
[b] Two bands eluted together.
[c] The total remaining pigment on the column was eluted as astacene with 15% KOH in 90% methanol.

TABLE 6

Carotenoid Distribution in Crustacea[a]

Group, genus, and species	Carotenoids	Reference
Branchiopoda		
Apus cancriformus	β-Carotene, γ-carotene, echinenone, canthaxanthin, lutein	(*69*)
Artemia salina	β-Carotene, isozeaxanthin, echinenone, canthaxanthin, isocryptoxanthin, astaxanthin (?)	(*75,83,137, 190,191*)
Brachinecta packardi	β-Carotene, lutein, echinenone, canthaxanthin, astaxanthin	(*118*)
B. paludosa	β-Carotene, echinenone, isocryptoxanthin, canthaxanthin, astaxanthin, lutein	(*51*)
Branchipus stagnalis	β-Carotene, lutein, echinenone, canthaxanthin, astaxanthin	(*118*)
Chirocephalus diaphanus	β-Carotene, lutein, isozeaxanthin, echinenone, canthaxanthin, astaxanthin	(*118*)
Daphnia magna	β-Carotene, echinenone, canthaxanthin, astaxanthin, crustaxanthin	(*146,147*)
D. pulex	β-Carotene, echinenone, canthaxanthin	(*75,300*)
Tanymastix lucunae	β-Carotene, lutein, echinenone, canthaxanthin, astaxanthin	(*118*)
Copepoda		
Artodaptomus salinus	β-Carotene, astaxanthin, crustaxanthin	(*15,16*)
Argulus foliaccus (L.)	β-Carotene, isocryptoxanthin, lutein, neoxanthin, isozeaxanthin, astacene	(*44*)

(*Continued*)

TABLE 6—*Continued*

Group, genus, and species	Carotenoids	Reference
Cyclops stenuus stenuus	β-Carotene, lutein, echinenone, canthaxanthin, astaxanthin	(*65*)
Diaptomus castor	β-Carotene, γ-carotene, violaxanthin, astaxanthin	(*63*)
Hemidiaptomus amblyodon	β-Carotene, γ-carotene, violaxanthin, astaxanthin	(*63*)
Labidocera acutifrons	Astaxanthin	(*324*)
Cirripeda (barnacles)		
Chthamalus fragilis	β-Carotene, zeaxanthin, astaxanthin	(*148*)
Conchoderma virgatum	Astaxanthin, phoenicoxanthin	(*148*)
Elminius modestus	β-Carotene, zeaxanthin, astaxanthin	(*148*)
Lepas anatifera	β-Carotene, zeaxanthin, astaxanthin, fucoxanthin, phoenicoxanthin	(*148*)
L. fascicularis	β-Carotene, zeaxanthin, astaxanthin, peridinin fucoxanthin	(*111,148*)
L. pectinata	Zeaxanthin, astaxanthin, phoenicoxanthin	(*148*)
Pollicipes polymerus	β-Carotene, lutein, zeaxanthin, isozeaxanthin (?), astaxanthin	(*156*)
Malacostraca		
Aristeus antennatus	Astaxanthin	(*29*)
Astacus astacus	β-Carotene, lutein, canthaxanthin, astaxanthin	(*67*)
A. leptodactylus	β-Carotene, echinenone, canthaxanthin, cryptoxanthin, lutein, zeaxanthin, neoxanthin, violaxanthin, astaxanthin	(*46*)
Betaeus harfordi	β-Carotene, echinenone, astaxanthin	(*210*)
Cambarus bartonii tenebrosus	β-Carotene, lutein, astaxanthin	(*314*)
Carcinus manenas	β-Carotene, γ-carotene, lutein, zeaxanthin, echinenone, canthaxanthin, astaxanthin	(*119*)
C. mediterraneus	β-Carotene, lutein, canthaxanthin, astaxanthin	(*244*)
Clibanarius erythropus	β-Carotene, echinenone, canthaxanthin, phoenicoxanthin, astaxanthin, lutein, α-doradexanthin	(*27*)
C. misanthropus	β-Carotene, lutein, canthaxanthin, astaxanthin	(*68*)
Crangon crangon	β-Carotene, lutein, zeaxanthin, canthaxanthin, astaxanthin	(*74*)
C. vulgaris Fabr.	β-Carotene, astaxanthin	(*278*)
Emerita analoga	α-Carotene, β-carotene, echinenone, canthaxanthin, zeaxanthin, diatoxanthin, alloxanthin, astaxanthin	(*120*)
Eupagurus barnhardus	Astaxanthin	(*32*)
Homarus gramarus	Astaxanthin	(*116*)
Hyale perieri	β-Carotene, lutein, canthaxanthin, astaxanthin, neoxanthin	(*43*)
Idotea granulosa	β-Carotene, lutein, isozeaxanthin, echinenone, canthaxanthin, isocryptoxanthin, 4-hydroxy-4′-keto-β-carotene, 2-hydroxy-β-carotene	(*201,203*)
I. metallica	β-Carotene, zeaxanthin, astaxanthin, crustaxanthin, isocryptoxanthin, idoxanthin	(*145*)

(*Continued*)

TABLE 6—*Continued*

Group, genus, and species	Carotenoids	Reference
I. montereyensis	β-Carotene, lutein, echinenone, crustaxanthin, 2-hydroxy-β-carotene, isocryptoxanthin	(*200,203*)
I. resecata (Stimpson)	α-Carotene, β-carotene, echienenone, canthaxanthin, 4-hydroxy-4′-keto-β-carotene, lutein, zeaxanthin, flavoxanthin, violaxanthin, 2-hydroxy-β-carotene, isocryptoxanthin	(*202,203*)
Leander serratus	β-Carotene, echinenone, isocryptoxanthin, canthaxanthin, lutein, zeaxanthin, astaxanthin, cryptoxanthin	(*52*)
Macropipus holsatus	β-Carotene, canthaxanthin, astaxanthin, lutein (?), zeaxanthin (?)	(*13*)
Nephrops norvegicus	β-Carotene, astaxanthin, lutein, zeaxanthin	(*52*)
Niphargus aquilex schellenbergi	Canthaxanthin, astaxanthin, isozeaxanthin, 4-keto-4′-hydroxy-β-carotene, rubixanthin, celaxanthin	(*73*)
N. tatrensis	β-Carotene, astaxanthin, rubixanthin, celaxanthin	(*73*)
Orconectes limosus	β-Carotene, lutein, echinenone, neoxanthin, astaxanthin, canthaxanthin, zeaxanthin	(*49,66;* cf. *58*)
O. pellucidus pellucidus	β-Carotene, lutein	(*314*)
O. rusticus	β-Carotene, lutein, cryptoxanthin, astaxanthin	(*317*)
Pachygrapus marmoratus	β-Carotene, lutein, zeaxanthin, canthaxanthin, astaxanthin	(*14*)
Panulirus japonicus	β-Carotene, β-zeacarotene, isocryptoxanthin, 4-hydroxyechinenone, echinenone, astaxanthin, 3-hydroxycanthaxanthin, β-doradexanthin	(*181,216*)
Penaeus japonicus	β-Carotene, echinenone, canthaxanthin, astaxanthin, phoenicoxanthin, tunaxanthin, lutein, zeaxanthin, dihydroxypiradixanthin	(*171,174*)
P. setiferus	Cryptoxanthin, astaxanthin	(*196*)
Pinnotheres pisum	β-Carotene, lutein, echinenone, canthaxanthin, astaxanthin, alloxanthin	(*24*)
Pleuroncodes planipes (Stimpson)	β-Carotene, astaxanthin	(*315*)
Portunus trituberculatus	Canthaxanthin, 4-hydroxyechinenone, 3-hydroxycanthaxanthin, astaxanthin, β-carotene, echinenone, isocryptoxanthin, β-carotene, lutein, cryptoxanthin, astaxanthin	(*176*)
Procambarus clarkii	β-Carotene, echinenone, astaxanthin, phoenicoxanthin	(*234*)
Rhithropanopeus harrisi tridentatus	β-Carotene, γ-carotene, lutein, zeaxanthin, canthaxanthin, astaxanthin	(*70*)

[a] See also Czeczuga (*57*), Czeczuga and Ktyszejko (*72*), and Tanaka *et al.* (*296*).

pigment is a common algal pigment, and the amphipod was isolated from the algae growing on a seawall.

Alloxanthin (**XXX**) was isolated from the commensal crab *Pinnotheres pisum* (*24*) and may well have been assimilated directly from the algal food of the crab. Fucoxanthin (**VIII**), a common algal pigment, was isolated from the barnacles *Lepas fascicularis* and *L. anatifera* and was also isolated from fecal pellets produced by living animals (*148*). Peridinin (5,6′-epoxy-3,5,3′-trihydroxy-6,7-didehydro-5,6,5′,6′-tetrahydro-10,11,12-trinor-β,β-caroten-19′,11′-olide) (**XXXIV**), a characteristic dinoflagellate carotenoid (*288*), was also isolated from *L. fascicularis* (*111*). Phoenicoxanthin (**XXXVI**) was isolated from species of *Lepas*, from *Conchoderma virgatum* (*148*), from the spider mite *Tetranychus cinnabarins* (*308*), and from the flamingo (*109*).

Idoxanthin (3,3′,4′-trihydroxy-β,β-caroten-4-one) (**XXXV**) was isolated from *Idotea metallica* (*145*) and would appear to be a pigment peculiar to Crustacea. Crustaxanthin (β,β-carotene-3,4,3′,4′-tetrol) (**XXXVI**), originally isolated from the copepod *Arctodiaptomus salinus* (*15,16*) and other Crustacea (*72, 145, 147, 212*), has also been isolated from the green alga *Haematococcus pluvialis* (*77*).

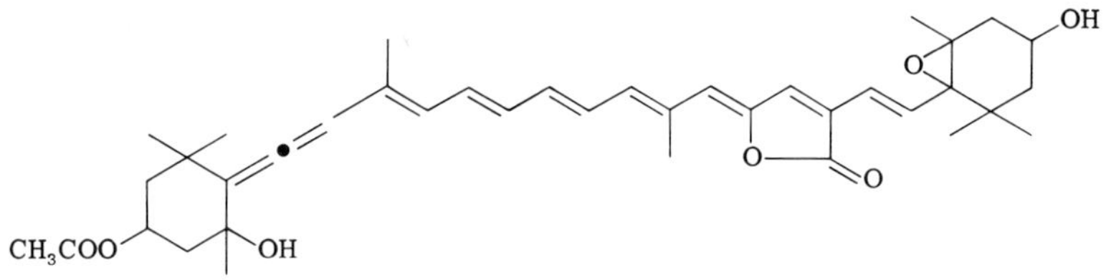

(XXXIV)

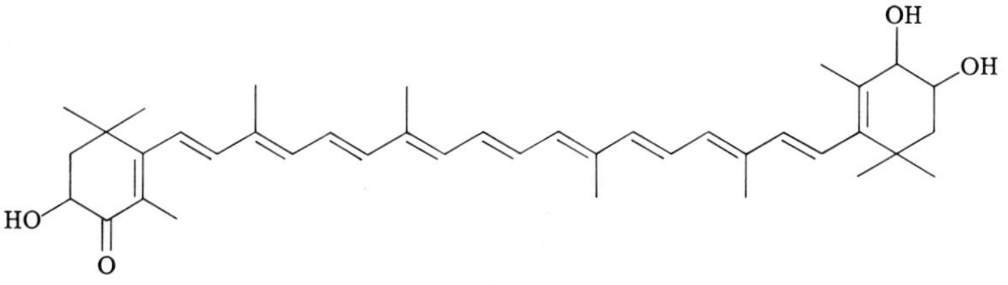

(XXXV)

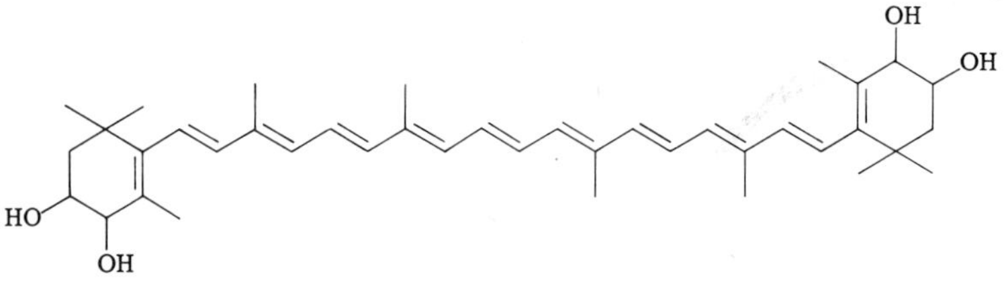

(XXXVI)

It has been suggested that the lack of carotenoids in some cavernicolous crustaceans is due to the absence of light in their environment. However, cave-dwelling crayfish (*316*) and benthic marine crustaceans, which live in total darkness, contain carotenoids absorbed from the diet.

The Atlantic deep-sea red crab *Geryon quinquedens* is found in water depths greater than 300 meters. At this depth the crab is in nearly total darkness, yet it is intensely red. Herring (*149*) compared the pigment concentration and depth of some 47 decapod species. Between 250 and 1250 meters, the range in values among individual species was such that Herring concluded that there was no consistent relationship between depth and pigment concentration in the species examined.

Geryon quinquedens from the west Atlantic was found to contain astaxanthin, ester-free astaxanthin, zeaxanthin ester, and β-doradexanthin ester (*194*). The latter pigment has been reported in only one other crustacean, the spiny lobster *Panulirus japonicus* (*216*). The same crab isolated from the east Atlantic contained, in addition, canthaxanthin, lutein, epoxide, and hydroxyechinenone. β-Doradexanthin ester was not reported (*72*).

H. Fish

The carotenoids in fish were reviewed by Goodwin in 1951 (*121*) and by Fox in 1957 (*103*). Crozier more recently (1974) covered carotenoids briefly in his chapter on the pigments of fish (*40*). Love (*209*) indexed carotenoids along with other chemical substances, and extensive use was made of this book in equating the various common and scientific names used in Table 7.

Fish, along with other animals, are thought not to be able to perform a *de novo* synthesis of carotenoids. Whereas other lower forms appear able to make extensive alterations in their dietary carotenoids, fish seem limited in the changes that they can make. Some species are able only to deposit a single carotenoid pigment. When carotenoids do occur in fish, they are found in the skin, flesh, eggs, gonads, milt, liver, and eyes. The xanthophylls, often as esters, predominate greatly over the carotenes.

Table 7 lists the species from which carotenoids have been reported since the reviews by Goodwin and Fox. It can be seen that astaxanthin is the most common pigment isolated, followed by lutein and tunaxanthin. It should be noted that prior reviews mention a carotenoid of then unknown structure, taraxanthin. Taraxanthin has since been shown to be 5,6-lutein epoxide (5,6- epoxy- 5,6- dihydro- β,ϵ-carotene-3,3′-diol) (**XXXVII**) and tunaxanthin (3,3′-dihydroxy-ϵ-carotene) (**IX**).

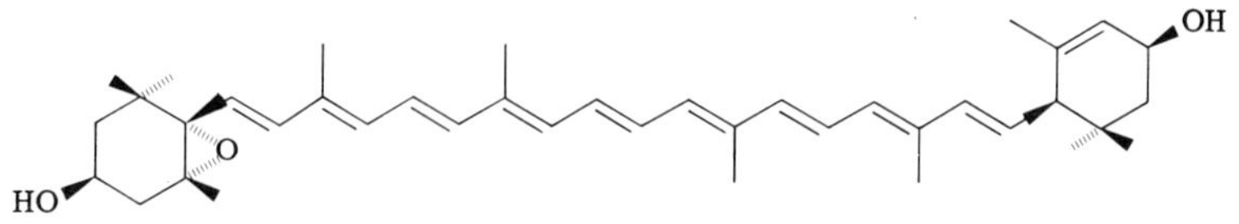

(XXXVII)

TABLE 7

Carotenoid Pigments in Fin Fish[a]

Genus and species	Common name	Carotenoids	Tissue	Reference
Acanthopagrus schlegeli		Tunaxanthin, lutein, zeaxanthin, cynthiaxanthin, diatoxanthin, α-cryptoxanthin	Skin, fin	(*223*)
Ambloplites rupestris	Sunfish	Lutein	Skin, fin	(*264*)
Apodichthys flavidus	Marine gunnel	ϵ-Carotene, tunaxanthin, lutein, isozeaxanthin, zeaxanthin	Skin	(*41*)
Auxis thazard	Frigate mackerel	Tunaxanthin, zeaxanthin, lutein, α-cryptoxanthin, diatoxanthin, cynthiaxanthin	Skin, fin	(*223*)
Beryx splendens	Alfoncino	Astaxanthin	Skin	(*154,303*)
Blennius sanguinolentus		β-Carotene, tunaxanthin, astaxanthin		(*53*)
B. sphinx		Lutein, tunaxanthin, taraxanthin, astaxanthin		(*53*)
Branchiostegus japonicus	"Tile fish"	Astaxanthin, tunaxanthin	Skin	(*303*)
Callionymus punctatus	"Nezumigochi"	Cynthiaxanthin, zeaxanthin, tunaxanthin, lutein, diatoxanthin	Skin, fin	(*223*)
Carassius auratus	Goldfish	β-Carotene, lutein, zeaxanthin, α-doradexanthin, astaxanthin, β-doradexanthin	Skin	(*138,139,152, 185,186*)
		Zeaxanthin, lutein, 4-keto-hydroxy-β-carotene, canthaxanthin, astaxanthin	Egg	(*138*)
		Zeaxanthin, β-carotene, lutein	Liver	(*138*)
C. carassius	Crucian carp	Lutein, astaxanthin	Skin	(*155*)
Careharinus glaucus	Blue shark	Carotenoids present	Liver	(*99*)
Centrophorus squamosus	Shark	Carotenoids present	Liver	(*99*)
Centroscyllium fabricii	Black dogfish	Carotenoids present	Liver	(*99*)
Centroscymnus coelolepis	Portuguese shark	Carotenoids present	Liver	(*99*)
Cetorhinus maximus	Basking shark	Carotenoids present	Liver	(*99*)
Chaenobryttus gulosus	Sunfish	Lutein	Skin, fin	(*98*)
Chaenogobius isaza	"Isaza"	Cynthiaxanthin, tunaxanthin, β-doradecin, cryptoxanthin, zeaxanthin, lutein, diatoxanthin, astacene	Skin, fin	(*221*)
Chelidonichthys kumu	Gurnard	Astaxanthin, violaxanthin (?), tunaxanthin, lutein	Skin	(*155,303*)

Chimaera monstrosa	Rabbit fish	Carotenoids present	Liver	(*99*)
Chrysophrys major	Sea bream	Zeaxanthin, α-carotene, tunaxanthin, lutein, α-doradecin, astaxanthin	Skin	(*169,172,173 182*)
		Echinenone, tunaxanthin, β-carotene, zeaxanthin, lutein, astaxanthin	Organs	(*182*)
Clupea harengus	Atlantic herring	Astacene, astaxanthin, canthaxanthin, lutein	Muscle	(*59*)
		Astaxanthin, astacene, isozeaxanthin, canthaxanthin	Liver	(*59*)
		Astacene, astaxanthin, lutein, canthaxanthin	Eggs	(*59*)
		Canthaxanthin, tunaxanthin, lutein epoxide, isozeaxanthin, astaxanthin	Milt	(*54*)
Clunpanodon punctatus	"Konoshiro"	Lutein, diatoxanthin, zeaxanthin, tunaxanthin, cynthiaxanthin	Skin, fin	(*226*)
Clupea pallasii	Pacific herring	Tunaxanthin, lutein, zeaxanthin, α-cryptoxanthin	Skin, fin	(*226*)
Coilia mystus	"Estu"	Zeaxanthin, lutein	Skin, fin	(*226*)
Cololabis saira	Skipper	Tunaxanthin, lutein, zeaxanthin, diatoxanthin, cynthiaxanthin	Skin, fin	(*225*)
Coregonus lavaretus	Lavaret	Canthaxanthin, 4-hydroxy-4′-keto-β-carotene, isozeaxanthin, zeaxanthin, astaxanthin	Milt	(*54*)
		4-Hydroxy-α-carotene, lutein epoxide, lutein, zeaxanthin		(*55*)
C. peled		Echinenone, tunaxanthin, astaxanthin, cryptoxanthin	Fin (male)	(*55*)
		Astaxanthin	Fin (female)	(*60*)
		Astaxanthin, canthaxanthin, echinenone, isozeaxanthin	Skin (male)	(*60*)
		Astaxanthin	Skin (female)	(*60*)
		Canthaxanthin, tunaxanthin, astaxanthin	Muscle (male)	(*60*)
		Astaxanthin, tunaxanthin, canthaxanthin	Muscle (female)	(*60*)
		Canthaxanthin, astaxanthin, isozeaxanthin,	Liver (male)	(*60*)
		4-Hydroxy-α-carotene, astaxanthin, canthaxanthin	Liver (female)	(*60*)
		Astaxanthin, lutein epoxide, canthaxanthin	Roe	(*60*)
Coreoperca kawamebari	Japanese perch	Tunaxanthin, lutein, zeaxanthin, cynthiaxanthin, cryptoxanthin, diatoxanthin	Skin, fin	(*224*)
Coryphaena hippurus	Dolphin fish	Tunaxanthin, α-cryptoxanthin, lutein, zeaxanthin	Skin, fin	(*223*)

(*Continued*)

TABLE 7—*Continued*

Genus and species	Common name	Carotenoids	Tissue	Reference
Cottus pollus	"Kajika"	Tunaxanthin, lutein, zeaxanthin, cryptoxanthin, diatoxanthin, cynthiaxanthin, β-doradecin	Skin, fin	(*221*)
C. reinii	Utsusemikajika	Tunaxanthin, cynthiaxanthin, zeaxanthin, lutein, cryptoxanthin, diatoxanthin, astacene, β-doradecin	Skin, fin	(*221*)
Crenilabrus tinca		Tunaxanthin, canthaxanthin, lutein, astaxanthin		(*53*)
Cyprinus carpio	Red carp	Lutein, astaxanthin	"Body"	(*99*)
Diodon holacanthus	Porcupinefish	Tunaxanthin, β-carotene, α-cryptoxanthin, lutein, zeaxanthin, astacene, α-doradecin	Fin, skin	(*217*)
Diplodus annularis		β-Carotene, canthaxanthin, tunaxanthin, lutein, zeaxanthin, astacene		(*53*)
D. sargus		β-Carotene, tunaxanthin, lutein, taraxanthin, zeaxanthin, astaxanthin		(*53*)
Ditrema temmincki	Sea chub	Tunaxanthin, astaxanthin, α-doradexanthin, lutein, α-cryptoxanthin, zeaxanthin	Skin, fin	(*223*)
Doderleinia berycoides	"Serranid fish"	Astaxanthin, tunaxanthin	Skin	(*303*)
Engraulis japonica	Anchovy	Zeaxanthin, diatoxanthin, cynthiaxanthin, lutein	Skin, fin	(*226*)
Entosphenus japonicus	Lamprey	Lutein, astaxanthin	"Body"	(*155*)
Epinephelus akaara	"Grouper"	Astaxanthin, tunaxanthin	Skin	(*303*)
E. fasciatus	Red hata	Astaxanthin, tunaxanthin	Skin	(*303*)
Erythrichthys schlegelt	"Hachibiki"	Astaxanthin, tunaxanthin	Skin	(*303*)
Etmopterus spinax	Shark	Carotenoids present	Liver	(*99*)
Etrumeus micropus	Sardine	Tunaxanthin, lutein	Skin, fin	(*155,226*)
Eugaleus galeus	Tope	Carotenoids present	Liver	(*99*)
Evynnis japonica	Red sea bream	Canthaxanthin, lutein, astaxanthin	Organs	(*180*)
		β-Cryptoxanthin, tunaxanthin, zeaxanthin, lutein, 3-hydroxycanthaxanthin, astaxanthin	Skin	(*180*)
Exocoetus volitans	"Idatentobiuo"	Tunaxanthin, lutein, zeaxanthin, astaxanthin, α-doradexanthin	Skin, fin	(*225*)
Fistularia petimba	Flute-mouth	Astaxanthin, tunaxanthin	Skin	(*303*)

Gasterosteus aculeatus	Three-spined sticle-back fish	Tunaxanthin, letein, astaxanthin, zeaxanthin, cynthiaxanthin	Skin, fin	(*19,225*)
Gephyroberyx japonicus	"Berycoid fish"	Astaxanthin	Skin	(*303*)
Gobius melanostomus		β-Carotene, astacene		(*53*)
Harengula tunasi	"Sappa"	Tunaxanthin, lutein, zeaxanthin, diatoxanthin, α-cryptoxanthin	Skin, fin	(*226*)
Helicolenus hilgendorfi	"Rockfish"	Astaxanthin, violaxanthin, tunaxanthin	Skin	(*303*)
Heterodontus francisci	Horned shark	Zeaxanthin	Skin, liver	(*106*)
Hexagrammox otakii	"Ainame"	Tunaxanthin, α-doradexanthin, α-cryptoxanthin, lutein, diatoxanthin	Skin, fin	(*223*)
Hexanchus griseus	Six-gilled shark	Carotenoids present	Liver	(*99*)
Hypomesus japonicus	"Chika"	Zeaxanthin, cryptoxanthin, cynthiaxanthin, lutein, diatoxanthin, β-carotene	Skin, fin	(*221*)
H. olidus	Pond smelt	Zeaxanthin, cryptoxanthin, cynthiaxanthin, lutein, diatoxanthin, β-carotene	Skin, fin	(*221*)
Ilisha elongata	"Hira"	Tunaxanthin, lutein, zeaxanthin, cynthiaxanthin, α-cryptoxanthin, diatoxanthin	Skin, fin	(*226*)
Ischikauia steenackeri	Wataka	Lutein, zeaxanthin, diatoxanthin, cynthiaxanthin, α-doradexanthin	Skin, fin	(*227*)
Lamna cornubica	Porbeagle	Carotenoids present	Liver	(*99*)
Lateolabrax japonicus	"Seigo"	Astaxanthin, tunaxanthin	Skin, fin	(*223*)
Lepidotrigla guntheri	"Sea robin"	Astaxanthin, tunaxanthin	Skin	(*303*)
Lepomis cyanellus	Sunfish	Lutein	Skin, fin	(*264*)
L. macrochirus	Sunfish	Lutein	Skin, fin	(*264*)
L. megalotis	Sunfish	Lutein, astaxanthin	Skin, fin	(*264*)
L. microlophus	Sunfish	Lutein	Skin, fin	(*264*)
Limander yokohamae		Astaxanthin		(*155*)
Micropterus salmoides	Sunfish	Lutein	Skin, fin	(*264*)
M. punctulatus	Sunfish	Lutein	Skin, fin	(*264*)
Misgurnus fossilis	Loach	Lutein	"Body"	(*155*)
Moroco steindachneri	"Aburahaya"	Lutein, zeaxanthin, diatoxanthin, α-doradexanthin	Skin, fin	(*227*)
Mugil auratus		β-Carotene, canthaxanthin, lutein, zeaxanthin, astaxanthin		(*53*)

(*Continued*)

TABLE 7—*Continued*

Genus and species	Common name	Carotenoids	Tissue	Reference
M. cephalus		Zeaxanthin, lutein, diatoxanthin, tunaxanthin, isocryptoxanthin, cryptoxanthin, β-carotene, cynthiaxanthin	Skin, fin	(*217,220*)
Mustelus mustelus	Smooth hound	Carotenoids present	Liver	(*99*)
Myliobatis aquila	Eagle ray	Carotenoids present	Liver	(*99*)
Navodon modestus	Black scraper	Tunaxanthin, cynthiaxanthin, lutein, zeaxanthin, α-cryptoxanthin, astaxanthin	Skin, fin	(*223*)
Oncorhynchus gorbuscha	Pink salmon	Astaxanthin	Muscle	(*166*)
O. keta	Chum or dog salmon	Astaxanthin	Muscle	(*166*)
O. kisutch	Coho or silver salmon	Astaxanthin	Muscle	(*166*)
O. masou	Cherry salmon	Astaxanthin	Muscle	(*3*)
O. nerka	Sockeye salmon	Astaxanthin	Muscle	(*166*)
O. tschawytscha	Chinook salmon	Astaxanthin	Muscle	(*33,166*)
Opsariichthys uncirostris		Lutein, zeaxanthin, diatoxanthin, cynthiaxanthin, astaxanthin, α-doradexanthin	Skin, fin	(*227*)
Oryzias latipes	Telost fish	β-Carotene, zeaxanthin, tunaxanthin, lutein, cynthiaxanthin	Skin, fin	(*151,225*)
Osmerus dentex	Japanese smelt	Zeaxanthin, diatoxanthin, lutein, cynthiaxanthin, cryptoxanthin	Skin, fin	(*222*)
Parapristipoma trilineatum	"Isaki"	Tunaxanthin, zeaxanthin, lutein, α-cryptoxanthin, diatoxanthin, cynthiaxanthin	Skin, fin	(*223*)
Parasilurus asotus	Catfish	Lutein, astaxanthin	"Body"	(*155*)
P. asotus		β-Zeacarotene-3,17′-diol, zeacarotene, β-zeacaroten-3-ol, 9,10′-didehydrozeacaroten-3,17′-diol, βzeacarotenetriol, β-carotene, cryptoxanthin, lutein, β-zeacarotene-3,17′-diol, zeaxanthin, diatoxanthin, cynthiaxanthin	Skin, fin	(*218*)
Pimelometopon pulchrum	Teleost fish	Tunaxanthin, zeaxanthin, astaxanthin	Skin	(*36,38*)
Platyrhinoidis triseriata	Thornback ray	Zeaxanthin	Skin, liver	(*106*)
Plecoglossus altirelis	Sweet smelt	Lutein, astaxanthin	"Body"	(*155*)
		Zeaxanthin, cynthiaxanthin, lutein, cryptoxanthin	Skin, fin	(*221*)

Pomoxis nigromaculatus	Sunfish	Lutein	Skin, fin	(*264*)
P. annularis	Sunfish	Lutein	Skin, fish	(*264*)
Priacanthus boops	"Big eye"	Astaxanthin, tunaxanthin	Skin	(*303*)
Pristipomoides amoenus		Tunaxanthin, astaxanthin	Fin	(*155*)
P. filamentosus	"Chibiki"	Astaxanthin, tunaxanthin	Skin, fin	(*223*)
Pristiurus melanostomus	Black-mouth dogfish	Carotenoids present	Liver	(*99*)
Prognichthys agoo	Flying fish	Tunaxanthin, lutein, zeaxanthin	Skin, fin	(*225*)
Pungitius sinensis	"Tomiyo"	Tunaxanthin, lutein, zeaxanthin, cynthiaxanthin, α-cryptoxanthin, diatoxanthin	Skin, fin	(*225*)
Raja batis	Blue skate	Lutein	Skin	(*103*)
		Carotenoids present	Liver	(*99*)
R. Brachura	Blonde ray	Carotenoids present	Liver	(*99*)
R. circularis	Sandy ray	Carotenoids present	Liver	(*99*)
R. clavata	Thornback ray	Carotenoids present	Liver	(*99*)
		Lutein	Skin	(*103*)
R. fullonica	Ray	Carotenoids present	Liver	(*99*)
R. lintea	Ray	Carotenoids present	Liver	(*99*)
R. montagui	Spotted ray	Carotenoids present	Liver	(*99*)
R. naevus	Cuckoo ray	Carotenoids present	Liver	(*99*)
R. oxyrhynchus	Long-nosed skate	Carotenoids present	Liver	(*99*)
R. rădiata	Starry skate	Carotenoids present	Liver	(*99*)
Rhinoplagusia japonica	Black cow tongue	Astaxanthin	Skin	(*155*)
Roccus saxatilis	Striped bass	Lutein, tunaxanthin, zeaxanthin, diatoxanthin, cynthiaxanthin	Skin, fin	(*223*)
Sabastes inermis	"Mebaru"	Tunaxanthin, astaxanthin, α-doradexanthin, lutein, cynthiaxanthin, zeaxanthin, α-cryptoxanthin	Skin, fin	(*223*)
Salangichthys microdon	Whitefish, "shiravo"	Zeaxanthin, cynthiaxanthin, lutein, β-carotene, cryptoxanthin	Skin, fin	(*221*)
Salmo gairdneri (= *S. irideus*)	Rainbow trout	Astaxanthin, lutein, β-carotene, taraxanthin	Skin	(*17,249*)
		Canthaxanthin	Muscle	(*268*)
		β-Carotene, canthaxanthin, 4-hydroxy-4-keto-β-carotene, lutein, astaxanthin	Milt	(*54*)
		Astaxanthin, lutein	Eggs	(*117*)

(*Continued*)

TABLE 7—*Continued*

Genus and species	Common name	Carotenoids	Tissue	Reference
S. trutta (= *S. fario*)	Sea trout, brown trout	β-Carotene, Canthaxanthin, astaxanthin, lutein	Skin, muscle	(*17,287,299, 323*)
S. salar	Atlantic salmon	Astaxanthin, canthaxanthin, lutein, zeaxanthin	Muscle	(*55,188,260*)
Salvelinus fontinalis	Brook trout	Lutein, astaxanthin	Muscle	(*260*)
		Canthaxanthin	Skin	(*262*)
Sardinops melanosticta	Sardine	Tunaxanthin, lutein, zeaxanthin		(*155,226*)
Scomber japonicus	Japanese mackerel	Tunaxanthin, lutein	Skin, fin	(*223*)
Scombrops boops	"Big eye"	Lutein, violaxanthin	"Body"	(*155*)
		Tunaxanthin, lutein, α-doradexanthin	Skin, fin	(*223*)
Scorpaenodes littoralis	"Isokasago"	Astaxanthin, tunaxanthin, α-doradexanthin, lutein, α-cryptoxanthin, zeaxanthin	Skin, fin	(*223*)
Scymnodon jonsonii	Shark	Carotenoids present	Liver	(*99*)
Scymnorhinus licha	"Darkie Charlie"	Carotenoids present	Liver	(*99*)
Sebastes baramenuke (*S. maramenuke*)	"Rockfish"	Astaxanthin	Skin	(*303*)
S. flammeus	"Rockfish"	Astaxanthin	Skin	(*303*)
S. inermis	"Rockfish"	Astaxanthin, tunaxanthin	Skin	(*154*)
S. iracundus	"Rockfish"	Astaxanthin	Skin	(*154*)
S. tompsoni	"Rockfish"	Astaxanthin, tunaxanthin	Skin	(*303*)
Sebastiscus marmoratus	"Rockfish"	Astaxanthin, α-doradexanthin, tunaxanthin, lutein, α-cryptoxanthin	Skin, fin	(*223,303*)
Sebastodes atrovirens	Sebastodes	Zeaxanthin, tunaxanthin	Flesh	(*37*)
S. carnatus	Sebastodes	Zeaxanthin, tunaxanthin, astaxanthin	Flesh	(*37*)
S. constellatus	Sebastodes	Astaxanthin, zeaxanthin, tunaxanthin	Flesh	(*37*)
S. eos	Sebastodes	Astaxanthin, zeaxanthin, tunaxanthin	Flesh	(*37*)
S. flavidus	Sebastodes	Astaxanthin, zeaxanthin, tunaxanthin	Flesh	(*37*)
S. miniatus	Sebastodes	Astaxanthin, zeaxanthin, tunaxanthin	Flesh	(*37*)
S. umbrosus	Sebastodes	Astaxanthin, zeaxanthin, tunaxanthin	Flesh	(*37*)
Sebastolobus macrochir	Channel rockfish	Astaxanthin, tunaxanthin	Skin	(*154,303*)
Somniosus microcephalus	Greenland shark	Carotenoids present	Liver	(*99*)

Sparus sarba (Rhabdosargus sarba)	Silver bream	Tunaxanthin, lutein, zeaxanthin, diatoxanthin	Skin, fin	(*223*)
Spheroides niphobles				(*217*)
Spirinchus lanceolatus	Smelt	Zeaxanthin, cryptoxanthin, diatoxanthin, lutein	Skin, fin	(*222*)
Sprattus sprattus	Sprat	Astacene, astaxanthin, canthaxanthin, tunaxanthin, lutein	Muscle	(*59*)
		Astaxanthin, β-carotene, isozeaxanthin, taraxanthin	Liver	(*59*)
		Astaxanthin, astacene, tunaxanthin	Intestine	(*59*)
		Canthaxanthin, isozeaxanthin, astaxanthin, 4-hydroxy-4-keto-β-carotene	Milt	(*54*)
Squalus acanthias	Piked dogfish	Carotenoids present	Liver	(*99*)
Squatina squatina	Angelfish, angel ray	Carotenoids present	Liver	(*99*)
Stephanolepis cirrhifer	Filefish	Tunaxanthin, cynthiaxanthin, lutein, α-cryptoxanthin, zeaxanthin, β-carotene, diatoxanthin	Skin, fin	(*223*)
Stolephorus japonicus	"Kibinago"	Tunaxanthin, lutein, zeaxanthin, α-cryptoxanthin	Skin, fin	(*226*)
Theragra chalcogramma	Alaska pollack	Tunaxanthin, lutein, zeaxanthin	Skin, fin	(*223*)
Thunnus albacares (Neothunnus macropterus)	Yellow fin tuna	Tunaxanthin	Fin	(*155*)
T. thynnus orientalis	Pacific blue fin tuna	Tunaxanthin	"Body"	(*154,155*)
Torpedo nobiliana	Electric ray	Carotenoids present	Liver	(*99*)
Trachurus japonicus	Jack mackerel	Tunaxanthin		(*155*)
Trygon pastinaca	Stingray	Carotenoids present	Liver	(*99*)
		Lutein	"Body"	(*155*)
Tribolodon hakonensis	"Ugua"	Lutein, zeaxanthin, diatoxanthin, cynthiaxanthin, astaxanthin, α-doradexanthin	Skin, fin	(*227*)
Zacco platypus	Minnow	Lutein, zeaxanthin, diatoxanthin, cynthiaxanthin, α-doradexanthin	Skin, fin	(*227*)
Z. temmincki	"Kawamutsu"	Lutein, zeaxanthin, diatoxanthin, cynthiaxanthin, α-doradexanthin	Skin, fin	(*227*)

[a] See Czeczuga (*61*) for carotenoids in the brain of fin fish species.

Since the isolation of tunaxanthin from the Pacific blue fin tuna in 1957 by Hirao *et al.* (*154*), this pigment has been shown to be widely distributed in various marine species. Reports of the isolation of taraxanthin from marine species should be questioned until verified. Whereas taraxanthin was thought to be the major companion pigment to astaxanthin, Table 7 shows that tunaxanthin is probably the second most important carotenoid pigment in marine fish.

Taraxanthin has been reported in freshwater salmonids [*Salmo irideus* (*249*), *S. trutta* eggs (*55*), *Hucho hucho* eggs (*55*), *Coregonus albula* eggs (*55*), *C. lavaretus* eggs (*55*), *C. peled* eggs (*55*)]. Czeczuga (*55*) also reported the isolation of tunaxanthin from some of the eggs of the above species. It should be noted that the identifications cited above and most of those listed in Table 7 were based on spectral, chromatographic, and, in some cases, partition data only.

Lutein is a dominant pigment in freshwater fish and is generally of minor significance in marine fish. On the other hand, tunaxanthin is a pigment characteristic of marine fishes. Exceptions to this rule are *Oryzias latipes* (*151*), in which the opposite is true, and some of the Salmonidae (*55*), in which both pigments have been reported. Hirao *et al.* (*151*) suggested that *Oryzias* is a marine rather than freshwater fish. Their suggestion is based on the fact that this teleost can survive a highly saline environment and does not form vitamin A_2, which is common to freshwater fish.

Matsuno *et al.* (*220*) cultured striped mullets in both fresh and salt water and found zeaxanthin (**XI**) in both groups, but lutein (**II**) was the other main pigment in fresh water, and diatoxanthin (7,8-didehydro-β,β-carotene-3,3′-diol) was isolated from saltwater mullets. While the pigment makeup of the skin and fins of these fish was the same, the distribution appeared to change relative to a salt- or freshwater environment. On the other hand, mature sweet smelt (*Plecoglossos altivelis*) showed little difference between marine and freshwater origins (*221*), although a larger difference was seen for young and cultured fish.

Matsuno *et al.* (*218*) reported the isolation of a number of β-zeacarotenoids from the fins and skin of the common Japanese catfish *Parasilurus asotus*. These remarkable pigments included β-zeacarotenediol (3,17′-diol) (30.4%) and a β-zeacarotenetriol. The hydroxylated compounds were found as the free forms rather than the more common esterified forms.

Matsuno *et al.* (*219*) examined six species of gobioninaeous fish and found that cynthiaxanthin (34–60%) and zeaxanthin (10–31%) represented the major pigments. Lutein was found in lesser amounts in these freshwater fishes.

IV. BIOSYNTHESIS OF CAROTENOIDS

Since animals are thought not capable of the *de novo* synthesis of carotenoids, these compounds must be supplied in the diet. Thus, a discussion of biosynthesis must be limited to alterations of the ingested carotenoids. It can be assumed that the carotenoids in fungi, bacteria, and plants are there because of a complete synthesis. No such assumption can be made on the isolation of a pigment from animals because the pigment either may have originated directly from the diet or may be the result of a transformation of a dietary carotenoid. It is also possible that the isolated pigments were in the gut when the animal was killed and were totally unrelated to the animals' absorptive or biosynthetic capabilities. It is tempting to construct pathways based on the isolation of a series of pigments, and it would seem that the literature contains such proposals. It is important that certain precautions be taken: (a) The animal should be placed on a carotenoid-free diet; (b) pure carotenoids should be fed in a physical form that can be absorbed by the animal; (c) the animal should again be placed on a carotenoid-free diet.

The most conclusive evidence can be obtained only with the use of radioactive tracers and by the isolation and characterization of key enzymes. The biosynthetic ability of a species is important in the construction of fish food. For example, both trout and goldfish can absorb lutein from plant sources. Goldfish are reported to convert lutein to the characteristic goldfish pigments. Trout apparently can absorb only lutein, and a noncharacteristic yellow-colored trout is produced.

A. Shellfish

Fox (*104*) listed four chief classes of carotenoid-metabolizing animals and stated that the gastropods of predominantly herbivorous diet, tend to be nonselective carotenoid assimilators, storing both oxygenated and hydrocarbon types of carotenoids.

McBeth (*211,212*) studied the carotenoids of a nudibranch and proposed that members of this class of organisms lack the ability to alter carotenoids from one structure to another. Isorenieratene was isolated, but it is also contained in the sponge on which the nudibranch feeds. These mollusks seem able to make carotenoproteins and esters of their dietary pigments.

Astaxanthin is not a common molluscan pigment (*129*) yet has been reported in herbivorous gastropods (*104*). It remains to be established whether the organisms are able to oxidize plant carotenoids to astaxanthin or to absorb the pigment from zooplankton or algae.

Jensen and Sakshaug (*163*) studied the relationship between the total carotenoids in the sea mussel (*Mytilus edulis*) and the occurrence of the phytoplankton in the open fjord. High concentrations of carotenoids in the mussel were observed in the spring and late summer. Minimal amounts were detected in the summer and mid-winter samples. The high concentration of carotenoids corresponded to the phytoplankton bloom in May dominated by *Skeletonema* and to the autumn bloom of *Chaetoceros* and *Ceratoim* species. The carotenoids of the phytoplankton and mussel were not compared, so that a food chain could not be established. There is, however, a good correlation between the phytoplankton blooms and mussel pigmentation.

The bay scallop *Aequipecten irradians* was maintained on phytoplankton harvested from the sea (*266*) and on the flagellate *Monochrysis lutheri* (*267*) in laboratory studies.

Oysters are being raised in heated hatcheries on public grounds in the United States and in hanging cultures in Japan (*10*). Mussels are also being raised in hanging cultures in Spain (*10*).

The most promising method of culturing shellfish under intensive care is the fertilization of the bays, fjords, or enclosed ponds that produce the phytoplankton and zooplankton on which shellfish live. The runoff of nitrates and other agricultural chemicals and sewage into bays, however, should make fertilization unnecessary in many locations.

Carotenoids are responsible for the yellow colors of shellfish species of economic importance, but at present it does not appear to be necessary to add them to feed. These pigments are contained in adequate amounts in the natural food of these species. If under intensive care a synthetic diet were constructed, some provision would have to be made for pigmentation.

Certain mollusks have been shown to have an extreme anoxic tolerance, and Karnaukhov (*167*) suggested that the carotenoids could serve as terminal electron acceptors. Zs.-Nagy (*326*), however, estimated that the carotenoids present could supply the oxygen demand for a few minutes but could not be responsible for the survival of these organisms without oxygen. Zs.-Nagy suggested that the cytosomes are responsible for maintaining oxidative phosphorylation during anoxia.

B. Crustaceans

1. Branchipods

The biosynthesis of the carotenoids of the brine shrimp *Artemia salina* (L.) has been studied more extensively than that of any other crustacean. This is not surprising since brine shrimp eggs are readily available and the organisms are small enough to be maintained in sufficient numbers in the laboratory. The brine shrimp is also important since it serves as food for other organisms both in the laboratory and in natural habitats. Whereas most crustaceans contain astaxanthin as their main pigment, the usual brine shrimp strains cannot biosynthesize this pigment (*75,78,83,136,158*). Czygan (*75,78*) reported the isolation of a small amount of astacene from a Canadian strain. Hsu *et al.* (*158*) found echinenone and canthaxanthin as the only pigments present in eggs of the brine shrimp obtained from California, whereas Czeczuga (*47*) isolated β-carotene, γ-carotene, echinenone, canthaxanthin, lutein, violaxanthin, and astaxanthin from eggs obtained from Bordeaux, France (*298*).

Davies *et al.* (*82*) fed β-carotene to brine shrimp and isolated β-carotene, echinenone, and canthaxanthin after 5 days of feeding. After the fish returned to a carotenoid-free diet, only canthaxanthin and echinenone were isolated. Echinenone, when fed, was converted to canthaxanthin. Zeaxanthin (*158*) and isozeaxanthin (*83*) are absorbed but not converted to ketocarotenoids. Isocryptoxanthin can be converted to 4-hydroxy-4′-keto-β-carotene but not to either echinenone or canthaxanthin (*83*). Davies *et al.* (*83*) proposed the following scheme for the biosynthesis of carotenoids in *Artemia*:

β-Carotene
↓
echinenone
↓
canthaxanthin

Isocryptoxanthin
(4-hydroxy-β-carotene)
↓
4-hydroxy-4′-keto-β-carotene

The lack of conversion of hydroxylated derivatives of β-carotene would cast doubt on their being free intermediates (*78*).

Herring (*147*) raised *Daphnia magna* on a diet of either algae or yeast and various pure carotenoids, none of which contained astaxanthin. The overall pathway, β-carotene → echinenone → canthaxanthin → astaxanthin, was confirmed by the feeding of intermediates. Isocryptoxanthin is converted to echinenone and canthaxanthin, and isozeazanthin is converted to canthaxanthin. *Chlorella pyrenoidosa* cells were fed, and the pigmentation in *Daphnia* was seen as coming from β-carotene rather than from lutein, violaxanthin, or neoxanthin.

The role of the hydroxycarotenoids in the metabolism of *Daphnia* is not clear. Alhtough these compounds can be converted to ketocarotenoids, they were not isolated when β-carotene or echineone was fed. A similar situation may be present in other species of *Anostraca* [see Gilchrist (*118*)] in which astaxanthin, isocryptoxanthin, isozeaxanthin, and unidentified xanthophylls were isolated. Phoenicoxanthin (3-hydroxycanthaxanthin) was isolated (*118*) in this study and may represent an intermediate between canthaxanthin and astaxanthin.

2. *Malacostraca*

Lee (*201*) identified the pigments from the marine isopod *Idotea granulosa* and proposed the following pathway on the basis of the pigments isolated:

β-Carotene → isocryptoxanthin → echinenone →
4-hydroxy-4′-keto-β-carotene → isozeaxanthin
→ canthaxanthin

Lee also proposed alternative routes for isozeaxanthin formation from β-carotene. The reduction of xanthophylls to β-carotene was proposed on the basis of the relative amounts of carotenoids isolated from the three varieties of *Idotea*.

Czeczuga and Czerpak (*71*) fed *Daphnia magna* with *Chlorella vulgaris, Scenedesmus quadricuada,* and *Saccharomyces cerevisiae* and also found astaxanthin in *D. magna* when it was not contained in the feed. The pigments from the live food were not consistently incorporated into the *Daphnia*.

Herring (*145*) studied the carotenoids of the marine isopod *Idotea metallica*. Under starved conditions β-carotene, isocryptoxanthin, astaxanthin, and zeaxanthin increased and idoxanthin (3,3′,4-trihydroxy-4′-keto-β-carotene) and crustaxanthin (3,3′,4,4-tetrahydroxy-β-carotene) decreased. Herring proposed that astaxanthin was reduced to idoxanthin and crustaxanthin from idoxanthin and crustaxanthin. Controlled feeding experiments would be necessary to prove the latter proposal. Gilchrist and Lee (*120*) fed green algae (*Ulva*) labeled from $^{14}CO_2$ to the sand crab *Emerita analoga*. The change in the ratio of carotenes and xanthophylls and the increase in ketocarotenoids indicated that the sand crab converted carotenes to echinenone, canthaxanthin, and astaxanthin. Katayama *et al.* (*171,174*) proposed, on the basis of pigment isolation, the following metabolic pathway for the prawn *Penaeus japonicus:*

β-Carotene → echinenone → canthaxanthin → phoenicoxanthin (3-hydroxycanthaxanthin) → astaxanthin

This pathway was supported by the feeding of ^{14}C-labeled β-carotene (*175*) and canthaxanthin (*295*) to the prawn with the isolation of astaxanthin. The feeding of *Spirulina* and crab waste also resulted in the deposition of astaxanthin. A second pathway to astaxanthin in the prawn was discovered in feeding zeaxanthin. Tanaka *et al.* (*295*) proposed the second pathway as follows:

Zeaxanthin → β-doradexanthin → astaxanthin

In these studies the carotenoid composition of the carapace was found to be somewhat different from that of the internal organs (*174*).

The crab *Portunus trituberculatus* (*177*) was fed a special diet that included radioactive β-carotene. Labeled β-carotene, isocryptoxanthin, and echinenone were isolated from the internal organs, whereas 4-hydroxyechinenone, canthaxanthin, 3-ketocanthaxanthin, and astaxanthin were isolated from the carapace. A similar metabolism exists in the Japanese lobster *Panulirus japonicus* (*181*), in which radioactivity from ingested β-carotene was found in isocryptoxanthin and echinenone in the internal organs, and canthaxanthin, 3-hydroxycanthaxanthin, and astaxanthin were labeled in the carapace.

The isolation of β,β-caroten-2-ol from three species of *Idotea* (*203*) together with isocryptoxanthin suggests that these monohydroxycarotenes could also be present in other crustaceans, with the 2-hydroxy derivative not detected. Lee and Gilchrist suggested that the β,β-caroten-2-ol is a metabolism product of the isopod rather than an ingested pigment. This suggestion is based on the lack of isolation of the compound from the main food of the isopod (kelp—*Macrocystis pyrifera*). Controlled feeding experiments are necessary to prove this point.

C. Fish

Although, as can be seen from Table 7, there are many recent papers on the carotenoid composition of fish, studies on the metabolism or biosynthesis of carotenoids are relatively few. With few exceptions, papers on metabolism have been concerned with fish, such as goldfish, trout, and salmon, in which economic factors are involved. One has only to look at the price of prawn, trout, salmon, etc., to realize the economic pressure to develop efficient means of raising and coloring these species.

1. Salmonids

Steven (*286,287*) showed that the pigment in trout is of alimentary origin. Trout fed a diet of chopped meat and earthworms soon lost their yellow and red pigmentation. Lutein, a yellow carotenoid, was the only remaining carotenoid. When the fish were reared in the laboratory on a diet of *Simocephalus, Daphnia* species, and the larvae of *Corethra,* the normal coloration was obtained. Steven experienced difficulty in obtaining coloration with the use of pure carotenoids such as β-carotene, lutein, or astacene. These results suggest that either the physical form of the pigment was such that the fish could not absorb it, or else the pigments could not be converted to the normal fish pigments.

Thommen and Gloor (*299*) reported the presence of canthaxanthin in addition to astaxanthin and β-carotene in the sea trout *Salmo trutta.*

Peterson *et al.* (*249*) found that lutein, when fed to trout, was deposited in the flesh without modification. β-Carotene fed at a level of 200 mg of pigment per 1 lb of trout food resulted in no deposition of color after 4 weeks of feeding. Brook trout fed either ground crayfish or crayfish extract contained the same crayfish pigments, although in different amounts.

Deufel (*93*), Saito and Regier (*262*), and Schmidt and Baker (*269*) showed that canthaxanthin can be used to color the skin, flesh, and eggs of trout. Hata and Hata (*142*) administered pure, ^{14}C-labeled preparations of β-carotene, lutein, and zeaxanthin directly into the stomach of anesthetized rainbow trout. After 48 hr at 25°C, the three pigments were found distributed among the skin, flesh, liver, and digestive tract, but no radioactivity was found associated with the astaxanthin fraction. The recovery of radioactivity for each pigment was 6.0% or less, and in the case of β-carotene most of the activity was associated with the digestive tract. These results seem to show a lack of conversion of these pigments to astaxanthin in the trout. The present authors would have preferred an experimental design in which the fish were not subjected to as much stress, i.e., pigment incorporated into normal diets and a lower and more normal temperature maintained.

Savolainen and Gyllenberg (*268*) fed the carotenoids obtained from *Rhodotorula sanneii* to rainbow trout. The major pigments of this yeast are the monocyclic acid torularhodin, the monocylic carotenes torulene and γ-carotene, and β-carotene (about 8%). On feeding these pigments, lutein and a small amount of canthaxanthin were isolated. These results could indicate that the trout is able to make carotenoid transformations unknown in any other system and thus would bear repeating.

Thommen and Gloor (*299*) found only esterified astaxanthin in sea trout, and

Kuo *et al.* (*193*) found both free and esterified astaxanthin in the skin and flesh of rainbow trout that had been fed on a diet of red crab (*Geryon quinquedens*) waste (*194*).

A number of authors (*34,143,275*) have shown that astaxanthin is deposited in the skin as the mono- and diesters. Other xanthophylls, such as lutein, zeaxanthin, and cynthiaxanthin, are also deposited in the esterified form. Astaxanthin and other xanthophylls are deposited in the flesh, liver, ovary, and digestive organs in the free form (*143*).

Torrissen and Braekkan (*301*) fed purified astaxanthin and its mono- and diester to 250-gm rainbow trout at a level of 12 mg/kg feed for 33 days. The level of astaxanthin was 0.84 μg/gm from fed astaxanthin, 0.63 μg/gm from the monoester, and 0.33 μg/gm from the diester of astaxanthin. The pigments from pure astaxanthin and its diester (*275*) and from capelin oil were deposited preferentially in the skin. Astacene (*275*) was incorporated in both the skin and flesh as the ester and free form at an efficiency of incorporation similar to that of astaxanthin.

The pigment of the salmon is astaxanthin (*3,9,39,166*) and is deposited in the skin as esters and in the eggs as free astaxanthin (*39*). Evelyn (*98*) reported an unusually yellow-colored sockeye salmon that contained xanthophylls, none of which were astaxanthin. Canthaxanthin colors salmon skin and flesh but seems to be deposited rather than converted to astaxanthin.

2. *Goldfish*

Hirao *et al.* (*153*) fed three groups of goldfish (*Carassius auratus*) the following diets: (a) rich in carotene and lutein, (b) free of carotenoids, (c) rich in astaxanthin. The mean amount of carotenoids per individual fish greatly increased with diet a, remained unchanged with diet b, and decreased with diet c.

Katayama *et al.* (*185,186*) isolated lutein, β-carotene, astaxanthin, and a compound named α-doradexanthin (3-hydroxy-3′,4′-diketo-α-carotene) from goldfish. On the basis of structural considerations of the carotenoids isolated, these authors proposed that astaxanthin was formed from lutein via α-doradexanthin and β-doradexanthin by oxidation and isomerization (α-ionone ring $\rightarrow$ β-ionone ring) (Scheme 3).

Experimental proof for the conversion of lutein to astaxanthin was obtained by labeling experiments of Hsu *et al.* (*159*). Goldfish were fed [^{14}C]lutein for 4 days and returned to a regular diet for 4 days. Analysis of the pigments showed that 63% of the label was in astaxanthin, 5% in α-doradexanthin, and 0.8% in β-doradexanthin.

Hata and Hata (*140*), on the other hand, obtained very poor incorporation of the carbon from lutein into astaxanthin in the goldfish. These authors obtained 4-ketolutein (α-doradexanthin) as the major product after feeding [^{14}C]lutein for 30 days.

Clearly there is no obvious explanation for the divergent results obtained by Hsu *et al.* and Hata and Hata. Possible answers might lie in a strain difference in the

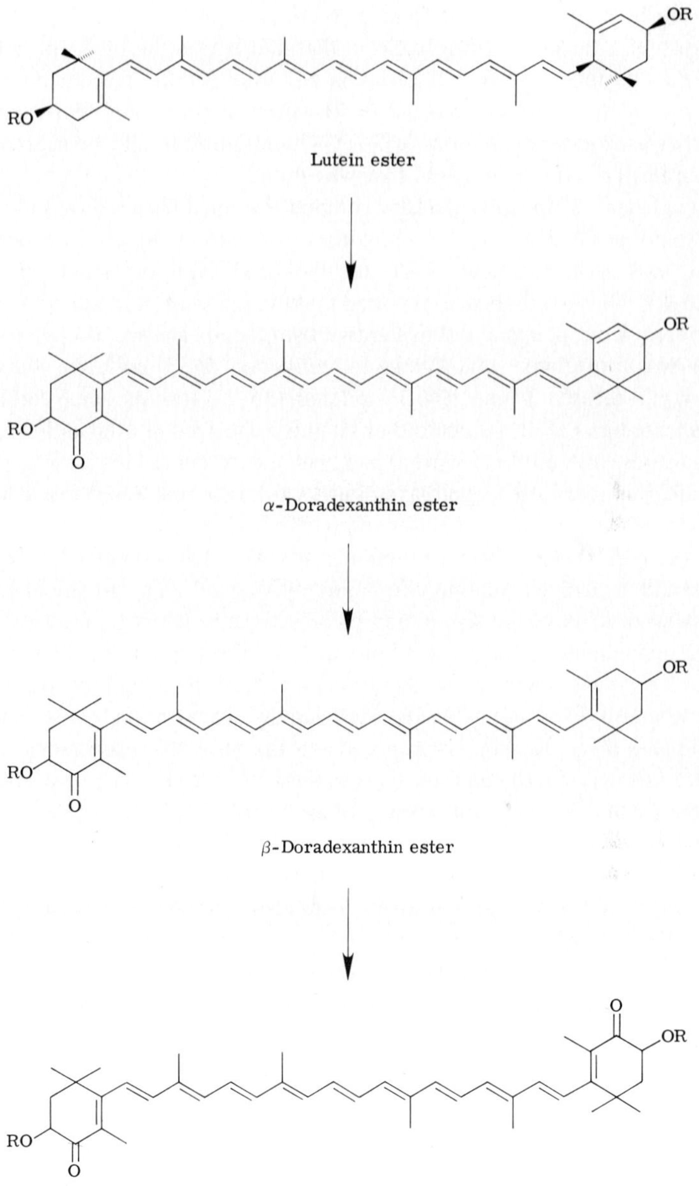

Scheme 3. Metabolism of lutein in goldfish (*186*).

organisms or in the different feeding times used. Andrewes *et al.* (*4*) questioned the biosynthesis of astaxanthin from lutein in the goldfish on the basis of their results showing the chiroptical properties of lutein and astaxanthin. These authors argued that, "if astaxanthin were synthesized *in vivo* from lutein without epimerization on the enzyme, astaxanthin should have 3*S*,3′*R* configuration and be achiral" rather than the optically active compound that was found.

Buchecker *et al.* (*20*) isolated a new compound named fritschiellaxanthin, which was shown to be 3*S*,3′*R*,6′*R*,3,3′-dihydroxy-β,ε-caroten-4-one. This compound, however, was not identical with the α-doradexanthin found in goldfish (3*S*,3′*S*,6′*R*). These authors also reported that the lutein of goldfish is not identical with the lutein from plants and thus the two compounds are not 3′ epimers of each other. It was postulated that lutein is converted to 3′-epilutein through the 3-hydroxy-β,ε-caroten-3′-one. The 3′-epilutein (3*S*,3′*S*), of the same configuration as α-doradexanthin (3*S*,3′*S*), could thus be biosynthesized to astaxanthin (3*S*,3′*S*). This elegant research eliminated what had been a perplexing biosynthetic problem.

Hata and Hata (*137*) fed cynthiaxanthin to goldfish and found that it was converted to the 4-keto and 4,4′-diketo compounds (Scheme 4).

Rodriguez (*255*) fed pure carotenoids to fish that contained only lutein, α-doradexanthin, and astaxanthin after being maintained on a carotenoid-free food. The astaxanthin level of the fish increased significantly when β-carotene, isocryptoxanthin, echinenone, or canthaxanthin was fed. When an intermediate was fed, it caused a temporary increase in the more oxidized members of the pathway. β-Carotene and its derivatives disappeared rapidly when the fish were returned to carotenoid-free diets. Rodriguez proposed, on the basis of experimental findings, the pathway shown in Scheme 5 for the metabolism of β-carotene in goldfish. This is consistent with the significant labeling of astaxanthin with ^{14}C-labeled β-carotene in the goldfish (*159*).

In other experiments β-carotene was converted to astaxanthin in the red carp (*179*) at very low levels, and low levels were also reported for red carp, goldfish, and fancy red carp (*175*).

Again it appears that there is no consensus among various laboratories as to the major precursor (i.e., lutein or β-carotene) of astaxanthin. It is now known (*258*) that β-carotene and echinenone are formed from β-carotene on MicroCel C columns, but the purification of β-carotene before feeding was on MgO (*27*). Purification of the β-carotene incorporated into the fish on MicroCel C columns would not be expected to yield astaxanthin.

Rodriguez *et al.* (*256,257*) did not detect phoenicoxanthin when canthaxanthin was fed. In contrast, Hata and Hata (*137*) reported the isolation of the pigment, but from goldfish fed on natural diets. Isozeaxanthin was slowly converted to 4-hydroxy-4′-keto-β-carotene but did not raise the levels of canthaxanthin or astaxanthin.

Hata and Hata (*140,141*) fed [^{14}C]zeaxanthin to goldfish and found significant conversion of zeaxanthin to astaxanthin through 4-ketozeaxanthin.

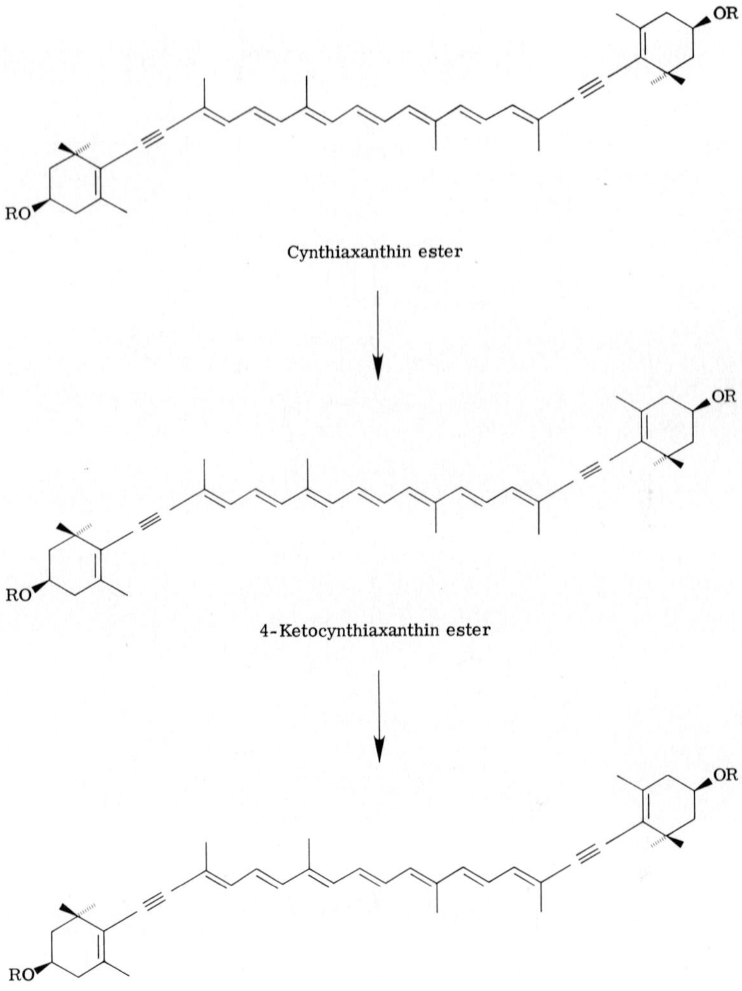

Scheme 4. The metabolic pathways of ingested cynthiaxanthin in goldfish (*137*).

3. *Other Fish*

Katayama and co-workers fed labeled β-carotene to sea bream (*Chrysophrys major*) (*183*) and red sea bream (*Evynnis japonica*) (*180*) and found that the incorporation of label into astaxanthin was poor. When prawn were fed radioactive β-carotene and the prawn were in turn fed to either of these fish, the pigments from the prawn were readily transferred to the fish (*171,174*).

Further work with the feeding of pure lutein, canthaxanthin, and zeaxanthin

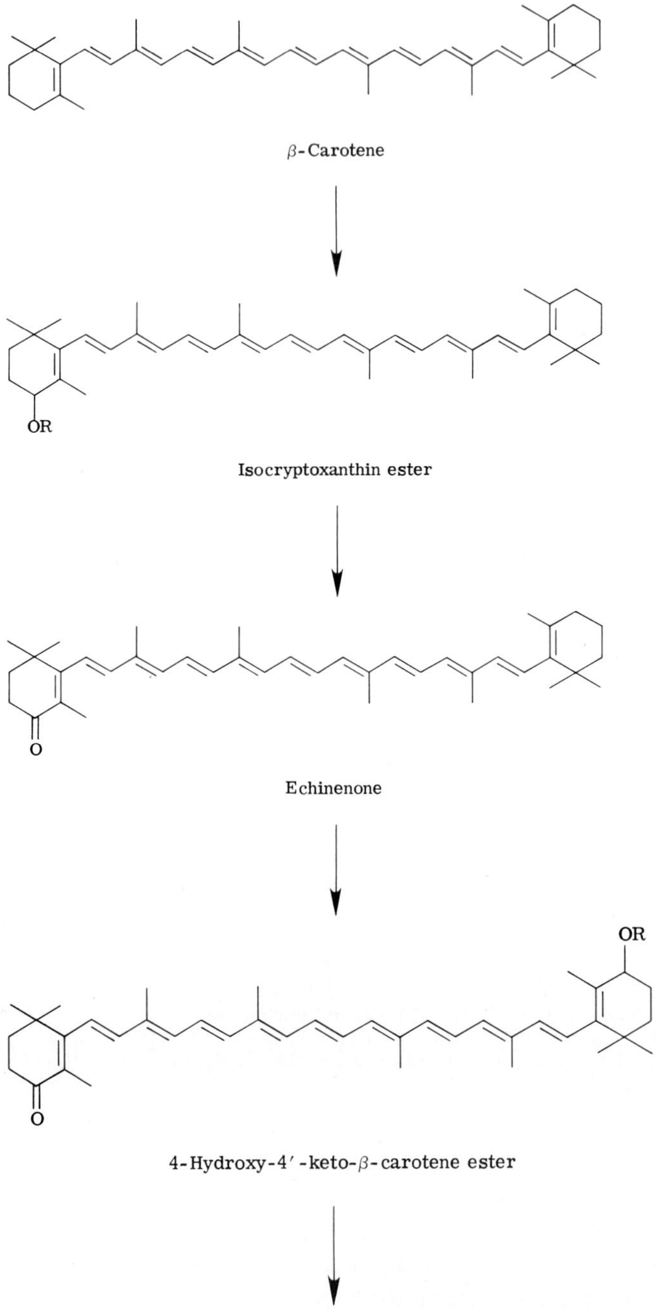

Scheme 5. Metabolism of β-carotene to astaxanthin in the goldfish (*255*).

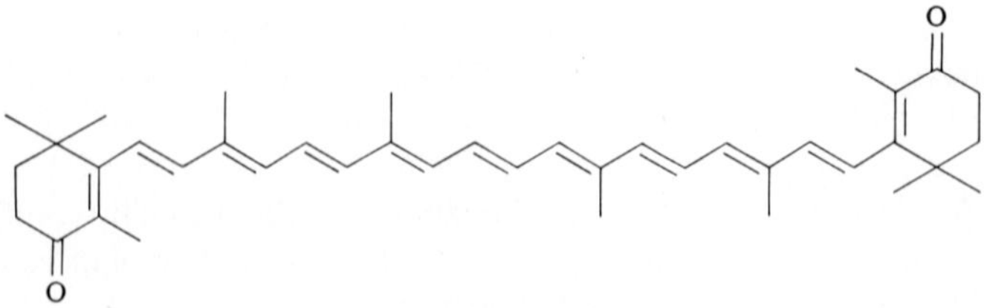

Canthaxanthin

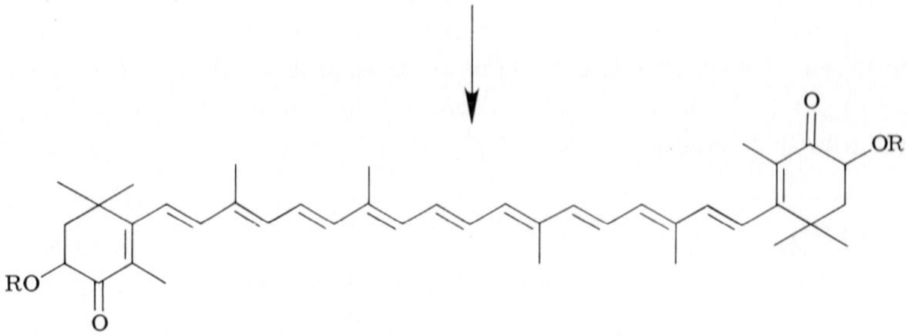

Astaxanthin ester

Scheme 5. (*Continued*)

(*290*) showed that sea bream could deposit their pigments in the flesh without modification but could not convert them to astaxanthin. This result suggests that these marine red fish cannot oxidize the 3,3′ and 4,4′ positions of the β-ionone ring. If these red fish are to be cultured in aquaculture programs, red carotenoids will have to be included in the diets (*291*).

The biosynthesis of tunaxanthin in marine fish presents a rather curious problem. This pigment has not been isolated as yet from any lower form, i.e., plant or animal. Thus, one must assume that it is converted from some preformed pigment. It is generally assumed that ε-carotene is the precursor of tunaxanthin. The assumption is based on the isolation of ε-carotene and tunaxanthin from some fish (*41,169,172*). Tunaxanthin is often the main pigment of marine fish and may represent almost 90% of the carotenoids in some species (*41*). Although ε-carotene is not a common pigment, it was recently isolated from the Siphonalean green alga *Codium fragile* (*313*). A compound tentatively identified as a monohydroxy-ε-carotene was also isolated from *C. fragile*.

Crozier (*40*) fed a marine teleost a diet deficient in carotenoids and found that the level of tunaxanthin decreased and that of astaxanthin increased. Crozier (*36,38*) suggested that tunaxanthin might be converted to zeaxanthin, and this pigment was isolated with tunaxanthin. Such a conversion together with a similar one in the formation of β- from α-doradexanthin esters represents a departure from the well-documented lack of conversion of the α and β rings in carotenes and xanthophylls in higher plants (*124*).

The information to date suggests that the metabolism of carotenoids involves oxidative changes. However, Gross and Budowski (*134*) reported an increase in vitamins A_1 and A_2 on feeding astaxanthin, isozeaxanthin, canthaxanthin, β-carotene, but not lutein. Barua *et al.* (*11*) fed β-carotene and zeaxanthin to vitamin A-depleted fish and found no increase in dehydroretinol (vitamin A_2) in the liver, as measured by color reaction with $SbCl_3$. The liver oils of fish fed lutein, however, produced an intense color with $SbCl_3$. These authors claimed that lutein is transformed into dehydroretinol by *Saccobranchos fossilis;* however, no direct conversion was demonstrated. It remains to be seen whether an actual transfer of carbon into purified vitamin A can be demonstrated or whether other compounds substitute for the retinoids.

D. Overview

It is clear that animals differ greatly in their ability to ingest and transform various carotenoid pigments. When the feeding experiments and the distribution of carotenoids in aquatic animals are taken together, consistent patterns are seen. Thus aquatic animals, and the main carotenoid of which is astaxanthin, might be divided into three classes on the basis of where astaxanthin is biosynthesized (*177*).

Group I. These animals can convert zeaxanthin or lutein to astaxanthin, but β-carotene is not the major precursor of astaxanthin. They can transfer astaxanthin in the diet to their body astaxanthin. Goldfish, red carp, and fancy red carp belong to this group. They can oxidize the 4 and 4′ positions of β-ionone rings but do not oxidize the 3,3′ position as easily.

Group II. These animals can convert β-carotene and zeaxanthin to astaxanthin. Almost all crustaceans in which the main carotenoid is astaxanthin belong to this group. The 4, 4′, 3, and 3′ positions of β-ionone rings are oxidized.

Group III. These animals cannot convert β-carotene, lutein, or zeaxanthin to astaxanthin but can transfer astaxanthin, lutein, zeaxanthin, tunaxanthin, etc., to their body pigments, often as esters.

V. USE OF CAROTENOIDS IN COMMERCIAL FISH FEEDS

In recent years there has been an upsurge of interest in the commercial culture of marine and freshwater animals. In some countries the need is for a good protein source, whereas in others the demand for luxury items, such as trout, salmon, and shrimp, exceeds the supply. It is a mistake to think of fish farming or aquaculture as a recent invention. Fish farming has been practiced for many generations, and carp raising, in particular, has been known for centuries and is now practiced in Europe, Israel, Asia, and other countries (*10*). In addition, the farming of milk fish and that of catfish, trout, prawn, and shrimp are in various stages of progress.

It has been estimated that warmwater fish ponds produce enough natural food to support 100–300 lb of fish per acre (*207*). If aquaculture is to be an economic reality, conditions must be such that this figure approaches 1500–2000 lb/acre. Careful water management, protection from disease and predators, and correct nutrition are areas that must be considered if efficient use of ponds is to be achieved.

As seen above, animals differ in their requirements, from those that require a specific pigment such as astaxanthin to those that can make a pigment from a number of other pigments.

In Japan there are some very impressive prawn culture and research facilities. Since 1963 more than 15 farms have been established using over 100 ha of discarded salt beds or sandy inlets (*92*). It has been estimated that more than 100,000 tons of aquatic animals are produced each year, including trout, sea bream, prawn, red sea bream, crab, lobster, goldfish, carp, and fancy red carp.

Pan-sized salmon are now being produced in floating pens off the coast of the northwestern United States, Maine, and elsewhere.

It is important in evaluating the effect of carotenoids in the diets of fish that all of the nutritional and environmental factors be considered. One could work with imbalanced diets or stressed fish and get results there were different from the results of studies in which more normal conditions were used. A classic example of results obtained under different sets of conditions was reported by Poston. In an initial study (*251*) the conversion of β-carotene to liver vitamin A was investigated in fingerling brook trout at a temperature of 9°C. Whereas vitamin A was absorbed and stored in the liver, dietary β-carotene was not stored in the liver as vitamin A. It was concluded that fingerling brook trout apparently cannot convert β-carotene to vitamin A. However, in a later study (*252*) run at 12.4°C, brook trout fingerlings were found to be able to convert β-carotene to vitamin A.

A. Crustaceans

Growing shrimp and crayfish in ponds is now practiced in some countries, and shrimp are grown from egg to market size (*114*). Intensive lobster culture has not yet become a reality because of the problems associated with high-density rearing. The shrimping industry in the United States commands about a quarter of the value of the catch of the fishing industry, and the consumption in the United States has averaged about 1 million lb/day.

The U.S. crayfish industry, located mainly in southern Louisiana, produces about 10 million lb valued in excess of $5 million (*8*). Although it is probable that pond-related crustaceans will not replace the natural harvest, they certainly will be a valuable supplement, particularly during certain times of the year.

1. *Natural Crustacean Food*

Studies of the feeding habits of crustaceans usually involve the identification of the plant and animal contents of the stomach. It is perhaps the best method since crustaceans are selective in what they consume. However, different foods may be digested at different rates.

Flint (*100*) found that the intake of the larval shrimp was limited to filamentous blue-green algae and diatoms. Allen (*2*) compared the stomach contents of inshore and offshore populations of *Crangon allmani* and concluded that mollusks were consumed inshore and polychaetes offshore by these dacapods.

Eldred *et al.* (*96*) found leaves, algae, diatoms, flagellates, polychaetes, nematodes, shrimps, mysids, copepods, isopods, amphipods, ostracods, and mollusks in the stomachs of *Penaeus duorarum* collected in the fall in Florida. Sastrakusumah (*265*) studied the food of juvenile pink shrimp migrating out of an estuary to offshore grounds. Observations were made throughout the year. It was found that the highest feeding activity was in September. There appeared to be little difference in the diets on the basis of size of the animals or time of year. Crustaceans and polychaetes were preferred, whereas seagrasses, diatoms, and foraminifers were avoided.

Richards (*254*) studied the use of vertical substrates in the production of pink shrimp. Although chopped squid was the main food source used, it was noted that the shrimp showed a preference for filamentous green algae and copepods and avoided blue-green algae and nematodes.

The crayfish lives on plant and animal matter, and the use of commercial fertilizers has had some benefit (*231*). The deep-sea red crab taken off the coast of New England and Canada appears to feed on sea anemones (*229*).

2. *Crustacean Feeds*

Little is known of the nutritional requirements of crustaceans (*231,237*). Existing fish chows lack the desired physical characteristics, i.e., proper durability (*231*). Because of the lack of nutritional information on shrimp, artificial foods that will support normal growth rates for any of the stages of development are not available. Neal (*237*) found that the growth of postlarval shrimp on artificial diets did not exceed 0.3–0.4 mm/day, whereas the growth of those fed freshly hatched brine shrimp nauplii can average 1.3 mm/day (Figs. 4 and 5).

Forster (*101*) experimented with a series of diet formulations for juvenile prawn, *Palaemon serratus,* in which nutrients as well as binding agents were investigated. The diets consisted of whitefish meal, dried skimmed milk, distillers' solids, wheat germ, yeast, shrimp meal, seaweed meal, extracted soya, cotton seed, and peanut meal, corn gluten, and wheat meal. Although a number of the ingredients contain carotenoids, only the shrimp and seaweed meals (57%) would be expected to contain carotenoids of value to the prawn. Growth data were described as inferior to those obtained using fresh mussel gonads.

Meyers *et al.* (*231*) investigated the use of fish protein concentrate (FPC), marine protein concentrate (MPC), or and single-cell protein (SCP) (*Torulopsis* yeast) in pellet and microcapsulated form (Table 8). Growth was described as favorable. It is not clear whether carotenoids were added to the diet.

Fig. 4. A penaeid shrimp consumes semipurified pellets, one of a series of diets being tested in a sea grant shrimp nutrition project at Skidway Institute of Oceanography, Savannah, Georgia. [Used by permission from the Center for Marine Resources/Texas A & M University.]

Fig. 5. A University of Miami laboratory technician succeeds in coaxing a shrimp to take experimental food from his hand. [Used by permission from the Center for Marine Resources/Texas A & M University.]

TABLE 8

Pelleted Ration for Crustaceans[a]

Ingredient	Concentration (%)
FPC, MPC, or SCP[b]	39.1
Rice bran	45.6
Alphacel (a substituted cellulose)	10.0
Mineral mix (NBC 23)	1.0
Soy oil	4.3

[a] From Meyers *et al.* (*231*).
[b] FPC, fish protein concentrate; MPC, marine protein concentrate; SCP, single-cell protein.

A shrimp feed must have special water stability if it is to be utilized. The average hard pellet either leaches its water-soluble components or breaks down within a few minutes after immersion in water (*230*). Hastings *et al.* (*135*) described a basic shrimp feed that is water stable for 24 hr due to gelatinization of starch in its processing. The ration (Table 9) contains 10% shrimp meal.

Meyers and Rutledge (*233*) discussed the value of using shrimp meal as a crustacean food and stressed the importance of quality control in its production if the pigments are to be utilized by crustaceans.

Experimental diets for the prawn (*175*), Japanese lobster (*181*), and crab (*180*) have been published. In these artificial diets it was necessary to add clam meat or cabbage, both of which are sources of sulfides, as attractants.

Deshimaru and Shigeno (*92*) investigated the protein requirement for prawn by including protein sources that approximated the amino acid pattern of the prawn.

TABLE 9

Formula of Basic Shrimp Feed That May Be Further Modified by Special Processing within a Feed Mill[a]

Ingredient	Concentration (%)
Soybean flour (50% protein)	20
Shrimp meal	10
Fish meal	10
Wheat flour	20
Rice by-product fractions	20
Alfalfa meal	10
Feather meal	10

[a] From Hastings *et al.* (*135*).

Fish meal was found to be a poor protein source because of its deficiency in basic amino acids.

Consumer acceptance of the prawn on the Japanese market is based largely on color. The red prawn is preferred, whereas the blue prawn, generally artificially cultured, is not. The blue prawn contains about half the total pigment of the red prawn. The blue pigment is a carotenoprotein and is a nutritional rather than a genetic problem. Prawn fed crab waste are red rather than blue (*170*).

It is obvious that there is a need for nutritional data on the requirements of crustaceans at various stages of development. Intensive cultivation of crustaceans will mean that "natural" sources of carotenoids will not be available, and these pigments will have to be added.

B. Pigmented Carp

The carotenoids of the common goldfish (*Carassius auratus*), crucian carp (*C. carassius*), and fancy red carp (*Cyprinus carpio*) have been intensively studied. There are perhaps 100 fancy varieties of goldfish, but only about 20 kinds, such as comet, fantail, veiltail, and nymph, are sold commercially. The fancy carps are commercially bred in Japan, where they may command very high prices. The goldfish may be scaleless because of their thin, transparent scales. These fish may be gold or red with spots of blue, purple, or lavender or they may be calico. Some golden carp are known to have lived for 50 years, but most live for many fewer years. They are popular as pets because of their color and easy care. These fish probably originated from the plain-colored carp of China and Japan and lost their striking colors on being returned to a body of water devoid of feed containing carotenoids.

In common with other animals, these fish must be supplied with carotenoids in the diet. However, goldfish are unique in the biochemical changes that they are able to make. Goldfish are able to make the rare conversion of the ϵ ring to the β ring and the even rarer conversion of lutein to 3′-epilutein. Thus, lutein and to a lesser extent β-carotene can be converted to astaxanthin. Most fish, on the other hand, can only store ingested carotenoids or perhaps esterify carotenoid alcohols.

A number of recent studies have resulted in postulated pathways for carotenoid conversions (see Schemes 3, 4, and 5). These studies have shown the conversion to astaxanthin of β-carotene (*159,256,257*) and questioned its role as a major precursor (*175,179,292*); of lutein (*159,162,178,184*) and questioned its role (*143*); of zeaxanthin (*143,214,295*); and of isocryptoxanthin, echinenone, and canthaxanthin (*257*). Canthaxanthin was not converted to astaxanthin in another study (*292*).

Because of the conflicting reports, it is difficult to draw many conclusions. Isuahashi and Wakui (*162*) found that astaxanthin was better than lutein for the pigmentation of fancy carp (*162*). These authors found that cholesterol had no effect on the level of accumulated carotenoids. Hata and Hata found that zeaxanthin and astaxanthin were also better than lutein (*143*). Fish fed zeaxanthin were redder than those fed lutein (*292*). Thus, the conflicting results may reflect the reporting of relative significance of incorporation. It is also possible that in some cases the crystal red pigments had not dissolved.

C. Salmonids

Recent surveys (*7*) show that some 16 million lb of trout are produced annually in the United States on commercial trout farms. In addition, sizable amounts of trout have been shipped to the United States from Japan and Denmark. It is estimated that trout production in France and Italy exceeds 20 million lb annually. Salmon culture has centered on the production of fingerling or smolt-stage fish, which are released to make their way to the sea. Present hatchery feeding programs understandably make very little effort to pigment these fish. Significant production of 5- to 7-lb and pan-sized salmon is a reality, and thus salmon pigmentation is very important.

1. Natural Salmonid Food

Manzer (*214*) examined the stomach contents of 1189 pink, chum, sockeye, and coho salmon caught off the northern British Columbia coast from June through August. The pink (*Oncorhynchus gorbuscha*) and chum (*O. keta*) were mainly planktophagous. The coho (*O. kitutch*) were mainly piscivorous, whereas sockeye (*O. nerka*) were mainly planktophagous but consumed fish as well. Table 10 lists the various organisms recovered from the stomachs of salmon. It can be seen that the natural diet of adult salmon is rich in crustaceans and other organisms that contain large amounts of carotenoid pigments.

Needham (*241*) conducted an extensive study of the food consumed by various species of trout. Table 11 lists the classes of food consumed by rainbow trout. Although algae are listed as being present, they occurred in 20 of the 80 stomachs and averaged 36% by volume of the stomach contents of the fish in which they were found. Plant food was not found in the stomachs of any brook trout or brown trout. The diets of brown, brook, and rainbow trout are similar except that rainbow trout also eat algae. Table 12 shows that trout select their food on the basis of availability of a given food, with some exceptions.

Needham studied the food eaten by brook trout during a year. As might be expected, land foods were less plentiful during the winter months (Table 13). It is interesting that larvae, pupae, and nymphs make up such a large proportion of the diet and yet the crustaceans are thought to provide the bulk of trout pigments. None of the insects mentioned by Needham were listed in a recent review (*127*) as containing carotenoids. This does not necessarily mean that these insects do not contain carotenoids but may simply point out our lack of knowledge of insect pigments. The main pigments of insects are the leaf pigments (β-carotene and lutein), although ketocarotenoids have been isolated.

TABLE 10

Relative Importance of Taxonomic Groups Consumed by Juvenile Salmon in Chatham Sound and Adjacent Waters, Early June to Late August 1955[a,b]

Taxonomic group	Pink		Chum		Sockeye		Coho	
	F	V	F	V	F	V	F	V
Chaetognatha	+	+	+	+	—	—	1	+
Cladocera	—	—	—	—	—	—	1	+
Gastropoda	2	+	+	+	2	+	—	—
Polychaeta	—	—	1	+	—	—	—	—
Ostracoda	2	+	3	+	1	1	—	—
Copepoda	48	31	29	19	45	48	7	6
Cirripedia	13	6	5	2	3	1	—	—
Mysidacea	+	+	1	+	—	—	1	+
Cumacea	1	+	2	+	1	+	—	—
Isopoda	1	+	1	+	—	—	1	+
Amphipoda	7	1	8	2	3	3	11	3
Euphausiacea	7	3	7	4	9	1	4	3
Decapoda	14	4	20	5	7	5	14	8
Crustacean remains	—	—	—	—	—	—	4	1
Insecta	8	1	23	5	10	4	19	5
Larvacea	41	40	62	51	25	14	1	+
Fish	2	5	5	4	19	24	52	70
Eggs	1	+	+	+	—	—	—	—
Unidentifiable	18	7	13	4	5	2	8	2
No. stomachs examined	537		410		98		144	
Percentage with food	95		95		91		98	
No. fish measured	470		329		50		144	
Size range of fish measured (mm)	35–114		32–106		57–122		62–147	

[a] From Manzer (*214*). Reproduced by permission of Information Canada.
[b] F, percent occurrence; V, percentage of volume; +, <0.5%.

Algal carotenoids have been reviewed by Goodwin (*128*) and can vary from leaf-type carotenoids to ketocarotenoids depending on the organism and conditions of growth.

Nilsson and Andersson (*245*) studied the food of the allopatric brown trout in the lakes of northern Sweden. Although brown trout prefer *Trichoptera* larvae, the consumption of the small crustaceans *Bythotrephes*, *Eurycereus*, and *Daphnia* was quite high in the summer. Even though crustaceans were consumed by all sizes of trout, the flesh of those weighing less than 100 gm was white; fish heavier than 150 gm were mainly red-fleshed. In lakes where crustaceans are abundant and are found in the stomach contents, the trout are pink to red and are considered to be of the best

TABLE 11

Foods Consumed by 80 Rainbow Trout[a,b]

Class of food	Number found in 80 stomachs	Percentage of total
Mayflies	490	37.1
Caddisflies	247	18.7
Two-winged flies	234	17.8
Beetles	105	7.9
Ants, bees, and wasps	88	6.6
Stoneflies	44	3.3
Moth larvae	17	1.2
Snails	14	1.1
Leafhoppers	13	1.0
Crayfish and scuds	13	1.0
Alderfly larvae	11	0.8
True bugs	11	0.8
Grasshoppers	7	0.5
Fish and salamanders	6	0.5
Algae (*Cladophora*)	+	+
Miscellaneous	23	1.71
	1323	

[a] From Needham (*241*).

[b] Average length 6 inches, ranging from 3 to 12 inches. Most of these were taken in May, June, and July. All were taken from streams, none from lakes.

TABLE 12

Comparison of Available Bottom Foods with Those Consumed by Trout[a]

Food	Available Aquatic foods (%)	Aquatic foods eaten by trout (%)[b]
Mayfly nymphs	36.9	30.1
Caddisfly larvae and pupae	21.3	44.7
Stonefly nymphs	14.7	3.5
Fly larvae and pupae	13.8	15.8
Beetle larvae	7.6	2.8
Crayfish and scuds	3.7	1.1
Miscellaneous	1.9	1.9

[a] From Needham (*241*).

[b] Percentage based on number of each eaten. From stomach examination of 147 trout; 32 rainbow, 6 brown, and 109 eastern brook trout taken in streams of central New York State. Length from 4 to 10 inches.

TABLE 13

Foods Eaten by Eastern Brook Trout Each Month of the Year[a]

Month	Number of trout exam-ined	Food eaten[b] First choice	Food eaten[b] Second choice	Land food (total number eaten)	Water food (total number eaten)
January	13	Caddisfly larvae and pupae	Aquatic fly larvae and pupae	2	158
February	9	Caddisfly larvae and pupae	Mayfly nymphs	37	122
March	4	Caddisfly larvae and pupae	Aquatic fly larvae and pupae	2	48
April	21	Caddisfly larvae and pupae	Aquatic fly larvae and pupae	181	738
May	10	Mayfly nymphs	Adult mayflies	119	111
June	25	Caddisfly larvae and pupae	Mayfly nymphs	122	165
July	62	Caddisfly larvae and pupae	Mayfly nymphs	304	591
August	40	Mayfly nymphs	Adult flies	302	276
September	18	Leafhoppers	Grasshoppers	89	50
October	18	Leafhoppers	Land beetles	94	53
November	19	Caddisfly larvae and pupae	Land beetles	31	46
December	12	Springtails	Caddisfly larvae and pupae	97	340
	251			1380	2698

[a] From Needham (*241*).
[b] Choice determined by numbers consumed.

quality (*279*). Tusa (*306*) investigated the food consumed by brown trout in a creek over a 2-year period. The most abundant aquatic organisms were primarily mayfly larvae and caddis worms. Crustaceans represented only a small part of the diet.

2. Carotenoids in Salmonid Feed

Numerous diets for trout and salmon have appeared in the literature, and fish feeds are available from companies all over the world. Generally a company does not produce a set formulation but alters it, depending on the price and availability of ingredients, keeping the analysis relatively constant. Salmonid nutrition is beyond the scope of this chapter except as it relates to pigmentation. "Trout Feeds and Feeding," by Phillips *et al.* (*250*), "The Nutrition of Fish," by Pearson (*248*), and "Topics in Nutrition," in *Fish in Research,* edited by Neuhaus and Halver (*242*), treat the subject in a practical manner. The Tunison Laboratory of Fish Nutrition,

Cortland, New York, has published a long series of research bulletins and is an excellent source of nutrition information on salmonids. The Oregon test diet (*198*) has been used as a vehicle for pigment incorporation studies, and its development is contained in a series of progress reports (*160*). The Oregon moist pellet consists of a meal mix: 22% herring meal, 22% cottonseed meal, 4% shrimp meal, 3% wheat germ meal, 3% corn distillers' dried solubles, and a vitamin premix. The wet mix consists of two fish products, including salmon and tuna viscera, turbot, herring, dogfish, and hake.

If the fresh fish in the moist pellet contains the enzyme lipoxygenase, decoloration of the diet may result. That is, the astaxanthin from shrimp, crab, etc., may be lost due to the activity of this enzyme.

a. Crustacean Meal. In terms of dollars, shrimp is the most valuable fishery in the United States. Current domestic landings and imported shrimp amount to more than a 0.5 billion lb/year (*233*). In processing shrimp, the head and the hard carapace are removed during semimechanized peeling operations. Waste material accounts for approximately 70% of the whole shrimp, and thus the use of shrimp by-products becomes an economic as well as an environmental pollution problem. The value of shrimp meal as a feeding supplement as well as a pigment source is affected by processing and storage temperatures, light, oxygen, moisture, and possible mineral imbalances in the fish.

Studies in Denmark described by Meyers and Rutledge (*233*) showed that vacuum-dried meal or fresh raw shrimp impart a good flavor and color to pond-raised trout. Improved color was not obtained when shrimp meal, dried at high temperatures, was used.

Lambertsen and Braekkan (*195*) determined the astaxanthin content of industrial, foreign (to Norway), and vacuum-dried shrimp meals. The results (Table 14A)

TABLE 14A

Astaxanthin Content of Some Shrimp Meals[a]

Sample	Astaxanthin (μg/gm)	% as diester
Industrial meals		
A	8.9	82.0
B	3.9	82.8
C	0.0	
Unknown, of foreign origin		
A	24.5	84.9
B	16.3	89.9
C	12.8	86.8
Vacuum-dried		
A	76.2	77.0

[a] From Lambertsen and Braekkan (*195*).

TABLE 14B

Composition of Carotenoids in Shrimp Meal before and after Drying

Carotenoid	Before drying (μg/gm dry basis)	After tray-drying (μg/gm dry basis)
3,3′-Dihydroxy-ϵ-carotene	1.36	
Echinenone	0.25	
Isocryptoxanthin	0.38	
Canthaxanthin	0.39	
4-Keto-4′-hydroxy-β-carotene	0.74	
Dihydroxypiradixanthin (?)	4.66	
Lutein	Trace	
Zeaxanthin	Trace	
Astaxanthin ester	66.1	10.3
Astacene	55.4	6.26
Astaxanthin	7.19	Trace

showed that the astaxanthin content is generally low or even absent in commercial shrimp meals. The vacuum-dried meal contained three times more astaxanthin than the best industrial sample. Table 14B shows the analysis of shrimp protein before and after tray-drying. It can be seen that the minor carotenoids are lost entirely and that astaxanthin, its ester, and astacene are all greatly reduced.

Studies by Spinelli (*281*) show the high losses experienced in drying pelagic red crab (*Pleuroncodes planipes*) meal. These crabs are of interest because of their intense red color (*315*). Spinelli *et al.* (*280*) prepared a pelagic red crab meal and added it at concentrations of 10 and 25% of an Oregon moist pellet diet. The diet was fed to 7-month-old rainbow trout for 60 days. The fish fed on both diets as well as a 25% shrimp diet had a hue similar to the reddish cast of ocean-caught salmon. In this study the muscle of fish fed canthaxanthin had an orange cast.

Saito and Regier (*262*) fed diets containing 20 and 30% shrimp waste and 20% crab waste (vacuum-dried below 50°C for 48 hr). After 8 weeks on the 30% diet, brook trout were similar in color to those fed for 12 weeks on 20% shrimp meal. The crab-fed trout did not show a rapid coloration, and the control fish fed a commercial diet were only slightly colored. Peterson *et al.* (*249*) fed raw crayfish and a crayfish extract to rainbow and brown trout. Although the amount of pigment was adjusted to be equal, the extract provided better coloration. Tables 15 and 16 show the effect of feeding the crayfish extract to brown and rainbow trout, respectively.

Steel (*285*) fed rainbow trout a diet containing 15% shrimp waste meal over a 24-week period. Sensory evaluations were performed on the trout at the end of the feeding trials. The meal-fed fish received a significantly higher score for color as well as for flavor. In a 34-week study with several diets, shrimp waste meal or a pigment extract was most effective in achieving muscular pigmentation. The most rapid pigmentation was obtained with the shrimp meal extract.

TABLE 15

Effect of Crayfish Extract on the Xanthophyll Content of Brown Trout[a,b]

Fish	Fins	Skin	Flesh	Total	Weeks fed	Preparation
A	4.67	3.87	0	0.48	2	Extract
B	5.88	3.74	0.70	0.44	2	Control
C	24.00	8.63	0.70	1.50	4	Extract
D	5.13	2.44	0	0.34	4	Control

[a] From Peterson *et al.* (*249*).
[b] Data expressed as grams pigment per gram wet weight tissue.

Chinook and coho salmon and rainbow trout were fed formulations containing as much as 40% shrimp scrap over a 22-week period (*35*). It was shown that levels of less than 5% produced very little pigmentation (*285*), whereas levels of greater than 40% were impractical from both a nutritional and manufacturing point of view (*35*).

Ellis (*97*) described the feeding of coho salmon for 8–12 weeks on a semimoist pellet containing 20% shrimp waste. At the end of 12 weeks the pigmentation was described as similar to the color of wild salmon.

Spinelli (*282*) described a dry euphausid feedstuff that was added to other test diets. This 20% part consisted of 6 parts nondried *Euphausia pacifica,* 1 part soy oil (with 0.025% ethoxyquin), and 4 parts dry shrimp waste. The total mixture was dried under vacuum (to 6%) and passed through a hammer mill. Good stability of the carotenoids was obtained when the oil and antioxidant were added. Rainbow trout fed (60 days) either the dry diet or euphausids in an Oregon moist pellet gave good organoleptic scores and good pigmentation. The fish fed the dry diet with 15% oil pigmented at a faster rate than those with 10% oil.

Kuo *et al.* (*193*) fed rainbow trout an artificial diet containing 20% freeze-dried crab waste meal from a plant processing Atlantic red crab, *Geryon quinquedens.*

TABLE 16

Effect of Crayfish Extract on the Xanthophyll Content of Rainbow Trout[a,b]

Fish	Fins	Skin	Flesh	Total tissue	Weeks fed	Preparation
A	6.50	3.13	0.43	0.67	2	Extract
B	3.33	3.10	0	0.43	2	Control
C	24.44	5.80	0.99	1.58	4	Extract
D	5.83	2.47	0	0.31	4	Control
E	48.63	14.83	1.36	2.61	7	Extract
F	3.04	3.09	0	0.09	7	Control

[a] From Peterson *et al.* (*249*).
[b] Data expressed as grams pigment per gram wet weight tissue.

Fish fed the crab meal diet grew as fast as the controls, but there was no significant coloring effect over the controls during 15- and 23-week trials. On the other hand, fish fed a carotenoid extract from the crab meal showed an excellent red color after 6 weeks of feeding. It was concluded that *G. quinquedens* meal, even under the best of processing conditions, was not a good pigment source for trout. This is in contrast to shrimp meals and pelagic crab meals, from which trout are able to ''extract'' the pigment.

The limiting factor in the use of crustacean meals as protein supplements has been the large proportion of exoskeleton material in the meal. Although the protein content on a dry basis varies from 25 to 40%, the high levels of calcium have a deleterious effect on the overall nutrition (*232*). Red crab waste (*193*) contained 27% protein but also contained 36% ash, of which 30% was calcium.

Rutledge (*259*) described a decalcification procedure for crustacean meals based on a differential screening effect between the protein and the skeletal material. The protein level of the meals nearly doubled in this procedure.

Spinelli (Fig. 6) described a separation scheme for *Pleuroncodes planipes* in which the protein is solubilized with NaOH, separated from the chitinous residue, and then precipitated at its isoelectric point with HCl. It is anticipated that the astaxanthin content of this preparation might be low.

A similar procedure was used by Marine Commodites International, Brownsville, Texas, for the separation of the shrimp protein from the chitinous residue. Base was not used, and the amount of acetic acid used was carefully controlled. The resulting

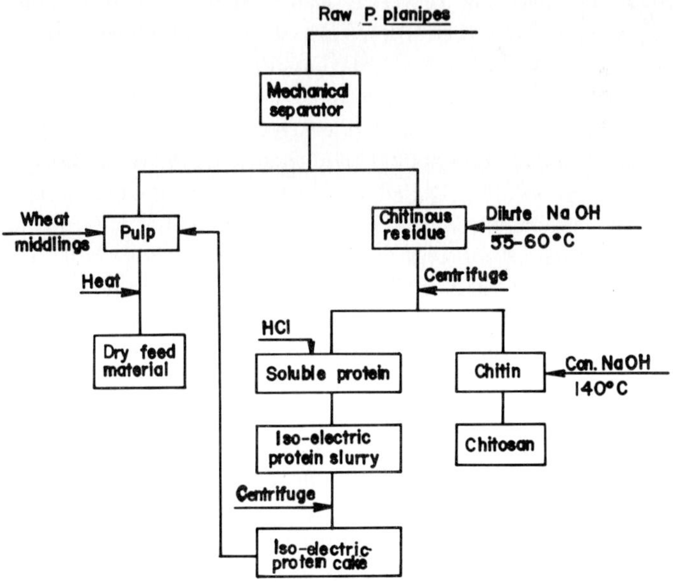

Fig. 6. Processing of red crab into a feed material and chitinous residue (*284*).

protein cake was deep red. This protein was incorporated into a fish diet, and coloration of trout was achieved (*276*).

b. Carotenoid Extracts. The use of shrimp (*285*) and crayfish (*249*) extracts is described above. Extracts of paprika pepper have been added as a supplement to many animal diets including fish. The main pigment in paprika is capsanthin, which has never been isolated from wild fish. De la Mar and Francis (*84*) isolated 54 and 37 pigments from bleached and unbleached paprika, respectively. The hydrocarbon, monoal, and diol fractions amounted to 25% of the total and included β-carotene, cryptoxanthin, and zeaxanthin. Peterson *et al.* (*249*) added paprika pepper extract to a commercial diet at the rate of 107 mg per pound of trout food. Coloration was detected at 2 weeks and was fairly prominent at 4 weeks. The color was described as natural except for a small amount of an undesirable yellow. Lutein obtained from dried marigold petals was fed to brook trout at the rate of 200 mg per pound of feed and produced a "natural" yellow color in 2–4 weeks. Lutein is a common leaf and algal pigment and is probably responsible for the yellow color in pond-reared trout.

Neamtu *et al.* (*238*) fed rainbow trout diets containing 4–5 mg % carotenoid pigment extracts from the floral parts of *Aesculus hippocastanum* and *A. parviflora*. The *Aesculus* pollen contained pentacyclic carotenes and xanthophylls, and these were incorporated into the flesh of the trout. The incorporation of pigment from *A. hippocastanum* was low and the extract was toxic. Fish fed *A. parviflora* were slightly pink at 4 weeks and slightly smaller than the controls.

Lambertsen and Braekkan (*195*) compared the astaxanthin content of various crustacean products (Table 17). The results showed that prawn, krill, and calanus oil are particularly good sources of astaxanthin.

Spinelli and Mahnken (Fig. 7) described a procedure in which wet shrimp or pelagic crab waste is extracted with hot soybean oil. The resulting oil is highly pigmented and contains high-quality pigments. The oil was rapidly incorporated into the flesh of test salmon. The advantages of the procedure are that the destructive drying step is eliminated and no recovery of solvent is needed. The oil can be fed either as a pigmented oil or as part of the protein solids (*283*).

Extracts containing 155 mg per 100 gm oil were made from red crab and incorporated into Oregon moist pellet diets at 3, 6, and 9 mg carotenoid per 100 gm feed. Rainbow trout were fed for 120 days on the three diets. Diets containing either 6 or 9 mg were judged to give good to excellent pigmentation (*284*).

c. Pure Carotenoid Preparations. The literature contains conflicting reports on the use of crystalline carotenoids. Steven (*287*) injected into earthworms carotenoids dissolved in *Arachis* oil and found that lutein, β-carotene, and astacene were not deposited in the skin or fins of trout fed the worm. Steel (*285*) fed canthaxanthin in a crystalline form and obtained no incorporation, whereas a water-dispersible canthaxanthin produced trout rated significantly higher in color

TABLE 17

Astaxanthin Content of Different Crustacean Products[a]

Sample	Astaxanthin (μg/gm)	% astaxanthin as Diester	Monoester	Free
Shrimp (*Pasiphaea* sp.)	14.9	69	17	14
Krill (euphausiids), mixed samples				
A	77.4	48	49	3
B	22.4	33	36	31
Calanus finniarchicus[b]	46.5	37	38	25
Stomach content[c] of herring, mainly *Calanus*	90.8	55	26	19
Lobster shells, boiled	35.3	34	26	40
Prawn offal, boiled	66.0	67	28	5
Prawn oil (*Pandalus borealis*)	1095	79	18	3
Krill oil (euphausiids)	727	51	43	6
Calanus oil (*C. finmarchicus*)	520	27	33	40

[a] From Lambertsen and Braekkan (*195*).
[b] 3.5% fat.
[c] Represents 4% of whole herring.

than control fish. In order to obtain a water-dispersible form, the carotenoid can be incorporated into a 40- to 100-mesh beadlet containing 10% canthaxanthin in a matrix of gelatin vegetable oil, sugar, and starch (Roxanthin Red 10).

Several studies have shown that canthaxanthin can rapidly be taken up by trout and salmon in flesh and skin. Schmidt and Baker (*269*) fed 190 mg/kg of a beadlet form of stabilized canthaxanthin feed to rainbow trout for 31 weeks and 450 mg/kg of feed to cutthroat trout for 11 weeks [see Bauernfeind (*12*)]. At the end of the test periods the fish had a flesh color similar to that of red coho salmon. These authors reported that the induced pigmentation may be more stable to processing than the natural pigmentation of coho salmon.

Lambertsen and Braekkan (*195*), Steel (*285*), Saito and Regier (*262*), Deufel (*93*), Carlsson (*26*), Malak *et al.* (*213*), and Spinelli *et al.* (*280*) reported good coloration of salmonids with the feeding of canthaxanin.

Duefel (*93*) fed rainbow trout 40 mg/kg of canthaxanthin for 24 weeks and described the trout as having a salmonlike coloration in both the flesh and skin (Table 18). He noted in this report as well as in a subsequent paper (*94*) that

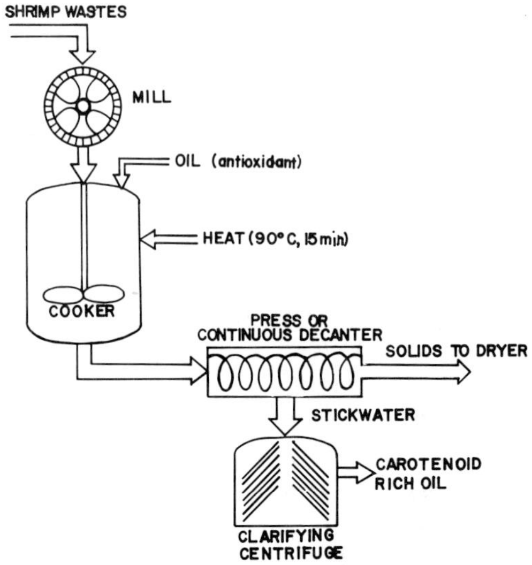

Fig. 7. Preparation of carotenoid extract from shrimp wastes (*284*).

carotenoids, especially ketocarotenoids, provide beneficial effects on spawning (fertilization, sexual maturity, and embryonic development).

The hue of the flesh of salmonids is slightly different when astaxanthin or canthaxanthin is fed (*262*). Canthaxanin gives a slightly more orange color than salmon color. A number of authors have reported that canthaxanin tends to fade more than expected on cooking (*282*), although Carlsson (*26*) noted that an acceptable color was obtained after cooking if the feeding program was greater than 2 months. He

TABLE 18

Tissue Carotenoid Content of Control and Canthaxanthin-fed Fish[a]

Feeding period (weeks)	Carotenoid content (mg/100 gm)	
	Canthaxanthin	Total carotenoids
Control series		
8	0.0	0.067
16	Trace	0.078
24	Trace	0.078
Canthaxanthin series		
8	0.034	0.126
16	0.080	0.209
24	0.124	0.238

[a] From Deufel (*93*); see also Bauernfeind (*12*).

further noted that the color intensity decreased rapidly if the feeding program was interrupted. Tess (Stavanger, Norway) adds canthaxanthin and red capelin oil to its trout and salmon feeds no. 7 and 8. This company recommends a 2-month feeding period for trout and 6–8 months for salmon (4–5 kg). A shorter feeding results in a mottled appearance.

Zeaxanthin has been fed to a number of fish and crustaceans in an effort to establish its place in the biosynthesis of astaxanthin. It appears that zeaxanthin, although a yellow pigment, is converted to astaxanthin, and thus an intensification of the red colors results from its feeding. Tanaka *et al.* (*295*) fed pure zeaxanthin to prawn (*Penaeus japonicus*) and found that it was converted to astaxanthin. Zeaxanthin is also converted to astaxanthin in fancy red carp, golden yellow carp, and goldfish, and an intensification of color was noted on feeding zeaxanthin (*144, 292*).

Torrissen and Braekkan (*301*) and Simpson and Kamata (*275*) reported on the feeding of pure preparations of astaxanthin, its mono- and diester, and astacene to rainbow trout. These reports taken together show that the skin preferentially takes up the esters and deposits all fed forms as the esters. Astaxanthin is found only in the flesh as the free form, and astaxanthin (*301*) is the most efficient form for pigmenting the flesh. Regardless of the form fed the fish, either esterified or deesterfied, the esters are deposited in the skin and the free form deposited in the flesh. Since the reports by Steven (*286,287*), astacene has been considered an artifact of the isolation procedure and a breakdown product of astaxanthin. When astacene was prepared by chromatography rather than saponification and fed to trout, it was found to be deposited in the flesh and skin at a rate similar to astaxanthin. A compound tentatively identified as "astacene ester" was also obtained. Other authors also have reported the isolation of astacene from aquatic animals [e.g., herring and sprat in one sample, 42% (*59*); crab, 16% (*62*); and brine shrimp, trace (*75*)].

Nakazoe and Hata (*235*) extracted from the internal organs of trout, carp, and goldfish enzyme fractions that hydrolyzed the monoester of astaxanthin to free astaxanthin. Astacene was not reported as a by-product even though the pH optimum was given as 8–12.

The use of β-carotene in trout has not been very successful (*249*), and lutein results in yellow coloration.

d. Other Ingredients Containing Carotenoids. A number of items found in commercial diets contain carotenoids. Pigments may be found in fish meals made from viscera of fish or from "trash" fish.

Salmon oil or eggs have been used when available. Plant preparations from seeds, including cottonseed, oats, wheat, and corn, and leaf preparations, such as alfalfa and kelp, all contain carotenoids. However, unless an ingredient contained xanthophylls, it would not be expected to be of significant value as a pigment source.

Some red-colored fish oils were found to be a good source of astaxanthin (*195*).

These included oil samples from capelin, mackerel, ocean perch, and polar cod. One sample of the four tested from capelin contained 94.3 μg/gm of astaxanthin in one sample but as little as 5.7 μg/gm in another. The astaxanthin contents of crustacean oils are given in Table 17. These red fish and crustacean oils should be good sources of astaxanthin for shrimp and salmonids.

Joseph and Williams (*165*) extracted shrimp heads (*Penaeus setiferus*) and impregnated a commercial feed with the shrimp oil. The augmented diet was fed to the Malaysian prawn *Macrobrachium rosenbergii*. It was found that the prawn on the augmented diet had a higher survival rate, mean weight, biomass, total lipid and total pigment content, and a better feed conversion efficiency. The total pigment levels in the experimental prawn were 15 times higher than in the control, and the experimental animals were visibly more red than the controls. Thus, an oil extract of crustacean waste could supply pigments as well as needed growth factors for crustaceans and salmonids.

Pigmented preparations from corn gluten, alfalfa, and *Spirulina* (blue-green alga) were found to increase the body astaxanthin of prawn (*Penaeus japonicus*) to various degrees (*295*).

The green alga *Haematococcus* contains astaxanthin, and broken cells of this alga were fed to sea bream. Astaxanthin was deposited in the skin and flesh (*236*). *Scenedesmus* and *Spirulina* cells raised the xanthophyll level but did not increase the astaxanthin level.

Andrewes *et al.* (*5*) reported the isolation of astaxanthin from the fermenting yeast *Phaffia rhodozyma*. Astaxanthin occurred at levels greater than 80%, with small amounts of β-carotene, echinenone, 3-hydroxyechinenone, and phoenicoxanthin also present. The chiroptical property of yeast astaxanthin was 3*R*,3′*R* (*6*) rather than the 3*S*,3′*S* reported for *Haematococcus pluvialis,* the lobster *Homarus gammarus* (*4*), and the spider mite *Schizonobia sycophanta* (*309*).

Johnson *et al.* (*164*) reported on the use of broken *Phaffia rhodozyma* cells at a level of 15% in the diets of lobster and trout. Pigmentation of trout was reported (10 μg/gm in 43 days), but levels lower than those obtained with the control brine shrimp diet were found for lobster.

It would appear that a major problem in the industrialization of algae (*Haematococcus*) or red yeast (*Phaffia*) is the method of cell rupture. The French press or Braun tissue disintegrator are good for small samples but are not applicable to large samples. The drying of cells would also cause them to rupture but would probably also result in pigment loss. One application of enzyme disintegration (*236*) was not successful.

The flowers of *Adonis aestivalis* contain a large amount of the esters of astaxanthin (*239*). This plant may be a potential source of astaxanthin since the petal can be dried without significant loss of the pigment. It remains to be seen whether this "weed" can be cultivated and harvested in a non-labor-intensive operation. It also must be determined whether a low-alkaloid plant results from cultivation.

VI. CONCLUSIONS

Aquaculture can be defined as the cultivation and harvest of fresh and marine aquatic animals. It more closely resembles agriculture than it does commercial fishing. Fish farming, like farming, is based on the assumption that proper management of conditions should yield more and better products.

Unlike commercial fishing, aquaculture does not rely on common property such as the sea but on controlled ponds, bays, or specific water systems. Aquaculture fish, even though raised under different conditions, must nevertheless compete in price and quality characteristics with the caught product. With few exceptions, the aquaculture product has not set its own standard of identity but is compared to the fishery product.

Since the purpose of aquaculture is to prepare a product for food, the characteristics of food quality must be considered to be as important as the biology of the animal, the system, and the disease vectors. The term "flavor" is very complex and involves more than a strict definition based on the taste and aroma of the product. Certainly texture, size and shape, and appearance are also involved in consumer acceptance [see Spinelli (*281*), Meyers and Rutledge (*232*), and Simpson (*274*)].

Ostrander *et al.* (*247*) showed that color was listed most frequently, by itself or in conjunction with other factors such as size and shape, as a distinguishing characteristic between salmon and trout. Indeed, without such clues as color and size, the trained panel could not distinguish salmon from trout. It is well known that the price of salmon is directly related to the intensity of hue. It is not difficult to find many other examples of color being related to consumer acceptance.

Thus, aquaculturalists must feed the correct pigment in order to obtain an animal that meets consumers' expectations. It is also necessary to avoid adding with the pigment adventitious compounds that could adversely affect the flavor or safety of the aquaculture product.

The kinds of carotenoids added to the diet of aquatic animals should be selected according to the species because of differing biosynthetic capabilities and expected final pigment content. Katayama *et al.* (*177*) proposed three groups of aquatic organisms of economic interest. Specific recommendations within those groups can be made:

Group 1. Goldfish, red carp, and fancy red carp: Color can be improved by adding a number of pure carotenoids, such as 1–4% zeaxanthin, lutein, canthaxanthin, and astaxanthin. *Spirulina* (zeaxanthin main carotenoid), marigold flowers (lutein), and other plant preparations can be fed.

Group 2. Prawn, crab, lobster, and other crustaceans in which astaxanthin is the main carotenoid. Color can be improved by feeding 1–3% β-carotene, astaxanthin, canthaxanthin, zeaxanthin, or *Spirulina*.

Group 3. Sea breams and salmonids: A tunaxanthin–lutein–zeaxanthin–astaxanthin mixture (2 : 1 : 1 : 6) would have to be fed to sea bream to equal the pigmentation of wild sea bream. Trout and salmon would require 20 mg per 100 gm of astaxanthin for 2–4 months and 40 mg per 100 gm of canthaxanthin for a similar period of time.

These are suggested as guides since age and species of fish, temperature and other environmental factors, and fat content of the diet would all have to be considered.

ACKNOWLEDGMENTS

This chapter was first written while one of us (K. L. S.) was on sabbatical leave at the Department of Biochemistry, University of Liverpool, England.

The Sea Grant Program at the University of Rhode Island, the Department of Biochemistry, University of Liverpool, and the F. Hoffmann-La Roche Company, Basel, contributed support in the gathering of data for the chapter, and their help is gratefully acknowledged.

Rhode Island Agricultural Experiment Station Contribution Number 1972.

REFERENCES

1. Aguilar-Martinez, M., and Liaaen-Jensen, S. (1974). *Acta Chem. Scand., Ser. B* **28,** 1247–1248.
2. Allen, J. A. (1960). *J. Mar. Biol. Assos. U.K.* **39,** 481–508.
3. Amano, K., Tsukuda, N., Bito, M., Sasayama, S., Yamanaka, H., and Fujii, Y. (1968). *Tokaiku Suisan Kenkyusho Kenkyu Hokoku* **55,** 237–244.
4. Andrewes, A. G., Borch, G., Liaaen-Jensen, S., and Snatzke, G. (1974). *Acta Chem. Scand., Ser. B* **28**, 730–736.
5. Andrewes, A. G., Phaff, H. J., and Starr, M. P. (1976). *Phytochemistry* **15,** 1003–1007.
6. Andrewes, A. G., and Starr, M. P. (1976). *Phytochemistry* **15,** 1009–1011.
7. Anonymous (1971). *Am. Fish. U.S. Trout News* **16** (1), 14.
8. Avault, J. W., Jr., de la Bretonne, L., Jr., and Jaspers, E. J. (1970). *Am. Fish Farmer* **1** (10), 8–12, 26–27.
9. Bailey, B. E. (1937). *J. Biol. Board Can.* **3,** 469–472.
10. Bardach, J. E. (1968). *Science* **161,** 1098–1106.
11. Barua, A. B., Singh, H. T., and Das, R. C. (1973). *Br. J. Nutr.* **30,** 1–12.
12. Bauernfeind, J. C. (1976). *Prog. Fish Cult.* **38,** 180–183.
13. Bilaus, C., Nicoara, E., Bodea, C., and Petrescu, I. (1970). *Stud. Cercet. Biochim.* **13,** 15–18.
14. Bilaus, C., Nicoara, E., Bodea, C., and Petrescu, I. (1971). *Stud, Cercet. Biochim.* **14,** 123–125.
15. Bodea, C., Nicoara, E., Illyes, G., and Serban, M. (1965). *Rev. Roum. Biochim.* **2,** 205–211.
16. Bodea, C., Nicoara, E., Illyes, G., and Suteu, M. (1966). *Vol. Omagial Acad. R. Ripan (Bucharest)* pp. 153–156.
17. Bretthauer, R. (1967). *Z. Naturforsch. Teil B* **22,** 858–864.
18. Britton, G., and Goodwin, T. W. (1971). *In* "Vitamins and Coenzymes," Part C (D. B. McCormick and L. D. Wright, eds.), Methods in Enzymology, Vol. 18, pp. 654–701. Academic Press, New York.
19. Brush, A. H., and Reisman, H. M. (1965). *Comp. Biochem. Physiol.* **14,** 121–125.
20. Buchecker, R., Eugster, H., and Weber, A. (1968). *Helv. Chim. Acta* **61,** 1962–1968.
21. Buchwald, M., and Jencks, W. P. (1968). *Biochemistry* **7,** 834–844.
22. Buchwald, M., and Jencks, W. P. (1968). *Biochemistry* **7,** 844–859.

23. Bullock, E., and Dawson, C. J. (1970). *Comp. Biochem. Physiol.* **34,** 799–804.
24. Campbell, S. A. (1969). *Comp. Biochem. Physiol.* **30,** 803–812.
25. Campbell, S. A., Mallams, A. K., Waight, E. S., Weedon, B. C. L., Barbier, M., Lederer, E., and Salaque, A. (1967). *Chem. Commun.* pp. 941–942.
26. Carlsson, H. (1978). *FAO Symp. Finfish Nutr. Feed Technol., Hamburg.*
27. Castillo, R., and Lenel, R. (1978). *Comp. Biochem. Physiol. B* **59,** 67–73.
28. Ceccaldi, H. J., and Allemand, B. H. (1964). *Recl. Trav. Stn. Mar. Endoume-Marseille Fasc. Hors Ser. Suppl.* pp. 59–64.
29. Ceccaldi, H. J., and Zagalsky, P. F. (1967). *Comp. Biochem. Physiol.* **21,** 435–438.
30. Chang, C. W. J., and Moore, J. C. (1971). *J. Chem. Educ.* **48,** 408–409.
31. Cheesman, D. F., Lee, W. L., and Zagalsky, P. F. (1967). *Biol. Rev. Cambridge Philos. Soc.* **42,** 131–160.
32. Cheesman, D. F., and Prebble, J. (1966). *Comp. Biochem. Physiol.* **17,** 929–935.
33. Chichester, C. O., and Nakayama, T. O. M. (1965). *In* "Chemistry and Biochemistry of Plant Pigments" (T. W. Goodwin, ed.), pp. 439–457. Academic Press, New York.
34. Choubert, G., Jr., and Luquet, P. (1975). *Ann. Hydrobiol.* **6,** 123–130.
35. Crawford, D. L. (1971). Personal communication.
36. Crozier, G. F. (1967). *Am. Zool.* **7,** 202.
37. Crozier, G. F. (1967). *Comp. Biochem. Physiol.* **23,** 179–184.
38. Crozier, G. F. (1968). *J. Exp. Mar. Biol. Ecol.* **4,** 1–8.
39. Crozier, G. F. (1969). *J. Fish Res. Board Can.* **27,** 973–975.
40. Crozier, G. F. (1974). Pigments of fishes. *In* "Oeuterostomians, Cyclostomes and Fishes" (M. Florkin and B. T. Scheer, eds.), Chemical Zoology, Vol. 8, pp. 509–521. Academic Press, New York.
41. Crozier, G. F., and Wilkie, D. W. (1966). *Comp. Biochem. Physiol.* **18,** 801–804.
42. Czeczuga, B. (1970). *Mar. Biol.* **5,** 141–144.
43. Czeczuga, B. (1970). *Mar. Biol.* **6,** 117–120.
44. Czeczuga, B. (1971). *Acta Parasitol. Pol.* **19,** 185–194.
45. Czeczuga, B. (1971). *Mar. Biol.* **10,** 254–257.
46. Czeczuga, B. (1971). *Comp. Biochem. Physiol. B* **39,** 945–953.
47. Czeczuga, B. (1971). *Comp. Biochem. Physiol. B* **40,** 47–52.
48. Czeczuga, B. (1971). *Hydrobiologia* **37,** 301–307.
49. Czeczuga, B. (1971). *Hydrobiologia* **38,** 185–191.
50. Czeczuga, B. (1973). *Mar. Biol.* **19,** 206–209.
51. Czeczuga, B. (1973). *Bull. Acad. Pol. Sci., Ser. Sci. Biol.* **21,** 365–368.
52. Czeczuga, B. (1973). *Mar. Biol.* **21,** 139–143.
53. Czeczuga, B. (1973). *Hydrobiologia* **41,** 113–125.
54. Czeczuga, B. (1974). *Bull. Acad. Pol. Sci., Ser. Sci. Biol.* **22,** 211–214.
55. Czeczuga, B. (1975). *Hydrobiologia* **46,** 223–239.
56. Czeczuga, B. (1976). *Bull. Acad. Pol. Sci., Ser. Sci. Biol.* **24,** 53–56.
57. Czeczuga, B. (1976). *Bull. Acad. Pol. Sci., Ser. Sci. Biol.* **24,** 349–353.
58. Czeczuga, B. (1976). *Bull. Acad. Pol. Sci., Ser. Sci. Biol.* **24,** 175–180.
59. Czeczuga, B. (1976). *Pol. Arch. Hydrobiol.* **23,** 123–129.
60. Czeczuga, B. (1977). *Acta Hydrobiol.* **19,** 183–190.
61. Czeczuga, B. (1977). *Folia Histochem. Cytochem.* **14,** 343–346.
62. Czeczuga, B. (1975). *Comp. Biochem. Physiol. B* **51,** 309–311.
63. Czeczuga, B., and Czerpak, R. (1966). *Comp. Biochem. Physiol.* **17,** 523–534.
64. Czeczuga, B., and Czerpak, R. (1968). *Comp. Biochem. Physiol.* **25,** 547–552.
65. Czeczuga, B., and Czerpak, R. (1968). *Comp. Biochem. Physiol.* **26,** 101–110.
66. Czeczuga, B., and Czerpak, R. (1968). *Eur. J. Biochem.* **5,** 429–432.
67. Czeczuga, B., and Czerpak, R. (1969). *Hydrobiologia* **33,** 497–506.
68. Czeczuga, B., and Czerpak, R. (1969). *Mar. Biol.* **4,** 24–27.

69. Czeczuga, B., and Czerpak, R. (1970). *Hydrobiologia* **35,** 163–173.
70. Czeczuga, B., and Czerpak, R. (1970). *Int. Rev. Gesamten Hydrobiol.* **55,** 213–220.
71. Czeczuga, B., and Czerpak, R. (1976). *Acta Hydrobiol.* **18,** 1–21.
72. Czeczuga, B., and Ktyszejko, B. (1977). *Bull. Acad. Pol. Sci., Ser. Sci. Biol.* **43,** 767–772.
73. Czeczuga, B., and Skalski, A. (1973). *Hydrobiologia* **42,** 355–362.
74. Czerpak, R., and Czeczuga, B. (1969). *Mar. Biol.* **3,** 122–125.
75. Czygan, F.-C. (1966). *Z. Naturforsch. Teil B* **21,** 801–805.
76. Czygan, F.-C. (1968). *Arch. Microbiol.* **61,** 81–102.
77. Cygan, F.-C. (1968). *Flora (Jena), Abt. A* **159,** 339–345.
78. Czygan, F.-C. (1968). *Z. Naturforsch. Teil B* **23,** 1367–1368.
79. Czygan, F.-C., and Seefried, H. (1970). *Z. Naturforsch., Teil B* **25,** 761–762.
80. Dales, R. P. (1962). *J. Mar. Biol. Assos. U.K.* **42,** 259–274.
81. Davies, B. H. (1965). *In* "Chemistry and Biochemistry of Plant Pigments" (T. W. Goodwin, ed.), pp. 489–532. Academic Press, New York.
82. Davies, B. H., Hsu, W.-J., and Chichester, C. O. (1965). *Biochem. J.* **94,** 26P.
83. Davies, B. H., Hsu, W.-J., and Chichester, C. O. (1970). *Comp. Biochem. Physiol.* **33,** 601–615.
84. De la Mar, R. R., and Francis, F. J. (1969). *J. Food Sci.* **34,** 287–290.
85. De Nicola, M. (1954). *Biochem. J.* **56,** 555–558.
86. De Nicola, M. (1956). *Exp. Cell Res.* **10,** 441–446.
87. De Nicola, M. (1958). *Ric. Sci.* **28,** 2076–2083.
88. De Nicola, M. (1959). *Ric. Sci.* **29,** 78–84.
89. De Nicola, M. G. (1959). *Boll. Sedute Accad. Gioenia Sci. Nat. Catania* **5,** 201–214; *C. A.* **55,** 12684 (1961).
90. De Nicola, M., and Goodwin, T. W. (1954). *Pubbl. Staz. Zool. Napoli* **25,** 145–160.
91. De Nicola, M., and Goodwin, T. W. (1954). *Exp. Cell Res.* **7,** 23–31.
92. Deshimaru, O., and Shigeno, K. (1972). *Aquaculture* **1,** 115–133.
93. Deufel, J. (1965). *Arch. Fischereiwiss.* **16,** 125–132.
94. Deufel, J. (1975). *Schweiz. Z. Hydrol.* **37,** 244–248.
95. Eimhjellen, K. E. (1967). *Acta Chem. Scand.* **21,** 2280–2303.
96. Eldred, B., Ingle, R. M., Woodburn, K. D., Hutton, R. F., and Jones, H. (1961). *Fla. State Board Conserv., Prof. Pap., Ser 3* pp. 1–139.
97. Ellis, J. N. (1979). *In* "Proceedings of World Symposium on Finish Nutrition and Fish feed Technology" (J. E. Halver and K. Tiews eds.), Vol. 2, pp. 353–364. Heenemann Verlagsgesellschaft M5H, Berlin.
98. Evelyn, T. P. T. (1967). *J. Fish. Res. Board Can.* **24,** 2195–2199.
99. Fisher, L. R. (1964). "Vitamin A, Carotenoids and Total Lipids in the Livers of some Elasmobranches," pp. 1–18. Mar. Res. Ser. Scotl. Home Dep., Edinburgh, U.K.
100. Flint, L. H. (1956-). *Proc. La. Acad. Sci.* **19,** 11–14.
101. Forster, G. R. (1951). *J. Mar. Biol. Assoc. U.K.* **30,** 333–360.
102. Fox, D. L. (1953). "Animal Biochromes and Structural Colours." Cambridge Univ. Press, London and New York.
103. Fox, D. L. (1957). *In* "Physiology of Fishes" (M. E. Brown, ed.), Vol. 2, pp. 367–385. Academic Press, New York.
104. Fox, D. L. (1966). *In* "Physiology of Mollusca" (K. M. Wilbur and C. M. Yonge, eds.), Vol. 2, pp. 249–274. Academic Press, New York.
105. Fox, D. L. (1972). *Comp. Biochem. Physiol. B* **43,** 919–927.
106. Fox, D. L., and Crozier, G. F. (1965). *Science* **150,** 771–773.
107. Fox, D. L., Crozier, G. F., and Smith, V. E. (1967). *Comp. Biochem. Physiol.* **22,** 177–188.
108. Fox, D. L., and Hopkins, T. S. (1966). *In* "Physiology of Echinodermata" (R. A. Boolootian, ed.), pp. 277–300. Wiley (Interscience), New York.
109. Fox, D. L., and Hopkins, T. S. (1966). *Comp. Biochem. Physiol.* **17,** 841–856.

110. Fox, D. L., and Pantin, C. F. A. (1951). *Philos. Trans. R. Soc. London, Ser. B* **230,** 415–450.
111. Fox, D. L., Smith, V. E., and Wolfson, A. (1967). *Experientia* **23,** 965–967.
112. Fox, D. L., and Wilkie, D. W. (1970). *Comp. Biochem. Physiol.* **36,** 49–60.
113. Francis, G. W., Upadhyay, R. R., and Liaaen-Jensen, S. (1970). *Acta Chem. Scand.* **24,** 3050–3052.
114. Fujinaga, M. (1969). U.S. Patent 3,477,406.
115. Galasko, G., Hora, J., Toube, T. P., Weedon, B. C. L., André, D., Barbier, M., Lederer, E., and Villaneuva, V. R. (1969). *J. Chem. Soc. C* pp. 1264–1265.
116. Gammack, D. B., Raper, J. H., Zagalsky, P. F., and Quarmby, R. (1971). *Comp. Biochem. Physiol. B* **40,** 295–300.
117. Georgiev, G. S. (1971). *Folia Balc.* **2,** 1–10.
118. Gilchrist, B. M. (1968). *Comp. Biochem. Physiol.* **24,** 123–147.
119. Gilchrist, B. M., and Lee, W. L. (1967). *J. Zool.* **151,** 171–180.
120. Gilchrist, B. M., and Lee, W. L. (1972). *Comp. Biochem. Physiol. B* **42,** 263–294.
121. Goodwin, T. W. (1951). *Biochem. Soc. Symp.* **6,** 63–82.
122. Goodwin, T. W. (1952). "The Comparative Biochemistry of Carotenoids." Chapman & Hall, London.
123. Goodwin, T. W. (1960). *In* "Physiology of Crustacea" (T. Waterman, ed.), pp. 101–140. Academic Press, New York.
124. Goodwin, T. W. (1965). *In* "The Chemistry and Biochemistry of Plant Pigments" (T. W. Goodwin, ed.), pp. 127–142, 143–173. Academic Press, New York.
125. Goodwin, T. W. (1968). *In* "Porifera, Coelenterata, and Platyhelminthes" (M. Florkin and B. T. Scheer, eds.), Chemical Zoology, Vol. 2, pp. 37–41. Academic Press, New York.
126. Goodwin, T. W. (1968). *In* "Porifera, Coelenterata, and Platyhelminthes" (M. Florkin and B. T. Scheer, eds.), Chemical Zoology, Vol. 2, pp. 149–155. Academic Press, New York.
127. Goodwin, T. W. (1971). *In* "Arthropoda," Part B (M. Florkin and B. T. Scheer, eds.), Chemical Zoology, Vol. 6, pp. 279–306. Academic Press, New York.
128. Goodwin, T. W. (1971). *In* "Aspects of Terpenoid Chemistry and Biochemistry" (T. W. Goodwin, ed.), pp. 315–356. Academic Press, New York.
129. Goodwin, T. W. (1972). *In* "Mollusca" (M. Florkin and B. T. Scheer, eds.), Chemical Zoology, Vol. 7, pp. 187–199. Academic Press, New York.
130. Goodwin, T. W., and Fox, D. L. (1955). *Nature (London)* **175,** 1086–1088.
131. Griffiths, M. (1966). *Dev. Biol.* **13,** 296–309.
132. Griffiths, M., and Perrott, P. (1976). *Comp. Biochem. Physiol. B* **55,** 435–441.
133. Gross, J., Camon, M., Lifshitz, A., and Sklarz, B. (1975). *Comp. Biochem. Physiol. B* **52,** 459–464.
134. Gross, J., and Budowski, P. (1966). *Biochem. J.* **101,** 747–754.
135. Hastings, W. H., Meyers, S. P., and Butler, D. P. (1971). *Feedstuffs* **43**(47), 38.
136. Hata, M., and Hata, M. (1969). *Comp. Biochem. Physiol.* **29,** 985–994.
137. Hata, M., and Hata, M. (1970). *Tohoku J. Agric. Res.* **21,** 183–188.
138. Hata, M., and Hata, M. (1971). *Int. J. Biochem.* **2,** 11–19.
139. Hata, M., and Hata, M. (1971). *Int. J. Biochem.* **2,** 182–184.
140. Hata, M., and Hata, M. (1972). *Nippon Suisan Gakkaishi* **38,** 331–338.
141. Hata, M., and Hata, M. (1972). *Nippon Suisan Gakkaishi* **38,** 339–343.
142. Hata, M., and Hata, M. (1973). *Tohoku J. Agric. Res.* **24,** 192–196.
143. Hata, M., and Hata, M. (1975). *Tohoku J. Agric. Res.* **26,** 35–40.
144. Hata, M., and Hata, M. (1975). *Nippon Suisan Gakkaishi* **41,** 653–655.
145. Herring, P. J. (1959). *J. Mar. Biol. Assoc. U.K.* **49,** 766–779.
146. Herring, P. J. (1968). *Comp. Biochem. Physiol.* **24,** 205–221.
147. Herring, P. J. (1968). *Comp. Biochem. Physiol.* **24,** 187–203.
148. Herring, P. J. (1971). *Experientia* **27,** 1027–1028.

149. Herring, P. J. (1973). *J. Mar. Biol. Assoc. U.K.* **53,** 539–562.
150. Hertzberg, S., Liaaen-Jensen, S., Enzell, C. R., and Francis, G. W. (1969). *Acta Chem. Scand.* **23,** 3290–3312.
151. Hirao, S., Kikuchi, R., and Hama, T. (1969). *Nippon Suisan Gakkaishi* **35,** 187–198.
152. Hirao, S., Kikuchi, R., and Taguchi, H. (1963). *Nippon Suisan Gakkaishi* **29,** 371–381.
153. Hirao, S., Kikuchi, R., and Taguchi, H. (1963). *Nippon Suisan Gakkaishi* **29,** 382–386.
154. Hirao, S., Yamada, J., and Kikuchi, R. (1957). *Nippon Suisan Gakkaishi* **16,** 53–58.
155. Hirao, S., Yamada, J., and Kituchi, R. (1959). *Nippon Suisan Gakkaishi* **16,** 53–58.
156. Holter, A. R. (1969). *Comp. Biochem. Physiol.* **28,** 675–684.
157. Hora, J., Toube, T. P., and Weedon, B. C. L. (1970). *J. Chem. Soc. C* pp. 241–242.
158. Hsu, W.-J., Chichester, C. O., and Davies, B. H. (1970). *Comp. Biochem. Physiol.* **32,** 69–79.
159. Hsu, W.-J., Rodriguez, D. B., and Chichester, C. O. (1972). *Int. J. Biochem.* **3,** 333–338.
160. Hublou, W. F., McKee, T. B., Law, D. K., Sinnhuber, R. O., and Yu, T. C. (1963). *Res. Briefs* **9**(1), 57–62.
161. Isler, O., ed. (1971). "Carotenoids." Birkhaeuser, Basel.
162. Iwahashi, M., and Wakui, H. (1976). *Nippon Suisan Gakkaishi* **42,** 1339–1344.
163. Jensen, A., and Sakshaug, E. (1970). *J. Exp. Mar. Biol. Ecol.* **5,** 180–186.
164. Johnson, E. A., Conklin, D. E., and Lewis, M. J. (1977). *J. Fish. Res. Board Can.* **34,** 2717–2421.
165. Joseph, J. D., and Williams, J. E. (1975). *Proc World Maricult. Soc.* **6,** 147–155.
166. Kanemitsu, T., and Aoe, H. (1958). *Nippon Suisan Gakkaishi* **24,** 555–558.
167. Karnaukhov, V. N. (1971). *Exp. Cell Res.* **64,** 301–306.
168. Karrer, P., and Jucker, E. (1948). "Carotenoids." Birkhaeuser, Basel; Engl. transl. by E. A. Braude, Elsevier, Amsterdam, 1950.
169. Katayama, T. (1969). *Kagoshima Diagaku Suisan Gakubu Kiyo* **18,** 79–86.
170. Katayama, T. (1975). Personal communication.
171. Katayama, T., Hirata, K., and Chichester, C. O. (1971). *Nippon Suisan Gakkaishi* **37,** 614–620.
172. Katayama, T., Hirata, K., Yokoyama, H., and Chichester, C. O. (1970). *Nippon Suisan Gakkaishi* **36,** 709–714.
173. Katayama, T., Ikeda, N., and Harada, K. (1965). *Nippon Suisan Gakkaishi* **31,** 947–952.
174. Katayama, T., Kamata, T., and Chichester, C. O. (1972). *Int. J. Biochem.* **3,** 363–368.
175. Katayama, T., Kamata, T., Shimaya, M., Deshimaru, O., and Chichester, C. O. (1972). *Nippon Suisan Gakkaishi* **38,** 1171–1175.
176. Katayama, T., Kunisaki, Y., Shimaya, M., Sameshima, M., and Chichester, C. O. (1973). *Nippon Suisan Gakkaishi* **39,** 283–287.
177. Katayama, T., Kunisaki, Y., Shimaya, M., Simpson, K. L., and Chichester, C. O. (1973). *Comp. Biochem. Physiol. B* **46,** 269–272.
178. Katayama, T., Miyahara, T., and Sameshima, M. (1973). *Kagoshima Daigaku Suisan Gakubu Kiyo* **22,** 39–45.
179. Katayama, T., Miyahara, T., Shimaya, M., and Chichester, C. O. (1972). *Int. J. Biochem.* **3,** 569–572.
180. Katayama, T., Miyahara, T., Tanaka, Y., Sameshima, M., Simpson, K. L., and Chichester, C. O. (1974). *Nippon Suisan Gakkaishi* **40,** 97–103.
181. Katayama, T., Shimaya, M., Sameshima, M., and Chichester, C. O. (1973). *Int. J. Biochem.* **4,** 223–226.
182. Katayama, T., Shintani, K., and Chichester, C. O. (1973). *Comp. Biochem. Physiol. B* **44,** 253–257.
183. Katayama, T., Shintani, K., Shimaya, M., Imai, S., and Chichester, C. O. (1972). *Nippon Suisan Gakkaishi* **38,** 1399–1403.
184. Katayama, T., Tsuchiya, H., and Chichester, C. O. (1971). *Kagoshima Daigaku Suisan Gakubu Kiyo* **20,** 173–184.
185. Katayama, T., Yokoyama, H., and Chichester, C. O. (1970). *Int. J. Biochem.* **1,** 438–444.

186. Katayama, T., Yokoyama, H., and Chichester, C. O. (1970). *Nippon Suisan Gakkaishi* **36,** 702–708.
187. Khare, A., Moss, G. P., and Weedon, B. C. L. (1973). *Tetrahedron Lett.* **40,** 3921–3924.
188. Khare, A., Moss, G. P., Weedon, B. C. L., and Matthews, A. D. (1973). *Comp. Biochem. Physiol. B* **45,** 971–973.
189. Knowles, R. E., Livingston, A. L., Nelson, J. W., and Kohler, G. O. (1968). *J. Agric. Food Chem.* **16,** 654–658.
190. Krinsky, N. I. (1965). *Comp. Biochem. Physiol.* **16,** 181–187.
191. Krinsky, N. I., and Lenhoff, H. M. (1965). *Comp. Biochem. Physiol.* **16,** 189–198.
192. Kuhn, R., and Kühn, H. (1967). *Eur. J. Biochem.* **2,** 349–360.
193. Kuo, H.-C., Lee, T.-C., Kamata, T., and Simpson, K. L. (1976). *Alimenta* **15,** 47–51.
194. Kuo, H.-C., Lee, T.-C., Chichester, C. O., and Simpson, K. L. (1976); *Comp. Biochem. Physiol.* B **54,** 387–390.
195. Lambertsen, G., and Braekkan, O. R. (1971). *J. Sci. Food Agric.* **22,** 99–101.
196. Larry, D., and Salwin, H. (1966). *J. Assoc. Off. Anal, Chem.* **49,** 681–683.
197. Lederer, E. (1938). *Bull. Soc. Chim. Biol.* **20,** 567–610.
198. Lee, D. J., Roehm. J. N., Yu, T. C., and Sinnhuber, R. O. (1967). *J. Nutr.* **92,** 93–98.
199. Lee, K.-H., and Kim, C.-B. (1971). *Pusan Susan Taehak Yon'gu Pogo* **11,** 57–62.
200. Lee, W. L. (1966). *Comp. Biochem. Physiol.* **18,** 17–36.
201. Lee, W. L. (1966). *Comp. Biochem. Physiol.* **19,** 13–27.
202. Lee, W. L., and Gilchrist, B. M. (1972). *J. Exp. Mar. Biol. Ecol.* **10,** 1–27.
203. Lee, W. L., and Gilchrist, B. M. (1975). *Comp. Biochem. Physiol. B* **51,** 247–253.
204. Lee, W. L., Gilchrist, B. M., and Dales, R. P. (1967). *J. Mar. Biol. Assoc. U.K.* **47,** 33–37.
205. Lenel, R. (1967). *Bull. Soc. Zool. Fr.* **92,** 109–113.
206. Lewis, R. W. (1970). *Lipids* **5,** 947–949.
207. Lewis, W. M., and Heidinger, R. (1971). *Am. Fish Farmer* **2**(5), 14–16.
208. Liaaen-Jensen, S. (1971). *In* "Carotenoids" (O. Isler, ed.), pp. 61–188. Birkhaeuser, Basel.
209. Love, R. M. (1970). "The Chemical Biology of Fishes." Academic Press, New York.
210. McBeth, J. W. (1970). *Experientia* **26,** 144.
211. McBeth, J. W. (1972). *Comp. Biochem. Physiol. B* **41,** 55–68.
212. McBeth, J. W. (1972). *Comp. Biochem. Physiol. B* **41,** 69–77.
213. Malak, N. A., Zwingelstein, G., Jouanneteau, J., and Koenig, J. (1975). *Ann. Nutr. Aliment.* **29,** 459–475.
214. Manzer, J. I. (1968). *J. Fish. Res. Board Can.* **26,** 2219–2223.
215. Matsumoto, J., and Obika, M. (1968). *J. Cell Biol.* **39,** 233–250.
216. Matsuno, T., Kusumoto, T., Watanabe, T., and Ishihara, Y. (1973). *Nippon Suisan Gakkaishi* **39,** 43–50.
217. Matsuno, T., Nagata, S., Sato, Y., and Watanabe, T. (1974). *Nippon Suisan Gakkaishi* **40,** 579–584.
218. Matsuno, T., Nagata, S., and Vemura, M. (1975). *Nippon Suisan Gakkaishi* **41,** 343–349.
219. Matsuno, T., Katsuyama, M., and Iwasaki, N. (1975). *Nippon Suisan Gakkaishi* **41,** 351–355.
220. Matsuno, T., Nagata, S., and Chiba, K. (1975). *Nippon Suisan Gakkaishi* **41,** 459–464.
221. Matsuno, T., Katsuyama, M., and Vemura, M. (1975). *Nippon Suisan Gakkaishi* **41,** 681–684.
222. Matsuno, T., Katsuyama, M., and Kashizaki, M. (1976). *Nippon Suisan Gakkaishi* **42,** 465–467.
223. Matsuno, T., and Katsuyama, M. (1976). *Nippon Suisan Gakkaishi* **42,** 645–649.
224. Matsuno, T., Katsuyama, M., and Ishida, T. (1976). *Nippon Suisan Gakkaishi* **42,** 651–654.
225. Matsuno, T., and Katsuyama, M. (1976). *Nippon Suisan Gakkaishi* **42,** 761–763.
226. Matsuno, T., and Katsuyama, M. (1976). *Nippon Suisan Gakkaishi* **42,** 765–768.
227. Matsuno, T., and Katsuyama, M. (1976). *Nippon Suisan Gakkaishi* **42,** 847–850.
228. Mayer, H., and Isler, O. (1971). *In* "Carotenoids" (O. Isler, ed.), pp. 325–575. Birkhaeuser, Basel.
229. Meade, T. L. (1971). Personal communication.

230. Meyers, S. P. (1971). *Feedstuffs* **43**(38), 27.
231. Meyers, S.P., Avault, J. W., Jr., Rhee, J. S., and Butler, D. (1970). *La. State Univ., Coastal Stud., Bull.* No. 5, 157–172.
232. Meyers, S. P., and Rutledge, J. E. (1971). *Feedstuffs* **43,** 16.
233. Meyers, S. P., and Rutledge, J. E. (1971). *Feedstuffs* **43**(49), 31–32.
234. Nakagawa, H., Kayama, M., Yamada, H., and Asakawa, S. (1974). *Hiroshima Diagaku Sui-Chikusangakubu Kiyo* **13,** 1–13.
235. Nakazoe, J., and Hata, M. (1978). *Proc. Jpn. Soc. Sci. Fish., 53rd Meet.,* Tokyo Abstr. No. 557.
236. Nakazoe, J., and Hata, M. (1978). *Proc. Jpn. Soc. Sci. Fish., 53rd Meet.,* Tokyo Abstr. No. 558.
237. Neal, R. A. (1971). *Am. Fish Farmer* **2**(6), 10–11, 18–20.
238. Neamtu, G., Weaver, C. M., Wolke, R. E., and Simpson, K. L. (1976). *Rev. Roum. Biochim.* **13,** 25–30.
239. Neamtu, G., Tamas, V., and Bodea, C. (1966). *Rev. Roum. Biochim.* **3,** 305–310.
240. Needham, A. E. (1970). *Comp. Biochem. Physiol.* **35,** 509–534.
241. Needham, P. R. (1969). "Trout Streams" (revised by C. F. Bond). Holden-Day, San Francisco, California.
242. Neuhaus, O. W., and Halver, J. E., eds. (1969). "Fish in Research," pp. 209-306. Academic Press, New York.
243. Neve, R. A., and Howard, G. A. (1970). *Alaska Univ. Inst. Mar. Sci., Publ.* No. 1, 549–562.
244. Nicoara, E., Bilaus, C., Bodea, C., and Petrescu, I. (1970). *Stud, Cercet. Biochim.* **13,** 65–68.
245. Nilsson, N.-A., and Andersson, G. (1967). *Inst. Freshwater Res. Rep.* No. 47, 120–125.
246. Nishibori, K. (1960). *Publ. Seto Mar. Biol. Lab.* **8,** 95–105.
247. Ostrander, J., Martinsen, C., Liston, J., and McCullough, J. (1976). *J. Food Sci.* **41,** 386–390.
248. Pearson, W. E. (1968). "The Nutrition of Fish." Hoffmann-La Roche, Basel.
249. Peterson, D. H., Jäger, H. K., Savage, G. M., Washburn, G. N., and Westers, H. (1966). *Trans. Am. Fish. Soc.* **95,** 408–414.
250. Phillips, A. M., Jr., Tunison, A. V., d Balzer, G. C. (1963). *U.S. Dept. Fish Wildl. Serv.,* No. 159.
251. Poston, H. A. (1969). *Fish. Res. Bull.* No. 32, 41–43.
252. Poston, H. A. (1975). Personal communication.
253. Rajagopal, M. V., and Sohonie, K. (1956). *Curr. Sci.* **25,** 327.
254. Richards, W. L. (1971). *Miami Univ. Sea Grant Tech. Bull.* No. 10.
255. Rodriguez, D. B. (1971). Ph.D. Thesis, Univ. of California, Davis.
256. Rodriguez, D. B., Simpson, K. L., and Chichester, C. O. (1973). *Int. J. Biochem.* **4,** 213–222.
257. Rodriguez, D. B., Simpson, K. L., and Chichester, C. O. (1974). *Int. J. Biochem.* **5,** 157–166.
258. Rodriguez, D. B., Tanaka, Y., Katayama, T., Simpson, K. L., Lee, T.-C., and Chichester, C. O. (1976). *Agric. Food Chem.* **24,** 819–822.
259. Rutledge, J. E. (1971). *J. Agric. Food Chem.* **19,** 236–237.
260. Saito, A. (1969). *J. Fish. Res. Board Can.* **26,** 2234–2236.
261. Saito, A.,and Regier, L. W. (1970). *Fish. Res. Board Can., New Ser. Circ.* No. 36.
262. Saito, A., and Regier, L. W. (1971). *J. Fish. Res. Board Can.* **28,** 509–512.
263. Sakurai, H., Kato, K., Sakai, T., Masuda, Y., and Kuriyama, T. (1971). *Bull. Chem. Soc. Jpn.* **44,** 481–484.
264. Sarphie, T. G., and Crozier, G. (1970). *Comp. Biochem. Physiol.* **34,** 963–965.
265. Sastrakusumah, S. (1971). *Miami Univ. Sea Grant Tech. Bull.* No. 9.
266. Sastry, A. N. (1968). *Physiol. Zool.* **41,** 44–53.
267. Sastry, A. N., and Blake, N. J. (1971). *Biol. Bull. (Woods Hole, Mass.)* **140,** 274–283.
268. Savolainen, J. E. T., and Gyllenberg, H. G. (1970). *Lebensm.-Wiss. Technol.* **3,** 18–20.
269. Schmidt, P. J., and Baker, E. G. (1969). *J. Fish. Res. Board Can.* **26,** 357–360.
270. Shimizu, T., and Monma, R. (1968). *Nippon Suisan Gakkaishi* **34,** 159–162.
271. Shimizu, T., and Oda, A. (1968). *Nippon Suisan Gakkaishi* **34,** 627–632.

272. Shimizu, T., and Ohta, Y. (1968). *Nippon Suisan Gakkaishi* **34,** 210–213.
273. Shimizu, T., and Uchida, K. (1968). *Nippon Suisan Gakkaishi* **34,** 154–158.
274. Simpson, K. L. (1978). *R.I. Resour.* **24**(2), 1–3.
275. Simpson, K. L., and Kamata, T. (1979). *In* "Proceedings of World Symposium on Finfish Nutrition and Fishfeed." (J. E. Halver and K. Tiews, eds.), Vol. 2, pp. 415–424. Heenemann Verlugsgesellschaft MbH., Berlin.
276. Simpson, K. L., Kamata, T., Collins, J. G., and Collins, J. H. (1976). *Proc. Trop. Subtrop. Fish Tech. Conf., Corpus Christi, ex* pp. 395–411.
277. Simpson, K. L., Lee, T.-C., Rodriguez, D. B., and Chichester, C. O. (1976). *In* "Chemistry and Biochemistry of Plant Pigments" (T. W. Goodwin, ed.), Vol. 2, pp. 780–842. Academic Press, New York.
278. Snauwaert, F., Tobback, P. P., and Maes, E. (1973). *Lebensm.-Wiss. Technol.* **6,** 43–47.
279. Sømme, I. D. (1948). "Ørretboka." Jacob Dybwads Forlag, Oslo.
280. Spinelli, J., Lehman, L., and Wieg, D. (1974). *J. Fish. Res. Board Can.* **31,** 1025–1029.
281. Spinelli, J. (1975). *Improv. Fish Feed, Pac. Technol. Meet., Portland, Ore.*
282. Spinelli, J. (1979). *In* "Proceedings of World Symposium on Finfish Nutrition and Fishfeed." (J. E. Halver and K. Tiews, eds.), Vol. 2, pp. 345–352. Heenemann Verlagsgelsellschaft MbH, Berlin.
283. Spinelli, J. (1979). *In* "Proceedings of World Symposium on Finfish Nutrition and Fishfeed." (J. E. Halver and K. Tiews, eds.), Vol. 2, pp. 383–392. Heenemann Verlagsgesellschaft MbH., Berlin.
284. Spinelli, J., and Mahnken, C. (1978). *Aquaculture* **13,** 213–223.
285. Steel, R. E. (1971). M.S. Thesis, Oregon State Univ., Corvallis.
286. Steven, D. M. (1947). *Nature (London)* **160,** 507–508.
287. Steven, D. M. (1947). *Nature (London)* **160,** 540–541.
288. Strain, H. H., Svec, W. A., Aitzetmüller, K., Grandolfo, M. C., Katz, J. J., Kjøsen, H., Norgard, S., Liaaen-Jensen, S., Haxo, F. T., Wegfahrt, P., and Rapoport, H. (1971). *J. Am. Chem. Soc.* **93,** 1823–1825.
289. Tanaka, T., and Katayama, T. (1976). *Nippon Suisan Gakkaishi* **42,** 801–805.
290. Tanaka, T., and Katayama, T. (1976). *Nippon Suisan Gakkaishi* **42,** 807–812.
291. Tanaka, Y., Katayama, T., Simpson, K. L., and Chichester, C. O. (1976). *Nippon Suisan Gakkaishi* **42,** 1177–1182.
292. Tanaka, Y., Katayama, T., Simpson, K. L., and Chichester, C. O. (1976). *Nippon Suisan Gakkaishi* **42,** 885–891.
293. Tanaka, Y., Fujita, Y., and Katayama, T. (1977). *Nippon Suisan Gakkaishi* **43,** 767–772.
294. Tanaka, Y., Fujita, Y., and Katayama, T. (1977). *Nippon Suisan Gakkaishi* **43,** 761–765.
295. Tanaka, Y., Matsuguchi, H., Katayama, T., Simpson, K. L., and Chichester, C. O. (1976). *Nippon Suisan Gakkaishi* **42,** 197–202.
296. Tanaka, Y., Matsuguchi, H., Katayama, T., Simpson, K. L., and Chichester, C. O. (1976). *Comp. Biochem. Physiol. B* **54,** 391–393.
297. Tappel, A. L. (1953). *Food Res.* **18,** 104–108.
298. Thommen, H. (1971). *In* "Carotenoids" (O. Isler, ed.), p. 645. Birkhaeuser, Basel.
299. Thommen, H., and Gloor, U. (1965). *Naturwissenschaften* **52,** 161–162.
300. Thommen, H., and Wackernagel, H. (1964). *Naturwissenschaften* **51,** 87–89.
301. Torrissen, O., and Braekkan, O. R. (1979). *In* "Proceedings of World Symposium on Fishfish Nutrition and Fishfeed." (J. E. Halver and K. Tiews, eds.), Vol. 2, pp. 377–382. Heenemann Verlagsgesellschaft MbH., Berlin.
302. Tsukuda, N. (1970). *Nippon Suisan Gakkaishi* **36,** 725–733.
303. Tsukuda, N., and Amano, K. (1966). *Nippon Suisan Gakkaishi* **32,** 334–345.
304. Tsukuda, N., and Amano, K. (1967). *Nippon Suisan Gakkaishi* **33,** 962–969.
305. Tsukuda, N., and Amano, K. (1968). *Nippon Suisan Gakkaishi* **34,** 633–639.

306. Tusa, I. (1968). *Zool. Listy* **17,** 379–395.
307. Upadhyay, R. R., and Liaaen-Jensen, S. (1970). *Acta Chem. Scand.* **24,** 3055–3057.
308. Veerman, A. (1970). *Comp. Biochem. Physiol.* **36,** 749–763.
309. Veerman, A., Borch, G., Pedersen, R., and Liaaen-Jensen, S. (1975). *Acta Chem. Scand., Ser. B* **29,** 525.
310. Vetter, W., Englert, G., Rigassi, N., and Schwieter, V. (1971). *In* "Carotenoids" (O. Isler, ed.), pp. 189–266. Birkhaeuser, Basel.
311. Vevers, H. G. (1966). *In* "Physiology of Echinodermata" (R. A. Boolootian, ed.), pp. 267–275. Wiley (Interscience), New York.
312. Wald, G., Nathanson, N., Jencks, W. R., and Tarr, E. (1948). *Biol. Bull. (Woods Hole, Mass.)* **95,** 249–250.
313. Walton, T. J. (1970). Thesis, Univ. of Liverpool, Liverpool.
314. Weedon, B. C. L. (1971). *In* "Carotenoids" (O. Isler, ed.), pp. 29–59. Birkhaeuser, Basel.
315. Wilkie, D. W. (1972). *Comp. Biochem. Physiol. B* **42,** 731–734.
316. Wolfe, D. A., and Cornwell, D. G. (1964). *Science* **144,** 1467–1469.
317. Wolfe, D. A., and Cornwell, D. G. (1965). *Comp. Biochem. Physiol.* **16,** 205–213.
318. Yamaguchi, M. (1957). *Bull. Chem. Soc. Jpn.* **30,** 111–114.
319. Yamaguchi, M. (1957). *Bull. Chem. Soc. Jpn.* **30,** 979–983.
320. Yamaguchi, M. (1958). *Bull. Chem. Soc. Jpn.* **31,** 51–55.
321. Yamaguchi, M. (1958). *Bull. Chem. Soc. Jpn.* **31,** 739–742.
322. Yamaguchi, M. (1960). *Bull. Chem. Soc. Jpn.* **33,** 1560–1562.
323. Zagalsky, P. F., Ceccaldi, H. J., and Daumas, R. (1970). *Comp. Biochem. Physiol.* **34,** 579–607.
324. Zagalsky, P. F., Cheesman, D. F., and Ceccaldi, H. J. (1967). *Comp. Biochem. Physiol.* **22,** 851–871.
325. Zagalsky, P. F., and Herring, P. J. (1972). *Comp. Biochem. Physiol. B* **41,** 397–415.
326. Zs.-Nagy, I. (1974). *Comp. Biochem. Physiol. A* **49,** 399–405.

5

CAROTENOIDS IN WILD AND CAPTIVE BIRDS

Alan H. Brush

I. INTRODUCTION

Birds have a unique place among pets and zoo animals. Their bright colors and complex plumage patterns provide an attractiveness perhaps unmatched in the vertebrates. For the most part, they are easily tamed, do well in fairly confined quarters, and can be kept with a minimum of labor. One need only recall the number of zoos, parks, and other public places where large birds are an attraction, and consider the number of hobbyists who keep and breed cage birds, to realize the extent of public and private interest in these organisms. Over 560,000 birds were declared for import into the United States in 1969, and over 1.2 million were declared from 1967 to 1970 (*3*). This is even more impressive when one considers that canaries and psittacines are excluded from this tally. Presumably, the greatest part of these went to zoos and private individuals.

Collectors and breeders are not the only individuals interested in the colors and patterns of birds. Early evolutionists (*50*) had surprisingly deep insights into the evolutionary history and biological role of colors. Even Teddy Roosevelt, the twenty-sixth president, was concerned with the role that colors and patterns played

ISBN 0-12-082850-2

in animal biology (*74*). In birds, as well as higher primates, the senses of sight and hearing are highly developed and provide roughly similar systems of identification and communication. This applies at both the individual and specific levels. Because of the inherent similarities in the sensory system, coupled with other factors, such as bipedalism, complex behavioral repertories, and the conspicuous nature and gregariousness of many species, birds have been interesting and profitable organisms for many types of studies by ornithologists, ethologists, and ecologists. By the same token, some physiological and biochemical parameters that are responsible for these features have become the foci of study.

Carotenoids are an interesting group of molecules in that they are practically ubiquitous in plants and animals and play a variety of biological roles. These include their function in the retina for photoreception; in various aspects of lipid metabolism and the maintenance of membranes; as vitamins; in reproduction; and reputedly in olfaction. In addition to these physiological roles, in many groups (including bony fish, echinoderms, and crustaceans), and especially in birds (*90*), they are displayed prominently in portions of the integument. Thus, they are intimately associated with the visual system, species recognition, mating systems, and various behavioral patterns of many species. It is their function in the dramatic and conspicious displays and plumage colors of a wide variety of avian species that gives carotenoids their special attractiveness.

Carotenoid chemistry lends itself to detailed analysis. Investigation of the biosynthetic metabolic pathways and the end products is possible [see Thommen, cited in Isler (*47*)]. This facilitates a variety of studies whereby the chemical basis for numerous morphological and behavioral features may be investigated. In some cases (e.g., displays and integumentary color patterns), relatively simple compounds such as carotenoids can produce complex and important characters and are thus ideally suited for detailed analysis. One hopes that such analyses will lead to understanding of the mechanisms involved, their role in the biology of the species, and the use of this information to derive generalities on the evolution of the particular combination of color, pattern, and behavior. Information on carotenoid biochemistry and metabolism of carotenoids is useful to commercial poultry producers [Bauernfeind *et al.*, cited in Isler (*47*)].

Carotenoids are distributed widely, but irregularly in higher vertebrates. They are conspicuous in lower vertebrates, especially teleost fish [see Chapter 4 by Simpson *et al.* and Fox (*29,31*)]. Recently, a considerable body of information has developed on the distribution and metabolism of carotenoids in the invertebrates. This material is not discussed extensively here. Nevertheless, it is essential in reconstructing metabolic pathways of carotenoids.

II. FUNCTIONS OF AVIAN CAROTENOIDS

The importance of carotenoids in the visual process was demonstrated in birds by Wald and Zussman (*99*). Carotenoids form an important source of yolk pigments and meat color in chickens (see Chapter 3 by Marusich and Bauernfeind) and are also found in the egg yolk of noncommercial species (*38,41*). The functions of yolk carotenoids are not known, but their deposition can be controlled by diet (*60*).

The most interesting loci of deposition here are the integumentary structures, including skin, feathers, scales, and fleshy structures usually associated with the head and neck, the reason being that captive animals are more likely to be valued for their visual appearance than for the color quality of their flesh or eggs. Investigations of these structures in several species have led to the most interesting models and questions on the metabolism, control, and evolution of these pigments. In addition to establishing accurate documentation of the carotenoids in animals, workers have attempted to understand the biochemical basis of the colors and their metabolic and genetic controls.

III. DISTRIBUTION OF CAROTENOIDS

No recent systematic survey of integumentary carotenoids exists. Carotenoids have been identified in integumentary structures (feathers, bills, feet, etc.) of the following orders:

Pelecaniformes—*Phaethon*
Ardeiformes (= Ciconiformes)
  Plataleidae
Phoenicopteriformes
  Phoenicopteridae
Anseriformes
Galliformes
  Phasianidae
Charadriiformes
Columbiformes
  Megaloprepia, Ptilinopus
Trogoniformes
Piciformes
  Picadae, Capitonidae, Ramphastidae
Passeriformes
  Pipridae, Codtingidae, Tyrannidae, Sylviidae, Oriolidae, Bombycillidae, Laniidae, Icteridae, Fringillidae, Thraupidae, Carduelidae, Estrildidae, Ploceidae

The first surveys were concerned mainly with simply establishing the presence or absence of carotenoids (*24*) and the relationship between pigment and feather structure (*39*). Two generalities that emerged from these and subsequent studies were that carotenoids are generally not present in young birds (juvenile plumage) or in birds regarded as phylogenetically more primitive. Correlations with physiological conditions and aging were noticed (*58*). As more investigators, zookeepers, and breeders become aware of the potential dietary involvement of carotenoids, these questions may be reexamined. For example, yellow washes and yellow feathers in the Sphenisciformes and feathers of the more obscure and less accessible passerine families (Oxyruncidae, Drepanididae, Campephagidae, Meliphagidae, Cyclarthidae) should be investigated.

The pigments of juvenile plumages are very poorly known. Melanins appear to be the most common pigments and account for the pale yellows in most hatchlings. However, comparative data are rare. In the coot (*Fulica atra*), ethanolic extracts of the bill and the feathers of the head, face, and neck produced a yellow-orange solution with no distinct visible spectral bands (Brush and Zamlauski, unpublished). The extracted pigment was insoluble in common organic solvents and gave a green fluorescence in UV light. There was no reaction to concentrated HCl, peroxide, or sodium hydrosulfite. Thus, both the chemistry and identity of these pigments remain obscure. A proposed function (*5*) and the relation to feather structure (*77*) have been discussed.

The range of carotenoids found in birds is rather broad. The list tabulated below was compiled from numerous sources.

α-Carotene	β-Carotene
Lutein (3,3′-dihydroxy-α-carotene)	Isocryptoxanthin (4-hydroxy-β-carotene)
Canary xanthophyll (definite identity unknown; possibly lutein epoxide)	Echinenone (4-keto-β-carotene)
Zeinoxanthin (3-hydroxy-α-carotene)	Canthaxanthin (4,4′-diketo-β-carotene)
Phoenicopterone (4-keto-α-carotene)	Zeaxanthin (3,3′-dihydroxy-β-carotene)
Picoverdin (picofulvin) (structure unknown; possibly a carotenoid	Isozeaxanthin (4,4′-dihydroxy-β-carotene)
	Rhodoxanthin (3,3′-diketoretrodehydro-β-carotene)
	Phoenicoxanthin (3-hydroxy-4,4′-diketo-β-carotene)
	Astaxanthin (3,3′-dihydroxy-4,4′-diketo-β-carotene)
	Guaraxanthin (7,8-dihydroxyastaxanthin)

The unsubstituted carotenoids (carotenes) do not appear abundantly in integumentary structures. The α-, β-, and ϵ-carotenes have been reported in internal tissues, especially liver, serum and body fat.

There appears to be no relationship between taxonomic position and carotenoid metabolism. Thus, the presence of feather pigment is not a derived or advanced character. In those taxa in which carotenoid pathways are active, apparently nothing precludes a bird from assimilating an unsubstituted carotene and oxidizing it completely. The implication is that, once the ability to deposit carotenoids arose phylogenetically, the taxa evolved the ability to carry out additional oxidations. This explains the broad taxonomic convergence in the presence of these pathways.

Most authorities agree with Searle (*76*) that, except for the corpora lutea, adrenals, body fat, plasma, and other minor internal sites, carotenoids are not deposited to any significant extent by mammals. Little mammalian skin and no hair coloration are due to carotenoids. To some extent, plasma coloration by carotenoids is a function of the diet. The relationship is far from exact, however, and demonstrable differences in absolute or relative carotenoid levels, plasma half-life, and transport mechanisms are inconsistent (*31*).

Carotenoids are rarely, if ever, added to the feed of mammals with the idea of enhancing integumentary coloration. Most diets prescribed for mammals contain some natural carotenoids and are universally supplemented with vitamin A. In an evolutionary sense, the reason for the lack of integumentary carotenoids relates to mammalian biology. The factors that would reduce the usefulness of integumentary carotenoids are numerous. Many mammals are nocturnal. Small mammals tend to rely heavily on the olfactory senses for territory maintenance and in reproduction rather than on visual display. Larger, herding mammals use audible signals in addition to vision for many social purposes. Many birds have developed visual displays and signals that involve carotenoids, whereas mammals have developed alternative systems.

IV. TRANSPORTATION

Vertebrates, or for that matter animals in general, do not synthesize carotenoids *de novo* but must absorb them from the gut. In simple animals such as hydra, the ingested carotenoids are deposited in an unaltered form (*40*). In higher animals, some unmodified carotenoids are deposited in body tissues, but considerable modification may occur. Some holotherians (Echinodermata), for example, carry carotenes and hydroxycarotenoids in the gut but preferentially deposit the ketocarotenoids in the body wall. Thus, most animal groups are faced with the problems of absorption, transportation, and conversion of carotenoids. Possible mechanisms for these conversions in birds are discussed below.

Carotenoids, in general, and provitamin A, in particular, are believed to be absorbed from the gut along with unsaturated fats and oils (*4*). Relatively few details are available on the requirements and assimilation of carotenoids in birds. The most readily available source is through the ingestion of carotene precursors (*27,35,37*). Mammals differ in the intestinal metabolism of β-carotenes. Human beings absorb carotenoids in an unchanged state, whereas rats (*Rattus* sp.) and the domestic pig do not absorb β-carotenes across the intestinal mucosa (*46*). In mammals, the lack of carotenoid deposition may be due in some measure to an inadequate transport system. It was early noted (*69*) that animals that do not store carotenoids internally contain none in their blood. However, a hybrid ibis (*Guara alba* × *rubra*) maintained high plasma carotenoid levels but was unable to transfer it to the feather follicles (*31*). Animals that store carotenoids in their tissues usually transport it in the form of a carotenoid–protein complex. Among the higher vertebrates, the carotenoid–protein complex usually involves mostly substituted carotenes. Mammalian and avian plasma tends to have a yellow or orange cast, whereas invertebrate plasmas and integuments can be blue, green, violet, or other colors. The carotenoid moiety may bind to simple proteins or glycolipoproteins or more complex protein combined with carbohydrates. The best characterized carotenoids are those of invertebrates, and some insights into their comparative biochemistry are available (*19,81*).

The carotenoid–proteins in birds have been studied inadequately. In the ring-necked pheasant (*Phasianus colchicus*), one is present in the serum of both sexes. In starch-gel electrophoresis under slightly alkaline conditions, it migrates as a broad band behind the serum albumin. Baker *et al.* (*2*). reported some individual variation in band width and location in a population of birds from Illinois. A presumed carotenoid–protein (fraction 8) of the plasma of hens has been associated with the cessation of laying (*59*). The protein appears in detectable amounts during periods of nonlaying, supposedly because it is no longer being transferred to the yolk. Lush cited only unpublished evidence that the fraction binds carotenoid pigments. In addition to the roles in the egg yolk, the carotenoid–protein may function in electron transport, as a membrane-stabilizing mechanism, or in molecular transport.

In the scarlet ibis (*Guara ruber*) a single, large (MW 2×10^5 or more) protein apparently acts as the carotenoid carrier (*83a*). Even though five to seven presumed carotenoids can be separated by thin-layer chromatography (TLC) from plasma extracts of the scarlet ibis, only a single carrier was present in acrylamide electrophoresis. By comparison, serum of the closely related white ibis (*Guara alba*) produced no more than three carotenoids in TLC and had plasma carotenoid–proteins present in only barely detectable amounts.

The impression gained from these data is that birds have a high molecular weight serum component capable of binding carotenoids. Chemical evidence indicates that it is a lipoprotein. This complex is instrumental in the transport of carotenoids. However, there are no data on how the lipoprotein binds the carotenoid, the specificity or selectivity of the carotenoids, or the kinetics of the relationship. In human beings, the transport of hydrocarbon carotenes was associated predominantly with low-density lipoproteins (LDL), whereas the hydroxycarotene lutein was more widely distributed among high-density lipoproteins (HDL) and low-density plasma proteins (*52*). On the other hand, in birds preliminary data indicate that an LDL ($d = 1.063$) was involved with the carotenoids as they were absorbed from the gut, whereas the HDL fraction ($d = 1.21$) was associated with the various metabolically modified carotenoids of hepatic origin (*83a*).

In the roseate spoonbill *Ajaia ajaja,* the HDL contained both canthaxanthin and astaxanthin probably dissolved in the lipid moiety, whereas the LDL contained much less of the neutral fraction (*34*). An oriole (*Icterus* sp.) had two orange bands upon electrophoresis of plasma on cellulose acetate strips (Brush, unpublished). It is not known whether both proteins contained the same or different carotenoids. Although the exact mechanisms of transport are unknown and the molecular basis of protein selectivity remains to be elucidated, preliminary evidence indicates that the carotenoids are associated with the lipoidal fraction rather than the protein portion of the serum lipoproteins (*34*). A retinol-transporting protein has been isolated and characterized from human beings (*70*). The fraction contains four subunits (MW 21,400), which are identical in double immunodiffusion but differ in their isoelectric point and electrophoretic mobility on polyacrylamide-gel electrophoresis. There are no neutral sugar or hexosamine residues, and approximately 1 mole of retinol is bound per mole of protein. The nature and occurrence of this protein in other mammals and birds remains to be investigated.

Although direct chemical studies are lacking, several authors have speculated on the binding mechanism of carotenoid–proteins in which the two elements bind in stoichiometric proportions (*19,31,81*). Specific sites are undoubtedly involved since spectral shifts occur in the carotenoid. The carbonyl groups, especially the 4-keto group of the carotenoid, are crucial. Both single oxo and keto radicals can combine. It is thought that the binding occurs first by carbonyl–amino linkages. Then, if sufficient dehydration follows, a Schiff's base linkage is formed. No direct experimental evidence is available for the pigmented feather keratins. However, the isolation of so many mono- and diketocarotenoids provides circumstantial evidence for the existence of this type of mechanism in birds.

Few measurements of the turnover rates of carotenoids in the plasma are available. Trams (*83a*) estimated the half-life of plasma carotenoids in *Guara* on a carotene diet at 3.5–4.5 days. On the basis of a series of prolonged feeding experiments with American flamingos, Fox (*31*) gave some estimates of "whole-body" turnover rates. In one case it took 85 days for the blood carotenoids (mainly canthaxanthin) to drop from 1 mg to 0.25 mg per 100 gm and 128 days for restoration in animals on a diet supplemented with β-carotene. In another experiment it took 6 months for carotenoid levels in the blood to fall 86% in animals on a carotene-free diet. These figures, when compared to Tram's estimate, suggest that there is a considerable reservoir of carotenoids, especially oxycarotenoids in the liver and other tissues of this species.

Additional information of this type is needed urgently because it relates directly to feeding schedules of captive birds. Details of the metabolism and storage of vitamin A are available only for chickens (*25*).

V. METABOLIC FRACTIONATION

The most thorough studies on metabolic fractionation of carotenoids in birds are those of Denis Fox and associates on the Phoenicopteridae and various Ciconiiform birds. Because of their size, adaptability to captivity, and attractive coloration, these waterbirds are suitable for study in addition to their common use as display birds. The work on the distribution, fractionation, metabolism, and identification of the pigments in this group has been reviewed (*30,31*). No data as comprehensive as these are available for any other group. There are several patterns discernible, but it is difficult to determine how general the picture may be. For example, canthaxanthin is the major plasma carotenoid and is also widely distributed in feathers.

Monoketocarotenoids are almost entirely absent from integumentary structures and are found in the plasma and liver. These molecules may have a rather transient existence. Astaxanthin is more prominent in the skin than elsewhere. Phoenicoxanthin, another potentially intermediate compound (see Fig. 1), tends to be found in the skin rather than in the liver or blood. Few studies are available on the tissue distribution of pigments in other bird species. In the scarlet tanager, the predominant integumentary pigment is also found in the liver of nonmolting individuals (*9*).

The next largest body of data is a group of extensive observations on the integumentary carotenoids of birds. There are two reasons for this. First, the integumentary structures are of primary importance for zoo curators and gamekeepers and for commercial reasons. Colors also provide potentially important information in taxonomic studies. Second, feathers are extremely stable structures and can be procured easily from either living or prepared specimens. Carotenoids have been recovered from feathers in museum skins 80–100 years old. This stability and availability of feathers make them attractive sources of carotenoids for the laboratory investigator.

A variety of carotenoids have been isolated from the feathers of a wide range of species. In many cases, the pigment is pure, or nearly pure, canthaxanthin (*9,12,91*). A variety of other oxygen-containing carotenoids have been isolated from birds (*92,94*). The *in vivo* color frequently resembles the *in vitro* color of the pigment, but this is not always the case. In the cotinga (*Xipholena punicea*), a combination of red carotenoids and blue structural colors gives a deep purple coloration to the feathers (*11,85,87*). Direct *in vivo* and *in vitro* comparisons are of increasing importance in pigment studies (*48*).

Rather pure deposits of pigments are sometimes associated with unusual feather structures. In the cedar waxwing (*Bombycilla cedrorum*), the waxy structure on the tips of the wing feather contains chromatographically pure deposits of astaxanthin (*14*). Superficially, similar structures are also seen in some melanin-containing feathers (*8,10*). In other passerines, complex color patterns that contain elements of essentially pure carotenoids have been observed, e.g., the Gouldian finch. In the suboscines, although only one visible feather color may be involved, the depositions often consist of more than one pigment (*84*). Unfortunately, many studies on avian coloration were done before the introduction of various chromatographic techniques. Thus, a body of literature exists in which the pigments are identified only as lipochromes. There is a trend in more advanced species toward having only single pigments in the feathers. The taxonomic and metabolic implications of this and the relation to the overall pattern remain to be resolved.

In summary, birds with conspicuous deposits of carotenoids have the ability to transport and fractionate a variety of carotenoids. In the plasma, carotenoids are associated with lipid and HDL fractions. A large number of the carotenoids fractionated in the avian body are essentially unmodified from those found in the diet. Other molecules, especially ketocarotenoids, are found almost exclusively in animals. In some cases, their distribution is limited to certain tissues, often the reproductive system. This raises two questions regarding the metabolic modification of carotenoids. First, where are carotenoids metabolized and what types of pathways and control mechanisms are involved? Second, how is the selectivity in certain tissues established and maintained? This applies not only to the selectivity of certain tissues, to the binding mechanisms of transport and deposition, but also to the ability to produce patterns of carotenoid displays in both time and space.

VI. METABOLIC PATHWAYS

Some question exists with regard to the preliminary catabolism of even simple carotenoids. Current evidence suggests that, in mammals, β-carotene is cleaved at the central double bond by soluble enzymes in the gut and liver (*42,67*). The products appear to be retinal. On the other hand, the metabolic fate of α-carotene is less clear. McAnally and Szymanski (*62*) suggested that feeding α-carotene to mammals resulted in at least two different end products. Thus, they could not determine whether cleavage occurred principally at the central double bond or at one end of the molecule.

The carotenoids in the digestive tract presumably enter the body and are picked up in the circulation. Unlike the larger fatty compounds and vitamin A, carotenoids have not been found in the lymphatic system. Rather, they appear to be carried by plasma lipoproteins, presumably to the liver. Once the carotenoids are ingested, they can undergo any of several fates. In many mammals, a good deal of the carotenoid material appears unaltered in the feces. In birds, some pigments may be absorbed and deposited essentially unmodified (e.g., lutein and rhodoxanthin). This is generally interpreted as simply a reflection of the plant or algal material in the diet (*32*). Other pigments may undergo a series of metabolic changes. These usually involve the adding of various side groups, which change the absorption characteristics and therefore the color of the molecule. It appears that animals are incapable of changing the form of the backbone or long-chain chromophore or of interconverting α-carotenes and β-carotenes (*43*). The relatively simple chemical changes of which animals are capable may cause dramatic changes in the color, solubility, and partitioning behavior of the molecule and therefore greatly affect the biochemical behavior both intra- and extracellularly (*81*). These pigments usually undergo a series of oxidative steps and are not commonly found in plants (*91,92*). Presumably, they are derived by metabolic modification of unoxygenated precursors.

In the evolution of carotenoid usage in animals, it is parsimonious to assume that the unoxidized hydrocarbons were used first as active pigments. Unsubstituted carotenes are widely distributed in organisms. During the course of evolution, pressure for a more efficient usage or distinctive role of these molecules arose. Both α- and β-carotenes have a yellow color, and there might be little from which to choose between the deposition of these two, although intestinal transport may have played a role at one time. Efficiency, in an evolutionary sense, could be improved if oxycarotenoids were used in displays. Because of the intensity of their spectral absorption, the large variety of chemical combinations possible, and the possible tissue-specific functions, this appears to be a reasonable hypothesis. Two lines of evidence support this supposition. First, in a highly colored, but taxonomically diverse group such as the birds, the predominant integumentary pigments are xanthophyllic. The unoxygenated forms tend to be limited to such tissues as the retina and liver. Second, the unsubstituted dietary carotenes tend to disappear quickly (*35*). This is due in part to rapid conversion to vitamin A and presumably in part to their conversion by oxidation to modified tissue and integumentary pigments. These metabolic steps are initiated in the gut but are also important in the liver and perhaps in the integument itself. Finally, it should be pointed out that many of the intermediate forms can function as pigments (e.g., echinenone and phoenicopterone). In this case, intermediate forms are those that are only partly substituted. The simplest substitutions seem to be the end product in many species of both vertebrates and invertebrates. Thus, the available evidence indicates the presence of both a mechanism and selective pressures for the evolution of metabolic pathways leading to more complex pigments.

The organic chemistry of carotenoids has been reviewed (*55,56,61*). The ultimate biosynthetic origin of carotenoids lies in plants, fungi, and bacteria. Nevertheless, certain aspects of carotenoid biosynthesis are relevant to animal metabolism, specifically with regard to the precursor molecules and the pathways of subsequent metabolic modification. Lycopene, a C_{40} product of phytoene dehydrogenation, appears to be the biological precursor of both α- and β-carotene [references in Goodwin (*43*)]. However, the process of cyclization appears to be different for the α and β rings. That is, there is no interconversion from one ring structure to the other since it is generally held that both rings are formed separately but by the same or similar mechanisms (*43*).

Pathways of carotenoid modification have been elucidated in several invertebrates (*18,23,54*), but no complete pathways have been published for birds. It is possible to suggest one based on current data. Mutants of the Gouldian finch suggest that birds are incapable genetically of converting α and β chromophores, as previously proposed on chemical grounds (*15*). In experiments in which chickens were fed canthaxanthin, it was deposited in the skin (*53*) (A. L. Livingston, personal communication). Therefore, it was absorbed, transported, and deposited in the skin but not in the feathers. However, when fed a diet containing β-carotene, chickens did not produce canthaxanthin. This implies that *Gallus* lacks the metabolic pathway necessary for the conversion of β-carotene to canthaxanthin and does not deposit it in the feathers even when available, unless fed at very high levels (see Chapter 3 by Marusich and Bauernfeind). The information for carotenoid modification is apparently under relatively strict genetic control. For example, carotenoids are present in the red but not the white ibis and are absent in the closely related black-headed ibis, *Threskiornis melanocephala* (*33*). Dietary changes can eliminate color differences in caged birds but do not account wholly for such differences in native populations (*16*). Furthermore, there is some indication that the genetic systems, in turn, may be under hormonal influence (*9,16*) (Power and Brush, unpublished). Thus, it is possible by using information on the presence of various pigments in an assortment of tissues to attempt to reconstruct pathways for carotenoid metabolism in birds (Fig. 1).

It is apparent from Fig. 1 that the α-carotene pathway is shorter than the β-carotene one. From the viewpoint of structural chemistry, the asymmetric nature of the α-carotene molecule must limit the degree of possible oxidative substitutions. The yellow lutein molecule is probably widespread in birds. The occurrence of

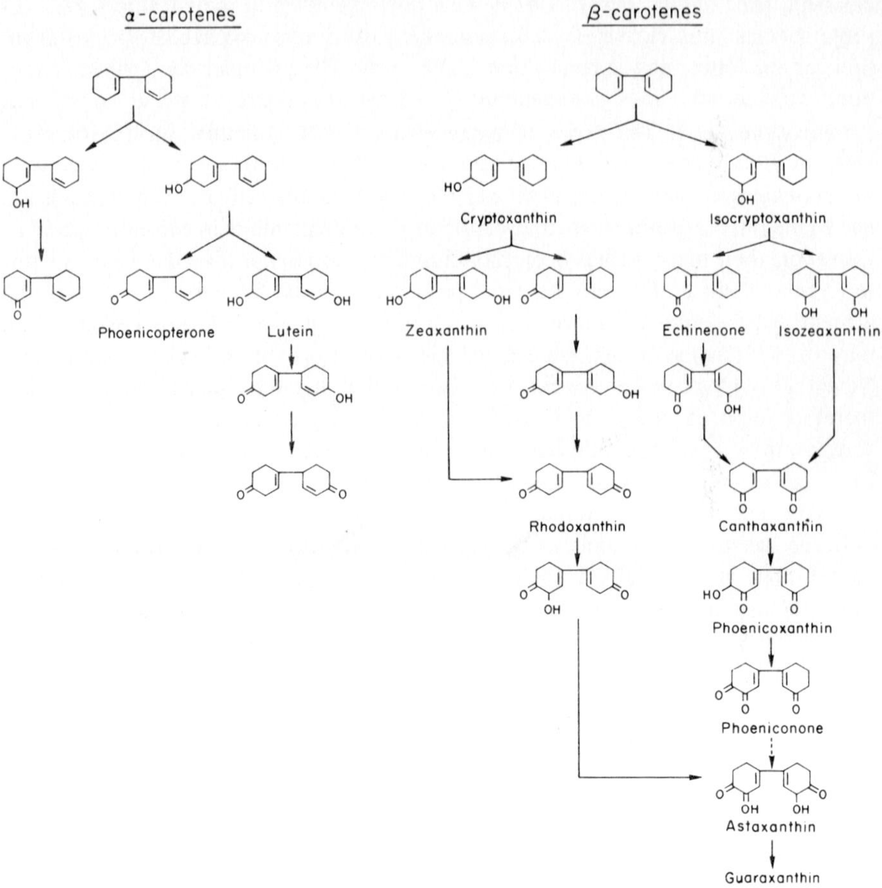

Fig. 1. Proposed intermediary pathways for carotenoids in birds. Parallel mechanisms begin with either α- or β-carotene backbone. Nomenclature after Foppen (*28*). Only ring portions of the molecules are shown. Only those molecules currently known from avian material are labeled. For further details see text.

phoenicopterone, which is orange, has been reported only in flamingos. Such molecules as 4-keto-α-carotene and 3,3′-diketo-α-carotene have not been reported in birds and are included only to illustrate potential homologies in the pathways. In spite of the limitations associated with α-carotenes, they occur widely. In addition, birds use the epoxy-α-carotenes in some unusual conditions. An example is the rare orange-faced mutant of the Gouldian finch, in which the β-carotene pathway has been deleted (*15*). One might interpret this as a case of "metabolic necessity."

The β-carotene series has frequently been used as a dietary color source in birds. These compounds, especially those that transmit in the red end of the spectra, occur repeatedly in the tissues of many avian species. They range from yellow to intense red, and one might suggest that every known "chemical trick" to increase redness has been tried. Thus, many birds deposit canthaxanthin in their feathers (*91*). In some species, this derivative is accompanied by a more oxidized form, astaxanthin, or the latter may appear alone. Among the Phoenicopteridae, phoenicoxanthin, astaxanthin, and guaraxanthin (7,8-dihydroastaxanthin) occur in various combinations (*31*). The series of papers from Fox's laboratory indicate that the hydroxy-β-carotenes (isocryptoxanthin and isozeaxanthin) are relatively rare in the Phoenicopteridae. Davies *et al.* (*23*) reported that the 4-hydroxycarotenoids are not found in the conversion of β-carotene to canthaxanthin in the brine shrimp. However, their presence has been established in many other invertebrates. In order to explain this, Davies *et al.* proposed a mechanism in which the 4-hydroxycarotenoid occurs only as a transient enzyme-bound intermediate. This may be widespread in animals (*81*). This may also explain Fox's observations. Nevertheless, hydroxycarotenoids have been found in avian material (*9,13*) and are therefore included in Fig. 1. These observations suggest an intense selective pressure toward ketonization of the integumentary carotenes, thus producing "redness" from the original yellow pigments at minor metabolic costs. The ultimate survival value of this strategy is axiomatic.

In the substitution sequence of β-carotenes, one of two symmetric pathways is chosen. The scheme calls for substitution at either the 3,3′ or 4,4′ ring position. Thus, the asymmetric series, e.g., 3,4′ structure, do not appear to be used by birds. This is based on the fact that molecules such as 3′-hydroxyechinenone or euglenanone (possibly 3,4′-diketo-β-carotene) have not yet been reported in birds. The evidence is that, at the level of echinenone, birds find it expedient to add the oxo group symmetrically on the opposite ring and continue toward canthaxanthin rather than substitute asymetrically. This conserves steps and presumably material since isozeaxanthin can then also act as a canthaxanthin precursor. Note that euglenanone would not necessarily be a dead end because it could be converted to phoeniconone (3,4,4′-triketo-β-carotene). Rhodoxanthin (3,3′-diketoretrodehydro-β-carotene) is red and has been recovered from both Passeriformes and Columbiformes. Nothing is known regarding its subsequent metabolism. In Fig. 1, the step from phoeniconone to astaxanthin is shown as a dashed arrow to indicate its questionable nature. It is possible, but admittedly unlikely that these conversions would occur *in vivo*. They would require several steps, probably including an isomerase. Since all these molecules are intensely red, an isomerase may not even be necessary; nevertheless, its potential existence should be mentioned.

It should be noted that essentially nothing is known about the physical or kinetic characteristics of the enzymes that catalyze these reactions. Fox *et al.* (*36*) speculated that the oxidation of β-carotenes to the mono and diketo derivatives might occur in the liver. The conversion of the more highly oxidized xanthophylls to astaxanthin (which is absent from the blood) might occur in the feather follicles or fluids in the pen feathers (*30*). There are no hard data to support either suggestion. Nevertheless, one hopes that ultimately we can combine information on metabolism, transportation, selectivity, histochemistry, and enzymology to obtain an integrated picture of carotenoid biochemistry.

VII. SOURCES OF INDIVIDUAL VARIATION IN WILD BIRDS

Carotenoids in solution are bleached rapidly by atmospheric oxygen, especially in the presence of light. Generally the carotenoids of feathers are rather stable. However, controlled exposure to sunlight produced rather rapid fading in *Colaptes* feathers (*78*). As shown below, diet has an important influence on feather color, especially in captive birds. Thus, both physical and biological parameters must be appreciated when one is considering the natural variation of pigments.

One consideration is age. In the red crossbill (*Loxia curvirostra*), there are pigmentary differences in the immature (first winter) and adult birds (*83*). In many other species, young birds tend to be less intensely colored than older ones. Juveniles also have plumage patterns that are different from those of adults. It is well known that seasonal differences in carotenoids also occur, probably in part due to dietary changes (availability of pigmented foods). A classic example is the bill coloration of the starling (*Sturnus vulgaris*) (*103*). In the scarlet tanager (*Piranga olivacea*), both males and females deposit isozeaxanthin in the feathers during the basic molt in the fall. The spring breeding (alternate) plumage of males contains canthaxanthin. Presumably, this change during both molts is mediated by a single enzyme induced by the male sex hormone (*9*). Other cases of sexual dichromatism are fairly common in birds, and the behavioral and communicatory aspects of these changes are obvious even if the biochemistry remains obscure.

Many species show polymorphisms with regard to carotenoid colors. One of the most spectacular of these is the bush shrikes of the genus *Malaconotus* (*44*). With no direct information on either the genetics or biochemistry of this complicated situation, Hall *et al.* (*44*) presented an astute theoretical analysis. The hypotheses still require testing. The situation is different in the case of the Gouldian finch (*Poephila gouldiae*). Because of its popularity as a cage bird, a wealth of data is available on the genetics of the color phases. The pigments and feather structure are also well known (*15,95*). It is obvious from this work that color, especially from carotenoids, is under genetic control, separate from that of the feather structure, and that both must be considered in interpretations of color in birds. In the sooty-capped tanager (*Chlorospingus pileatus*), color morphs based on quantitative distribution of carotenoids have been described by both reflectance spectrophometry and chemical methods (*48*). A similar difference has been demonstrated in two parulid warblers of the genus *Vermivora* (*17*).

A final factor that may affect the color of birds in the wild is the existence of hybrids. Test (*78*) reported on the pigments of interspecific hybrids in flickers (*Colaptes* spp.). In tanagers of the genus *Ramphocelus,* studies on parental and hybrid populations indicate the presence of only a single carotenoid, the relative concentration of which is responsible for the variation in color from scarlet (*R. flammigerus*) to yellow (*R. icteronotus*). In nature, there must be genetic control of the pigment concentration because colors are population specific, there is no noticeable variation in the color of "pure" forms, and hybrids are of intermediate shades depending on their distance from the "pure" populations (*12*). The biochemical studies on these hybrids have been supplemented by direct spectrophotometric analysis of the pigments within the feather (*48*). Additional observations on various mutations support this hypothesis and are valuable in understanding both the intraspecific variation and the few reported cases of melanism in carotenoid-containing feathers. (*12*).

VIII. DIETS, FEEDING, AND COLOR MANIPULATION

Certainly diet is important for the successful maintenance of any animal species in captivity (*98*), from the individual caged bird in one's home to the most elaborate zoo or commercially kept species. General diet becomes especially important as increasing numbers of rare or endangered species are entrusted to zoos. It must be realized, however, that diet is not the only parameter that determines success in animal keeping or breeding. Proper handling, adequate space, and many environmental considerations must also be considered.

Balanced diets (i.e., those containing adequate amounts and proportions of fat, carbohydrates, vitamins, and proteins) have been published for many zoo animals (*73*). Diets are designed to conform as closely as possible to the normal diets of the species. For example, among mammals, diets are available for herbivorous, omnivorous, carnivorous, and picivorous species. In many cases, dietary constituents take the form of selected additives that are supplemental to hay, raw meat, or fish. Dietary additives are available in a variety of physical forms but are usually fed as pellets or cake. Food concentrates such as these usually contain 18-23% crude protein, 2-7% crude fat, 1-3% calcium, and about 1% phosphorus. In addition, 3-8% of the total consists of fibrous binding. Salts of various elements, e.g., Fe, Cu, Mn, I, Zn, Co, may be added in trace amounts. Depending on the diet, various amounts of the vitamins deemed necessary for human nutrition are also added.

Since vitamin A is inevitably included (in amounts of 8000–20,000 IU per kilogram feed), a carotenoid precursor is available.

Dietary constituents for a variety of birds are given in Table 1. Notice that in each case there are several sources of carotenoids. The actual utilization and pigmentary potential of the individual sources are unknown. The inclusion of various materials was developed rather empirically, and synthetic carotenoids were not added until the mid-1950's (*96*). Diet A is from Wackernagel (*97*). It is fed to many small to medium-sized birds and may be supplemented with seeds, grain, meat, fruit, or lettuce. The overall composition is different from that of commercial poultry feed (crude protein, 21.9%; digestible protein, 18.7%; crude fiber, 5.2%). Ratcliffe (*73*) suggested a similar diet with slightly different proportions. In addition, oyster shell flour was substituted for bone meal and ground meat, and eggs were included in significant amounts.

There are many published and unpublished diet formulas for flamingos, ibises, and spoonbills (*30,49*). All diets take into consideration the feeding habits of these birds and are usually fed as a slurry. Diet B (Table 1) is the formula used in the San Diego Zoo (supplied courtesy of Mr. K. C. Lint, Curator of Birds). Quantities are "per bird." This diet may be considered in view of diets proposed by others. Shrimp and lobster are relatively expensive, and Conway (*20*) pointed out the economic and spoilage problems associated with them.

It appears that many feeders are more secure with several potential carotenoid sources than simply one (*72*). Carotenoid supplements are fed to a variety of smaller and taxonomically quite diverse birds. Bruning [(*7*), and personal communication] reported successful plumage color control in cotingas, barbets, quetzals, tanagers, sapsuckers, and a variety of fringillid finches. An extensive literature exists for gallinaceous bird diets (*1*). Carotenoids have not yet been reported in some of these species; nevertheless, they are listed in Section III.

In addition to the general diets presented here, Lint (*57*) discussed insect diets for birds, especially with regard to their high protein and vitamin content. The problems of keeping insect-eating species and those that eat soft plant material were considered extensively by Roots (*75*). Ratcliffe (*73*), in his comprehensive listing, also included at least two diets for hummingbirds.

Diets for birds with a wide variety of other food habits also are available [e.g., parrots, in Hauser (*45*)]. In many cases these are similar to those fed to mammals. Generally, diets are designed to accommodate picivorous, carnivorous, frugivorous, insectivorus, seed-eating, and omnivorous species. Details on these diets are readily available and need not be discussed here (see, for example, such series as the "International Zoo Yearbook" and *Aviculture* magazine). The principles of feeding zoo animals were the subject of a symposium (*71*). Since dietary carotenoids have a much more direct relation to integumentary colors in birds, diets and dietary supplements will be considered in greater detail.

It is possible to produce a carotenoid-free state in birds. In flamingos deprived of adequate dietary carotenoids, the tendency is toward a paling of feathers produced

TABLE 1

Three Basic Diets Fed to Zoo Birds[a]

Diet A[b]

Basic mixture		Composition	
Ground plate maize	18%	Crude protein	22.5%
Ground wheat	11%	Crude fat	7%
Ground barley	10%	Crude fiber	3%
Rolled oats	10%	Calcium	1.0%
Groundnut oil meal	8%	Phosphorus	0.8%
Soya bean oil meal	10%		
Nettle leaf meal	6%		
Dried yeast	8%		
Skimmed-milk powder	10%		
Stabilized fat	5%		
Bone meal	2%		
Salt	0.9%		
Trace element mixture[c]	0.1%		
Vitamin mixture[d]	1%		
	100%		

Final mixture		Composition	
Basic mixture	50%	Crude protein ca.	19%
Minced cooked meat	20%	Crude fat	6%
Ground carrots	20%	Crude fiber	2%
Ground hard-boiled eggs with shells	7%	Calcium	0.6%
		Phosphorus	0.4%
Supplement[e]	3%	Aureomycin	20 mg/kg food
	100%		

Diet B[f]

Basic mixture	
Cooked brown rice	2 oz
Fresh carrot pulp	2 oz
Purina kibbled dog food	2 oz
Fresh shrimp	1 oz
Spiny lobster	2 oz
Fresh ground meat	2 oz
Table salt	2% by wt
Bread	$\frac{1}{2}$ slice
Caradee	$\frac{1}{5}$ oz
Red salmon	1 oz
Paprika	1 tsp
Canthaxanthin	1%

(*Continued*)

TABLE 1—*Continued*

Diet C[g]	
Basic mixture	
Crushed limestone	0.275 lb
Brewer's yeast	0.650 lb
Vitamin mix	0.275 lb
Super laying mash	7.0 lb
Caradee (carrot oil)	0.5 lb
	8.7 lb
Final mixture	
Basic mixture	6 lb
Vitamin A-D_3-E	2 oz
Crushed limestone	3 tbs
Ground carrot	3 lb
Cooked ground meat	3 lb
Ground liver	2 lb
Cooked, ground whole egg with shell	2 lb
Brown rice	1.75%
Raisins	1.75%

[a] In actual feeding practice, rigid dietary programs may not always be adhered to in the maintenance of red feather pigmentation. Dietary supplements used are influenced by experience. The Bronx Zoo uses Caradee and carrots for flamingos, carrot juice and canthaxanthin for the tufted puffin, and canthaxanthin for tanagers and barbets. The Philadelphia Zoo uses canthaxanthin in a zoo cake–water mix for flamingos. The Washington National Zoo uses shrimp meal and canthaxanthin in a feed mix for its scarlet ibis and roseate spoonbills and shrimp meal, carrot juice, carotene, and canthaxanthin in a feed mix for its flamingos. The St. Louis Zoo feeds a mixture containing dried shrimp, ground raw carrots, carrot oil, and fresh carrot juice to its flamingos.

[b] Diet A from Wackernagel (*97*).

[c] Micronutrients added per kilogram of food: Fe, 20 mg; Cu, 2 mg; Mn, 50 mg; I, 2 mg; Zn, 10 mg; Co, 1 mg.

[d] Added per kg of food: Vitamin A, 40,000 IU; D_3, 6000 IU; B_2, 40 mg.

[e] The supplement contains 1 part Aurofac and 1 part fat-soluble extract of lucerne (Pancosma AG, Geneva).

[f] Diet B from K. C. Lint, San Diego Zoo (personal communication).

[g] Diet C from D. Bruning, Bronx Zoo (personal communication). The final mixture in this case is a generalized feed for insectivorous birds.

in response to either plucking or normal molt. In adult quail (*Coturnix c. japonica*) maintained on a carotenoid-free diet, there was a tendency toward feather loss and a matted condition of those that remained. Skin color did not change noticeably (*64*). The oil droplets of the cones (*100*) in the retina retained normal coloration for 11 months (*64*). Thus, the rate of disappearance was considerably slower than that reported for plasma carotenoids of the same bird and implies differential rates of catabolism and exchange between the oil droplets and the general circulation.

The absence of colored dietary carotenoids in quail did not affect egg productivity, fertility, or hatchability; the yolk was pale but not colorless. Young from these eggs were normal externally but lacked carotenoids in the oil droplets of the eye. Their general development was not affected adversely (*64*).

The observation that feather coloration and diet are related is an old one. As early as the late 1930's, zookeepers added ground carrots to the diets of many species. This was necessary despite the fact that some natural carotenoids were available in the diet. These sources included such items as nettle leaf meal and alfalfa meal. In addition to carrots, early supplements included various crustaceans and capsicum and insect larvae. Flamingos and the scarlet ibis, because of their size and color, have been more extensively studied with regard to diet (*20,21,66,72*) and color manipulation (series of papers by Fox *et al.*) than any other avian groups. It is now well established that carotenoids, especially canthaxanthin or its precursor, β-carotene, are necessary to produce and maintain good colors in the feathers. This is emphasized by the fact that most of the diets proposed in the last decade include an artificial carotenoid source, usually canthaxanthin prepared by Hoffmann-La Roche. Generally, the canthaxanthin is added to the diet at a level of about 1% [Neiboer (*66*), and personal communication from several zoo directors]. There appears to be no experimental basis on which this figure is based. It is necessary to consider the timing of diet relative to molt, specific feeding behavior, quantities of components, and the best vehicle for the foodstuff [e.g., Comben, in Phillipson (*71*)]. Because of its lower cost and relatively greater distribution as both a pigment and pigment precursor, one would expect β-carotene to be used rather than canthaxanthin. Given the metabolic pathways and fractionation of carotenoids in these birds, this is not an unwise choice. The fractionation of carotenoids in one flamingo species on a specific regimen was described by Fox *et al.* (*35*). It appears, as one might expect, that these species can use canthaxanthin directly and can modify such metabolic precursors as β-carotene to produce canthaxanthin. Similar data are available for the house finch *Carpodacus mexicanus* (*16*).

Manipulation of integumentary colors by dietary control is known in other species, the classic work on canaries being one of the first examples (*6,88,93*). This was followed by similar work on other species. In the goldfinch (*Carduelis carduelis*), when the natural carotenoids were replaced by rhodoxanthin, the regenerated feathers were red instead of the normal yellow (*89*). These experiments do not explain the color changes observed by Middleton (*65*) in native birds of this species. In an examination of his birds we found no qualitative difference in the pigment content of the red and yellow facial feathers, but quantitative differences existed, and only a single major xanthophyll was present (Brush and Zamlauski, unpublished). Individual differences in facial color were due to differential distribution of yellow and deep orange feathers. Each of the extreme examples had at least some of the feathers of the opposite color. Since there was no obvious relation to diet, breeding condition, age, or sex, the variation in the population was difficult to explain. All feathers had reduced barbules and the flattened, expanded, coalescing barbs typical of carotenoid-containing feathers, but we did not find the structural difference between red and yellow feathers mentioned by Desselberger (*24*). Another documented case of color manipulation in birds was reported for woodpeckers of the genus *Colaptes* (*79,80*). The typical color loss of feathers in birds on diets relatively free of carotenoids was observed. When fed a common diet (including carotenoid sources), the two species (*C. auratus* and *C. cafer*) produced the same color as wild conspecific individuals. This suggests genetic rather than simple dietary control. By varying the carotenoid levels of the diet during molt, Test was able to manipulate pigment deposition to the point of producing barring of feathers.

Finally, Weber (*101,102*) suggested that the feather colors of birds molting in captivity were determined not by diet alone but by the degree of freedom of movement. Both redpolls (*Carduelis flammea*) and crossbills (*Loxia curvirostra*) produced redder feathers when kept in larger cages than control birds kept in a confined space. No physiologically based reason for this was proposed, and the observation, to my knowledge, has not been verified independently. In a related study on the pine grosbeak (*Pinicola enucleator*) we found a complex situation in captive birds. Samples of this sexually dichromatic species molted under both wild and captive conditions were supplied by C. S. Adkisson, Virginia Polytechnic Institute. The situation overall was reminiscent of that of the scarlet tanager (*9*). The main pigment in red males was canthaxanthin, whereas isozeaxanthin predominated in the females. The situation differed in *Pinicola,* however, because of the presence of minor amounts of hypophasic, yellow pigments in both sexes. In yellow males, produced by molting in captivity, no canthaxanthin was found. These individuals had both isozeaxanthin and other yellow pigments in the plumage. The minor pigments differed from those in females. When these birds were allowed to molt in larger cages situated in a botanical garden, the male plumage was red. We concluded that the yellow plumage in captivity was a result of the lack of sufficient carotenoid precursor and that the diet of captive birds was supplemented adequately in the garden aviaries by insects or other means to produce the red plumage. The dichromatism in this case must be due to the female's inability to produce canthaxanthin from isozeaxanthin. The possibility of aptochromatic changes in the plumage of both wild and captive birds still has adherents [e.g., Meier (*63*)] but lacks adequate observational and chemical support.

Management of wild animal populations has only recently gained the interest of a wide spectrum of biologists [see, e.g., Boyd, in Crawford (*22*)]. Unlike the situation in zoological gardens or with pet animals, feeding managed wild animals is not a simple matter. Dietary control can be obtained by hand feeding prepared diets or by management of the vegetation through the use of fertilizers, herbicides, or burning. The first is costly in both labor and materials, and the latter lack specificity. Obviously, these practices apply most easily to large, herding herbivorous forms. Presumably, there are lessons to be learned from the management of domesticated forms such as sheep or cattle. In addition to herbivorous mammals, managed populations of flocking or grazing birds could also be developed. Ecologically, it follows that increases in the number and stability of herbivorous forms could result in elevated, stable populations of both avian and mammalian carnivores. The symposium edited by Crawford (*22*) considered problems of total nutrition in several species and specific problems, such as water balance, vitamin deficiencies, and other pathology, in both wild and caged populations of a variety of mammals. Management of any natural population is important from the viewpoint of both species survival and the potential use of such populations as food for human consumption.

IX. STRUCTURAL CORRELATIONS

Desselberg (*24*) was one of the first to use modern histochemical techniques to investigate the relationship between lipochromes and feather structure. These and other studies were done before the reapplication of Tswett's chromatographic techniques. As a result, in none of these early studies were the pigments fractionated or identified. It is now established that there is a direct correlation between feather structure and carotenoid pigments [see Brush and Allen (*14*), Brush and Seifried (*15*), and discussion in this report]. Specific carotenoids can also be combined with structurally produced colors to provide colors in feathers (*11,68,87*).

Functionally, the modifications in feather structure associated with carotenoids appear to produce an increase in the colored surface to maximize the exposure of pigment. This is conspicuously so in those species in which the pigmented feathers are limited to relatively small or isolated patches (*15*). These structures increase the brightness component of the reflected spectra (*48*). The display color of individual carotenoids can be affected also by binding to the feather proteins themselves, as seems to be the case in the fruit pigeons *Ptilinopus* (*86*). As more becomes known about the chemistry of feather proteins and the ultrastructural relationship between pigments and proteins, we shall achieve greater insight into color phenomena and presumably better methods for their control. For example, the situation with the complex of green, blue, and yellow feathers in which colors traditionally were presumed to be the result of the interaction between physical parameters and pigments has been reinvestigated (*26,26a*). In addition to proposing a different mechanism for the physical production of color, Dyck observed spectral changes of pigments within the feather as compared to the spectra in solution. The origin of these shifts as well as the contribution of fluorescent pigments to feather color require investigation.

X. SYSTEMATIC AND EVOLUTIONARY CONSIDERATIONS

Carotenoids are only one class of compounds that are of relatively low molecular weight compared to proteins but that are biologically important. Other such classes include amino acids, phosphonucleotides, vitamins, and coenzymes. For example, small carbohydrate moieties can profoundly affect the behavior of proteins. Much of the biological significance of these molecules is based on the information contained in their biological organization or their synthesis rather than on their intrinsic chemical information [for discussion, see Throckmorton (*82*)]. Generally, these chemicals are ubiquitous in plants and animals and may play a variety of chemical or physiological roles. The carotenoids, for example, in addition to their role in integumentary structures are involved in a number of physiological processes (*51*).

Their small size, relatively broad distribution, and dietary origin reduce the usefulness of carotenoids as carriers of taxonomic information (*104*). This is reflected in their distribution in birds. Carotenoids do not appear to be distributed in a fashion that reflects ancestral relationships or indicates unique biochemical homologies. Thus, the occurrence of individual carotenoids has not been used to establish biochemical relationships among any groups of animals. However, cladistic analysis of entire pathways in well-defined groups may prove to be informative (Brush, in preparation).

In spite of considerable recent progress, much more must be learned about the metabolism of carotenoids. One important question involves the existence of species-specific carotenoids. Frequently, molecules were given trivial names to show their origin, e.g., taraxanthin, echinenone, and phoenicopterone. As additional comparative evidence accumulates, it appears that we were not always dealing with species-specific pathways. Thus, the specificity of the nomenclature was not valid. Nevertheless, the information now available on genetic control of pathways and the specificity and selectivity of various tissues implies that carotenoid metabolism is an intricate mechanism with survival value. Integumentary carotenoids appear in many invertebrates, especially those with complex life cycles and some social interaction (i.e., crustaceans and echinoderms). Bright colors and patterning play numerous biological roles. Among vertebrates, fish and birds are the brightest groups. As mentioned earlier, mammals rarely rely on integumentary carotenoids. In birds, the ability to metabolize and deposit integumentary carotenoids has arisen several times. Within the groups acquiring this ability, the enzymes producing some degree of ketonization are almost always present, which allows the production of a range of colors from yellow to deep orange or red. The distinction between those taxa with and without the ability to metabolize carotenoids is presumably more significant evolutionarily than the distinction among those with various degrees of ketonization.

Probably of equal importance is the problem of selectivity in metabolic fractionation or deposition of individual carotenoids. The genetic and biochemical evidence indicates some degree of selectivity of both these parameters. Areas that require future study are the transport system of the blood and the metabolic pathways of both the liver and the feather papillae. This would help to clarify questions of specificity. In an overall evolutionary sense, the carotenoids seem to represent a series of metabolic pathways that have been redesigned numerous times to meet the specific needs of individual groups of organisms.

ACKNOWLEDGMENTS

This chapter was prepared while I was a fellow of the National Institutes of Health (GM 5085-01) in the Department of Biochemistry, University of California, Berkeley. Carotenoid and feather structure studies in my laboratory have been supported generously by the National Science Foundation and the University of Connecticut Research Foundation.

I take great pleasure in dedicating this paper to Professor D. L. Fox on the occasion of his seventy-sixth birthday.

REFERENCES

1. Anonymous (1976). "Aliments Composes pour Gibier a Plumes," Information Technique 685D. Bur. Nutr. Anim. Elevage, Marseille.
2. Baker, C. M. A., Manwell, C., Labisky, R. F., and Harper, J. A. (1966). *Comp. Biochem. Physiol.* **17,** 467–499.
3. Banks, R. C. (1970). *U.S. Bur. Sport Fish. Wildl., Spec. Rep.* No. 136.
4. Bell, G. H., Davidson, J. N., and Scarborogh, H. (1965). "Textbook of Physiology and Biochemistry." Williams & Wilkins, Baltimore, Maryland.
5. Boyd, H. J., and Alley, R. (1948). *Ibis* **90,** 582–593.
6. Brockmann, H., and Völker, O. (1934). *Hoppe-Seyler's Z. Physiol. Chem.* **224,** 193–215.
7. Bruning, D. (1971). *Int. Zoo Yearb.* **11,** 215–218.
8. Brush, A. H. (1965). *Auk* **82,** 155–160.
9. Brush, A. H. (1967). *Condor* **69,** 549–559.
10. Brush, A. H. (1967). *Wilson Bull.* **79,** 322–328.
11. Brush, A. H. (1969). *Condor* **71,** 431–433.
12. Brush, A. H. (1970). *Comp. Biochem. Physiol.* **36,** 785–793.
13. Brush, A. H. (1978). *In* "Aves" (A. H. Brush, ed.), Chemical Zoology, Vol. 10, pp. 141–164. Academic Press, New York.
14. Brush, A. H., and Allen, K. (1963). *Science* **142,** 47–48.
15. Brush, A. H., and Seifried, H. (1968). *Auk* **85,** 416–430.
16. Brush, A. H., and Power, D. M. (1976). *Auk* **93,** 725–739.
17. Brush, A. H., and Johnson, N. K. (1976). *Condor* **78,** 412–414.
18. Castillo, R., and Lenel, R. (1978). *Comp. Biochem. Physiol.* **59,** 67–73.
19. Cheesman, D. F., Lee, W. L., and Zagalsky, P. F. (1967). *Biol. Rev. Cambridge Philos. Soc.* **42,** 131–160.
20. Conway, W. G. (1960). *Avicult. Mag.* **65,** 108–112.
21. Conway, W. G. (1965). *Int. Zoo Yearb.* **5,** 162–164.
22. Crawford, M. A., ed. (1968). "Comparative Nutrition of Wild Animals," Symposia of the Zoological Society of London, No. 21. Academic Press, New York.
23. Davies, B. H., Hsu, W. J., and Chichester, C. O. (1970). *Comp. Biochem. Physiol.* **33,** 601–615.
24. Desselberger, H. (1930). *J. Ornithol.* **78,** 228–376.
25. Deuel, H. J. (1957). "The Lipids," Vol. 3. Wiley (Interscience), New York.
26. Dyck, J. (1971). *Z. Zellforsch. Mikrosk. Anat.* **115,** 17–29.
26a. Dyck, J. (1973). *Zool. Jb. Anat. Bd.* **90,** 550–566.
27. Fisher, H. (1972). *In* "Avian Biology" (D. S. Farner and J. R. King, eds.), Vol. 2, pp. 431–471. Academic Press, New York.
28. Foppen, F. H. (1971). *Chromatogr. Rev.* **14,** 133–298.

29. Fox, D. L. (1957). *In* "The Physiology of Fishes" (M. E. Brown, ed.), Vol. 2, pp. 367-385. Academic Press, New York.
30. Fox, D. L. (1975). *In* "Flamingos" (J. Kear and N. Duplaix-Hall, eds.), pp. 162-182. Poyser, Berkhamsted, England.
31. Fox, D. L. (1976). "Animal Biochromes and Structural Colors," 2nd ed. Univ. of California Press, Berkeley.
32. Fox, D. L., and Hopkins, T. S. (1966). *Comp. Biochem. Physiol.* **17,** 841-856.
33. Fox, D. L., and Hopkins, T. S. (1966). *Comp. Biochem. Physiol.* **19,** 267-278.
34. Fox, D. L., Hopkins, T. S., and Zilversmit, D. B. (1965). *Comp. Biochem. Physiol.* **14,** 641-649.
35. Fox, D. L., McBeth, J. W., and MacKinney, G. (1970). *Comp. Biochem. Physiol.* **36,** 253-262.
36. Fox, D. L., Smith, V. E., and Wolfson, A. A. (1967). *Comp. Biochem. Physiol.* **23,** 225-232.
37. Fox, D. L., Wolfson, A. A., and McBeth, J. W. (1969). *Comp. Biochem. Physiol.* **29,** 1223-1229.
38. Fox, H. M., and Vevers, G. (1960). "The Nature of Animal Colors." Univ. of Washington Press, Seattle.
39. Frank, F. (1939). *J. Ornithol.* **87,** 426-523.
40. Frigg, M. (1970). *Z. Vgl. Physiol.* **69,** 186-224.
41. Goldsmith, T. H. (1965). *Peabody Mus. Nat. Hist. Yale Univ., Postilla* No. 91.
42. Goodman, D. S., and Huang, H. S. (1965). *Science* **149,** 879-880.
43. Goodwin, T. W. (1971). *Biochem. J.* **123,** 293-329.
44. Hall, B. P., Moreau, R. E., and Galbraith, I. C. J. (1966). *Ibis* **108,** 161-182.
45. Hauser, K. W. (1976). *Prakt. Tierarzt* **57,** 17.
46. Huang, H. S., and Goodman, D. S. (1965). *J. Biol. Chem.* **240,** 2839-2844.
47. Isler, O., ed. (1971). "Carotenoids." Birkhaeuser, Basel.
48. Johnson, N. K., and Brush, A. H. (1972). *Syst. Zool.* **21,** 245-262.
49. Kear, J. (1974). *Int. Zoo Yearb.* **14,** 142-144.
50. Keeler, C. A. (1893). "Evolution of the Colors of North American Land Birds." Calif. Acad. Sci., San Francisco, California.
51. Krinsky, N. I. (1971). *In* "Carotenoids" (O. Isler, ed.), pp. 669-716. Birkhaeuser, Basel.
52. Krinsky, N. I., Cromwell, D. G., and Oncely, J. L. (1958). *Arch. Biochem. Biophys.* **73,** 233-246.
53. Kuzmicky, D. D., Kohler, G. O., Livingston, A. L., Knowles, R. E., and Nelson, J. W. (1968). *Poult. Sci.* **47,** 389-397.
54. Lee, W. L. (1966). *Comp. Biochem. Physiol.* **18,** 17-36.
55. Liaaen-Jensen, S. (1970). *Experientia* **26,** 697-710.
56. Liaaen-Jensen, S. (1971). *In* "Carotenoids" (O. Isler, ed.), pp. 63-188. Birkhaeuser, Basel.
57. Lint, K. C. (1961). *Int. Zoo Yearb.* **3,** 104.
58. Lönnberg, E. (1934). *Proc. Int. Ornithol. Congr., 8th,* pp. 410-424.
59. Lush, I. E. (1963). *Br. Poult. Sci.* **4,** 255-260.
60. Marusich, W. L., and Bauernfeind, J. C. (1970). *Poult. Sci.* **49,** 1555-1566.
61. Mayer, H., and Isler, O. (1971). *In* "Carotenoids" (O. Isler, ed.), pp. 325-575. Birkhaeuser, Basel.
62. McAnally, J. S., and Szymanski, C. D. (1966). *Nature (London)* **210,** 1366.
63. Meier, G. (1957). *Gefiederte Welt* **81,** 122-124.
64. Meyer, D. B., Stuckey, S. R., and Hudson, R. A. (1971). *Comp. Biochem. Physiol.* **40,** 61-70.
65. Middleton, A. L. A. (1969). *Emu* **69,** 145-154.
66. Neiboer, E. (1965). *Int. Zoo. Yearb.* **5,** 164-165.
67. Olson, J. A., and Hayaishi, O. (1965). *Proc. Natl. Acad. Sci. U.S.A.* **54,** 1364-1370.
68. Olson, S. L. (1970). *Condor* **72,** 424-430.
69. Palmer, L. S. (1922). "Carotenoids and Related Pigments." Chem. Catalog Co. (Tudor), New York.

70. Peterson, P. A., and Berggard, I. (1971). *J. Biol. Chem.* **246,** 25–33.
71. Phillipson, A. T. (1976). *Int. Zoo Yearb.* **16,** 1–32.
72. Poulsen, H. (1960). *Avicult. Mag.* **66,** 48–51.
73. Ratcliffe, H. L. (1966). *Int. Zoo Yearb.* **6,** 5–23.
74. Roosevelt, T. R. (1911). *Bull. Am. Mus. Nat. Hist.* **30,** 120–231.
75. Roots, C. (1970). "Softbilled Birds." Gifford, London.
76. Searle, A. G. (1968). "Comparative Genetics of Coat Color in Mammals." Academic Press, New York.
77. Steinbacher, G. (1930). *J. Ornithol.* **78,** 53–58.
78. Test, F. H. (1940). *Condor* **42,** 76–80.
79. Test, F. H. (1942). *Univ. Calif., Berkeley, Publ. Zool.* **46,** 371–390.
80. Test, F. H. (1969). *Condor* **71,** 206–211.
81. Thommen, H. (1971). *In* "Carotenoids" (O. Isler, ed.), pp. 637–668. Birkhaeuser, Basel.
82. Throckmorton, L. H. (1968). *Annu. Rev. Entomol.* **13,** 99–114.
83. Tordoff, H. B. (1952). *Condor* **54,** 200–203.
83a. Trans, E. G. (1969). *Comp. biochem. Physiol.* **28,** 1177–1184.
84. Völker, O. (1951). *J. Ornithol.* **93,** 20–26.
85. Völker, O. (1952). *J. Ornithol.* **93,** 122–129.
86. Völker, O. (1953). *J. Ornithol.* **94,** 263–273.
87. Völker, O. (1955). *Naturwissenschaften* **12,** 612–613.
88. Völker, O. (1955). *J. Ornithol.* **96,** 54–57.
89. Völker, O. (1958). *J. Ornithol.* **99,** 422–425.
90. Völker, O. (1960). *Fortschr. Chem. Org. Naturst.* **18,** 177–222.
91. Völker, O. (1961). *Naturwissenschaften* **17,** 581–582.
92. Völker, O. (1961). *J. Ornithol.* **102,** 430–437.
93. Völker, O. (1962). *J. Ornithol.* **103,** 276–286.
94. Völker, O. (1963). *Carotine Carotenoide* **9,** 282–291.
95. Völker, O. (1964). *J. Ornithol.* **105,** 186–189.
96. Wackernagel, H. (1959). *Avicult. Mag.* **65,** 20–21.
97. Wackernagel, H. (1960). *Int. Zoo Yearb.* **2,** 95–102.
98. Wackernagel, H. (1968). *In* "Comparative Nutrition of Wild Animals" (M. A. Crawford, ed.), Symposia of the Zoological Society of London, No. 21, pp. 1–12. Academic Press, New York.
99. Wald, G., and Zussman, H. (1938). *J. Biol. Chem.* **122,** 449–460.
100. Walls, G. L., and Judd, H. D. (1933). *Br. J. Ophthalmol.* **17,** 641–675, 705–725.
101. Weber, H. (1953). *J. Ornithol.* **94,** 342–346.
102. Weber, H. (1961). *J. Ornithol.* **102,** 159–163.
103. Witschi, E. (1961). *In* "Biology and Comparative Physiology of Birds" (A. J. Marshall, ed.), Vol. 2, pp. 115–168. Academic Press, New York.
104. Zuckerkandl, E., and Pauling, L. (1965). *J. Theor. Biol.* **8,** 357–366.

6

CAROTENES AND OTHER VITAMIN A PRECURSORS IN ANIMAL FEED

J. C. BAUERNFEIND, C. R. ADAMS, AND W. L. MARUSICH

ISBN 0-12-082850-2

I. INTRODUCTION

The carotenoids, a class of fat-soluble pigments occurring naturally (*901*) in groups of living organisms, were recognized as chemical entities for about 100 years before the existence of vitamin A was known (Table 1). Wachenroder (*891*) isolated

TABLE 1

Some Events in the History of Vitamin A, Carotene, and Provitamin A[a]

Date	Event	Reference[b]
1831	Carotene isolated from carrots	1
1906–1911	Column chromatography developed for carotenoid separation	2
1907	Empirical formula of carotene established	3
1913–1915	Fat-soluble growth factor "A" recognized in cod liver oil and butter	4, 5
1917	Xerophthalmia recognized as vitamin A deficiency	6
1919	Growth factor observed in poultry liver	7
1919–1920	Vitamin A activity related to yellow carotenoids of corn	8
1921	Quantitative rat growth assay method developed for vitamin A	9
1923	Poultry shown to require vitamin A	10

(*Continued*)

TABLE 1—*Continued*

Date	Event	Reference[b]
1920–1925	Vitamin activity in fish oil associated with purple color reaction of sulfuric acid or arsenic trichloride	11
1924–1926	Ruminants shown to require vitamin A	12, 13
1926	Stable color with antimony trichloride in chloroform	14
1928	Carotene gives Carr–Price reaction and cures vitamin A deficiency in rats	15
1928–1931	Structure of carotene and vitamin A established	16
1930	Carotene demonstrated to be converted to vitamin A	17
1931	First vitamin A standard developed: 0.6 μg β-carotene = 1 IU vitamin A (League of Nations)	
1937–1940	Vitamin A crystallized	18, 19
1947	Vitamin A synthesized	20, 21
1948	Vitamin A_2 isolated	22
1949–1950	Commercial production of vitamin A	
1951–1952	Development of gelatin beadlet stablized vitamin A for aminal applications	
1949–1953	Modification of vitamin A in vision revealed	23, 24
1950	Present vitamin A standard adopted: 0.0344 μg vitamin A acetate = 1 IU (World Health Organization)	
1950	β-Carotene synthesized	25–27
1951	Vitamin A_2 synthesized	28
1954–1956	Industrial synthesis of β-carotene; commercial production of β-carotene	29
1958–1959	β-Apo-8′-carotenal synthesized	30, 31
1970	Several quadrillion (10^{15}) IU of vitamin A are produced by chemical synthesis annually	

[a] From Bauernfeind and Cort (*50*) and Bauernfeind and DeRitter (*51*).

[b] Key to references: 1. Wackenroder, H., *Mag. Pharm.* **33,** 144 (1831); 2. Tswett, M., *Ber. Dtsch. Bot. Ges.* **24,** 316, 384 (1906); **26,** 88 (1908); **29,** 630 (1911); 3. Willstätter, R., and Mieg, W., *Justus Liebigs Ann. Chem.* **355,** 1 (1907); 4. McCollum, E.V., and Davis, M., *J. Biol. Chem.* **15,** 167 (1913); **23,** 181 (1915); 5. Osborne, T.B., and Mendel, L.B., *J. Biol. Chem.* **15,** 311 (1913); **20,** 379 (1915); 6. McCollum, E.V., and Simmonds, N., *J. Biol. Chem.* **32,** 181 (1917); 7. Palmer, L.S., and Kempster, H.L., *J. Biol. Chem.* **39,** 299 (1919); 8. Steenbock, H., *et al., Science* **50,** 352 (1919); *J. Biol. Chem.* **41,** 81 (1920); 9. Zilva, S.S., and Miura, M., *Biochem. J.* **15,** 654 (1921); 10. Emmett, A.D., and Peacock, G., *J. Biol. Chem.* **56,** 679 (1923); 11. Rosenheim, O., and Drummond, J.C., *Lancet* **1,** No. 198, 862 (1920); **1,** No. 19, 753 (1925); 12. Hart, E.B., *et al., J. Biol. Chem.* **62,** 316 (1924); 13. Jones, R.I., *et al., J. Dairy Sci.* **9,** 119 (1926); 14. Carr, F.H., and Price, E.A., *Biochem. J.* **20,** 497 (1926); 15. von Euler, H., von Euler, B., and Hellstrom, H., *Biochem. Z.* **203,** 370 (1928); 16. Karrer, P., *et al., Helv. Chim. Acta* **12,** 1142 (1929); **13,** 1084 (1930); **14,** 614, 1431 (1931); 17. Moore, T., *Biochem. J.* **24,** 692 (1930); 18. Holmes, J.G., and Corbet, R.E., *J. Am. Chem. Soc.* **59,** 2042 (1937); 19. Baxter, J.G., and Robeson, C.D., *Science* **92,** 202 (1940); 20. Isler, O., *et al., Helv. Chim. Acta* **30,** 1911 (1947); **32,** 489 (1949); 21. Arens, J.F., and vanDorf, D.A., *Nature (London)* **160,** 187 (1947); 22. Salah, M.K., and Morton, R.A., *Biochem. J.* **43,** lvi (1948); 23. Wald, G., *Annu. Rev. Biochem.* **22,** 497 (1953); *Doc. Opthalmol.* **3,** 94 (1949); 24. Ball, S., and Morton, R.A., *Biochem. J.* **45,** 298 (1949); 25. Karrer, P., and Eugster, C.H., *Helv. Chim. Acta* **33,** 1172 (1950); 26. Inhoffen, H.H., *et al., Justus Liebigs Ann. Chem.* **570,** 54 (1950); 27. Milas, N.A., *et al. J. Am. Chem. Soc.* **72,** 4844 (1950); 28. Farrar, K.R., *et al., Chem. Ind.* **70,** 49 (1951); 29. Isler, O., *et al., Helv. Chim. Acta* **39,** 249 (1956); 30. Isler, O., *et al. Chimia* **12,** 89 (1958); 31. Rüegg, R., *et al., Helv. Chim. Acta* **42,** 854 (1959).

carotene from carrots as early as 1831, but it was not until about 1931 that Karrer *et al.* (*440*) prepared the first concentrate of vitamin A from natural sources. Goodwin (*324*) concluded that carotenogenesis proceeds primarily in the plant world. The biosynthesized plant pigments (Fig. 1) are ingested by animals and are metabolized, resulting in significant pigment storage in tissues in some animals, in others to a lesser degree, and in some practically not at all.

All the mechanisms that control the absorption and assimilation of some carotenoids or that may modify others during ingestion, according to species differences, are not fully understood, but useful data continue to accumulate. A great deal of species specificity has been noted regarding the ability to absorb dietary carotenoids. Thus, certain species, such as man and bovines, can absorb both vitamin A and carotenoids; both species convert provitamin A carotenoids to

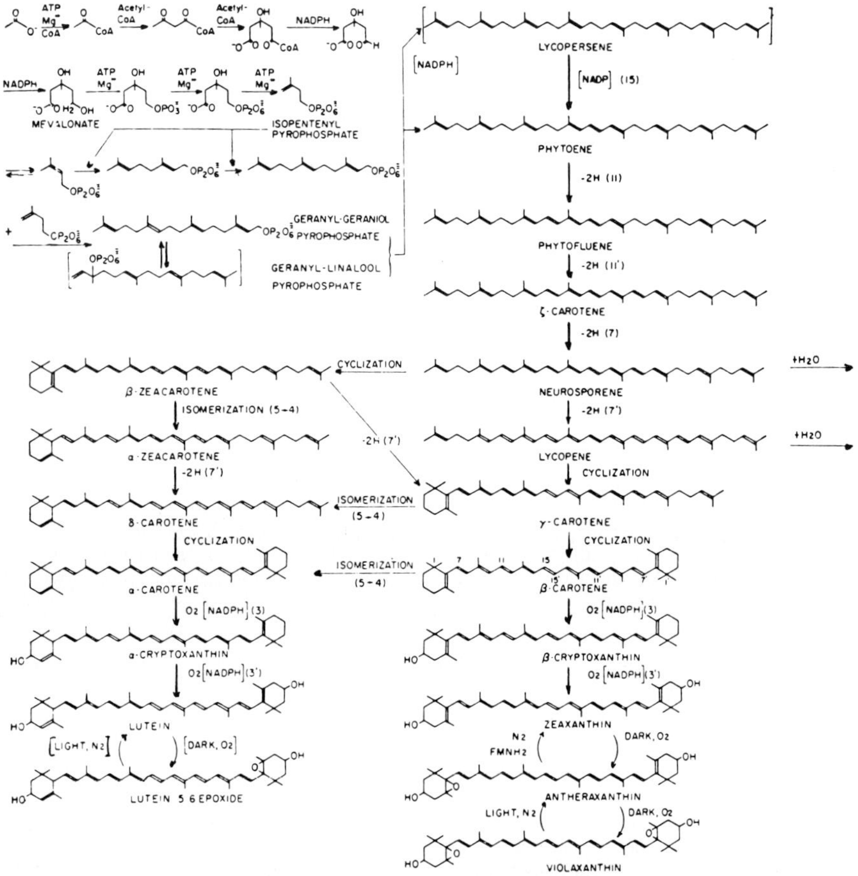

Fig. 1. Composite of postulated pathways for carotenoid biosynthesis. [From Olson (*656*).]

vitamin A. Whereas β-carotene is the chief tissue carotenoid in the dairy cow, little is deposited in the goat. In contrast, such species as the rat and pig do not appear to deposit significant amounts of carotenoid pigments in their tissues whether they are provitamins or nonprovitamin oxycarotenoids, although they have the ability to convert provitamins A to vitamin A. Birds, including the chicken, absorb and deposit oxycarotenoids in their body tissues and convert precursors to vitamin A. Early reviews are those of Palmer (*666*) and Peterson *et al.* (*694*).

Ganguly *et al.* (*290,291*) studied the species specificity of the absorption mechanism and put forward the hypothesis that suitable receptor proteins in an organism determine its ability to absorb or reject a particular carotenoid. Their findings were on the intracellular distribution of β-carotene, lycopene, and lutein in the mucosal cells of rats and chicks following the feeding of a single carotenoid or mixtures of carotenoids. In the chicken, significant amounts of lutein were found in the intestinal mucosa and muscle. The other two carotenoids were present in trace amounts in both rat and chick tissues. The bulk of the lutein in the intestinal mucosa of the chick was associated with the mitochondrial and microsomal fractions of the mucosal cell homogenate. In view of present concepts (*539*) of lipid absorption in the form of micelles, carotenoid absorption might well be due to one species being able to form suitable micellar solutions in the intestinal lumen with one carotenoid but not with others. In a 1975 study (*890*) Verma and Wallach, observing resonance-enhanced Raman spectra of erythrocyte ghosts, proposed the presence of membrane-bound β-carotene. Hence, the affinity of tissues of species for specific carotenoid structures may influence species metabolism of these compounds.

II. VITAMIN A ACTIVITY

The naturally occurring pigments, as components principally of plants, are ingested by animals and metabolized by one or more pathways: (a) unselective absorption, (b) selective absorption, (c) chemical alteration en route through the digestive tract or during the absorption process, or before tissue storage or *in vivo* function. One transformation of practical significance to mammals, birds, amphibians, and fish is the conversion of certain structural carotenoids to vitamin A. Vitamin A is essentially an animal compound and is present in practically all animal species.

The activity of carotenoid vitamin A precursors can be illustrated in two ways: (a) by feeding the carotenoid to an animal under prescribed bioassay conditions and noting tissue responses, such as vitamin A levels in blood and liver, or noting an effect on growth; and (b) by using a purified or partially purified carotenoid-cleavage enzyme in contact with the carotenoid under *in vitro* conditions and measuring retinal (vitamin A aldehyde) formation. With the *in vitro* enzyme approach, such carotenoids as α-carotene, β-carotene, bis-3,3′,4,4′-dehydro-β-carotene, 3′,4′-dehydro-17′-oxo-γ-carotene, and β-apo-10′-carotenol have shown activity (*499*).

For general vitamin A activity in mammals, a carotenoid provitamin A compound must have at least an unsubstituted β-ionone ring and a polyene side chain attached. The other end of the molecule may be cyclic or acyclic and may be lengthened but not shortened to less than an 11-carbon polyene fragment. The symmetric isoprenoid chain containing an all-trans conjugated double-bond system is necessary for maximal activity. Moore (*608*), over 50 years ago, unequivocally showed by biological criteria that β-carotene is converted to vitamin A, although Steenbock (*824*), over 60 years ago, first conceived that yellow color might be associated with biological activity. The relation of carotene to vitamin A was suggested by Steenbock on the basis of the intensity of color due to the yellow pigments in butter, peas, and maize being roughly proportional to their vitamin A biopotency. Later it was shown that extremely potent vitamin A preparations were colorless and that some highly colored yellow compounds were without vitamin A activity. These observations challenged the concept that vitamin A and carotene are identical. When von Euler and co-workers (*256*) demonstrated that purified β-carotene administered in small amounts could cure vitamin A deficiency in rats, it was evident that both a colorless compound (an absorption band at 328 nm) and a highly colored yellow compound (no absorption band at 328 nm) had vitamin A activity. Moore (*605*) demonstrated that colorless vitamin A appeared in the liver fat of vitamin A-deficient rats after the administration of either vitamin A or β-carotene, and in both cases the product had the characteristic absorption band at 328 nm. Colorless vitamin A is therefore the active principle, and β-carotene is a precursor of vitamin A, or a "provitamin A."

β-Carotene possesses two β-ionone rings, one at either end of a long polyene chain, and is a provitamin A with high activity. α-Carotene possesses both a β-ionone and an α-ionone ring, whereas γ-carotene is a carotenoid containing one β-ionone ring and an opened ring. α-Carotene and γ-carotene are also biologically active but at approximately half the β-carotene value. Some of the carotenoids possessing vitamin A activity (*47*) and their occurrence are listed in Table 2. The activity values are presented in terms of β-carotene valued at 100%.

Oxygen may be introduced into the carotene molecule in certain locations and oxidation states. The β-apocarotenals are postulated to be degradation products, formed if terminal oxidation of the β-carotene molecule is one pathway for conversion to vitamin A (*311*). They are biologically active when given orally (*441,538*) or parenterally (*558*). The associated carotenals, carotenoic acids, and some acid esters are also active. A linear relationship was found between loss of β-carotene exposed to molecular oxygen at 60°C and time (*239*). The products of the reactions were β-carotene 5,6-monoepoxide, β-carotene 5,6,5′,6′-diepoxide, β-carotene 5,8-monoepoxide, and β-carotene 5,8,5′8′-diepoxide and isomers. The predominant formation of epoxides is further evidence that the initial site of oxygen attachment on the molecule can be the terminal double bond. The 5,6-monoepoxide and the 5,8-monofuranoxide are believed to be vitamin A precursors despite the altera-

TABLE 2

Occurrence and Vitamin A Activity of Some Carotenoids[a]

Carotenoid[b]	Activity (%)	Occurrence
β-Carotene	100	Green plants, vegetables, carrots, sweet potatoes, squash, spinach, tomatoes, green peppers, pineapples, paprika, oranges, cranberries, figs, grapes, berries, apricots, peaches, prunes, apples, pears, strawberries, watermelons, wheat, corn, pasta products, palm oil, sorghum, hay, silage, algae, lichens, crustaceans, bivalves, eggs, fish, trout, alfalfa, *Tagetes* meal
α-Carotene	50–54	Green plants, carrots, squash, corn, green peppers, watermelons, potatoes, apples, peaches, oranges, cherries, figs, berries, grapes, bananas, pineapples, pasta, bleached paprika, *Tagetes* meal, trout, palm oil, chestnuts
3,4-Dehydro-β-carotene	75	Microorganisms
3,4,3′,4-Bisdehydro β-carotene	38	
2,2′-Dimethyl-β-carotene	50	
1,1′-Bisdemethyl-1,1′-bisethyl-β-carotene	41	
γ-Carotene	42–50	Carrots, sweet potatoes, corn, tomatoes, algae, some fruits including apricots and watermelons, microorganisms, palm oil
7′,8′-Dihydro-γ-carotene, β-zeacarotene	20–40	Corn, tomatoes, yeast, cherries
β-Carotene 5′,6′-monoepoxide	21	Plants, potatoes, red peppers
α-Carotene 5,6-monoepoxide	25	Plants, flowers, bleached paprika
β-Carotene 5,6:5′,6′-diepoxide	Active?	Plants
β-Carotene 5′,8′-monofuranoxide, mutatochrome, citroxanthin, flavacin	50	Orange peel, red peppers, tomatoes, sweet potatoes, cranberries, algae, bleached paprika

(*Continued*)

TABLE 2—*Continued*

Carotenoid[b]	Activity (*4*)	Occurrence
β-Carotene 5,6-monoepoxide 5′,8′-monofuranoxide, luteochrome	Active?	
β-Carotene 5,8:5′,8′-difuranoxide, aurochrome	Active?	
4-Keto-β-carotene, 4-oxo-β-carotene, echinenone, aphanin, myxoxanthin	44–54	Algae, sea urchins, *Daphnia, Hydra,* red sponges, brine shrimp, crustaceans
3-Keto-β-carotene, 3-oxo-β-carotene	52	
3,4-Dehydro-4′-keto-β-carotene	Active	
β-Semicarotenone	Active	Citrus fruit, *M. exotica, T. trifolia*
3,4-Diketo-β-carotene, euglenanone	Probably active	Algae
3-Hydroxy-β-carotene, cryptoxanthin	50–60	Yellow corn, green peppers, lichens, persimmons, papayas, lemons, oranges, prunes, apples, apricots, peaches, strawberries, cranberries, pineapples, paprika, pasta, eggs, poultry, *Tagetes* meal
3-Hydroxy-β-carotene 5′6-monoepoxide, cryptoxanthin monoepoxide	Probably active	Potatoes, oranges, lemons, prunes, flowers
4-Hydroxy-β-carotene, isocryptoxanthin	48	Brine shrimp
Isocryptoxanthin methyl ether	52	
3,4-Dehydro-3′-hydroxy-β-carotene, anhydrolutein, deoxylutein	Active, ca. 10	Alfalfa meal, acidulated soybean soapstock
5,6-Dihydroxy-β-carotene	Active	
3-Hydroxy-4-keto-β-carotene, hydroxyechinenone	Probably active	Algae, bacteria, flowers
β-Apo-2′-carotenal	Active	Citrus fruit
β-Apo-8′-carotenal	36–72	Citrus fruit, green plants, animal tissue, alfalfa meal, grass

(*Continued*)

TABLE 2—*Continued*

Carotenoid[b]	Activity (%)	Occurrence
β-Apo-10′-carotenal	Active	Citrus fruit, green plants, alfalfa meal
β-Apo-12′-carotenal	120	Alfalfa meal
β-Apo-8′-carotenoic acid	Active	Corn, animal tissue
3′,4′-Dehydro-17′-oxo-γ-carotene, oxotorulene	Active	Microorganisms, yeast
β-Apo-8′-carotenoic acid ethyl ester	25, 78	
Citranaxanthin	44	Citrus, *S. citrangequat*
Torularhodin	Active, <50	Yeast, microorganisms, fungi, plants
Cryptocapsin	Probably active	Red peppers
Lycopene	Inactive	Tomatoes, carrots, green peppers, pink citrus fruit, apricots, watermelons, microorganisms
3,3′-Dihydroxy-β-carotene, zeaxanthin	Inactive	Spinach, paprika, yellow corn, green peppers, fruits, pasta, poultry, eggs, bivalves, brine shrimp, algae
3,3′-Dihydroxy-α-carotene, lutein	Inactive	Green leaves, corn, potatoes, spinach, green peppers, carrots, tomatoes, fruits, apples, pears, apricots, peaches, prunes, oranges, strawberries, cranberries, figs, grapes, blackberries, pineapples, alfalfa meal, *Tagetes* meal, trout, algae, grass, poultry, eggs, pasta
3,3′-Dihydroxy-4,4′-diketo-β-carotene, astaxanthin	Inactive	Oranges, crustaceans, lobster, fish, algae, *Daphnia,* trout, salmon
4,4′-Diketo-β-carotene, canthaxanthin aphanicin	Inactive	Mushrooms, trout, *Daphnia, Hydra,* microorganisms, algae, crustaceans, brine shrimp
Capsanthin	Inactive	Red peppers, paprika
Capsorubin	Inactive	Red peppers, paprika
Bixin	Inactive	Annatto seeds

[a] Adapted from Bauernfeind (*47*).

[b] Other potential provitamin A compounds believed to exist in nature are the following: β,γ-carotene (5,18-didehydro-5,6-dihydro-β,β-carotene) in fungi; torulene (3′,4′-dehydro-γ-carotene) in yeasts and fungi; 7′,8′,11′,12′tetrahydro-γ-carotene; 1′,2′-dihydro-1′-hydroxytorulene in bacteria; 1′,2′-dihydro-1′-hydroxy-γ-carotene in bacteria; aleuriaxanthin in mushrooms and yeasts; 4′-hydroxy-γ-carotene; 3′4-didehydro-18′-hydroxy-γ-carotene in yeasts; 3,4-dihydroxy-β-carotene; plectaniaxanthin (1′,2′-dihydroxy-1′,2′-dihydrotorulene) in bacteria; 1′-methoxy-1′,2′-dihydro-3′,4′-dehydro-γ-carotene in bacteria; 2′-dehydroplectaniaxanthin (1′-hydroxy-2′-keto-1′,2′-dihydrotorulene); β-apo-4′-carotenal; β-apo-2′-carotenol in yeasts; β-apo-8′-carotenol and β-apo-12′-carotenol; neurosporaxanthin (β-apo-4′-carotenoic acid) in microorganisms; sintaxanthin (7′,8′-dihydro-7′-apo-β-caroten-8′-one); β-apo-8′-carotenoic acid in eggs and animal tissue; β-apo-10′-carotenoic acid in yolk; and β-apo-12′-carotenoic acid in yolk and animal tissue.

tion of one of the β-ionone rings. It has been assumed that, *in vivo,* these compounds are converted back to β-carotene. Activity of the diepoxides is not supported by the data of Subbarayan *et al.* (*830*) and Mallia *et al.* (*531*). Kishore *et al.* (*455*) also reported that the 5,6-monoepoxide of α-carotene was inactive. Other epoxides, such as 5,6- and 5,8-epoxy-β-apo-8′-carotenyl acetates, were tested by Singh and Cama (*800*), but only the 5,6 compound was active. Other oxygen-containing carotenoids are biologically active, such as cryptoxanthin, isocryptoxanthin, 5,6-dihydroxy-β-carotene, citranaxanthin, and torularhodin. Torularhodin is considered to be a minor vitamin A precursor. Surmatis *et al.* (*831*) reported that adding a methyl group to both rings of β-carotene in the 2,2′ positions does not destroy activity, nor does the replacement of methyl with an ethyl in the 1 or 1,1′ positions. However, the replacement of the methyl with an isobutyl group in the 1,1′ positions destroys biological activity. Oxidation that opens both rings of β-carotene destroys activity, but when only one ring is attacked, as in the case of β-semicarotenone, for example, activity remains. Dehydrogenation of the β-ionone ring, as in the case of 3,4-dehydro-β-carotene, still allows the compound to be active. Budowski *et al.* (*127,128*) showed that 3-hydroxy-3,4-dehydro-β-carotene (anhydrolutein) acts as a precursor of vitamin A_2 in the chick and mouse in addition to fish. This is the first known instance of the formation of vitamin A_2 from a carotenoid by terrestrial animals. Vitamin A_2 is 3-dehydroretinol (2 H less than vitamin A_1) and has about 40% of the activity of vitamin A. Activity is lost when hydroxyls and ketones are attached to both rings (positions 3 and 3′,4 and 4′) or when both β-ionone rings are removed and the molecule becomes acyclic. Hydrogenation of both rings of β-carotene or removal of methyl groups (positions 13 and 13′) completely destroys its activity.

Exceptions to the above are the potential vitamin A value of astaxanthin and canthaxanthin in fish. Astaxanthin and canthaxanthin have been claimed to have vitamin A activity in fish, which, if true, would mean that these function as provitamin A compounds in the marine food chain. When the fish *Gambosia holbrookii* was depleted of vitamin A and then given astaxanthin (3,3′-dihydroxy-4,4′-diketo-β-carotene), significant amounts of retinol (vitamin A) were found in the intestine and liver (*330,342*). Conversion also occurs in the intestines (*541*) of fish *in vitro*. In vitamin A-depleted guppies and platies, astaxanthin, canthaxanthin (4,4′-diketo-β-carotene), isozeaxanthin (4,4′-dihydroxy-β-carotene), and β-carotene increase the vitamin A content of the liver (*329,342*). Since the β-carotene content usually increases in the tissues during the conversion process of astaxanthin and canthaxanthin, these carotenoids are presumably reduced to β-carotene before cleavage to vitamin A. Astaxanthin appears to have an antixerophthalmic action in the albino rat (*141a*). It was reported that the diacetate of astaxanthin was converted to retinol by rats' eyes incubated *in vitro* (*540*). The report claimed that astaxanthin can replace vitamin A in the diet of the rat in the maintenance of the electroetinogram, a limited role.

III. PRINCIPAL VITAMIN A PRECURSORS

Certainly, β-carotene is the most important vitamin A precursor naturally present in animal rations. Its concentration in ingredients of plant origin, particularly leafy plants, greatly exceeds that of other carotenes. Due to its high degree of vitamin A activity and frequent occurrence in nature, β-carotene is best known among the carotenoid vitamin A precursors. α-Carotene accompanies β-carotene to a lesser degree in plant sources, and γ-carotene rarely comprises more than a small percentage of the total carotenes. In yellow corn (maize), cryptoxanthin is the primary precursor and contributes significant vitamin A activity when high levels of yellow corn are used in animal rations. β-Zeacarotene, also present in yellow corn, has about one-fourth the biological value of β-carotene and therefore is of less value than cryptoxanthin. Although cryptoxanthin is a relatively important precursor in mammals and birds, another vitamin A precursor, namely, echinenone (myxoxanthin), is regarded as an important precursor in fish. The latter is of some importance when crustacean by-products are fed to farm animals. Another compound, torularhodin, is present in certain yeasts. The carotenals, such as β-apo-8′-carotenal, comprise a relatively recently recognized vitamin A precursor group. Finally, epoxides deserve some attention.

In β-carotene conversion to vitamin A one must assume that (a) half of each molecule of pigment is wasted during conversion or (b) one out of every two molecules is wasted and one is converted to two molecules of vitamin A. Because β-carotene has two β-ionone rings, it would seem that two molecules of vitamin A should be derived from each β-carotene molecule during conversion. Presumably, the remaining molecules either are not formed and absorbed, or are destroyed without vitamin A function. Since β-carotene is the best known carotenoid, more information is available concerning its digestion, absorption, and deposition in tissues. In man, β-carotene is a normal constituent of blood, and the amount present is related to the level of intake of the provitamin. When exceedingly large amounts are ingested over a sufficient time interval, a yellow pigmentation of the body tissue may occur, which is described as carotenemia or hypercarotenosis. The yellowing disappears when excessive levels of carotenoids are no longer consumed. Hypercarotenosis in man from ingestion of carotenoids is not generally regarded as a serious physiological problem. Hypercarotenosis has been associated with anorexia nervosa (*738*) and the rare genetic inability to convert carotene to vitamin A (*570*).

The biological activity or effectiveness of β-carotene has been examined in great detail. The Agricultural Research Council (ARC) of Britain and the National Research Council (NRC) of the United States make similar recommendations in terms of retinol activity. The NRC considers 8 μg β-carotene and the ARC 5 μg β-carotene equivalent in activity to 1 μg retinol. Branion and Emslie (*111*) tabulated all the ratios, on a weight basis, of the biological effectiveness of β-carotene and retinol for animals and man and found them to vary from 2:1 to as high as 12:1.

In most domestic animals and man the ratio varied from 4:1 to 10:1, with a mean of 7:1. In view of these observations the practice of converting carotene values to vitamin A activity by using the relationship of 0.6 μg β-carotene equivalent to 1 IU of vitamin A activity (1 mg β-carotene equals 1667 IU) is in error on the basis of experimental data and practical feeding experience. A more realistic relationship would be about 2 μg of β-carotene equivalent to 1 IU of vitamin A activity or 1 mg β-carotene equivalent to about 500 IU of vitamin A activity.

The absorptions of α- and γ-carotenes are in a range similar to that of β-carotene, according to Deuel *et al.* (*198*), Zechmeister *et al.* (*946*), and Johnson *et al.* (*433*), and they have a biopotency about 50% of that of β-carotene (*293,334,494,495, 565,918*). As the amount of administered α-carotene is increased in the diet its digentibility decreases (*280*). The presence of α-carotene has been demonstrated in animal tissues such as the eggs and liver of chickens and the liver of rats (*294*). Ganguly *et al.* (*293*) demonstrated that α-carotene is converted to vitamin A in the intestinal wall of the chicken. McNally and Szymanski (*571*), upon feeding α-carotene to rats, found α-vitamin A in addition to vitamin A stored in the livers. α-Carotene makes up 5–40% of the total carotenes of the carrot. Orange varieties contain more than red varieties. α-Carotene also constitutes a high percentage of the total carotenoid content of palm oil and certain fruits. γ-Carotene is widely distributed in nature but is usually present in small amounts. It is a constituents of carrots and palm oil, berries, fruits, salt-marsh dodder, and fermentation products. Cis isomers of γ-carotene are more widely distributed in nature than cis isomers of some other carotenes. Using chicks as experimental animals, Greenberg *et al.* (*334*) found that pro-γ-carotene has 51% of the potency of all-*trans*-β-carotene, whereas all-*trans*-γ-carotene (prepared by stereoisomerization of pro-γ-carotene) has 42%. β-Zeacarotene (*trans*-7′,8′-dihydro-γ-carotene) obtained from *Rhodotorula glutinis* grown at 5°C and from frozen yellow corn (*797*) was found to be spectroscopically and chromatographically identical to synthetic β-zeacarotene and probably identical with the β-zeacarotene previously isolated from corn gluten (*696*). It has approximately 25–30% of the biological activity of β-carotene.

Cryptoxanthin is an oxycarotenoid with a smaller proportion of oxygen than some other oxycarotenoid; as an active carotenol, it is quite similar to β-carotene. Because of a hydroxyl grouping it often occurs in ester form. Cryptoxanthin has practical application in the vitamin A nutrition of livestock and poultry. It occurs naturally in yellow corn in substantial amounts and, with chick growth as the criterion of activity, has been shown to have about 55% (*333*) of the biological activity of β-carotene at relatively low levels of feeding, similar to findings obtained with the rat. Other biological activity studies have confirmed these findings (*197,678*). However, John *et al.* (*428*) in 1970 reported a biological potency of 22% of that for β-carotene. Cryptoxanthin is present in the skin and eggs (*307*) of chickens, and trace amounts are found in butter (*306*) and the blood serum (*305*) of cattle. The conversion of cryptoxanthin to vitamin A occurs in the intestinal wall, as evidenced by research with the rat (*678*) and other animals. The biopotency (*292*) of another carotenoid, echinenone, a component of algae, crustaceans, and other marine life, is about 50%.

The β-apocarotenals are another source of provitamin A and occur widely in biological materials (*799*). In 1938 von Euler *et al.* (*258*) reported that β-apo-8′-carotenal was the main product resulting from the partial oxidation of β-carotene with permanganate and that daily supplements of β-apo-8′-carotenal fed to vitamin A-deficient rats improved growth. Glover and Redfearn (*311*) prepared β-apo-8′-, β-apo-10′-, and β-apo-12′-carotenals and fed them to vitamin A-deficient rats. All were transformed into vitamin A. According to Thompson (*857*) the β-apocarotenals have approximately 50% of the activity of β-carotene when given orally. The vitamin A activity of β-apo-8′-carotenal was determined quantitatively in 1960 by Marusich *et al.* (*538*) using the classic U.S.P. rat curative growth assay. An oil solution of β-apo-8′-carotenal was assayed in two separate tests. With oral dosing, an average potency of 1,200,000 IU of vitamin A per gram of β-apo-8′-carotenal (72% of the activity of β-carotene) with 95% confidence limits of 1,070,000–1,330,000 IU was observed. Stabilized gelatin–sugar beadlets of β-apo-8′-carotenal added to the diet were assayed in three separate growth assays and exhibited vitamin A activity equal to 1,070,000 IU per gram, with a range of 950,000–1,190,000. The differences in results obtained with the oil solution and with the beadlet preparation were not statistically significant. It was of interest to compare the performance of β-apo-8′-carotenal with β-carotene at levels of feeding high enough to elicit liver stores of vitamin A. Levels of 20–40 μg were fed to rats daily for a 2-week period. The animals were sacrificed, and liver and kidney tissue levels of vitamin A were measured. The vitamin A activity of β-apo-8′-carotenal was found to be lower (about 600,000 IU/gm), about 36% of the activity of β-carotene, when the liver and kidney storage assay was used than when the previously mentioned U.S.P. rat curative growth test was applied. In chick liver storage tests carried out in 1963 by Brüggemann *et al.* (*124*) β-apo-8′-carotenal was shown to have 41% of the activity of β-carotene (Fig. 2). In liver trials the level of

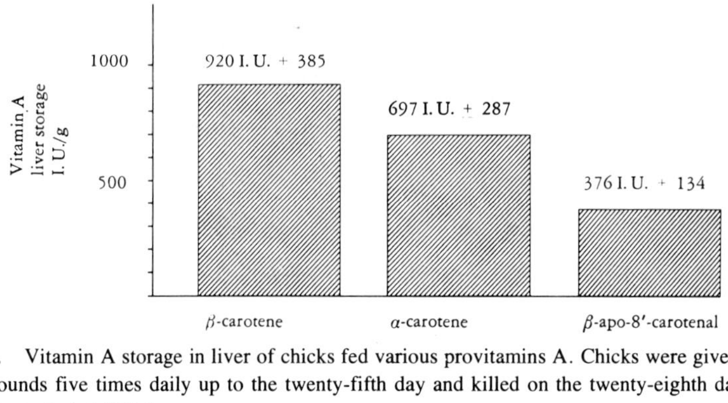

Fig. 2. Vitamin A storage in liver of chicks fed various provitamins A. Chicks were given 1 mg of the compounds five times daily up to the twenty-fifth day and killed on the twenty-eighth day. [From Brüggemann *et al.* (*124*).]

feeding is substantially higher than in growth trials. Discrepancies between the two test procedures are not unusual, as the literature indicates (*431*). In 1968 Al-Hasani and Parrish (*17*) reported on the vitamin A activity of the β-apocarotenals in the Japanese quail, *Coturnix coturnix japonica*. β-Apo-8′-carotenal, β-apo-10′-carotenal, and β-apo-12′-carotenal prepared from β-carotene were used. Growth, viability, egg production, fertility, hatchability of eggs, and viability and growth of progeny demonstrated that these β-apocarotenals are vitamin A-active compounds. When they were fed, substantial quantities of vitamin A were found in blood serum, liver, and egg yolk, which is further evidence of the vitamin activity of β-apocarotenals. In 1972 these studies were extended (*16*). Sharma *et al.* (781) observed 8′-, 10′-, and 12′-apo-β-carotenals in the intestines of chickens fed β-carotene, thus suggesting enzymatic attack at several double-bond locations. A year earlier Sharma *et al.* (*782*) showed that the β-apocarotenals have almost comparable growth-promoting activity on a molar basis but give more variable results in liver storage tests.

In view of the use of β-apocarotenals as coloring agents for food and feeds, the metabolism of these compounds has evoked interest. The major product of their metabolism, other than retinol, is the corresponding carotenoic acid (*309,849*). In oral administration trials with laying hens, β-apo-8′-carotenal was retained in egg yolk partly in the form of β-apo-8′-carotenoic acid. Carotenals have been found in intestinal tissue and a number of higher plants (*919,920*). Thommen and Wiss (*850*)

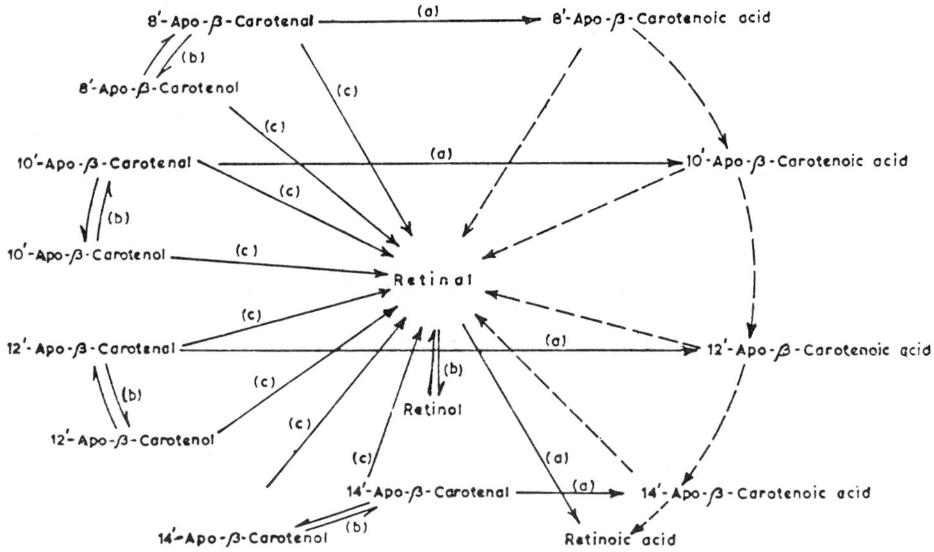

Fig. 3. Schematic representation of metabolism of apocarotenoids in Animals. Solid arrows, major route; dashed arrows, unknown route; (a) aldehyde oxidase; (b) aldehyde reductase; (c) carotene 15,15′-dioxygenase. [From Singh and Cama (*799*).]

reported the isolation from alfalfa meal, by means of column and thin-layer chromatography, of two carotene aldehydes: β-apo-8′-carotenal and β-apo-10′-carotenal. Both aldehydes can be regarded as intermediates in the degradation of β-carotene to vitamin A. Carotenoic acids, which presumably are products of carotenoid metabolism, have also been demonstrated in mushrooms, carrots, and spinach (*799*). A schematic pathway (*799*) for the metabolism of apocarotenoids to retinol was proposed in 1974 by Singh and Cama (Fig. 3).

Over a decade ago Subbarayan *et al.* (*830*) orally administered 5,6-mono-epoxy-β-carotene to rats and observed some conversion to vitamin A. The biological activity was estimated to be 21% of that for β-carotene. The monoepoxy compound was not effective when administered intraperitoneally, and the diepoxy compound was not effective when administered orally. A few years later Kishore *et al.* (*455*) reported 5,6-mono-epoxy-α-carotene to be poorly converted to α-vitamin A, a nonbiologically active compound. Orally administered 5,8-epoxy-β-carotene (citroxanthin) was converted to vitamin A and judged by John and Cama (*427*) to have 31% of the biological activity of β-carotene. The authors suggested that, when the mono-, epoxy-, and hydroxy-β-carotene structures are administered orally, the mechanism for biological activity may be eccentric cleavage of the molecule.

IV. ACTIVITY OF ISOMERS

The provitamin A activity of the all-trans structure of these carotenoids is usually markedly reduced by isomerization to a cis form at one or more of the double bonds. Hence, the molecular arrangement influences interaction with enzymatic conversion to vitamin A. There are a great number of possible geometric isomers in the case of the carotenoids. With β-carotene, only 20 unhindered isomers are expected, with 12 having been observed out of a theoretical 272 possibilities (*945*). The all-trans structure, the more stable geometric form, can undergo cis-trans isomerization in solution; therefore, it is sometimes difficult to decide whether a cis isomer occurs in nature or whether it is formed during its isolation due to the ease of formation of cis-trans equilibrium mixtures under exposure to light, heat, and acids. Whether these cis stereoisomers are active themselves or must first be transformed in the digestive tract to the all-trans form is not known. One explanation for the biological activity of the *cis*-carotene steroisomers is that during digestion and/or absorption they are rotated, at least partially, at their asymmetric C atoms to the linear all-trans form.

The most complete study of the provitamin A activity of carotenoid stereoisomers was made by Zechmeister (*945*) in collaboration with Deuel (*193*). With the exception of pro-γ-carotene, cis isomers of the provitamin A structures have a lower biopotency than do the naturally occurring all-trans forms. The following provitamin A activities have been recorded for stereoisomers in rats, all-*trans*-β-carotene being assigned a potency of 100%; 3- or 9-mono-*cis*-(neo-α-carotene U),

13% 3,6- or 2,7-di-*cis*-(neo-α-carotene B), 16%; 3-mono-*cis*-(neo-β-carotene U), 38%; 3,6-di-*cis*-(neo-β-carotene B), 53%; 15,15′-mono-*cis*-(β-carotene), 30–50%; 11,11′-di-*cis*-(β-carotene), 30%; 3-mon-*cis*-(neo-γ-carotene P), 19%, mixed neo-γ-carotenes, 16%; poly- or penta-*cis*-(pro-γ-carotene), 43% 2-mono-*cis*-(neocryptoxanthin U), 27%; and central mono-*cis*-(neocryptoxanthin A), 42%. The provitamin activity of some carotene isomers is slightly higher in the chick than in the rat.

Different biological procedures have been tried for estimating the vitamin A value of carotene isomers. Carotenes investigated by Sweeney and Marsh (*835*) in 1973 were all-*trans*-β-carotene, neo-β-carotene B (9,13-di-*cis*), neo-β-carotene U (9-mono-*cis*), all-*trans*-α-carotene, neo-α-carotene B (13,9′-di-*cis*), and neo-α-carotene U (9-mono-*cis*). Assays for vitamin A activity were made by measurement of vitamin A storage in tissues of rats fed the isomers. Liver vitamin storage values for cis isomers were consistently higher than values reported by others using, as a criterion, growth response assays. These higher values would indicate the formation of isomers at the expense of all-*trans*-carotenes, which would result in smaller decreases in provitamin A activity than those calculated using the growth response values. For all-*trans*-α-carotene, liver vitamin storage values were lower than those obtained by the growth assay. Vitamin A storage in the kidneys was small compared to that in the liver. The change from an all-trans to a cis, trans configuration of the provitamin A molecule usually causes a decrease in the biological potency of the molecule, although this may not necessarily occur. Deuel (*193,194*) mentioned that some cis isomers are changed back to the all-trans form in the digestive tract. Cis isomers of β-carotene, such as neo-β-carotene and neo-β-carotene U have been observed in large amounts in fresh and preserved plant tissue. Bickoff (*72*) reported that commercial dehydration of alfalfa greatly promotes the formation of cis isomers. It is likewise an important phenomenon in the curing and storing of hay, ensilage production, and pelleting. Isomerism also may be induced by refluxing the pigment in a solvent, by illumination, by treatment with acids or iodine, by melting carotenoid crystals under high heat, or in the normal cooking of carotene-containing foods.

V. METABOLIC PATHWAY

The biosynthesis of vitamin A from β-carotene takes place mainly in the intestinal mucosa during absorption by the following reaction sequence: (a) the cleavage of β-carotene to form retinal and (b) the reduction of retinal to retinol (Fig. 4). Conversion can be either by (a) symmetric or asymmetric fission or (b) terminal oxidation. The finding of apocarotenals in animal and plant tissue supports the terminal oxidation scheme. A diagram (*659*) of the overall metabolism of vitamin A-active structures is shown in Fig. 4.

In 1954, Glover and Redfearn (*311*) made the unique suggestion that β-carotene and other carotenoids might be cleaved by a continuous stepwise process, leading

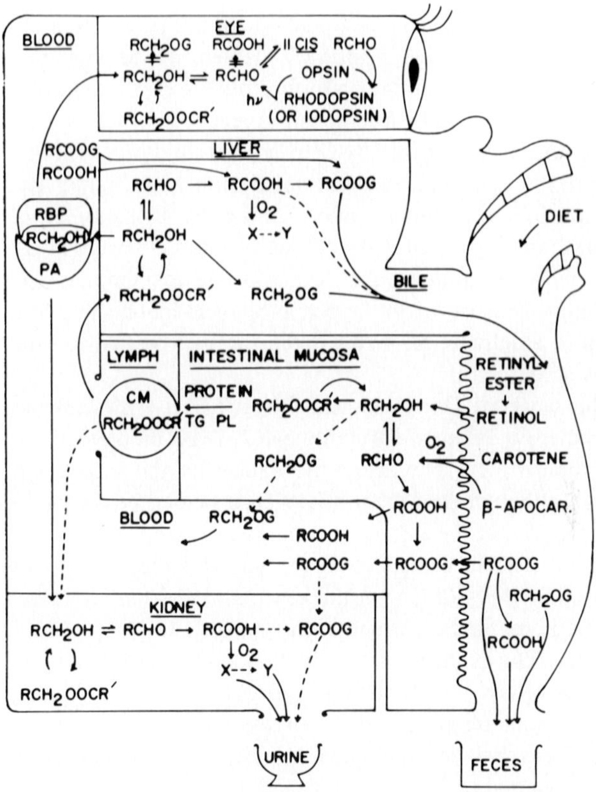

Fig. 4. Carotene and vitamin A metabolism. Abbreviations: RCH_2OH, retinol; RCHO, retinal; RCOOH, retinoic acid; RCH_2OG, retinyl β-glucosiduronate; RCOOG, retinoyl β-glucuronide; RCH_2OOCR', retinyl ester, mainly palmitate; RBP, retinol-binding protein; PA, prealbumin acidic protein; CM, chylomicron; TG, triglyceride. [From Olson (*659*).]

ultimately to 1 mole of vitamin A per mole of carotene consumed. This concept is supported by the fact that β-apo-8′-, β-apo-10′-, β-apo-12′-, and β-apo-14′-carotenals are biologically active. Furthermore, β-apo-12′-carotenal has a higher vitamin activity than β-carotene or β-apo-8′-carotenal. Yet animals given large amounts of β-carotene do not yield chemically measurable amounts of the anticipated intermediates, presumably because they may exist only as a fleeting stage, are protein bound, etc. Several radioactive β-apocarotenals were readily converted in animals to vitamin A ester with the release of small radioactive fragments. Although apocarotenals exist in plants, the above reaction apparently does not occur there. McGillivray (*558*) suggested that β-carotene might be cleaved into β-apocarotenals in the blood. No evidence supporting this suggestion was obtained with a perfused liver system (*944*). Glover implied that oxidative attack probably takes place at more than one position in the chain and that oxidase enzymes are probably responsible for carotene degradation. Initial penetration of the β-carotene molecule probably takes place nearer the center, at the site of the conjugated double-bond system rather than the end. If terminal oxidation occurs, further degradation follows until the vitamin A molecule is reached. Much of the earlier work on this problem has been summarized by Glover (*309*) and Moore (*605*). Glover (*309*) and Goodwin (*322*) concluded that central fission is the major pathway and that terminal degradation is a minor route. Studies by Koizumi *et al.* (*468*) using isolated rat intestines suggested that, although central cleavage of β-carotene might take place *in vivo,* β-carotene also seems to be metabolized to vitamin A by terminal oxidation because of the appearance of β-ionone in addition to protein-bound vitamin A aldehyde; the latter is the form converted to vitamin A alcohol and vitamin A acid.

Whereas the work of Goodman and Olson (*320*) with enzymes supports the central fission theory, Pullman and Pullman (*717*) on theoretical grounds favor the terminal oxidation theory. The actual mechanism of the conversion process is far from clear, and, although the weight of evidence favors the central fission theory, it must be recognized that β-carotene can be metabolized or degraded by processes other than central fission; actually both mechanisms may be operating.

Karrer and co-workers (*440*), in the elucidation of the structure of vitamin A, considered that it might be formed from β-carotene by central fission and the addition of water. The central fission hypothesis was supported by the early studies of Kuhn and co-workers (*495*) and von Euler *et al.* (*259*) and later by Hunter (*416*) on the basis of the biological activity of the carotene isomers. α-Carotene and γ-carotene had only half the biological activity of the β isomer; however, when vitamin A was obtained in a pure crystalline form and compared with β-carotene on a weight basis, its biological potency in the growth test was unexpectedly found to be twice as great. Apparently one molecule of vitamin A is formed from one molecule of β-carotene. Many subsequent biological assays confirmed these results. The fact that only one molecule of vitamin A is formed from one molecule of β-carotene in biological assays has led more than one investigator to consider that, instead of occurring centrally, fission occurs asymmetrically, at least part of the time (*614,799*). If asymmetric fission operated totally, the assumption would have to be made that the two ends of the conjugated system of double bonds in the carotenoids were equally susceptible to attack to fit the foregoing biological evidence. An alternate proposal would be that half of the β-carotene molecules were split centrally, whereas the other half were completely degraded by oxidative enzymes.

In 1965, the existence of a cell-free and particle-free cleavage enzyme was independently reported by two laboratories: Goodman and Huang in New York from rat intestine (*318*) and Olson and Hayaishi in Kyoto from rat liver (*658*). The enzyme is a dioxygenase which requires no cofactors and is referred to as β-carotene 15,15′-dioxygenase. It requires molecular oxygen and yields retinal (retinene or vitamin A aldehyde) as its sole product.

The carotenoid-cleavage enzyme 15,15′-dioxygenase has been demonstrated in a number of species of animals (*315,316,798*) including the rat, pig, rabbit, and guinea pig. In the future, the *in vitro* enzyme approach may deserve higher priority in metabolism studies, considering the variables encountered in the intact animal bioassay and the need for greater volume of the compound for testing. On the other hand, (a) all vitamin A precursor compounds active in the intact animal, such as β-carotene epoxides, do not respond (*798*) to the enzyme *in vitro* (Table 3), and (b) the central cleavage approach excludes terminal oxidation as an activity-producing mechanism. In the marine food chain, β-carotene, in addition to being converted to vitamin A, forms oxycarotenoid structures. When adult brine shimp, *Artemia salina,* are fed ^{14}C-labeled β-carotene, a conversion takes place, first to echinenone and then to canthaxanthin and/or astaxanthin (*182,488,490*). Isocryptoxanthin and isozeaxanthin also may be intermediates in some species. The metabolism of carotene and/or vitamin A has been reviewed by Olson (*653–655,659*), Thommen (*848*), Krinsky (*487*), Goodman (*316*), and others (*904,927*).

TABLE 3

Enzymatic Cleavage of Carotenoids[a,b]

Substrate	Guinea pig enzyme: Specific activity	Guinea pig enzyme: Relative activity	Rabbit enzyme: Specific activity	Rabbit enzyme: Relative activity
β,β-Carotene	5.93	1.00	8.41	1.00
β,ε-Carotene	4.23	0.72	4.67	0.56
5,6-Epoxy-5,6-dihydro-β,β-carotene	1.92	0.32	2.12	0.25
5,8-Epoxy-5,8-dihydro-β,β-carotene	1.11	0.19	0.91	0.11
5,8,5′,8′-Diepoxy-5,8,5′,8′-tetrahydro-β,β-carotene	0.10	0.02	0.12	0.02
5,6-Epoxy-5,6-dihydro-β,ε-carotene	1.46	0.25	1.81	0.22
5,8-Epoxy-5,8-dihydro-β,ε-carotene-	1.55	0.09	0.71	0.09
β,β-Caroten-3′-ol	0.22	0.04	0.22	0.03
3′,4′-Didehydro-β,β-caroten-3-ol	0.18	0.03	0.16	0.02
8′-Apo-β-carotenal	0.56	0.09	0.28	0.03
10′-Apo-β-carotenal	3.37	0.57	4.65	0.55
12′-Apo-β-carotenal	0.68	0.12	0.57	0.07
8′-Apo-β-caroten-8′-yl acetate	0.96	0.16	0.80	0.10
8′-Apo-β-carotenol	0.41	0.07	0.15	0.02
10′-Apo-β-carotenol	2.98	0.50	3.60	0.43
12′-Apo-β-carotenol	0.59	0.08	0.51	0.07

[a] From Singh and Cama (*798*).

[b] Substrate specificity of carotene 15,15′-dioxygenase from guinea pig and rabbit intestinal mucosa. Different substrates were incubated with the final preparation of the enzyme from guinea pig and rabbit intestine (15.0 mg per reaction mixture) under the standard assay conditions. The activity was estimated by determining the amount of formation of products.

VI. SITE OF CONVERSION

The biosynthesis of vitamin A from β-carotene takes place principally in the intestinal mucosa during the absorption of dietary β-carotene and other carotenoid vitamin A precursors. In the normal *in vivo* reaction discussed earlier, retinol is formed and then mainly esterified with long-chain fatty acids; these retinyl esters are incorporated into lymph chylomicrons and transported from the intestine via the intestinal lymphatics (*315*) to various tissues.

Over 30 years ago Sexton *et al.* (*779*) proposed the concept that the intestinal wall is an important conversion site. Deuel and co-workers (*549,912*) demonstrated in rats that, with the feeding of β-carotene, vitamin A appears in the intestinal wall before an appreciable amount is found in the liver. Chickens were also shown to convert α-carotene and cryptoxanthin in the intestinal mucosa (*292*). Other investigations during this period revealing the intestine as the site of transformation of β-carotene to vitamin A were conducted by Goodwin and Gregory (*327*) in rabbits and goats; Glover *et al.* (*310*) and Mattson *et al.* (*549*) in rats, and Thompson *et al.* (*860,862*) in pigs; Cheng and Deuel (*153*), and Sibbald and Olsen (*795*) in chicks; Thompson *et al.* (*860*) and Klosterman *et al.* (*462*) in sheep; and Stallcup and Herman (*823*) and Elliott (*235*) in cattle. Chicks convert carotene to vitamin A as early as the first day of life (*371*) and support physiological functions involving vitamin A. When β-carotene is introduced into the intestinal tract, conversion to vitamin A occurs in the intestinal wall and mucosa, and vitamin A storage in the liver rises. The capacity of the intestinal wall to accomplish this is significantly greater than the need (*944*).

When β-carotene was fed to vitamin A-depleted chicks, vitamin A first appeared in the intestinal wall after 1 hr and in the liver after 3 hr in studies by Cheng and Deuel (*153*). Goodwin and Gregory (*327*) proved the absence of carotene in the portal blood of rabbits, goats, and sheep after giving carotene by the oral, abomasal, or duodenal routes; moreover, an increased vitamin A content was observed in the thoracic lymph after carotene was fed. ^{3}H-Labeled β-carotene administered orally to three human subjects was present in lymph mainly as retinyl esters, primarily of palmitic, stearic, and some oleic and linoleic acids, but some unconverted β-carotene was also present (*317,657*).

The greater conversion efficiency of the intestine (*559*) over that of other tissues is probably due to its excellent system for emulsifing lipids and thus making carotene more accessible to the enzyme system. Carotenoids not transformed into vitamin A in the intestinal mucosa are probably cleaved mainly in the liver (*658,944*). The formation of retinol esters from β-carotene has been demonstrated in isolated perfused rat liver. Other organs, such as the kidney, may be active as well. The kidney is considerably less active *in vitro*. Vitamin A is not formed from β-carotene in heart or lung preparations (*655*) or in isolated blood. In the rat, the conversion rate by the intestine is approximately twice as great as that by the liver on the basis of whole organs but four to seven times as great per gram wet weight of tissue (*944*). Both organs convert carotene into retinol at a rate much faster than that required to satisfy the nutritional requirements.

The possibility of absorption of labeled β-carotene from the colon or cecum of sheep was studied using four mature wethers considered to have a normal vitamin A status (*875*). Ligated intestinal loops were injected with ^{14}C-labeled β-carotene suspended in aqueous polysorbate 80. At the termination of the experimental absorption period, the liver, kidneys, and intestinal loops were excised for determination of radioactivity and were found to be free of radioactivity, which tends to confirm the belief that carotene cannot be absorbed from the cecum or large intestine.

In the presence of an adequate concentration of bile salts in the intestines the uptake of β-carotene seems to be the rate-limiting step in the overall mechanism rather than the intracellular processes leading to vitamin A ester formation. The total amount of β-carotene in the intestinal wall is small when bile or a bile substitute is absent (*657*). Vitamin A formation (*657*) in the lower third of the small intestine is slower than that in the upper and middle intestine.

VII. VITAMIN A STANDARD

A summary of vitamin A standards is presented in Table 4. The international unit (or U.S.P. unit) of vitamin A is currently equivalent to 0.6 μg of all-*trans*-β-carotene or 0.344 μg of all-*trans*-vitamin A (acetate) (0.3 μg of retinol).

The 1931 standard, of which 1 μg was to be 1 IU, was a product obtained by recrystallization of a mixture of samples of carotene prepared in eight different laboratories in seven different countries (*734*). However, this standard was found to be impure and to contain a certain amount of inactive material. P. Karrer then prepared, for the National Institute for Medical Research, about 1 gm of pure crystalline β-carotene. The 1931 standard was then compared in biological tests with this new preparation by a number of workers. The dose of β-carotene equivalent to 1 μg of the 1931 standard carotene was found to be 0.57, 0.66, 0.50, and 0.66 μg with an average value of 0.6 μg of β-carotene. This is equivalent to 1 unit of vitamin A and was accepted by the 1934 international conference (*774*). A decision was also reached at the time to issue the standard as a solution of carotene in coconut oil, with the addition of hydroquinone. The 1934 conference, recognizing the usefulness of the spectrophotometric method for determining vitamin A, recommended that a sample of cod liver oil, the potency of which had been accurately determined, be provided as a subsidiary standard of reference. The U.S.P. reference cod liver oil was selected. The spectrophotometric method of determining vitamin A (absorption maximum near 328 nm) was regarded as trustworthy. The intensity of absorption $E_{1\,\text{cm}}^{1\%}$ could be measured to within ±2.5%, and, as a means of converting the $E_{1\text{cm}}^{1\%}$ value to a figure representing international units per gram, the factor 1600 was recommended.

TABLE 4

Vitamin A Standards

Date	Compound	Reference
	β-Carotene (0.6 μg = 1 IU)	
1935–1956[a]	Crystalline β-carotene, 300 μg (500 IU)/gm in coconut oil; after 1949 in cottonseed oil	415
1956 to present	No material standard; pure β-carotene commercially available for use if needed; potency expressed as IU of provitamin A	419
	Vitamin A acetate (0.344 μg = 1 IU)	
1948–1957	Crystalline vitamin A acetate in cottonseed oil; 10,000 IU/gm in gelatin capsules	934,935
1957 to present	Crystalline vitamin A (trans) acetate in cottonseed oil; 100,000 IU/gm in gelatin capsules[b,c]	932,933
1959 to present	Crystalline vitamin A (trans) acetate in gelatin beadlets; 10,000 IU/gm[b]	24

[a] The International Biological Standard for provitamin A was discontinued in 1956.

[b] Currently available from U.S. Pharmacopeia Reference Standards, 4630 Montgomery Avenue, Bethesda, Maryland, as liquid and dry products.

[c] Currently available from Center for Authentic Chemical Substances, Apotekens Kontrollaboratorium, Stockholm, Sweden.

Later the international standard was changed to vitamin A acetate, at which time 0.344 μg (also 0.3 μg of vitamin A alcohol) became equivalent to 1 IU of vitamin A activity, the presently accepted standard. These values were established by the Expert Committee on Biological Standardization of the World Health Organization (*935*) after careful statistical evaluation of collaborative rat curative growth bioassays of β-carotene and vitamin A. Thus, 1 μg of vitamin A (alcohol) is generally recognized as virtually equivalent to 2 μg of β-carotene in biological activity. This gives a theoretical effective carotene to vitamin A ratio of 2 : 1 on a weight basis. Vitamin A (alcohol), vitamin A acetate, and vitamin A palmitate contain 3,333,000, 2,907,000, and 1,818,000 IU of vitamin A activity per gram, respectively, as compared to 1,667,000 IU of vitamin A activity for 1 gm of all-*trans*-β-carotene.

VIII. SPECIES DIFFERENCES

It is known that animals can differ in their intestinal metabolism of β-carotene. Some species, such as man, absorb unchanged carotenoids to some extent. For example, the presence of β-carotene, lycopene, and lutein has been determined in human plasma (*489*). The rat and pig absorb virtually no carotene beyond the intestinal mucosa (*656,862*). The quantitative difference in the ability of man and the rat to absorb intact β-carotene is not great since most of the radioactivity absorbed into human lymph after the feeding of labeled β-carotene was found in retinyl esters and not in unconverted β-carotene (*413*).

Branion and Emslie (*111*) summarized studies on the utilization of carotene by the fox, mink, dog, rabbit, cat, guinea pig, hamster, monkey, mouse, and rat. A similar summary was made for poultry by Evans (*261*) and for cattle, calves, sheep, swine, and horses by Phillips (*699*). The role of carotene and vitamin A in animal feeding with emphasis on dairy cattle was studied by McGillivray (*560*). Mammalian species that accumulate carotenes or carotenoids in their tissues are man, cattle, and horses. Those that do not or usually have only traces include sheep, goats, deer, rabbits, swine, dogs, foxes, cats, mink, guinea pigs, hamsters, rat, and mice. Of the avian species, certain breeds of chickens, pheasants, flamingos, and others and of the marine species, salmon, trout, lobsters, crabs, shrimp, prawns, crayfish, and others accumulate oxycarotenoids in their tissues.

In the past years a number of very valuable reviews have appeared dealing with various aspects of carotenoid metabolism. Much of the material covered relates directly to the role of carotene and vitamin A in animal feeding, and particular reference is made to Deuel (*193*), Goodwin (*325*), and Moore (*605*). The present review deals mainly with animal studies involving factors influencing the activity and utilization of carotene, other vitamin A precursors, factors influencing content in natural sources, the stability of carotenoids, their practical significance in animal nutrition, and the practical utility of vitamin A.

A. Fish

Hypovitaminosis A in fish (*361*) is characterized by poor vision, poor growth, keratinization of epithelial tissue, and tissue hemorrhages. More specifically, signs of deficiency (*450*) of vitamin A in carp (*30*) and catfish and trout (*456*) include reduced growth, poor feed conversion, anemia, anorexia, discoloration of skin, hemorrhages, hepatic atrophy, bent opercles, xerophthalmia, ocular opacity, and increased mortality. β-Carotene is present in the ovaries and ova of a variety of fish (*257*) and also in the muscle tissue of some fish such as the salmon. Czeczuga (*176*) reported the presence of β-carotene, echinenone, and cryptoxanthin in the tissues of crayfish. β-Carotene is predominant in the hepatopancreas and ovary of the crab (*303*). Ketola (*450*) indicated a vitamin A requirement of 7500–20,000 IU per kilogram dry ration; requirements given by the NAS–NRC committee (*641*) are 2000–4500 IU per kilogram dry ration. α-Carotene has been observed in the liver of the sunfish (*815*). The Atlantic cod converts carotene to vitamin A. According to Deuel (*193*) it seems reasonable to postulate that much vitamin A in fish livers originates from dietary carotenes. The Atlantic cod (*624a*) and perch and dace

(*615*) convert dietary β-carotene to vitamin A, but more trials are needed on other types of fish to demonstrate clearly the efficiency of this conversion process. Brook trout utilize carotene poorly (*712*), and dietary supplemental β-carotene prevented ocular lesions in swim-up trout held in warm water (12°C) but not in cold water (9°C), according to Poston *et al.* (*711*). The dietary requirements of fish for vitamin A acetate and β-carotene are discussed by Dupree (*216*).

B. Mouse

Vitamin A or a vitamin A precursor is a required nutrient (*175*) for the mouse. Deficiency symptoms recorded by Fenton *et al.* (*268*) and McCarthy and Cerecedo (*551*) and summarized by the NAS-NRC committee (*636*) are xerophthalmia, diarrhea, tremors, unkempt fur, decreased growth, and rectal hemorrhages; in the male, they also include permanent sterility and, in the female, vaginal hemorrhages and abortion of young. Epithelium in various tissues and organs becomes keratinized as the dificiency progresses.

Morris (*613*) estimated the daily needs of the mouse to be about 5 μg β-carotene, or 0.3–0.6 μg vitamin A. The NAS-NRC committee (*636*) estimates that 250 IU of vitamin A per kilogram diet or 1–2 IU/day should provide the requirements of growing, pregnant, and lactating mice, but without a margin of safety. Some workers have used β-carotene in their studies; others have used vitamin A. Although there is evidence that the mouse converts carotene to vitamin A, there is no accurate information on the degree of efficiency of carotene conversion to vitamin A.

C. Rat

The dietary requirements of vitamin A for the rat have been reviewed (*175,636*), and deficiency signs have been detailed by Wolbach (*926*) and McCoy (*554*). Vitamin A is essential for vision; a deficiency results in night blindness. Xerophthalmia may also occur and appears as a reddish exudate on the lids, opacity of the cornea, and distortion of the shape of the eye. There is usually growth depression. Failure of implantation with the production of aborted or short-lived litters is common in the female, whereas in the male testicular degeneration is characteristic of the deficiency. The deficiency (*636*) is characterized particularly by malformations of the epithelial structures and epiphyseal cartilage. Epithelial tissues become keratinized. Disorganization of the tooth structure causes an impairment of its growth, a distortion of the incisors, and loss of the normal inner pink-orange color. Retardation of skeletal growth results in a compression of the brain, spinal cord, and roots and herniation of the brain into the foramen magnum with consequent mechanical injury to the brain and nerve roots, with incoordination occurring in about 5–6 weeks in weanling rats. The minimal requirement of vitamin A for growth is approximately 200 IU per kilogram of rat daily (2000 IU/kg diet); for reproduction the daily requirement is 1200 IU per kilogram of rat daily (12,000 IU/kg diet) (*636*).

The most exhaustive and quantitative biological work with β-carotene has been done with rats. There are numerous reports that the vitamin A activity of β-carotene is dependent on the specific conditions of the rat assay. The addition of vitamin E (*131,376,387,387a,467,511*) at low levels in the growth assay enhances the vitamin A activity of β-carotene; on the other hand, larger quantities of vitamin E may be antagonistic to carotene utilization (*376,387,387a,566,568*). The addition of vitamin C (*550*) to the ration, B vitamins (*393*), the protein source (*335, 341,411,424,842*), the quantity and nature of the dietary fat (*221,509,791*), emulsifiers (*254,803*), antioxidants (*388,392,394*), the presence of thyroxine (*193*), bile (*657*), the carrier and food source (*874*), and the age (*888*) of the animal are some of the variables that can affect the conversion and utilization of β-carotene by the rat. The rat is efficient in converting carotene (*413*) to vitamin A only at low levels of intake. For example, rats given 100 μg ^{14}C-labeled β-carotene or 1.3 mg [^{14}C]retinol (with α-tocopherol) in a fat-base carrier or aqueous dispersion showed a similar vitamin concentration in the lymph (Table 5). Retinyl palmitate predominates in the intestinal lymph whether β-carotene or retinol is administered (*413*). Carotene is less digestible and less efficiently converted to vitamin A in the rat at high levels of intake.

On the basis of rat assays, the World Health Organization (WHO) in 1950 defined 1 IU as equivalent to 0.3 μg of vitamin A alcohol. In a collaborative assay, the U.S. Pharmacopeia Vitamin Advisory Board (*878*) reported a potency of 1,667,000 IU/gm for β-carotene and 3,300,000 IU/gm for vitamin A alcohol. Marusich *et al.* (*537*) observed, by rat growth assay, that synthetic β-carotene contained 1,730,000 IU/gm; that is, 0.58 μg of β-carotene was equivalent to 1 IU of vitamin A activity. The weight of evidence supports a biological value of 0.6 μg being equivalent to 1 IU of vitamin A for low levels of supplementation in the rat. When carotene is fed at higher levels sufficient to elicit liver stores of vitamin A, the relative efficiency of conversion to, and tissue storage as, vitamin A are markedly reduced (*66,101, 424,543,605,874*). Data from this laboratory indicate that dietary β-carotene provides one-third to one-fifth the liver storage of vitamin A as compared to an equivalent dietary unitage of vitamin A. this is an effective carotene to vitamin A ratio of 6:1 to 10:1 on a weight basis. Given the above-described variability of the vitamin A activity of β-carotene in rats, it is not surprising that variable carotene to vitamin A activity ratios have been reported in other species of animals by other investigators.

D. Rabbit

According to Branion and Emslie (*111*) symptoms of vitamin A deficiency in rabbits include xerophthalmia (*154,532,574,698*), retarded growth (*698*), ataxia and spastic paralysis, incoordination and disturbance of equilibrium, reproductive failure (*501*), degeneration of nerves, sometimes blindness and deafness, corneal lesions and lowered ascorbic acid content of the aqueous humor, hydrocephalus,

TABLE 5

Intestinal Absorption and Metabolism of Radioactive β-Carotene and Vitamin A in the Rat[a,b]

Labeled compound fed	Test meal	No. of samples[c]	Percent distribution of labeled retinyl esters[d]			
			Palmitate	Stearate	Oleate	Linoleate
^{14}C-Labeled β-carotene, preparation 1	Trilinolein	3	47 ± 0[e]	21 ± 0[e]	13 ±1	17 ± 4
	Triolein	2	45 ± 2	23 ± 5	25 ± 4	8 ± 1
	Triolein	1 (chylos)	46	22	24	8
	Triolein	1 (bottom)	47	26	19	8
	Sodium taurocholate (aq)	1	50	25	15	10
^{14}C-Labeled β-carotene, preparation 2	Olive oil	2	45 ± 3	24 ± 2	21 ± 3	10 ± 1
[15-^{14}C]Retinol	Trilinolein	2	48 ± 8	23 ± 0	14 ± 4	16 ± 5
	Triolein	1	48	19	22	11
	Corn–olive oil (1:1)	1	40	33	17	10
	Corn–olive oil (1:1)	1 (chylos)	43	21	23	13
	Corn–olive oil (1:1)	1 (bottom)	37	30	21	12

[a] From Huang and Goodman (*413*).

[b] Labeled retinyl ester compositions in lymph after test meals containing ^{14}C-labeled β-carotene or ^{14}C-labeled retinol.

[c] Samples consisted of whole lymph unless otherwise indicated.

[d] Mean plus or minus standard error.

[e] $n = 2$. In the third sample palmitate + stearate comprised 75% of the labeled retinyl esters, but these were not individually estimated quantitatively by reversed-phase paper chromatography.

and death. Lack of appetite is characteristic of a vitamin A deficiency in rabbits, as cited by Saksena *et al.* (*758*); these authors also noted abnormal elasticity of the lungs and aorta associated with a decreased elastin content. The requirement of rabbits for vitamin A has been included in the 1977 NAS–NRC committee report on nutrient allowances (*638*).

Investigators have administered carotene in studies of vitamin A deficiency. Mellanby (*575*) prevented the symptoms of vitamin A deficiency by feeding young rabbits 1–3 mg of carotene daily. Phillips and Bohstedt (*698*) found that 50 μg of carotene per kilogram body weight prevented the deficiency syndrome and maintained health, but a 6- to 20-fold increase in this dose was required as a remedial dose. Pirie and Wood (*704*) concluded that rabbits responded well to carotene, as judged by plasma and liver levels of the vitamin. Payne *et al.* (*681*) reported that 8 μg of vitamin A per kilogram body weight daily was adequate for growth of female and breeding male rabbits. However, reproducing females required somewhat more than the highest level fed, which was 14 μg per kilogram body weight. The authors suggested that this requirement may approximate 20 μg (1660 IU/kg diet). Normally, rabbits obtain their vitamin A as provitamin A, principally carotene. No attempt has been made to determine quantitatively the efficiency of conversion of β-carotene to vitamin A, although qualitatively it appears effective in meeting the requirements for growth, health, and reproduction. Rabbits injected with carotene formed appreciable amounts of vitamin A that were distributed between the liver and kidneys (*81*).

E. Cat

The studies of Ahmad (*10*), Rea and Drummond (*725*), and Gershoff *et al.* (*301*) demonstrated that for all practical purposes β-carotene is not utilized by the cat as a source of vitamin A activity, whether administered orally or parenterally. Vitamin A from fish oil or that prepared by chemical synthesis is effective in meeting the requirement (*180,301,302*). It has been reported that the enzyme carotene 15,15′-dioxygenase, which converts certain carotenoid vitamin A precursors to retinal, is absent in the cat (*500*).

Vitamin A deficiency was produced in kittens by Gershoff *et al.* (*301*). Symptoms are emaciation, weakness of the hind legs with some signs of rigidity, and reproductive failure (*636,773*); also, a decrease in food consumption has been observed. According to the NAS–NRC committee (*636*) estimates made using various adequate rations, 9,000 IU of vitamin A per 100 gm of diet meets the nutritional requirements of growing cats.

F. Mink

The NAS–NRC committee (*634*) suggested that the vitamin A allowance for the mink be the same as that for the fox. Vitamin A deficiency symptoms are night blindness, diarrhea, eye disorders, incoordination (*42,896*), decreased growth, respiratory tract infections, and inferior pelt quality (*43*).

Evans and Czarnocki (*262*) reported that, at relatively high levels of feeding, carotene was 1–4% as effective as an equivalent unitage of vitamin A in promoting liver storage of vitamin A in mink. It was observed by Warner (*896*) that essentially no difference in liver and plasma vitamin A levels existed in mink receiving carotene at moderate levels compared to those on a purified diet deficient in vitamin A. Experiments (*898*) in which plasma and liver vitamin A levels were measured after the feeding of carotene or alfalfa meal showed that mink are inefficient in converting carotene to vitamin A. This work demonstrated that alfalfa meal and probably other plant sources of carotene are poorly utilized by mink. Performed vitamin A, fed at the same unitage level, produced significantly higher liver storage and plasma levels. As with cats, carotene probably should not be considered a satisfactory vitamin A precursor source for the rations of mink. Mink require about 200 IU of vitamin A per kilogram live weight daily (*634*). The NAS–NRC publication (*634*) and current literature should be consulted for further details on carotene utilization by mink.

G. Fox

According to the Branion and Emslie review (*111*) vitamin A deficiency symptoms in the fox (*810*) are nervous disturbances (trembling and cocking of the head, coma), xerophthalmia, widespread epithelial metaplasia, demyelinization of many nerve fibers, and abortion in the female. Deficiency symptoms (*44,810*) also include unsteadiness of gait, weaving, whirling or running in circles, emaciation, and death. Autopsy of deficient animals revealed urinary calculi, gastritis, enteritis, pyelitis, and general inflammation of the urinary tract. At times, the milk teeth and adult teeth were present in the same socket, and often the small discolored adult incisors were chipped or broken at the upper corners. There are variations in vitamin A content of the liver of foxes, but there has been no consistent relation of vitamin liver levels to fur quality (*408,702*).

Foxes can utilize carotene as a dietary source of vitamin A, but the conversion is inefficient (*44,166*). When carotene is used to meet the vitamin A requirement of foxes, a conversion factor of 6 should be applied because foxes do not convert carotene to vitamin A as efficiently as do rats. The NAS–NRC committee recommended in 1953 and again in 1968 (*634*) that growing foxes be fed 100 IU of performed vitamin or 600 IU (360 μg) of β-carotene daily per kilogram body weight daily. For dry feed, the daily vitamin A allowance would be 2410 IU per kilogram dry feed, assuming a body weight of 4.7 kg and a feed intake of 195 gm.

H. Dog

The nutrition of dogs has been studied by Steenbock (*824*), Stimson and Hedley (*828*), Crimm and Short (*170*), Branion and Emslie (*111*), Michaud and Elvehjem

(*581*), Singh *et al.* (*801*), and the NAS–NRC committee (*633*). Early in the history of nutrition, vitamin A deficiency was recognized in the dog. Deficiency signs, such as anorexia, weight loss, ataxia, xerophthalmia, conjunctivitis, corneal opacity and ulceration, skin lesions, metaplasia of the bronchiolar epithelium, pneumonitis, increased susceptibility to infection, faulty bone remodeling, failure of the neural foramina to enlarge, and impaired nerve functioning, have been noted. Magendie (*528*) in 1816 reported the first evidence associating the development of sore eyes in dogs with a vitamin A-deficient diet. Steenbock *et al.* (*825*) clarified these earlier observations by noting that dogs fed cod liver oil were normal but those deprived of ''fat-soluble vitamins'' developed xerophthalmia, which could be cured by the administration of cod liver oil. The NAS–NRC committee (*633*) recommended in 1962 that puppies be fed 200 IU vitamin A per kilogram body weight daily (90 IU/lb). Adult dogs can be maintained on 100 IU of vitamin A per kilogram daily (45 IU/lb). In 1974, allowances of 100 IU per kilogram body weight for adult dog maintenance and 220 IU/kg for growing puppies were established. These amounts can be provided by a dietary concentration of 5000 IU per kilogram feed on a dry matter basis.

The administration of large doses of carotene does not produce high serum and liver levels in the dog (*286,612,825*). Although the dog can utilize carotene, its ability to transform carotene to vitamin A and to store the surplus vitamin A appears to be limited (*133,877*). Nevertheless, in the past, carotene was judged to be an effective source of vitamin A for the dog (*109*). Bradfield and Smith (*109*) found that 200 U.S.P. units of vitamin A per kilogram body weight supplied by either cod liver oil, carotene in oil, or carrots were effective for growing pups. Michaud and Elvehjem (*581*) reported that the weight of carotene required was two or three times that of vitamin A, and the NAS–NRC committee (*633*) concluded that, for dogs, carotene is approximately one-half as valuable as vitamin A; converted to an international unit basis for carotene to vitamin A, the factor would be 4 on this basis.

I. Horse

A deficiency of vitamin A in horses (*352,635*) is characterized by night blindness, lachrymation, keratinization of the cornea, respiratory symptoms, abscesses of the sublingual gland, incoordination, convulsive seizures, reproductive difficulties in both the mare and stallion, anorexia or capricious appetite, polyuria, and progressive weakness. Garton *et al.* (*296*) and Goodwin in his review (*324*) cited the early finding of carotene in horse plasma and depot fat and stated that the carotene content of plasma followed a seasonal fluctuation, the probable result of the alternating influence of fresh pasture and dry pasture or cured hay.

A survey of seasonal variations in the concentration of carotene and vitamin A in the blood plasma of brood mares was conducted to obtain information directly related to the horse (*883*). Approximately 24 brood mares on six farms in six different sections of New Jersey were involved. The breeds represented were Shetland ponies, standardbreds, thoroughbreds, and grades. For a period of 18 months, blood samples were taken at approximately 6-week intervals. No effort was made to influence the owners in their feeding or management programs since the object of the survey was to evaluate the mares under typical managerial practices. The average seasonal changes in the carotene and vitamin A concentrations (*883*) in the blood plasma of the mares are shown in Fig. 5. The ranges for carotene and vitamin A concentrations found in this study were from 0 to 336 and from 8.3 to 36.2 μg per 100 ml plasma, respectively. The means for the total collection period were 50.4 and 19.8 μg per 100 ml plasma, respectively. This range reflects seasonal variations as well as variations among individual horses and management practices.

Although vitamin A is required in the ration of horses, few studies (*276–278*) have been undertaken on the relative efficiency of conversion of carotene to vitamin A in this species. Levels of vitamin A intake required by the horse indicate that it needs 6 times as much carotene as vitamin A for growth and maintenance and 10 times as much at the higher requirement level for pregnancy and lactation. Therefore, 1 mg of carotene in the rations of horses (Table 11) would be equivalent to 333–556 IU of vitamin A. Experiments (*278*) with mature geldings indicated that an average carotene intake of 198 mg/day was insufficient to maintain plasma carotene or vitamin A concentrations. Forages fed were Lincoln bromegrass, timothy, Atlanta fescue, Reed Canary grass, Atlantic alfalfa, Pennscot red clover, orchard grass, and Midland Bermuda grass. Plasma vitamin A concentrations declined to <10 μg per 100 ml in horses receiving as much as 172 mg of carotene per day from grass hay, but vitamin A deficiency symptoms were not observed. A lower intake of

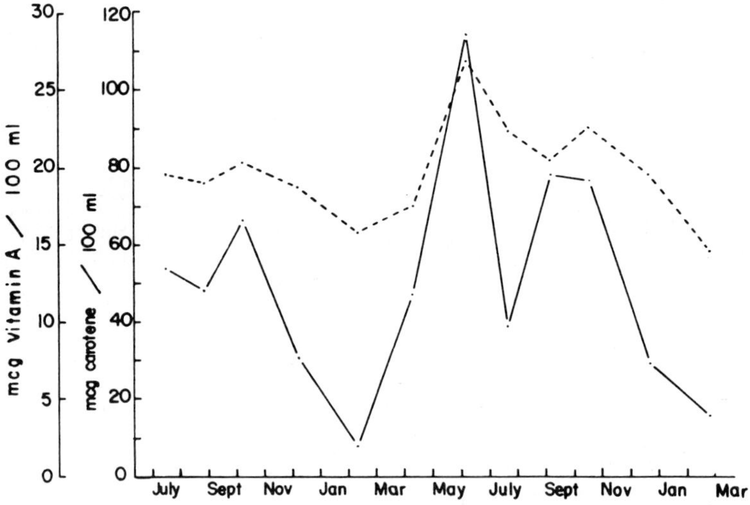

Fig. 5. Seasonal variation in concentration of carotene and vitamin A in plasma of brood mares of New Jersey (—, vitamin A; -.-, carotene). [From VanderNoot *et al.* (*883*).]

carotene from alfalfa maintained plasma vitamin A concentrations more effectively than the intake from grass hays. In establishing a vitamin A requirement for horses according to the NAS–NRC guide (*635*), it is estimated that 25 IU of vitamin A per kilogram body weight are adequate for maintenance, 40 IU/kg are adequate for growth, and 50 IU/kg are adequate for gestation and lactation. In the NAS–NRC guide it is assumed that 1 mg of carotene is equivalent to 400 IU of vitamin A.

The vitamin A requirement of horses can be met either by carotene contained in feeds or by vitamin A supplements. Fresh green forages and good-quality hays are satisfactory sources of carotene. Pasture quality varies with the change of season. Normally, early spring growth produces a lush green pasture high in carotene content. A decrease in carotene content occurs during the summer months, when the pastures are dormant; some dry off and become poor sources of carotene. Rejuvenation occurs during the autumn with cooler temperatures and increased rainfall. Over the winter months, when pastures are not available, the horse is dependent on hay. Even though hay quality may be high at the beginning of the winter season, storage causes a loss of carotene content. Even hay of the best quality cannot compare with lush, growing pasture.

The intake of carotene by horses fed rations devoid of good-quality forage is likely to be below their requirements. No ill effects may be apparent for several months, provided that the animals have good stores of vitamin A. Because carotene is unstable at higher temperatures or when exposed to light, carotene content decreases with storage and eventually becomes too low to meet the vitamin A requirement. The conversion of carotene from dried forages to vitamin A is inefficient in the horse (*278*). It is sound nutritional practice to provide horses with vitamin A-containing feeds or a vitamin A supplement that will supply at least the minimal requirements. Larger amounts of vitamin A supplementation should be provided for gestating–lactating mares and for breeding stallions.

J. Poultry

Vitamin A deficiency symptoms in poultry are described in detail by Moore (*605*), Wolbach (*926*), and the NAS–NRC subcommittee (*637*) on poultry nutrition. Vitamin A is a dietary requirement of poultry. In growing chickens on a diet severely deficient in vitamin A, growth is markedly retarded. Other symptoms are general weakness, emaciation, staggering gait, ruffled plumage, reduced resistance to infection, and increased mortality. The secretory mechanisms of the intestinal mucous glands, the tear glands, and the salivary glands fail and keratinization occurs. In mature chickens, the symptoms develop much more slowly. Eye disorder becomes acute, and there is often a cheesy exudate around the eyes as well as a sticky discharge from the nostrils. Egg production and hatchability are markedly reduced.

Although true vitamin A has been fed in commercial feeds for decades, β-carotene can be an important provitamin in poultry under simulated natural conditions. The chicken, like the rat, is able to convert β-carotene to vitamin A, although the chicken can be raised to maturity without carotenoid pigments in the diet, provided that supplementary vitamin A is fed (*668*). By 1931, Capper *et al.* (*142*) had already shown that the chicken can convert β-carotene to vitamin A. That this conversion takes place in the intestines was demonstrated by Cheng and Deuel (*153*). Sibbald and Olsen (*795*) demonstrated that the conversion can take place in the duodenum. Most of the carotene consumed by poultry is metabolized, some is converted to vitamin A (*351*). Small amounts are found in the liver and the yolk of the egg (*293*). In young chicks, β-carotene can be converted to vitamin A by the first week of life or earlier (*98,371*).

The vitamin A activity of carotene in the chicken, as in the case of the rat, must be defined in relation to dosage. Table 6 summarizes various carotene to vitamin A activity ratios obtained with different sources of carotene and at different levels of feeding. The relative effectiveness of vitamin A and carotene in promoting liver stores was calculated directly by comparing the absolute vitamin A liver stores resulting from feeding an equivalent international unitage of vitamin A or carotene. Many of these studies were based on one level of dietary supplementation. At lower levels of feeding, when the criterion is growth, the vitamin A activity of carotene is essentially equal to the activity found in rats, namely, 0.6 μg equal to 1 IU of vitamin A (*20,71,144,245,536,651,672,673,731,735,752,787,867,916*). On the basis of liver storage of vitamin A when higher than minimal maintenance levels are fed, a marked reduction in the efficiency of carotene conversion is noted (*240,245,283,308,371,505,536,651,726*).

At a low β-carotene feeding level of 500 IU vitamin A per pound, the data from another broiler chick trial (*536*) indicated that β-carotene functions equally as well as vitamin A when compared on an international unit basis; hence, an effective weight ratio of 2 : 1 was obtained under the conditions of the trial. Likewise, in another study (*273*) on 1-day-old broiler-type cockerels, β-carotene from a fermentation source compared favorably with crystalline all-*trans*-β-carotene in its influence on growth, feed efficiency, and plasma and liver vitamin A concentration. The finding of a 2 : 1 ratio at a lower feeding level (400 or 800 IU) was supported by data of Parrish *et al.* (*673*). β-Carotene of microbial origin was also found to be active in several Russian studies (*706,871*). At levels of vitamin A activity that supported chick growth, β-carotene appeared to be as effective as vitamin A acetate on an international unit basis. Similar data were reported by Tiews *et al.* (*867*). Olsen *et al.* (*651*) found preformed vitamin A to be more effective than carotene even at levels close to the requirement for growth. Krieg (*484*), using liver stores of vitamin A as the criterion, found that this was also the case with hens.

In Russian reports (*870,949*) grass meal with added antioxidants was declared to inhibit the rate of carotene oxidation and to have a positive effect on chick growth and feed efficiency to a greater extent than grass meal without anti-oxidants added. Ascarelli *et al.* (*33*) observed the influence of several sources of carotene on growth and vitamin A liver storage (Table 7). Carotene from clover and oatmeals, added to

TABLE 6

Relative Carotene to Vitamin A Equivalency for Poultry

Criterion for comparison	Feed level[a] (IU/lb)	Source of carotene; source of vitamin A	Conversion ratio,[b] carotene to vitamin A	μg β-carotene ≎ IU vitamin A
Growth	340–680	Alfalfa leaf meal; U.S.P. cod liver oil	2:1	0.6
	200	Dehydrated alfalfa meal; fish oil	2:1	0.6
	500–1000	Crystalline carotene[c]; crystalline vitamin A acetate	2:1	0.6
	100–900	Dehydrated cereal grass; stabilized dry vitamin A	2:1	0.6
	400–700	Dehydrated alfalfa meal; stabilized dry vitamin A	2:1	0.6
	500	Stabilized β-carotene[d]; stabilized dry vitamin A	2:1	0.6
Liver storage	2,000	Fresh alfalfa; fish oil	7:1	2.1
	4,000	Crystalline carotene[c];	4:1	0.6
	8,000	crystalline vitamin A acetate	10:1	1.5
	700	Crystalline carotene[c]; vitamin A natural esters	4:1	1.2
	1,000	Dehydrated alfalfa meal; stabilized dry vitamin A	4:1	1.2
	10,000	Dehydrated grass meal; crystalline vitamin A acetate	5:1	1.5
	1,200	Dehydrated cereal grass; stabilized dry vitamin A	5:1	1.5
	4,800	Dehydrated alfalfa meal; stabilized dry vitamin A	10:1	3.0
	2,500 5,000	Dehydrated alfalfa meal; stabilized dry vitamin A	5:1	1.5
	2,500 5,000	Stabilized dry β-carotene[d]; stabilized dry vitamin A	5:1	1.5

[a] Calculated on basis of 1.0 gm of crystalline β-carotene = 1,667,000 IU vitamin A.
[b] Weight basis.
[c] From natural sources: 90% β-carotene; 10% α-carotene.
[d] Synthetic crystalline β-carotene dissolved in oil in emulsified beadlet form.

mash feed and fed to chickens, was utilized at the same rate as that from lucerne meal, the criterion being the amount of vitamin stored in the liver. The highest vitamin A storage was obtained when chemically synthesized carotene was added to the feed.

Studies by Evans showed that published reports on *(261)* a variety of vitamin A-active sources do not, for practical purposes, confirm the generally accepted view that carotene and vitamin A are equally effective in terms of biological

TABLE 7

Comparison of Utilization of Carotene from Various Dried Plants[a,b]

Carotene source	Body weight at 6 weeks (gm)	Vitamin A (IU/liver)	Mean carotene intake (μg/chick)
Synthetic carotene[c]	505	524	3836
Commercial lucerne meal[d]	531	402	4164
Lucerne meal	515	384	4158
Clover meal	523	486	4126
Oatmeal	501	357	4098

[a] From Ascarelli *et al.* (*33*).
[b] Twenty chicks per group.
[c] Gelatin-coated water-soluble concentrate containing 15% chemically synthesized carotene and added tocopherol; 10 chicks per treatment.
[d] Containing 0.0125% ethoxyquin; 10 chicks per treatment.

activity in feeding poultry. Rather, the evidence points to a definite superiority of vitamin A over carotene. Evaluation may also be affected, according to Evans, by such factors as age and breed of birds; exposure to disease; concentrations of other nutrients in the ration; nutritional history of the birds, vitamin A status or storage, particularly depletion; level of intake of the vitamin or provitamin; source or carrier of the vitamin or provitamin and its history, including processing and storage; antioxidants, antibiotics, bentonite, or other additives or medicaments in the ration; and natural or incidental inhibitors or stimulants in ration components. One of the most important factors affecting utilization is the level of intake of the vitamin or provitamin. In the reports examined, Evans correctly noted that perhaps the most obvious feature was the heterogeneity of intakes. However, although the relative efficacies may have differed for different functions, vitamin A was more effectively used than carotene in more than half of the 43 studies reviewed and in 76% of 25 studies reported between 1950 and 1966. No differences were suggested between males and females, among breeds, or among age groups in the response to the two sources of vitamin A activity. These comparisons, taking into consideration physical and chemical forms of pure β-carotene and pure vitamin A (palmitate), demonstrated greater efficacy for vitamin A as a vitamin source for poultry at current levels of use in commercial production practices. The Agricultural Research Council of Great Britain (*642*) declared that the activity of β-carotene, on a weight basis, is defined as half that of vitamin A, but its biological value in terms of vitamin A may be reduced by a variety of factors affecting its utilization in the diet. It was therefore suggested that, in practice, β-carotene may have, weight for weight, about one-sixth the value of vitamin A, although opinion is still divided on this matter.

In turkeys (*355*) crystalline vitamin A acetate was compared to crystalline carotene using both growth and liver stores of vitamin A as criteria of activity after 8 weeks of feeding starting with 1-day-old poults. On the basis of international units, crystalline vitamin A acetate was about four times as efficient in supporting growth as corresponding levels of crystalline carotene. This is an effective carotene to vitamin A ratio of 8 : 1 based on weight. In the liver storage test, approximately 30,000 IU/lb from carotene were equivalent to feeding 4000 IU/lb as crystalline vitamin A acetate, a ratio of about 14 : 1 on a weight basis. Creger *et al.* (*169*), on the basis of blood and liver vitamin A values after oral administration of β-carotene in an oil solution, concluded that the young turkey is not an efficient converter of carotene to vitamin A.

Comparative studies have been made on other avian species. In the bobwhite quail, Nestler *et al.* (*625*) found that carotene was utilized only one-third to one-seventh as efficiently as vitamin A, one-half to one-tenth as effectively as natural vitamin A esters, and only one-fourth to one-seventeenth as effectively as vitamin A acetate, based on liver vitamin A storage values. Vitamin A was superior to carotene in other studies (*368*) on bobwhite quail. Various carotenals have been found to be effective vitamin A precursors for the quail (*17*).

K. Swine

Swine, like poultry, require an exogenous source of vitamin A. Deficiency symptoms involving epithelial tissues of the respiratory, reproductive, nerve, and genitourinary systems are similar to those described for other species. Under natural conditions, a partial posterior paralysis may develop before other symptoms are apparent. Diarrhea, muscular incoordination, unthriftiness, poor growth, infertility or congenital malformations, and birth of dead pigs make this dietary nutrient extremely important to swine producers. Low liver storage and reduced plasma levels of vitamin A and elevated cerebrospinal fluid pressure are criteria of inadequate vitamin A status in the pig.

The vitamin A requirements of swine can be met by either vitamin A or β-carotene (*640*). Vitamin A is critical for breeding swine during the reproductive phases, with the requirements being higher than those during growing and finishing. In 1941, Braude *et al.* (*112*) estimated that about 40 μg of β-carotene per kilogram body weight (0.7 mg/kg ration) were adequate for good growth in pigs but allowed limited liver storage. A survey by Parrish *et al.* (*670*) showed that the vitamin A concentration in the colostrum and blood serum of gilts and in the blood serum of young pigs was one-half to three-fourths as high in those fed carotene as those fed vitamin A. In the livers of newborn pigs, carotene produced only one-fourth as much vitamin A, unit for unit fed, as did vitamin A. Hentges *et al.* (*380*) in 1952 suggested that about 25 μg of purified β-carotene per kilogram body weight (about 0.4 mg/kg ration) was the minimal daily requirement for depleted 8-week-old pigs to restore initial blood plasma vitamin A levels and provide some vitamin A liver storage. Ullrey *et al.* (*881*) in 1965 reported that 0.5 mg of β-carotene per kilogram ration gave optimal weight gains in the 50- to 100-kg weight category.

Season of feeding can influence the vitamin A nutrition of the pig. In one German study (*179*) assays were made of vitamin A of colostrum and milk from sows during lactation. In the fall, when the feed contained sufficient carotene, the levels of vitamin A in the colostrum and milk were about 60% higher than the values found in the spring, when the rations were deficient in carotene. Levels of vitamin A in the fall and spring were, respectively: colostrum (first day of lactation), 328, 154; milk (eleventh to sixteenth day of lactation), 60, 30 IU/liter. The spring rations provided only about 40% of the sows' vitamin A requirements.

In 1970–1975 many reports appeared in the Russian literature on the value of adding antioxidants (ethoxyquin, dilutin, etc.) to grass meals to delay oxidation of carotene and hence give some improvement in biological response in growing-fattening swine (*818,943*), namely, higher blood or liver vitamin A values or better production indices. However, a supplement of stabilized vitamin A was more effective and more economical than the corresponding amount of stabilized carotene given in that form or as grass meal (*887*).

Yellow corn (maize) is the primary, and frequently the only, vitamin A precursor ingredient in pig rations. The estimation of the carotenoid content in corn presents a challenge to the chemist, for it contains three biologically active carotenoids—β-carotene, β-zeacarotene, and cryptoxanthin—approximately in the proportions 0.25, 0.25, and 0.5. Because β-carotene has twice the activity of the other two, they should be determined separately. Ullrey *et al.* (*879–881*) determined the vitamin A activity of corn by feeding it to pigs and measuring the liver store of vitamin A. He observed that 1 μg of carotene from corn had the same activity as 0.08 μg of retinol, whereas according to the NAS–NRC (*640*) relationship its activity should have been equivalent to 0.15 μg of retinol.

Compared with the rat, the pig is an inefficient converter of β-carotene to vitamin A. It has been shown that in swine a decreasing efficiency of carotene utilization occurs with increasing intake (*620*). Graded levels of 40, 80, 160, and 320 μg of carotene from alfalfa were compared to 8, 16, 24, and 32 μg of vitamin A per pound live weight each day in a 6-week feeding test with swine using plasma vitamin A concentration and liver stores of vitamin A as the two criteria of relative performance. On the basis of plasma vitamin A values, the relative effectiveness of carotene to vitamin A on a weight basis is as follows: at the 40 μg level, 6 : 1; at 80 μg, 7 : 1; at 160 μg, 9 : 1; and at 320 μg, 11 : 1. On the basis of liver vitamin A stores, the following relationships were found: at the 40 μg level, 4 : 1; at 80 μg, 5 : 1, at 160 and 320 μg, 9 : 1. At lower levels of feeding adequate for growth and reproduction, an average ratio of 5 to 6 : 1 is found; however, as the feeding levels are increased to provide for adequate liver stores, the β-carotene to vitamin A ratio widens to 8 to 10 : 1 on a weight basis. The same effect of increasing intake on the efficiency of utilization of carotene has been reported for dairy cattle and lambs.

The NAS–NRC (*640*) accepts a relationship of 1 mg carotene equal to 500 IU vitamin A for swine. On a weight basis, this is an effective ratio of 6 mg to 1 mg (6th: 1). some reports are at variance with the NAS–NRC statement on the biological value of β-carotene for the pig. In a study on swine by Wellenreiter *et al.* (*906*), conversion efficiencies were calculated by dividing the regression coefficients for carotene by the regression coefficients for retinyl palmitate. At higher levels of carotene intake, when corn gluten meal was included in the repletion diets, 1 mg of carotene had a vitamin A potency of 123–174 IU (Table 8). When corn was the only source of carotenes and the concentration was more typical of corn–soy swine diets, 1 mg of carotene had a vitamin A potency of 261 IU. In still another study, in which the provitamin A was supplied in the form of alfalfa meal or yellow corn, provitamin A was used less efficiently than stabilized vitamin A, according to Parrish and Aubel (*669*). It would appear that the NAS–NRC relationship of 1 mg of dietary carotene being equal to 500 IU of vitamin A activity exaggerates the usefulness of carotenes in corn for swine when liver vitamin A storage is used as the criterion. Table 9 records the results of some studies on the biological value of β-carotene and vitamin A for swine. Using night blindness as the criterion of effectiveness, a ratio of 6 μg β-carotene to 1 μg of vitamin A (6 : 1) was established for swine (*352*). On the basis of reproduction and lactation, this weight ratio rises to about 10 : 1 (*352*). With growth and reproduction as criteria (*353*), a ratio of 5 : 1 was reported. Liver stores of newborn and 4-day-old pigs from gilts fed crystalline carotene or vitamin A concentrate indicated that vitamin A produces four to five times the storage of vitamin A as an equivalent amount of carotene on an international unit basis (*670*).

TABLE 8

Vitamin A Activity of Corn Carotenes and Efficiency of Conversion by the Pig[a]

Trial	Conversion efficiency (%)[b]	Activity (IU/mg)[c]
1	6.76	123
2	9.56	174
3	14.33	261

[a] Three trials with weanling pigs whose vitamin A-deficient diet was supplemented with three levels of the carotenes (1–10 mg/kg diet) from corn and corn gluten or both and compared to vitamin A palmitate (73–654 μg/kg diet). From Wellenreiter *et al.* (*906*).

[b] Calculated by dividing the regression coefficient of the carotene source by the regression coefficient of all-*trans*-retinyl palmitate and multiplying by 100.

[c] Calculated by multiplying the vitamin A activity of all-*trans*-retinyl palmitate (1818 IU/mg) by the relative efficiency of conversion of crude corn carotenes.

TABLE 9

Relative Carotene to Vitamin A Equivalency for Swine

Criterion for comparison	Level fed/lb body wt (μg carotene : IU vitamin A	Source of carotene; source of vitamin A	Conversion ratio, carotene to vitamin A[a]	μg β-carotene ≎ 1 IU vitamin a
Growth	18:10	Dehydrated alfalfa meal; cod liver oil	5:1	1.5
Growth and finishing	26–120:14–64	All natural feed sources; vitamin A supplement	6:1	1.8
Growth and reproduction plus safety factor	40:26	Dehydrated alfalfa meal; cod liver oil	5:1	1.5
Plasma level	40:26	Dehydrated alfalfameal;	6:1	1.8
	80:53	stabilized dry vitamin A	7:1	2.1
	160:79		9:1	2.7
	320:106		11:1	3.3
Reproduction and lactation	73:49	Dehydrated alfalfa meal; cod liver oil	10:1	3.0
Night blindness	15:10	Dehydrated alfalfa meal; cod liver oil	6:1	1.8
Liver storage	13:22	Crystalline carotene[b]; vitamin A oil	8–10:1	2.4–3.0
	40:26	Dehydrated alfalfa meal;	4:1	1.2
	80:53	stabilized dry vitamin A	5:1	1.5
	160:79		7:1	2.1
	320:106		9:1	2.7

[a] Weight basis.

[b] From natural sources: 90% β-carotene, 10% α-carotene.

L. Ruminants

Cattle (*560*) normally circulate a relatively large quantity of β-carotene and other carotenoids in the blood, and substantial amounts of carotenoid pigment may be present in liver and kidney tissues and in fat deposits. Sheep and goats circulate small or no appreciable quantities of carotene in their blood. Vitamin A is a nutrient of prime importance in ruminant nutrition. It is required for growth, reproduction, adequate milk production, and support of many normal body functions, examples of which are maintenance of cerebrospinal fluid pressure (CFP) and prevention of night blindness (nyctalopia). Under practical conditions a deficiency can show up as impaired fertility, a smaller calf or lamb crop, offspring of lowered viability, impaired growth and feed efficiency, poor hair coat, running noses and eyes, impaired vision, and edema (anasarca). Unfortunately, under practical management conditions, a vitamin A deficiency is difficult to diagnose and marginal deficiency may exist without obvious deficiency symptoms. Nevertheless, a subclinical deficiency can be diagnosed by determination of the vitamin A and carotene levels in body tissues (blood plasma or, preferably, from a liver biopsy) by measuring the pressure of the spinal fluid or, in an advanced state of deficiency, by examining the eyes for papilloedema. Cerebrospinal fluid pressure appears to be accurate at severely low plasma levels, 10–15 μg per 100 ml, but is of questionable accuracy with higher plasma levels. In one study (*228*) CSP dropped from a high of 355 mm of saline with plasma levels of 3.5 μg vitamin A per 100 ml to a constant pressure of 74 mm at concentrations of 14 μg per 100 ml and above. Lack of growth may be a sign, but often in the subclinical or marginal deficiency state the animals may show satisfactory growth with no outward deficiency signs for a long time. Under drought conditions, Schmidt (*766*) noted vitamin A deficiency signs of stringy nasal discharge, staggering gait, and cloudy cornea, which is often mistaken for pinkeye among ruminant animals. Whether it is a marginal or more severe vitamin A deficiency state, these animals show increased sensitivity to solar heat. It was observed by some investigators that with vitamin A deficiency a high incidence of urinary calculi may occur.

Phillips (*699*) reminds us that at birth the young animal normally has a very low level of blood plasma vitamin A, and concentrations of 10 μg or more per 100 ml must be quickly attained within a few days for a favorable chance of survival (*379*). This may be accomplished by suitable supplementation, namely, colostrum feeding or vitamin A-active concentrate feeding to the young, vitamin A-active concentrate in the diet of the dam, or both. In ruminants, particularly under grazing conditions, carotene is relied on to a greater extent to provide the necessary vitamin A than in most other farm animals. In the young ruminant a critical evaluation of the intake of carotene as a vitamin A source should not be overlooked.

1. Goats, Sheep

Vitamin A or its provitamin is required for adequate nutrition of sheep, as with other mammalian species. The first recognized sign of vitamin A deficiency in sheep is nyctalopia, and this criterion has been used often to determine the requirement of sheep for vitamin A or carotene (*352,354*). Australian reports describe vitamin A deficiency of sheep during prolonged drought, the chief sign being the production of weak or dead lambs, prematurely or at full term. Night blindness has been observed, but not total blindness (*140,552,687*). Urinary calculi have been found quite often in vitamin A-deficient sheep but less frequently in cattle and goats.

Several investigators reported normal plasma levels of vitamin A for sheep. Pope *et al.* (*708*) cited 37–48 μg %, Hoefer and Gallup (*405*) gave a wider range, 20–45 μg %, whereas Peirce (*686*) believed the normal range to be 25–30 μg %. Moore and Payne (*609*) were in close agreement with Harms (*367*) in reporting normal liver vitamin A concentrations in sheep to be from 460 to 503 IU/gm. Harms also reported a vitamin A range in grazing sheep of 230–810 μg per gram of liver but only 2–5 μg of carotene per gram. Peirce (*686*) also reported that lower plasma carotene levels, 5–15 μg %, were found in practice. Early reports (*324*) cited the presence of some carotenes in plasma, colostrum, and liver of sheep, particularly when they are on fresh pasture or receiving dried forage still containing carotenoids. Goats, on the other hand, do not contain carotenes in their body tissues, or, if they do, they are present only in trace amounts. The milk of sheep and goats contains only traces of carotenes. Peirce (*685*) reported about 8–10 μg % in colostrum and milk of sheep; however, colostrum is a rich source of vitamin A (800 μg %). Chanda (*147*) determined that the partition of carotenoids and vitamin A in the colostrum and milk of cows and goats was affected by diet and stage of lactation. Goat colostrum contained β-carotene, whereas goat milk did not. A parabolic relationship between the stages of lactation and the vitamin A content of milk fat was found in both species. Sheep and goats can subsist apparently for long periods on poor pasture if scrub brushes and trees are available as a source of carotene. Dikshit and Ranganathan (*200*) conducted carotene metabolism studies in goats. Gartner and Anson (*295*) found during a 16-month drought period in Australia that the liver of their sheep contained reserves of vitamin A despite the lack of green pasture. The animals consumed sufficient carotene (43–90 mg/kg dry matter) from the leaves of the mulga (*Acacia aneura*). Peirce (*686*) determined the concentration of carotene in the edible portion of native fodder plants in Australia. In a New Zealand survey (*453*) about 1 in 1000 lamb carcasses were rejected for export because of excessively yellow fat; incidences of 5–10% were occasionally reported on individual farms. Perirenal fat usually contained more pigment than omental fat. A study with 562 lambs showed that the fat of the progeny of three Romney rams and Romney dams was more yellow than the fat of the progeny of 31 other sires of five different breeds mated to Romney ewes. Lutein and flavoxanthin were the chief carotenoids (*168*) isolated from the yellow fat of sheep. A third oxycarotenoid was probably auroxanthin, and small amounts of β-carotene and flavochrome were present. The estimated heritability of yellow fat based on visual scores was 0.18. In general, lamb fat is much whiter than beef fat, with corresponding lower concentrations of carotenoids.

The feeding of silage to sheep is not deleterious to the utilization of the carotene it contains, although it may be so with lambs (*373*). Brune and Zaddach (*125*) fed equal amounts of carotene (11 mg/day) from silage or hay made from grass from the same field to sheep and found, from liver vitamin A values, that silage carotene was better utilized than hay carotene. Pcelinceva (*683*) found that carotene (22–45 mg/day) was better utilized from hay than from silage, as judged by the level of vitamin A in ewe's milk. Martin *et al.* (*535*) reported that carotene from corn silage produced adequate liver stores in lambs. Phillips (*699*) mentioned that factors influencing the utilization of dietary carotene and vitamin A by cattle, such as previous nutritional state (*202*), are in some cases applicable to sheep.

Regarding recommended allowances of carotene for sheep, Peirce (*687*) administered 2.5 mg of carotene per 100 lb body weight and noted that night blindness was cured, but no appreciable effect on vitamin A reserves in the liver was observed. The classic studies of Guilbert *et al.* (*352*) demonstrated the minimal requirement of sheep to be 200 μg vitamin A or 1.2–1.6 mg carotene daily per 100 lb body weight.

A unit of carotene as normally measured has different feeding values, depending on the circumstances under which it is fed. For sheep fed in normal fashion for usual productivity, this value seems to vary between 400 and 550 IU of vitamin A activity per milligram of carotenes. This range of values was verified by the studies of Martin *et al.* (*535*) on growing lambs, in which the vitamin A activity of corn silage carotenes was compared to a vitamin A standard, all-*trans*-retinyl palmitate. Blood serum and livers were analyzed for vitamin A content. From the liver storage data, 1 mg of corn silage carotenes was found to be equivalent to 436 IU of vitamin A activity under the conditions of this trial.

In the NAS–NRC recommendation (*639*) 4.3 μg of vitamin A or 25 μg of carotene per kilogram body weight were used to calculate the minimal requirements. The carotene requirements for late pregnancy and lactation were calculated at 500% of the minimal level necessary to prevent nyctalopia (25 μg/kg body weight). The recommendation for replacement lambs and yearlings was calculated at 250% of this minimal level. For fattening lambs and for ewes during the first 15 weeks of gestation, 150% of this minimum was used. Because sheep do not convert carotene to vitamin A in the usual quoted ratio of 1 mg of β-carotene equivalent to 1667 IU vitamin A, the NAS–NRC (*639*) assumed 400–700 IU vitamin A per milligram carotene in calculating the numerous requirements when expressed as international units of vitamin A activity.

Variation in carotene conversion values has been observed in feeding trials with sheep. Diven and Erwin (*202*) noted different conversion values for vitamin A-deficient and healthy sheep, indicating that previous nutritional state of the animal may affect carotene utilization. For normal sheep, preformed vitamin A was 3.3 times as effective as an equal weight of β-carotene, and for deficient sheep it was 4.6 times as effective. On the basis of plasma and liver values of vitamin A in sheep after feeding carotene from alfalfa and vitamin A from fish oil (*405*) or vitamin A in a dry carrier (*620*), it was determined that the relative value on a weight basis varies with the level of intake, which is similar to results obtained with dairy calves (*339,748*). Graded levels of 40, 80, 160, and 320 μg of carotene were fed daily to sheep for 6 weeks compared to 8, 16, 24, and 32 μg of vitamin A per pound live weight (*620*). On the basis of plasma vitamin A levels, the following carotene to vitamin A (alcohol) ratios on a weight basis were found: at the 40 μg level, 8 : 1; at 80 μg, 9 : 1; at 160 μg, 11 : 1; and at 320 μg, 13 : 1. Similar values (*620,748*) based on liver stores of vitamin A were found: at the 40 μg level, 6 : 1; at 80 μg, 8 : 1; at 160 μg, 10 : 1; and at 320 μg, 13 : 1 (Table 10). A survey of data on sheep shown in Table 11 indicates that essentially similar activity responses occurr in cattle for β-carotene and vitamin A. For growth and at marginal levels of feeding, a 5 : 1 to 6 : 1 ratio of carotene to vitamin A is observed, and at higher feeding levels, to provide a safety factor, a ratio of 10 : 1 is more frequently reported. These relationships have been accepted in calculating the relative effectiveness of the two vitamin A sources.

2. *Dairy Cattle*

In dairy cattle (*632*) vitamin A deficiency may be observed as degeneration of the mucosa of the respiratory tract, mouth, salivary glands, eyes, tear glands, intestinal tract, urethra, kidneys, and vagina and keratinization of epithelial tissue. Histopathological changes in the parotid ducts, with thickened, stratified squamous epithelium replacing the normal pseudostratified columnar epithelium, appear to be specific for avitaminosis in cattle. Diarrhea, loss of appetite, and emaciation are commonly observed in a well-developed deficienty state, and greater susceptibility to infections, colds, and pneumonia may occur. In later stages, characteristic changes in the eye may take place, and sometimes blindness results. Blindness is generally permanent and is due to constriction of the optic nerves caused by stenosis of the optic foramen. As vitamin A deficiency develops, adaptation to the dark is reduced, and night blindness occurs. Staggering gait, convulsive seizures, and papilloedema are usually present as a result of elevated CFP. Loss of appetite is of economic importance since it soon leads to slowing of growth, with papilloedema as an early objective sign of vitamin A deficiency in growing cattle.

More specific symptoms of vitamin A deficiency are those affecting reproductive performance, such as decreased sexual activity and lowered sperm count in bulls and reduced implantation rate, abortion, premature births, and retained placenta in cows (*49,242*). In many instances, these conditions occur with marginal vitamin A intake that often goes undetected. The first signs of vitamin A deficiency in pregnant cows are altered gestation periods, abortions, high incidence of retained placenta, stillbirths, and uncoordinated or blind calves. Retention of the placenta is common. Other effects of vitamin A deficiency on reproduction are cessation of estrus, prolonged gestation period, and difficult parturition (*246a*). In the bull, there is inability to perform and lack of normal spermatogenesis (*527*). If symptoms of deficiency occur before the breeding age, bulls fail to breed but usually show improvement with time when given a vitamin A supplement. If bulls reach breeding

TABLE 10

Relative Value of Carotene from Artificially Dehydrated Alfalfa and Vitamin A from a Dry Carrier[a]

Amount of carotene from alfalfa fed (μg/lb live wt daily)	Amount of vitamin A from a dry carrier necessary to obtain equivalent responses to carotene (μg/lb live wt daily)	μg carotene/μg vitamin A
Based on plasma vitamin A		
Calves[b]		
40	9.9	4.0
80	12.5	6.4
160	16.6	9.6
320	21.9	14.6
Lambs		
40	6.1 (5.2)[c]	6.6 (7.7)[c]
80	9.7 (8.8)	8.2 (9.1)
160	15.4 (14.9)	10.4 (10.7)
320	24.4 (25.2)	13.1 (12.7)
Pigs		
40	7.1	5.6
80	11.5	7.0
160	18.5	8.6
320	29.9	10.7
Based on log liver vitamin A		
Calves[b]		
40	7.6	5.3
80	10.2	7.8
160	13.7	11.7
320	18.4	17.4
Lambs		
40	7.3 (6.2)[c]	5.5 (6.4)[c]
80	10.8 (9.8)	7.4 (8.2)
160	16.0 (15.5)	10.0 (10.3)
320	23.8 (24.4)	13.4 (13.1)
Pigs		
40	10.7	3.7
80	15.7	5.1
160	23.1	6.9
320	34.0	9.4

[a] From Myers *et al.* (*620*) and Rousseau *et al.* (*748*).

[b] Calculated from the data of Rousseau *et al.* (*748*).

[c] Values calculated from the plasma and liver vitamin A concentrations adjusted by covariance for the plasma vitamin A values determined the day before vitamin supplementation.

TABLE 11

Relative Carotene to Vitamin A Equivalency for Ruminants and Horses

Animal	Criterion for comparison	Level fed/lb body wt (μg carotene : IU vitamin A)	Source of carotene; source of vitamin A	Conversion wt ratio, carotene to vitamin A	μg β-carotene ≏ 1 IU vitamin A	IU vitamin A ≏ 1 mg β-carotene
Sheep	Growth, reproduction, and lactation	60:30	Dehydrated alfalfa meal; cod liver oil	6.6:1	2.0	500
	Growth, reproduction, and lactation	17–58:9–23	Natural feed sources: fish liver oil	6:1	1.8	556
	Plasma level	40:26	Dehydrated alfalfa meal; stabilized dry vitamin A	8:1	2.4	417
		80:53		9:1	2.7	380
		160:79		11:1	3.3	303
		320:106		13:1	3.9	256
	Night blindness	13:10	Dehydrated alfalfa meal; cod liver oil	6:1	1.8	556
	Liver storage	40:26	Dehydrated alfalfa meal; stabilized dry vitamin A	6:1	1.8	556
		80:53		8:1	2.4	417
		160:79		10:1	3.0	333
		320:106		13:1	3.9	256
	Liver storage	136:230	Dehydrated alfalfa meal; fish liver oil	6:1	1.8	556
	Liver storage	507:270	Stabilized dry β-carotene; stabilized dry vitamin A	12:1	3.6	278
Dairy cattle	Growth	60:30	Dehydrated alfalfa meal; cod liver oil	6.6:1	2.0	500
	Plasma levels	34:7	Crystalline carotene and dehydrated alfalfa; shark liver oil	5:1[a]	1.5	266
				8:1[b]	2.4	417
	Plasma levels	31:12.5	Dehydrated alfalfa meal; stabilized dry vitamin A	8:1	2.4	417

	Plasma levels	60:18	Dehydrated alfalfa meal; stabilized dry vitamin A	5.4:1	1.6	625
		180:36		10.4:1	3.2	313
		540:69		20.0:1	6.0	167
	Plasma levels	40:30	Stabilized dry β-carotene; stabilized dry vitamin A	7.7:1	2.3	435
		80:40		11.1:1	3.4	313
		160:50		16.2:1	4.9	204
	Spinal fluid pressure	33:21	Dehydrated alfalfa meal; stabilized dry vitamin A	5:1	1.5	666
	Prevention of papillary edema	36:26	Dehydrated alfalfa meal; stabilized dry vitamin A	4.5:1	1.4	714
	Liver storage	60:18	Dehydrated alfalfa meal; stabilized dry vitamin A	6.1:1	1.8	556
		180:36		12.5:1	3.8	263
		540:69		23.4:1	7.1	141
	Liver storage	30:10	Dehydrated alfalfa meal; stabilized dry vitamin A	10:1	3.1	223
Beef cattle	Growth	60:30	Dehydrated alfalfa meal; cod liver oil	6.6:1	2.0	500
	Growth and reproduction	17–48:3–19	All feed sources; vitamin A oil	8:1	2.4	417
	Liver storage	12–36:7–33	Dehydrated alfalfa meal; cod liver oil	10:1	3.0	333
	Liver storage	62:25	Stabilized dry β-carotene; stabilized dry vitamin A	8.3:1	2.5	400
	Liver storage	1–4 × NRC	Stabilized dry β-carotene; stabilized dry vitamin A	8.3:1	2.5	400
Horse	Growth, night blindness		Dehydrated alfalfa	6:1	1.8	556
	Pregnancy, lactation		Dehydrated alfalfa	10:1	3.0	333

[a] For Holstein yearlings.
[b] For Guernsey yearlings.

age before developing symptoms, they try to breed but usually show improvement with time when given a vitamin A supplement. If bulls reach breeding age before developing symptoms, they try to breed but may be too weak to mount and perform. Semen obtained by artificial stimulation may be fertile, but it is usually scanty and high in number of abnormal spermatozoa (*404*).

A preferential accumulation of carotene in the tissues of cows was first noted in the pioneering work of Palmer and Eckles (*667*). Carotene constitutes the major portion of carotenoids present (*294,675*). Small amounts of α-carotene, lycopene, cryptoxanthin, and other oxycarotenoids are also present (*304,328,429*). The carotenoid content of cows' plasma shows a marked seasonal variation, and the plasma levels directly and the levels in the butterfat of the milk indirectly reflect the dietary intake of carotenoids. Marked increases in plasma carotenoid values are obtained after animals are put out to pasture in the spring, and high levels usually continue during the summer. With cows grazing all year round, the monthly variation in blood plasma levels corresponds to the condition of the pasture grasses (*384*). There are marked variations in plasma levels among different breeds of cattle, and these are reflected in the yellow color of the milk and butter (*189,384,491,674*). β-Carotene is also present in the erythrocytes (*629*). Goodwin and Wilson (*328*) showed that the levels of carotene and vitamin A in cattle blood remained fairly constant while the animals were on grass pasture from May to October. The levels dropped after parturition, when large amounts of both carotene and vitamin A were transferred to the colostrum; 95% of the colostral vitamin A was esterified. Cattle blood, and that of other species, was fractionated by Ganguly *et al.* (*291*). The colostrum of cows, as in the case of human beings, is rich in carotene (*328,675,760*). At parturition, a drop occurs in plasma carotene levels (*328,493*). Sutton *et al.* (*833*) investigated the carotene and vitamin A content of blood plasma of dairy cows with reference to variations associated with parturition and lactation. They found that plasma carotene decreased markedly at the time of parturition and during the first week of lactation. The needs of the nursing or milk-fed calf for vitamin A are normally met by vitamin A and carotene in the colostrum and milk. The importance of colostrum as a vehicle for carotene in the nursing newborn calf was demonstrated by Spielman *et al.* (*816*) and Wise *et al.* (*921*). Young calves receiving colostrum and then milk from birth to a few weeks of age showed an average apparent absorption of vitamin A of 81–95% and 38–65% absorption of carotenoids (*671*).

In attempts to estimate the vitamin A depletion time in calves, Teichman *et al.* (*844*) realized that, within age group and breed, seasonal or yearly variations did not significantly affect the rate of decrease of plasma vitamin A when a depletion diet was fed. It did, however, appear that with increase in age the rate of decrease of plasma vitamin A became more rapid. Eaton *et al.* (*223,224*) ascertained when the plasma vitamin A level of calves reached 5 μg per 100 ml, they were essentially depleted of their vitamin A liver stores. The vitamin A level in the blood decreased in a uniform manner regardless of the initial concentration. Jones *et al.* (*436*), reviewing a large number of case histories of vitamin A deficiencies in young calves, reported an average mortality rate of 15.5% over a 2-year period; in most cases diarrhea and pneumonia were evident. Blood carotene and vitamin A concentrations were 2 and 7.2 μg per 100 ml, respectively. Plasma vitamin A levels below 10 μg per 100 ml were judged not to be adequate. Eye conditions and poor hair coat occurred frequently but appeared not to be specifically related to plasma level of vitamin A. It is particularly noteworthy that the authors observed severe vitamin A deficiency symptoms in calves with low plasma vitamin A levels of 2.6 μg per 100 ml but with a fairly high plasma carotene concentration of 20.3 μg per 100 ml. It was suggested that many calves die from pneumonia and diarrhea precipitated by vitamin A deficiency before symptoms of a deficiency are observed. Eaton *et al.* (*224,227*) demonstrated that a linear relationship between plasma vitamin A concentration and the log of liver vitamin A concentration exists in dairy calves but that the linear regressions of plasma concentration have limits in the estimation of vitamin A concentration in the liver. Plasma vitamin A and liver vitamin A were not entirely dependent on one another.

The efficiency with which carotene can meet the vitamin A requirement depends on the amount absorbed from the feed, the destruction of carotene before it reaches the intestine, losses by oxidation in the small intestine, and any increase in carotenoids due to microbial synthesis in the lower gut (*562*). The administration of high levels of vitamin A and carotene to dairy calves (*422*) resulted in higher vitamin A levels in plasma, as would be expected. In addition, the feeding of vitamin A also depressed the level of plasma carotene. With carotene administration the rise in plasma carotene as well as vitamin A was only slight. Dolge *et al.* (*204*) also observed that increasing the alfalfa meal content of a dairy calf ration to 25, 50, and 75% increased the plasma carotene level but had only a slight effect on the plasma vitamin A concentration. In a later study Dolge *et al.* (*205*) reported on the interrelationships between carotene from artificially dehydrated alfalfa and vitamin A from a dry carrier when fed to Holstein calves. McGillivray (*561*) showed close quantitative relationships among protein, vitamin A, carotene, and tocopherol levels and their rapid decrease upon the onset of lactation in the dairy cow. As other feedstuffs are introduced (*921*) after the period of colostrum feeding, it is important to ensure that the intake of vitamin A or provitamin A is adequate, particularly if the calf is raised for veal or fattened on concentrates, without access to pasture grasses. Abrams *et al.* (*2*) observed a depression of feed consumption in growing calves as the first objective sign of vitamin A deficiency. It is interesting that, when the feed intake of normal calves was restricted to that of vitamin A-deficient animals, the weight gains were the same until the deprived calves began to have respiratory and intestinal disorders. The economic significance of an insufficient intake of vitamin A was shown by Ritzman *et al.* (*737*), who demonstrated that calves receiving a vitamin A-deficient diet consumed more feed but gained 50% less weight than those receiving an adequate amount of vitamin A. Protein utilization decreased about 25% owing to lack of sufficient vitamin A.

Carotenoids, mainly β-carotene, accumulate in cattle in body organs and tissues other than the liver, such as adrenals, bile, body fat, corpus luteum, corpus rubrum, iris, kidney, muscle, ovaries, epithelium, pituitary, placenta, testes, and thymus. The most visible sites in which carotenoids are deposited in beef animals are the fat deposits and the butterfat of milk. The highest concentration occurs in the corpus luteum and corpus rubrum (*324*). Performed vitamin A is likewise present in the blood, liver, and milk of cattle as well as other tissues.

Parallel feeding trials with β-carotene and vitamin A are needed to obtain valid comparisons and so develop conversion factors for the vitamin A activity of β-carotene in dairy cattle as well as different breeds of dairy cattle. Palmer and Eckles (*667*) were the first to study the pigments of cow's milk quantitatively. Later it was recognized that under similar feeding conditions Guernseys accumulate considerable greater quantities of carotene in their tissues than do Holsteins (*324*). Conversely, Holsteins have been reported to have greater concentrations of vitamin A in their blood plasma, liver, and milk than Guernseys. Moore and co-workers (*603*) reported that, to prevent increased CFP, Guernseys required 1.13 times as much, and Jerseys 1.07 times as much, carotene as Holsteins. To maintain marginal plasma vitamin A concentrations with slight liver storage, Elliott (*235*) observed that Guernsey calves needed a daily intake of 100 μg of carotene per pound live weight and Holsteins only 60 μg, or a requirement for Guernseys 1.7 times that for Holsteins. Baumann *et al.* (*58*) also compared breeds and found that Guernsey milk is richer in carotenoids than Friesian or Ayrshire milk. This has been confirmed by Thompson (*858*). Variations in the vitamin A activity of milk are due to variations in the carotene content of the cow's ration (*105,520,863*). All countries in the Northern Hemisphere show a similar seasonal fluctuation, with higher values in summer than in winter, depending on the carotene content of the feed. Moore and Berry (*601*) investigated the vitamin A and carotene content of blood plasma of three breeds of dairy calves from birth to 4 months of age. Eaton *et al.* (*226*) showed that in dairy cattle, depending on the criteria used, Holsteins converted carotene to vitamin A from 1.4 to 1.8 times as efficiently as Guernseys. Boyer *et al.* (*106*) concluded that 18 μg of vitamin A was required per kilogram body weight for Holsteins and Guernseys or 75–85 μg of carotene for Holsteins and 125–135 μg for Guernseys. On a weight basis the above requirements would mean that the ratio of vitamin A to carotene utilization was between 4 : 1 and 5 : 1 for Holsteins and 7 : 1 to 8 : 1 for Guernseys. In a 1974 Indian study (*438*), plasma carotenoids and vitamin A were established in two groups of Holstein, Holstein × Hariana crossbred, and Hariana calves for a 9-month period. Both plasma carotenoids and plasma vitamin A concentrations were related to carotene content in all breeds. The study indicated that the absorption of carotene in Holsteins and crossbreds was better than that in Hariana breeds. In general, although breed differences are clearly defined, great variability may exist among animals of the same breed kept under the same conditions. In contrast, monozygous twins are extremely uniform in their blood carotenoid pictures. The question of variability in blood constituents among animals could form the basis of a genetic study.

Levels of carotene intake sufficient for normal reproduction seem to be more than adequate for lactation (*834*). Higher levels of intake influence the vitamin A potency of the milk, in that a rather high threshold for secretion of the vitamin and its precursor into milk has been demonstrated (*756*). Some workers attempted to find out how much carotene is required to maintain milk vitamin A and carotene at the level typical of summer pasture. Wilbur *et al.* (*913*) found that 330 mg of carotene from dried lucerne (alfalfa) hay maintained vitamin A at summer levels. Atkeson *et al.* (*35*) noted that 600 mg daily was needed to achieve pasture levels, but increasing intake to 3.6 gm had no further effect. Thompson *et al.* (*858*) found that 600–800 mg carotene from kale was needed to maintain the vitamin A and carotene levels in milk through the months of October, November, and December. Hjarde and Larsen (*398*) reported that, when cows were fed freshly cut lucerne, a daily intake of 670 mg carotene from this source gave milk fat normal vitamin A values for summer Danish butter, and these values were not elevated when the carotene intake was increased to 2.9 gm. They also observed that 700 mg carotene from silage was not as effective as 670 mg from lucerne, and that maize-germ meal, although raising the fat intake of cows from 170 to 560 gm daily, reduced markedly the vitamin A and carotene content of the butterfat. McGillivray (*561a*) in New Zealand showed that when Jersey cows were given additional tocopherol the vitamin A level of milk was raised from summer levels to typical winter levels. Apparently, seasonal fluctuations in the carotene and vitamin content of milk are not all due to the stage of growth of the pasture grasses. McGillivray (*561a*) found that, despite a nearly constant carotene intake from pasture, the vitamin A potency of the milk of cows in New Zealand showed a pronounced downward trend in midsummer, which could be prevented by feeding additional tocopherol. Worker and McGillivray (*931*), in experiments with monozygous twins, found marked differences between the utilization of equal amounts of carotene from either clover or ryegrass. Subsequent work indicated that the botanical composition of the plant life changed at the time of the decrease in vitamin A potency from a predominance of ryegrass to clover and that other factors, such as the composition of pasture grass fat, may decrease the utilization of carotene and tocopherol. Pasture grass fats are highly unsaturated, but a considerable amount of hydrogenation may occur in the summer, according to Shorland *et al.* (*793,794*).

The relationship between blood and milk vitamin A and carotene levels has been studied. Blood carotene levels show diminishing increments as carotene dietary intake increases; milk carotene is also diminished. Owens (*663*) demonstrated clearly that β-carotene, at all levels of intake, does not readily pass unchanged into the blood from the feed and that there is similar difficulty in passing from blood into the milk (Fig. 6). Thompson and Ascarelli (*859*) studied the interesting difference in behavior of vitamin A and carotene in milk and in blood and concluded that the

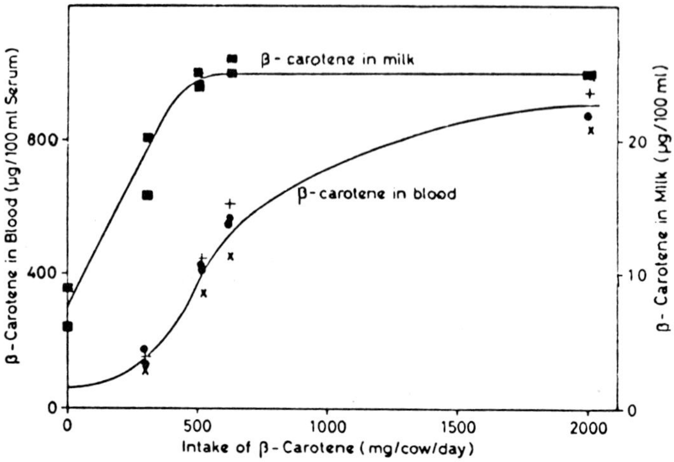

Fig. 6. β-Carotene in milk and blood of Ayrshire cows as a function of intake of carotene in food. β-Carotene in blood increases rapidly when the food intake of carotene is low, but there is a diminishing increment at higher intakes. β-Carotene in milk, even at the highest intake, is much lower than β-carotene in blood but reaches its maximal value at a lower intake. [From Owens (*663*).]

cow, after a rapid change from a low to a high carotene intake, requires a few days to adjust its vitamin A metabolism, with the vitamin A in the milk always first rising to high values independently of the initial level and then returning to those typical of pasture feeding. The studies of Thompson and Ascarelli also eliminate the probability of changes in the composition of the fat of green herbage influencing the utilization of carotene. McGillivray and Worker (*566*) found that this factor may be of importance under conditions of normal feeding in which seasonal or stage of growth variations in the nature and proportion of plant waxes, sterols, and triglycerides could well affect the utilization of carotene. Owen (*663*) found the partition of vitamin A between ester and alchohol form in milk to be a function of the carotene content in the milk (Fig. 7). Thompson and Ascarelli (*859*) reported that vitamin A in milk was closely related to the vitamin A ester concentration in blood but that surprising fluctuations in the latter could not be related to carotene intake. An explanation of the slow rise in carotene content and rapid rise in vitamin A content is that at high levels of carotene intake vitamin A in milk is derived directly from vitamin A ester in the chylomicrons of blood, which are of immediate dietary origin, and that carotenoids in the milk are derived from the lipoprotein-bound forms of carotenoids in blood. Byers *et al.* (*135*) reported the carotene requirement for pregnancy in Friesian and Jersey cattle to be about 130 μg per kilogram body weight. It can be calculated that a high-yielding cow on pasture producing 700 gm fat daily will secrete into her milk about 5 mg carotene and 7 mg vitamin A (equivalent to about 35 mg carotene). Hence, the total daily requirement of carotene to replace the vitamin A and carotene secreted in the milk is about 40

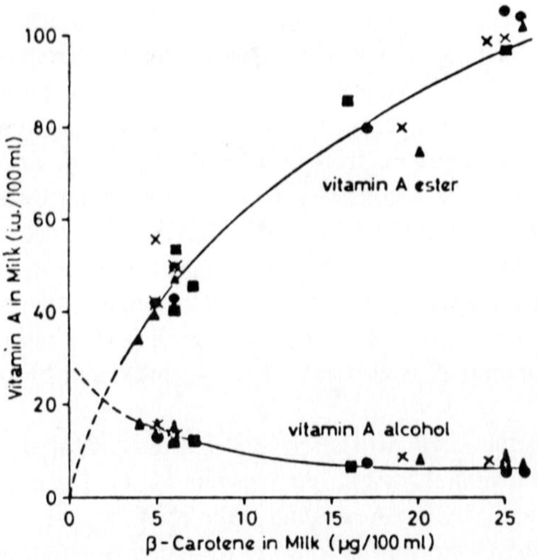

Fig. 7. Partition of vitamin A between ester and alcohol in milk as a function of β-carotene in the milk. The changes in β-carotene reflect the carotene content of the diet. As carotene intake decreases, vitamin A intake increases, while vitamin A ester decreases. [From Owen (*663*).]

mg. Guilbert *et al.* (*354*) found that a 500-kg cow needs about 15 mg carotene, so the total requirement may be considered to be 55 mg (40 + 15). Such a level of intake is about normal in winter, when the only source of carotene is hay. This level of intake is clearly insufficient to keep milk fat vitamin A and carotene at normal levels in summer months, at which time a cow will eat enough grass to supply about 2–6 gm of ingested carotene daily.

In reviewing his studies, Thompson (*858*) pointed out that, because of the difficulty in interpreting results of digestibility trials, it may be more meaningful with lactating animals to determine the biological effectiveness of carotene in feeds by measuring the level of vitamin A and carotene in plasma, liver, and milk, particularly in the latter since the level of vitamin A in milk rapidly reflects changes in carotene or retinol intake; moreover, milk is readily available and easily analyzed. According to Thompson, in the interpretation of blood levels it must be considered that it is not the alcohol but the ester form of the vitamin that reflects the immediate dietary intake and it is the concentration of the ester that determines the level of vitamin A in milk fat. Rapid alterations in plasma vitamin A levels and milk vitamin A and carotene levels are often seen when cows change from one source of carotene to another. When cows given a (*859*) ration providing 20–60 mg carotene daily were put on pasture supplying 2000–6000 mg daily, the plasma and milk vitamin A increased to a maximum within 2–5 days, whereas carotene took 3 weeks to reach maximal values; reverting to a low carotene intake caused a slow drop in carotene and a rapid drop in vitamin A. In another instance in which a parallelism between carotene intake and vitamin A and carotene levels did not occur, cows were given

2000 mg carotene from silage daily (about 4 mg/kg live weight) as part of a winter ration, which increased milk carotene levels only slightly, from 4.6 to 5.8 μg per gram fat but increased milk fat vitamin A from 7.2 to 11.1 μg/gm. When the same cows were turned out to pasture supplying 4000–6000 mg daily, the carotene content of the milk fat increased from 5.8 to 10.9 μg/gm, and surprisingly the vitamin A decreased from 11.1 to 8.1 μg per gram fat. Other instances of a lack of parallelism have been reported (*585,805*), with several explanations being offered. Perhaps, beyond a certain level of carotene intake, the absorption of vitamin A formed in the gut is depressed by the presence of carotene in the same way that the absorption of carotene is depressed by feeding high levels of retinol. Chanda *et al.* (*149*) reported on vitamin A esterification in the milk and blood of cows given dietary carotene.

Thompson (*856*) gave cows artificially dried grass prepared at three stages of maturity to provide between 2.5 and 3 gm carotene daily. During the 20-week trial period, the carotene and vitamin A content of the milk fat was 8.5 and 7.4 μg/gm, respectively. These values are similar to those from cows on pasture, demonstrating the value of dried grass as a milk colorant and vitamin A source. In these trials it was also shown that carotene from the more mature grass was not utilized as efficiently as that of the grass cut earlier, data in agreement with earlier findings of McDowall and McGillivray (*555*). The relation of carotene and vitamin A content of summer milk to the carotenoid content of pasture grasses or forage was also studied by Hibbs *et al.* (*384*).

Requirements for vitamin A were assessed in the past in terms of the provitamin A carotenes because this carotenoid group represented the more common dietary source of vitamin A for cattle (*632*) and had a longer history of assay methodology. On the basis of a reassessment by Eaton *et al.* (*227,229*) using the most sensitive criterion (maintenance of normal CFP), 10.6 mg carotene per 100 kg live weight appeared to be the minimal rerequirement for growing calves. In the absence of more definitive data, it seems justifiable to assume that a similar level is required for the maintenance of normal CFP in mature animals. There was successful reproduction for extended periods with an intake of 19 mg carotene per 100 kg live weight, but carotene intake appeared to be most critical during the last 2–3 months of gestation (*740*). In a series of papers (*350–354*) Guilbert and co-workers reported that differences between the relative efficiencies of vitamin A and carotene widen with increasing levels of intake. For cattle, the daily requirement to prevent night blindness was found to be 25–33 μg carotene per kilogram body weight. Lewis and Wilson (*513*) found that calves require nearly twice as much vitamin A for maximal growth as is required to prevent night blindness and nearly 15 times as much to achieve high plasma levels of vitamin A. At minimal levels adequate to prevent any detectable degree of nyctalopia in cattle, sheep, swine, and horses, a ratio of efficiency of vitamin A to carotene by weight is about 6 : 1; at the minimal level that results in significant liver storage and successful reproduction, the ratio is about 10 : 1. In another study, in which plasma vitamin A levels were used as the criterion of effectiveness (*106*), a ratio of 4 : 1 to 7 : 1 by weight of carotene to vitamin A was required for equivalent plasma values. Using growth and reproduction in cattle as the criteria, Guilbert and Loosli (*353*) found 6.6 times as much carotene on a weight basis to be required as vitamin A.

Heifers have been reported to utilize vitamin A alcohol and the natural esters more efficiently than carotene concentrates (*745*). The calf can utilize carotene (*816*); however, the efficiency of conversion is not high (*171*). Preformed vitamin A is much more effective. The Connecticut group has been one of the most active and careful teams of investigators in assessing the relative merits of carotene and vitamin A, particularly in the calf (Fig. 8). In one study (*747*) they determined quantitatively the relative value of carotene from alfalfa and vitamin A from a dry carrier fed at minimal levels. The criteria were blood and liver values of vitamin A, CFP, occurrence of papillary edema, and tissue alterations in the Holstein calf. They found that, to maintain a plasma vitamin A value of 10 μg %, a daily intake of 31 μg of carotene or a daily intake of 3.8 μg of vitamin A per pound body weight was required (approximately 8 : 1). To achieve a liver concentration of 0.6 μg/gm, a daily carotene intake of 30 μg or a daily vitamin A intake of 3.0 μg per pound body weight was necessary (10 : 1), and to maintain a minimal CFP of 120 mm of water a daily intake of 33 μg of carotene or 6.4 μg of vitamin A was necessary (approximately 5 : 1). In addition to these relationships, the carotene and vitamin

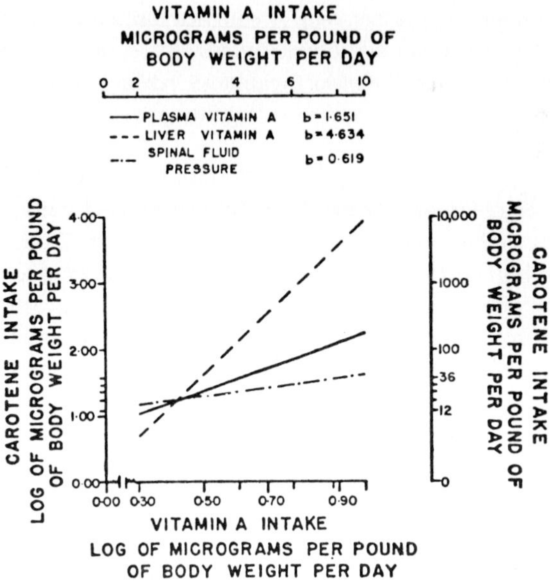

Fig. 8. Amount of carotene from alfalfa necessary to obtain responses of plasma vitamin A, liver vitamin A, and spinal fluid pressure in Holstein calves equivalent to specified levels of vitamin A from a dry stabilized form. [From Rousseau *et al.* (*747*).]

A intakes in micrograms per pound body weight required daily to prevent papillary edema were greater than 36 and 8, respectively; to prevent squamous metaplasia in the main duct of the parotid gland required 24 and 6, respectively; and to prevent similar lesions in the interlobular ducts of the same glands required 30 and 4, respectively. The average ratio of carotene to vitamin A with these three criteria is 5 : 1 on a weight basis. The same workers (*748*) evaluated carotene from alfalfa and vitamin A in a dry carrier fed at medium to high levels to Holstein calves. They found a variable ratio of effectiveness that was dependent on level of intake when plasma and liver vitamin A concentrations and depletion times were used as the criteria. From 5 to 7 times as much carotene as vitamin A was required (*748*) at the level of 60 μg/day per pound body weight; at 180 μg of carotene, the ratio rose to 10 : 1 to 13 : 1; and at the highest level, 540 μg of carotene, it rose to 20 : 1 to 24 : 1. Other workers, such as Blaxter (*86*) and Ross *et al.* (*745*), have also investigated the relative amounts of carotene and vitamin A required to meet the vitamin A needs of cattle. From summary Table 11, one notes that the ratio of effectiveness varies with the levels fed and the criteria used to measure performance. From a practical consideration, the ratio varies on a weight basis from a minimum of 5 : 1 or 6 : 1 to 10 : 1 for carotene to vitamin A to provide a safety margin as indicated by measurable storage in the liver and adequate plasma A levels. An examination of the literature indicates that the relative value of carotene to vitamin A can vary from 4 : 1 to 10 : 1 depending on the criteria used and other experimental conditions (*599*). With practical dairy cattle rations, it appears to be advisable to consider 1 mg of carotene as equivalent to only 400 IU of vitamin A.

In Section XII, B, some additional, relatively recent data (*9,165a,521,522, 578,761*) are discussed in relation to β-carotene, not as a vitamin source, but as a dietary ingredient with specific action on fertility parameters.

3. *Beef Cattle*

A lack of β-carotene or vitamin A in the diet of beef cattle causes the same deficiency signs as those mentioned in the section on dairy cattle. Cattle suffering from mild vitamin A deficiency reduce their feed intake and fail to make satisfactory weight gains, an effect of great economic significance. This is especially noticeable in feedlot cattle receiving high-concentrate rations. The lowered feed intake may result in other deficiencies when the ration barely meets other nutrient requirements (*631*). In finishing cattle, vitamin A deficiency results in generalized edema or anasarca, with clinical symptoms of lameness in the hock and knee joints and swelling in the brisket area. Also, pulmonary complications culminating in pneumonia have been reported. Some other more common signs of vitamin A deficiency are dull, lusterless eyes or cloudy corneas, excessive lacrimation, [profuse discharge of tears] or mucus from eyes, rough, dry haircoat, diarrhea, and night blindness. The last condition can be observed at dusk, with deficient cattle bumping into each other, fence posts, water troughs, or other objects. This sign is often overlooked because of the difficulty of recognizing it in animals not subjected to specific tests.

The hepatic reserves and plasma level of vitamin A are important indicators of vitamin A status. In the feeding of baby-beef animals, according to Bridge and Spratling (*113*) of England, very young calves are usually bought from other herds and are reared on milk or milk substitutes until they are about 7 weeks old. They are then fed *ad libitum* on a ration high in cereal grains with protein, mineral, and vitamin supplements. They usually gain from 2.5 to 3.5 lb/day and are slaughtered at 800–900 lb live weight when 10–12 months of age. Because the ration usually consists of feed ingredients of low or variable carotene content, synthetic vitamin A is usually added at the rate of 4,000–6,000 IU per kilogram ration. The high energy level of the ration is obtained by including a large proportion of cereal grains, omitting roughage, and, in many cases, adding small quantities of fat. When vitamin A is not supplied in some form (*113*), vitamin A deficiency results (Fig. 9). Although Hale and co-workers (*357*), Erwin and co-workers (*250*), and Beaumont and Parr (*59*) found liver vitamin A values to be borderline under their feeding regimes, low hepatic levels can be avoided, as demonstrated by Foll (*275*), who found mean values of vitamin A to be 374 IU per gram of liver in steers fed an all-concentrate ration supplemented at the rate of 10,000 IU vitamin A per kilogram ration. Levels of vitamin A supplementation should be related to the energy content of the ration as well as to the weight of the animals consuming it.

Bridge and Spratling (*113*) stated that, during vitamin A deprivation, a previously adequately fed animal is able to survive because of a liver reserve of vitamin A, the time of survival being roughly dependent on the magnitude of the reserve. Cattle after grazing have survived up to 6 months on carotene- or vitamin A-free rations (*264*), lambs up to 6 months (*354*), and ewes over 1 year (*684*). Frey and Jensen (*282*) found that in yearling steers the depletion of liver vitamin A stores resulting from the consumption of a ration low in carotene occurred at the rate of approximately half the liver stores every 40 days. Smith and co-workers (*805*) observed that steers on a deficient ration were also depleted at almost exactly this rate, whereas steers receiving maize silages were depleted about half as slowly.

Vitamin A deficiency in bulls of breeding age leads to a decline in sexual activity. Spermatozoa decrease in number and motility, and there is a marked increase in abnormal forms. Some German studies (*905*) led to the conclusion that β-carotene had a specific effect on spermatogenesis independent of the vitamin A supply, but these results need confirmation. In breeding cows, estrus may continue when vitamin A deficiency has advanced to the point where convulsions are common, but the ability to become pregnant is impaired. Deficiency in pregnant animals, if sufficiently severe, occasionally results in abortion or birth at term of dead, weak, or blind calves. Retained placentas are more common in cows deficient in vitamin A. The vitamin A requirement of breeding, growing, and finishing beef cattle can be met by provitamin A compounds in green pasture or forage and in

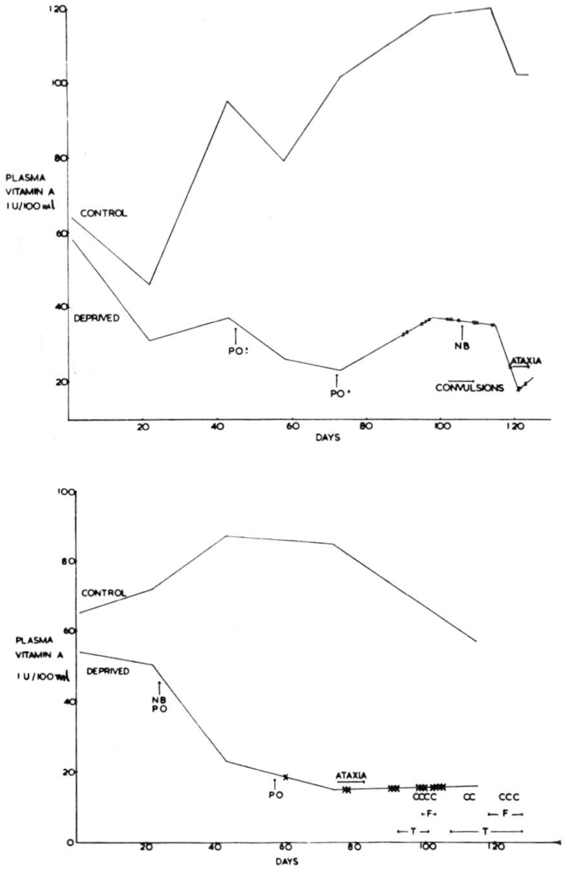

Fig. 9. Onset of night blindness and other abnormalities in vitamin A-deficient calves in relation to plasma vitamin A levels. Twin calves, one deficient and one control in each of two experiments. Key: X, diarrhea; PO, papilloedema; NB, night blindness; F, fever; C, coughing; T, tachypnea. [From Bridge and Spratling (*113*).]

ration concentrates, or by oral or injected vitamin A supplements or combinations of the two (*631*). It is realized, however, that the conversion rate for cattle is different under different conditions. Some of the factors that influence this conversion rate are breed of animals, individual differences among animals, level of feed or carotene intake, and ration antioxidants. Stress conditions have also been suggested as causing reduced conversion of carotene. Not only does β-carotene have a role (a) as a provitamin A source for the cow and (b) as an attractive yellow food colorant indirectly applied to butter, but evidence is now accumulating that it has (c) a specific action on fertility parameters in female cattle, influencing ovulation, conception, and gestation (*9,165a,521,522,578,761*).

Brood cows, bulls, and replacement stock raised under range or pasture conditions more often than is realized do not receive enough or make efficient use of the carotenoid vitamin A precursors present in forage. With dry or drought conditions, the limited grass available is usually low in carotene. Also, the higher nitrate concentration in forages from pastures treated with heavy applications of nitrate fertilizers has been associated with a reduction in the animals' ability to convert carotene to vitamin A efficiently. Under many range conditions, the vitamin A activity actually available to the animals is often not sufficient for optimal breeding performance, normal growth, or maximal disease resistance.

Meyer (*579*) showed that there was considerable variation in the vitamin A concentration of livers obtained from cattle shipped from different parts of the country into California feedlots. Two lots of cattle from the same area and wintered on the same range showed wide differences in vitamin A stores. One group had an average of 40 μg per gram liver, with a range of 24–48 μg/gm. The other group averaged 14 μg per gram liver, with a range of 2–41 μg/gm. Kolari *et al.* (*470*), upon observing vitamin A deficiency in beef cattle, considered that 10–12 μg per 100 ml plasma was indicative of a deficiency and that 30–44 μg of vitamin A per 100 ml suggested adequate vitamin A stores. Swingle (*838*) stated that, under range conditions, when plasma carotene or vitamin A levels drop to 25 or 16 μg per 100 ml plasma, respectively, clinical symptoms may soon appear. With pregnant cows, if blood vitamin A levels drop significantly to very low levels, one can expect the birth of weak or dead calves.

Although the animal may not show any outward symptoms of a vitamin A deficiency, pregnant cows frequently do not have adequate vitamin A body stores to carry them through the entire gestation period. The vitamin A requirements of bred cows are greatest during the last 2 or 3 months (the last trimester) of the gestation period. These requirements are usually three to five times the level needed for maintenance, yet it is during this critical period of gestation that vitamin A body stores may be dangerously low or entirely depleted. Adequate vitamin A reserves in the dam are thus vital for maintaining pregnancy and for providing the calf with adequate levels of vitamin A in the milk during early life. However, the amount of vitamin A transferred (*38*) from the mother to the fetus may be very small. Hence, most calves are born with low liver concentrations of vitamin A, even if the dam has had an adequate intake of this vitamin. The newborn calf thus becomes dependent on colostrum and subsequent milk as a source of vitamin A. Cows in a marginal or depleted vitamin A state may not provide sufficient vitamin A to their young for normal growth or life itself. The amount of vitamin A in the colostrum and milk depends on adequate liver reserves and/or current vitamin A intake of the dam. When cattle arrive at the feedlot from the range, they usually have been off feed and water, have been subjected to the stress of handling and shipping, have been exposed to infectious disease, parasitism, and environmental conditions conducive to lowered vitality, and have reduced resistance and are thus drawing heavily on their vitamin A reserves.

Under practical feeding conditions, feeders should consider (a) liver stores of vitamin A as influenced by previous feeding and (b) carotene destruction in feeds during storage. Cattle store vitamin A and carotene in the liver and body fat during times of abundant intake from pasture and other feeds. There reserves may be large enough to reduce or meet the needs of older animals for as long as 6 months. Another possibility is that the needs of cattle for vitamin A may appear to be increased because of the destruction of vitamin A or carotene in feeds during storage or even in the digestive tract. In steers, the efficiency of conversion of carotene to vitamin A decreases as the level of feeding of carotene is increased.

There are a number of trials in which the relative vitamin A value of carotenes has been studied for the past 40 years. Guilbert and Loosli reported that on the average 6.6 μg of carotene is equivalent to 1.0 μg of vitamin A (6.6 : 1) for normal growth and reproduction in beef cattle (*353*). The NAS–NRC committee (*631*) accepts 1 mg of carotene as equivalent to 400 IU of vitamin A for all beef cattle. This is an effective weight ratio of 8 mg of carotene to 1 mg of vitamin A (8 : 1). This relationship was derived from the comparative results found in the literature (*61,62,350,434*) on the basis of quantities of carotene per se required for various functions (*526*) and extrapolated from data obtained with dairy cattle (*183,443,492,600,604,741*). Brocklesby (*118*) pointed out that experimental cattle kept on graded dosages of vitamin A or carotene utilize carotene less efficiently at higher levels of intake. At intakes of 60 μg per pound body weight, carotene was equal to one-seventh the weight of vitamin A; at 180 μg, one-thirteenth; and at 540 μg, one-twenty-fourth. Comparative studies on relative carotene to vitamin A equivalency are summarized in Table 11. Perry *et al.* (*692*), in studies involving added vitamin A supplements to rations containing different levels of carotene, observed little vitamin A increases as judged by plasma or liver values due to its carotene contribution. Hence, one may question the availability of the carotene of rations to meet the vitamin A needs of growing–fattening cattle under all management conditions.

M. Farm Animals

Table 11 summarizes relative carotene to vitamin A equivalency for ruminants and horses. From this table and Table 9 it can be concluded that, at practical levels of feeding to provide for growth and reproduction, ratios of 5 : 1 to 7 : 1 on a weight basis for β-carotene and vitamin A are required for swine, ruminants, and horses. When higher levels of supplementation are required to provide a margin of safety as measured by plasma A levels or liver stores, the ratio is 10 : 1 on a weight basis. Poultry utilization (Table 6) of β-carotene is relatively much more efficient, so that an average feeding ratio of only 2 : 1 is required for growth. However, poultry require five times or more the amount of β-carotene as vitamin A on a weight basis at higher feeding levels to provide liver vitamin A storage.

Branion and Emslie (*111*) cited the needless confusion caused by authors reporting the biological value of provitamins A in international units without giving the relationship used to convert mass or weight units of the provitamins to their corresponding biological value, "international" units. Much of the confusion would be eliminated if evaluation in units were discarded and requirements were expressed as weight of the pure substances used in the reported study. Ames (*23*) urged that "all β-carotene sources be expressed solely in terms of milligrams of all-*trans*-β-carotene" and suggested that a partial solution might be "to express all β-carotene levels used in a study solely on a weight basis." The committee working with

TABLE 12

Ratios of Biological Effectiveness of β-Carotene and Vitamin A for Various Animals[a]

Animal	Authority[b]	mg β-carotene[b] equivalent to 1 mg of vitamin A alcohol[c]	IU preformed vitamin A per mg β-carotene[d]
Rat	International	2	1667
Mink	NAS-NRC	12	277
Fox	NAS-NRC	12	277
Dog	NAS-NRC	4 or 8	418-834
Poultry	NAS-NRC	2	1667
	NCAN	3	1112
	ARC	6	556
Horse	NAS-NRC	6-10	333-556
	NCAN	7	476
Dairy cattle	NAS-NRC	8-10	333-418
	ANRC	8	418
	NCAN	7	476
Beef cattle	NAS-NRC	8.3	400
	ANRC	8	418
	NCAN	7	476
Sheep	NAS-NRC	5.8-8.3	400-578.
	ANRC	8	418
	NCAN	7	476
Swine	NAS-NRC	6.2	533
	NCAN	7	476
Man	MRC	6	556
	NAS-NRC	4	834
	BMA	6	556
	CCN	8	417
	NCAN	7	476

[a] From Branion and Emslie (*111*).

[b] ANRC, Animal Nutrition Research Council, North America; ARC, Agricultural Research Council of Great Britain; BMA, British Medical Association; CCN, Canadian Council on Nutrition; MRC, Medical Research Council, Great Britain; NAS-NRC, National Academy of Sciences, National Research Council, United States; NCAN, National Committee on Animal Nutrition, Canada.

[c] On the basis of all-*trans*-β-carotene.

[d] 1 IU of vitamin A activity is usually regarded as ≎ to 0.6 μg of β-carotene or 0.3 μg of vitamin A alcohol or 0.344 μg of vitamin A acetate or 0.55 μg of vitamin A palmitate.

Branion and Emslie recommended that, with studies in animal nutrition, quantitative measures of performed vitamin A and of provitamins A be expressed in mass units of vitamin A acetate, palmitate, alcohol, and β-carotene or other carotenoid provitamin A compounds. When rations are formulated it is desirable or even necessary to equate a known mass or quantity of provitamin A with the appropriate amount of preformed vitamin A to know its contributing vitamin value. To be meaningful, a conversion factor should be the ratio of the weight of a pure preparation of the carotenoid vitamin A precursor to a standard preparation of vitamin A acetate or vitamin A alcohol that would provide an equal biological effect when consumed under clearly defined conditions by a specific class of animal of stated physiological development and condition. From the literature Branion and Emslie (*111*) compiled such a relationship (Table 12). Other reviews by Tiews (*865*), Beeson (*61,62*), and Ullrey (*879*) may be consulted.

IX. FEED SOURCES OF CAROTENES AND VITAMIN A PRECURSORS

Carotenoids are present in tissues of green plants and are usually associated with chlorophyll in the chloroplasts, probably as protein complexes, and in lipid formations (*324*). The amount of carotene and other carotenoid vitamin A precursors in grasses varies with plant species and stage of growth. Plant species that produce little leaf, or mature very quickly, or are very susceptible to drought tend to contain little carotene because the level of carotene seems to parallel that of chlorophyll and protein (*892*). This variation is less marked in the leafier green legumes.

For large farm animals, green grass, immature and early-bloom legumes, and grasses cut and quickly dried into hay provide the major sources of provitamin A (*167*). The most common carotene in all green leaves is β-carotene. The β-apocarotenals have been identified in grass and alfalfa (*850*). More than 40 different carotenoids are present in dehydrated alfalfa meal (*75*). Best known of the seed products are yellow corn and corn gluten meal, which contain β-carotene, cryptoxanthin, some α-carotene, γ-carotene, and carotene isomers (*88*). In a chromatographic study by Moster *et al.* (*616*) of the carotenoids of a corn seeding, 18 individual structures were cited. The relative amounts differ with varieties and strains. *Tagetes* meal (*13*) contains α-cryptoxanthin, antheraxanthin, α-carotene, β-carotene, and lutein. Sorghums contain considerably smaller amounts of carotenoids (*832*) than does yellow corn, but β-carotene content can vary from a trace to 25% of the total pigment in yellow sorghum varieties. Tabulations (for example, Table 13) of carotene-containing feedstuffs have been prepared and appear in various publications (*11,449,622*). All grazing animals are usually dependent on plant synthesis of carotenoid vitamin A precursors in natural feed sources to meet their vitamin A needs.

TABLE 13

Carotene Content of Major Feedstuffs[a]

Feedstuff	Carotene, mg/kg (as fed)
Alfalfa, dehydrated 17%	161
Alfalfa, sun-cured	96
Alfalfa, ensiled	27
Bermuda grass, hay	117
Bermuda grass, fresh	103
Bluegrass, Kentucky, fresh	116
Bluestem grass, fresh	69
Bromegrass, hay	4.4
Bromegrass, fresh	189.0
Buffalo grass, fresh	45
Canary grass, fresh	82.5
Cattle milk, whole dried	7.0
Clover ladino, hay	147
Clover, red, hay	32.3
Clover, red, fresh	43.5
Corn cobs, ground	0.6
Corn silage	11.7
Corn grain	1.8
Corn, hominy	9.2
Corn, distilled grain and solids	3.7
Linseed meal	0.2
Grama grass, fresh	19.3
Oat, hay	88.9
Orchard grass, fresh	80.3
Peanut, shell grind	0.8
Red top grass, fresh	40.1
Sagebrush	12.2
Saltbush	15.7
Sorghum stover	5.8
Sorghum, silage	9.7
Soybean, hay	31.8
Soybean, silage	21.7
Timothy, hay	47.2
Timothy, silage	29.6
Wheat, hay	95.9
Wheat, straw	2.0
Wheat, fresh	11.8
Wheatgrass, fresh	41.4

[a] From NAS–NRC (*622*).

A. Pasture, Hay, Silage

Cattle are reared almost exclusively on feed stuffs of plant origin; thus, they depend on the carotene contained in fresh plants, hay, or silage for their supply of vitamin A activity. Although the carotene contained in green plants is a product of various endogenous and exogenous factors (*314,460,594*), plant sources may be generally presumed to contain abundant amounts of it. β-Carotene was present in 59 species of which also contained α-carotene. Healthy, mature cattle and other ruminants grazing on fresh, green pasture have no difficulty in meeting their vitamin A needs as long as the fresh condition of the pasture is maintained. The level of carotene in pasture grass, clover, alfalfa, etc., is highest in the young plant but decreases as the plant becomes older (*457*). Old pasture cannot be guaranteed to supply sufficient carotenoid vitamin A precursors. Their availability to animals is impaired because of progressive lignification of plant tissues. Kramer and Tarjan (*481*) reported on the biological activity of carotene in different plants. Year-round grazing is practiced mainly in countries that have extensive cattle production on pasture, for example, South America, South Africa, New Zealand, and the southern parts of the United States. The vitamin A status of pasture-fed animals fluctuates with the season (*802*); it is best during periods of fresh vegetation, deteriorating during the rest of the year, when the quantity of pasture decreases.

The carotene content of forage plants (Table 14) has been measured, but, since this is quite variable, the figures cited in studies can be used only as guides (*287*) to actual situations. In temperate climates, where pasturing is possible only during certain times of the year, supplemental vitamin A feeding practices are required. The use of hay, silage, and grain concentrates, supplemented with limited periods of grazing on pasture, is the general practice. The carotene level in these dry-lot or stall-fed rations depends on their composition and can fluctuate greatly if not carefully controlled.

Many studies continue to be made on the carotene yield from fodder or forage crops, the method of harvesting and conservation, the type of equipment involved, and the influence of temperature and other environmental factors (*228,383,710, 893,940*). Data on the carotene content of roughage assembled by Aitken and Hankin (*11*) are shown in Table 15. The carotene in forage plants is most fully available to animals on fresh pasture; immediately after harvesting, it begins to be reduced (*70,349,895*). Degradation occurs rapidly immediately after cutting and gradually becomes slower with time (*786*). The wilting of cut grass and forage is accompanied by a fall in plant carotene content (*269*). In artificially dried grass the carotene level depends on the maturity of the grass when cut, the temperature in the drier, and conditions and duration of storage before it is fed. High-temperature, short-time drying seems to cause the least destruction. If hay or dried grass is not handled and stored properly, there may be losses of 30–50% of the carotenes during storage. The carotene content of freshly mowed green forage may fall by more than half its content within 24 hr (*822,900*). The total loss depends to a

TABLE 14

Carotene Content of Various Forage Plants[a]

Plant	Dry substance (%)	mg carotene/kg fresh substance		
		Minimal value	Mean value	Maximal value
Perennial ryegrass, *Lolium perenne*	26.0	46	104	204
Orchard grass, *Dactylis glomerata*	21.1	13	67	183
Timothy, *Phlenm pratense*	23.5	18	53	111
Smooth-stalked meadow grass, *Poa pratensis*	22.1	9	44	168
Meadow fescue, *Festuca pratensis*	22.0	18	74	138
Lucerne, *Medicago sativa*	23.9	17	47	93
Red clover, *Trifolium pratense*	21.9	19	40	59
White clover, *Trifolium repens*	16.6	23	53	92
Sugar beet leaves and tops	14.8	0	5	14

[a] From Futterwerttabellen (*287*).

great extent on the drying methods used in curing. During curing in the field, the plants are exposed to the elements for several days, resulting in the greatest loss of carotene. Carotene losses in hay may be from 73 to 84% during drying on the ground and from 39 to 81% when various types of racks are used. Hay made by the tripod method contained more carotene than ground-cured hay (*177*). Barn-dried hay may contain more carotene than hay dried in the field (*403*), but barn losses are still considerable. Zubricky (*952*) reported on the carotene content in 17 samples of alfalfa hay, alfalfa–grass hay, oat hay, and other cereal hay. There was a direct correlation between the color of the hay and its carotene content. Two samples with a green color had carotene contents above 20 mg/kg. All other samples having a yellowish color were very low in carotene, containing from 2 to 7 mg/kg. In another study (*769*) carotenoids in the form of β-carotene and oxycarotenoids were examined in seven grass species, six clovers, four lucerne species, and eight meadow hays at different ages, places, seasons, and storage periods. Toward the end of the growing stage, the carotenoid content fell. Fertilization, season, and soil all affected the carotenoid content, but on the whole the difference among species was usually larger than that among sites of the same species. Hay can then be a relatively poor source of carotene because of many factors, such as overmaturity of the grass when

TABLE 15

Carotene Content of Roughage[a]

Feed	Source	Average carotene content of aerial parts of plants (mg/kg dry matter)[b]										
		Early growth			Flowering			Seed heads developed (mature)			Drying in the field	Artificially dehydrated
		G	S	D	G	S	D	G	S	D		
Agropyron cristatum, crested wheatgrass	US	280			180			140			45–0	
Agropyron smithii, western wheatgrass	US	320			215			120			50	
Andropogon furcatus, big bluestem	US	350			245			140			23–8	
Arrhenatherum elatius, tall oatgrass	UK	316			176					24		
Beta vulgaris, fodder beet tops	Sweden							238				
Beta vulgaris, fodder beet leaves	CSSR							236				
	Rumania							133				
Beta vulgaris, sugar beet tops	France							56	13			
	USSR							129				
Bouteloua gracilis, blue grama	US	265			185			100			65–20	
Brassica oleracea, cabbage white	CSSR							161				
Brassica oleracea, cauliflower leaves	Palestine							400				
	CSSR							333				
Brassica oleracea, marrowstem kale	CSSR							288				
	Sweden							70				
Brassica, turnip leaves	CSSR							315				

Brassica, swede turnip leaves	Germany					190			
Bromus inermis, bromegrass	US	344		88	95	47	11		
Buchloë dactyloides, buffalo grass	US	260		143		90			33
Calmagrostis montanensis, plains reedgrass	US	255		165		100			60
Cenchrus ciliaris,	India	374		194		63			
buffels grass/killukaltai grass	South Africa							6–2	
Chloris gayana,	India	408–370				245–164			
Rhodes grass	S. Africa					140–40			6–4
Cyndon dactylon,	US	386		226		263			
Bermuda grass/dhub grass	India	398							
Dactylis glomerata cocksfoot/orchard grass	US	410	198–238	195		20	13–2		
	CSSR			136					
	Sweden	242		175					
	UK	346		144		50			
	USSR			67		31			
Daucus carota,	India					174			
carrot tops	CSSR					174			
	Sweden					350			
Festuca pratensis,	Sweden	128		118					
meadow fescue	Poland						21		
	UK	314		142					
	USSR			79		12			
Festuca rubra, red fescue	Sweden	277							
Lolium multiflorum,	CSSR	350		221					
Italian ryegrass	Netherlands		288						
	UK	209		100					

(*Continued*)

TABLE 15—(*Continued*)

Feed	Source	Average carotene content of aerial parts of plants (mg/kg dry matter)[b]										
		Early growth			Flowering			Seed heads developed (mature)			Drying in the field	Artificially dehydrated
		G	S	D	G	S	D	G	S	D		
Lolium perenne, perennial ryegrass	Sweden	136			75							
	UK	259			67							
	Australia				370			76				
	Poland	384			313							
Medicago sativa, *Medicago media* alfalfa, lucerne	US	340			143	155	13–33					180
	Netherlands											185
	India	240										
	S. Africa	400										
	CSSR	315			160		37–0	88				
	Poland	280			260		41–1					
	Sweden	225			200	70–48						148
	Australia						72–18					
Panicum maximum, Guinea/hairy buffels grass	West Indies	232			125			91				
	India	254			124			118				
	S. Africa	308						78			2–1	
Panicum purpurascens, para grass	West Indies	242						63				
Pennisetum clandestinum, Kiku/kikuyu grass	India				166							
	Kenya							78				
Pennisetum purpureum, Napier grass	India	221			106						3	
	US						45					
Phleum pratense, timothy	US						28	56				
	Sweden	342			184							
	UK	274			73			42				
	USSR				85			23				

Poa pratensis, Kentucky bluegrass	US	406		223						
	Poland								38	
	Sweden	245		210			126			
Sorghum sudanense, Sudan grass	US	317		198			58		3	
	India	117					97			
	Palestine	142		72						
	S. Africa			135						
	Rumania	286		155			57			
Trifolium pratense, red clover	US			233		35–12				
	Germany			152	84					
	Hungary	312		165						
	Poland	348		130						
	Sweden	349		146						277
	Rumania	530		240						
	Australia			700						
	Poland	542		703						
Trifolium repens, white clover	US	294							5	40
	Sweden			222						
	UK	509								
Zea mays, maize	India	360								
	Palestine			72						
	Poland	350	78[c]	175	56[d]		65[e]			
	Germany						44[d]	33[e]		
	Rumania	391								
	USSR						521[e]			
	US				94–75					

[a] From Aitken and Hankin (*11*).
[b] G, green feed; S, silage; D, dried (hay).
[c] Sweet corn.
[d] Flint varieties.
[e] Seed in wax stage.

cut, field weathering, heating, and long storage, all of which tend to destroy provitamin A. Much can be gained from the proper handling of a forage crop, or its value may be greatly decreased through neglect or failure to take proper precautions. These factors then influence the nutritive value of the product at the time of feeding. A comparative study of the feed value of alfalfa hays produced by different curing methods and nutrient yield was made by Hashizume *et al.* (*372*).

Carotene losses occur not only during silage making, but again during storage. The rate of breakdown depends on the initial carotene content, so that the greatest losses are usually sustained when the original content has been least impaired during silage making. Further losses occur during storage. When silage is fed to animals, it contains only a part of the carotene originally contained in the fresh plant.

Silage production from sugar beet, sunflower, dent corn, red clover, and alfalfa under different storage conditions was examined for maximal carotene retention by Wierzchowski and Basaj (*910*). The pH of the silage, proportions of acid and ammonia compounds, and growth stage of the plant influenced fermentation losses of carotene and other nutrients in the ensilaging of meadow grass, red clover, and maize (*624*). After the silage has been stored, uniformity of carotene content may not exist with the passage of time. Bahtin (*37*) found that the carotene content of maize silage at the top, bottom, and middle of a silo showed wide variations in individual values, although average values for each zone or layer were similar. Carotene balance studies in heifers fed about 550 mg carotene per day in the form of maize silage containing 27 mg carotene per kilogram and hay showed high digestibility coefficients, but the author reported low utilization of the carotene as judged by blood carotene values. Kovalcik *et al.* (*476*) stated that the β-carotene content of haylage is, in most cases, insufficient to maintain cattle. In a 1970 study by Miller *et al.* (*585*), carotene and vitamin A levels were determined in liver and blood in Holstein bullocks. These animals were then paired on the basis of liver vitamin A concentration and given a concentrate mixture plus either maize silage or lucerne pellets, which were adjusted weekly to equalize estimated net energy and carotene intakes within the pairs. The bullocks given pellets had higher plasma carotene and liver vitamin A concentrations than the bullocks given silage. Plasma vitamin A values were higher for the bullocks given silage.

Hence, the carotene content of silage depends (*662*) on (a) carotene content of original material, (b) possible losses resulting from intense feed fertilization practices, (c) losses that occur during filling of the silo, (d) losses due to wilting, (e) losses during the fermentation period, and (f) storage losses. Well-made silages may be closer to freshly cut hay than cured hay and hence can be richer in carotene (*469*). Usually corn (maize) silage retains carotenoids much better than drying (*770*). This is also reflected in studies (*662*) of vitamin A and carotene content of milk, which can be higher with silage feeding than with hay feeding without silage. Ensiling jowar (*677*) resulted in a lower carotene loss (52%) than did making it into hay (88%).

Maize (corn) plants were significantly higher in carotene content when harvested before frost than after it (*395,534,770*). Coarsely ground plants contained a higher content than finely ground plants after standing. Fermentation and silo storage resulted in a carotene loss ranging from 19 to 40% of ingoing levels. Storage losses tended to be greater in finely chopped silage than in more coarsely ground silage. Treatment of corn silage having high levels of moisture with organic acids produces significantly lower levels of carotene and vitamin E. Treatment of maize that had been artificially dried (having 12% moisture) with acid did not significantly reduce the values of vitamin E or carotene (*425*). Corn ensiled with $NaHSO_4$ (5 gm/kg) contained 40% more carotene and 35% more lactate but 28% less acetate than a similar silage without the preservative (*939*). Soil-covered plastic sheets laid over silage help preserve it and ensure anaerobic conditions (*46*).

Carotene losses can vary considerably during the preparation of silage, as mentioned by Friesecke (*285*). The proportion of the content in the original forage found to remain in turnip leaf silage was 32–81%, and in lucerne (alfalfa) silage it was 21–97% (*763*). Some 10% is usually destroyed during the actual fermenting process, but losses of as much as 30% have also been recorded (*314,899*). In open silos, the destruction of carotene often continues at a rapid rate (*314*). The carotene content (Table 16) of various fresh silages has been determined according to the German Agricultural Society's feed value (*287*). Wilted silage usually has a lower carotene content, because 20–70% can be lost during the wilting process. Depending on the extent of wilting, losses of carotene of 52–64% from lucerne grass silage have been reported (*618*). In other cases, wilted silage contained only 17%, or 3.6% (*786a*) of the original carotene. In practice, the quality of wilted silage, as a feeding stuff, can be rated above that of fresh silage (*618*), but, whereas the quality rises as the proportion of dry matter increases, the carotene content falls (*68*) because it is adversely affected by wilting (*786*). Much more carotene is lost during hay and silage making than in artificial drying, but the high cost of the latter process makes it

TABLE 16

Carotene Content of Various Fresh Silages[a]

Silage	Dry substance (%)	mg carotene/kg fresh substance		
		Minimal value	Mean value	Maximal value
Meadow grass silage	21.0	4	9	41
Clover silage	20.0	20	39	43
Lucerne silage	22.7	0	20	67
Red clover silage	18.1	9	38	85
Maize silage, green ripe	12.2	0	6	19
Silage of sugar beet leaves and tops	15.9	1	5	11

[a] From Futterwerttabellen (*287*).

of less practical use in the preservation of feeds for cattle than the other two methods. The quality of the original material and the method of preparation affect silage or haylage, which may be a good source of carotene, but if wilting takes place or overheating occurs in the silo the amount of carotene can be reduced substantially. Hay older than one year or excessively weathered or haylage overheated in storage may contain little or no carotene. When cows are fed rations without carotene, such as corn cobs, bleached hay, overheated silage or straw, with or without grain for several months or more, vitamin A deficiency can occur.

A better understanding of the carotenoid content of silage products would probably result if total carotenoid analyses of the forage crop to be preserved by ensilaging were made all the way through the process of preservation and during its feeding to farm animals. Almendinger and Hinds (*18*) explored carotenoid extraction procedures of haylage in an attempt to gain more knowledge of the physiological plant and animal processes and the carotenoid content of the product.

Silage prepared from well-eared corn is used extensively for feeding cattle in the United States. There have been many reports of suboptimal vitamin A nutrition associated with the feeding of corn silage in amounts presumed adequate to meet vitamin A needs. On the other hand, silage made from well-eared, well-matured corn from normally fertilized fields has been reported to supply the daily vitamin A requirements of beef cattle (*461*). These findings suggest four possibilities that may apply in some silage conditions: (a) the carotenes of silage were being poorly utilized by the livestock, (b) the ingestion of such silage increased the cattle's rate of usage of vitamin A, (c) the ingestion of silage adversely affected the ability of the liver to store the vitamin, or (d) the carotene content of the silage was low and became a more unreliable source of the provitamin as a sole vitamin A source. The latter possibility probably prevails most frequently. The adequacy of silage (*100*) as a source of vitamin A activity for cattle has been of practical concern for some time. Much depends on the factors previously discussed and the percentage of the total ration it makes up. When it is used as a complete feed with additives supplying nitrogen, minerals, and energy, it may be important to also add vitamin A. Whether to add vitamin A to silage, whether the fermentation of silage would promote the destruction of added vitamin A, or whether to use direct oral or intramuscular forms of vitamin A are some of the decisions the feeder must make.

B. Alfalfa (Luçerne)

Alfalfa has traditionally been a prime source of nutrients for farm animals. Dehydrated alfalfa meal has been one of the major sources of carotenes and xanthophylls used in poultry rations and is usually sold with a guaranteed carotene (vitamin A activity) content. The β-carotene content of the alfalfa plant shows a parabolic development and reaches a maximum generally just before blooming. About 85–95% of the β-carotene is present in the leaves (*597*). After the maximal carotenoid content is reached, at or before the first flowering stage, it gradually declines. The carotene and xanthophyll contents of alfalfa were higher during the September–November period than during the June–July period in a study (*519*) involving 11 trials.

An understanding of the influence of handling and processing on the carotene content of alfalfa should lead to the production of improved products from alfalfa, an intrinsically very valuable hay and forage plant for livestock. An extensive comparison of the influences of variety and certain environmental factors on the carotene and protein contents of alfalfa was made (*435*). No variety was found to be consistently higher than others in either carotene or protein content. Any differences could be attributed to differences in leafiness. A single third-cutting comparison of 23 varieties and 15 selections presented ranges of 190–274 ppm carotene and 16.0–22.6% protein. In general, protein values were parallel to carotene content.

As plant cells are injured or die during the drying process, carotene destruction begins. Leaves were held at given moisture contents during enzymatic studies (*70*), and the greatest carotenoid losses occurred when 80–95% of the water content had been lost, whereas in completely dehydrated leaves no enzymatic losses occurred. Atmospheric oxygen is necessary for enzymatic oxidation of carotenoids. Photodestruction rates for carotene increase with decreasing moisture content until 70–80% of the water is gone and then decrease to a lower rate when complete dehydration is reached (*70*). Raking alfalfa hay at the wrong time may cause loss of leaves, reducing the yield by as much as 20–30%. The quality suffers even more because most of the nutrients are found in the leaves. Loss by shattering is pronounced under conditions of low humidity when the hay is handled at a moisture content below 55%. Maximal carotene and protein yield may be obtained by raking when the hay reaches a dryness of between 55 and 65% moisture (*70,676*).

Various methods of small-scale drying of lucerne (alfalfa) under the sun in the field and under a shed were examined (*676*) in India. Carotene loss progressed exponentially, most of the loss occurring during the first 2½ months. Ultimate retention was 20–30%. The preferred method was short exposure to the sun and further drying under a shed in small bundles. In another investigation in India (*780*) alfalfa was dehydrated by sun, air, or hot air drying processes with and without blanch pretreatments to determine the effect on carotene and xanthophylls. Blanching followed by hot air drying gave the best retention of carotene. Light did not appear to affect losses, and low temperature storage reduced the losses. In a U.S. study (*572*) the effect of cutting treatments in the handling of irrigated Ladak alfalfa in the Yakima Valley influenced the yield of dry matter, an important factor associated with carotene content. Highest total seasonal carotene with adequate carotene content for the market was produced when the yield per cutting was 0.5–2 tons/acre. Spring clipping, previous fall cutting, and interval between cuttings had little direct effect on carotene content. It is common practice to cut alfalfa for dehydration in the bud stage because any further increase in dry matter leads to poor quality.

Alfalfa products contain a wide variety (*76*) of carotenoids (Fig. 10). In the

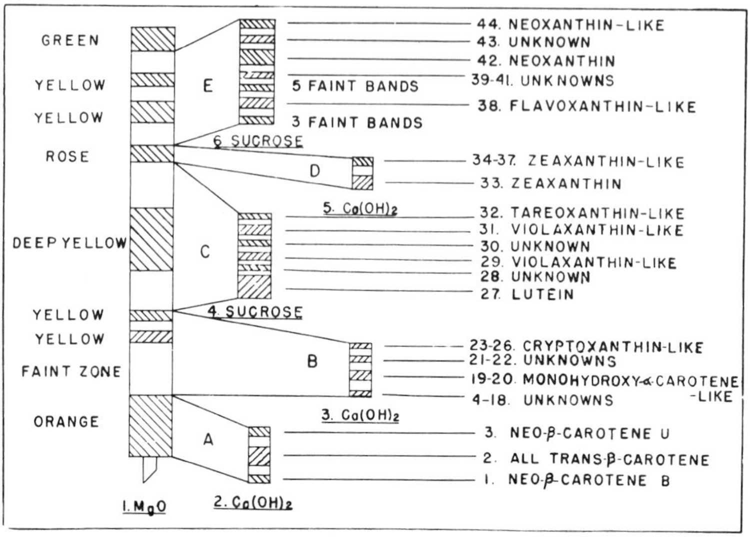

Fig. 10. Chromatographic separation of the carotenoids of dehydrated alfalfa. 1, Developed with hexane–acetone–alcohol (82.5:15:2.5); 2, with hexane–*p*-cresyl methyl ether (98.5:1.5); 3 and 5, with hexane–acetone (9:1); 4 and 6, with hexane–alcohol (99.5:0.5). [From Bickoff *et al.* (*75*).]

processing of alfalfa (*516*) the content of xanthophylls in dehydrated alfalfa meal was greatly reduced as the outlet drying temperature was increased and the moisture content of the meal decreased. Although total carotene content was relatively unaffected by the temperature of dehydration or the moisture of the meal, the β-carotene isomers increased with higher outlet temperature and lower moisture levels. Livingston *et al.* (*518*) observed losses of xanthophylls to be from 28 to 73% and carotene losses from none to 33% during pilot- and industrial-scale alfalfa dehydration. The effects of these dehydration conditions on the stability of xanthophylls, carotene and α-tocopherol in alfalfa meal were studied by using industrial-scale Arnold and Sterns–Roger dryers and a pilot-scale modified Arnold dryer. Carotenoid losses during dehydration depended on moisture content, outlet temperature of the dehydrator, and dehydrator retention time. Accordingly, during dehydration plus storage the total xanthophyll and carotene retention (*517*) of the medium-moisture meals was greater than that of the high-moisture meals (Table 17). Carotene and α-tocopherol were more stable than xanthophylls during dehydration. An industrial process has been developed for alfalfa products with higher protein and carotenoid contents (*466*).

C. Corn (Maize), Sorghum

Two general classes of carotenoid pigments, carotenes and xanthophylls (or oxycarotenoids), are primarily responsible for the yellow color of maize. Although

TABLE 17

Stability of Carotene during Pilot Alfalfa Dehydration and Storage[a,b]

Dryer temperature				Retained in meal (%)	
At inlet (°F)	At outlet (°F)	Moisture of meal (%)	Initial carotene content in fresh alfalfa (mg/kg)	After dehydration	After dehydration and storage
1600	220	11.0	225	100	28
1400	220	9.0	267	100	35
1200	220	8.6	205	100	35
1600	270	4.0	238	100	49
1400	270	3.2	275	100	58
1200	270	1.6	258	100	50
1600	320	1.2	282	100	52
1400	320	0.7	269	100	49
1200	320	0.3	284	88	43

[a] From Livingston *et al.* (*517*).
[b] Stored 12 weeks at 90°F.

the content of xanthophylls is considerably higher than that of carotene in kernels and leaves, the concentration of both classes of pigments varies appreciably with the genotype. The xanthophyll content of average commercial maize is approximately 10 times greater than the carotene content (*89*). Major carotenoids were determined in 15 lines, five high-carotenoid exotic strains, and five crosses of maize. In general, zeaxanthin and lutein were the major carotenoids. In two flint types of maize, zeaxanthin accounted for 67% of the total carotenoids. However, with certain color separations, cryptoxanthin was the major component. It appears to be possible to select for any one of the major carotenoids (*340*). Quackenbush *et al.* (*721*) in 1961 reported on five corn samples from different parentage; β-carotene values ranged from 0.6 to 4.4, β-zeacarotene from 0.1 to 4.0, and cryptoxanthin from 1.0 to 3.2 μg/gm Quackenbush *et al.* (*720*) conducted a study dealing with the carotenoid, oil, and tecopherol content of some 125 corn inbreds. They found that the provitamin A content of yellow inbreds varied from a trace to 7.3 μg/gm, including β-zeacarotene, cryptoxanthin, and β-carotene as vitamin A precursor forms. The content of pigmenting xanthophylls also varied widely (2–33 μg per gram of corn) and bore no close relationship to the provitamin A pigments. On the basis of analyses of several hybrid samples of inbred corn, Blessin *et al.* (*89*) indicated that xanthophylls and carotenes may be independently inherited. In a study (*582*) of corn in Yugoslavia, 13 single-cross hybrids analyzed contained provitamin A, 3.3–9.8; carotene, 1.4–8.1; and cryptoxanthin, 2.5–7.5 mg/kg. The 17 double-cross hybrids contained provitamin A, 3.7–7.4; carotene, 2.2–4.0; and cryptoxanthin, 2.5–7.2

mg/kg. The 3 double-cross white hybrids contained only traces of carotenoids. Thus, these authors stated that the provitamin A content of maize is subject to genetic influences. Likewise, sorghum, although lower in provitamin A content, is also subject to genetic influence. In an Indian study 57 varieties were analyzed and a wide range of carotene values observed, from traces to 26% of the total pigment content (*832*). Gross and Heller (*343*) also reported on the carotene content of grain sorghums.

Temperature of storage and moisture content of the corn grain influence carotenoid content. Quackenbush (*718*) reported on the effects of temperature and moisture on corn carotenoid losses in a high-carotenoid, single-cross dent corn (carotenes, 4.8; cryptoxanthin and esters, 5.5 μg per gram corn) over a 3-year period. Two moisture levels of the grain (3 and 11%) and two storage temperatures (7°C and 25°C) were used (Table 18). The carotene fraction was the least stable of all the pigments. Storage temperature was more important than moisture content; over the 3-year period, one-half of the carotene was lost at 7°C, whereas three-quarters was lost at 25°C. Cryptoxanthin and its esters were quite stable under storage at the lower temperature at both moisture levels. In an Italian study (*502*), the content of carotene in sorghum was found to be dependent on the variety and varied with the stage of growth and the harvesting period. Blessin *et al.* (*90*) observed that exposure of four strains of yellow-endosperm sorghum grain to weathering produced a continuous loss of both carotenes and xanthophylls. Grain exposed for 41 days retained only 47–58% of the carotenes and 54–66% of the xanthophylls present at zero exposure time. A trend toward higher ratios of xanthophylls to carotenes was indicated for the exposed samples, but the significance is questionable because of the low level of carotenes and the presence of xanthophyll esters in the carotene fraction. The highest contents of carotenes (1.2 μg/gm) and xanthophylls (9 μg/gm) in the samples under the most favorable conditions (zero exposure) were low compared with those of Nebraska 701 yellow corn (carotene, 2.4, xanthophylls, 28 μg/gm) grown under the same conditions.

More recently attention has been paid to correlating chemical analyses of feed ingredients with their biological value. Such a study was undertaken by Wellenreiter *et al.* (*906*) on the vitamin value of corn and corn gluten meal for pigs. Carotenoid analyses (Table 19 and 20) were performed on the two corn products. A calculation of the theoretical biopotencies of the vitamin A-active carotenoid pigments in corn produced some interesting figures for comparison with the potencies determined experimentally. If one arbitrarily assigns a relative potency of 100 to all-*trans*-β-carotene and assumes that β-zeacarotene has 25% and cryptoxanthin about 50% of this value, then the β-carotene equivalent of the active carotenoids is about 50% of the AOAC carotene values determined on the three samples of corn and two samples of corn gluten meal (Table 20). Assuming that the pig is only 30% as efficient as the rat in converting β-carotene to vitamin A, one would expect about 15% of the AOAC carotene to be converted and stored as liver vitamin A. When these percentages are multipled by 1667 IU, which represents the theoretical vitamin A potency

TABLE 18

Changes in Carotenoid Fractions of Corn during Storage[a,b]

Moisture (%)	Storage temp (°C)	Period of storage (months)	Carotenes	Zeinoxanthin	Cryptoxanthin and esters	Lutein	Zeaxanthin	Polyoxy pigments	Sum of pigments
11	25	0	4.8	3.4	5.5	19.8	16.9	3.4	53.8
		4	3.6	2.1	5.1	17.5	16.5	2.0	46.8
		8	2.5	2.2	3.3	16.0	10.4	2.8	37.2
		12	1.8	1.9	2.4	10.8	6.7	1.6	25.2
		24	1.7	1.8	3.2	11.3	7.7	1.9	27.6
		36	1.0	1.3	2.8	7.6	5.2	1.8	19.5
11	7	0	4.8	3.4	5.5	19.8	16.9	3.4	53.8
		4	3.1	2.2	5.0	20.5	12.5	2.6	45.9
		8	3.5	2.4	4.5	23.2	15.1	2.7	51.4
		12	3.4	2.9	4.0	14.6	14.2	2.3	41.4
		24	3.4	2.4	6.1	19.3	12.4	2.5	46.1
		36	2.6	2.5	5.6	15.8	12.6	2.3	41.1
3	25	0	3.9	2.3	5.0	15.7	9.9	2.8	39.6
		4	3.3	1.8	5.0	14.4	10.1	2.3	36.9
		8	2.6	1.7	3.4	12.0	7.8	2.0	29.5
		12	2.7	2.1	3.1	8.4	6.3	1.7	24.2
		24	1.4	1.4	3.1	8.6	5.8	2.0	22.3
		36	1.1	1.3	2.5	6.5	4.4	1.6	17.2
3	7	0	3.9	2.3	5.0	15.7	9.9	2.8	39.6
		4	3.1	1.4	5.0	13.2	10.7	2.5	35.9
		8	3.5	2.1	4.0	17.9	12.4	2.2	42.1
		12	3.5	2.5	4.0	11.5	8.8	2.5	32.8
		24	3.0	2.3	5.5	15.0	10.2	2.0	38.0
		36	2.2	2.1	4.5	10.3	8.6	2.0	29.5

[a] From Quackenbush (*718*).
[b] All values are averages of two analyses, expressed as micrograms per gram of corn.

TABLE 19

Carotenes in Corn and Corn Gluten Meal[a,b]

Component	Corn 1[c]	Corn 2[c]	Corn 3[c]	CGM1[d]	CGM2[d]
Phytoene	3.7 ± 0.4	5.0 ± 0.2	8.7 ± 0.6	64.7 ± 2.4	53.2 ± 1.5
Phytofluene	2.0 ± 0.1	6.1 ± 1.4	4.0 ± 0.2	69.7 ± 4.6	44.0 ± 1.6
β-Carotene	0.5 ± 0.0	1.3 ± 0.0	2.0 ± 0.0	13.8 ± 0.4	20.1 ± 0.1
β-Zeacarotene	0.4 ± 0.0	1.2 ± 0.0	1.9 ± 0.0	12.6 ± 0.4	20.9 ± 0.1
ζ-Carotene	0.2 ± 0.0	0.7 ± 0.0	1.1 ± 0.0	7.4 ± 0.1	13.7 ± 0.1
Zeinoxanthin	0.9 ± 0.1	2.1 ± 0.1	4.0 ± 0.1	42.5 ± 4.0	40.5 ± 1.4
Cryptoxanthin	0.9 ± 0.1	1.9 ± 0.1	3.7 ± 0.1	35.6 ± 2.6	36.3 ± 0.8
Esters	2.5 ± 0.1	11.8 ± 2.1	16.0 ± 0.3	81.2 ± 3.5	176.9 ± 1.0
Lutein	2.6 ± 0.1	12.2 ± 2.1	16.2 ± 0.2	84.5 ± 3.4	184.8 ± 0.3
Zeaxanthin	0.7 ± 0.0	4.2 ± 0.8	6.5 ± 0.2	31.5 ± 1.2	80.7 ± 0.2
Polyoxy pigments	0.9 ± 0.0	1.9 ± 0.0	1.4 ± 0.1	7.7 ± 0.2	22.4 ± 0.9
Total	15.3	48.4	65.5	451.2	693.5

[a] From Wellenreiter *et al.* (*906*).
[b] Data expressed as milligrams per kilogram plus or minus standard error of analyses.
[c] Corn used in trials 1, 2, and 3, respectively.
[d] Corn gluten meal used in trials 1 and 2, respectively.

of 1 mg of all-*trans*-β-carotene, then 1 mg of AOAC carotene from corn or corn gluten meal would have a calculated potency ranging from 254 to 267 IU. These values are very close to the experimentally determined value of 261 IU/mg when corn was the source of carotenes.

The results of these studies reemphasize the need for more careful appraisal,

TABLE 20

Provitamin A-Active Components of Corn and Corn Gluten Meal[a]

Component	Corn 1[b]	Corn 2[b]	Corn 3[b]	CGM1[c]	CGM2[c]
			% active pigments		
β-Carotene	27.8	29.5	26.3	22.3	26.0
β-Zeacarotene	22.2	27.3	25.0	20.3	27.0
Cryptoxanthin	50.0	43.2	48.7	57.4	47.0
			mg/kg		
Total active	8.8	4.4	7.6	62.0	77.3
Crude carotene	2.1 ± 0.1[d]	5.1 ± 0.2[d]	8.6 ± 0.1[d]	69.7 ± 1.7[d]	90.8 ± 1.4[d]

[a] From Wellenreiter *et al.* (*906*).
[b] Corn used in trials 1,2, and 3, respectively.
[c] Corn gluten meal used in trials 1 and 2, respectively.
[d] Standard error of analyses.

during genetic selection, of the nutritional properties of grains widely used as animal feeds, and illustrate that even high-quality feedstuffs can be rendered poor by faulty handling or storage practices.

X. FACTORS INFLUENCING CAROTENE CONTENT

Carotenoid and carotene content are influenced by the genetics and environment of the plant (*89*). In parasitic-type plants, α structures predominate over β structures; carotenoid epoxides are favored in high altitudes (*94*). Special corn varieties with high carotenoid content can be bred (*89,126*). The content of xanthophylls and carotenes of 11 lucerne clones was higher during September and November than during June and July, according to Livingston *et al.* (*518*). Both types of carotenoids reached a maximum at or before first flower and then declined as blossoming progressed. Lutein decreased in November, whereas violaxanthin increased. Values have been tabulated for carotene, lutein, neoxanthin, and violaxanthin in six grasses and four legumes. A diurnal pattern exists in the carotene content of the leaf (*852*). Variations also exist in the leaves, with the oldest having the lowest content. Interplant variation can be considerable. Maximal carotene and oxycarotenoid content occurs after germination and in alfalfa (lucerne), clovers, and others reaches a maximum before flowering and then decreases (*519*). In alfalfa leaves there is sunlight-sensitized and enzymatic destruction after cutting (*336*). Livingston *et al.* (*517*) reported oxycarotenoid losses from 28 to 73% and carotene losses from 0 to 33% during pilot- and industrial-scale alfalfa dehydration.

Carotenoid content is usually correlated with the green color of hays and silage. The rate of decrease depends on original content, temperature, moisture content, amount of exposure to air and sunshine, type of equipment used in processing, and other factors (*134*). When the ambient temperature rises, the rate of loss is accelerated. The carotene content of natural feed ingredients shows a large variation among feeds and a large range of assay values for any particular forage or feed concentrate. Among the grain products common in ruminant nutrition, only yellow corn and its by-products contain measurable amounts of carotenes and cryptoxanthin. The individual and total carotenoid contents of corn are not constant, nor is the ratio of the individual carotenoids (*209*). None of the individual oxycarotenoids are equally available (*41*).

A. Assay

Bickoff (*72*) pointed out that most of the older chemical methods for determining β-carotene content measured only total carotene, no allowance being made for the difference in activity of the various provitamin A carotenoids (α-, β-, γ-carotenes and cryptoxanthin) that might be present. The problem is further complicated by the knowledge that each of the carotenoids can exist in a number of stereoisomeric forms, some having widely different biological activities. The all-trans form of β-carotene and other carotenoids is usually found and maintained only in fresh growing plants or in the crystalline state when protected from oxygen and heat. Since carotenoids contain unsaturated bonds, they are subject to isomerization. Cis–trans isomerism can occur at those double bonds in the central chain adjacent to the methyl groups. The all-trans or straight form is fully active; the cis or bent forms are usually less active.

The carotenoids and their isomers can be separated by chromomatographic procedures and the individual carotenoids determined spectrophotometrically. However, a widely used chromatographic procedure for the determination of the β-carotene in feed ingredients, such as the AOAC method (*29,74*), fails to separate the isomers of lower biological potency from all-*trans*-β-carotene, and thus the biological activity may be overestimated by as much as 30% and possibly more. These isomeric changes assume increasing importance depending on the processing or pretreatment to which the carotenoid-containing feeds have been subjected. Isomerism may be induced as follows: by refluxing the pigment in a solvent, by illumination, by treatment with acids or iodine, or by melting the crystals at high heat.

Workers who have compared the biological response of a feed containing a mixture of carotenoids and their isomers point out that, although it is possible to work out a factor for each feedstuff converting the chemical value to that obtained biologically, the biological evaluation is currently the only method that indicates the actual vitamin A activity of such a feed ingredient, although excellent progress has been made in assessing values by the chemical method in certain feed ingredients.

B. Lability

During the curing and storage of the hay crop, the carotene content decreases at a rapid rate. The mechanisms responsible for the destruction of carotene in alfalfa are enzymatic, photochemical, and autoxidative. The rate of carotene loss depends on original content, particle size, temperature, moisture, exposure to sunshine or heat, rain leaching, and other factors. Drying may be done by the sun or in dehydration equipment. In studies of factors affecting loss of carotene in alfalfa during harvesting, Griffith and Thompson (*336*) observed that only 7–8% was caused by sunlight and 27–28% by enzymes. In the stems the sunlight caused about 5% loss, and enzymes 70%. Thompson and co-workers (*853*) also showed that increased temperature during storage promoted the formation of cis isomers in alfalfa, whereas irradiation with visible light decreased the amount of cis isomers. Irradiation of alfalfa meal produced up to a 10% increase in the vitamin A activity.

Mills and Hart (*586*) reported that the most effective method of retaining carotene in dried material was either to store at low temperatures or to remove the air and store in vacuum or under inert gas. Oil sprays and antioxidant addition also slow oxidation (*591,648,851,854*). Bolin (*96*) observed the stability of carotene in green grasses and alfalfa stored at cold temperatures. Under conditions of low temperature, the carotene content of hay can be expected to decrease about 6–7% per month. When the temperature is about 66°F the loss may be over 18% per month. The loss of carotenoids from dehydrated alfalfa hays and calf starters was investigated by Dolge *et al.* (*203*). Losses were found to occur essentially at a linear rate, but the latter was influenced by the initial carotenoid concentration. The intermediate concentrations could be predicted from the initial carotenoid concentration and length of storage period. In one study (*676*), lucerne was collected at one time and dried by nine methods; the variables studied included drying in the sun or shade, loose or in bundles, on the ground or in suspension. The highest value was for lucerne wilted for 1 day on the ground and then dried and stored under a shed. Only slightly lower was that for material dried by hanging bundles under a shed. The lowest content was for lucerne dried outside in sunlight in loose heaps. Losses of carotene during the storage of lucerne by five methods up to 100 days were reported. Most of the loss occurred during the first 70 days; after 100 days 13.2–33.2 μg carotene per gram dried lucerne remained (Table 21). In another study, considerably more carotene was lost from the middle of a stack of dry lucerne than from its edges. Pelleting reduces the loss of carotenes. The loss of carotenes (gradient loss through a stack) from dried green crop during storage was observed by Booth (*102*). In one investigation (*346*), pelleting dried green feeds prepared from alfalfa, alfalfa-grass mixture, red clover, and pasture grass reduced carotene losses after a 224-day storage period by 4–14%. However, under the most favorable conditions, savings in carotene in legumes amounted to only 17% of the cost of the pelleting operation. Storage in plastic bags improved only slightly the preservation of the carotene. Storage at low temperatures (– 10°F) was more effective in preserving the carotenoid pigments during storage of dried citrus flavedo than the addition of an antioxidant before the drying phase (*107*). Bethke *et al.* (*71*) and Halverson and

TABLE 21

Loss of Carotene in Lucerne on Storage[a,b]

Period of storage (days)	Perennial lucerne: Dried under shed	1 day wilted	2 days wilted	Dried in bundles under shed	Annual lucerne, dried on *mandap* under shed
10	82.80	71.85	23.60	71.63	65.08
25	51.58	48.52	20.65	33.18	48.75
40	22.37	41.55	17.95	29.85	41.62
70	21.40	30.45	17.55	19.25	39.57
100	21.25	28.47	13.25	18.37	33.17

[a] From Patel *et al.* (*676*).
[b] Loss expressed as micrograms carotene per gram lucerne.

Hendrick (*363*) reviewed the literature on the instability of carotene in mixed feeds and reported that losses increased markedly in rations containing added minerals in quantity and in those containing meat scraps. Antioxidant-treated animal fats increase the stability of carotene when added to commercial poultry feeds (*57,796*).

All carotenoids, whether in synthetic or isolated form or as components of natural feeds or forages, are sensitive to oxidative destruction. The pure crystalline carotenoids produced by chemical synthesis lend themselves ideally to physiochemical stabilization when embedded in a carrier matrix of varied composition such as gelatin, starch, and gums. Dry products of excellent stability can be obtained following the principle for stabilization of fat-soluble vitamins (*459*). The stability of carotene added to solid carriers was studied by Bickoff and Williams (*80*). The addition of an effective antioxidant to alfalfa meal helps to stabilize the carotenes and carotenoids. The use of a vacuum or an inert gas such as nitrogen can aid in stabilizing the carotene content of alfalfa. There is an interesting record of this (*951*). Alfalfa stems and leaves were steam-blanched, dried, and ground to a meal containing 374 ppm carotene. Samples stored *in vacuo* in darkness at 90°F for 27, 35, and 55 weeks retained 98% of the carotene. A final sample, kept 27 years at room temperature, retained 96% of it original carotene. Isomerization of carotene was negligible in all samples.

C. Antioxidant Use

The stabilization of carotenes and oxycarotenoids in feed ingredients consumed by domestic animals has been a long-time objective of a number of investigators around the world. Mitchell and Silker (*590*) observed that the addition of natural products such as cottonseed meal or soybean meal increases the stability of carotene in alfalfa. The addition of 5% of animal fats or animal oils to dehydrated alfalfa meal decreased the loss of carotene during storage, despite their content of free fatty acids (*79*). The addition of oil or fat is also useful in reducing the dustiness of alfalfa meal (*882*).

In later studies, Mitchell and Silker (*591*) reported that various vegetable or animal oils, applied as antidusting agents at concentrations of 0.8%, do not protect carotene in alfalfa meal substantially unless an antioxidant, such as ethoxy-1,2-dihydro-2,2,4-trimethylquinoline (ethoxyquin), was added. Similar experiences with mixed feeds were reported by Livingston *et al.* (*515*). Evidently, the protective effect of fats depends on the amount applied and on the type of fat used, since fats under some conditions may even accelerate the destruction of carotene. The relationship between the structure of quinoline derivatives and their antioxidant activity was investigated (*76*). The 6-alkoxy derivatives were most effective for alfalfa meal, whereas the 6-hydroxy derivative best protected the carotene in mineral oil (*76*). Thompson (*851*) reported that carotene in dehydrated alfalfa meal was stabilized by the addition of antioxidants dissolved in ethylene glycol monoethyl ether and applied in the form of a spray to dehydrated meal in a rotating drum.

Fifty-four compounds were tested as antioxidants. Disubstituted hydroquinone in the 2,5 positions, para-substituted phenylenediamines, and derivatives of 2,2,4-trimethyl-1,2-dihydroquinoline were the most active compounds. A synergistic effect of certain combinations was observed; crude cottonseed oil and acetone proved to be the best solvents. The effect of increasing concentrations of antioxidant on the stability of carotene was determined. Samples of meal were sprayed with increasing concentrations (from 0 to 0.5%) of 2,5-di-*tert*-butylhydroquinone. When the percentage of carotene remaining was plotted against the amount of antioxidant added, it was apparent that increasing amounts up to 0.125% were increasingly effective. Above this level the additional antioxidant gave proportionately smaller effects. In a confirming study, the antioxidant 2,5-di-*tert*-butylhydroquinone (about 0.1%) applied to alfalfa meal in an oil was found to be effective in delaying the oxidation of carotene. In other studies, 2,5-disubstituted hydroquinone, para-substituted phenylenediamines, derivatives of 2,2,4-trimethyl-1,2-dihydroquinoline, ureas, phenothiazine, and naphthylamine were observed to be active (*73,77,78*).

Mitchell *et al.* (*592*) noted that the application of oil alone had some effect on carotene stability in certain instances. The untreated meal lost 49% of its carotene in 3 months, compared to losses of 54 and 41% for the meals sprayed with 16 and 80 lb of oil per ton, respectively. Heating the meal after spraying with 80 lb of oil greatly improved carotene retention, the combination of oil and heat reducing the loss to 29%. Some reduction in carotene loss was obtained in all cases when 0.01% antioxidant was added along with 16 lb of oil, although the effect at this level of antioxidant was not appreciable unless heat was applied also. Even when heated, the most stable sample at this oil level and antioxidant concentration (substituted dihydroquinoline) still lost 36% of its carotene. Marked reduction in carotene loss was obtained when a combination of 80 lb of oil, 0.1% antioxidant, and heat was used. Under these conditions (*592*) the dihydroquinoline-treated meal lost only 9% of its carotene, whereas losses in the presence of the substituted hydroquinone and the substituted hexamethylenediamine were 17 and 15%, respectively (Fig. 11). The suggestion is offered that the stabilizing effect of 0.01% antioxidant would have been greater under commercial conditions, particularly if applied before grinding, because some heating of the meal occurs during processing. When alfalfa meal was sprayed with 0.1% diphenyl-*p*-phenylenediamine (DPPD), isoproxydiphenylamine, or *p*-hydroxydiphenylamine in oil, a marked improvement in stability was observed during storage, according to Kephart (*877a*).

The preservation of carotene and vitamin A continues to be investigated. Booth (*103*) published an account of the problem of preventing loss of carotene during storage of dried green crops. Scandinavian workers have reported extensively on the preservation, formation, and utilization of carotene (*650*). The American studies of the early 1950's were continued, and reports by workers in other countries began to appear in the literature during the late 1960's and early 1970's. Stabilization of carotenoids in harvested fresh alfalfa was accomplished with ethoxyquin by Knowles *et al.* (*465*). Alfalfa commercially dehydrated to yield meals with dif-

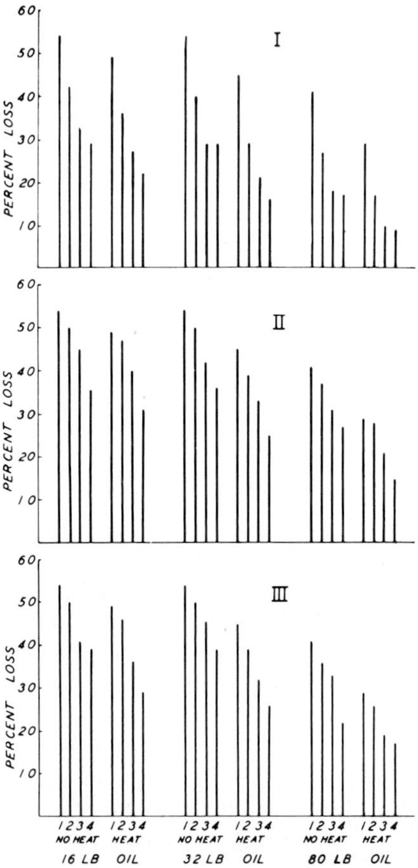

Fig. 11. Effect of oil, antioxidant, and heat on loss of carotene in dehydrated alfalfa meal stored at 25°C for 3 months. I, β-Ethoxy-1,2-dihydro-2,2,4-trimethylquinoline; II, *N,N'*-diphenylhexanethylenediamine; III, 2.5-di-*tert*-butylhydroquinone; 1, 2, 3, and 4 are 0, 0.01, 0.05, and 0.1% antioxidant, respectively. [From Mitchell *et al.* (*592*).]

ferent moisture contents and freeze-dried alfalfa were subjected to storage conditions (*464*). The meals that had undergone the smallest losses during dehydration had the greatest losses during storage. Freeze-dried samples were the most stable. Ethoxyquin treatment reduced oxidative carotenoid losses but isomerization losses of neoxanthin and violaxanthin continued. Dvinskaja (*219*), in a study on the preservation of carotenes in grass meal, noted losses of carotene after storage for 4, 6, and 8 months as 31, 52, and 62%, respectively. Of the antioxidants tried, ethoxyquin was the most effective, with corresponding losses of 7–13, 15–21, and 21–29%. The most effective concentration of ethoxyquin was 0.091%, this preserved 78% of the carotene after 8 months. Meals of clover, vetch, peas, oats, and timothy, of vetch and oats, and of clover and timothy were treated with 0.02% ethoxyquin. Preservation of carotene was higher in all treated meals than in control samples. When given to livestock and poultry, the treated meal increased the vitamin A content of liver, fat, blood plasma, milk, and eggs and improved productivity. In Russian trials, the addition of 10% beet molasses with 0.02% ethoxyquin retarded the decline in carotene content five- to eightfold. Molasses itself has some stabilizing influence, probably by sealing off oxygen, as shown in grass and alfalfa experiments (*178*). Animal fat could replace the molasses. Other antioxidant studies (*271,272,347,941*) confirmed or extended these findings.

Antioxidants other than ethoxyquin that are effective in stabilizing the carotenes include kurasan (*507*), 3,5-disubstituted 2,6-dimethyl-1,4-dihydropyridines (*214*), diphenyl-*p*-phenylenediamine, ionol (*947*), diludin (*819,869,948*), digisan (2,2,4-trimethyl-6-ethoxy-1,2,3,4-tetrahydroquinoline) (*948,950*), *p*-hydroxydiphenylamine derivatives (*950*), tanan (2,2,6,6-tetramethylpiperidine-1-oxyl) (*950*), and butylated hydroxyanisole (*207,942*). Diludin was an effective as ethoxyquin in its carotenes stabilizing ability in applications of dried grass or alfalfa meal (*818*). In Poland, the highest β-carotene retention was obtained in granulated, dried alfalfa, packaged in polyethylene bags and stored at low temperature (4°C), with ethoxyquin added (*121*). Alfalfa leaf protein concentrates have been prepared (*583*). In reports from California the combined addition of ethoxyquin and vegetable oil was stated to have a favorable effect on carotene stability (*952*).

Thus, it is well documented in the literature that the use of antioxidants in the processing of alfalfa or lucerne is beneficial in retarding the loss of vitamin A precursors and carotenoid pigments. Likewise, in the processing of various grass meals from clover, oats, timothy, vetch, wheat, and others, antioxidant addition retards losses of valuable provitamin A compounds such as the carotenes. Since this has been an accepted fact for some time around the world, investigators might turn to other subjects of investigation.

XI. FACTORS INFLUENCING CAROTENE ACTIVITY

Some factors affecting the biological activity of carotenoid vitamin A precursors are the physical form of the carotenoid; its solubility; the state of isomerization; the dietary level of the carotenoid administered; the dietary fat level, the degree of unsaturation and state of oxidation of the fat; the presence of dietary antioxidants (e.g., tocopherol, ascorbic acid) and prooxidants (e.g., nitrites); the influence of dietary minerals; the presence of absorption inhibitors (e.g., mineral oil); the presence of adequate bile in the tract or added dietary emulsifiers; the type and adequacy of dietary protein; the role of thyroid-active compounds; the surface area of the digestive tract and the enzymatic digestion efficiency; the presence of disease and/or parasites in the organism; age; species; breed; sex; and environmental temperature. This subject has been reviewed in detail elsewhere (*111,193,194,564,753,855,857*).

The carotenes are best utilized when consumed at relatively low levels in the ration (*111,753*). This fact was first recognized in the feeding of natural feedstuffs containing carotenes and other carotenoid vitamin A precursors (*193,194*). A fairly well-developed mechanism exists in the absence of deterrents to convert limited dietary intakes of β-carotene to vitamin A. As the ingested amounts are increased, however, the efficiency of conversion decreases markedly. This phenomenon has been noted with both natural crystalline and chemically synthesized all-*trans*-β-carotene, as well as stabilized dry forms of β-carotene (*55*). In view of the multiplicty of factors influencing the utilization of carotene, it is extremely difficult, if not impossible, to affirm the biological activity of carotene for any animal. There is no single conversion ratio (*865*) for the transformation of β-carotene to vitamin A. Rather, it depends on the conditions of use. Ratios of the biological effectiveness of β-carotene to that of vitamin A for various species have been recommended (*111*) by different authorities in the United States, Canada, and Britain. Rather than the theoretical ratio of 2 : 1 most often cited in nutrition textbooks, ratios of 4 : 1 to 8 : 1 and higher are encountered in actual feeding practices.

A. Source

A number of conditions alter the rate at which carotene can be absorbed. Carotene is absorbed with relative difficulty due to its low solubility and because it is a hydrocarbon without functional groups. There is no possibility that β-carotene can yield smaller and more easily absorbed molecules by hydrolysis or oxidation and retain its characteristics other than by its conversion to vitamin A. Carotenoid provitamins A are present in the cells of plants as solutions in fats, as protein complexes, or in crystalline forms. During passage through the digestive tract, the plant cells must be broken down. In the small intestine the provitamins can be (a) absorbed as such, (b) converted to vitamin A, or (c) destroyed by enzymes, bacteria, or oxidation. The digestibility or freeing of carotenoids from other plant constituents is usually not complete, even when they are given in minimal quantities, and some unabsorbed carotenoid is usually present in the feces. Because of the limited efficiency in digestive and absorptive processes, the apparent coefficient of digestibility decreases with an increased intake of the carotenoids. Kemmerer and Fraps (*446*) fed rats β-carotene in dehydrated alfalfa at levels of 1, 10.5, and 20 ppm and observed apparent digestibilities of 43, 22, and 18–23%, respectively. They also determined digestibility figures of 51 and 22%, respectively, for β-carotene when administered in oil or in dried alfalfa fed to rats on a fat-free regimen. According to the work of Drummond and McWalter (*208*), one explanation of the difficulty with the absorption of β-carotene and other carotenoids may be their low solubility. A concentration of 0.1% mixed carotenoids found naturally in red palm oil is about the highest stable solution that can be achieved unless the carotenoids are complexed with protein such as occurs in carrot juice (*829*). It is difficult to make a stable solution of crystalline β-carotene in vegetable oil much greater than 0.05–0.1% (*55, 608*). Vegetable sources of carotene are not usually absorbed as well as oily solutions, but fat is only partially necessary for proper utilization; colloidal dispersions and emulsions can be utilized in the absence of fat.

The source of carotene in the diet is an important factor with regard to its utilization by the animal and is consequently also an important consideration in determining dietary requirements. The relative efficiency of utilization of carotenoids in various foods varies considerably. One factor responsible for such differences is the different nature of the carotenoids in different foods and feeds. For example, in rat liver and kidney storage tests, Johnson and Baumann (*431*) reported the relative efficiencies of all-*trans*-β-carotene, neo-β-carotene B, and neo-β-carotene U to be 100 : 48 : 33; these are in excellent agreement with the results obtained by Deuel *et al.* (*196*) in rat growth studies. Carotene is liberated in the stomach or rumen from plant tissue during digestion. However, the release of carotene from plant tissue depends to some extent on the type of plant and stage of growth. The biological value of carotene from different kinds of fresh forage varies considerably. Ascarelli and Bondi (*32*) observed that when plant materials were autoclaved before use there was no difference among sources. Therefore, it was believed that differences in the activity of plant enzymes were responsible for the differences found with fresh forage. When alfalfa was autoclaved and dried before feeding, it had higher biological value, as judged by liver stores, than fresh forage. In studies of chick liver storage, Frey and Wilgus (*283*) found the carotene in fresh alfalfa to be utilized more efficiently than that in dehydrated alfalfa meal. The variable response of carotene from different kinds of grasses is particularly well illustrated by data of Thompson (*858*). He demonstrated the relative efficiency of different grasses compared with crystalline carotene in gelatin beadlet form of increasing the vitamin A level in milk fat. The increase in vitamin A per gram of milk fat that was derived from 3 gm of carotene in cocksfoot grass was seven times that produced by 4 gm of carotene from rye grass. The availability of the crystalline β-carotene in this trial was always several times greater than that of the carotene derived from plants. Wing (*917*) found that grasses, legumes, and alfalfa hay had apparent digestibilities of 55% for dry matter and 78% for carotene. Lower digestibilities of 31% for silage and 25% for dried forage were observed by Nehring and Hoffman (*623*), who found that digestibility decreased with advancing maturity of the forage, was affected by its lignin content, and was generally lower for carotene than for some of the other nutrients. The physical form of the plant material ingested by the animal plays an essential role in determining the availability of carotene. Yeast carotenoids proved to be a good source of carotene for pigs (*661*). Ullrey *et al.* (*881*) found that fermentation carotene from mycelia of *Blakeslea trispora* was effective; as little as 0.5 mg carotene per kilogram feed supported maximal weight gain in depleted pigs but 3.5 mg was needed to restore liver vitamin A stores.

The method of preparing alfalfa forage can influence carotene bioavailability. Hansen *et al.* (*365*) fed Guernsey cows three different forms of alfalfa: (a) artificially dried and chopped, (b) artificially dried and pelleted, and (c) field-cured and field-baled. Of the three types of alfalfa the artificially dried and chopped form produced the highest intake of carotenoids per cow daily, the highest levels of carotenoids and vitamin A per unit of milk secreted, the highest concentrations of carotene and vitamin A in the milk fat, and the highest total of these substances secreted in the milk. Except for the intake of carotenoids, there was little difference among values for the pelleted and field-baled alfalfa.

The efficiency of conversion of carotene to vitamin A is related to the size of the dose. Absorption varies with dosage rate, apparent digestibility and absorption coefficients decreasing at higher dosage rates. At very high levels of intake, as little as about 1% absorption was reported (*58*). An example of the poor utilization of carotene at high dosage levels was shown by Thompson (*858*) in the cow. Increasing the dose of β-carotene 4-fold produced a 2½-fold increase in milk carotene and vitamin A content, and a 16-fold increase in the dose produced only a 4-fold increase. Booth (*101*) showed that high food intakes and faster passage of food through the gut decrease the apparent absorption of carotene, measured as the difference between the amount ingested and that excreted. Hence, although a number of nutritional factors can affect the utilization of carotene, a decisive one is the amount of carotene administered because the utilization of carotene follows the law of diminishing returns: The more carotene the animal ingests the lower is its vitamin A activity per unit of weight (*253,338,620*). Not only by the administration of large quantities of carotene is the utilization reduced, but numerous trials have shown that after periods of deficiency carotene is poorly utilized (*135,136,253,742*). The longer a deficiency of carotene is maintained, the more impaired becomes the capacity of the animal to subsequently utilize carotene (*338,589*). Under these conditions vitamin A deficiency can be relieved more effectively by the administration of preformed vitamin A rather than a carotene source (*28,242,742*).

B. Sex, Age, Environmental Factors

The sex of the animal plays a role in the metabolism of lipids in general, so it would not be surprising if it had some influence on the metabolism of carotene or vitamin A. Sex differences exist in the plasma levels of carotene and vitamin A in man. β-Carotene values in the blood are usually higher in females than in males, but plasma vitamin A levels show exactly the reverse relationship between sexes (*193*). Vitamin A values for hepatic vitamin A storage have been reported to be higher in females than in males, under normal feeding conditions, not only in rats (*104,137*) but also in cattle (*244*). Goodwin (*326*) reported that the concentrations of carotenoids were more abundant in the organs of females than in those of males in species in which they were absorbed. Booth (*101*), however, believes that the absorption efficiency of carotene and the conversion to vitamin A are not functions of sex.

McGillivray pointed out that a number of workers (*304,833*) have found the carotenoid level of cow blood plasma to be up to five times greater than that of bull plasma. Although sex differences are common in carotenoid and vitamin A metabolism, it seems that the explanation might lie in the different management given the animals or the different role for carotene in the bovine female. There is a difference in feeding practice or intake; in the male there is only a maintenance requirement, whereas in the female there are maintenance plus milk production and reproduction requirements. Valuable (and sometime ferocious) bulls are frequently kept penned indoors and not usually run on pasture. When bulls are run on pasture throughout the year, their blood plasma carotenoid values are more similar to those of cows of the same breed and under the same management practices (*560*). Differences, of course, could be anticipated in the pregnant and lactating female, depending on the stage of the pregnancy cycle. Normally, the blood level of carotenoids of the newborn calf is quite low (*602*).

Once the animals are on pasture, carotenoid levels begin to increase with age. The blood carotenoid content of one-year-old pasture-fed heifers becomes about 70% of adult cows. In their first lactation young cows have slightly lower carotenoid levels than more mature cows, but thereafter levels are more comparable (*560*). Vavich and Kemmerer (*888*) studied the influence of the size of rats on liver storage of vitamin A after feeding a given dose of β-carotene. Storage was higher in the smaller animals. Older male rats are less efficient in converting carotene to vitamin A, according to studies by Murray and Erdody (*619*).

Interest in the possible effects of environmental temperature on the metabolism of carotene and vitamin A has developed in recent years. The relative biopotency of β-carotene was observed by Smith and Borchers (*809*) to decrease slightly with each increment of increase in environmental temperature (Table 22). This apparent decrease may have resulted from the destruction of β-carotene in the diet, particularly at 35°C, during the 6- to 10-hr period required for consumption of the 5-gm meal by the animal. No reduction in the level due to vitamin A was detected after a similar exposure. Environmental temperature may affect vitamin A depletion or the utilization of carotenes as a source of vitamin A or both. Increased requirements for vitamin A have been reported for animals maintained at both abnormally high (*498*) and abnormally low temperatures (*247*). Ershoff (*247*), in his trials with rats, showed a more rapid depletion and shorter survival time in animals on a vitamin A-deficient diet. Phillips (*701*) observed, in trials conducted in Canada, the effects of low environmental temperature versus room temperature on vitamin A and carotene metabolism in the rat. He found that utilization of carotene for storage as vitamin A in the liver was greater in rats kept at 2° than at 22°C, but hepatic storage of orally administered vitamin A and rate of metabolism from storage were unaffected by these environmental temperatures. Environmental effects reported for different species are not always in agreement.

Animals with insufficient body stores or inadequate intakes of vitamin A do not appear to tolerate thermal stress. A series of experiments to determine the effect of heat as well as solar radiation on the utilization of vitamin A by cattle was conducted by Page *et al.* (*664*). In one study, cattle outside and exposed to temperatures of

TABLE 22

Effect of Environmental Temperature on the Utilization of β-Carotene Relative to Retinyl Acetate When Fed in the Diet[a,b]

Treatment	Temperature (°C) 5	25	35
	μg vitamin A per liver ± SEM		
Retinyl acetate			
1.52 mg	906 ± 44	910 ± 17	828 ± 32
0.73 mg	147 ± 10	162 ± 9	165 ± 6
β-Carotene			
1.03 mg	117 ± 10[c]	94 ± 4	58 ± 4[d]
	% molar biopotency relative to retinyl acetate		
Biopotency of β-carotene[e]	111	104	94

[a] From Smith and Borchers (*809*).

[b] Ten animals at each temperature were fed 1.03 mg β-carotene and five animals 0.73 mg and 1.52 mg retinyl acetate as a component of the diet.

[c] Significantly different from 25°C treatment, $p < 0.05$.

[d] Significantly different from 5° and 25°C treatments, $P < 0.01$.

[e] The regression lines from which the biopotency values were calculated are $Y_{5°C} = 0.961\ X - 554$, $Y_{25°C} = 0.947\ X - 530$, and $Y_{35°C} = 0.784\ X - 407$, where X = retinyl acetate intake (μg) and Y = vitamin A (μg) (expressed as retinol) per liver.

72°–110°F lost three times as much liver vitamin A as those remaining inside at temperatures of 75°–89°F. A second study, in which shade was provided for one group of cattle but not for the other, showed that at the same temperature the cattle exposed to solar radiation lost about 2½ times as much of their vitamin A liver stores as those provided with shade. Subsequent intraruminal injections of vitamin A in animals previously exposed to solar radiation resulted in a greater absorption of vitamin A than-unexposed animals. Attempting to determine the influence of location and season on vitamin A status, Ralston and Dyer (*723*) obtained blood and liver samples for vitamin A determination during the four seasons at nine different geographic locations in the state of Washington. They ascertained that plasma vitamin A levels were significantly higher in spring and summer than in fall or winter. Liver concentration of vitamin A was also related to the condition of the animal; however, plasma vitamin A and plasma or liver carotenoid levels were not. Higher plasma and liver vitamin A levels were probably related more to quantity and quality of forage than to ambient temperature. Smithcors (*811*), after reviewing studies of vitamin A with beef cattle, concluded that fattening cattle need supplemental vitamin A when under physiological stress, including hot weather and the use of feed additives for growth stimulation.

There is a seasonal influence not affected by temperature. The carotenoid content of the ration is the main factor affecting the level of these pigments in blood and milk. This is particularly apparent in countries where cows are transferred from low-carotenoid winter rations to spring pasture. Average carotene figures can vary from as low as 130 μg during the dry ration feeding period to as high as 1680 μg per 100 ml of plasma for cows on green pasture (*520*). There are thus clearly defined seasonal trends, in most countries, in the carotenoid content of cows' blood associated with periods of dry rations and pasture feeding.

C. Intestinal Loss, Synthesis

Carotenoids, like vitamin A, suffer some destruction in the rumen. King and co-workers (*451*), by means of *in vitro* and *in vivo* studies, showed that about 40% of ingested vitamin A and carotene is destroyed in the reticulorumen of sheep. Of the carotene reaching the intestine the major part is converted to retinol, inappreciable amounts are absorbed unchanged, and some is excreted unchanged in the feces. Little work has been carried out on the mode of destruction of carotene in the intestinal tract. Hove (*412*) reported that clear stomach extracts, in the presence of methyl linoleate, oxidize carotene very much in the same way as soybean lipoxidase. Extracts from the small intestine were observed to be much less active. Dudley (*215*) noted soy protein inhibition of intestinal β-carotene absorption. In one trial (*713*) mature wethers with abomasal cannulas were fed isocaloric and isonitrogeneous diets, with alfalfa meal or carotene in stabilized beadlet form as the sole source of carotene. Preintestinal losses of 25–33% of the dietary intakes were noted. Mean losses were significantly greater with alfalfa than with the beadlets. Others have described losses from preintestinal destruction (*442,716*). Fernandez *et al.* (*270*) pursued the study of the stability of carotene in the rumen of ruminants by *in vitro* and *in vivo* trials and concluded that little carotene is normally lost (about 10%) during its passage through the preintestinal part of the alimentary tract.

McGillivray (*562*), at the University of New Zealand, presented evidence to indicate that there may be an appreciable synthesis of carotene in the cecum of sheep. In digestion trials designed to give an indication of the effect of composition of feed on the availability of carotene, it was noted that some sheep were excreting, via the feces, more carotene than they were consuming. Carotene-synthesizing bacteria, mainly gram-positive, nonmotile rods, polymorphic or cocci, were isolated by Pivnyak and Konyakhin (*707*) from the digestive tract of sheep. These were facultative anaerobes developing at an optimal temperature of 37°C. About half the bacterial cultures, when incubated in a carotene-free medium, formed 5–7 μg carotene per 100 ml medium, and the amounts were increased by adding kerosene. In India, Majumdar and Gupta (*530*), working with goats maintained for 2 years on a virtually carotene-free diet (70 μg/kg feed or about 3 μg carotene/kg body weight daily), observed that more carotene was excreted than was consumed and that excreted carotene was biologically active in rat tests. It was postulated that microbial synthesis took place in the cecum and ileum. In another trial with sheep, Tucker *et al.* (*876*) fed a ration supplying 202 μg of carotene daily to sheep (3 μg/kg body weight) and observed the excretion of 465 μg of a pigment that was chromatographically and spectrophotometrically very similar to β-carotene but biologically inactive in chick tests. Almendinger and Hinds (*18*) reported apparent carotenoid increases in the digestive tract of beef cattle. In careful and detailed investigations, they gave cattle hay and haylage made from the same cut and found that the fecal output of β-carotene was four times greater than the input when haylage was fed, whereas when hay was consumed the fecal output was slightly less than the input. Although the reported trials seemed to be well controlled and analytical concepts of the day applied, this potentiality warrants reexamination with current technology and modern instrumentation.

D. Vitamin A, Oxycarotenoids

Usually, the administration of massive amounts of vitamin A has been reported to lower (*206,210,422,548,751*) serum carotenoid values in poultry and cattle; liver, skin, and egg yolk carotenoid values in poultry; and milk fat carotene content (*87*) in cattle. It is still believed that vitamin A interferes with the absorption of carotene and other carotenoids from the intestinal tract. Since vitamin A and carotenoids are structurally related, this may indicate a competition between the two for absorption from the intestinal tract. Whether vitamin A (more soluble than carotene) at high levels exerts a direct or indirect effect, involving some physiological or biochemical mechanism, on the utilization of carotenoids cannot be explained from present data. It has been postulated that many of the effects of excessive vitamin A intake appear to result indirectly from vitamin A actions on the lipoprotein membranes of the cell (*201*). The effect of thiouracil and high vitamin A levels (*211*) on carotenoids appears to be somewhat similar. However, these data do not suggest a possible indirect effect of vitamin A on thyroid in depressing carotenoid utilization. It would still appear that vitamin A interferes mainly with the absorption of these pigments, possibly by its direct effect in or on the intestinal tract. Obviously, there may be other mechanisms by which fat-soluble vitamins influence the utilization of carotenoids. The data reported by Hendricks *et al.* (*378*) also suggest that the utilization of β-carotene is depressed when the diet contains high levels of vitamin D.

The presence of oxycarotenoids in the diet can influence the absorption of carotenes, but apparently the effect depends on the relative amount fed. It has been claimed that lutein or oxycarotenoids in general tend to interfere with the utilization of carotene (*447*), but Sherman (*789*) reported that the presence of lutein actually protects carotene from intestinal destruction. Others have not noted this protective action; in fact, more investigators have noted an opposite effect. High and Day (*390*) found that small doses of lutein increased, and large amounts of squalene and phytol decreased, the utilization of carotene. Kemmerer *et al.* (*447*) observed that the addition of oxycarotenoids to the ration decreased the utilization of carotene by

20%, as judged by liver stores. Added chlorophyll had the same effect. Kelley and Day (*445*) found that large amounts of xanthophylls administered with β-carotene decreased carotene utilization, as measured by tissue storage of vitamin A.

E. Vitamin E

Much work has been carried out on the practical aspects of protecting carotenoids in the intestinal tract. Moore (*606*) made the original observation that α-tocopherol (vitamin E) when administered with vitamin A or carotene to rats results in a greater storage of vitamin A in the liver owing to improved stability of the vitamin A and carotene in the gut. This and subsequent observations by Davies and Moore (*181*) confirm that ingestion of vitamin E increases the storage of vitamin A in the liver of the rat, increases the time taken to deplete a rat of vitamin A, and delays the onset of dental symptoms attributable to vitamin A deficiency. A series of investigations by many researchers, stimulated by the original Moore observations, has repeatedly confirmed his early work under various experimental conditions (*21,313,348,369, 387,387a,467,511*). The effect of added tocopherols on the utilization of vitamin A was tested by Hickman *et al.* (*387,387a*) over a wide logarithmic range covering all practical dosages. From these and other results it can be deduced that vitamin E exerts a sparing action on vitamin A and to a greater extent on β-carotene at certain dietary intakes. β-Carotene fed with an optimal vitamin E intake becomes a more efficient source of vitamin A for the rat. It was found that the apparent potency of a 0.1-μg dose of carotene was multiplied by 2–2.5, whereas the effect of a 4-μg dose was increased by a factor of 1.25–1.5, and the response of rats to 10 μg was not improved by added tocopherol. The apparent response to 1 μg of vitamin A was increased 1.5–2.0 times by the optimal dose of tocopherol, but the response produced by 30 μg was not improved by the administration of tocopherol. The influence of various levels of tocopherol intake on growth induced by carotene was nicely demonstrated graphically by Harris *et al.* (*369*) (Fig. 12). Hebert and Morgan (*376*), at certain levels of carotene feeding, observed an improved response from tocopherol (0.5 mg daily).

Quackenbush and associates (*719*), from their experiments on the activity of very small quantities of vitamin E, believe that, in promoting a greater biological response to carotene, tocopherol functions as an antioxidant in the gastrointestinal tract rather than regulates some phase of metabolism in the tissues. Dubouloz *et al.* (*213*) are also of the opinion that tocopherols act as antioxidants in the gastrointestinal tract, protecting vitamin A from destruction. Sherman (*790*) suggested that the enhanced utilization of vitamin A in the presence of tocopherols can be ascribed to the prevention of a physiological antagonism between carotene and unsaturated fatty acids. Hickman and Harris (*386*) concluded that the action of tocopherols in augmenting the biological activity of both carotene and vitamin A may be the result of a positive action in the gastrointestinal tract. The results of these workers suggested that tocopherols act synergistically with carotene and vitamin A and that the intes-

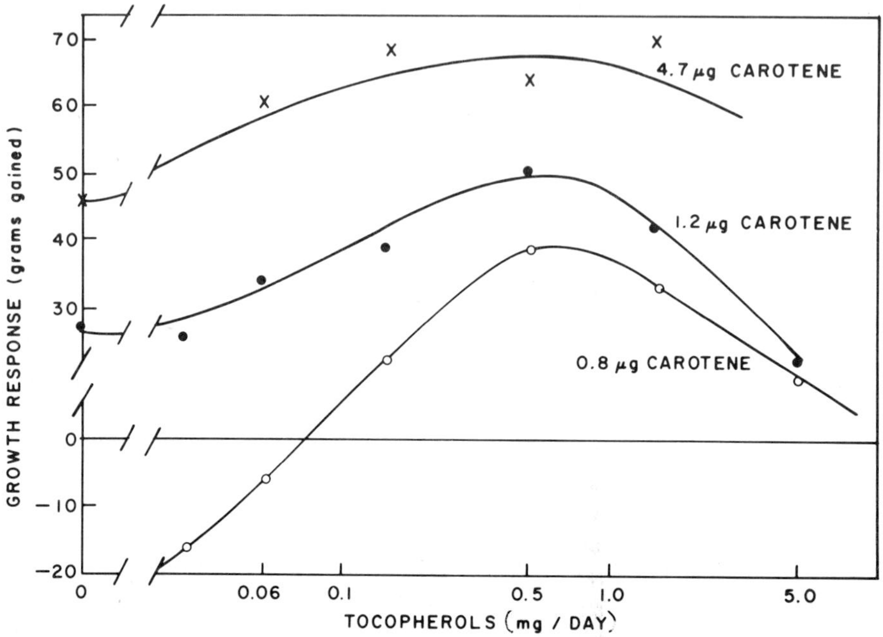

Fig. 12. Influence of tocopherol intake on the growth response induced in vitamin A-depleted rats by carotene. [From Harris *et al.* (*369*).]

tine is probably the site of action of the tocopherols, inasmuch as the effect was greatly diminished when the supplements of vitamins A and E were given on alternate days, as compared with the results when they were given together. Baumann and co-workers (*430,836*) reported that large doses of α-tocopherol actually may decrease the amount of vitamin A stored in the liver. Guerrant (*345*) failed to get a response from α-tocopherol (3 mg). According to postulates of Moore (*605*), high dose levels of tocopherol inhibit the oxidation phase of retinol formation and, at lower levels, prevent the complete oxidative destruction of carotene. Hence, conditions, timing, and quantitative relationships are critical factors under which α-tocopherol best augments the vitamin A activity of the ingested carotenes.

The observation of Moore (*606*), verified often in the literature, was extended by studies of Ames and associates (*21,22,25*), which indicated that vitamin E adequacy in the ration is a prerequisite not only for improved intestinal absorption of vitamin A but also for *in vivo* utilization of vitamin A. Rao (*724a*) found that the difference in growth response produced by carotene dissolved in various vegetable oils could be reduced by equalizing the tocopherol content of the oils, in agreement with the findings of Guggenheim (*348*) that the vitamin A efficacy of the carotenes in different vegetable fat sources was related to their tocopherol content as well as to their melting point, indicated by Brown and Bloor (*119*). The last few reports are of interest since it was Mattill (*547*) in 1927 who first showed that materials containing vitamin E delayed the *in vitro* oxidative destruction of carotene and vitamin A in fatty food mixtures before tocopherols were known to have antioxidant properties. Ascorbic acid (vitamin C) and derivatives such as vitamin E also have an effect on utilization of carotene. According to Hickman (*385*) and associates, when rats were fed 0.8 μg/day of carotene, the following increases in body weight occurred over a period of 36 days as a result of the addition of the following supplements: without vitamin E, 2 gm; with 0.15 mg mixed tocopherols, 31 gm; with an equal amount of pure α-tocopherol, 27 gm; with pure β-tocopherol, 30 gm; with pure γ-tocopherol, 30 gm; with 0.1 mg pure ascorbic acid (vitamin C), 7.5 gm; with 1 mg ascorbic acid, 11 gm; with 10 mg ascorbic acid, 20 gm; and with 0.5 mg palmitoylascorbic acid, 30 gm.

F. Antioxidants

Naturally occurring and intentionally added antioxidants in carriers of carotene and vitamin A have received attention as factors influencing the conversion of carotenoid vitamin A precursors to vitamin A. These substances also are thought to retard destruction of the vitamin and provitamins within the digestive tract and possibly within body tissues, as well as in stored feeds before ingestion. These aspects merit review because they are of practical importance. Indeed, some of the variability in observations on vitamin A sources presumably resulted from different antioxidant influences in feed sources and rations, whether by design or happenstance. The amount of antioxidant or the ratio of antioxidant to carotenoid may have a modifying influence on the results obtained. Dietary antioxidants appear to have an important effect on carotene utilization and perhaps on the absorption of dietary provitamin A. It is not clear, however, whether the antioxidants contribute directly to the efficient absorption of the provitamins A, or whether they merely protect either the provitamins or the vitamin A from oxidative breakdown during ingestion, absorption, and transport to the tissues.

In 1952 it was observed by Kramke *et al.* (*482*) that, when a suboptimal level of vitamin A activity was furnished in the ration as alfalfa meal, chick growth was stimulated by the addition of either crude soybean lecithin, defatted lecithin, or crude soybean oil. Slanetz and Scharf (*803*) reported that the addition of lecithin (crude soybeans, phosphatides) greatly increased the accumulation of vitamin A in the liver of rats fed carotene (50 IU) daily. Because lecithin has both antioxidant and emulsifing properties, it is not clear whether both properties or only one are functioning. The effect may have been due to the tocopherols contained in the crude lecithin product. A subsequent study in which purified lecithin or phosphatides was employed, Jensen *et al.* (*426*) indicated that the enhancing property was due principally to the tocopherols naturally present in the crude product. Subsequently, Potter *et al.* (*714*) showed that DPPD increased the vitamin A and carotenoid content of the blood plasma and liver of chicks fed either a purified or a practical-type ration. Within a year Monson *et al.* (*596*) confirmed earlier reports that the inclusion of

DPPD (0.025%) in the ration increased liver storage of vitamin A in chickens fed rations containing vitamin A in oil or graded levels of alfalfa meal. In 1955, DPPD addition to the ration of chickens was reported by Matterson *et al.* (*545*) to improve the utilization of carotene and vitamin A. Another antioxidant, 2,6-di-*tert*-butyl-14-methylphenol (DBMP), was found to be somewhat less effective than DPPD. In a previous report by these workers (*546*) favorable action by antioxidants such as by DPPD on increasing vitamin A concentration in the liver was observed.

Studies with antioxidants were also carried out on ruminants. Teichman *et al.* (*845*) found that the addition of DPPD (0.01%) to rations increased the utilization of carotene in dairy cows. These studies were extended by Rousseau *et al.* (*746*) in a publication dealing with the influence of some antioxidants on the utilization of carotene by Holstein calves. Three antioxidants, DPPD, ethoxyquin, and 2,5-di-*tert*-butylhydroquinone, as separate ration additives resulted in increases in the utilization of carotene from alfalfa as judged by higher plasma carotenoid levels and longer vitamin A depletion times. The main location of activity of these antioxidants was postulated to be in the digestive tract of the calves, with limited evidence of activity within the body proper. A slight sparing action of these antioxidants on the total tocopherol concentration in the liver was also noted. The favorable effect of DPPD (250 ppm) in the ration of the dairy calf in improving the utilization of carotene and vitamin A was further confirmed by Pirchner *et al.* (*703*). High *et al.* (*388,391,392,394*), over a period of several years, studied added antioxidants and observed that low dietary levels of an effective antioxidant increased carotene utilization in the laboratory rat. High intakes (10 mg) of such antioxidants as DPPD, DBMP, and tocopherol (vitamin E) were also observed for their influence on the metabolism of carotene (34 μg) and vitamin A (13 μg) in the rat. The DPPD, which is relatively insoluble, had little or no enhancing effect, whereas DBMP, at the level fed, decreased vitamin A deposition. Feeding of a large amount (*388*) of antioxidant decreased the deposition of vitamin A when carotene was fed but not when vitamin A was fed (Table 23). The authors suggested that antioxidants in large amounts may suppress oxidative steps necessary for conversion of carotene to vitamin A.

G. Protein

Vitamin A and protein nutrition have been closely correlated both clinically and biochemically. Qualitative or quantitative protein deficiency appears to impair the utilization of carotene. Berger *et al.* (*67*) discovered that feeding a protein-free diet decreased but did not prevent carotene conversion to vitamin A. The type and adequacy of dietary protein are considered to affect the biological value of carotene. Fraps (*279*) found that the substitution of casein for corn meal in the ration of albino rats increased the apparent digestibility of carotene and storage of vitamin A in the liver. James and El Gindi (*424*) were among the early investigators to study the effect of the protein constituents of the diet on the utilization of carotene by rats. They showed that, with equal carotene ingestion, the liver stores of vitamin A were

TABLE 23

Influence of 2,6-Di-*tert*-butyl-4-methylphenol on Carotene and Vitamin A Metabolism[a]

Series	Supplement	No. of rats[b] and sex	Average weight gain (gm)	Vitamin A present (μg) Liver	Kidneys	Total
1	34 μg carotene	10 (5 males)	57 ± 6	85.9 ± 8.1	2.5	88.4 ± 8.1[c]
	34 μg carotene + 10 mg DBMP	7 (5 males)	67 ± 6	25.8 ± 5.0	7.2	33.0 ± 5.2
2	13 μg vitamin A	9 (4 males)	67 ± 6	45.0 ± 5.3	14.0	59.0 ± 7.0
	13 μg vitamin A + 10 mg DBMP	7 (3 males)	69 ± 6	45.0 ± 5.5	7.4	52.4 ± 6.3

[a] From High (*388*).
[b] Wistar strain.
[c] Standard error of the mean.

1690, 990, 940, and 230 IU when the rats consumed casein, lactalbumin, gluten, and zein in the diet, respectively. Dietary carotene was much more efficiently utilized by the rats when high-quality protein (casein) was the dietary protein source than when low-quality protein (zein or gluten) was used. Tarjan (*843*) confirmed these general findings. By studying the vitamin A content of blood and liver in animals fed simultaneously β-carotene and various levels of protein, a number of workers (*192,423,542*) concluded that either the absorption of β-carotene or its conversion to vitamin A is directly dependent on an adequate protein intake. For example, Arnich and Pederson (*31*) observed that, when carotene was fed, the amount of vitamin A stored in the liver increased with increasing levels of protein in the diet. After reviewing the literature, Campbell (*141*) suggested that protein depletion of an animal may result in reduced absorption of carotene and retarded mobilization of hepatic reserves of retinol. Although it has been known since 1953 that dietary intake of protein affects tissue storage of vitamin A when β-carotene is fed, the mechanism of this effect remained relatively obscure until the last decade. Deshmukh and Ganguly (*192*) showed that in rats fed β-carotene the vitamin A content of the intestinal wall varied with dietary protein; it therefore seems most likely that the conversion process is affected by the dietary protein level.

Olsen *et al.* (*652*) observed with poultry that, when the protein level was changed from high to low with the same energy level, an inverse relationship occurred between protein level of the diet and carotene and vitamin A storage. Friend and co-workers (*284*), studying pigs on a diet low in protein, observed the stores of vitamin A in the liver to be low; this occurrence they believed, as had previous workers, to be due to impaired conversion of carotenoids to retinol and to reduced subsequent intestinal absorption of the formed vitamin. Sheep grazing on poor pasture may be receiving a diet not only low in carotene but also low in protein. Anderson *et al.* (*27*) gave ewes vitamin A-deficient diets containing either 6 or 10% protein. The low-protein diet caused loss of body weight and decreases in serum albumin, albumin to globulin ratio, and plasma retinol. However, the loss of liver retinol was less with the low-protein diet, indicating that protein is concerned with the mobilization of liver stores or with the transport of vitamin A in plasma. In lambs previously adequate in vitamin A and given urea-containing rations plus carotene or a vitamin A source, there was no adverse effect on vitamin A status or carotene utilization (*217*).

Vitamin A deficiency affects the serum protein patterns in large ruminants. Erwin *et al.* (*249*) demonstrated with vitamin A-deficient steers that serum albumin was increased, α-globulin reduced, and β-globulin reduced when oral dosages of carotene (15 mg/100 lb body weight) were given for 10 days. Liver vitamin A stores dropped in the vitamin A-deficient steers but rose in the normal steers, suggesting an impaired conversion of carotene in vitamin A deficiency. Further work by Erwin *et al.* (*252*) using identical twin heifers supported these results. When one twin was subjected to vitamin A deficiency, serum albumin was progressively lowered. Intravenous injections of carotene were ineffective in raising the serum albumin content. Serum albumin did return to normal in the deficient co-twin within 10 days following an intravenous injection of vitamin A. The apparent inability of the bovine under these conditions to convert metabolic carotene to vitamin A is in agreement with earlier findings of Church *et al.* (*160*) and Kon *et al.* (*472*).

Since Goodman *et al.* (*319*) and Olson and Hayaishi (*658*) have developed workable assays for the enzyme (carotene dioxygenase) that converts β-carotene to retinal, it is possible to test directly the hypothesis that dietary protein influences this carotene to vitamin A enzymatic process. In these studies dietary protein was necessary for the conversion of carotene to retinol at both steps: the conversion to retinaldehyde (*341*) and the conversion of retinaldehyde to retinol (*191*). Olson (*653*) pointed out that the synthesis of a number of key proteins in the utilization of provitamin A carotenoids and in the absorption and transport of vitamin A is depressed by protein deficiency. They include β-carotene 15,15′-dioxygenase, pancreatic retinyl ester hydrolase, chylomicron protein, and the retinol-binding protein involved in blood transport. These findings are of particular significance to human nutritionists as well as animal nutritionists since protein–calorie malnutrition and vitamin A deficiency are considered to be the two common nutritional deficiency diseases of man along with iron and iodine deficiency. More recently Kamath and Arnich (*437*), utilizing ^{14}C-labeled β-carotene in a study on the influence of dietary protein on intestinal biosynthesis of retinol, found, with rats, that dietary protein enhances carotene conversion in diets up to at least 40% protein (casein).

H. Choline, Betaine, Methionine

Few data exist in the literature on the influence of added choline, betaine, and methionine in rations relative to their influence on carotene conversion or utilization of vitamin A. Thorbjarnarson and Drummond (*864*) reported that the addition of choline to a choline-deficient diet fed to rats brought about a decrease in fat and vitamin A content of the livers. Popper and Chinn (*709*) observed in an experiment with rats that liver stores of vitamin A were dependent on the intake of choline. A deficiency of choline produced fatty livers that were low in vitamin A. In 1946 Bentley and Morgan (*66*) investigated the effect of choline on liver storage of vitamin A by rats. Storage by depleted rats fed vitamin A with a high-fat, low-protein diet was not affected by choline deficiency. With a low-fat, low-protein basal diet, the addition of choline increased the liver vitamin A. When carotene was fed with the high-fat diet, the vitamin A deposits were very small, and somewhat less fatty livers than normal livers were observed. With the low-fat diet, carotene caused better vitamin A deposition, which was little affected by choline. It was noted by Esh and Sutton (*254*) that the feeding of lecithin plus vitamin A or a carotene supplement stimulated greater weight gains and an increased liver storage of vitamin A in laboratory rats than did vitamin A or carotene fed alone. Clayton and Baumann (*163*) observed that even during choline deficiency vitamin A was distributed normally in the liver of rats.

Camp *et al.* (*139*) found that choline supplements enhanced the utilization by chickens of carotene from alfalfa, but Kramke *et al.* (*482*) found no effect. The latter group did observe enhanced utilization with supplements of soybean lecithin (crude or defatted) and of soybean oil. In 1958 Edwards *et al.* (*230*) added choline (0.25%) and betaine (0.25%) separately to chicken rations containing variable levels of alfalfa meal. The basal ration was judged to be marginal or deficient in choline content. No increase in the utilization of carotene from alfalfa meal was obtained by supplementing the ration with choline or betaine, as judged by chick growth responses.

Davis *et al.* (*184*) conducted a study on the effect of three lipotrophic agents, lecithin, choline, and methionine, on the absorption and utilization of vitamin A and carotene by young dairy calves. Fifty-four 2-day-old male Jersey calves were used in the study. The feeding of these three lipotrophic agents plus vitamin A did not materially increase the plasma levels or liver stores of vitamin A above those of a group receiving vitamin A alone. However, the addition of either choline (2 gm daily) or methionine (3 gm daily) plus carotene to the ration of calves caused a significant increase in the absorption of carotene, as measured by plasma levels of carotene. All treated groups showed significantly higher plasma carotene levels than did the untreated controls. None of the lipotrophic agents had any significant effect on the liver stores of carotene when carotene was fed.

I. Fat

Most of the early data on carotene absorption indicate that carotene must be dissolved in fat or oil in order to be utilized effectively. When provitamin A is dissolved in fat or occurs naturally in a fat carrier, utilization follows, whereas, in the absence of fat, carotene may not be absorbed from the intestine. Rats on a low-fat diet were shown by Ahmad (*10*) to be unable to absorb much carotene, although almost complete utilization of the provitamin was obtained when 10% of fat was incorporated in the diet. Burns *et al.* (*132*) found that the fat content of the diet influenced the utilization of vitamin A and β-carotene by the rat. Both were more effective when fed with a diet containing 5% lard than with a diet that contained no added fat. Animals were all supplemented with 0.1 gm of refined corn oil daily. Shaw and Deuel (*785*) observed a correlation between the quantities of fat and carotene absorbed by rats. The addition of fat to the ration of chickens also influenced the absorption of carotene but had no effect on the absorption of vitamin A, according to Russell *et al.* (*757*). Hens absorbed less carotene when consuming a 0.1% fat ration than when fed a 4% fat ration. These general conclusions have been confirmed, with both animal and human subjects, by a number of investigators (*45,861,884*). According to the work of McGillivray (*563*), it was shown that carotene in arachis oil was relatively more effective in maintaining the vitamin A potency of milk fat than summer pastures.

In various ways the nature of the fat may affect the absorption of the provitamins.

The type of fat influences the biological activity of β-carotene. Dyer *et al.* (*221*) found that β-carotene was five to six times more effective for rats when dissolved in arachis oil than when dissolved in hardened cottonseed oil or ethyl laurate. Other findings were observed in vegetable oils by workers (*566*), including Lease *et al.* (*509*) and Kraybill and Shrewsbury (*483*). Sherman (*791*) found that soybean oil was a better solvent for carotene than were cottonseed oil, linseed oil, and corn (maize) oil, and the worst were butterfat and coconut oil. On the other hand, Aldersberg and associations (*6*) found butter to be a better vehicle for promoting carotene absorption than cottonseed oil. The inclusion in diets of oils and fats that are poorly digested, such as paraffin oil or beeswax, lowers utilization (*566*) of carotene. An interesting study was made by Brown and Bloor (*119*) on the influence of the chain length of dietary fatty acids on the absorption of carotenoids from carrots by the rat. Various fatty acid fractions of butter were fed to rats; these fatty acids represented 10% of the total dietary intake of the animals. The amount of carotene absorbed from raw carrots was highest when low molecular weight fatty acids were given, and the percentage of absorbed carotenoids fell as the chain length of the fatty acids increased. McGillivray (*560*) has made some observations and speculations on fats. Fats found in pasture grasses are highly unsaturated, but recent work has shown that a considerable amount of hydrogenation may occur during passage through the rumen. This modification of pasture fats may have a profound influence on carotene absorption, since the degree of unsaturation of the oil vehicle has a marked effect on the absorption of carotene at high intake levels. Furthermore, although this aspect does not appear to have been investigated, it seems possible that carotenoids themselves might be susceptible to hydrogenation in the rumen, with a resultant loss of provitamin activity.

Ability to assist in the preservation of carotenes by tocopherols and other antioxidants is possibly an important qualification of an effective dietary fat. An antagonism between unsaturated fatty acids and vitamin E affects carotene metabolism. If dietary fats are not stabilized and rancidity develops, destruction of carotene and retinol may occur, which tends to give rise to deficiency signs in animals, as reported by Matsushima and Dowe (*544*) after they had fed a fat-supplemented ration to yearling steers. Siedler *et al.* (*796*) observed that antioxidants (BHA, BHA plus BHT, DPPD, and ethoxyquin) added to animal fats in animal feeds were effective in increasing the carotene stability over that observed with nonstabilized fats. Another factor is the influence of emulsifying the fat. For example, Deuel *et al.* (*195*) reported that feeding β-carotene in margarine enhanced its activity for rat growth by as much as 30%. This enhanced effect of margarine was examined by Marusich *et al.* (*537*) and could not be confirmed. Differences in the findings may be due to variations among laboratories in dosing schedules, tocopherol content of margarine fed, etc.

Randoin *et al.* (*724*) reported that rats on a fat-free diet can utilize β-carotene as well as rats on a diet containing 10% of peanut oil. A greater efficiency was found in the utilization of carotene when it was mixed in the diet than when it was given as a supplement in an oil solution. This result may mean that some dietary component has a protective action on carotene in the gastrointestinal tract. Also, carotene can be used in the absence of fat, either in aqueous colloidal dispersions or as dispersions prepared with surface-active agents (*472*). Particle size is very important; the smaller the particle, the better the utilization. Thompson (*858*) found that if the particles were less than 1 μm the provitamin was very effectively utilized. Molander (*595*) noted that carotene dissolved in corn oil could be demonstrated in human serum when the particle size was 0.5 μm.

J. Emulsifiers

In contrast to vitamin A, which is well absorbed by animals, carotene is poorly absorbed, and hence most of it can be excreted in the feces. Bile secretion is essential for carotene to be biologically effective. According to the work of Drummond and McWalter (*208*), one explanation of the difficulty in the absorption of β-carotene and other carotenoids may be their low solubility. β-Carotene or fat-soluble vitamins are chiefly absorbed from the small intestine. According to Shaw and Deuel (*785*) the rate of absorption is a function of the surface area of the animal. Bile action is required for the normal utilization of β-carotene consumed in a fat carrier. Thus, in hepatic dysfunction, little or no absorption of carotene takes place (*332*). Greaves and Schmidt (*332*) and Irvin *et al.*, (*420*) demonstrated that bile salts, glycodeoxycholic and deoxycholic acids, could affect the absorption of carotene in animals deprived of natural bile flow or hepatic function. In isolated studies of intestinal loops of dogs reported by Irvin *et al.* (*420*), insignificant amounts of the provitamin were absorbed when carotene was introduced into the intestinal loops in cottonseed oil solution. When either pancreatic lipase or a combination with bile salts was given concomitantly, considerable amounts of carotene were absorbed. The favorable action of bile salts is related to their action in bringing about emulsification of the fat and thus assisting in the digestion and absorption of it and the provitamin contained therein.

Emulsifying agents other than bile salts improve the absorption of carotene (*233*). This was observed some time ago by Adlersberg and Sobotka (*7*) and Adlersberg and associates (*6*). When lecithin was added to cottonseed oil, the amount of carotene absorbed from this modified carrier was greater. Esh and co-workers (*255*) showed that the same improvement in utilization of vitamin A and carotene occurs in the rat and particularly when soybean lecithin is fed. Similar results with lecithin were obtained by Slanetz and Scharf (*804*); also, favorable data were collected by Patrick and Morgan (*679*) with chickens. Polyoxyethylene sorbitan monopalmitate or monooleate has been used for producing water-dispersible solutions of carotene for oral and parenteral injection. A number of workers (*131,861,922*) observed that carotene was as well or even better absorbed when administered in an aqueous colloidal dispersion solubilized by dispersing agents than when given in an oil solution. Jacobson and associates (*422*) studied the efficiency of various modes of administering carotene and vitamin A to dairy calves. Feeding by nipple of aqueous dispersions of the substance in milk resulted in particularly fast absorption. Vavich and co-workers (*889*) found that carotene was better utilized when fed as a homogenate in nonfat milk than when it was homogenized in water.

K. Paraffin, Mineral Oil

The presence of poorly absorbed hydrocarbons such as mineral oil has a deleterious effect on carotene utilization. An early investigator, Rowntree (*750*), reported that the minimal quantities of vitamin A required for the growth of rats were inadequate when mineral oil was present in the diet. Dutcher and associations (*218*) found that carotene absorption was depressed by this hydrocarbon to a far greater extent than was that of vitamin A. The observation that plasma carotene and liver vitamin A values are lower after the ingestion of carotene in the presence of mineral oil than when mineral oil is absent has been confirmed by a number of workers (*14,26,174,529,593,924*). The deleterious effect of mineral oil on carotene absorption also applies to man. Curtis and Kline (*174*) and Alexander *et al.* (*14*) showed that paraffin medicinally given to man markedly lowers the absorption of carotene using plasma carotenoid level as the response criterion. However, Mahle and Patton (*529*) reported that certain hydrophilic mucilloids, which are used as cathartics in place of mineral oils, do not decrease the utilization of carotene. Alumina gel, which is often administered over a protracted period in the treatment of peptic ulcer, is without a depressing action on carotene utilization (*407*). When mineral oil in the form of imitation mayonnaise replaced the usual vegetable oil product, studies showed that the use of this product decreased plasma carotene concentrations.

The depressing effect of paraffin on carotene may be of concern in ruminant nutrition when one is considering the use of liquid paraffin to control bloat in grazing cattle. Bloat is a major problem with ruminant animals. Reid and Johns (*732*) found fats and oils to act as antifoaming agents, and a current antibloat practice is to drench the animals, or spray the pastures, with suitable oils. Animal and vegetable oils have little or no effect on carotene absorption (*557*), but mineral oils of the heavy liquid paraffin type, which are equally effective in the control of bloat and considerably cheaper, are known to reduce carotene uptake. Burns and co-workers (*131,132*) noted that as little as 0.08% of mineral oil, sufficient to prevent dustiness of lucerne meal, had a deterious effect on the utilization of carotene. The dose of paraffin oil recommended (*556*) for bloat control (about 150 ml per cow daily) has been found to reduce carotene absorption by 40–50%. There is a corresponding decrease in intestinal vitamin A formation, and these effects of paraffin ingestion are reflected in marked reductions in the carotene and vitamin A ester levels in the blood and milk fat. If the duration of this practice is short, detrimental effects on the vitamin A nutrition of the animal may not become significant. Likewise, the single dose that man may take before retiring at night causes no significant interference (*826*).

L. Thyroxine, Thiouracil

In addition to the observations of von Euler and Klussman (*260*) of an antagonism between thyroxine and carotene affecting the growth of rats deficient in vitamin A, Kunde (*496*) in 1926 noted vitamin A deficiency in thyroidectomized rabbits that were fed carotene. Fellenberg and Greuter (*267*) and Fasold and Heidemann (*266*) claimed that carotene appeared in the milk of thyroidectomized goats, and Abelin (*1*) reported its metabolic influence on carotene and vitamin A. All these findings helped to develop a lively interest in the relationship between carotene and thyroxine. The thyroid gland is believed to have some role in the absorption of carotene. Johnson and Baumann (*432*) demonstrated unequivocally that less vitamin A accumulated in the livers and kidneys of rats rendered hypothyroid with thiouracil administration than in those of normal rats receiving identical amounts of carotene. Also, the livers of hyperthyroid rats contained more vitamin A than did those of the normal controls. These observations have been confirmed by some researchers but not by others. Cama and Goodwin (*138*) demonstrated that, in the case of the rat, the absorption of carotene is increased when thyroxine is given and is decreased after the thyroid antagonist, thiouracil, is administered. Also, it was shown that the fecal excretion of carotene was increased after thiouracil was given. Kowalweski *et al.* (*477*) injected carotene into the portal vein of thyroidectomized dogs and reported no conversion to vitamin A. The metabolism of carotene administered intravenously to rats was not influenced by changes in thyroid function, as shown by McGillivray *et al.* (*565*), thus supporting earlier findings of Cama and Goodwin (*138*). Barrick *et al.* (*40*) consider that very high doses of thiouracil inhibit the carotene conversion in feeder lambs, and this appears also to be true with mature sheep. Increased conversion of dietary carotene to vitamin A in the intestine was found by Chanda (*146*) to be responsible for the increase in blood vitamin A that occurs when chickens are given thyroxine.

Chanda and Owen (*148*) reported that thyroxine increased and thiouracil decreased the carotene and vitamin A content of cow's milk. Chanda and co-workers (*150,151*) observed an improved utilization of carotene in cows and goats after an injection of thyroxine, as well as a decreased utilization when thiouracil was given. This would indicate that the beneficial effect of thyroxine must be exerted on the absorption of carotene rather than on its subsequent metabolism. Normal goats absorb carotene more efficiently than do normal cows.

Schultze and Turner (*772*) demonstrated that the goat thyroid is more active than the cow thyroid. The conversion of the provitamin A in milk is increased by the administration of the thyroid hormone to lactating cows. If a functioning thyroid gland is necessary for carotene conversion, it is conceivable that the vitamin A insufficency caused by nitrates and nitrites present in the ration at troublesome levels may be an indirect result of an induced abnormal thyroid function. Bloomfield *et al.* (*92*) showed that dietary nitrate (0.5%) can adversely affect the iodine uptake of the thyroid gland of the rat.

Remington *et al.* (*733*) stated that eye symptoms of vitamin A deficiency are cured by the oral administration of carotene to thyroidectomized animals. DiBella (*199*) found carotene to be effective, but at a reduced efficiency. In vitamin A-deficient thyroidectomized albino rats, symptoms of avitaminosis A disappeared after administration of carotene; hence, the conversion of β-carotene to vitamin A is independent of the thyroid, according to studies by Boguth and Sari (*95*). Wiese and collaborators (*911*) were unable to demonstrate variations in carotene metabolism in rats rendered hypothyroid by the feeding of thiouracil. Morgan and Arnrich (*611*) are of the opinion that the normally functioning thyroid gland is not essential for carotene utilization in rats and dogs.

Viewing all studies, one must come to the conclusion that the thyroid gland is active in improving the availability of dietary carotene. The question to be answered is whether this effect occurs as a result of an increased absorption of carotene or whether it might be ascribed to a more efficient conversion of carotene to vitamin A. It would appear from the work of Chanda and co-workers (*150,151*) that subcutaneous injections of thyroxine increase and of thiouracil decrease the digestibility or absorbability of carotene. This was indicated by the fact that fecal carotene was slower to appear with thyroxine treatment than with thiouracil treatment. Thyroxine was also reported to increase the carotene and vitamin A content of milk fat in cows, and the effect was greater with oral administration (*146*). Observations that the amount of vitamin A ester in the blood was increased in cows fed carotene when they were treated with either thyroxine or a thyrotropin preparation can be interpreted as evidence that the extra vitamin A originates from the carotene in the intestine. Goodwin (*323*) believed that the beneficial effect of the thyroid hormone in reducing the loss of carotene in the feces is due to better absorption of carotene. If the effect of the thyroid hormone were on the dioxgenase in the gastrointestinal tract, one might assume that an accelerated removal of carotene from the tract could be due to conversion to vitamin A. Goodwin concluded that, although the bulk of evidence suggests that the thyroid functions in carotene metabolism by increasing the absorption of the provitamin, one cannot completely rule out the possibility that the thyroid hormone has some influence on dioxygenase activity. At least, all the evidence is against improved stability in the lumen of the intestines as an explanation.

M. Nitrites, Nitrates

Several investigators have reported that carotene and vitamin A can be destroyed *in vitro* by nitrite but not by nitrate (*186,660,716,728,739*). The *in vitro* reaction is dependent on pH, and certain compounds related to carotene are less severely attacked than β-carotene itself. Bloomfield *et al.* (*91*), by their rat and sheep trials, showed that dietary nitrate can adversely influence the normal functioning of the thyroid gland. This was also observed by Wyngaarden *et al.* (*938*) and was regarded as traceable to an impaired functioning of the gland, the correction of which would require supplemental iodine. Smith *et al.* (*807*) and Holst *et al.*, (*409*) reported that the feeding of rations with a high nitrate content has an adverse effect on the vitamin A status of the animal. Microbial reduction, principally in the ruminant, can convert nitrates to nitrites.

Phillips (*700*) conducted investigations in an attempt to answer the controversial question of whether nitrite acts on vitamin A metabolism before or after the vitamin is absorbed from the diet. Female rats were started at 21 days of age on a semipurified ration free of vitamin A and supplemented with different forms of vitamin A. In these experiments the feeding of 1% potassium nitrite concomitantly with either carotene or vitamin A decreased the amount of vitamin A in the liver, but this did not result from the inability of the liver to store vitamin A since increased concentrations of vitamin A in the liver are observed after intraperitoneal injection of the vitamin when nitrite is fed. Phillips concluded in the rat that reduction of liver storage of vitamin A from carotene or vitamin A is the direct effect of oxidation of the vitamin or provitamin by nitrite. It does not involve interference with absorption, enzymatic processes associated with conversion of carotene, or liver function. Previously, Emerick and Olson (*243*) had done similar rat experiments but fed orally either 3% sodium nitrate or 0.5% sodium nitrite. Liver storage of vitamin A after oral administration of β-carotene was lowered, more due to nitrite than to nitrate feeding. Experimenting with rats, O'Dell *et al.* (*643*) found that a level of 0.3% potassium nitrite in the diet caused a rapid depletion of vitamin A body stores. Smith *et al.* (*807*), also experimenting with rats, demonstrated that chronic nitrite toxicity could be overcome with vitamin A or carotene supplementation. Equivalent doses of carotene were not as effective as vitamin A in alleviating this condition. Greathouse *et al.* (*331*), studying the disappearance of carotene from ligated intestines of rats fed nitrate, found that, with nitrate feeding, carotene disappeared more rapidly from the intestine.

Dietary nitrite depressed vitamin A storage in chickens whether vitamin or carotene was fed but less so when vitamin A was injected, according to studies (Table 24) by Sell and Roberts (*777*). An equivalent quantity of vitamin A administered in the ration failed to increase liver vitamin A to the same extent as that injected. This would suggest an increase in the destruction of oral vitamin A (or carotene) before absorption due to nitrite or to interference by nitrite in the absorption of vitamin A from the digestive tract or both. Similar observations on the carotene-depressing effect of nitrite were made by Brüggemann and Tiews (*123*), Adams *et al.* (*3*), and Lazar *et al.* (*508*). In poultry studies by Marrett and Sunde (*533*) vitamin A always overcame the toxicity of nitrate or nitrite. When marginal levels of vitamin A were used and high levels of nitrate or nitrite were fed, mortality increased, growth was reduced, and rapid breathing developed. Nitrite was more toxic than nitrate.

Wood *et al.* (*928*) fed nitrate (0.3%) or nitrite (0.08%) in drinking water to growing pigs. The diet contained either 3520 IU from β-carotene or 1173 IU from vitamin A palmitate per kilogram. In pigs given β-carotene, nitrate significantly

TABLE 24

Influence of Nitrite on Weight Gain, Liver Vitamin A and Carotene, and Thyroid Gland Weight of Chicks[a,b]

Treatment[c]	Weight gain[d] (gm)	Liver vitamin A (μg/gm fresh wt)	Liver carotene (μg/gm fresh wt)	Thyroid gland weight (mg/100 gm body wt)	Mortality (%)	Vitamin A value (IU/kg feed)
1	72 ± 3	0.88 ± 0.27	1.05 ± 0.19	7.67 ± 0.65	7	0
1 + N	51 ± 3	0.19 ± 0.14	1.14 ± 0.07	16.97 ± 2.62	33	0
2_0	88 ± 2	24.13 ± 7.20	1.03 ± 0.07	6.66 ± 0.89	0	4,080
2_0 + N	86 ± 2	1.26 ± 0.34	1.14 ± 0.15	13.59 ± 2.83	7	4,080
3_0	94 ± 6	87.25 ± 4.67	1.16 ± 0.21	5.67 ± 0.88	0	16,320
3_0 + N	80 ± 3	15.87 ± 5.98	1.05 ± 0.50	10.21 ± 0.78	16	16,320
2_i	91 ± 1	35.39 ± 7.70	1.47 ± 0.10	8.13 ± 0.59	9	4,080
2_i + N	85 ± 2	8.20 ± 1.78	1.27 ± 0.13	12.98 ± 1.20	2	4,080
3_i	85 ± 4	69.80 ± 4.27	1.54 ± 0.14	6.74 ± 0.91	11	16,320
3_i + N	86 ± 1	55.58 ± 5.78	1.16 ± 0.10	10.92 ± 0.92	2	16,320
2_c	92 ± 3	4.55 ± 2.03	1.41 ± 0.17	5.93 ± 1.03	4	4,080
2_c + N	79 ± 6	1.35 ± 1.27	1.05 ± 0.27	12.31 ± 2.55	13	4,080
3_c	92 ± 6	36.66 ± 4.31	3.05 ± 0.25	6.32 ± 1.08	4	16,320
3_c + N	85 ± 6	2.55 ± 0.49	1.04 ± 0.15	10.01 ± 2.55	2	16,320

[a] From Sell and Roberts (777).
[b] Data represent means ± SE.
[c] The subscript to each treatment number denotes the following: (0) vitamin A fed, (i) vitamin A injected, and (c) carotene fed as the vitamin A source. Dietary nitrite (N) as potassium nitrite, 0.4% ration.
[d] Weight gain per cockerel from 1 to 4 weeks of age.

decreased liver vitamin A levels, but in pigs given vitamin A the decrease was not as great. Nitrite significantly decreases liver vitamin A levels in pigs given either β-carotene or vitamin A. Hutagalung *et al.* (*417*) determined the effects of adding potassium nitrate or nitrite to the diets of pigs on the utilization of dietary carotene. Feeding of 3% nitrate significantly ($p < 0.01$) depressed gains. A reduction in average daily feed consumption was observed only in pigs receiving 3% nitrate supplemented with β-carotene. No adverse effect of the nitrate treatments was observed on liver vitamin A stores, serum vitamin A values, or methemoglobin levels. Serum vitamin A values were significantly ($p < 0.01$) reduced when 0.3% nitrite was included in the diet, regardless of the source of supplemental vitamin A. Potassium nitrite in the diet gave measurable increases in methemoglobin at the higher levels of nitrite. A definite trend toward a decrease in liver vitamin A stores (*417*) was observed when the nitrite level in the diet increased (Table 25).

Sheep and cattle losses due to nitrate poisoning were reported in the midwestern and western United States decades ago (*110,908*). The high nitrate content in certain forages has been known for some time to be toxic to ruminants and to exert an adverse effect on carotene and vitamin A metabolism. A toxic forage condition observed in Missouri during a drought was revealed by Muhrer *et al.* (*617*) to be nitrite toxicity. Nitrate content in forage exceeding 0.5% often leads to difficulty. In one feeding trial with nitrogen-fertilized forages (*172*), five uniform groups of eight steer calves each were randomly assigned to the following treatments: (1) alfalfa hay, (2) low-N coastal Bermuda grass hay, (3) high-N hay, (4) high-N hay plus

TABLE 25

Effect of Nitrite in Feed on Pig Performance, Serum Vitamin A, and Liver Vitamin A[a,b]

	Level of NO_2 (%)				
	0	0.075	0.150	0.300	0.300[c]
Daily gain (kg)	0.68	0.65	0.63	0.55**	0.51**
Feed/gain	3.90	3.77	3.73	3.65	3.90
Serum vitamin A (μg/100 ml)[d]					
0 days	28.7 ± 2.6	31.2 ± 6.6	30.9 ± 9.0	38.8 ± 9.7	33.3 ± 3.4
42 days	17.9 ± 1.7	16.0 ± 6.4	15.1 ± 4.0	12.8 ± 1.9	15.1 ± 4.3
87 days	20.3 ± 2.4	19.7 ± 5.9	17.3 ± 2.4	16.0 ± 2.4	11.7 ± 0.4*
Liver vitamin A (μg/gm, 87 days)[d]	21.1 ± 5.9	17.2 ± 7.4	11.9 ± 5.3	13.3 ± 5.5	9.5 ± 5.4

[a] Eight pigs per treatment. Average initial weight was 26.1 kg. From Hutagalung *et al.* (*417*).

[b] Single asterisk denotes significantly ($p < 0.05$) different from control; double asterisk, significantly ($p < 0.01$) different from control.

[c] Source of supplemental vitamin A activity for this group was vitamin A palmitate; source of supplemental vitamin A activity for the other groups was β-carotene.

[d] Plus or minus standard deviation.

TABLE 26

Effect of Forage Species, Level of Nitrogen Fertilization, and Carotene and Vitamin A Supplementation on Blood and Liver Carotene and Vitamin A Levels in Steers[a,b]

	Treatment[c]				
	Alfalfa hay	Low-N CBG hay	High-N CBG hay	High-N CBG hay +carotene	High-N CBG hay+ vitamin A
Blood carotene (μg/100 ml plasma)					
Initial	15.97	19.12	12.30	18.90	18.33
After 127 days on treatment	67.96*	73.02*	65.71*	94.65*	69.82*
At 11 days post-treatment[d]	30.00*	38.15*	29.08*	53.06*	34.75*
Blood vitamin A (μg/100 ml plasma)					
Initial	33.91	35.87	38.48	34.13	32.55
After 127 days on treatment	19.12*	19.36*	24.81*	19.92*	33.36
After 11 days post-treatment[d]	15.56*	16.90*	18.15*	18.36*	29.94
Liver carotene (μg/gm)					
Initial	0.99	1.11	1.22	1.25	1.36
At 11 days post-treatment[d]	3.32*	2.72*	2.94*	3.42*	2.90*
Liver vitamin A (μg/gm)					
Initial	16.03	22.39	20.73	17.20	19.93
At 11 days post-treatment[d]	8.55	12.88	7.20*	5.94*	21.98

[a] From Cullison and Ward (*172*).
[b] Asterisk denotes significantly ($p < 0.05$) different from initial level.
[c] CBG, coastal Bermuda grass.
[d] On a low-carotene diet for 11 days following treatment.

β-carotene, and (5) high-N hay plus vitamin A (50,000 IU per head daily). All groups received in addition a low-carotene concentrate mixture over a 127-day feeding period. Supplementary carotene administration (*172*) had no influence on blood or liver vitamin A, as contrasted with the positive influence of supplementary vitamin A (Table 26).

The main toxic effect of nitrate is produced by conversion to nitrites as a result of the reducing action of rumen microflora. South Dakota researchers (*241,660*) have clearly shown in laboratory trials that nitrate accelerates the destruction of carotene in corn silage during the fermentation phase (Fig. 13). This is perhaps an explanation for some of the vitamin A deficiencies reported in cattle fed corn silage. In the blood of nitrite-fed ruminants, nitrite reacts with hemoglobin, inducing methemoglobin, which in sufficient concentration causes death.

Studies with dietary nitrates and nitrites have also been conducted on sheep. From 0.1 to 0.75% sodium and potassium nitrite (equimolar) fed to sheep reduced the vitamin A liver stores, according to Holst *et al.* (*409*). After 56 days (*321*) sheep fed sodium nitrate (3%) in a corn–alfalfa ration had liver vitamin A concentration 46% lower than control animals. Similar findings were reported in subsequent trials

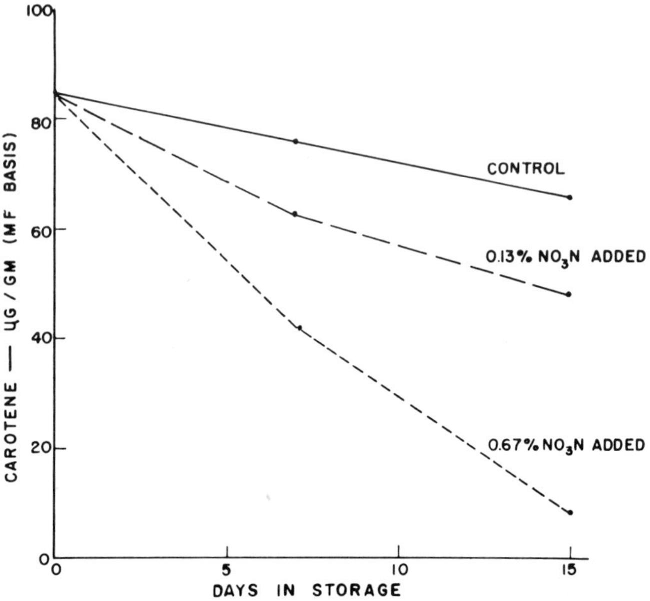

Fig. 13. Effect of nitrate on carotene stability in corn silage. [From Embry and Emerick (*241*).]

(*402*). On the other hand, the feeding of 4% potassium nitrate in rations to sheep by Cline *et al.* (*164*) had no effect on weight gain or vitamin A storage. The application of calcium ammonium nitrate to ryegrass–clover pasture during the grazing season did not appear to influence vitamin A liver storage in sheep (*647*).

Nitrates in feeds have been suggested as causes of vitamin A deficiency in cattle under feedlot conditions. One percent nitrate (KNO_3) in a ration for fattening cattle was reported by Hale *et al.* (*359*) to reduce the vitamin A content of the liver and by Smith *et al.* (*806*) and Weichental *et al.* (*903*) to reduce plasma vitamin A levels. Case (*143*) reported that the addition of carbohydrate and vitamin A helped reduce the toxicity if the ration did not exceed 1.5% nitrate. McIlwain and Schipper (*569*), in trials with dairy calves, found nitrates or nitrites to interfere with normal carotene utilization. Keating *et al.* (*442*) showed vitamin A degradation by nitrate-containing rumen liquor. Dietary nitrate (1–2%) failed to accelerate depletion of vitamin A stores in a beef cattle fattening trial by Lichtenwalner *et al.* (*514*). In Hereford and Angus steer calves, high- and low-nitrogen fertilized coastal Bermuda grass did not show significant effects on blood or liver vitamin A levels. According to Petrova (*695*), nitrate, which accumulated in plants as a result of high-nitrogen fertilization or which was added during ensilage up to 5% of dry mass, did not have any deleterious effect on the carotene content of the silage. Addition of 0.1% sodium nitrite to silage sharply decreased the carotene, but 0.25–0.30% completely destroyed it. Although some investigators have shown that nitrate has little effect on vitamin A nutrition in cattle (*185,584,894*), the influence of nitrate and nitrite on the nutrition and health of ruminants remains controversial. Cattle rations containing carotene in excess of or meeting recommendations have failed to meet vitamin A needs, and one of the suspected reasons has been excessive nitrate levels. Dietary variables, such as roughage, iodine content, and energy level, have differed among studies, and these, too, may have influenced the results.

One variable not always appreciated (*660*) in the destruction of β-carotene by nitrite is degree of acidity. Nitrite solutions of low pH caused destruction of carotene and vitamin A in *in vitro* trials, confirming earlier reports by Wilson (*914*). As pH decreases (*660*), the rate of carotene destruction increases (Table 27). Furthermore, carotenoid vitamin A precursors (*716*) other than β-carotene are also sensitive to nitrite destruction (Fig. 14). Two carotenals are more stable in the presence of nitrite than is β-carotene, β-apo-12′-carotenal being more stable than β-apo-8′-carotenal. Vitamin A alcohol is more susceptible to degradation by nitrite than the other three substances. In feeds, vitamin A is usually present or added in the ester form.

In summary, dietary plant nitrates are converted in the rumen by microbial action to nitrites, which are toxic at high levels and probably interfere in some fashion with the stability or conversion of carotene to vitamin A in the intestinal wall. Absorbed nitrite ion deranges hemoglobin by oxidizing the iron atom from the ferrous to the ferric state, forming methemoglobin, which lowers the total oxygen-carrying capacity of the blood markedly. In addition, nitrites are involved in many oxidative processes, such as formation of nitrosoprotein compounds. The oxidative property of nitrite suggests that it may exert its effect through oxidation of vitamin A, carotene, or other oxidation-labile compounds. Reviews on nitrates and nitrites are available (*60,115,471,868,937*). In more recent years, the deleterious effects of nitrosamines formed by the reaction of nitrites with tertiary amines have become more significant causing some concern for the health of animals.

TABLE 27

Destruction of β-Carotene by Addition of Nitrite Solutions of Various pH[a]

pH of buffer	pH of mixture (final)	β-Carotene remaining, (% of control)[b]
7.50	7.65	99
6.05	6.25	95
5.40	5.55	65
5.00	5.15	27
4.55	4.75	0
4.05	4.30	0

[a] From Olson *et al.* (*660*).

[b] Percentage of amount remaining in tube to which no nitrite was added.

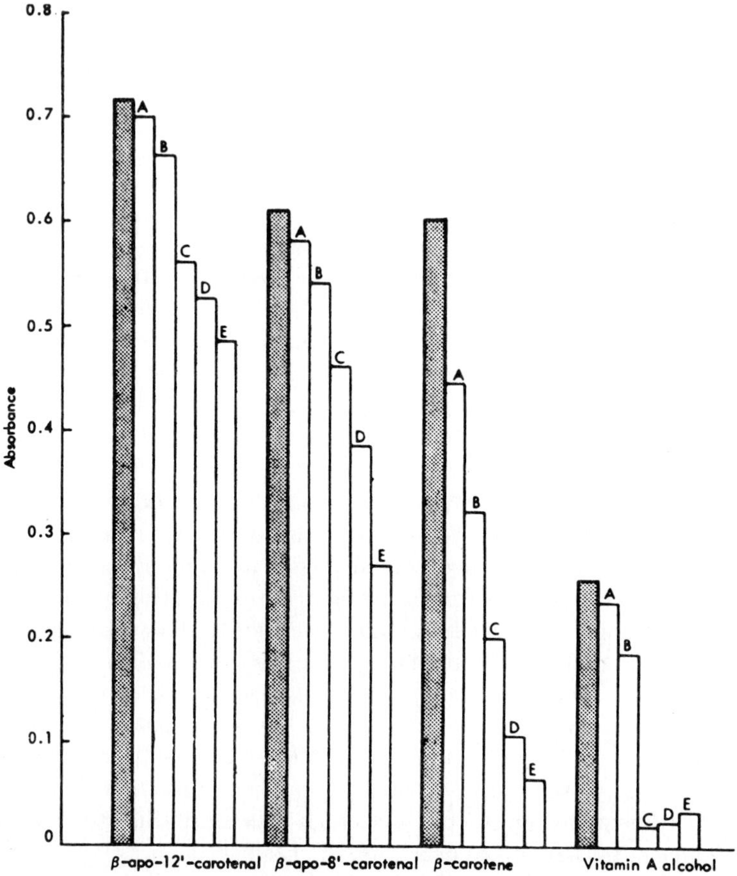

Fig. 14. Comparison of the destruction of three provitamin A compounds and vitamin A alcohol by nitrite at pH 2. Molar nitrite to carotenoid ratios: solid bar, control; A, 0.5:1; B, 1:1; C, 4:1; D, 10:1; E, 20:1. [From Pugh and Garner (*716*).]

N. Disease

Considerable evidence exists showing a relationship between vitamin A and disease. In a review of vitamin A metabolism in sick animals, Tagwerker (*839*) cited many instances in which vitamin A utilization is impaired by various diseases. The vitamin A status of a diseased animal is generally affected and the requirement for vitamin A increased whenever demands are made on the defense mechanism of the animal. Numerous stress factors, such as liver disease, intestinal disease, and parasitic infections, can adversely affect vitamin A metabolism.

A number of researchers have reported a diminished conversion of carotene because of ill health. Since the major site of conversion of carotene to retinol is the intestinal wall, diseases that damage the intestinal epithelium and/or are accompanied by diarrhea may interfere with the vitamin A status of the animal. Such damage occurs in diseases like coccidiosis, parasitic gastroenteritis, and salmonellosis. In 1960, Erasmus *et al.* (*246*) showed that both carotene and preformed vitamin A were used less for growth and liver storage in coccidiosis-infected than in normal chickens. The absorption of vitamin A and of carotenoids was disturbed in the upper two-thirds of the intestine of chickens infected with *Eimeria acervulina* (*410,754,755*). In infected birds the concentration of carotene in the blood had begun to fall by the second day after the infection, reached its lowest level (one-fifth of the original concentration) on the fourth to the sixth day, and then slowly returned to normal after the ninth day. The concentration of vitamin A in the blood, however, did not decrease until the third or fourth day after the infection, and returned to normal at a later date.

Some experimental evidence exists incriminating parasitic gastroenteritis in sheep. The absorption of nutrients and presumably of carotene and retinol is diminished (*263*). Hepatic changes caused by the liver fluke *Fasciola hepatica* in parasitized stock may be associated with lowered liver concentrations of vitamin A. Davtyan and Akopjan (*187*), working with sheep, observed lower vitamin A levels in infected animals. Sheep with (30,000 IU/head daily) and without vitamin A supplementation were infected with lungworm by Akopjan (*12*). After several weeks, the number of larvae in the feces of the sheep receiving supplemental vitamin A was noticeably lower. After the animals were killed, it was observed that the number and size of the worm nodules on the lungs were greatly reduced in the viatmin A-supplemented group. In cases of hyperkeratosis or X-disease described by Olafson (*649*), cattle exhibited symptoms similar to those of vitamin A deficiency. In most cases, animals respond to doses of vitamin A but not to carotene.

Another view should be entertained here, as vitamin A is known to be responsible for the integrity of cells. The effects of both clinical and subclinical deficiencies of carotene or vitamin A on the production of antibodies and on the resistance of tissues against microbiological infection or parasitic infestation have frequently been demonstrated. The addition of vitamin A to a diet reduced the mortality rate in chickens suffering from fowl typhoid. The production of antibodies against typhoid is reduced in rats deficient in vitamin A, and the resistance to *Escherichia coli* and *Salmonella* is impaired in deficient sheep. A study of numerous case histories of diseases in suckling calves by Dvorak (*220*) showed that 66% of the animals had a low plasma level of vitamin A. Low plasma vitamin A levels were found in all cases of bronchopneumonia and in 86% of the cases of dyspepsia simplex. Vitamin A levels were always reduced in calves under 30 days of age, with diarrhea that lasted more than 2 days. There was little or no carotene in plasma of diseased calves. Although the incidence of viral infections (such as foot-and-mouth disease) may be reduced by general malnutrition, a deficiency of vitamin A may impair resistance and intensify the degree of infection. This has been clearly demonstrated in the case of viral hepatitis and of scours in calves delivered by cows deficient in vitamin A. The same observation has been made in chicks suffering from Newcastle disease and chronic respiratory disease. Uncomplicated cases of acetonemia (ketosis) in cattle were observed by Patton (*680*) to respond promptly and recover completely with vitamin A treatment. Vitamin A treatment was not as effective when this disease was complicated with milk fever or metritis, but, when given as supportive therapy with other medication, the vitamin A helped correct the ketosis. Herrick (*382*) briefly reviewed the influence of vitamin A on disease states. Worm infestation in dogs and chicks and the occurrence of lungworms in sheep were aggravated by a deficiency of vitamin A. Supplementation with vitamin A improved the level of health of animals infected with roundworms, of hens infected with capillaria, and of rats infected with hookworms. Vitamin A is believed to act against most intestinal worm infestations by improving the state of the mucosa.

O. Drugs

Responses to antibiotics, although variable, usually show a positive result in influencing carotene conversion to vitamin A. As additives to rations, antibiotics in common use have influenced the utilization of carotene and vitamin A. Burgess *et al.* (*130*) reported that penicillin in chicken rations increased liver stores of vitamin A and levels of carotenoids in blood serum. Coates *et al.* (*165*) also obtained a positive effect with penicillin. On the other hand, Swick *et al.* (*837*) observed no effect of penicillin addition. Chlorotetracycline increased the conversion of carotene to vitamin A in chickens, as measured by the vitamin content of livers. As observed by Almquist and Maurer (*19*), the combination of an antibiotic (chlorotetracycline and diamine penicillin mixture) and an antioxidant or an antibiotic and a fat source was somewhat better than the antibiotic alone in improving vitamin A utilization.

According to Guerrant (*344*), chlorotetracycline added to a vitamin A-deficient diet markedly influenced the growth-stimulating effect of supplementary β-carotene, the magnitude of influence depending on the amount of antibiotic incorporated. In subsequent rat trials other antibiotics gave comparable results, whereas others gave little stimulation. Oxytetracycline added to a feed in which alfalfa supplied the vitamin A activity increased liver stores of vitamin A in steers in two out of three experiments (*222*); similar effects have been observed with stilbestrol. Neither chlorotetracycline nor stilbestrol had any influence on liver carotene or vitamin A storage in ruminants, according to Erwin *et al.* (*248*). It was also shown by Perry *et al.* (*692*) that neither the feeding of chlorotetracycline nor the implantation of stilbestrol had any effect on the liver or plasma concentration of vitamin A. They did show, however, that growth responses to the antibiotic or stilbestrol were obtained only with supplemental vitamin A in the ration. Sulfonamide therapy has no influence on normal vitamin A and carotene metabolism of young calves, according to Ronning and Knodt (*743*).

When carotene was given to rats in studies by High (*389*), chlorotetracycline increased the deposition of vitamin A in liver but penicillin was ineffective. Chlorotetracycline increases the biological effectiveness of vitamin A by a mechanism other than that of increased absorption from the gut (*184*). One suggestion is that the antibiotics have different effects on the intestinal microbial synthesis of vitamin B_{12}. This is based on the observation that, when carotene was given to rats, added vitamin B_{12} increased the deposition of vitamin A in liver and kidneys. In another study (*370*) neither chlorotetracycline nor vitamin B_{12} appeared to affect the utilization of preformed vitamin A in rats. Vitamin B_{12} does, however, increase tissue deposition of vitamin A when carotene is fed, as shown by the research of High and Wilson (*393*). The evidence that vitamin B_{12} improves the efficiency of utilization of dietary carotene for liver storage of vitamin A in rats raises the question regarding the desirability of including vitamin B_{12} or liver extract in the classic vitamin A bioassay diet. The results of High and co-workers suggest that, in the bioassay of carotene, the assay diet fortified with vitamin B_{12} would probably yield a truer picture of carotene as a vitamin A source under practical conditions than similar diets low in or devoid of vitamin B_{12}.

P. Minerals

Other nutritional factors can have an adverse effect on the utilization of carotene. Conversion to vitamin A is slowed down in phosphorus deficiency. This is an important finding in view of the usually inadequate phosphorus content of pasture grasses and hay. Ross and Gallup (*744*) observed an inverse relationship between blood plasma inorganic phosphorus and plasma carotene levels in beef cattle fed phosphorus-deficient rations. In determining the effect of phosphorus deficiency on vitamin A and carotene metabolism, Gallup *et al.* (*289*), working with sheep and cattle, reported that the level of carotene in plasma increased with a phosphorus deficiency. The vitamin A concentration in plasma and liver was, however, greater in animals fed adequate phosphorus rations. A phosphorus deficiency seemed to interfere with the conversion of carotene to vitamin A. The feeding of supplements to cattle on winter range with vitamin A, phosphorus, and a combination of the two by Thomas *et al.* (*847*) improved the rate of weight gain. The addition of vitamin A and phosphorus separately improved weight gain to about the same degree, whereas the combination gave the greatest growth response. It has been reported that "soft phosphate" interferes with the utilization of carotene and vitamin A, presumably through the adsorptive action (*251*) of the clay in the intestinal tract. In other studies it has been observed that, if maize silage is treated with limestone or urea, the availability of the carotene is enhanced (*627*).

Brocklesby (*118*) noted that sulfate is reduced to sulfide in the rumen, and enough hydrogen sulfide may be present to interfere with the conversion of carotene to vitamin A. A direct relationship (*481,610*) was found between the concentration of copper and that of vitamin A in blood of normal ewes at pasture. Also, in lambs that had been experimentally poisoned with copper and in one animal poisoned accidentally, the levels of vitamin A in the plasma were reduced. As treatment was given and the blood levels of copper fell, there was an accompanying rise in the plasma levels of vitamin A. An interrelationship between zinc nutrition and vitamin A utilization has been indicated by several investigators (*759,808*).

Q. Miscellaneous

When peanuts are badly handled during harvesting or are allowed to remain damp, they may be attacked by the fungus *Aspergillus flavus,* which then produces a powerful hepatotoxin, aflatoxin. Calves suffering from aflatoxicosis have been observed to have livers with a very low vitamin A content compared to those of control calves; thus, interference in the carotene utilization process is suspected. Experimentally, Clark and Colburn (*162*) found low concentrations of vitamin A in the liver of rats given large doses of cortisone, which impairs the conversion of carotene to vitamin A. Alloxan-diabetic rats stored only one-fourth as much vitamin A in the liver after a fixed dose of carotene as did nondiabetic controls (*812*). Octachloronaphthalene decreased markedly the utilization of carotene when given orally to rats, but not when injected (*188*). Chlorinated hydrocarbons depress the conversion of carotene to vitamin A. Various carcinogenic drugs reduce liver vitamin A values (*312*).

Some ingredients added to carotene-containing rations can modify the nutritive value of the carotenoid. For example, Hove (*411*) reported that the addition of dried yeast (8%) to the ration of rats lowered the conversion of β-carotene to vitamin A. The lipid fraction of yeast appears to cause this reduction. Cholestyramine, an ion-exchange resin, when added to the ration of chickens, decreased carotenoid utilization (*212*) either indirectly by intestinal binding of bile salts required for carotenoid absorption or by a direct effect on the carotenoids themselves. Many pelleted rations contain bentonite, a substance that absorbs carotene and vitamin A and reduces the utilization of carotene from a purified diet (*506*). Briggs and Spievy (*114*) reported that bentonite produced vitamin deficiency in chicks when added to a synthetic diet but not when included in practical chick rations. Inclusion of raw soybeans in the diet decreases the biological value of carotene (*237,821*). The feeding of a concentrate ration containing 30% soybeans increases the carotene requirement for maintenance of adequate plasma carotene and vitamin A levels in dairy calves (*784*). Reports have appeared which indicate that the feeding of raw soybeans or soybean hay produces symptoms of vitamin deficiency in dairy cattle (*396,783*) as well as calves (*237*). Reproductive disturbances also have been reported from the feeding of soybean hay to rabbits (*448*). Raw soybean contains an enzyme that oxidizes fat and destroys carotene (*236*). A vitamin A shortage (*448*), brought about by enzymatic destruction of carotene in the soybean hay by lipoxidase, may be an explanation for the reproductive disturbances reported in rabbits and ruminants and lowered plasma vitamin A values (*236,857*).

XII. PURE CAROTENE BY ORAL OR PARENTERAL ADMINISTRATION

A. Injectable Carotene

Kon and Thompson (*473*), in a 1951 summary of their work and that of others, showed that the conversion of orally administered β-carotene to vitamin A occurs in the intestinal wall. Reports of utilization of carotene by several species of animals when it is injected intravenously (*81,160,472,567,568*) indicate that there are sites of conversion of carotene to vitamin A other than the intestinal wall. In the rat, intramuscular, intraperitoneal, or subcutaneous injections of various aqueous suspensions of carotene were utilized for growth restoration and the eradication of xerophthalmia (*84,872*). Tomarelli and associates (*872*) and Bieri and Sandman (*85*) reported that β-carotene could be readily utilized by rats when given parenterally, provided that an aqueous suspension in polysorbate 80 or 40 were used. Oil solutions were less efficiently utilized. Additional studies of Bieri and Pollard (*83*) demonstrated that carotene in aqueous dispersions when administered intravenously to rats is rapidly converted in part to vitamin A even if the bile duct is ligated or the small intestine, kidneys, or 60–75% of the liver are removed. McGillivray *et al.* (*567*) and Worker (*929,930*) reported similar results with rats whose liver or liver and viscera (comprising the stomach, large and small intestines, pancreas, adrenals, kidneys, and gonads) or lung tissues were removed. It has not been possible to identify the tissue or tissues responsible for this process. However, carotene introduced in the form of an aqueous dispersion is continuously carried by circulating blood to all the tissues of the animal body, so that the possibility exists for the oxidative cleavage of carotene in many of the tissues. The conversion of injected carotene to vitamin A was not influenced by the daily administration of tocopherol (*82*). McGillivray and Worker (*568*) demonstrated that in the rat the conversion of intravenously administered carotene causes an increase in plasma vitamin A within 15 min. Subsequent studies by McGillivray (*558*) revealed that intravenously administered C_{30}, C_{27}, and C_{25} carotenals can also be converted to vitamin A but apparently not as efficiently as carotene. Sexton *et al.* (*779*) showed that vitamin A deficiency and death can occur when considerable amounts of carotene are still in the liver.

Injections of carotene emulsions into the residual yolks of newly hatched chicks resulted in increased liver vitamin A contents at 4 days of age and prolonged survival times (*371*). Solubilized, aqueous β-carotene injected intravenously into vitamin A-deficient chicks was rapidly converted to vitamin A, which appeared in the liver and serum. Injected cryptoxanthin did not form significant amounts of vitamin A and was not as active as β-carotene in promoting growth (*81*). The utilization of injected cryptoxanthin was less than that of cryptoxanthin given by the oral route. Rabbits injected with carotene formed appreciable amounts of vitamin A, which was about equally distributed between the liver and kidneys (*81*). Similar data have been obtained in the dog (*478*). With pigs, Hentges *et al.* (*381*) found that water-miscible preparations of carotene administered intravenously or intramuscularly resulted in complete remission of vitamin A deficiency symptoms.

Church *et al.* (*161*) and Kirschman and MacVicar (*452*) demonstrated the conversion of carotene to vitamin A in sheep given intravenous injections of carotene solubilized in aqueous solution of polysorbate 40. However, they were unable to demonstrate a similar conversion in Hereford, Guernsey, or Holstein calves. Menzies *et al.* (*577*) agreed that sheep can utilize intravenously administered carotene. Related studies by Boling *et al.* (*99*) showed that aqueous dispersions of carotene administered intravenously result in increases in plasma vitamin A values in sheep. Physiological quantities of ^{3}H-labeled vitamin A and ^{14}C-labeled β-carotene (in 20% polyoxyethylene sorbitan monooleate, polysorbate 80) were administered intravenously to bile-duct-cannulated sheep. Data obtained from this trial established the secretion of vitamin A and β-carotene metabolites in the bile of sheep and suggested that the enterohepatic circulation plays a positive role in the vitamin A metabolism of sheep.

Warner and Maynard (*897*) reported that intravenously injected carotene in coconut oil did not cure vitamin A-deficient calves but that similar injections of aqueous colloidal suspensions of carotene significantly increased blood levels of vitamin A. The data of Schuh *et al.* (*768*) indicated that a portion of intravenously injected carotene was utilized by calves, since plasma vitamin A levels increased, signs of avitaminosis A became less intense or entirely disappeared, and longer survival occurred after the injection of carotene. Failure to observe high mean levels of plasma vitamin A in ruminants following carotene injection might be explained by a rapid absorption of vitamin A from the bloodstream or by slower rates of utilization of the carotene suspension, as suggested by McGillivray and Worker (*567*). Eaton and co-workers (*225*) found that calves depleted of vitamin A had higher plasma levels of β-carotene and greater amounts of vitamin A in the liver and lung after an aqueous suspension of carotene had been given intravenously than after it had been administered orally.

The utilization of carotene by different species was clarified by Kon *et al.* (*472*). They showed that with a normal diet the state of dispersion was important, since carotene given to rats, calves, or rabbits by mouth was better absorbed and more efficiently converted to vitamin A from an aqueous dispersion in polysorbate 40 than from an oily solution. This difference in favor of aqueous dispersions is more obvious when carotene is given by injection. The conversion of intravenously administered aqueous dispersions of carotenoids to vitamin A was reviewed by McGillivray (*560*).

B. Oral Carotene

The relative performance of orally administered carotene and vitamin A in several species has been investigated. In general, aqueous solutions or dispersions of vitamin A or provitamin A (carotene) were utilized equally well or better for survival and for liver storage of vitamin A than were oily solutions. Preformed vitamin A was always utilized more effectively than carotene in similar preparations given orally or as ration components, as measured by liver stores and survival time.

Gurcay *et al.* (*355*) and Castano *et al.* (*144*) compared vitamin A acetate and crystalline carotene using as criteria the growth and liver stores of vitamin A in turkey poults and chicks (Table 28), respectively. On the basis of international units, the vitamin A compound was several times as efficient as crystalline carotene, judging by liver vitamin A stores. Laughland and Phillips (*505*) also studied the utilization of vitamin A and carotene by normal deutectomized chicks. Liver stores after 8, 15, 22, and 29 days were from 5 to 10 times higher when vitamin A acetate was fed at 10,000 IU per pound feed as when 5.6 mg β-carotene (9333 IU) were fed

TABLE 28

Utilization by the Chick of Vitamin A from Different Sources[a,b]

Vitamin A per pound of diet (IU)	Total weight gain (gm)			Vitamin A in blood plasma (IU/100 ml plasma)		Vitamin A in liver (IU/gm liver)	
	Male	Female	Average	5th week	8th week	5th week	8th week
			Crystalline carotene				
500	626	530	578	56	61	1	2
6,000	679	601	640	169	234	44	74
12,000	682	592	637	219	253	126	243
18,000	681	547	614	231	258	212	377
24,000	615	581	598	270	271	303	446
			Crystalline vitamin A acetate				
500	605	478	542	54	54	2	2
6,000	660	635	648	238	255	183	316
12,000	657	623	640	263	283	482	728
18,000	620	609	614	278	288	685	1,163
24,000	679	569	624	450	356	1,139	2,291
			"Black cod" liver oil				
500	597	520	558	52	48	2	3
6,000	654	557	606	224	220	119	219
12,000	702	528	615	255	255	425	685
18,000	639	552	596	293	258	636	1,065
24,000	629	547	588	355	355	795	1,645

[a] From Castano *et al.* (*144*).

[b] Gain in body weight from the fourteenth day to the eighth week of age and vitamin A concentration in blood plasma and liver. Concentration at hatching: 1038 IU/100 ml plasma, 73 IU/gm liver. Concentration at end of 14-day depletion period: 105 IU/100 ml plasma, 9 IU/gm liver.

per pound. It was acknowledged by Ely (*240*) that dry stabilized vitamin A was about 2.6 times more effective in promoting chick liver storage than provitamin A. Similar results were found by Olsen *et al.* (*651*) in comparing several dry vitamin A sources and vitamin A feeding oil with carotene from cereal grass. At levels of 100–1200 IU per pound feed no differences were seen in growth for carotene from cereal grass or dry vitamin A beadlets. However, on the basis of liver stores of vitamin A and survival time, the cereal grass was decidedly inferior. The authors concluded that, in short-term experiments, growth may not be a reliable criterion for the evaluation of vitamin A availability. Ribarova (*736*) observed that synthetic β-carotene had a limited effect on the growth of chicks and the utilization of food and was not effective as a source of vitamin A. Erasmus *et al.* (*246*) reported that a level of 800 IU of stabilized vitamin A per pound of diet was as effective in maintaining feed consumption and growth rate in chicks artificially infected with coccidiosis as was 2400 IU/lb from dry β-carotene beadlets.

Trials in which known compound purity, isomer content, and stabilized form of comparable physical forms of provitamin A (such as β-carotene) and vitamin A (such as vitamin A palmitate), if studied at increasing levels of intake and at a varied ratio to each other, ought to minimize some of the variability in past studies and provide for more reliable comparison. Such a trial was run by Marusich and Bauernfeind (*536*). Synthetic β-carotene (gelatin beadlets) and synthetic vitamin A palmitate (gelatin beadlets) in water-dispersible form were compared for both growth and liver storage of vitamin A in the chicken. Over a feeding range, the vitamin A product consistently provided 2.5 times the liver stores of vitamin A as the equivalent unitage from β-carotene. Expressed on a weight basis, 5 mg β-carotene provided liver stores of vitamin A equivalent to those found from feeding 1 mg of vitamin A, or a ratio of 5 : 1. Based on chick growth at a low level of feeding in this trial, a weight ratio of 2 : 1 was obtained. Hence, for poultry, when growth and liver storage were the criteria for activity, ratios of 2 : 1 to 5 : 1 (Fig. 15) or more were observed when comparisons were made on a weight basis of pure β-carotene to pure vitamin A (*536*).

In studies of 1-day-old broiler-type cockerel by Flegal *et al.* (*273*) β-carotene of fermentation origin compared favorably with a commercial crystalline all-*trans*-β-carotene in its influence on growth, feed efficiency, and plasma and liver vitamin A concentration. At low levels of vitamin A intake that support optimal chick growth, β-carotene appeared to be as effective as the ANRC Vitamin A Reference Standard on an international unit basis. However, both β-carotene products, when fed at levels high enough to induce plasma concentration and liver storage, were inferior to the ANRC Vitamin A Reference Standard (vitamin A acetate). According to Semin *et al.* (*778*) microbial carotene (*Blakeslea trispora*) in chicken experiments did not have the biological value of carrot carotene; at 50% higher doses, it gave results close to those for vitamin A acetate and was equivalent to lucerne meal carotene.

Relative values of vitamin A (distilled concentrate) and carotene (crystalline) for

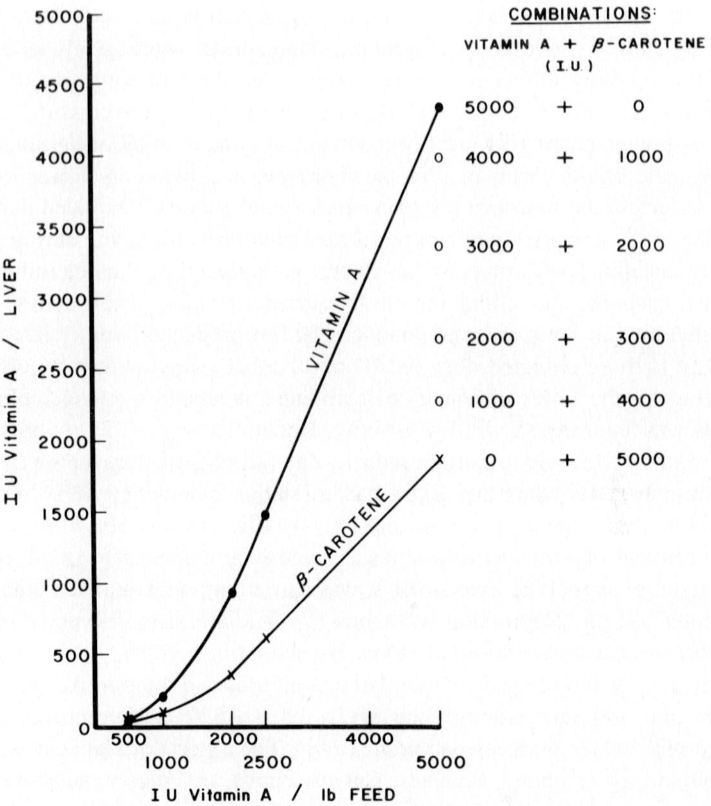

Fig. 15. Chick liver vitamin A values after the feeding of vitamin A or β-carotene. [From Marusich and Bauernfeind (*536*).] Stabilized vitamin A palmitate beadlets, stabilized β-carotene beadlets, or combinations were incorporated in a range of quantities in the diet and fed to broiler-type chicks for 4 weeks. Based on liver vitamin A values the relative effectiveness of β-carotene to vitamin A is as follows (assuming 0.6 mg β-carotene = 1000 IU = 0.3 mg vitamin A):

β-Carotene fed/lb feed		Weight ratio effectiveness, β-carotene to vitamin A
mg	IU equivalents	
0.3	500	2:1
0.6	1000	4.4:1
1.2	2000	5.1:1
4.8	5000	4.6:1

meeting the vitamin A needs of swine during gestation and early lactation were studied by Parrish *et al.* (*670*). Vitamin supplements were given at a level of 6500–7100 IU daily, either in capsule form or dissolved in corn oil and added to soybean concentrate. Both groups consumed a ration of low vitamin A content. Carotene-supplemented gilts had lower vitamin A content in blood serum, in colostrum, and the blood serum of their newborn pig and markedly lower vitamin A storage values in the newborn pig liver. Hence, the authors concluded that unit for unit, carotene is less effective than preformed vitamin A for swine during gestation and early lactation. β-Carotene, 95% all-*trans*, produced by the heterothallic mold *Blakeslea trispora* and contained in the dried mycelia, when fed to vitamin A-depleted pigs in 1-mg amounts (about 1600 IU) produced liver vitamin A levels equivalent to those obtained with 192 IU of all-*trans*-retinyl palmitate (*881*). Even accounting for the lower efficiency of carotene conversion by swine compared to rats, this was an unexpectedly low activity. Hendricks *et al.* (*378*) published similarly low values. In a study starting with 10-day-old piglets, they reported that 1 mg of β-carotene of fermentation origin had a relative biopotency of 90–132 IU, as judged by liver storage data. The swine study by Ullrey *et al.* (*880,881*), in which 1 mg of β-carotene of fermentation origin was found to be equivalent to 192 IU of vitamin A palmitate for support of liver storage, was carried out with depleted pigs, 50–100 kg in body weight. Regression equations (*881*) relating total liver vitamin A to dietary β-carotene and vitamin A intake are shown in Fig. 16.

The relative value of carotene (dehydrated alfalfa) and vitamin A (dry form) fed to lambs and pigs was examined by Myers *et al.* (*620*) and compared to results obtained with calves by Rousseau *et al.* (*748*). The criteria of response were based on blood and liver vitamin A values. Calves, lambs, and pigs were about equal in

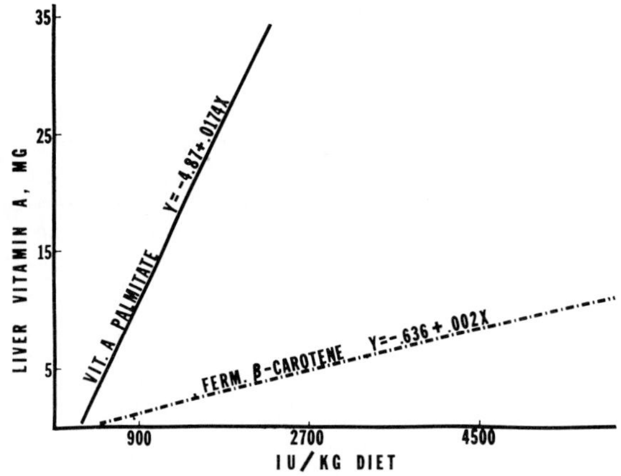

Fig. 16. Vitamin A activity of fermentation β-carotene for swine. Regression equations relate total liver vitamin A and dietary concentration of all-*trans*-vitamin A palmitate or fermentation β-carotene. [From Ullrey *et al.* (*881*).]

the conversion of carotene to vitamin A at the lower levels of carotene intake, but, at higher levels of intake, the pigs appeared to be more efficient than either the calves or lambs, and the lambs more efficient than the calves. Farugue and Walker (*265*), in trials with young lambs, fed β-carotene in the form of water-miscible all-*trans*-β-carotene dispersed in milk in amounts of from 68.5 to 2,200 μg β-carotene per kilogram live weight per 24-hr. The relative weights of β-carotene and retinyl palmitate (expressed as retinol) that were required to produce equivalent concentrations of retinol in the serum varied from 5 : 1 to 25 : 1, and, for liver storage, they varied from 3 : 1 to 9 : 1 (Table 29).

Ross *et al.* (*745*) compared the relative efficiency of β-carotene, vitamin A, and vitamin A natural esters when fed in capsule form at the rate of 100 IU per kilogram body weight to Holstein heifers receiving a low-vitamin A ration. Vitamin A and its esters were equivalent and more efficiently utilized, as judged by blood plasma concentration of vitamin A. Blood carotene or vitamin A values are widely used as measures of the carotene and vitamin A nutritional status of livestock. It is generally agreed that liver vitamin A concentration is a more sensitive indicator of vitamin A

TABLE 29

Relative Biological Potencies of β-Carotene and Retinyl Palmitate (Expressed as Retinol) in Maintaining Equivalent Concentrations of Retinol in Blood and Liver of Milk-Fed Lambs as Compared with Ruminant Lambs and Young of Other Species[a]

Amount of β-carotene given (μg/kg 24 Hr)	Ratio of β-carotene to retinol			
	Plasma[b]	Liver	Plasma	Liver
	Calves		Pigs	
88.2	4.0	5.3	5.6	3.7
176	6.4	7.8	7.0	5.1
353	9.6	11.7	8.6	6.9
705	14.6	17.4	10.7	9.4
	Milk-fed lambs[c]		Ruminant lambs	
68.5	5.4	3.1	—	—
88.2	(6.2)	(3.6)	6.6	5.5
176	(8.4)	(5.2)	8.2	7.4
275	10.2	6.2	—	—
353	(11.4)	(6.8)	10.4	10.0
705	(15.4)	(7.9)	13.1	13.4
1100	18.7	8.4	—	—
2200	25.3	8.9	—	—

[a] From Farugue and Walker (*265*).
[b] Serum was used in milk-fed lambs.
[c] Values in parentheses were calculated from published reports.

status than is blood vitamin A levels (*281*). The development of the liver biopsy technique in cattle has facilitated the use of liver vitamin A concentration in assessing the nutritional status of animals with respect to this vitamin (*775,907*). Carotene in alfalfa leaf meal was found to be efficiently utilized, carotene in carrot oil least efficient, and synthetic β-carotene intermediate in a three-source carotene dietary study with dairy calves (*129*). Synthetic vitamin A was more efficiently utilized, as judged by vitamin A storage data. Vitamin A also produced milk fat with the highest vitamin A value. A high correlation existed among milk fat vitamin A, blood vitamin A, and liver vitamin A.

In a calf study (*922*) carotene and vitamin A, respectively, were dispersed in reconsituted milk (10% nonfat milk solids) and soybean oil (3%) and fed at the rate of 100,000 IU per 100 lb body weight. When the nipple system of feeding calves was employed, the uptake of carotene from an oil concentrate in the milk was greater when dispersion was accomplished by homogenizing rather than be stirring. A similar comparison of dispersions of a natural vitamin A ester in oil revealed no difference between homogenization and stirring. The rate of absorption of vitamin A, however, was greater than that of carotene. Vitamin A oil dispersed by means of an emulsifying agent was absorbed more rapidly and apparently to a greater degree than the unemulsified oil. The differences were more pronounced when given via nipple then when delivered by stomach tube into the rumenoreticular cavity. In a subsequent study, Jacobson and co-workers (*422*) confirmed that, when carotene was dispersed in milk and fed to calves from a bottle with a nipple, the supplement was more rapidly absorbed than when administered by stomach tube. Moreover, the rate of absorption of β-carotene and vitamin A was somewhat less rapid when the concentrations were fed in gelatin capsules than when the foregoing procedures were employed.

Wise *et al.* (*923*) continued to study different dispersions of β-carotene (crystalline) and vitamin A (synthetic and natural ester concentrate) and methods of administration to dairy calves. Emulsification of the pure carotenoid supplement by homogenization enhanced its passage into the bloodstream of calves. Kon *et al.* (*472*) also found that, when calves received carotene in an emulsion (homogenized), the resulting concentrations in the blood were as high as those from the consumption of an aqueous dispersion of carotene and were maintained at a high level longer. Hence, the nature of dispersion of the carotenoid evidently plays an important role in absorption by calves.

Earlier studies had shown that emulsification of an oil solution of vitamin A or carotene is accomplished by the use of dispersing agents or homogenization. Aqueous dispersions or collodial suspensions have resulted in more rapid utilization of carotene or vitamin A but do such dispersion forms influence the relative value of the two vitamin A sources? Grifo *et al.* (*339*) placed male Holstein calves on a milk replacer and compared a water-dispersible form of carotene (gelatin beadlets) with water-dispersible vitamin A palmitate (gelatin beadlets). A third group was given vitamin A in oil. An amount of carotene 3 to 16 times as great as vitamin A in

TABLE 30

Effect of Carotene and Vitamin A on Plasma Carotenoids and Plasma Vitamin A of Holstein Calves Fed a Milk Replacer[a,b]

	Plasma carotenoids (μg/100 ml)		Plasma vitamin A (μg/100 ml)		
Intake (μg)[c]	Initial	Comparison period average	Initial	Comparison period average	Adjusted comparison period average[d]
Carotene (beadlets)					
40	9	25	11.1	11.0	9.9
80	2	46	3.7	10.1	11.7
160	10	81	8.2	12.9	12.9
Vitamin A (beadlets)					
9	6	6	8.0	12.2	12.3
12	10	7	7.8	14.3	14.4
15	4	5	9.0	15.0	14.6
Vitamin A (oil)					
9	12	9	11.6	9.8	8.5
15	6	6	5.2	9.9	10.9
Standard deviation per calf	—	—	4.1	2.6	2.2

[a] From Grifo *et al.* (*339*).
[b] Represents data from three calves per treatment group.
[c] Per pound of live weight per day.
[d] Adjusted by covariance for initial plasma vitamin A values.

dispersion form was required to produce comparable plasma vitamin A concentrations. The data (Table 30) also confirmed a widening ratio of carotene to vitamin A as intakes are increased, and the superiority of a water-dispersible form of vitamin A over the vitamin A oil solution was confirmed as well. The relative effect of vitamin A and β-carotene, both administered in the emulsion beadlet form (*339*), is illustrated graphically in Fig. 17. The same workers (*337*) investigated the conversion of carotene from alfalfa and synthetic carotene in dry beadlets to vitamin A and found the dry beadlets to be 1.4 times as effectively converted as the carotene from alfalfa, as indicated by plasma and liver vitamin A stores.

Thompson *et al.* (*860*), in studies on the effectiveness of different states of dispersion of β-carotene in the rat, found that finely powdered carotene mixed with a fat-free diet was converted to a slight extent to vitamin A and that the addition of fat (10%) improved the conversion. Further improvement in absorption was obtained by dissolving carotene in diethyl ether and then pouring it over the fat-free diet and letting the ether evaporate. With colloidal carotene (*860*), the conversion was the most efficient (Fig. 18). Under the same comparative conditions, the uptake

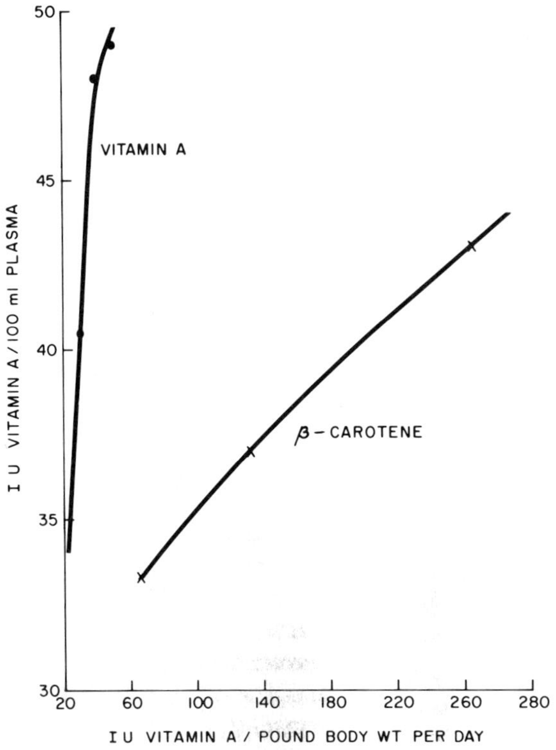

Fig. 17. Calf plasma vitamin A values after the feeding of vitamin A or β-carotene. [Adapted from Grifo *et al.* (*339*).] Stabilized vitamin A palmitate beadlets or stabilized β-carotene beadlets were incorporated in a range of quantities in a milk replacer and fed to standardized Holstein male calves. Based on linear regressions of increase in plasma vitamin A values relative to log dose, the relative effectiveness of β-carotene to vitamin A is as follows (assuming 0.6 mg β-carotene = 1000 IU = 0.3 mg vitamin A):

β-Carotene fed/lb body wt		Weight ratio effectiveness, β-carotene to vitamin A
mg	IU equivalent	
0.04	66	8:1
0.08	133	11:1
0.16	266	16:1

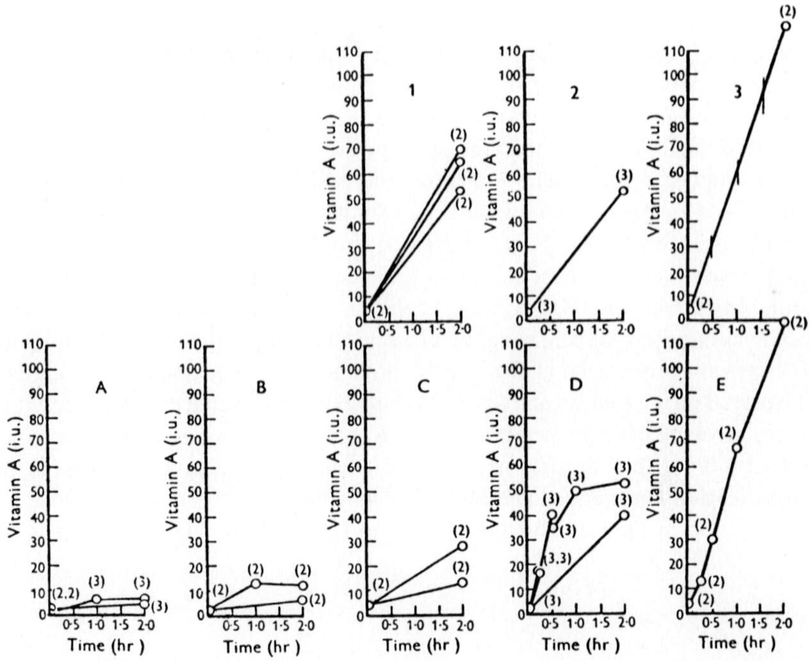

Fig. 18. Effect of the state of dispersion on the conversion of β-carotene to vitamin A and on the absorption of preformed vitamin A in the small intestine of vitamin A-deficient rats. Rate of increase in hours in the concentration of total vitamin A in international units in the wall of the small intestine. (A) 2.5 mg powdered β-carotene on 1.5-gm fat-free vitamin A-deficient diet; (B) 2.5 mg powdered β-carotene on 1.5-gm vitamin A-deficient diet containing 10% fat: (C) 2.6 mg β-carotene, dissolved in ether, on 1-gm fat-free diet; ether removed by evaporation; (D) 3.2–5 mg β-carotene as 2- to 5-ml colloidal solution mixed with 1-gm fat-free diet; (E) 4 mg β-carotene in 400 mg arachis oil, or 400 mg arachis oil alone, either mixed with 1-gm diet; (1) 600 IU vitamin A acetate dissolved in *n*-hexane on 1-gm fat-free diet; hexane removed by evaporation; (2) 1200 IU vitamin A as 2-ml colloidal solution of vitamin A acetate, and 1-gm fat-free diet; (3) 600 IU vitamin A acetate in 100 mg fat and 1-gm fat-free diet. Figures in parentheses indicate the number of rats on which values are based. [From Thompson *et al.* (*860*).]

of the preformed vitamin was less related to the state of dispersion than was carotene.

Thompson (*858*) continued to employ cows to study the effectiveness of carotene. When β-carotene was in a microcrystalline or a colloidal preparation, with a particle size of 1 μm or less, it was very well utilized. Thompson described a study in which 14.5% β-carotene beadlets were fed to six sets of Friesian twins for a 21-day period. A level of 4 gm of carotene daily was given to represent the level normally ingested by a cow on pasture. A 1-gm level of carotene was used as an intermediate level, and a 0.25 gm level was chosen as a low level of carotene intake representing the use of carotene-depleted feeds in winter. The results were in line with the levels fed. A rapid rise in vitamin A contrasted sharply with the slow rise in carotene. As little as 0.25 gm carotene raised both the carotene and the vitamin A levels. The carotene content of the milk fat increased steadily to the end of the feeding period. After dosing was stopped, the vitamin A decreased nearly as rapidly as they had risen. When 4 gm carotene were given, the carotene level (*858*) in the milk fat attained a high value typical of the milk of cows of the Channel Island breeds (Fig. 19). About 1 gm of carotene from gelatin beadlets (*858*) was judged to be equally as effective as 3–4 gm of carotene from pasture in increasing the vitamin A and carotene content of the milk fat (Table 31). Moon *et al.* (*598*), in trials with Ayshire cows, showed that synthetic colloidal β-carotene (beadlet form) was as effective as cut grass in increasing the vitamin A content of the secreted fluid milk.

The previous discussion of the role of carotene in the nutrition of dairy cattle has concentrated on its vitamin activity and relative nutritive value compared to vitamin A. Does carotene have a specific function in cattle other than as a source of vitamin A activity? Published reports from Germany and England over the past 5 years reinforce earlier correlations and observations which indicate that β-carotene has a specific action in the reproductive process in female cattle and suggest that, even

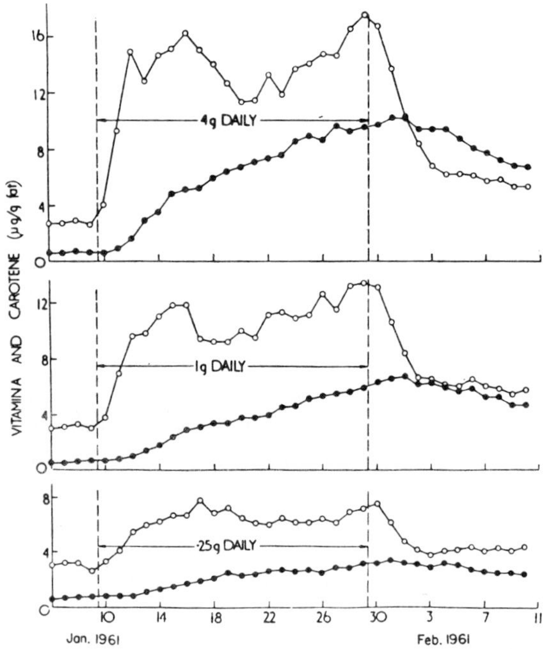

Fig. 19. Vitamin A and carotene in the milk fat of 12 Fresian cows on a low-carotene diet supplemented for 21 days with 0.25, 1.0, or 4.0 gm carotene daily as a 14.5% microcrystalline gelatin beadlet preparation. Each point represents the mean value for four cows (○-○, vitamin A; ●-●, carotene). [From Thompson (*857,858*).]

TABLE 31

Levels of Vitamin A and Carotene in Milk Fat of Cows under Various Conditions[a,b]

Source and amount of β-carotene	No. of cows used	Initial control period (hay and concentrates mean for 3 days): Vitamin A	Initial control period: Carotene	Carotene feeding period: Vitamin A: Level at 15th day	Vitamin A: Increase	Carotene: Level at 15th day	Carotene: Increase
Gelatin beadlets							
1 gm	4	3.2	0.5	11.0	7.8	5.2	4.7
4 gm		2.7	0.6	13.5	10.8	8.6	8.0
Ryegrass, 4 gm	9	9.5	5.6	8.6	−0.9	8.9	3.3
Mixed pasture, 4 gm	24	4.2	1.4	7.0	2.8	6.0	4.6
Cocksfoot grass,[c] 3 gm	4	5.5	1.5	10.9	5.4	6.5	5.0

[a] From Thompson (*858*).
[b] Measured in cows during the stall feeding period and after 15 days of grass feeding or 15 days on a winter diet supplemented with 1–4 gm daily of synthetic β-carotene as the 14.5% microcrystalline gelatin beadlet preparation. Values in micrograms per gram fat.
[c] Cut and fed in the stall.

though vitamin A is supplied at optimal dietary intakes, β-carotene adequacy must be considered also.

Brüggemann and Niesar (*122*), Schultz (*771*), and Goodwin (*324*) had earlier observed a high β-carotene content in the corpora lutea of cattle but no vitamin A; furthermore, there was a correlation between the carotene content of corpora lutea and dietary β-carotene. Blood carotene levels and reproductive performances in range Hereford cattle were studied by Payne and King (*682*). Alexsson (*15*) in 1947 was able to lower the insemination index by feeding β-carotene sources within a time period before the insemination period. A correlation of improved fertility with a higher intake of β-carotene but not with vitamin A was reported by Konermann (*474*), and Seitaridis (*776*) observed lower blood carotene values of cows with ovarian cysts. Since the practical problem of fertility exists in high-yielding dairy herds and since the carotene content of rations is variable, studies were initiated in Germany to gain direct evidence on the significance of dietary β-carotene in the reproduction process.

Meyer *et al.* (*578*) in 1975 conducted two feeding trials with two groups of black-pied heifers, 12 control animals ingesting a low-carotene ration and 18 animals on the same ration supplemented with β-carotene (30 mg per 100 kg body weight). Both groups were adequately supplemented with vitamin A. In the control (carotene-deficient) group, ovulation was retarded, cystic degeneration appeared on the ovaries, and a slower development of the corpora lutea was observed. In a subsequent report by Schams *et al.* (*761*), delayed ovulation and a decrease in conception were reported. The suggestion was put forward that β-carotene has a specific action on fertility.

In additional studies an experimental design included two groups of black-pied heifers, one group receiving β-carotene (stabilized gelatin form) added to a low-carotene ration and a nonsupplemented group consuming the low-carotene ration. Both groups received supplementary vitamin A. The compiled data from trials run in Germany by Lotthammer and Ahlswede (*521*), Lotthammer *et al.* (*522*), Schams *et al.* (*762*), and Ahlswede and Lotthammer (*9*) show for the β-carotene group (30 mg per 100 kg body weight) a more timely ovulation, a more pronounced estrus, a lower incidence of corpora luteal and follicular cysts, and a higher percentage of pregnancies after one or two insemination periods. A subsequent field trial was set up by Cooke and Comben (*165a*) in the United Kingdom wherein a commercial dairy herd of 182 cows were divided into two groups, one group receiving grass silage and a feed concentrate, the other receiving corn (maize) silage and a feed concentrate. Both groups were given supplementary vitamins A and E. Fertility records monitored over 1 year revealed a higher percentage of pergnancies in the corn silage group (740 μg carotene per 100 ml plasma), having the higher carotene intake, than on the grass silage group (290 μg per 100 ml). In 1979 Lotthammer (*520a*) summarized the experiments which indicate that β-carotene plays a specific role in the fertility of cattle that is not performed by vitamin A.

There is probably little danger of a deficiency of β-carotene in cows fed high-quality hay, grass silage, grass cubes, etc. However, there will be instances of low-carotene silages and roughages, dried-up pastures and drought conditions, or concentrates without added dehydrated green meals or yellow maize products, in which feeding programs may not supply ample β-carotene. Fortunately, qualitative and quantitative estimation of carotene in blood plasma is an easy analysis. Contrary to vitamin A, body tissue studies of β-carotene need to be carried out; low plasma carotene values may be cause for concern when low fertility of cows is experienced.

If the β-carotene content of rations fed to cows is judged to be inadequate, it can be supplemented with natural feed concentrates or by the addition of synthesized β-carotene in stabilized dry form. The β-carotene in these products is in a fine emulsion or colloidal form. When the dry product is consumed and brought into contact with the digestive juices, the product readily disperses, allowing for full bioavailability of the β-carotene.

XIII. BASIS OF ROLE FOR PREFORMED VITAMIN A

It is well recognized that, if domestic animals have access to growing or fresh green pasture of grasses, legumes, etc., either through grazing or unlimited foraging practices, vitamin A deficiency is not a primary problem. However, when pastures fail during the dry season or during prolonged drought, or when animals are stall-fed roughage and/or concentrate feeds of inadequate vitamin A activity for a sufficiently long period, deficiency symptoms occur. In animal management practices of the past involving a green pasture summer grazing program followed by winter indoor or dry-lot feeding, the liver stores of vitamin A accumulated during the summer played an important role in getting the animals through the following winter. Many feeds used during the winter were devoid of carotenoid precursors of vitamin A and feeds believed to be rich in this substance may have been poor because of losses during preparation and storage. At the end of the winter the animals' reserves frequently were exhausted if a vitamin A supplement had not been provided. The function (*705*) of vitamin A and its importance in the feeding of farm animals (*113,855*) has been reviewed periodically.

In semiarid parts of the world where green roughage is available only for a very limited period of the year and at other times conserved roughage, such as hay, is the only source of vitamin A precursors, it is important for animals to store sufficient vitamin A to carry them through to the next period of adequate intake. Indoor feeding of deficient rations, such as cereal by-products (corn cobs, etc.), sugar beet pulp, oilseed cake or meal, and straw, requires supplementation with carotene or vitamin A. Carotene or other precursors may be supplied as dried grass or lucerne (alfalfa) meal, or stabilized synthetic vitamin A may be added. If the indoor ration is of the high-energy, low-roughage type or has corn silage as a major constituent, stabilized preformed vitamin A may be required, as mentioned by Bridge and Spratling (*113*).

Natural carotenoid vitamin A precursors have not been a dependable source of vitamin A activity. The efficiency of conversion of carotene to vitamin A is related to the level of intake. Earlier sections of this chapter have discussed the relative instability of carotenes and other vitamin A precursors in feeds of plant origin and attempts to preserve them by processing controls, the addition of chemical additives, and the avoidance of prolonged storage before consumption by the animal. Stabilized vitamin A esters have the advantage of ease of stabilization since pure compound availability permits their incorporation into physical matrices with antioxidants to yield improved products as ration additives. As Bridge and Spratling (*113*) indicated, however, although synthetic vitamin A is protected by gelatin or wax with the addition of antioxidants, protection or stability still must be considered relative.

It is generally recognized that an animal must store vitamin A in its liver if it is to maintain an adequate level of vitamin A in the serum in order to supply all of the target tissues and organs. In the past animals accumulated substantial stores of vitamin A when grazing pastures and forage crops. Some experiments (*885*) completed at the Institute for Animal Feeding and Nutrition Research, Hoorn, Holland, with dairy cows showed that, when the level of vitamin A in 1 gm of liver is below 25 IU, the udder fails to develop properly. Calving occurs days too soon. When the calves are born, they are weak, and the production of milk is poor. Colostrum and milk are the primary sources of vitamin A for calves and lambs from which liver stores can begin to accumulate, thus initiating the cycle in the new generation. Koopman *et al.* (*475*) studied vitamin A and carotene in the livers of the calf, the cow, and the bovine fetus. At all stages of development the level of vitamin A in the fetal liver was low, but the level of carotene was very similar to that in the liver of the parent. The livers of calves that died shortly after birth tended to show normal fetal levels of carotene but subnormal fetal levels of vitamin A. The vitamin A nutrition of the newborn animal was reviewed by Phillips (*697*).

The degree of vitamin A deficiency may affect an animal's ability to utilize vitamin A and/or carotene upon subsequent treatment. Helmboldt *et al.* (*377*) showed that the administration of vitamin A but not carotene completely reversed metaplastic changes in the parotid ducts and markedly lowered raised CFP in cattle. Byers *et al.* (*135*), upon studying the effect of suboptimal carotene intake on successive generations of dairy cattle, realized that low carotene intake over extended periods adversely affects the animals' ability to convert the carotene present in the diet. Byers *et al.* (*136*) observed that animals of the Jersey and Holstein breeds from dams fed low-carotene rations in the latter part of their pregnancies and maintained for nearly 5 years at 50, 90, and 130 μg of carotene per kilogram of body weight had extremely low liver carotene and vitamin A values. These workers also indicated that, in normal Holstein cows, alfalfa hay does not maintain the liver vitamin A value as well as grass–legume silage and that neither maintains it as well as does irrigated fresh growing pasture.

In 1954, O'Donoghue (*646*) reported a field condition occurring in sheep and cattle in Alberta that was suspected of being a vitamin A deficiency. Animals presumably were on pasture or receiving alfalfa in the ration. Recovery was so dramatic following vitamin A therapy that it was concluded to be vitamin A deficiency despite the presence of atypical symptoms. Symptoms involving the nervous system that were most commonly observed were extension of head and neck, convulsions, staggering, muscular spasms, and often scouring. There was no indication of eye disorders or loss of appetite. Later (*645*), after reviewing similar case histories in beef cattle responding to vitamin A treatment, O'Donoghue concluded that frequently cases of vitamin A deficiency go undetected because typical symptoms are not always present. Brocklesby (*117*) suggested that marginal vitamin A deficiencies without marked clinical symptoms are fairly widespread. He observed that deficient cattle were unable to utilize carotene efficiently and recommended treatment with high levels of vitamin A. The need of cattle for vitamin A has been reviewed by McGillivray (*560*), Nadai (*621*), and others.

The use of liver storage values as a criterion for evaluating vitamin A supplements was shown by Harms *et al.* (*366*) to have tangible significance. These workers demonstrated that the onset of deficiency symptoms and survival time among chicks ingesting an essentially vitamin A-free diet are directly related to the amount of vitamin A stored in the liver before the time of feeding such a diet. Birds with high liver stores of vitamin A are more resistant to the onset of deficiency symptoms and also survive longer. Section XII,B indicates that true vitamin A is a more effective supplement to ensure liver stores than is orally administered carotene. Dry preparations of synthetic vitamin A esters in a gelatin matrix, according to Ascarelli and Senger (*34*) and Hvidsten *et al.* (*418*), are most effectively utilized, as judged by liver storage data. Hale and co-workers (*358*) fattened steers of 731 lb average live weight for 168 days on a ration containing little or no carotene but various levels of stabilized retinyl palmitate. It was found by liver biopsy and blood sampling that a daily intake of 40,000 IU of vitamin A was necessary to maintain the initial liver values.

Vitamin A activity of milk and dairy products is due not only to vitamin A but also to provitamin A carotenoids, the latter contributing color as well. Variations in the carotene intake of cows under normal husbandry practices are primarily responsible for the greatest variations in the vitamin A potency of milk (*863*). It is now well recognized that, for the production of milk containing fat of high vitamin A potency, cows are more dependent on a daily carotene and vitamin A intake than on liver reserves of vitamin A (*399*). Both the carotene and the vitamin A contents of the milk fat respond very rapidly to changes in provitamin A or vitamin A intakes, irrespective of body reserves under normal feeding operations, but, during periods of severe vitamin A deprivation, there is some transfer from liver and body depots of carotene and vitamin A to maintain milk fat levels. Since the intake of carotenoids is generally higher during the summer than during the winter, the fraction of total vitamin A activity due to the provitamin is greater in summer milk. Summer milk contains from 1.5 to as much as 15 times the amount of total vitamin A contained in winter milk, with small differences being more common. When cows are transferred from pasture to winter or stall-fed rations, a decline in the vitamin A content of the milk begins and continues to fall slowly through the dry feeding season until the cows are once more put on pasture in the spring.

The vitamin A in the milk of most species is mainly in the esterified form (*863*). The amount present appears to be fairly closely correlated with blood vitamin A ester levels. Large increases in the vitamin A content of milk can be obtained by supplementing the ration of the cow with a supplement form of vitamin A. When cows were fed large amounts of vitamin A, the concentration in the milk produced was from 3 to over 15 times that in normal pasture milk (*630*). Along with an increase in the vitamin A content of the milk, there is a simultaneous decrease in the carotene content, being greater with higher levels of vitamin A administered.

The efficiency of utilization of ingested vitamin A or carotene in the diet of the cow with regard to secretion in milk is lower when large amounts are fed but higher when smaller amounts are fed. Hjarde *et al.* (*399*) estimated that under normal conditions pasture-fed cows in Denmark consumed between 2 and 5 gm of carotene per head daily. Only about 1% of the total carotene intake was utilized as vitamin A excreted in the milk. With large amounts of carotene, there is an apparent transfer from the diet to the milk of between <1 and 6%, whereas, with low or very low intakes, values of from 10 to 47% have been observed. About 700,000 IU or more of vitamin A are required as a daily supplement to raise the total vitamin A value of winter milk to that of summer milk. If synthetic β-carotene were added, it would take 2.5–4.0 gm daily.

Placental transfer of vitamin A from the bovine dam and other pregnant animals to the fetus is very limited, and fetal liver and blood vitamin levels are normally extremely low. Only when pregnant animals receive massive doses of vitamin A does the passage of vitamin A to the fetus via the placenta become significant. Feeding high levels of vitamin A or carotene to dairy cows during late pregnancy (*817*) resulted in a 75 times greater liver vitamin A reserve in the newborn calves of cows receiving vitamin A. The newborn are dependent on colostrum for their vitamin A supply. Colostrum may be 6–20 times richer in carotene and vitamin A than milk (*560,760,921*). Since the uptake of carotene from milk by calves seems limited, their blood containing little carotene until after they have access to pasture, the intake of vitamin A in colostrum has high significance. Later, the vitamin A secreted in the milk of the dairy cow continues to meet, in part, the dietary need of the young animal. It is also desirable to attain a substantial level of vitamin A in fluid milk, as a component of its index of quality, when it serves as food for man. It should not be difficult to understand why a dairy cow producing a large volume of fluid milk daily has a higher vitamin A dietary requirement than a low-volume producer or a nonmilking animal. If colostrum and subsequent fluid milk are insufficient sources of vitamin A for the calf, another recourse is possible. Jones *et al.* (*436*), from field observations, noted that high calf mortality, particularly from scours and pneumonia, was correlated with vitamin A inadequacy in the calf. Prohaszka (*715*) found a correlation between prophylaxis of white scours in calves and early dosing with vitamin A. The administration of supplementary vitamin A to calves within the first week after birth has become a standard prophylactic measure, particularly during the months of dry pasture and/or the feeding of poor-quality dry forage.

With milk cows, a highly productive life, including high milk yield and production of healthy calves, is the ideal to be attained. To ensure satisfactory fertility among dairy cattle, a continuing adequate intake of vitamin A is required. Infertility is one of the most important economic losses in beef and dairy cattle herds. Nutritional inadequacy of the ration should not be overlooked. Vitamin A inadequacy can result in impaired reproduction. As has been shown in the bovine female, β-carotene also plays a specific role (*521,522,578,761*). Ehrlich (*231*) and Hoelzer (*406*) observed a close correlation between dietary vitamin A intake and herd fertility and suggested dietary supplementation or intramuscular (IM) administration of vitamin A as protection against infertility. The high vitamin A requirement of the breeding bull was estimated to be 45,000–120,000 IU by Karg (*439*). Roussel *et al.* (*749*) administered a daily dose of 45,000 IU vitamin A to breeding bulls in one group versus a group of bulls receiving their normal ration. Compared to the group on the normal ration, the supplemental group had a greater voume of ejaculate and ejaculate density and lower numbers of abnormal spermatoza. Other workers have also observed favorable results with vitamin A supplementation of breeding bulls.

Between 1960 and 1970 there was great concern about the adequacy of practical rations for supplying sufficient quantities of vitamin A to ruminants, particularly beef cattle. There were considerable discrepancies in the literature concerning the quantity of vitamin A required by feedlot cattle. Early requirement studies were conducted with carotene, and low requirement values were postulated. Some of these discrepancies were probably due to criteria used in assessing vitamin A adequacy. There have been a number of reports on beef cattle describing vitamin A deficiency symptoms, even though rations were thought to contain adequate quantities of carotene. Various factors, such as nitrates in forage and depressed thyroid activity, have been suggested as the cause. In two experiments at Purdue University by Beeson *et al.* (*62,63*), steers required a supplemental source of vitamin A if fed a ration containing 1.0 or 2.3 mg of carotene per pound or a ration supplying a daily intake of 18–46 mg of carotene. The ration furnishing the higher level of carotene contained 10% of sun-cured alfalfa pellets. Fortification of either ration with 20,000 IU of dry stabilized vitamin A palmitate per head daily increased daily gain 15–30% and improved feed efficiency 6–10%.

Why is there a need for higher vitamin A levels? Early studies on the carotene and vitamin A requirement of cattle were based on the minimal amount necssary to prevent night blindness and not on the need for optimal performance. Also, many changes have been made in cattle production in recent years, which influence the vitamin A requirement: (a) increased growth rate of animals, (b) improved beef cattle genotypes, (c) intensive management practices, and (d) introduction of all-concentrate or high-energy feeding practices. Several workers have provided clear indications that, to maintain liver stores of vitamin A in animals given all-concentrate rations, it is necessary to supply larger amounts of vitamin A than to comparable animals receiving traditional diets to maintain the higher rate of gain. It is generally recognized that feeding high-energy rations increases the requirement for vitamin A and carotene (*274,722*). Moreover, rations that contain high proportions of grain contain less roughage and are therefore lower in carotene content, and their supplementation with vitamin A is generally recommended (*250,274*).

For the past three decades various application forms (*51,56*) of vitamin A have been developed (Table 32), a tribute to developmental chemists and technologists, to meet the wide range of animal husbandry practices (Table 33). In practice, vitamin A can be given as a daily supplement in the feed or drinking water, or it can be administered at intervals of the animal's life by oral or parenteral administration. An experiment (*48*) illustrates the performance of in-diet and parenterally administered vitamin A in growing animals.

Sprague–Dawley male weanling rats were maintained on the U.S.P. XIV vitamin A-free ration supplemented with 1 IU vitamin A per gram feed until they reached a weight of 150–200 gm. They were then grouped according to weight distribution. There were four experimental variables: (a) vitamin A dry stabilized beadlets fed as a component of the diet (20 IU/gm diet) for 12 weeks, or a total intake of

TABLE 32

Application Forms of Vitamin A[a]

Product	Potency (IU/gm)
Liquid	
Fish oil concentrates	10,000–200,000
Vitamin A acetate in oil	500,000–2,500,000
Vitamin A palmitate in oil	500,000–1,500,000
Vitamin A ester emulsions of A, AD, ADE	50,000A/5000 D/30 E to 400,000 A/100,000 D/200 E
Dry	
Beadlets or related particles	
Vitamin A acetate ± D_3	A 500,000–750,000; D_3, 125,000–325,000
Vitamin A palmitate ± D_3	A, 250,000–325,000; D_3, 250,000–325,000
Water-dispersible vitamins A, AD_3	A, 250,000–500,000; D_3, 100,000–250,000
Parenteral	
Vitamins A, AD, ADE	50,000 A/25,000 D/20 E to 500,000 A/75,000 D/50E[b]

[a] From Bauernfeind and De Ritter (*51*).
[b] International units per milliliter.

TABLE 33

Mode of Administration of Vitamin A[a]

Route	Comments	Animals
Ingredients of feed or daily ration	Usually incorporated in the complete dry feed or food by the manufacturer	All animals
	In moist or semisolid canned or film packages	Dogs, cats
Diet or ration supplements	In tablet, treats, or award items	Dogs, cats
	In dry feed supplements fed with farm-grown cereal and plant protein feeds	Swine, ruminants, horses, more rarely poultry
	In liquid feed supplements	Cattle, some sheep
	In salt or protein field blocks	Cattle, sheep, goats
Drinking water	Usually in a multinutrient water-dispersible dry formulation under conditions of disease or other stress	Poultry, some swine
Parenteral	In water-base emulsion	Cats, dogs, horses, young farm animals
	In water-emulsifiable liquids	Older and mature swine and ruminants

[a] From Bauernfeind and De Ritter (*51*).

30,050 IU vitamin A per rat; (b) a single IM dose of vitamin A of 30,300 IU per rat in the muscle of the tibial area on the first day of the 12-week experiment, the animals being maintained on the vitamin A-free ration; (c) a combination of in-diet feeding plus single IM dosing (30,330 IU) [(a) plus (b)], or a total intake of 59,700 IU of vitamin A per rat for the 12-week period; and (d) the same combination (total dose 61,500 IU) as in (c) but the parenteral portion given in thirds at 0, 4, and 8 weeks of testing. Some animals in each group were allowed to continue to 16 weeks before sacrifice. Storage of vitamin A in the livers of rat sacrificed by CO_2 asphyxiation served as the criterion for biological efficacy.

Vitamin A administered at levels beyond the growth and maintenance requirement was stored in the liver of animals when either properly formulated dry stabilized products for addition to the diet or properly formulated parenteral prod-

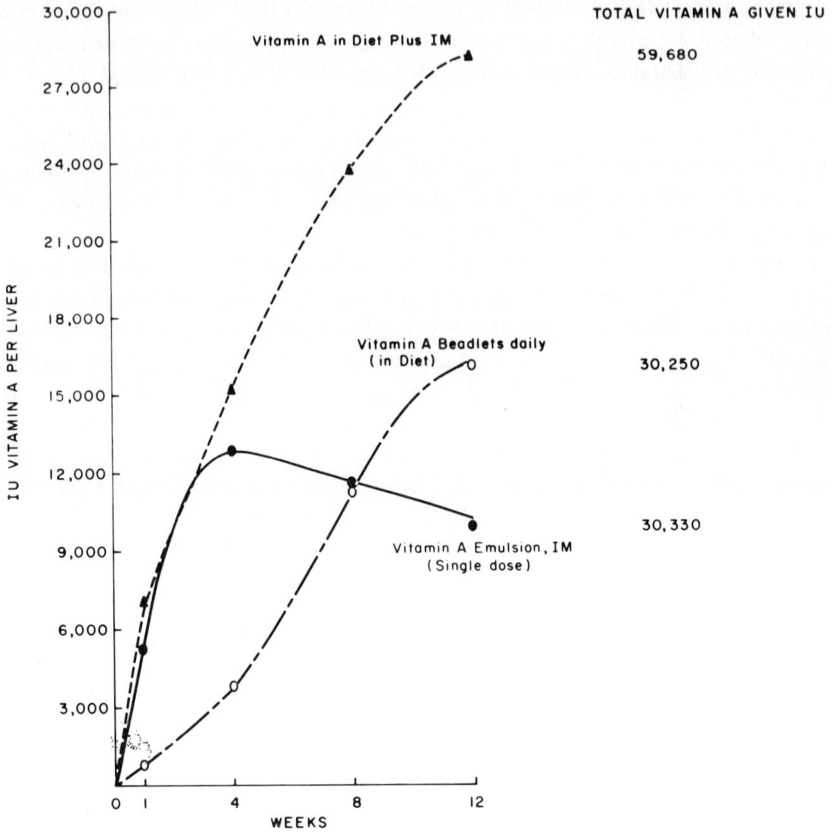

Fig. 20. Response of dietary and parenteral vitamin A. Stabilized vitamin A palmitate beadlets were fed in the diet to rats at a level of 20 IU/gm for 12 weeks (30,250 IU total) compared to an initial single IM injection of 30,330 IU of vitamin A in an emulsifiable parenteral composition and the combined dietary and parenteral application. [From Bauernfeind (*48*).]

ucts (*48*) injected IM were used (Fig. 20). When vitamin A was supplied in the diet at a level beyond the minimal needs, the liver concentration gradually increased, reaching a maximal value at the end of the feeding period. When administered parenterally, markedly increased liver concentrations were reached within a week after administration. When vitamin A was supplied by both routes simultaneously, high liver concentrations were quickly obtained, which increased further over the feeding period. One should bear in mind these patterns of vitamin deposition in the liver when weighing the merits of administering vitamin A to livestock by either of the two routes in field practice, the choice depending on the objective.

A. Oral Vitamin A

Before the recognition of vitamin A as an essential nutrient and the inception of the practice of adding a vitamin A source to animal feeds, domestic animals obtained vitamin A through the metabolic conversion of ingested carotenes and other precursors from plant concentrates and forages. The initial practice of adding fish oil and fish oil concentrates to feed gradually changed to the addition of chemically synthesized vitamin A in dry stabilized form (*48*). The addition of vitamin A to the ration is still the most economic and efficient mode of administering the vitamin. Occasionally, when prepared feeds or rations are not consumed readily, water-dispersible forms of the vitamin may be added to drinking water, but this method is more applicable to avian species. There are animal management situations in which dietary supplements and drinking water applications are not feasible or controlled and a real need exists for a parenteral form of vitamin A which is usually administered IM. Thompson (*855*) indicated that, when the intake of vitamin A is too low or too uncertain to yield optimal responses for health and economic goals, supplementary preformed vitamin A should be given because it is stable, cheap, and readily available. Synthetic β-carotene can be used commercially if the ration is insufficient in carotene content to meet the needs of the reproducing female bovine or possibly to improve milk color (*165a*).

1. Dry Rations

Greater amounts of vitamin A are used today in animal applications as an ingredient blended into the dry feed or ration. Although used to some extent in certain parts of the world, fish oil has been on the decline since the advent of synthetic vitamin A in 1950. In the manufacture of feed, liquids such as oils are sprayed onto the feed in the batch or continuous blending operation, the amount being controlled volumetrically. In some such instances, synthetic vitamin A with antioxidants in oil dilutions continues to be used; the addition of oil plays a secondary role of (a) reducing the dustiness of the feed and/or (b) supplying an added concentrated source of energy or calories. Although the spreading of oil solutions over the large surface area of the dry feed ingredient particles exposes the vitamin A more readily to oxidation by oxygen of the air, adequate short-term vitamin A stability can be achieved if antioxidants are incorporated in the oil solutions and if the feed is fed to the animals shortly after the feed is manufactured.

To counteract the stress factors involving vitamin A in feed manufacturing such as exposure to oxygen, trace mineral elements, pelleting, and feed storage, the feed industry has readily accepted the dry stabilized forms of the vitamin. One of the most effective forms developed to date (Fig. 21) is the dry gelatin beadlet in which the vitamin A ester (palmitate or acetate), emulsified into a gelatin–plasticizer–antioxidant viscous liquid formulation, is spray-dried (Fig. 22) into discrete dry particles. Such products have characteristics of good chemical stability, good physical stability (*51*), excellent biological availability, and very satisfactory mixing performance (Table 34) in mash and pelleted feeds. Dry stabilized vitamin A products of variable potencies, with or without vitamin D and/or vitamin E, are available on the market depending on the specifications of use, the blending equipment involved, and the market handling, storage, and use pattern of the final feed. Most frequently, the dry vitamin A form is incorporated into a premix with other small-volume nutrients before the premix is blended uniformly into the complete feed. Dry stabilized vitamin A is also used in milk replacers for calves and lambs and for dry pet foods.

Once inside the alimentary tract of the animal, the gelatin beadlet softens and releases the vitamin A, which is so finely dispersed that it is readily absorbed through the intestinal wall (*69,120,375,463*), with rapid increases in blood vitamin A levels and liver reserves. Substances other than gelatin have been used as carriers

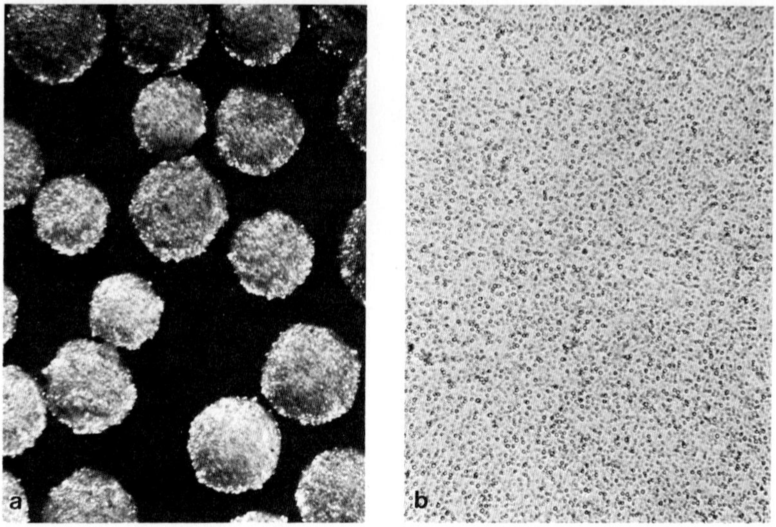

Fig. 21. Physical structure of stabilized vitamin A gelatin beadlets. (a) Outer view showing spherical shape and rough surface. (b) Inner view showing matrix emulsion of active ingredient.

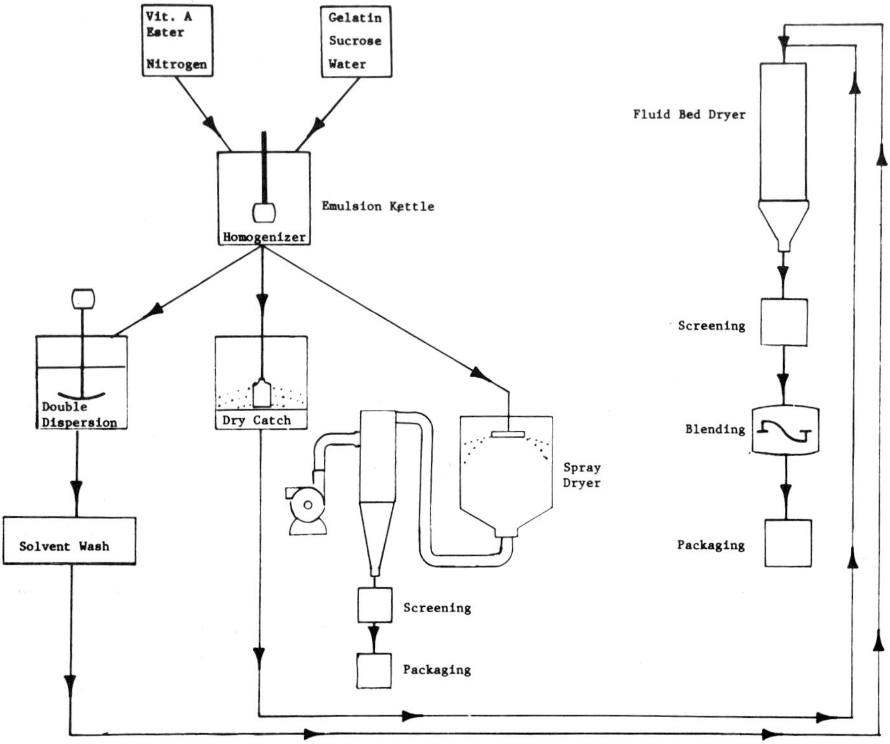

Fig. 22. Schematic layout for production of dry coated carotenoid or vitamin A products. (Courtesy of M. Cannalonga.)

in dry preparations of vitamin A (such as wax, fat, gums, and pectin), but gelatin-type formulations have been most widely used. To demonstrate beadlet digestion and vitamin absorption in the body (*120,841*), tritium-labeled vitamin A ([11,12-^{3}H] vitamin A acetate) incorporated into gelatin beadlets was fed to chickens as a component of the ration for a 30-min period (Fig. 23). About 755 IU of labeled vitamin A were consumed per bird. One hour after feeding 93% had already passed from the tract into the body. The majority of the unabsorbed portion, not having yet completely passed the crop, still had not reached the upper segment of the small intestine, where vitamin A is absorbed. Hence, vitamin A in gelatin beadlet form is biologically available. Even though biological availability and stability in feed have been demonstrated for the gelatin beadlet (*486,651*), questions arise as to whether there is an adequate number of dry vitamin A-active particles in the feed to provide a small animal (a 1-day-old chick, for example) with ample vitamin A. Particle counts, chemical assays, and growth studies have given assurance of satisfactory distribution of dry vitamin A in feeds.

TABLE 34

Stability of Several Forms of Vitamin A[a,b]

A. Dry vitamin A acetate, 650,000 IU/gm, in eight commercial broiler premixes

Storage test	Average retention (%) ± *SD*
3 weeks at 45°C	94 ± 6.3
4 weeks at RT	97 ± 5.3
8 weeks at RT	95 ± 3.4
16 weeks at RT	95 ± 4.1

B. Dry vitamin A palmitate, 325,000 IU/gm, in mash and pelleted broiler feed

	Retention of vitamin A (%)	
Storage test	Mash feed	Pelleted feed
30 days at 37°	94	72
30 days at RT	95	86
60 days at RT	98	78
90 days at RT	95	71

C. Dry vitamin A acetate, 650,000 and 325,000 IU/gm, under various conditions of use

		Retention of vitamin A (%)	
Condition	Storage test	Type 650,000 IU/gm	Type 325,000 IU/gm
Pelleted feed	30 days at RT	92	84
	60 days at RT	84	80
	30 days at 37°C	67	64
Mineral mix	120 days at RT	96	97
	42 days at 45°C	87	87

D. Water-dispersible liquid vitamin A in liquid feed supplement

Vitamin A form:	Acetate	Palmitate	Acetate + palmitate
Potency (IU/lb):	250,000	300,000	400,000
Liquid feed supplement, initial potency (IU/lb):	34,000	33.700	34.500
Storage test	Retention of vitamin A (%)		
2 weeks at RT			
Top half	90	91	97
Bottom half	102	115	114
Average	96	100	100
4 weeks at RT			
Top half	86	86	94
Bottom half	106	108	101
Average	96	97	98

(*Continued*)

TABLE 34—*Continued*

E. Dry vitamin A acetate, assayed, 325,000 IU/gm, in range mineral blocks (vitamin A potency: 345,000 IU/lb)

Storage test	Retention of vitamin A (%)
30 days at 45°C	90
60 days at 45°C	89
30 days at RT	92
90 days at RT	95
120 days at RT	96

F. Parenteral vitamin A, 100,000 and 500,000 IU/ml

Storage test	Retention of vitamin A (%)
Type 100,000 IU/ml	
3 weeks at 45°C	95
6 weeks at 45°C	91
3 months at 37°C	94
12 months at RT	94
Type 500,000 IU/ml	
3 weeks at 45°C	98
6 weeks at 45°C	93
6 months at 37°C	96
12 months at RT	98

[a] From Bauernfeind and De Ritter (*51*).
[b] RT, room temperature (23°C).

2. *Liquid Feed Supplements*

The term "supplement" in the feed industry has been usually applied to a combination of vitamins and minerals with added protein or nitrogen sources to be fed free-choice, blended with other ingredients grown on the farm, or purchased locally to form a complete animal ration. "Liquid feed supplement" is a term applied to a variety of products prepared from liquid or semisolid feed ingredients and used to provide supplementary nutrients to the roughage or dry portion of the animal ration. Liquid supplements are used mainly in ruminant rations, and they represent an expansion of the old practice of adding molasses to dry cattle feed as a source of energy and to render the dry feed more palatable to the animal. Many liquid supplements contain, in addition to molasses, fat, urea, a phosphorus source, trace minerals, salt, a calcium source, and frequently vitamins, A, D, and E. Because the viscosity, pH, and solids content of liquid feed supplements vary considerably, the development of vitamin A forms that would blend uniformly and be stable in such products represented a new challenge in the technology of vitamin A appplication forms. The products of choice are liquid emulsions of fat-soluble vitamins A, D, and E in formulations of emulsifiers, antioxidants, synergists, preservatives, and carriers. In addition to stability, homogeneity of distribution of the fat-soluble

Feed:

Vitamin A content:
100 000 I.U./kg

Radioactivity:
320 400 c/min g

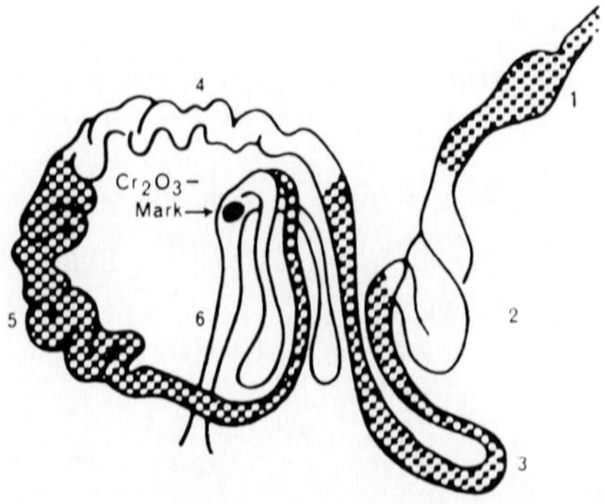

Total amount of Vitamin A consumed:
630 I.U.

Chicks killed one hour after administration

Radioactivity found in the intestinal tract				
Location	Wall	Content	Quotient	% of absorbed activity
	ipm	ipm		
1. Crop	0	6000	0	0·3
2. Proventriculus stomach and gizzard	4000	4000	1	0·4
3. Duodenum	6000	8000	0·75	0·69
4. Anterior part of the small intestine with bile and pancreas ducts	5900	800	7·4	0·33
5. Rest	14000	53000	0·26	3·32
6. Colon and caeca	0	0	—	0
	33000	71800		5·04

Fig. 23. Absorption of tritium-labeled vitamin A (in the form of gelatin beadlet) in chicks. [From Brubacher *et al.* (*120*) and Tagwerker (*841*).]

vitamins in the supplement, after standing, must be achieved if all animals consuming it are to receive an adequate supply of nutrients.

3. *Self-Feeding Situations*

There are special situations in which ruminants do not regularly eat a dry formulated ration or receive a liquid supplement; nor are they rounded up periodically to be injected with a vitamin A parenteral form. In such cases, supplementary vitamin A must be given for good reproductive or growth performance if the animals are on dry pasture or consuming poor-quality roughage. One method of providing some supplementary vitamin A is to mix a dry stabilizied vitamin A form into a salt or complete mineral supplement to be self-fed. The self-feeding stations should protect the supplement from rain. Frequent mixing and serving of the supplement are advisable, as the stress of certain weather conditions, moisture, etc., lowers the stability of this type of vitamin A product (*4,362*). In some instances, instead of a dry, granular mineral formulation, a self-feeding product, such as a range protein mineral or range mineral block or cubes, may contain vitamin A.

TABLE 35

Stability of Dry Vitamin A Mixed with Salt[a]

Week	Vitamin A (IU/gm)	Potency retained (%)
Trial 1[b]		
0	1000	100
1	740	74
2	790	79
3	620	62
4	630	63
5	100	10
6	13	1
Trial 2[c]		
0	970	100
1	790	81
2	783	81
3	788	81
4	754	78
5	791	82
6	819	84

[a] From Mitchell (*588*).
[b] Salt + vitamin A beadlets.
[c] Salt + vitamin A beadlets + 0.25% calcium stearate.

A test mixture in one trial (*588*) was a simple mixture of gelatinized vitamin A ester and salt. Wetting of the mixture by rain was not observed, but enough moisture seemed to be picked up from humid air and perhaps saliva to induce caking. This may have dissolved some of the gelatin coating, resulting in a loss of vitamin A potency. Calcium stearate added in a second trial prevented caking, greatly increased stability (*588*), and provided more uniform dispersion (Table 35). Moisture can dissolve the gelatin and permit contact between the salt and vitamin A, thereby accelerating loss of potency. Destruction in trace mineral mixtures under conditions of moisture and high temperature can also occur.

4. Drinking Water Preparations

In today's intensive animal production, situations develop in which animals "go off the feed" or eat sparingly. When animals are switched abruptly to new facilities and feed, they may not readily adjust. Often the cause of the decreased feed intake may be an acute microbial or viral infection. Animals that cease eating or eat very little continue to drink water. Stress vitamin and drug formulations exist for mass application to groups of animals, such as turkeys, chickens, ducks, and game birds.

Such preparations could also be used for larger animals. However, more frequently animals such as sheep are "drenched" or injected rather than allowed to drink the medicated water voluntarily. These stress formulations are usually dry products dispersed in the drinking water at the time of use, either on a batch scale or metered from a liquid concentrate into the flow of drinking water. The vitamin A product best suited for this application must possess (a) rapid dispersibility in cold water, (b) no residue of particulate matter that might foul or plug valves and metering devices, and (c) acceptable stability, both as a liquid concentrate and in the final drinking water.

Preparations of the vitamin in water-dispersible forms are similar in composition to the dry stabilized preparations, as they, too, consist of the vitamin in a very fine emulsion together with antioxidants, emulsifiers, and other stabilizing substances. They provide excellent availability of the vitamin to the animal (*792,814,923*) because the minute droplets of the very finely dispersed vitamin are easily absorbed through the intestinal wall.

5. *Oral Drench*

In 1950–1960, a series of pioneering vitamin A drenching investigations was performed by Australian scientists on lambs and weaners under conditions of protracted drought. In controlled experiments with weaned lambs and ewes, a vitamin A oral drench significantly increased live-weight gains, food consumption, and wool production as compared with untreated controls (*140*). Also, oral doses of vitamin A greatly reduced mortality among lambs (Table 36). These results confirmed earlier observations of Briggs *et al.* (*116*) and Kehar (*444*) that greater wool production is associated with higher carotene intakes and higher blood vitamin A levels. Kehar reported "that the total fleece of lambs held on a suboptimal level of carotene for a period of about 10 months yielded in two clips 559 gm. An identical group of animals whose ration was fortified with carotene produced 1189 gm. The maximal lengths of the fibers in the two groups were 2.8 and 8.4 in., respectively. From these trials it was concluded that a single oral dose of 10,000 IU of vitamin A per pound body weight (approximately 500,000 IU per animal) confered a high degree of protection against vitamin A deficiency for a period of up to 6 months for weaned lambs. The restoration of fertility in rams adversely affected by vitamin A deficiency under drought conditions was favorably influenced by vitamin A supplementation. Even after relatively short droughts, which usually do not have harmful effects on mature sheep, vitamin A supplements should be administered to rams to protect them against infertility. Ewes to be mated after long periods of low carotene intake may also benefit from vitamin A supplementation.

Evidence is therefore available that vitamin A drenching is beneficial to sheep under drought or semidrought conditions. It is therefore recommended that prophylactic dosing be carried out on lambs and weaners before and during the dry season. Vitamin A drenching of rams 6–8 weeks before breeding is also recommended. The reproduction performance of ewes as well as the health of future

TABLE 36

Prophylactic and Therapeutic Vitamin A Supplementation in Drought-Affected Sheep

Animals, investigator	Diet composition	Duration of experiment	Experimental group	Vitamin A supplementation[a]	Mortality (%)
Lambs, McClymont *(552)*	Wheat, 90%; wheaten chaff, 10%; ground limestone, 1.5%; restricted intake	250 days	Test group	515,000 IU (given in 3 doses orally)	18
			Control group	None	56
Weaners, Franklin *et al. (278a)*	Six subgroups receiving different percentages of wheat and wheaten chaff; fed daily or weekly	243 days	Test group	500,000 IU (one initial dose) orally	37.5[b]
			Control group	None	63[b]
Lambs, Campbell *(140)*	Wheaten chaff; 45%; oats, 25%; bran, 25%; linseed meal, 5%; limestone supplement	240 days	Test group	345,000 IU at 9, 131,000 IU at 18, 150,000 IU at 36 weeks of age, orally	8
			Control group	None	48
Weaners Campbell *(140)*	Five subgroups receiving different percentages of wheat and wheaten chaff; intake was restricted at drought maintenance levels; others were fed *ad libitum*	240 days	Test group	500,000 IU 21 days after beginning of experiment; one-half of weaners also received 345,000 IU by mouth 132 days after start of experiment	14.5
			Control group	None	55.5

[a] In oily solutions.
[b] Averages from all subgroups.

progeny may be seriously affected by poor vitamin A reserves. Under field conditions it is often difficult to predict the state of vitamin A reserves in animals or, in some countries, to predict when the next rain will come to renew pastures. The cost of vitamin A drenching is small; hence, prophylactic dosing of pregnant ewes seems advisable in areas that are likely to be afflicted with prolonged periods of severe drought.

B. Injectable Vitamin A

Past observations indicate that vitamin A injected IM from properly formulated parenteral products is transported and deposited efficiently in the liver, and from this depot the needs of the animal are served for an extended period. Thus, another mode of administration has been provided for one of the most critical vitamins for growth and reproduction in household and farm animals.

Vitamin A storage in the liver has been the main criterion for the effectiveness of vitamin A supplementation, regardless of the mode of administration. In mammals, about 95% of the total absorbed vitamin A is found in the liver (*159,553,607,788*), which releases the vitamin A back into the bloodstream when required, even during a deficiency of dietary intake (*605*). Aqueous emulsions or dispersions of vitamin A have been shown generally to be more effective than oil solutions when administered by the oral route in human beings (*512,902*), rats (*813,814*), chickens (*524*), calves (*421*) cows (*813*), sheep (*234*), and swine (*155,158,401*). Neumann *et al.* (*626*) reported that the IM injection of vitamin A in an oil solution was relatively ineffective in a 76-day trial with Hereford steers, as evidenced by plasma values, weight gains, and symptoms of foot rot and shipping fever. As early as 1942, little or no liver and blood response to IM injection of vitamin A in oil solution was noted in the rat (*39*), and hence the oral route (*510*) was favored as a mode of administration.

Aqueous vitamin A preparations given parenterally have been more effective than plain oil solutions in human beings (*480*), rats (*190,485,587,764*), chicks (*764,820*), goats (*568*), and swine (*155,156,400,485*). The IM administration of aqueous vitamin A dispersions or water-miscible preparations has been shown to be equal to or better than oral (or intraruminal) dosing of oil solutions in lambs (*97*), rats (*190*), swine (*156,159*), and poultry (*820*). More data from trials with rats, chicks, rabbits, and calves have been reported (Fig. 24). The IM administration of an aqueous dispersion was also highly effective in the buffalo and camel (*238*). The IM injection of vitamin A in aqueous dispersion or oral administration in oil produced equivalent liver stores in rats and pigs, according to Christensen *et al.* (*156*). In later work (*157*), in which tritium-labeled vitamin A was administered either in groundnut oil or as a polysorbate 80 aqueous dispersion, the two preparations were of equal value when given orally to rats. When injected, the vitamin A in aqueous dispersions had, after a 10-day period, been transferred from the muscle to the blood and liver, whereas the vitamin in the oil solution still remained in the muscle. The

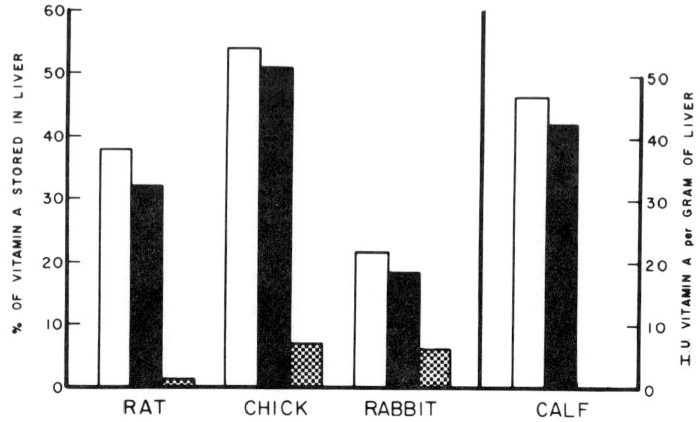

Fig. 24. Liver vitamin A values in four animal species after oral and IM administration of vitamin A preparations. Seven-day liver data for the rat and chick, 28-day data for the rabbit, and 100-day data for the calf; 50,000 IU were given to the rat, 25,000 IU to the chick, 100,000 IU to the rabbit, and 1,000,000 IU IM to the calf in one group and 10,000 IU/day in the diet for 100 days in another group. Key: rat, chick, and rabbit: open bars, water emulsifiable, single dose, oral; closed bars, water emulsifiable, single dose, IM; dotted bars, oil solution, single dose, IM. Calf: open bar, water emulsifiable, in-diet, daily; closed bar, water emulsifiable, single dose, IM.

weight of evidence favors the water-miscible form of vitamin A over oil solutions on the basis of rapidity and magnitude of liver storage of vitamin A in various species. Studies with tritium-labeled vitamin A injected intravenously (IV) as a polysorbate 80 aqueous dispersion in beef cattle (*374*) demonstrate that vitamin A efficiently stored in the liver is in continuous turnover; however, additional studies are needed to confirm and extend this pattern in liver performance. Not all parenteral water-dispersible vitamin A products perform (*53*) in an identical manner. Specific formulation factors, such as the type and level of emulsifier in the preparation, influence efficacy.

One of the special features of vitamin A administered parenterally to ruminants is that it is not subject to the environmental hazards of the rumen or stomach; ruminal destruction of vitamin A has been demonstrated in bovine species (*458*). Data have already been cited showing that nitrite in the diet interacts with dietary vitamin A but not with vitamin A given subcutaneously (SC) (*243*). The favorable influence of vitamin E on the biological efficacy of orally administered vitamin A is generally accepted. Vitamin A absorption is definitely impaired in vitamin E-deficient animals. Even when vitamin A is administered IM in emulsified form, utilization of the vitamin in vitamin E-deficient animals is impaired. With the simultaneous injection of vitamin E, the utilization of vitamin A is markedly increased in rats. These data indicate that the presence of an adequate amount of vitamin E is a prerequisite not only for improved intestinal absorption but also for *in vivo* utilization of vitamin A (*23,25*).

1. Parenteral Solutions

In the last decade parenteral administration of vitamin A has become a means of giving vitamin A either as a therapeutic measure or more frequently as a prophylactic approach when oral or dietary administration is either inconvenient or impossible (*48,52,840*). Frequently, parenteral vitamin A administration accompanies other individual animal treatments such as in roundups for dehorning, deworming, branding, or tagging. In other instances, such as with feedlot cattle, animals of unknown history brought in from range are given an IM injection of vitamin A as part of the adaptation or preconditioning process to help the animals adjust to the new environment. Parenteral administration of vitamin A may be indicated (*49,53*) when conditions for in-diet feeding of the vitamin are unfavorable, as (a) when animals are off feed, (b) when animals are on dry pasture, (c) during drought, (d) when manifest avitaminosis is evident, (e) when new animals are received with an unknown history, or (f) when extremely large quantities are to be given to protect animals over a prolonged period. Parenteral administration ensures that each animal receives its intended vitamin A allotment, whether or not it is consuming sufficient quantities of feed.

Parenteral solutions of vitamin A were originally used until they were recognized to be relatively ineffective biologically. The preparations of choice are (a) water-free emulsifiable vitamin A (sometimes with vitamin E) and (b) aqueous, water-dispersible vitamin A. The emulsifiable preparations, which can be prepared in a vitamin A potency up to 1,000,000 IU/ml, have very good biological performance, particularly in cattle. Usually these products (*53*) contain 500,000 IU/ml and an emulsifier that readily speeds the distribution of the fat-soluble vitamins by an emulsifying action in the tissue fluids following IM injection. Aqueous base-dispersible preparations (those containing ester) give excellent biological response (*54*) and are better tolerated by more sensitive animal species, such as horses, cats, and dogs. The translocation pattern (*54*) of an effective water-dispersible vitamin A formulation (100,000 IU/ml) from the muscle injection site to the liver depot is illustrated in Fig. 25. The potency of aqueous injectable vitamin A is more limited; most formulations of wide application usually contain 100,000 IU/ml.

Although IM injection deep into the muscle is favored, other techniques have been used. In the past, intraruminal (IR) administration in cattle was practiced; however, the IR approach does not differ significantly from "drenching" (both being oral variations), and currently it is rarely used. Comparisons of IM and in-diet administration have been made with several species of animals.

2. Weight Gain and Feed Efficiency

The daily needs of an animal for vitamin A must be met for maximal performance. Hale (*356*) reported that an oral intake of 40,000 IU of vitamin A daily is needed in the bovine to maintain liver vitamin A storage levels of 100 μg (333 IU) of vitamin A per gram and recommended vitamin A administration regardless of the carotene content of the ration.

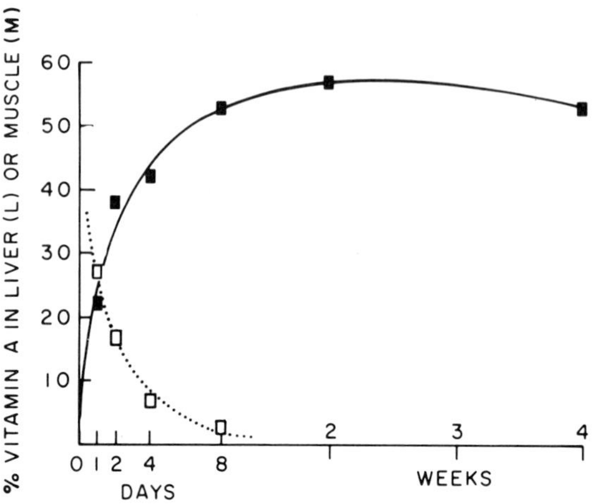

Fig. 25. Pattern of translocation of an effective water-dispersible vitamin A palmitate formulation injected IM to the liver in the chicken. Open squares, muscle; closed squares, liver. [From Bauernfeind *et al.* (*54*).]

Hinds *et al.* (*397*), noting that injected vitamin A is apparently useful under specific management practices, set up a trial with 70 head of cattle to determine how efficiently IM-injected vitamin A (1,000,000 IU) is stored, when the liver concentration reaches its peak, and how long a single injection of vitamin A lasts. Liver values peaked in 2 weeks and began to drop in 4 weeks. Steers maintained on a haylage ration stored a higher percentage of the injected vitamin A than those on silage. The question of injectable vitamin A in cattle was also considered by Meyer *et al.* (*580*), who concluded that a dose of 500,000–1,000,000 IU of vitamin A (IM or IR) in an effective vehicle would provide adequate vitamin protection for 60–90 days, as judged by liver stores.

For several years, Beeson *et al.* (*63*) and Perry *et al.* (*688–691*) at Purdue University compared single parenteral injections of vitamin A with oral vitamin A fed daily as a means of administering the vitamin to growing–fattening Hereford steers on a low-carotene diet (1.5 mg carotene per kilogram). A dry stabilized vitamin A palmitate gelatin beadlet product (325,000 IU/gm) and an IM-injectible vitamin A formulation (500,000 IU/ml) were employed. Individual blood and liver vitamin A levels were measured in animals in two trials for 210 days, as were weight gains and feed consumption values. The data (*690*) are presented in Table 37. Vitamin A administered orally (20,000 IU per head daily) initially appeared to be equally effective as that administered IM (1 million, 4 million or 6 million IU) in stimulating the rate of gain over that of control cattle receiving no supplemental vitamin A. However, in a feeding trial of 210 days, the two higher levels of parenteral vitamin A (4 million and 6 million IU) were more effective in maintaining serum and liver vitamin A concentrations than the lower level (1 million IU).

This indicated that the lower level should be administered several times during such a long feeding period, if this parenteral form is to be the vitamin A source. Both methods of administering vitamin A resulted in significant ($p < 0.05$) increases in daily gain and improved feed conversions. No effect of treatment on carcass grades was detected. Record *et al.* (*727*) reviewed the Purdue experiments.

Cunha (*173*) recommended that steers coming into feedlots, vitamin A-depleted steers, or steers suffering from stress (such as shipping stress, excessive handling, subclinical disease level, and cold weather) be given 500,000–1,000,000 IU of vitamin A shortly after arrival, either in the feed, in the water, or by injection. Trials at Purdue University (Table 37), in Illinois, and elsewhere have demonstrated that parenteral administration is effective in beef cattle. Workers at Purdue University (*915*) suggested an IM injection of 1,000,000 IU of vitamin A or more per head of growing–fattening cattle on arrival at the feedlot. Range beef cattle arriving at feedlots for fattening with relatively unknown nutriture and subjected to stress of handling and shipment are often given an IM injection of 1,000,000 IU of vitamin A on arrival as a protective measure against shipping fever and to ensure vitamin A adequacy. Whether vitamin A is administered by daily feeding, massive oral doses, or IM or IR injections, supplementary vitamin A should be viewed as an insurance program, according to Lane (*503*). Perry (*688*), reviewing his experience with water-emulsifiable parenteral vitamin A in cattle and published reports on vitamin

TABLE 37

Response of Beef Cattle[a] to Dietary and Injectable Vitamin A[b]

Trial	Daily gain (kg)	Weight increase (%)	Feed/ kg gain (kg)	Feed saving (%)	Carcass grade
1a. Corn, cob, soy ration[c]	0.80	—	9.4	—	5.7
1b. 20,000 IU vitamin A[d]/head/daily ration	0.94[e]	18	8.5	9	6.4
1c. 1,000,000 IU vitamin A injection[f]	0.91[e]	14	8.5	9	7.0
1d. Oral vitamin A (1b) plus injection (1c)	0.94[e]	18	8.5	9	6.0
1e. 6,000,000 IU vitamin A injection	0.88[e]	11	8.8	6	5.7
1f. Oral vitamin A (1b) plus injection (1e)	0.93[e]	16	8.3	12	5.5
2a. Corn, cob, soy ration[c]	0.88	—	9.9	—	6.4
2b. 20,000 IU vitamin A[d]/head/daily ration	0.98[e]	11	9.4	5	6.4
2c. 4,000,000 IU vitamin A injection[f]	0.94[e]	7	9.7	2	6.3

[a] Hereford steer calves; average initial weight 245 kg (trial 1), 275 kg (trial 2); 72 steers in each trial; 210-day period.

[b] From Perry *et al.* (*690*).

[c] All cattle fed rations of rolled yellow corn, 69%; ground corn cobs, 20%; soybean meal (44% protein), 10%; bone meal, 0.7%; salt with cobalt 0.3% (trial 1), same with 3% molasses (trial 2).

[d] Stabilized vitamin A palmitate beadlet.

[e] Significant at 5% level over lot 1a (trial 1), 2a (trial 2).

[f] Vitamin A alcohol emulsifiable parenteral.

A, cites (a) improved reproductive performance, (b) heavier calves, and (c) efficient liver vitamin A storage as advantages of the treatment.

Animals grazing during prolonged droughts or throughout the winter season may become deficient in vitamin A, and, if no suitable supplementary feed can be given, vitamin A is probably best provided by a single large dose as an oral drench or by injection, according to Bridge and Spratling (*113*). Weight gain responses to parenteral vitamin A in beef cattle receiving a limited intake from concentrate while on pasture were observed by Chapman *et al.* (*152*).

Vitamins A and E were administered parenterally to lambs and sheep by Perry *et al.* (*693*) and, although some improvement in weight gain and lambing performance was obtained, the degree of improvement was variable and, according to the authors, did not justify routine use. Hanke and Jordan (*364*), administering vitamin A (500,000 IU) IM, failed to obtain significant weight gain in fattening lambs in field trials. Bell *et al.* (*65*) reported a variable response to supplementary injectable vitamin A in growing lambs. In some flocks, increased gains were obtained and wool length tended to be greater; in others, little difference was noted. Injection of vitamin A (500,000 IU) into pregnant ewes about 4 weeks before lambing tended to increase the vitamin content of the colostrum, and the liver values were higher in lambs from ewes receiving the injection. Of interest is the work of Varnell and Erwin (*886*) in which lambs were injected with vitamin A by IR, IM, SC, and intraperitoneal routes. Increased blood levels and liver stores of vitamin A were obtained with all modes of administration.

3. *Improved Reproduction*

The dependency of the fertility and reproduction of farm animals on vitamin A or vitamins A and E has been observed and reviewed (*5,8,49,93,605,841*). Grazing cattle, especially those in arid or drought areas, may be low in the calf crop. One-half of a Texan herd of 182 previously bred cows was injected IM with 2,000,000 IU of water-emulsifiable vitamin A in the fall and again in the spring. The injected group produced 12% more calves than the control group. The authors (*298,299*) concluded from this study that during the prebreeding and breeding period the injection of 500,000 IU of vitamin A per 250 lb of body weight increases liver stores for stress periods and improves breeding performance by increasing the percentage of conceptions and the number of healthy calves. Although Totusek (*873*), in cattle trials in Oklahoma, did not find IR injection of vitamin A (1,000,000 IU) before calving to have an influence on cow weight or calf survival, he noted that fewer injected cows remained unbred the following year than untreated cows and that calves from treated cows were heavier. In a desert range environment in Arizona, Lane (*504*) observed that vitamin A injected IM in yearlings and heifers yielded an increase in calf crop and heavier weaned calves. In his subsequent vitamin A studies, the number of cows reached several thousand in eight trials, one-half of which received injections of vitamin A (1,000,000 IU). Vitamin A administration to yearling heifers that calved as 2-year-olds produced calves averaging 20 lb heavier than calves from untreated animals. When treatment was before the breeding season, the calf crop increased 10–11%. Melton (*576*) and Meacham *et al.* (*573*) also showed improved reproductive performance in beef cows with vitamin A supplementation.

Bradfield and Behrens (*108*) conducted a 4-year study on nine ranches in the desert of southern Nevada involving 2417 beef cows on the range and obtained a 12–14% increase in pregnancy among animals administered vitamins A, D, and E IM compared to nontreated animals. Comparing three age groups of cows, a higher percentage of pregnancy was found for the injected group in each case (Table 38). Data comparing pasture and range breeding in the presence and absence of treatment again revealed an improvement in the percentages of pregnancy for the treated group. As a result of these studies, parenteral vitamin A administration before breeding has been extended to practical field use.

Kupferschmied and Rehm (*497*) studied fertility disturbances in bovine males due to vitamin A deficiency. At the Insemination Station in Neuenberg, Switzerland, they tested the effect of parenterally administered vitamin A on the breeding performance of bulls receiving a ration adequate in vitamin A (carotene) content. Control bulls were kept on the regular ration. Treated bulls were given an IM injection of 2–3 million IU of vitamin A. Observations were made before the experiment, several months after the parenteral treatment, and following associated

TABLE 38

Pregnancy Improvement with Parenterally Administered Vitamins A, D, and E in Cows[a]

	Treated group		Control group	
Age group	Animals (no.)	Pregnant (%)	Animals (no.)	Pregnant (%)
Mature cows	1097	84.5	582	70.1
First-calf heifers	241	83.0	129	74.9
Replacement heifers	261	79.3	107	64.5

	Control group				Treated group			
	Pasture breeding		Range breeding		Pasture breeding		Range breeding	
Year	Animals (no.)	Pregnant (%)	Animals (no.)	Pregnant (%)	Animals (no.)	Pregnant (%)	Animals (no.)	Pregnant (%)
1964	49	81.7	13	61.5	89	90.1	12	75.0
1965	116	80.6	159	74.6	214	92.7	233	91.2
1966	109	65.6	108	50.0	305	88.5	212	63.5
1967	207	75.7	57	66.8	387	84.8	147	69.5
Total	481	75.2	337	64.9	995	88.1	604	75.9

[a] From Bradfield and Behrens (*108*).

pregnancies. In a number of comparative trials, a lower incidence of stillbirths (Table 39) and an increased volume and density of sperm were observed as resulting from vitamin A treatment. The authors pointed out the economic significance of these findings. Later studies involved tritium-labeled vitamin A (*730*). In a study of Aberdeen Angus breeding bulls (*298*,*299*) involving two trials of 50 and 40 animals, respectively, the effect of IM injections of water-emulsifiable vitamin A (3,000,000 IU) on the fertility of breeding bulls was determined. One to two months after treatment a significant number of animals had improved spermatogenesis and breeding capabilities. In a study (*454*) at a Pennsylvania breeders' cooperative, dairy and beef bulls receiving corn silage, alfalfa hay, and a 14% protein concentrate responded to supplementary vitamin A, which was given as a 2,000,000 IU IM injection initially and again 90 days later. Semen characteristics and volume, sperm concentration and motility seemed to be improved by the supplementary parenteral vitamin A administration. Ehrlich (*232*) also stressed the close connection between the fertility of the herd and the carotene supply and recommended administering vitamin A (1,000,000 IU) to provide protection against infertility under conditions of low carotene intake. Lunca and Musetescu (*523*) injected cattle with vitamin A (500,000 IU) and vitamin E (250 mg) three or four times and observed an improvement in fertility. Nicholson and Cunningham (*628*) attempted to correlate the level of vitamin A blood plasma or liver with the incidence of retained placenta, abortion, or stillbirth but found that no correlation exists. Thomas *et al.* (*854*) found that supplemental vitamins did not improve sperm output of bulls or production by cows.

Maintenance of spermatogenesis in the boar was influenced by injected vitamin A, according to Palludan (*665*). Parenteral administration of vitamin A or vitamins A and E improved the reproductive performance of the horse (*5*). In a pilot study (*36*) conducted at Trangie Agricultural Research Station, Australia, merino rams grazing on pastures were given a single injection of 1,000,000 IU of vitamin A. Eight weeks later, the rams had significantly higher densities of sperm in their ejaculates than untreated rams. Sperm production may be reduced if rams graze on dry pasture for longer than 9 months, and if sperm production falls below a critical level the fertility of the flocks attached to those rams will be reduced. In most

TABLE 39

Influence of IM Administration of Vitamin A on Stillbirths in Cattle[a]

Group	Offspring (no.)	Stillbirths (no.)	Stillbirths (%)
Control	1614	109	6.75
Treated	959	34	3.54
Difference			3.21

[a] From Kupferschmied and Rehm (*497*).

pasture conditions, sufficient quantities of carotene should be available to meet the needs of rams.

4. *Support for Disease Therapy*

Vitamin A is essential for the maintenance of normal mucosal tissue of the respiratory, digestive, reproductive, and urinary tracts and the mucous membranes of the eyes. There are abundant reports that an insufficiency of vitamin A increases an animal's susceptibility to microbial diseases and internal parasitic infections (*414,644,767,839,841,866*) and that in a number of diseases there is a disturbed absorption and/or mobilization of the vitamin. Greater resistance to disease has also been observed with orally administered vitamin A in poultry. Attention has been paid to the influence of vitamin A status on disease, in particular coccidiosis in poultry. Increased demand for vitamin A by birds with parasitic infections is probably a reflection of poorer absorption due to damage to the mucosa of the digestive tract. Increased susceptibility to infection in deficient birds may also be due to the replacement of secretory epithelia by keratinized membranes. As reviewed by Thompson (*855*), it has long been known that chicks with a marginal vitamin A intake are more susceptible to coccidiosis than chicks with a normal intake. Erasmus *et al.* (*246*) found that the severity of the disease was the same regardless of the level of vitamin A but that dosing with vitamin A improved the recovery. β-Carotene was not as effective as vitamin A, and infected birds stored less of the vitamin in their livers. Gerriets (*300*), in experiments with some 40,000 pullets, found that increasing vitamin A intakes from 3000 to 10,000 IU per kilogram was more effective against coccidiosis than giving 800 ppm of the coccidiostat 3,5-dinitro-*o*-toluamide. In turkeys artificially infected with *Histomonas meleagridis*, mortality and lesions were reduced by additional vitamin A (*909*). Direct treatment of young animals with water-emulsifiable vitamins A and E is recommended by Kraft (*479*) as supportive therapy from a nutritional and disease control standpoint.

Hale *et al.* (*360*) reported that the IM injection of vitamin A (1,000,000 IU) in incoming steers was very effective in the recovery of weight lost in shipment when compared to control animals. A reduction in the number of steers requiring medication was not observed. The authors also noted a smaller incidence of ringworm in cattle in the field that were administered supplementary vitamin A IM. In east central Texas, veterinarians (*827*) studying eye pathology found corneal opacity, marked conjunctivitis, profuse lacrimation (symptoms characteristic of vitamin A deficiency), and pinkeye in Hereford bulls. They noted a marked reduction in the afflicted animals 3 weeks after injection of an emulsifiable vitamin A preparation (2.5 million or 5 million IU). Pasture-grazing beef cattle with diarrhea but free of intestinal parasites and with a diagnosis of high nitrate intake responded to IM vitamin A injection (2,500,000 IU) 12 days after the injection. Although the authors deny parenteral vitamin A is a cure-all, they acknowledge its therapeutic value in disease conditions involving epithelial tissue and suggest its use in concomitant chemotherapy.

When calf scours have been a problem in U.S. dairy and beef herds on poor roughage in late winter, IM administration of vitamin A to the dam several weeks before parturition has been helpful in alleviating the conditions. German (*299*) and German and Adams (*298*) reported observations made on animals treated with water-emulsifiable vitamin A injections. In one study, 775 grazing yearling steers were losing weight and had rough hair coats and watery eyes. Twelve of the steers were injected (IM) with vitamin A (1,000,000 IU), and within 48 hr the animals showed marked improvement. The remainder of the animals were then treated and responded similarly. In another case, 20 dairy herd replacement calves were injected IM with 500,000 IU of emulsifiable vitamin A, and 20 were left untreated. At 3 months of age, and following the first cold weather, all the injected calves were normal, whereas all the untreated calves developed scours and two calves in this group died. A beneficial effect of vitamin A administration on calf diarrhea in Nebraska trials was noted by Schleicher and Clanton (*765*). Calving difficulties in heifers wintered on low-quality roughages might be avoided by vitamin injections, according to Gay (*297*) at Iowa State University. Heifers originating from a drought area in New Mexico and injected with a combination of vitamins A, D, and E showed a much lower incidence of calving problems. The number of normal births was higher in the group receiving vitamins, and such complications as retained placenta and uterine prolapse were almost eliminated. Vitamin injections had no effect on liver weight gains.

Emulsifiable, parenteral vitamin A, D, and E preparations were field-tested with concomitant chemotherapy employing a sulfonamide (sulforthomidine) to treat infectious cattle diseases such as respiratory infections (pneumonia) and alimentary tract infections (gastroenteritis) (*64*). Among the animals treated were 382 cattle, 293 calves, and 819 pigs. The cattle received 1,000,000–2,000,000 IU, calves 500,000 IU, pigs 1,000,000 IU, and piglets 250,000 IU of vitamin A IM in addition to sulfa drug therapy. Data (*64*) from these trials are summarized in Table 40 and support the use of combined parenteral vitamin A administration and sulfa drug therapy for these infectious diseases. Chaffee (*145*) reported favorably on the combined use of parenteral vitamin A and sulfadimethoxine.

In certain situations with cattle (*8*), massive doses of vitamins (1,000,000 IU of vitamin A and 250–1000 IU of vitamin E) administered orally or parenterally may be indicated. Treatment may be repeated, depending on the circumstances. In dairy cattle, such situations include retained placenta, weak or dead calves at birth, unexplained abortion, acetonemia or ketosis (*936*), downer cows, vague appetite problems and nitrate poisoning, and cows' deprived of pasture or green-chop because of drought conditions. O'Donoghue (*644*) and Tagwerker (*839,841*) recommend supplementary vitamin A for pregnant beef cattle and feedlot cattle. Symptoms in cattle that may respond to vitamin supplementation are abortion, weak calves, scours, central nervous system derangement, impaired vision, pneumonia, and subcutaneous edema (anasarca).

TABLE 40

Vitamin ADE and Sulfonamide Therapy for Animal Infections[a]

Species	Treatment	Total animals (no.)	Cured (%)	Improved (%)	Efficacy tally	
					Cured or improved (%)	Doubtful or negative (%)
Cattle	Sulforthomidine	336	63	22	85	15
	Sulforthomidine plus vitamin	46	76	17	93	7
Calves	Sulforthomidine	110	66	13	79	21
	Sulforthomidine plus vitamin	183	84	9	93	7
Pigs	Sulforthomidine	295	49	37	86	14
(>30 kg)	sulforthomidine plus vitamin	35	94	0	94	6
Pigs	Sulforthomidine	397	54	34	88	12
(<30 kg)	Sulforthomidine plus vitamin	92	64	33	97	3
All animals	Sulforthomidine	1138	56	29	85	15[b]
	Sulforthomidine plus vitamins	356	78	16	94	6[b]

[a] From Behrens *et al.* *(64)*.

[b] Highly significant difference in the chi-square test.

C. Requirements

National Research Council estimated requirements of various farm animals for vitamin A as well as feeding levels administered to animals under field conditions are shown in Table 41. The relative needs (*753*) of vitamin A activity for various farm animals are illustrated in Fig. 26.

TABLE 41

Vitamin A Requirements of Domestic Animals

Animal[a]	NRC[b]	Practical use levels
	IU/kg diet (90% dry matter)	
Cats	9,000	12,000
Chickens		
Start-grow	1,500	5,000
Lay-breed	4,000	6,000
Dogs	4,500	5,800
Ducks	4,000	5,000
Fish	2,000	4,000
Foxes	2,410	4,000
Mink	3,500	6,000
Pheasants	3,000	5,000
Quail	5,000	8,000
Rabbits		
Growth	580	2,000
Gestation-lactation	1,160	3,000
Turkeys		
Start	4,000	8,000
Grow	4,000	8,000
Breed	4,000	8,000
	IU/head/daily	
Cattle, beef		
Fattening (350)	18,000	30,000
Gestation (500)	24,000	40,000
Lactation (500)	38,000	50,000
Bulls (600)	48,000	60,000
Cattle, dairy		
Veal calves (75)	3,200	10,000
Growing stock (250)	10,000	20,000
Lactating cows (500)	38,000	60,000
Bulls (600)	25,000	50,000
Horses		
Growing (400)	14,000	30,000
At rest (500)	12,500	25,000

(Continued)

TABLE 41—*Continued*

Animal[a]	NRC[b]	Practical use levels
Med. work (500)	14,000	25,000
Gestation (500)	30,000	40,000
Lactation (500)	33,000	50,000
Sheep		
Fattening (35)	990	2,000
Replacements (50)	2,350	3,000
Gestation (60)	5,660	6,000
Lactation (60)	5,650	6,000
Swine		
Growing		
5-10 kg	1,100	2,000
60-100 kg	3,900	5,000
Gestation	7,000	10,000
Lactation	8,000	10,000
Boars	9,500	12,000

[a] Numbers in parentheses indicate weight of animal in kilograms.
[b] NRC, National Research Council-National Academy of Sciences: Nutrient requirements of domestic animals.

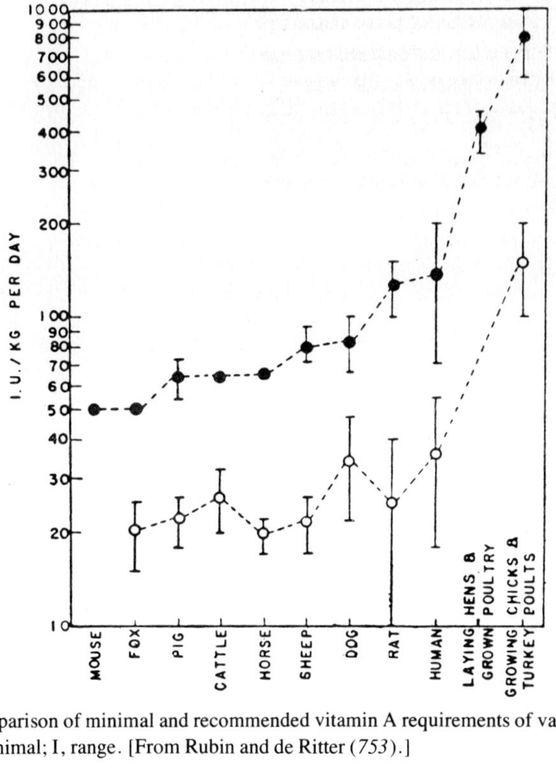

Fig. 26. Comparison of minimal and recommended vitamin A requirements of various species. Key: ●, optimal; ○, minimal; I, range. [From Rubin and de Ritter (*753*).]

REFERENCES

1. Abelin, I. (1933). *Hoppe-Seyler's Z. Physiol. Chem.* **217,** 109–114.
2. Abrams, J. T., Barnett, K. C., Bridge, P. S., Palmer, A. C., Spratling, F. R., and Sharman, I. M. (1969). *Int. Z. Vitaminforsch.* **39,** 416–425.
3. Adams, A. W., Emerick, R. J., and Carlson, C. W. (1966). *Poult. Sci.* **45,** 1215–1222.
4. Adams, C. R. (1974). *Feedstuffs* **46**(34), August 26.
5. Adams, C. R. (1972). "Stud Managers Handbook," Vol. 8, pp. 37–41. Agriserv. Found., Clovis, California.
6. Adlersberg, D. S., Kann, A. P., Mauer, K., Newerly, W., Winterernitz, W., and Sobotka, H. (1949). *Am. J. Dig. Dis.* **16,** 333–337.
7. Adlersberg, D. S., and Sobotka, H. (1943). *J. Nutr.* **25,** 255–263.
8. Aehnelt, E., and Konermann, H. (1961). *Proc. Int. Congr. Anim. Reprod., The Hague* **3,** 437–439.
9. Ahlswede, L., and Lotthammer, K. H. (1978). *Dsch. Tieraerztl. Wochenschr.* **85,** 7–12.
10. Ahmad, B. (1937). *Biochem. J.* **25,** 1195–1204.
11. Aitken, F. C., and Hankin, R. G. (1970). "Vitamins in Feeds for Livestock," Tech. Commun. No. 28. Common. Bur. Anim. Nutr., Commonw. Agric. Bur., Farnham Royal, England.
12. Akopjan, W. E. (1956). *Veterinariya (Moscow)* **1,** 129.
13. Alam, A. U., Couch, J. R., and Creger, C. R. (1968). *Can. J. Bot.* **46,** 1539–1541.
14. Alexander, B., Lorenzen, E., Hoffmann, R., and Garfinkel, A. (1947). *Proc. Soc. Exp. Biol. Med.* **65,** 275–278.
15. Alexsson, J. (1949). *J. Dairy Sci.* **32,** 60–70.
16. Al-Hasani, S. M., and Parrish, D. B. (1972). *J. Nutr.* **102,** 1437–1440.
17. Al-Hasani, S. M., and Parrish, D. B. (1968). *J. Nutr.* **94,** 402–406.
18. Almendinger, R., and Hinds, F. C. (1969). *J. Nutr.* **97,** 13–18.
19. Almquist, H. J., and Maurer, S. (1955). *Arch. Biochem. Biophys.* **55,** 297–298.
20. Almquist, H. J., Jukes, T. H., and Newton, W. E. (1938). *Calif. Agric. Exp. Stn., Circ. No.* 108.
21. Ames, S. R. (1969). *Am. J. Clin. Nutr.* **22,** 934–935.
22. Ames, S. R. (1968). *Proc. Symp. Assoc. Vitam. Chem., Chicago* p. 17.
23. Ames, S. R. (1962). *Feedstuffs* **35**(6), 52.
24. Ames, S. R. (1960). *Feedstuffs* **32**(10), 38.
25. Ames, S. R., and O'Rourke, W. F. (1965). *J. Anim. Sci.* **24**(3), 871.
26. Anderson, O. (1939). *Klin. Wochenschr.* **18,** 499–502.
27. Anderson, T. A., Hubbert, F., Jr., Roubicek, C. B., and Taylor, R. E. (1962). *J. Nutr.* **78,** 341–347.
28. Anderson, T. A., Hubbert, F., Roubicek, C. B., and Taylor, R. E. (1961). *J. Anim. Sci.* **20,** 923–924.
29. AOAC Methods of Analysis (1965). 10th ed. (1975). 12th ed. Assoc. Off. Anal. Chem., Washington, D.C.
30. Aoe, H., Masuda, I., Saito, T., and Komo, A. (1967). *Nippon Suisan Gakkaishi* **33,** 959; **33**(4), 355–360; **33**(10), 970–974.
31. Arnich, L., and Pederson, D. J. (1956). *Fed. Proc., Fed. Am. Soc. Exp. Biol.* **15,** 212.
32. Ascarelli, I., and Bondi, A. (1957). *J. Agric. Sci.* **49,** 113–119.
33. Ascarelli, I., Minkov, U., and Nir, I. (1969). *J. Sci. Food Agric.* **20**(6), 336–339.
34. Ascarelli, I., and Senger, M. (1962). *J. Sci. Food Agric.* **13,** 332–338.
35. Atkeson, F. W., Hughes, J. S., Kunerth, B. L., Peterson, W. J., and Kramer, M. (1937). *J. Nutr.* **14,** 621–629.
36. Australian Country (1975). Mar.
37. Bahtin, S. N. (1968). *Vestn. Sk. Nauki (Moscow)* **12,** 70–71.
38. Baker, F. H., MacVicar, R., Pope, L. S., and Whitehair, C. K. (1953). *Proc. Soc. Exp. Biol. Med.* **83,** 571–574.

39. Barlow, O. W., and Kocher, A. (1942). *Am. J. Physiol.* **137,** 213–216.
40. Barrick, E. R., Andrews, F. N., and Bullard, J. F. (1948). *J. Anim. Sci.* **7,** 539.
41. Bartov, I., and Bornstein, S. (1967). *Poult. Sci.* **46,** 796–805.
42. Bassett, C. F. (1962). *Dan. Pelsdyravl.* **25,** 136–137.
43. Bassett, C. F. (1962). *Am. Fur Breeder* **34**(12), 18–20.
44. Bassett, C. F., Harris, L. E., Smith, S. E., and Yeoman, E. D. (1946). *Cornell Vet.* **36,** 5–16.
45. Basu, N. K. (1937). *Z. Vitaminforsch.* **6,** 106–110.
46. Bat'kaev, R. Y., and Roslyakov, L. A. (1974). *Vestn. Sk. Nauki Kaz.* **17**(3), 47–50.
47. Bauernfeind, J. C. (1972). *J. Agric. Food Chem.* **20**(3), 456–473.
48. Bauernfeind, J. C. (1969). *World Rev. Anim. Prod.* **5**(21), 20–50.
49. Bauernfeind, J. C., and Adams, C. R. (1970). *Proc. Int. Conf. Cattle Dis. 6th, Philadelphia, Pa.* pp. 411–416.
50. Bauernfeind, J. C., and Cort, W. M. (1974). *Crit. Rev. Food Technol.* **4,** 337–375.
51. Bauernfeind, J. C., and De Ritter, E. (1972). *Feedstuffs* **44**(5), 34–41.
52. Bauernfeind, J. C., and Marusich, W. L. (1971). *News Rev., Roche, Basel* pp. 11–17.
53. Bauernfeind, J. C., Marusich, W. L., and Newmark, H. L. (1970). *Bull. Parenter. Drug Assoc.* **24**(4), 169–175.
54. Bauernfeind, J. C., Marusich, W. L., and Roncalli, R. A. (1972). *Can. Vet. Med. Assoc. Meet., Quebec.*
55. Bauernfeind, J. C., Osadca, M., and Bunnell, R. H. (1962). *Food Technol.* **16,** 101–107.
56. Bauernfeind, J. C., Rubin, S. H., Surmatis, J. D., and Ofner, A. (1970). *Int. J. Vitam. Res.* **40**(3), 391–416.
57. Baumann, C. A., Foster, E. G., and Lavik, P. S. (1941). *J. Nutr.* **21,** 431–444.
58. Baumann, C. A., Steenbock, J., Beeson, W. M., and Rupel, I. W. (1934). *J. Biol. Chem.* **105,** 167–176.
59. Beaumont, W. H., and Parr, W. H. (1963). *Vet. Rec.* **75,** 1344.
60. Becker, M. (1967). *Qual. Plant. Mater. Veg.* **15,** 48–64.
61. Beeson, W. M. (1965). *Fed. Proc., Fed. Am. Soc. Exp. Biol.* **24,** 924–926.
62. Beeson, W. M. (1962), *Distill. Feed Res. Counc. Conf., Proc.* pp. 43–47; also *Nat. Beef Cattle Conf., Ames, Iowa.*
63. Beeson, W. M., Perry, T. W., Smith, H. W., and Mohler, M. T. (1963). *J. Anim. Sci.* **22**(4), 1117; also *Purdue Univ. Agric. Prog. Rep.* No. 88 (1963); *Feedstuffs* **36**(13), 42–43 (1964).
64. Behrens, H., Kupferschmid, H., Nadai, J., Rehm, W., Scholl, E., and Weiser, H. (1967). *Tieraerztl. Umsch.* **22,** 91–96.
65. Bell, T. D., Dahmen, J. J., Baker, J. P., and Sharp, G. D. (1965). *Univ. Idaho Agric. Res. Bull.* No. 67.
66. Bentley, L. S., and Morgan, A. F. (1946). *J. Nutr.* **31,** 333–346.
67. Berger, S., Rechcigl, M., Jr., Loosli, J. K., and Williams, H. H. (1962). *J. Nutr.* **77,** 174–178.
68. Bergner, H., Lehmann, H., and Münchmeyer, R. (1962). *Arch. Tierernaehr.* **12,** 153–160.
69. Bernhard, K., and Heinzelmann, F. (1955). *Int. Z. Vitaminforsch.* **26**(1/2), 91–95.
70. Bernstein, L., and Thompson, J. F. (1947). *Bot. Gaz.* **109,** 204–219.
71. Bethke, R. M., Record, P. R., and Wilder, O. H. M. (1939). *Poult. Sci.* **18,** 179–187.
72. Bickoff, E. M. (1957). *Methods Biochem. Anal.* **4,** 1–423.
73. Bickoff, E. M. (1957). *J. Am. Oil Chem. Soc.* **28,** 65–68.
74. Bickoff, E. M., Livingston, A. L., Bailey, G. F., and Thompson, C. R. (1954). *J. Assoc. Off. Agric. Chem.* **37,** 894–902.
75. Bickoff, E. M., Livingston, A. L., Bailey, G. F., and Thompson, C. R. (1954). *J. Agric. Food Chem.* **2,** 563–567.
76. Bickoff, E. M., Livingston, A. L., Guggolz, J., and Thompson, C. R. (1954). *J. Agric. Food Chem.* **2,** 1229–1231.
77. Bickoff, E. M., Livingston, A. A., Guggolz, J., and Thompson, C. R. (1952). *J. Am. Oil Chem. Soc.* **29,** 51–53 and 445–446.

78. Bickoff, E. M., Livingston, A. L., and Thompson, C. R. (1955). *J. Am. Oil Chem. Soc.* **32,** 64–67.
79. Bickoff, E. M., Thompson, C. R., Livingston, A. L., Van Atta, G. R., and Guggolz, J. (1955). *J. Agric. Food Chem.* **3,** 67–69.
80. Bickoff, E. M., and Williams, K. T. (1944). *Ind. Eng. Chem.* **36,** 320–323.
81. Bieri, J. G. (1955). *Arch. Biochem. Biophys.* **56,** 90–96.
82. Bieri, J. G. (1955). *Proc. Soc. Exp. Biol. Med.* **88,** 482–484.
83. Bieri, J. G., and Pollard, C. J. (1954). *Br. J. Nutr.* **8,** 32–44.
84. Bieri, J. G. and Sandman, R. P. (1951). *Fed. Proc., Fed. Am. Soc. Exp. Biol.* **10,** 376.
85. Bieri, J. G., and Sandman, R. P. (1951). *Proc. Soc. Exp. Biol. Med.* **77**, 617–619.
86. Blaxter, K. L. (1951). *J. R. Agric. Soc. Engl.* **112,** 25–35.
87. Blaxter, K. L., Kon, S. K., and Thompson, S. Y. (1946). *J. Dairy Res.* **14,** 225–230.
88. Blessin, C. W., Brecher, J. D., and Dimler, R. J. (1964). *Cereal Chem.* **41,** 543–548.
89. Blessin, C. W., Brecher, J. D., Dimler, R. J., Grogan, C. O., and Campbell, C. M. (1963). *Cereal Chem.* **40,** 436–442.
90. Blessin, C. W., Dimler, R. J., and Webster, O. J. (1962). *Cereal Chem.* **39,** 389–392.
91. Bloomfield, R. A., Welsch, C. W., and Muhrer, M. E. (1964). *J. Anim. Sci.* **23,** 1207.
92. Bloomfield, R. A., Welsch, C. W., Garner, G. B., and Murer, M. E. (1961). *Science* **134,** 1690; also *J. Anim. Sci.* **20,** 981.
93. Boch, J., and Bronsch, K. (1966). *Tieraerztl. Umsch.* **21,** 62–69.
94. Bodea, C. (1969). *Pure Appl. Chem.* **20,** 517–530.
95. Boguth, W., and Sari, O. (1962). *Int. Z. Vitaminforsch.* **32,** 132–134.
96. Bolin, D. W. (1939). *J. Dairy Sci.* **22,** 111–113.
97. Bolin, F. M., Eveleth, D. F., and Bolin, D. W. (1950). *Vet. Med.* **45,** 451–452.
98. Bolin, D. W., Lampman, C. E., and Berg, L. R. (1943). *Poult. Sci.* **22,** 348–353.
99. Boling, J. A., Mitchell, G. E., Jr., Little, C. O., Fields, C. L., and Call, J. L. (1969). *J. Anim. Sci.* **29,** 504–508.
100. Bolton, W., and Common, R. H. 1942. *J. Soc. Chem. Ind.* **61,** 50–51.
101. Booth, V. H. (1956). *Br. J. Nutr.* **10,** 241–250.
102. Booth, V. H. (1956). *J. Sci. Food Agric.* **7,** 114–119.
103. Booth, V. H. (1955). "Preservation of Carotene in Dried Green Crops." Assoc. Green Crop Driers, England.
104. Booth, V. H. (1952). *J. Nutr.* **48,** 13–30.
105. Booth, R. G., Kon, S. K., Dann, W. F., and Moore. T. (1933). *Biochem. J.* **27,** 1189–1196.
106. Boyer, P. D., Phillips, P. H., Lundquist, N. S., Jensen, C. W., and Rupel, I. W. (1942). *J. Dairy Sci.* **25,** 433–440.
107. Braddock, R. J., and Kesterson, J. W. (1974). *J. Food Sci.* **39,** 712–714.
108. Bradfield, D., and Behrens, W. C. (1968). *Proc. West. Sect. Am. Soc. Anim. Sci.* **19,** 1–5.
109. Bradfield, D., and Smith, M. C. (1938). *Am. J. Physiol.* **124,** 168–173.
110. Bradley, W. B., Eppson, H. R., and Beath, O. A. (1940). *Wyo. Agric. Exp. Stn., Bull.* No. 241.
111. Branion, H. D., and Emslie, A. R. G. (1966). *Can. Dep. Agric., Publ.* No. 1238.
112. Braude, R., Foot, A. S., Henry, K. M., Kon, S. K., Thompson, S. Y., and Mead, T. H. (1941). *Biochem. J.* **35,** 693–770.
113. Bridge, P. S., and Spratling, F. R. (1966). *In* "Recent Advances in Animal Nutrition" (J. T. Abrams, ed.), PP. 143–187. Churchill, London.
114. Briggs, G. M., and Spivey, M. R. (1954). *Poult. Sci.* **33,** 1044–1045.
115. Briggs, J. E. (1964). *AFMA Nutr. Counc., Chicago, Ill.*
116. Briggs, P. K., Franklin, M. C., and McClymont, G. L. (1956). *Aust. Vet. J.* **32,** 299.
117. Brocklesby, H. N. (1961). *Merck Agric. Livestock Memo* No. 3.
118. Brocklesby, H. N. (1960). *Merck Agric. Livestock Memo* No. 5.
119. Brown, E. F., and Bloor, W. R. (1945). *J. Nutr.* **29,** 239–360.

120. Brubacher, G., Gloor, U., Wiss, O., Würsch, J., and Tagwerker, F. (1962). *Proc. World Poult. Congr. 12th, Sydney,* 164–167.
121. Brudzynski, A., Dowhyluk-Ochnio, A., and Roginski, H. (1971). *Przem. Ferment. Rolny* **15**(11), 15–20.
122. Brüggemann, J., and Niesar, K. H. (1957). *Vitam. Horm. (N.Y.)* **7,** 10–25.
123. Brüggemann, J., and Tiews, J. (1964). *Int. Z. Vitaminforsch.* **34,** 233–240.
124. Brüggemann, J., Tiews, J., and Zucker, H. (1963). Unpublished observations.
125. Brune, H., and Zaddach, M. (1965). *Z. Tierphysiol., Tierernaehr. Futtermittelkd.* **20,** 224–234.
126. Brunson, A. M., and Quackenbush, F. W. (1962). *Crop Sci.* **2,** 344–347.
127. Budowski, P., and Gross, J. (1965). *Nature (London)* **206,** 1254–1255.
128. Budowski, P., Ascarelli, I., Gross, J., and Nir, I. (1963). *Science* **142,** 969–971.
129. Bullis, D. D., Weswig, P. H., Jones, I. R., and Bone, J. E. (1955). *Ohio Agric. Exp. Stn., Techn. Pap.* No. 1159.
130. Burgess, R. C., Gluck, M., Brisson, G., and Laughland, D. H. (1951). *Arch. Biochem. Biophys.* **33,** 339–340.
131. Burns, M. J., Hauge, S. M., and Quackenbush, F. W. (1951). *Arch. Biochem.* **30,** 341–346.
132. Burns, M. J., Hauge, S. M., and Quackenbush, F. W. (1951). *Arch. Biochem.* **30,** 347–350.
133. Busson, A., and Simonnett, H. (1932). *C. R. Soc. Biol.* **109,** 716, 1253.
134. Butler, J. L., Wilkinson, W. S., Hellwig, R. E., Barbee, C., and Know, F. E. (1969). *Am. Soc. Agric. Eng.* **12,** 552–555.
135. Byers, J. H., Jones, I. R., and Bone, J. R. (1956). *J. Dairy Sci.* **39,** 1556–1564.
136. Byers, J. H., Weswig, P. H., Bone, J. F., and Jones, I. R. (1955). *J. Dairy Sci.* **38,** 657–664.
137. Callison, E. C., and Knowles, V. H. (1945). *Am. J. Physiol.* **143,** 444–452.
138. Cama, H. R., and Goodwin, T. W. (1949). *Biochem. J.* **45,** 236–241.
139. Camp, A. A., Cartrite, H. T., Quisenberry, J. H., and Couch, J. R. (1955). *Poult. Sci.* **34,** 1134–1138.
140. Campbell, B. L. (1956). *Proc. Aust. Soc. Anim. Prod.* **1,** 46–51.
141. Campbell, J. A. (1964). *Proc. Nutr. Soc.* **23,** 31–38.
141a. Carricaburu, P., Tayebi, B., and Chardenot, P. (1970). *C. R. Acad. Sci., Ser. D* **270,** 1282.
142. Capper, N. S., McKibbin, I. M. W., and Prentics, J. H. (1931). *Biochem. J.* **25,** 265–274.
143. Case, A. A. (1957). *J. Am. Vet. Med. Assoc.* **130,** 323–329.
144. Castano, F. F., Boucher, R. V., and Callenbach, E. W. (1951). *J. Nutr.* **45,** 131–141.
145. Chaffee, P. S. (1968). *J. Small Anim. Pract.* **9,** 34–136.
146. Chanda, R. (1956). *Nature (London)* **178,** 541–542.
147. Chanda, R. (1953). *J. Agric. Sci.* **43,** 54–71.
148. Chanda, R., and Owen, E. C. (1952). *Biochem. J.* **51,** 404–417.
149. Chanda, R., Clapham, H. M., and Owen, E. C. (1954). *Biochem. J.* **56,** 453–458.
150. Chanda, R., Clapham, H. M., McNaught, M. L., and Owen, E. C. (1951). *Biochem. J.* **50,** 95–99.
151. Chanda, R., Clapham, H. M., McNaught, M. L., and Owen, E. C. (1951). *J. Agric. Sci.* **41,** 179–186.
152. Chapman, H. L., Shirley, R. L., Palmer, A. Z., Haines, C. E., Carpenter, J. W., and Cunha, T. J. (1964). *J. Anim. Sci.* **23,** 669–673.
153. Cheng, A. L. S., and Deuel, H. J., Jr. (1950). *J. Nutr.* **41,** 619–628.
154. Chevrel, M. L., and Cornier, M. (1948). *C. R. Acad. Sci.* **226,** 1854–1855.
155. Christensen, N. O. (1959). *Proc. Int. Vet. Congr., 16th, Madrid* **2**(41), 21–27.
156. Christensen, N. O., Elkjaer, R., Engelund, A., Munch, T., and Terp, P. (1958). *Nord Vet. Med.* **10,** 49–62.
157. Christensen, N. O., Engelund, A., Handler, J. A., and Terp, P. (1962). *Nord. Vet. Med.* **14,** 270–279.
158. Christensen, N. O., Engelund, A., Munch, T., and Terp, P. (1958). *Nord. Vet. Med.* **10,** 393–406.

159. Christensen, N. O., Engelund, A., Munch, T., and Terp, P. (1957). *Nord. Vet. Med.* **9,** 455–463.
160. Church, D. C., MacVicar, R., Bieri, J. G., Baker, F. H., and Pope, L. S. (1954). *J. Anim. Sci.* **13,** 677–683.
161. Church, D. C., MacVicar, R., Bieri, J. G., Baker, F. H., and Pope, L. S., Jr. (1953). *J. Anim. Sci.* **12,** 910–911.
162. Clark, I., and Colburn, R. W. (1953). *Fed. Proc., Fed. Am. Soc. Exp. Biol.* **12,** 190.
163. Clayton, C. C., and Baumann, C. A. (1944). *J. Nutr.* **27,** 155–163.
164. Cline, T. R., Hatfield, E. E., and Garrigus, U. S. (1962). *J. Anim. Sci.* **21,** 991.
165. Coates, M. E., Harrison, G. F., Kon, S. K., Porter, J. W. G. and Thompson, S. Y. (1952). *Chem. Ind. (London)* p. 149.
165a. Cooke, B. C., and Comben, N. (1978). *Br. Soc. Anim. Prod.,* **26**(3), 356–357.
166. Coombes, A. I., Ott, G. L., and Wisnicky, W. (1940). *North Am. Vet.* **21,** 601–605.
167. Crampton, E. W., and Harris, L. E. (1969). "Applied Animal Nutrition," 2nd ed. Freeman, San Francisco, California.
168. Crane, B., and Clare, N. T. (1975). *N.Z. J. Agric. Res.* **18,** 273–275.
169. Creger, C. R., Couch, J. R., and Chavez, R. (1971). *Nutr. Rep. Int.* **3,** 191–195.
170. Crimm, P. D., and Short, D. M. (1937). *Am. J. Physiol.* **118,** 477–482.
171. Crowley, J. W., and Allen, N. N. (1950). *J. Dairy Sci.* **33,** 380–381.
172. Cullison, A. E., and Ward, C. S. (1965). *J. Anim. Sci.* **24,** 969–972.
173. Cunha, T. J. (1963). *Feedlot* **5**(3), 28.
174. Curtis, A. C., and Kline, E. M. (1939). *Arch. Intern. Med.* **63,** 54–93.
175. Cuthbertson, W. F. J. (1957). *Proc. Nutr. Soc.* **16,** 70–76.
176. Czeczuga, B. (1971). *Comp. Biochem. Physiol. B* **39,** 945–953.
177. Danasoury, M. S., El Noubi, H. M., and Makky, A. (1971). *U.A.R. J. Anim. Prod.* **11,** 307–308.
178. Danilenko, I. A., Maksakov, V. J., Dolgorucenko, L. E., Mirosnicenko, N. A., and Alemov, T. K. (1968). *Zivotnovodstvo* **9,** 45–48.
179. Dannenberg, D. (1969). *Arch. Exp. Veterinaermed.* **23,** 1063–1078.
180. Da Silva, A. C., Fried, R., and De Angelis, R. C. (1952). *J. Nutr.* **46,** 399–409.
181. Davies, A. W., and Moore, T. (1941). *Nature (London)* **147,** 794–796.
182. Davies, B. H., Hsu, W. J., and Chichester, C. O. (1970). *Comp. Biochem. Physiol.* **32,** 69–79; **33,** 601–615.
183. Davies, R. E., and Madsen, L. L. (1941). *J. Nutr.* **21,** 135–146.
184. Davis, C. L., Elliott, R. F., and Lassiter, C. A. (1956). *J. Dairy Sci.* **39,** 440–447.
185. Davison, K. L., Hansel, W., Krook, L., McEntee, K., and Wright, M. J. (1964). *J. Dairy Sci.* **47,** 1065–1073.
186. Davison, K. L., and Seo, J. (1963). *J. Dairy Sci.* **46,** 862–863.
187. Davtyan, E. A., and Akopjan, W. D. (1959). *Tr. Helminthol. Lab.* **9,** 79.
188. Deadrick, R. E., Bieri, J. G., and Cardenas, R. R., Jr. (1955). *J. Nutr.* **57,** 287–295.
189. DeHaas, J. H., and Meulemans, O. (1937). *Geneeskd. Tijdschr. Ned.-Indie* **77,** 279–288.
190. DeMan, T. J., Roborgh, J. R., and Ten Ham, E. J. (1968). *Diergeneeskd. Memo.* **83,** 380–387.
191. Deshmukh, D. S., and Ganguly, J. (1967). *Indian J. Biochem.* **4,** 18–21.
192. Deshmukh, D. S.,and Ganguly, J. (1964). *Indian J. Biochem.* **1,** 204–207.
193. Deuel, H. J., Jr. (1957). "The Lipids, Their Chemistry and Biochemistry," Vol. 3. Wiley (Interscience), New York.
194. Deuel, H. J., Jr. (1955). "The Lipids," Vol. 2, Wiley (Interscience), New York.
195. Deuel, H. J., Jr., Greenberg, S. M., Savage, E. E., and Melnick, D. (1951). *J. Nutr.* **43,** 371–388.
196. Deuel, H. J., Jr., Johnston, C., Meserve, E. R., Polgar, A., and Zechmeister, L. (1945). *Arch. Biochem.* **7,** 247–255.
197. Deuel, H. J., Jr., Meserve, E. R., Johnston, C. H., Polgar, A., and Zechmeister, L. (1945). *Arch. Biochem.* **7,** 447–450.

198. Deuel, H. J., Jr., Sumner, E., Johnston, C. H., Polgar, A., and Zechmeister, L. (1945). *Arch. Biochem.* **6,** 157–161.
199. DiBella, L. (1940). *Arch. Sci. Biol. (Bologna)* **26,** 469–492; **30,** 1–8 (1944).
200. Dikshit, P. K., and Ranganathan, S. (1955). *Indian J. Med. Res.* **43,** 489–496.
201. Dingle, J. T., and Lucy, J. A. (1965). *Biol. Rev. Cambridge Philos. Soc.* **40,** 422–461.
202. Diven, R. H., and Erwin, E. S. (1958). *Proc. Soc. Exp. Biol. Med.* **97,** 601–603.
203. Dolge, K. L., Dembeczak, M., Rousseau, J. E., Jr., Eaton, H. D., Beall, G., and Moore, L. A. (1955). *Conn., Storrs Agric. Exp. Stn., Bull.* No. 314.
204. Dolge, K. L., Eaton, H. D., Avampato, J. E., Mochrie, R. D., Elliott, F. I., and Beall, G. (1952). *J. Dairy Sci.* **35,** 424–435.
205. Dolge, K. L., Rousseau, J. E., Jr., Teichman, R., Eaton, H. D., and Beall, G. (1956). *J. Dairy Sci.* **39,** 1660–1670.
206. Donovan, G. A., Henderson, D. C., and Luneau, R. N. (1961). *Poult. Sci.* **40,** 1395–1396.
207. Dorogush, A. G., and Novikova, E. N. (1971). *Biokhim. Patochim. Obmena Veshch. Mekh. Ego Regul.* pp. 146–151.
208. Drummond, J. C., and McWalter, R. J. (1934). *J. Physiol. (London)* **83,** 236–242.
209. Dua, P. N., Day, E. J., Hill, J. E., and Grogan, C. O. (1967). *J. Agric. Food Chem.* **15,** 324–328.
210. Dua, P. N., Day, E. J., Tipton, H. C., and Hill, J. E. (1966). *J. Nutr.* **90,** 117–122.
211. Dua, P. N., Dubal, V. L., Day, E. J., and Hill, J. E. (1968). *Proc. Soc. Exp. Biol. Med.* **128,** 262–265.
212. Dua, P. N., Dubal, V. L., Day, E. J., and Hill, J. E. (1967). *Poult. Sci.* **46,** 1599–1601.
213. Dubouloz, P., Pistone, G., and Marville, R. (1949). *Bull. Soc. Chim. Biol.* **31,** 1273–1278.
214. Duburs, G., Uldrikis, J., Valdmanis, A., Spruys, J., Trizitis, G., Petersone, I., and Melbergs, J. (1973). *Otkryt., Izobret., Prom. Obraztsy. Tovarnye Znaki* **50**(7), 8.
215. Dudley, R. P. (1971). *Diss. Abstr. Int. B* **32,** 2574–2575.
216. Dupree, H. K. (1970). *U.S. Bur. Sport Fish. Wildl., Prog. Sport Fish. Res. 1969, Publ.* No. 88, 148–150.
217. Durdle, W. M., Love, S. B., Smith, G. S., and Hatfield, E. E. (1962). *J. Anim. Sci.* **21,** 993–994.
218. Dutcher, R. A., Harris, P. L., Hartzler, E. L., and Guerrant, N. B. (1934). *J. Nutr.* **8,** 269–283.
219. Dvinskaja, L. M. (1968). *Svinovodstvo (Moscow)* No. 4, 19–20.
220. Dvorak, M. (1960). *Vet. Med.* **5,** 521–536.
221. Dyer, F. J., Kay, K. M., and Coward, K. H. (1934). *Biochem. J.* **28,** 875–881.
222. Dyer, I. A., Ensminger, M. E., and Blue, R. L. (1957). *J. Anim. Sci.* **16,** 828–833.
223. Eaton, H. D., Carpenter, C. A., Caverno, R. J., Johnson, R. E., Elliot, F. I., and Moore, L. A. (1951). *J. Dairy Sci.* **34,** 124–135.
224. Eaton, H. D., Helmbolt, C. F., Jungherr, E. L., Carpenter, C. A., and Moore, L. A. (1951). *J. Dairy Sci.* **34,** 386–395.
225. Eaton, H. D., Matterson-, L. D., Helmboldt, C. F., and Jungherr, E. L. (1951). *J. Dairy Sci.* **34,** 1073–1080.
226. Eaton, H. D., Myers, G. S., Dicks, M. W., Dehority, B. A., Grifo, A. P., Jr., Teichman, R., Hemboldt, C. F., Jungherr, E. L., and Gosslee, D. G. (1959). *J. Dairy Sci.* **42,** 642–650.
227. Eaton, H. D., Rousseau, J. E., Jr., Dicks, M. W., Teichman, R., and Grifo, A. P., Jr. (1958). *J. Dairy Sci.* **41,** 1456–1459.
228. Eaton, H. D., Rousseau, J. E., Jr., and Norton, H. W. (1961). *J. Anim. Sci.* **44,** 1368–1370.
229. Eaton, H. D., Rousseau, J. E., Jr., Woelfel, C. G., Calhoun, M. C., Nielsen, S. W., and Lucas, J. J. (1964). *Conn., Storrs Agric. Exp. Stn., Bull.* No. 383.
230. Edwards, H. M., Dunahoo, W. S., and Fuller, H. L. (1958). *Poult. Sci.* **37,** 1141–1443.
231. Ehrlich, C. (1962). *Prakt. Tieraerzt* No. 4, 134, 136, 138.
232. Ehrlich, C. (1962). *Prakt. Tieraerzt* No. 3, 387–389.
233. El-Gorab, M. I., Underwood, B. A., and Loerch, J. D. (1975). *Biochim. Biophys. Acta* **401,** 265–277.

234. Elhatw, I. A. E. (1961). *Acta Vet. Hung.* **11,** 339-343.
235. Elliott, R. F. (1949), *J. Dairy Sci.* **32,** 710-712.
236. Ellmore, M. F., and Shaw, J. C. (1954). *J. Dairy Sci.* **37,** 1269-1272.
237. Ellmore, M. F., Shaw, J. C., Hatziolos, B. C., Moore, L. A., and Sykes, F. J. (1948). *J. Dairy Sci.* **31,** 665.
238. Elmoty, I. A., El Mulla, A., and Elmosallami, S. (1967). *Acta Vet. Hung.* **17,** 87-90.
239. El-Tinay, A. H., and Chichester, C. O. (1970). *J. Org. Chem.* **35,** 2290-2293.
240. Ely, C. M. (1959). *Poult. Sci.* **38,** 1316-3124.
241. Embry, L. B., and Emerick, R. J. (1963). *S. D. Farm Home Res.* **14**(3), 10-13.
242. Embry, L. B., and Kortan, L. J. (1962). *S. D. Agric. Exp. Stn., Pamphl.* No. 3.
243. Emerick, R. J., and Olson, D. E. (1962). *J. Nutr.* **78,** 73-77.
244. Ender, F. (1934). *Z. Vitaminforsch.* **3,** 247-253.
245. Erasmas, J. (1961). *Meet. Soc. Adv. Anim. Nutr., South. Afr., Pretoria.*
246. Erasmus, J., Scott, M. L., and Levine, P. P. (1960). *Poult. Sci.* **39,** 565-571.
246a. Erb, R. E., Andrews, F. W., Hauge, S. M., and King, W. A. (1947). *J. Dairy Sci.* **30,** 687-702.
247. Ershoff, B. H. (1950). *Proc. Soc. Exp. Biol. Med.* **74,** 586-587.
248. Erwin, E. S., Dyer, I. A., and Ensminger, M. E. (1956). *J. Anim. Sci.* **15,** 1147-1153.
249. Erwin, E. S., Elam, C. J., and Dyer, I. A. (1957). *Science* **126,** 702.
250. Erwin, E. S., Gordon, R. S., and Algeo, J. W. (1963). *J. Anim. Sci.* **22,** 341-345.
251. Erwin, E. S., and Page, H. M. (1958). *J. Anim. Sci.* **17,** 662-666.
252. Erwin, E. S., Varnell, T. R., and Page, H. M. (1959). *Proc. Soc. Exp. Biol. Med.* **100,** 373-375.
253. Erwin, E. S., Varnell, T. R., and Roubicek, C. B. (1957). *J. Anim. Sci.* **16,** 1022.
254. Esh, G. C., and Sutton, T. S. (1948). *J. Nutr.* **36,** 391-404.
255. Esh, G. C., Sutton, T. S., Hibbs, J. W., and Krauss, W. E. (1948). *J. Dairy Sci.* **31,** 461-478.
256. von Euler, B., von Euler, H., and Hellström, H. (1928). *Biochem. Z.* **203,** 370-384.
257. von Euler, H., Hellström, H., and Malmberg, M. (1933). *Sven. Kem. Tidskr.* **45,** 151-152.
258. von Euler, H., Karrer, P., and Solmsson, U. (1938). *Helv. Chim. Acta* **21,** 211-222.
259. von Euler, H., Karrer, P., and Zubrys, P. (1934). *Helv. Chim. Acta* **17,** 24-29.
260. von Euler, H., and Klussman, E. (1932). *Hoppe-Seyler's Z. Physiol. Chem.* **213,** 21-34.
261. Evans, E. V. (1966). *Can. DeP. Agric., Publ.* No. 1238
262. Evans, E. V., and Czarnocki, J. (1966). *Can. Dep. Agric., Publ.* No. 1238.
263. Eveleth, D. F., Goldsby, A. I., Bolin, F. M., and Bolin, D. W. (1953). *Vet. Med.* **48,** 441-442.
264. Everhart, E. L. (1958). *J. Am. Vet. Med. Assoc.* **132,** 344-347.
265. Farugue, O., and Walker, D. M. (1970). *Br. J. Nutr.* **24,** 23-28.
266. Fasold, H., and Heidemann, E. R. (1933). *Z. Dsch. Gesamte Exp. Med.* **92,** 53-56.
267. von Fellenberg, T., and Greuter, F. (1932). *Biochem. Z.* **253,** 42-63.
268. Fenton, P. F., Shrader, R. E., and Cowgill, G. R. (1949). *Fed. Proc., Fed. Am. Soc. Exp. Biol.* **8,** 382.
269. Ferguson, W. S. (1963). *In* "Animal Health, Production and Pasture" (A. N. Worden, K. C. Sellers, and D. E. Tribe, eds.), p. 119. Longmans, Green, New York.
270. Fernandez, G., Albertazi, C., and Bressani, R. (1971). *Turrialba* **21,** 157-164.
271. Ferrer, F., Valdes, F., Pantoja, A. M., and Boado, A. (1974). *Cienc. Agropecu. Ing. Pecu.* No. 8, 19-24.
272. Fischerova, J. (1970). *Biol. Chem. Vyz. Zvirat* **6,** 27-32.
273. Flegal, C. J., Schaible, P. J., and Hall, H. H. (1971). *Poult. Sci.* **50,** 349-358.
274. Foll, N. J. (1964). *Fert. Feed. Stuffs J.* **60,** 406-407.
275. Foll, N. J. (1964). *Vet. Rec.* **76,** 460-461.
276. Fonnesbeck, P. V., Vander Noot, G. W., and Symons, L. D. (1966). *J. Anim. Sci.* **25,** 890-891.
277. Fonnesbeck, P. V., and Vander Noot, G. W. (1964). *J. Anim. Sci.* **23,** 1232.
278. Fonnesbeck, P. V., and Symons, L. D. (1967). *J. Anim. Sci.* **26,** 1030-1038.
278a. Franklin, M. C., McClymont, G. L., Briggs, P. K., and Campbell, R. L. (1955). *Aust. J. Agric. Res.* **6,** 324-342.

279. Fraps, G. S. (1946). *Arch. Biochem.* **10,** 485–489.
280. Fraps, G. S., and Meinke, W. W. (1945). *Arch. Biochem.* **6,** 323–327.
281. Frey, P. R., Jensen, R., and Connell, W. E. (1947). *J. Nutr.* **34,** 421–430.
282. Frey, P. R., and Jensen, R. (1947). *Science* **105,** 313.
283. Frey, P. R., and Wilgus, H. S. (1949). *J. Nutr.* **39,** 517–528.
284. Friend, C. J., Heard, C. R. C., Platt, B. S., Stewart, R. J. C., and Turner, M. R. (1961). *Br. J. Nutr.* **15,** 231–240.
285. Friesecke, H. (1964). "Fertility and Performance of Cattle in Relation to Vitamin A Metabolism." Roche Inf. Serv., Basel.
286. Frohring, W. O. (1935). *Proc. Soc. Exp. Biol. Med.* **33,** 280–782; also *Vet. Med.* **32,** 190–195 (1937).
287. Futterwerttabellen der Deutschen Landwirtschafts-Gessellschaft (1961–1963). Frankfurt am Main.
288. Galgan, M. W., Bell, J. M., and Spencer, E. Y. (1950). *Sci. Agric.* **30,** 316–323.
289. Gallup, W. D., Thomas, O. O., Ross, O. B., and Whitehari, C. L. (1953). *J. Anim. Sci.* **12,** 715–721.
290. Ganguly, J. (1960). *Vitam. Horm. (N.Y.)* **18,** 387–402.
291. Ganguly, J., Krinsky, N. I., Mehl, J. W., and Deuel, H. J., Jr. (1952). *Arch. Biochem. Biophys.* **38,** 275–282.
292. Ganguly, J., Krinsky, N. I., Pinckard, J. H.,and Deuel, H. J., Jr. (1953). *Hoppe-Seyler's Z. Physiol. Chem.* **295,** 61–66.
293. Ganguly, J., Mehl, J. W., and Deuel, H. J., Jr. (1953). *J. Nutr.* **50,** 59–72.
294. Ganguly, J., Mehl, J. W., and Deuel, H. J., Jr. (1953). *J. Nutr.* **50,** 73–83.
295. Gartner, R. J. W., and Anson, R. J. (1966). *Aust. J. Exp. Agric. Anim. Husb.* **6,** 321–325.
296. Garton, C. L., Vander Noot, G. W., and Fonnesbeck, P. V. (1964). *J. Anim. Sci.* **23,** 1233.
297. Gay, N. (1965). *Cattlemen* Oct.
298. German, H. L., and Adams, C. R. (1962). *J. Anim. Sci.* **21**, 966; also *Feedstuffs* **35**(43), 92 (1962); *Proc. Tex. Nutr. Feed Conf.* pp. 63–72 (1962).
299. German, H. L. (1964). *Feed Illus.* **15**(2), 10–12.
300. Gerriets, E. (1966). *Berl. Muench. Tieraerztl. Wochenschr.* **79,** 271–275.
301. Gershoff, S. N., Andrus, S. B., Hegsted, D. M., and Lentine, E. A. (1957). *Lab. Invest.* **6,** 227–240.
302. Gershoff, S. N. (1959). *J. Am. Vet. Med. Assoc.* **134,** 139–141.
303. Gilchrist, B. M., and Lee, W. L. (1967). *J. Zool.* **151,** 171–180.
304. Gillam, A. E., and El-Ridi, M. S. (1945). *Biochem. J.* **31,** 251–253.
305. Gillam, A. E., and El-Ridi, M. S. (1935). *Biochem. J.* **29,** 2465–2468.
306. Gillam, A. E., and Heilbron, I. M. (1935). *Biochem. J.* **29,** 834–836.
307. Gillam, A. E., and Heilbron, I. M. (1935). *Biochem. J.* **29,** 1064–1067.
308. Gledhill, R. H., and Smith, S. B. (1955). *Poult. Sci.* **34,** 942–948.
309. Glover, J. (1960). *Vitam. Horm. (N.Y.)* **18,** 371–386.
310. Glover, J., Goodwin, T. W., and Morton, R. A. (1948). *Biochem. J.* **43,** 512–518.
311. Glover, J., and Redfearn, E. R. (1954). *Biochem. J.* **58,** 15P.
312. Goerner, A., and Goerner, M. M. (1939). *Am. J. Cancer* **37,** 518–520.
313. Goldharber, P., Zacharias, L., and Kinsey, V. E. (1950). *J. Nutr.* **42,** 453–462.
314. Gondos-Gromicov, M., Maxim, V., Bejan, G., and Calatoiu, A. (1959). *Lucr. Stiint. Inst. Cercet. Zooteh., Bucharest* **17,** 33–47.
315. Goodman, D. S. (1969). *Am. J. Clin. Nutr.* **22,** 963–965.
316. Goodman, D. S. (1967). *Prog. Biochem. Pharmacol.* **3,** 487–497.
317. Goodman, D. S., Bloomstrand, R., Werner, B., Huang, H. S., and Shiratori, T. (1966). *J. Clin. Invest.* **45,** 1615–1623.
318. Goodman, D. S., and Huang, H. S. (1965). *Science* **149,** 879–880.
319. Goodman, D. S., Huang, H. S., Kanai, M., and Shiratori, T. (1967). *J. Biol. Chem.* **242,** 3543–3554.

320. Goodman, D. S., and Olson, J. A. (1969). *In* "Steroids and Terpenoids" (R. B. Clayton, ed.), Methods in Enzymology, Vol. 15, pp. 462-475. Academic Press, New York.
321. Goodrich, R. D., Emerick, R. J., and Emery, L. B. (1964). *J. Anim. Sci.* **23,** 100-104.
322. Goodwin, T. W. (1963). *Wiss. Veroeff. Dsch. Ges. Ernaehr.* **9,** 1-25.
323. Goodwin, T. W. (1954). *In* "Chemical Pathology of Animal Pigments" (R. T. Williams, ed.), pp. 71-84. Cambridge Univ. Press, London and New York.
324. Goodwin, T. W. (1954). "Carotenoids, Their Comparative Biochemistry," Chemical Publ. Co., New York.
325. Goodwin, T. W. (1952). "Carotenoids, Their Comparative Biochemistry," Chapman & Hall, London.
326. Goodwin, T. W. (1950). *Biol. Rev. Cambridge Philos. Soc.* **25,** 391-413.
327. Goodwin, T. W., and Gregory, R. A. (1948). *Biochem. J.* **43,** 505-512.
328. Goodwin, T. W., and Wilson, A. A. (1951). *Biochem. J.* **49,** 499-502.
329. Grangaud, R., Massonet, R., Conguy, T., and Ridolfo, J. (1962). *C. R. Acad. Sci.* **254,** 579-581.
330. Grangaud, R., Nicol, M., LeGall, J. Y., and Soussy, A. (1964). *Arch. Sci. Physiol.* **18,** 225-251.
331. Greathouse, T., Little, C. O., and Mitchell, G. E., Jr. (1962). *J. Anim. Sci.* **21,** 383.
332. Greaves, J. D., and Schmidt, C. L. A. (1935). *Am. J. Physiol.* **111,** 492-506.
333. Greenberg, S. M., Chatterjee, A., Calbert, C. E., Deuel, H. J., Jr., and Zechmeister, L. (1950). *Arch. Biochem.* **25,** 61-65.
334. Greenberg, S. M., Calbert, C. E., Pinckard, J. H., Deuel, H. J., J., and Zechmeister, L. (1949). *Arch. Biochem.* **24,** 31-39.
335. Gridgemann, N. T., Lees, H., and Wilkinson, H. (1940). *J. Soc. Chem. Ind.* **59,** 120-125.
336. Griffith, R. B., and Thompson, C. R. (1949). *Bot. Gaz.* **111,** 165-175.
337. Grifo, A. P., Jr., Rousseau, J. E., Jr., Eaton, H. D., and Gosslee, D. G. (1961). *J. Dairy Sci.* **44,** 556-558.
338. Grifo, A. P., Rousseau, J. E., Jr., Eaton, H. D., Dehority, B. A., Hazzard, D. G., Helmboldt, C. F., and Gossee, D. G. (1960). *J. Dairy Sci.* **43,** 1809-1819.
339. Grifo, A. P., Jr., Rousseau, J. E., Jr., Eaton, H. D., and Gosslee, D. G. (1960). *J. Dairy Sci.* **43,** 1003-1006.
340. Grogan, C. O., and Blessin, C. W. (1968). *Crop Sci.* **8,** 730-732.
341. Gronowska-Senger, A., and Wolf, G. (1970). J. *Nutr.* **100,** 300-308.
342. Gross, J., and Budowski, P. (1966). *Biochem. J.* **101,** 747-754.
343. Gross, W., and Heller, V. G. (1944). *Proc. Okla. Acad. Sci.* **24,** 97.
344. Guerrant, N. B. (1960). *Proc. Soc. Exp. Biol. Med.* **105,** 400-403; **111,** 34-37 (1962).
345. Guerrant, N. B. (1949). *J. Nutr.* **37,** 37-51.
346. Guether, G. (1972). *Arch. Tierernaehr.* **22,** 555-559.
347. Guether, G., and Hennig, A. (1974). *Arch. Tierernaehr.* **24,** 263-271.
348. Guggenheim, K. (1944). *Biochem. J.* **38,** 260-264.
349. Guilbert, H. R. (1935). *J. Nutr.* **10,** 45-62.
350. Guilbert, H. R., and Hart, G. H. (1935). *J. Nutr.* **10,** 409-427.
351. Guilbert, H. R., and Hinshaw, W. R. (1934). *J. Nutr.* **8,** 45-56.
352. Guilbert, H. R., Howell, C. E., and Hart, G. H. (1940). *J. Nutr.* **19,** 91-103.
353. Guilbert, H. R., and Loosli, J. K. (1951). *J. Anim. Sci.* **10,** 22-41.
354. Guilbert, H. R., Miller, R. F., and Hughes, E. H. (1937). *J. Nutr.* **13,** 543-564.
355. Gurcay, R., Boucher, R. V., and Callenbach, E. W. (1950). *J. Nutr.* **41,** 565-582.
356. Hale, W. H. (1968). *Proc. Symp. Assoc. Vitam. Chem., Chicago, Ill.*, pp. 57-58.
357. Hale, W. H., Hubbert, F., Jr., and Taylor, R. E. (1962). *Proc. Soc. Exp. Biol. Med.* **109,** 289-290.
358. Hale, W. H., Hubbert, F., Jr., Taylor, R. E., Anderson, T. A., and Taylor, B. (1962). *Am. J. Vet. Res.* **23,** 992-996.
359. Hale, W. H., Hubbert, F., Jr., and Taylor, R. E. (1961). *J. Anim. Sci.* **20,** 934.

360. Hale, W. H., Taylor, B., Theuer, B., and Essig, H. (1967). *Ariz. Cattle Feeder's Day Rep. Univ. Ariz.*
361. Halver, J. E. (1972). *In* "Fish Nutrition" (J. E. Halver, ed.), pp. 79–82. Academic Press, New York.
362. Halverson, A. W., and Hendrick, C. M. (1955). *Poult. Sci.* **34,** 355–360.
363. Halverson, A. W., and Hendrick, C. M. (1955). *S.D. Agric. ExP. Stn., Bull.* No. 114.
364. Hanke, H. E., and Jordan, R. M. (1963). *Proc. Annu. Sheep Lamb Feeders' Day, 35th, Univ. Minn.,* Feb. 8, pp. 26–27.
365. Hansen, H. M., Porter, G. H., Johnson, R. E., and Eaton, H. D. (1954). *J. Dairy Sci.* **37,** 311–316.
366. Harms, R. H., Reid, B. L., and Couch, J. R. (1955). *Poult. Sci.* **34,** 1125–1133.
367. Harms, F. (1942). *Vitam. Horm. (N.Y.)* **2,** 151–158.
368. Harper, T. A., Boucher, R. V., and Callenbach, E. W. (1952). *Poult. Sci.* **31,** 273–283.
369. Harris, P. L., Kaley, M. W., and Hickman, K. C. D. (1944). *J. Biol. Chem.* **152,** 313–320.
370. Hartsook, E. W., Batchelor, E., and Johnson, B. C. (1953). *Proc. Soc. Exp. Biol. Med.* **83,** 43–45.
371. Harvey, J. D., Parrish, D. B., Sanford, P. E., and Hughes, J. S. (1955). *Poult. Sci.* **34,** 1348–1357.
372. Hashizume, T., Fujita, H., Matsuoka, S., Tanaka, K., and Urakami, K. (1976). *Obihiro Chikusan Daigaku Gakujutsu Kenkyu Hokoku, Dai-1-Bu* **10,** 117–128.
373. Hatfield, F. E., Smith, G. S., Neuman, A. L., Forbes, R. M., Garrigus, U.S., and Ross, O. B. (1961). *J. Anim. Sci.* **20,** 676.
374. Hayes, B. W., Mitchell, G. E., Jr., Little, C. O., and Sewell, H. B. (1967). *J. Anim. Sci.* **26,** 855–857.
375. Heaney, D. P., and Thomas, O. O. (1956). *J. Anim. Sci.* **15,** 1252.
376. Hebert, J. W., and Morgan, A. F. (1953). *J. Nutr.* **50,** 175–190.
377. Helmboldt, C. F., Jungherr, E. L., Eaton, H. D., and Moore, L. A. (1953). *Am. J. Vet. Res.* **14,** 343–354.
378. Hendricks, D. G., Miller, E. R., Ullrey, D. E., Struthers, R. D., Baltzer, B. V., Hoefer, J. A., and Luecke, R. W. (1967). *J. Nutr.* **93,** 37–43.
379. Hensen, R. G., Phillips, P. H., and Rupel, I. W. (1946). *J. Dairy Sci.* **29,** 761–766.
380. Hentges, J. F., Jr., Grummer, R. H., Phillips, P. H., and Bohstedt, G. (1952). *J. Anim. Sci.* **11,** 721–726.
381. Hentges, J. F., Jr., Grummer, R. H., and Sorensen, D. K. (1952). *J. Anim. Sci.* **11,** 794–795.
382. Herrick, J. B. (1972). *Vet. Mod.* **67,** 908–909.
383. Hibbs, J. W., and Evans, M. W. (1950). *Agron. J.* **42,** 94–95.
384. Hibbs, J. W., Krauss, W. E., and Monroe, C. F. (1949). *J. Dairy Sci.* **32,** 955–960.
385. Hickman, K. (1943). *Annu. Rev. Biochem.* **12,** 353–396.
386. Hickman, K. C. D., and Harris, P. L. (1946). *Adv. Enzymol. Relat. Subj. Biochem.* **6,** 469–524.
387. Hickman, K. C. D., Harris, P. L., and Woodside, M. R. (1942). *Nature (London)* **150,** 91–396.
387a. Hickman, K. C. D., Kaley, M. W., and Harris, P. L. (1944). *J. Biol. Chem.* **152,** 303–311; 321–328.
388. High, E. G. (1956). *Arch. Biochem. Biophys.* **60,** 456–462.
389. High, E. G. (1955). *Fed. Proc., Fed. Am. Soc. Exp. Biol.* **14,** 437.
390. High, E. G., and Day, H. G. (1952). *J. Nutr.* **48,** 369–376.
391. High, E. G., and Day, H. G. (1951). *J. Nutr.* **43,** 245–260.
392. High, E. G., Smith, H. C., Jr., Taylor, H. H., and Wilson, S. S. (1954). *J. Biol. Chem.* **201,** 681–686.
393. High, E. G., and Wilson, S. S. (1953). *J. Nutr.* **50,** 203–212.
394. High, E. G., Wood, L. A., Jr., and Wilson, S. S. (1952). *J. Biol. Chem.* **195,** 787–793.
395. Hillman, D., Huber, J. T., Emery, R. S., Thomas, J. W., and Cook, R. M. (1973). *Mich. Agric. Ext. Bull.* E-702.

396. Hilton, J. H., Hauge, S. M., and Wilbur, J. W. (1935). *J. Dairy Sci.* **18,** 795-800.
397. Hinds, F. C., Zimmerman, J. E., Hatfield, E. E., Smith, G. S., and Neumann, A. L. (1965). *Proc. Ill Cattle Feeders' Day Univ. Ill,* Apr. 15, pp. 50-52.
398. Hjarde, W., and Larsen, J. B. (1956). *Proc. Int. Dairy Congr., 14th, Rome* **2**(1), 552.
399. Hjarde, W., Nielsen, J., and Porontnikoff, O. (1954). *Acta Agric. Scand.* **4,** 3-16.
400. Hjarde, W. A., Sorensen, A. N., Palludan, B., and Sorensen, P. H. (1961). *Acta Agric. Scand.* **11,** 13-54.
401. Hjarde, W. A., Neimann-Sorensen, A., Palludan, B., and Sorensen, P. H. (1958). *Aarsberet. Inst. Sterilitets forsk. K. Vet. Landbohoejsk.* **143,** 165-193; **144,** 144-167 (1959).
402. Hoar, D. W., Embry, L. B., and Emerick, R. J. (1968). *J. Anim. Sci.* **27,** 1727-1733.
403. Hodgson, R. E., Davis, R. E., Hosterman, W. H., and Heinton, T. E. (1948). *U.S. Dep. Agric. Yearb.* pp. 161-167.
404. Hodgson, R. E., Hall, S. R., Sweetman, W. J., Wiseman, H. G., and Converse, H. T. (1946). *J. Dairy Sci.* **29,** 669-687.
405. Hoefer, J. A., and Gallup, W. D. (1947). *J. Anim. Sci.* **6,** 325-333.
406. Hoelzer, H. (1960). *Tieraerztl. Umsch.* **15,** 408-412.
407. Hoffman, W. S., and Dyniewicz, H. A. (1945). *Gastroenterology* **5,** 512-522.
408. Holmes, A. D., Tripp, F., Ashbrook, G., and Kellogg, C. E. (1941). *U.S. Wildl. Res. Bull.* No. 3.
409. Holst, W. O., Flynn, L. M., Garner, G. B., and Pfander, W. H. (1961). *J. Anim. Sci.* **20,** 936.
410. van der Horst, C. J. G., and Kouwenhoven, B. (1970). *J. Parasitol.* **56,** 351.
411. Hove, E. L. (1952). *Fed. Proc., Fed. Am. Soc. Exp. Biol.* **11,** 446-447.
412. Hove, E. L. (1943). *Science* **98,** 433-434.
413. Huang, H. S., and Goodman, D. S. (1965). *J. Biol. Chem.* **240,** 2839-2844.
414. Huber, W. G. (1963). *Vet. Med.* **57,** 311-313; **58,** 875-880.
415. Hume, E. M., and Chick, H. (1935). *Med. Res. Counc. (G. B.), Spec. Rep. Ser.* No. 202, iv.
416. Hunter, R. F. (1946). *Nature (London)* **158,** 257-260.
417. Hutagalung, R. I., Chaney, C. H., Wood, R. D., and Waddil, D. G. (1968). *J. Anim. Sci.* **27,** 79-82.
418. Hvidsten, H., Sannan, F., and Osuldsen, T. (1963). *Meld. Norg. Landbrukshoegsk.* **42**(18), 1-19.
419. International Union of Pure and Applied Chemistry (1959). "The Vitamin A Potency of β-Carotene." IUPAC, London.
420. Irvin, J. L., Kopala, J., and Johnstone, C. G. (1941). *Am. J. Physiol.* **132,** 202-210.
421. Jacobson, N. L., Allen, R. S., Blake, J. T., and Homeyer, P. G. (1954). *J. Nutr.* **54,** 143-153.
422. Jacobson, N. L., Wise, G. H., Allen, R. S., and Kempthorne, O. (1950). *J. Dairy Sci.* **33,** 645-656.
423. Jagannathan, S. N., and Patwardhan, V. N. (1960). *Indian J. Med. Res.* **48,** 775-784.
424. James, W. H., and El Gindi, I. M. (1953). *J. Nutr.* **51,** 97-108.
425. Jensen, A. H., Baker, D. H., Lynch, P. B., and Harmon, B. G. (1973). *Proc. Ill. Pork Ind. Day, Univ. Ill.* Dec. 4.
426. Jensen, J. L., Hickman, K. C. D., and Harris, P. L. (1943). *Proc. Soc. Exp. Biol. Med.* **54,** 294-296.
427. John, J. and Cama, H. R. (1972). *Int. J. Vitam. Nutr. Res.* **42,** 278-284.
428. John, J., Kishore, G. S., Subbarayan, C., and Cama, H. R. (1970). *Indian J. Biochem.* **7,** 222-225.
429. Johnson, B. C., Peterson, W. H., and Steenbock, H. (1941). *J. Dairy Sci.* **24,** 813-817.
430. Johnson, R. M., and Baumann, C. A. (1948). *J. Biol. Chem.* **175,** 811-816.
431. Johnson, R. M., and Baumann, C. A. (1947). *Arch. Biochem.* **14,** 361-367.
432. Johnson, R. M., and Baumann, C. A. (1947). *J. Biol. Chem.* **171,** 513-521.
433. Johnson, R. M., Swick, R. W., and Baumann, C. A. (1949). *Arch. Biochem.* **22,** 122-131.
434. Jones, J. H., Riggs, J. K., Fraps, G. S., Jones, J. M., Schmidt, H., Dickson, R. E., Howe, P. E., and Black, W. H. (1938). *Proc. Am. Soc. Anim. Prod.* **31,** 94-102.

435. Jones, L. G., Zscheile, F. P., and Griffith, R. B. (1953). *Hilgardia* **22**(6), 179–202.
436. Jones, T. J., Hough, R. M., and Gerard, W. T. (1962). *Can. Vet. J.* **3**, 248–256.
437. Kamath, S. K., and Arnich, L. (1973). *J. Nutr.* **103**, 202–206.
438. Kapoor, U. R., and Ranjhan, S. K. (1974). *Indian J. Anim. Sci.* **44**, 22–31.
439. Karg, H. (1964). *Tierzuechter* **16**(10), 349–350, 353.
440. Karrer, P., Morf. R., and Schopp, K. (1931). *Helv. Chim. Acta* **14**, 1036, 1431.
441. Karrer, P., and Solmssen, U. (1937). *Helv. Chim. Acta* **20**, 682–690.
442. Keating, E. K., Hale, W. H., and Hubbert, F., Jr. (1964). *J. Anim. Sci.* **23**, 111–117.
443. Keener, H. A., Bechdel, S. I., Guerrant, N. B., and Thorp, W. T. S. (1942). *J. Dairy Sci.* **25**, 571–583.
444. Kehar, N. D. (1956). *Proc. Int. Congr. Anim. Husb., 7th, Madrid* Final Rep. Part 8, p. 105.
445. Kelly, B., and Day, H. G. (1950). *J. Nutr.* **40**, 159–168.
446. Kemmerer, A. R., and Fraps, G. S. (1938). *J. Nutr.* **16**, 309–315.
447. Kemmerer, A. R., Fraps, G. S., and De Mottier, J. (1947). *Arch. Biochem.* **12**, 135–138.
448. Kendall, K. A., Salisbury, G. W., and Vandemark, N. L. (1950). *J. Nutr.* **42**, 487–500.
449. Kent-Jones, D. W., Amos, A. J., and Thackery, G. B. 1958. *Analyst.* **83**, 479–480.
450. Ketola, H. G. (1976). *Feedstuffs* Feb. 16, pp. 42–45.
450. Ketola, H. G. (1976). *Feedstuffs* Feb. 16, pp. 42–45.
451. King, T. B., Lohman, T. G., and Smith, G. S. (1962). *J. Anim. Sci.* **21**, 1002.
452. Kirschman, J. C., and MacVicar, R. (1955). *Fed. Proc. Fed. Am. Soc. Exp. Biol.* **14**, 439.
453. Kirton, H. H., Crane, B., Paterson, D. J., and Clare, N. T. (1975). *N. Z. J. Agric. Res.* **18**, 267–272.
454. Kish, A., Weidler, F., and Adams, C. R. (1962). Unpublished observations.
455. Kishore, G. S., Subbarayan, C., and Cama, H. R. (1970). *Indian J. Biochem.* **7**, 98–101.
456. Kitamura, S., Suwa, T., Ohara, S., and Nakagawa (1967). *Nippon Suisan Gakkaishi* **33**, 1120, 1126.
457. Kivimae, A. (1959). *Acta Agric. Scand., Suppl.* **5**, 142.
458. Klatte, F. J., Mitchell, G. E., Jr., and Little, C. O. (1964). *Anim. Sci. Res. Prog. Rep.* No. 138, pp. 37–38.
459. Klaui, H., Hausheer, W., and Huschke, G. (1970). *In* "Encyclopedia of Food and Nutrition" (R. A. Morton, ed.), pp. 113–159. Pergamon, Oxford.
460. Klimes, I. (1960). *Zivocisna Vyroba* **5**, 543–558; **6**, 627–638.
461. Klosterman, E. W., Johnson, L. J., Moxon, A. L., and Grifo, A. P., Jr. (1964). *J. Anim. Sci.* **23**, 723–728.
462. Klosterman, E. W., Bolin, D. W., and Light, M. R. (1949). *J. Anim. Sci.* **8**, 624.
463. Knapstein, H. (1959). *Int. Z. Vitaminforsch.* **30**, 59–77.
464. Knowles, R. E., Livingston, A. L., Nelson, J. W., and Kohler, G. O. (1968). *J. Agric. Food Chem.* **16**, 654–658.
465. Knowles, R. E., Livingston, A. L., Nelson, J. W., and Kohler, G. O. (1968). *J. Agric. Food Chem.* **16**, 985–989.
466. Knuckles, B. E., Bickoff, E. M., and Kohler, G. O. (1972). *J. Agric. Food Chem.* **20**, 1055–1057.
467. Koehn, C. J. (1948). *Arch. Biochem.* **17**, 337–344.
468. Koizumi, I., Suzuki, T., and Sahaski, Y. (1963). *J. Vitaminol.* **9**, 154–159.
469. Kolari, O. E. (1964). *Minn., Feed Serv. Agric. Ext. Release* Spring.
470. Kolari, O. E., Harvey, A. L., and Meiske, J. C. (1960). *Minn., Beef Cattle Grassl. Field Day Rep.* B-28.
471. Kolari, O. E., and Jordan, R. M. (1961). *Feedstuffs* **33**(25), 28–31.
472. Kon, S. K., McGillivray, W. A., and Thompson, S. Y. (1955). *Br. J. Nutr.* **9**, 244–267.
473. Kon, S. K., and Thompson, S. Y. (1951). *Br. J. Nutr.* **5**, 114–119.
474. Konermann, H. (1967). *Z. Tierphysiol., Tierenaehr. Futtermittelkd.* **33**, 212–214.

475. Koopman, J. J., Wijbenga, A., and Bloem, A. C. M. (1971). *Tijdschr. Diergeneeskd.* **96,** 1019-1027.
476. Kovalcik, T., Zubricky, J., Lazar, J., Kohut, J., Valocek, J., and Palkoci, L. (1975). *Folia Vet.* **18,** 21-26.
477. Kowalewski, K., Henrotin, E., and van Geertruyden, J. (1951). *Acta Gastro-Enterol. Belg.* **14,** 607-616.
478. Kowalewski, K., and Henrotin, E. (1950). *Acta Gastro-Enterol. Belg.* **13,** 864-871.
479. Kraft, H. (1960). *Tieraeztl. Umsch.* **15,** 246-249.
480. Kramer, B., Sobel, A. E., and Gottfried, S. P. (1947). *Am. J. Dis. Child.* **73,** 543-553.
481. Kramer, M., and Tarjan, R. (1959). *Int. Z. Vitaminforsch.* **30,** 49-59; 271-278.
482. Kramke, E. H., Lloyd, M. D., and Fritz, J. C. (1952). *Poult. Sci.* **31,** 49-53.
483. Kraybill, K. R., and Shewsbury, C. L. (1936). *J. Nutr.* **11,** 103-110.
484. Krieg, R. (1963). *Arch. Gefluegelkd.* **27,** 215-234.
485. Kring, P. L., and Lund, A. (1958). *Acta Pharmacol. Toxicol.* **15,** 189-196.
486. Kring, P. L., and Rekling, E. (1958). *Dan. Tidsskr. Farm.* **32,** 19-25.
487. Krinsky, N. I. (1971). *In* "Carotenoids" (O. Isler, ed.), pp. 669-716. Birkhaeuser, Basel.
488. Krinsky, N. I. (1965). *Comp. Biochem. Physiol.* **16,** 181-188.
489. Krinsky, N. I., Cornwell, D. G., and Onclay, J. L. (1958). *Arch. Biochem. Biophys.* **13,** 233-246.
490. Krinsky, N. I., and Lenhoff, H. M. (1965). *Comp. Biochem. Physiol.* **16,** 189-198.
491. Krukowsky, V. N., Whiting, F., and Loosli, J. K. (1950). *J. Dairy Sci.* **33,** 791-796.
492. Kuhlman, A. H., and Gallup, W. D. (1940). *Proc. Am. Soc. Anim. Prod.* **33,** 67-73.
493. Kuhlman, A. H., and Gallup, W. D. (1944). *J. Dairy Sci.* **27,** 633-634.
494. Kuhn, R., and Brockmann, H. (1933). *Klin. Wochenschr.* **12,** 972-973.
495. Kuhn, R., Brockmann, H., Scheunert, A., and Schieblich, M. (1933). *Hoppe-Seyler's Z. Physiol. Chem.* **221,** 129-136.
496. Kunde, M. M. (1926). *Proc. Soc. Exp. Biol. Med.* **23,** 812.
497. Kupferschmied, H., and Rehm, W. F. (1967). *Conv. Insem. Veterinarians, Fed. Inst. Artif. Insem. Domest. Anim. Wels, Austria;* also *Proc. Int. Symp. Biol. Found. Reprod. Domest. Anim., Max-Planck-Inst.* (1967); *Dsch. Tieraerztl. Wochenschr.* **76**(6), 142-143 (1968).
498. Kurokawa, G. (1941). *Kokumin Eisei* **18,** 123-135.
499. Lakshmanan, M. R., Olson, J. A., and Pope, J. L. (1969). *Fed. Proc., Fed. Am. Soc. Exp. Biol.* **28,** 490.
500. Lakshamanan, M. R., Chansang, H., and Olson, J. A. (1972). *J. Lipid Res.* **13,** 477-483.
501. Lamming, G. E., Salisbury, G. W., Hays, R. L., and Kendall, K. A. (1954). *J. Nutr.* **52,** 217-239.
502. Landi, R., and Antongiovanni, M. (1973). *Maydica* **18**(½), 50-62.
503. Lane, A. M. (1962). *Proc. Ariz. Cattle Feeders' Day* May 5, pp. 22-30.
504. Lane, A. M. (1964). *Proc. Ariz. Cattle Feeders' Day* May 7, pp. 28-29 (1964).
505. Laughland, D. R., and Phillips, W. E. J. (1955). *Poult. Sci.* **34,** 1359-1362.
506. Laughland, D. H., and Phillips, W. E. J. (1954). *Can. J. Biochem. Physiol.* **32,** 593-599, 610-620.
507. Lautner, V. (1968). *Biol. Chem. Vyz. Zvirat* **4,** 379-383.
508. Lazar, J., Kaliska, A., and Kovac, J. (1972). *Folia Vet.* **16**(1/2), 13-19.
509. Lease, E. J., Lease, J. G., Steenbock, H., and Baumann, C. A. (1939). *J. Nutr.* **17,** 91-102.
510. Lemley, J. M., Brown, R. A., Bird, O. E., and Emmett, A. D. (1947). *J. Nutr.* **33,** 53-64.
511. Lemley, J., Brown, R., Bird, O., and Emmett, A. (1947). *J. Nutr.* **34,** 205-218.
512. Lewis, J. M., Bodansky, O., Birmingham, J., and Cohlan, S. Q. (1947). *J. Pediatr.* **31,** 496-508.
513. Lewis, J. M., and Wilson, L. T. (1945). *J. Nutr.* **30,** 467-475.
514. Lichtenwalner, R. E., Fontenot, J. P., and Tucker, R. E. (1973). *J. Anim. Sci.* **37,** 837-847.
515. Livingston, A. L., Bickoff, E. M., and Thompson, C. R. (1955). *J. Agric. Food Chem.* **3,** 439-441.

516. Livingston, A. L., Knowles, R. E., Israelsen, M., Nelson, J. W., Mottola, A. C., and Kohler, G. O. (1966). *J. Agric. Food Chem.* **14,** 643–644.
517. Livingston, A. L., Knowles, R. E., and Kohler, G. O. (1970). *U.S. Dep. Agric., Agric. Res. Serv., Tech. Bull.* No. 1414.
518. Livingston, A. L., Knowles, R. E., Nelson, J. W., and Kohler, G. O. (1968). *J. Agric. Food Chem.* **16,**(1), 84–87.
519. Livingston, A. L., Smith, D., Carnahan, H. L., Knowles, R. E. Nelson, J. W., and Kohler, G. (1968). *J. Sci. Food Agric.* **19,** 632–636.
520. Lord, J. W. (1945). *Biochem. J.* **39,** 372–374.
520a. Lotthammer, K. H. (1979). *Feedstuffs,* **51**(43), 16–50.
521. Lotthammer, K. H., and Ahlswede, L. (1977). *Dsch. Tieraerztl. Wochenschr.* **84,** 220–226.
522. Lotthammer, K. H., Ahlswede, L., and Meyer, H. (1976). *Dsch. Tieraerztl. Wochenschr.* **83,** 353–356.
523. Lunca, N., and Musetescu, M. (1966). *Zooteh. Med. Vet.* **16**(4), 26–29.
524. Luther, H. G., Goett, E. J., and Cragwell, G. O. (1952). *Int. J. Vitam. Res.* **23,** 362–380.
525. Mackinney, G. (1935). *J. Biol. Chem.* **108,** 45–49.
526. Madsen, L. L., and Davis, R. E. (1949). *J. Anim. Sci.* **8,** 625–626.
527. Madsen, L. L., Eaton, O. N., Heemstra, L., Davis, R. E., Cabell, C. A., and Knapp, B., Jr. (1948). *J. Anim. Sci.* **7,** 60–69.
528. Magendie, F. (1816). *Ann. Chim. Phys.* **1**(3), 66.
529. Mahle, A. E., and Patton, H. M. (1947). *Gastroenterology* **9,** 44–53.
530. Majumdar, B. N., and Gupta, B. N. (1960). *Indian J. Med. Res.* **48,** 388–393.
531. Mallia, A. K., John, J., John, K. V., Lakshmanan, M. R., Jungalwala, F. B., and Cama, H. R. (1970). *Indian J. Biochem.* **7,** 102–103.
532. Mann, I., Pirie, A., Tansley, K., and Wood, C. (1946). *Am. J. Ophthalmol.* **29,** 801–815.
533. Marrett, L. E., and Sunde, M. L. (1968). *Poult. Sci.* **47,** 511–519.
534. Martin, F. H., Ullrey, D. E., Miller, E. R., Kemp, K. E., Geasler, M. R., and Henderson, H. E. (1971). *J. Anim. Sci.* **32,** 1233–1238.
535. Martin, F. H., Ullrey, D. E., Newland, H. W., and Miller, E. R. (1968). *J. Nutr.* **96,** 269–274.
536. Marusich, W. L., and Bauernfeind, J. C. (1963). *Poult. Sci.* **42,** 949–957.
537. Marusich, W., DeRitter, E., and Bauernfeind, J. C. (1957). *J. Am. Oil Chem. Soc.* **34**(5), 217–221.
538. Marusich, W. L., DeRitter, E., Vreeland, J., and Krukar, R. J. (1960). *J. Agric. Food Chem.* **8,** 390–393.
539. Masoro, E. J. (1968). "Physiological Chemistry of Lipids in Mammals." Saunders, Philadelphia, Pennsylvania.
540. Massonet, R., Conguy, T., and Grangaud, R. (1965). *Ann. Nutr. Aliment.* **19,** 655–659.
541. Massonet, R., Conguy, T., and Ridolfo, J. (1961). *C. R. Acad. Sci.* **253,** 336–338.
542. Mathews, J., and Beaton, G. H. (1963). *Can. J. Biochem. Physiol.* **41,** 543–549.
543. Matsumuro, H., Kitaguwa, S., and Kaga, A. (1959). *Annu. Rep. Natl. Inst. Nutr., Tokyo* pp. 74–75.
544. Matsushima, J., and Dowe, T. W. (1954). *J. Am. Oil Chem. Soc.* **31,** 54–56.
545. Matterson, L. D., Bunnell, R. H., Potter, L. M., and Singsen, E. P. (1955). *Poult. Sci.* **34,** 1209–1210.
546. Matterson, L. D., Bunnell, R. H., Stinson, L. D., Singsen, E. P., and Potter, L. M. (1955). *Poult. Sci.* **34,** 1080–1084.
547. Mattill, H. A. (1927). *J. Am. Med. Assoc.* **89,** 1505–1508.
548. Mattson, F. H., and Deuel, H. J., Jr. (1943). *J. Nutr.* **25,** 103–112.
549. Mattson, F. H., Mehl, J. W., and Deuel, H. J., Jr. (1947). *Arch. Biochem. Biophys.* **15,** 65–73.
550. Mayfield, H. L., and Roehm, R. R. (1956). *J. Nutr.* **58,** 203–217, 483–493.
551. McCarthy, P. T., and Cerecedo, L. R. (1952). *J. Nutr.* **46,** 361–376; **49,** 357–367 (1953).
552. McClymont, G. L. (1956). *Proc. Aust. Soc. Anim. Prod.* **1,** 45.

553. McCoord, A. B., and Luce-Clausen, E. M. (1934). *J. Nutr.* **7,** 557-572.
554. McCoy, R. H. (1942). *In* "The Rat in Laboratory Investigation" (J. Griffith and C. Farris, eds.), pp. 67-101. Lippencott, Philadelphia, Pennsylvania.
555. McDowall, F. H., and McGillivray, W. H. (1963). J. *Dairy Res.* **30,** 59-66.
556. McDowall, F. H., McGillivray, W. A., and Reid, C. S. W. (1957). *N. Z. J. Sci. Technol.* **38,** 839-852.
557. McDowall, F. H., Patchell, R. M., and Reid, C. S. W. (1957). *N. Z. J. Sci. Technol. Sect. A* **38,** 1036-1053.
558. McGillivray, W. A. (1961). *Br. J. Nutr.* **15,** 313-325.
559. McGillivray, W. A. (1960). *N. Z. J. Agric. Res.* **3,** 276-278.
560. McGillivray, W. A. (1960). *World Rev. Nutr. Diet.* **2,** 131-155.
561. McGillivray, W. A. (1959). *N. Z. J. Agric. Res.* **2,** 694-701.
561a. McGillivray, W. A. (1952). *J. Dairy Sci.* **19,** 119-126.
562. McGillivray, W. A. (1951). *Br. J. Nutr.* **5,** 223-228.
563. McGillivray, W. A. (1957). *J. Dairy Res.* **24,** 95-101 and 346-351.
564. McGillivray, W. A., Thompson, S. Y., and Worker, N. A. (1958). *J. Dairy Res.* **25,** 439-446.
565. McGillivray, W. A., Thompson, S. Y., and Worker, N. A. (1956). *Br. J. Nutr.* **10,** 126-134.
566. McGillivray, W. A., and Worker, N. A. (1958). *N. Z. J. Agric. Res.* **1,** 267-272.
567. McGillivray, W. A., and Worker, N. A. (1958). *N. Z. J. Agric. Res.* **1,** 273-280.
568. McGillivray, W. A., and Worker, N. A. (1957). *Br. J. Nutr.* **11,** 47-56.
569. McIlwain, P. K., and Schipper, I. A. (1963). *J. Am. Vet. Med. Assoc.* **142,** 502-505.
570. McLaren, D. S., and Zehian, B. (1971). *Am. J. Dis. Child.* **121,** 278-280.
571. McNally, J. S., and Szymanski, C. D. (1966). *Nature (London)* **210,** 1366.
572. McQuire, W. S., and Jackobs, J. A. (1951). *Agron. J.* **43,** 379-381.
573. Meacham, T. N., Bovard, K. P., and Priode, B. M. (1964). *J. Anim. Sci.* **23,** 308.
574. Mellanby, E. (1934). *J. Pathol. Bacteriol.* **38,** 391-407.
575. Mellanby, E. (1935). *Brain* **58,** 141-173.
576. Melton, A. A. (1964). *Feedlot* Mar.; also *Proc. Tex. A&M Nutr. Conf.* Oct., 1964.
577. Menzies, C. S., Mitchell, G. E., Jr., and Little, C. O. (1967). *Int. Z. Vitaminforsch.* **37,** 443-448.
578. Meyer, H., Ahlswede, L., and Lotthammer, K. H. (1975). *Dsch. Tieraerztl. Wochenschr.* **82,** 444-449.
579. Meyer, J. (1961). *Calif. Cattle Feeders' Day Rep., Univ. Calif., Davis.*
580. Meyer, J. H., Strong, H. T., Kromann, R., Hughes, J. P., and Garett, W. N. (1962). *Proc. Univ. Calif. Cattle Feeders' Day* pp. 33-41.
581. Michaud, L., and Elvehjem, C. A. (1944). *North Am. Vet.* **25,** 657-666.
582. Mihajlovic, M., Hadzi-Taskovic, V., Sukalovic, V., and Markovic, G. (1971). *Proc. Meet. Maize Sorghum Sect., EUCARPIA, 5th, Milan, 1969* pp. 82-84.
583. Miller, R. E., Edwards, R. H., Lazar, M. E., Bickoff, E. M., and Kohler, G. O. (1972). *J. Agric. Food Chem.* **20,** 1151-1154.
584. Miller, R. E., Moore, L. A., Waldo, D. R., and Wrenn, T. R. (1967). *J. Anim. Sci.* **26,** 624-627.
585. Miller, R. W., Hemken, R. W., Waldo, D. R., and Moore, L. A. (1970). *J. Anim. Sci.* **30,** 984-987.
586. Mills, R. C., and Hart, E. B. (1945). *J. Dairy Sci.* **28,** 1-13.
587. Misra, U. K. (1968). *Agric. Biol. Chem.* **32,** 707-710.
588. Mitchell, G. E. (1967). *J. Am. Vet. Med. Assn.* **151,** 430-436.
589. Mitchell, G. E., Neumann, A. L., Garrigus, R. R., Durdle, W. M., and Kendall, K. A., (1960). *Proc. Ill. Cattle Feeders' Day, Univ. Ill., Champaign-Urbana.*
590. Mitchell, H. L., and Silker, R. E. (1950). *Ind. Eng. Chem.* **42,** 2325-2327.
591. Mitchell, H. L., and Silker, R. E. (1955). *J. Agric. Food Chem.* **1,** 1163-1164.
592. Mitchell, H. L., Beauchene, R. E., and Silker, R. E. (1954). *J. Agric. Food Chem.* **2,** 939-941.
593. Mitchell, H. S. (1933). *Proc. Soc. Exp. Biol. Med.* **31,** 231-233.

594. Mitchell, J. H., and Wise, G. H. (1944). *J. Dairy Sci.* **27,** 189–196.
595. Molander, D. W. (1949). *Yale J. Biol. Med.* **21,** 201–210.
596. Monson, W. J., Gade, E. T., Lloyd, M. D., and Groschke, A. C. (1957). *Poult. Sci.* **36,** 166–170.
597. Montemuno, O., Cianci, D., Bufano, G., and Ciruzzi, B. (1966). *Ann. Fac. Agric. Univ. Bari* **20,** 101–134.
598. Moon, F. E., Munro, P., and Ferns, H. R. (1966). *Proc. Nutr. Soc.* **25,** 3–4.
599. Moore, L. A. (1961). *Proc. Meet. Anim. Nutr. Res. Counc., Washington, D.C.* pp. 8–9.
600. Moore, L. A. (1939). *J. Dairy Sci.* **22,** 803–812.
601. Moore, L. A., and Berry, M. H. (1945). *J. Dairy Sci.* **28,** 821–826.
602. Moore, L. A., and Berry, M. H. (1944). *J. Dairy Sci.* **27,** 867–873.
603. Moore, L. A., Berry, M. H., and Sykes, J. F. (1943). *J. Nutr.* **26,** 649–658.
604. Moore, L. A., Sykes, J. F., Jacobson, W. C., and Wiseman, H. G. (1948). *J. Dairy Sci.* **31,** 533–538.
605. Moore, T. (1957). "Vitamin A." Am. Elsevier, New York.
606. Moore, T. (1940). *Biochem. J.* **34,** 1321–1328.
607. Moore, T. (1931). *Biochem. J.* **25,** 275–286.
608. Moore, T. (1930). *Biochemistry* **24,** 692–702.
609. Moore, T., and Payne, J. E. (1942). *Biochem. J.* **36,** 34–36.
610. Moore, T., Sharman, I. M., Todd, J. R., and Thompson, R. H. (1972). *Br. J. Nutr.* **28,** 23–30.
611. Morgan, A. F., and Arnrich, L. (1953). *Fed. Proc., Fed. Am. Soc. Exp. Biol.* **12,** 424–425.
612. Morgan, A. F., and Bentley, L. S. (1943). *Fed. Proc., Fed. Am. Soc. Exp. Biol.* **2,** 67.
613. Morris, H. P. (1947). *Vitam. Horm. (N.Y.)* **5,** 175–195.
614. Morton, R. A. (1940). *Chem. Ind.* **59,** 301–307.
615. Morton, R. A., and Creed, R. H. (1939). *Biochem. J.* **33,** 318.
616. Moster, J. B., Quackenbush, F. W., and Porter, J. W. (1952). *Arch. Biochem. Biophys.* **38,** 287–296.
617. Muhrer, M. E., Case, A. A., Garner, G. B., and Pfander, W. H. (1955). *J. Anim. Sci.* **14,** 1251.
618. Munzberg, H., and Aperdannier, R. (1962). *Mitt. DLG* **77,** 20.
619. Murray, T. K., and Erdody, P. (1977). *Nutr. Rep. Int.* **3,** 129–134.
620. Myers, G. S., Eaton, H. D., and Rousseau, J. E., Jr. (1959). *J. Anim. Sci.* **18,** 288–297.
621. Nadai, J. (1966). *World Congr. Anim. Feed., Madrid.* Oct. 2, 1966.
622. *NAS-NRC* (1969). *Publ.* No. 1684.
623. Nehring, K., and Hoffmann, M. (1967). *Arch. Tierernaehr.* **17,** 263–279.
624. Nehring, K., and Hoffman, M. (1967). *Arch. Tierernaehr.* **16,** 553–568.
624a. Neilands, J. B. (1947). *Arch. Biochem. Biophys.* **13,** 415–419.
625. Nestler, R. B., Derby, J. V., and DeWitt, J. B. (1948). *J. Nutr.* **36,** 323–329.
626. Neumann, A. L., Smith, G. S., Lamb, P. E., Albert, W. W., and Zimmerman, J. E. (1962). *Proc. Ill. Cattle Feeders' Day* pp. 28–29.
627. Newland, H. W. (1964). *Mich. State Univ. Exp. Stn., Bull.* No. 465.
628. Nicholson, J. W. G., and Cuningham, H. M. (1965). *Can. Vet. J.* **6,** 275–281.
629. Nilsson, B. (1971). *Acta Physiol. Scand.* **83,** 1–3.
630. *Nutr. Rev.* (1947). **5**(9), 265–266.
631. Nutrient Requirements of Beef Cattle (1958). *NAS-N.R.C., Publ.* No. 579; *Publ.* No. 0-309-01754-8 (1970).
632. Nutrient Requirements of Dairy Cattle (1958). *NAS-N.R.C., Publ.* No. 464; *Publ.* No. 1349 (1966); *Publ.* No. 0-309-01916-8 (1971).
633. Nutrient Requirement of Dogs (1953). *NAS-N.R.C., Publ.* No. 300; *Publ.* No. 989 (1962); *Publ.* No. 0-309-02315-7 (1974).
634. Nutrient Requirement of Foxes and Mink (1953). *NAS-N.R.C., Publ.* No. 296; *Publ.* No. 1676 (1968).

635. Nutrient Requirements of Horses (1961). *NAS-N.R.C., Publ.* No. 912; *Publ.* No. 0-309-02045-X (1973); *Publ.* No. 0-309-02760-8 (1978).
636. Nutrient Requirements of Laboratory Animals (1962). *NAS-N.R.C., Publ.* No. 990; *Publ.* No. 0-309-02028-X (1972); *Publ.* No. 0-309-02767-5 (1978).
637. Nutrient Requirements of Poultry (1960). *NAS-N.R.C., Publ.* No. 827; *Publ.* No. 1345 (1966); *Publ.* No. 0-309-01861-7 (1971); *Publ.* No. 0-309-02725-X (1977).
638. Nutrient Requirements of Rabbits (1954). *NAS-N.R.S., Publ.* No. 331; *Publ.* No. 1194 (1966); *Publ.* No. 0-309-02607-5 (1977).
639. Nutrient Requirements of Sheep (1957). *NAS-N.R.C., Publ.* No. 504; *Publ.* No. 1693 (1968); *Publ.* No. 0-309-02212-6 (1975).
640. Nutrient Requirements of Swine (1959). *NAS-N.R.C., Publ.* No. 648; *Publ.* No. 1599 (1968); *Publ.* No. 0-309-02140-5 (1973).
641. Nutrient Requirements of Trout, Salmon and Catfish (1973). *NAS-N.R.C., Publ.* No. 0-309-02141-3.
642. Nutritive Requirements of Farm Livestock (1960). *Publ.* No. 1 (Poultry). Agric. Res. Counc. G. B., London.
643. O'Dell, B. L., Erek, Z., Flynn, L., Garner, G. B., and Muhrer, M. E. (1960). *J. Anim. Sci.* **19,** 1280.
644. O'Donoghue, J. G. (1962). *Proc. Annu. Feed Ind. Conf., 2nd, Alberta.*
645. O'Donoghue, J. G. (1955). *Can. J. Comp. Med.* **19,** 367–372.
646. O'Donoghue, J. G. (1954). *Can. J. Comp. Med.* **18,** 141–147.
647. O'Donovan, P. B., and Conway, A. (1967). *Anim. Prod.* **9,** 283.
648. Ogden, R. L. (1956). *J. Agric. Food Chem.* **4,** 428–431.
649. Olafson, P. (1947). *Cornell Vet.* **37,** 279–291.
650. Olesson, N., Akerberg, E., and Blixt, B. (1955). *Acta Agric. Scand.* **5,** 113–184.
651. Olsen, E. M., Harvey, J. D., Hill, D. C., and Branion, H. D. (1959). *Poult. Sci.* **38,** 929–942.
652. Olsen, E. M., Hill, D. C., and Branion, H. D. (1964). *Poult. Sci.* **43,** 1488–1501.
653. Olson, J. A. (1969). *Am. J. Clin. Nutr.* **22,** 953–962.
654. Olson, J. A. (1968). *Vitam. Horm. (N.Y.)* **26,** 1–63.
655. Olson, J. A. (1967). *Pharmacol. Rev.* **19,** 559–596.
656. Olson, J. A. (1964). *J. Lipid Res.* **5,** 281–299.
657. Olson, J. A. (1961). *J. Biol. Chem.* **226,** 349–356.
658. Olson, J. A., and Hayaishi, O. (1965). *Proc. Natl. Acad. Sci. U.S.A.* **54,** 1364–1370.
659. Olson, J. A. (1969). *Fed. Proc., Fed. Am. Soc. Exp. Biol.* **28,** 1670–1677.
660. Olson, O. E., Nelson, D. L., and Emerick, R. J. (1963). *J. Agric. Food Chem.* **11,** 140–143.
661. Onufriev, P. U., and Cikova, A. P. (1970). *Svinovodstvo (Moscow)* No. 9, pp. 34–35.
662. Orth, A., and Koch, G. (1963). *Wiss. Veroeff. Dsch. Ges. Ernaehr.* **9,** 363–376.
663. Owen, E. C. (1965). *World Rev. Nutr. Diet.* **5,** 132–208.
664. Page, H. M., Erwin, E. S., and Nelms, G. E. (1959). *Am. J. Physiol.* **196,** 917–918.
665. Palludan, B. (1966). *Nature (London)* **21,** 639–640.
666. Palmer, L. S. (1922). "Carotenoids and Related Pigments," ACS Monograph Series. Chem. Catalog Co. (Tudor), New York.
667. Palmer, L. S., and Eckles, C. H. (1914). *J. Biol. Chem.* **17,** 191–210.
668. Palmer, L. S., and Kempster, H. L. (1919). *J. Biol. Chem.* **39,** 299–313.
669. Parrish, D. B., and Aubel, C. E. (1957). *Kans. Agric. Exp. Stn., Circ.,* No. 349.
670. Parrish, D. B., Aubel, C. E., Hughes, J. S., and Wheat, J. D. (1951). *J. Anim. Sci.* **10,** 551–560.
671. Parrish, D. B., Bartley, E. E., Burris, D. U., and McIntyre, R. T. (1953). *J. Dairy Sci.* **36,** 489–494.
672. Parrish, D. B., and Sanford, P. E.(1962). *Informal Poult. Nutr. Conf., Atlantic City, N.J.*
673. Parrish, D. B., Zimmerman, R. A., Stanford, P. E., and Hung, E. (1963). *J. Nutr.* **79,** 9–17.
674. Parrish, D. B., Wise, G. H., Atkeson, F. W., and Hughes, J. S. (1949). *J. Dairy Sci.* **32,** 209–221.

675. Parrish, D. B., Wise, G. H., and Hughes, J. S. (1947). *J. Biol. Chem.* **167,** 673–678.
676. Patel, B. M., Patel, C. A., and Shukla, P. C. (1966). *Indian J. Vet. Sci.* **36**(3), 124–129.
677. Patel, L. G., and Dave, A. D. (1972). *Indian J. Dairy Sci.* **25**(1), 35–38.
678. Patel, S. M., Mehl, J. W., and Deuel, H. J., Jr. (1951). *Arch. Biochem.* **30,** 103–109.
679. Patrick, H., and Morgan, C. L. (1943). *Science* **98,** 434–435.
680. Patton, J. W. (1944). *Vet. Med.* **39,** 150–153.
681. Payne, A. S., Donefer, E., and Baker, R. D. (1972). *Can. J. Animal Sci.* **52**(1), 125–136.
682. Payne, M. G., and King, H. E. (1947). *J. Anim. Sci.* **6,** 50–55.
683. Pcelinceva, E. M. (1966). *Ovcevodstvo* No. 2, pp. 31–32.
684. Peirce, A. W. (1954). *Aust. J. Agric. Res.* **5,** 470–483.
685. Peirce, A. W. (1947). *Aust. J. Exp. Biol. Med. Sci.* **25,** 111–117.
686. Peirce, A. W. (1946). *Aust. J. Exp. Biol. Med. Sci.* **24,** 231–240.
687. Perice, A. W. (1945). *Aust. J. Exp. Biol. Med. Sci.* **23,** 295–303.
688. Perry, T. W. (1966). *Merck Agric. Livestock Memo* No. 11(1), pp. 3, 7.
689. Perry, T. W., Beeson, W. M., Smith, W. H., Harrington, R. B., and Mohler, M. T. (1968). *J. Anim. Sci.* **27,** 190–194.
690. Perry, T. W., Beeson, W. M., Smith, W. H., and Mohler, M. T. (1967). *J. Anim. Sci.* **26,** 115–118.
691. Perry, T. W., Beeson, W. M., Smith, W. H., and Mohler, M. T. (1964). *J. Anim. Sci.* **23,** 888.
692. Perry, T. W., Beeson, W. M., Mohler, M. T., and Smith, W. H. (1962). *J. Anim. Sci.* **21,** 333–339.
693. Perry, T. W., Outhouse, J. B., and Osborn, J. A. (1964). *Purdue Univ. Agric. Res. Prog. Rep.* Nos. 137 and 138.
694. Peterson, W. J., Hughes, J. S., and Payne, L. F. (1939). *Kans. State Coll. Agric. Exp. Stn., Tech. Bull.* No. 46.
695. Petrova, M. V. (1969). *Sb. Nauchn. Rab., Vses. Nauchno-Issled. Inst. Zhivotnovod.* No. 15, pp. 25–27.
696. Petzold, E. N., Quackenbush, F. W., and McQuistan, M. (1959). *Arch. Biochem. Biophys.* **82,** 117–124.
697. Phillips, P. H. (1953). *In* "Symposium on Nutrition" (R. M. Herriott, ed.), pp. 3–20. Johns Hopkins Press, Baltimore, Maryland.
698. Phillips, P. H., and Bohstedt, G. (1938). *J. Nutr.* **15,** 209–219.
699. Phillips, W. E. J. (1966). *Can. Dep. Agric., Publ.* No. 1238.
700. Phillips, W. E. J. (1966). *Can. J. Biochem.* **44,** 1–7.
701. Phillips, W. E. J. (1962). *Can. J. Biochem. Physiol.* **40,** 491.
702. Picard, F. H. J., and Bakker, T. (1939). *Tijdschr. Diergeneeskd.* **66,** 353–356; also *Dsch. Pelstierz.* **14,** 342–345.
703. Pirchner, F., Allen, R. S., and Jacobson, N. L. (1957). *J. Dairy Sci.* **40,** 1448–1456.
704. Pirie, A., and Wood, C. (1946). *Biochem. J.* **40,** 557–560.
705. Pitt, G. A. J. (1971). *In* "Carotenoids" (O. Isler, ed.), pp. 717–742. Birkhaeuser, Basel.
706. Pivnyak, I. G., and Konyakhin, A. N. (1972). *Dok. Vses. Ordena Lenina Akad. Skh. Nauk* No. 7, pp. 30–32.
707. Pivnyak, I. G., and Konyakhin, A. N. (1972). *Skh. Biol.* **7,** 416–419.
708. Pope, A. L., Phillips, P. H., and Bohstedt, G. (1949). *J. Anim. Sci.* **8,** 57–66.
709. Popper, H., and Chinn, H. (1942). *Proc. Soc. Exp. Biol. Med.* **49,** 202–204.
710. Portal, N., Davila, F., Hernandez, R., and Boado, A. (1974). *Cienc. Agropec. Ing. Pec.* No. 8, pp. 1–6.
711. Poston, H. A., Riis, R. C., Rumsey, G. L., and Ketola, H. G. (1977). *Cornell Vet.* **67,** 472–509.
712. Poston, H. A. (1968). *Fish. Res. Bull.* No. 32, pp. 41–43.
713. Potkanski, A. A., Tucker, R. E., and Mitchell, G. E., Jr. (1974). *Int. J. Vitam. Res.* **44,** 3–7.
714. Potter, L. M., Bunnell, R. H., Matterson, L. D., and Singsen, E. P. (1956). *Poult. Sci.* **35,** 452–456.

715. Prohaszka, L. (1965). *Monatsh. Veterinaermed.* **20,** 656.
716. Pugh, D. L., and Garner, G. B. (1963). J. *Agric. Food Chem.* **11,** 528–529.
717. Pullman, B., and Pullman, A. (1963). "Quantum Biochemistry," pp. 435–470. Wiley (Interscience), New York.
718. Quackenbush, F. W. (1963). *Cereal Chem.* **40,** 266–269.
719. Quackenbush, F. W., Cox, R. P., and Steenbock, H. (1942). *J. Biol. Chem.* **145,** 169–177.
720. Quackenbush, F. W., Firch, J. G., Brunson, A. M., and House, L. R. (1963). *Cereal Chem.* **40,** 250–259.
721. Quackenbush, F. W., Firch, J. G., Rabourn, W. J., McQuistan, M., Petzold, E. N., and Kargl, T. E. (1961). *J. Agric. Food Chem.* **9,** 132–135.
722. Quarterman, J., and Mills, C. F. (1964). *Proc. Nut. Soc.* **23,** 10–11.
723. Ralston, A. T., and Dyer, I. A. (1959). *J. Anim. Sci.* **18,** 874–879.
724. Randoin, L., Hugot, D., and Causeret, J. (1951). *C. R. Soc. Biol.* **145,** 65–70.
724a. Rao, S. D. (1945). *Nature (London)* **156,** 449.
725. Rea, J. L., and Drummond, J. C. (1932). *Int. Z. Vitaminforsch.* **1,** 177–183.
726. Record, P. R., Bethke, R. M., and Wilder, O. H. M. (1937). *Poult. Sci.* **16,** 25–33.
727. Record, P. R., Beeson, W. M., and Smith, W. H. (1963). *J. Anim. Sci.* **22,** 845.
728. Reddy, B. S., and Thomas, J. W. (1962), *J. Anim. Sci.* **21,** 1010.
729. Rehm, W., and Kupferschmied, H. (1966). *Schweiz. Arch. Tierheilkd.* **108,** 428–438.
730. Rehm, W., Kupferschmied, H., and Rietz, P. (1970). *Schweiz. Arch. Tierheilkd.* **112,** 500–511.
731. Reid, B. L., Daugherty, H. K., and Couch, J. R. (1955). *Poult. Sci.* **34,** 603–608.
732. Reid, C. S. W., and Johns, A. T. (1957). *N. Z. J. Sci. Technol. Sect. A* **38,** 908–927.
733. Remington, R. E., Harris, P. L., and Smith, C. L. (1942). *J. Nutr.* **24,** 597–606.
734. Report of the Conference on Vitamin Standards (1931). *League Nations Publ., Geneva* iii, 10 (C.H.1055/1).
735. Reynolds, J. B., Parrish, D. B., Mitchell, H. L., and Payne, L. F. (1948). *Poult. Sci.* **27,** 182–185.
736. Ribarova, M. (1969). *Zhivotnovad. Kauki* **6**(1), 67–76.
737. Ritzman, E. G., Colovos, N. F., Keener, H. A., and Teeri, A. E. (1945). *N.H. Agric. Exp. Stn., Tech. Bull.* No. 87, p. 28.
738. Robboy, M. S., Sato, A. S., and Schwabe, A. D. (1974). *Am. J. Clin. Nutr.* **27,** 362–367.
739. Roberts, W. K., and Sell, J. L. (1963). *J. Anim. Sci.* **22,** 1081–1085.
740. Ronning, M. E., Berousek, E. R., Griffith, J. R., and Gallup, W. D. (1959). *Okla. Agric. Exp. Stn., Bull.* T-76.
741. Ronning, M., Berousek, E. R., Kuhlman, A. H., and Gallup, W. D. (1953). *J. Dairy Sci.* **36,** 52–56.
742. Ronning, M., and Knodt, C. B. (1952). *J. Dairy Sci.* **35,** 320–323.
743. Ronning, M., and Knodt, C. B. (1950). *J. Dairy Sci.* **33,** 424–429.
744. Ross, O. B., and Gallup, W. D. (1949). *J. Anim. Sci.* **8,** 628–629.
745. Ross, R. H., Knodt, C. B., and Guerrant, N. B. (1948). *J. Dairy Sci.* **31,** 111–117.
746. Rousseau, J. E., Jr., Eaton, H. D., Teichman, R., Helmboldt, C. F., Jungherr, E. L., Bacon, E. L., Dolge, K. L., Beall, G., and Moore, L. A. (1956). *J. Dairy Sci.* **39,** 1671–1682.
747. Rousseau, J. E., Jr., Eaton, H. D., Helmboldt, C. F., Jungherr, E. L., Robrish, S. A., Beall, G., and Moore, L. A. (1954). *J. Dairy Sci.* **37,** 887–899.
748. Rousseau, J. E., Jr., Eaton, H. D., Teichman, R., Helmboldt, C. F., Jungherr, E. L., Bacon, D. L., and Beall, G. (1956). *J. Dairy Sci.* **39,** 1565–1573.
749. Roussel, J. D., Patrick, T. E., Kellgren, H. C., Randel, P. F., and Rusoff, L. L. (1963). *J. Dairy Sci.* **46,** 583–585.
750. Rowntree, J. I. (1931). *J. Nutr.* **3,** 345–351.
751. Rubin, M., and Bird, H. R. (1946). *Science* **103,** 584.
752. Rubin, M., and Bird, H. R. (1941). *Poult. Sci.* **20,** 291–297.

753. Rubin, S. H., and De Ritter, E. (1954). *Vitam. Horm. (N.Y.)* **12,** 101–135.
754. Ruff, M. D., and Fuller, H. L. (1975). *J. Nutr.* **105,** 1447–1456.
755. Ruff, M. D., Reid, W. M., and Johnson, J. K. (1974). *Poult. Sci.* **53,** 1801–1809.
756. Rusoff, L. L., Skipper, H. E., and Dix-Arnold, P. T. (1942). *J. Dairy Sci.* **25,** 807–813.
757. Russell, W. C., Taylor, M. W., Walker, H. A., and Polskin, L. J. (1942). *J. Nutr.* **24,** 199–211.
758. Saksena, J. S., Mehta, J. A., and Naimark, A. (1971). *Can. J. Physiol. Pharmacol.* **49,** 127–133.
759. Saraswat, R. C., and Arora, S. P. (1972). *Indian J. Anim. Sci.* **42,** 358–362.
760. Sarkar, B. C. R. (1948). *J. Dairy Sci.* **31,** 479–487.
761. Schams, D., Hoffmann, B., Lotthammer, K. H., and Ahlswede, L. (1978). *Acta Endocrinol. (Copenhagen), Suppl.* **215,** 101.
762. Schams, D., Hoffmann, B., Lotthammer, K. H., and Ahlswede, L. (1977). *Dsch. Tieraerztl. Wochenschr.* **84,** 307–310.
763. Scharrer, K., and Raker, K. (1955). *Z. Tierernaehr Futtermittelkd.* **10,** 275–300.
764. Schenck, F., and Pilz, W. (1956). *Med. Chem.* **5,** 399; also *Tieraerztl. Umsch.* **12**(6), 199–200.
765. Schleicher, E., and Clanton, D. (1964). *Nebr. Cattlemen* **21**(1), 30, 34.
766. Schmidt, H. (1952). *Sheep Goat Raiser* **33,** 20.
767. Schoop, G. (1956). *Tieraerztl. Umsch.* **11,** 281–284.
768. Schuh, J. D., Ronning, M., and Gallup, W. D. (1959). *J. Dairy Sci.* **42,** 159–165.
769. Schultz, G. (1972). *Wirtschaftseigene Futter* **18,** 222–236.
770. Schultz, G., and Reupke, B. (1973). *Wirtschaftseigene Futter* **19,** 1–4.
771. Schultz, H. (1956). Thesis, Muenchen Univ. Diss. Med. Vet.
772. Schultze, A. B., and Turner, C. W. (1945). *Univ. M. Agric. Exp. Stn., Res. Bull.* No. 392.
773. Scott, P. P., and Scott, M. G. (1964). *J. Reprod. Fertil.* **8,** 270–271.
774. Second Conference on Vitamin Standardization (1934). *League Nations Publ., Geneva* No. 3, Extract 15, p. 428.
775. Seghetti, L., and Marsh, H. (1953). *Am. J. Vet. Res.* **14,** 9–11.
776. Seitaridis, K. (1963). Thesis, Hannover Diss. Med. Vet.
777. Sell, J. L., and Roberts, W. K. (1963). *J. Nutr.* **79,** 171–178.
778. Semin, V. N., Tkachev, I. F., and Bukhtiyarova, O. N. (1973). *Vitam. Pitan. Skh. Zivotz.* pp. 286–398.
779. Sexton, E. L., Mehl, J. W., and Deuel, H. J., Jr. (1946). *J. Nutr.* **31,** 299–319.
780. Shah, M. A., Qureshi, A. K., and Elahi, M. (1970). *Sci. Ind. (Karachi)* **7**(3), 153–159.
781. Sharma, R. V., Mathur, S. N., Dmitrovoskii, A. A., Das, R. C., and Ganguly, J. (1977). *Biochim. Biophys. Acta* **486,** 183–194.
782. Sharma, R. V., Mathur, S. N., and Ganguly, J. (1976). *Biochem. J. Cell. Aspects* **158,** 377–383.
783. Shaw, J. C. (1950). *J. Dairy Sci.* **33,** 486–495.
784. Shaw, J. C., Moore, L. A., and Sykes, J. F. (1951). *J. Dairy Sci.* **34,** 176–180.
785. Shaw, R. J., and Deuel, H. J. (1944). *J. Nutr.* **27,** 395–401.
786. Shepherd, J. B., Wiseman, H. G., Ely, R. E., Melin, C. G., Sweetman, W. J., Gordon, C. H., Schoenleber, L. G., Wagner, R. E., Campbell, L. E., Roane, G. D., and Hosterman, W. H. (1954). *U.S. Dep. Agric., Tech. Bull.* No. 1079.
786a. Shepherd, J. B., Schoenleber, L. G., Wisseman, H. G., Melin, C. G., Sweetman, W. J., Hosterman, W. H., and Tysdal, H. M. (1948). *J. Dairy Sci.* **31,** 688–699.
787. Sherman, H. C., and Trupp, H. Y. (1949). *J. Nutr.* **37,** 467–474.
788. Sherman, H. C., and Boynton, K. C. (1925). *J. Am. Chem. Soc.* **47,** 1646–1653.
789. Sherman, W. C. (1947). *Proc. Soc. Exp. Biol. Med.* **65,** 207–210.
790. Sherman, W. C. (1941). *Proc. Soc. Exp. Biol. Med.* **47,** 199–200.
791. Sherman, W. C. (1941). *J. Nutr.* **22,** 153–165.
792. Sherman, W. C., Kuhn, G. O., White, E. A., Lewis, W. D., Reynolds, W. M., and Luther, H. G. (1958). *J. Anim. Sci.* **17,** 586–592.
793. Shorland, F. B. (1944). *Nature (London)* **153,** 168.

794. Shorland, F. B., Weenink, R. D., and Johns, A. T. (1955). *Nature (London)* **175,** 1129–1130.
795. Sibbald, I. R., and Olsen, E. (1958). *Poult. Sci.* **37,** 1465–1467.
796. Siedler, A. J., Enzer, E., and Schweigert, B. S. (1956). *J. Agric. Food Chem.* **4,** 1023–1029.
797. Simpson, K. L., and Goodwin, T. W. (1965), *Phytochemistry* **4,** 193–195.
798. Singh, H., and Cama, H. R. (1974). *Biochim. Biophys. Acta* **370,** 49–61.
799. Singh, H., and Cama, H. R. (1974). *Biochem. Rev.* **45,** 77–88.
800. Singh, H., and Cama, H. R. (1972). *Int. Z. Vitaminforsch.* **42**(3), 378–388.
801. Singh, M. M., Sinha, G. K., and Gupta, B. N. (1965). *Indian Vet. J.* **42,** 879–880.
802. Skinner, J. D. (1963). *Proc. S. Afr. Soc. Anim. Prod.* **2,** 143.
803. Slanetz, C. A., and Scharf, A. (1945). *J. Nutr.* **30,** 239–243.
804. Slanetz, C. A., and Scharf, A. (1943). *Proc. Soc. Exp. Biol. Med.* **53,** 17–19.
805. Smith, G. S., Durdle, W. M., Zimmerman, J. E., and Neumann, A. L. (1964). *J. Anim. Sci.* **23,** 625–632.
806. Smith, G. S., Hatfield, E. E., Durdle, W. M., and Neumann, A. L. (1962). *J. Anim. Sci.* **21,** 1013–1014.
807. Smith, G. S., Neumann, A. L., and Hatfield, E. E. (1961). *J. Anim. Sci.* **20,** 683.
808. Smith, J. C., Jr., McDaniel, E. G., Fan, F. F., and Halsted, J. A. (1973). *Science* **181,** 954–955.
809. Smith, J. E., and Borchers, R. (1972). *J. Nutr.* **102,** 1017–1024.
810. Smith, S. E. (1941). *Am. Fur Breeder* **14**(3), 10, 12; also *J. Nutr.* **24,** 97–107 (1942).
811. Smithcors. J. F. (1962). *Mod. Vet. Pract.* **43**(12), 47–49, 52–53.
812. Sobel, A. E., Rosenberg, A., and Adelson, H. (1953). *Arch. Biochem. Biophys.* **44,** 176–180.
813. Sobel, A. E., and Rosenberg, A. A. (1950). *J. Nutr.* **42,** 557–562.
814. Sobel, A. E., Rosenberg, A. A., and Engel, E. (1952). *J. Nutr.* **48,** 183–192.
815. Sørenson, N. A. (1933). *K. Nor. Vidensk. Selsk., Forh.* **6**(40), 154–156.
816. Spielman, A. A., Loosli, J. K., Thomas, J. W., Norton, C. L., and Turk, K. L. (1946). *J. Dairy Sci.* **29,** 381–391.
817. Spielman, A. A., Thomas, J. W., Loosli, J. K., Norton, C. L., and Turk, K. L. (1946). *J. Dairy Sci.* **29,** 707–715.
818. Spruzs, J. (1971). *Latv. Lopkopibas Vet. Zinat. Petnieciska Inst. Raksti* **26,** 109–117.
819. Spruzs, J. (1971). *Latv. Lopkopibas Vet. Zinat. Petieciska Inst. Raksti* **26,** 119–124.
820. Squibb, R. L., Braham, J. E., and Scrimshaw, N. S. (1956). *Poult. Sci.* **35,** 73–76.
821. Squibb, R. L., Cannon, C. Y., and Allen, R. S. (1948). *J. Dairy Sci.* **31,** 421–427; **32,** 565–569.
822. Stahl, W., Steger, H., Kasdorff, K., and Püschel, F. (1957). *Z. Tierernaehr. Futtermittelkd.* **12**(6), 333–339.
823. Stallcup, O. T., and Herman, H. A. (1950). *J. Dairy Sci.* **33,** 237–242.
824. Steenbock, H. (1919). *Science* **50,** 352–353.
825. Steenbock, H. E., Nelson, E. M., and Hart, E. B. (1921). *Am. J. Physiol.* **58,** 14–19.
826. Steigmann, F., Popper, H., Dyniewicz, H., and Maxwell, I. (1952). *Gastroenterology* **20,** 587–594.
827. Stevens, W. R., and Williams, C. D. (1962). Unpublished observations.
828. Stimson, A. M., and Hedley, O. W. (1933). *U.S. Public Health Rep.* **48**(17), 445–449.
829. Strauss, W. (1942). *Helv. Chim. Acta* **25,** 705–717.
830. Subbarayan, C., Lakshmanan, M. R., and Cama, H. R. (1966). *Biochem. J.* **99,** 308–311.
831. Surmatis, J. D., Maricq, J., and Ofner, A. (1958). *J. Org. Chem.* **23,** 157–162.
832. Suryanarayana, R. K., Rukmini, C., and Mohan, V. S. (1968). *Indian J. Agric. Sci.* **38,** 368–372.
833. Sutton, T. S., Kaeser, H. E., and Soldner, P. A. (1945). *J. Dairy Sci.* **28,** 933–939.
834. Swanson, E. W., Martin, G. G., Pardue, F. E., and Gorman, G. M. (1968). *J. Anim. Sci.* **27,** 541–548.
835. Sweeney, J. P., and Marsh, A. C. (1973). *J. Nutr.* **103,** 20–25.
836. Swick, R. W., and Baumann, C. A. (1952). *Arch. Biochem. Biophys.* **36,** 120–126.
837. Swick, R. W., Lih, H., and Baumann, C. A. (1952). *Fed. Proc., Fed. Am. Soc. Exp. Biol.* **10,** 395.

838. Swingle, K. F. (1950). *Proc. Annu. Mont. Nutr. Conf., 1st, Montana State Coll.;* also *Feedstuffs* **22**(35), 37–40.
839. Tagwerker, F. J. (1961). *Wien. Tieraerztl. Monatsschr.* **48,** 661–690.
840. Tagwerker, F. J. (1968). *Proc. Conf. Yugosl. Feed Ind., Optija.*
841. Tagwerker, F. J. (1962). "Technical Service Brochure." Roche, Basel.
842. Tainsh, P., and Wilkinson, H. (1939). *Chem. Ind.* **58,** 1051–1052.
843. Tarjan, R. (1958). *Vopr. Pitan.* **17**(4), 3–8.
844. Teichman, R., Helmboldt, C. F., Rousseau, J. E., Jr., Jungherr, E. L., and Eaton, H. D. (1953). *Conn. Storrs Agric. Exp. Stn., Bull.* No. 301.
845. Teichman, R., Morgan, M. E., Eaton, H. D., and MacLeod, P. (1955). *J. Dairy Sci.* **38,** 693–695.
846. Thomas, J. W., Hillman, D., and Hafs, H. D. (1968). *J. Dairy Sci.* **51,** 979.
847. Thomas, O. O., Woodward, R. R., Payne, G. F., Quisenberry, J. R., and Willson, F. S. (1957). *J. Anim. Sci.* **16,** 1093–1094.
848. Thommen, H. (1971). *In* "Carotenoids" (O. Isler, ed.), pp. 637–668. Birkhaeuser, Basel.
849. Thommen, H. (1961). *Chimia* **15,** 433–434.
850. Thommen, H., and Wiss, O. (1963). *Z. Ernaehrungswiss.* **3,** 18–23.
851. Thompson, C. R. (1950). *Ind. Eng. Chem.* **42,** 922–925.
852. Thompson, C. R. (1949). *Agron. J.* **41,** 294–297.
853. Thompson, C. R., Bickoff, E. M., and Maclay, W. D. (1951). *Ind. Eng. Chem.* **43,** 126–129.
854. Thompson, C. R., Bickoff, E. M., Van Atta, G. R., Kohler, G. O., Guggolz, J., and Livingston, A. L. (1960). *U.S. Dep. Agric., Agric. Res. Serv., Tech. Bull.* No. 1232.
855. Thompson, S. Y. (1975). *World Rev. Nutr. Dietet.* **21,** 224–280.
856. Thompson, S. Y. (1970). *Proc. Int. Dairy Congr., 18th* **1E,** 569.
857. Thompson, S. Y. (1965). *Proc. Nutr. Soc.* **24,** 136–146.
858. Thompson, S. Y. (1963). *Wiss. Veroeff. Dsch. Ges. Ernaehr.* **9,** 262–281.
859. Thompson, S. Y., and Ascarelli, I. (1962). *Proc. Int. Dairy Congr., 16th, Copenhagen* Sect. A, pp. 81–88.
860. Thompson, S. Y., Braude, R., Coates, M. E., Cowie, A. T., Ganguly, J., and Kon, S. K. (1950). *Br. J. Nutr.* **4,** 398–421.
861. Thompson, S. Y., Braude, R., Cowie, A. T., Ganguly, J., and Kon, S. K. (1949). *Biochem. J.* **44,** 9P-10P.
862. Thompson, S. Y., Ganguly, J., and Kon, S. K. (1947). *Br. J. Nutr.* **1,** 5P-6P; **3,** 50–78·(1949).
863. Thompson, S. Y., Ganguly, J., Mawson, E. H., and Kon, S. K. (1949). *Proc. Int. Dairy Congr., 12th, Stockholm* **2,** 238.
864. Thorbjarnarson, T., and Drummond, J. C. (1938). *Biochem. J.* **32,** 5–9.
865. Tiews, J. (1963). *Wiss. Veroeff. Dsch. Ges. Ernaehr.* **9,** 235–262.
866. Tiews, J. (1962). *Tieraerztl. Umsch.* **17**(10), 357–362.
867. Tiews, J., Zucker, H., and Wendel, E. (1969). *Int. Z. Vitaminforsch.* **39,** 141–156.
868. Tillman, A. D. (1965). *Feed Age* **15**(6), 20–23.
869. Tkachev, I. F., Malyshev, A. N., Bukhtiyarova, O. N., and Malysheva, L. F. (1972). *Khim. Sel'sk. Khoz.* **10**(4), 301–304.
870. Tkachev, I. F., Malyshev, A. N., Bukhtiyarova, O. N., and Malysheva, L. F. (1971). *Zhivotnovodstvo* **31**(6), 41–43.
871. Tkachev, I., Semin, V., and Solodukin, A. (1968). *Ptitsevodstvo (Moscow)* No. 4, pp. 14–15.
872. Tomarelli, R. M., Charney, J., and Bernhardt, F. W. (1946). *Proc. Soc. Exp. Biol. Med.* **63,** 108–110.
873. Totusek, R. (1965). *Feedstuffs* **37**(32), 57.
874. Treichler, R., Kemmerer, A. R., and Fraps, G. S. (1942). *J. Nutr.* **24,** 57–64.
875. Tucker, R. E., Mitchell, G. E., and Little, C. O. (1967). *J. Anim. Sci.* **26,** 225–226.
876. Tucker, R. E., Mitchell, G. E., Jr., and Little, C. O. (1967). *J. Nutr.* **93,** 518–522.
877. Turner, R. G. (1934). *Proc. Soc. Exp. Biol. Med.* **31,** 866–868.

877a. U.S. Patent 2,474,182 (1949).
878. U.S. Pharmacopeia Vitamin Advisory Board (1947). "Summary of Collaborative Biological Assay Study," Letter No. 172. E. Fullerton Cook, Phila. Pa.
879. Ullrey, D. E. (1972). *J. Anim. Sci.* **35,** 648–657.
880. Ullrey, D. E., Miller, E. R., Struthers, R. D., Hoeffer, J. A., and Hall, H. H. (1963). *6th Internat. Congr. Nutr. Edinburgh.* p. 219.
881. Ullrey, D. E., Miller, E. R., Struthers, R. D., Peterson, R. E., Hoefer, J. A., and Hall, H. H. (1965). *J. Nutr.* **85,** 375–385.
882. Van Atta, G. R., Thompson, C. R., Bickoff, E. M., and Maclay, W. D. (1953). *Feed Age* **3**(7), 30.
883. Vander Noot, G. W., Fonnesbeck, P. V., and Garton, C. L. (1965). *N.J. Agric.* **47**(2), 3–5.
884. Van Eckelen, M., and Pannevis, W. (1938). *Nature (London)* **141,** 203–204.
885. Van Leeuwen, J. M. (1973). *Tijdschr. Diergeneeskd.* **98,** 65–71.
886. Varnell, T. R., and Erwin, E. S. (1960), *J. Anim. Sci.* **19,** 961–962.
887. Vasyutenkova, N. S. (1973). *Khim. Sel'sk. Khoz.* **10**(5), 383–384.
888. Vavich, M. G., and Kemmerer, A. R. (1950). *J. Nutr.* **40,** 605–610.
889. Vavich, M. G., Stull, J. W., Raica, N., and Kemmerer, A. R.(1955). *Arch. Biochem. Biophys.* **55,** 310–314.
890. Verma, S. P., and Wallach, D. F. H. (1975). *Biochim. Biophys. Acta* **401,** 168–176.
891. Wachenroder, H. (1831). *Mag. Pharm.* **33,** 144.
892. Waite, R., and Sastry, K. N. S. (1949). *Emp. J. Exp. Agric.* **17,** 179–187.
893. Waite, R., and Sastry, K. N. S. (1949). *J. Agric. Sci.* **39,** 174–182.
894. Wallace, J. D., Raleigh, R. J., and Westwig, P. H. (1964). *J. Anim. Sci.* **23,** 1042–1045.
895. Walsh, K. A., and Hauge, S. M. (1953). *J. Agric. Food Chem.* **1,** 1001–1004.
896. Warner, R. G. (1961). *Proc. Cornell Nutr. Conf.* pp. 96–100.
897. Warner, R. G., and Maynard, L. A. (1952). *J. Anim. Sci.* **11,** 780.
898. Warner, R. G., Travis, H. F., Bassett, C. F., Krook, L., and McCarthy, B. (1963). *Mink Farmer Res. Found. Prog. Rep.*
899. Washburn, R. G., Kraus, W. E., and Monroe, C. F. (1947). *J. Dairy Sci.* **30,** 568.
900. Waugh, R. K., Hauge, S. M., and Hilton, J. H. (1944). *J. Dairy Sci.* **27,** 585–590.
901. Weedon, B. C. L. (1971). *In* "Carotenoids" (O. Isler, ed.), pp. 29–59. Birkhaeuser, Basel.
902. Week, E. F., and Sevigne, F. J. (1950). *J. Nutr.* **40,** 563–576.
903. Weichental, B. A., Emerick, R. J., Rmbry, L. B., and Whetzal, F. W. (1961). *J. Anim. Sci.* **20,** 955.
904. Wiss., O., and Thommen, H. (1963). *Wiss. Veroeff. Dsch. Ges. Ernaehr.* **9,** 179–192.
905. Weiss, R. R. (1975). Thesis, Tieraerztliche Hochschule, Hannover.
906. Wellenreiter, R. H., Ullrey, D. E., Miller, E. R., and Magee, W. T. (1969). *J. Nutr.* **99,** 129–136.
907. Whitehair, C. K., Peterson, D. R., Van Arsdell, W. J., and Thomas, O. O. (1952). *J. Am. Vet. Med. Assoc.* **121,** 285–287
908. Whitehead, E. I., and Moxon, A. L. (1952). *S.D. Agric. Exp. Stn., Bull.* No. 424.
909. Whitmore, J. H., Sullivan, T. W., and Grace, O. D. (1968). *Poult. Sci.* **47,** 159–164.
910. Wierzchowski, Z., and Basaj, J. (1966). *Rocz. Nauk Roln., Ser. B* **88**(1), 79–92.
911. Wiese, C. E., Mehl, J. W., and Deuel, H. J. (1948). *J. Biol. Chem.* **175,** 21–28.
912. Wiese, C. E., Mehl, J. W., and Deuel, H. J., Jr. (1947). *Arch. Biochem.* **15,** 75–79.
913. Wilbur, J. W., Hilton, J. H., and Hauge, J. M. (1940). *J. Dairy Sci.* **23,** 765–769.
914. Wilson, J. K. (1943–1945). *Cornell Univ. Agric. Exp. Stn., Mem.* Nos. 243 and 271.
915. Wilson, L. L., MacDonald, K. G., and Mayo, H. H. (1966). *Purdue Univ. Agric. Serv. Release* AS-341, March.
916. Wilson, W. O., Schroeder, C. H., and Higgins, W. A. (1936). *Poult. Sci.* **15,** 426–436.
917. Wing, J. M. (1969). *J. Dairy Sci.* **52,** 479–483.

918. Winterstein, A. (1933). *Hoppe-Seyler's Z. Physiol. Chem.* **215,** 51–58.
919. Winterstein, A., and Hegedüs, B. (1960). *Chimia* **14,** 18–19.
920. Winterstein, A., Studer, A., and Rüegg, A. (1960). *Chem. Ber.* **93,** 2951.
921. Wise, G. H., Caldwell, M. J., Parrish, D. B., Atkeson, F. W., and Hughes, J. S. (1948). *J. Anim. Sci.* **7,** 70–77.
922. Wise, G. H., Jacobson, N. L., Allen, R. S., and Yang, S. P. (1949). *J. Dairy Sci.* **32,** 711.
923. Wise, G. W., Jacobson, N. L., Allen, R. S., and Yang, S. P. (1958). *J. Dairy Sci.* **41,** 143–151.
924. With, T. K. (1942). *Vitam. Horm. (N.Y.)* **2,** 369–399.
925. Witt, S. C., Bickoff, E. M., and Kohler, G. O. (1972). *Feedstuffs* **44**(41), 26–27.
926. Wolbach, S. B. (1954). *In* "The Vitamins" (H. Sebrell and R. Harris, eds.), Vol. 1, pp. 106–137. Academic Press, New York.
927. Wolf, G. (1969). *Am. J. Clin. Nutr.* **22,** 898–968.
928. Wood, R. D., Chaney, C. H., Waddill, D. G., and Garrison, G. (1967). *J. Anim. Sci.* **26,** 510–513.
929. Worker, N. A. (1957). *Br. J. Nutr.* **11,** 44–46.
930. Worker, N. A. (1956). *Br. J. Nutr.* **10,** 169–175.
931. Worker, N. A., and McGillivray, W. A. (1957). *J. Dairy Res.* **24,** 85–114.
932. World Health Organization (1960). *WHO Tech. Rep. Ser.* No. 187.
933. World Health Organization (1959). *WHO Pharmacopeia, Int. Suppl.* **955,** 178.
934. World Health Organization (1953). *WHO Tech. Rep. Ser.* No. 68.
935. World Health Organization (1950). *WHO Tech. Rep. Ser.* No. 3.
936. Wright, A. W., and Roncalli, R. (1964). *Vet. Med. Small Anim. Clin.* **59,** 1248–1252.
937. Wright, M. J., and Davison, K. L. (1964). *Adv. Agron.* **16,** 197–247.
938. Wyngaarden, J. B., Wright, B. M., and Ways, P. (1952). *Endocrinology* **50,** 537–549.
939. Yanchilin, L. V., Serykh, V., and Ogui, V. G. (1974). *Khim. Sel'sk. Khoz.* **12,** 611–612.
940. Yoshida, N., Oshara, Y., and Okamoto, A. (1971). *Obihiro Chijusan Daigaku Gakujutsu Kenkyu Hokoku, Dai-1-Bu* **7,** 340–350.
941. Zababa, V. (1972). *Svinovodstvo (Moscow)* No. 9, pp. 16–17.
942. Zabara, V. N. (1971). *Svinovodstvo (Moscow)* No. 14, pp. 38–47.
943. Zabara, V. N. (1971). *Svinovodstvo (Moscow)* No. 15, pp. 35–40.
944. Zachman, R. D., and Olson, J. A. (1965). *J. Biol. Chem.* **238,** 541–546.
945. Zechmeister, L. (1962). "Cis-Trans Isometric Carotenoids." Academic Press, New York.
946. Zechmeister, L., Pinckard, J. H., Greenberg, S. M., Straub, E., Fukui, T., and Deuel, H. J., Jr. (1949). *Arch. Biochem.* **23,** 242–245.
947. Zhedek, M. S., and Anderson, T. P. (1968). *Khim. Sel'sk. Khoz.* **6,** 695–698.
948. Zhedek, M. S., Anderson, T. P., Rozantsev, E. G., and Shapiro, A. B. (1970). *Sin. Issled. Eff. Khim. Polim. Mater.* No. 4, 283–288.
949. Zhedek, M. S., Grigoryan, G. L., and Rogantsev., E. G. (1970). *Sin. Issled. Eff. Khim. Polim. Mater.* No. 4, 277–283.
950. Zhedek, M. S., and Rozantsev, E. G. (1971). *Fiziol. Biokhim. Osn. Povysh. Prod. Skh. Zhivotn.* pp. 232–241.
951. Zscheile, F. P., Jr. (1973). *J. Agric. Food Chem.* **21,** 1117–1118.
952. Zubricky, J. (1973). *Folia Vet.* **17,** 375–378.

7

CAROTENOIDS IN PHARMACEUTICAL AND COSMETIC PRODUCTS

K. Münzel

I. INTRODUCTION: COLOR AND FLAVOR

Medicinal preparations that have been appropriately colored and flavored may not have any added therapeutic advantage, but they do have considerable psychological impact. A water-white medicine is unattractive to many patients, who usually believe such a preparation to be quite useless. Even medicine that itself is disagreeable can be made to look attractive and to be pleasant to the taste if a careful selection of an appropriate color, flavor, and diluent is made. The judicious use of these substances is therefore important if the cooperation of the patient in taking or using some prescribed medication is to be secured.

Coloring agents are added to pharmaceutical products in order both to provide a form of control during the manufacturing processes and also to distinguish one product from another. Colors so used to characterize or identify products play as

ISBN 0-12-082850-2

important a part in pharmaceutical manufacture as do the coloring agents used for aesthetic or psychological reasons.

Synthetic carotenoids, which are identical to those that occur in nature, are included in the group of accepted food colors. They are therefore generally admitted automatically as colors for drugs in those countries where there is no special list of acceptable colors for drugs (such as the "D&C dyes" and the "external D&C dyes" listed in the United States).

Although carotenoids are classified as oil-soluble pigments, their solubility in vegetable oils (and many organic solvents) is limited. As a rule, because of this low solubility, the pure crystalline carotenoids are not commercially practical forms. Through certain technological developments, some convenient forms have become available which can be used both in fat- and water-based pharmaceuticals. Water-dispersible forms of carotenoids are being successfully applied in the coloring of sugar-coated tablets and soft and hard gelatin capsules; oily suspensions of the finely dispersed carotenoids serve as coloring agents for fatty vehicles, such as fatty suppositories.

II. COLORING SUGAR-COATED TABLETS

Dry water-dispersible beadlets of carotenoids have been tested as coloring agents for sugar-coated tablets (1–4). Colored sugar syrups were prepared by the simple addition of an aqueous solution (or a highly concentrated aqueous suspension) of these beadlets to plain sugar syrup.

A. Example 1: Preparation of a Colored Syrup with a Low Concentration of a Carotenoid (0.01–0.05%, w/w) for Color Coating

1. Aqueous Carotenoid Solution (About 0.1–0.5% of the Carotenoid)

One to five parts of the water-soluble beadlets (which contain 10% of the appropriate carotenoid) are dissolved in sufficient boiling water and made up to 100 parts.

2. Sugar Syrup

One hundred parts of sucrose are heated with 40 parts of distilled water until the boiling point reaches 108°C. [The concentration of the sugar rises, as a consequence, from 71.5% (w/w) to 77.5% (w/w).]

3. Colored Syrup

Ten parts of the hot solution 1 are mixed with 90 parts of the hot syrup 2, thus yielding a syrup having 0.01–0.05% (w/w) of the carotenoid.

B. Example 2: Preparation of a Colored Syrup with a Higher Concentration of a Carotenoid for Color Coating

1. Carotenoid Beadlet Suspension (20%)

One part of the carotenoid beadlets (10%, water soluble) is triturated in a pestle and mortar with four parts of distilled water at 80°C.

2. Sugar Syrup

One hundred parts of sucrose are heated with 40 parts of distilled water until the boiling point reaches 108°C. [The concentration of the sugar rises, as a consequence, from 71.5% (w/w) to 77.5% (w/w).]

3. Colored Syrup

Depending on the shade and intensity of the final color that is desired, different amounts of suspension 1 may be mixed with sugar syrup 2, calculated as grams of carotenoid for every 100 gm of sugar in the coloring coat. A coating of "100%" is usually required; that is, the total weight the sugar coating is the same as that of the kernel. For example:

Weight of kernel	200 mg	
Weight of subcoat	About 150 mg	200 mg
Weight of coloring coat	About 50 mg	

The coloring coat of 2000 sugar-coated (s.c.) tablets will contain about 100 gm of sugar. Four-tenths of a gram of the pure carotenoid (or 4 gm of the water-soluble beadlets, 10%) per 100 gm of sugar will yield a very deep color; 0.1 gm (or 1 gm of the beadlets) a strong color; and 0.04 gm (or 0.4 gm of the beadlets) a "normal" color (see Fig. 1). Since 100 gm of sugar are present in 129 gm of syrup boiling at 108°C, beadlet suspension 1 (20%) will have to be added to 129 gm of syrup in the amounts (depending on the amount of carotenoid desired) shown in the following tabulation.

Desired weight (gm) of pure carotenoid per 100 gm sugar (= 129 gm syrup, bp 108°C)	Beadlets (gm)	20% slurry 1 (gm)
0.4	4	20
0.2	2	10
0.1	1	5
0.05	0.5	2.5
0.025	0.25	1.25
0.0125	0.125	0.525

Water-dispersible carotenoid (10%)	Amount of pure carotenoid per 40 mg sugar	Apocarotenoic ester	β-Carotene	Apocarotenal	Canthaxanthin
1.5 mg	0.15 mg				
0.45 mg	0.045 mg				
0.15 mg	0.015 mg				
0.045 mg	0.0045 mg				
0.015 mg	0.0015 mg				

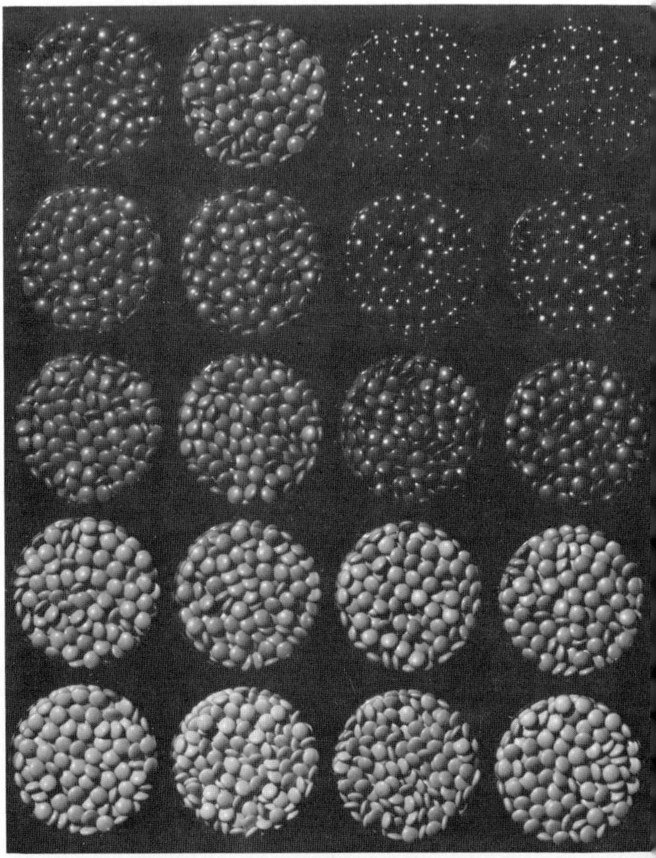

Fig. 1. Sugar-coated tablets colored with carotenoids. The colored coat of one tablet contains 40 mg sucrose. (Se color insert for color representation.)

The appropriate amount of suspension 1 (see the tabulation) is added to 129 gm of the nearly boiling syrup 2, which is then stirred until a homogeneously colored syrup is obtained (and no undissolved aggregates of the beadlets remain). After the tablets have been sealed and subcoated in the usual way, the (still hot) syrup is used to apply the color coat by pouring and spraying; warm air is also blown into the pan if this is the usual coating practice. The sugar-coated tablets are then waxed and polished.

Surrounding the carotenoids with a heavy coating of sugar helps to maintain good stability of color. The carotenoids that were used in these tests and the colors of the sugar-coated tablets obtained are illustrated in Fig. 1. The colors that can be obtained range from pure yellow through orange and red to reddish-brown. Tablets coated with these preparations of synthetic carotenoids show excellent stability of color both during storage at raised temperatures and during prolonged storage at room temperatures, and they show good stability toward light except when very pale shades of a color are used. Because the coloring power of carotenoids is so high, the amounts involved are exceedingly small and the slightest trace of oxidation may cause fading. Deeper colors, however, have proved to be very stable. New forms of canthaxanthin have been developed which, even at low concentrations, show good color stability.

The tablets colored with β-carotene showed essentially no color change after several months at elevated temperatures or 3 months of exposure to direct light. β-Apo-8′-carotenal-colored tablets were likewise unchanged in color after several months of storage at elevated temperatures or 2 weeks in a light chamber. Samples of both the β-carotene- and β-apo-8′-carotenal-colored tablets stored at room temperature for more than 3 years are as they were originally. Canthaxanthin-colored tablets are also very stable on storage at elevated temperatures. After 4 weeks in a light chamber, the tablets show only a slight change in color depth (*3*).

Carotenoids in the "water-soluble" form have excellent covering power (because, being colloidal substances, they occupy a position somewhere between water-soluble dyes and pigments). They therefore provide coatings that are only some 25–50% of the total finished weight, that is, a total weight of the coat that is only one-quarter to one-half of the weight of the core, and this yields a colored coating that is entirely uniform.

The cost of coloring sugar-coated tablets with carotenoids is in the range of about 0.2–5 cents per thousand tablets, depending on the shade and depth of color. This is in the range of the cost of the color of many commercially marketed tablets.

III. COLORING SUPPOSITORIES

The coloring of fat-based suppositories (*5*) in order to characterize, distinguish, or identify them is best accomplished using fat-soluble dyestuffs. Colored fat-based suppositories are especially required by pharmacies in hospitals that have their own facilities for the manufacture of suppositories, where these forms of rectal medication are often not distinguishable by the packing; they are dispensed and then dispatched to the medical departments in bulk. A color can serve as a safeguard against mistakes. The therapeutic use of this form of rectal medication (and its popularity among medical practitioners) varies significantly from one country to another. It is used more on the European continent (especially in the Latin countries, such as France, Italy, and Spain) than in the Scandinavian and Anglo-Saxon countries (Sweden, Great Britain, and the United States).

The use of so-called permitted *fat-soluble* coloring agents for this purpose springs readily to mind. Synthetic fat-soluble dyes are allowed, however, in only a few countries, and in still fewer as coloring agents for food and, hence, for pharmaceutical products. The "permitted" *natural* coloring agents are mainly crude and ill-defined vegetable or animal extracts. The pure carotenoids, on the other hand, are accepted almost universally for use in foods and are already widely used in food industries for the coloring of fats, butter, and oils. To color fats, carotenoids are available as crystals or as suspensions of microcrystals in oil. Carotenoids are sensitive to light and to oxygen, and an antioxidant must be added to the preparation; the addition of 0.01–0.03% of α-tocopherol is recommended since this greatly enhances the stability of the color. A level of 0.1% of the carotenoid usually imparts a rich color to suppositories; paler shades can be obtained by appropriate dilution of the fat base.

The simplest method is the use of an oily suspension of the microcrystalline carotenoid. Although any of the carotenoids can in principle be produced in the form of such a suspension, only two concentrated suspensions each of β-carotene, apocarotenal, and apocarotenoic ester are available, since at present these are the only ones being used in large quantities for coloring margarine and butter.

The other carotenoids must be dissolved in the basic fat, using a concentrated solution in chloroform as an intermediate stage. (The α-tocopherol that is used as an antioxidant may also be dissolved in the same chloroform.) This solution is mixed with as little as possible of the basic fat previously warmed to 40°–50°C. After gentle stirring the chloroform evaporates at this temperature. The mixture is then made up to the desired weight by the addition of the melted basic fat and allowed to cool (see Chart).

When a drug is to be included, its compatibility with the carotenoid and the stability of the combination during storage must first be thoroughly investigated. If the suppository is wrapped or heat-welded in foil packs, the tests for stability should include any possible interaction between the wrapping material and the carotenoid (such as the diffusion of the carotenoid into the foil or the bleaching of the carotenoid by oxidizing substances present in the foil).

In summary, carotenoid colors offer interesting possibilities for the safe coloring of suppositories (Fig. 2).

IV. COLORING HARD- AND SOFT-SHELLED GELATIN CAPSULES

Since the matrix of water-soluble carotenoid preparations contains gelatin (type B, derived from an alkali-treated precursor, which exhibits an isoelectric point between pH 4.7 and 5.0), these products are entirely compatible with the gelatin in capsules, which is also type B. Recently developed forms of canthaxanthin allow successful coloration of hard-shelled capsules in shades ranging from pink (combined with titanium dioxide) to dark red. When the carotenoids are mixed with other

Flow Chart for Production of 100 gm of Suppository Mass Containing 0.1% of Carotenoid

I. Dissolving the carotenoid
α-Tocopherol
10–30 mg
Carotenoid
100 mg
Dissolved in
Chloroform q.s. (approx 500–1000 μl)

II. Basic fat (at 40°–50°C)
10 gm

Mix

Evaporate $CHCl_3$ at 40°–50°C

III. Concentrated fat solution
containing the carotenoid

IV. Basic fat (at 40°–50°C)
q.s. up to 100 gm

Mix

Allow to cool

V. Finished basic fat
colored with 0.1%
of carotenoid

Keep in a cool, dark place

Fig. 2. Fat-based suppositories colored with carotenoids. The wreath consists of nine pairs of differently colored suppositories containing the following compounds and concentrations (% w/w): 1) crocetin diethyl ester (0.025); 2) β-carotene (0.1); 3) β-apo-8′-carotenic acid ethyl ester (0.1); 4) canthaxanthin (0.1); 5) β-apo-8′-carotenal (0.1); 6) torularhodin diethyl ester (0.02); 7) torularhodin diethyl ester (0.001); 8) P-518,2,2′-diketospirilloxanthin (0.002); and dodecaprenocarotene (0.01). (See color insert for color representation.)

coloring compounds, such as riboflavin or riboflavin 5′-phosphate, special effects can be obtained, but because of the sensitivity of riboflavin to light any such capsules have to be stored where they are protected from light to avoid changes in color. Water-soluble carotenoid preparations have been successfully incorporated in soft shelled gelatin capsules.

V. COLORING OTHER FORMS OF MEDICINAL PREPARATIONS

The use of both the crystalline carotenoids and the above-mentioned special forms of application in the coloring of pharmaceutical preparations not encased in sugar, oil, fat, or wax has not been very successful because of the ease with which they become oxidized in air, especially when they are exposed to light. The coloring of plain tablets is possible if the water-dispersible carotenoids are dispersed in concentrated sugar syrups, which serve as the granulating fluid for the tablet powders. Sugar syrups, however, are rarely used as the granulating fluid because they retard the disintegration of the tablet. It might be possible that with additional antioxidant protection tablets may be colored with water-dispersible carotenoid preparations, but this has not been studied.

The coloring of fat-based or hydrocarbon-based ointments and oil/water (o/w) or water/oil (w/o) creams is to be recommended. The coloring may be applied as with suppositories. Microcrystalline suspensions are added to the melted fat-based ointment, which is then cooled with gentle stirring until it becomes a semisolid gel. The addition of an antioxidant such as α-tocopherol, butylated hydroxytoluene, or butylated hydroxyanisole is recommended. If this colored basic fat is to be the disperse phase of an O/W cream containing a W/O emulsifier (such as cetyl alcohol), it is heated to 70°–80°C; the aqueous phase, at the same temperature and containing the complexing O/W emulsifier, is then poured into it with stirring.

Liquid pharmaceutical oral O/W emulsions (such as cod liver oil emulsion or vitamin A emulsion) containing mainly acacia gum and tragacanth as the emulsifiers may have the disperse oil phase colored with carotenoids.

The coloring of pharmaceutical elixirs and syrups (as distinct from foods and drinks) is made more difficult by the fact that the degree of stability required of these preparations is much higher than that sufficient for fruit juices and beverages. A minimal shelf life of 2 years at room temperature generally has to be guaranteed, but the use of an inert gas or the addition of stabilizing agents such as hesperidin or ascorbic acid is not usually feasible. Recently developed beadlets with an improved antioxidant system give clear, stable solutions. These beadlets show promise for use in pharmaceutical elixirs and syrups outside of the pH range of 3.7–4.7. Within this pH range, the matrix of the beadlets, which contains gelatin, tends to precipitate.

VI. CAROTENOIDS IN COSMETIC PRODUCTS

Carotenoids are already being widely used to color edible fats and oils; it is obvious, therefore, that they can be used just as well to color fat-based cosmetics or cosmetics that contain a fat as a vehicle, such as in liquid emulsions and creams. The procedure used to color fat-based suppositories can similarly be used for fat-based cosmetics.

Technologically, suppositories closely resemble lipsticks. The main point of interest, therefore, rests in the use of carotenoids to replace synthetic lipstick colors, since the safety of the certified colors now being used in lipsticks has been a matter of great concern to the cosmetic industry for many years. In principle, carotenoids are entirely suitable for use as colors for lipsticks. The shades and tints of color that can be obtained from the present commercial carotenoids in lipsticks are, however, often not fashionable and cannot therefore fully satisfy the requirements of the cosmetic industry. Nevertheless, a strong red carotenoid such as canthaxanthin or lycopene can help to vary or improve a desirable lipstick color if it is added as a color supplementary to the usual permitted certified D&C colors. Incorporation of suitable antioxidants is required to maintain color stability of the carotenoids in lipsticks.

A patent by Roche Products Limited* provides an example of how red carotenoids can be used to color cosmetic powders. A solution of the carotenoids in chloroform is blended with rice powder, and this colored concentrate is then mixed with a powder base consisting of talc and other common powder ingredients, yielding a powder that has an attractive shade of pink. Shades of orange and beige can also be produced.

To color emulsions according to the same patent the carotenoids can be dissolved in a nonionic emulsifier such as Cremophor EL† and then incorporated in the oil phase of the emulsion. When carotenoids that contain free hydroxyl groups are used, a liquid emulsion can be colored in shades of pink, yellow, or orange if the concentration of the carotenoid is varied between 3 and 10 mg per kilogram of emulsion.

REFERENCES

1. Kläui, H., Capeder, A., and Münzel, K. (1963). *Pharm. Ind.* **25,** 173.
2. Kläui, H., and Münzel, K. (1965). *Pharm. Acta Helv.* **40,** 153.
3. Magid, L. (1966). *Drug Cosmet. Ind.* **99**(5), 64.
4. Münzel, K. (1967). *Dtsch. Apoth.-Ztg.* **107,** 277.
5. Münzel, K., and Füller, W. (1961). *Pharm. Acta Helv.* **44,** 208.

*Austrian Patent 233'931, June 1964.

†Manufacturer: Badische Anilin- und Sodafabrik (BASF), Ludwigshafen, West Germany. Similar emulsifier: Emulphor EL, GAF Corporation, Dye Stuff & Chemical Division, Chattanooga, Tennessee 37402.

8

CAROTENOIDS IN MEDICAL APPLICATIONS

Micheline M. Mathews-Roth

I. INTRODUCTION

Until recently, the primary medical use in man of carotenoid pigments, whether taken in as carotenoid-containing foods or administered as dietary supplements, has been the prevention or correction of vitamin A deficiency. The discovery that the carotenoid pigments ubiquitous in the plant kingdom play an important role as protective agents against photosensitization by the plants' own chlorophyll has led to another medical use for carotenoids in man: as protective agents in certain skin diseases aggravated by light. In this chapter we will examine in detail this most recent use of carotenoid pigments and also briefly discuss the studies in bacteria, plants, and animals that led to the use of carotenoids in the treatment of human photosensitivity.

CAROTENOIDS AS COLORANTS
AND VITAMIN A PRECURSORS

ISBN 0-12-082850-2

II. ROLE OF CAROTENOIDS IN PLANTS

Carotenoid pigments are found in every green plant. A variety of functions has been postulated for these pigments, such as roles in phototaxis, oxygen evolution, oxygen transport, as accessory pigments in photosynthesis, and as protective agents against photosensitization by the plants' own chlorophyll. At present, only the latter two have received sufficient unequivocal experimental substantiation to suggest that they are indeed significant functions of carotenoids in green plants (*88,89*). There exist comprehensive reviews on both the well-documented and speculative functions of carotenoids (*60,88,89*). We will discuss here briefly only the two well-established functions of carotenoids in green plants.

A. Role in Photosynthesis

The chlorophyll of plants plays the central role in photosynthesis, but it has long been noted that other pigments also participate to some degree in the process. Engelmann (*41,42*) was the first to observe that carotenoids could act as accessory light-absorbing pigments in photosynthesis in green plants. Carotenoids serve a useful role for the plant because, by their absorption at wavelengths lower than that absorbed by chlorophyll, they extend the wavelengths of light that can be used in photosynthesis. Chlorophyll normally absorbs light above 600 nm. Carotenoids, which absorb light between 400 and 600 nm, transfer the light energy they have absorbed in this region to the chlorophyll, thus providing it with additional energy, which it would not have had if the carotenoids were not present.

Engelmann's work has been confirmed and extended by many workers in many species of green plants as well as in photosynthetic bacteria (*37*). The role of accessory pigments, carotenoids as well as phycobilins, in photosynthesis has been thoroughly reviewed by Blinks (*14*).

B. Role in Protection against Photosensitization

Sistrom *et al.* first suggested that carotenoids might be acting as protective agents against photodynamic action in bacteria as a result of observations they made on the wild type and a certain mutant of *Rhodopseudomonas spheroides* (*138*). This mutant, which they called "blue-green," did not produce colored carotenoids, but accumulated the colorless carotenoid precursor, phytoene. They found that when the blue-green mutant was grown in the presence of light and air, growth stopped, chlorophyll was destroyed, and the cells were killed. On the other hand, the wild type, with the normal component of carotenoid pigments, is not injured by growth in light and air; in the presence of air, chlorophyll synthesis stops, and the organism grows by aerobic metabolism. It should be noted that the deletion of the carotenoid pigments did not markedly affect the ability of the blue-green mutant to grow photosynthetically in the presence of light in an atmosphere of nitrogen (bacterial photosynthesis is anaerobic); only when light and air were present did the lack of carotenoids give rise to problems.

These workers were able to show that it was the bacteriochlorophyll which was responsible for the lethal photosensitization of the blue-green mutant and that both oxygen and light were necessary for the destructive reaction to occur (a photodynamic action). They concluded that the carotenoid pigments were functioning as protective agents against this photodynamic killing and that the bacteriochlorophyll was the endogenous photosensitizer. These findings, that carotenoids can protect against chlorophyll photosensitization, have been confirmed in other photosynthetic bacteria and algae [see Krinsky (*88*) for a detailed review].

In their report (*138*), Stanier and his group suggested that the carotenoid pigments of all green plants function as protective agents to prevent photosensitization by chlorophyll. Observations on carotenoidless mutants of higher plants suggest that this indeed may be the case. Koski and Smith (*87*) and Smith *et al.* (*139*) reported studies on a white mutant of corn that lacked colored carotenoids and that, when exposed to light, would die when the food supply in the cotyledons was exhausted. Examination of the plants showed that the chloroplasts were destroyed. Anderson and Robertson (*4*) showed that destruction of chlorophyll was occurring in light in these colorless mutants and that the presence of carotenoids prevented this destruction. Wallace and Schwarting (*150*) studied a mutant of sunflowers (*HA*) which did not contain colored carotenoids. This plant also died when grown in air and light once the food supply in the cotyledons was used up.

Thus, in view of all the data now available, it can be concluded that the carotenoid pigments in all photosynthetic organisms, bacteria, algae, and higher plants, play an important role in protecting these organisms against the seriously damaging effects of photooxidation by their own endogenous photosensitizer, chlorophyll.

With these data as a background, it was not surprising that Stanier and his associates also suggested that the carotenoids of nonphotosynthetic bacteria might also play a protective role (*138*), even though these organisms did not seem to contain large amounts of an obvious endogenous photosensitizer, as did the photosynthetic organism. Kunisawa and Stanier (*92*) induced a colorless mutant of *Corynebacterium poinsettiae,* a nonphotosynthetic bacterium that contains carotenoids. They found that the mutant was killed in the presence of an exogenous photosensitizer (toluidine blue), light, and air, whereas the wild type, which contained colored carotenoids, was not affected by this exposure. Again, they found that this was a photodynamic phenomenon, as air was necessary for the phenomenon to occur; the cells were not killed in the presence of dye and light in a nitrogen atmosphere. The authors attempted to demonstrate the presence of an endogenous photosensitizer in this organism by exposing the wild type and colorless mutant to light [4000 foot-candles (fc) for 4 hr] in the presence of air but with no added photosensitizer. However, this treatment did not kill the cells. They concluded that carotenes indeed could protect the cells against photosensitization with an exogenous photosensitizer, but they could not determine whether this ability was of any use to the cells. The author and Sistrom (*102,103*) were able to show that nonphotosynthetic bacteria might indeed contain endogenous photosensitizers by exposing wild type *Sarcina lutea,* another nonphotosynthetic carotene-containing organism, and its colorless mutant to natural sunlight in air for 4 hr. We found that, at these high light intensities (12,000 fc at the start of the experiment and 5000 fc at the end), the mutant was killed and the wild type was not. We showed that this was also a photodynamic effect, as the mutant was not killed after exposure to sun in an atmosphere of nitrogen. Since these studies were done (*92,102,103*), many other workers have confirmed the protective function of carotenoid pigments in nonphotosynthetic bacteria [see Krinsky (*88*) for a detailed review].

C. Ecological Importance of the Protective Function of Carotenoids

Stanier and Cohen-Bazire have pointed out the important role played by carotenoid pigments in the evolution of photosynthetic organisms, both bacteria and higher plants, during the development of porphyrin-containing photosynthetic pigments in these organisms (*141,142*). Since porphyrins are powerful photosensitizers (*15*), it was crucial for photosynthetic organisms to evolve protective mechanisms against photosensitization by their chlorophyll. Hence, carotenoids have proved to be important not only as trappers of additional energy for photosynthesis, but also as the mechanism whereby the plant protects itself against photosensitization by its own chlorophyll in our aerobic atmosphere. As mentioned above, Stanier and Cohen-Bazire also suggested that a similar protective function could be postulated in nonphotosynthetic carotenoid-containing bacteria. They cited the fact that carotenogenesis in fungi and mycobacteria is stimulated by light and that there is a negative correlation between anaerobiosis and the presence of carotenoids (*142*). In fact, organisms that are normally exposed to both light and air contain carotenoid pigments. For example, many aquatic bacteria, such as the flavobacteria, which are normally exposed to light and air contain carotenoid pigments (*60,80,156*). Aquatic organisms of the genera *Vibrio, Pseudomonas,* and *Corynebacterium* also contain these pigments (*28,70*). On the other hand, most aerobic terrestrial bacteria probably are not exposed to high light intensities while in the soil; for example, carotenoids are rarely found in coliform organisms. Bacteria commonly found in air would be subject to intense light. Spores are probably resistant to photodynamic killing, and carotenoids are seldom, if ever, found in the bacilli. Non-spore-forming bacteria commonly found in air will, on the Stanier hypothesis, contain carotenoids. Examples of these organisms are *Mycobacterium, Corynebacterium, Staphylococcus,* and *Sarcina,* all of which contain carotenoids (*28,70,80*). Thus, both in photosynthetic organisms and in nonphotosynthetic bacteria, carotenoid pigments play an important role in providing protective mechanisms against photosensitization by endogenous photosensitizers that would cause damage to the organisms in the presence of the light and air normally present in their environment.

III. ROLE OF CAROTENOIDS IN ANIMALS

A major part of this volume is concerned with the important role of carotenoids as constituents of animal feeds. Hence, the discussion on the role of carotenoids in animals will be brief. Carotenoid pigments are found in many animals, both vertebrate and invertebrate (*60,80*). It is important to note that animals do not synthesize these compounds; they are obtained from the animals' food. In herbivores the pigments come from the plants that the animals eat, and in carnivores they come from the animals' prey, which in turn obtained the pigments from their own food. In fact, most carnivores get their vitamin A preformed from prey. Some animals are thought to be unable to convert carotenoids efficiently to vitamin A.

Although experimental evidence indicates that animals cannot synthesize the entire carotene molecule, many, such as birds and echinoderms, have enzymes that can effect changes in the molecular structure of the pigments absorbed from the diet. These modified pigments very often are used by the animals for coloration (*49,60*).

Many invertebrates contain carotenoids, a large part of which are present as carotenoproteins (*20,144*). Several functions have been ascribed to these interesting complexes, such as stabilizing molecular configurations, protective coloration, and involvement in embryonic development. In fact, it has been suggested that carotenoids may play a protective function in the case of the eggs of some mollusks. It has been reported that only species whose eggs are pigmented lay their eggs in positions out of water exposed to light or in shallow water and that carotenoprotein content is increased under conditions of increased light (*20*).

Photosensitization does indeed occur in animals, mainly from the ingestion of feed containing photosensitizers and, in rare instances, from genetic disease, such as "congenital porphyria" in cattle (*21,22*). Nothing is known about the presence of endogenous photosensitizers in animals, and no definitive experiments have been done to show whether carotenoids have a protective function. The pigmented mollusks mentioned above indeed may prove to be interesting subjects for a study of endogenous photosensitization in an animal model. Much work needs to be done in investigating the function (other than nutritional) of carotenoids in animals.

IV. ROLE OF CAROTENOIDS IN MAN

The main function of carotenoid pigments in man is a nutritional one: that of providing a source of vitamin A. Man, like other animals, cannot synthesize this vitamin. Carotenoids also play some part in the coloration of human skin (*38,39*).

This property, of accumulating in human skin, has led to another use of carotenoids in man, as light-protective agents in human photosensitivity diseases. We will look at these two uses of carotenoids in man and also consider the evidence for lack of toxicity of carotene on oral administration.

A. Use in Nutrition

In spite of the current availability and use of synthetic vitamin A either in pure or concentrate form for the nutrification of food, or in single or multiple vitamin supplements, foods containing carotenoid vitamin A precursors are still a very significant source of this vitamin in the diets of human beings. Diets in various parts of the world vary greatly in their content of vitamin A, available either directly in meat, fish, eggs, or dairy foods, or as the various provitamins A in vegetable sources. β-Carotene is the most important provitamin A, mainly because of its prevalence in plants widely consumed by man and livestock and because it is the provitamin A with the greatest activity (*146*). About 50% of the apparent vitamin A intake in the American diet is in the form of provitamin A (*146*). In Britain, the contribution of dietary vitamin A needs from provitamin A is one-third to nearly one-half from fruits and vegetables (*146*). In India, attempts are being made to increase the consumption of leafy vegetables as one method of improving vitamin A intake. It is obvious, then, that the carotenoid vitamin A precursors such as β-carotene are still an important source of vitamin A in the human diet.

In addition to availability in the diet, several factors may influence the absorption and utilization of β-carotene and other provitamins. For example, studies in animals have shown that dietary protein deficiency may have a deleterious effect on the utilization of provitamin A (*13,33,72*). In severe protein deficiency, kwashiorkor, serum and liver vitamin A is low (*6*), but in milder cases of protein deficiency there seems to be no problem of provitamin A utilization (*40,94*). Studies on isolated carotene dioxygenase, the enzyme responsible for central cleavage of the β-carotene molecule, from rat and human intestine suggest that the activity of this enzyme is dependent on dietary protein intake (*65*).

The amount, nature, and dispersion of dietary fat and the amount and digestibility of the carotenoid sources in the diet are also important factors in the maximal absorption and utilization of provitamins A (*146*). Interestingly, a requirement for emulsifiers (bile salts or synthetic detergents) and lipids for the activity of carotene dioxygenase has been found (*59,93*).

The transformation of β-carotene to vitamin A (retinol) in animals (plants are thought not to contain vitamin A as such) is considered to occur by cleavage of the carotene molecule at the central double bond (*58,119*) or by cleavage of the conjugated chain adjacent to one β-ionone ring followed by sequential oxidative removal of fragments containing two to five carbon atoms (*55,56*). The activity of carotene dioxygenase isolated from rat and human intestinal mucosa and also from liver seems to suggest that the cleavage of the molecule at the central double bond is the more important mechanism for metabolism of carotene in mammals (*56,57*).

Many studies have shown that the main point of conversion of ingested carotenoids is the intestinal mucosa. Conversion also occurs to a fairly significant degree in the liver. It was suggested that the lungs might also be a site of conversion, but data from heart–lung preparations have not confirmed this (*120*). Greenberg *et al.* suggested that sebaceous glands may convert β-carotene to vitamin A (*27,62,63*).

Traditionally, it has been thought that only plants and bacteria can synthesize carotenoids from simple precursors (*60*). Recently, some evidence has been presented that bovine corpus luteum tissue could synthesize β-carotene from acetate *in vitro* (*7*). Further confirmation of this finding must be obtained before the present view that there is no carotenoid synthesis in animals can be abandoned.

Carotenoids have been found in human milk (*86,96*), and there are at least two reported instances of breast-fed infants becoming carotenemic on their carotenemic mothers' milk (*2,147*). It also seems that carotenoids can cross the placental barrier. DeBuys reported that a child of a carotenemic mother was born carotenemic (*32*). Claussen and McCoord suggested that 1 mg of carotenoid per kilogram of fetus crosses the placenta daily (*24*). Claussen reported the presence of some carotenoid in the cord blood at birth (*23*). After birth, the amount of carotenoid present is dependent on the child's diet.

The manner in which carotenoids are excreted is not clear. Hashimoto (*67*) claimed that none is excreted in the urine. Hess and Meyers found that urine got yellower within 45 min after concentrated carrot extract was given (*68*). Unfortunately, they did not chemically extract the urine to determine the presence of carotenoids. Almond and Logan (*2*) made the same observation and attempted to extract the urine chemically. They extracted a yellow pigment, but this was found not to be carotene. We looked for carotenoids in the urine of some erythropoietic protoporphyria patients taking carotene and also found that the yellow pigment contained in the urine was not a carotenoid on spectrophotometric or chromatographic analysis. More work certainly must be done on the question of the excretion of carotenoids.

B. Development of the Idea of Using Carotenoids for the Treatment of Photosensitivity

The data presented in Section II,B indicated that carotenoid pigments have a well-documented ability to protect plants and bacteria against photosensitivity. From these data, it seemed sensible to determine whether the administration of carotenoid pigments could prevent photosensitization in patients with photosensitivity diseases in which the photosensitizer had some resemblance to the endogenous photosensitizer in plants. Such a disease was light-sensitive porphyria, in which the porphyrins produced are similar in structure to the porphyrin ring of chlorophyll. Preliminary experiments done in the summer of 1961 with L. C. Harber at New

York University Medical School suggested that the onset of erythema to artificial light could be delayed by the oral administration of β-carotene. A search of the literature up to that time revealed that Kesten had delayed the onset of erythema in a patient with urticaria solare by the use of β-carotene (*83*).

Because of this promising clinical information, it was thought worthwhile to develop an animal model in order to test more thoroughly the hypothesis that carotenoids could protect against photosensitization in an animal system. Lipson and Baldes had developed a model of lethal and sublethal porphyrin photosensitivity in mice, in which they injected the animals intraperitoneally with a derivative of hematoporphyrin and exposed the animals to light (*99*). At about the same time, Forssberg *et al.* (*48*) had devised a Tween 80 suspension of β-carotene, which they administered intraperitoneally to mice 18–24 hr before X irradiation to determine whether β-carotene could protect the animals against the lethal effects of the X rays. The author used the photosensitization method of Lipson and Baldes and the carotene suspension of Forssberg *et al.* in an experiment to determine whether carotene could prevent the lethal photosensitization of injected hematoporphyrin and subsequent light exposure. Suspensions of β-carotene in Tween 80 were prepared according to the method of Forssberg *et al.* (*48*), and 3 mg of β-carotene or the equivalent volume of Tween 80 alone was administered intraperitoneally to mice 18–24 hr before they were injected with the hematoporphyrin derivative prepared by the method of Lipson and Baldes (*99*). Before exposure to light, these mice received 1 mg of the hematoporphyrin derivative intraperitoneally. Another group of mice received 1 mg of hematoporphyrin alone. In addition, dark controls, animals receiving either hematoporphyrin alone or β-carotene in Tween 80 alone, were included. It can be seen from Table 1 that significantly more animals that received β-carotene survived the treatment with hematoporphyrin and light than did those that did not receive the β-carotene. Thus, β-carotene was effective in mice in preventing the lethal photosensitization induced by injection of hematoporphyrin and exposure to visible light (*101*).

On the basis of these findings and those mentioned above on bacteria and plants, it seemed feasible to administer β-carotene orally to patients with photosensitivity. For this study, the author entered into collaboration with Dr. T. B. Fitzpatrick, Dr. L. C. Harber, Dr. E. H. Kass, and Dr. M. A. Pathak (*107,108*) and later with a number of physicians in different parts of the country [see Mathews-Roth *et al.* (*110,110a*) for a list]. The disease we chose to study was erythropoietic protoporphyria (EPP), a disorder characterized by abnormally elevated levels of protoporphyrin in red blood cells, plasma, feces, and skin and by sensitivity to visible light (380–560 nm). The photosensitivity is manifested by a burning sensation in the exposed skin followed by various degrees of erythema and edema (*32a,148*). The erythema and edema produced by exposure to natural sunlight can be elicited in the laboratory under controlled conditions by exposing the patient's skin to a strong artificial light source, such as a carbon arc or a xenon arc lamp.

The first patient we treated, a 10-year-old girl, was initially seen in the winter of

TABLE 1

Protective Effect of β-Carotene on Photosensitization of Mice by Hematoporphyrin and Visible Light[a]

Group	Light exposure	Animals: Total number	Animals: Number dead	Animals: Number alive
Hematoporphyrin[b]	Yes	27	21	6
Hematoporphyrin[b] and carotene[c]	Yes	27	9	18
Hematoporphyrin[b] and Tween 80–saline	Yes	8	7	1
Hematoporphyrin[b]	No	8	0	8
Carotene[c]	No	8	0	8

[a] Data from Mathews (*101*).
[b] Animals received 1 mg intraperitoneally before exposure to light.
[c] Carotene was suspended in Tween 80–saline. Animals received 3 mg of β-carotene intraperitoneally 18–24 hr before exposure to light.

1967. At that time she could tolerate only brief exposures to sunlight (30 min or less). Exposure to a carbon arc lamp produced erythema in 2 min. In June 1968, she was given a preparation of concentrated carrot oil in doses approximately equivalent to 30 mg of β-carotene per day. After a month of carrot oil ingestion, she could tolerate at least 30 min of carbon arc light and more than 1 hr of sunlight. By the middle of the summer, she could play outdoors in the afternoon without experiencing any symptoms of photosensitivity. As is characteristic of the disease, her porphyrin levels remained elevated throughout the period of treatment. Because the carrot oil was unpleasant to take and caused diarrhea, its use was discontinued. In the summer of 1969, the patient (and all subsequent patients) were given β-carotene in the form of 10% β-carotene "beadlets" (Hoffmann-La Roche), after FDA approval of the use of this preparation was obtained. She and two other patients were successfully treated with this preparation during the summer of 1969 (*107,108*). In 1970, since the beadlet preparation had proved effective in our three patients, we enlarged the study to include all the patients of Dr. L. C. Harber and those of other physicians who had contacted us concerning the use of β-carotene since the publication of our first three case histories. By July 1975, we had treated 133 patients with EPP with β-carotene. The results of the collaborative study and a description of the protocol that all collaborating physicians adhered to have been published (*110,110a*).

Table 2 shows the effect of treatment with β-carotene on the tolerance to sun exposure of the 133 patients treated. Of these patients, 22 have reported little or no improvement, and the remaining 111 have reported significant improvement in their

TABLE 2

Effect of β-Carotene on Tolerance to Sunlight of Patients with Erythropoietic Protoporphyria[b]

	Number of patients	Percentage	Protection index[b]
	10	16	1[c]
	12		2[c]
	27	84	3-5
	39		6–10
	15		11-15
	6		16–20
	24		>21
Totals	133	100	

[a] Data from Mathews-Roth *et al.* (*110, 110a*).

[b] Protection index is defined as the number of minutes of summer sunlight tolerated without the development of symptoms after therapy, divided by the number of minutes of summer sunlight tolerated without the development of symptoms before therapy. An index of less than 3 represents no improvement.

[c] Patients showed little or no improvement despite blood carotene levels of at least 400 μg % following intake of largest dose appropriate for their ages.

ability to tolerate sunlight without developing symptoms of photosensitivity. Over 75% have reported a fourfold or more increase in tolerance to sunlight. All of these patients received the beadlet preparation in individualized doses, which were increased until the patient reported either improvement in sun tolerance or no improvement after taking the beadlets at a dose that resulted in blood carotene levels of 800 μg/dl or higher for a period of 3 months (*110,110a*).

On the average, it took between 1 and 3 months for the patients who received benefit from carotene therapy to notice increased tolerance to sun exposure. The majority are now engaging in outdoor activities which they were unable to do before therapy started, most stating that they spend more time in the sun since taking β-carotene. This seems to be especially gratifying to the children, who before carotene therapy were unable to play outdoors to any great extent; now the majority can spend hours outside with their friends.

Many patients stated that they were able to develop a suntan for the first time in their lives. Indeed, the results of tests with artificial light sources indicate that patients with EPP react normally to sunburn radiation when tested with that radiation alone. However, because of the presence of light of 380 nm and above in natural sunlight, which light causes a patient's symptoms, patients with EPP have never been able to tolerate sun exposure long enough to develop a tan. It was the impression of several of the patients that the acquisition of the tan, plus the β-carotene, aided their protection from the sun's effects.

The majority of the patients noted that when they are taking β-carotene, reactions from the sun that do occur are less severe in intensity and duration than before therapy. Not only have the patients increased the time during which they can remain in the sun symptom free, but they do not develop significant numbers of cutaneous lesions during their increased exposure time.

In addition to the patients' subjective reactions of improved tolerance to sunlight, we used phototesting with a polychromatic xenon arc light before and during β-carotene therapy as an objective method of evaluating the effects of this therapy (*110*). Tolerance to the xenon arc light increased in those patients who reported benefitting from β-carotene therapy, but no increased tolerance to this radiation has been found in patients reporting no improvement (Table 3; note patient 8). How-

TABLE 3

Tolerance to Xenon Arc Light (Phototest) Measured before and During Therapy[a]

	Time to develop minimal erythema (min)	
Patient	Before therapy	After therapy
1	20	40
2	20	40
3	0.5	7
4	20	30
5	10	20
7	5	10
8	5	5
9	15	30
10	5	30
11	3	45
12	5	15
13	5	30
14	5	20
15	5	20
16	10	20
34	5	10
35	5	25
37	5	15
38	10	25
39	2	10
42	5	15

[a] Data are taken from Mathews-Roth *et al.* (*110*), where a description of the phototesting technique is also given. The data are the average for each patient of three tests done before treatment and the average during treatment of those tests done when blood carotenoid levels had reached at least 400 μg %.

ever, it should be noted that we did not find a definite relationship between the ability to tolerate a given amount of xenon arc light and the ability to tolerate a given amount of sunlight. Response to artificial light shows a qualitative but not a quantitative relationship to response to sunlight.

Treatment with β-carotene has had no effect on the abnormalities of porphyrin metabolism that are characteristic of EPP. All of the 133 patients had elevated erythrocyte, plasma, and stool porphyrin levels at the time they entered the study, and their porphyrins have remained elevated in spite of β-carotene treatment.

In July 1975, the FDA approved β-carotene (Solatene, Roche) for the treatment of the photosensitivity associated with EPP. Since our first publication in 1969, 11 reports of the successful use of β-carotene in the treatment of photosensitivity in EPP have appeared in the medical literature (*3,8,11,19,66,71,91,97,114,154,155*). In these reports, over three-fourths of the patients were described as having experienced significant improvement in their ability to tolerate light exposure, and no side effects from the use of β-carotene were observed. One study reporting little or no improvement has also appeared (*25a*); however, these workers used a much lower dosage of β-carotene than that recommended by use (*110*). Later, some of the patients from the unsuccessfully treated group were given higher doses of β-carotene by another investigator and the patients noted some increased tolerance to sun (*135a*). These results emphasize the importance, as we have mentioned, of individualizing the dose to each patient and increasing the dose until the patient reports some improvement.

The problem of individualizing treatment is also important in the conduct of a controlled trial and makes double-blind designs almost impossible. At a minimum, such trials should use a dose of β-carotene large enough to produce amelioration of symptoms in the majority of patients (a minimal period of 3 months of treatment at doses giving blood levels of at least 800 μg/dl; for adults at least 180 mg/day should be given).

From our results and those of the other workers listed above it can be concluded that β-carotene, when administered in sufficiently large doses, can be effective in ameliorating photosensitivity in most patients with EPP.

C. Use of β-Carotene in Other Photosensitive Conditions

Since β-carotene seemed to be effective in preventing photosensitivity in EPP, it seemed logical to see whether it could prevent photosensitivity in other photosensitivity disorders. Several workers have already investigated this possibility.

Sneddon (*140*) reported some success in treating patients suffering from congenital porphyria (Gunther's disease) with β-carotene, as did Seip *et al.* (*134*) using a β-carotene–canthaxanthin combination product (Phenoro). We also have treated with β-carotene two children having this condition and have noted some improvement. New lesions were significantly decreased in number and severity, and the patients have been able to increase their sun exposure somewhat (*105a*).

Swanbeck and Wennersten treated 8 patients with polymorphous light eruption (PMLE) with β-carotene and/or the β-carotene–canthaxanthin combination product and found an increase in the minimal erythema dose of all the patients after treatment (*143*). Jansen treated 66 patients with the β-carotene–canthaxanthin combination product, 55 of whom improved, as judged by patients' responses to questionnaires and physicians' examinations (*72a*). Norlund and Klaus, in collaboration with the author and Pathak found that a topical sunscreen had to be used in conjunction with oral β-carotene to get an effective therapeutic result in a severe case of PMLE (*118*). During the course of the collaborative study on EPP, we also treated 27 patients suffering from PMLE with large doses of β-carotene. Nine of these patients increased their ability to tolerate exposure to sunlight by more than a factor of 3 (Table 4). Only 2 of these 9 reported marked improvement in their photosensitivity with the use of β-carotene alone; the other 7 had to use a sunscreen in addition to the βcarotene. (None of the 27 patients with PMLE had received any relief from the use of sunscreens alone.) The remaining 18 patients reported no improvement with the use of both oral β-carotene and the topical sunscreen.

Thune (*147a*) administered the β-carotene–canthaxanthin combination product to 17 patients with PMLE and found "a significant increase in minimal erythema dose

TABLE 4

Effect of β-Carotene on Tolerance to Sunlight of 39 Patients with Various Forms of Photosensitivity Other Than EPP[a]

	Protection index[b] (number of patients)						
Diagnosis	1	2	3–5	6–10	11–15	16–20	>21
Polymorphous light eruption	12	6	5[c,d]	2[e,f]	1[e]	0	1[c,f]
Solar urticaria	5	0	1[h,f]	0	0	0	0
Hydroa aestivale	1	1	0	0	0	0	1[e,g]
Porphyria cutanea tarda	1	0	0	0	0	0	0
Actinic reticuloid	2	0	0	0	0	0	0

[a] Data taken from Mathews-Roth *et al.* (*110a*).

[b] Protection index is defined in Table 2. Patients with protection indices of 1 and 2 showed little or no improvement despite blood carotene levels of at least 400 μg % following intake of largest dose of β-carotene for their ages.

[c] One patient reporting a protection index of 4 and one patient reporting a protection index of 24 stated that they did not need to use a topical sunscreen in conjunction with β-carotene ingestion. All other PMLE patients needed both β-carotene and topical sunscreen to achieve the stated protection.

[d] Three patients were in the study for one summer, two patients for two summers.

[e] One summer in study.

[f] Two summers in study.

[g] Boy, 7 years old.

[h] Woman; 40 years old.

for unflltered (xenon arc) light at the 5% level of probability.'' However, only 5 of the 17 patients had a protection index of more than 3 with respect to sunlight exposure, a much lower rate than that obtained by the above two authors and more in line with our findings. Thune could not show a direct correlation between the protection index obtained for artificial light and that obtained for sunlight exposure. He also found, as we have, that some patients benefited from the combined therapy of oral β-carotene and a topical sunscreen preparation.

Porphyria cutanea tarda is successfully treated by phlebotomy, but this treatment is occasionally contraindicated. We treated one patient with this condition with β-carotene in the hope of decreasing the appearance of new cutaneous lesions, which in this condition are porphyrin-induced photoinjury. However, examination of the patient's skin and the patient's own report after 2 months of β-carotene therapy showed no appreciable decrease in the number of new lesions formed.

Kobza *et al.* reported that β-carotene alone had no effect in the treatment of two cases of actinic reticuloid and three cases of solar urticaria (*85*). We treated six cases of solar urticaria, three cases of hydroa aestivale, and two cases of actinic reticuloid with large doses of β-carotene. Only one patient with solar urticaria and one patient with hydroa aestivale reported noticeable relief (Table 4). Both stated that they obtained improvement without the use of a topical sun screen.

In summary, it would appear that, for diseases other than EPP, β-carotene may be of some use in the treatment of congenital porphyria if given in large doses starting when the patients are very young. However, this treatment seems to be of limited value in PMLE, solar urticaria, and hydroa aestivale; we would recommend its use only when other treatment modalities have failed. Carotene treatment seems to be of no use in either porphyria cutanea tarda or actinic reticuloid.

Two independent observations suggested that perhaps carotenemia might have an effect on a normal individual's response to sunlight. Bendes (*12*) had observed that the presence of carotenemia in children undergoing heliotherapy for tuberculosis prevented sunburn, and Sandler found that carotenemia facilitated tanning of the skin (*132*). These observations, plus our success with β-carotene in the treatment of EPP, led us to conduct a controlled trial to determine whether the administration of large doses of β-carotene would alter the response of a normal fair-skinned individual to sunlight. We found that large doses of β-carotene had a small but statistically significant effect in increasing the minimal erythema dose for eliciting erythema produced by sunlight. However, the observed effects were too small to recommend the use of β-carotene as a protective agent for sunburn (*109*). We also found that the men in the group taking carotene in this study developed more pigmentation (tanning) than did the men in the placebo group, thus confirming the findings of Bendes (*12*) and Sandler (*132*).

Anyan (*5*) reported that a young girl who had been carotenemic while hypothyroid had not been able to tan while in this condition. Once she became euthyroid after appropriate treatment, her carotenemia disappeared, and the ability to tan returned. These findings are not necessarily at variance with ours. Perhaps, as

Anyan suggested, marked, long-standing carotenodermia may protect the skin to such a degree that tanning may become difficult to induce. In addition, one does not know the effects of hypothyroidism itself on tanning. Unfortunately, serum carotene levels were not given. Also, it is difficult to compare the degree of sun exposure of this youngster to that of the patients in our study. As we mentioned above, many of our EPP patients report tanning for the first time because of their new-found ability to stay in the sun. Some actually sunbathe on the beach. None have reported not being able to tan. These observations support our findings and those of Bendes (*12*) and Sandler (*132*) that tanning increases and burning decreases in carotenemic individuals. However, more work must be done to elucidate the extent of the ability of carotenoids to alter the skin's reaction to sunburn.

Because of the finding that β-carotene has some, albeit slight, effect of increasing the minimal erythema dose of sunlight necessary to elicit erythema, we thought it of interest to investigate other carotenoids for their ability to affect sensitivity to the "sunburn range" (280–320 nm, UV–B). To date, we have investigated phytoene, the colorless triene precursor of β-carotene, the absorption spectrum of which partially overlaps the UV–B range and which has been found to inhibit the formation of free radicals (*111*). We found that guinea pigs receiving daily intraperitoneal injections of phytoene for 14 days before exposure to UV–B radiation developed significantly less erythema to this radiation than did animals receiving daily injections of the same amount of placebo for 14 days (*106a*). Studies with phytoene and other carotenoids are continuing.

Other therapeutic uses have also been reported for carotenoids. Karagezyan *et al.* (*75*) and Shabykin and Godorazhy (*135*) have used carotene in various skin diseases (not photosensitivity diseases) with some success. Some authors had suggested a role for carotenoids in olfaction. However, Moulton conclusively demonstrated that carotenoids are not involved in the olfactory mechanism (*113*). On the somewhat more speculative side, some work has been done suggesting that carotenoids may have some role in "oxidations." Karnaukov has done a series of studies in which he suggests that carotene and myoglobin function as an "intracellular oxygen stock" in lipofuscin granules in aging tissues and in mollusk neurons (*76–79*). In addition Campbell (*16,17*) and Nelson *et al.* (*116*) suggest that a diet rich in carrots protects rats against acute anoxia. Perhaps in view of these interesting preliminary observations, the possible role of carotenoids in oxidative metabolism ought to be thoroughly investigated.

D. Clinical–Pharmacological Considerations

1. Lack of Toxicity

Studies on carotene and sunburn radiation that we conducted on patients with EPP treated with carotene and on normal volunteers given similar doses of carotene for a 10-week period (*109,110*) have confirmed reports in the literature that the ingestion of large amounts of pure β-carotene has not produced toxic side effects (*117,122*). Some of our patients have been taking carotene for up to 11 years, with no appearance to date of toxicity. In addition, no toxicity has been noted by the other workers cited above using β-carotene therapy. We detected no abnormalities in blood sugar or blood urea nitrogen in either EPP patients or normal volunteers. All except one patient with EPP had normal total bilirubin and serum glutamate–oxaloacetate transaminase levels. This patient had fairly consistently elevated levels of these parameters before as well as during therapy but never complained of or gave a history of liver disease. He did not respond to carotene therapy and had sustained high levels of porphyrins. Although liver disease is seen in some cases of EPP (*133,145*), this patient is the only one in our group to demonstrate significant liver function test abnormalities. Another patient had his gallbladder removed, but his liver function tests were normal.

The complete blood count was not affected by carotene intake. Many of the patients with EPP had a mild anemia (*104*). This was not worsened or improved by carotene intake. Leukopenia has been reported in patients who ingested large amounts of carrots (*54,74*). This phenomenon reverses itself when the patient stops eating an abnormal amount of carrots. We found no evidence of leukopenia in any of our patients or volunteers. The absence of this finding in our study strengthens the suggestion of earlier workers that the leukopenia is probably due to substances present in the carrots other than β-carotene (*117,122*). Indeed, they reported the the administration of pure β-carotene did not induce leukopenia in a patient who had experienced leukopenia after eating large amounts of carrots (*117,122*).

Methemoglobinemia has been shown to result from the consumption of large amounts of carrot juice (*82*). This is another example of the toxicity of large amounts of vegetables and their products and a reason why we do not encourage photosensitive patients to become carotenemic by ingesting large amounts of these foods.

None of our patients developed abnormally high levels of serum vitamin A. This confirms what has been repeatedly reported in the literature: that individuals ingesting large amounts of β-carotene do not develop hypervitaminosis A (*69,117,122*).

Magnus and collaborators reported that one patient treated with β-carotene experienced an increase in hay fever attacks, and another swelling of the tongue and face (*85*). We found no such reactions in our patients. It should be kept in mind that an occasional patient could be allergic to some of the other components of the beadlet preparation (gelatin, antioxidants, etc.). Perhaps such patients should be given crystalline β-carotene in vegetable oil to determine if, indeed, the reaction was caused by beadlet components. If the adverse reaction disappears, the patient can continue to take the carotene in oil if carotene has benefited his photosensitivity. Thus, it would seem that the only "side effect" from the ingestion of large amounts of β-carotene is carotenodermia, the yellowish coloration of the skin observed in many people who ingest large amounts of the pigment.

2. *Carotenemia and Carotenodermia*

Hess and Myers (*68*) noted that the skin of people who consumed large amounts of carrots and other vegetables often developed a yellow color; they used the term "carotenemia" for this condition. The phenomenon of yellow skin color was probably first noted by Baeltz (*9*), who called it aurantiasis cutis. He did not, however, make the connection between a diet abnormally high in carotene-containing foods and the yellow skin color. Apparently, Hess and Myers were the first to do so. In actuality, carotenodermia is the more accurate term for the phenomenon of yellow skin coloration by the ingestion of large amounts of carotenoid pigments; carotenemia literally means carotene in the blood. Carotenemia has been thoroughly reviewed (*25,73,74*); we will note only highlights of this condition here.

Mammals can be divided into three groups according to their ability to accumulate carotenoids in blood and tissues as carotenoids (not vitamin A) after oral administration: (a) those that can accumulate a mixture of carotenoids, (b) those that accumulate predominantly carotenes as opposed to xanthophylls, and (c) those that cannot accumulate any carotenoids (*60*). Human beings, monkeys, foxes, fitcherts, and probably badgers and roedeer are examples of the first group; cattle, horses, dogs, sheep, buffalo, deer, antelope, caribou, and hedgehogs are examples of the second; and goats, swine, hares, and the common laboratory animals, mice, rats, rabbits, and guinea pigs, are examples of the third group. Castle (*18*) studied a strain of rabbit that carried a recessive gene for "yellow fat," or the ability to accumulate carotenoids. We have found that greatly increasing the carotenoid content of the diet of guinea pigs and mice can lead to the accumulation of high levels of carotenoids in skin (*104d,105b*). This would suggest that the difference in carotenoid metabolism between the carotenoid accumulators (group a) and the nonaccumulators (group c) is a quantitative rather than a qualitative one.

In man, the most common cause of carotenemia and carotenodermia is eating excessive amounts of carotenoid-containing vegetables and fruits. The predominant pigment found in the serum is the predominant pigment of the vegetable eaten. Usually this is β-carotene. There was a report of high levels of lycopene, the principal pigment of the tomato, in a patient whose dietary indiscretion was tomato juice (*129*). There have been a few cases reported in which carotenemia and carotenodermia developed in individuals whose vegetable intake was not abnormal. It seems that these patients have a genetic inability to split carotenoids into vitamin A to the same degree as normal individuals (*25,100,112,136*). These patients have low to normal serum vitamin A levels and may or may not show symptoms of vitamin A deficiency (*25,100,112,136*).

The development of carotenodermia seems to be quite variable; different individuals on essentially the same diet can vary in the degree of carotenodermia they develop (*74*). Factors that seem to influence the development of carotenodermia, besides the amount of carotenoid-containing foods ingested, are the amount, nature, and dispersion of lipid in the diet, the digestibility of the carotenoid-containing food, and the general adequacy of nutrition and level of protein in the diet (*146*). It has been found, for example, that the ingestion of mineral oil definitely decreases the absorption of carotenoids (and, for that matter, also that of vitamin A). The time of development of carotenemia and carotenodermia from the onset of increased carotenoid intake varies widely, depending on the amount and kind of carotenoid-containing foods ingested. In our studies of EPP patients and normal volunteers taking pure β-carotene and on otherwise normal diets, we found that the development of carotenodermia occurred between $1\frac{1}{2}$ and 3 months, but the development of serum carotene levels above normal values occurred within 1 month. Those taking the larger doses from the beginning (180 mg/day) developed carotenemia and carotenodermia somewhat earlier. Some development of carotene tolerance, manifested by a decrease in blood levels of carotene after several months of therapy, has been noted by some workers (*62,149*). We have not found any conclusive evidence of this kind of phenomenon in our patients or subjects, but we have noted that some individuals, even on fairly large doses, never achieve as high a level as do others on the same dose. This probably indicates the variation in the ability of different individuals to absorb carotene.

Clinical observation of patients with carotenodermia has long demonstrated that this coloration is most obvious in areas of the body with a prominent amount of horny layer, such as the palms of the hands, soles of the feet, knees, and other areas of skin that are subject to mechanical irritation. Gandy (*51,52*) also suggested that in addition to areas subjected to mechanical irritation, carotenoids are deposited in greater amounts in areas subjected to irritation by physical factors, including light. The observation that light may play a part in regulating the intensity of carotenodermia was suggested by Klose (*84*), who observed that carotenodermia was confined to areas exposed to light in infants fed carrots and suggested that this might be due to the influence of light. He found that, in a group of infants given the same food, carotenodermia developed only in those infants whose cribs were kept near the window. Interestingly, the distribution of carotene in the skin of the whole body found by Edwards and Duntley fairly well parallels those areas of the body exposed to light (*38*). These findings are interesting in view of the protective function of carotenoids. They might suggest a function of the carotenoid pigments normally present in skin; that is, a certain level of protective pigments is always present to "balance" the potential photosensitizing pigments such as cytochromes and other heme pigments which are also always present. Thus, in conditions of increased light exposure, the body might increase the deposition of carotenoids in light-exposed areas. If there were sufficient carotene present in the diet, this would be seen as some carotenodermia in the light-exposed areas, explaining the findings of Klose. It is known that visible light can penetrate the epidermis into the dermal layer (*43*). Since skin contains potential photosensitizers, as we pointed out above, it would indeed seem important that exposed skin have some kind of protective mechanism to guard against photosensitization. Perhaps this protective mechanism involves the normally occurring carotenoid pigments.

It is intriguing to think that there might be some increase of carotenoid pigments in light-exposed areas in view of the protective function of carotenoids. However, other than the observations of Klose (*84*), there are no data to suggest that light actually plays a role in increasing the deposition of carotenes in the skin of man. Interestingly, in certain bacteria and fungi, light indeed does play a role in the deposition of carotenoids (*10*). The possibility remains that carotenoids are more prevalent in light-exposed areas simply because there is more horny layer in these areas. In addition, those areas of the body exposed to light are also usually exposed to mechanical irritation, resulting in increased formation of horny layer, so it is difficult to say which factor could be more important. It would seem that certain physical and probably also anatomical factors influence the development of the horny layer. Because of the affinity of carotenoids for the horny layer, the pigments are predominant in those areas of marked horny layer deposit. If the diet is high in carotenoids, carotenodermia becomes most marked in these areas. Grof *et al.* suggested a histamine-mediated mechanism to explain the deposition of carotene in these areas (*64*). Perhaps such a mechanism is not necessary since the carotenoids naturally accumulate in skin cells. When physical irritation occurs, there is a more rapid division of cells and resultant formation of horny layer, and thus the degree of accumulation of carotenoids in the area is again related to the amount of carotenoid intake. However, the Grof suggestion is appealing if there were some mechanism in the skin for the regulation of the level of photoprotector; then some mechanism would be necessary to get more carotenoids into the area susceptible to photosensitization. Clearly, more work is needed to study the factors involved in the development of carotenodermia.

Although the presence of carotenoids in skin had been established by visual and reflectance spectrophotometric observation, until now no one had chemically isolated carotenoids from skin. Dohi and Ono observed yellow material in the horny layer of an unstained section of skin (*36*), and Grof extracted carotenoids from horny layer scraped from the soles and knees of a carotenemic patient (*64*). We isolated carotenoids from human skin (*95*). We obtained specimens of abdominal skin from noncarotenemic individuals at autopsy and analyzed whole skin, epidermis, and dermis for the presence of carotenoids. We found that the epidermis contained more carotenoids than did the dermis. Some authors have proposed that carotene, which they thought to be present in sebum, is excreted onto the surface of the skin and rubbed into the horny layer (*74*). Our findings that carotene is present in both layers of the skin, and also the fact that abdominal skin does not have a prominent horny layer and does not sweat profusely but does contain carotenoids, tend to militate against this mechanism, if indeed it exists, playing a prominent part in the accumulation of carotenoids in skin. In addition, carotenoids oxidize rapidly when applied to the surface of the skin (*110*). It is more likely that carotenoids are carried to all the skin cells via the circulation and extracellular fluid, and the pigments enter the cells and bind to membranes. A precedent for this entry into cells and binding to membranes is the presence of carotenoids in the membranes of liver cell mitochondria (*61*). Additional studies of the subcellular anatomy of skin are necessary to pinpoint exactly the localization of the carotenoid pigments in the cellular organelles of the various layers.

Another area of the body in which a localized deposit of carotenoids has been found is a preexisting arcus senilis of the cornea. Giorgio *et al.* (*53*) reported the case of a man in whose corneas were found golden-yellow rings. At first they were thought to be Kayser–Fleischer rings, but on more careful examination it was found that these rings did not have the characteristics of Kayser–Fleischer rings; nor did the patient have elevated serum copper and ceruloplasmin levels or disease of the liver or basal ganglia, factors pathognomonic of Wilson's disease, in which Kayser–Fleischer rings occur exclusively. A detailed history elicited from the patient revealed that he ate a diet prominent in vegetables and in addition consumed daily 1–2 qt of carrot juice. The serum carotene level at the time of examination was 900 μg %. Neither we nor any of our collaborating physicians have seen any patient or normal subject with a carotenoid-stained arcus senilis, but it is conceivable that, as the number of patients receiving high doses of carotenoids for the treatment of photosensitivity increases, some patients may develop this phenomenon or smaller amounts of carotenoid deposits in the cornea. Hence, the physician should be aware of the existence of this phenomenon. Another ophthalmological finding was noted in patients receiving large doses of β-carotene for EPP by DeVries-de Mol *et al.* (*34*). In a study or 39 patients they noted that treatment with β-carotene had no effect on the patients' visual acuity, dark adaptation, or color discrimination, all of these parameters being within the normal range. They found no yellow color in the lens. No mention was made of any pigmentation in the cornea. Most patients seemed to have "grains" in the macular region, which the authors considered not to be pathological.

Levels of carotenoids in blood and skin are influenced by several disease conditions. The blood carotene level may be markedly decreased in malabsorption diseases, such as steatorrhea and sprue. Workers have found that, although the actual level of carotenoids in blood is not an absolute diagnostic parameter for diseases of malabsorption, the rate at which carotene is absorbed can be used as an indication of the presence of these diseases (*44,81,121*). Hence, a carotene loading test has been developed for use in diagnosing malabsorption diseases.

Studies have been conducted on carotenoid levels in various infectious diseases. Carotenodermia has been reported in polio (*98*) and malaria (*74*), and lowered carotene absorption (or intake?) in several fevers, pneumonia, sepsis, and grippe (*60*). Claussen (*23*) suggested that high levels of carotene might prevent infections, after having observed that infants with high carotene levels develop less severe upper respiratory infections.

High carotene levels have also been reported in castrated males (*39*) and retarded children (*125*). In the latter it seems that the size of injested food particles is an important matter; the smaller the particles the greater the likelihood of carotenodermia developing. These children probably do not chew their food well, so large-particle food may pass rather unchanged through the digestive tract.

Jehgers (*73*) suggested that carotenemia may be present in liver disease. However, Ditlefsen and Stoa (*35*) found decreased carotene levels in alcoholics (perhaps this may be somewhat related to diet), but patients with cirrhosis and chronic hepatitis had either normal or only moderately decreased carotene levels, with much individual variation. Blood carotene levels were found to be below normal patients with obstructive jaundice, acute hepatitis and cholecystitis, and secondary carcinoma of the liver. Adlersberg *et al.* (*1*) found that carotene absorption was markedly decreased in severe hepatic damage.

Patients with conditions associated with hyperlipemia have in many cases been found to have concomitant carotenemia and, in some cases, carotenodermia. Examples are nephrosis, diabetes, and hypothyroidism (*24,73,74*). It is not clear, however, whether the hyperlipemia is the important factor in the increased carotene levels. In nephrosis there is some question about decreased excretion of carotenoids, but since, as noted above, there is disagreement as to whether carotenoids are excreted in the urine, this point is still not settled. Interestingly, and for reasons unknown, patients with nephrosis do not often develop carotenemia.

Carotenemia and carotenodermia have been observed for many years in diabetics, and the question has long been asked as to whether this phenomenon is due to some role of insulin on carotenoid metabolism. Animal experiments suggested such a role, as it was found that in alloxan-diabetic rats there was markedly less conversion of β-carotene to vitamin A in the isolated intestinal loop (*131*). Ralli *et al.* (*124,127*) suggested that diabetics had an impaired ability to convert β-carotene to vitamin A in the liver. However, since that time, studies of carotene tolerance curves in diabetics and normal individuals have revealed no difference between the two groups in absorption and conversion of carotene to vitamin A (*115,128*). This finding suggests that there is no direct role of insulin in carotenoid metabolism. The carotenemia and carotenodermia that result are probably due to the diet containing ample amounts of vegetables and fruits that most diabetics follow (*60*). It is interesting that, on the other hand, vegetarians do not often develop carotenodermia. This is because their diet in actuality does not contain an overabundance of carotenoid-containing vegetables, as these are low in protein. Vegetarians increase the concentration of high-protein vegetables, such as various seeds and beans, in their diet. These are not very high in carotenoid content.

Carotenemia and carotenodermia are often noted in cases of hypothyroidism. For many years it was suggested that thyroid hormone had some effect on the absorption of carotenoids and their conversion to vitamin A. Studies by Walton *et al.* (*151*) suggest that the absorption of carotenoids is unimpaired in hypothyroidism but that the mechanism for the disposal of the carotenoids once they are taken into the body as such (''intrinsic'' carotenoids, or those carotenoids not split into vitamin A by the intestine and absorbed into the circulation) is interfered with. They found that the amount of low-density lipoproteins (S_f 3–9) responsible for carriage of carotenoids in serum is significantly higher in patients with hypothyroidism than in euthyroid patients and also that this fraction is overloaded with respect to carotene as compared to the concentration of carotene in the S_f 3–9 fraction in euthyroid blood. Absorption and utilization of preformed vitamin A seem to be within normal limits. The authors suggest that serum lipoprotein levels may be influenced by the level of protein metabolism, which is under the influence of thyroid hormone. Conversion of high-S_f to low-S_f classes, however, does not seem to be affected by this hormone. The authors suggest that the conversion of absorbed and circulating carotenoids to vitamin A might be dependent on complete catabolism of the S_f 3–9 lipoprotein fraction. Thus, it would seem that once carotenoids are absorbed and bound to lipoprotein, there is a defect in the disposal of the carotenoids and their carrier lipoprotein and, as a result, a decrease in the conversion of some of this intrinsic carotene into vitamin A, probably in organs other than the intestinal wall. Clearly, careful studies of carotene absorption and conversion with concurrent studies of lipoprotein metabolism in a significant number of patients are needed to establish whether the effect of thyroid hormone is indeed limited to the fate of the "intrinsic" carotenoids and has no effect on the absorption of carotene and the conversion of some of this absorbed carotene to vitamin A in the intestinal wall.

Page (*123*) reported menstrual disorders and hypervitaminosis A in women ingesting large amounts of carotenoid-containing food. Two of the women had anorexia nervosa. Several authors have reported carotenemia in anorexia nervosa, which condition also results in menstrual disorders (*29,30,126,130,137*). Pops and Schwabe suggest that there might be decreased catabolism of β-lipoproteins in anorexia nervosa, as there is in hypothyroidism, as found by Walton (*126*). This may be true, as lipoprotein metabolism is so closely linked to protein metabolism, which may be altered in anorexia nervosa. More investigation of lipoprotein metabolism in anorexia nervosa is certainly in order. In all of these cases, Page's included, one cannot rule out the possibility of other components of the ingested vegetables causing some of the "symptoms," such as tiredness, nervousness, and depression, occasionally attributed to carotenemia due to excessive vegetable ingestion and also found in anorexia nervosa. In our patients taking pure β-carotene, no such symptoms have developed, and the women patients between 17 and 45 have developed no menstrual abnormalities. The unusual finding in the patients of Page (*123*) and Robboy *et al.* (*130*) is that of abnormally high serum vitamin A levels; all previously published reports that we are aware of indicate that this does not happen if ingested vegetables are the only source of carotenoids. It is not clear whether these patients were taking any supplemental vitamins. In studies of this kind, it is important to ascertain (and report) whether the patients are taking supplementary vitamins containing even a small amount of vitamin A and also to be sure that appropriate corrections are made in the vitamin A assays for the cross-reacting carotene, the latter being especially important when there is carotenemia.

We also examined the carriage of carotenoids in the lipoprotein fractions of normal individuals taking carotene and normal individuals on a regular diet (*105*). We found that the amount of carotene in each lipoprotein fraction had increased in the serum from carotenemic individuals but that the relative increase was significant only in the low-density and high-density lipoprotein fractions. There was no significant difference between the cholesterol levels of each fraction in the sera of the carotenemic and noncarotenemic individuals, indicating that there is no increase in serum β-lipoprotein in carotenemia. We also found, as Krinsky *et al.* had noted (*90*), that the majority of the pigments were in the β-lipoprotein fraction in both the carotenemic and noncarotenemic volunteers.

Human blood cells have been examined for the presence of carotenoid pigments. Stefanini *et al.* found that platelets contained carotenoids and that the amount of pigments in these cells represented a moderate concentration of the pigments in the platelets with respect to the level of the pigments in plasma (*142a*). We examined the erythrocytes from noncarotenemic individuals and carotenemic individuals (patients receiving β-carotene for the treatment of EPP). We found that the erythrocytes from both groups contained a small amount of carotenoids, the erythrocytes from the latter group containing more pigments than those from the former (*104a*). We found that the pigment is associated with the cell membrane. The principal pigment isolated from erythrocytes and erythrocyte membranes from carotenemic individuals was β-carotene; from those from noncarotenemic individuals, a mixture of carotenoids was isolated. We also found small amounts of carotenoids in the leukocytes from carotenemic and noncarotenemic individuals (*104b*). Here also, the pigment is associated with the cell membrane. As in the erythrocytes, β-carotene was the main pigment found in the leukocytes of the carotenemic individuals, whereas the cells of the noncarotenemic individuals contained a mixture of pigments. Neither leukocytes nor erythrocytes seemed to concentrate the pigments, as compared to plasma. From a technical standpoint, we found that we had to extract the pigments in the presence of antioxidants in order to demonstrate the presence of pigments.

We examined some organs taken from two patients ingesting β-carotene (*104c*). The first patient, a 4-year-old girl with Rothmund–Thompson syndrome, had been taking β-carotene (30 mg/day) to alleviate the photosensitivity associated with this disease. She had died suddenly of acute bacterial meningitis. At autopsy, specimens of brain and liver were obtained. The second patient, a 22-year-old woman with EPP who had been taking β-carotene (180 mg/day) for her photosensitivity, underwent a cholecystectomy because of a history of several attacks of right upper quadrant and midepigastric pain after fatty meals. Biopsy material of the liver, as well as gallbladder, gallbladder contents, and duodenal drainage, were obtained for carotenoid analysis.

Gross and microscopic examination of the liver of the first patient was within normal limits. The brain showed the characteristic gross and microscopic appearance of acute meningitis but no other abnormalities. The gallbladder of the second patient had the characteristic findings of cholecystitis on gross and microscopic examination. The liver appeared normal on gross examination. Light microscopic examination revealed normal overall hepatic architecture, with some intrahepatic and canalicular cholestasis. Red-fluorescent deposits were seen on fluorescent microscopy, findings characteristic of EPP. No fatty changes or marked pigmentation were seen in the liver of either patient. Electron microscopic observations of the liver of the second patient revealed the presence of semicrystalline inclusions of stacked lamellar character similar to those having been reported in EPP, in patients taking steroids, (this patient had taken steroids up until about 2 years before her operation), and in control subjects [references about these inclusions are given in Mathews-Roth *et al.* (*104c*)].

There have been several series of liver carotenoid and vitamin A levels reported in the literature—one done on diabetic individuals and the others done on apparently normal individuals. The liver carotenoid levels of our patients (first patient, 21.88 mg/kg wet wt; second patient, 11.92 mg/kg wet wt) were well within the ranges given for the normal individuals. Likewise, the liver vitamin A levels of our patients remained within the normal ranges for liver vitamin A given in those studies. We also found no detectable carotenoids in the gallbladder wall or in the gallbladder contents removed at operation. Only traces of carotenoids were found in the biliary drainage (180 μg/liter) and duodenal drainage (60 μg/liter) taken the day before the operation.

The carotenoid content of the cerebral tissue of the first patient was 13 μg per kilogram wet weight of brain. The cerebrum of an elderly patient who died of atherosclerosis (who was not taking β-carotene) contained 27 μg per kilogram wet weight of brain. Thus, β-carotene therapy does not cause increased accumulation of this pigment in brain. Ours is the first report in the literature of the detection of carotenoids in brain tissue in human beings.

Our findings suggest that, as in animal studies (*9a*), abnormal amounts of β-carotene and vitamin A do not accumulate in the liver as a result of the ingestion of crystalline β-carotene. This is probably true also of brain. In addition, it would appear that no pathological structural changes have occurred in liver as a result of high levels of carotenoid intake. Again, the same is probably also true for brain.

From the results of our studies and from the data found in the literature, it would seem that there is no marked direct toxicity from the ingestion of large amounts of pure (crystalline) β-carotene. On the other hand, individuals developing carotenemia and carotenodermia by increased intake of carotenoid-containing foods are at some risk for the development of some hematological abnormality. They also may encounter other difficulties because of the usually inadequate diet, in which protein sources are often neglected, and occasionally because of the misdiagnosis of their "jaundice" by a physician unaware of their dietary indiscretions and unfamiliar with carotenodermia.

Because of the potential dangers associated with excessive ingestion of carotenoid-containing foods, we do not recommend that photosensitive patients obtain relief from their photosensitivity by increasing the intake of such foods. Pure β-carotene, either in the beadlet (water dispersible) formulation or as crystalline β-carotene mixed with butter or margarine for those rare individuals allergic to the noncarotenoid beadlet components, is indeed the preparation to be recommended.

V. MODE OF ACTION OF THE LIGHT-PROTECTIVE FUNCTION OF CAROTENOID PIGMENTS

In this chapter we have presented data from experiments done in bacteria, plants, and animals and from a clinical study done with human subjects which show that carotenoid pigments can eliminate or ameliorate the photosensitivity caused by either endogenous or exogenous photosensitizers. It is thus of interest to examine by what mechanism the carotenoid pigments could be exerting their protective function.

In 1968, Krinsky outlined four possible mechanisms to explain this function of the carotenoids (*88*): (a) a filter system in the cell envelope to filter out potentially harmful light, (b) systems that can interact with and quench photosensitizer triplet states, (c) systems that can serve as preferred substrates for photosensitized oxidations, and (d) systems that can stabilize membranes or repair damaged membranes. Since that time additional experimental evidence has indicated that only the second suggested mechanism, now extended to include quenching of singlet oxygen by carotenoids, seems to be significantly associated with the pigments' protective function (*89*). Mechanism (a) may have some function in certain plants, although it may not be the sole explanation for the pigments' protective effects in these organisms. One could suggest some filtering role of carotenoids in the human or animal porphyrin model, as the absorption spectrum of heme molecules and that of β-carotene as well as many other colored carotenoids overlap to some degree. However, the amount of carotenoids that would have to be deposited in the skin to provide a sufficient level of density is much more than the amount actually deposited in the skin in carotenodermia. In addition, the photoprotective effect of carotenoids seems to be independent of the absorption spectrum. The pigments can protect against photosensitization by exogenous photosensitizers with absorption spectra bearing no relation to that of the carotenoids present. The findings in photosynthetic organisms that led to the suggestion of mechanism (c) by several groups of workers now are thought to be connected with reactions involved with photosynthesis rather than with the protective function of carotenoids (*89*). The author and Krinsky showed that the presence of carotenoids does not seem to be involved with membrane stability in *S. lutea* (*106*), thus suggesting that mechanism (d) may not be of great significance in the protective function of the carotenoids.

Let us consider the second proposed mechanism briefly. In 1957, Fujimori and Livingston first showed that carotenoids could quench the triplet state of chlorophyll. These observations have since been confirmed by other workers, including ourselves, who showed that the carotenoids of *S. lutea* could also quench the free radicals formed by illuminated chlorophyll *a* in the presence of hydroquinone, air, and light (*50,111*).

In 1964, it was shown that singlet oxygen ($^1O_2$) could carry out reactions similar to those of photosensitized oxidations (*26,47*). The participation of $^1O_2$ in many chemical reactions and some biological systems has now been well-established (*45,152*). The studies of McDonagh (*99a*) on bilirubin and of Dalton and McAuliffe (*31*) on protoporphyrin showed that $^1O_2$ is produced on illumination of these compounds in the presence of air or oxygen. These findings suggest that $^1O_2$ may be involved in biological systems containing these compounds, for example, the photosensitization by protoporphyrin in EPP.

In 1968, Foote and Denny showed that carotenoids could quench $^1O_2$ (*46*). They suggested that quenching of singlet oxygen rather than triplet quenching may be the more dominant mechanism in the protective action of carotenoids. However, triplet quenching cannot be ruled out completely; it would simply seem to be a minor component of the reactions involved in the protective action of carotenoids against photosensitization.

Although $^1O_2$ quenching is an extremely attractive mechanism to explain the protective function of the carotenoids in EPP, the definitive experiment of determining the actual occurrence of $^1O_2$ in an *in vivo* system and its subsequent quenching in the presence of carotenoids has yet to be done. Unfortunately, this kind of experiment is frought with technical difficulties (and probably cannot be done in man), so that is will probably be a long time before the question can be definitively resolved, namely, whether $^1O_2$ is produced in a case of photosensitivity and whether the administration of carotenoids quenches this species, thus ameliorating the photosensitivity.

VI. SUMMARY

In this chapter we have presented evidence that carotenoid pigments can exert a protective action against photosensitivity in man as well as in animal and plant systems. From current evidence, it seems that the administration of high levels of carotenoids is effective for the amelioration of photosensitivity in the majority of patients with EPP and does not lead to toxic side effects. A larger number of patients with other forms of photosensitivity (PMLE, congenital porphyria, etc.) must be studied before it can definitely be determined whether β-carotene or other carotenoids are effective in ameliorating photosensitivity in these conditions. Present *in vitro* evidence suggests tht carotenoids may exert their protective function by quenching $^1O_2$ and perhaps also free radicals. Much more work must be done to ascertain whether this is the mechanism actually operative *in vivo*.

REFERENCES

1. Adlersberg, D., Kann, S., Maurer, A. P., Newerly, K., Winternitz, W., and Sobotka, H. (1949). *Am. J. Dig. Dis.* **16,** 333–337.
2. Almond, S., and Logan, R. F. L. (1942). *Br. Med. J.* **ii,** 235–241.
3. Anderson, C. R. (1972). *Arch. Dermatol.* **106,** 412–413.
4. Anderson, I. C., and Robertson, D. S. (1960). *Plant Physiol.* **35,** 531–534.
5. Anyan, W. P. (1972). *Arch. Dermatol.* **105,** 130.
6. Arroyave, G., Wilson, D., Mendez, J., Behar, M., and Scrimshaw, N. (1961). *Am. J. Clin. Nutr.* **9,** 180–185.
7. Austern, B. M., and Gawienowski, A. M. (1969). *Lipids* **4,** 227–229.
8. Baart de la Faille, H., Suurmond, D., Went, L. N., van Steveninck, J., and Schorthorst, A. A. (1972). *Dermatologica* **145,** 389–394.
9. Baeltz, cited in Miyake, I. (1924). *Arch. Dermatol. Syph.* **147,** 184–195.
9a. Bagdon, R. E., Zbinden, G., and Studer, A. (1960). *Toxicol. Appl. Pharmacol.* **2,** 225–236.
10. Batra, P. (1971). *In* "Photophysiology" (A. C. Giese, ed.), Vol. 6, pp. 47–76. Academic Press, New York.
11. Beckert, E., and Metz, J. (1974). *Hautarzt* **25,** 195–196.
12. Bendes, J. H. (1926). *Minn. Med.* **9,** 112–114.
13. Berger, S., Recheigl, M., Loosli, J. K., and Williams, H. Y. (1962). *J. Nutr.* **77,** 174–178.
14. Blinks, L. R. (1964). *In* "Photophysiology" (A. C. Giese, ed.), Vol. 1, pp. 199–222. Academic Press, New York.
15. Blum, H. F. (1941). "Photodynamic Action and Diseases Caused by Light." Reinhold, New York.
16. Campbell, J. A. (1938). *Q. J. Exp. Physiol. Cogn. Med. Sci.* **28,** 231–241.
17. Campbell, J. A. (1939). *Q. J. Exp. Physiol. Cogn. Med. Sci.* **29,** 259–275.
18. Castle, W. E. (1933). *Proc. Natl. Acad. Sci. U.S.* **19,** 947–950.
19. Chapel, T. A., Stewart, R. H., and Webster, S. B. (1972). *Arch. Dermatol.* **105,** 572–573.
20. Cheesman, D. F., Lee, W. L., and Zagalsky, P. F. (1967). *Biol. Rev. Cambridge Philos. Soc.* **42,** 132–160.
21. Chu, T. C., and Chu, E. J.-H. (1962). *Biochem. J.* **83,** 318–325.
22. Clare, N. T. (1956). *Radiat. Biol.* **3,** 693–723.
23. Clausen, S. W. (1931). *Trans. Am. Pediatr. Soc.* **43,** 27–30.
24. Clausen, S. W., and McCoord, A. B. (1938). *J. Pediatr.* **13,** 635–650.
25. Cohen, Lord (1958). *Ann. Intern. Med.* **48,** 219–227.
25a. Corbett, M. F., Herxheimer, A., Magnus, I. A., Ramsay, C. A., and Kobza-Black, A. (1977). *Br. J. Dermatol.* **97,** 655–662.
26. Corey, E. J., and Taylor, W. C. (1964). *J. Am. Chem. Soc.* **86,** 3681–368.
27. Cornbleet, T., and Greenberg, R. (1957). *Arch. Dermatol.* **76,** 431–433.
28. Courington, D. P., and Goodwin, T. W. (1955). *J. Bacteriol.* **70,** 568–571.
29. Crisp, A. H., and Stonehill, E. (1967). *Postgrad. Med. J.* **43,** 721–725.
30. Dally, P. J. (1959). *Br. Med. J.* **ii,** 1333.
31. Dalton, J., and McAuliffe, C. A. (1972). *Nature (London)* **235,** 388.
32. DeBuys, L. R. (1923). *Trans. Miss. State Med Assoc.* pp. 133–138.
32a. DeLeo, V., Poh-Fitzpatrick, M., Mathews-Roth, M., and Herber, L. (1976). *Am. J. Med.* **60,** 8–22.
33. Deshmukh, D. S., and Ganguly, J. (1964). *Indian J. Biochem.* **1,** 204–207.
34. DeVries-de Mol, E. C., Went, L. M., and Volker-Dieben, H. J. (1974). *Med. Probl. Opthalmol.* **13,** 349–352.
35. Ditlefsen, E.-M. L., and Stoa, K. F. (1954). *Scand. J. Clin. Lab. Invest.* **6,** 210–216.
36. Dohi, D., and Ono, T. (1922). Cited in Hashimoto, reference 67.

37. Duysens, L. M. N. (1951). *Nature (London)* **168,** 548–550.
38. Edwards, E. A., and Duntly, S. Q. (1939). *Am. J. Anat.* **63,** 1–33.
39. Edwards, E. A., Hamilton, J. B., Duntly, S. Q., and Hubut, G. (1951). *Endocrinology* **28,** 119–128.
40. Ehrlich, S. T., Farthing, B. R., and Moschetti, D. (1954). *J. Nutr.* **84,** 389–394.
41. Engelmann, T. W. (1883). *Bot. Z.* **41,** 1–16.
42. Engelmann, T. W. (1884). *Bot. Z.* **42,** 81–93.
43. Everett, M. A., Yeargers, E., Sayre, R. M., and Olson, R. L. (1966). *Photochem. Photobiol.* **5,** 533–542.
44. Feldman, E., and Adlersberg, D. (1959). *Am. J. Med. Sci.* **238,** 733–739.
45. Foote, C. S. (1968). *Science* **162,** 963–970.
46. Foote, C. S., and Denny, R. W. (1968). *J. Am. Chem. Soc.* **90,** 6233–6235.
47. Foote, C. S., and Wexler, S. (1964). *J. Am. Chem. Soc.* **86,** 3874–3877.
48. Forssberg, A., Lingen, C., Ernster, L., and Lindberg, O. (1959). *Exp. Cell Res.* **16,** 7–14.
49. Fox, H. M., and Vever, G. (1960). "Nature of Animal Colors." Sidgwick & Jackson, London.
50. Fujimori, E., and Tavla, M. (1966). *Photobiol. Photochem.* **5,** 877–887.
51. Gandy, D. T. (1935). *South. Med. J.* **8,** 444–446.
52. Gandy, D. T. (1964). *Med. Rec. Ann. (Houston)* **47,** 274–275.
53. Giorgio, A. L., Cartwright, G. E., and Wintrobe, M. M. (1964). *Arch. Intern. Med.* **113,** 817–818.
54. Gjerlow, J. (1966). *Tidsskr. Nor. Laegeforen.* **86,** 33–34.
55. Glover, J., and Redfearn, E. R. (1954). *Biochem. J.* **58,** xv.
56. Glover, J., and Redfearn, E. R. (1960). *Vitam. Horm. (N.Y.)* **18,** 371–386.
57. Goodman, D. S. (1967). *Prog. Biochem. Pharmacol.* **3,** 487–497.
58. Goodman, D. S., and Huang, H. S. (1966). *Science* **149,** 879–880.
59. Goodman, D. S., Huang, H. S., Kanai, M., and Shiratori, T. (1967). *J. Biol. Chem.* **242,** 3543–3554.
60. Goodwin, T. W. (1952). "The Comparative Biochemistry of the Carotenoids." Chapman & Hall, London.
61. Green, D. E., and Hatefi, Y. (1961). *Science* **133,** 13–19.
62. Greenberg, R., Cornbleet, T., and Jeffay, A. L. (1959). *J. Invest. Dermatol.* **32,** 599–604.
63. Greenberg, R., Cornbleet, T., and Demovsky, R. (1957). *Arch. Dermatol.* **76,** 17–23.
64. Grof, P., Bodzay, J., and Kovaks, A. (1967). *Acta Med. Acad. Sci. Hung.* **24,** 129–139.
65. Gronowska-Senger, A., and Wolf, G. (1970). *J. Nutr.* **100,** 300–308.
66. Gschnait, F., and Wolff, K. (1974). *Hautarzt* **25,** 72–80.
67. Hashimoto, H. (1922). *J. Am. Med. Assoc.* **78,** 1111–1112.
68. Hess, A. F., and Myers, V. C. (1919). *J. Am. Med. Assoc.* **73,** 1743–1745.
69. Hoch, H. (1943). *Biochem. J.* **37,** 430–433.
70. Hodgkiss, W., Liston, J., Goodwin, T. W., and Jamikorn, M. (1954). *J. Gen. Microbiol.* **11,** 438–450.
71. Hopsu-Havu, V. K., Terho, P. E., and Hollmen, T. (1973). *Ann. Clin. Res.* **5,** 181–185.
72. Jagannathan, S. N., and Patwardhan, V. N. (1960). *Indian J. Med. Res.* **48,** 775–784.
72a. Jansen, C. T. (1974). *Dermatologica* **149,** 363–373.
73. Jehgers, H. (1943). *N. Engl. J. Med.* **228,** 678–686.
74. Josephs, H. W. (1944). *Am. J. Dis. Child.* **67,** 33–43.
75. Karagezyan, M. A., Antonik, N. N., and Ulitina, I. A. (1962). *Vestn. Dermatol. Venerol.* **36,** 30–33.
76. Karnaukhov, V. N. (1971). *Exp. Cell Res.* **64,** 301–306.
77. Karnaukhov, V. N. (1973). *Exp. Cell Res.* **80,** 479–483.
78. Karnaukhov, V. N. (1973). *Tsitologiia* **15,** 538–542.
79. Karnaukhov, V. N., Taturyunas, T. R., and Petrunyaka, V. V. (1972). *Mech. Ageing Dev.* **2,** 201–210.

80. Karrer, P., and Jucker, E. (1950). "Carotenoids" (E. A. Braude, transl.). Elsevier, Amsterdam.
81. Kasper, H., and Hospach, R. (1974). *Dtsch. Med. Wochenschr.* **99,** 198–200.
82. Keating, J. P., Lell, M. E., Strauss, A. M., Zarkowsky, H., and Smith, G. E. (1973). *N. Engl. J. Med.* **288,** 824–826.
83. Kesten, B. M. (1951). *Arch. Dermatol. Syph.* **64,** 221–228.
84. Klose, E. (1919). *Muench. Med. Wochenschr.* **66,** 419–420.
85. Kobza, A., Ramsay, C. A., and Magnus, I. A. (1973). *Br. J. Dermatol.* **88,** 157–166.
86. Kon, S., and Mawson, M. E. (1950). *Med. Res. Counc. (G.B.), Spec. Rep. Ser.* No. 269.
87. Koski, V. M., and Smith, J. H. C. (1951). *Arch. Biochem. Biophys.* **34,** 189–195.
88. Krinsky, N. I. (1968). *In* "Photophysiology" (A. C. Giese, ed.), Vol. 3, pp. 123–195. Academic Press, New York.
89. Krinsky, N. I. (1971). *In* "Carotenoids" (O. Isler, ed.), pp. 669–716. Birkhaeuser, Basel.
90. Krinsky, N. I., Cornwell, D. G., and Oncley, J. L. (1958). *Arch. Biochem. Biophys.* **73,** 233–246.
91. Krook. G., and Haeger-Aronson, B. (1974). *Acta Derm.-Venereol.* **54,** 39–44.
92. Kunisawa, R., and Stanier, R. Y. (1958). *Arch. Mikrobiol.* **31,** 146–156.
93. Lakshamanan, M. R., Pope, J. L., and Olsen, J. A. (1968). *Biochem. Biophys. Res. Commun.* **33,** 347–352.
94. Lala, V. R., and Reddy, V. (1970). *Am. J. Clin. Nutr.* **23,** 110–113.
95. Lee, R., Mathews-Roth, M. M., and Pathak, M. A. (1974). *J. Invest. Dermatol.* **64,** 175–177.
96. Lesher, M., Brody, J. K., Williams, H. H., and May, I. G. (1945). *Am. J. Dis. Child.* **70,** 182–192.
97. Lewis, M. B. (1972). *Aust. J. Dermatol.* **13,** 75–78.
98. Linn, H. W. (1954). *Med. J. Aust.* **I,** 581–588.
99. Lipson, R. L., and Baldes, E. J. (1960). *Arch. Dermatol.* **82,** 508–516.
99a. McDonagh, A. F. (1972). *Biochem. Biophys. Res. Commun.* **48,** 408–415.
100. McLaren, D. S., and Zekian, B. (1971). *Am. J. Dis. Child.* **121,** 278–280.
101. Mathews, M. M. (1964). *Nature (London)* **203,** 1092.
102. Mathews, M. M., and Sistrom, W. R. (1959). *Nature (London)* **184,** 1892.
103. Mathews, M. M., and Sistrom, W. R. (1960). *Arch. Mikrobiol.* **35,** 139–146.
104. Mathews-Roth, M. M. (1974). *J. Am. Med. Assoc.* **230,** 824.
104a. Mathews-Roth, M. M. (1975). *Clin. Chem. (N.Y.)* **21,** 258–259.
104b. Mathews-Roth, M. M. (1978). *Clin. Chem. (N.Y.)* **24,** 700–701.
104c. Mathews-Roth, M. M., Abraham, A., and Gabuzda, T. (1976). *Clin. Chem. (N.Y.)* **22,** 922–923.
104d. Mathews-Roth, M. M., Crean, C., and Clancy, M. (1978). *Nutr. Rep. Int.* **17,** 581–584.
105. Mathews-Roth, M. M., and Gulbrandsen, C. (1974). *Clin. Chem. (N.Y.)* **20,** 1578–1579.
105a. Mathews-Roth, M. M., Haining, R. G., and Kinney, T. R. (1977). *Am. J. Dis. Child.* **131,** 366.
105b. Mathews-Roth, M. M., Hummel, D., and Crean, C. (1977). *Nutr. Rep. Int.* **16,** 419–423.
106. Mathews-Roth, M. M., and Krinsky, N. I. (1970). *Biochim. Biophys. Acta* **203,** 357–359.
106a. Mathews-Roth, M. M., and Pathak, M. A. (1975). *Photochem. Photobiol.* **21,** 261–263.
107. Mathews-Roth, M. M., Pathak, M. A., Fitzpatrick, T. B., Harber, L. C., and Kass, E. H. (1970). *N. Engl. J. Med.* **282,** 1231–1234.
108. Mathews-Roth, M. M., Pathak, M. A., Fitzpatrick, T. B., Harber, L. C., and Kass, E. H. (1970). *Trans. Assoc. Am. Physicians* **83,** 176–184.
109. Mathews-Roth, M. M., Pathak, M. A., Parrish, J., Fitzpatrick, T. B., Kass, E. H., Toda,K., and Clemans, W. (1972). *J. Invest. Dermatol.* **59,** 349–353.
110. Mathews-Roth, M. M., Pathak, M. A., Fitzpatrick, T. B., Harber, L. C., and Kass, E. H. (1974). *J. Am. Med. Assoc.* **228,** 1004–1008.
110a. Mathews-Roth, M. M., Pathak, M. A. Fitzpatrick, T. B., Harber, L. C., and Kass, E. H. (1977). *Arch. Dermatol.* **113,** 1229–1232.
111. Mathews-Roth, M. M., Wilson, T., Fujimori, E., and Krinsky, N. I. (1974). *Photochem. Photobiol.* **19,** 217–222.
112. Molholm-Hansen, J. (1965). *Nord. Med.* **22,** 385–388.

113. Moulton, D. G. (1962). *Nature (London)* **195,** 1312–1313.
114. Moynahan, E. J., and Leppard, B. (1973). *Proc. R. Soc. Med.* **66,** 1085–1086.
115. Murrill, W. A., Horton, P. B., Leiberman, E., and Newburgh, L. H. (1941). *J. Clin. Invest.* **20,** 395–400.
116. Nelson, D., Goetzel, S., Robins, S., and Ivy, A. C. (1943). *Proc. Soc. Exp. Biol. Med.* **52,** 1–2.
117. Neimann, C., and Klein Obbink, H. J. (1954). *Vitam. Horm. (N.Y.)* **12,** 69–99.
118. Nordlund, J. J., Klaus, S. N., Mathews-Roth, M. M., and Pathak, M. A. (1973). *Arch. Dermatol.* **108,** 710–712.
119. Olson, J. A. (1961). *Am. J. Clin. Nutr.* **9**(4), Part 2, 1–12.
120. Olson, J. A. (1964). *J. Lipid Res.* **5,** 281–299.
121. Onstad, G. R. (1972). *J. Am. Med. Assoc.* **221,** 677–679.
122. Ostwald, R., and Briggs, G. M. (1967). *In* "Toxicants Occurring Naturally in Foods" (Food Protection Commission, ed.), pp. 183–220. Nat. Sci. Found., Washington, D.C.
123. Page, S. W. (1971). *Aust. N.Z. J. Obstet. Gynecol.* **11,** 32–36.
124. Pariente, A., Present, C., and Ralli, E. P. (1936). *Am. J. Med. Sci.* **192,** 365–371.
125. Patel, H., Dunn, H. G., Tischler, B., McBurney, A. K., and Hach, E. (1973). *Can. Med. Assoc. J.* **108,** 848–852.
126. Pops, M. A., and Schwabe, A. D. (1968). *J. Am. Med. Assoc.* **205,** 121–122.
127. Ralli, E. P., Brandaleone, H., and Mandelbaum, T. (1934–1935). *J. Lab. Clin. Med.* **20,** 1266–1275.
128. Ramachandran, K. (1973). *Indian J. Med. Res.* **61,** 1831–1834.
129. Reich, P., Swachman, H., and Craig, J. M. (1960). *N. Engl. J. Med.* **262,** 263–269.
130. Robboy, M. S., Sato, A. S., and Schwabe, A. D. (1974). *Am. J. Clin. Nutr.* **27,** 362–367.
131. Rosenberg, A., and Sobel, A. E. (1953). *Arch. Biochem. Biophys.* **44,** 326–329.
132. Sandler, A. S. (1935). *Arch. Pediatr.* **52,** 391–406.
133. Scott, A. J., Ansford, A. J., Webster, B. H., and Stinger, H. C. W. (1973). *Am. J. Med.* **54,** 251–259.
134. Seip, M., Thune, P. O., and Eriksen, L. (1974). *Acta Derm.-Venereol.* **54,** 239–240.
135. Shabykin, G. P., and Godorazhy, A. I. (1967). *Vestn. Dermatol. Venerol.* **41,** 71–73.
135a. Shaffrir, A. (1977). *Proc. R. Soc. Med.* **70,** 574.
136. Sharvill, D. E. (1970). *Proc. R. Soc. Med.* **63,** 605–606.
137. Silverman, J. A. (1974). *J. Pediatr.* **84,** 68–73.
138. Sistrom, W. R., Griffiths, M., and Stanier, R. Y. (1956). *J. Cell. Comp. Physiol.* **48,** 473–515.
139. Smith, J. H. C., Durham, L., and Warster, C. F. (1959). *Plant Physiol.* **34,** 340–345.
140. Sneddon, I. B. (1974). *Proc. R. Soc. Med.* **67,** 593–594.
141. Stanier, R. (1960). *Harvey Lect.* **54,** 219–255.
142. Stanier, R., and Cohen-Bazire, G. (1957). *In* "Microbial Ecology" (R. E. O. Williams and C. C. Spicer, eds.), Symposium of the Society for General Microbiology, Vol. 7, pp. 56–89. Cambridge Univ. Press, London and New York.
142a. Stefanini, M., Krinsky, N., and Magalini, S. (1957). *J. Lab. Clin. Med.* **50,** 225–228.
143. Swanbeck, G., and Wennersten, G. (1974). *Acta Derm.-Venereol.* **54,** 491–499.
144. Thommen, H. (1971). *In* "Carotenoids" (O. Isler, ed.), pp. 637–668. Birkhaeuser, Basel.
145. Thompson, R. P. H., Molland, E. A. Nicholson, D. C., and Gray, C. H. (1973). *Gut* **14,** 934–938.
146. Thompson, S. Y. (1965). *Proc. Nutr. Soc.* **24,** 136–146.
147. Thomson, M. L. (1973). *Arch. Dis. Child.* **18,** 112.
147a. Thune, P. (1976). *Acta Derm.-Venereol.* **56,** 127–133.
148. Tschudy, D. P., Magnus, I. A., and Kalivas, J. (1971). *In* "Dermatology in General Medicine" (T. B. Fitzpatrick, ed.), pp. 1143–1166. McGraw-Hill, New York.
149. Urbach, C. (1952). *Exp. Med. Surg.* **10,** 7–20.
150. Wallace, R. H., and Schwarthing, A. E. (1954). *Plant Physiol.* **29,** 431–436.

151. Walton, K. W., Campbell, D. A., and Tonks, D. (1965). *Clin. Sci.* **29,** 199–215.
152. Wilson, T., and Hastings, J. W. (1970). *In* "Photophysiology" (A. C. Giese, ed.), Vol. 5, pp. 49–95. Academic Press, New York.
153. Wiskemann, A. (1974). *In* "Sunlight and Man" (M. A. Pathak, L. C. Harber, M. Seiji, and A. Kukita, eds.), pp. 669–676. Univ. of Tokyo Press, Tokyo.
154. Wurker, I. (1972). *Z. Haut.-Geschlechtskr.* **47,** 1006–1007.
155. Zaynoun, S. T., Hunter, J. A. A., Darby, F. J., Zarembski, P., Johnson, B. E., and Frain-Bell, W. (1977). *Br. J. Dermatol.* **97,** 663–668.
156. ZoBell, C. E., and Feltham, C. B. (1934). *Bull. Scripps Inst. Oceanogr., Tech. Ser.* **3,** 279–296.

9

CAROTENOIDS AS PHOTOCONDUCTORS

J. O. Williams

I. INTRODUCTION

The now famous Koranyi lecture (*1,2*) delivered by Szent-Györgyi in 1941 did much to rekindle the interest in the phenomena of energy transfer and electron (or charge carrier) motion in the "ordered" structures found in biological systems. Since that time the study of semi- and photoconduction in organic materials has been extensive, the chief stimuli being the possible discovery of tailor-made highly conducting materials of technological importance and the solution of hitherto unexplained phenomena occurring in certain biophysical processes (*3,4*), namely, photosynthesis and vision. At the forefront of these studies have been the carotenoids (*5,6*), as a class of highly colored compounds present by themselves

ISBN 0-12-082850-2

and in combination with other macromolecules in certain loci of undisputed importance in controlling several of the complex processes so vital to life. In vertebrates the visual pigments rhodopsin and iodopsin, which are complexes of carotenoids with the protein opsin, are located in the outer segments of the rods and cones (*7*). However, the precise way in which optical excitation of the visual pigment is converted to a discharge of the optic nerve has led to controversy, and recently a physical (*8*), as opposed to a chemical (*9*), approach has been proposed invoking the photoconduction process. This model, put forward by Rosenberg, involves the direct generation of charge carriers upon light excitation followed by motion of the charge carriers in a concentration gradient of charge or in an electric field. Subsequently, much of the work on the photoconduction of carotenoids has involved seeking justification for this idea. Since neither of the protein–carotenoid complexes has yet been crystallized in pure form free of detergent micelles, it has been necessary to use model compounds to investigate the photoconduction theory of the photoreceptor mechanism. Foremost in this field has been the study of the various isomers of β-carotene and retinal derivatives both in the solid state (single crystals, powders, films) and as integral components of model, synthetic bilayer membranes. Of course, the serious disadvantage of employing these model carotenoids is that motion of the charge carriers through essentially molecularly homogeneous aggregates is probably very different from that through a medium involving the protein moiety to which the real chromophore would be attached. Nevertheless, certain invaluable information has become available which has furthered our understanding of the possible role of these materials in biological systems and has suggested possible technological utilization.

In this chapter we shall consider the essential theoretical background relevant to any discussion of the role of carotenoids as photoconductors, assess critically current experimental evidence, and suggest possible areas where further effort is needed either to seek further clarification of physical or biophysical points or in an effort to develop carotenoid photoconductive technology. Any discussion of photoconductivity would not be complete without consideration of the allied property of semiconductivity. We shall therefore also develop ideas on this aspect, considering it an integral part of the photoconduction process, namely, motion of the charge carriers *after* generation. However, it will become apparent throughout the discussion that semiconductivity per se in carotenoids, as with most other organic insulators, need not necessarily be an "intrinsic" process.

II. PHOTOCONDUCTIVITY IN ORGANIC SOLIDS: INTRODUCTORY REMARKS

The essential problem in the photoconduction of organic materials, particularly in those that, in general, are highly conjugated and bound together in the solid state by weak, physical forces, is to locate the conducting states relative to the molecular energy states (exciton states) (*10*). It has long been realized (*11,12*) that the adoption of the concepts relating to classical semiconductors (*13*) (namely, Si and Ge) is not wholly applicable to these molecular aggregates. The main reason for this is the invariable absence of a photoconduction band edge corresponding to the energy gap in organic materials and the creation of electronic charge carriers indirectly by excitation into low-lying, nonconducting exciton states. The conducting levels for most organic solids remain a mystery even though accurate determinations have been carried out for some materials, notably anthracene and tetracene (*14,15*). As we shall see later an estimation of the positions of these states is the best we can hope for in the carotenoids.

Following the generation of the electronic charge carriers after optical excitation, their motion in an electric field is considered to be controlled either by motion in delocalized levels—band conduction (*16,17*)—or by hopping (or tunnelling) between localized states—hopping (tunnelling) conduction (*18,19*). Consequently, band and hopping conduction may be differentiated by measurement of a relative activation energy for carrier transport. Therefore, the parameters $n_{\pm}$, the number of charge carriers, and $\mu_{\pm}$, their intrinsic mobility in an electric field E, usually related by the equations

$$J/E = \sigma = nq\mu = n_{+}q\mu_{+} + n_{-}q\mu_{-} \tag{1}$$

where J is the current density, q the electronic charge, and σ the conductivity, must be separated. It is generally found that the conductivity σ is thermally activated, being described by an equation of the form

$$\sigma(T) = \sigma_0 \exp(-E_C/kT) = \sigma_0 \exp(-E_G/2kT) \tag{2}$$

where T is the temperature, k is Boltzmann's constant, and σ_0, E_G, and E_C are constants. In the same way the photocurrent is controlled by the constants σ_{0p} and E_p in the equation

$$\sigma_p(T) = \sigma_{0p} \text{ and } (-E_p/kT) \tag{3}$$

Since σ is given by the product of the charge density and the mobility, the temperature dependence of σ may be due to the exponential temperature dependence of n or μ or both:

$$n(T) = n_0 \exp(-E_n/kT) \tag{4a}$$

$$\mu(T) = \mu_0 \exp(-E_\mu/kT) \tag{4b}$$

Consequently, to describe completely the conduction process we have to be in a position to describe the temperature dependence both of the charge carrier density and of the intrinsic mobility. The carotenoids, as we shall see, fall short of this basic requirement in that the temperature dependence of neither $n_{\pm}$ nor $\mu_{\pm}$ is known; all their electrical properties are described in terms of the temperature dependence of σ. Even though it is a major disadvantage in the complete understanding of processes occurring in this class of solids, the absence of this important information has not excluded the carotenoids as important organic photoconductors, and much additional and valuable information has emerged in the wake of the photoconduction studies.

III. MOLECULAR STRUCTURES AND AGGREGATES

β-Carotene is a typical member of the C_{40} carotenoid pigments and is considered a precursor of retinal (vitamin A aldehyde), the C_{20} carotenoid (Fig. 1c) considered to be an essential component of visual pigments. The interesting part of the molecule as far as photoconduction and spectroscopic studies are concerned is the linear chain of conjugated double bonds. The molecule may exist as a number of geometric isomers, that in which the linear chain is straight being the all-trans form (Fig. 1a). With rotation about a number of possible double bonds various mono-cis and di-cis stable forms occur, but the one of major interest to us is the 15,15′-cis shown in Fig. 1b. Even though most of our concern will be with the C_{40} members, reference will also be made to the C_{20} series, including the all-*trans*- (Fig. 1c), 13-*cis*- (Fig. 1d), 11-*cis*- (Fig. 1e), and 9-*cis*- retinals (Fig. 1f). Figure 1g shows the molecular structure of 13-*cis*-retinal-*m*-nitroaniline hydrogen chloride, which is a protonated Schiff's base of 13-*cis*-retinal and believed to contain a linkage similar to that existing between retinal and opsin in rhodopsin.

Because of the reluctance of these carotenoids to yield large single crystals from the melt, vapor, or solution their electrical and spectral properties have been examined using either polycrystalline powder, thin films deposited from solution, or transparent glasses prepared by melting. Material preparations employing the above techniques suffer gross disadvantages: (a) these carotenoids are extremely sensitive to a variety of adsorbed gases (see Sections IV and VI,A) the effects of which are prevalent in the large-surface-area polycrystalline solids, and (b) the transparent glasses formed by melting invariably contain a mixture of stereoisomers. For the spectral studies thin evaporated films deposited on Pyrex or quartz substrates are particularly well suited (*20*). Most of the electrical studies, however, have been carried out with the polycrystalline powders and transparent glasses sandwiched between contacts provided by conducting glass plates. Only one comprehensive study has been reported for single crystals of all-*trans*-*β*-carotene (*21*).

The purity of the carotenoids does not seem to have been of great concern to the early investigators. It has been argued (*22*) that there is a fundamental difference between "biological" and "inorganic" semiconductors in that the latter contain long-range order whereas the former do not. Any local alteration of this long-range order will have important repercussions, whereas the short-range order associated with biological semiconductors will not be modified until the impurity reaches high concentration. However, this argument may not be entirely applicable to photoconduction studies since it is known that single-crystalline samples are desirable for

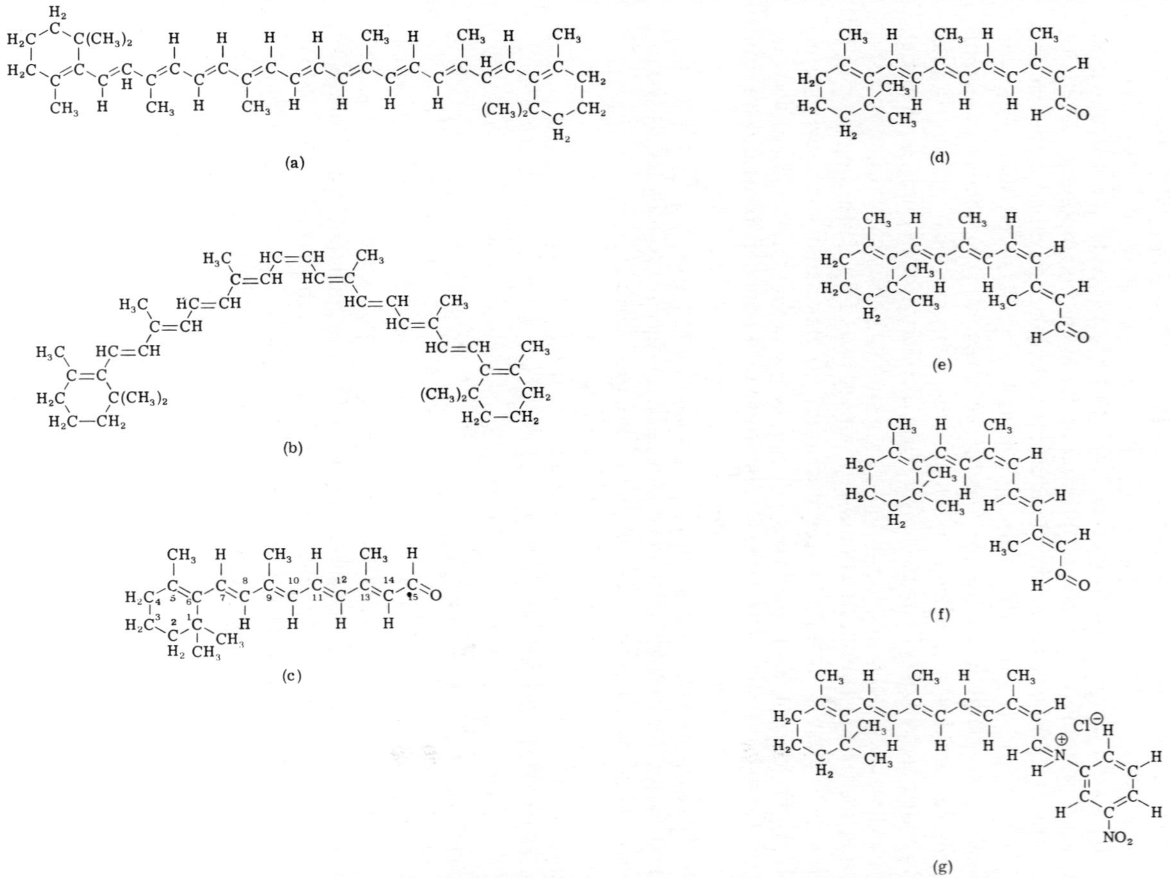

Fig. 1. Molecular structures of some typical photoconducting carotenoids. (a) All-*trans*-β-carotene, (b) 15,15′-*cis*-β-carotene, (c) all-*trans*-retinal, (d) 13-*cis*-retinal, (e) 11-*cis*-retinal, (f) 9-*cis*-retinal, (g) 13-*cis*-retinal-*m*-nitroaniline hydrogen chloride.

such studies, and even then the impurities and imperfections associated with real crystals (*23*) through their participation in such phenomena as surface states (*24*) and bulk trapping (*25*) can dominate the behavior. An excellent illustration of the role of imperfections (both chemical and physical) can be seen by considering the mobility of the charge carriers. As mentioned earlier (Section II) the determination of the carrier mobility may serve to differentiate between two proposed conductivity mechanisms. If the charge carrier is localized for a length of time on a particular site and then "hops" or "tunnels" to a neighboring site in a relatively short time, low values of the mobility ($<10^{-4} m^2 V^{-1} sec^{-1}$) are expected (*26*), whereas motion of the charge carrier within a band (to which it has been previously excited) would yield higher mobilities (*27*) ($>10^{-4} m^2 V^{-1} sec^{-1}$)). Furthermore, in the band model the temperature dependence of the mobility would be dependent only on acoustic phonon scattering, and the functional dependence would be of the form $\mu \propto T^{-n}$ with $n \simeq \frac{3}{2}$. Mobility in the hopping model would be expected to increase exponentially with increasing temperature according to $\mu = \mu_0 \exp(-E_\mu/kT)$ This picture is considered to be idealized in that no account is taken of trapping. To do this we may define two types of mobility; μ, the intrinsic mobility, and μ_D, the drift mobility, related by an expression of the form $\mu_D = \mu\Theta$, where $\Theta < 1$ (*28*). Consequently, without regard for trapping phenomena, measured values of mobility totally unrepresentative of the true or intrinsic mobility may be obtained.

IV. STUDIES ON β-CAROTENE

A. Photoconductivity in β-Carotene

The photoconductivity excitation spectra of organic crystals have in general been found to be closely related to the absorption spectrum of the crystal (*29*). The surface photoresponse is invariably identical to the absorption spectrum, whereas in the bulk an inverse relationship may be observed; i.e., high absorption leads to low photoconductivity. The situation with regard to β-carotene is summarized in Fig. 2, which gives the absorption spectra in solution for polycrystalline material, isomerized glass, and single crystals together with the corresponding photoconduction excitation spectra. The visible-region absorption spectrum of a solution of all-*trans*-β-carotene (1) consists of a single electronic band, which is a singlet–singlet π-electron transition covering the region 400–500 nm, with vibrational levels separated by 0.177 eV. The spectrum of the various cis forms is almost identical in this region, and a new peak appears at ca. 350 nm called the cis peak. This cis peak also corresponds to a singlet–singlet transition and presumably arises from a dipole moment perpendicular to the straight linear chain when the molecule is bent (*30*). This peak is also present in molten glasses of all-*trans*-β-carotene and shows that isomerization has occurred on cooling. The absorption spectra of the glass films are broadened relative to the dilute solution spectra, and there is a slight

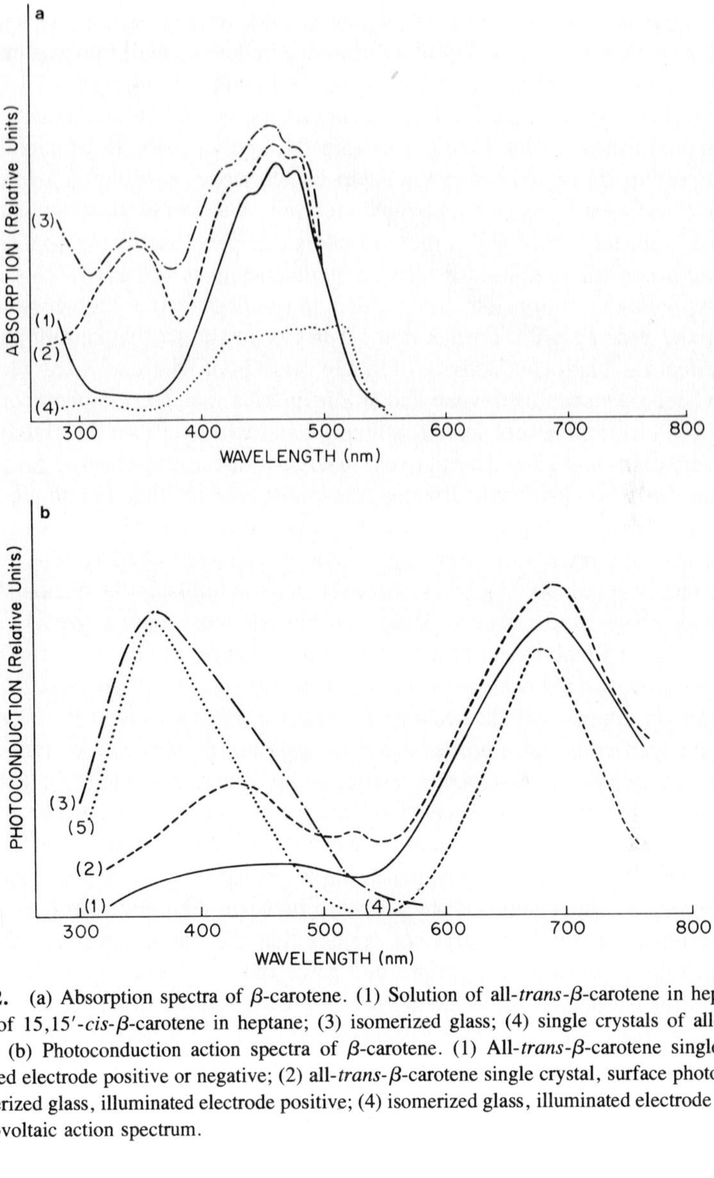

Fig. 2. (a) Absorption spectra of β-carotene. (1) Solution of all-*trans*-β-carotene in heptane; (2) solution of 15,15′-*cis*-β-carotene in heptane; (3) isomerized glass; (4) single crystals of all-*trans*-β-carotene. (b) Photoconduction action spectra of β-carotene. (1) All-*trans*-β-carotene single crystal, illuminated electrode positive or negative; (2) all-*trans*-β-carotene single crystal, surface photocurrents; (3) isomerized glass, illuminated electrode positive; (4) isomerized glass, illuminated electrode negative; (5) photovoltaic action spectrum.

red shift of the longest wavelength band. We are not concerned with polarized absorption spectra of β-carotene in this chapter, but the unpolarized spectrum of single crystals is included in Fig. 2. The photoconduction action spectra depend largely on the polarity of the illuminated electrode and the nature of the sample. It has been shown that in all but the single crystals the photocurrents are higher ($\times 10^5$) when the illuminated electrode is positive. This indicates that the contribution of positive holes to the currents is higher than that of the electrons. This may be a result of the difference in mobility or a difference in the extent of trapping of the two charge carriers. Furthermore, Rosenberg (*30*) found that in both 15,15′-*cis*-β-carotene and the isomeric glass the photoconduction due to positive holes is strong in the region of the cis peak [350 nm (3.45 eV)], and virtually no photoconduction was observed in the region of the lowest π–π^* transition at 450 nm (2.58 eV). (In all-*trans*-β-carotene samples the photoconductivity was too weak to be measured.) Light that is weakly absorbed in these systems can penetrate to the opposite electrode, and when this is biased positive a photocurrent is observed. This is again consistent with the notion that the major contribution to the photoconduction is made by the positive hole. Furthermore, it has been shown that certain gases and vapors affect the photoconductivity of single crystals of all-*trans*-β-carotene (*31*). Electron acceptors such as oxygen and sulfur dioxide increase the photoconductivity, whereas electron donors such as ammonia and iodine reduce the photoconductivity. Inert gases and gases having weak charge-transfer properties, e.g., nitrogen and argon, have no observable effect on the photoconductivity. For single crystals of all-*trans*-β-carotene bulk photoconduction appears in a region of the spectrum far removed from the main absorption bands, and the positive and negative photocurrents are nearly equal (see Fig. 2.) Consequently, it is unlikely that excitation to the first excited state of β-carotene is responsible for these photocurrents, and the possibility of the involvement of a low-lying triplet state, following earlier predictions by Rosenberg (*32*), was proposed by Chapman *et al.* (*21*). However, no direct transition from the ground state to any triplet level has been found for β-carotene, and the delayed emission observed from β-carotene samples at low temperatures cannot be unequivocally assigned to a triplet state of the molecule (*65*). The possibility that this band can arise because of impurities cannot, however, be ruled out even though the suggestion has been made (*33*) that impurity concentration as high as 1–3% is needed to generate new bands in the photoconductivity action spectra of organic crystals. The square-root relationship between the photocurrent at 750 nm and light intensity for single crystals implies that the photocurrent is limited by recombination of the charge carriers and rules out exciton–exciton interactions. However, for isomerized β-carotene glasses the photocurrent at 500 nm is linear with light intensity and could be representative of an exciton mechanism. If, indeed, triplet excitons are responsible for the photocurrents in β-carotene, as postulated by Rosenberg, then their generation via singlet to triplet intersystem crossing would still result in a photocurrent peaking at 350 nm.

The activation energy for photoconduction in β-carotene [see Eq. (3)] has been determined for a variety of isomeric modifications, and the values are summarized in Table 1.

The activation energies are multiples of the main vibrational spacing of the absorption band of β-carotene ($h\nu$ = 0.177 eV), and Rosenberg and Harder (*42*) have shown that for the unhindered cis isomer and the β-carotene glass the photocurrents are markedly larger (ca. 10^3) than for the all-trans isomer. However, as

TABLE 1

Activation Energies for Photoconduction in Various Isomers of β-Carotene

Isomer	E_p(eV)
All-*trans*-β-carotene	0.37 ± 0.04
15,15′-Mono-*cis*-β-carotene	0.20 ± 0.02
β-Carotene glass	0.19 ± 0.02

mentioned earlier it is difficult to assess whether it is n or μ that varies between the isomers.

B. Drift Mobility Measurements

An attempt has been made to measure the drift mobility of the charge carriers in β-carotene glasses (*34*). The glass was contained between two conducting glass electrodes (see Section III) and a light pulse of short duration (<10 μsec) generated charge carriers close to the illuminated electrode. Under an applied field E the charge carrier of appropriate sign drifted across the sample of thickness d to the opposite electrode. The resulting transients (*35,36*) were recorded on an oscilloscope (see Fig. 3a) and the mobility μ_D, given by the expression $\mu_D = d/ET_{tr}$, was evaluated from dependencies of t_{tr}, the transit time of the charge carrier, on the applied field E. A typical plot for positive holes is shown in Fig. 3b. The mobility at 298°K is 1.6×10^{-8} m^2 V^{-1} sec^{-1}. There is evidence, from the shape of the transients, of strong carrier trapping in these β-carotene samples, and, indeed, this may be the reason why electron transients have not been observed. Furthermore, we have no reliable data on the temperature dependence of the hole mobility, and so no conclusions can be made regarding the mechanism of charge transport. The absolute value at 298°K, however, does indicate a hopping model according to Glarum's criterion (*37*). Clearly, there is a need to develop more ways of growing large single crystals of β-carotene so that further drift mobility measurements can be carried out.

C. Rectification, Gain, and Quantum Efficiency for Photoconduction

It has been shown that for β-carotene glasses the photocurrent, when the illuminated electrode is positive, is ca. 10^5 times larger than when this electrode is negative. This is generally known as the rectification effect. The rectification ratio R is then given by

$$R = \frac{n_+\mu_+}{n_-\mu_-} \simeq 10^5 \qquad (5)$$

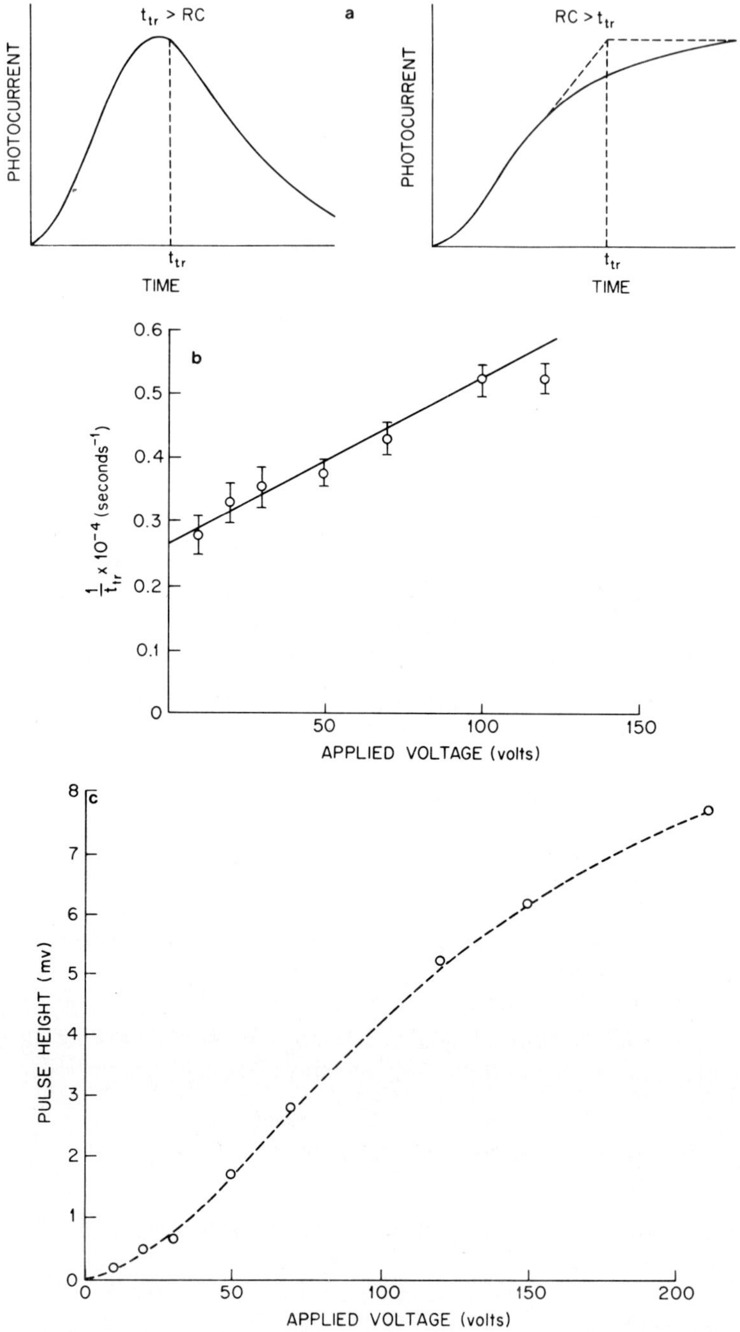

Fig. 3. (a) Typical transient signals for positive holes in a β-carotene glass at room temperature; (b) dependence of $1/t_{tr}$ versus E for positive holes (thickness, 2.5×10^{-5} m; drift mobility, 1.6×10^{-8} m^2 V^{-1} sec^{-1}); (c) dependence of pulse height versus E.

Most measurements of the ratio of mobilities μ_+/μ_- for organic solids show the value to be close to unity, so that we may write

$$R = n_+/n_- \simeq 10^5 \tag{6}$$

Another important property when one is considering photoconduction is the gain factor defined as the ratio τ/t_{tr}, where τ is the lifetime of the charge carrier and t_{tr} the transit time of the charge carrier between two electrodes separated by a thickness of material d. This is the factor by which the number of charge carriers traversing the circuit is larger than the number of photons absorbed and is synonymous with the quantum efficiency of generation, provided that all charge carriers generated by light absorption contribute to the current and are not trapped or suffer annihilation. As mentioned previously (Section III) it is likely that there is strong trapping in β-carotene glasses, and so any value for the quantum efficiency derived from experimental observations will be a lower limit.

In Fig. 3c we see that the plot of the photocurrent against the applied field tends to saturate at higher fields. This means that we are approaching the condition in which all the carriers generated by the light are being drawn across the sample by the applied field, i.e., $\tau \geqslant t_{tr}$. Taking 300 V as the voltage at which this commences in a β-carotene film 2×10^{-5} m thick and knowing the photon flux incident on the cell yields a value for the quantum efficiency of *ca*. 10^{-2}. Similarly, from typical transit time values of 50 μsec we must have lifetime values of the same magnitude to yield gain factors of unity. Although we have as yet no reliable values for τ in β-carotene, a study of anthracene yields free hole lifetimes of between 100 and 1000 μsec. Consequently, we see that in β-carotene the gain factor probably falls appreciably short of the value of 10^4 often found for good inorganic photoconductors.

D. Photovoltaic Effect in β-Carotene

In many sandwich cells of organic solids a photovoltaic voltage appears when the cell is nonuniformly excited, e.g., by strongly absorbed light (*38*). Positive charges diffuse from the illuminated layer toward the unilluminated part of the crystal, and the illuminated electrode becomes negative. Ultimately diffusion of the charge carriers is opposed by a counterfield. Kallmann and Pope (*38*) gave a general expression for the photovoltaic effect in terms of n_+, n_-, the number of positive and negative charge carriers of mobility μ_+ and μ_-, respectively, in a material between electrodes A and B as

$$V_{\text{ph}} = -\frac{kT}{q} \int \frac{n_+^B n_-^B}{n_+^A n_-^A} \quad \frac{(\mu_+ \, dn)^+ - (\mu_- \, dn)^-}{(\mu_+ n_+) - (\mu_- n_-)} \tag{7}$$

In most cases as in β-carotene only one carrier contributes to the conductivity, and the photovoltage is then given by

$$V_{ph} = -\frac{kT}{q} \log \frac{n_+^B}{n_+^A} \tag{8}$$

An important consequence of this expression is that only if $n_+{}^B \neq n_+{}^A$ will a photovoltage develop, and consequently for an observable photovoltage there must be an interaction between the light and the illuminated electrode. If this interaction leads to positive hole injection from the illuminated electrode, then the density $n_+{}^A$ becomes so large that as many positive holes move back to this electrode as are emitted by the impinging excitons, and $n_+{}^A$ becomes larger than $n_+{}^B$, giving rise to a positive photovoltage at B. A photovoltage could also develop if charge carriers were generated within the bulk of the material. In order to get a difference between $n_+{}^A$ and $n_+{}^B$ it must be assumed that positive and negative charges contribute to the conductivity. Since only a small negative current is observed in β-carotene, the photovoltage probably arises due to electrode effects. In β-carotene glasses the photovoltaic action spectrum closely resembles the photoconductivity action spectrum (see Fig. 2) and reaches a maximum in the cis peak region. The photovoltages developed in such samples are of the order of 200 mV.

E. Involvement of the Triplet State

Whereas there is little direct evidence (*65*) for the existence of and consequently the localization of the triplet state in β-carotene, triplet excitons have been considered to be essential to the understanding of the photoconduction of this material. [The triplet state of the related C_{40} carotenoid lycopene has been located at 3.1 eV (*66*).] The fact that the triplet state is an important kinetic intermediary step in the process leading to the generation of charge carriers in other organic solids was clearly demonstrated for the case of anthracene by Sharp and Schneider (*39*) and for copper phthalocyanine by Day and Williams (*40*). In both these cases the photoconduction action spectrum was found to follow closely the ground state–triplet (S_0–T_1) absorption spectrum and was linearly dependent on the exciting light intensity. This result is independent of whether triplet excitons are generated directly by absorption or indirectly by absorption and then intersystem crossing from a singlet state. Contrary to the anthracene case (*41*) phosphorescence has not been directly observed in β-carotene, but Rosenberg interprets the photoconductivity results on the basis of the energy diagram shown in Fig. 4.* The position of a triplet state T is considered crucial to any discussion of the photoconductivity, and for β-carotene one has been taken from the value of the activation energy for semiconduction in the material (see Section VI). (Some support for this assumption comes from the fact that the activation energy for semiconductivity in lycopene corresponds to the triplet state energy.)

*The observation of delayed emission from β-carotene in a rigid matrix at low temperatures by Cherry *et al.* (*65*) locates a triplet level at *ca.* 2.1 eV, provided that the emission is, indeed, from β-carotene and not some unidentified impurity. The triplet level found from semiconductivity data would then correspond to T_2.

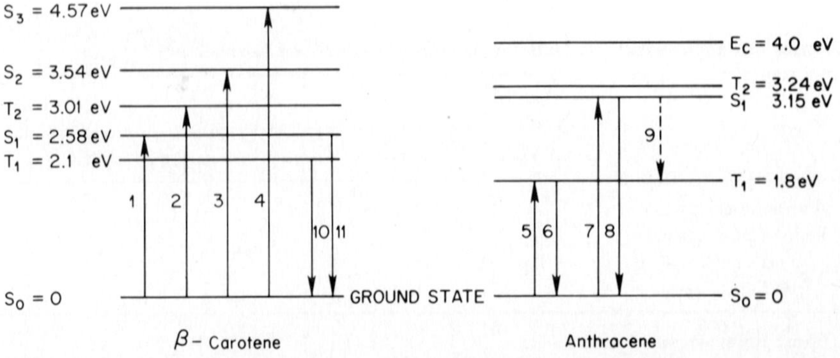

Fig. 4. Schematic energy level diagrams for β-carotene and anthracene. The processes indicated include the direct absorption from the ground state to the first $S_0 \rightarrow S_1$ (1), second $S_0 \rightarrow S_2$ (3), and third $S_0 \rightarrow S_3$ (4) excited states of β-carotene. The corresponding absorptions to the first singlet $S_0 \rightarrow S_1$ (7) and first triplet $S_0 \rightarrow T_1$ (5) states of anthracene are also shown. The $S_0 \rightarrow T_2$ (2) in β-carotene corresponds to twice the activation for semiconductivity. Whereas the fluorescence of anthracene (8) and its phosphorescence (6) are well established, these processes in β-carotene (11 and 10, respectively) have only been reported once (see text), and the origin of the delayed emission is in some doubt. Intersystem crossing, e.g., $S_1 \rightarrow T_1$ (9), occurs in anthracene.

This triplet level falls above the lowest excited singlet level and below the second excited singlet level, and since in the cis isomer the photocurrent peaks in the cis peak (350 nm region), the transition $S_2 \rightarrow T_2$ is more favored than $S_1 \rightarrow T_2$. Consequently, photocurrents in β-carotene samples containing the 15,15′-cis isomer are higher than in the all-trans compound. In the latter, since absorption to the cis peak is small, the only way to populate the T_2 state is by thermal excitation from the ground state. Furthermore, when the cis isomer is present the photoconduction activation energy (see Table 1) is smaller, indicating that intersystem crossing from this level will dominate over those from the lower singlet level and the photoconduction excitation spectrum will follow the absorption spectrum in the cis peak region.

V. PHOTOCONDUCTIVITY IN CAROTENOIDS OTHER THAN β-CAROTENE

Table 2 summarizes the information available on the photoconductivity of some other carotenoids. The study of these materials is far less exhaustive than that on β-carotene, and the quantities measured are limited to the absolute value of the photoconductivity and the activation energy (*42*). One striking fact is that all-*trans*-retinal is not included in the table; it exhibits no detectable photoconductivity. Furthermore, the photoconduction activation energy of the 13-*cis*-retinal Schiff's base is depressed to 0.15 eV from 0.25 eV of the corresponding 13-*cis*-retinal. This

TABLE 2

Summary of Photoconductivity Data for Isomers of Retinal and Derivatives[a]

	Retinal				13-*cis*-Retinal-*m*-nitroaniline HCl
	All-*trans*-	13-*cis*-	11-*cis*-	9-*cis*-	
Activation energy (eV)	—	0.25	0.24	—	0.15
Ratio of photocurrent to dark current at 300°K	1	12	10	2	52

[a] After Rosenberg and Harder (*42*).

suggests that attachment of retinal to opsin via a Schiff's base in rhodopsin may similarly result in lowering the photoconduction activation energy from that of the chromophore alone. The lack of further studies of photoconduction in these materials probably stems from their extreme sensitivity to both heat and light.

VI. SEMICONDUCTIVITY OF CAROTENOIDS

A. Dark dc Conductivity and the Effect of Gases

Some insight into the mode of transport of charge carriers through bulk samples can be sometimes obtained by studying the semiconductivity of the material in question. This, of course, assumes that such a property is intrinsic and is, indeed, a reflection of a process occurring in the bulk material. Organic materials in general and the carotenoids in particular are essentially insulators at room temperature, i.e., the band gap is sufficiently large compared to kT that few charge carriers are generated. Those that are created are nevertheless swamped by those originating at the electrodes or arising as a result of some impurities present in the parent material. Most of the studies involving carotenoids have been made employing dc methods, and the roles of electrode injection, trapping, and impurities have not been assessed. One phenomenon, however, that is fairly well understood is the effect of the adsorption of gases on the semiconductivity of all-*trans*-β-carotene. This has been studied in some detail by Misra *et al.* (*43*), Cherry and Chapman (*31*), and Williams and Rosenberg (*44*). The presence of a variety of adsorbed gases, such as oxygen, methanol, and sulfur dioxide, increases the semiconduction currents in crystalline powders of β-carotene by several orders of magnitude and decreases the activation energy [see Eq. (2)] for semiconduction. The current increase depends on the amount and nature of the adsorbed gas (see Table 3) and is slightly less for single crystalline specimens than for polycrystalline powder. Misra *et al.* (*43*) note that the adsorbed gases have no effect on the σ_0 term in the conductivity expression;

TABLE 3

Conductivity Increase for All-*trans*-β-carotene Powder in the Presence of Various Adsorbates[a]

Conductivity increase, $\sigma_A/\sigma V$	Gas or vapor adsorbed
0–10	N_2O
	Butanol
	Hexanol
	CO_2
	Ethyl acetate
	Butyl acetate
	Argon
	Helium
	H_2
	N_2
10^2–10^3	Isopropanol
	SO_2
	O_2
10^4–10^6	NH_3, ethanol
	H_2S, methanol
	NO_2, propanol
	Acetone, trimethylene, NO

[a] After Misra *et al.* (43).

i.e., the number of conducting states does not change. A theory is proposed to account for the change in activation energy in terms of a change in the effective, local dielectric constant κ'. In many cases the conductivity on adsorption of a quantity of gas m is given by the expression

$$\sigma_A(m) = \sigma_V \exp(\alpha_m) \tag{9}$$

where σ_V is the conductivity in the virgin state, and α is a constant.

Combining this expression with that for a semiconductor [see Eq. (2)],

$$\sigma_V(T) = \sigma_0 \exp(-E_G/2kT) \tag{10}$$

gives

$$\sigma_A(m,T) = \sigma_0 \exp\left(-\frac{E_G - \gamma m}{2kT}\right) = \sigma_0 \exp\left(-\frac{E_A}{2kT}\right) \tag{11a}$$

where

$$\gamma = 2kT\alpha \text{ and } E_A = E_G - \gamma m \tag{11b}$$

The E_G and E_A are written in terms of the static dielectric constant κ and κ^1 before and after gas adsorption, respectively, as

$$E_G = I_g - A_g - \frac{q^2}{R}\left(1 - \frac{1}{\kappa}\right) \tag{12}$$

$$E_A = I_g - A_g - \frac{q^2}{R}\left(1 - \frac{1}{\kappa'}\right) \tag{13}$$

where I_g and A_g are the ionization potential and electron affinity of the molecules, respectively, R is the radius of the polarization field, and q is the electronic charge.

Subtraction of Eq. (13) from Eq. (12) and combining with Eq. (11b) yields

$$\gamma m = \frac{q^2}{R}\left[\frac{1}{\kappa} - (1 - \kappa')\right] \tag{14a}$$

or

$$\kappa^1 = \kappa/(1 - \kappa X m) \tag{14b}$$

where

$$X = R\gamma/q^2$$

Equation (11a) then becomes

$$\sigma_A(T\kappa') = \sigma_0 \exp\left(-\frac{E_G}{2kT}\right)\exp\left[\frac{q^2}{2kTR}\left(\frac{1}{\kappa} - \frac{1}{\kappa'}\right)\right] \tag{15}$$

Rosenberg concluded that the changes in the effective dielectric constant [possibly due to the formation of donor–acceptor complexes (*20*)] and the resulting changes in the activation energy can satisfactorily account for the electrical behavior of all-*trans*-β-carotene crystals upon adsorption of various gases. These results prompted Rosenberg *et al.* (*45*) to propose a semiconduction theory of olfaction. Up to that time no satisfactory theory of olfaction had been put forward that could adequately explain the transduction part of the process, i.e., how the attachment of a gaseous molecule to the receptor sites could trigger off an impulse leading to odor perception. Carotenoid pigments, both in the free state and bound to proteins, have been identified in the olfactory region (*46–50*), although their precise location, distribution, and function are not accurately known. The semiconductive theory of olfaction is based on the formation of a weakly bound donor–acceptor type of complex between the adsorbed gas molecules and the carotenoid pigment. Williams and Rosenberg (*51*) have constructed an olfactometer incorporating a variety of carotenoids and operated on the basis of the semiconductive theory. It has been shown that the pattern of responses from individual carotenoids is in excellent agreement with existing electrophysiological data when aliphatic, homologous series of alcohols (Fig. 5), acetates (Fig. 6), and acids (Fig. 7) are used as stimulants. Furthermore, the behavior of certain isomeric compounds is consistent with such data; e.g., *n*-propanol is far more effective than isopropanol in stimulating a response. The current responses from a group of carotenoids adequately characterize

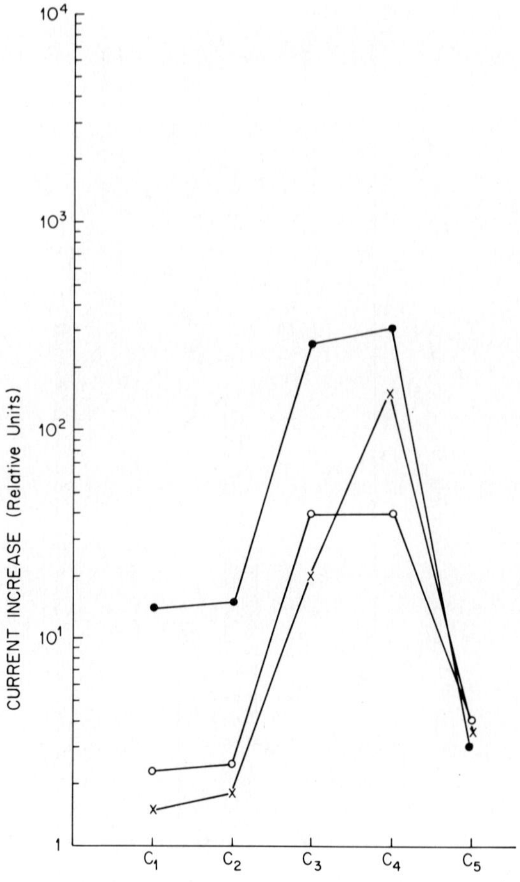

Fig. 5. Current responses obtained from three conductivity cells containing retinal (x), retinol (●), and all-*trans*-β-carotene (○) after exposure to aliphatic alcohols (C_1–C_5) at a partial pressure of 10 mm.

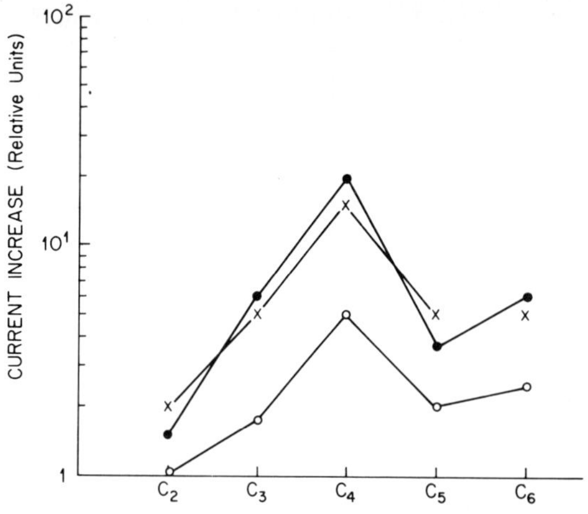

Fig. 6. Current responses obtained from three conductivity cells containing retinal (x), retinol (●), and all-*trans*-β-carotene (○) after exposure to aliphatic acetates (C_2–C_6) at a partial pressure of 10 mm.

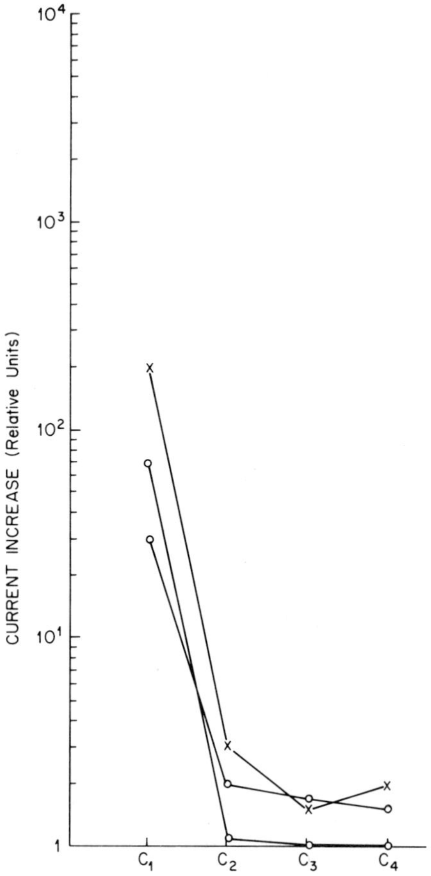

Fig. 7. Current responses obtained from three conductivity cells containing retinal (x), retinol (•) and all-*trans*-β-carotene (○) after exposure to carboxylic acids (C_1–C_4) at a partial pressure of 1 mm.

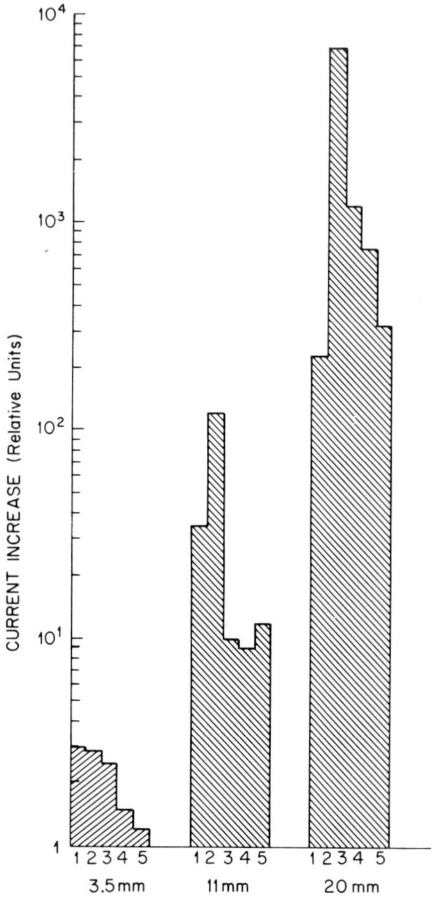

Fig. 8. Histograms illustrating the effects of an increasing *n*-propanol concentration on the current responses from five different carotenoids (1, retinal; 2, retinol; 3, norbixin; 4, retinoic acid; 5, all-*trans*-β-carotene). Note how the pattern of responses changes.

the quality of the odors, and it has been shown how odor quality can change with changing concentration of both *n*-propanol and methyl acetate (see Figs. 8 and 9, respectively). Finally, by employing the responses from five different carotenoids, namely, retinal, retinol, norbixin, retinoic acid, and all-*trans*-β-carotene, it is possible to identify any one of the given odors, provided that the vapor has been previously exposed to the carotenoids in the olfactometer at the same concentration.

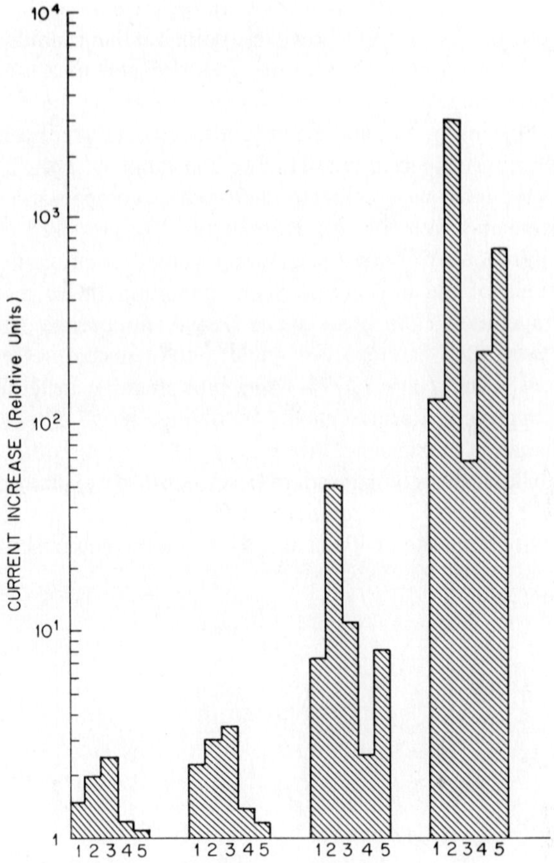

Fig. 9. Histograms illustrating the effect of an increasing methyl acetate concentration on the current responses from five different carotenoids (1, retinal; 2, retinol; 3, norbixin; 4, retinoic acid; 5, all-*trans*-β-carotene). Note how the pattern of responses changes.

B. Compensation Effect

The semiconduction characteristics of crystals and glasses of various cis and trans isomers of retinal (*42*) display a compensation effect (*52*) in that higher activation energies have correspondingly larger preexponential factors. These data have become part of a more extensive body of information that supports the applicability of an expression of the form

$$\sigma(T) = \sigma_0' \exp(E_0/2kT_0) \exp(-E_0/2kT) \tag{16}$$

to describe the semiconductivity of a range of organic and biological substances

(*52–54*). Here, σ_0' and T_0 are constants that are believed to be characteristic of the material. [Johnston and Lyons (*54*), however, point out that a similar dependence of σ_0 on E_0 could in many cases arise from experimental measurements within a narrow temperature range.] The information available for the retinals is plotted according to Eq. (10) in Fig. 10, and the dependence of the preexponential factor on the activation energy is shown in Fig. 11. The constants σ_0' and T_0 are also given.

There have been numerous attempts to interpret this compensation effect in terms of physical parameters. Kemeny and Rosenberg (*55*) proposed a mechanism of charge carrier migration involving the motion of small polarons in thermally activated energy levels of the molecules. Such tunnelling (band motion) gives the characteristic temperature T_0 in terms of the Debye temperature. Since the data on the semiconductivity of β-carotene also yield a compensation effect (with $\sigma_0' = 1 \times 10^{-3}\ \Omega^{-1}\ \text{cm}^{-1}$ and $T_0 \simeq 2900°\text{K}$), the interpretation in terms of the small-polaron theory suggests a transport model involving charge carrier motion within narrow energy bands in this material. Eley *et al.* (*53*), on the other hand, consider several ways in which this compensation effect could be explained: (a) electrode injection of charge carriers, (b) a temperature dependence of the activation energy for semiconductivity associated with an adsorbed film on the surface of the material,

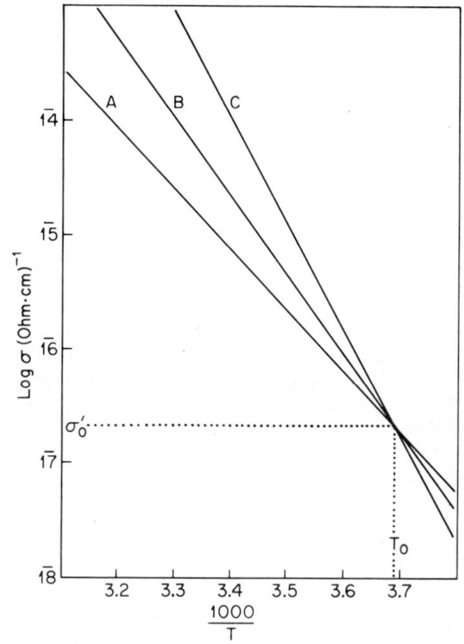

Fig. 10. Semiconductivity data for retinals and retinal complexes. (A) Protonated 13-*cis*-retinal *m*-nitroaniline Schiff base crystals, (b) 11-*cis*-retinal crystals, (c) 9-*cis*-retinal glass; $T_0 = 270°\text{K}$, $\sigma_0' = 2.2 \times 10^{-17}\ (\Omega\ \text{cm})^{-1}$. [After Rosenberg *et al.* (52).]

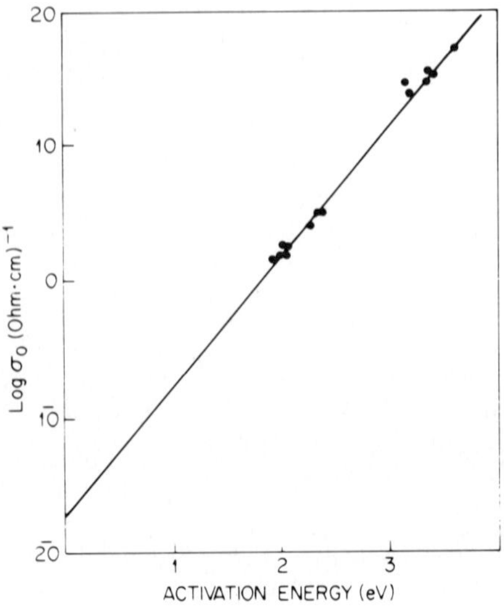

Fig. 11. Plot of log σ_o versus activation energy for the retinals. [After Rosenberg *et al.* (52).]

(c) bulk generation of charge carriers followed by hopping over or tunnelling through intermolecular barriers, and (d) narrow band theory. Roberts and Polanco (*19*) consider two alternative mechanisms, one that involves band motion and one that involves the hopping of the charge carriers between localized levels. Thus, we see that the occurrence of a compensation effect cannot unequivocally identify the conduction mechanism in one given material.

C. ac Studies

The possibility that conduction may occur by means of carrier hopping between localized states can be investigated by means of ac techniques (*56*). Such measurements have been performed on "dry" β-carotene samples by Pethig (*57*) and on β-carotene in the presence of a variety of adsorbed species by Williams and Leffler (*58*). Typical results are shown in Fig. 12. The functional dependence of the conductivity G on frequency ω is of the form

$$G(\omega) = A\omega^s \tag{17}$$

with s having values of 0.89 and 0.94 for the case of all-*trans*-β-carotene in the presence of methanol and ethanol, respectively, and 0.88 in the dry state at ca. 298°K. These values of s are similar to that predicted theoretically by Pollak and Geballe (*18*) for conduction occurring as a result of a thermally activated hopping process between pairs of localized states for which the difference in energy levels is of the order of kT. In such cases s is close to 0.8 with A complex. Thus, we see that in polycrystalline powders of all-*trans*-β-carotene, in which conduction along sur-

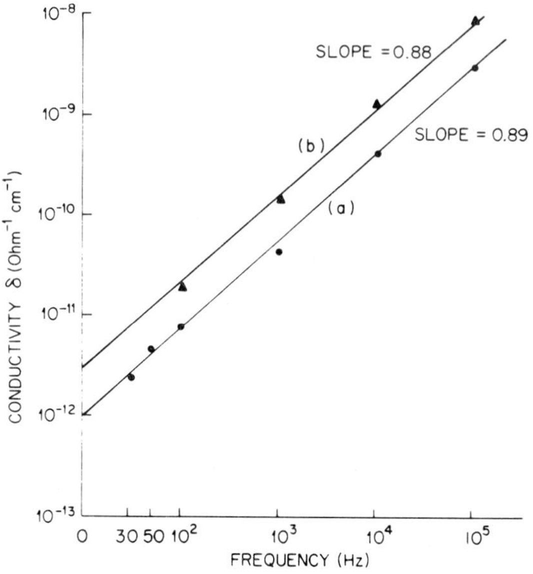

Fig. 12. The ac conductivity versus frequency dependence for (a) "dry" all-*trans*-β-carotene powder, (b) all-*trans*-β-carotene powder in the presence of 3.5% adsorbed methanol.

faces probably dominates, charge migration is an activated process. Consequently the dc and ac results, depending on which approach one takes to explain the compensation effect, can yield different conduction mechanisms, and further work must be done to resolve this problem. Clearly, the observation of a thermally activated drift mobility would support the ac measurements in favor of a hopping process.

VII. CAROTENOIDS IN MODEL BILAYER SYSTEMS

In recent years experimental bimolecular (bilayer) lipid membranes have been studied extensively as a model for a variety of biological membranes (*59–61*). Reviews (*62*) indicating the relevance of such model systems to the understanding of biological membrane structure and function have been published, and it has been shown that certain pigments, notably carotenoids and chlorophylls (*60,63,64*) can be incorporated into such bilayers. In particular it has been shown that the "pigmented membranes" exhibit photo-emf's upon irradiation with light in the chlorophyll absorption bands and photoconductivity with an externally applied electric field. Kobamoto and Tien (*64*) studied the photoresponses arising from the presence of all-*trans*-retinal in membranes of lecithin and oxidized cholesterol. The photoresponse consists of a fast and slow component—the slw component being opposite

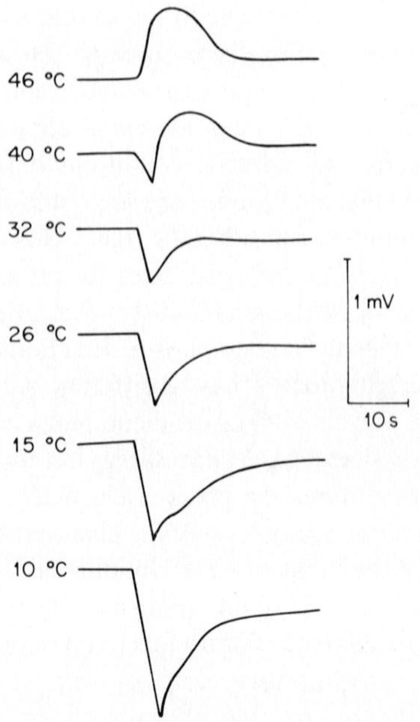

Fig. 13. Time dependence of the photoresponses of all-*trans*-retinal bimolecular lipid membrane as a function of temperature. The duration of the illumination was 1 sec. [After Kobamoto and Tien (64).]

in sign to the fast component—and has an action spectrum that follows closely the absorption spectrum of the all-*trans*-retinal. It was further shown that these membranes could generate, upon illumination, temperature-sensitive biphasic photopotentials (Fig. 13), which have been likened to the early receptor potentials of visual receptors. However, these studies tend to employ high light intensities, and the possibility of appreciable heating effects in the extremely thin bilayers (ca. 100 A°) cannot be totally ruled out. Furthermore, since a central dogma of photosynthesis and vision is that only processes of high quantum efficiency are relevant it appears that, as a result of the small magnitude of the photoresponses even with high light intensities, bilayers incorporating these pigments are not completely satisfactory model systems in which to study these phenomena.

VIII. CAROTENOIDS AS PHOTODETECTORS AND COLOR DISCRIMINATORS

As mentioned earlier (Section IV,C) the quantum efficiency for the generation of charge carriers by light in β-carotene glasses is quite low, and it is unlikely, therefore, that this system can be used to develop ideas that can be directly related to vision and photosynthesis. However, in studying such systems important insights have been gained which may be of value to the development of β-carotene glasses as efficient photodetectors and color discriminators. (However, it must be stressed at the outset that the long-term photodegradation and thermal degradation of carotenoids in general do not make them attractive commercial propositions.)

It has been shown that under specific conditions a β-carotene glass cell will develop a set of responses that are unique for each color of incident light regardless of the wavelength distribution constituting the light. This method of color detection is dependent on the fact that the two photoelectric effects—photoconductive and photovoltaic effects—can occur simultaneously in a single cell and that the two processes generate currents that flow in opposite directions. Under the conditions of opposing current flow each process has a different action spectrum. For more specific color differences one cell is not sufficient, and a combination of three cells, including one for intensity detection, is necessary. Let us consider Fig. 14. In Fig. 14a the cell is illuminated through the positive electrode, and light that is strongly absorbed (blue-green spectral region) causes a photocurrent due to positive holes flowing in the direction of the incident light. Simultaneously, a photovoltaic current flows in the same direction, and if the applied voltage is of the same order of magnitude as the photovoltage (ca. 100 mV) then a monophasic current having two

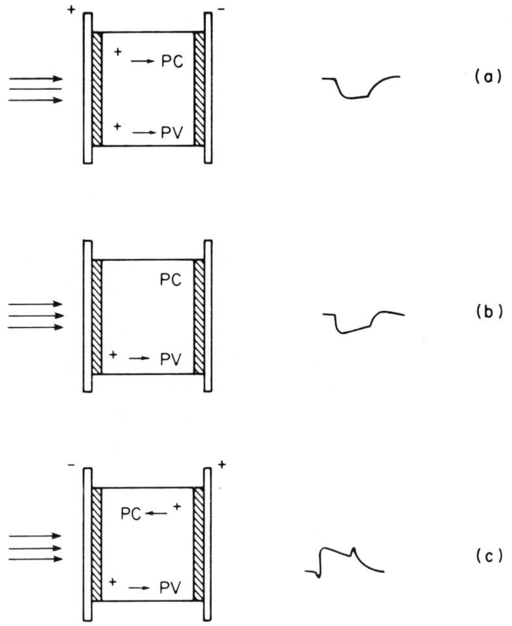

Fig. 14. Schematic diagram illustrating the two photoprocesses: photoconduction and photovoltaic effects. These can be superimposed to give a monophasic current curve consisting of both currents adding in phase (a) or a diphasic current curve when the two currents are in opposition (c). The case of no applied voltage is shown in (b). Note the "on" and "off" peaks in (c). [After Rosenberg (30).]

almost equal components will flow. The resultant transient when the light is switched on and off is also shown in Fig. 14. Light in the red region of the spectrum will not generate either of these two currents (see Section IV,A and Fig. 2). In Fig. 14b we show the case when no voltage is applied to the cell. Here, only the current due to the photovoltaic effect flows when short-wavelength light (blue-green region) impinges on the cell. The transient response is again shown. Figure 14c shows the case for irradiation through the negative electrode. In this case the photocurrents due to short-wavelength (blue-green) light are negligible, but this light generates a positive hole current in the direction of the light by the photovoltaic effect. Long-wavelength light does, however, generate a positive hole current when absorbed near the back electrode, but now the current flows so as to oppose that due to the photovoltaic effect. The shape of these two currents is now diphasic with two components. (It is found that the photovoltaic contribution to the current increases and decreases faster than the photocurrents.) Consequently, if short wavelengths predominate in the light, the steady state will be negative; if long wavelengths predominate in the light, the steady state will be positive; and if the two steady-state currents are of equal magnitude due to the balance of long and short wavelengths, the net current flow will be zero; only "on" and "off" peaks will occur. What then

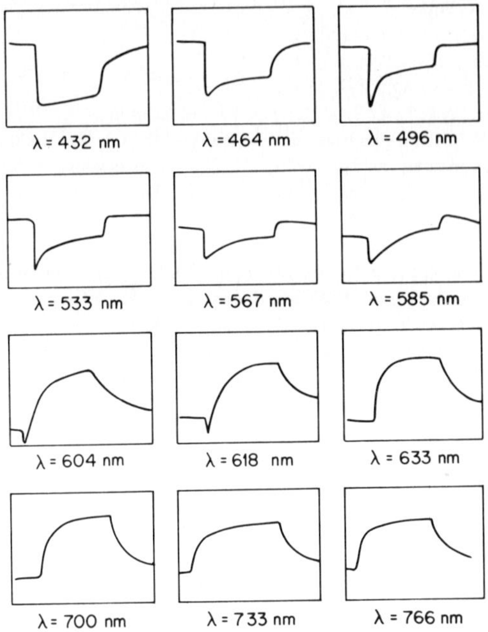

Fig. 15. The response curves for monochromatic light of different wavelengths in a β-carotene sandwich cell irradiated through the negative electrode. The responses are negative (photovoltaic effect) at short wavelengths, are positive (photoconduction) for long wavelengths, and display "on" and "off" peaks only in the green spectral region. [After Rosenberg *et al.* (*52*).]

happens when monochromatic radiation in any region of the spectrum is used? The answer is provided by Rosenberg's data, which are summarized in Fig. 15. Here, the cell has been irradiated with monochromatic light at different peak wavelengths from 432 to 766 nm. The response for green light ($\lambda = 585$ nm) is as predicted: a net current of zero but with "on" and "off" peaks superimposed. Rosenberg has also shown that the response to broad-band light is the same as that to monochromatic light and further demonstrated that three cells consisting of carotenoids having well-defined action spectra are capable of showing color matching and complete color discrimination.

REFERENCES

1. Szent-Györgyi, A. (1941). *Science* **93,** 609.
2. Szent-Györgyi, A. (1941). *Nature (London)* **148,** 157.
3. Gergely, J. (1961). *In* "Symposium on Electrical Conductivity in Organic Solids" (H. Kallmann and M. Silver, eds.), p. 369. Wiley, New York.
4. Szent-Györgyi, A. (1959). *Discuss. Faraday Soc.* **27,** 111.
5. Terenin, A. (1961). See reference 3, p. 39.
6. Rosenberg, B. (1961). See reference 3, p. 291.
7. Fernandez-Moran, H. (1959). *In* "Biophysical Science—A Study Program" (J. L. Oncley, ed.), p. 319. Wiley, New York.
8. Rosenberg, B. (1962). *Photochem. Photobiol.* **1,** 117.
9. Wald, G. (1958). *Exp. Cell Res., Suppl.* **5,** 398.
10. Kasha, M. (1959). See reference 7, p. 162.
11. Gutmann, F., and Lyons, L. E. (1967). "Organic Semiconductors." Wiley, New York.
12. Kommandeur, J. (1965). *In* "Physics and Chemistry of the Organic Solid State" (D. Fox, M. M. Labes, and A. Weissberger, eds.), Vol. 2, p. 1, Wiley, (Interscience), New York.
13. Garret, C. G. B. (1959). *In* "Semiconductors" (N. B. Hannay, ed.), p. 89. Reinhold, New York.
14. Pope, M. (1967). *Sci. Am.* **87,** 38.
15. Pope, M. (1968). *Mol. Cryst. Liq. Cryst.* **4,** 183.
16. Kittel, C. (1956). "Introduction to Solid State Physics." Wiley, New York.
17. Munn, R. W., and Siebrand, W. (1970). *J. Chem. Phys.* **52,** 47.
18. Pollack, M., and Geballe, T. H. (1961). *Phys. Rev.* **122,** 1742.
19. Roberts, G. F., and Polanco, J. I. (1970). *Phys. Status Solidii* **1,** 409.
20. Misra, T. N., and Rosenberg, B. (1968). *J. Chem. Phys.* **48,** 5734.
21. Chapman, D., Cherry, R. J., and Morrison, A. (1967). *Proc. R. Soc. London, Ser. A* **301,** 173.
22. Rosenberg, B., and Postow, E. (1969). *Ann. N.Y. Acad. Sci.* **158,** 161.
23. Thomas, J. M., and Williams, J. O. (1967). *Trans. Faraday Soc.* **63,** 1720.
24. Eley, D. D. (1967). *J. Polym. Sci. Part C* **17,** 73.
25. Bube, R. H. (1960). "Photoconductivity of Solids." Wiley, New York.
26. Joffe, A. F. (1959). *Sov. Phys.—Solid State* **1,** 139.
27. Eley, D. D. (1968). *In* "Organic Semiconducting Polymers" (J. E. Keaton, ed.), p. 259. Dekker, New York.
28. Rose, A. (1955). *Phys. Rev.* **97,** 1538.
29. Lyons, L. E. (1955). *J. Chem. Phys.* **23,** 220.
30. Rosenberg, B. (1966) *Adv. Radiat. Biol.* **2,** 193.
31. Cherry, R. J., and Chapman, D. (1967). *Nature (London)* **215,** 956.
32. Rosenberg, B. (1961). *J. Chem. Phys.* **34,** 63.

33. Almeleh, N., and Harrison, S. E. (1965). *J. Phys. Chem. Solids* **26,** 1571.
34. Edwards, J. B. (1971). M.S. Thesis, Univ. of Wales, Aberystwyth.
35. Kepler, R. C. (1960). *Phys. Rev.* **119,** 1226.
36. Spear, W. E. (1969). *J. Non-Cryst. Solids* **1,** 197.
37. Glarum, S. H. (1963). *J. Phys. Chem. Solids* **24,** 1577.
38. Kallmann, H., and Pope, M. (1961). See reference 3, p. 1.
39. Sharp, J. H., and Schneider, W. G. (1964). *J. Chem. Phys.* **41,** 3657.
40. Day, P., and Williams, R. J. P. (1965). *J. Chem. Phys.* **42,** 4049.
41. Williams, D. F. (1967). *J. Chem. Phys.* **47,** 344.
42. Rosenberg, B., and Harder, H. C. (1967). *Photochem. Photobiol.* **6,** 629.
43. Misra, T. N., Rosenberg, B., and Switzer, R. (1968). *J. Chem. Phys.* **48,** 2096.
44. Williams, J. O., and Rosenberg, B. (1968). Unpublished observations.
45. Rosenberg, B., Misra T. N., and Switzer, R. (1968). *Nature (London)* **217,** 423.
46. Milas, N. A., Postman, W. M., and Heggie, R. (1939). *J. Am. Chem. Soc.* **61,** 1929.
47. Briggs, M. H., and Duncan, R. B. (1961). *Nature (London)* **191,** 1310.
48. Gerebtzoff, M. A., and Shkapenko, A. (1952). *C. R. Assoc. Anat.* **68,** 511.
49. Briggs, M. H., and Duncan, R. B. (1962). *Nature (London)* **195,** 1313.
50. Kurihara, K. (1967). *Biochim. Biophys. Acta* **148,** 328.
51. Williams, J. O., and Rosenberg, B. (1969). *Umsch. Wiss. Tech.* **11,** 348.
52. Rosenberg, B., Bhowmik, B. B., Harder, H. C., and Postow, E. (1968). *J. Chem. Phys.* **49,** 4108.
53. Eley, D. D., Fawcett, A. S., and Willis, M. R. (1968). *Trans. Faraday. Soc.* **64,** 1513.
54. Johnston, G. R., and Lyons, L. E. (1970). *Phys. Status Solidii* **37,** 43.
55. Kemeny, G., and Rosenberg, B. (1970). *J. Chem. Phys.* **52,** 4151.
56. Anderson, J. C. (1972). *J. Vac. Sci. Technol.* **9,** 1.
57. Pethig, R. (1972). *Symp. Chem. Org. Solid State, 3rd, Glasgow.*
58. Williams, J. O., and Leffler, R. (1968). Unpublished observations.
59. Mueller, P., Rudin, D. O., Tien, H. T., and Wescott, W. C. (1964). *Recent Prog. Surf. Sci.* **1,** 379.
60. Tien, H. T. (1968). *Nature (London)* **219,** 272.
61. Hesketh, T. R. (1969). *Nature (London)* **224,** 1026.
62. Tien, H. T., and Diana, A. L. (1968). *Chem. Phys. Lipids* **2,** 55.
63. Tien, H. T., and Kobamoto, N. (1969). *Nature (London)* **224,** 1107.
64. Kobamoto, N., and Tien, H. T. (1972). *Biochim. Biophys. Acta* **266,** 56.
65. Cherry, R. J., Chapman, D., and Langelaar, J. (1968). *Trans. Faraday Soc.* **64,** 1.
66. Lewis, G. N., and Kasha, M. (1944). *J. Am. Chem. Soc.* **66,** 2100.

10

CAROTENOID ANALYTICAL METHODS

E. De Ritter and A. E. Purcell

I. GENERAL METHODS FOR DETERMINATION OF CAROTENOIDS IN NATURAL PRODUCTS

A. Introduction

Carotenoid analysis may appear to be formidable to someone contemplating research in this area. Since there are several thousand references compiled in

CAROTENOIDS AS COLORANTS
AND VITAMIN A PRECURSORS

ISBN 0-12-082850-2

"Chemical Abstracts," a novice is overwhelmed, not knowing where or how to start. However, after extensive literature review, it is found that there are only a few methods with many modifications adapting analytical methods to the convenience of a facility and preference of an individual.

Carotenoid analysis may be performed for a number of reasons. The objective may be merely to define the color of an agricultural or food product. Although such analysis may be simple and provide an objective figure, it may be less descriptive than color comparison. If carotenoid analysis is performed in an effort to explain differences in color, analysis may involve the most detailed isolation and identification procedures.

A major reason for carotenoid analysis is to determine vitamin A value. Although more than 50 carotenoids possess vitamin A activity only 10 are of real significance (*188*). Of these, β-carotene is the most abundant, and it theoretically yields two molecules of Vitamin A per molecule. The others yield only one for one.

Determination of vitamin A value usually requires separation and measurement of the vitamin A-active carotenes, but good estimates can sometimes be obtained by multiplying total carotenoids by factors calculated from previous analysis of the source under investigation. When the best possible estimate is needed, meticulous separation and measurements of isomers are required, Availability of carotenoids from foods and absorption by the intestine are sources of greater variation than isomerization in routine feeding experiments.

Carotenoid analysis to determine composition or in the course of studies on carotenogenesis requires all available methods of isolation and identification.

There is no attempt in this chapter to review all literature concerning carotenoid analytical methods since such a mass compilation of literature would be too cumbersome for easy use and would contribute little except as a historic guide to the development of acceptable methods. Methods recommended in this chapter have been used by the authors and found to be efficient and effective. References pertinent to understanding and applying these methods are given.

Development of analytical methods must be based on the information desired and the properties of the compound of concern. Types of information that may be desired have been discussed.

Properties of carotenoids that must be considered in analyses are the following:

1. Carotenoids are lipids and, therefore, are soluble in other lipids and nonpolar solvents.
2. The colors of carotenoids are due to conjugated double bonds. Most naturally occurring carotenoids possess a trans configuration for all conjugated double bonds. Free halogens, acid, excessive light, and high temperature cause isomerization to mixtures of cis–trans isomers.
3. The double bonds are easily oxidized and must be protected from conditions or compounds that cause oxidation.

4. Spectral absorption by the double-bond system provides information about the nature of the chromophore as an aid in identification of the molecule.
5. Carotenoids are adsorbed onto various oxides and other adsorbents according to the number and arrangement of double bonds and the number of hydroxyl or epoxide groups on the molecule.

Naturally occurring carotenoids are found inside cells and are surrounded by an aqueous protoplasmic medium. Carotenoids of some plants, such as carrots, tomatoes, and sweet potatoes, are deposited on a protein matrix and appear as crystals of the major pigment (*14,186,211,234*). Carotenoids of other plants, such as squash and pansies (*71*), and also of animal tissue occur in colored droplets, suggesting that the pigments are dissolved in other lipids. If the tissues are processed by heat, much of the cell structure is broken and chromoplasts disintegrate, releasing the carotenoids to be dissolved in other lipids. Lipid micelles containing dissolved carotenoids become embedded in carbohydrate or protein. Some cells are not ruptured, thus leaving some of the carotenoids inside intact cells, with the cells embedded in masses of carbohydrate or protein.

B. Materials

1. Solvents

Solvents must be free of oxidizing compounds, acids, and free halogens. Other quality factors are less stringent, permitting the use of many commercial hydrocarbon fractions. Since spectra of carotenoids change according to the solvent in which they are measured, the solvents should be reasonably free of other solvents. One of the authors once obtained a shipment of petroleum ether, bp 30°–60°C, which caused the longest-wavelength maximum of β-carotene to occur 2 nm higher than previous batches obtained from the same source. Presumably, more than reasonable amounts of aromatics were present.

Reasonably pure solvents have negligible optical absorbance in the visible range of the spectrum. If it is necessary to work in the ultraviolet range, solvent purity may become a major problem. Almost every nonpolar solvent has been used for carotene extraction, and yet the ideal solvent has not been found. Carbon disulfide is the best solvent for carotenoids, but its volatility, flammability, toxicity, and degradation to foul-smelling products limit its usefulness. Benzene and toluene are good solvents, but the mobility of electrons contributes to rapid autoxidation of carotenoids in these solvents. Chlorinated hydrocarbons are good solvents and may be used with due caution to toxicity. These solvents should be of a high quality, free of acid and halogen. Peroxide-free ether is an efficient solvent, but its use is inconvenient at times due to its volatility and flammability. Hydrocarbons, e.g., hexane, heptane, isooctane, are not efficient solvents for carotenoids, although other properties favor their use.

Acetone is a better solvent for hydroxycarotenoids, although it is not appreciably better for carotene. Furthermore, small amounts of water picked up by acetone force carotenes out of solution. The lower alcohols are useful in dissolving hydroxycarotenoids but are not adequate solvents for carotenes.

The choice of solvents is somewhat subjective, but commercial, high-purity hexane with a 3°C boiling range has many desirable properties. It is free of oxidizing and isomerizing agents, gives reproducible spectra without further purification, and can be cleaned up for work in the UV region by passage through silica gel (*173*).

2. *Filter Aids*

Filtration is a common part of carotenoid analysis and is frequently complicated by the presence of large hydrophilic molecules, which clog filters. Addition of diatomaceous earth filter aids, eliminates, or at least greatly reduces clogging. Celite preparations marketed by Johns-Manville Co. have been most frequently cited for use, but a number of other commercial filter aids have been used with equally good results. It is conceivable that some filter aids contain sufficient iron or other transition elements to catalyze the oxidation of carotenoids, but specific instances have not been documented.

Filter aids heated to very high temperatures acquire adsorptive capabilities for carotenoids, but these are quickly destroyed in the presence of water. Filter aids are frequently mixed with adsorbents to obtain more rapid and uniform flow through chromatographic columns. When they are used for this purpose, more care is required to obtain uniform flow characteristics and absence of adsorption.

3. *Adsorbents*

A great number of adsorbents have been used to separate carotenoids (*207*). Quality standards for adsorbents are more critical than for other materials used in carotenoid studies. Destruction of carotenoids during chromatography seems inevitable, and the degree of loss is one measure of the usefulness of an adsorbent. Loss with the best adsorbents is small enough to be tolerated. Uniformity from column to column and lot to lot is another criterion of usefulness and implies the need for stability in storage.

Various well-standardized adsorbents are commercially available and discussed in more detail in Section I,D,3,c.

C. Procedures for Selection, Preparation, and Storage of Samples

A sample for analysis must represent the material from which it is taken. Carotenoids are not uniformly distributed in tissues or present in all types of tissues that may be closely associated with each other. The nature and quantity of carotenoids in plant tissue are usually a function of maturity and may vary from one part of a tissue mass to another. When fruits and vegetables are to be analyzed, it must be determined whether the analysis is to represent the whole or just the edible portion. The color that carotenoids impart to tissues provides a visual index to the distribution of carotenoids and may serve as a guide to the selection and description of the samples analyzed.

Some tissues, e.g., alfalfa and soybean leaves, contain lipoxidase enzymes, which cause rapid destruction of carotenoids as soon as the tissue wilts (*144*). Until it has been established that a tissue does not contain lipoxidase, samples should be blanched immediately to inactivate it if it is present. Samples should be weighed before blanching to prevent error arising from the loss of solids.

Any sample that respires will lose dry matter in the form of carbohydrates. Carotenoids will be unaffected; consequently, extensive respiration will cause an increase in the concentration of carotenoids although no carotenoid synthesis occurs (*9*). The effect of respiration is slow compared to the action of lipoxidase, but, if respiring samples are to be held for more than a few hours, respiration should be eliminated by blanching or reduced by low temperature.

Carotenoid biosynthesis can occur after the selection of samples. An outstanding example of this is seen in tomato tissue. Synthesis also is known to occur in carrot slices and may occur in a number of other tissues.

All samples containing carotenoids are subject to autoxidation. Dehydrated samples and oils containing carotenoids are particularly susceptible. After an induction period, loss of carotenoids in oils can be detected in a matter of minutes, and procedures must be adjusted accordingly (*213*). Dehydrated solid samples are less susceptible, but changes can sometimes be measured in a day's time when the samples are stored in the dark. Light increases the rate of oxidation, suggesting the need to protect samples from light as much as possible.

It is not difficult to obtain representative samples from liquid or finely ground products, but tissue samples should be homogenized. Drying may cause losses of carotenoids. Vacuum-drying of blanched samples apparently causes negligible loss if the tissues are immediately ground and assayed after removal from vacuum. Another method that is useful for most tissues is blending a weighed amount of tissue with a weighed aliquot of solvent while the blender is running at low speed to keep the sample stirred (*127*).

Carotenoids suspended in liquids, e.g., fruit juices, sewage samples, or microbial cultures, can be prepared for analysis by filtration through a filter aid such as celite. If the filter has a tendency to clog due to slime and pectins, the addition of methanol to give a 50% concentration will flocculate larger molecules, resulting in better filtration. Centrifugation can also be used but is not as quantitative if carotenoids are finely dispersed.

D. Analysis

Complete carotenoid analysis consists of extraction of the pigments from tissues, gross purification by saponification, separation into class according to the number of hydroxyl groups, isolation by chromatography, identification and measurements by spectral absorption, and confirmation of identification by special tests. One or more of these steps may be eliminated depending on the nature of the system under study and the amount of information required (Table 1, Figs. 1 and 2).

The simplest form of carotenoid analysis consists of estimating pigment content by direct spectral measurement of the tissue. Reflectance measurements of grapefruit (*126,219*) and tomato products (*55,131,230*) have provided reasonable estimates of pigment content. Such measurements are subject to light scattering, nonuniform distribution of pigment, and interference from other pigments. Heat processing causes color changes by destroying the chromoplasts (*175*) as well as changing the state of other cellular constituents. Because of the many variables, direct spectral readings are useful only for a few simple systems.

The carotenoids must be extracted when more than tissue color is desired. If the pigment extract is free of turbidity and interference from noncarotenoid pigments, carotenoid content may be calculated from spectral absorbance of the extract. Extracted pigments from sweet potato, Centennial and Goldrush varieties, contain 85% β-carotene; thus, vitamin A potency may be estimated nearly as well by spectral absorption of the extract as by separation and measurement of isolated pigments. Lycopene and β-carotene, which constitute 90–95% of the carotenoids of tomato, can be measured with good accuracy by determining spectral absorbance at two wavelengths and calculating concentrations by simultaneous equations.

The second step of carotenoid analysis is saponification. Saponification is a gross purification since it removes neutral fats, fatty acids, and many esters, thus simplifying subsequent manipulations. The nonsaponifiable extract is sufficiently free of interfering substances to permit measurement of total carotenoid content by spectral absorbance. If a particular source has been previously analyzed in detail and found to be relatively constant in composition, factors may be computed for estimating nutritional value or the level of the major pigments by spectral absorbance. Saponification is required to obtain clean separation of carotenoids according to hydroxyl groups and sharp chromatographic separations.

1. Extraction

Extraction consists of removing hydrophobic carotenoids from hydrophilic media. Nonpolar solvents are of little use in extracting pigments because of limited penetration through the hydrophilic mass that surrounds much of the pigment. Slightly polar solvents dissolve little carotene when dry and much less when water is present. It may be postulated that carotenoid extraction can be accomplished by reducing the amount of water to a low level and extracting with slightly polar solvents in the presence of nonpolar solvents. The most successful methods of extraction are based on this principle.

A simple method of extracting carotenoids from fresh tissue involves blending or grinding the tissue in sequential portions of acetone, each portion of acetone being removed by decantation or filtration before a new portion is added. Relatively large amounts of acetone are required, and the extracted pigments must be transferred to a hydrophobic solvent before subsequent steps may be taken. The only way to extract carotenoids from some very tough tissues, such as certain molds or shrimp shells, is to grind the tissue with sand under acetone using mortar and pestle. Finely ground dried materials can also be extracted by soaking the sample in acetone at 4°C for 8–16 hr. Two or three soakings with fresh portions are required for quantitative extraction. Carotenoids of uncooked, finely ground, dried tissue can be extracted nearly quantitatively with ether in a Soxhlet or a Goldfinch extractor, but this method cannot be used on cooked tissue since gelatinized starch, sugars, and denatured proteins form an ether-resistant layer around the lipids.

Extraction in blenders is more rapid and complete than by macerating in various solvents, but blenders cause considerable splattering, and care must be exercised to obtain quantitative recovery (*147*). Foaming mixtures have been developed to reduce splattering (*147*). The mixture consists of ethanol, hydrocarbon, and water. Since the samples may contain water, size of sample and volume of mixture may become factors in the effectiveness of this mix. Carotenoids can be extracted from dried products by the addition of ¼ volume of water.

The most universally applicable method of extraction involves precipitation of carbohydrates with methanol followed by extraction with an acetone–hexane mixture as follows. Blend fresh plant material with sufficient water to make a slurry thin enough for smooth blending. Add methanol to make the final concentration of methanol 50–60%. Allow a few minutes for coagulation, and then stir in ca. 1 gm of filter aid per gram of dry tissue. Filter the mixture under vacuum through a filter paper coated with filter aid in a Büchner funnel. As the precipitate mat begins to dry, tamp it to seal cracks that develop as the mat shrinks. When the mat appears to have lost nearly all the liquid phase, scrape it out of the funnel into a beaker and cover it with a 1 : 1 mixture of acetone–hexane, and stir until the mat is broken up. After 4 or 5 min of standing with occasional stirring, filter through the same funnel with vacuum, allowing the filtrates to combine. Tamp the mat as it dries and wash it with small portions of the acetone–hexane mixture until the filtrate is nearly colorless. Allow the mat to dry, scrape it out, and repeat extraction with acetone and hexane until extraction is complete. Usually two extractions are sufficient. If dry material is to be extracted, it should be finely ground and reconstituted with enough water to make a thin slurry. From here on the sample may be handled as a fresh sample.

Whatever method has been used for extraction, transfer the filtrate to a separatory funnel and allow two distinct phases to form. Additional nonpolar solvent (i.e., hexane) may be desirable to increase the nonaqueous phase. If distinct phases are slow in forming, the addition of small amounts of water may hasten separation. Water should be added gently and mixed with gentle swirling, not shaking, since these systems form emulsions at this stage. When two distinct layers form, separate the aqueous layer, which may contain traces of the more oxidized carotenes, and pass it through ether with added water to recover dissolved xanthophylls and emul-

TABLE 1

Suggested Procedures for Routine Carotenoid Analysis of Various Types of Samples

Source	Size of sample	Method	Reference
Animal			
Blood and fluids	2–10 ml	Equal volume ETOH; extract with ether; direct optical absorbance 450 nm × 4 = concentration in extract	*40, 50, 227*
Fats	1 gm	Two milliliters 10% KOH in ethanol per gram fat, reflux, add 4 volumes ether, 4 volumes water; pass bottom phase through ether, combine ether extracts, add 1 volume hexane, wash away alkali, evaporate, take up in hexane for optical absorbance or chromatography (see Fig. 1)	*15, 161, 217*
Feathers Fish scales Invertebrate shells	1–10 gm dry powder	Sequential extraction with acetone while grinding with sand in a mortar; add 1 volume hexane and 1 volume water; pass bottom layer through ether; wash out acetone, saponify with 4 volumes MeOH saturated with KOH; wash out alkali; chromatograph	*73, 77, 129, 137, 242*
Liver and organs		Same procedure as for fats	*198*
Dairy products		Same procedure as for fats	
Fruits and vegetables			
High carotenoid content	1–10 gm		*6*
Apricot		Blend fresh tissue with enough water	*43, 150*
Canteloupe		to form a thin slurry; grind dry	*49, 203*
Carrot		material to fine grains, add water to reconsti-	*56, 110, 112, 155, 233, 238*
Mango		tute to a thin slurry; add methanol to a final	*10, 102, 183*
Palm oil fruits		concentration of 50%, filter with filter aid;	*91, 220*
Papaya		extract mat with acetone–hexane, 1 : 1;	*142, 212, 244*
Peppers		combine extracts; pass aqueous layer through	*35, 36, 37, 105*

Pumpkin		ether; combine ether and hexane extracts; estimate total carotenes by 4 × 450 nm; saponify by shaking with ¼ volume of methanol saturated with KOH; pass methanol layer through ether; combine ether and hexane extracts, wash, dry; chromatograph (see Fig. 2)	*110*
Rose hips			*95*
Squashes			*110, 125*
Sweet potato (orange)			*110, 141, 143, 156, 174*
Tomato			*45, 57, 136, 168, 230*
Watermelon			*152, 168*
Intermediate carotenoid content			
Avocado	25–100 gm fresh weight, 2.5–25 gm dry weight	Same procedure as above except that volumes of solvents must be adjusted to the greater bulk of the sample; concentrate the final extract to obtain absorbance into the best range for the spectrophotometer available	*122*
Banana peel			*228*
Citrus			*46, 116, 146, 216, 219, 226, 245, 247*
Peach			*134*
Pineapple			*76, 151*
Squash (summer)			*89*
Sweet potato (yellow)			*112*
Low carotenoid content			
Apple	100–1000 gm	Same procedure as above with appropriate changes of solvent volumes and concentrations	*72, 171*
Banana (edible portion)			*228*
Cherry			*218*
Corn (sweet)			*30, 100*
Dates			*202*
Pear			*72*
Melon (honeydew)			*203*
Potato			*162, 231*
Strawberry			*72*
Feeds			
Forages	20–50 gm	Same extraction procedure as for fruits and vegetables; must be saponified before spectral absorption is measured	*6, 21, 33, 74, 112, 132, 133, 170*
Green leaves			*208, 209a, 210, 240, 241*
Flowers	Content is so variable that sample	Blend with 25 ml acetone per gram fresh petals; add 2 gm filter aid per gram and	*22, 59, 225, 235, 246, 249*

(*continued*)

TABLE 1—*Continued*

Source	Size of sample	Method	Reference
	size must be determined	filter with vacuum; repeat extraction until the extract is colorless; after 3 extractions, add a few drops of water to the mixture; add ½ volume ether to the combined extracts, mix, add 1 volume water, mix, and let stand until layers form; pass the lower layer through ether; reflux 10 min with ¼ volume methanol saturated with KOH; add ¼ volume water, mix, pass bottom layer through ether; evaporate ether, take up in hexane, partition, and chromatograph	
Grains and seeds	20–100 gm	Grind dry grains to pass 40-mesh screen; refluxing acetone or ethanol will extract some grains completely; suggested method: blend ground grains or wet grains with water to make a thin slurry; precipitate; add methanol to make the final concentration to 50% and proceed as with fruits and vegetables	*17, 18, 67, 119, 153, 154, 166, 167, 196, 200, 236, 243*
Microorganisms			
Mold mycelia and spores	1–25 gm	Mix with an equal volume of sand and grind in a mortar with successive portions of acetone; repeat until acetone extracts are colorless; transfer to hexane, saponify, partition, chromatograph	*5, 34, 38, 121, 135, 140, 176, 222*
Vegetative bacteria, algae, yeasts	10–100 gm	Scrape colonies from solid cultures and suspend in methanol; filter liquid cultures through filter aid and suspend in methanol; proceed as with fruits and vegetables	*63, 66, 92, 94, 97, 98, 106, 107, 124, 159, 163, 189, 191, 192, 194, 214*

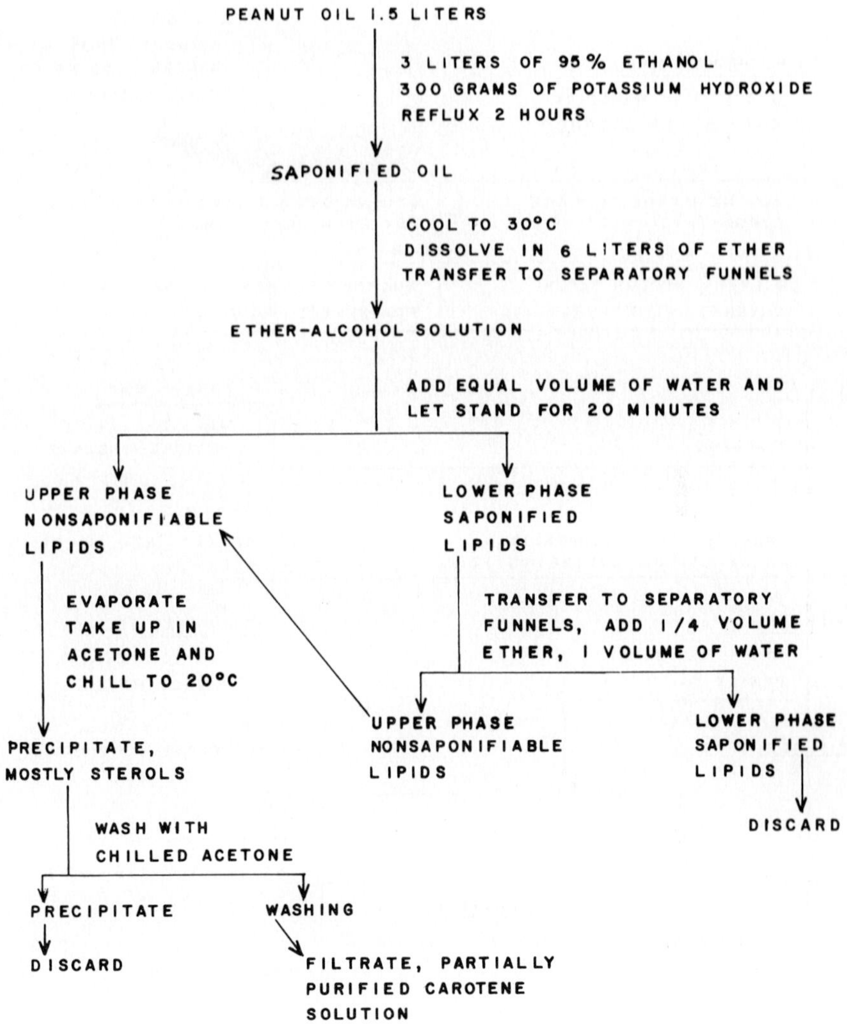

Fig. 1. Scheme for extracting carotenoids from peanut oil.

sified carotenes. Remove polar solvents, methanol and acetone, by gently washing the nonpolar phase two or three times with water. If chlorophylls are not present, a good estimate of total carotenes may be obtained by drying the nonpolar phase with sodium sulfate and determining spectral absorbance.

Animal tissues and body fluids sometimes require special handling. Body fluids can be extracted efficiently by coagulating proteins with addition of an equal volume of ethanol and extracting lipids with ether. Some samples, such as liver, egg, or animal fat, form emulsions so readily that they are most easily extracted by

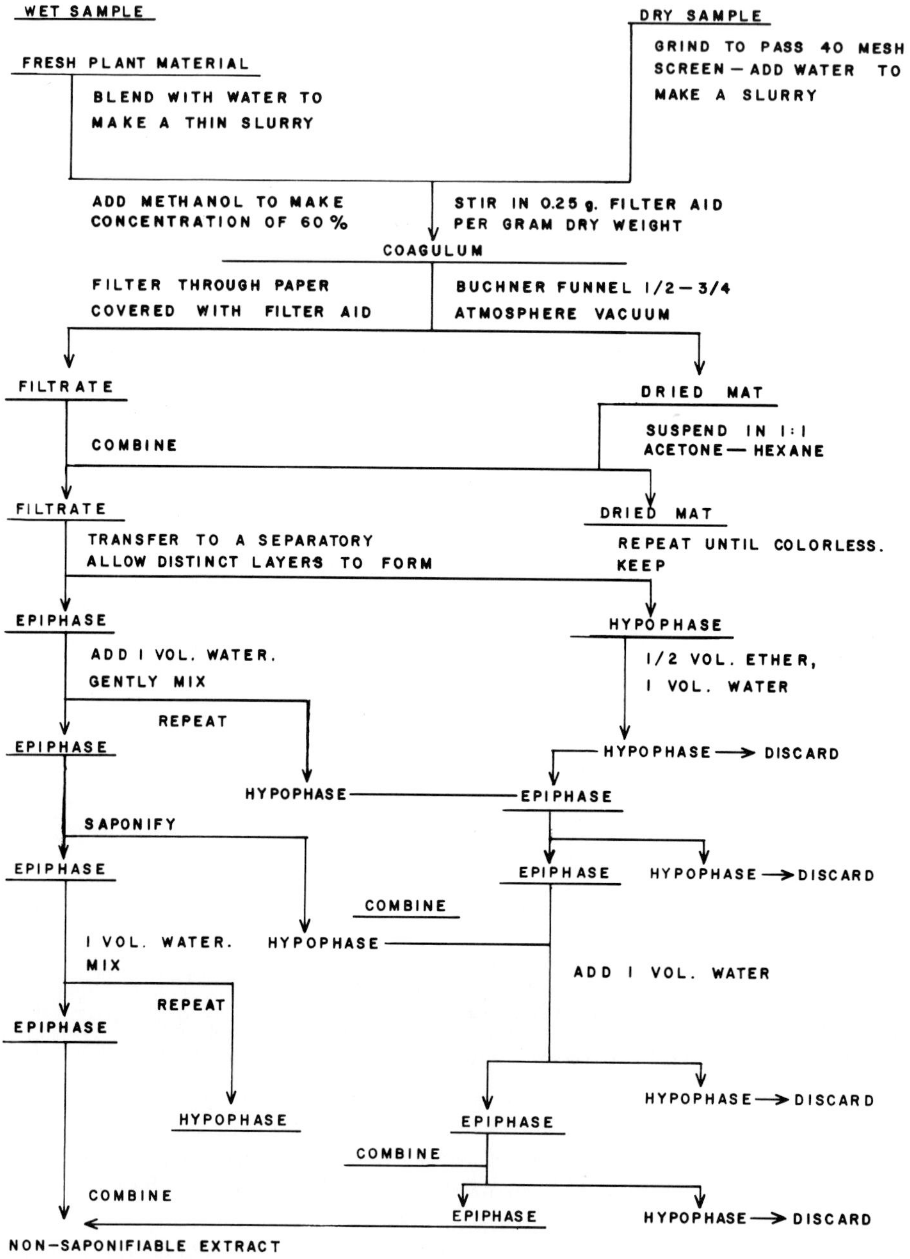

Fig. 2. Scheme for extracting carotenoids from low-lipid samples.

saponification (*7*). Refluxing the sample in methanol saturated with potassium hydroxide will hydrolyze proteins, carbohydrates, and lipids. Nonsaponifiable lipids, including carotenoids, can be easily recovered by gently shaking the hydrolyzate with 2–4 volumes of ether and adding 1 volume of water to cause ready formation of phases. After the ether phase is washed free of alkali, dried with sodium sulfate, and transferred to hexane, the extract is ready for direct spectral measurements or chromatography.

2. *Saponification*

Carotenoid analysis of fats and oils may be simple or very difficult, depending on the amount of carotenoids and other colored material. Carotenoids in fats and oils are in solution and may be measured by direct spectral measurement in the absence of interfering materials. It is possible to apply oil samples to chromatographic columns and isolate carotenoids (*217*). Better separation can be obtained by removing the neutral fat fraction by saponification (*161*). A useful procedure is as follows. Dissolve or disperse the sample in 4–6 volumes of ether, add 1 volume of methanol saturated with potassium hydroxide per gram of oil, and reflux for 30 min. Add 2 volumes of water per volume of methanol to cause formation of an ether epiphase. Pass the water phase through ether and add another volume of water to recover the last traces of carotenoids. Combine the ether extracts and wash three or four times with water. Great care must be taken in all these manipulations because the abundance of soaps creates a large amount of stable emulsions; however, with patience and an occasional pinch of sodium chloride, one may recover the carotenoids quantitatively. It may be necessary to saponify again with sodium ethylate to obtain a preparation that will chromatograph cleanly.

3. *Separation Procedures*

a. Partition Separation. Natural carotenoid mixtures are so complex that separation of minor pigments by chromatography alone would be nearly hopeless without some prior method of separating them into classes. Chromatography on sugar or starch columns can be used to separate the carotenoids according to the number of hydroxyl groups on the molecule, but other functional groups may prevent clean separations. Partitioning between solvents such as hexane–95% methanol gives better separation on the basis of hydroxyl groups. In its simplest form, partitioning can be accomplished by countercurrent washings with a series of separatory funnels. Three transfers are adequate to separate the carotenes from the polyhydroxy compounds. Separation of the hydroxy compounds into the two classes of monohydroxy and polyhydroxy compounds requires at least 12 transfers, suggesting the need for a countercurrent extraction apparatus. Large countercurrent partitioners can be used to separate carotenoids into at least five groups and provide measurements of partition coefficients (*41,42,79*). Partition chromatography has been used to separate carotenoids according to the number of hydroxyl groups. Silica gel–methanol columns give clean separation and high recovery (*172*). They work best when the hydroxy fractions are small compared to the hydrocarbon fraction.

Carotenoids can be separated into fractions by "reverse-phase" partition chromatography (*3,101*). Powdered polyethylene is the stationary phase and acetone in water is the mobile phase. More polar compounds, such as chlorophylls and polyhydroxycarotenoids, move more rapidly than less polar compounds. Increasing the amount of acetone in the water–acetone mixture increases the mobility of nonpolar carotenoids. This system works well when the hydrocarbon fraction is small compared to the hydroxy fraction.

Saponification. The simplest and most widely used method of saponification consists of shaking a hydrocarbon solution of extracted materials with ¼ volume of methanol saturated with potassium hydroxide for 2–3 min and allowing it to sit for 20–30 min. Distinct layers form. The upper layer or epiphase contains most of the carotenoids. The bottom layer, hypophase, contains some hydroxycarotenoids and some carotenes emulsified by soaps. These can be recovered by passing the hypophase into ether and adding 2 volumes of water. All carotenoids except those with acidic groups can be quantitatively recovered in the ether layer that forms. Excess alkali can be removed by washing several times with water. This saponification removes chlorophyll and hydrolyzes most of the neutral fats and lower esters of hydroxycarotenoids. Some higher esters of the carotenoids such as physalien (zeaxanthin dipalmitate), various waxes, and sterol esters may not be hydrolyzed. Chromatography usually proceeds well after this degree of saponification.

If more strenuous saponification is desired, it can be accomplished by transferring the extract to ether before adding methanolic potassium hydroxide to form a single phase, which may be refluxed for 10–20 min. This method hydrolyzes physalien. The addition of water causes the ether to form an epiphase with most of the carotenoids. Small amounts remaining in the ethanol–water phase can be recovered by passing the phase through ether again.

If extreme purity is required, for example, in radiotracer studies, it is necessary to use even stronger saponification. This can be done by refluxing an ether solution with sodium ethylate or lithium aluminum hydride (*182*). Lithium aluminum hydride reduces carboxy carotenes so its use is limited when such carotenes are being sought (*73,165*). Lithium aluminum hydride is a dangerous chemical and should not be used until safe handling and use have been studied.

b. Chromatography. Paper (*209*) and thin-layer chromatographic methods (*20,79a,193*) have been developed. Although not widely used for isolation, these methods are of some use in identification.

Gas chromatography is not of great value for carotenoid analysis. At temperatures necessary to achieve sufficient volatilization, cracking and polymerization occur. One of the authors once attempted to eliminate polymerization by saturating lycopene with bromine before chromatography, but cracking increased. Carotenoids saturated by catalytic hydrogenation can be cleanly separated by gas chromatography. This yields a "carbon skeleton" analysis and is of limited use in identification. It has been used to further purify isolated carotenoids in radiotracer studies (*4,13*). Because of the great complexity of natural carotenoid mixtures and the similarity of many carotenoids to each other, column chromatography is needed for isolation.

c. Adsorbent. A number of compounds have been studied for use as adsorbents for carotenoids (*207*), but magnesium oxide, aluminum oxide, and calcium hydroxide fill every need for chromatographic analysis of carotenoids. Micro-Cel C, long considered inert, has been shown to cause substantial hydroxylation of β-carotene (*186a*).

Magnesium oxide is the most useful and widely used adsorbent. It is readily available in standardized preparations designated for chromatographic use, is compatible with most solvents, and is stable. Chromatographic preparations are so finely ground that it must be mixed with a filter aid to obtain reasonable flow rates through columns. In the absence of hydroxyl groups its affinity for carotenes increases with the number of double bonds. Noncyclic molecules are bound more tightly than cyclic, the order of elution being phytoene, phytofluene, α-carotene, β-carotene, ζ-carotene, γ-carotene, neurosporene, and lycopene. In-chain episodes are split by MgO (*49a*).

A useful system for separating carotenes on magnesium oxide is as follows. Mix Fisher Sea Sorb and Hyflo Super-Cel (1 : 1, w/w) and pour the dry powder into a chromatographic tube. Settle the powder by lightly tapping the tube on all sides while applying a vacuum of up to 25 in. of mercury. Cover the top of the column with 1–2 cm anhydrous sodium sulfate, and then wet the column with hexane. Packing a column is as much of an art as it is a science. This method gives as low a rate of failure as any other method. Preparing a column of the right size is largely dependent on experience and intuition, since the capacity of a column is influenced by the composition of the carotenoid fraction and noncarotenoid constituents in the extract. A column 18 × 350 mm will separate carotenoid mixtures from low-lipid, high-carotenoid vegetables containing between 10 and 50 mg total carotene.

Put the pigments on the column in hexane and develop with hexane while monitoring the column with a long-wave UV lamp. Phytofluene can be seen as a blue-green fluorescent band moving ahead of the yellow pigments. As phytofluene nears the bottom of the column, it can be assumed that phytoene has been eluted. At this time, begin developing with 2% acetone in hexane to elute phytofluene and α-carotene in turn. Increase the acetone concentration to 5% to elute β-carotene and obtain good separation of the remaining pigments.

Allow the column to dry. Between 1 and 2 min after the top of the column has dried, the flow of solvent at the bottom of the column will increase rapidly and begin to decrease. The column will at this time have a good consistency for extruding and carving out the bands. Place the carved-out fractions into beakers and remove the pigments from the adsorbent with acetone–hexane (1 : 1).

The ζ-carotene appears as a pale canary yellow band immediately above β-carotene. It is contaminated with most of the sterols from the sample and does not form a compact band. A number of carotenoid epoxides are found between ζ- and γ-carotene. The latter occurs as a fairly compact, deep orange band. The next major band is neurosporene, which is not as deep orange as γ-carotene and occurs immediately below lycopene, which is a tightly adsorbed red band. Without clean partition separation the neurosporene and lycopene bands will be contaminated with various monohydroxycarotenoids.

The mono- and polyhydroxy fractions may be separated in about the same order of double-bond arrangement. The first solvent for monohydroxycarotenoids is 10% acetone in hexane followed by 0.5% methanol, then progressing to 1.0% methanol. The polyhydroxy fraction may start the same way and progress up to 2% methanol.

Various well-standardized aluminum oxide preparations are commercially available for carotenoid chromatography. Less expensive grades may be made into reproducible adsorbents by washing with water, drying at 350°C for 1–2 hr, and then deactivating to the desired level by the addition of water. We have found that 5% water gives a reasonable level of adsorption for separating the hydroxycarotenoids. In general, aluminum oxide binds more tightly and has a larger capacity than magnesium oxide. Open-end compounds are bound relatively more tightly, so that phytofluene may be eluted after α-carotene and ζ-carotene and nearer to γ-carotene than to β-carotene.

Hydroxyl and epoxy groups are also bound relatively more tightly, so it is possible to obtain cleaner fractions by alternate chromatography on magnesium oxide and aluminum oxide. Hexane solutions require 10% acetone and up to 2% ethanol to cause lycopene to migrate. Benzene with 2–10% ethanol is preferred for developing hydroxycarotenoids.

Aluminum oxide columns can be packed by pouring the dry powder, without filter aid, into a tube and settling it with vibration. If vacuum is used in developing, it should not exceed 5 in. of mercury.

In nutritional studies, it is often necessary to separate cis–trans isomers. Calcium hydroxide is the best adsorbent; it is used in the same way as magnesium oxide, and the order of elution is the same. Pigments are adsorbed less tightly, and developing systems must be modified accordingly. Because of better separation of the cis isomers, pigment bands often appear as a series of smaller bands above and below a central main band.

Chromatography may be used as a cleanup procedure rather than for detailed separation of carotenoids (*252*). The pigments are put on a short column of magnesium oxide–Hyflo Super-Cel (1 : 1) in 10% acetone–hexane. Colorless carotenes, α-carotene, and β-carotene come off the column with the solvent front while the more tightly adsorbed pigments including chlorophylls remain on the column. This method eliminates the need for saponification.

Liquid chromatography has been applied to carotenoid analysis (*54,145,169*). The art of liquid chromatography appears to be sufficiently undeveloped that modifications must be adapted to various sources (*206*). Stewart (*205a*) described a HPLC procedure for α-carotene, β-carotene, and β-cryptoxanthin, the provitamin carotenoids in orange juice, using a magnesia column with a gradient elution technique. Stewart (*205b*) also determined the provitamin A and carotenoid content of seven citrus cultivars at various stages of maturation using HPLC with a MgO column for separating the provitamin A compounds and a silica column (Pellosil, Reeve Angel) for the xanthophylls. Eskins *et al.* (*61a*) separated plant pigments by HPLC on a 37–75 μm Bondapak C_{18}-Porasil B column. Hajlbrahim *et al.* (*81a*) used a 5-μm silica (Partisil) column with gradient elution to separate pigments in algal extracts. In studies of model carotenoids, Fiksdahl *et al.* (*65a*) found HPLC with a silica column coupled with a UV-VIS scanning spectrophotometer to be especially advantageous for analysis of cis–trans isomerization mixtures. Zakaria *et al.* (*245a*) separated and determined the contents of β-carotene and lycopene in six tomato samples using a reversed-phase HPLC method with a column containing 5-μm particles of Whatman Partisil-PXS-5/ODS and a mobile phase containing 8% chloroform in acetonitrile. In addition to measuring retention times, which alone are not sufficient for positive identification, stopped-flow visible spectra of the chromatographic peaks were also obtained at the top of each peak. Once procedures have become standardized, complete carotenoid analysis will be simplified and chromatogrphic mobility will be measured objectively and accurately; consequently, appreciably more weight can be given to chromatographic mobility in identifying individual carotenoids.

4. *Spectrophotometry*

Some of the older published maxima seem to be in error by 5–6 nm. These maxima were probably obtained with spectroscopes, and the error of the longest-wavelength maximum is inherent in the nature of these instruments. When there is more absorption on one side of a maximum than on the other, the position of the band is shifted away from the greater absorbance. Spectrophotometers are not subject to this effect; consequently, the longest-wavelength maxima appeared to occur at wavelengths 5–6 nm lower than those reported from spectroscopic measurement. This led to the practice of reporting the highest possible wavelength of maxima consistent with interpretation. Many published maxima for pigments were obtained under the most rigid conditions to avoid isomerization and these maxima are 1–2 nm higher in wavelength than those obtained by routine analysis.

Another source of confusion is repetitious reporting of errors in review articles. Several review articles list the longest-wavelength maximum of β-carotene in carbon disulfide as 521 nm, but an errata published subsequent to the original reports 512 nm. Most American references report 506 nm.

Solvents affect location of maxima. The longest-wavelength maximum of β-carotene in benzene is variously reported as 483–486 nm, and there is no apparent difference in reports from America and Europe. European literature usually reports this maximum in petroleum as 480–484 nm, whereas American literature reports it as 477–478 nm. The latter is in good agreement with European maxima reported in hexane.

When attempting to identify a pigment, it is good policy to consult original reports rather than reviews or books and obtain assurance that the solvents are comparable. If other criteria suggest proper identification, do not reject such identification in deference to a 1–3 nm discrepancy between reported maxima and observed maxima.

Absorption maxima of some common carotenoids are given in Table 2 to illustrate differences in absorption spectra caused by changes in structure. Differences in the series phytoene, phytofluene, ζ-carotene, neurosporene, and lycopene are caused by the successive addition of two more conjugated double bonds in an open-chain chromophore. When one end of the chain is closed, as in γ-carotene, the main absorption maximum is shifted about 10 nm toward the blue. When both ends are closed, as in β-carotene, the maximum is shifted about 20 nm. If one of the cyclized double bonds is shifted out of conjugation, as in α-carotene, or eliminated by the formation of a 5,6-epoxide, as in antheraxanthin, the maximum is shifted an additional 5–6 nm toward the blue. The shift is doubled as both cyclized bonds are taken out of conjugation, as in ε-carotene, or eliminated by formation of a di-5,6-epoxide, as in violaxanthin. The conversion of a 5,6-epoxide to a 5,8-epoxide removes one conjugated double bond from the chain with a shift of about 20 nm toward the blue. This is illustrated by isomerization of antheraxanthin to mutatoxanthin. Isomerization of violaxanthin to auroxanthin forms two 5,8-epoxides, causing a shift of 40–45 nm. The resulting absorption spectrum becomes similar to that of ζ-carotene, suggesting that the conjugated system is responsible for the visible absorption spectrum and is little affected by substituents. The negligible influence of substituents on the spectrum is also illustrated by the similarities among spectra of β-carotene, cryptoxanthin with one hydroxyl group, and zeaxanthin with two.

The shape of the spectral absorption curve is of some value in suggesting the identity of the pigment, but this is still mostly subjective. When the conjugated chain ends in cyclic structures as in β-carotene, the difference in height of the maxima and minima decreases, and the third maximum characteristic of a noncyclic compound becomes an inflection. Each conjugated double bond ending in a cyclic structure decreases the wavelength of absorption by 7–9 nm. When a cyclic double bond is shifted out of conjugation, as in α-carotene, the structure of the spectral curve is restored, but the position of the longest-wavelength maximum decreases 3–4 nm. The same changes occur when a 5,6-epoxide is formed so that differentiation between α-carotene and β-carotene 5,6-epoxide must be made on the basis of chromatographic mobility and tests for epoxides.

Isomerization to cis forms may cause a 3–5 nm hypsochromic shift but does not restore the structure of the curve as does moving a cyclic double bond out of conjugation. Judgment verified by chromatographic behavior may be sufficient to

TABLE 2

Absorption Maxima of Selected Carotenoids[a]

Carotenoid	Solvent	Maxima[b]	Ref.	$E^{1\%}_{1\,cm}$	Maximum	Ref.
Phytoene	Hexane	(298), 286, 275	*(182)*	850	286	*(182)*
ε-Carotene	Hexane	467, 437, 410	*(48)*	2890	492	*(93)*
	Hexane	467, 437, 414	*(189)*	2900	437	*(32)*
Phytofluene	Hexane	366, 348, 330	*(174)*	1520	347	*(232)*
α-Carotene	Hexane	474, 446, 422	*(251)*	2730	446	*(251)*
	Hexane	471, 443, 419	*(174)*	2800	445	*(75)*
	Hexane	472, 443, 418	*(48)*			
β-Carotene	Hexane	478, 450	*(251)*	2580	450	*(251)*
	Hexane	478, 450, (425)	*(174)*	2505	451	*(75)*
	Hexane	477, 448, (423)	*(48)*			
ζ-Carotene	Hexane	425, 400, 300	*(174)*	2540	400	*(181)*
	Hexane	423, 397, 376	*(48)*	2270	400	*(74)*
δ-Carotene	Hexane	488, 456, 430	*(104)*	3210	456	*(214)*
γ-Carotene	Petroleum ether	494, 462, 437	*(93)*	3100	462	*(93)*
	Hexane	493, 461, 436	*(250)*	2720	461	*(250)*
Neurosporene	Hexane	468, 441, 416	*(104)*			
		470, 440, 416	*(93)*	2990	440	*(93)*
Lycopene	Hexane	504, 472, 445	*(104)*			
	Hexane	505, 472, 446	*(93)*	3450	472	*(93)*
	Hexane	503, 472, 445	*(250)*	3460	472	*(250)*
Cryptoxanthin	Hexane	472, 445, (422)	*(48)*			
		478, 452	*(251)*	2470	452	*(251)*
Lutein	Ethanol	477, 447, 423	*(251)*	2560	447	*(251)*
	Benzene	487, 455, 431	*(48)*			
Zeaxanthin	Ethanol	479, 452	*(251)*	2480	452	*(251)*
	Benzene	487, 458, (434)	*(48)*			
Antheraxanthin	Ethanol	477, 447, 424	*(53)*			
	Benzene	486, 457, (433)	*(44)*	2300	457	Estimation
Mutatoxanthin	Ethanol	456, 428, 410	*(120)*	2240	428	*(120)*
	Benzene	468, 439	*(108)*			
Violaxanthin	Hexane	472, 443	*(75)*			
	Ethanol	471, 442, 417	*(108)*			
	Benzene	483, 454, 428	*(64)*	2140	454	*(64)*
Auroxanthin	Ethanol	427, 402, 381	*(53)*	1850	402	*(75)*
	Benzene	434, 407, 385	*(44)*			
Neoxanthin	Ethanol	467, 438, 417	*(120)*			
	Benzene	477, 447	*(75)*	2270	447	*(75)*
	Benzene	479, 449, 425	*(44)*			
Capsanthin	Hexane	498, 468	*(169)*	1940	468	*(169)*
	Benzene	519, 486	*(75)*	2200	486	*(75)*

[a] Listed in approximate order of elution from magnesium oxide.

[b] Data in parentheses indicate inflection.

establish the nature of the pigment, but isomerization tests should be made for confirmation.

Carbonylcarotenoids usually have one broad absorption maximum and perhaps one or two inflections. The best way to complete identification of such a compound is to reduce it to the hydroxy compound.

5. *Special Tests*

Major pigments can usually be identified with reasonable certainty on the basis of source, chromatographic behavior, and spectral properties. Identification of minor pigments is uncertain without additional tests. The tests to be discussed lend strength to the characterization of various minor carotenoids, but it is the authors' view that even with these tests many of the minor pigments can not be positively identified. If positive identification or characterization of new pigments is required, more sophisticated procedures should be used. At times this becomes complicated because some of the carotenoids appear to be so labile that they are isolated as equilibrium mixtures of isomers or oxidation states. Many of the pigments are accompanied by unidentified contaminants that appear to have the same solubility and chromatographic characteristics as the pigment. In radiotracer studies, it has been observed that only by multiple crystallization can β-carotene be obtained with constant specific radioactivity. Pigments such as ζ-carotene cannot be crystallized but can presumably be obtained in pure form by multiple chromatography before and after treatment with $LiAlH_4$ and reesterification of the remaining contaminants (*182*).

a. Partition Coefficients. Partition of pigments between hexane and 90 or 95% methanol is reliable in determining the difference between hydrocarbons and monohydroxy- and polyhydroxycarotenoids (*164*). Hydrocarbons partition greater than 10 : 1 into the hexane layer in contact with 95% methanol. Monohydroxy pigments partition about equally between hexane and 95% methanol. Dihydroxy compounds partition 10 : 1 into 95% methanol and greater than 5 : 1 into 90% methanol. Epoxy groups have a much smaller effect on partition behavior than hydroxyl groups, making conclusions based on such differences unreliable. Movements in large countercurrent separators have greater reliability in determining partition coefficients. Constancy of partition coefficients is a measure of purity and can often be used to advantage in demonstrating that minor constituents with complex spectra are mixtures.

b. Iodine Isomerization. Iodine in pigment solutions catalyzes isomerization, resulting in an equilibrium mixture of cis–trans isomers (*248*). In practice, add 50–100 ppm iodine to a hexane solution of a pigment and expose the mixture to bright light, e.g., sunlight through a window for 1–2 min. If the pigment is an all-trans isomer, a hypsochromic shift of 3–4 nm, perhaps even 5 nm, will occur. If it is a cis isomer, a hyperchromic shift of 1–3 nm will occur. Many of the minor pigments from any source are cis isomers of other pigments.

c. Test for Epoxides. Epoxides of carotenoids are either 5,6- or 5,8-epoxides. Both types give a blue or green color in the presence of concentrated hydrochloric acid. The hydrochloric acid–ether test is based on this behavior. Isomerization of 5,6-epoxides to 5,8-epoxides is catalyzed in the presence of acid, causing a decrease of 18–24 nm in the position of the absorption maxima. The procedure recommended by Curl and Bailey (*47*) is as follows. Prepare two 10-ml aliquots of the pigment in ether with an absorbance of 1. Add 9 ml methanol to each; then 1 ml concentrated hydrochloric acid to one aliquot and 1 ml H_2O to the other. Shake and observe for 1 hr. Monoepoxides yield a yellow-green color and diepoxides a blue-green color, whereas nonepoxides remain relatively unchanged. To distinguish between 5,6- and 5,8-epoxides, prepare a third aliquot as before and add 10 ml 0.5% citric acid in methanol. Obtain the absorbance spectrum after 20–30 min. The position of the spectral absorption maxima of mono-5,6-epoxides will decrease about 20 nm, and no significant change will occur with 5,8-epoxides.

The epoxide test is sufficiently sensitive to give a color change when the epoxide is present as a contaminant. This can sometimes be detected by a scaled-down epoxide test on each phase of the partition test. If one phase gives a relatively stronger test than the other, the probability of the sample being a mixture is rather great.

d. Test for Allylic Hydroxyls. A hydroxyl allyl to a double bond, as in lutein, forms a methyl ether in the presence of concentrated hydrochloric acid in methanol. The methyl ether is less polar, so the new compound partitions as though it had one less hydroxyl group than the original compound.

Fieser (*65*) developed formulas for calculating the number of conjugated double bonds in chromophores in relation to the position of the highest absorption maximum of the pigment. These are often of considerable aid in identifying unknown pigments. The equation for carotenoids is as follows:

$$\lambda_{\max}^{\text{hexane}} = 114 + 5m + n(48.0 - 1.7n) - 16.5R_{\text{i}} - 10R_{\text{e}}$$

where m = number of C substituents, n = number of conjugated double bonds, R_{i} = number of rings with an internal double bond (type VI), R_{e} = number of rings with an external double bond (type VII).

e. Quantitation. Quantitation of carotenoids is based on spectral absorption and calculation of concentration using absorption coefficients. The number of chromophores found among carotenoids is smaller than the number of carotenoids. For example, β-carotene, cryptoxanthin, and zeaxanthin have the same chromophore; thus, they have the same absorption maxima and essentially the same molar absorption coefficient. Considering the differences in molecular weight, it is convenient to multiply the absorbance by 4.00 to find milligrams per liter of β-carotene, and 4.12 and 4.26, respectively, in the case of cryptoxanthin and zeaxanthin.

Extinction coefficients for 1% solutions are listed for the pigments in Table 2.

There are differences among various reports of extinction coefficients as there are about the position of maxima, but recent values for the major pigments are close. Extinction coefficients are difficult to obtain. Many of the pigments are present in parts per billion from the best sources; consequently, isolation and purification of enough pigment to obtain an accurately weighed sample may be a major project. Some carotenoids may firmly retain solvents of crystallization, which must be removed without isomerization or oxidation. All carotenoids are subject to oxidation in the crystalline state, creating problems of accurate weighing and redissolving the pigment without change.

Davies (*53*) presented an extensive catalogue of the carotenoids with absorption maxima and extinction coefficients. It is advised that workers consult original references as much as possible (see Appendix, pp. 883–916).

It is not always necessary to achieve separation before measuring pigments. Absorption spectra of lycopene and β-carotene are such that they can be estimated with good accuracy by reading absorbance at two wavelengths and calculating concentrations with binary equations. In hexane, lycopene has an absorption maximum at 503–504 nm, an area in which β-carotene has little absorbance; β-carotene has a maximum at 450 nm, which nearly coincides with a minimum of lycopene. The following equations were obtained using the spectra of freshly crystallized pigments obtained with Cary model 15 spectrophotometer:

$$\text{Lycopene (mg/liter)} = 3.59A_{503} - 0.69A_{450}$$

$$\beta\text{-Carotene (mg/liter)} = 4.59A_{450} - 3.08A_{503}$$

The same type of equations can be prepared for β-carotene and ζ-carotene. No source has been reported in which only these two constitute a preponderance of the pigments, but the following equation is useful in checking the purity and quantitating the amount of these two pigments, which are somewhat difficult to separate chromatographically:

$$\beta\text{-Carotene (mg/liter)} = 4.07A_{450} - 0.146A_{400}$$

$$\zeta\text{-Carotene (mg/liter)} = 5.34A_{400} - 2.21A_{450}$$

II. DETERMINATION OF CAROTENES AND OXYCAROTENOIDS (XANTHOPHYLLS) IN DRIED PLANT PRODUCTS

Methods for analysis of dry plant products and mixed feeds have been developed by Quackenbush *et al.* (*179–180a*) and adopted as official AOAC procedures. These methods utilize a mixture of hexane, acetone, ethanol, and toluene for extraction of the carotenoids. The main advantage of this system is that it is applicable to both corn and alfalfa products as well as other pigmenters found in normal poultry rations. Earlier methods of extraction have not been as generally applicable. A mixture of toluene, ethanol, ethyl acetate, and water was developed for corn and corn products (*185*), whereas a hexane–acetone mixture containing alkali was found to be effective for alfalfa meals (*118,128*). In addition to separating carotenes and oxycarotenoids, the column chromatographic procedure of Quackenbush *et al.* (*179*) separates the latter into monohydroxy pigments (MHP), dihydroxy pigments (DHP), which are more effective, and polyoxy pigments, which have little or no pigmenting potency (*178*). The procedures (*179,180*) are described in detail below.

A. Apparatus

Chromatographic tube: 12.5 mm id × 30 cm, Pyrex, with bottom capillary tube 2 mm id × ca. 10 cm to extend into neck of 25-ml volumetric flask.

Vacuum filtration device: For collection of eluate in volumetric flask (Fisher Scientific Co. Filtrator or equivalent). Attach rubber stopper to column to fit device.

B. Reagents

Acetone: Dry, alcohol-free. Distill over Zn (granular, ca. 10 mesh).

Hexane: Phillips Petroleum Co. "high purity" or equivalent.

Extractant: Hexane–acetone–absolute alcohol–toluene (10 + 7 + 6 + 7).

Adsorbent I: Mix in mechanical blender 1–2 hr 1 + 1 (w/w) silica gel G (according to Stahl, Brinkmann Instruments, Inc.) and diatomaceous earth (Hyflo Super-Cel, Fisher Scientific Co.).

Adsorbent II: Mix in mechanical blender 1–2 hr 1 + 1 (w/w) activated magnesia (Sea Sorb 43, Fisher Scientific Co.) and diatomaceous earth (Hyflo Super-Cel).

Methanolic potassium hydroxide, 40%. Dissolve 40gm KOH in MeOH, cool, and dilute to 100 ml with MeOH.

Sodium sulfate solution, 10%: Dissolve 10gm anhydrous Na_2SO_4 in 100 ml H_2O.

Eluants: 1. *Carotenes:* Hexane–acetone (96 + 4), 2. *MHP:* Hexane–acetone (90 + 10), 3. *DHP:* Hexane–acetone (80 + 20); and 4. *Total xanthophylls:* Hexane–acetone–MeOH (80 + 10 + 10).

1-(Phenylazo)-2-naphthol (CI Solvent Yellow 14; Sudan I) standard solutions: 1. *Stock solution,* 1.0 m*M*: Recrystallize standard (Aldrich Chemical Co., Milwaukee, Wisconsin 53210, or Matheson, Coleman and Bell) from hot absolute alcohol. Dry crystals to constant weight in 70°C vacuum oven. Dissolve 0.1241 gm in 500 ml acetone–isopropanol (1 + 1), and 2. *Working solution,* 0.04 m*M*: Dilute 20 ml stock solution to 500 ml with acetone–isopropanol (1 + 1). Store in dark.

C. Preparation of Sample

Grind the sample to pass a no. 40 sieve. Accurately weigh the sample (2 gm corn gluten or alfalfa meal; 50 mg marigold meal; 4 gm mixed feed) into a 100-ml volumetric flask. Pipet 30 ml extractant into the flask, stopper, and swirl 1 min. For low-moisture samples, e.g., marigold meal, dehydrated alfalfa, or corn gluten (not air-dried samples), also pipet 1 ml H_2O per 2 gm of sample into the flask, stopper, and swirl 1 min. For high-moisture (air-dried) samples, omit addition of H_2O.

1. Hot Saponification (for Rapid Extraction and for Samples Containing Xanthophyll Esters, e.g., *Marigold Meal and Citrus By-products)*

Pipet 2 ml (4 ml for 4-gm sample of mixed feed) of 40% methanolic KOH into the flask, swirl 1 min, and place the flask in a 56°C H_2O bath for 20 min. Attach an air condenser or cool the neck of the flask to prevent loss of solvent. Cool the sample, and let stand in the dark for 1 hr. Pipet 30 ml hexane into the flask, swirl 1 min, dilute to volume with 10% Na_2SO_4, and shake vigorously 1 min. Let stand in the dark for 1 hr before chromatography. Upper phase is 50 ml.

2. Cold (Overnight) Saponification (for Samples Not Containing Xanthophyll Esters)

Let the mixture stand in the dark for about 16 hr. Pipet 2 ml 40% methanolic KOH into the flask and swirl 1 min. Let stand in the dark for 1 hr; then proceed with hexane addition as in hot saponification.

D. Chromatography

With the column on a Filtrator, place absorbent cotton or a glass wool plug in the bottom and add ca. 12-cm layer of adsorbent I. Apply full vacuum and add more adsorbent to give a 7-cm layer. Use a flat instrument, such as an inverted cork on a glass rod, to press and flatten the surface of the adsorbent. Place a 2-cm layer of anhydrous Na_2SO_4 above the adsorbent and press firmly.

1. Total Carotenes

With a 25-ml volumetric flask in place to collect the eluate, pipet 5 ml (or 10 ml if low pigment) of the upper phase onto the column and adjust the vacuum for a flow of two or three drops per second. The needle valve in the vacuum line helps control the flow rate. Add the carotene eluant as the last of the solution enters the adsorbent and continue until the carotene band is collected in the flask. Keep the adsorbent covered with solvent at all times. Release the vacuum, place the carotene solution in the dark until it reaches room temperature, and dilute to volume with the carotene eluant. Invert the flask several times to mix; then determine the absorbance A immediately, as in Section II,E.

Xanthophylls remain on the column. For the separation of MHP from DHP (both free from epoxy and polyoxy pigments), or for total xanthophylls, proceed as in Section II,D, 2 or 3 below.

2. Separation of Xanthophylls

1. With a 25-ml volumetric flask in the Filtrator and a vacuum applied to the column, let the eluant level approach the adsorbent surface; then immediately add

MHP eluant. The band of MHP (zeinoxanthin, cryptoxanthin) and any persistent mono- or diesters should move down the column ahead of the other bands. When the elution of the MHP band is complete, place the flask in the dark to attain room temperature before diluting to volume with the MHP eluant and A determination.

2. Proceed as in step 1 using the DHP eluant to collect the pigments of the next band (lutein, zeaxanthin, and their isomers) in a 25- or 50-ml volumetric flask. Violaxanthin, neoxanthin, and other polyoxy pigments remain on column.

3. *Total Xanthophylls*

If the value for total xanthophylls is desired, pipet a fresh aliquot from the original extractant upper phase onto a 7-cm column of adsorbent II, and elute the carotenes with hexane–acetone (90 + 10) and the total xanthophylls with hexane–acetone–MeOH (80 + 10 + 10).

E. Determination

Measure A promptly to minimize isomerization and autoxidation losses. First, check the calibration of the spectrophotometer by reading the working standard solution at 1-nm intervals between 469 and 479 nm. If the maximal value is not at 474 nm, recalibrate the instrument. When the instrument shows maximum A at 474 nm and the slit width is 0.03, the working solution readings should be 0.561 (474 nm) and 0.460 (436 nm). Correct the calculations from the equations below for the instrument deviation factor. If the instrument lacks a controllable slit, estimate the concentration by assuming that the working standard solution of dye has the same A as 2.35 mg carotenes per liter at 436 nm and 2.38 mg xanthophylls per liter at 474 nm.

Determine A of the carotene fraction at 436 nm and of the MHP and DHP fractions at 474 nm. For highest accuracy, control the solution volumes to give A between 0.25 and 0.75.

F. Calculations

The following equations are applicable to absorbance data obtained from calibrated spectrophotometers that operate with a narrow slit width. Values of 196 and 236 are a for *trans*-β-carotene and *trans*-lutein at prescribed wavelengths; b is cell length in centimeters; d (dilution factor) = (grams sample × milliliter extractant on column)/(50 ml upper phase × milliliters final dilution); and f (instrument deviation factor) = 0.460/observed A_{436} or 0.561/observed A_{474}. Carotene fraction concentration (milligrams per pound) = $(A_{436} \times 454f)/(196bd)$. The MHP fraction concentration (milligrams per pound) or DHP fraction concentration (milligrams per pound) or total xanthophylls (milligrams per pound) = $(A_{474} \times 454f)/(236bd)$.

As an estimate of the pigmenting potency of the sample, Quackenbush *et al.* (*179*) proposed that a "DHP equivalent" be calculated as the sum of the calculated value for DHP plus one-half the MHP concentration. When cis and trans isomers of the individual pigments are present, both forms appear to behave similarly in the chromatography, and there is no evidence available to indicate that moderate cis–trans isomerization seriously diminishes the pigmenting activity of the oxycarotenoids. Since each of the measured fractions, carotenes, MHP, and DHP, consists of a mixture of pigments, the absorptivity values used in calculation are necessarily arbitrary. The reading of carotenes at 436 nm is subject to possible error because of the steepness of the curve at that point, and the absorptivity value of 196 will tend to overestimate in the presence of appreciable amounts of cis isomers.

The measurement of DHP equivalents by the above method was reported by Marusich (*139a*) to show good correlation for alfalfa meal and corn gluten meal with the biological effectiveness in skin pigmentation of broilers. Manz and Vuilleumier (*137a*), however, emphasized the importance of determining the ratio of lutein to zeaxanthin because their egg yolk pigmentation studies showed that these two dihydroxycarotenoids have different pigment shades and also different deposition rates. These authors reported a method of analysis in which the sample of feed mix or concentrate is saponified with alkali, and the carotenoids extracted with ether and separated from carotenes on partially deactivated aluminum oxide. Subsequently, the total xanthophylls are determined, and the lutein and zeaxanthin fractions are separated by means of a magnesium oxide–Dicalite column and measured by spectrophotometry. Methods are also described for other pigmenters, including canthaxanthin, β-apo-8′-carotenoic acid ethyl ester, citranaxanthin, capsanthin, capsorubin, and bixin.

Knuckles *et al.* (*114*) studied the application of the above method to dried protein concentrates prepared during the wet fractionation of alfalfa. The method proved satisfactory for analysis of the dried whole juice, but a modification of the method of Kohler *et al.* (*118*) yielded significantly higher values for carotenes and xanthophylls in a dry protein–xanthophyll concentrate prepared from the whole juice. In the modified method carotenoids are extracted by an overnight soaking procedure or a rapid refluxing procedure. In the former procedure a 1- to 2-gm sample is swirled for 1 min with 30 ml of hexane–acetone (7 + 3). After the addition of 0.5 ml of water and swirling for 1 min, the mixture is allowed to stand overnight in the dark. In the rapid procedure, a 1- to 2-gm sample is swirled 1 min with 20 ml of hexane–acetone (1 + 1), 0.5 ml of water is added and swirled 1 min, the mixture is refluxed 15 min at one or two drops per second, and 4–6 ml of hexane are rinsed through the condenser and heating stopped immediately. Following either extraction procedure, the extract is swirled for 1 min with 2 ml of 40% methanolic KOH and for another minute after 1.5 ml of water is added. After dilution to volume with hexane, mixing, and allowing to settle, carotenes and xanthophylls are separated on an 11-cm × 12.5-mm-diameter column of magnesia–Hyflo Super-Cel (1 + 1). Five milliliters of extract are put into the column. Carotenes are eluted with hexane–acetone (9 + 1) in a volume of ca. 21 ml of eluate, and xanthophylls are eluted with hexane–acetone–methanol (8 + 1 + 1) in a volume of ca. 40 ml of eluate. Carotenes are measured in a colorimeter with a 400-nm filter and xanthophylls with a 470-nm filter. Nonepoxide xanthophylls are measured after treating the xanthophyll eluate with 1 ml of 0.5 *N* ethanolic HCl before diluting to volume.

Knowles *et al.* (*113*) described a method for the rapid determination of carotene and xanthophyll in dried plant materials by stirring or reflux extraction followed by chromatography on a magnesia column. Modifications in fineness of sample, extracting solvent, and time of refluxing or stirring are made to obtain optimal recovery of carotene and xanthophyll from various plant materials. The conditions recommended adequately exclude chlorophyll from column eluates and yield good agreement with the results obtained by overnight extraction at room temperature.

Knuckles *et al.* (*115*) described a method for the analysis of xanthophyll as well as carotene in fresh alfalfa and low-fiber materials such as whole juice and protein–xanthophyll concentrates. The method involves cutting or chopping of alfalfa to obtain uniform mixing, followed by blending 25 gm for 2 min with 250 ml absolute ethanol. After filtration, a 5-ml aliquot is mixed with four drops of 40% methanolic KOH and evaporated just to dryness under vacuum (60°–65°C). The residue is transferred to a chromatographic column containing 11 cm of a 1 : 1 mixture of activated magnesia and diatomaceous earth with the aid of three 4-ml portions of hexane–acetone (9 + 1), and carotenes are eluted with the same mixture until ca. 20 ml of eluate are collected. Xanthophylls are eluted with hexane–acetone–methanol (8 + 1 + 1) until 40 ml of eluate are collected. After dilution of these solutions to volume, carotene is measured at 436 nm and total xanthophylls at 475 nm. Neoxanthin and violaxanthin, which are not effective pigmenters, are removed by shaking 20 ml of the xanthophyll solution with 1 ml of ethanolic HCl, diluting to 25 ml with acetone, and reading the absorbance at 475 nm after 5–15 min. Absorption coefficients used in calculating concentrations are 196 for carotene, 236 for total xanthophylls, and 210 for pigmenting or nonepoxide xanthophylls. The lower coefficient for the latter is used because acidification isomerizes and dehydrates a portion of the lutein (*113,115*).

III. DETERMINATION OF SYNTHETIC CAROTENOIDS

A. β-Carotene

Methods for determining β-carotene involve the measurement of its yellow-orange color at the appropriate wavelength (Fig. 3).

1. In Concentrates

β-Carotene is manufactured in pure, crystalline form, but crystals are not used commercially as such due to the difficulty of handling as well as the instability of the crystals on exposure to air. More practical, high-potency dosage forms are suspensions of carotene crystals in vegetable oil, liquid emulsions, and stabilized dry powders, usually in the form of gelatin–sugar beadlets. Analytical methods for the various forms are described below.

a. Crystals. Weigh accurately about 50 mg of sample into a 100-ml volumetric flask. Dissolve the carotene with 10 ml of acid-free chloroform and dilute to mark with cyclohexane previously purified by passage through a column of silica gel and

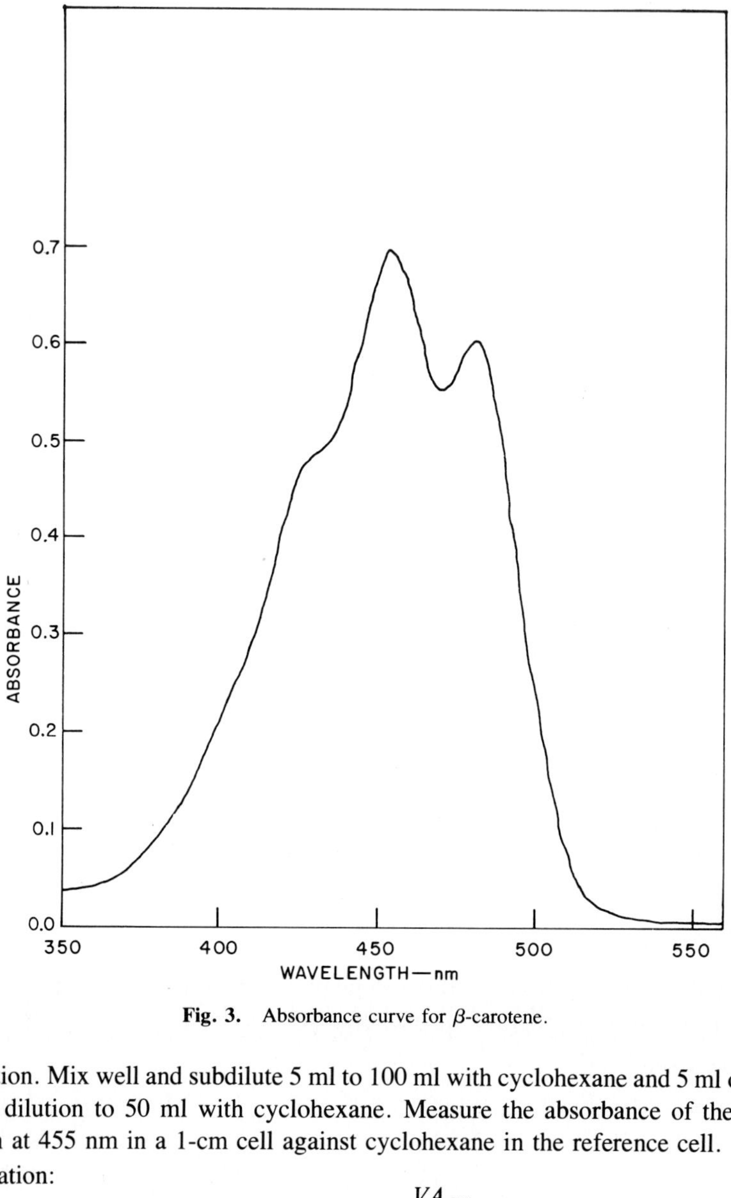

Fig. 3. Absorbance curve for β-carotene.

distillation. Mix well and subdilute 5 ml to 100 ml with cyclohexane and 5 ml of this second dilution to 50 ml with cyclohexane. Measure the absorbance of the final dilution at 455 nm in a 1-cm cell against cyclohexane in the reference cell.

Calculation:

$$\%\beta\text{-carotene} = \frac{VA_{455}}{W2500}$$

where A_{455} = absorbance at the maximum at 455 nm, W = weight of sample (gm), V = total volume (ml) containing W gm of sample (20,000), and 2500 = $E_{1\,\text{cm}}^{1\%}$ of reference *trans*-β-carotene at 455 nm in cyclohexane.

The identity of β-carotene can be checked by determining the ratio of the absorbance A_{455} to the absorbance at 340 and 483 nm. Since A_{340} is normally low for *trans*-β-carotene, it can be measured more easily by reading the second dilution above and dividing by 10. The ratio A_{455}/A_{340} should be greater than 15, and A_{455}/A_{483} should be in the range 1.14–1.18 for pure, all-*trans*-β-carotene.

An additional check on the identity of β-carotene can be made by thin-layer chromatography. Twenty-five milligrams of *trans*-β-carotene are dissolved in 5 ml of chloroform and 0.01 ml applied to a silica gel G plate. After development to a height of 10–12 cm with 20% benzene in *n*-heptane, the plate is sprayed with a freshly prepared mixture of equal volumes of a 10% aqueous solution of $FeCl_3 \cdot 6H_2O$ and a 5% aqueous solution of $K_3Fe(CN)_6$. The R_f value for *trans*-β-carotene should be 0.5–0.6.

b. Suspensions (e.g., 22% β-Carotene). Weigh accurately about 110 mg of sample into a 100-ml volumetric flask, add 10 ml of acid-free chloroform, and dilute to volume with cyclohexane. Mix well until a clear solution is obtained. Subdilute 10 ml to 100 ml and again 5 ml to 50 ml in cyclohexane. Read and calculate as described above for β-carotene crystals.

c. Emulsions (e.g., 3.6% β-Carotene). In market forms of β-carotene, such as beadlets and emulsions, which are prepared by heating oil solutions, the β-carotene occurs as an equilibrium mixture of cis and trans stereoisomers. Consequently, special methods of assay had to be developed to arrive at a correct estimate of the total β-carotene content of the market forms themselves and of food products in which they were used.

Since the absorbancy of cis stereoisomers of a carotenoid is not the same as that of the trans stereoisomer, the $E^{1\%}_{1\,\mathrm{cm}}$ value for the all-trans form cannot be used to give a true estimate of the total carotenoid present in a given stereoisomeric mixture. For a correct estimate either (a) the stereoisomers must be separated by some means and individually determined, or (b) a corrected $E^{1\%}_{1\,\mathrm{cm}}$ must be derived which will give a true measure of the total carotenoid present in the cis–trans mixture. Since the separation of cis–trans isomers is quite difficult and not readily accomplished in all cases, this approach is not practical for routine assay. The second approach is therefore used.

The addition of catalytic amounts of iodine to a solution of a carotenoid stereoisomer or mixture of stereoisomers results in a reproducible isomeric mixture of fairly constant spectrophotometric properties. The same stereoisomeric pattern develops regardless of the starting stereoisomer. To determine the absorbancy at the absorption maximum of the stereoisomeric equilibrium mixture, a solution of all-*trans*-carotenoid of known concentration is treated with iodine. With the $E^{1\%}_{1\,\mathrm{cm}}$ value thus known, an unknown stereoisomeric mixture can be treated with iodine, the absorbancy measured, and the total concentration of carotenoid calculated.

A stock solution of iodine of a concentration of 2 mg % (10 mg of crystalline iodine in 500 ml of cyclohexane) is prepared. According to the procedure previously described for the assay of β-carotene, 2 ml of this iodine solution are added before the final dilution is made up. The $E_{1\,\text{cm}}^{1\%}$ value for the equilibrium mixture of stereoisomers of β-carotene is determined by measuring the absorbancy of this solution after isomerization with iodine by heating in a water bath for 30 min at 70°C. The 30-min heating period was found to be sufficient since heating for longer periods produced no further change in absorbancy. The isomerization is conducted in amber or low-actinic flasks since carotenoids are quite sensitive to light, especially in the presence of iodine. The average value for $E_{1\,\text{cm}}^{1\%}$ as determined by repeated measurements is 2230 at 452 ± 2 nm (*23*). The stereoisomeric mixture extracted from β-carotene emulsions and beadlets was treated with iodine under the same conditions, but the change in absorbancy was negligible. It is therefore apparent that the stereoisomeric mixture produced by heating *trans*-β-carotene in oil solutions during processing for these market forms is the same as that produced by iodine isomerization. The use of $E_{1\,\text{cm}}^{1\%} = 2230$ for the assay of these stereoisomeric mixtures of β-carotene in emulsions or beadlet forms is therefore valid for routine assay purposes.

The procedure for assay of a 3.6% β-carotene emulsion is as follows. Accurately weigh a sample of about 200 mg and quantitatively transfer the sample into a 250-ml separatory funnel with the aid of 50 ml of water. Add 50 ml of denatured 3A alcohol and extract with 50-ml portions of petroleum ether (PE), shaking for 3 min each time until the aqueous layer is colorless. Two to three extractions are generally sufficient. If an emulsion forms during the extraction procedure, add approximately 5 gm of sodium sulfate or sodium chloride to the separator. Combine the extracts into a 250-ml low-actinic volumetric flask and dilute to volume with PE. Add 2–3 gm of granular anhydrous sodium sulfate and shake for approximately 3 min. Allow the sulfate to settle. Accurately pipet 5 ml of the clear solution into a 50-ml low-actinic volumetric flask and evaporate to dryness with a stream of nitrogen without application of heat. Dissolve the residue and dilute to volume with cyclohexane. Measure the absorbance of this solution in 1-cm cells at 452 nm against cyclohexane.

Calculation:

$$\frac{VA_{452}}{W2230} = \%\ \text{total}\ \beta\text{-carotene}$$

where V = total volume (ml) containing W gm of sample (2500), W = weight of sample (gm), A_{452} = absorbance at 452 nm, and 2230 = reference $E_{1\,\text{cm}}^{1\%}$ of isomeric mixture of β-carotene.

d. Beadlets (e.g., 2.4 or 10% β-Carotene). Accurately weigh a sample of ca. 250 mg of 2.4% beadlets or 60 mg of 10% beadlets and transfer into a 250-ml Erlenmeyer flask containing 50 ml of distilled water. Swirl and warm, if necessary, to disperse the beadlets completely. Transfer to a 250-ml low-actinic separatory funnel with the aid of 50 ml of 3A alcohol and proceed as described above for 3.6% β-carotene emulsion.

2. *In Foods*

Procedures are described below for the quantitative determination of β-carotene added to a number of food products representing the most important types in which β-carotene has found application. These include margarine, shortening, orange beverage products, dry mixes, soups, and baked goods. Methods of separating β-carotene from natural pigments in foods and from other food coloring agents and of identifying the various pigments are given also.

The procedure used to extract β-carotene varies with the nature of the food product to be assayed. For fatty products such as margarine or shortening a direct extraction with PE will suffice. For dry products a preliminary blending or shaking with water may be necessary before extraction with the organic solvent. In some cases the use of diethyl ether may be necessary since it is a more effective solvent than PE. The β-carotene extract is clarified by shaking with Hyflo Super-Cel and/or anhydrous sodium sulfate. The β-carotene is measured in a photoelectric colorimeter with a 455-nm filter and the potency of the solution determined from a calibration curve prepared by reading graded, known levels of pure β-carotene under the same conditions. Throughout the procedures care must be taken to avoid undue exposure to air, actinic light, or oxidizing agents that destroy carotene. The use of nonactinic glassware is advantageous.

Reagents

Alcohol: Denatured ethanol (3A alcohol).

Petroleum ether: Boiling range 30°–60°C; purified by passage through activated silica gel.

Skellysolve B: Boiling range 60°–71°C.

Methanol: Reagent grade.

Sulfuric acid, concentrated: Reagent grade.

Hydrochloric acid, concentrated: Reagent grade.

Potassium hydroxide: Reagent grade, 50 or 5% (w/w) in water.

Ammonium hydroxide: Reagent grade, 5% in 55% ethanol.

Diethyl ether: Mallinckrodt's Black Label, peroxide-free.

Sodium sulfate: Anhydrous, granular, AR grade.

Magnesium oxide: Westvaco no. 2642 or Sea Sorb 43.

Hyflo Super-Cel: Johns-Manville.

Alumina: Merck reagent grade, suitable for chromatographic absorption, no. 71707.

n-Propanol: Reagent grade.

Calcium carbonate: AR grade.

Stannous chloride: AR grade.

Acetone: Reagent grade, dry, alcohol-free.

Apparatus

Blender: Waring Blendor, Sorvall Omnimixer, or similar blender that can be tightly closed and cooled in water bath so that solvents can be used for blending.

Rotating vacuum evaporator.

Chromatographic tubes: 10–15 mm id, 250–300 mm high.

Photoelectric colorimeter: Evelyn or similar colorimeter with filter having peak close to 455 nm.

Preparation of standard calibration curve for β-carotene in skellysolve B: Weigh precisely a quantity of about 50 mg of pure all-*trans*-β-carotene into a 500-ml volumetric flask. Dissolve in Skellysolve B, dilute to mark with Skellysolve B, and mix. Subdilute 10 ml to 100 ml with Skellysolve B to obtain approximately 10 μg/ml. Make further dilutions to contain about 3, 2, 1, 0.5, 0.2, and 0.1 μg of β-carotene per milliliter. In each case before dilution to volume, add a solution of free tocopherol in Skellysolve B to provide 5 ml of tocopherol per 100 ml of β-carotene solution in order to stabilize the dilute solutions. Using the 455-nm filter, set the photoelectric colorimeter to 100% transmission with Skellysolve B. At this setting read each of the above standard solutions of β-carotene. Plot the absorbancies against concentrations to obtain the standard calibration curve.

a. Margarine. The following simple procedure can be used for margarine in which pure β-carotene is the only added coloring agent. For margarine containing about 5000 U.S.P. units of β-carotene per pound, weigh a 10-gm sample and transfer quantitatively to a glass-stoppered 100-ml mixing cylinder. For higher-potency margarine a proportionately smaller sample is taken. Add 50–60 ml of Skellysolve B and dissolve the margarine by shaking at room temperature. Dilute to volume with additional Skellysolve B. After allowing to settle for at least 5 min, measure the volume of the aqueous layer in the bottom. Subtract this from 100 ml to get the volume of the solvent layer. If the extract is sparkling clear, an aliquot may be read directly with the 455-nm filter. If, as is generally the case, a slight cloudiness persists, the extract is clarified by treating a portion in a glass-stoppered centrifuge tube with Hyflo Super-Cel and centrifuging. If a sample of the uncolored margarine is available, it (or the oil) may be diluted exactly as the colored sample and read as a blank. Many samples of uncolored margarines yield little or no color blank at this dilution.

Calculation:

Determine the concentration of β-carotene in the tube by comparing with the standard curve.

$$\text{IU/lb} = \frac{\mu\text{g}}{\text{ml}} \times \frac{V}{10} \times \frac{454}{1000} \times \frac{1667}{1}$$

where V is the volume of Skellysolve B layer (100 ml minus the volume of the aqueous layer).

A similar procedure for direct spectrophotometric reading of β-carotene in margarine at 455 nm, using a 40% solution of the margarine oil in cyclohexane and an uncolored oil solution as a blank, has been described by Luckmann *et al.* (*130*). Espoy and Barnett (*62*) described a procedure for carotene in butter or margarine in which a solution of the oil in hexane is chromatographed without saponification on a magnesia–Hyflo Super Cel column. The eluted carotene is read in a spectrophotometer at 450 nm. Background absorption from the uncolored margarine is not completely removed but amounts to only about 0.4 ppm calculated as carotene. These authors also described a column chromatographic technique for determining bixin in butter or margarine colored with annato extract.

Margarine colored with natural carotene concentrates or butter may contain cis and trans isomers of α- and β-carotene as well as γ-carotene. These can be separated by the following procedure. Weigh a sample containing ca. 1 mg of carotene (maximum 10 gm; if larger sample is needed, increase quantities of solvents accordingly) into a saponification flask. Add 2 ml of hydroquinone solution (20 gm in 100 ml of absolute ethanol) and 15 ml of methanolic KOH (18 gm KOH dissolved in 10 ml of H_2O and diluted to 100 ml with absolute methanol). Reflux 15–20 min at 60°–70°C with occasional agitation. Pour 10 ml of absolute methanol through condenser and transfer quantitatively to a 250-ml separatory funnel with the aid of 40 ml of methanol. Cool and extract with 100 ml PE (30°–40°C). Add 10 ml of water and let stand for 10–20 min. Separate the PE extract, pour into another separatory funnel, and wash with 10 ml of water. Evaporate an aliquot of the PE extract containing ca. 200 μg of β-carotene to dryness under vacuum and take up the residue in 10 ml PE.

Prepare a 10 × 1.4-cm chromatographic column with Al_2O_3 activated by heating at 100°C in vacuum for 16–20 hr (remains suitable for use for several days if stored in tightly closed bottle). Pour PE through the column and close the outlet stopcock when a 1-cm layer covers the surface. Pour the above 10 ml of extract onto the column, followed by 40 ml PE added portionwise (do not let the surface run dry). Change the receiver and elute the carotenes as follows:

Fraction 1: Elute with ca. 100 ml of 5% diethyl ether in PE; will elute *cis*-β-carotene and then *cis*-α-carotene.

Fraction 2: Elute with ca. 50 ml of 25% diethyl ether in PE; will elute *trans*-β-carotene and then *trans*-α-carotene.

Fraction 3: Elute with 50 ml of 50 : 50 mixture of diethyl ether and PE; will elute γ-carotene.

Evaporate fraction 2 containing *trans*-β-carotene to dryness under vacuum and take up the residue in sufficient PE so that 100-ml contains ca. 0.2 mg of carotene. Measure the absorption *A* at the maximum at 452–454 nm.

Calculation:

$$\frac{1000A}{2560} = \text{mg } \beta\text{-carotene per 100 ml solution}$$

Identification of the carotene is carried out by measuring the absorption curve from 300 to 500 nm. Maxima should be found at 452–454 nm and at 480–482 nm and a minimum at 468–470 nm. In order to differentiate between α- and β-carotene, an aliquot of the PE solution containing about 50 μg of carotene is evaporated to dryness under vacuum. The residue is dissolved in 5 ml of CS_2, and the position of the absorption maximum is measured in the spectrophotometer. Pure synthetic β-carotene has a characteristic absorption maximum at 484–485 nm, whereas the maximum for a mixture of equal parts of α- and β-carotene is found at 481 nm.

b. Shortening Colored with Pure β-Carotene. Weigh 10 gm of shortening containing about 5 mg of β-carotene (8333 U.S.P. units) per pound on a small piece of glassine weighing paper. Fold the paper cylindrically and insert into a 100-ml glass-stoppered cylinder. Add Skellysolve B almost to the 100-ml mark. Moisten the stopper with a drop of water so that an effective seal is obtained. Shake until the shortening is dissolved, add Skellysolve B to the 100-ml mark, and mix well. Read in a colorimeter with a 455-nm filter against a blank of unfortified shortening similarly dissolved. Determine the concentration of β-carotene in micrograms per milliliter from the standard calibration curve.
Calculation:

$$\text{mg of carotene per lb} = \frac{\mu\text{g}}{\text{ml}}\ 4.54$$

c. β-Carotene Added to Orange Juice Concentrates. Two procedures are described. Procedure 1 is used when a parallel sample of unfortified beverage is available so that the amount of added β-carotene can be determined by difference. Procedure 2 is used when an unfortified sample is not available and the added β-carotene must be separated by column chromatography before measurement. In the latter procedure α-, β-, and ζ-carotenes naturally present in orange juice to the extent of 2–5% of the total carotenoids are eluted with the added β-carotene, but these natural carotenes are normally present at a total concentration of one-tenth or less of that of the added β-carotene.

i. Procedure 1. To a 40-ml glass-stopped centrifuge tube add 0.5 gm of $CaCO_3$ and 0.25 gm of Hyflo Super-Cel. Add 2 ml of a 1 : 1 : 1 mixture of water, methanol, and *n*-propanol, thoroughly wetting the adsorbent mixture. Using a 2-ml blowout volumetric pipette, transfer 2 ml of thoroughly mixed orange juice concentrate into the tube. Follow with 20 ml of Skellysolve B, stopper, and shake 5 min in a horizontal mechanical shaker. Centrifuge briefly and add 2 gm of anhydrous Na_2SO_4. Shake and centrifuge briefly and again add 4 gm of anhydrous Na_2SO_4. Shake 5 min and centrifuge 5 min or until the Skellysolve B supernatant is clear. Although only 20 ml of Skellysolve B are added to the tube, 0.5 ml is added to this volume by a salting-out from the aqueous phase of much of the *n*-propanol. Using a dilution of the supernatant suitable for reading in a colorimeter or spectrophotomer, determine the concentration of total pigment expressed as β-carotene.

For a reading in an Evelyn colorimeter with a 440-nm filter, when the expected total pigment is about 40 mg/liter, pipet 5 ml of the supernatant into a 25-ml volumetric flask. Dilute to mark with Skellysolve B and read in the colorimeter set to 100% transmission with Skellysolve B. From the absorbancy reading, determine the concentration of the pigment (expressed as β-carotene in micrograms per milliliter) from the standard curve or by the use of a Beer's law constant if the concentration is on the linear portion of the curve.
Calculation:

$$\frac{\mu g}{ml} \times \frac{20.5}{2} \times \frac{25}{5} = \frac{\text{mg of added } \beta\text{-carotene}}{\text{liter of concentrate}}$$

By subtraction of the total pigment obtained with the unfortified sample from that obtained with the fortified sample, the amount of added β-carotene is obtained.

ii. Procedure 2. Ten milliliters of the extract, obtained as described in Procedure 1, are evaporated to dryness in a water bath (40°C) under a stream of nitrogen in a 125-ml Erlenmeyer flask, and the residue is dissolved in 5 ml of Skellysolve B. Pack a chromatographic column of about 10 mm id with 8 cm of Merck reagent aluminum oxide suitable for chromatographic adsorption (no. 71707) under a vacuum of about 15–20 in. of Hg. The vacuum is diminished before the sample is applied so that the elution rate is about two drops per second. Transfer the residue dissolved in Skellysolve B into the alumina column. Rinse the flask two times with 5 ml of Skellysolve B and transfer the rinsings into the column. Elute the column with 35 ml of 2% acetone in Skellysolve B into a glass-stoppered, graduated cylinder so that the final volume is 40 ml. The colorless initial eluate is discarded to keep the volume in this range. The carotenes (α-, β-, and ζ-carotene) are eluted as a red-orange band; a deep yellow band that follows should be at least 2 cm from the bottom of the column after all the carotenes have been eluted. The eluate is read in an Evelyn colorimeter as described above.
Calculation:

$$\frac{\mu g}{ml} \times \frac{20.5}{2} \times \frac{40}{10} = \frac{\text{mg of total pigment}}{\text{liter of concentrate}}$$

d. Total Carotenoids and Carotenes in Orange Juice. Higby (*87*) described a simplified method for determining total carotenoids and carotenes in natural and carotene-fortified orange juice. The method involves extraction with a foaming mixture of isopropanol and hexane. For total carotenoids, the extract is dried, diluted to volume, and measured with a spectrophotometer. Carotenes are separated by column chromatography on magnesia–diatomaceous earth after saponification.

Normally, carotenes are present in orange juice at a level of less than 10% of the total carotenoids. An increase to 20% or more is readily detected and is indicative of the addition of β-carotene to the juice. Reeder and Park (*185a*) and Stewart (*205a*) described specific, rapid methods for determining provitamin A carotenoids in orange juice via high-speed liquid chromatography.

e. Dry Mixes. Weigh 10 gm into a blender, add 50 ml of H_2O previously warmed to about 50°C, and blend for several minutes. Add 50 ml of 3A alcohol and reblend for several minutes. Transfer quantitatively to a 250-ml glass-stoppered separatory funnel with the aid of a little alcohol–water (50 : 50), add about 80 ml of Skellysolve B, shake well, and allow to separate. Draw off the water–alcohol layer into a second separatory funnel and repeat the extraction twice more with 80-ml portions of Skellysolve B (additional extractions may be necessary if the extracts continue to be colored). Combine all Skellysolve B extracts in a 500-ml separatory funnel and wash with three 50-ml portions of water. After discarding the last water wash, shake with about 10 gm of anhydrous sodium sulfate and transfer to a 250-ml (or 500-ml, if necessary) volumetric flask. Dilute to volume with Skellysolve B, mix, and read in a colorimeter with a 455-nm filter. If the concentration is too low, concentrate to a suitable carotene level by evaporation under vacuum with gentle warming (in a water bath at 45°–50°C).
Calculation:

$$\text{mg } \beta\text{-carotene per lb} = \frac{\mu g}{ml} \times \frac{250}{10} \times \frac{454}{1000} = \frac{\mu g}{ml} \times 11.34$$

f. Liquid Soups. Mix sample well and weigh 100 gm into a blender. Add 100 ml 3A alcohol and blend for several minutes. Transfer quantitatively to a 500-ml glass-stoppered separatory funnel with the aid of alcohol–water (50 : 50), add about 80 ml of Skellysolve B, shake well, and allow to separate. Continue as described for dry mixes above.
Calculation:

$$\text{mg } \beta\text{-carotene per lb} = \frac{\mu g}{ml} \times \frac{250}{100} \times \frac{454}{1000} = \frac{\mu g}{ml} \times 1.134$$

g. Baked Goods. Weigh a 20-gm sample and transfer to a blender. Add 100 ml H_2O and blend for several minutes. Add 100 ml 3A alcohol and blend again for several minutes. Transfer quantitatively to a 500-ml glass-stoppered separator with the aid of alcohol–water (50 : 50), add about 80 ml of Skellysolve B, shake well, and allow to separate. Continue as described for dry mixes above.
Calculation:

$$\text{mg } \beta\text{-carotene per lb} = \frac{\mu g}{ml} \times \frac{250}{20} \times \frac{454}{1000} = \frac{\mu g}{ml} \times 5.67$$

h. β-Carotene Added to Egg Products. A rapid method is described for determining β-carotene added to spray-dried or frozen egg products. The assay conditions have been designed to be as simple and rapid as possible. For example, the use of Skellysolve B for extraction rather than acetone gives essentially complete extraction of unwanted oxycarotenoid pigments. Water washing to remove acetone and the occurrence of troublesome emulsions are eliminated. The alumina for removal of oxycarotenoid pigments is used directly as received; no activation by heat or moisture is employed. The addition of ethyl ether to the Skellysolve B at a level of 16% permits complete adsorption of the oxycarotenoids and leaves at least 97% of the β-carotene unadsorbed on the alumina.

Weigh a 4-gm sample containing a total of about 800–1000 μg of β-carotene either on glassine weighing paper or on a small watch glass (allow frozen whole egg to defrost and shake well before sampling). Transfer the sample with the aid of 40 gm of anhydrous sodium sulfate powder and a rubber policeman to a mortar 4-in. in diameter. Add about 30 ml of Skellysolve B to just cover the mixture. Grind thoroughly with a pestle until all large, dried egg particles or lumps are reduced and mixed intimately with the sodium sulfate. Transfer the ground mixture with the aid of Skellysolve B from a wash bottle into a 500-ml glass-stoppered cylinder. Fill the cylinder to about 250 ml with Skellysolve B. Shake vigorously for 2 min. Fill the cylinder to the 500-ml mark (volume of solvent, 470 ml) and shake vigorously for 1 min. Allow the sodium sulfate to settle.

Into a glass-stoppered 200-ml volumetric flask pipet 100 ml of the Skellysolve B extract of the egg followed by 20 ml of diethyl ether. Weigh 2 gm of the Merck reagent aluminum oxide suitable for chromatographic adsorption and transfer to the volumetric flask. Shake for 1 min. Fill the volumetric flask to mark with a solution of 16% diethyl ether in Skellysolve B and mix. Allow the alumina to settle. Pour about 30 ml into a 45-ml glass- or cork-stoppered centrifuge vessel containing about ½ in. of anhydrous sodium sulfate. Shake and centrifuge for 3 min or until clear or filter through a 1-in. layer of anhydrous sodium sulfate held in a funnel by a plug of glass wool. Pipet at least 10 ml of the clear supernatant into an Evelyn colorimeter tube. Read in the Evelyn colorimeter with a 440-nm filter against a Skellysolve B blank. Obtain the concentration of β-carotene from a previously prepared standard curve for β-carotene in Skellysolve B.

Calculation:

$$\mu\text{g carotene per gm} = \frac{\mu\text{g}}{\text{ml}} \times \frac{198}{100} \times \frac{470}{4}$$

3. *Separation of β-Carotene from Other Food Colors*

A number of procedures are described below for the separation and identification of β-carotene. Specific directions are given for margarine and orange beverages, but similar tests can be applied to Skellysolve B extracts of any food obtained by the extraction procedures detailed above.

a. Margarine. *i. Procedure 1.* A cube of butter or margarine is melted in a warm-water bath, and the oil is centrifuged off. About 30–35 gm of the oil is weighed and diluted to 100 ml in hexane. This solution is used for the various tests. Although the oil produced by a simple centrifuging is often turbid, this does not interfere with the determination.

(a) By color tests. A portion of the oil solution is shaken with a small amount of 5% ammonium hydroxide in 55% ethyl alcohol. A straw color in the aqueous phase indicates that annatto has been added. If yellow color remains in the hexane phase after this test, the presence of carotene is probable.

(b) By chromatographic behavior. The behavior of the pigments on a magnesium oxide column is also very revealing as to their identity. A chromatographic tube, 15 × 150 mm, is fitted into a 50-ml filter flask. The column is packed to a depth of 4–5 cm using a moderate vacuum with an adsorbent consisting of 50% magnesium oxide (Westvaco no. 2642) and 50% filter aid (Johns-Manville Hyflo Super-Cel). Then 15 ml of the oil solution are added. Without being allowed to dry, the column is rinsed with a total of 30 ml of hexane–acetone (9 : 1), which is added in successive small portions. Annatto behaves characteristically in coloring the very top of the column with a brick color, which remains fixed throughout the elutions. Carotene passes through the column with very little resistance, whereas other natural carotenoids, such as xanthophylls, are held back until eluted with more polar solvents.

ii. Procedure 2. *(a) By spot tests.* Dissolve about 5 gm of margarine in 30 ml PE or Skellysolve B. Allow 30 min for settling. Spot a strip of filter paper with the extract and dry. Repeat until a yellow spot becomes visible. Two such spots are needed. Cut out one spot, place on spot plate, and add several drops of PE. Press the spot repeatedly with a glass rod. If the pigment washes off, carotene may be present. To determine which pigment is present, a fresh spot is treated with one drop of concentrated sulfuric acid. A blue-gray spot indicates carotene.

If the pigment cannot be washed off with PE, annatto or turmeric may be present. A fresh spot is treated with one drop of concentrated H_2SO_4. Annatto (bixin) turns blue and turmeric red and finally pink. These spot tests are summarized in Table 3.

TABLE 3

Spot Test on Filter Paper

Pigment or coloring agent	Color of spot	Petroleum ether	One drop of conc. H_2SO_4	One drop of 50% KOH
Carotene	Yellow	Washes off	Blue-gray	No significant change
Annatto	Orange	No change	Blue	Yellow
Turmeric	Yellow	No change	Red, turns pink	Red-orange-brown

TABLE 4

Color Tests in Solution

Pigment or coloring agent	Conc. HCl	Conc. H_2SO_4	5% KOH
Carotene	No apparent change	Initially bluish; turns gray to brown	No apparent change
Annatto (bixin)	Orange color	Blue	Orange—little change
Turmeric (cucumin)	Pink-red color	Red	Deep orange-red-brown
Mixture of turmeric and annatto	Pink-red color	Greenish	Orange-red-brown

(b) By phase separation and color tests. Dissolve 5 gm of margarine in 30 ml PE. Allow 30 min for settling. Transfer 10 ml of PE extract into separatory funnel and add 10 ml of methanol (90% methanol with 10% water saturated with petroleum ether) and 1 ml of 50% KOH. Shake and let the PE and alcohol phases separate. Drain off the alcohol phase. To the PE phase add again 10 ml of 90% methanol and 1 ml 50% KOH. Shake. Drain off the alcohol phase and combine with the first one. The PE phase will contain carotene (epiphasic), and the alcohol phase will contain annatto (bixin) and tumeric (curcumin), if present. In an acidic PE–90% methanol mixture (1 ml concentrated HCl added), carotene and bixin are epiphasic (PE phase), but curcumin is hypophasic (alcohol–water phase).

To 10 ml of PE extract add 10 ml of indicated reagent and mix. The color changes of the reagent phase are listed in Table 4.

(c) Other confirmatory tests. Annatto (bixin): A purple color appears when stannous chloride ($SnCl_2$) is allowed to react with the dried aqueous extract on filter paper. Turmeric (curcumin): An alcoholic spot dried and treated with boric acid solution turns light purple when heated at 100°C.

b. Orange Beverages. The present yellow-orange beverage colors can be classified in two groups: oil soluble and water soluble. The oil-soluble agents considered below are β-carotene, β-apo-8′-carotenal, and annatto extracts (bixin). In beverages, these are used in a water-dispersible form. The water-soluble colors are Yellow No. 5 (tartrazine) and Yellow No. 6 (Sunset Yellow).

To a 50-ml glass-stoppered centrifuge tube add 1 gm $CaCO_3$, 5 ml of orange beverage (orange juice concentrate should be reconstituted with water in the normal manner before testing), 10 ml acetone, and 20 ml PE (boiling range 30°–60°C). Shake 3 min and centrifuge until the phases are separated and the solution is clear. Examine both the water and the PE phase. A yellow-orange color in the lower water phase (solution A) indicates the presence of water-soluble compounds, probably Yellow No. 5 and/or Yellow No. 6. A pale yellow color indicates the absence of added color.

Examine the upper PE phase. Yellow-orange or orange color indicates added carotenoids. A pale yellow color very likely indicates the presence of only the original (natural) carotenoid. If added carotenoids are present, proceed as follows. Transfer most of the PE phase to a small separatory funnel and wash three or four times with water. Let the phases separate each time, and drain off the water phase. To the washed PE extract add about 1 gm of anhydrous Na_2SO_4, shake, and let settle. Decant the clear extract (solution B) and save for identification of oil-soluble color.

Solution B is spotted on a 3 × 20 cm strip of Whatman no. 1 chromatography paper about 2 cm from the bottom of the strip. More solution B is added until the spot becomes visible (about 0.2 ml). Into a cylinder about 4 cm in diameter and 15–20 cm high, pour 20 ml of a solution of 2% acetone in PE. Immerse the strip of paper in the solvent with the spot very close to the solvent surface but not touching it. Observe the ascending solvent front and the position of the spot or spots. Develop the chromatogram for 5–10 min, or until the solvent front moves upward about 7 cm. Remove the paper strip from the cylinder and mark the position of the solvent front. Measure the distance from the starting (spotting) point to the new position of the spot and the distance traveled by the liquid front.

$$R_f = \frac{\text{distance traveled by pigment}}{\text{distance traveled by solvent}}$$

Carotene travels with the solvent front and is yellow. β-Apo-8′-carotenal follows carotene (a little slower), and the spot is orange-red. Natural xanthophylls present in orange juice travel at about the same rate as bixin esters (orange), but they are bright yellow and are present in very small amounts. Bixin scarcely moves under these conditions. Table 5 summarizes the R_f values of the carotenoids and the color of the spots.

4. Carotene In Blood Serum

A simple method of measuring β-carotene in serum has been described by Roels and Trout (*187*). To a 3-ml portion of serum in a 15-ml centrifuge tube are added slowly with shaking 3 ml of freshly prepared 1 *N* KOH in 90% ethanol. After refluxing for 20 min at 50°–60°C in a water bath, the tube is cooled and shaken vigorously for 10 min after 6 ml PE (40°–60°C) are added. After centrifugation for 1 min at 500 rpm, 5 ml of the PE extract are removed for measurement of color at 452 nm. Carotene concentration is determined from a standard calibration curve.

Bayfield *et al.* (*12*) described a simple and relatively rapid method for determining carotenoids in serum or plasma. Two-dimensional paper chromatography is used, and separation of carotenoids in human plasma or animal serum into at least

TABLE 5

Qualitative Color Test Data

Carotenoid	R_f	Color of spot
β-Carotene	0.95–1.00	Yellow
β-Apo-8′-carotenal	0.70–0.85	Orange (reddish)
Bixin esters	0.25–0.38	Orange-yellow
Bixin	0.00–0.10	Orange-yellow

four components is achieved. α-Tocopherol can be determined from the same paper chromatogram.

B. β-Apo-8′-carotenal

1. *In Concentrates*

Concentrates of apocarotenal, including crystals, suspensions, emulsions, or beadlets, are extracted and diluted in the same manner as described above for the corresponding β-carotene product except that all-*trans*-apocarotenal is measured at the maximum at 460–462 nm (Fig. 4) and the iodine-isomerized equilibrium mixture of isomers at 457 ± 2 nm. The $E^{1\%}_{1\,cm}$ values used for calculating potencies are 2640 and 2400 for the all-trans isomer and the equilibrium mixture of isomers, respectively.

2. *Identity and Purity Test by Thin-Layer Chromatography*

Prepare a 5 × 20 cm or 10 × 20 cm TLC plate using a thin paste consisting of 20 ml of water and 10 gm of secondary magnesium phosphate previously screened through a U.S.P. no. 200 sieve. Dry plates for 3 hr at 120°C and store with protection from humidity. Evaporate a portion of the apocarotenal extract under N_2 at 40°C and dissolve the residue in $CHCl_3$ to obtain a concentration of 10–50 μg/ml. Apply a volume containing about 5 μg apocarotenal to the plate with a micropipet and develop by ascending chromatography for 20–30 min with 5% diethyl ether in low-boiling PE (35°–45°C). Examine spots visually under a long-wave UV light after spraying with 20% antimony trichloride in $CHCl_3$ (Carr–Price reagent) or with concentrated H_2SO_4. The all-trans isomer of apocarotenal will have an R_f value of about 0.5 and the cis isomer about 0.7.

3. *In Foods*

Applications of apocarotenal as a food color have been studied by Bauernfeind and Bunnell (*11*). The choice of procedure for extraction and purification of apocarotenal in foods depends on the nature of the sample. For solid food samples, blending with the appropriate solvent mixture is an effective extraction method. The

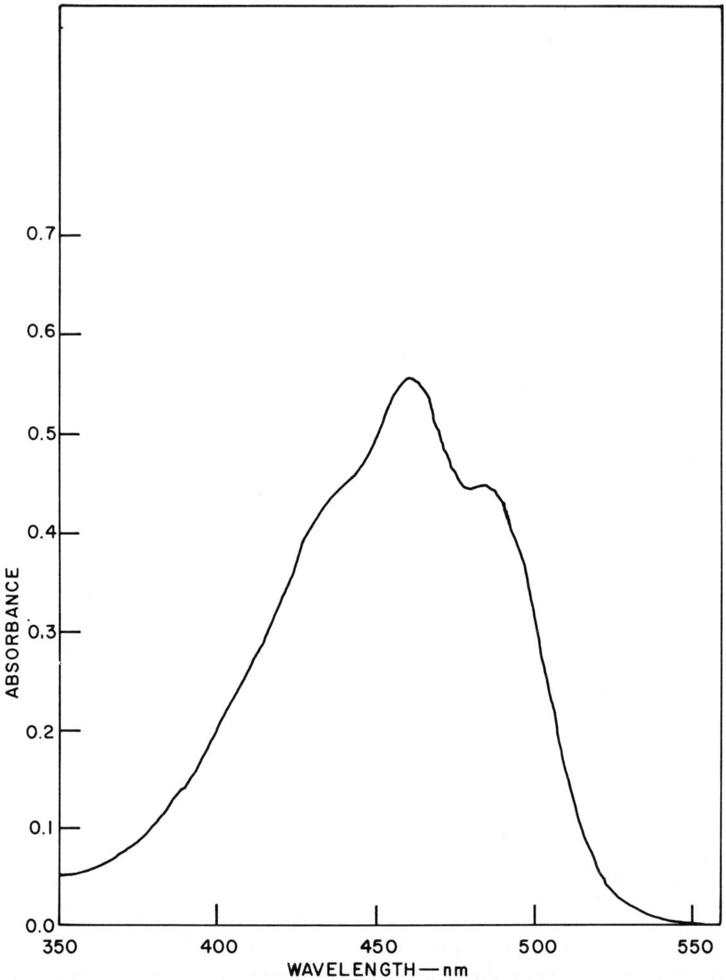

Fig. 4. Absorbance curve for β-apo-8′-carotenal.

techniques for separating apocarotenal from other food colors include solvent partition and column chromatography on alumina and/or magnesia. Analytical procedures have been reported by Osadca *et al.* (*157*).

Reagents

Petroleum ether: Boiling range 30°–60°C. Purify by silica gel column filtration.

Cyclohexane: Purify by silica gel column filtration.

Magnesium oxide: Sea Sorb 43, Westvaco, or Baker's AR grade.

Alumina: Woelm, neutral, activity grade I. To prepare with 5, 10, 15, or 17% water, pipet, respectively, 5, 10, 15, or 17 ml water into a 250-ml glass jar and roll to distribute the water on the walls. Add 100 gm alumina, close jar tightly, and shake in a mechanical shaker for 30 min to break up all lumps. Do not use until at least 2 hr after the water is added. Keep in a tightly closed jar and use within 24 hr.

Iodine solution, 10 μg/ml: Dissolve 0.5 gm iodine in 50 ml 3A alcohol. Transfer 1 ml into a 100-ml volumetric flask and make to volume with cyclohexane. Mix and transfer 5 ml into a 50-ml volumetric flask and make to mark with cyclohexane.

Apparatus

Blender: Waring Blendor, Servall OmniMixer, or similar blender that can be closed tightly and cooled in water so that solvents can be used for blending.

Chromatographic tubes: 1.2–1.5 cm id, 25–30 cm long with capillary tube at bottom.

Procedure

Extraction: For solid samples such as cheese or cake containing 2–5 mg apocarotenal per pound, blend a 5-gm sample with 20 ml 3A alcohol for 3 min. Add 40 ml PE and blend again for 3 min. Transfer quantitatively to a 250-ml separatory funnel, using 10 ml water to rinse the blender. Add water to a separator, swirl, and allow to separate. For semisolids such as frosting or sherbet (2–5 mg/lb), shake a 10-gm sample in a separatory funnel with 40 ml 3A alcohol for 3 min, add 40 ml PE, and shake vigorously for 5 min; add 20 ml water, swirl, and let separate. For liquids (2–5 mg/qt) take 10-ml samples and extract as for semisolids. For lower-potency samples, increase the sample size and the volume of each solvent proportionately.

Separate the lower water–alcohol phase into another separatory funnel and extract again with 25 ml PE, repeating if necessary to extract all solvent-soluble color. Combine the PE extracts, which contain the apocarotenal, carotenes, most natural carotenoids, and part of any annatto or bixin present. To remove annatto or bixin, shake the combined PE extracts with 25 ml of a methanol–water–60% ammonia mixture (87:10:3, v/v/v). Let settle, draw off the alcohol phase, reextract with 25 ml PE, and combine the PE with the previous PE extract. Wash the PE extract three times with 50-ml portions of water and separate from the water. Evaporate the entire PE extract under nitrogen at about 40°C and dissolve the residue in 10 ml PE.

Column chromatography: For cheese or cake, use a 12-cm column of alumina with 10% water; for frosting or sherbet, an 8-cm column of alumina with 15% water; and for orange juice or drink, a 12-cm column of alumina with 15% water. Pack the column under slight suction by adding the alumina in small portions with tapping. Transfer the 10-ml PE extract into the column, using small portions of PE to rinse the flask. Continue to elute with PE at a rate of about two drops per second until the forerunning, yellow band of carotene is eluted. Replace the collecting cylinder and continue to elute with a 4 : 6 (v/v) benzene–PE mixture until the brick red band of apocarotenal, which spreads slightly as it moves down the column, is eluted. In the case of orange juice or drink, elute a second yellow band following the carotene with a 2 : 8 (v/v) benzene–PE mixture before using the 4 : 6 mixture to elute the apocarotenal. If separation of the apocarotenal band on the column is not complete due to the high fat content of extracts of samples such as cheese or cake, evaporate the eluate containing the apocarotenal under nitrogen at 40°C, dissolve the residue in 10 ml PE, and repeat the column chromatographic step.

Spectrophotometric measurement: Evaporate the eluate containing the apocarotenal just to dryness under nitrogen in a water bath at 40°C and dissolve the residue in sufficient cyclohexane to yield a concentration of 1–2 μg/ml. Read the maximal absorption, which for foods is usually in the range of 455–459 nm due to the presence of cis isomers. Calculate the potency in this case with an $E^{1\%}_{1\,\text{cm}}$ value of 2400.

The TLC system described above for use on concentrates can be used also to check the identity of the apocarotenal in the column eluate as well as the completeness of separation from other coloring agents. A portion of the eluate is evaporated and the residue taken up in $CHCl_3$ as described. About 5 μg of apocarotenal in this solution are applied to the plate and compared with a control chromatogram of about the same amount of pure apocarotenal.

4. *In Feeds*

The isolation and determination of small amounts of β-apo-8′-carotenal (1–5 mg/lb) in feedstuffs requires separation from chlorophylls, carotenes, oxycarotenoids, and other colored ingredients. Total carotenoids in certain feeds (fortified with alfalfa, corn gluten meal, or both) may amount to 100 mg/lb, or 20–100 times the amount of β-apo-8′-carotenal added.

The method for the determination of β-apo-8′-carotenal in feedstuffs is based on spectrophotometric measurement in cyclohexane after purification by column chromatography. Recoveries of β-apo-8′-carotenal are very close to 100%, and interferences from common feed sources of pigmentation such as alfalfa, yellow corn, and corn gluten meal are negligible. The identity of the separated β-apo-8′-carotenal can be checked by TLC.

Preparation of Extracting Mixtures. Into a 3-liter separatory funnel add the following: 600 ml freshly boiled and cooled distilled water, 50 ml concentrated ammonia (ca. 58%), 600 ml ethanol (3A), 500 ml ethyl ether (peroxide-free), 1000 ml PE (boiling range 60°–80°C, purified by silica gel column filtration). Mix well and allow to separate. Draw off the aqueous phase and keep separate from the ether phase.

Extraction (Level of β-Apocarotenal = 5 mg/lb): Weigh a 50-gm sample into a 1000-ml amber volumetric flask; then add 200 ml of aqueous phase, swirl for 2 min, and follow with exactly 300 ml of ether–PE phase. Replace the air in the flask with nitrogen, stopper the flask, and shake for 15 min. Let the phases separate. Transfer a 25-ml aliquot of the upper phase into a 50-ml flask or cylinder, add about 3 gm of anhydrous sodium sulfate, shake well, and let settle. Decant most of the dried extract into a 50-ml glass-stoppered centrifuge tube, rinse the flask containing sodium sulfate with three small portions of PE (to ensure quantitative transfer), adding the rinsings to the centrifuge tube, and evaporate under a stream of nitrogen in a water bath (40°C). Dissolve the residue immediately in 20 ml PE containing 2.5% Wesson oil and use 10 ml of this "final solution" for the column chromatography. (See Note 1 below if level is 1 mg β-apocarotenal per pound).

Column Chromatography: To 100 gm of alumina (Woelm, neutral), add 15 ml of distilled water and mix well to break down all lumps. Continue mixing on mechanical shaker for 30 min. Do not use this alumina for at least 1–2 hr after addition of water. Store in a tightly closed jar.

Pack a chromatographic column with alumina to a height of 9.5–10 cm, adding it in small portions and tapping the column. Place a ¼-in. layer of anhydrous sodium sulfate on top of the column. Transfer 10 ml of "final solution" (Note 2) into the column; follow this with PE (in small portions) to remove carotenes and other colored pigments (20–30 ml). Replace the collecting cylinder and elute the β-apo-8′-carotenal, which is located below chlorophylls, oxycarotenoids, and other pigments, with a 6:4 mixture (v/v) of PE–benzene. As the spreading orange-red β-apo-8′-carotenal band moves down, discard the first few milliliters of eluate, continue eluting, and collect the β-apo-8′-carotenal band with an additional 25–40 ml of the PE–benzene mixture.

Measurement of β-Apo-8′-carotenal: Evaporate the eluted β-apo-8′-carotenal fraction under a stream of nitrogen in a water bath (40°C). Dissolve the residue immediately in 15 ml of cyclohexane and locate the peak of maximal absorption in the region of 455–462 nm. If maximal absorption occurs at 457 ± 2 nm, a stereoisomeric mixture of β-apo-8′-carotenal is indicated, and calculation is made with an $E^{1\%}_{1\,\text{cm}}$ value of 2400. If the maximal absorption occurs at 460–462 nm, all-*trans*-β-apo-8′-carotenal is indicated, and calculation is made with an $E^{1\%}_{1\,\text{cm}}$ value of 2640.

Calculation:

$$\frac{A_{\max} D \times 454 \times 10}{E^{1\%}_{1\,\text{cm}}} = \text{mg } \beta\text{-apocarotenal per lb feed}$$

where $A_{\max}$ = maximal absorption at the peak, $E^{1\%}_{1\,\text{cm}}$ = 2400 for a stereoisomeric mixture or 2640 for all-*trans*-β-apo-8′-carotenal, and D = dilution factor = (300/50) (20/25) (15/10) = 7.2.

Note 1. For a level of 1 mg of β-apocarotenal per pound of feed, take 100 gm of sample and double the amount of extracting mixture and double the aliquot of the initial extract used for evaporation. After column chromatography, evaporate the eluate and dissolve the residue in 10 ml of cyclohexane. For other concentrations follow these rules: (a) For column chromatography take an aliquot containing about 25–50 μg of β-apo-8′-carotenal; (b) the aliquot used for column chromatography should contain 0.25 ml of Wesson oil in PE; and (c) the concentration of β-apo-8′-carotenal in cyclohexane used for absorption measurements should not be less than 1 μg/ml.

Note 2. In some feeds containing a very high level of chlorophylls, magnesia columns may be used to separate chlorophylls from β-apocarotenal. β-Apocarotenal with other carotenoids is eluted with a mixture of acetone–chloroform (1 : 1, v/v). The residue is evaporated, dissolved in PE containing 2.5% of Wesson oil, and used for alumina column chromatography as described above. Confirmation of the identity of apocarotenal in the column eluate can be made by the TLC procedure described above for foods.

C. Canthaxanthin

Canthaxanthin is determined spectrophotometrically by measuring the absorbance at the maximum at 470 ± 2 nm. Figure 5 shows the absorption curve for all-*trans*-canthaxanthin. Isomerization to an equilibrium mixture of cis and trans isomers causes no significant shift in the wavelength of maximal absorption. The values of $E^{1\%}_{1\,cm}$ at 470 ± 2 nm in cyclohexane are 2200 for all-*trans*-canthaxanthin (Fig. 5) and 1970 for the cis–trans equilibrium mixture.

1. In Concentrates

a. Crystals. The procedure for assay of crystalline canthaxanthin is essentially the same as that described above for crystalline β-carotene. It has been observed, however, that there is significant adsorption of canthaxanthin on the glass surface of pipets when very dilute solutions are transferred. Hence, the following procedure is recommended for pipetting the aliquot of the second dilution of canthaxanthin in cyclohexane into the volumetric flask in which the final dilution is made.

With a pipet manipulator, withdraw the solution into a 5-ml Ostwald-Folin (blowout) pipet to slightly above the mark and then adjust to the exact amount. Transfer the solution into the appropriate volumetric flask and rinse the pipet several times by introducing cyclohexane from the top of the pipet. Collect all the rinsings in the volumetric flask and dilute to volume with cyclohexane.

b. In 10% Beadlets. Weigh accurately ca. 100 mg of beadlets in a 200-ml volumetric flask. Add ca. 100 ml of hot water (60°–70°C) and swirl mix for 2–3 min or until all beadlets are dispersed. Cool to room temperature, dilute to volume with distilled water, and mix. Transfer a 10-ml aliquot into a 50-ml glass-stoppered centrifuge tube. Add 2 ml of 1 *N* HCl and 25 ml of $CHCl_3$ and shake for 5 min. Break any emulsion that forms with two or three drops of 3A alcohol. Centrifuge at ca. 2500 rpm for 3 min. If the aqueous phase is not colorless, add ca. 1 gm of NaCl, shake for an additional 5 min, and centrifuge again.

Transfer by pipet ca. 15 ml of the chloroform extract into another glass-stoppered centrifuge tube containing ca. 2 gm of anhydrous sodium sulfate. Mix well and centrifuge for ca. 3 min. Pipet a 4-ml aliquot of the clear chloroform extract into a 50-ml volumetric flask. Place the flask in a water bath at ca. 40°C and evaporate the $CHCl_3$ under nitrogen almost to dryness. Dissolve the residue in ca. 30 ml of

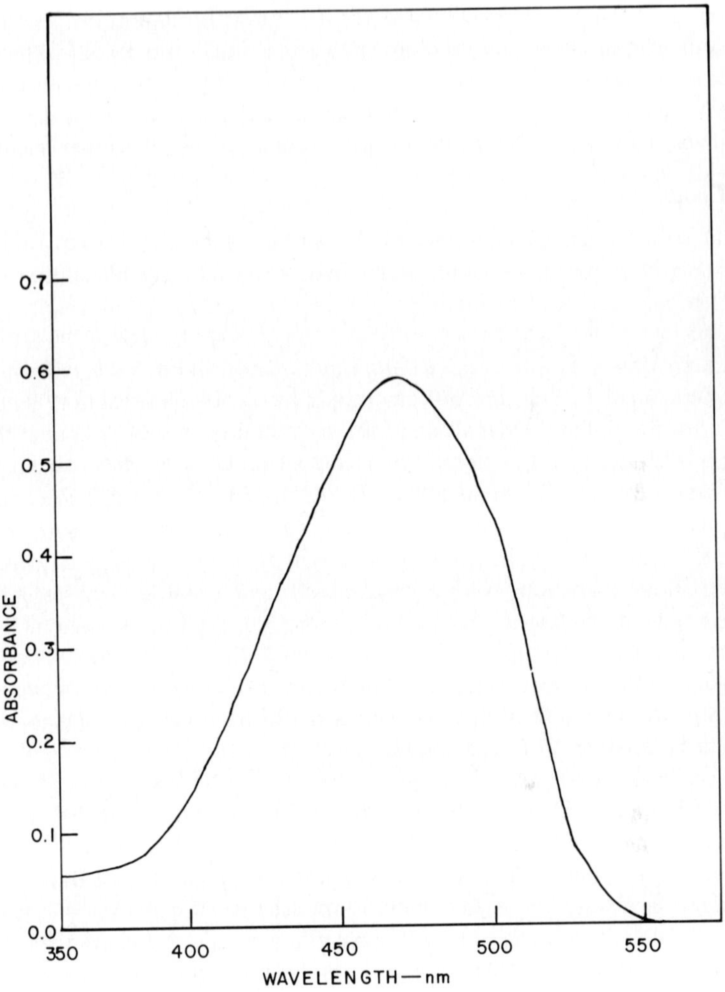

Fig. 5. Absorbance curve for canthaxanthin.

cyclohexane, add 2.5 ml of iodine solution (25 μg of iodine) in cyclohexane, and dilute to volume with cyclohexane. Mix and keep in the dark for 2 hr at room temperature. Measure the absorbance of this solution against cyclohexane in the reference cell at the maximum at 470 ± 2 nm.

Calculation:

$$\frac{6250A}{1970W} = \%\ \text{canthaxanthin}$$

where A = absorbance at 470 ± 2 nm and W = initial weight of sample (gm).

Identity and Purity Test by TLC: Apply 0.1 ml of the chloroform extract of the 10% beadlets after drying with sodium sulfate to a TLC plate coated with silica gel G. Chromatograph by the ascending technique with a 9 : 1 (v/v) methylene chloride–ether mixture until the solvent front has ascended ca. 10 cm. The R_f value for all-*trans*-canthaxanthin will be about 0.56 and for the cis isomer about 0.46.

2. *In Foods*

Applications of canthaxanthin as a food color have been described by Bunnell and Borenstein (*24*). Procedures for the determination of canthaxanthin added to various types of foods have been reported by Osadca *et al.* (*157*) as follows:

a. Extraction. *i. For Foods Containing Little or No Fat Such as Tomato Soup or Spaghetti Sauce.* Transfer a 20-gm sample into a 500-ml separatory funnel, add 100 ml water and 80 ml ethanol, and mix. Extract three times (or four if the third extract is still appreciably colored) with 100-ml portions of ether, shaking for at least 2 min each time. Combine all ether extracts, add 100 ml PE, mix, and wash three times with 100 ml water by swirling (do not shake). Separate the water completely, transfer the ether extract to a 500-ml cylinder, and bring to volume with PE. Add 50 gm anhydrous Na_2SO_4, shake well, and let settle. Transfer an aliquot containing about 50–100 μg canthaxanthin into a 125-ml round-bottom flask and evaporate to dryness (complete dryness is essential for good separations in column chromatography) in a 45°C water bath under nitrogen. Dissolve the residue in 10 ml PE with gentle warming in a water bath and use for column chromatography as described in Sections III,C,2,b,i and ii below.

ii. For Sausage Products. Blend a 40- to 50-gm sample with 100 ml acetone for 3 min using a blender with a gasketed screw cover. Add 50 ml $CHCl_3$ and blend again for 3 min. Filter through a 6-cm Whatman no. 1 paper on a Büchner funnel with suction and wash the residue with 50 ml acetone. Return the filter paper and solids to the blender and repeat the same blending and filtering operations. Transfer the combined extracts to a 500-ml cylinder and bring to volume with acetone. Add 50 gm anhydrous Na_2SO_4 and shake well. Proceed as described for the ether extract in Section III,C,2,a,i. Purify by column chromatography as described in Sections III,C,2,b,ii and iii below.

iii. For Processed Cheese. Blend 10 gm cheese with 100 ml acetone for 3–5 min. Filter through a 6-cm Whatman no. 1 paper on a Büchner funnel with suction, using about 40 ml acetone to wash the blender and funnel. Transfer the filter paper and residue back to the blender. If the filtrate turns turbid on standing, filter through Whatman no. 1 paper on a regular funnel, wash the flask and paper with a little acetone, collect the filtrate in a 500-ml cylinder, and return this filter paper to the blender. Repeat the blending with 100 ml acetone and repeat the filtration. Combine the acetone filtrates and bring to a measured volume with acetone. Add about 50 gm anhydrous Na_2SO_4 and shake well. Proceed as described for the ether extract in Section III,C,2,a,i. Purify by column chromatography as described in Sections III,C,2,b,ii and iii below.

b. Column Chromatography. Extracts of foods containing little or no fat prepared as described in Section III,C,2,a,i above are chromatographed first on a column of alumina with 17% water and then on a column of a 50 : 50 mixture of magnesia and Hyflo Super-Cel.

i. Alumina Column, 17% Water. Place a pledget of cotton in the bottom of a chromatographic tube and pack with alumina to a height of 8 cm. Transfer the 10 ml canthaxanthin extract in PE into the column. Rinse the flask several times with a few milliliters of PE and add rinsings to the column. Continue development with PE at a rate of about two drops per second until ca. 25 ml are collected. Then develop with a 4 : 6 mixture of benzene–PE, using small portions to rinse the flask and remove all traces of canthaxanthin that may be left in the flask. Discard all eluate containing any colored bands, including the violet-red band of apocarotenal (if present) moving ahead of the orange-red canthaxanthin band. Usually about 60 ± 20 ml benzene–PE mixture are required. At this point the spreading canthaxanthin band may tend to form two bands due to the presence of both cis and trans isomers. Elute and collect all the canthaxanthin, which will include a small amount of other tomato carotenoids, using ca. 50 ml 4 : 6 (v/v) mixture of acetone–PE. Evaporate the canthaxanthin eluate almost to dryness under a stream of nitrogen, dissolve the residue in 10 ml PE by warming in water, and use for magnesia column chromatography.

ii. Magnesia Column. Transfer 100 gm magnesia and 100 gm Hyflo Super-Cel to a 500-ml jar, close tightly, and shake in mechanical shaker for 1 hr. Store mixture in a tightly closed jar. Place a cotton pledget in the bottom of a chromatographic tube and pack with magnesia mixture to a height of 10 cm under a vacuum of ca. 30 mm, adding magnesia mixture in small portions and leveling each time with a blunt glass rod. Add a ¾-inch layer of anhydrous Na_2SO_4 on top of the column. Transfer the PE solution of the alumina column eluate into the column. Rinse the flask several times with small portions of PE and add rinsings to column. Adjust the vacuum to obtain an elution rate of two or two and one-half drops per second. Continue to develop with PE until ca. 15 ml are collected. Discard this eluate and continue development with 100 ml 10% acetone in PE, again using small portions to rinse the flask and dissolve the last traces of canthaxanthin. Discard this eluate and develop further with 100 ml 15% acetone in PE. Chlorophylls, bixin, and other carotenoids will stay at the top of the column. A deep violet band of canthaxanthin will move slightly down the column; if the canthaxanthin moves to the bottom of the column, reduce the volume of 10 and 15% acetone in PE as needed. Discard this eluate and elute the canthaxanthin band with about 80 ± 20 ml 40% acetone in PE, discarding any colored eluate that may precede the spreading violet band of canthaxanthin. If the eluate contains about 100 μg canthaxanthin, the entire volume may be taken for evaporation in preparation for measurement. If the canthaxanthin level is higher, the eluate should be brought to volume with PE and an aliquot containing about 100 μg taken for isomerization and measurement as described below.

Extracts containing fats as prepared in Sections III,C,2,a,ii and iii above are chromatographed first on an alumina column with 5% water as described in Section III,C,2,b,iii below, followed, if necessary, by chromatography on a magnesia column as in Section III,C,2,a,ii above. If TLC as described below indicates satisfactory separation of canthaxanthin by the alumina column, then the chromatography on magnesia may be omitted.

iii. Alumina Column, 5% Water. Proceed as described for the alumina column, 17% water (Section III,C,2,b,i) with the following changes. Use 25 ml of a 4:6 benzene–PE mixture and ca. 50 ml of a 6:4 mixture. Elute the canthaxanthin with ca. 35 ml of a 2:8 acetone–PE mixture, discarding any colored eluate that may precede the canthaxanthin. Evaporate the eluate almost to dryness under nitrogen in a water bath at 45°C and take up in 10 ml PE for magnesia column chromatography, if required. If the magnesia column is not necessary, proceed with isomerization as described below.

c. Isomerization and Spectrophotometric Measurement. Transfer a portion of the eluate containing about 100 μg canthaxanthin to a 50-ml volumetric flask; evaporate almost to dryness in a water bath at 45°C under a stream of nitrogen. Dissolve the residue in about 30 ml cyclohexane, warm in water to dissolve completely, and cool to room temperature. Add 2 ml iodine solution (20 μg iodine), bring to volume with cyclohexane, mix, and keep in the dark at room temperature for 2 hr to obtain an equilibrium mixture of stereoisomers. Measure absorbance of this solution at the maximum of 468 ± 1 nm against a cyclohexane blank. For calculation of potency use $E^{1\%}_{1\,cm}$ for canthaxanthin of 1970.

d. TLC for Purity Check. To check the identity and purity of the canthaxanthin in the final eluate, evaporate to dryness an aliquot containing about 50 μg canthaxanthin and dissolve the residue in 1.0 ml $CHCl_3$. Apply 50 μl (2.5 μg) to a thin-layer plate coated with silica gel G and apply a similar quantity of a solution of crystalline canthaxanthin to another spot. Chromatograph by the ascending technique with 9:1 (v/v) methylene chloride–ether mixture until the solvent front has ascended about 10 cm. Approximate R_f values will be 0.56 for all-*trans*-canthaxanthin and 0.46 for the cis isomer. If the plate shows significant spots not seen in the standard chromatogram, scrape the canthaxanthin spots into two centrifuge tubes, extract with 5 ml $CHCl_3$, filter through cotton, wash filter with a small amount of $CHCl_3$, evaporate $CHCl_3$ almost to dryness, and dissolve each residue in 3 ml cyclohexane. Measure the absorbance at maximum of the all-trans isomer at

468–472 nm and of the cis isomer at 464 nm. For calculation use $E_{1\,cm}^{1\%} = 2200$ for all-*trans*-canthaxanthin and 1420 for the cis isomer. As an alternative procedure, the noncanthaxanthin spot or spots can be separated, eluted, and measured in a similar manner.

3. *In Feeds*

For the pigmenting of broilers canthaxanthin is added to feeds in the form of 10% gelatin–sugar beadlets at levels of 1–4 gm of canthaxanthin per ton. These levels are low compared to the amounts of natural carotenoids contained in feedstuffs and necessitate purification of feed extracts by column chromatography followed by TLC. The analytical method for feeds developed by Osadca *et al.* (*158*) is as follows:

Reagents (*Use analytical grade throughout*)

Ammonium hydroxide solution, 7% with 0.5% propyl gallate: Mix 3 volumes of water with 1 volume of ammonium hydroxide (ca. 28%); add and dissolve 0.5 gm propyl gallate per 100 ml of mixture.

Petroleum ether: Merck (Benzin) boiling range 30°–60°C, purified on a silica gel column (desiccant-activated, Davison Chemical).

Ethanol: 3A alcohol may be used.

KOH solution, ca. 50%: Using a magnetic stirrer, dissolve 50 gm of KOH pellets in 60 ml of a mixture of methanol–water (2:1, v/v) contained in a 100-ml volumetric flask. Dilute to volume with the methanol–water mixture.

Iodine solution: Dissolve 25 mg of pure iodine in 100 ml of alcohol. Prepare fresh weekly and store in refrigerator. Pipet 5 ml into a 50-ml volumetric flask and dilute to volume with benzene. Prepare this working solution fresh daily.

Diethyl ether, peroxide-free: Mallinckrodt Black Label has been found suitable.

Florisil, 100 to 200 mesh, adsorbent: Fisher Scientific Co.

Silica gel G: For TLC, according to Stahl, EM reagents.

Celite: Analytical filter aid. Johns-Manville Products Corp., Celite Division.

Extracting solution: Mix PE and diethyl ether in the ratio 2:1 (v/v) and add 1 gm butylated hydroxytoluene (BHT) and 1 gm butylated hydroxyanisole (BHA) per 100 ml of mixture.

30% Acetone solution: Mix PE and acetone in the ratio 7:3 (v/v).

Thin-layer chromatography developing solution: Mix 160 ml benzene, 20 ml diethyl ether, 10 ml methanol, 1 ml pyridine, 0.25 gm BHT, and 0.25 gm BHA.

Apparatus

Spectrophotometer: Beckman DU, Gilford 240, or equivalent instrument equipped with 5-cm cells.

Chromatography tubes: 18 × 200 mm (ca. 15 mm id) sealed to 5 × 100 mm tube at one end and to reservoir ca. 35 × 90 mm at the other end.

Chromatographic chamber: All-glass (clear) tank approximately 27 cm long, 7 cm wide, 27 cm high with cover.

Glass plates: 20 × 20 cm.

Blender: Waring Blendor or similar blender with jar of about 1000-ml capacity that can be closed tightly.

Powerstat transformer: To control speed of blender.

Preparation of Adsorption Column

Place a small pledget of cotton at the bottom of a chromatographic tube and pack with Florisil to a height of 8 cm by adding in several portions and tamping lightly with a rod. Add a 1-cm layer of anhydrous Na_2SO_4 on the top of the column. Prepare columns just before use.

Preparation of TLC Plates

Blend two parts of water and one part of silica gel G (60 gm silica + 120 ml water are sufficient to coat five plates) for 2 min. Coat the plates with the mixture at 0.75 mm thickness. Let the plates dry at room temperature for at least 4 hr (leaving the plates at room temperature for 24 hr will not change their performance). Heat at least 4 hr at 60°C and use within a few minutes after removal from a 60°C oven.

Procedure

All steps should be carried out protected from daylight or fluorescent light. Red or amber glassware or yellow safe lights should be used.

Sample preparation: Grind pellets or crumbles to pass a 30-mesh screen. Use mash as received.

Sample extraction and saponification. Transfer 45.0 gm feed sample into a 250-ml Erlenmeyer flask. Add 100 ml of 7% ammonium hydroxide solution that has been warmed to 60°C. Mix carefully, using a glass rod, if necessary, to ensure that the sample is thoroughly mixed and wetted. Place the flask in a water bath at ca. 65° for 15 min. Transfer the contents of the flask quantitatively into a 1000-ml blender container using 150 ml of alcohol to rinse the Erlenmeyer flask. Add rinsings to the blender. Cover tightly with the cap and blend for 5 min with the speed of the blender adjusted so that the mixture remains well below the cap. Add to the blender exactly 450 ml of extracting solution, and cover with the cap tightly (*caution:* flammable liquid; blender must not leak at bottom, and cap must fit very snugly) and blend for a total of about 5 min with two short stops (ca. 10 sec) after 1.5 and 3 min. Adjust to slow speed to keep liquid level below the cap. (*Caution:* If the blender is equipped with a screw cap, stop blending after 3–4 sec and unscrew the cap momentarily to release pressure.) Let settle briefly, decant most of the liquid into a 1000-ml separatory funnel, and let the phases separate. Drain off and discard the lower aqueous phase. For a 1 gm/ton level of canthaxanthin in the feed transfer 100 ml (for 2 gm/ton take 50 ml; for 4 gm/ton take 25 ml) of clear upper phase (if the extract is not clear, filter rapidly through a filter funnel containing a pledget of glass wool) into an amber 125-ml, flat-bottom, round flask and evaporate in a water bath at 45°C under a stream of nitrogen until no smell of diethyl ether or petroleum ether is detectable (about 5 ml of liquid will remain in the flask). Add to the flask 10 ml of PE, 10 ml of 50% KOH solution, and 0.5 gm of propyl gallate. Swirl carefully. Keep at room temperature for 15 min with occasional swirling. Transfer the contents into a 125-ml separatory funnel using two 5-ml portions of water and two 5-ml portions of alcohol to rinse the flask and make the transfer quantitative. Finally, rinse the flask twice with 10-ml portions of PE and add these rinses to the separatory funnel. Add to the separator 25 ml water (do not shake) and let the phases separate (if necessary, use a small amount of alcohol to break emulsion). Discard the aqueous phase and retain the entire PE phase in the separator. Wash three times more, adding 50 ml of water each time and swirling gently. Discard the lower aqueous phase as completely as possible without losing any of the solvent phase. Add to the washed PE extract about 3 gm of anhydrous, granular Na_2SO_4 and mix carefully. Insert a filter funnel with a glass wool pledget at the bottom into a 100-ml flask and pour PE extract into the funnel. Rinse the sodium sulfate in the separator with three successive 10-ml portions of PE, pouring each portion into the filter funnel. Use the combined PE extract and rinsings for column chromatography.

Column chromatography: Wash the packed Florisil column with 10 ml of PE. Before the top surface of the column runs dry, transfer the above PE solution into the column. Make the transfer quantitative by using two 10-ml portions of PE to rinse the flask and wash the column. Add each rinsing to the column just before the previous solution disappears into the top of the column. Develop the column by adding 40–50 ml of diethyl ether in small portions until a broad yellow band is removed. The brown-red band of canthaxanthin should stay close to the top of the column. Replace collecting vessel with a 40- or 50-ml conical centrifuge tube and elute the canthaxanthin band with 30% acetone in PE (ca. 30 ml). To keep the volume of eluate down, discard any eluate collected before the canthaxanthin band approaches the bottom of the column. (*Caution:* During chromatography do not let the column run dry but add each portion of solvent just before the previous portion disappears into top of column.) Evaporate the eluate containing the canthaxanthin band and several other smaller bands almost to dryness in a water bath at 45°C under a stream of nitrogen. Dissolve the residue in 0.5 ml of benzene and use the entire 0.5 ml for TLC.

Thin-layer chromatography: Using a micropipet or other spotting device, apply the benzene solution to the plate about 2 cm from the edges and the bottom as a uniform streak about 15 cm long and not more than 0.5 cm wide. To keep the streak narrow use a gentle stream of nitrogen but avoid complete evaporation of benzene. Use several additional portions of ca. 0.2 ml of benzene to make the transfer and spotting quantitative. Pour TLC developing solution into the chromatographic chamber and place the plate into the chamber and cover. Allow to develop (ascending) until the height of the solvent front is about 12 cm or less if the orange-red band of canthaxanthin is already separated from other minor bands. (Examine the plate under white light for only a few seconds to determine if the separation is satisfactory. Interfering bands should be 0.25–0.5 cm from the canthaxanthin band, which has an R_f value of ca. 0.6 in this system. Development usually takes 30–45 min.) Remove the plate, mark the canthaxanthin band, remove all silica gel around the canthaxanthin band, and immediately scrape off the canthaxanthin band through a small funnel into a 50-ml glass-stoppered centrifuge tube containing 5 ml of alcohol. Swirl by hand for 10 sec. Pipet 20 ml benzene and rinse through the funnel into the tube. Remove the funnel and shake the tube on a mechanical shaker for about 7 min. Add to the tube about 1 gm of celite filter aid, shake for a few seconds, and centrifuge at 2000 rpm for 3 min. Remove the centrifuge tube, swirl once by hand to wash down small particles adhering to the walls, and centrifuge again at 2000 rpm for 10 min. Decant carefully most of the clear solution (do not decant any turbid solution to avoid repeated centrifuging) into another glass-stoppered centrifuge tube. Pipet 20 ml of clear solution (centrifuge if not clear) into a 25-ml amber volumetric flask, add 0.1 ml of iodine working solution, and mix. Stopper the flask loosely and immerse in a water bath at 65°C for 15 min. Cool to room temperature in a water bath. Mix and use for spectrophotometric reading.

Spectrophotometry: Determine the absorbance A of the canthaxanthin solution at the maximum of 470 ± 2 nm in a 5-cm cell using an alcohol–benzene mixture (1:4, v/v) as a reagent blank.

Calculation:

$$\frac{AD \times 9070}{E_{1\,\mathrm{cm}}^{1\%} \times 5 \times 0.85} = \text{gm canthaxanthin per ton}$$

where A = absorbance of sample, D = dilution factor = $(450 \times 25)/45V = 250/V$, V = volume of original extraction solution used (100, 50, or 25 ml), $E_{1\,\mathrm{cm}}^{1\%} = 1900$, 0.85 = recovery factor, and 9070 = factor for converting result to gm/ton.

D. β-Apo-8′-carotenoic acid ethyl ester

1. In Concentrates

This synthetic carotenoid (Carophyll) is an effective egg yolk pigmenter when fed to laying hens (*204,205*) and is also effective as a broiler pigmenter (*139*). A stabilized 10% beadlet form has been adopted as a reference standard by the Animal Nutrition Research Council (ANRC) for use in broiler pigmentation studies (*139*). Crystalline β-apo-8′-carotenoic acid ethyl ester and concentrates can be assayed by the methods described above for the corresponding β-carotene products. For the all-trans form of the ester, the value of $E_{1\,\mathrm{cm}}^{1\%}$ in cyclohexane at 448–450 nm (Fig. 6) is 2550. For the equilibrium mixture of isomers of the ester, the $E_{1\,\mathrm{cm}}^{1\%}$ value in cyclohexane at 448–450 nm is 2320.

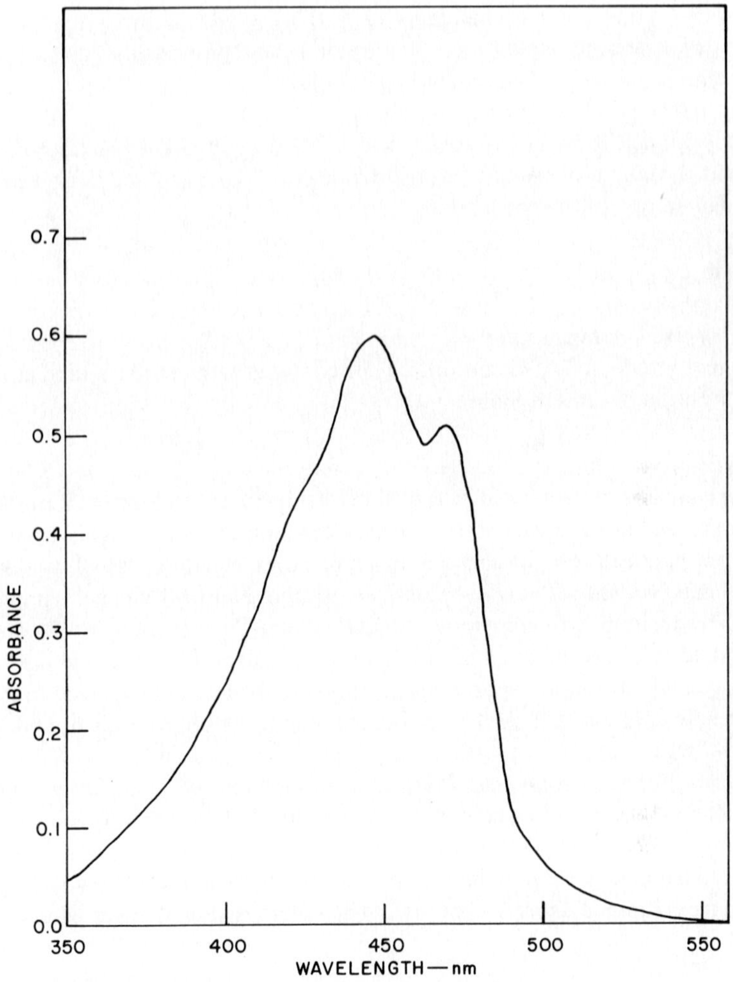

Fig. 6. Absorbance curve for β-apo-8′-carotenoic acid ethyl ester.

2. *In Feeds*

The following procedure is used for the determination of the ester in feeds. After grinding and extraction, the extract is chromatographed on a column of alumina to separate the apocarotenoic acid ester from carotenes, chlorophylls, and oxycarotenoids.

Reagents

Petroleum ether (60°–90°C): Filtered through aluminum oxide.

Diethyl ether: Peroxide-free, filtered through aluminum oxide.

Aluminum oxide for chromatography: Activity grade I, Brockman or almost neutral; deactivated by addition of 11% water 1 day before use.

Magnesium oxide for column chromatography.

Celite filtering aid: Hyflo Super-Cel.

Ammonium hydroxide solution, 7% with 0.5% propyl gallate: Mix 3 volumes of water with 1 volume of ammonium hydroxide (ca. 28%); add and dissolve 0.5 gm propyl gallate per 100 ml of mixture.

Apparatus

Spectrophotometer.

Rotating vacuum evaporator.

Chromatography tube: 22 mm id, 25–30 cm long with capillary tube at bottom with 0.5 mm id, 9–10 cm long.

Procedure

Extraction: For extraction of β-apo-8′-carotenoic acid ethyl ester from the feed sample, proceed as described above for the extraction of canthaxanthin from feeds, except use duplicate 60-gm samples of feed. After blending, allow to stand for about 15 min, withdraw two 150-ml aliquots of ether extract from each sample, and dry each by shaking with anhydrous sodium sulfate. To one of these aliquots from each feed sample, add an amount of ethyl ester of β-apo-8′-carotenoid acid in the form of a standard solution approximately equal to the amount expected in the 150 ml of sample solution (e.g., for a feed containing 2 mg of the ester per kilogram, add ca. 40 μg).

Chromatography on alumina: Plug the restricted end of four chromatographic tubes with cotton and fill to a height of 10 cm with PE. Adjust the flow rate to about two drops per second by pressing the cotton plugs with a glass rod. Pour a slurry of deactivated alumina in PE into the column and let it settle in layers by vibrating until the alumina column is about 13 cm high. The columns should never be allowed to run dry.

Filter the four aliquots of sample extract through sintered-glass filters (porosity 3) into 250-ml round-bottom flasks that can be fitted to a rotary vacuum evaporator. Evaporate the solvent at a temperature not exceeding 40°C, take up the residue in a few milliliters of PE, and transfer quantitatively into the column, using a few rinses with small portions of PE. Elute the first yellow bands containing the carotenes with PE. Then elute with 8 : 2 and 6 : 4 PE–benzene mixtures to remove the orange band of the ester of β-apo-8′-carotenoic acid. Evaporate this eluate (30–60 ml) to dryness in a rotary evaporator after flushing with nitrogen. Take the residues of sample up in cyclohexane to a volume of 25 or 50 ml, and the residues of sample plus standard up to 50 or 100 ml in cyclohexane. Measure the absorbance of each solution at the maximum at 448–450 nm.

Calculation:

$$\frac{A_{\mathrm{Spl}}}{2A_{\mathrm{Spl+Std}} - A_{\mathrm{Spl}}} S = \text{mg ethyl ester of } \beta\text{-apo-8}'\text{-carotenoic acid per kg feed}$$

where A_{Spl} = mean absorbance of the two sample solutions, $A_{\mathrm{Spl+Std}}$ = mean absorbance of the two sample solutions to which standard has been added, S = added standard, calculated as mg/kg feed (= mg standard per aliquot × 50).

IV. MEASUREMENT OF BIOLOGICAL ACTIVITY OF CAROTENOIDS

The conversion of carotenoids, in particular β-carotene, by animals to vitamin A is of major importance in animal nutrition. The evaluation of the provitamin A activity of carotenoids is made by some of the methods normally used for bioassay of vitamin A. The rat and chick have been most commonly used for such bioassays (*78*). Methods of assessing vitamin A potency have been based on such specific responses as night blindness, xerophthalmia, cornification of vaginal epithelium, blood levels, and liver storage and on such nonspecific responses as growth during prophylactic or curative dosing or survival after a single dose.

A. Procedures

Detailed directions for conducting the various bioassays for vitamin A activity have been summarized in reviews by Bliss and György (*19*), Guerrant (*80*), Embree *et al.* (*61*), and Harris (*83*).

B. Reference Standards

In 1934 the international unit of vitamin A was defined by the Permanent Commission of the Health Organization of the League of Nations as the vitamin A activity of 0.6 μg of β-carotene. Following the availability of pure synthetic vitamin A, the U.S. Pharmacopeia in 1948 established the U.S.P. unit of vitamin A as equal to 0.3 μg of vitamin A alcohol or 0.344 μg of vitamin A acetate. The same definition of an international unit was made in 1949 by the Committee on Biological Standardization of the World Health Organization, which also retained the definition that 1 IU of vitamin A equals 0.6 μg β-carotene.

Two reference standards are available for use in vitamin A bioassays. The U.S.P. Vitamin A Reference Solution (*224*) is a solution of vitamin A acetate in cottonseed oil at a defined potency of 100,000 U.S.P. units per gram and is supplied in the form of gelatin capsules containing 250 mg of this solution. The ANRC Vitamin A

Reference Standard (*2*) is in the form of gelatin beadlets derived from the U.S.P. Vitamin A Reference Solution and containing about 10,000 U.S.P. units of vitamin A per gram. The U.S.P. standard is suitable for direct oral dosing but is not stable when added to the test diet. The ANRC standard is stable in diets and suitable for use in bioassays when the vitamin A or carotenoid is fed in the diet.

When β-carotene is fed in the diet or administered orally to test animals at levels in the range of the daily requirement, the β-carotene is absorbed efficiently and utilized. When progressively higher doses of β-carotene are administered, the vitamin A activity decreases due to incomplete absorption (*51,148*). This is illustrated by the data in Table 6. With 500 U.S.P. units of vitamin A or β-carotene per pound of feed, the same level of storage of vitamin A in the chick liver is found, but, at higher feeding levels, storage of vitamin A in the liver is considerably less when β-carotene is fed than with vitamin A feeding. Similar patterns of liver storage as well as of levels of vitamin A in blood plasma of chickens have been reported by Castano *et al.* (*31*) after feeding crystalline β-carotene and crystalline vitamin A acetate at levels ranging from 500 to 24,000 IU per pound of diet for 8 weeks.

Factors Affecting the Utilization of Carotene by the Rat or Chick

The provitamin A activity of β-carotene is dependent on the specific conditions of the bioassay. The nature of the source of carotene and the ingredients of the test diet play a significant role in this regard, but the results from different laboratories have not always been in agreement on some of the effects noted. Wilson *et al.* (*237*) reported that carotene present in the tissues of green vegetables was better absorbed than carotene in oil by vitamin A-deficient rats. On the other hand, Fraps and Meinke (*68*) and Treichler *et al.* (*221*) found that the carotene of vegetables is utilized to a lesser extent than carotene dissolved in cottonseed oil for storage of vitamin A in livers of rats. One reason for this difference was found by Kemmerer *et al.* (*111*), who observed that xanthophylls and chlorophylls, when fed with

TABLE 6

Storage of Vitamin A in the Liver of the Chick after Feeding Beadlet Forms of Vitamin A and β-Carotene

U.S.P. units vitamin A or provitamin A per kg feed	U.S.P. units vitamin A per liver	
	Vitamin A fed	β-Carotene fed
500	30	30
2000	875	350
2500	1460	580
4000	3100	1350
5000	4250	1840

carotene dissolved in cottonseed oil, decreased the utilization of carotene for liver storage of vitamin A about 20%. Vavich and Kemmerer (*229*) studied this interaction of xanthophylls and carotene in the chick. They found that 300 and 600 μg of xanthophylls did not reduce the storage of vitamin A in the livers of chicks fed 65 μg of β-carotene daily, but 100, 300, and 600 μg of xanthophylls markedly reduced the storage of vitamin A when 130 μg of β-carotene were fed. This occurred even when the daily supplement of 130 μg of carotene was given in three doses at 3-hr intervals. Frey and Wilgus (*70*) reported that carotene in cottonseed oil was utilized less efficiently than that in alfalfa as measured by liver stores of vitamin A in laying pullets and by the vitamin A and carotene content of the eggs laid. Fresh alfalfa was utilized somewhat more efficiently than dehydrated alfalfa meal. Vitamin A from fish liver oil fed at the same level of vitamin A activity was more effective in promoting liver stores of vitamin A but less effective than alfalfa or carotene in oil in causing storage of total potential vitamin A activity in eggs. Parrish *et al.* (*160*) compared vitamin A with alfalfa meal in chick growth tests and found them to be equally effective when fed at levels of 400 and 800 units of vitamin A per pound of diet.

The nature of the fat used in bioassay diets can also affect the response to carotene. Lease *et al.* (*123*) found varying responses with various common edible fats. Sherman (*195*) obtained poor utilization of β-carotene in methyl linoleate but found soy bean oil to have a salutary effect. Soybean phosphatides were reported by Slanetz and Scharf (*199*) to influence the activity of doses of β-carotene in the assay range. Kemmerer and Fraps (*109*) found increased absorption of carotene by rats when fed in cottonseed oil. Deuel *et al.* (*58*) obtained as much as 30% greater than theoretical response when carotene in margarine was fed to vitamin A-depleted rats. In contrast to these results, Marusich *et al.* (*138*) found no significant increase in the utilization of carotene when fed in a series of commercial margarines in rat curative growth assays conducted according to the U.S.P. method (*223*) modified by inclusion of vitamin B_{12} and tocopherols in the diet, both of which have been claimed to promote the utilization of carotene. Mineral oil, which was reported by Rountree (*190*) to have a deleterious effect on the absorption of vitamin A by growing rats, has an even greater effect in lowering the absorption of carotene (*60*).

The nature of the vegetable or fruit has a marked influence on the availability of its carotenoids to the animal body (*28,239*). Methods of processing, such as blanching and dehydration, can also affect the ratio of bioassay to chemical assay value as shown by Wilson *et al.* (*239*) for carrots. Callison *et al.* (*29*) pointed out the need for appropriate chemical assay methodology in evaluating relative biological and chemical assays. These authors found a biologically inactive pigment in corn meal to be unseparated from active carotenoids on a MgO–Hyflo Super-Cel adsorbent, which resulted in an apparent bioavailability of only 53%, which was in agreement with the value of 52% reported by Wilson *et al.* (*239*). However, when this inactive pigment was removed using lime as the adsorbent, the biological availability of the active carotenoids in corn meal was 82% (*29*).

The presence as well as the level of vitamin E in the diet have been reported to have a marked influence on the utilization of β-carotene. Moore (*149*) was the first to report the effect of vitamin E deficiency on the vitamin A reserves of the rat. Davies and Moore (*52*) showed that vitamin E increased the time required for depletion of liver stores of vitamin A and concluded, as did Quackenbush *et al.* (*177*), that tocopherols act as antioxidants in the gastrointestinal tract. Sherman (*197*) postulated an antagonism between unsaturated fatty acids and carotene in the absence of tocopherols, which results in the inefficient use of carotene, probably through enzymatic destruction. Hickman *et al.* (*85*) demonstrated the enhancement by vitamin E of the growth-promoting effect of vitamin A in the rat. The time required for depleting young rats of vitamin A and the survival time after vitamin A supplementation were increased by vitamin E. These workers (*82*) extended these studies to carotene and found the vitamin activity of the latter to be influenced markedly by the tocopherol intake of test rats. A daily dose of about 0.5 mg of mixed tocopherols yielded the optimal sparing action on carotene. The free tocopherols were more effective in this regard than their esters, and the individual natural tocopherols (α, β, and γ) were essentially equally active (*86*). A mixture of tocopherols, as distilled from edible oils, was reported (*86*) to be slightly more active than any of the individual tocopherols, and the combination of palmitylascorbic acid to the tocopherol further enhanced the synergistic effect on carotene utilization.

Johnson and Baumann (*99*) reported a depressing effect of high levels of tocopherol on the liver stores and growth of rats. Koehn (*117*), using a low-fat diet, found as good a growth response of rats to 1 μg of carotene daily as to the same weight of vitamin A when 1 mg of α-tocopherol was given daily. Burns *et al.* (*27*) fed depleted rats 0, 0.5, 1.0, or 2.0 mg of α-tocopherol per day with 1 μg of vitamin A acetate or of β-carotene. At the 1 mg level of tocopherol the growth response was as good with the carotene as with the vitamin A. The tocopherol did not influence the growth of the rats fed vitamin A.

Hebert and Morgan (*84*) found that 0.5 mg of α-tocopherol given daily in the diet of partially depleted rats caused a significant increase in liver stores of vitamin A when 87–174 μg of carotene were given in oil daily for 14 days. Above and below these levels no effect was found. The tocopherol supplement had no significant effect when vitamin A was fed at levels of 35–129 μg daily for 14–28 days, except at the 35 μg level fed for 28 days. The depressing effect of feeding sodium selenite to rats on liver plus kidney storage of vitamin A after feeding of vitamin A or carotene was reported by Bieri *et al.* (*16*). The addition of 0.3 mg of *d,l*-α-tocopherol to the vitamin A supplement failed to prevent the decrease in vitamin A storage in selenium-fed animals.

Almquist and Maurer (*1*), Burgess *et al.* (*25,26*), and Coates *et al.* (*39*) reported that penicillin and/or chlorotetracycline feeding increased blood levels and liver storage of vitamin A in growing chicks receiving vitamin A or carotene. High (*88*) found that feeding of chlorotetracycline increased liver and kidney storage of vitamin A in rats when carotene was fed but not when vitamin A was fed. The addition of chlorotetracycline to a vitamin A-deficient diet was found by Guerrant (*81*) to stimulate additional growth in vitamin A-deficient rats receiving restricted intakes of β-carotene or vitamin A. The extent of the additional growth depended on the amount of antibiotic added to the basal diet and on the amount of supplemental vitamin fed. The greatest effect was found with 1 gm of chlorotetracycline per kilogram of diet in rats subsisting on a minimal maintenance intake of vitamin A or β-carotene.

The utilization of carotene by growing rats is affected also by the nature and amount of the protein fed. Fraps (*69*) found that substitution of casein for corn meal in the diet of albino rats increased the apparent digestibility of carotene and the storage of vitamin A in the liver. James and El Gindi (*96*) compared the effects of casein, lactalbumin, gluten, and zein in isocaloric, isonitrogenous, and essentially isophosphoric rations on the nutritional utilization of β-carotene by growing albino rats. Fecal excretion of carotene was greater for rats receiving lactalbumin or gluten in the diet than for those receiving casein or zein. The livers of rats fed casein were about 20% smaller, relative to the size of the body, than the livers of animals fed the other three proteins, but at the same carotene intake for all groups, the smaller livers of rats receiving casein contained two to five times as much vitamin A as the livers of the other three groups. Tainish and Wilkinson (*215*) reported that 30% coconut-cake meal + 10% casein added to the British Pharmacopeia diet in place of casein and part of the starch yielded a higher carotene–vitamin A response ratio in the rat growth bioassay. Ramasarma (*184*) and Bacharach (*8*), on the other hand, reported that the introduction of coconut-cake meal into their normal vitamin A-deficient diets retarded the performance of test animals. Hove (*90*) found that a factor in the lipid fraction of yeast diminished the growth response to β-carotene. Kamath *et al.* (*103*) fed adequate, moderately high, and high levels of carotene to rats along with various amounts of protein. Usually, the utilization of carotene, when evaluated by the combined hepatic and renal deposits of retinol, was linearly dependent on dietary protein.

Smith and Borchers (*201*) studied the effect of environmental temperature on the utilization of β-carotene by the rat using liver storage of vitamin A for evaluation. Exposure of rats to 5°C consistently elevated the biopotency of β-carotene relative to those of retinyl acetate when compared to environments of 25° or 35°C. The method of administration influenced the magnitude of the results. Techniques that resulted in poor utilization of vitamin A and β-carotene increased the differences observed at different temperatures. When utilization was high with feeding of β-carotene in the diet, the differences in temperature had only a minor effect on the conversion of β-carotene to vitamin A.

It is obvious from the above discussion that the biochemical relation between β-carotene and vitamin A is a variable dependent on the criterion of response and on such dietary factors as the level and nature of fat and protein, vitamin E and other vitamins, antibiotics, oxycarotenoids, and minerals. Thus, as stated by Gridgeman

(*78*), if the object of bioassays is to obtain representative relative potency, then the more varied the conditions between comparisons, the more acceptable would be the average results. Although other carotenoids have been studied much less intensively than β-carotene, the various factors that influence the determination of the vitamin A potency of carotene would be expected to influence the bioassay of other carotenoids possessing vitamin A activity. The need expressed by Gridgeman (*78*) still exists for a new criterion that, unlike rat growth, is specific to vitamin A.

REFERENCES

1. Almquist, H. J., and Maurer, S. (1955). *Arch. Biochem.* **55,** 297–298.
2. Ames, S. R. (1962). *Feedstuffs* **32,** 38–39.
3. Anderson, A. F. H., and Calvin, M. (1962). *Nature (London)* **194,** 285–286.
4. Anderson, D. G., and Porter, J. W. (1962). *Arch. Biochem. Biophys.* **97,** 509–528.
5. Anderson, R. F., Arnold, M., Nelson, G. E. N., and Ciegler, A. (1958). *J. Agric. Food Chem.* **6,** 543–545.
6. AOAC Methods of Analysis (1970). "Carotenes (Determination of); Carotenes in Fresh Plant Material and Silages," AOAC Method 39.014-39.017, 11th ed., p. 769. Assoc. Off. Agric. Chem., Washington, D.C.
7. Austin, C. R., and Shipton, J. (1944). *J. Counc. Sci. Ind. Res.* **17,** 115–126.
8. Bacharach, A. L. (1940). *Biochem. J.* **34,** 542–550.
9. Bailey, G. F., and Button, H. J. (1945). *Fruit Prod. J.* **24,** 138, 142, 155.
10. Banarjee, B. N., and Ramasarma, G. B. (1938). *Agric. Live-Stock India* **8,** 253–258.
11. Bauernfeind, J. C., and Bunnell, R. H. (1962). *Food Technol.* **16,** 76–82.
12. Bayfield, R. F., Falk, R. H., and Barrett, J. D. (1968). *J. Chromatogr.* **36,** 54–62.
13. Beeler, D. A., and Porter, J. W. (1962). *Biochem. Biophys. Res. Commun.* **8,** 367–371.
14. Ben-Shaul, Y., Teffry, T., and Klein, S. (1968). *J. Microsc. (Paris)* **7,** 265–274.
15. Bickoff, E., and Williams, K. T. (1943). *Ind. Eng. Chem., Anal. Ed.* **15,** 266–268.
16. Bieri, J. G., Pollard, C. J., and Cardenas, R. R., Jr. (1957). *Proc. Soc. Exp. Biol. Med.* **94,** 318–320.
17. Blessin, C. W., Dimler, R. J., and Webster, O. J. (1962). *Cereal Chem.* **39,** 389–392.
18. Blessin, C. W., Brecher, J. D., and Dimler, R. J. (1964). *Cereal Chem.* **41,** 543–550.
19. Bliss, C. I., and György, P. (1951). *In* "Vitamin Methods," (P. György, ed.), Vol. 2, pp. 41–275. Academic Press, New York.
20. Bolliger, H. R., König, A., and Schwieter, U. (1964). *Chimia* **181,** 136.
21. Booth, V. H. (1957). *J. Sci. Food Agric.* **8,** 371–376.
22. Booth, V. H. (1964). *Phytochemistry* **3,** 229–234.
23. Bunnell, R. H., Driscoll, W., and Bauernfeind, J. C. (1958). *Food Technol.* **12,** 536–541.
24. Bunnell, R. H., and Borenstein, B. (1967). *Food Technol.* **21,** 13A–16A.
25. Burgess, R. C., Gluck, M., Brisson, G., and Laughland, D. H. (1951). *Arch. Biochem.* **33,** 339–340.
26. Burgess, R. C., Gluck, M., and Laughland, D. H. (1953). *Poult. Sci.* **32,** 444–449.
27. Burns, M. J., Hauge, S. M., and Quackenbush, F. W. (1951). *Arch. Biochem.* **30,** 341–350.
28. Callison, E. C., Orent-Keiles, E., Frenchman, R., and Zook, E. G. (1949). *J. Nutr.* **37,** 139–152.
29. Callison, E. C., Hallman, L. F., Martin, W. F., and Orent-Keiles, E. (1953). *J. Nutr.* **50,** 85–100.
30. Camillo, L. J., Hoppert, C. A., and Fabian, F. W. (1942). *Food Res.* **7,** 339–352.
31. Castano, F. F., Boucher, R. V., and Callenbach, E. W. (1951). *J. Nutr.* **45,** 131–142.
32. Chapman, D. J., and Hano, F. T. (1963). *Plant Cell Physiol.* **4,** 57–63.
33. Charkey, L. W., and Wilgus, H. S., Jr. (1944). *Ind. Eng. Chem., Anal. Ed.* **16,** 184–187.

34. Chichester, C. O., Nakayama, T., Mackinney, G., and Goodwin, T. W. (1955). *J. Biol. Chem.* **214,** 515–517.
35. Cholnocky, L., Györgyfy, K., Nagy, E., and Panczel, M. (1955). *Acta Chim. Hung.* **6,** 143–171.
36. Cholnoky, L. V. (1937). *Kiserletugyi Kozl.* **40,** 173–186.
37. Cholnoky, L. V. (1939). *Z. Unters. Lebensm.* **78,** 157–161.
38. Ciegler, A., Arnold, M., and Anderson, R. F. (1959). *Appl. Microbiol.* **7,** 94–98.
39. Coates, M. E., Harrison, G. F., Kon, S. K., Porter, J. W. G., and Thompson, S. Y. (1952). *Chem. Ind. (London)* **7,** 149.
40. Conner, C. L. (1928). *J. Biol. Chem.* **77,** 619–626.
41. Curl, A. L. (1953). *J. Agric. Food Chem.* **1,** 456–460.
42. Curl, A. L. (1960). *J. Agric. Food Chem.* **8,** 356–358.
43. Curl, A. L. (1960). *Food Res.* **25,** 190–196.
44. Curl, A. L. (1960). *Food Res.* **25,** 670–674.
45. Curl, A. L. (1961). *J. Food Sci.* **26,** 106–111.
46. Curl, A. L. (1962). *J. Food Sci.* **27,** 171–176.
47. Curl, A. L., and Bailey, G. F. (1962). *J. Agric. Food Chem.* **9,** 403–405.
48. Curl, A. L. (1964). *J. Agric. Food Chem.* **12,** 522–524.
49. Curl, A. L. (1966). *J. Food Sci.* **31,** 759–761.
49a. Cyronak, M. J., Osianu, D., Britton, G., and Simpson, K. L. (1978). *J. Agric. Food Chem.* **26,** 712–715.
50. Daniel, E. V., and Beres, T. (1936). *Hoppe-Seyler's Z. Physiol. Chem.* **238,** 160–162.
51. Davies, A. W., and Moore, T. (1934). *Biochem. J.* **28,** 288–295.
52. Davies, A. W., and Moore, T. (1941). *Nature (London)* **147,** 794–796.
53. Davies, B. H. (1965). *In* "Chemistry and Biochemistry of Plant Pigments" (T. W. Goodwin, ed.), pp. 489–532. Academic Press, New York.
54. Davies, B. H. (1967). *Biochem. J.* **103,** 51P.
55. Davis, W. B. (1949). *Anal. Chem.* **21,** 1500–1503.
56. Della Monica, E. S., and McDowell, P. E. (1965). *Food Technol.* **19,** 1597–1599.
57. Denisen, E. L. (1951). *Iowa State Coll. J. Sci.* **25,** 549–564.
58. Deuel, H. J., Greenberg, S. M., and Savage, E. E. (1951). *J. Nutr.* **43,** 371–387.
59. Drumm, J., and O'Conner, W. F. (1945). *Biochem. J.* **39,** 211–212.
60. Dutcher, R. A., Harris, P. L., Hartzler, E. R., and Guerrant, N. B. (1934). *J. Nutr.* **8,** 269–283.
61. Embree, N. D., Ames, S. R., Lehman, R. W., and Harris, P. L. (1957). *Methods Biochem. Anal.* **4,** 43–98.
61a. Eskins, K., Scholfield, C. R., and Dutton, H. J. (1977). *J. Chromatogr.* **135,** 217–220.
62. Espoy, H. M., and Barnett, H. M. (1955). *Food Technol.* **9,** 367–372.
63. Etchells, J. L., Bell, T. A., and Jones, I. D. (1953). *Farlovia* **4,** 265–304.
64. Eugster, C. H., and Karrer, P. (1957). *Helv. Chim. Acta* **40,** 69–79.
65. Fieser, L. F. (1949). *J. Org. Chem.* **15,** 930–943.
65a. Fiksdahl, A., Mortensen, J. T., and Liaaen-Jensen, S. (1978). *J. Chromatogr.* **157,** 111–117.
66. Fox, D. L., and Lewin, R. A. (1963). *Can. J. Microbiol.* **9,** 753–768.
67. Fraps, G. S., and Kemmerer, A. R. (1941). *Ind. Eng. Chem., Anal. Ed.* **13,** 806–809.
68. Fraps, G. S., and Meinke, W. W. (1945). *Food Res.* **10,** 187–196.
69. Fraps, G. S. (1946). *Arch. Biochem.* **10,** 485–489.
70. Frey, P. R., and Wilgus, H. S. (1949). *J. Nutr.* **39,** 517–528.
71. Frey-Wyssling, A., and Kreutzer, E. (1958). *Planta* **51,** 104–114.
72. Galler, M., and MacKinney, G. (1965). *J. Food Sci.* **30,** 393–395.
73. Ganguly, J., Krinksy, N. I., and Pinckard, J. H. (1956). *Arch. Biochem. Biophys.* **60,** 345–351.
74. Glein, E. G., Tressler, D. K., and Fenton, F. (1944). *Food Res.* **9,** 471–490.
75. Goodwin, T. W. (1955). *In* "Modern Methods of Plant Analysis" (K. Paech and M. V. Tracey, eds.), Vol. III, p. 272. Springer-Verlag, Berlin and New York.

76. Gortner, W. A. (1965). *J. Food Sci.* **30,** 30–32.
77. Grayson, J. M., and Tauber, O. E. (1943). *Iowa State Coll. J. Sci.* **17,** 191–196.
78. Gridgeman, N. T. (1945). "The Estimation of Vitamin A." Lever Brothers and Unilever Ltd., Port Sunlight, Cheshire, England.
79. Grob, E. C., Pfauder, H., Leuenberger, U., and Signer, R. (1971). *Chimia* **25,** 332–333.
79a. Gross, J. (1980). *Chromatographia* **13,** 572–573.
80. Guerrant, N. B. (1951). *In* "Vitamin Methods," (P. György, ed.), Vol. 2, pp. 1–40. Academic Press, New York.
81. Guerrant, N. B. (1960). *Proc. Soc. Exp. Biol. Med.* **105,** 400–403.
81a. Hajlbrahim, S. K., Tibbets, P. J. C., Watts, C. D., Maxwell, J. R., Eglenton, G., Colén, H., and Guiochon, G. (1978). *Anal. Chem.* **50,** 549–553.
82. Harris, P. L., Kaley, M. W., and Hickman, K. C. D. (1944). *J. Biol. Chem.* **152,** 313–320.
83. Harris, P. L. (1960). *Vitam. Horm. (N.Y.)* **18,** 341–370.
84. Hebert, J. W., and Morgan, A. F. (1953). *J. Nutr.* **50,** 175–190.
85. Hickman, K. C. D., Kaley, M. W., and Harris, P. L. (1944). *J. Biol. Chem.* **152,** 303–311.
86. Hickman, K. C. D., Kaley, M. W., and Harris, P. L. (1944). *J. Biol. Chem.* **152,** 321–328.
87. Higby, W. K. (1962). *J. Food Sci.* **27,** 42–49.
88. High, E. G. (1955). *Fed. Proc., Fed. Am. Soc. Exp. Biol.* **14,** 437.
89. Holmes, A. D., Spelman, A. F., and Jones, C. P. (1945). *Food Res.* **10,** 489–496.
90. Hove, E. L. (1952). *Fed. Proc., Fed. Am. Soc. Exp. Biol.* **11,** 446–447.
91. Hunter, R. F., Scott, A. D., and Williams, N. W. (1944). *Biochem. J.* **38,** 209–211.
92. Ingraham, M. A., and Baumann, C. A. (1934). *J. Bacteriol* **28,** 31–40.
93. Isler, O., and Schuder, P. (1963). *Wiss. Veroeff. Dtsch. Ges. Ernaehr.* **9,** 54–103.
94. Jackman, L. M., and Jensen, S. L. (1964). *Acta Chem. Scand.* **18,** 1403–1411.
95. Jacoby, F. C., and Wokes, F. (1944). *Biochem. J.* **38,** 279–282.
96. James, W. H., and El Gindi, I. M. (1953). *J. Nutr.* **51,** 97–108.
97. Jensen, S. L. (1964). *Acta Chem. Scand.* **18,** 1462–1464.
98. Jensen, S. L., Hegge, E., and Jackman, L. M. (1964). *Acta Chem. Scand.* **18,** 1703–1718.
99. Johnson, R. M., and Baumann, C. A. (1948). *J. Biol. Chem.* **175,** 811–816.
100. Jones, I. D., White, R. C., and Gibbs, E. (1962). *Food Technol.* **16,** 96, 99–102.
101. Jones, I. D., White, R. C., Gibbs, E., and Denard, C. D. (1968). *J. Agric. Food. Chem.* **16,** 80–83.
102. Jungalwala, F. B., and Cama, H. R. (1963). *Indian J. Chem.* **1,** 36–40.
103. Kamath, S. K., MacMillan, J. B., and Arnrich, L. (1972). *J. Nutr.* **102,** 1579–1584.
104. Kargl, T. E., Quackenbush, F. W., and Tomes, M. L. (1960). *Proc. Am. Soc. Hortic. Sci.* **75,** 574–578.
105. Karrer, P., and Schlientz, W. (1934). *Helv. Chem. Acta* **17,** 7–8.
106. Karrer, P., and Sollmsen, U. (1936). *Helv. Chem. Acta* **19,** 1019–1024.
107. Karrer, P., and Rutschmann, J. (1943). *Helv. Chem. Acta* **26,** 2109–2114.
108. Karrer, P., and Jucker, E. (1950). "Carotenoids," p. 384. Elsevier, Amsterdam.
109. Kemmerer, A. R., and Fraps, G. S. (1938). *J. Nutr.* **16,** 309–315.
110. Kemmerer, A. R., Fraps, G. S., and Meinke, W. W. (1945). *Food Res.* **10,** 66–71.
111. Kemmerer, A. R., Fraps, G. S., and DeMottier, J. (1947). *Arch. Biochem.* **12,** 135–138.
112. Kemmerer, A. R. (1956). *J. Assoc. Off. Agric. Chem.* **29,** 18–24.
113. Knowles, R. E., Livingston, A. L., and Kohler, G. O. (1972). *J. Agric. Food Chem.* **20,** 1127–1129.
114. Knuckles, B. E., Witt, S. C., Miller, R. E., and Bickoff, E. M. (1971). *J. Assoc. Off. Anal. Chem.* **54,** 769–772.
115. Knuckles, B. E., Bickoff, E. M., and Kohler, G. O. (1972). *J. Assoc. Off. Anal. Chem.* **55,** 1202–1205.
116. Koch, J., and Sajak, E. (1965). *Z. Lebensm.—Unters. Forsch.* **26,** 260–271.
117. Koehn, C. J. (1948). *Arch. Biochem.* **17,** 337–344.

118. Kohler, G. O., Knowles, R. E., and Livingston, A. L. (1967). *J. Assoc. Off. Anal. Chem.* **50,** 707–711.
119. Kramer, A., Maloney, C. H., Stephenson, E., and Marks, A. L. (1945). *Md. Agric. Exp. Stn., Bull.* **A39,** 67–85.
120. Krinsky, N. I., and Goldsmith, T. H. (1960). *Arch. Biochem. Biophys.* **91,** 271–279.
121. Kryeminski, L. F., and Quackenbush, F. W. (1960). *Arch. Biochem. Biophys.* **88,** 64–67.
122. Lassen, S., Bacon, K., and Sutherland, J. (1944). *Food Res.* **9,** 427–433.
123. Lease, E. J., Lease, J. G., Steenbock, H., and Baumann, C. A. (1939). *J. Nutr.* **17,** 91–102.
124. Lederer, E. (1938). *Bull. Soc. Chim. Biol.* **20,** 611–634.
125. Lewis, E. P., and Merrow, S. B. (1962). *J. Agric. Food Chem.* **10,** 53–56.
126. Lime, B. J., Stephens, T. S., and Griffiths, F. P. (1954). *Food Technol.* **8,** 566–569.
127. Lime, B. J., Griffiths, F. P., O'Conner, R. T., Heinzelman, D. C., and McCall, E. R. (1957). *J. Agric. Food Chem.* **5,** 941–944.
128. Livingston, A. L., Nelson, J. W., and Kohler, G. O. (1969). *J. Assoc. Off. Anal. Chem.* **52,** 617–622.
129. Lonnber, I. E. (1933). *Ark. Zool.* **25A** 1, 1–17.
130. Luckmann, F. H., Melnick, D., and Vahlteich, H. W. (1952). *J. Am. Oil Chem. Soc.* **29,** 321–330.
131. MacGillivray, J. H. (1935). *Canner* **81,** 7–8.
132. MacKinney, G., and Milner, H. W. (1933). *J. Am. Chem. Soc.* **55,** 4728–4729.
133. MacKinney, G. (1935). *J. Biol. Chem.* **111,** 75–84.
134. MacKinney, G. (1937). *Plant Physiol.* **12,** 216–218.
135. MacKinney, G., Chichester, C. O., and Wong, P. (1954). *Arch. Biochem. Biophys.* **53,** 479–483.
136. MacKinney, G., Rich, C. M., and Jenkins, J. A. (1954). *Proc. Natl. Acad. Sci. U.S.A.* **40,** 695–699.
137. Manunta, C. (1937). *Bull. Soc. Ital. Biol. Sper.* **12,** 33–34.
137a. Manz, U., and Vuilleumier, J. P. (1977). *Z. Lebensm.-Unters. Forsch.* **163,** 21–24.
138. Marusich, W., De Ritter, E., and Bauernfeind, J. C. (1957). *J. Am. Oil Chem. Soc.* **34,** 217–221.
139. Marusich, W. L. (1970). *Feedstuffs* **42**(8), 30–31.
139a. Marusich, W. L. (1971). *Feedstuffs* **43**(8), 32–34.
140. Mase, Y., Rabourne, W. J., and Quackenbush, F. W. (1957). *Arch. Biochem. Biophys.* **68,** 150–156.
141. Matlach, M. B. (1937). *J. Wash. Acad. Sci.* **27,** 493–495.
142. Miller, C. D., and Robbins, R. C. (1936). *Hawaii Agric. Exp. Stn., Annu. Rep.* pp. 58–59.
143. Miller, J. C., and Covington, H. M. (1942). *Proc. Am. Soc. Hortic. Sci.* **40,** 519–522.
144. Mitchell, H. L., and Hauge, S. M. (1946). *J. Biol. Chem.* **164,** 543–550.
145. Moneger, R. (1968). *Physiol. Veg.* **6,** 367–402.
146. Monselise, S. P., and Halevy, A. H. (1961). *Science* **113,** 1478.
147. Moore, L. A., and Ely, R. (1941). *Ind. Eng. Chem., Anal. Ed.* **13,** 600–601.
148. Moore, T. (1931). *Biochem. J.* **25,** 275–286.
149. Moore, T. (1940). *Biochem. J.* **34,** 1321–1328.
150. Morgan, A. F., and Madren, E. O. (1933). *J. Nutr.* **6,** 83–93.
151. Morgan, R. C. (1966). *J. Food Sci.* **31,** 213–217.
152. Morgan, R. C. (1967). *J. Food Sci.* **32,** 275–278.
153. Munsey, V. E. (1938). *J. Assoc. Off. Agric. Chem.* **21,** 331–351.
154. Murri, I. K. (1940). *Sov. Plant Ind. Rec.* **4,** 125–132.
155. Naef, J., and Turian, G. (1963). *Phytochemistry* **2,** 173–177.
156. O'Conner, R. T., Heinzelman, D. C., and Jefferson, M. T. (1946). *Ind. Eng. Chem., Anal. Ed.* **18,** 557–562.
157. Osadca, M., De Ritter, E., and Bunnell, R. H. (1966). *J. Assoc. Off. Anal. Chem.* **49,** 1078–1083.
158. Osadca, M., Araujo, M., and De Ritter, E. (1972). *J. Assoc. Off. Anal. Chem.* **55,** 110–113.
159. Owen, E. C. (1954). *J. Sci. Food Agric.* **5,** 449–453.

160. Parrish, D. B., Zimmerman, R. A., Sanford, P. E., and Hung, E. (1963). *J. Nutr.* **79,** 9–17.
161. Pattes, H. E., and Purcell, A. E. (1966). *J. Am. Oil Chem. Soc.* **44,** 328–330.
162. Pendlington, S., Dupont, M. S., and Trussell, F. J. (1965). *Biochem. J.* **94,** 25P–26P.
163. Peterson, W. J., Evans, W. R., Leich, E., and Etchells, J. L. (1958). *J. Bacteriol.* **75,** 586–591.
164. Petracek, F. J., and Zechmeister, L. (1956). *Anal. Chem.* **28,** 1484–1485.
165. Petracek, F. J., and Zechmeister, L. (1956). *Arch. Biochem. Biophys.* **61,** 137–139.
166. Petzold, E. N., Quackenbush, F. W., and McQuistan, M. (1959). *Arch. Biochem. Biophys.* **82,** 117–124.
167. Petzold, E. N., and Quackenbush, F. W. (1960). *Arch. Biochem. Biophys.* **86,** 163–165.
168. Poe, C. F., Gant, O. K., and Griffin, E. (1939). *Fruit Prod. J.* **19,** 73–74.
169. Polgar, A., and Zechmeister, L. (1944). *J. Am. Chem. Soc.* **66,** 186–190.
170. Porter, T., Wharton, M. A., and Beltz, R. M. (1944). *Food Res.* **9,** 434–441.
171. Potter, M. T. (1933). *J. Nutr.* **6,** 99–102.
172. Purcell, A. E. (1958). *Anal. Chem.* **30,** 1049–1051.
173. Purcell, A. E., Thompson, G. A., and Bonner, J. (1958). *J. Biol. Chem.* **234,** 1081–1084.
174. Purcell, A. E., and Walter, W. M., Jr. (1968). *J. Agric. Food Chem.* **16,** 769–770.
175. Purcell, A. E., Walter, W. M., Jr., and Thompkins, W. T. (1969). *J. Agric. Food Chem.* **17,** 41–42.
176. Purcell, A. E., and Walter, W. M., Jr. (1971). *In* "Vitamins and Coenzymes" (D. B. McCormick and L. D. Wright, eds.), Methods in Enzymology, Vol. 18C, pp. 701–706. Academic Press, New York.
177. Quackenbush, F. W., Cox, R. P., and Steenbock, H. (1942). *J. Biol. Chem.* **145,** 169–177.
178. Quackenbush, F. W., Kvakovszky, S., Hoover, T., and Rogler, J. C. (1965). *J. Assoc. Off. Anal. Chem.* **48,** 1241–1244.
179. Quackenbush, F. W., Dyer, M. A., and Smallidge, R. L. (1970). *J. Assoc. Off. Anal. Chem.* **53,** 181–185.
180. Quackenbush, F. W., and Miller, S. (1972). *J. Assoc. Off. Anal. Chem.* **55,** 617–621.
180a. Quackenbush, F. W. (1973). *J. Assoc. Off. Anal. Chem.* **56,** 748–753.
181. Rabourn, W. J., and Quackenbush, F. W. (1954). Unpublished observations.
182. Rabourn, W. J., Quackenbush, F. W., and Porter, J. W. (1954). *Arch. Biochem. Biophys.* **48,** 267–274.
183. Ramasarma, G. B., and Banerjee, B. N. (1940). *J. Indian Inst. Sci., Sect. A* **23,** 1–10.
184. Ramasarma, G. B. (1942). *Ann. Biochem. Exp. Med.* **2,** 103–112.
185. Rathmann, D. M., and Daggy, E. E. (1961). "Corn Xanthophylls in Poultry Nutrition." Corn Products Co., 717 Fifth Avenue, New York, New York.
185a. Reeder, S. K., and Park, G. L. (1975). *J. Assoc. Off. Anal. Chem.* **58,** 595–598.
186. Reeve, R. M. (1943). *Food Res.* **8,** 137–145.
186a. Rodriguez, D. B., Tanaka, Y., Katayama, T., Simpson, K. L., Lee, T.-C., and Chichester, C. O. (1976). *J. Agric. Food Chem.* **24,** 819–822.
187. Roels, O. A., and Trout, M. (1959). *Am. J. Clin. Nutr.* **7,** 197–202.
188. Roels, O. A. (1967). *In* "The Vitamins" (W. H. Sebrell and R. S. Harris, eds.), Vol. 1, pp. 113–121. Academic Press, New York.
189. Rothblatt, G. H., Ellis, D. S., and Kritchevsky, D. (1964). *Biochim. Biophys. Acta* **84,** 340–347.
190. Rountree, J. I. (1931). *J. Nutr.* **3,** 345–351.
191. Ryvarden, L., and Jensen, S. L. (1964). *Acta Chim. Scand.* **18,** 643–654.
192. Saperstein, S., and Starr, M. P. (1954). *Biochem. J.* **57,** 273–275.
193. Schalteggar, K. H. (1964). *J. Chromatogr.* **19,** 75–80.
194. Seybold, A., and Hulsbruch, W. (1943). *Bot. Arch.* **44,** 336–341.
195. Sherman, H. C. (1941). *J. Nutr.* **22,** 153–165.
196. Sherman, W. C., and Salmon, W. D. (1939). *Food Res.* **4,** 371–380.
197. Sherman, W. C. (1941). *Proc. Soc. Exp. Biol. Med.* **47,** 199–200.
198. Skurnik, L., and Suhonen, P. (1939). *Z. Vitaminforsch.* **8,** 316–323.

199. Slanetz, C. A., and Scharf, A. (1943). *Proc. Soc. Exp. Biol. Med.* **53,** 17–19.
200. Smith, A. M., and Robb, W. (1943). *J. Agric. Sci.* **33,** 119–121.
201. Smith, J. E., and Borchers, R. (1972). *J. Nutr.* **102,** 1017–1024.
202. Misth, M. C., and Farrankop, H. (1945). *Ariz. Agric. Exp. Stn., Mimeo. Rep.* **69,** 611.
203. Smith, M. C., Farrankop, H., Caldwell, E., and Wood, M. A. (1944). *Ariz. Agric. Exp. Stn., Mimeo. Rep.* **67,** 1–11.
204. Steinegger, P., Menzi, M., and Berger, R. (1956). *Gefluegelhof* **19,** 215–220.
205. Steinegger, P., and Zanetti, G. (1957). *Arch. Gefluegelkd.* **21,** 236–245; **23,** 166–173 (1959).
205a. Stewart, I. (1977). *J. Assoc. Off. Anal. Chem.* **60,** 132–136.
205b. Stewart, I. (1977). *J. Agric. Food Chem.* **25,** 1132–1137.
206. Stewart, I., and Wheaton, T. A. (1971). *J. Chromatogr.* **55,** 325–336.
207. Strain, H. H. (1934). *J. Biol. Chem.* **105,** 523–535.
208. Strain, H. H. (1935). *J. Biol. Chem.* **111,** 85–93.
209. Strain, H. H. (1953). *J. Phys. Chem.* **57,** 638–640.
209a. Strain, H. H., Sherma, J., Benton, F. L., and Katz, J. J. (1965). *Biochim. Biophys. Acta* **109,** 1–15; 16–22.
210. Strain, H. H. (1954). *Arch. Biochem. Biophys.* **48,** 458–468.
211. Straus, W. (1942). *Helv. Chim. Acta* **9,** 705–717.
212. Subbara, C., and Cama, H. R. (1964). *Indian J. Sci.* **2,** 451–453.
213. Sumner, J. B., and Sumner, R. J. (1940). *J. Biol. Chem.* **134,** 531–533.
214. Suzere, G. (1959). *J. Biol. Chem. Jpn.* **46,** 1497–1504.
215. Tainish, P., and Wilkinson, H. (1939). *J. Soc. Chem. Ind., London,* **58,** 1051–1052.
216. Taylor, A. L., and Witte, J. P. (1938). *Ind. Eng. Chem.* **30,** 110–111.
217. Thaler, H. (1937). *Fette Seifen* **44,** 38–42.
218. Tichenor, D. A., Martin, B. C., and Wells, C. E. (1965). *Food Technol.* **19,** 406–409.
219. Ting, S. V., and Deszyck, E. J. (1958). *Proc. Am. Soc. Hortic. Sci.* **71,** 271–277.
220. Tobler, F.,and Tobler, G. (1921). *Ber. Bot. Ges.* **39,** 213–218.
221. Treichler, R., Kemmerer, A. R., and Fraps, G. S. (1942). *J. Nutr.* **24,** 57–64.
222. Turian, G. (1957). *Physiol. Plant.* **10,** 667–680.
223. U.S. Pharmacopeia XIV (1950), p. 789. Mack Publ. Co., Easton, Pa.
224. U.S. Pharmacopeia XVI (1960). p. 833. Mack Publ. Co., Easton, Pa.
225. Valadon, L. R. G., and Mummery, R. S. (1968). *Biochem. J.* **106,** 479–484.
226. Vandercook, C. E., and Yokoyama, H. (1965). *J. Food Sci.* **30,** 865–868.
227. Van Eekelen, M., and Emmerie, A. (1935). *Acta Brevia Neerl. Physiol., Pharmacol., Microbiol.* **4,** 171–172.
228. Van Loeseke, H. (1929). *J. Am. Chem. Soc.* **51,** 2439–2443.
229. Vavich, M. G., and Kemmerer, A. R. (1950). *Arch. Biochem.* **28,** 295–298.
230. Vogele, A. C. (1937). *Plant Physiol.* **12,** 929–955.
231. Von Euler, I. H., Ahestrom, L., Hogbert, B., and Tengstom, S. (1943). *Chem. Zentralbl.* **11,** 41.
232. Wallace, V., and Porter, J. W. (1952). *Arch. Biochem. Biophys.* **36,** 468–472.
233. Weckel, K. G., Santon, B., Herman, E., Laferriere, L., and Gabelmann, W. H. (1962). *Food Technol.* **16,** 91–94.
234. Weier, T. E. (1942). *Am. J. Bot.* **29,** 35–44.
235. Werchmeister, P. (1960). *Zuechter* **30,** 164–168.
236. White, J. W., Jr., Zscheile, F. P., and Brunson, A. M. (1942). *J. Am. Chem. Soc.* **64,** 2603–2606.
237. Wilson, H. E. C., Ahmad, B., and Majumdar, B. N. (1936). *Indian J. Med. Res.* **24,** 399–409.
238. Wilson, R. H., Thomas, J. O., and De Eds, F. (1942). *Fruit Prod. J.* **22,** 15–17.
239. Wilson, R. H., Ambrose, A. M., De Eds, F., Dutton, H. J., and Bailey, G. F. (1946). *Arch. Biochem.* **10,** 131–140.
240. Wilstold, H., and Jensen, H. B. (1937). *Sven. Kem. Tidskr.* **49,** 258–260.
241. Wineman, H. G., Kane, E. A., Shinn, L. A., and Cary, C. A. (1938). *J. Agric. Res.* **57,** 635–669.
242. Wolf, D. A., and Cornwell, D. G. (1964). *Science* **144,** 1467–1468.

243. Wood, G. R. (1941). *Sci. Agric.* **22,** 93–103.
244. Yamamoto, H. Y. (1964). *Nature (London)* **201,** 1049–1050.
245. Yokoyama, H., White, M. J., and Vandercook, C. E. (1965). *J. Org. Chem.* **30,** 2482–2483.
245a. Zakaria, M., Simpson, K., Brown, P. R., and Krstulovic, A. (1979). *J. Chromatogr.* **176,** 109–117.
246. Zechmeister, L., and Cholnoky, L. V. (1932). *Hoppe-Seyler's Z. Physiol. Chem.* **208,** 26–32.
247. Zechmeister, L., and Tuzson, P. (1936). *Hoppe-Seyler's Z. Physiol. Chem.* **240,** 191–194.
248. Zechmeister, L., and Tuzson, P. (1939). *Ber. Dtsch. Chem. Ges.* **72,** 1340–1346.
249. Zechmeister, L., and Polgar, A. (1941). *J. Biol. Chem.* **140,** 1–3.
250. Zechmeister, L., LeRosen, A. L., Schroeder, W. A., Polgar, A., and Pauling, L. (1943). *J. Am. Chem. Soc.* **65,** 1940–1951.
251. Zscheile, F. P., White, J. W., Jr., Beadle, B. W., and Roach, J. R. (1942). *Plant Physiol.* **17,** 331–346.
252. Zscheile, F. P., and Whitmore, R. A. (1947). *Anal. Chem.* **19,** 170–172.

Appendix: ABSORPTION MAXIMA AND $E^{1\%}_{1\,cm}$ VALUES FOR CAROTENOIDS[a]

Carotenoid	Solvent[b]	Absorption maxima, nm[c]			$E^{1\%}_{1\,cm}$, λ (nm)	Ref.[d]
Actinioerythrin	PE	(470)	496	529		*107*
	Ac	(480)	508	538		*107*
	C			518	(550)	*107*
Adonirubin	Hex		462			*(84)*
	EtOH		468			*154*
	C		479			*84*
Adonixanthin	PE		460			*73*
Aleuriaxanthin	PE	434	460	491	2600 (460)	*177*
	Ac	(440)	463	493	2440 (463)	*16*
Alloxanthin	PE	(422–424)	450–451	478–479		*34, 97, 190*
(cynthiaxanthin,	Hex		451	480		*37*
pectenoxanthin)	EtOH	(425–430)	450–454	478–483		*34,95,97*
	C	436	460	489		*95*
	B	443	463	492		*33,34,97*
	CS		481	513		*97*
Anhydrocryptocapsol	B	440	466	498	1950 (466)	*40*
Anhydroescholtzxanthin	PE	470	495–500	528–531		*46,141,267*
	Hex	471	499	532–534	3018 (499)	*131,267*
	EtOH	474	497	530		*45*
	Cy	474	503	538		*141*
	C	484–487	510–516	542–549		*45,141*
	B	486	512	545		*45*
	CS	503–506	528–539	566–579		*45,141*
Anhydrorhodovibrin	PE	453–455	480–483	512–516	2700 (483)	*170,186*
(P-481)	Ac	459	488	522		*169*
	C	471	499	533		*169*
	B	471	500	534		*169*
	CS	493	521	559		*169*
Anhydrowarmingol	PE	445	462	498		*203*
(lycopenol)						

(*Continued*)

Appendix—*Continued*

Carotenoid	Solvent[b]	Absorption maxima, nm[c]			$E^{1\%}_{1\,cm}$, λ (nm)	Ref.[d]
Anhydrowarmingone	PE		459–496			*6, 212*
(lycopenal)	Ac		496			*6*
Antheraxanthin	Hex	(421)	445	470		*238*
	PE	422–423	444–445	472–477		*64,70,153*
	EtOH	422	444	472		*228*
	C	430	456–461	484–491		*91,228*
	B	(431)	456	487		*53*
	CS		478	510		*135*
Aphanicin, *see* Canthaxanthin						
Aphanin, *see* Echinenone						
Aphanizophyll	MeOH	444	475	507		*108, 109*
	EtOH	445	472	502		*229*
	PE	446	472	502		*105*
	Ac	450	476	507		*105*
	C	453	482	510		*229*
	Pyr	462	494	531		*140*
Aphanol, *see* Isocryptoxanthin						
β-Apo-2′-carotenal	PE		494		2300 (494)	*215*
	B		515	556		*248*
	CS	505	544	579		*248*
β-Apo-2′-carotenol	PE	447	473	503		*208*
β-Apo-2′-carotenyl-acetate	PE	448	474	504	2905 (474)	*208*
β-Apo-4′-carotenal	PE		487		2500 (487)	*215*
β-Apo-6′-carotenal	PE		472		2120 (472)	*215*
β-Apo-8′-carotenal	PE		457		2640 (457)	*208,215*
	Cy		461	488		*208*
	EtOH		463			*259*
	B		470	504		*248*
	CS		488	526		*248*
β-Apo-4′-carotenoic acid ethyl ester	Hex		472			*264*
	PE		474		2835 (474)	*215*
	B		486			*264*

β-Apo-8′-carotenoic acid ethyl ester	PE		445	470	2500 (445)	*215*
	PE		448	(472)	2515 (448)	*125*
	Cy		449	475		*125*
β-Apo-10′carotenal	PE		435		2190 (435)	*208*
	PE		432	(458)		*259*
	C		453	487		*248*
β-Apo-10′carotenoic acid	PE		430		2235 (430)	*125*
β-Apo-12′carotenal	PE		414		2160 (414)	*207,208*
	PE		412	435		*28*
Apo-10′-fucoxanthinal	PE	389	410	435		*28*
Apo-12′-fucoxanthinal (C_{25})	PE	368	384	408		*28*
Apo-12′-fucoxanthinal (C_{27})	PE	389	408	431		*28*
Apo-6′-lycopenal	PE	(462)	479	(515)		*149*
	Ac	(448)	473			*149*
	EtOH		475			*149*
	MeOH		477			*149*
	B		502	540		*249*
	CS	494	525	569		*249*
Apo-6′-lycopenoic acid	MeOH	(440)	460	488		*149*
	Ac	(447)	468	(493)		*149*
Apo-6′-lycopenoic acid methyl ester	PE	(448)	471	503	2600 (471)	*149*
	Ac	445	469	(495)		*149*
Apo-6′-lycopenol	PE	426	451	483		*149*
Apo-8′-lycopenal	PE		468			*149*
	Ac		468			*149*
	EtOH		465			*149*
	MeOH		468			*149*
	Cy		475	505	2839 (475)	*231*
	B		487–488	(517–518)		*27,249*
	CS	473	508	546		*249*

(*Continued*)

Appendix—*Continued*

Carotenoid	Solvent[b]	Absorption maxima, nm[c]			$E^{1\%}_{1\,cm}$, λ (nm)	Ref.[d]
Apo-1-norbixinal (*cis*) methyl ester	B		456	485		*20*
Apo-10′-violaxanthal	Hex	403	424	448		*59*
	EtOH		440			*59*
Apo-12′violaxanthal	B		412	(430)		*58*
Astacene	PE		473			*145*
	EtOH		478			*145*
	Ac		482			*89*
	Pyr		498		1690 (498)	*126*
	CS		500			*92*
Astaxanthin	Hex		467–468			*60,85*
	PE		468–470			*143,144*
	MeOH		472			*168*
	EtOH		478			*144*
	B		478			*33*
	Ac		480			*89*
	C		483–485			*85,168*
	CS		502–503			*92,168*
Astaxanthin diacetate	EtOH		474			*17*
	B		482			*17*
Asteroidenone	PE		455–457			*75,91*
	Hex		461			*91*
	EtOH		462			*229*
	C		471			*229*
	B		472			*168*
	CS		494			*91*
Aurochrome (ζ-carotene)	PE	380	400	425		*130*
	Hex	380–381	401	426		*79,236*
	B	387	409	434	2042 (409)	*20*
	CS		426–428	454–457		*135,140,236*

Auroxanthin	Hex	380	400	425		*239*
	PE	382	402	427		*246*
	EtOH	379–382	400–403	425–428	1850 (402)	*48,49,224,242,260*
	C	385	413	438		*246*
	B	385	410	(433)		*73*
	CS		423	454		*135*
Azafrin	PE		409		2200 (409)	*65*
	C		428	458		*140*
	Pyr		428	458		*140*
Azafrinaldehyde	CS		457	489		*249*
α-Bacterioruberin	PE	464	494	528	2350 (494)	*170*
	Ac	468–470	498–500	532–534	2540 (498)	*147,170*
	C	475	506	544		*170*
	CS	501	534	572		*170*
Bisanhydrobacterioruberin	PE	465	493	527		*198*
	Ac	471	498	533		*198*
Bisanhydrocapsanthol	B	461	489	522	2530 (489)	*40*
3,4,3′,4′-Bisdehydro-β-carotene	PE		471		2400 (471)	*115,117*
	Hex		471		2370 (471)	*201*
Bisdehydrolycopene	Hex	476	504	542		*140*
	PE	480	510	540		*230*
	C	493	528	567		*140*
	B	493	531	570		*140*
	Pyr	492–496	528–536	568–575		*230,249*
	CS	519	558	601		*249*
Bisdihydrocanthaxanthin	Hex		(406)	422		*85*
Bixin	PE	432	456	490	4200 (456)	*126*
	C	443	475	510		*102*
	CS	457	489	524		*140*
Caloxanthin	EtOH	(426)	449	475		*229*
	C	(432)	458	484		*229*

(*Continued*)

Appendix—*Continued*

Carotenoid	Solvent[b]	Absorption maxima, nm[c]			$E^{1\%}_{1\,cm}$, λ (nm)	Ref.[d]
Canthaxanthin (aphanicin)	PE		465–467		2200 (466)	*126,268*
	Hex		466		2200 (466)	*100,126*
	Cy		469		2200 (469)	*232*
	EtOH		474–478			*94,172,201,229*
	C		482			*92*
	B		480–485		2092 (480)	*65,201,244*
	CS		500			*92,100*
Capsachrome	B	(431)	456	483		*53*
	CS		482	515		*136*
Capsanthin	Hex		475	504		*92*
	PE	(450)	475	505		*70*
	B		481–484	(510)		*55,73*
	B	(460)	483	518	2072 (483)	*70*
	CS		503	542		*140*
Capsanthin monoepoxide	PE		472			*70*
	C		481	511		*136*
	B	(455)	478–479	509–513		*53,70*
	B		483	516		*136*
	CS		499	534		*136*
Capsanthinone (capsanthone, ketocapsanthin diacetate)	B	454	487	524		*140*
	CS	468	503	541		*140*
	B	443	473	507		*265*
Capsanthol	B	(430)	456	487		*55*
Capsochrome	B	(431)	456	(483)		*53*
	CS		482	515		*140*
Capsolutein	B	(430)	455	486		*53*
Capsolutein 5,6-epoxide	B	425	451	483		*53*
Capsolutein 5,8-epoxide	B	406	430	459		*53*
Capsorubin	PE	444	474	506		*92*
	PE	(455)	479	510		*70*
	B	(460)	487–489	522–523	2200 (489)	*53,73*
	CS	467–470	502–503	541–542		*92,140,243a*

Capsorubol	B	424	449	481		*53*
Capsorubone	B		450	(518)		*73*
α-Carotene	Hex	418–421	441–445	471–474		*57,192,239*
	PE	422–423	444–447	473–475	2800 (444)	*49,126,146,214, 259,260*
	EtOH	423	444	473		*96*
	Ac	424	448	476		*110*
	C		454–455	484–485		*92,146*
	CS		477	509	2180 (477)	*92*
α-Carotene 5,6-epoxide	Hex	(399)	441–442	471–472		*73,92*
	C		454	483		*140*
	B		455	484		*140*
	CS		471	503		*140*
β-Carotene	Hex	423	446	475	2592 (453)	*79*
	PE	425	448–453	475–482	2592 (453)	*92,105,116,146, 174,214,259,260*
	AC	(429)	452	478		*88*
	EtOH		453	480	2620 (453)	*116*
	Cy		457	485	2505 (457)	*116*
	C	(438)	462	485–487		*146,228*
	C		465–466	491–497	2396 (465)	*92,116*
	B	(435)	462	487		*18*
	B		465–466	494	2337 (465)	*65,116*
	CS	(450)	484–485	512	2008 (484)	*92,116,140*
β-Carotene 5,6-epoxide	Hex	423	444–446	473–475	2590 (446)	*79,236*
	PE		447	478		*135*
	C		459	492		*135*
	B		460	492		*135*
	CS		479	511		*135*
β-Carotene 5,6,5′,6′-diepoxide	Hex	417	440	468–470	2690 (440)	*79,236*
	EtOH	418	442	471		*228*
	C	242	448	477		*228*
	B	426	451	481	2394 (451)	*20*
	CS		472	502		*177*

(*Continued*)

Appendix—*Continued*

Carotenoid	Solvent[b]	Absorption maxima, nm[c]			$E^{1\%}_{1\,cm}$, λ (nm)	Ref.[d]
β-Carotene 5,8,5′,8′-diepoxide	Hex	381	401	426		*79*
β,γ-Carotene	PE	420–421	444–448	472–478		*9,15*
γ,γ-Carotene	Hex	437	462	492	2760 (462)	*23*
	PE	433–438	457–462	487–494	3100 (462)	*177,205,214,259*
	EtOH	(437)	460	490		*95*
	Ac	(439)	461	491		*88*
	C	446	475	509		*140*
	B	447	477	510		*140*
	CS	463	496	533		*92*
γ-Carotene	PE	414	438	468		*9*
δ-Carotene	Hex	428	458	490		*92*
	PE	431	456	489	3290 (456)	*214*
	C	440	470	503		*92*
	CS	457	490	526		*248*
ϵ-Carotene	Hex	414–415	438–439	468–470		*36,84*
	PE	416	440	470	3120 (440)	*214*
	EtOH	417	440	470		*36*
	B	425	451	482		*36*
ζ-Carotene	PE	376–378	398–400	418–426		*57,72,177,205, 259,260*
	Hex	377–380	397–400	423–426	2555 (400)	*53,72,192*
	EtOH	377	399	425		*205*
Asymmetric	PE	376	396	418		*66*
ζ-Carotene 1,2-epoxide	PE	377	400	425		*22*
η-Carotene	PE	376	397	422		*259*
	Hex	382	405	429		*85*
β-Carotenone	Hex	417	440	468–470		*79,236*
	PE		443	471		*135,163*
	C		456	484		*135,163*
	C		(460)	490	1626 (490)	*261*
	B		453–456	485–486		*135*
	CS		466–470	499–502		*135*

β-Carotenonealdehyde-like	Hex	399	422	448		*250*
Celaxanthin	Hex	453	481	516		*26*
	EtOH	455	488	521		*140*
	CS	487	521	562		*169*
Chlorobactene	PE	435	461	491		*27*
	B	450	476	508		*27*
	CS	(470)	496	530		*27*
Chloroxanthin	Hex	417	440	470		*172*
Chrysanthemaxanthin	Hex	400	420–421	446–450		*92,239*
	EtOH	400	421	448	2100 (421)	*130,133,140*
	C		430	459		*140*
	CS		451	481		*92*
Citranaxanthin	Hex		463	495		*259*
	PE		463	495	2145 (463)	*255*
	EtOH		475	(488)		*255*
β-Citraurin	Hex		450	476		*259*
	B		464	(493)		*58*
	CS	457	488	523		*266*
Citroxanthin	PE	402–404	424–426	450–456		*247,260*
(mutatochrome)	Hex	403–404	427–428	450–456	2440 (427)	*25,73,79,92,236*
	EtOH	405	427	454		*247*
	DE	415	428	452	2260 (428)	*105*
	C		435	469		*92*
	B		437–440	470	1989 (437)	*20,92*
	CS		456–459	489–490		*92,135,139,236*
Compound X	Hex	(280)	288	(298)		*233*
Corynebacterium						
poinsettiae	PE	410	435	463		*198*
	Ac	414	439	468		*198*
-450-monool	Ac	434	455	484		*198*
-473-monool	Ac	450	475	507		*198*
Corynexanthin	EtOH	415	437	467		*92*
	C	423	447	478		*92*
	CS	435	466	495		*92*

(*Continued*)

Appendix—*Continued*

Carotenoid	Solvent[b]	Absorption maxima, nm[c]			$E^{1\%}_{1\,cm}$, λ (nm)	Ref.[d]
Crocetin	Hex	400	420	445		*140*
	PE	400	422	450	4320 (450)	*126*
	C		435	463		*140*
	Pyr	411	436	464		*140*
	CS	426	453	482		*140*
Crocetin dimethyl ester	PE	400	422	450	3980 (450)	*124,125*
Crocetin dialdehyde	PE	408–411	430–434	458–460		*28,120*
	B	422	445	473	3970 (445)	*83*
Crocetin semialdehyde	B	422	445	473		*83*
Crocoxanthin	Hex	422	445	475		*37*
	EtOH	421	443	472		*96*
	C	430	454	482		*96*
Crustaxanthin	Hex	423	448–450	475–480		*85,194*
	EtOH		450	477		*102*
	C		457	(482)		*237*
	CS		471			*237*
Cryptocapsin	Hex	(445)	470	(497)		*53*
	PE		475	505		*70*
	B		483	518		*70*
	B		486	(520)	1972 (486)	*40*
Cryptocapsol	B	(428)–432	454–457	483–487	2240 (457)	*40,53*
Cryptocapsone	B		491		1905 (491)	*40*
Cryptochrome	Hex		398	424		*54*
	CS		424	456		*137*
Cryptoflavin	Hex		424	452		*56*
	PE	403	426	453		*49*
	EtOH		430	460		*140*
	C		438	468		*92*
	B		439	470		*140*
	CS		459	490		*140*

α-Cryptoxanthin	Hex	421–422	445–446	473–475	2636 (445)	*37,39,202*
(zeinoxanthin)	C	434	456	485		*182*
	B	433	457	488	2355 (457)	*39*
	CS	449	474	505		*182*
β-Cryptoxanthin	PE	425	449	475–476		*70,104*
(cryptoxanthin)	PE	(425)	450–452	479–480	2386 (452)	*123,260*
	Hex	425	451	483		*125*
	EtOH	(428)	449–452	473–478		*95,123*
	CS	453	483	518		*92*
Cryptoxanthin 5,6-epoxide	Hex	(419)	443	472		*56*
	PE	(422)	444–445	473–474		*52,54,260*
	EtOH	424	445–449	477–481		*96,137*
	C		456	488		*137*
	C	(434)	456	483		*228*
	B		461	494		*137*
	CS		479	512		*137*
Cryptoxanthin 5,8-epoxide	Hex	424	452			*56*
Cryptoxanthin 5,6,5′,6′-di-	PE	415–416	439	469–470		*52,260*
epoxide	Hex	416	439	470		*54*
	EtOH	423	442	472–473		*94,137*
	C	432	452	480		*94*
	B		455	486		*137*
	CS		473	503		*137*
Cryptoxanthin 5,6,5′,8′-di-	PE	400	422	448		*260*
epoxide	Hex	400	423	448		*54*
Cryptoxanthin 5,8,5′,8′-di-	PE	399	424			*260*
epoxide						
Decaprenoxanthin	Hex	(394)	417	439		*245*
(dehydrogenans P-439)	Ac	(400)	419	443		*180*
Deepoxyneoxanthin	EtOH	(423–426)	445–447	472–475	2350 (445)	*49,159,260*
(trollein)	B	432	456	485		*48*
3,4-Dehydro-β-carotene	PE		461		2330 (461)	*117*
	Hex		462			*131*

(*Continued*)

Appendix—*Continued*

Carotenoid	Solvent[b]	Absorption maxima, nm[c]			$E^{1\%}_{1\,cm}$, λ (nm)	Ref.[d]
Dehydroflexixanthin	Ac		485	(505)		*3*
Dehydrogenans-ζ-carotene						
Monohydroxy	Hex	378	400	425		*245*
Dihydroxy	Hex	378	400	423		*245*
3,4-Dehydrolycopene	PE	462–465	491–493	524–527	3000 (492)	*8,177,205*
	EtOH	465	492	525		*205*
	B	481	514	551		*249*
	Pyr	492	528	568		*230*
	CS	501	536	574		*249*
11′,12′-Dehydrononaprenoxanthin	Hex	399	422	449		*245a*
2′-Dehydroplectaniaxanthin (monoester)	PE		495	(522)		*17*
3,4-Dehydrorhodopin	PE	451–455	479–483	511–517		*170,205*
	EtOH	454	481	513		*205*
3′-4′-Dehydrotorulene	PE		490	520		*3*
	Ac		494	(523)		*3*
Dehydroxydecaprenoxanthin	Hex	414	438	437		*245a*
Deoxyflexixanthin	PE		477	503	≧2650 (477)	*3*
	MeOH		477	(500)		*3*
	Ac		481	508		*3*
Diadinochrome	EtOH	(409)	427	456		*195*
Diadinoxanthin	Hex	(421)	445	475		*183*
	EtOH	(424)	445–448	474–478		*92,228*
	C	432	455	482		*228*
Cis/trans	CS	449	474	506		*183,189*
Diatoxanthin	Hex		439	467		*37*
	EtOH	(425–428)	449–453	475–481		*79,92,195,228*
	C	(433)	458	486		*228*

7,8-Didehydroastaxanthin	DE		473			*87*
(may contain 7, 8, 7′,8′-	Pyr	(478)	495	(522)		*219*
tetradehydroastaxanthin)	CS	(480)	518	541		*219*
3′,4′-Didehydro-18′-hydroxy-γ-carotene	PE		481		2790 (481)	*208*
9,10,9′,10′-Diepoxycanthaxanthin	EtOH		336			*200*
	B		344			*200*
3,4-Dihydroanhydrorhodovibrin	PE	444–446	470–473	501–504		*68,186*
7,7′-Dihydro-β-carotene	PE	382	405	429	2990 (405)	*119*
	Cy	386	407	433	2690 (407)	*113*
Dihydrocrocetin	MeOH	336	355	377		*78*
1,2-Dihydro-3,4-dehydrolycopene	PE	457	483	518		*184*
1′,2′-Dihydro-1′-2′-dihydroxy-4-ketotorulene	EtOH	(455)	478	(501)		*205*
1′,2′-Dihydro-4,2′-diketotorulene	PE	(469)	494	526		*205*
	EtOH		502			*205*
1′,2′-Dihydro-1′-hydroxy-γ-carotene	PE	(438)	459	488		*103*
	PE	444	462	494		*27*
1′,2′-Dihydro-1′-hydroxy-4-keto-γ-carotene	PE		462			*167*
	PE		465	(490)		*103*
	Ac		471	(491)		*103*
1′,2′-Dihydro-1′-hydroxytorulene	PE	440	465	497		*205*
	EtOH	444	468	499		*205*
1,2-Dihydrolycopene	PE	444	470	502		*184*
1,2-Dihydroneurosporene	PE	414	439	468		*184*
2′-Dihydrophillipsiaxanthin	PE	482	510	544		*17*
1′,2′-Dihydrophytoene	PE	275	285	296		*187*
1′,2′-Dihydrophytofluene	PE	331	348	367		*187*
3,4-Dihydrospheroidene	PE	414	438	468		*186*
11′,12′-Dihydrospheroidene	PE	392	413	439		*186*
3,4-Dihydrospirilloxanthin	PE	453	481	512		*186*

(*Continued*)

Appendix—*Continued*

Carotenoid	Solvent[b]	Absorption maxima, nm[c]			$E^{1\%}_{1\,cm}$, λ (nm)	Ref.[d]
2,2′-Dihydroxy-β-carotene	Ac	425	452	480	2060 (452)	*151*
3,4-Dihydroxy-γ-carotene	—	443	463	490		*193*
x,y-Dihydroxy-3,4-dehydro-α-carotene	Hex	(418–430)	451	(469–484)		*237*
	C	(420–445)	465	(480–490)		*237*
	CS	(436–462)	484	(501–520)		*237*
5,6-Dihydroxy-5,6-dihydrolycopene	Hex	431	456	468	2820 (456)	*32*
3,3′-Dihydroxy-4′-oxo-β-carotene	PE		458			*53*
	EtOH		464			*53*
2-(Dihydroxyisopentenyl)-2′-isopentenyl-β-carotene	MeOH	(420)	448	473		*198*
	PE	(425)	451	480		*198*
	Ac	(427)	454	481		*198*
3,3′-Dihydroxyisorenieratene	PE	(425)	445	475		*10,11*
	MeOH		453			*199*
	EtOH	(430)	451	487		*199*
	B	438	463	493		*11*
Dihydroxylycopene	PE	444–446	470–473	503–504		*210,230*
	Ac	446–448	474	507	1585 (474)	*210*
	C	456–458	482–486	515–521		*210,230*
	B	458	485–486	520–521		*210*
	CS	475–479				*210*
Dihydroxypirardixanthin	PE	413	436	467		*160*
3,4-Diketo-α-carotene	Hex		462			*237*
	C		476			*237*
	CS		500			*237*
4,4′-Diketo-γ-carotene	EtOH		493	(513)		*205*
4,4′-Diketocynthiaxanthin	PE	(446)	474	504		*98*
4,4′-Diketolycopene	EtOH	(477)	498	(526)		*205*

2,2′-Diketospirilloxanthin	PE	488	518	555	2000 (518)	*171*
	DE		522			*171*
	Ac	495	528	559		*171*
	B	510	539	575		*171*
	C		543		2210 (543)	*216*
	CS	528–530	561–562	601–603		*127,216*
Dimethoxy-β-carotene	Hex		450–451			*267*
1,1′-Dimethoxy-1,2,1′2′-tetrahydrolycopene	PE	443	471	503	2990 (471)	*230*
Dinoxanthin	Hex	416	439	469		*183*
	MeOH	416	438	467		*183*
	EtOH	419	441	470–471		*92,183*
	CS	441	467	498		*183*
α-Doradecin	PE		455	471		*144*
	EtOH		(457)	473		*144*
β-Doradecin	PE		461	475		*144*
	EtOH		465			*144*
α-Doradexanthin	PE		455	470		*146*
	EtOH		458	478		*146*
Echinenone	DE		455			*105*
(myxothanin)	PE		456–458	(482)	2158 (458)	*65,112,146*
(aphanin)	Hex	(432)	459	(483)		*93*
	EtOH		457–461			*94,146*
	Ac		460			*88*
	Cy		461		2110 (461)	*232*
	C		471–473			*92,95*
	B		470–472	(490)	2091 (472)	*69,244*
	Pyr		477	508		*140*
	CS		488–494			*92,244*
Eschscholtzxanthin	Hex	442–444	472	502	3269 (472)	*80,141*
	EtOH	446	472	503		*140*
	Ac	448	474	505		*106*
	C	457	481	511		*106*
	B	459	486	520		*141*
	Pyr	463–464	489	521		*106,140*
	CS	474	507	542		*141*

(*Continued*)

Appendix—*Continued*

Carotenoid	Solvent[b]	Absorption maxima, nm[c]			$E^{1\%}_{1\,cm}$, λ (nm)	Ref.[d]
Eugorgianenoic acid ethyl ester	Hex	376	399	421		*86*
Flavochrome	Hex		422	450		*140*
	C		433	461		*140*
	B		434	462		*140*
	CS		451	482		*140*
Flavoxanthin	Hex	400	420	447		*239*
	PE		421	450		*140*
	EtOH	400	421	448		*130*
	C		430	459		*92*
	B	406–408	430–432	456–459	2550 (432)	*51,164,192*
	CS		449	479		*140*
Flexixanthin	Ac		483	510		*3*
Foliachrome, *see* Neochrome						
Foliaxanthin, *see* Neoxanthin						
Fucoxanthin	Hex	425–427	450	476–478		*28,92*
	PE	425	446	473		*152*
	EtOH	(426)	449	(465)		*96*
	C		460	(478)		*92*
	B	(443)	461	485		*28*
	CS	(450)	478	508	2036 (478)	*28*
Fucoxanthinol	PE	425	448	476		*28*
	EtOH		452		1453 (452)	*28*
	B	(436)	460	(487)		*90*
Galloxanthin	Hex	378	398	420		*7*
Gazaniaxanthin	PE	434	460	490		*31*
	Hex	435	463	495		*92*
	EtOH	435	462	495		*140*
	Ac	439	462	492	2580 (462)	*14*
	B	448	476	509		*92*
	CS	461	495	531		*92*
Guaraxanthin	Hex		(436)	455		*85*

Helenien (lutein dipalmitate)	Hex	420	445	475	1394 (445)	*39*
	B	433	457	487	1254 (457)	*39*
Heteroxanthin	EtOH	(423)	445–448	474–478		*225,226*
	C	434	456	485		*228*
1,2,7,8,11,12-Hexahydrolycopene (85%) + 15% of 7,8,1′,2′,7′,8′ isomer	PE	375	395	419		*187*
3,4,3′,4′,7′,8′-Hexahyrospirilloxanthin (tentative)	PE	414	438	467		*186*
Hopkinsiaxanthin	PE		466	497		*222*
Hydroxyanhydrorhodovibrin (hydroxy-P-481)	Hex	455	483	516		*170*
	C	469	498	532		*170*
	CS	491	522	559		*170*
3-Hydroxy-β-apo-10′-carotenal	Hex		431	(456)		*259*
Hydroxycapsolutein	B	(430)	456	486		*53*
Hydroxycapsolutein 5,6-epoxide	B	421	449	482		*53*
Hydroxy-α-carotene	Hex	416–421	440–445	471–475	2620 (445)	*32,57,84,192*
	Ac	(425)	447	475	2330 (447)	*151*
Hydroxy-α-carotene 5,6-epoxide	PE	417	441	469		*52*
Hydroxy-α-carotene 5,8-epoxide	PE	396	419	446		*52*
2-Hydroxy-β-carotene	Ac	(430)	452	479	2290 (452)	*151*
Hydroxy-ζ-carotene	Hex	380	401	425		*54*
x-Hydroxy-β,β-caroten-*y*-one	PE		453–457			*34,91*
	Hex		461			*91*
	CS		494			*91*
Hydroxychlorobactene	PE	(435)	461	491		*27*
	Ac	(440)	464	494		*175*
	B	(446)	467	495		*27*
4′-Hydroxycryptoxanthin	Hex	424	452	483		*142a*
	CS	451	481	516		*142a*

(*Continued*)

Appendix—*Continued*

Carotenoid	Solvent[b]	Absorption maxima, nm[c]			$E^{1\%}_{1\,cm}$, λ (nm)	Ref.[d]
4-Hydroxy-3′,4′-dehydro-β-carotene	Hex		461		2250 (461)	*201*
1-Hydroxy-1,2-dihydro-apo-3-lycopenal	Cy	449	474	505	2420 (474)	*231*
8-Hydroxy-7′,8′-dihydro-citranaxanthin	Hex	402	423	449		*258*
1′-Hydroxy-1′,2′-dihydro-2-isopentenyl-2′-(hydroxy-isopentenyl)torulene	PE	446	473	504		*198*
(c.p. 473)	Ac	451	478	509		*198*
1-Hydroxy-1,2-dihydro-phytoene	PE	275	285	298		*185*
1-Hydroxy-1,2-dihydro-phytofluene	PE	331	348	367		*185*
8-Hydroxy-7′,8′-dihydro-reticulataxanthin	Hex	401	422	448		*259*
5-Hydroxy-5,6-dihydro-zeaxanthin	Hex		449	477		*61*
	EtOH	(425)	450	478		*61*
	C		460	490		*61*
	B	(438)	460	490		*61*
	Pyr		466	495		*61*
	CS	(455)	478	508		*61*
3-Hydroxy-3′,4′-diketo-α-carotene, *see* α-Doradecin						
3-Hydroxy-3′,4′-diketo-β-carotene, *see* β-Doradecin						
3-Hydroxyechinenone	PE		457–458			*158,168*
	EtOH		460			*168*
	C		472			*168*
	B		472			*168*

3′-Hydroxyechinenone	PE		457			*34,91,106,168*
	C		472			*168*
4′-Hydroxyechinenone	PE		452–453	472		*166*
	PE		453–454	(472–474)	2250 (454)	*34,176*
	Hex		456–458		2155 (458)	*155,166,201*
	EtOH		460			*94*
	B		466			*34*
	C		470			*229*
	CS		490			*166*
3-Hydroxyisorenieratene	PE	425	446	475		*11*
	DE		449	(472)		*199*
	MeOH		450	(474)		*199*
	B	438	463	493		*11*
3-Hydroxy-3′-keto-α-carotene	EtOH	425	445	473		*94*
3-Hydroxy-4′-keto-β-carotene	PE		455			*53*
	EtOH		461			*53*
Hydroxyokenone (okenol)	PE		470	501		*203*
4′-Hydroxy-3′-oxo-echinenone	Hex		458–462			*155*
2-Hydroxyplectaniaxanthin	PE	445	471	502		*181*
	Ac	450	476	507	2445 (476)	*181*
	C	458	486	519		*181*
	B	460	488	521		*181*
3-Hydroxysintaxanthin	Hex	(426)	446	474		*259*
Hydroxyspheroidene (hydroxy-Y)	Hex	429	454	486		*201*
	C	440	466	500		*201*
	B	439	467	501		*201*
	CS	457	486	522		*201*
Hydroxyspheroidenone	Hex	460	483	516		*173*
	EtOH		487			*173*
	C		501			*173*
	B	(475)	501	530	2070 (501)	*127,173*
	CS	(490)	520	555		*173*

(*Continued*)

Appendix—*Continued*

Carotenoid	Solvent[b]	Absorption maxima, nm[c]			$E_{1\,cm}^{1\,\%}$, λ (nm)	Ref.[d]
3-Hydroxy-3,4,4′-triketo-β-carotene	CS		502			*155*
Idoxanthin	PE		457–461			*102*
Isocarotene	Hex	448	474	499		*132*
	Cy	447	473	504		*232*
Isocryptoxanthin (aphanol)	PE	(425)	448–449	474–475		*69,105*
	Hex	(427)	450–451	(479)		*93,267*
	EtOH	427	450	475		*94*
	EtOH	(425)	450	477		*101*
	Ac	(425)	449	472		*88*
	CS		482	509		*166*
Isofucoxanthin	PE	(430)	453	482	1600 (453)	*129*
	MeOH		454			*28*
	C		467			*28*
Isofucoxanthinol	MeOH		454			*28*
	C		467			*28*
Isolutein	Hex		442	471		*92*
	B		453	482		*92*
	CS		472	502		*92*
2-Isopentenyl-3,4-dehydrorhodopin	PE	454	482	514		*198*
	Ac	458	486	518		*197*
	B	470	499	533		*198*
Isorenieratene (leprotene)	PE	(426)	448	475	2080 (448)	*150*
	Hex	425	452	484		*92*
	C	428	460	495		*92*
	B	440–443	464–465	492–493		*44,126*
	CS	452	484	520		*253*
β-Isorenieratene	B	(440)	464	492		*148*
	CS	456	487	508		*44*

Isozeaxanthin	Hex	(427)	450	475	2360 (450)	*93,166,201*
	PE	(428)	451	478–479	2400 (451)	*118,176*
	EtOH	430	451	478		*94*
	CS		479	508		*166*
4-Keto-α-carotene (phoenicopterone)	Hex		449–454	(470)	2280 (451)	*32,81,84*
4-Keto-γ-carotene	PE		462			*167*
	PE		465	(490)		*103*
	PE	(437)	464	(486)		*205*
	Hex		468			*32*
	EtOH	(438)	468	(490)		*205*
	Ac		471	(491)	2386 (471)	*103*
4-Keto-3′,4′-dehydro-β-carotene	Hex		470			*201*
4-Keto-3′-hydroxy-β-carotene	PE		452			*104*
4-Keto-4′-hydroxy-β-carotene	Hex		458			*201*
4-Ketophleixanthophyll	Ac		480	(507)	1160 (480)	*106*
3-Ketomutatochrome	B		(437)	458		*73*
4-Ketotorulene	PE	(465)	489	519		*205*
	EtOH	(470)	491	516		*205*
Laetiporxanthin	MeOH (acid)		448	474		*241*
Loroxanthin (19-hydroxylutein)	EtOH	(425)	446	473		*156*
	C	431	455	482		*95*
Lutein	Hex	420–423	445–447	474–477		*37,92,94*
	PE	421–423	445–448	474–476		*70,146*
	EtOH	420–422	444–447	474–476	2550 (445)	*140,153,221,260*
	C	428–435	456–458	485–487		*92,143,146*
	B	431–433	455–458	487	2236 (458)	*42,51*
	CS	445	475	508	2160 (475)	*140,161*
Lutein 5,6-epoxide, *see* Taraxanthin						

(*Continued*)

Appendix—*Continued*

Carotenoid	Solvent[b]	Absorption maxima, nm[c]			$E^{1\%}_{1\,cm}$, λ (nm)	Ref.[d]
Lutein 5,8-epoxide	B	404	428	457		*57*
Luteochrome	Hex	398–404	422–427	448–450	2520 (422)	*79,236*
	CS		451	480–487		*135,236*
Lutoexanthin	EtOH	395–396	420–424	446–450		*96,224,260*
	B	406–408	431–432	460–461		*48,51,53*
Lycopenal, *see* Anhydro warmingone						
Lycopene						
Trans	PE	442–446	468–472	499–504	3450 (472)	*17,150,177,205, 214*
	Hex	446–448	472–473	504–505		*23,126*
	EtOH	443–444	469–472	500–502		*140,205*
	Ac	447–448	474–475	505–506		*106,170*
	C	458	484	518		*106*
	B	455	487	522	3370 (487)	*140,230*
	Pyr	462	496	525		*106*
	Cs	477	508	548		*140*
Neo A	Ac	443	469	500		*149*
Neo B	Ac	440	465	493		*149*
Lycopene 1,2-epoxide	PE	443	469	500		*22*
Lycopene 5,6-epoxide	PE	431	454	483		*22*
Lycophyll	PE	444	473	504		*140*
	EtOH	444	474	505		*140*
	Ac	447	474	504–508	2340 (508)	*151,190*
	B	456	487	521		*92*
	CS	473	507	547		*92*
Lycoxanthin	PE	443–445	469–473	500–504		*51,151,170,220*
	EtOH	444–445	470–474	502–505		*140,149*
	Ac	448	474–476	505–507	3080 (474)	*151,170*
	B	463	489	520		*220*
	CS	473–475	505–507	545–547		*92,220*

Methylbixin	PE	430–432	455–456	490	4050 (456)	*114,124*
Trans 10,10′-Cis	PE	444	470	502	2445 (470)	*114*
Micronone	PE	(416)	440	467		*206*
	EtOH		~445			*206*
	C		~453			*206*
Microxanthin	C	403	427	455		*206*
Mimulaxanthin	EtOH	425	448	476		*196*
Monadoxanthin	Hex	422	445	475		*37*
	EtOH	425	448	476		*196*
Monodemethylated	PE	462	494	528		*170*
spirilloxanthin (hydroxyspirilloxanthin, dehydrorhodovibrin)	Ac	467	499	533		*170*
Mutatochrome, *see* Citroxanthin						
Mutatoxanthin	PE		426	456		*140*
	EtOH	409	430	458		*140*
	C		437	468		*140*
	B		439	468		*140*
	B	410	434	462		*192*
	Pyr		443	473		*140*
	CS		459	488		*135,140*
Mutatoxanthin a	B	414	437	464		*48*
Mutatoxanthin b	B	414	436	465		*48*
Mytiloxanthin	EtOH		470			*35*
	B		480			*35*
Myxobactin	PE	(450)	473	504		*205*
	EtOH	(452)	475	505		*205*
Myxobactor	PE	(454)	475	(500)		*205*
	EtOH	(458)	479	(502)		*205*
Myxoxanthophyll	EtOH	445–448	471–473	503		*92,229*
	MeOH	(449)	473	503		*211*
	DE	(447)	474	504		*211*
	Ac	450–453	476–478	506–510	2160 (478)	*104,105,107,211*

(Continued)

Appendix—*Continued*

Carotenoid	Solvent[b]	Absorption maxima, nm[c]			$E^{1\%}_{1\,cm}$, λ (nm)	Ref.[d]
	THF	456	482	514	2030 (482)	*211*
	C	(460)	488	520–522		*107,211*
	B	462	488	522		*107*
	Pyr	465	493	528		*107*
	CS	479	508	544		*140*
Neochrome	EtOH	401	422–424	448–451	2270 (424)	*41,43,224*
	B	407–408	431–434	457–460		*43,140,192*
Neomethylbixin C	B	440	465	498		*20*
Neoxanthin	Hex	415	435–437	462–466		*92,239*
	EtOH	415–417	438–439	467–468	2243 (439)	*41,153,158,188*
	C	421–423	447–448	476–477		*228,246*
	B	421–425	447–450	477–480		*51,57,92,192*
	CS		463	493		*92*
Neoxanthin a	EtOH	417	437	465		*260*
	B	(421)	447	480		*56*
Neoxanthin b	EtOH	416	436	464		*260*
	B	423	447	479		*56*
Neurosporaxanthin	MeOH	432	465	500		*238*
	Hex	(440)	472	(505)	2450 (472)	*238*
	Ac		472	(496)		*1*
	PE		477	(504)		*1*
	PE		479	507	2500 (479)	*125*
	B	(460)	486	(515)	2270 (486)	*238,264*
	CS	(463)	496	(523)		*238*
Neurosporaxanthin methyl ester	PE		476	505		*125*
Neurosporene	PE	413–416	436–440	466–470		*67,126,177,205 259*
	Hex	415–416	438–440	468–470	2918 (440)	*72,99*
	EtOH	416	440	469		*205*
	CS	440–443	470–471	502–503		*99,142*

Nonaprenoxanthin (dehydrogenans P-373)	Hex	354	373	394		*245a*
2′-Norastaxanthin ester	Ac		490			*89*
Nostoxanthin	EtOH	(426)	448	475		*229*
	C	(432)	457	485		*229*
3,4,3′,4′,7′,8′,11′,12′-Octa-hydrospirilloxanthin (tentative)	PE	375	395	420		*186*
Okenol	PE	444	470	500		*179*
	Ac	445	472	501	2560 (472)	*179*
Okenone	PE		484	517		*203*
	PE	(460)	484	516	2320 (484)	*179*
	Ac	(465)	487	518		*179*
Oscilloxanthin	MeOH	464	496	531		*134*
	EtOH	468	492	526		*229*
	Ac	470	499	534		*108*
	C	480	510	548		*108*
	Pyr	483	514	552		*134,140*
	CS	494	528	568		*134,140*
3-Oxoechinenone	EtOH		466			*76*
Paracentrone	Hex	417	439	468		*111*
	B	440	456	480		*111*
Pectenolone diacetate	EtOH		463			*33*
	B		470			*33*
Pectenoxanthin, *see* Cyn-thiaxanthin						
Peridinin (sulcatoxanthin)	Hex	(431)	454–455	484–485		*183,227*
	MeOH		467			*183*
	EtOH		475			*186*
	B		465	502		*183*
	CS	450–454	480–482	512–516		*92,140,183*
Persicaxanthin	Hex	353	371	392		*50*
Pheixanthophyll	Ac	454	478	509		*106*
	C	(465)	489	522		*106*
	Pyr	(465)	493	526		*106*

(*Continued*)

Appendix—*Continued*

Carotenoid	Solvent[b]	Absorption maxima, nm[c]			$E^{1\%}_{1\,cm}$, λ (nm)	Ref.[d]
Phillipsiaxanthin (2,2′-diketobacterioruberin)	Ac	(495)	528	559		*17*
	C		543		2430 (543)	*216*
Phillipsiaxanthin diester	PE	488	518	554		*17*
	Ac	(495)	528	559		*17*
Philosamiaxanthin	EtOH	424	446	474	2555 (446)	*38*
	Ac	(425)	449	477		*178*
Phoeniconone (3-oxocanthaxanthin)	Hex		~467			*62*
	PE		468			*145*
	EtOH		477			*76*
	C		486			*62*
	Pyr		498		1700 (498)	*155*
	CS		498			*62*
Phoenicopterone, *see* 4-Keto-α-carotene						
Phoenicoxanthin (3-hydroxycanthaxanthin)	Hex		462–468			*62,84*
	C		486			*62,191*
	Pyr		490			*62*
	CS		498			*62*
Physalien (zeaxanthin dipalmitate)	PE		452	480	1340 (452)	*121*
Physoxanthin	Hex		445	474	2400 (445)	*24*
Phytoene	PE	275–276	285–286	296–297	1250 (285)	*12,64,66,205*
	Hex	(276)	286	(298)		*53*
	Hex	276	286	297	915 (286)	*72*
	EtOH	276	286	297		*72*
Phytoene 1,2-oxide	PE	275	285	296		*30*
Phytofluene	PE	330–333	347–348	366–368	1350 (348)	*49,64,66,177,205*
	Hex	331–332	347–348	366	1577 (347)	*49,72*
	EtOH	331	348	366		*205*
Phytofluene epoxide	PE	331	348	367		*22*

Phytofluenol	Hex	332	348	368		*92*
Plectaniaxanthin	PE	445–450	471–475	502–505		*13,19*
	Ac	450–454	474–478	507–509	2505 (474)	*13,19,181*
Prolycopene	PE		444	470	1920 (444)	*92*
	C		454	484		*92*
	B		456	485		*92*
	CS		470	501		*92*
Pyrenoxanthin	PE	420	448	472		*254*
	C	(432)	454	482	1763 (454)	*254*
Pyrroxanthin	Hex		459	487		*183*
	EtOH		471			*183*
	CS		512	586		*183*
Renierapurpurin	PE	447	468	497	2500 (468)	*150*
	B	464	487	519		*44*
	CS	475–477	504	544	2065 (504)	*150,201,251*
γ-Renierapurpurin	PE	(448)	474	504		*27*
	B	(460)	485	522		*27*
Renieratene	B	(457)	476	(507)		*44*
	CS	463	496	532		*252*
Reticulataxanthin	PE		463	490		*256*
	Hex		465	495		*55*
	B	(457)	478	(507)		*55*
Reticulataxanthol	B	431	456	486		*55*
Retrodehydro-β-carotene	PE	446	472	502	3205 (472)	*126*
	Hex		474	504		*267*
Retrodehydro-γ-carotene	Ac	(462)	487	519		*97*
Rhodopin	PE	443–445	470–473	503–504		*170,205*
	EtOH	445	470	501		*201*
	Cy	447	474	507	2989 (474)	*231*
	Ac	448	477	508		*170*
	Ac	(450)	475	508		*175*
	C	455	482	516		*170*
	B	456	483–484	515–519		*170,205*
	CS	476	505	544		*170*

(Continued)

Carotenoid	Solvent[b]	Absorption maxima, nm[c]			$E^{1\%}_{1\,cm}$, λ (nm)	Ref.[d]
Rhodopinal	PE			498		*203*
(warmingone)	Ac	448	472	503		*6,213*
	EtOH		(495)	512		*174,213*
Rhodopinol	PE	445	471	495		*203*
(warmingol)	Ac			503–505		*6,174*
Rhodovibrin	PE	455	483	516–517		*170,231*
	Ac	460	488	522		*170*
	C	469	498	532		*170*
	B	473	502	535		*170*
	CS	491	522	559		*170*
Rhodoxanthin	Hex	458	489–490	524	2500 (490)	*92,126*
	Cy	466	491	521	2386 (491)	*232*
	EtOH		496	538		*140*
	C	482	510	546		*140*
	B	474	504	542		*140*
	CS	491	525	564		*92,128*
Rosaxanthin	MeOH (acid)	359	379	407		*240*
Roserythrin	Ac		515			*89*
Rubichrome	PE		447	475		*138*
	C		455	486		*138*
	B		455	486		*138*
	CS		472–476	501–506		*92,138*
Rubixanthin	PE	434–435	460	490	2750 (460)	*31,175*
	Hex	432	462	494		*92*
	EtOH	433	463	496		*140*
	Ac	435	462	492	2909 (462)	*14*
	C	439	474	509		*140*
	CS	461	494	533		*140*

Rubixanthin 5,6-epoxide	PE		461	491		*138*
	C		474	504		*138*
	B		475	504		*138*
	CS	461	491	526		*138*
Rubixanthone	Hex	456	479	508		*193*
Saprospira grandis 434	PE	410	434	462		*2*
	MeOH	410	433	460		*2*
	Ac	411	437	465		*2*
Saprospira grandis 500	MeOH		490–505			*2*
	Ac		490–505			*2*
Saprospira thermalis 483	Ac		483	510		*4*
Saproxanthin	Ac	451	479	509	2920 (479)	*2*
Sarcinaxanthin	Hex	415	440	462		*92*
	EtOH	415	441	470		*140*
	C	423	451	480		*92*
	B	423	451	480		*92*
	CS		469	494		*92*
Sarcinene	MeOH	414	438	469		*234*
Semi-α-carotenone	Hex	438	462	490		*263*
	C		480	500	1431 (500)	*263*
	CS		499	533		*140*
Semi-β-carotenone	Hex	440	467	495	1850 (467)	*261*
	PE	446	470	501		*140*
	B	458	486	518		*140*
	C		487	(519)		*140*
	CS		499	538		*140*
Sinensiachrome	EtOH	353	374	396		*49*
Sinensiaxanthin	EtOH	372	392	416		*59*
	B	380	402	427		*50*
Sintaxanthin	Hex	(425)	448	475		*257*
	B	(436)	462	(485)	2588 (462)	*257*
Sintaxanthol	Hex	404	425	451		*257*
	B	415	437	463	1657 (437)	*257*

(*Continued*)

Appendix—*Continued*

Carotenoid	Solvent[b]	Absorption maxima, nm[c]			$E^{1\%}_{1\,cm}$, λ (nm)	Ref.[d]
Siphonaxanthin	PE		446	(468)		*152*
	EtOH		448–455			*223,243*
	C		466			*206*
Siphonaxanthol	EtOH	397	421	450		*243*
Siphonein	PE		450	(473)		*152*
	EtOH		456			*243*
	C		466			*206*
Spheroidene	Hex	429	455	487		*201*
(pigment Y)	EtOH	429	454	486		*201*
	C	440	466	500		*201*
	B	441	468	502	2785 (468)	*201*
	CS	457	486	522		*201*
Spheroidenone	Hex	(460)	482	513		*165*
(pigment R)	PE	460	483	515		*171*
	Ac	(455)	484	(505)		*171*
	EtOH		488			*165*
	C		499			*165*
	B	(475)	499–501	530–533		*165,171*
	CS	(495)	520	553		*165*
Spirilloxanthin	PE	461–467	490–494	523–529		*173,186,204,220*
	Hex	462	493	526		*204*
	EtOH	465	491	526		*170*
	Ac	468	499	533–534		*170,230*
	C	(475)	505	543		*170*
	B	479–80	510	546–549	2470 (510)	*170,220*
	CS	495–500	532	567–572		*170,220*
Tangeraxanthin	B	(463)	484	(520)		*55*
Tangeraxanthol	B	437	463	494		*55*
Tarachrome	B	408	432	460	2045 (432)	*82*

Taraxanthin	Hex	418–420	442–445	470–472		*92,239*
(lutein 5,6-epoxide)	EtOH	416–420	442	470–471	2800 (442)	*29,140,153,162, 235*
	B	425–429	452–455	483–485	2373 (455)	*51,82*
	C	433	453	483		*95*
	CS	441	469	501		*92*
Taraxanthin dipalmitate	Hex	420	442	471		*29*
3,4,3′,4′-Tetradehydro-lycopene	PE	480	510	540		*230*
	Pyr	492	528	568		*230*
7,8,7′8′-Tetrahydro-astaxanthin	DE		475	(496)		*87*
	B		492	(522)		*87*
	Pyr		496	526		*87*
	CS		507	541		*87*
3,4,3′,4′-Tetrahydrobis-anhydrobacterioruberin	Ac	448	473	504		*198*
7,8,7′,8′-Tetrahydro-capsorubin (P-441)	B	392	414	441		*53*
7′,8′,11′,12′-Tetrahydro-γ-carotene	PE	(358)	378	397	1800 (378)	*71*
1,2,1′,2′-Tetrahydro-lycopene	PE	444	470	502		*187*
5,6,5′,6′(?)-Tetrahydro-lycopene	Hex	418	441	473		*32*
7,8,11,12-Tetrahydro-lycopene	PE	374	395	419		*67*
	Hex	374	395	420	2519 (395)	*72*
1,2,1′,2′-Tetrahydro-neurosporene	PE	414	439	468		*187*
Tetrahydrophytoene	MeOH		215			*157*
	PE		220			*157*
3,4,11′,12′-Tetrahydro-spheroidene	PE	374	394	418		*186*
3,4,3′,4′-Tetrahydro-spirilloxanthin	PE	444	470	501		*150,186*
	Ac	447	473	503	2935 (473)	*5*

(*Continued*)

Appendix—*Continued*

Carotenoid	Solvent[b]	Absorption maxima, nm[c]			$E^{1\%}_{1\,cm}$, λ (nm)	Ref.[d]
3,4,7,8-Tetrahydro-spirilloxanthin	PE	425	453	483		*186*
3,4,3′,4′-Tetrahydro-spirilloxanthin-20-al	Ac	368	398	527		*77*
Thiothece 478	PE		478	508		*203*
Thiothece 484	PE	458	484	515		*203*
Thiothece polar 484	PE		484	515		*203*
Torularhodin	PE	467	501	537		*92*
	PE		507	(535)	2040 (507)	*125,126*
	MeOH	(460)	493	529		*140*
	EtOH	463	493	529		*140*
	C	(483)	515	554	1932 (515)	*140,217*
	B	485	519	557		*140*
	Pyr	485	518	558		*140*
	CS	502	541	582		*140*
Torularhodinaldehyde	PE	480	514	548	2840 (514)	*215*
	Cy		513	544		*208*
	B	491	525	562		*249*
	CS	524	555	594		*249*
Torularhodin ethyl ester	PE	476	502	534	2850 (502)	*125*
	Cy	478	505	538		*125*
	B		519	549		*125*
	Pyr	485	518	558		*140*
Torularhodin methyl ester	PE	(475)	497	529	2950 (497)	*125*
Torulene	PE	454–460	480–484	513–518	3240 (480)	*177,205,209*
	Hex	458–460	484–487	518–519		*23,126*
	EtOH	456	485–486	517–520		*140,205*
	C	467–469	501	537–539		*92,249*
	B	470	503	541		*249*
	Pyr	475	508	545		*140*
	Cs	488–491	522–525	563		*92,249*

2,3,3′-Trihydroxy-β-carotene	EtOH		450	478		*218*
3,4,4′-Trihydroxy-β-carotene	Hex		448	478		*63*
	Pyr		470	496		*63*
Triphasiaxanthin	Hex	440	467	495		*262*
	C		480	510	1654 (480)	*262*
Trollein, *see* Deepoxyneoxanthin						
Trollichrome	EtOH		424	451		*140*
	C		430	458		*92*
	B		432	458		*92*
	CS		450	479		*92*
Trollixanthin	EtOH		447	474		*140*
	C		455	482		*92*
	B	427	454	482		*82*
	CS		473	501		*92*
Tunaxanthin	Hex	436	(454)	466		*47*
	PE	438		468		*146*
	C	449		479		*146*
	B	447	(464)	478		*47*
	CS	463	(481)	495		*47*
Valenciachrome	B	337	354	375		*48*
Valenciaxanthin	Hex	351	369	390		*50*
	B	358	377	399		*48*
Vaucherixanthin	EtOH	419	441–443	469–471		*153,225,228*
	C	426	451	480		*228*
Violaxanthin	PE	418	442	466		*68*
	Hex		443	472		*92*
	MeOH	415	440	469		*140*
	EtOH	419–420	440–443	470–471	2550 (443)	*45,140,153,260*
	C	426	449	478		*228*
	B	422–428	447–454	479–484	2240 (454)	*21,51,57,82*
	CS	440	470	501		*140*

(Continued)

Appendix—*Continued*

Carotenoid	Solvent[b]	Absorption maxima, nm[c]			$E^{1\%}_{1\,cm}$, λ (nm)	Ref.[d]
Violerythrin	Ac		556			*107*
	CS	540	576	625		*140*
Warmingone, *see* Rhodo-pinal						
α-Zeacarotene	Hex	399	421	449	1850 (421)	*202*
β-Zeacarotene	PE	(404)	425	452		*259*
	PE	406	428	454	2520 (428)	*209*
	Hex	406–407	427–428	454	1940 (427)	*126,202*
Zeaxanthin	PE	424–427	449–452	476–481	2348 (452)	*62,122,123,146, 259*
	Hex	423–426	450–451	480–483		*92,239*
	MeOH	422	450	481		*140*
	EtOH	424–426	450–451	481–483	2540 (450)	*140,221,260*
	Ac	(430)	452–453	479	2340 (452)	*4,104*
	C	429–433	462	492–494		*92,146*
	C	(434)	459	488		*228*
	B	(433–440)	459–463	488–491		*43,57*
	CS	450	482–483	518		*92,135*
Zeinoxanthin, *see* α-Cryp-toxanthin						

[a] Appendix compiled by E. DeRitter.

[b] Abbreviations: Ac, acetone; B, benzene; C, chloroform; CS, carbon disulfide; Cy, cyclohexane; DE, diethyl ether; EtOH, ethanol; Hex, hexane; MeOH, methanol; PE, petroleum ether; Pyr, pyridine; THF, tetrahydrofuran.

[c] Values in parentheses represent points of inflection.

[d] Key to references in Appendix (see p. 917).

Key to References in Appendix

1. Aasen, A. J., and Liaaen-Jensen, S., *Acta Chem. Scand.* **19,** 1843–1853 (1965).
2. Aasen, A. J., and Liaaen-Jensen, S., *Acta Chem. Scand.* **20,** 811–819 (1966).
3. Aasen, A. J., and Liaaen-Jensen, S., *Acta Chem. Scand.* **20,** 1970–1988 (1966).
4. Aasen, A. J., and Liaaen-Jensen, S., *Acta Chem. Scand.* **20,** 2322–2324 (1966).
5. Aasen, A. J., and Liaaen-Jensen, S., *Acta Chem. Scand.* **21,** 371–377 (1967).
6. Aasen, A. J., and Liaaen-Jensen, S., *Acta Chem. Scand.* **21,** 2185–2204 (1967).
7. Akiya, S., *Nippon Ganka Gakkei Zasshi* **67,** 1168–1181 (1963); *C. A.* **62,** 8045.
8. Andrewes, A. G., Kjøsen, H., Liaaen-Jensen, S., Weisgraber, K. H., Lousberg, R. J. J. C., and Weiss, U., *Acta Chem. Scand.* **25,** 3878–3880 (1971).
9. Andrewes, A. G., and Liaaen-Jensen, S., *Acta Chem. Scand.* **27,** 1401–1409 (1973).
10. Arcamone, F., Camerino, B., Cotta, E., Franceschi, G., Grein, A., Penco, S., and Spalla, C., *Experientia* **25,** 241–242 (1969).
11. Arcamone, F., Camerino, B., Franceschi, G., and Penco, S., *Gazz. Chim. Ital.* **100,** 581–590 (1970).
12. Aring, T., Bramley, P. M., Davies, B. H., and Rees, A. F., *Phytochemistry* **11,** 3187–3192 (1972).
13. Arpin, N., and Liaaen-Jensen, S., *Phytochemistry* **6,** 995–1005 (1967).
14. Arpin, N., and Liaaen-Jensen, S., *Phytochemistry* **8,** 185–193 (1969).
15. Arpin, N., Fiasson, J. L., Bouchez-Dangye-Caye, M. P., Francis, G. W., and Liaaen-Jensen, S., *Phytochemistry* **10,** 1595–1601 (1971).
16. Arpin, N., Kjøsen, H., Francis, G. W., and Liaaen-Jensen, S., *Phytochemistry* **12,** 2751–2758 (1973).
17. Arpin, N., and Liaaen-Jensen, S., *Bull. Soc. Chim. Biol.* **49,** 527–536 (1967).
18. Attwood, M. A., *Phytochemistry* **9,** 2415–2416 (1970).
19. Bae, M., Lee, T. H., Yokoyama, H., Boettger, H. G., and Chichester, C. O., *Phytochemistry* **10,** 625–629 (1971).
20. Barber, M. S., Davis, J. B., Jackman, L. M., and Weedon, B. C. L., *J. Chem. Soc.* 2870–2881 (1960).
21. Bartlett, L., Klyne, W., Mose, W. P., Scopes, P. M., Galasko, G., Mallams, A. K., Weedon, B. C. L., Szabolcs, J., and Tóth, Gy., *J. Chem. Soc. C* 2527 (1969).
22. Ben-Aziz, A., Britton, G., and Goodwin, T. W., *Phytochemistry* **12,** 2759–2764 (1973).
23. Bindl, E., Lang, W., and Rau, W., *Planta* **94,** 156–174 (1970).
24. Bodea, C., and Nicoară, E., *Justus Liebigs Ann. Chem.* **609,** 181–185 (1957).
25. Bodea, C., Nicoară, E., and Salontai, T., *Justus Liebigs Ann. Chem.* **648,** 147–150 (1961).
26. Bodea, C., Neamtu, G., and Tămas, V., *Rev. Roum. Chim.* **11,** 1123–1126 (1966).
27. Bonnett, R., Spark, A. A., and Weedon, B. C. L., *Acta Chem. Scand.* **18,** 1739–1744 (1964).
28. Bonnett, R., Mallams, A. K., Spark, A. A., Tee, J. L., Weedon, B. C. L., and McCormick, A., *J. Chem. Soc. C* 429–454 (1969).
29. Booth, V. H., *Phytochemistry* **3,** 229–234 (1964).
30. Britton, G., and Goodwin, T. W., *Phytochemistry* **8,** 2257–2258 (1969).
31. Brown, B. O., and Weedon, B. C. L., *Chem. Commun.* 382–384 (1968).
32. Bush, W. V., and Zechmeister, L., *J. Am. Chem. Soc.* **80,** 2991–2999 (1958).
33. Campbell, S. A., Mallams, A. K., Waight, E. S., Weedon, B. C. L., Barbier, M., Lederer, E., and Salaque, A., *Chem. Commun.* 941–942 (1967).
34. Campbell, S. A., *Comp. Biochem. Physiol.* **30,** 803–812 (1969).
35. Campbell, S. A., *Comp. Biochem. Physiol.* **32,** 97–115 (1970).
36. Chapman, D. J., and Haxo, F. T., *Plant Cell. Physiol.* **4,** 57–63 (1963).
37. Chapman, D. J., *Phytochemistry* **5,** 1331–1333 (1966).

38. Chino, H., Murakami, S., and Harashima, K., *Biochim. Biophys. Acta* **176,** 1-26 (1969).
39. Cholnoky, L., Szabolcs, J., and Nagy, E., *Justus Liebigs Ann. Chem.* **616,** 207-218 (1958).
40. Cholnoky, L., Szabolcs, J., Cooper, R. D. G., and Weedon, B. C. L., *Tetrahedron Lett.* 1257-1259 (1963).
41. Cholnoky, L., Györgyfy, K., Szabolcs, J., Weedon, B. C. L., and Waight, E. S., *Chem. Commun.* 404-405 (1966).
42. Cholnoky, L., Szabolcs, J., and Tóth, G., *Justus Liebigs Ann. Chem.* **708,** 218-223 (1967).
43. Cholnoky, L., Györgyfy, K., Ronai, A., Szabolcs, J., Tóth, G., Galasko, G., Mallams, A. K., Waight, E. S., and Weedon, B. C. L., *J. Chem. Soc. C* 1256-1263 (1969).
44. Cooper, R. D. G., Davis, J. B., and Weedon, B. C. L., *J. Chem. Soc.* 5637-5641 (1963).
45. Costes, C., and Thomas, C. *Ann. Physiol. Veg.* **10,** 171-197 (1968).
46. Costes, C., Laurence, J., and Thomas, C., *Bull. Soc. Fr. Physiol. Veg.* **15**, 55-70 (1969).
47. Crozier, G. F., and Wilkie, D. W., *Comp. Biochem. Physiol.* **18,** 801-804 (1966).
48. Curl, A. L., and Bailey, G. F., *J. Agric. Food Chem.* **2,** 685-690 (1954).
49. Curl, A. L., and Bailey, G. F., *J. Agric. Food Chem.* **4,** 156-162 (1956).
50. Curl, A. L., and Bailey, G. F., *J. Agric. Food Chem.* **9,** 403-405 (1961).
51. Curl, A. L., *J. Food Sci.* **26,** 106-111 (1961).
52. Curl, A. L., and Bailey, G. F., *J. Food Sci.* **26,** 442-447 (1961).
53. Curl, A. L., *J. Agric. Food Chem.* **10,** 504-509 (1962).
54. Curl, A. L., *J. Food Sci.* **27,** 171-176 (1962).
55. Curl, A. L., *J. Food Sci.* **27,** 537-543 (1962).
56. Curl, A. L., *J. Food Sci.* **28,** 623-626 (1963).
57. Curl, A. L., *J. Food Sci.* **29,** 241-245 (1964).
58. Curl, A. L., *J. Food Sci.* **30,** 13-18 (1965).
59. Curl, A. L., *J. Food Sci.* **32,** 141-143 (1967).
60. Czeczuga, B., and Czerpak, R., *Eur. J. Biochem.* **5,** 429-432 (1968).
61. Czygan, F.-C., and Heumann, W., *Arch. Mikrobiol.* **57,** 123-134 (1967).
62. Czygan, F.-C., and Kessler, E., *Z. Naturforsch., Teil B* **22,** 1085-1086 (1967).
63. Czygan, F.-C., *Planta* **85,** 35-41 (1969).
64. Davies, B. H., Ph.D. Thesis, Univ. of Wales (1961).
65. Davies, B. H., *In* "Chemistry and Biochemistry of Plant Pigments" (T. W. Goodwin, ed.), pp. 518-532. Academic Press, New York, 1965.
66. Davies, B. H., and Holmes, E. A., *Biochem. J.* **113,** 33P-34P (1969).
67. Davies, B. H., *Biochem. J.* **116,** 93-99 (1970).
68. Davies, B. H., *Biochem. J.* **116,** 101-110 (1970).
69. Davies, B. H., Hsu, W.-J., and Chichester, C. O., *Comp. Biochem. Physiol.* **33,** 601-615 (1970).
70. Davies, B. H., Matthews, S., and Kirk, J. T. O., *Phytochemistry* **9,** 797-805 (1970).
71. Davies, B. H., and Rees, A. F., *Phytochemistry* **12,** 2745-2750 (1973).
72. Davis, J. B., Jackman, L. M., Siddons, P. T., and Weedon, B. C. L., *J. Chem. Soc. C* 2154-2165 (1966).
73. De la Mar, R. R., and Francis, F. J., *J. Food Sci.* **34,** 287-290 (1969).
74. De Nicola, M., *Exp. Cell Res.* **10,** 441-446 (1956).
75. De Nicola, M. G., *Boll. Sedute Accad. Gioenia Sci. Nat. Catania* **5,** 201-214 (1959).
76. Egger, K., and Kleinig, H., *Phytochemistry* **6,** 903-905 (1967).
77. Eimhjellen, K. E., Steensland, H., and Traetteberg, J., *Arch. Mikrobiol.* **59,** 82-92 (1967).
78. Ellinghausen, H. C., Jr., and Pelczar, M. J., Jr., *J. Bacteriol.* **70,** 448-453 (1955).
79. El-Tiney, A. H., and Chichester, C. O., *J. Org. Chem.* **35,** 2290-2293 (1970).
80. Entschel, R., and Karrer, P., *Helv. Chim. Acta* **40,** 1809-1811 (1957).
81. Entschel, R., and Karrer, P., *Helv. Chim. Acta* **41,** 983-987 (1958).
82. Eugster, C. H., and Karrer, P., *Helv. Chim. Acta* **40,** 69-79 (1957).
83. Eugster, C. H., Hürlimann, H., and Leuenberger, H. J., *Helv. Chim. Acta* **52,** 806-807 (1969).

84. Fox, D. L., and Hopkins, T. S., *Comp. Biochem. Physiol.* **17,** 841–856 (1966).
85. Fox, D. L., and Hopkins, T. S., *Comp. Biochem. Physiol.* **19,** 267–278 (1966).
86. Fox, D. L., Smith, V. E., Grigg, R. W., and McLeod, W. D., *Comp. Biochem. Physiol.* **28,** 1103–1114 (1969).
87. Francis, G. W., Upadhyay, R. R., and Liaaen-Jensen, S., *Acta Chem. Scand.* **24,** 3050–3052 (1970).
88. Francis, G. W., and Halfen, L. N., *Phytochemistry* **11,** 2347–2348 (1972).
89. Francis, G. W., Hertzberg, S., Upadhyay, R. R., and Liaaen-Jensen, S., *Acta Chem. Scand.* **26,** 1097–1104 (1972).
90. Galasko, G., Hora, J., Toube, T. P., Weedon, B. C. L., André, D., Barbier, M., Lederer, E., and Villaneuva, V. R., *J. Chem. Soc. C* 1264–1265 (1969).
91. Gilchrist, B. M., *Comp. Biochem. Physiol.* **24,** 123–147 (1968).
92. Goodwin, T. W., *in* "Modern Methods of Plant Analysis" (K. Paech and M. V. Tracy, eds.), Vol. 3, pp. 272–311. Springer-Verlag, Berlin and New York, 1955.
93. Grob, E. C., and Pflugshaupt, R. P., *Helv. Chim. Acta* **48,** 930–939 (1965).
94. Hager, A., and Stransky, H., *Arch. Mikrobiol.* **71,** 132–163 (1970).
95. Hager, A., and Stransky, H., *Arch. Mikrobiol.* **72,** 68–83 (1970).
96. Hager, A., and Stransky, H., *Arch. Mikrobiol.* **73,** 77–89 (1970).
97. Halfen, L. N., Pierson, B. K., and Francis, G. W., *Arch. Mikrobiol.* **82,** 240–246 (1972).
98. Hata, M., and Hata, M., *Tohoku J. Agric. Res.* **21,** 183–188 (1970).
99. Haxo, F., *Arch. Biochem.* **20,** 400–421 (1949).
100. Haxo, F., *Bot. Gaz. (Chicago)* **112,** 228–232 (1950).
101. Healey, F. P., *J. Phycol.* **4,** 126–129 (1968).
102. Herring, P. J., *J. Mar. Biol. Assoc. U.K.* **49,** 767–779 (1969).
103. Hertzberg, S., and Liaaen-Jensen, S., *Acta Chem. Scand.* **20,** 1187–1194 (1966).
104. Hertzberg, S., and Liaaen-Jensen, S., *Phytochemistry* **5,** 557–563 (1966).
105. Hertzberg, S., and Liaaen-Jensen, S., *Phytochemistry* **5,** 565–570 (1966).
106. Hertzberg, S., and Liaaen-Jensen, S., *Acta Chem. Scand.* **21,** 15–41 (1967).
107. Hertzberg, S., Liaaen-Jensen, S., Enzell, C. R., and Francis, G. W., *Acta Chem. Scand.* **23,** 3290–3312 (1969).
108. Hertzberg, S., and Liaaen-Jensen, S., *Phytochemistry* **8,** 1281–1292 (1969).
109. Hertzberg, S., and Liaaen-Jensen, S., *Phytochemistry* **10,** 3251–3252 (1971).
110. Hiyama, T., Nishimura, M., and Chance, B., *Anal. Biochem.* **29,** 339–342 (1969).
111. Hora, J., Toube, T. P., and Weedon, B. C. L., *J. Chem. Soc. C* 241–242 (1970).
112. Hsu, W.-J., Chichester, C. O., and Davies, B. H., *Comp. Biochem. Physiol.* **32,** 69–79 (1970).
113. Inhoffen, H. H., Pommer, H., and Meth, E.-G., *Justus Liebigs Ann. Chem.* **572,** 151–160 (1951).
114. Inhoffen, H. H., and Raspé, G., *Justus Liebigs Ann. Chem.* **592,** 214–221 (1955).
115. Inhoffen, H. H., and Raspé, G., *Justus Liebigs Ann. Chem.* **594,** 165–176 (1955).
116. Isler, O., Lindlar, H., Montavon, M., Rüegg, R., and Zeller, P., *Helv. Chim. Acta* **39,** 249–259 (1956).
117. Isler, O., Lindlar, H., Montavon, M., Rüegg, R., and Zeller, P., *Helv. Chim. Acta* **39,** 274–282 (1956).
118. Isler, O., Lindlar, H., Montavon, M., Rüegg, R., and Zeller, P., *Helv. Chim. Acta* **39,** 449–454 (1956).
119. Isler, O., Montavon, M., Rüegg, R., and Zeller, P., *Helv. Chim. Acta* **39,** 454–462 (1956).
120. Isler, O., Gutmann, H., Lindlar, H., Montavon, M., Rüegg, R., Ryser, G., and Zeller, P., *Helv. Chim. Acta* **39,** 463–473 (1965).
121. Isler, O., Lindlar, H., Montavon, M., Rüegg, R., Saucy, G., and Zeller, P., *Helv. Chim. Acta* **39,** 2041–2053 (1956).
122. Isler, O., Montavon, M., Rüegg, R., and Zeller, P., *Justus Liebigs Ann. Chem.* **603,** 129–144 (1957).

123. Isler, O., Lindlar, H., Montavon, M., Rüegg, R., Saucy, G., and Zeller, P., *Helv. Chim. Acta* **40,** 456–467 (1957).
124. Isler, O., Gutmann, H., Montavon, M., Rüegg, R., Ryser, G., and Zeller, P., *Helv. Chim. Acta* **40,** 1242–1249 (1957).
125. Isler, O., Guex, W., Rüegg, R., Ryser, G., Saucy, G., Schwieter, U., Walter, M., and Winterstein, A., *Helv. Chim. Acta* **42,** 864–871 (1959).
126. Isler, O., and Schudel, P., *Wiss. Veroeff. Dtsch. Ges. Ernaehr.* **9,** 54–104 (1963).
127. Jackman, L. M., and Liaaen-Jensen, S., *Acta Chem. Scand.* **18,** 1403–1411 (1964).
128. Jaeger, L., and Karrer, P., *Helv. Chim. Acta* **46,** 683–687 (1963).
129. Jensen, A., *Acta Chem. Scand.* **20,** 1728–1730 (1966).
130. Jungalwala, F. B., and Cama, H. R., *Biochem. J.* **85,** 1–8 (1962).
131. Karmakar, G., and Zechmeister, L., *J. Am. Chem. Soc.* **77,** 55–60 (1955).
132. Karrer, P., and Schwab, G., *Helv. Chim. Acta* **23,** 578–581 (1940).
133. Karrer, P., and Jucker, E., *Helv. Chim. Acta* **26,** 626–630 (1943).
134. Karrer, P., and Rutschmann, J., *Helv. Chim. Acta* **27,** 1691–1695 (1944).
135. Karrer, P., and Jucker, E., *Helv. Chim. Acta* **28,** 427–436 (1945).
136. Karrer, P., and Jucker, E., *Helv. Chim. Acta* **28,** 1143–1146 (1945).
137. Karrer, P., and Jucker, E., *Helv. Chim. Acta* **29,** 229–233 (1946).
138. Karrer, P., Jucker, E., and Steinlin, K., *Helv. Chim. Acta* **30,** 531–535 (1947).
139. Karrer, P., and Jucker, E., *Helv. Chim. Acta* **30,** 536–537 (1947).
140. Karrer, P., and Jucker, E., "Carotenoids." Elsevier, Amsterdam, 1950.
141. Karrer, P., and Leumann, E., *Helv. Chim. Acta* **34,** 445–453 (1951).
142. Karrer, P., and Leumann, E., *Helv. Chim. Acta* **34,** 1412–1414 (1951).
142a. Karrer, P., and Jaeger, L., *Helv. Chim. Acta* **46,** 688–690 (1963).
143. Katayama, T., Hirata, K., Yokoyama, H., and Chichester, C. O., *Nippon Suisan Gakkaishi* **36,** 709–714 (1970).
144. Katayama, T., Yokoyama, H., and Chichester, C. O., *Int. J. Biochem.* **1,** 438–444 (1970).
145. Katayama, T., Katama, T., and Chichester, C. O., *Int. J. Biochem.* **3,** 363–368 (1972).
146. Katayama, T., Shintani, K., and Chichester, C. O., *Comp. Biochem. Physiol. B* **44,** 253–257 (1973).
147. Kelly, M., Norgård, S., and Liaaen-Jensen, S., *Acta Chem. Scand.* **24,** 2169–2182 (1970).
148. Khosla, M. C., and Karrer, P., *Helv. Chim. Acta* **43,** 453–457 (1960).
149. Kjøsen, H., and Liaaen-Jensen, S., *Phytochemistry* **8,** 483–491 (1969).
150. Kjøsen, H., Liaaen-Jensen, S., and Enzell, C. R., *Acta Chem. Scand.* **25,** 85–93 (1971).
151. Kjøsen, H., and Liaaen-Jensen, S., *Acta Chem. Scand.* **26,** 4121–4129 (1972).
152. Kleinig, H., and Egger, K., *Phytochemistry* **6,** 1681–1686 (1967).
153. Kleinig, H., and Egger, K., *Z. Naturforsch. Teil B* **22,** 868–872 (1967).
154. Kleinig, H., *Z. Naturforsch. Teil B* **22,** 977–979 (1967).
155. Kleinig, H., and Czygan, F.-C., *Z. Naturforsch., Teil B* **24,** 927–930 (1969).
156. Kleinig, H., Nitsche, H., and Egger, K., *Tetrahedron Lett.* 5139–5142 (1969).
157. Kleinkauf, H., *Arch. Mikrobiol.* **53,** 154–158 (1966).
158. Krinsky, N. I., and Goldsmith, T. H., *Arch. Biochem. Biophys.* **91,** 271–279 (1960).
159. Krinsky, N. I., and Levine, R. P., *Plant Physiol.* **39,** 680–687 (1964).
160. Krinsky, N. I., and Lenhoff, H. M., *Comp. Biochem. Physiol.* **16,** 189–198 (1965).
161. Kuhn, R., and Smakula, A., *Hoppe-Seyler's Z. Physiol. Chem.* **197,** 161–166 (1931).
162. Kuhn, R., and Lederer, E., *Hoppe-Seyler's Z. Physiol. Chem.* **200,** 108–114 (1931).
163. Kuhn, R., and Brockmann, H., *Ber. Dtsch. Chem. Ges.* **65,** 894–898 (1932).
164. Kuhn, R., and Brockmann, H., *Hoppe-Seyler's Z. Physiol. Chem.* **213,** 192–198 (1932).
165. Land, D. G., Ph.D. Thesis, Univ. of Liverpool (1955).
166. Lee, W. L., *Comp. Biochem. Physiol.* **19,** 13–27 (1966).
167. Leftwick, A. P., and Weedon, B. C. L., *Acta Chem. Scand.* **20,** 1195–1199 (1966).
168. Leftwick, A. P., and Weedon, B. C. L., *Chem. Commun.* 49–50 (1967).

169. LeRosen, A. L., and Zechmeister, L., *Arch. Biochem.* **1,** 17–26 (1943).
170. Liaaen-Jensen, S., "The Constitution of Some Bacterial Carotenoids and their Bearings on Biosynthetic Problems." Bruns, Trondheim, 1962.
171. Liaaen-Jensen, S., *Acta Chem. Scand.* **17,** 303–312 (1963).
172. Liaaen-Jensen, S., *Acta Chem. Scand.* **17,** 489–499 (1963).
173. Liaaen-Jensen, S., *Acta Chem. Scand.* **17,** 500–508 (1963).
174. Liaaen-Jensen, S., and Schmidt, K., *Arch. Mikrobiol.* **46,** 138–149 (1963).
175. Liaaen-Jensen, S., Hegge, E., and Jackman, L. M., *Acta Chem. Scand.* **18,** 1703–1718 (1964).
176. Liaaen-Jensen, S., *Acta Chem. Scand.* **19,** 1166–1174 (1965).
177. Liaaen-Jensen, S., *Phytochemistry* **4,** 925–931 (1965).
178. Liaaen-Jensen, S., and Hertzberg, S., *Acta Chem. Scand.* **20,** 1703–1709 (1966).
179. Liaaen-Jensen, S., *Acta Chem. Scand.* **21,** 961–969 (1967).
180. Liaaen-Jensen, S., Hertzberg, S., Weeks, O. B., and Schwieter, U., *Acta Chem. Scand.* **22,** 1171–1186 (1968).
181. Liu, I.-S., Lee, T. H., Yokoyama, H., Simpson, K. L., and Chichester, C. O., *Phytochemistry* **12,** 2953–2956 (1973).
182. Livingston, A. L., and Knowles, R. E., *Phytochemistry* **8,** 1311–1312 (1969).
183. Loeblich, A. R., and Smith, E., *Lipids* **3,** 5–13 (1968).
184. Malhotra, H. C., Britton, G., and Goodwin, T. W., *Chem. Commun.* 127–128 (1970).
185. Malhotra, H. C., Britton, G., and Goodwin, T. W., *FEBS Lett.* **6,** 334–336 (1970).
186. Malhotra, H. C., Britton, G., and Goodwin, T. W., *Phytochemistry* **9,** 2369–2375 (1970).
187. Malhotra, H. C., Britton, G., and Goodwin, T. W., *Int. Z. Vitaminforsch.* **40,** 315–322 (1970).
188. Mallams, A. K., Waight, E. S., Weedon, B. C. L., Cholnoky, L., Györgyfy, K., Szabolcs, J., Krinsky, N. I., Schimmer, B. P., Chichester, C. O., Katayama, T., Lowry, L., and Yokoyama, H., *Chem. Commun.* 484–485 (1967).
189. Mandelli, E. F., *J. Phycol.* **4,** 347–348 (1968).
190. Markham, M. C., and Liaaen-Jensen, S., *Phytochemistry* **7,** 839–844 (1968).
191. Matsuno, T., Kusumoto, T., Watanabe, T., and Ishihara, Y., *Nippon Suisan Gakkaishi* **39,** 43–50 (1973).
192. Morgan, R. C., *J. Food Sci.* **31,** 213–217 (1966).
193. Neamtu, G., Illyes, G., and Bodea, C., *Rev. Roum. Biochim.* **5,** 215–221 (1968).
194. Nicoară, E., Illyes, G., Suteu, M., and Bodea, C., *Rev. Roum. Chim.* **12,** 547–552 (1967).
195. Nitsche, H., *Tetrahedron Lett.* 4913–4916 (1970).
196. Nitsche, H., *Phytochemistry* **11,** 401–404 (1972).
197. Norgård, S., and Liaaen-Jensen, S., *Acta Chem. Scand.* **23,** 1463 (1969).
198. Norgård, S., Aasen, A. J., and Liaaen-Jensen, S., *Acta Chem. Scand.* **24,** 2183–2197 (1970).
199. Nybraaten, G., and Liaaen-Jensen, S., *Acta Chem. Scand.* **25,** 370–372 (1971).
200. Osianu, D., Nicoară, E., and Bodea, C., *Rev. Roum. Chim.* **6,** 925–927 (1971).
201. Petracek, F. J., and Zechmeister, L., *J. Am. Chem. Soc.* **78,** 1427–1434 (1956).
202. Petzold, E. N., Quackenbush, F. W., and McQuistan, M., *Arch. Biochem. Biophys.* **82,** 117–124 (1959).
203. Pfennig, N., Markham, M. C., and Liaaen-Jensen, S., *Arch. Mikrobiol.* **62,** 178–191 (1968).
204. Polgar, A., van Niel, C. B., and Zechmeister, L., *Arch. Biochem.* **5,** 243–264 (1944).
205. Reichenbach, H., and Kleinig, H., *Arch. Mikrobiol.* **76,** 364–380 (1971).
206. Ricketts, T. R., *Phytochemistry* **6,** 1375–1386 (1967).
207. Rüegg, R., Lindlar, H., Montavon, M., Saucy, G., Schaeren, S. F., Schwieter, U., and Isler, O., *Helv. Chim. Acta* **42,** 847–853 (1959).
208. Rüegg, R., Montavon, M., Ryser, G., Saucy, G., Schwieter, U., and Isler, O., *Helv. Chim. Acta* **42,** 854–864 (1959).
209. Rüegg, R., Schwieter, U., Ryser, G., Schudel, P., and Isler, O., *Helv. Chim. Acta* **44,** 994–999 (1961).
210. Ryvarden, L., and Liaaen-Jensen, S., *Acta Chem. Scand.* **18,** 643–654 (1964).

211. Schlegel, H. G., *Arch. Mikrobiol.* **31,** 231–239 (1958).
212. Schmidt, K., Pfennig, N., and Liaaen-Jensen, S., *Arch. Mikrobiol.* **52,** 132–146 (1965).
213. Schmidt, K., *Arch. Mikrobiol.* **77,** 231–238 (1971).
214. Schwieter, U., Bolliger, H. R., Chopard-dit-Jean, L. H., Englert, G., Kofler, M., König, A., von Planta, C., Rüegg, R., Vetter, W., and Isler, O., *Chimia* **19,** 294–302 (1965).
215. Schwieter, U., Gutmann, H., Lindlar, H., Marbet, R., Rigassi, N., Rüegg, R., Schaeren, S. F., and Isler, O., *Helv. Chim. Acta* **49,** 369–389 (1966).
216. Schwieter, U., Rüegg, R., and Isler, O., *Helv. Chim. Acta* **49,** 992–996 (1966).
217. Simpson, K. L., Nakayama, T. O. M., and Chichester, C. O., *J. Bacteriol.* **88,** 1688–1694 (1964).
218. Smallidge, R. L., and Quackenbush, F. W., *Phytochemistry* **12,** 2481–2482 (1973).
219. Sørensen, N. Z., Liaaen-Jensen, S., Børdalen, B., Haug, A., Enzell, C., and Francis, G., *Acta Chem. Scand.* **22,** 344–347 (1968).
220. Starr, M. P., and Saperstein, S., *Arch. Biochem. Biophys.* **43,** 157–168 (1953).
221. Strain, H. H., "Leaf Xanthophylls." Carnegie Inst., Washington, D.C., 1938.
222. Strain, H. H., *Biol. Bull. (Woods Hole, Mass.)* **97,** 206–209 (1949).
223. Strain, H. H., "Chloroplast Pigments and Chromatographic Analysis." Penn. State Univ. Press, University Park, Pennsylvania, 1958.
224. Strain, H. H., Sherma, J., and Grandolfo, M., *Anal. Chem.* **39,** 926–932 (1967).
225. Strain, H. H., Svec, W. A., Aitzetmüller, K., Grandolfo, M., and Katz, J. J., *Phytochemistry* **7,** 1417–1418 (1968).
226. Strain, H. H., Aitzetmüller, K., Svec, W. A., and Katz, J. J., *Chem. Commun.* 876–877 (1970).
227. Strain, H. H., Svec, W. A., Aitzetmüller, K., Grandolfo, M. C., Katz, J. J., Kjøsen, H., Norgård, S., Liaaen-Jensen, S., Haxo, F. T., Wegfahrt, P., and Rapoport, H., *J. Am. Chem. Soc.* **93,** 1823–1825 (1971).
228. Stransky, H., and Hager, A., *Arch. Mikrobiol.* **71,** 164–190 (1970).
229. Stransky, H., and Hager, A., *Arch. Mikrobiol.* **72,** 84–96 (1970).
230. Surmatis, J. D., and Ofner, A., *J. Org. Chem.* **28,** 2735–2739 (1963).
231. Surmatis, J. D., Ofner, A., Gibas, J., and Thommen, R., *J. Org. Chem.* **31,** 186–188 (1966).
232. Surmatis, J. D., Walser, A., Gibas, J., Schwieter, U., and Thommen, R., *Helv. Chim. Acta* **53,** 974–990 (1970).
233. Suzue, G., Tsukada, K., and Tanaka, S., *Biochim. Biophys. Acta* **144,** 186–188 (1967).
234. Thirkell, D., Strang, R. H. C., and Chapman, J. R., *J. Gen. Microbiol.* **49,** 157–164 (1967).
235. Thomas, D. M., M.S. Thesis, Univ. of Wales (1962).
236. Tsukida, K., and Zechmeister, L., *Arch. Biochem. Biophys.* **74,** 408–426 (1958).
237. Ungers, G. E., and Cooney, J. J., *J. Bacteriol.* **96,** 234–241 (1968).
238. Valadon, L. R. G., and Cooke, R. C., *Phytochemistry* **2,** 103–105 (1963).
239. Valadon, L. R. G., and Mummery, R. S., *Phytochemistry* **6,** 983–988 (1967).
240. Valadon, L. R. G., *Phyton (Buenos Aires)* **25,** 151–157 (1968).
241. Valadon, L. R. G., and Mummery, R. S., *Ann. Bot. (London)* **33,** 879–882 (1969).
242. Walton, T. J., Ph.D. Thesis, Univ. of Liverpool (1970).
243. Walton, T. J., Britton, G., Goodwin, T. W., Diner, B., and Moshier, S., *Phytochemistry* **9,** 2545–2552 (1970).
243a. Warren, C. K., and Weedon, B. C. L., *J. Chem. Soc.* 3972–3986 (1958).
244. Warren, C. K., and Weedon, B. C. L., *J. Chem. Soc.* 3986–3993 (1958).
245. Weeks, O. B., and Garner, R. J., *Arch. Biochem. Biophys.* **121,** 35–49 (1967).
245a. Weeks, O. B., Andrewes, A. G., Brown, B. O., and Weedon, B. C. L., *Nature (London)* **224,** 879–882 (1969).
246. Williams, B. L., M.S. Thesis, Univ. of Wales (1963).
247. Williams, R. J. H., and Goodwin, T. W., *Phytochemistry* **6,** 1037–1039 (1967).
248. Winterstein, A., *Hoppe-Seyler's Z. Physiol. Chem.* **219,** 249–252 (1933).

249. Winterstein, A., Studer, A., and Rüegg, R., *Chem. Ber.* **93,** 2951–2965 (1960).
250. Wright, H. E., Jr., Burton, W. W., and Berry, R. C., Jr., *Arch. Biochem. Biophys.* **82,** 107–116 (1959).
251. Yamaguchi, M., *Bull. Chem. Soc. Jpn.* **30,** 111–114 (1957).
252. Yamaguchi, M., *Bull. Chem. Soc. Jpn.* **30,** 979–983 (1957).
253. Yamaguchi, M., *Bull. Chem. Soc. Jpn.* **31,** 51–55 (1958).
254. Yamamoto, H. R., Yokoyama, H., and Boettger, H., *J. Org. Chem.* **34,** 4207–4208 (1969).
255. Yokoyama, H., and White, M. J., *J. Org. Chem.* **30,** 2481–2482 (1965).
256. Yokoyama, H., White, M. J., and Vandercook, C. E., *J. Org. Chem.* **30,** 2482–2483 (1965).
257. Yokoyama, H., and White, M. J., *J. Org. Chem.* **30,** 3994–3996 (1965).
258. Yokoyama, H., and White, M. J., *J. Org. Chem.* **31,** 3452–3454 (1966).
259. Yokoyama, H., and White, M. J., *Phytochemistry* **5,** 1159–1173 (1966).
260. Yokoyama, H., and Vandercook, C. E., *J. Food Sci.* **32,** 42–48 (1967).
261. Yokoyama, H., and White, M. J., *Phytochemistry* **7,** 1031–1034 (1968).
262. Yokoyama, H., Guerrero, H. C., and Boettger, H., *J. Org. Chem.* **35,** 2080–2082 (1970).
263. Yokoyama, H., and Guerrero, H. C., *Phytochemistry* **9,** 231–232 (1970).
264. Zalokar, M., *Arch. Biochem. Biophys.* **70,** 568–571 (1957).
265. Zechmeister, L., and Cholnoky, L. V., *Justus Liebigs Ann. Chem.* **523,** 101–118 (1936).
266. Zechmeister, L., and Tuzson, P., *Ber Dtsch. Chem. Ges.* **69,** 1878–1884 (1936).
267. Zechmeister, L., and Wallcave, L., *J. Am. Chem. Soc.* **75,** 4493–4495 (1953).
268. Zeller, P., Bader, F., Lindlar, H., Montavon, M., Müller, P., Rüegg, R., Ryser, G., Saucy, G., Schaeren, S. F., Schwieter, U., Stricker, K., Tamm, R., Zürcher, P., and Isler, O., *Helv. Chim. Acta* **42,** 841–847 (1959).

Index

A

B

D

E

P

R

Z

FOOD SCIENCE AND TECHNOLOGY

A SERIES OF MONOGRAPHS

Maynard A. Amerine, Rose Marie Pangborn, and Edward B. Roessler, PRINCIPLES OF SENSORY EVALUATION OF FOOD. 1965.

S. M. Herschdoerfer, QUALITY CONTROL IN THE FOOD INDUSTRY. Volume I — 1967. Volume II — 1968. Volume III — 1972.

Hans Reimann, FOOD-BORNE INFECTIONS AND INTOXICATIONS. 1969.

Irvin E. Leiner, TOXIC CONSTITUENTS OF PLANT FOODSTUFFS. 1969.

Martin Glicksman, GUM TECHNOLOGY IN THE FOOD INDUSTRY. 1970.

L. A. Goldblatt, AFLATOXIN. 1970.

Maynard A. Joslyn, METHODS IN FOOD ANALYSIS, second edition. 1970.

A. C. Hulme (ed.), THE BIOCHEMISTRY OF FRUITS AND THEIR PRODUCTS. Volume 1 — 1970. Volume 2 — 1971.

G. Ohloff and A. F. Thomas, GUSTATION AND OLFACTION. 1971.

George F. Stewart and Maynard A. Amerine, INTRODUCTION TO FOOD SCIENCE AND TECHNOLOGY. 1973.

C. R. Stumbo, THERMOBACTERIOLOGY IN FOOD PROCESSING, second edition. 1973.

Irvin E. Liener (ed.), TOXIC CONSTITUENTS OF ANIMAL FOODSTUFFS. 1974.

Aaron M. Altschul (ed.), NEW PROTEIN FOODS: Volume 1, TECHNOLOGY, PART A — 1974. Volume 2, TECHNOLOGY, PART B — 1976. Volume 3, ANIMAL PROTEIN SUPPLIES, PART A — 1978.

S. A. Goldblith, L. Rey, and W. W. Rothmayr, FREEZE DRYING AND ADVANCED FOOD TECHNOLOGY. 1975.

R. B. Duckworth (ed.), WATER RELATIONS OF FOOD. 1975.

Gerald Reed (ed.), ENZYMES IN FOOD PROCESSING, second edition. 1975.

A. G. Ward and A. Courts (eds.), THE SCIENCE AND TECHNOLOGY OF GELATIN. 1976.

John A. Troller and J. H. B. Christian, WATER ACTIVITY AND FOOD. 1978.

A. E. Bender, FOOD PROCESSING AND NUTRITION. 1978.

D. R. Osborne and P. Voogt, THE ANALYSIS OF NUTRIENTS IN FOODS. 1978.

Marcel Loncin and R. L. Merson, FOOD ENGINEERING: PRINCIPLES AND SELECTED APPLICATIONS. 1979.

Hans Riemann and Frank L. Bryan (eds.), FOOD-BORNE INFECTIONS AND INTOXICATIONS, second edition. 1979.

N. A. Michael Eskin, PLANT PIGMENTS, FLAVORS AND TEXTURES: THE CHEMISTRY AND BIOCHEMISTRY OF SELECTED COMPOUNDS. 1979.

J. G. Vaughan (ed.), FOOD MICROSCOPY. 1979.

J. R. A. Pollock (ed.), BREWING SCIENCE, Volume 1 — 1979. Volume 2 — 1980.

Irvin E. Liener (ed.), TOXIC CONSTITUENTS OF PLANT FOODSTUFFS, second edition. 1980.

J. Christopher Bauernfeind (ed.), CAROTENOIDS AS COLORANTS AND VITAMIN A PRECURSORS:
TECHNOLOGICAL AND NUTRITIONAL APPLICATIONS. 1981.

In preparation

Aaron M. Altschul (ed.), NEW PROTEIN FOODS: Volume 4, ANIMAL PROTEIN SUPPLIES, PART B — 1981.